Encyclopedia of Database Systems

Ling Liu • M. Tamer Özsu

Editors

Encyclopedia of Database Systems

Second Edition

Volume 5

Q–S

With 1374 Figures and 143 Tables

 Springer

Editors
Ling Liu
Georgia Institute of Technology College
of Computing
Atlanta, GA, USA

M. Tamer Özsu
University of Waterloo School of Computer Science
Waterloo, ON, Canada

ISBN 978-1-4614-8266-6   ISBN 978-1-4614-8265-9 (eBook)
ISBN 978-1-4614-8264-2 (print and electronic bundle)
https://doi.org/10.1007/978-1-4614-8265-9

Library of Congress Control Number: 2018938558

Printed on acid-free paper

This Springer imprint is published by the registered company Springer Science+Business
Media, LLC part of Springer Nature.
The registered company address is: 233 Spring Street, New York, NY 10013, U.S.A.

Preface to the Second Edition

Since the release of the first volume of this Encyclopedia, big data has emerged as a central feature of information technology innovation in many business, science, and engineering fields. Databases are one of the fundamental core technologies for big data systems and big data analytics. In order to extract features and derive values from big data, it must be stored, processed, and analyzed in a timely manner. Not surprisingly, big data not only fuels the development and deployment of database systems and database technologies, it also opens doors to new opportunities and new challenges in the field of databases. As data grows in volume, velocity, variety, and with the attendant veracity issues, there is a growing demand for volume-scalable databases, velocity-adaptive databases, and variety-capable databases that can handle data quality issues properly. As machine learning and artificial intelligence renew their momentum with the power of big data, there is an increasing demand for new generation of database systems that are built for extracting features from databases as efficient and effective as conventional database systems are capable of for querying databases.

The first edition of the *Encyclopedia of Database Systems* is a comprehensive, multivolume collection of over 1,250 in-depth entries (3,067 including synonyms), covering important concepts on all aspects of database systems, including areas of current interest and research results of historical significance. This second edition of *Encyclopedia of Database Systems* expands the first edition by enriching the content of existing entries, expanding existing topic areas with new entries, adding a set of cutting-edge topic areas, including cloud data management, crowdsourcing, data analytics, data provenance management, graph data management, social networks, and uncertain data management to name a few. The new entries and the new topic areas were determined through discussions and consultations with the Advisory Board of the *Encyclopedia of Database Systems*. Each of the new topic areas was managed by a new Area Editor who, together with the editor-in-chief, further developed the content for each area, soliciting experts in the field as contributors to write the entries, and performed the necessary technical editing. We also reviewed the entries from the first edition and revised them as needed to bring them up-to-date.

We would like to thank the members of the Advisory Board, the Editorial Board, and all of the authors for their contributions to this second edition. We would also like to thank Springer's editors and staff, including Susan Lagerstrom-Fife, Michael Hermann, and Sonja Peterson for their assistance and support throughout the project, and Annalea Manalili for her involvement in the early period of this project.

In closing, we trust the Encyclopedia can serve as a valuable source for students, researchers, and practitioners who need a quick and authoritative reference to the subject on database systems. Suggestions and feedbacks to further improve the Encyclopedia are welcome from readers and from the community.

Preface to the First Edition

We are in an information era where generating and storing large amounts of data are commonplace. A growing number of organizations routinely handle terabytes and exabytes of data, and individual digital data collections easily reach multiple gigabytes. Along with the increases in volume, the modality of digitized data that requires efficient management and the access modes to these data have become more varied. It is increasingly common for business and personal data collections to include images, video, voice, and unstructured text; the retrieval of these data comprises various forms, including structured queries, keyword search, and visual access. Data have become a highly valued asset for governments, industries and individuals, and the management of these data collections remains a critical technical challenge.

Database technology has matured over the past four decades and is now quite ubiquitous in many applications that deal with more traditional business data. The challenges of expanding data management to include other data modalities while maintaining the fundamental tenets of database management (data independence, data integrity, data consistency, etc.) are issues that the community continues to work on. The lines between database management and other fields such as information retrieval, multimedia retrieval, and data visualization are increasingly blurred.

This multi-volume *Encyclopedia of Database Systems* provides easy access to important concepts on all aspects of database systems, including areas of current interest and research results of historical significance. It is a comprehensive collection of over 1,250 in-depth entries (3,067 including synonyms) that present coverage of the important concepts, issues, emerging technology and future trends in the field of database technologies, systems, and applications. The content of the *Encyclopedia* was determined through wide consultations. We were assisted by an Advisory Board in coming up with the overall structure and content. Each of these areas were put under the control of Area Editors (70 in total) who further developed the content for each area, soliciting experts in the field as contributors to write the entries, and performed the necessary technical editing. Some of them even wrote entries themselves. Nearly 1,000 authors were involved in writing entries.

The intended audience for the *Encyclopedia* is technically broad and diverse. It includes anyone concerned with database system technology and its applications. Specifically, the *Encyclopedia* can serve as a valuable and authoritative reference for students, researchers and practitioners who need a quick and authoritative reference to the subject of databases, data management, and database systems. We anticipate that many people will benefit from this reference work, including database specialists, software developers, scientists and engineers who need to deal with (structured, semi-structured or unstructured) large datasets. In addition, database and data mining researchers and scholars in the many areas that apply database technologies, such as artificial intelligence, software engineering, robotics and computer vision, machine learning, finance and marketing are expected to benefit from the *Encyclopedia*.

We would like to thank the members of the Advisory Board, the Editorial Board, and the individual contributors for their help in creating this *Encyclopedia*. The success of the *Encyclopedia* could not have been achieved without the expertise and the effort of the many contributors. Our sincere thanks also go to Springer's editors and staff, including Jennifer Carlson, Susan Lagerstrom-Fife, Oona Schmid, and Susan Bednarczyk for their support throughout the project.

Finally, we would very much like to hear from readers for any suggestions regarding the *Encyclopedia's* content. With a project of this size and scope, it is quite possible that we may have missed some concepts. It is also possible that some entries may benefit from revisions and clarifications. We are committed to issuing periodic updates and we look forward to the feedback from the community to improve the *Encyclopedia*.

Ling Liu
M. Tamer Özsu

List of Topics

Database Security and Privacy

Section Editor: *Elena Ferrari*

Access Control Administration Policies
Access Control Policy Languages
Access Control
Administration Model for RBAC
Anonymity
ANSI/INCITS RBAC Standard
Asymmetric Encryption
Auditing and Forensic Analysis
Authentication
Blind Signatures
Data Encryption
Data Rank/Swapping
Database Security
Digital Rights Management
Digital Signatures
Disclosure Risk
Discretionary Access Control
GEO-RBAC Model
Hash Functions
Homomorphic Encryption
Inference Control in Statistical
    Databases
Information Loss Measures
Intrusion Detection Technology
k-Anonymity
Mandatory Access Control
Merkle Trees
Message Authentication Codes
Microaggregation
Microdata Rounding
Microdata
Multilevel Secure Database Management
    System
Noise Addition
Nonperturbative Masking Methods
PRAM
Privacy Policies and Preferences
Privacy Through Accountability
Privacy-Enhancing Technologies
Privacy-Preserving DBMSs
Privacy
Private Information Retrieval
Protection from Insider Threats
Pseudonymity
Record Linkage

Regulatory Compliance in Data
    Management
Role-Based Access Control
SDC Score
Secure Data Outsourcing
Secure Database Development
Secure Transaction Processing
Security Services
Steganography
Symmetric Encryption
Synthetic Microdata
Tabular Data
Trusted Hardware
Unobservability

Semantic Web and Ontologies

Section Editor: *Avigdor Gal*

Description Logics
Emergent Semantics
Integration of Rules and Ontologies
Lightweight Ontologies
Linked Open Data
Ontology Elicitation
Ontology Engineering
Ontology
OWL: Web Ontology Language
Resource Description Framework (RDF)
    Schema (RDFS)
Semantic Crowdsourcing
Semantic Matching
Semantic Social Web
Semantic Streams
Semantic Web Query Languages
Semantic Web Services
Semantic Web

Data Cleaning

Section Editor: *Venkatesh Ganti*

Column Segmentation
Constraint-Driven Database Repair
Data Cleaning
Data Profiling
Deduplication in Data Cleaning
Record Matching

Database Design

Section Editor: *Alexander Borgida*

Conceptual Schema Design
Database Design
Database Reverse Engineering
Design for Data Quality
Logical Database Design: From Conceptual to
 Logical Schema
Physical Database Design for Relational
 Databases

Text Indexing Techniques

Section Editor: *Mario A. Nascimento*

Indexing Compressed Text
Indexing the Web
Inverted Files
Signature Files
Suffix Tree
Text Compression
Text Indexing Techniques
Trie

Data Quality

Section Editor: *Felix Naumann*

Data Conflicts
Data Fusion
Data Quality Assessment
Data Quality Dimensions
Data Quality Models
Data Scrubbing
Probabilistic Databases

Web Search and Crawl

Section Editor: *Cong Yu*

Deep-Web Search
Faceted Search
Focused Web Crawling
Geo-Targeted Web Search
Incremental Crawling
Metasearch Engines
Peer-to-Peer Web Search
Personalized Web Search
Precision and Recall
Search Engine Metrics

Test Collection
Web Advertising
Web Characteristics and Evolution
Web Crawler Architecture
Web Information Extraction
Web Page Quality Metrics
Web Question Answering
Web Search Query Rewriting
Web Search Relevance Feedback
Web Search Relevance Ranking
Web Search Result Caching and Prefetching
Web Search Result De-duplication and
 Clustering
Web Spam Detection

Multimedia Databases

Section Editor: *Vincent Oria,
Shin'ichi Satoh*

Audio Classification
Audio Content Analysis
Audio Metadata
Audio Representation
Audio Segmentation
Audio
Automatic Image Annotation
Computational Media Aesthetics
Content-Based Video Retrieval
Human-Centered Computing: Application
 to Multimedia
Image Content Modeling
Image Database
Image Metadata
Image Querying
Image Representation
Image Retrieval and Relevance Feedback
Image Segmentation
Image Similarity
Image
Multimedia Data Indexing
Multimedia Data Querying
Multimedia Databases
Multimedia Data
Multimedia Metadata
Object Recognition
Semantic Modeling and Knowledge
 Representation for Multimedia Data
Video Content Analysis

Association Rule Mining

Section Editor: *Jian Pei*

Workflow Management

Section Editor: *Barbara Pernici*

Stream Mining

Section Editor: *Divesh Srivastava*

Distributed Database Systems

Section Editor: *Kian-Lee Tan*

Logics and Databases

Section Editor: *Val Tannen*

Bag Semantics
Conjunctive Query
Datalog
First-Order Logic: Semantics
First-Order Logic: Syntax
FOL Modeling of Integrity Constraints
    (Dependencies)
Relational Algebra
Relational Calculus

Structured and Semi-structured Document Databases

Section Editor: *Frank Tompa*

Document Databases
Document Representations (Inclusive Native and
    Relational)
Electronic Dictionary
Electronic Encyclopedia
Electronic Newspapers
Enterprise Content Management
Functional Dependencies for Semistructured
    Data
Grammar Inference
Hypertexts
Markup Language
Normal Form ORA-SS Schema Diagrams
Object Relationship Attribute Data Model for
    Semistructured Data
Path Query
Region Algebra
Semi-structured Data Model
Semi-structured Database Design
Semi-structured Query Languages
Structural Indexing
Unicode

Indexing

Section Editor: *Vassilis J. Tsotras*

B+-Tree
Bitmap Index
Bloom Filters

Covering Index
Dense Index
Extendible Hashing
Generalized Search Tree
Hash-Based Indexing
I/O Model of Computation
Indexed Sequential Access Method
Linear Hashing
Primary Index
Range Query
Secondary Index
Sparse Index
Tree-Based Indexing

Parallel Database Systems

Section Editor: *Patrick Valduriez*

Data Skew
Database Clusters
Inter-operator Parallelism
Intra-operator Parallelism
Online Recovery in Parallel Database Systems
Parallel Data Placement
Parallel Database Management
Parallel Hash Join, Parallel Merge Join, Parallel
    Nested Loops Join
Parallel Query Execution Algorithms
Parallel Query Optimization
Parallel Query Processing
Query Load Balancing in Parallel Database
    Systems
Shared-Disk Architecture
Shared-Memory Architecture
Shared-Nothing Architecture
Virtual Partitioning

Advanced Storage Systems

Section Editor: *Kaladhar Voruganti*

Initiator
Logical Unit Number Mapping
Logical Unit Number
Multi-pathing
Object Storage Protocol
SCSI Target
Software-Defined Storage

Approximation and Data Reduction Techniques

Section Editor: *Xiaofang Zhou*

Approximate Query Processing
Data Reduction
Data Sampling
Data Sketch/Synopsis
Database Clustering Methods
Dimensionality Reduction
Discrete Wavelet Transform and Wavelet
    Synopses
Fractal
Hierarchical Data Summarization
Histogram
K-Means and K-Medoids
Linear Regression
Log-Linear Regression
Multidimensional Scaling
Nonparametric Data Reduction Techniques
Parametric Data Reduction Techniques
Principal Component Analysis
Singular Value Decomposition
Two-Dimensional Shape Retrieval

Social Networks

Section Editor: *Nick Koudas*

Collaborative Filtering
Recommender Systems
Social Influence
Social Media Analysis

Cloud Data Management

Section Editor: *Amr El Abbadi*

CAP Theorem
Cloud Computing
Data Center Energy Efficiency
Data Management in Data Centers
Data Migration Management
Elasticity
Infrastructure-as-a-Service (IaaS)
Multi-data Center Replication Protocols
Multi-datacenter Consistency Properties
Multitenancy

Platform-as-a-Service (PaaS)
Software-as-a-Service (SaaS)

Data Analytics

Section Editor: *Fatma Özcan*

Big Data Platforms for Data Analytics
Distributed File Systems
Distributed Machine Learning
Interactive Analytics in Social Media
Social Media Analytics
SQL Analytics on Big Data
Streaming Analytics
Structure Analytics in Social Media
Temporal Analytics in Social Media
Text Analytics in Social Media
Text Analytics

Data Management Fundamentals

Section Editor: *Ramez Elmasri*

Data Definition Language (DDL)
Data Definition
Data Management Fundamentals: Database
    Management System
Data Manipulation Language (DML)
Database Administrator (DBA)
Database Schema
Database
Logical and Physical Data Independence

NoSQL Databases

Section Editor: *Ling Liu* and *M. Tamer Özsu*

Column Stores
Document
MapReduce
NoSQL Stores

Graph Data Management

Section Editor: *Lei Chen*

Graph Database
Graph Mining

RDF Stores
RDF Technology
SPARQL

Data Provenance Management

Section Editor: *Juliana Freire*

Provenance and Reproducibility
Provenance in Databases
Provenance in Scientific Databases
Provenance in Workflows
Provenance Standards
Provenance Storage
Provenance: Privacy and Security

Ranking Queries

Section Editor: *Ihab F. Ilyas*

Preference Queries
Preference Specification
Probabilistic Skylines
Rank-Aware Query Processing
Rank-Join Indices
Rank-Join
Ranking Views
Reverse Top-k Queries
Score Aggregation
Skyline Queries and Pareto Optimality
Top-k Queries
Uncertain Top-k Queries

Uncertain Data Management

Section Editor: *Minos Garofalakis*

Graphical Models for Uncertain Data
   Management
Indexing Uncertain Data
Karp-Luby Sampling
Managing Data Integration Uncertainty
Managing Probabilistic Entity Extraction
Monte Carlo Methods for Uncertain Data
Probabilistic Entity Resolution
Query Processing over Uncertain Data
Uncertain Data Lineage
Uncertain Data Mining
Uncertain Data Models
Uncertain Data Streams
Uncertain Graph Data Management
Uncertain Spatial Data Management

Crowd Sourcing

Section Editor: *Reynold Cheng*

Cost and Quality Trade-Offs in Crowdsourcing
Crowd Database Operators
Crowd Database Systems
Crowd Mining and Analysis
Human Factors Modeling in Crowdsourcing
Indexing with Crowds

About the Editors

Ling Liu Georgia Institute of Technology College of Computing, Atlanta, GA, USA

Ling Liu is Professor of Computer Science in the College of Computing at Georgia Institute of Technology. She holds a Ph.D. (1992) in Computer and Information Science from Tilburg University, The Netherlands. Dr. Liu directs the research programs in the Distributed Data Intensive Systems Lab (DiSL), examining various aspects of data intensive systems, ranging from big data systems, cloud computing, databases, Internet and mobile systems and services, machine learning, to social and crowd computing, with the focus on performance, availability, security, privacy, and trust. Prof. Liu is an elected IEEE Fellow and a recipient of IEEE Computer Society Technical Achievement Award (2012). She has published over 300 international journal and conference articles and is a recipient of the best paper award from numerous top venues, including ICDCS, WWW, IEEE Cloud, IEEE ICWS, and ACM/IEEE CCGrid. In addition to serving as general chair and PC chairs of numerous IEEE and ACM conferences in big data, distributed computing, cloud computing, data engineering, and very large databases fields, Prof. Liu served as the Editor-in-Chief of *IEEE Transactions on Service Computing* (2013–2016) and also served on editorial boards of over a dozen international journals. Her current research is sponsored primarily by NSF and IBM.

M. Tamer Özsu University of Waterloo School of Computer Science, Waterloo, ON, Canada

M. Tamer Özsu is Professor of Computer Science at the David R. Cheriton School of Computer Science and the Associate Dean of Research of the Faculty of Mathematics at the University of Waterloo. He was the Director of the Cheriton School of Computer Science from January 2007 to June 2010.

His research is in data management focusing on large-scale data distribution and management of nontraditional data, currently focusing on graph and RDF data. His publications include the book *Principles of Distributed Database Systems* (with Patrick Valduriez), which is now in its third edition. He was the Founding Series Editor of *Synthesis Lectures on Data Management* (Morgan & Claypool) and is now the Editor-in-Chief of *ACM Books*. He serves on the editorial boards of three journals and two book series.

He is a Fellow of the Royal Society of Canada, American Association for the Advancement of Science (AAAS), Association for Computing Machinery (ACM), and the Institute of Electrical and Electronics Engineers (IEEE). He is an elected member of the Science Academy, Turkey, and a member of Sigma Xi. He was awarded the ACM SIGMOD Test-of-Time Award in 2015, the ACM SIGMOD Contributions Award in 2006, and the Ohio State University College of Engineering Distinguished Alumnus Award in 2008.

Advisory Board

Ramesh Jain Department of Computer Science, School of Information and Computer Sciences, University of California Irvine, Irvine, CA, USA

Peter MG Apers Centre for Telematics and Information Technology, University of Twente, Enschede, The Netherlands

Timos Sellis Data Science Research Institute, Swinburne University of Technology, Hawthorn, VIC, Australia

Matthias Jarke Informatik 5 Information Systems, RWTH-Aachen University, Aachen, Germany

Ricardo Baeza-Yate Department of Information and Communication Technologies, University of Pompeu Fabra, Barcelona, Spain

Jai Menon Cloudistics, Reston, VA, USA

Beng Chin Ooi School of Computing, National University of Singapore, Singapore, Singapore

Elisa Bertino Department of Computer Science, Purdue University, West Lafeyette, IN, USA

Erhard Rahm Fakultät für Mathematik und Informatik, Institut für Informatik, Universität Leipzig, Leipzig, Germany

Gerhard Weikum Department 5: Databases and Information Systems, Max-Planck-Institut für Informatik, Saarbrücken, Germany

Stefano Ceri Department of Electronics, Information and Bioengineering, Politecnico di Milano, Milano, Italy

Asuman Dogac SRDC Software Research and Development and Consultancy Ltd., Cankaya/Ankara, Turkey

Hans-Joerg Schek Department of Computer Science, ETH Zürich, Zürich, Switzerland

Alon Halevy Recruit Institute of Technology, Mountain View, CA, USA

Jennifer Widom Frederick Emmons Terman School of Engineering, Stanford University, Stanford, CA, USA

John Mylopoulos Department of Computer Science, University of Toronto, Toronto, ON, Canada

Serge Abiteboul INRIA and ENS, Paris, France

Frank Tompa David R. Cheriton School of Computer Science, University of Waterloo, Waterloo, ON, Canada

Patrick Valduriez INRIA and LIRMM, Montpellier, France

Gustavo Alonso Department of Computer Science, ETH Zürich, Zürich, Switzerland

Krithi Ramamritham Department of Computer Science and Engineering, Indian Institute of Technology Bombay, Mumbai, India

Area Editors

Peer-to-Peer Data Management

Karl Aberer Department of Computer Science, École Polytechnique Fédérale de Lausanne (EPFL), Lausanne, Switzerland

Database Management System Architectures

Anastasia Ailamaki Department of Computer Science, Ecole Polytechnique Fédérale de Lausanne, Lausanne, Switzerland

Information Retrieval Models

Giambattista Amati Fondazione Ugo Bordoni, Rome, Italy

XML Data Management

Sihem Amer-Yahia CNRS, University Grenoble Alpes, Saint Martin D'Hères, France

Database Middleware

Cristiana Amza Electrical and Computer Engineering, University of Toronto, Toronto, ON, Canada

Database Management Utilities

Philippe Bonnet Department of Computer Science, IT University of Copenhagen, Copenhagen, Denmark

Visual Interfaces

Tiziana Catarci Department of Computer Engineering, Automation and Management, Sapienza – Università di Roma, Rome, Italy

Stream Data Management

Ugur Cetintemel Department of Computer Science, Brown University, Providence, RI, USA

Querying over Data Integration Systems

Kevin Chang Department of Computer Science, University of Illinois at Urbana-Champaign, Urbana-Champaign, IL, USA

Self Management

Surajit Chaudhuri Microsoft Corporation, Redmond, CA, USA

Text Mining

Zheng Chen Microsoft Corporation, Beijing, China

Extended Transaction Models

Panos K. Chrysanthis Department of Computer Science, School of Computing and Information, University of Pittsburgh, Pittsburgh, PA, USA

Privacy-Preserving Data Mining

Chris Clifton Department of Computer Science, Purdue University, West Lafayette, IN, USA

Digital Libraries

Amr El Abbadi Department of Computer Science, UC Santa Barbara, Santa Barbara, CA, USA

Data Models

David W. Embley Department of Computer Science, Brigham Young University, Provo, UT, USA

Complex Event Processing

Opher Etzion Department of Information Systems, Yezreel Valley College, Jezreel Valley, Israel

Database Security and Privacy

Elena Ferrari Department of Computer Science, Università degli Studi dell'Insubria, Varese, Italy

Semantic Web and Ontologies

Avigdor Gal Industrial Engineering and Management, Technion – Israel Institute of Technology, Haifa, Israel

Data Cleaning

Venkatesh Ganti Alation, Redwood City, CA, USA

Web Data Extraction

Georg Gottlob Computing Lab, Oxford University, Oxford, UK

Sensor Networks

Le Gruenwald School of Computer Science, University of Oklahoma, Norman, OK, USA

Data Clustering

Dimitrios Gunopulos Department of Informatics and Telecommunications, National and Kapodistrian University of Athens, Athens, Greece

Scientific Databases

Amarnath Gupta San Diego Supercomputer Center, University of California San Diego, La Jolla, CA, USA

Geographical Information Systems

Ralf Hartmut Güting Department of Computer Science, FernUniversität in Hagen, Hagen, Germany

Data Visualization

Hans Hinterberger Department of Computer Science, ETH Zurich, Zurich, Switzerland

Web Services and Service Oriented Architectures

Hans-Arno Jacobsen Department of Electrical and Computer Engineering, University of Toronto, Toronto, ON, Canada

Metadata Management

Manfred Jeusfeld IIT, University of Skövde, Skövde, Sweden

Health Informatics Databases

Vipul Kashyap CIGNA Healthcare, Bloomfield, CT, USA

Visual Data Mining

Daniel A. Keim Computer Science Department, University of Konstanz, Konstanz, Germany

Data Replication

Bettina Kemme School of Computer Science, McGill University, Montreal, QC, Canada

Storage Structures and Systems

Masaru Kitsuregawa Institute of Industrial Science, University of Tokyo, Tokyo, Japan

Views and View Management

Yannis Kotidis Department of Informatics, Athens University of Economics and Business, Athens, Greece

Structured Text Retrieval

Jaap Kamps Faculty of Humanities, University of Amsterdam, Amsterdam, The Netherlands

Information Quality

Yang W. Lee School of Business, Northeastern University, Boston, MA, USA

Relational Theory

Leonid Libkin School of Informatics, University of Edinburgh, Edinburgh, UK

Information Retrieval Evaluation Measures

Weiyi Meng Department of Computer Science, State University of New York at Binghamton, Binghamton, NY, USA

Logical Data Integration

Renée J. Miller Department of Computer Science, University of Toronto, Toronto, ON, Canada

Database Design

Alexander Borgida Department of Computer Science, Rutgers University, New Brunswick, NJ, USA

Text Indexing Techniques

Mario A. Nascimento Department of Computing Science, University of Alberta, Edmonton, AB, Canada

Data Quality

Felix Naumann Hasso Plattner Institute, University of Potsdam, Potsdam, Germany

Web Search and Crawl

Cong Yu Google Research, New York, NY, USA

Multimedia Databases

Vincent Oria Department of Computer Science, New Jersey Institute of Technology, Newark, NJ, USA

Shin'ichi Satoh Digital Content and Media Sciences ReseaMultimedia Information Research Division, National Institute of Informatics, Tokyo, Japan

Active Databases

M. Tamer Özsu Cheriton School of Computer Science, University of Waterloo, Waterloo, ON, Canada

Spatial, Spatiotemporal, and Multidimensional Databases

Dimitris Papadias Department of Computer Science and Engineering, Hong Kong University of Science and Technology, Kowloon, China

Data Warehouse

Torben Bach Pedersen Department of Computer Science, Aalborg University, Aalborg, Denmark

Stefano Rizzi DISI – University of Bologna, Bologna, Italy

Association Rule Mining

Jian Pei School of Computing Science, Simon Fraser University, Burnaby, BC, Canada

Workflow Management

Barbara Pernici Department di Elettronica e Informazione, Politecnico di Milano, Milan, Italy

Query Processing and Optimization

Evaggelia Pitoura Department of Computer Science and Engineering, University of Ioannina, Ioannina, Greece

Data Management for the Life Sciences

Louiqa Raschid Robert H. Smith School of Business, University of Maryland, College Park, MD, USA

Information Retrieval Operations

Edie Rasmussen Library, Archival and Information Studies, The University of British Columbia, VC, Canada

Query Languages

Tore Risch Department of Information Technology, Uppsala University, Uppsala, Sweden

Database Tuning and Performance

Dennis Shasha Department of Computer Science, New York University, New York, NY, USA

Classification and Decision Trees

Kyuseok Shim School of Electrical Engineering and Computer Science, Seoul National University, Seoul, Republic of Korea

Temporal Databases

Christian S. Jensen Department of Computer Science, Aalborg University, Aalborg, Denmark

Richard T. Snodgrass Department of Computer Science, University of Arizona, Tucson, AZ, USA

Stream Mining

Divesh Srivastava AT&T Labs-Research, Bedminster, NJ, USA

Distributed Database Systems

Kian-Lee Tan Department of Computer Science, National University of Singapore, Singapore, Singapore

Logics and Databases

Val Tannen Department of Computer and Information Science, University of Pennsylvania, Philadelphia, PA, USA

Structured and Semi-structured Document Databases

Frank Tompa David R. Cheriton School of Computer Science, University of Waterloo, Waterloo, ON, Canada

Indexing

Vassilis J. Tsotras Department of Computer Science and Engineering, University of California-Riverside, Riverside, CA, USA

Parallel Database Systems

Patrick Valduriez INRIA and LIRMM, Montpellier, France

Advanced Storage Systems

Kaladhar Voruganti Equinix, San Francisco, CA, USA

Transaction Management

Gottfried Vossen Department of Information Systems, Westfälische
Wilhelms-Universität, Münster, Germany

Mobile and Ubiquitous Data Management

Ouri Wolfson Department of Computer Science, University of Illinois at Chicago, Chicago, IL, USA

Multimedia Information Retrieval

Jeffrey Xu Yu Department of Systems Engineering and Engineering Management, The Chinese University of Hong Kong, Hong Kong, China

Approximation and Data Reduction Techniques

Xiaofang Zhou School of Information Technology and Electrical Engineering, University of Queensland, Brisbane, Australia

Social Networks

Nick Koudas Department of Computer Science, University of Toronto, Toronto, ON, Canada

Cloud Data Management

Amr El Abbadi Department of Computer Science, UC Santa Barbara, Santa Barbara, CA, USA

Data Analytics

Fatma Özcan IBM Research – Almaden, San Jose, CA, USA

Data Management Fundamentals

Ramez Elmasri Department of Computer Science and Engineering, The
University of Texas at Arlington, Arlington, TX, USA

NoSQL Databases

M. Tamer Özsu Cheriton School of Computer Science, University of
Waterloo, Waterloo, ON, Canada

Ling Liu College of Computing, Georgia Institute of Technology, Atlanta,
GA, USA

Graph Data Management

Lei Chen Department of Computer Science and Engineering, The Hong Kong University of Science and Technology, Hong Kong, China

Data Provenance Management

Juliana Freire Computer Science and Engineering, New York University, New York, NY, USA

Ranking Queries

Ihab F. Ilyas Cheriton School of Computer Science, University of Waterloo, Waterloo, ON, Canada

Uncertain Data Management

Minos Garofalakis Technical University of Crete, Chania, Greece

Crowd Sourcing

Reynold Cheng Computer Science, The University of Hong Kong, Hong Kong, China

List of Contributors

Daniel Abadi Yale University, New Haven, CT, USA

Sofiane Abbar Qatar Computing Research Institute, Doha, Qatar

Alberto Abelló Polytechnic University of Catalonia, Barcelona, Spain

Serge Abiteboul Inria, Paris, France

Maribel Acosta Institute AIFB, Karlsruhe Institute of Technology, Karlsruhe, Germany

Ioannis Aekaterinidis University of Patras, Rio, Patras, Greece

Nitin Agarwal University of Arkansas, Little Rock, AR, USA

Charu C. Aggarwal IBM T. J. Watson Research Center, Yorktown Heights, NY, USA

Lalitha Agnihotri McGraw-Hill Education, New York, NY, USA

Marcos K. Aguilera VMware Research, Palo Alto, CA, USA

Yanif Ahmad Department of Computer Science, Brown University, Providence, RI, USA

Gail-Joon Ahn Arizona State University, Tempe, AZ, USA

Anastasia Ailamaki Informatique et Communications, Ecole Polytechnique Fédérale de Lausanne, Lausanne, Switzerland

Ablimit Aji Analytics Lab, Hewlett Packard, Palo Alto, CA, USA

Alexander Alexandrov Database and Information Management (DIMA), Institute of Software Engineering and Theoretical Computer Science, Berlin, Germany

Yousef J. Al-Houmaily Institute of Public Administration, Riyadh, Saudi Arabia

Mohammed Eunus Ali Department of Computer Science and Engineering, Bangladesh University of Engineering and Technology (BUET), Dhaka, Bangladesh

Robert B. Allen Drexel University, Philadelphia, PA, USA

Gustavo Alonso ETH Zürich, Zurich, Switzerland

Omar Alonso Microsoft Silicon Valley, Mountain View, CA, USA

Bernd Amann Pierre & Marie Curie University (UPMC), Paris, France

Giambattista Amati Fondazione Ugo Bordoni, Rome, Italy

Sihem Amer-Yahia CNRS, Univ. Grenoble Alps, Grenoble, France

Laboratoire d'Informatique de Grenoble, CNRS-LIG, Saint Martin-d'Hères, Grenoble, France

Rainer von Ammon Center for Information Technology Transfer GmbH (CITT), Regensburg, Germany

Robert A. Amsler CSC, Falls Church, VA, USA

Yael Amsterdamer Department of Computer Science, Bar Ilan University, Ramat Gan, Israel

Cristiana Amza Department of Electrical and Computer Engineering, University of Toronto, Toronto, ON, Canada

George Anadiotis VU University Amsterdam, Amsterdam, The Netherlands

Mihael Ankerst Ludwig-Maximilians-Universität München, Munich, Germany

Sameer Antani National Institutes of Health, Bethesda, MD, USA

Grigoris Antoniou Foundation for Research and Technology-Hellas (FORTH), Heraklion, Greece

Arvind Arasu Microsoft Research, Redmond, WA, USA

Danilo Ardagna Politechnico di Milano University, Milan, Italy

Walid G. Aref Purdue University, West Lafayette, IN, USA

Marcelo Arenas Pontifical Catholic University of Chile, Santiago, Chile

Nikos Armenatzoglou Department of Computer Science and Engineering, Hong Kong University of Science and Technology, Kowloon, Hong Kong, Hong Kong

Samuel Aronson Harvard Medical School – Partners Healthcare Center for Genetics and Genomics, Boston, MA, USA

Paavo Arvola University of Tampere, Tampere, Finland

Colin Atkinson Software Engineering, University of Mannheim, Mannheim, Germany

Noboru Babaguchi Osaka University, Osaka, Japan

Shivnath Babu Duke University, Durham, NC, USA

Nathan Backman Computer Science, Buena Vista University, Storm Lake, IA, USA

Kenneth Paul Baclawski Northeastern University, Boston, MA, USA

Ricardo Baeza-Yates NTENT, USA - Univ. Pompeu Fabra, Spain - Univ. de Chile, Chile

James Bailey University of Melbourne, Melbourne, VIC, Australia

Peter Bailis Department of Computer Science, Stanford University, Palo Alto, CA, USA

Sumeet Bajaj Stony Brook University, Stony Brook, NY, USA

Peter Bak IBM Watson Health, Foundational Innovation, Haifa, Israel

Magdalena Balazinska University of Washington, Seattle, WA, USA

Krisztian Balog University of Stavanger, Stavanger, Norway

Farnoush Banaei-Kashani Computer Science and Engineering, University of Colorado Denver, Denver, CO, USA

Jie Bao Data Management, Analytics and Services (DMAS) and Ubiquitous Computing Group (Ubicomp), Microsoft Research Asia, Beijing, China

Stefano Baraldi University of Florence, Florence, Italy

Mauro Barbieri Phillips Research Europe, Eindhoven, The Netherlands

Denilson Barbosa University of Alberta, Edmonton, AL, Canada

Pablo Barceló University of Chile, Santiago, Chile

Luciano Baresi Dipartimento di Elettronica, Informazione e Bioingegneria – Politecnico di Milano, Milano, Italy

Ilaria Bartolini Department of Computer Science and Engineering (DISI), University of Bologna, Bologna, Italy

Saleh Basalamah Computer Science, Umm Al-Qura University, Mecca, Makkah Province, Saudi Arabia

Sugato Basu Google Inc, Mountain View, CA, USA

Carlo Batini University of Milano-Bicocca, Milan, Italy

Michal Batko Masaryk University, Brno, Czech Republic

Peter Baumann Jacobs University, Bremen, Germany

Robert Baumgartner Vienna University of Technology, Vienna, Austria

Sean Bechhofer University of Manchester, Manchester, UK

Steven M. Beitzel Telcordia Technologies, Piscataway, NJ, USA

Ladjel Bellatreche LIAS/ISAE-ENSMA, Poitiers University, Futuroscope, France

Omar Benjelloun Google Inc., New York, NY, USA

Véronique Benzaken University Paris 11, Orsay Cedex, France

Rafael Berlanga Department of Computer Languages and Systems, Universitat Jaume I, Castellón, Spain

Mikael Berndtsson University of Skövde, The Informatics Research Centre, Skövde, Sweden

University of Skövde, School of Informatics, Skövde, Sweden

Philip A. Bernstein Microsoft Corporation, Redmond, WA, USA

Damon Andrew Berry University of Massachusetts, Lowell, MA, USA

Leopoldo Bertossi Carleton University, Ottawa, ON, Canada

Claudio Bettini Dipartimento di Informatica, Università degli Studi di Milano, Milan, Italy

Nigel Bevan Professional Usability Services, London, UK

Bharat Bhargava Purdue University, West Lafayette, IN, USA

Arnab Bhattacharya Indian Institute of Technology, Kanpur, India

Ernst Biersack Eurecom, Sophia Antipolis, France

Alberto Del Bimbo University of Florence, Florence, Italy

Carsten Binnig Computer Science-Database Systems, Brown University, Providence, RI, USA

Christian Bizer Web-based Systems Group, University of Mannheim, Mannheim, Germany

Alan F. Blackwell University of Cambridge, Cambridge, UK

Carlos Blanco GSyA and ISTR Research Groups, Department of Computer Science and Electronics, Faculty of Sciences, University of Cantabria, Santander, Spain

Marina Blanton University of Notre Dame, Notre Dame, IN, USA

Toine Bogers Department of Communication and Psychology, Aalborg University Copenhagen, Copenhagen, Denmark

Philip Bohannon Yahoo! Research, Santa Clara, CA, USA

Michael H. Böhlen Free University of Bozen-Bolzano, Bozen-Bolzano, Italy

University of Zurich, Zürich, Switzerland

Christian Böhm University of Munich, Munich, Germany

Peter Boncz CWI, Amsterdam, The Netherlands

Philippe Bonnet Department of Computer Science, IT University of Copenhagen, Copenhagen, Denmark

Alexander Borgida Rutgers University, New Brunswick, NJ, USA

Vineyak Borkar CTO and VP of Engineering, X15 Software, San Francisco, CA, USA

Chavdar Botev Yahoo Research!, Cornell University, Ithaca, NY, USA

Sara Bouchenak University of Grenoble I – INRIA, Grenoble, France

Luc Bouganim INRIA Saclay and UVSQ, Le Chesnay, France

Nozha Boujemaa INRIA Paris-Rocquencourt, Le Chesnay, France

Shawn Bowers University of California-Davis, Davis, CA, USA

Stéphane Bressan National University of Singapore, School of Computing, Department of Computer Science, Singapore, Singapore

Martin Breunig University of Osnabrueck, Osnabrueck, Germany

Scott A. Bridwell University of Utah, Salt Lake City, UT, USA

Thomas Brinkhoff Institute for Applied Photogrammetry and Geoinformatics (IAPG), Oldenburg, Germany

Nieves R. Brisaboa Database Laboratory, Department of Computer Science, University of A Coruña, A Coruña, Spain

Andrei Broder Yahoo! Research, Santa Clara, CA, USA

Nicolas Bruno Microsoft Corporation, Redmond, WA, USA

François Bry University of Munich, Munich, Germany

Yingyi Bu Chinese University of Hong Kong, Hong Kong, China

Alejandro Buchmann Darmstadt University of Technology, Darmstadt, Germany

Thilina Buddhika Colorado State University, Fort Collins, CO, USA

Chiranjeeb Buragohain Amazon.com, Seattle, WA, USA

Thorsten Büring Ludwig-Maximilians-University Munich, Munich, Germany

Benjamin Bustos Department of Computer Science, University of Chile, Santiago, Chile

David J. Buttler Lawrence Livermore National Laboratory, Livermore, CA, USA

Yanli Cai Shanghai Jiao Tong University, Shanghai, China

Diego Calvanese Research Centre for Knowledge and Data (KRDB), Free University of Bozen-Bolzano, Bolzano, Italy

Guadalupe Canahuate The Ohio State University, Columbus, OH, USA

K. Selcuk Candan Arizona State University, Tempe, AZ, USA

Turkmen Canli University of Illinois at Chicago, Chicago, IL, USA

Alan Cannon Napier University, Edinburgh, UK

Cornelia Caragea Computer Science and Engineering, University of North Texas, Denton, TX, USA

Barbara Carminati Department of Theoretical and Applied Science, University of Insubria, Varese, Italy

Sheelagh Carpendale University of Calgary, Calgary, AB, Canada

Michael W. Carroll Villanova University School of Law, Villanova, PA, USA

Ben Carterette University of Massachusetts Amherst, Amherst, MA, USA

Marco A. Casanova Pontifical Catholic University of Rio de Janeiro, Rio de Janeiro, Brazil

Giuseppe Castagna C.N.R.S. and University Paris 7, Paris, France

Tiziana Catarci Dipartimento di Ingegneria Informatica, Automatica e Gestionale "A.Ruberti", Sapienza – Università di Roma, Rome, Italy

James Caverlee Department of Computer Science, Texas A&M University, College Station, TX, USA

Emmanuel Cecchet EPFL, Lausanne, Switzerland

Wojciech Cellary Department of Information Technology, Poznan University of Economics, Poznan, Poland

Ana Cerdeira-Pena Database Laboratory, Department of Computer Science, University of A Coruña, A Coruña, Spain

Michal Ceresna Lixto Software GmbH, Vienna, Austria

Ugur Cetintemel Department of Computer Science, Brown University, Providence, RI, USA

Soumen Chakrabarti Indian Institute of Technology of Bombay, Mumbai, India

Don Chamberlin IBM Almaden Research Center, San Jose, CA, USA

Allen Chan IBM Toronto Software Lab, Markham, ON, Canada

Chee-Yong Chan National University of Singapore, Singapore, Singapore

K. Mani Chandy California Institute of Technology, Pasadena, CA, USA

Edward Y. Chang Google Research, Mountain View, CA, USA

Kevin Chang Department of Computer Science, University of Illinois at Urbana-Champaign, Urbana, IL, USA

Adriane Chapman University of Southampton, Southampton, UK

Surajit Chaudhuri Microsoft Research, Microsoft Corporation, Redmond, WA, USA

Elizabeth S. Chen Partners HealthCare System, Boston, MA, USA

James L. Chen University of Chicago, Chicago, IL, USA

Jin Chen Computer Engineering Research Group, University of Toronto, Toronto, ON, Canada

Jinjun Chen Swinburne University of Technology, Melbourne, VIC, Australia

Jinchuan Chen Key Laboratory of Data Engineering and Knowledge Engineering, Ministry of Education, Renmin University of China, Beijing

Lei Chen Hong Kong University of Science and Technology, Hong Kong, China

Peter P. Chen Louisiana State University, Baton Rouge, LA, USA

James Cheney University of Edinburgh, Edinburgh, UK

Hong Cheng Department of Systems Engineering and Engineering Management, The Chinese University of Hong Kong, Hong Kong, China

Reynold Cheng Computer Science, The University of Hong Kong, Hong Kong, China

Vivying S. Y. Cheng Hong Kong University of Science and Technology, Hong Kong, China

InduShobha N. Chengalur-Smith University at Albany – SUNY, Albany, NY, USA

Mitch Cherniack Brandeis University, Wattham, MA, USA

Yun Chi NEC Laboratories America, Cupertino, CA, USA

Fernando Chirigati NYU Tandon School of Engineering, Brooklyn, NY, USA

Rada Chirkova North Carolina State University, Raleigh, NC, USA

Laura Chiticariu Scalable Natural Language Processing, IBM Research – Almaden, San Jose, CA, USA

Jan Chomicki Department of Computer Science and Engineering, State University of New York at Buffalo, Buffalo, NY, USA

Fred Chong Computer Science, University of Chicago, Chicago, IL, USA

Stephanie Chow University of Ontario Institute of Technology, Oshawa, ON, Canada

Peter Christen Research School of Computer Science, The Australian National University, Canberra, Australia

Vassilis Christophides INRIA Paris-Roquencourt, Paris, France

Panos K. Chrysanthis Department of Computer Science, University of Pittsburgh, Pittsburgh, PA, USA

Paolo Ciaccia Computer Science and Engineering, University of Bologna, Bologna, Italy

John Cieslewicz Google Inc., Mountain View, CA, USA

Gianluigi Ciocca University of Milano-Bicocca, Milan, Italy

Eugene Clark Harvard Medical School – Partners Healthcare Center for Genetics and Genomics, Boston, MA, USA

Charles L. A. Clarke University of Waterloo, Waterloo, ON, Canada

William R. Claycomb CERT Insider Threat Center, Software Engineering Institute, Carnegie Mellon University, Pittsburgh, PA, USA

Eliseo Clementini University of L'Aguila, L'Aguila, Italy

Chris Clifton Department of Computer Science, Purdue University, West Lafayette, IN, USA

Edith Cohen AT&T Labs-Research, Florham Park, NJ, USA

Sara Cohen The Rachel and Selim Benin School of Computer Science and Engineering, The Hebrew University of Jerusalem, Jerusalem, Israel

Sarah Cohen-Boulakia University Paris-Sud, Orsay Cedex, France

Carlo Combi Department of Computer Science, University of Verona, Verona, VR, Italy

Mariano P. Consens University of Toronto, Toronto, ON, Canada

Dianne Cook Iowa State University, Ames, IA, USA

Graham Cormode Computer Science, University of Warwick, Warwick, UK

Antonio Corral University of Almeria, Almeria, Spain

Maria Francesca Costabile Department of Computer Science, University of Bari, Bari, Italy

Nick Craswell Microsoft Research Cambridge, Cambridge, UK

Fabio Crestani University of Lugano, Lugano, Switzerland

Marco Antonio Cristo FUCAPI, Manaus, Brazil

Maxime Crochemore King's College London, London, UK

Université Paris-Est, Paris, France

Andrew Crotty Database Group, Brown University, Providence, RI, USA

Matthew G. Crowson University of Chicago, Chicago, IL, USA

Michel Crucianu Conservatoire National des Arts et Métiers, Paris, France

Philippe Cudré-Mauroux Massachusetts Institute of Technology, Cambridge, MA, USA

Sonia Leila Da Silva Cerveteri, Italy

Peter Dadam University of Ulm, Ulm, Germany

Mehmet M. Dalkiliç Indiana University, Bloomington, IN, USA

Nilesh Dalvi Airbnb, San Francisco, CA, USA

Marina Danilevsky IBM Almaden Research Center, San Jose, CA, USA

Minh Dao-Tran Institute of Information Systems, Vienna University of Technology, Vienna, Austria

Gautam Das Department of Computer Science and Engineering, University of Texas at Arlington, Arlington, TX, USA

Mahashweta Das Visa Research, Palo Alto, CA, USA

Sudipto Das Microsoft Research, Redmond, WA, USA

Manoranjan Dash Nanyang Technological University, Singapore, Singapore

Anupam Datta Computer Science Department and Electrical and Computer Engineering Department, Carnegie Mellon University, Pittsburgh, PA, USA

Anwitaman Datta Nanyang Technological University, Singapore, Singapore

Ian Davidson University of California-Davis, Davis, CA, USA

Susan B. Davidson Department of Computer and Information Science, University of Pennsylvania, Philadelphia, PA, USA

Todd Davis Department of Computer Science and Software Engineering, Concordia University, Montreal, QC, Canada

Maria De Marsico Sapienza University of Rome, Rome, Italy

Edleno Silva De Moura Federal University of Amazonas, Manaus, Brazil

Antonios Deligiannakis University of Athens, Athens, Greece

Alex Delis University of Athens, Athens, Greece

Alan Demers Cornell University, Ithaca, NY, USA

Jennifer Dempsey University of Arizona, Tucson, AZ, USA

Raytheon Missile Systems, Tucson, AZ, USA

Ke Deng University of Queensland, Brisbane, QLD, Australia

Amol Deshpande University of Maryland, College Park, MD, USA

Zoran Despotovic NTT DoCoMo Communications Laboratories Europe, Munich, Germany

Alin Deutsch University of California-San Diego, La Jolla, CA, USA

Yanlei Diao University of Massachusetts Amherst, Amherst, MA, USA

Suzanne W. Dietrich Arizona State University, Phoenix, AZ, USA

Nevenka Dimitrova Philips Research, Briarcliff Manor, New York, USA

Bolin Ding University of Illinois at Urbana-Champaign, Urbana, IL, USA

Chris Ding University of Texas at Arlington, Arlington, TX, USA

Alan Dix Lancaster University, Lancaster, UK

Belayadi Djahida National High School for Computer Science (ESI), Algiers, Algeria

Hong-Hai Do SAP AG, Dresden, Germany

Gillian Dobbie University of Auckland, Auckland, New Zealand

Alin Dobra University of Florida, Gainesville, FL, USA

Vlastislav Dohnal Masaryk University, Brno, Czech Republic

Mario Döller University of Applied Science Kufstein, Kufstein, Austria

Carlotta Domeniconi George Mason University, Fairfax, VA, USA

Josep Domingo-Ferrer Universitat Rovira i Virgili, Tarragona, Catalonia, Spain

Guozhu Dong Wright State University, Dayton, OH, USA

Xin Luna Dong Amazon, Seattle, WA, USA

Chitra Dorai IBM T. J. Watson Research Center, Hawthorne, NY, USA

Zhicheng Dou Nankai University, Tianjin, China

Ahlame Douzal CNRS, Univ. Grenoble Alps, Grenoble, France

Yang Du Northeastern University, Boston, MA, USA

Susan Dumais Microsoft Research, Redmond, WA, USA

Marlon Dumas University of Tartu, Tartu, Estonia

Schahram Dustdar Technical University of Vienna, Vienna, Austria

Curtis E. Dyreson Utah State University, Logan, UT, USA

Johann Eder Department of Informatics-Systems, Alpen-Adria-Universität Klagenfurt, Klagenfurt, Austria

Milad Eftekhar University of Toronto, Toronto, ON, Canada

Thomas Eiter Institute of Information Systems, Vienna University of Technology, Vienna, Austria

Ibrahim Abu El-Khair Information Science Department, School of Social Sciences, Umm Al-Qura University, Mecca, Saudi Arabia

Ahmed K. Elmagarmid Purdue University, West Lafayette, IN, USA

Qatar Computing Research Institute, HBKU, Doha, Qatar

Ramez Elmasri Computer Science, The University of Texas at Arlington, Arlington, TX, USA

Aaron J. Elmore Department of Computer Science, University of Chicago, Chicago, IL, USA

Sameh Elnikety Microsoft Research, Redmond, WA, USA

David W. Embley Brigham Young University, Provo, UT, USA

Vincent Englebert University of Namur, Namur, Belgium

AnnMarie Ericsson University of Skövde, Skövde, Sweden

Martin Ester Simon Fraser University, Burnaby, BC, Canada

Opher Etzion IBM Software Group, IBM Haifa Labs, Haifa University Campus, Haifa, Israel

Patrick Eugster Purdue University, West Lafayette, IN, USA

Ronald Fagin IBM Almaden Research Center, San Jose, CA, USA

Ju Fan DEKE Lab and School of Information, Renmin University of China, Beijing, China

Wei Fan IBM T.J. Watson Research, Hawthorne, NY, USA

Wenfei Fan University of Edinburgh, Edinburgh, UK

Beihang University, Beijing, China

Hui Fang University of Delaware, Newark, DE, USA

Alan Fekete University of Sydney, Sydney, NSW, Australia

Jean-Daniel Fekete INRIA, LRI University Paris Sud, Orsay Cedex, France

Pascal Felber University of Neuchatel, Neuchatel, Switzerland

Paolino Di Felice University of L'Aguila, L'Aguila, Italy

Hakan Ferhatosmanoglu The Ohio State University, Columbus, OH, USA

Eduardo B. Fernandez Florida Atlantic University, Boca Raton, FL, USA

Eduardo Fernández-Medina GSyA Research Group, Department of Information Technologies and Systems, Institute of Information Technologies and Systems, Escuela Superior de Informática, University of Castilla-La Mancha, Ciudad Real, Spain

Paolo Ferragina Department of Computer Science, University of Pisa, Pisa, Italy

Elena Ferrari DiSTA, University of Insubria, Varese, Italy

Dennis Fetterly Google, Inc., Mountain View, CA, USA

Stephen E. Fienberg Carnegie Mellon University, Pittsburgh, PA, USA

Michael Fink Institute of Information Systems, Vienna University of Technology, Vienna, Austria

Peter M. Fischer Computer Science Department, University of Freiburg, Freiburg, Germany

Simone Fischer-Hübner Karlstad University, Karlstad, Sweden

Fabian Flöck GESIS – Leibniz Institute for the Social Sciences, Köln, Germany

Avrilia Floratou Microsoft, Sunnyvale, CA, USA

Leila De Floriani University of Genova, Genoa, Italy

Christian Fluhr CEA LIST, Fontenay-aux, Roses, France

Greg Flurry IBM SOA Advanced Technology, Armonk, NY, USA

Edward A. Fox Virginia Tech, Blacksburg, VA, USA

Chiara Francalanci Politecnico di Milano University, Milan, Italy

Andrew U. Frank Vienna University of Technology, Vienna, Austria

Michael J. Franklin University of California-Berkeley, Berkeley, CA, USA

Keir Fraser University of Cambridge, Cambridge, UK

Juliana Freire NYU Tandon School of Engineering, Brooklyn, NY, USA

NYU Center for Data Science, New York, NY, USA

New York University, New York, NY, USA

Elias Frentzos University of Piraeus, Piraeus, Greece

Johann-Christoph Freytag Humboldt University of Berlin, Berlin, Germany

Ophir Frieder Georgetown University, Washington, DC, USA

Oliver Frölich Lixto Software GmbH, Vienna, Austria

Ada Wai-Chee Fu Chinese University of Hong Kong, Hong Kong, China

Xiang Fu University of Southern California, Los Angeles, CA, USA

Kazuhisa Fujimoto Hitachi Ltd., Tokyo, Japan

Tim Furche University of Munich, Munich, Germany

Ariel Fuxman Microsoft Research, Mountain View, CA, USA

Silvia Gabrielli Bruno Kessler Foundation, Trento, Italy

Isabella Gagliardi National Research Council (CNR), Milan, Italy

Avigdor Gal Faculty of Industrial Engineering and Management, Technion–Israel Institute of Technology, Haifa, Israel

Alex Galakatos Database Group, Brown University, Providence, RI, USA

Department of Computer Science, Brown University, Providence, RI, USA

Wojciech Galuba EPFL, Lausanne, Switzerland

Johann Gamper Free University of Bozen-Bolzano, Bolzano, Italy

Weihao Gan University of Southern California, Los Angeles, CA, USA

Vijay Gandhi University of Minnesota, Minneapolis, MN, USA

Venkatesh Ganti Microsoft Research, Microsoft Corporation, Redmond, WA, USA

Dengfeng Gao IBM Silicon Valley Lab, San Jose, CA, USA

Like Gao Teradata Corporation, San Diego, CA, USA

Wei Gao Qatar Computing Research Institute, Doha, Qatar

Minos Garofalakis Technical University of Crete, Chania, Greece

Wolfgang Gatterbauer University of Washington, Seattle, WA, USA

Buğra Gedik Department of Computer Engineering, Bilkent University, Ankara, Turkey

IBM T.J. Watson Research Center, Hawthorne, NY, USA

Floris Geerts University of Antwerp, Antwerp, Belgium

Johannes Gehrke Cornell University, Ithaca, NY, USA

Betsy George Oracle (America), Nashua, NH, USA

Lawrence Gerstley PSMI Consulting, San Francisco, CA, USA

Michael Gertz Heidelberg University, Heidelberg, Germany

Giorgio Ghelli Dipartimento di Informatica, Università di Pisa, Pisa, Italy

Gabriel Ghinita National University of Singapore, Singapore, Singapore

Giuseppe De Giacomo Dip. di Ingegneria Informatica Automatica e Gestionale Antonio Ruberti, Sapienza Università di Roma, Rome, Italy

Phillip B. Gibbons Computer Science Department and the Electrical and Computer Engineering Department, Carnegie Mellon University, Pittsburgh, PA, USA

Sarunas Girdzijauskas EPFL, Lausanne, Switzerland

Fausto Giunchiglia University of Trento, Trento, Italy

Kazuo Goda The University of Tokyo, Tokyo, Japan

Max Goebel Vienna University of Technology, Vienna, Austria

Bart Goethals University of Antwerp, Antwerp, Belgium

Martin Gogolla University of Bremen, Bremen, Germany

Aniruddha Gokhale Vanderbilt University, Nashville, TN, USA

Lukasz Golab University of Waterloo, Waterloo, ON, Canada

Matteo Golfarelli DISI – University of Bologna, Bologna, Italy

Arturo González-Ferrer Innovation Unit, Instituto de Investigación Sanitaria del Hospital Clínico San Carlos (IdISSC), Madrid, Spain

Michael F. Goodchild University of California-Santa Barbara, Santa Barbara, CA, USA

Georg Gottlob Computing Laboratory, Oxford University, Oxford, UK

Valerie Gouet-Brunet CNAM Paris, Paris, France

Ramesh Govindan University of Southern California, Los Angeles, CA, USA

Tyrone Gradison Proficiency Labs, Ashland, OR, USA

Goetz Graefe Google, Inc., Mountain View, CA, USA

Gösta Grahne Concordia University, Montreal, QC, Canada

Fabio Grandi Alma Mater Studiorum Università di Bologna, Bologna, Italy

Tyrone Grandison Proficiency Labs, Ashland, OR, USA

Peter M. D. Gray University of Aberdeen, Aberdeen, UK

Todd J. Green University of Pennsylvania, Philadelphia, PA, USA

Georges Grinstein University of Massachusetts, Lowell, MA, USA

Tom Gruber RealTravel, Emerald Hills, CA, USA

Le Gruenwald School of Computer Science, University of Oklahoma, Norman, OK, USA

Torsten Grust University of Tübingen, Tübingen, Germany

Dirk Van Gucht Indiana University, Bloomington, IN, USA

Carlos Guestrin Carnegie Mellon University, Pittsburgh, PA, USA

Dimitrios Gunopulos Department of Computer Science and Engineering, The University of California at Riverside, Bourns College of Engineering, Riverside, CA, USA

Amarnath Gupta San Diego Supercomputer Center, University of California San Diego, La Jolla, CA, USA

Himanshu Gupta Stony Brook University, Stony Brook, NY, USA

Cathal Gurrin Dublin City University, Dublin, Ireland

Ralf Hartmut Güting Fakultät für Mathematik und Informatik, Fernuniversität Hagen, Hagen, Germany

Computer Science, University of Hagen, Hagen, Germany

Marc Gyssens Hasselt University, Hasselt, Belgium

Peter J. Haas IBM Almaden Research Center, San Jose, CA, USA

Karl Hahn BMW AG, Munich, Germany

Jean-Luc Hainaut University of Namur, Namur, Belgium

Alon Halevy The Recruit Institute of Technology, Mountain View, CA, USA

Google Inc., Mountain View, CA, USA

Maria Halkidi University of Piraeus, Piraeus, Greece

Terry Halpin Neumont University, South Jordan, UT, USA

Jiawei Han University of Illinois at Urbana-Champaign, Urbana, IL, USA

Alan Hanjalic Delft University of Technology, Delft, The Netherlands

David Hansen The Australian e-Health Research Centre, Brisbane, QLD, Australia

Jörgen Hansson University of Skövde, Skövde, Sweden

Nikos Hardavellas Carnegie Mellon University, Pittsburgh, PA, USA

Theo Härder University of Kaiserslautern, Kaiserslautern, Germany

David Harel The Weizmann Institute of Science, Rehovot, Israel

Jayant R. Haritsa Indian Institute of Science, Bangalore, India

Stavros Harizopoulos HP Labs, Palo Alto, CA, USA

Per F. V. Hasle Royal School of Library and Information Science, University of Copenhagen, Copenhagen S, Denmark

Jordan T. Hastings Department of Geography, University of California-Santa Barbara, Santa Barbara, CA, USA

Alexander Hauptmann Carnegie Mellon University, Pittsburgh, PA, USA

Helwig Hauser University of Bergen, Bergen, Norway

Manfred Hauswirth Open Distributed Systems, Technical University of Berlin, Berlin, Germany

Fraunhofer FOKUS, Galway, Germany

Ben He University of Glasgow, Glasgow, UK

Thomas Heinis Imperial College London, London, UK

Pat Helland Microsoft Corporation, Redmond, WA, USA

Joseph M. Hellerstein University of California-Berkeley, Berkeley, CA, USA

Jean Henrard University of Namur, Namur, Belgium

John Herring Oracle USA Inc, Nashua, NH, USA

Nicolas Hervé INRIA Paris-Rocquencourt, Le Chesnay, France

Marcus Herzog Vienna University of Technology, Vienna, Austria

Jean-Marc Hick University of Namur, Namur, Belgium

Jan Hidders University of Antwerp, Antwerpen, Belgium

Djoerd Hiemstra University of Twente, Enschede, The Netherlands

Linda L. Hill University of California-Santa Barbara, Santa Barbara, CA, USA

Alexander Hinneburg Institute of Computer Science, Martin-Luther-University Halle-Wittenberg, Halle/Saale, Germany

Hans Hinterberger Department of Computer Science, ETH Zurich, Zurich, Switzerland

Howard Ho IBM Almaden Research Center, San Jose, CA, USA

Erik Hoel Environmental Systems Research Institute, Redlands, CA, USA

Vasant Honavar Iowa State University, Ames, IA, USA

Mingsheng Hong Cornell University, Ithaca, NY, USA

Katja Hose Department of Computer Science, Aalborg University, Aalborg, Denmark

Haruo Hosoya The University of Tokyo, Tokyo, Japan

Vagelis Hristidis Department of Computer Science and Engineering, University of California, Riverside, Riverside, CA, USA

Wynne Hsu National University of Singapore, Singapore, Singapore

Yu-Ling Hsueh Computer Science and Information Engineering Department, National Chung Cheng University, Taiwan, Republic of China

Jian Hu Microsoft Research Asia, Haidian, China

Kien A. Hua University of Central Florida, Orlando, FL, USA

Xian-Sheng Hua Microsoft Research Asia, Beijing, China

Jun Huan University of Kansas, Lawrence, KS, USA

Haoda Huang Microsoft Research Asia, Beijing, China

Michael Huggett University of British Columbia, Vancouver, BC, Canada

Patrick C. K. Hung University of Ontario Institute of Technology, Oshawa, ON, Canada

Jeong-Hyon Hwang Department of Computer Science, University at Albany – State University of New York, Albany, NY, USA

Noha Ibrahim Grenoble Informatics Laboratory (LIG), Grenoble, France

Ichiro Ide Graduate School of Informatics, Nagoya University, Nagoya, Aichi, Japan

Sergio Ilarri University of Zaragoza, Zaragoza, Spain

Ihab F. Ilyas Cheriton School of Computer Science, University of Waterloo, Waterloo, ON, Canada

Alfred Inselberg Tel Aviv University, Tel Aviv, Israel

Yannis Ioannidis University of Athens, Athens, Greece

Ekaterini Ioannou Faculty of Pure and Applied Sciences, Open University of Cyprus, Nicosia, Cyprus

Panagiotis G. Ipeirotis New York University, New York, NY, USA

Zachary G. Ives Computer and Information Science Department, University of Pennsylvania, Philadelphia, PA, USA

Hans-Arno Jacobsen Department of Electrical and Computer Engineering, University of Toronto, Toronto, ON, Canada

H. V. Jagadish University of Michigan, Ann Arbor, MI, USA

Alejandro Jaimes Telefonica R&D, Madrid, Spain

Ramesh Jain University of California, Irvine, CA, USA

Sushil Jajodia George Mason University, Fairfax, VA, USA

Greg Janée University of California-Santa Barbara, Santa Barbara, CA, USA

Kalervo Järvelin University of Tampere, Tampere, Finland

Christian S. Jensen Department of Computer Science, Aalborg University, Aalborg, Denmark

Eric C. Jensen Twitter, Inc., San Francisco, CA, USA

Manfred Jeusfeld IIT, University of Skövde, Skövde, Sweden

Aura Frames, New York City, NY, USA

Heng Ji New York University, New York, NY, USA

Zhe Jiang University of Alabama, Tuscaloosa, AL, USA

Ricardo Jiménez-Peris Distributed Systems Lab, Universidad Politecnica de Madrid, Madrid, Spain

Hai Jin Service Computing Technology and System Lab, Cluster and Grid Computing Lab, School of Computer Science and Technology, Huazhong University of Science and Technology, Wuhan, China

Jiashun Jin Carnegie Mellon University, Pittsburgh, PA, USA

Ruoming Jin Department of Computer Science, Kent State University, Kent, OH, USA

Ryan Johnson Carnegie Mellon University, Pittsburg, PA, USA

Theodore Johnson AT&T Labs – Research, Florham Park, NJ, USA

Christopher B. Jones Cardiff University, Cardiff, UK

Rosie Jones Yahoo! Research, Burbank, CA, USA

James B. D. Joshi University of Pittsburgh, Pittsburgh, PA, USA

Vanja Josifovski Uppsala University, Uppsala, Sweden

Marko Junkkari University of Tampere, Tampere, Finland

Jan Jurjens The Open University, Buckinghamshire, UK

Mouna Kacimi Max-Planck Institute for Informatics, Saarbrücken, Germany

Tamer Kahveci University of Florida, Gainesville, FL, USA

Panos Kalnis National University of Singapore, Singapore, Singapore

Jaap Kamps University of Amsterdam, Amsterdam, The Netherlands

James Kang University of Minnesota, Minneapolis, MN, USA

Carl-Christian Kanne University of Mannheim, Mannheim, Germany

Aman Kansal Microsoft Research, Redmond, WA, USA

Murat Kantarcıoğlu University of Texas at Dallas, Richardson, TX, USA

Ben Kao Department of Computer Science, The University of Hong Kong, Hong Kong, China

George Karabatis University of Maryland, Baltimore Country (UMBC), Baltimore, MD, USA

Grigoris Karvounarakis LogicBlox, Atlanta, GA, USA

George Karypis University of Minnesota, Minneapolis, MN, USA

Vipul Kashyap Clinical Programs, CIGNA Healthcare, Bloomfield, CT, USA

Yannis Katsis University of California-San Diego, La Jolla, CA, USA

Raghav Kaushik Microsoft Research, Redmond, WA, USA

Gabriella Kazai Microsoft Research Cambridge, Cambridge, UK

Daniel A. Keim Computer Science Department, University of Konstanz, Konstanz, Germany

Jaana Kekäläinen University of Tampere, Tampere, Finland

Anastasios Kementsietsidis IBM T.J. Watson Research Center, Hawthorne, NY, USA

Bettina Kemme School of Computer Science, McGill University, Montreal, QC, Canada

Jessie Kennedy Napier University, Edinburgh, UK

Vijay Khatri Operations and Decision Technologies Department, Kelley School of Business, Indiana University, Bloomington, IN, USA

Ashfaq Khokhar University of Illinois at Chicago, Chicago, IL, USA

Daniel Kifer Yahoo! Research, Santa Clara, CA, USA

Stephen Kimani Director ICSIT, Jomo Kenyatta University of Agriculture and Technology (JKUAT), Juja, Kenya

Sofia Kleisarchaki CNRS, Univ. Grenoble Alps, Grenoble, France

Craig A. Knoblock University of Southern California, Marina del Rey, Los Angeles, CA, USA

Christoph Koch Cornell University, Ithaca, New York, NY, USA

EPFL, Lausanne, Switzerland

Solmaz Kolahi University of British Columbia, Vancouver, BC, Canada

George Kollios Boston University, Boston, MA, USA

Christian Koncilia Institute of Informatics-Systems, University of Klagenfurt, Klagenfurt, Austria

Roberto Konow Department of Computer Science, University of Chile, Santiago, Chile

Marijn Koolen Research and Development, Huygens ING, Royal Netherlands Academy of Arts and Sciences, Amsterdam, The Netherlands

David Koop University of Massachusetts Dartmouth, Dartmouth, MA, USA

Poon Wei Koot Nanyang Technological University, Singapore, Singapore

Julius Köpke Department of Informatics-Systems, Alpen-Adria-Universität Klagenfurt, Klagenfurt, Austria

Flip R. Korn AT&T Labs–Research, Florham Park, NJ, USA

Harald Kosch University of Passau, Passau, Germany

Cartik R. Kothari Biomedical Informatics, Ohio State University, College of Medicine, Columbus, OH, USA

Yannis Kotidis Department of Informatics, Athens University of Economics and Business, Athens, Greece

Spyros Kotoulas VU University Amsterdam, Amsterdam, The Netherlands

Manolis Koubarakis University of Athens, Athens, Greece

Konstantinos Koutroumbas Institute for Space Applications and Remote Sensing, Athens, Greece

Bernd J. Krämer University of Hagen, Hagen, Germany

Tim Kraska Department of Computer Science, Brown University, Providence, RI, USA

Werner Kriechbaum IBM Development Lab, Böblingen, Germany

Hans-Peter Kriegel Ludwig-Maximilians-University, Munich, Germany

Chandra Krintz Department of Computer Science, University of California, Santa Barbara, CA, USA

Rajasekar Krishnamurthy IBM Almaden Research Center, San Jose, CA, USA

Peer Kröger Ludwig-Maximilians-Universität München, Munich, Germany

Thomas Kühne School of Engineering and Computer Science, Victoria University of Wellington, Wellington, New Zealand

Krishna Kulkarni Independent Consultant, San Jose, CA, USA

Ravi Kumar Yahoo Research, Santa Clara, CA, USA

Nicholas Kushmerick VMWare, Seattle, WA, USA

Alan G. Labouseur School of Computer Science and Mathematics, Marist College, Poughkeepsie, NY, USA

Alexandros Labrinidis Department of Computer Science, University of Pittsburgh, Pittsburgh, PA, USA

Zoé Lacroix Arizona State University, Tempe, AZ, USA

Alberto H. F. Laender Federal University of Minas Gerais, Belo Horizonte, Brazil

Bibudh Lahiri Iowa State University, Ames, IA, USA

Laks V. S. Lakshmanan University of British Columbia, Vancouver, BC, Canada

Mounia Lalmas Yahoo! Inc., London, UK

Lea Landucci University of Florence, Florence, Italy

Birger Larsen Royal School of Library and Information Science, Copenhagen, Denmark

Mary Lynette Larsgaard University of California-Santa Barbara, Santa Barbara, CA, USA

Per-Åke Larson Microsoft Corporation, Redmond, WA, USA

Robert Laurini INSA-Lyon, University of Lyon, Lyon, France

LIRIS, INSA-Lyon, Lyon, France

Georg Lausen University of Freiburg, Freiburg, Germany

Jens Lechtenbörger University of Münster, Münster, Germany

Thierry Lecroq Université de Rouen, Rouen, France

Dongwon Lee The Pennsylvania State University, Park, PA, USA

Victor E. Lee John Carroll University, University Heights, OH, USA

Yang W. Lee College of Business Administration, Northeastern University, Boston, MA, USA

Pieter De Leenheer Vrije Universiteit Brussel, Collibra NV, Brussels, Belgium

Wolfgang Lehner Dresden University of Technology, Dresden, Germany

Domenico Lembo Dip. di Ingegneria Informatica Automatica e Gestionale Antonio Ruberti, Sapienza Università di Roma, Rome, Italy

Ronny Lempel Yahoo! Research, Haifa, Israel

Maurizio Lenzerini Dip. di Ingegneria Informatica Automatica e Gestionale Antonio Ruberti, Sapienza Università di Roma, Rome, Italy

Kristina Lerman University of Southern California, Marina del Rey, Los Angeles, CA, USA

Ulf Leser Humboldt University of Berlin, Berlin, Germany

Carson Kai-Sang Leung Department of Computer Science, University of Manitoba, Winnipeg, MB, Canada

Mariano Leva Dipartimento di Ingegneria Informatica, Automatica e Gestionale "A.Ruberti", Sapienza – Università di Roma, Roma, Italy

Stefano Levialdi Sapienza University of Rome, Rome, Italy

Brian Levine University of Massachusetts, Amherst, MA, USA

Changqing Li Duke University, Durham, NC, USA

Chen Li University of California – Irvine, School of Information and Computer Sciences, Irvine, CA, USA

Chengkai Li University of Texas at Arlington, Arlington, TX, USA

Hua Li Microsoft Research Asia, Beijing, China

Jinyan Li Nanyang Technological University, Singapore, Singapore

Ninghui Li Purdue University, West Lafayette, IN, USA

Ping Li Cornell University, Ithaca, NY, USA

Qing Li City University of Hong Kong, Hong Kong, China

Xue Li The University of Queensland, Brisbane, QLD, Australia

Yunyao Li IBM Almaden Research Center, San Jose, CA, USA

Ying Li Cognitive People Solutions, IBM Human Resources, Armonk, NY, USA

Xiang Lian Department of Computer Science, Kent State University, Kent, OH, USA

Leonid Libkin School of Informatics, University of Edinburgh, Edinburgh, Scotland, UK

Sam S. Lightstone IBM Canada Ltd, Markham, ON, Canada

Jimmy Lin University of Maryland, College Park, MD, USA

Tsau Young Lin Department of Computer Science, San Jose State University, San Jose, CA, USA

Xuemin Lin University of New South Wales, Sydney, NSW, Australia

Tok Wang Ling National University of Singapore, Singapore, Singapore

Bing Liu University of Illinois at Chicago, Chicago, IL, USA

Danzhou Liu University of Central Florida, Orlando, FL, USA

Guimei Liu Institute for Infocomm Research, Singapore, Singapore

Huan Liu Data Mining and Machine Learning Lab, School of Computing, Informatics, and Decision Systems Engineering, Arizona State University, Tempe, AZ, USA

Jinze Liu University of Kentucky, Lexington, KY, USA

Lin Liu Department of Computer Science, Kent State University, Kent, OH, USA

Ning Liu Microsoft Research Asia, Beijing, China

Qing Liu CSIRO, Hobart, TAS, Australia

Xiangyu Liu Xiamen University, Xiamen, China

Vebjorn Ljosa Broad Institute of MIT and Harvard, Cambridge, MA, USA

David Lomet Microsoft Research, Redmond, WA, USA

Cheng Long School of Electronics, Electrical Engineering and Computer Science, Queen's University Belfast, Kowloon, Hong Kong

Boon Thau Loo ETH Zurich, Zurich, Switzerland

Phillip Lord Newcastle University, Newcastle-Upon-Tyne, UK

Nikos A. Lorentzos Informatics Laboratory, Department of Agricultural Economics and Rural Development, Agricultural University of Athens, Athens, Greece

Lie Lu Microsoft Research Asia, Beijing, China

Bertram Ludäscher University of California-Davis, Davis, CA, USA

Yan Luo University of Illinois at Chicago, Chicago, IL, USA

Yves A. Lussier University of Chicago, Chicago, IL, USA

Ioanna Lykourentzou CRP Henri Tudor, Esch-sur-Alzette, Luxembourg

Craig MacDonald University of Glasgow, Glasgow, UK

Ashwin Machanavajjhala Cornell University, Ithaca, NY, USA

Samuel Madden Massachusetts Institute of Technology, Cambridge, MA, USA

Paola Magillo University of Genova, Genoa, Italy

Ahmed R. Mahmood Computer Science, Purdue University, West Lafayette, IN, USA

David Maier Portland State University, Portland, OR, USA

Ratul kr. Majumdar Department of Computer Science and Engineering, Indian Institute of Technology Bombay, Mumbai, India

Jan Małuszyński Linköping University, Linköping, Sweden

Nikos Mamoulis University of Hong Kong, Hong Kong, China

Stefan Manegold CWI, Amsterdam, The Netherlands

Murali Mani Worcester Polytechnic, Worcester, MA, USA

Serge Mankovski CA Labs, CA Inc., Thornhill, ON, Canada

Ioana Manolescu INRIA Saclay–Île de France, Orsay, France

Yannis Manolopoulos Aristotle University of Thessaloniki, Thessaloniki, Greece

Florian Mansmann University of Konstanz, Konstanz, Germany

Svetlana Mansmann University of Konstanz, Konstanz, Germany

Shahar Maoz The Weizmann Institute of Science, Rehovot, Israel

Patrick Marcel Département Informatique, Laboratoire d'Informatique, Université François Rabelais Tours, Blois, France

Amélie Marian Computer Science Department, Rutgers University, New Brunswick, NJ, USA

Volker Markl IBM Almaden Research Center, San Jose, CA, USA

David Martin Nuance Communications, Sunnyvale, CA, USA

Maria Vanina Martinez University of Maryland, College Park, MD, USA

Maristella Matera Politecnico di Milano, Milan, Italy

Michael Mathioudakis Université de Lyon, CNRS, INSA-Lyon, LIRIS, UMR5205, F-69621, France

Marta Mattoso Federal University of Rio de Janeiro, Rio de Janeiro, Brazil

Andrea Maurino University of Milano-Bicocca, Milan, Italy

Jose-Norberto Mazón University of Alicante, Alicante, Spain

John McCloud CERT Insider Threat Center, Software Engineering Institute, Carnegie Mellon University, Pittsburgh, PA, USA

Kevin S. McCurley Google Research, Mountain View, CA, USA

Andrew McGregor Microsoft Research, Silicon Valley, Mountain View, CA, USA

Timothy McPhillips University of California-Davis, Davis, CA, USA

Massimo Mecella Dipartimento di Ingegneria Informatica, Automatica e Gestionale "A.Ruberti", Sapienza – Università di Roma, Roma, Italy

Brahim Medjahed The University of Michigan–Dearborn, Dearborn, MI, USA

Carlo Meghini The Italian National Research Council, Pisa, Italy

Tao Mei Microsoft Research Asia, Beijing, China

Jonas Mellin University of Skövde, The Informatics Research Centre, Skövde, Sweden

University of Skövde, School of Informatics, Skövde, Sweden

Massimo Melucci University of Padua, Padua, Italy

Niccolò Meneghetti Computer Science and Engineering Department, University at Buffalo, Buffalo, NY, USA

Weiyi Meng Department of Computer Science, State University of New York at Binghamton, Binghamton, NY, USA

Ahmed Metwally LinkedIn Corp., Mountain View, CA, USA

Jan Michels Oracle Corporation, Redwood Shores, CA, USA

Gerome Miklau University of Massachusetts, Amherst, MA, USA

Alessandra Mileo Insight Centre for Data Analytics, Dublin City University, Dublin, Ireland

Harvey J. Miller University of Utah, Salt Lake City, UT, USA

Renée J. Miller Department of Computer Science, University of Toronto, Toronto, ON, Canada

Tova Milo School of Computer Science, Tel Aviv University, Tel Aviv, Israel

Umar Farooq Minhas Microsoft Research, Redmond, WA, USA

Paolo Missier School of Computing Science, Newcastle University, Newcastle upon Tyne, UK

Prasenjit Mitra The Pennsylvania State University, University Park, PA, USA

Michael Mitzenmacher Harvard University, Boston, MA, USA

Mukesh Mohania IBM Research, Melbourne, VIC, Australia

Mohamed F. Mokbel Department of Computer Science and Engineering, University of Minnesota-Twin Cities, Minneapolis, MN, USA

Angelo Montanari University of Udine, Udine, Italy

Reagan W. Moore School of Information and Library Science, University of North Carolina at Chapel Hill, Chapel Hill, NC, USA

Konstantinos Morfonios Oracle, Redwood City, CA, USA

Peter Mork The MITRE Corporation, McLean, VA, USA

Mirella M. Moro Departamento de Ciencia da Computaçao, Universidade Federal de Minas Gerais – UFMG, Belo Horizonte, MG, Brazil

Kyriakos Mouratidis Singapore Management University, Singapore, Singapore

Kamesh Munagala Duke University, Durham, NC, USA

Ethan V. Munson Department of EECS, University of Wisconsin-Milwaukee, Milwaukee, WI, USA

Shawn Murphy Massachusetts General Hospital, Boston, MA, USA

John Mylopoulos Department of Computer Science, University of Toronto, Toronto, ON, Canada

Marta Patiño-Martínez Distributed Systems Lab, Universidad Politecnica de Madrid, Madrid, Spain

ETSI Informáticos, Universidad Politécnica de Madrid (UPM), Madrid, Spain

Frank Nack University of Amsterdam, Amsterdam, The Netherlands

Marc Najork Google, Inc., Mountain View, CA, USA

Ullas Nambiar Zensar Technologies Ltd, Pune, India

Alexandros Nanopoulos Aristotle University, Thessaloniki, Greece

Vivek Narasayya Microsoft Corporation, Redmond, WA, USA

Mario A. Nascimento Department of Computing Science, University of Alberta, Edmonton, AB, Canada

Alan Nash Aleph One LLC, La Jolla, CA, USA

Harald Naumann Vienna University of Technology, Vienna, Austria

Gonzalo Navarro Department of Computer Science, University of Chile, Santiago, Chile

Wolfgang Nejdl L3S Research Center, University of Hannover, Hannover, Germany

Thomas Neumann Max-Planck Institute for Informatics, Saarbrücken, Germany

Bernd Neumayr Department for Business Informatics – Data and Knowledge Engineering, Johannes Kepler University Linz, Linz, Austria

Frank Neven Hasselt University and Transnational University of Limburg, Diepenbeek, Belgium

Chong-Wah Ngo City University of Hong Kong, Hong Kong, China

Peter Niblett IBM United Kingdom Limited, Winchester, UK

Naoko Nitta Osaka University, Osaka, Japan

Igor Nitto Department of Computer Science, University of Pisa, Pisa, Italy

Cheng Niu Microsoft Research Asia, Beijing, China

Vilém Novák Institute for Research and Applications of Fuzzy Modeling, University of Ostrava, Ostrava, Czech Republic

Chimezie Ogbuji Cleveland Clinic Foundation, Cleveland, OH, USA

Peter Øhrstrøm Aalborg University, Aalborg, Denmark

Christine M. O'Keefe CSIRO Preventative Health National Research Flagship, Acton, ACT, Australia

Paul W. Olsen Department of Computer Science, The College of Saint Rose, Albany, NY, USA

Dan Olteanu Department of Computer Science, University of Oxford, Oxford, UK

Behrooz Omidvar-Tehrani Interactive Data Systems Group, Ohio State University, Columbus, OH, USA

Patrick O'Neil University of Massachusetts, Boston, MA, USA

Beng Chin Ooi School of Computing, National University of Singapore, Singapore, Singapore

Iadh Ounis University of Glasgow, Glasgow, UK

Mourad Ouzzani Qatar Computing Research Institute, HBKU, Doha, Qatar

Fatma Özcan IBM Research – Almaden, San Jose, CA, USA

M. Tamer Özsu Cheriton School of Computer Science, University of Waterloo, Waterloo, ON, Canada

Esther Pacitti INRIA and LINA, University of Nantes, Nantes, France

Chris D. Paice Lancaster University, Lancaster, UK

Noël de Palma INPG – INRIA, Grenoble, France

Nathaniel Palmer Workflow Management Coalition, Hingham, MA, USA

Themis Palpanas Paris Descartes University, Paris, France

Biswanath Panda Cornell University, Ithaca, NY, USA

Ippokratis Pandis Carnegie Mellon University, Pittsburgh, PA, USA

Amazon Web Services, Seattle, WA, USA

Dimitris Papadias Department of Computer Science and Engineering, Hong Kong University of Science and Technology, Kowloon, Hong Kong, Hong Kong

Spiros Papadimitriou IBM T.J. Watson Research Center, Hawthorne, NY, USA

Apostolos N. Papadopoulos Aristotle University of Thessaloniki, Thessaloniki, Greece

Yannis Papakonstantinou University of California-San Diego, La Jolla, CA, USA

Jan Paredaens University of Antwerp, Antwerpen, Belgium

Christine Parent University of Lausanne, Lausanne, Switzerland

Josiane Xavier Parreira Siemens AG, Galway, Austria

Gabriella Pasi Department of Informatics, Systems and Communication, University of Milano-Bicocca, Milan, Italy

Chintan Patel Columbia University, New York, NY, USA

Jignesh M. Patel University of Wisconsin-Madison, Madison, WI, USA

Norman W. Paton University of Manchester, Manchester, UK

Cesare Pautasso University of Lugano, Lugano, Switzerland

Torben Bach Pedersen Department of Computer Science, Aalborg University, Aalborg, Denmark

Fernando Pedone Università della Svizzera Italiana (USI), Lugano, Switzerland

Jovan Pehcevski INRIA Paris-Rocquencourt, Le Chesnay Cedex, France

Jian Pei School of Computing Science, Simon Fraser University, Burnaby, BC, Canada

Ronald Peikert ETH Zurich, Zurich, Switzerland

Mor Peleg Department of Information Systems, University of Haifa, Haifa, Israel

Fuchun Peng Yahoo! Inc., Sunnyvale, CA, USA

Peng Peng Alibaba, Yu Hang District, Hangzhou, China

Liam Peyton University of Ottawa, Ottawa, ON, Canada

Dieter Pfoser Department of Geography and Geoinformation Science, George Mason University, Fairfax, VA, USA

Danh Le Phuoc Open Distributed Systems, Technical University of Berlin, Berlin, Germany

Mario Piattini University of Castilla-La Mancha, Ciudad Real, Spain

Benjamin C. Pierce University of Pennsylvania, Philadelphia, PA, USA

Karen Pinel-Sauvagnat IRIT laboratory, University of Toulouse, Toulouse, France

Leo L. Pipino University of Massachusetts, Lowell, MA, USA

Peter Pirolli Palo Alto Research Center, Palo Alto, CA, USA

Evaggelia Pitoura Department of Computer Science and Engineering, University of Ioannina, Ioannina, Greece

Benjamin Piwowarski University of Glasgow, Glasgow, UK

Vassilis Plachouras Yahoo! Research, Barcelona, Spain

Catherine Plaisant University of Maryland, College Park, MD, USA

Claudia Plant University of Vienna, Vienna, Austria

Christian Platzer Technical University of Vienna, Vienna, Austria

Dimitris Plexousakis Foundation for Research and Technology-Hellas (FORTH), Heraklion, Greece

Neoklis Polyzotis University of California Santa Cruz, Santa Cruz, CA, USA

Raymond K. Pon University of California, Los Angeles, CA, USA

Lucian Popa IBM Almaden Research Center, San Jose, CA, USA

Alexandra Poulovassilis University of London, London, UK

Sunil Prabhakar Purdue University, West Lafayette, IN, USA

Cecilia M. Procopiuc AT&T Labs, Florham Park, NJ, USA

Enrico Puppo Department of Informatics, Bioengineering, Robotics and Systems Engineering, University of Genova, Genoa, Italy

Ross S. Purves University of Zurich, Zurich, Switzerland

Vivien Quéma CNRS, INRIA, Saint-Ismier Cedex, France

Christoph Quix RWTH Aachen University, Aachen, Germany

Sriram Raghavan IBM Almaden Research Center, San Jose, CA, USA

Erhard Rahm University of Leipzig, Leipzig, Germany

Habibur Rahman Department of Computer Science and Engineering, University of Texas at Arlington, Arlington, TX, USA

Krithi Ramamritham Department of Computer Science and Engineering, Indian Institute of Technology Bombay, Mumbai, India

Maya Ramanath Max-Planck Institute for Informatics, Saarbrücken, Germany

Georgina Ramírez Yahoo! Research Barcelona, Barcelona, Spain

Edie Rasmussen Library, Archival and Information Studies, The University of British Columbia, Vancouver, BC, Canada

Indrakshi Ray Colorado State University, Fort Collins, CO, USA

Colin R. Reeves Coventry University, Coventry, UK

Payam Refaeilzadeh Google Inc., Los Angeles, CA, USA

D. R. Reforgiato University of Maryland, College Park, MD, USA

Bernd Reiner Technical University of Munich, Munich, Germany

Frederick Reiss IBM Almaden Research Center, San Jose, CA, USA

Harald Reiterer University of Konstanz, Constance, Germany

Matthias Renz Ludwig-Maximilians-Universität München, Munich, Germany

Andreas Reuter Heidelberg Laureate Forum Foundation, Schloss-Wolfsbrunnenweg 33, Heidelberg, Germany

Peter Revesz University of Nebraska-Lincoln, Lincoln, NE, USA

Mirek Riedewald Cornell University, Ithaca, NY, USA

Rami Rifaieh University of California-San Diego, San Diego, CA, USA

Stefanie Rinderle-Ma University of Vienna, Vienna, Austria

Tore Risch Department of Information Technology, Uppsala University, Uppsala, Sweden

Thomas Rist University of Applied Sciences, Augsburg, Germany

Stefano Rizzi DISI, University of Bologna, Bologna, Italy

Stephen Robertson Microsoft Research Cambridge, Cambridge, UK

Roberto A. Rocha Partners eCare, Partners HealthCare System, Wellesley, MA, USA

John F. Roddick Flinders University, Adelaide, SA, Australia

Thomas Roelleke Queen Mary University of London, London, UK

Didier Roland University of Namur, Namur, Belgium

Oscar Romero Polytechnic University of Catalonia, Barcelona, Spain

Rafael Romero University of Alicante, Alicante, Spain

Riccardo Rosati Dip. di Ingegneria Informatica Automatica e Gestionale Antonio Ruberti, Sapienza Università di Roma, Rome, Italy

Timothy Roscoe ETH Zurich, Zurich, Switzerland

Kenneth A. Ross Columbia University, New York, NY, USA

Prasan Roy Sclera, Inc., Walnut, CA, USA

Senjuti Basu Roy Department of Computer Science, New Jersey Institute of Technology, Tacoma, WA, USA

Sudeepa Roy Department of Computer Science, Duke University, Durham, NC, USA

Yong Rui Microsoft China R&D Group, Redmond, WA, USA

Dan Russler Oracle Health Sciences, Redwood Shores, CA, USA

Georgia Tech Research Institute, Atlanta, Georgia, USA

Michael Rys Microsoft Corporation, Sammamish, WA, USA

Giovanni Maria Sacco Dipartimento di Informatica, Università di Torino, Torino, Italy

Tetsuya Sakai Waseda University, Tokyo, Japan

Kenneth Salem University of Waterloo, Waterloo, ON, Canada

Simonas Šaltenis Aalborg University, Aalborg, Denmark

George Samaras University of Cyprus, Nicosia, Cyprus

Giuseppe Santucci University of Rome, Rome, Italy

Maria Luisa Sapino University of Turin, Turin, Italy

Sunita Sarawagi IIT Bombay, Mumbai, India

Anatol Sargin University of Augsburg, Augsburg, Germany

Mohamed Sarwat School of Computing, Informatics, and Decision Systems Engineering, Arizona State University, Tempe, AZ, USA

Kai-Uwe Sattler Technische Universität Ilmenau, Ilmenau, Germany

Monica Scannapieco University of Rome, Rome, Italy

Matthias Schäfer University of Konstanz, Konstanz, Germany

Sebastian Schaffert Salzburg Research, Salzburg, Austria

Ralf Schenkel Campus II Department IV – Computer Science, Professorship for databases and information systems, University of Trier, Trier, Germany

Raimondo Schettini University of Milano-Bicocca, Milan, Italy

Peter Scheuermann Department of ECpE, Iowa State University, Ames, IA, USA

Ulrich Schiel Federal University of Campina Grande, Campina Grande, Brazil

Markus Schneider University of Florida, Gainesville, FL, USA

Marc H. Scholl University of Konstanz, Konstanz, Germany

Michel Scholl Cedric-CNAM, Paris, France

Tobias Schreck Department of Computer Science and Biomedical Engineering, Institute of Computer Graphics and Knowledge Visualization, Graz University of Technology, Graz, Austria

Michael Schrefl University of Linz, Linz, Austria

Erich Schubert Heidelberg University, Heidelberg, Germany

Matthias Schubert Ludwig-Maximilians-University, Munich, Germany

Christoph G. Schuetz Department for Business Informatics – Data and Knowledge Engineering, Johannes Kepler University Linz, Linz, Austria

Heiko Schuldt Department of Mathematics and Computer Science, Databases and Information Systems Research Group, University of Basel, Basel, Switzerland

Heidrun Schumann University of Rostock, Rostock, Germany

Felix Schwagereit University of Koblenz-Landau, Koblenz, Germany

Nicole Schweikardt Johann Wolfgang Goethe-University, Frankfurt am Main, Frankfurt, Germany

Fabrizio Sebastiani Qatar Computing Research Institute, Doha, Qatar

Nicu Sebe University of Amsterdam, Amsterdam, Netherlands

Monica Sebillo University of Salerno, Salerno, Italy

Thomas Seidl RWTH Aachen University, Aachen, Germany

Manuel Serrano University of Alicante, Alicante, Spain

Amnon Shabo (Shvo) University of Haifa, Haifa, Israel

Mehul A. Shah Amazon Web Services (AWS), Seattle, WA, USA

Nigam Shah Stanford University, Stanford, CA, USA

Cyrus Shahabi University of Southern California, Los Angeles, CA, USA

Jayavel Shanmugasundaram Yahoo Research!, Santa Clara, NY, USA

Marc Shapiro Inria Paris, Paris, France

Sorbonne-Universités-UPMC-LIP6, Paris, France

Mohamed Sharaf Electrical and Computer Engineering, University of Toronto, Toronto, ON, Canada

Mehdi Sharifzadeh Google, Santa Monica, CA, USA

Jayant Sharma Oracle USA Inc, Nashua, NH, USA

Guy Sharon IBM Research Labs-Haifa, Haifa, Israel

Dennis Shasha Department of Computer Science, New York University, New York, NY, USA

Shashi Shekhar Department of Computer Science, University of Minnesota, Minneapolis, MN, USA

Jialie Shen Singapore Management University, Singapore, Singapore

Xuehua Shen Google, Inc., Mountain View, CA, USA

Dou Shen Microsoft Corporation, Redmond, WA, USA

Baidu, Inc., Beijing City, China

Heng Tao Shen School of Information Technology and Electrical Engineering, The University of Queensland, Brisbane, QLD, Australia

University of Electronic Science and Technology of China, Chengdu, Sichuan Sheng, China

Rao Shen Yahoo!, Sunnyvale, CA, USA

Frank Y. Shih New Jersey Institute of Technology, Newark, NJ, USA

Arie Shoshani Lawrence Berkeley National Laboratory, Berkeley, CA, USA

Pavel Shvaiko University of Trento, Trento, Italy

Wolf Siberski L3S Research Center, University of Hannover, Hannover, Germany

Ronny Siebes VU University Amsterdam, Amsterdam, The Netherlands

Laurynas Šikšnys Department of Computer Science, Aalborg University, Aalborg, Denmark

Adam Silberstein Yahoo! Research Silicon Valley, Santa Clara, CA, USA

Fabrizio Silvestri Yahoo Inc, London, UK

Alkis Simitsis HP Labs, Palo Alto, CA, USA

Simeon J. Simoff University of Western Sydney, Sydney, NSW, Australia

Elena Simperl Electronics and Computer Science, University of Southampton, Southampton, UK

Radu Sion Stony Brook University, Stony Brook, NY, USA

Mike Sips Stanford University, Stanford, CA, USA

Cristina Sirangelo IRIF, Paris Diderot University, Paris, France

Yannis Sismanis IBM Almaden Research Center, Almaden, CA, USA

Hala Skaf-Molli Computer Science, University of Nantes, Nantes, France

Spiros Skiadopoulos University of Peloponnese, Tripoli, Greece

Richard T. Snodgrass Department of Computer Science, University of Arizona, Tucson, AZ, USA

Dataware Ventures, Tucson, AZ, USA

Cees Snoek University of Amsterdam, Amsterdam, The Netherlands

Mohamed A. Soliman Datometry Inc., San Francisco, CA, USA

Il-Yeol Song College of Computing and Informatics, Drexel University, Philadelphia, PA, USA

Ruihua Song Microsoft Research Asia, Beijing, China

Jingkuan Song Columbia University, New York, NY, USA

Stefano Spaccapietra EPFL, Lausanne, Switzerland

Greg Speegle Department of Computer Science, Baylor University, Waco, TX, USA

Padmini Srinivasan The University of Iowa, Iowa City, IA, USA

Venkat Srinivasan Virginia Tech, Blacksburg, VA, USA

Divesh Srivastava AT&T Labs – Research, AT&T, Bedminster, NJ, USA

Steffen Staab Institute for Web Science and Technologies – WeST, University of Koblenz-Landau, Koblenz, Germany

Constantine Stephanidis Foundation for Research and Technology-Hellas (FORTH), Heraklion, Greece

University of Crete, Heraklion, Greece

Robert Stevens University of Manchester, Manchester, UK

Andreas Stoffel University of Konstanz, Konstanz, Germany

Michael Stonebraker Massachusetts Institute of Technology, Cambridge, MA, USA

Umberto Straccia The Italian National Research Council, Pisa, Italy

Martin J. Strauss University of Michigan, Ann Arbor, MI, USA

Diane M. Strong Worcester Polytechnic Institute, Worcester, MA, USA

Jianwen Su University of California-Santa Barbara, Santa Barbara, CA, USA

Kazimierz Subieta Polish-Japanese Institute of Information Technology, Warsaw, Poland

V. S. Subrahmanian University of Maryland, College Park, MD, USA

Dan Suciu University of Washington, Seattle, WA, USA

S. Sudarshan Indian Institute of Technology, Bombay, India

Torsten Suel Yahoo! Research, Sunnyvale, CA, USA

Jian-Tao Sun Microsoft Research Asia, Beijing, China

Subhash Suri University of California-Santa Barbara, Santa Barbara, CA, USA

Jaroslaw Szlichta University of Ontario Institute of Technology, Oshawa, ON, Canada

Stefan Tai University of Karlsruhe, Karlsruhe, Germany

Kian-Lee Tan Department of Computer Science, National University of Singapore, Singapore, Singapore

Pang-Ning Tan Michigan State University, East Lansing, MI, USA

Wang-Chiew Tan University of California-Santa Cruz, Santa Cruz, CA, USA

Letizia Tanca Computer Science, Politecnico di Milano, Milan, Italy

Lei Tang Chief Data Scientist, Clari Inc., Sunnyvale, CA, USA

Wei Tang Teradata Corporation, El Segundo, CA, USA

Egemen Tanin Computing and Information Systems, University of Melbourne, Melbourne, VIC, Australia

Val Tannen Department of Computer and Information Science, University of Pennsylvania, Philadelphia, PA, USA

Abdullah Uz Tansel Baruch College, CUNY, New York, NY, USA

Yufei Tao Chinese University of Hong Kong, Hong Kong, China

Sandeep Tata IBM Almaden Research Center, San Jose, CA, USA

Nesime Tatbul Intel Labs and MIT, Cambridge, MA, USA

Christophe Taton INPG – INRIA, Grenoble, France

Behrooz Omidvar Tehrani Laboratoire d'Informatique de Grenoble, Saint-Martin d'Hères, France

Paolo Terenziani Dipartimento di Scienze e Innovazione Tecnologica (DiSIT), Università del Piemonte Orientale "Amedeo Avogadro", Alessandria, Italy

Alexandre Termier LIG (Laboratoire d'Informatique de Grenoble), HADAS team, Université Joseph Fourier, Saint Martin d'Hères, France

Evimaria Terzi Computer Science Department, Boston University, Boston, MA, USA

IBM Almaden Research Center, San Jose, CA, USA

Bernhard Thalheim Christian-Albrechts University, Kiel, Germany

Martin Theobald Institute of Databases and Information Systems (DBIS), Ulm University, Ulm, Germany

Stanford University, Stanford, CA, USA

Sergios Theodoridis University of Athens, Athens, Greece

Yannis Theodoridis University of Piraeus, Piraeus, Greece

Saravanan Thirumuruganathan Department of Computer Science and Engineering, University of Texas at Arlington, Arlington, TX, USA

Qatar Computing Research Institute, Hamad Bin Khalifa University, Doha, Qatar

Stephen W. Thomas Dataware Ventures, Kingston, ON, Canada

Alexander Thomasian Thomasian and Associates, Pleasantville, NY, USA

Christian Thomsen Department of Computer Science, Aalborg University, Aalborg, Denmark

Bhavani Thuraisingham The University of Texas at Dallas, Richardson, TX, USA

Srikanta Tirthapura Iowa State University, Ames, IA, USA

Wee Hyong Tok National University of Singapore, Singapore, Singapore

David Toman University of Waterloo, Waterloo, ON, Canada

Frank Tompa David R. Cheriton School of Computer Science, University of Waterloo, Waterloo, ON, Canada

Alejandro Z. Tomsic Sorbonne-Universités-UPMC-LIP6, Paris, France

Inria Paris, Paris, France

Rodney Topor Griffith University, Nathan, Australia

Riccardo Torlone University of Rome, Rome, Italy

Kristian Torp Aalborg University, Aalborg, Denmark

Nicola Torpei University of Florence, Florence, Italy

Nerius Tradišauskas Aalborg University, Aalborg, Denmark

Goce Trajcevski Department of ECpE, Iowa State University, Ames, IA, USA

Peter Triantafillou University of Patras, Rio, Patras, Greece

Silke Trißl Humboldt University of Berlin, Berlin, Germany

Andrew Trotman University of Otago, Dunedin, New Zealand

Juan Trujillo Lucentia Research Group, Department of Information Languages and Systems, Facultad de Informática, University of Alicante, Alicante, Spain

Beth Trushkowsky Department of Computer Science, Harvey Mudd College, Claremont, CA, USA

Panayiotis Tsaparas Department of Computer Science and Engineering, University of Ioannina, Ioannina, Greece

Theodora Tsikrika Center for Mathematics and Computer Science, Amsterdam, The Netherlands

Vassilis J. Tsotras University of California-Riverside, Riverside, CA, USA

Mikalai Tsytsarau University of Trento, Povo, Italy

Peter A. Tucker Whitworth University, Spokane, WA, USA

Anthony K. H. Tung National University of Singapore, Singapore, Singapore

Deepak Turaga IBM Research, San Francisco, CA, USA

Theodoros Tzouramanis University of the Aegean, Samos, Greece

Antti Ukkonen Helsinki University of Technology, Helsinki, Finland

Mollie Ullman-Cullere Harvard Medical School – Partners Healthcare Center for Genetics and Genomics, Boston, MA, USA

Ali Ünlü University of Augsburg, Augsburg, Germany

Antony Unwin Augsburg University, Augsburg, Germany

Susan D. Urban Arizona State University, Phoenix, AZ, USA

Jaideep Vaidya Rutgers University, Newark, NJ, USA

Alejandro A. Vaisman Instituto Tecnológico de Buenos Aires, Buenos Aires, Argentina

Shivakumar Vaithyanathan IBM Almaden Research Center, San Jose, CA, USA

Athena Vakali Aristotle University, Thessaloniki, Greece

Patrick Valduriez INRIA, LINA, Nantes, France

Maarten van Steen VU University, Amsterdam, The Netherlands

W. M. P. van der Aalst Eindhoven University of Technology, Eindhoven, The Netherlands

Christelle Vangenot EPFL, Lausanne, Switzerland

Stijn Vansummeren Hasselt University and Transnational University of Limburg, Diepenbeek, Belgium

Vasilis Vassalos Athens University of Economics and Business, Athens, Greece

Michael Vassilakopoulos University of Thessaly, Volos, Greece

Panos Vassiliadis University of Ioannina, Ioannina, Greece

Michalis Vazirgiannis Athens University of Economics and Business, Athens, Greece

Olga Vechtomova University of Waterloo, Waterloo, ON, Canada

Erik Vee Yahoo! Research, Silicon Valley, CA, USA

Jari Veijalainen University of Jyvaskyla, Jyvaskyla, Finland

Yannis Velegrakis Department of Information Engineering and Computer Science, University of Trento, Trento, Italy

Suresh Venkatasubramanian University of Utah, Salt Lake City, UT, USA

Rossano Venturini Department of Computer Science, University of Pisa, Pisa, Italy

Victor Vianu University of California-San Diego, La Jolla, CA, USA

Maria-Esther Vidal Computer Science, Universidad Simon Bolivar, Caracas, Venezuela

Millist Vincent University of South Australia, Adelaide, SA, Australia

Giuliana Vitiello University of Salerno, Salerno, Italy

Michail Vlachos IBM T.J. Watson Research Center, Hawthorne, NY, USA

Akrivi Vlachou Athena Research and Innovation Center, Institute for the Management of Information Systems, Athens, Greece

Hoang Vo Computer Science, Stony Brook University, Stony Brook, NY, USA

Hoang Tam Vo IBM Research, Melbourne, VIC, Australia

Agnès Voisard Fraunhofer Institute for Software and Systems Engineering (ISST), Berlin, Germany

Kaladhar Voruganti Advanced Development Group, Network Appliance, Sunnyvale, CA, USA

Gottfried Vossen Department of Information Systems, Westfälische Wilhelms-Universität, Münster, Germany

Daisy Zhe Wang Computer and Information Science and Engineering (CISE), University of Florida, Gainesville, FL, USA

Feng Wang City University of Hong Kong, Hong Kong, China

Fusheng Wang Stony Brook University, Stony Brook, NY, USA

Jianyong Wang Tsinghua University, Beijing, China

Jun Wang Queen Mary University of London, London, UK

Meng Wang Microsoft Research Asia, Beijing, China

X. Sean Wang School of Computer Science, Fudan University, Shanghai, China

Xin-Jing Wang Microsoft Research Asia, Beijing, China

Micros Facebook, CA, USA

Zhengkui Wang InfoComm Technology, Singapore Institute of Technology, Singapore, Singapore

Matthew O. Ward Worcester Polytechnic Institute, Worcester, MA, USA

Segev Wasserkrug IBM Research Labs-Haifa, Haifa, Israel

Hans Weda Phillips Research Europe, Eindhoven, The Netherlands

Gerhard Weikum Department 5: Databases and Information Systems, Max-Planck-Institut für Informatik, Saarbrücken, Germany

Michael Weiner Regenstrief Institute, Inc., Indiana University School of Medicine, Indianapolis, IN, USA

Michael Weiss Carleton University, Ottawa, ON, Canada

Ji-Rong Wen Microsoft Research Asia, Beijing, China

Chunhua Weng Columbia University, New York, NY, USA

Mathias Weske University of Potsdam, Potsdam, Germany

Thijs Westerveld Teezir Search Solutions, Ede, Netherlands

Till Westmann Oracle Labs, Redwood City, CA, USA

Karl Wiggisser Institute of Informatics-Systems, University of Klagenfurt, Klagenfurt, Austria

Jef Wijsen University of Mons, Mons, Belgium

Mark D. Wilkinson University of British Columbia, Vancouver, BC, Canada

Graham Wills SPSS Inc., Chicago, IL, USA

Ian H. Witten University of Waikato, Hamilton, New Zealand

Kent Wittenburg Mitsubishi Electric Research Laboratories, Inc., Cambridge, MA, USA

Eric Wohlstadter University of British Columbia, Vancouver, BC, Canada

Dietmar Wolfram University of Wisconsin-Milwaukee, Milwaukee, WI, USA

Ouri Wolfson Mobile Information Systems Center (MOBIS), The University of Illinois at Chicago, Chicago, IL, USA

Department of CS, University of Illinois at Chicago, Chicago, IL, USA

Janette Wong IBM Canada Ltd, Markham, ON, Canada

Raymond Chi-Wing Wong Department of Computer Science and Engineering, The Hong Kong University of Science and Technology, Clear Water Bay, Kowloon, Hong Kong

Peter T. Wood Birkbeck, University of London, London, UK

David Woodruff IBM Almaden Research Center, San Jose, CA, USA

Marcel Worring University of Amsterdam, Amsterdam, The Netherlands

Adam Wright Partners HealthCare, Boston, MA, USA

Sai Wu Zhejiang University, Hangzhou, Zhejiang, People's Republic of China

Yuqing Wu Indiana University, Bloomington, IN, USA

Alex Wun University of Toronto, Toronto, ON, Canada

Ming Xiong Bell Labs, Murray Hill, NJ, USA

Google, Inc., New York, NY, USA

Guandong Xu University of Technology Sydney, Sydney, Australia

Hua Xu Columbia University, New York, NY, USA

Jun Yan Microsoft Research Asia, Haidian, China

Xifeng Yan IBM T. J. Watson Research Center, Hawthorne, NY, USA

Jun Yang Duke University, Durham, NC, USA

Li Yang Western Michigan University, Kalamazoo, MI, USA

Ming-Hsuan Yang University of California at Merced, Merced, CA, USA

Seungwon Yang Virginia Tech, Blacksburg, VA, USA

Yang Yang Center for Future Media and School of Computer Science and Engineering, University of Electronic Science and Technology of China, Chengdu, Sichuan, China

Yun Yang Swinburne University of Technology, Melbourne, VIC, Australia

Yu Yang City University of Hong Kong, Hong Kong, China

Yong Yao Cornell University, Ithaca, NY, USA

Mikalai Yatskevich University of Trento, Trento, Italy

Xun Yi Computer Science and Info Tech, RMIT University, Melbourne, VIC, Australia

Hiroshi Yoshida VLSI Design and Education Center, University of Tokyo, Tokyo, Japan

Fujitsu Limited, Yokohama, Japan

Masatoshi Yoshikawa University of Kyoto, Kyoto, Japan

Matthew Young-Lai Sybase iAnywhere, Waterloo, ON, Canada

Google, Inc., Mountain View, CA, USA

Hwanjo Yu University of Iowa, Iowa City, IA, USA

Ting Yu North Carolina State University, Raleigh, NC, USA

Cong Yu Google Research, New York, NY, USA

Philip S. Yu Computer Science Department, University of Illinois at Chicago, Chicago, IL, USA

Jeffrey Xu Yu Department of Systems Engineering and Engineering Management, The Chinese University of Hong Kong, Hong Kong, China

Pingpeng Yuan Service Computing Technology and System Lab, Cluster and Grid Computing Lab, School of Computer Science and Technology, Huazhong University of Science and Technology, Wuhan, China

Vladimir Zadorozhny University of Pittsburgh, Pittsburgh, PA, USA

Matei Zaharia Douglas T. Ross Career Development Professor of Software Technology, MIT CSAIL, Cambridge, MA, USA

Ilya Zaihrayeu University of Trento, Trento, Italy

Mohammed J. Zaki Rensselaer Polytechnic Institute, Troy, NY, USA

Carlo Zaniolo University of California-Los Angeles, Los Angeles, CA, USA

Hugo Zaragoza Yahoo! Research, Barcelona, Spain

Stan Zdonik Brown University, Providence, RI, USA

Demetrios Zeinalipour-Yazti Department of Computer Science, Nicosia, Cyprus

Hans Zeller Hewlett-Packard Laboratories, Palo Alto, CA, USA

Pavel Zezula Masaryk University, Brno, Czech Republic

Cheng Xiang Zhai University of Illinois at Urbana-Champaign, Urbana, IL, USA

Aidong Zhang State University of New York, Buffalo, NY, USA

Benyu Zhang Microsoft Research Asia, Beijing, China

Donghui Zhang Paradigm4, Inc., Waltham, MA, USA

Dongxiang Zhang School of Computer Science and Engineering, University of Electronic Science and Technology of China, Sichuan, China

Ethan Zhang University of California, Santa Cruz, CA, USA

Jin Zhang University of Wisconsin Milwaukee, Milwaukee, WI, USA

Kun Zhang Xavier University of Louisiana, New Orleans, LA, USA

Lei Zhang Microsoft Research Asia, Beijing, China

Lei Zhang Microsoft Research, Redmond, WA, USA

Li Zhang Peking University, Beijing, China

Meihui Zhang Information Systems Technology and Design, Singapore University of Technology and Design, Singapore, Singapore

Qing Zhang The Australian e-health Research Center, Brisbane, Australia

Rui Zhang University of Melbourne, Melbourne, VIC, Australia

Dataware Ventures, Tucson, AZ, USA

Dataware Ventures, Redondo Beach, CA, USA

Yanchun Zhang Victoria University, Melbourne, VIC, Australia

Yi Zhang Yahoo! Inc., Santa Clara, CA, USA

Yue Zhang University of Pittsburgh, Pittsburgh, PA, USA

Zhen Zhang University of Illinois at Urbana-Champaign, Urbana, IL, USA

Feng Zhao Microsoft Research, Redmond, WA, USA

Ying Zhao Tsinghua University, Beijing, China

Baihua Zheng Singapore Management University, Singapore, Singapore

Yi Zheng University of Ontario Institute of Technology, Oshawa, ON, Canada

Yu Zheng Data Management, Analytics and Services (DMAS) and Ubiquitous Computing Group (Ubicomp), Microsoft Research Asia, Beijing, China

Zhi-Hua Zhou National Key Lab for Novel Software Technology, Nanjing University, Nanjing, China

Jingren Zhou Alibaba Group, Hangzhou, China

Li Zhou Partners HealthCare System Inc., Boston, MA, USA

Xiaofang Zhou School of Information Technology and Electrical Engineering, University of Queensland, Brisbane, QLD, Australia

Huaiyu Zhu IBM Almaden Research Center, San Jose, CA, USA

Xiaofeng Zhu Guangxi Normal University, Guilin, Guangxi, People's Republic of China

Xingquan Zhu Florida Atlantic University, Boca Raton, FL, USA

Cai-Nicolas Ziegler Siemens AG, Munich, Germany

Hartmut Ziegler University of Konstanz, Konstanz, Germany

Esteban Zimányi CoDE, Université Libre de Bruxelles, Brussels, Belgium

Arthur Zimek Ludwig-Maximilians-Universität München, Munich, Germany

Department of Mathematics and Computer Science, University of Southern Denmark, Odense, Denmark

Roger Zimmermann Department of Computer Science, School of Computing, National University of Singapore, Singapore, Republic of Singapore

Lei Zou Institute of Computer Science and Technology, Peking University, Beijing, China

Q

Q-Measure

Tetsuya Sakai
Waseda University, Tokyo, Japan

Synonyms

None

Definition

Q-measure is a graded-relevance version of the well-known Average Precision. Let R denote the number of known relevant documents for a topic; let $gain_l$ denote the *gain value* for a document of relevance level l: for example, let it be 3 for a highly relevant document, 2 for a relevant document, and 1 for a partially relevant document. For a given ranked list of documents, let $I(r) = 0$ if the document at rank r is nonrelevant, and $I(r) = 1$ otherwise; then $C(r) = \sum_{k=1}^{r} I(k)$: the number of relevant documents within top r. Note that Precision at r is given by $C(r)/r$. Let $g(r) = gain_l$ if the document at rank r is l-relevant and let $g(r) = 0$ otherwise; let the *cumulative gain* be $cg(r) = \sum_{k=1}^{r} g(k)$. Define an *ideal ranked list* for the topic by sorting the R relevant documents by relevance level; let $cg^*(r)$ denote the cumulative gain at r for the ideal list. Then Q-measure is defined as:

$$Q\text{-}measure = \frac{1}{R} \sum_{r} I(r) \frac{C(r) + \beta cg(r)}{r + \beta cg^*(r)}$$

$$= \frac{1}{R} \sum_{r} I(r) BR(r)$$

where $\beta (\geq 0)$ is called the *persistence parameter* and $BR(r)$ is the *blended ratio* which combines Precision and Normalised Cumulative Gain. For a shallow *measurement depth* d (i.e., *document cutoff*), the R in the above definition may be replaced by $\min(d, R)$ to ensure that the maximum possible value is 1 for every topic.

Main Text

Q-measure reduces to Average Precision if $\beta = 0$; a large β (e.g., $\beta = 10$) represents a patient user who is willing to dig deep down the ranked list; the default is $\beta = 1$. In a binary relevance environment, Q-measure equals Average Precision for any ranked list if the list does not contain any relevant documents below rank R; otherwise, Q-measure is greater than Average Precision. It is known to be a statistically highly stable measure.

Q-measure is a member of the *Normalised Cumulative Utility* (NCU) family, which is defined by combining a *stopping probability distribution* over the document ranks for the user population and the *utility* at a given rank. For Q-measure, the probability distribution is uniform across all

© Springer Science+Business Media, LLC, part of Springer Nature 2018
L. Liu, M. T. Özsu (eds.), *Encyclopedia of Database Systems*,
https://doi.org/10.1007/978-1-4614-8265-9

relevant documents (just like Average Precision), and the utility is given by $BR(r)$ (whereas Average Precision uses Precision as the utility).

Variants of Q-measure include *R-measure* (a blended-ratio version of R-precision) and P^+ (a blended-ratio version of Reciprocal Rank). These also belong to the NCU family: R-measure (just like R-precision) assumes that all users stop at rank R; P^+ assumes that the stopping probability distribution is uniform across relevant documents retrieved at or above rank r_p, the rank of the most relevant document in the list that is nearest to the top.

Cross-References

▶ Advanced Information Retrieval Measures
▶ Average Precision
▶ Discounted Cumulated Gain
▶ D-Measure
▶ Expected Reciprocal Rank
▶ R-Precision

Quadtrees (and Family)

Michael Vassilakopoulos[1] and
Theodoros Tzouramanis[2]
[1]University of Thessaly, Volos, Greece
[2]University of the Aegean, Samos, Greece

Synonyms

Hierarchical regular-decomposition structures; Hierarchical spatial indexes; Quadtree variations

Definition

In general, the term quadtree refers to a class of representations of geometric entities (such as points, line segments, polygons, regions) in a space of two (or more) dimensions that recursively decompose the space containing these entities into blocks until the data in each block satisfy some condition (with respect, for example, to the block size, the number of block entities, the characteristics of the block entities, etc.).

In a more restricted sense, the term quadtree (octree) refers to a tree data structure in which each internal node has four (eight) children and is used for the representation of geometric entities in a two (three) dimensional space. The root of the tree represents the whole space/region. Each child of a node represents a subregion of the subregion of its parent. The subregions of the siblings constitute a partition of the parent's regions.

Several variations of quadtrees are possible, according to the dimensionality of the space represented, the criterion guiding the subdivision of space, the type of data represented, the type of memory (internal or external) used for storing the structure, the shape, position and size of the subregions, etc. However, the term quadtree usually refers to tree structures that divide space in a hierarchical and regular (decomposing to equal parts on each level) fashion. Since final subregions/blocks (that are no further subdivided, i.e., the blocks of the tree leaves) do not overlap, the underlying space is partitioned to a set of regions in favor of the efficient processing of spatial queries.

Historical Background

As computer science evolved, the need for representing and manipulating spatial data (data expressing geometric properties of entities, conceptually expressed by points, line segments, regions, geometric shapes, etc.) arose in several applications areas, like Computer Graphics, Multimedia, Geographical Information Systems, or VLSI Design. The recursive decomposition of space was naturally identified as a means for organizing spatial data. The term "quadtree" was used by Finkel and Bentley [1] to express an extension of the Binary Search Tree in two dimensions that was able to index points (point quadtree). Since then, several quadtree variations for a multitude of spatial data types that were used for almost all sorts of spatial data manipulations have appeared in the literature.

The term quadtree has taken a generic meaning and is used to describe a class of hierarchical data structures whose common property is that they are based on the principle of recursive decomposition of space. The quadtree, a variable resolution structure, is often confused with the Pyramid [2], a multi-resolution representation consisting of a hierarchy of arrays. quadtrees and family are space-driven methods (follow the embedding space hierarchy): the division to subregions obeys a predefined way. On the contrary, R-trees and family are data-driven methods (follow the data space hierarchy): the division to subregions depends on the data.

The region quadtree is the most famous such structure. As its name implies, it is used for representing regions and was initially termed Q-tree by Klinger and Dyer [3]. This is a main memory tree. A secondary memory implementation, the linear quadtree, was proposed by Gargantini [4]. The MX and PR quadtrees are adaptations of the region quadtree for representing point data [2]. The MX-CIF quadtree is a structure able to represent a collection of rectangles [2]. The PM family of quadtrees is used for representing polygonal maps and collections of line segments. In general, extensions to three or more dimensions of each quadtree-like structure are possible. For example, the extension of the region quadtree for three-dimensional volume data is called Region Octree. More details about the history of these and other quadtree-like structures can be found at [2, 5]. There have also been proposed quadtree variations for storing a pictorial database, like the DI-quadtree, for binary images and the generic quadtree, for grayscale and color images. These and other quadtree-based methods that have been proposed for representation and querying in image database applications are reviewed in [6]. Quadtree variations have also been proposed for evolving regional data, like overlapping quadtrees [7], Overlapping linear quadtrees, the multiversion linear quadtree, the multiversion access structure for evolving raster images, and the time-split linear quadtree [8]. The XBR tree [8] is a quadtree-like structure for indexing points or line segments especially designed for external memory. The skip quadtree

[9] is based on region quadtrees and skip lists. It indexes points and can be used for efficiently answering point location, approximate range, and approximate nearest neighbor queries. Quadtrees have also been used in commercial database management systems [10]. Demos of several quadtree structures can be found at [11].

Foundations

This section briefly presents some key quadtree structures among the numerous structures that have appeared in the literature.

Point Quadtree

The point quadtree [2] is an indexing mechanism for points. Each tree node corresponds to a point that subdivides the region of the node in four parts defined by two lines (parallel to the coordinate axes) on which this point lies. Thus, the shape and position of subregions depend on the coordinates of the point (data-driven subdivision). However, each region is always subdivided in four parts. A region is considered closed in relation to its lower and left border. If multiple points with the same coordinates are allowed, each node contains a list of the points it stores. In Fig. 1, a collection of points and the resulting partitioning of space (a) and the corresponding point quadtree (b) are depicted. By convention, in this and other quadtree variations, the children of a node correspond in order to the North-West, North-East, South-West, and South-East subregions of the node. Note that the partitioning of space and the tree shape depend on the order of insertion of the points. Balanced versions of this tree have been proposed in the literature [2].

A point quadtree is not only suitable for point indexing (e.g., discovering if there is and which is the gas station that lies at a given pair of coordinates) but for answering range queries, as well (e.g., finding all the cities that reside within a specified distance from a given pair of coordinates). The efficiency of the point quadtree during this operation comes from pruning the search space only to nodes that may contain part of the answer.

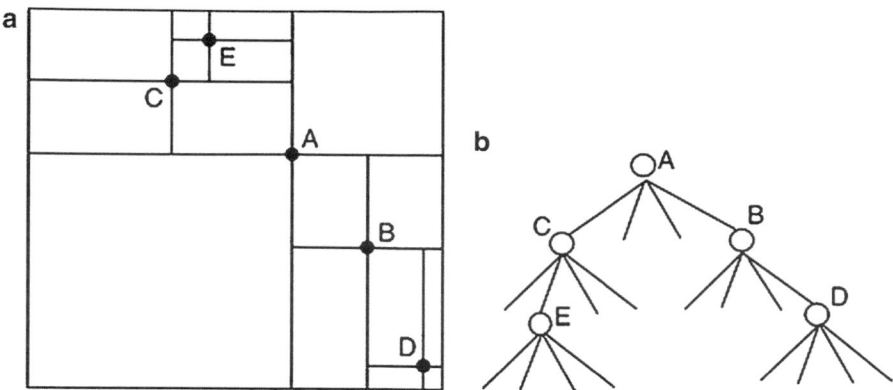

Quadtrees (and Family), Fig. 1 A collection of points (**a**) and the corresponding point quadtree (**b**)

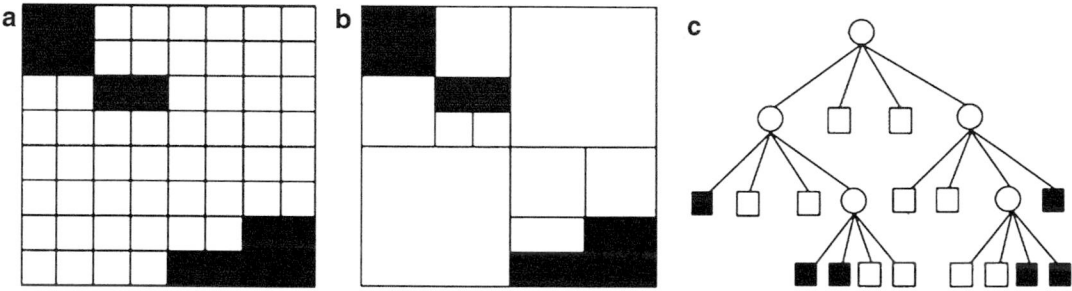

Quadtrees (and Family), Fig. 2 A binary image (**a**), its partition to blocks (**b**), and its region quadtree (**c**)

Region Quadtree

The region quadtree [2] (or Q-tree [3]) is a popular hierarchical data structure for the representation of binary images or regional data. Such an image can be represented as a $2^n \times 2^n$ binary array, for some natural number n, where an entry equal to 0 stands for a white pixel and an entry equal to 1 stands for a black pixel. The region quadtree for this array is made up either of a single white (black) node if every pixel of the image is white (black) or of a gray root, which points to four subquadtrees, one for every quadrant of the original image. Each region is always subdivided in four equal parts (space-driven subdivision). An example of an 8×8 binary image, its region quadtree, and the unicolor blocks to which it is partitioned by the quadtree external nodes are shown in Fig. 2a, b, and c, respectively.

The region quadtree, depending on the distribution of data, may result in considerable space savings. However, it can be used for several operations on regional data. One of them is the

determination of the color of a specific pixel or the determination of the block where this pixel resides (a specialized point-location query). Set theoretic operations are also well adapted to region quadtrees (recoloring, or dithering of a single image, or overlaying, union, intersection of sets of images). Connected component labeling (that is grouping pixels according to their color) is performed by utilizing region quadtrees. Other operations that are well adapted to region quadtrees include window clipping and certain transformations (like scaling by a power of two or rotating by 90°). For more operations and details, see [12].

The pointer-based implementation of region quadtrees is a main memory implementation with one node type that consists of pointer fields and one color field. This node type is used for internal and external nodes according to the value of the color field. Although such an implementation is memory consuming it simplifies several operations. A secondary memory implementation of a region quadtree, called a linear quadtree [4], consists of a list of values where there is one value for each black node of the pointer-based quadtree. The value of a node is an address describing the position and size of the corresponding block in the image. These addresses can be stored in an efficient structure for secondary memory (such as a B-tree or one of its variations). Evidently, this representation is very space efficient, although it is not suited to many useful algorithms that are designed for pointer-based quadtrees. The most popular linear implementations are the FL (Fixed Length) and the FD (Fixed length – Depth) linear implementations. In the former implementation, the address of a black quadtree node is a code word that consists of n base 5 digits. Codes 0, 1, 2, and 3 denote directions NW, NE, SW, and SE, respectively, while code 4 denotes a do-not-care direction. If the black node resides on level i, where $n >= i >= 0$, then the first $n-i$ digits express the directions that constitute the path from the root to this node and the last i digits are all equal to 4. In the latter implementation, the address of a black quadtree node has two parts: the first part is a code word that consists of n base 4 digits. Codes 0, 1, 2, and 3 denote directions NW, NE, SW, and SE, respectively. This code word is formed in a similar way to the code word of the FL-linear implementation with the difference that the last i digits are all equal to 0. The second part of the address has $[\log_2(n+1)]$ bits and denotes the depth of the black node or,

in other words, the number of digits of the first part that express the path to this node. Another interesting secondary memory implementation of the region quadtree is the paged-pointer quadtree [13] that partitions the tree nodes into pages and manages these pages using B-tree techniques.

PR Quadtree

The PR quadtree [2] (P comes from point and R from region) is an indexing technique for points (with similar functionality to point quadtrees) that is based on the region quadtree. Points are associated with quadrants that are formed according to the region quadtree rules. A leaf node may be white (without any points residing in its region) or black (with one point residing in its region). In Fig. 3, the collection of points of Fig. 1 and the resulting partitioning of space (a) and the corresponding PR quadtree (b) are depicted.

The final shape of the PR quadtree is independent to the order of insertion of the points. A problem with this structure is that the maximum depth of recursive decomposition depends on the minimum distance between two points (if there are two points very close to each other, the decomposition can be very deep). This effect is reduced by allowing leaf nodes to hold up to C points. When this capacity is exceeded, the node is split in four. By storing leaf nodes on secondary memory and setting C according to the

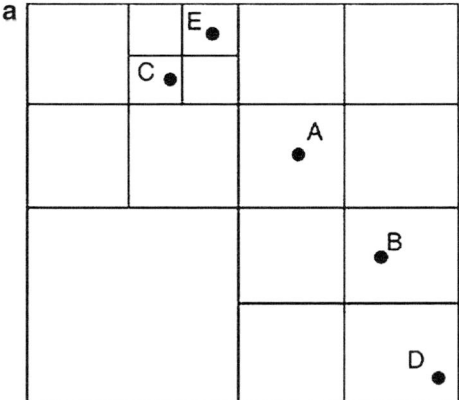

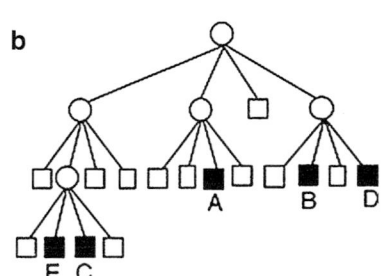

Quadtrees (and Family), Fig. 3 A collection of points (**a**) and the corresponding PR quadtree (**b**)

disk-page size, a structure that partially resides on disk is created.

PMR Quadtree

The PMR quadtree [2] is capable of indexing line segments and answering window queries (e.g., find the line segments that intersect a given window/area in the plane). The internal part of the tree consists of an ordinary region quadtree. The leaf nodes of this quadtree are bucket nodes that hold the actual line segments. Each line segment is stored in every bucket whose quadrant (region) it crosses. A line segment can cross the region of a bucket either fully or partially. Each bucket has a maximum capacity. When this capacity is exceeded due to an insertion of a line segment, the bucket is split in four equal quadrants. However, it is possible that one (or more) of these quadrants holds a number of line segments that still exceeds the bucket capacity. Since this is not occurring very often in practical applications (e.g., when line segments represent a road network) a bucket is split only once in four and overflow buckets are created when needed. The PMR quadtree can be implemented with bucket nodes residing on disk.

In Fig. 4, an example of the splitting of regions during the creation of a PMR quadtree by the successive insertion of line segments is depicted. The bucket capacity is two (just for demonstration purposes). Figures 4a, b, and c show the subdivision of space and the buckets created as line segments are inserted. Overflow buckets do not result from the insertions of Fig. 4. Note that the shape of the PMR quadtree depends on the order of insertion of the line segments.

XBR Tree

The XBR tree (XBR stands for eXternal Balanced Regular) is an indexing method suitable for indexing points (like a bucket PR quadtree) or line segments (like a PMR quadtree). It totally resides in secondary memory. Its hierarchical decomposition of space is the same as the one in region quadtrees. There are two types of nodes in an XBR tree. The first are the internal nodes that constitute the index. The second are the leaves containing the data items. Both the leaves and the internal nodes correspond to disk pages.

In an internal node, a number of pairs of the form < address, pointer > are contained. The number of these pairs is nonpredefined because the addresses being used are of variable size. An address expresses a child node region and is paired with the pointer to this child node. Both the size of an address and the total space occupied by all pairs within a node must not exceed the node size. The addresses in these pairs are used to represent certain subquadrants that result from the repetitive subdivision of the initial space. This is done by assigning the numbers 0, 1, 2, and 3 to NW, NE, SW, and SE quadrants, respectively. For example, the address 1 is used to represent the NE quadrant of the initial space, while the address 10 to represent the NW subquadrant of the NE quadrant of the initial space.

In the XBR tree, the region of a child is the subquadrant specified by the address in its pair minus the subquadrants corresponding to all the previous pairs of the internal node to which it

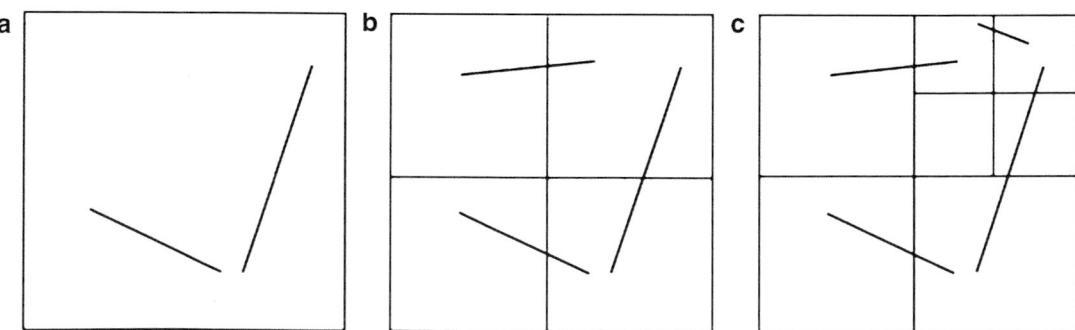

Quadtrees (and Family), Fig. 4 Splitting of PMR-quadtree regions by the successive insertion of line segments

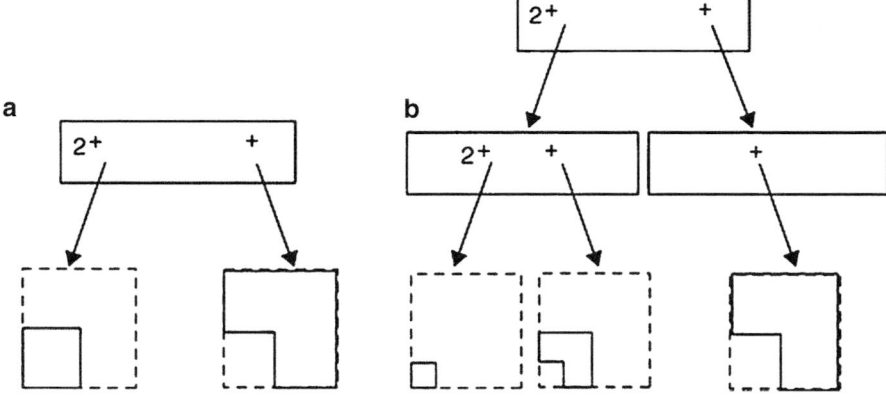

Quadtrees (and Family), Fig. 5 XBR trees with one level (**a**) and two levels (**b**) of internal nodes

belongs. Figure 5 presents XBR trees of one (a) and two (b) levels of internal nodes. The $^+$ is used to denote the end of each variable size address. The address 2^+ in the root denotes the SW quadrant of the initial space. On the other hand, the address $^+$ in the root specifies the initial space minus the SW quadrant.

Each leaf node in the XBR tree may contain a number of data items, which is limited by a predefined capacity C. When an insertion causes the number of data items of a leaf to exceed C, the leaf is split following a hierarchical decomposition analogous to the quadtree decomposition. In case line segments are stored (like PMR quadtrees), a leaf is split only once in four and overflow buckets are created when needed.

Due to the incremental (level-by-level) formation of absolute addresses and the variable length coding of them, XBR trees are very compact structures. Thus, I/O during query processing is reduced, favoring processing efficiency.

Quadtree and Time-Evolving Regional Data

In Tzouramanis et al. [8] and previous papers by the same authors, four temporal extensions of the linear region quadtree are presented: the time-split linear quadtree, the multiversion linear quadtree, the multiversion access structure for evolving raster images, and the overlapping linear quadtrees. These methods comprise a collec-

tion of specialized quadtree-based access methods that can efficiently store and manipulate consecutive raster images. Through these methods, efficient support for spatio-temporal queries referring to the past is provided. An extensive experimental space and time performance comparison of all the above-mentioned access methods, presented in Tzouramanis et al. [8], has shown that the overlapping linear quadtrees is the best performing method. For more details, see [8] and its references.

Key Applications

Quadtree-based access methods speed up access and queries in database systems that support spatial data. Some common uses of quadtrees include representation and indexing of images, spatial indexing for several spatial types (points, regions, line segments, polygonal maps), several set operations, point location, range, and nearest neighbor queries and temporal queries on series of evolving images. Quadtrees have been employed in numerous application areas that require efficient retrieval of complex objects, such as computer graphics and animation, computer vision, robotics, geographical information systems (GIS), image processing, image and multimedia databases, content-based image retrieval, medical imaging, urban planning, computer-aided design (CAD), or even in recent novel database applications such as P2P networks. Furthermore, together with the R-tree family, quadtrees serve as an important bridge for extending spatial databases to applications of several scientific areas, such as agriculture, oceanography, atmospheric physics, geology, astronomy, molecular biology, etc. Commercial database vendors like IBM and Oracle [10] have implemented the quadtree and the linear quadtree to cater for the large and diverse above application markets.

Future Directions

Since quadtrees were mainly introduced as main memory structures, the development of further external memory versions of several quadtree variations and the study of their performance for several queries remain a target. Papers, like [14], show that the comparative performance study for several query types between space-driven and data-driven indexing techniques can lead to interesting conclusions. Algorithms for queries based on the joining of data (e.g., image data) traditionally stored in quadtree structures and other types of spatial data stored in data-driven structures (e.g., point data stored in R-tree family structures) are worth developing and studying, especially when the evolution of the data is considered (spatio-temporal data).

Cross-References

▶ Indexing Historical Spatiotemporal Data
▶ Main Memory
▶ Query Processing and Optimization in Object Relational Databases
▶ R-Tree (and Family)
▶ Spatial Indexing Techniques
▶ Tree-Based Indexing

Recommended Reading

1. Finkel R, Bentley JL. Quad trees: a data structure for retrieval on composite keys. Acta Informatica. 1974;4(1):1–9.
2. Samet H. The design and analysis of spatial data structures. Reading: Addison Wesley; 1990.
3. Klinger A, Dyer C. Experiments on picture representation using regular decomposition. Comput Graphics Image Process. 1976;5(1):68–105.
4. Gargantini I. An effective way to represent quadtrees. Commun ACM. 1982;25(12):905–10.
5. Samet H. Foundations of multidimensional and metric data structures. Amsterdam: Morgan Kaufmann; 2006.
6. Manouvrier M, Rukoz M, Jomier G. Quadtree-based image representation and retrieval. In: Spatial databases: technologies, techniques and trends. Hershey: Idea Group Publishing; 2005. p. 81–106.
7. Vassilakopoulos M, Manolopoulos Y, Economou K. Overlapping quadtrees for the representation of similar images. Image Vis Comput. 1993;11(5):257–62.
8. Tzouramanis T, Vassilakopoulos M, Manolopoulos Y. Benchmarking access methods for time-evolving regional data. Data Knowl Eng. 2004;49(3):243–86.
9. Eppstein D, Goodrich MT, Sun JZ. The skip quadtree: a simple dynamic data structure for multidimensional data. In: Proceedings of the 21st Annual Symposium on Computational Geometry; 2005. p. 296–305.
10. Kothuri R, Ravada S, Abugov D. Quadtree and r-tree. indexes in oracle spatial: a comparison using gis data. In: Proceedings of the ACM SIGMOD International Conference on Management of Data; 2002. p. 546–57.
11. Brabec F, Samet H. Spatial index demos. http://donar.umiacs.umd.edu/quadtree/index.html. Last accessed in Dec 2016.
12. Samet H. Applications of spatial data structures. Reading: Addison Wesley; 1990.
13. Shaffer CA, Brown PR. A paging scheme for pointer-based quadtrees. In: Abel D, Chin Ooi B, editors. Advances in Spatial Databases, Proceedings of the 3rd International Symposium on Large Spatial Databases; 1993. p. 89–104.
14. Kim YJ, Patel JM. Rethinking choices for multidimensional point indexing: making the case for the often ignored quadtree. In: Proceedings of the 3rd Biennial Conference on Innovative Data Systems Research; 2007. p. 281–91. http://cidrdb.org/2007Proceedings.zip
15. Vassilakopoulos M, Manolopoulos Y. External balanced regular (x-BR) trees: new structures for very large spatial databases. In: Fotiadis DI, Nikolopoulos SD, editors. Advances in informatics. Singapore: World Scientific; 2000. p. 324–33.

Qualitative Temporal Reasoning

Paolo Terenziani
Dipartimento di Scienze e Innovazione
Tecnologica (DiSIT), Università del Piemonte
Orientale "Amedeo Avogadro", Alessandria,
Italy

Synonyms

Nonmetric temporal reasoning; Reasoning with qualitative temporal constraints

Definition

Qualitative temporal constraints are nonmetric *temporal constraints* stating the *relative* temporal position of facts that happen in time (e.g., fact F_1 is *before* or *during* fact F_2). Different types of qualitative constraints can be defined, depending on whether facts are instantaneous, durative, or repeated. *Qualitative temporal reasoning* is the process of reasoning with such temporal constraints. Given a set of qualitative temporal constraints, qualitative temporal reasoning can be used for different purposes, including checking their *consistency*, determining the *strictest constraints* between pairs of facts (e.g., for query-answering purposes), and pointing out a *consistent scenario* (i.e., a possible instantiation of all the facts on the timeline in such a way that all temporal constraints are satisfied).

Historical Background

In several domains and/or application areas, *temporal indeterminacy* has to be coped with. In such domains, the *absolute time* when facts hold (i.e., the exact temporal location of facts) is generally unknown. On the other hand, in many of such domains, *qualitative* temporal constraints about the *relative* temporal location of facts are available, and reasoning about such constraints is an important task. As a consequence, there is a long tradition for qualitative temporal reasoning within *philosophical logic* (consider, e.g., Prior's seminal branching time logics [12]).

In particular, qualitative temporal reasoning is important in several artificial intelligence areas, including planning, scheduling, and natural language understanding. Therefore, starting from the beginning of the 1980s, several *specialized* approaches (as opposed to *logical* approaches, which are usually general purpose) to the *representation* of *qualitative* temporal constraints and to temporal *reasoning* about them have been developed, especially within the artificial intelligence area.

The first milestone in the specialized approaches to qualitative temporal reasoning dates back to Allen's interval algebra (IA) [1], coping with qualitative temporal constraints between intervals (called *periods* by the database community and henceforth in this entry), to cope with durative facts. Further on, several other algebras of qualitative temporal constraints have been developed, to cope with instantaneous [15] or repeated/periodic [9, 13] facts, and several temporal reasoning systems have been implemented and used in practical applications (see, e.g., some comparisons in Delgrande et al. [4]).

Significant effort in the area has been devoted to the analysis of the *trade-off* between the *expressiveness* of the representation formalisms and the *complexity* of *correct* and *complete* temporal reasoning algorithms operating on them (see, e.g., the survey by Van Beek [14]). Since consistency checking in Allen's interval algebra is NP-complete, several approaches, starting from Nebel and Burkert [11], have focused on the identification of *tractable fragments* of it. The *integration* of qualitative and metric constraints has also been analyzed (see, e.g., Jonsson and Backstrom [7]). Recent developments also include *incremental* [6] and *fuzzy* [2] qualitative temporal reasoning. The recent book by Ligozat [10] contains a description of the main (both temporal and spatial) qualitative calculi which have been developed over the last three decades. Finally, starting in the 1990s, some approaches

Q

also started to investigate the adoption of qualitative temporal constraints and temporal reasoning within the temporal database context [3, 8].

Scientific Fundamentals

Qualitative temporal constraints concern the relative temporal location of facts on the timeline. A significant and milestone example is Allen's interval algebra (IA). Allen pointed out the 13 primitive qualitative relations between time periods: before, after, meets, met-by, overlaps, overlapped-by, starts, started-by, during, contains, ends, ended-by, and equal. These relations are *exhaustive* and *mutually exclusive* and can be combined in order to represent disjunctive relations. For example, the constraints in (Ex.1) and (Ex.2) state that F_1 is before or during F_2, which, in turn, is before F_3.

- (Ex.1) F_1 (BEFORE,DURING) F_2
- (Ex.2) F_2 (BEFORE) F_3

In Allen's approach, qualitative temporal reasoning is based on two algebraic operations over relations on time periods: intersection and composition. Given two possibly disjunctive relations R_1 and R_2 between two facts F_1 and F_2, temporal intersection (henceforth $\cap$) determines the most constraining relation R between F_1 and F_2. For example, the temporal intersection between (Ex.2) and (Ex.3) is (Ex.4). On the other hand, given a relation R_1 between F_1 and F_2 and a relation R_2 between F_2 and F_3, composition (@) gives the resulting relation between F_1 and F_3. For example, (Ex.5) is the composition of (Ex.1) and (Ex.2) above:

- (Ex.3) F_2 (BEFORE,MEETS,OVERLAPS) F_3
- (Ex.4) F_2 (BEFORE) F_3
- (Ex.5) F_1 (BEFORE) F_3

In Allen's approach, temporal reasoning is performed by a *path consistency* algorithm that basically computes the *transitive closure* of the constraints by repeatedly applying intersection

and composition. Abstracting from many optimizations, such an algorithm can be schematized as follows:

Repeat
  For all triples of facts$<F_i,F_k,F_j>$.
    Let R_{ij} denote the (possibly ambiguous)
    relation between F_i and F_j.
    $R_{ij} \leftarrow R_{ij} \cap (R_{ik} @ R_{kj})$.
Until quiescence.

Allen's algorithm operates in a time *cubic* in the number of periods. However, such an algorithm is *not complete* for the interval algebra (in fact, checking the consistency of a set of temporal constraints in the interval algebra is NP-hard [15]).

While in many approaches researchers chose to adopt Allen's algorithm, in other approaches they tried to design less expressive but *tractable* formalisms. For example, the point algebra is defined in the same way as the interval algebra, but the temporal elements are time points. Thus, there are only three primitive relations between time points (i.e., $<$, $=$ and $>$) and four disjunctive relations (i.e., $(<,=)$, $(>,=)$, $(<,>)$, and $(<,=,>)$). In the point algebra, sound and complete constraint propagation algorithms operate in polynomial time (namely, in $O(n^4)$, where n is the number of points [14]). Obviously, the price to be paid for tractability is expressive power: not all (disjunctive) relations between periods can be mapped onto relations between their end points. For instance, F_1 (*BEFORE,AFTER*) F_2 cannot be mapped into a set of (possibly disjunctive) pairwise relations between time points; indeed, an explicit disjunction between two different pairs of time points is needed (i.e., $(end(F_1)<start(F_2))$ and $(end(F_2)<start(F_1))$. The continuous point algebra restricts the point algebra excluding inequality (i.e., $(<,>)$). Allen's path consistency algorithm is both sound and complete for such an algebra and operates in $O(n^3)$ time, where n is the number of time points (for more details, see, e.g., the survey by Van Beek [14]).

A different simplification of Allen's interval algebra has been provided by Freksa [5]. Freksa has identified coarser qualitative temporal relations than Allen's ones, based on the notion of *semi-intervals* (i.e., beginning and ending points of durative events). As an example of relation on semi-intervals, Freksa has introduced the "older" relation (F_1 is *older* than F_2 if F_1's starting point is before F_2's starting point, with no constraint on the ending points). Notice that Freksa's *older* relation corresponds to a disjunction of five Allen's relations (i.e., *before, meets, overlaps, finished-by, contains*). Freksa has also shown that relations between semi-intervals result in a possible more compact notation and more efficient reasoning mechanisms, in particular if the initial knowledge is, at least in part, coarse knowledge.

Another mainstream of research about qualitative temporal reasoning focused on the identification of *tractable fragments* of Allen's interval algebra. The milestone work by Nebel and Burkert [11] first pointed out the "*ORD-Horn subclass*," showing that reasoning in such a class is a polynomial time problem and that it constitutes a maximal tractable subclass of Allen's interval algebra.

Starting in the early 1990s, some *integrated* temporal reasoning approaches were devised in order to deal with both qualitative and quantitative (i.e., metric) temporal constraints. For instance, Jonsson and Backstrom [7] proposed a framework, based on *linear programming*, that deals with both qualitative and metric constraints and that also allows one to express constraints on the relative duration of events (see, e.g., (Ex.6)):

- (Ex.6) John's drive to work is at least 30 min more than Fred's.

Many other important issues must be taken into account when considering qualitative temporal reasoning. For example, starting from Ladkin's seminal work [9], qualitative constraints between *repeated* facts have been considered. In the same mainstream, Terenziani has proposed an extension of Allen's interval algebra to consider qualitative relations between periodic facts [13]. Terenziani's approach deals with constraints such as (Ex.7):

- (Ex.7) Between January 1, 1999, and December 31, 1999, on the first Monday of each

month, Andrea went to the post office *before* going to work.

In Terenziani's approach, temporal reasoning over such constraints is performed by a path consistency algorithm which extends Allen's one. Such an algorithm is sound but not complete and operates in cubic time with respect to the number of periodic facts.

As concerns more strictly in the area of (temporal) databases, starting in the mid-1990s, some researchers started to investigate the treatment of qualitative temporal constraints within temporal (relational) databases (see, e.g., [3, 8]). In such approaches, the *valid time* of facts (tuples) is represented by symbols denoting time periods, and *qualitative* and *quantitative temporal constraints* are used in order to express constraints on their relative location in time, on their duration, and so on.

Koubarakis [8] first extended the constraint database model to include indefinite (or uncertain) temporal information (including qualitative temporal constraints). Koubarakis proposed an explicit representation of temporal constraints on data; moreover, the local temporal constraints on tuples are stored into a dedicated attribute. He also defined the algebraic operators and theoretically analyzed their complexity. On the other hand, the work by Brusoni et al. [3] mainly focused on defining an integrated approach in which "standard" artificial intelligence temporal reasoning capabilities (such as the ones sketched above in this entry) are suitably extended and paired with an (extended) relational temporal model. First, the data model is extended in such a way that each temporal tuple can be associated with a set of identifiers, each one referring to a time period. A separate relation is used in order to store the qualitative (and quantitative) temporal constraints of such periods. The algebraic operations of intersection, union, and difference are defined over such sets of periods, and *indeterminacy* (e.g., about the existence of the intersection between two periods) is coped with through the adoption of *conditional intervals*. Algebraic relational operators are defined on such a data model, and their complexity is analyzed. Finally, an integrated and modular architecture combining a temporal reasoner with an extended temporal database is described, as well as a practical application to the management of temporal constraints in clinical protocols and guidelines.

Key Applications

Qualitative temporal constraints are pervasive in many application domains, in which the absolute and exact time when facts occur is generally unknown, while there are constraints on their relative order (or temporal location). Such domains include the "classical" domains of planning and scheduling but also more recent ones such as managing multimedia presentations or clinical guidelines.

As a consequence, temporal reasoning is already a well-consolidated area of research, especially within the artificial intelligence community, in which a large deal of works aimed at building *application-independent* and *domain-independent* managers of temporal constraints. Such managers are intended to be *specialized knowledge servers* that represent and reason with temporal constraints and that *cooperate* with other software modules in order to solve problems in different applications. For instance, in planning problems, a temporal manager could cooperate with a planner, in order to check incrementally the temporal consistency of the plan being built. In general, the adoption of a specialized temporal manager is advantageous from the computational point of view (e.g., with respect to general logical approaches based on theorem proving), and it allows programmers to focus on their domain-specific and application-specific problems and to design modular architectures for their systems.

On the other hand, the impact and potentiality of extensively exploiting qualitative temporal reasoning in temporal databases have only been minimally explored by the database community, possibly due to the computational complexity that it necessarily involves. However, (temporal) databases will be increasingly applied to new application domains, in which the structure and the interdependencies of facts (including the temporal dependencies) play a major role, while the assumption that the absolute temporal location of facts is known does no longer hold. Significant application areas include database applications to store workflows, protocols, guidelines (see, e.g., the example in [3]), and so on. To cope with such applications, "hybrid" approaches in which qualitative (and/or quantitative) temporal reasoning mechanisms are paired with classical temporal database frameworks (e.g., along the lines suggested in [3]) are likely to play a significant role in a near future. The role of qualitative temporal constraints (and temporal reasoning) may be even more relevant at the *conceptual* level. Several temporal extensions to conceptual formalisms (such as the entity relationship) have been proposed in recent years, and there is an increasing awareness that, in many domains, qualitative (and/or quantitative) temporal constraints between conceptual objects are an intrinsic part of the conceptual model. As a consequence, qualitative temporal reasoning techniques such as the ones discussed above, may in the near future, play a relevant role at the conceptual modeling level.

Future Directions

One of several possible future research directions of qualitative temporal reasoning, which may be particularly interesting for the database community, is its application to "Active Conceptual Modeling." In his keynote talk at ER'2007, Prof. P. Chen, the creator of the entity-relationship model, has stressed the importance of extending traditional conceptual modeling to "Active Conceptual Modeling." Roughly speaking, the term "active" denotes the need for coping with evolving models having learning and prediction capabilities. Such an extension is needed in order to adequately cope with a new range of phenomena, including disaster prevention and management. Chen has stressed that "Active Conceptual Modeling" requires, besides the others, an explicit treatment of time. The extension and integration of qualitative temporal reasoning techniques into the "Active Conceptual Modeling" context are likely to give a major contribution to the achievement of predictive and learning capabilities and to become a potentially fruitful line of research.

Cross-References

▶ Absolute Time
▶ Allen's Relations
▶ Relative Time
▶ Temporal Constraints
▶ Temporal Indeterminacy
▶ Temporal Integrity Constraints
▶ Temporal Periodicity
▶ Time in Philosophical Logic
▶ Time Period
▶ Valid Time

Recommended Reading

1. Allen JF. Maintaining knowledge about temporal intervals. Commun ACM. 1983;26(11):832–43.
2. Badaloni S, Giacomin M. The algebra IA ^{fuz}: a framework for qualitative fuzzy temporal reasoning. Artif Intell. 2006;170(10):872–908.
3. Brusoni V, Console L, Pernici B, Terenziani P. Qualitative and quantitative temporal constraints and relational databases: theory, architecture, and applications. IEEE Trans Knowl Data Eng. 1999;11(6): 948–68.
4. Delgrande J, Gupta A, Van Allen T. A comparison of point-based approaches to qualitative temporal reasoning. Artif Intell. 2001;131(1–2):135–70.
5. Freksa C. Temporal reasoning based on semi-intervals. Artif Intell. 1992;54(1–2):199–227.
6. Gerevini A. Incremental qualitative temporal reasoning: algorithms for the point algebra and the ORD-Horn class. Artif Intell. 2005;166(1–2):37–80.
7. Jonsson P, Backstrom C. A unifying approach to temporal constraint reasoning. Artif Intell. 1998;102(1):143–55.
8. Koubarakis M. Database models for infinite and indefinite temporal information. Inf Syst. 1994;19(2):141–73.
9. Ladkin P. Time representation: a taxonomy of interval relations. In: Proceeding of the 5th National Conference on AI; 1986. p. 360–6.
10. Ligozat G. Qualitative spatial and temporal reasoning. Wiley; 2013. 539p. ISSBN:978-1-84821-252-7.
11. Nebel B, Burkert HJ. Reasoning about temporal relations: a maximal tractable subclass of Allen's interval algebra. J ACM. 1995;42(1):43–66.
12. Prior AN. Past, present and future. Oxford: Oxford University Press; 1967.
13. Terenziani P. Integrating calendar-dates and qualitative temporal constraints in the treatment of periodic events. IEEE Trans Knowl Data Eng. 1997;9(5): 763–83.
14. Van Beek P. Reasoning about qualitative temporal information. Artif Intell. 1992;58(1–3):297–326.
15. Vilain M, Kautz H. Constraint propagation algorithms for temporal reasoning. In: Proceedings of the 5th National Conference on Artificial Intelligence; 1986. p. 377–382.

Quality and Trust of Information Content and Credentialing

Chintan Patel and Chunhua Weng
Columbia University, New York, NY, USA

Definition

The quality and reliability of biomedical information is critical for practitioners (clinicians and biomedical scientists) to make important decisions about patient conditions and to draw key scientific conclusions towards developing new drugs, therapies and procedures. Evaluating the quality and trustworthiness of biomedical information [1] requires answering questions such as, *where did the data come from, under what conditions was the data generated, how accurate and complete is the data,* and so on.

Key Points

The quality and reliability of biomedical information is dependent on the task or the context of the application. There is a basic set of domain-independent features that can be used to characterize the quality and trustworthiness of the information:

Accuracy: The correctness of the information. Inherent noisiness in the underlying data generating clinical processes or biological experiments often leads to various errors in the resulting data. It is critical to quantify the frequency and source of such errors using the measure of accuracy to enable the clinicians and researchers for making informed decisions using the data.

Completeness: In biomedical settings, it is often necessary to have all the pertinent (complete) information at hand while making important clinical decisions or performing complex biological experiments. There are various constraints due to limited technology and the nature of clinical practice that leads to incomplete biomedical data:

1. In healthcare settings, only partial data gets recorded in electronic form and vast amount of data is still only available on paper [4]. For example, various clinical notes such as admit, progress and discharges notes containing valuable clinical information are generally written on paper charts and not entered electronically.
2. In several instances, despite the availability of electronic data, the information is not usable or accessible due to differences in underlying data standards, information models, terminology, etc. Consider for example, two biological databases using different ontologies for protein annotation, which will not be able to support data reuse or sharing across.

Transparency: An important aspect for determining the trustworthiness of information is the understanding or knowledge of the underlying processes or devices generating the data. Various tools or resources that provide transparency to data sources are more likely to be trusted by clinicians or biologists [2].

Credentialing: The trustworthiness of information in turn depends on the level of trust in data originators. Hence, it is important to attribute the information to its source in order to allow consumers of information to make appropriate decisions. Consider for example that a biolog-

ical annotation reviewed by human curators will be trusted more than an annotation automatically extracted from literature using text-mining [3].

As the biomedical domain becomes more and more information driven, the methods and techniques to characterize the quality and trustworthiness of information will gain more prominence.

Cross-References

► Clinical Data Quality and Validation

Recommended Reading

1. Black N. High-quality clinical databases: breaking down barriers. Lancet. 1999;353(9160):1205–11.
2. Buza TJ, McCarthy FM, Wang N, Bridges SM, Burgess SC. Gene Ontology annotation quality analysis in model eukaryotes. Nucleic Acids Res. 2008;36(2):e12.
3. D'Ascenzo MD, Collmer A, Martin GB. PeerGAD: a peer-review-based and community-centric web application for viewing and annotating prokaryotic genome sequences. Nucleic Acids Res. 2004;32(10):3124–59.
4. Thiru K, Hassey A, Sullivan F. Systematic review of scope and quality of electronic patient record data in primary care. BMJ. 2003;26(7398):1070.

Quality of Data Warehouses

Rafael Romero[1], Jose-Norberto Mazón[1], Juan Trujillo[3], Manuel Serrano[1], and Mario Piattini[2]
[1]University of Alicante, Alicante, Spain
[2]University of Castilla-La Mancha, Ciudad Real, Spain
[3]Lucentia Research Group, Department of Information Languages and Systems, Facultad de Informática, University of Alicante, Alicante, Spain

Definition

Quality is an abstract and subjective aspect for which there is no universal definition. It is usually said that there is a quality definition for each person. Perhaps the most abstract definition for this topic is that the data warehouse quality means the data is suitable for the intended application by all users. In this way, it is very complex to measure or assess the quality of a data warehouse system. Normally, the *data warehouse* quality is determined by (i) the quality of the data presentation and (ii) the quality of the *data warehouse* itself. The latter is determined by the quality of the database management system (DBMS), the data quality, and the quality of the underlying data models used to design it. A good design may (or may not) lead to a good data warehouse, but a bad design will surely render a bad data warehouse of low quality. In order to measure the quality of a data warehouse, a key issue is defining and validating a set of metrics to help to assess the quality of a data warehouse in an objective way, thus guaranteeing the success of designing a good data warehouse.

Historical Background

Few works have been presented in the area of objective indicators or metrics for data warehouses. Instead, most of the current proposals for data warehouses still delegate the quality of the models to the experience of the designer.

Only the model proposed by Jarke et al. [6], which is described in more depth in Vassiladis' Ph.D. thesis [15], explicitly considers the quality of conceptual models for data warehouses. Nevertheless, these approaches only consider quality as intuitive notions. In this way, it is difficult to guarantee the quality of data warehouse conceptual models, a problem which has initially been addressed by Jeusfeld et al. [7] in the context of the DWQ (data warehouse quality) project. This line of research addresses the definition of metrics that allows designers to replace the intuitive notions of quality of conceptual models of the data warehouse with formal and quantitative metrics. Sample research in this direction includes normal forms for data warehouse design as originally proposed in [9] and generalized in [8]. These normal forms represent a first step

toward objective quality metrics for conceptual schemata.

Following the idea of assessing the quality of data warehouses in an objective way, several metrics for evaluating the quality of data warehouse logical models have been proposed in recent years and validated both formally and empirically [12, 13].

Lately, Si-Saïd and Prat [14] have proposed some metrics for multidimensional schemas analyzability and simplicity, and on [3], several objective metrics for both conceptual and logical design are listed. Nevertheless, none of the metrics proposed so far has been empirically validated, and therefore, their practical utility has not been proven.

Foundations

The information quality of a data warehouse is determined by (i) the quality of the system itself and (ii) the quality of the data presentation (see Fig. 1). In fact, it is important that the data of the data warehouse not only correctly reflects the real world but also that the data are correctly interpreted. Regarding data warehouse quality, three aspects must be considered: the quality of the DBMS (database management system) that supports it, the quality of the data models used in their design (conceptual, logical, and physical), and the quality of the data contained in the data warehouse. The presentation quality of data warehouses is more related to the data presentation according to front-end tools such as OLAP (online analytical processing), data reporting, or data mining. For this reason, the following issues pertaining to data warehouse quality are described next: (i) DBMS quality, (ii) data model quality, and (iii) data quality.

Quality of DBMS

The quality of the DBMS in which the data warehouse is implemented is important, since the database engine is the core of the data warehouse system and has a deep impact on the performance of the whole system. A good DBMS could improve the performance of the system and the quality of the data by implementing constraints and integrity rules. In order to assess the quality of a DBMS, several international standards can be used, such as ISO/IEC 25010 [4] and ISO/IEC 9075 [5] or even information from database benchmarks [10].

Quality of Data Warehouse Data Models

The quality of the data models used in the design of data warehouses relies on the quality of the conceptual, logical, and physical models used for its design. A first step toward obtaining high-quality data models is the definition of development methodologies. However, a methodology, though necessary, may not be sufficient to guarantee the quality of a data warehouse. Indeed, a good methodology may (or may not) lead to a good product, but a bad methodology will surely render a bad product of low quality. Furthermore, many other factors could influence the quality of the products, such as human decisions. Therefore, it is necessary to complete specific methodologies with metrics and techniques for product quality assessment.

Structural properties (such as structural complexity) of a software artifact have an impact on its cognitive complexity as shown in Fig. 2. Cognitive complexity means the mental burden on those who have to deal with the artifact (e.g., developers, testers, maintainers). High cognitive complexity of an artifact reduces its understandability and leads to undesirable external quality attributes as defined in the standard ISO/IEC 25010 [4]. The model presented in Fig. 2 is an adaptation of the general model for software artifacts proposed in Briand et al. [2].

Quality of Data Warehouses, Fig. 1 Quality of the information and the data warehouse

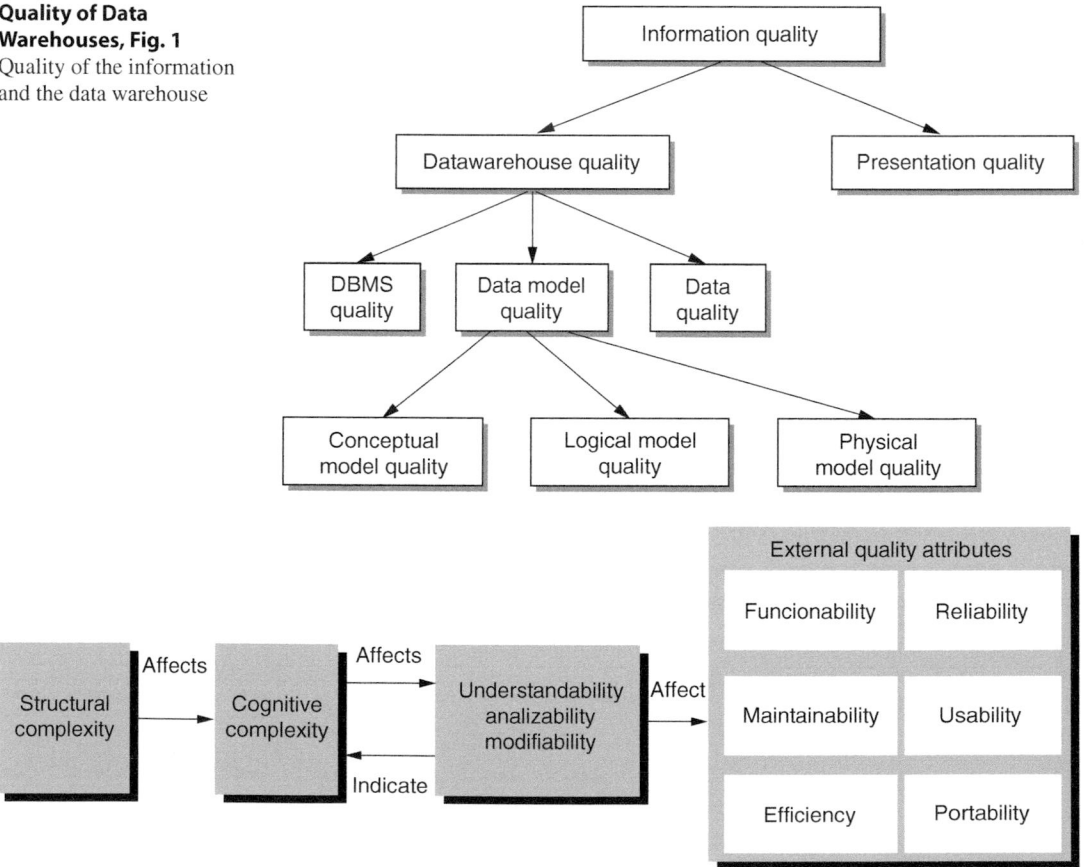

Quality of Data Warehouses, Fig. 2 Relationship between structural properties, cognitive complexity, understandability, and external quality attributes – based on the work described in Briand et al. [13]

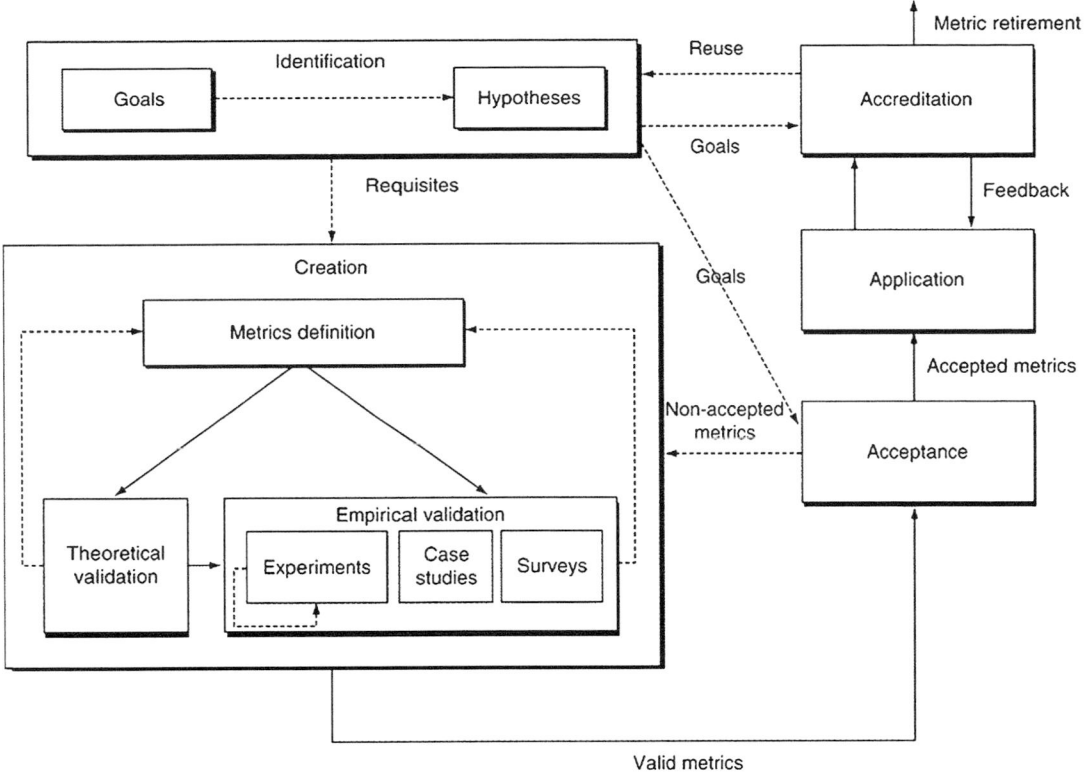

Quality of Data Warehouses, Fig. 3 Metrics creation process

Indeed, as data warehouse models are software artifacts, it is reasonable to consider that they follow the same pattern. It is thus important to investigate the potential relationships that can exist between the structural properties of these schemas and their quality factors.

In order to get a valid set of data warehouse metrics, the definition of metrics should be based on clear measurement goals, and the metrics should be defined following the organization's needs related to external quality attributes. In defining metrics, it is also advisable to take into account the expert's knowledge. Figure 3 presents a method (based on the method followed in [12, 13]) for obtaining valid and useful metrics. In this figure, continuous lines show metric flow and dotted lines show information flow.

This method has five main phases starting at the identification of goals and hypotheses and leading to the metric application, accreditation, and retirement:

1. *Identification*: Goals of the metrics are defined and hypotheses are formulated. All of the subsequent phases will be based upon these goals and hypotheses.
2. *Creation*: This is the main phase, in which metrics are defined and validated. This phase is divided into three subphases:

(a) *Metrics definition*. Metric definition is made by taking into account the specific characteristics of the system to be measured, the experience of the designers of these systems, and the work hypotheses. A goal-oriented approach as GQM (goal-question-metric [1]) can also be very useful in this step.
(b) *Theoretical validation*. The formal (or theoretical) validation helps identify when and how to apply the metrics. There are two main tendencies in measuring formal validation: the frameworks based on axiomatic approaches [8] and the ones

Practical Considerations

A detailed study of the empirical performance of quantile algorithms was carried out by Cormode et al. [5] on IP stream datasets. They concluded that with careful implementation, a commodity hardware machine (dual Pentium 2.8 GHz CPU and 4 GB RAM) can keep up with a 2 Gbit/s stream (310,000 packets per second). Performance numbers can depend also on the input distribution. For example, the deterministic algorithms presented above can have different memory usage and accuracy depending on the order in which the input values are presented, but sketching techniques such as the Count-Min sketch are not affected by the order of the input. The input value distribution can also impact perceived accuracy of the approximate quantiles. For example, for skewed distributions, the *numeric value* of the exact φ-th quantile can be arbitrarily far from the numeric values of the $(\varphi \pm \varepsilon)$-th quantile.

Extensions

Given the fundamental nature of quantiles and their widespread applications in data processing, it is no surprise that there are multiple extensions of the basic setting that have been considered so far. There are many interesting and practically-motivated applications, such as the latency of the web site mentioned earlier, where quantiles must be computed over distributed data, or over a sliding window portion of the stream etc. In the following, the current state of algorithms for these variants are briefly discussed.

Quantiles in Distributed Streams

In many settings, data of interest are naturally distributed across multiple sources, such as servers in a web application and measurement devices in a sensor network. In these applications, it is necessary to compute the quantile summary of the entire data, but *without* creating a centralized repository of data, which could be undesirable because of the additional latency, communication overhead, or energy constraints of untethered sensors. The efficiency of an algorithm in this distributed setting is measured by the amount of information each node in the system must transmit during the computation.

One natural approach for distributed approximation of quantiles is for each node (server, sensor, etc.) to compute a local summary of its data, and arrange the nodes in a virtual hierarchy that guides them to merge these summaries into a final structure computed at the root of the hierarchy. The tree-based Q-digest [18] algorithm extends rather easily to the distributed setting, as the histogram boundaries of the Q-Digest are aligned to binary partition of the original value space U. The space complexity of the distributed version remains the same as the stream version, namely, $O\left(\frac{1}{\epsilon} \log U\right)$. The GK algorithm is little more complicated to extend to distributed streams, but Greenwald and Khanna themselves developed such an extension in [9]. However, the space complexity of their distributed data structure grows to $O\left(\frac{1}{\epsilon}\log^3 n\right)$ [9]. The Bloom filter based Count-Min sketch [3] also extends easily to the distributed setting also without any increase in the space complexity.

Quantiles in Sliding Windows

In many applications, the user is primarily interested in the most recent portion of the data stream. This poses the *sliding window* extension of the quantiles problem, in which the desired quantile summary for the most recent N data elements - the window slides with the arrival of each new element, as the oldest element of the window is discarded and the new arrival added. In [12], Lin et al. presented such a sliding window scheme for quantile summaries, however, the space requirement for their algorithm is $O\left(\frac{1}{\epsilon^2} + \frac{1}{\epsilon}\log \epsilon^2 N\right)$. This was soon improved by Arasu and Manku [1] to $O\left(\frac{1}{\epsilon}\log\frac{1}{\epsilon}\log N\right)$.

Biased Estimate of Quantiles

The absolute measure of approximation precision is quite reasonable as long as the error εn is quite small compared to the rank of the quantile sought, namely, φn. This holds as long as the quantiles of interest are in the middle of the distribution. But if φ is either close to 0 or 1, one may prefer a *relative* error, so that the estimated quantile is in the range $[(1 - \varepsilon)\varphi, (1 + \varepsilon)\varphi]$. This variant was solved by Gupta and Zane [13] using random sampling techniques with a $O\left(\frac{1}{\varepsilon^3}\log n \log \frac{1}{\delta}\right)$ size data structure. The space bound has since been improved by Cormode et al. [4] to $O\left(\frac{1}{\varepsilon}\log U \log(\epsilon n)\right)$ using a deterministic algorithm.

Duplicate Insensitive Quantiles
In some distributed settings, a single event can be observed multiple times. For example in the Internet, a single packet is observed at multiple routers. In wireless sensor networks, due to the broadcast nature of the medium, and to add fault-tolerance, data can be routed along multiple paths. Summaries such as quantiles or the number of distinct items are clearly not robust against duplication of data; on the other hand, simpler statistics such as minimum and maximum are not affected by duplication. Flajolet and Martin's distinct counting algorithm [8] is the seminal work in this direction. Cormode et al. have introduced algorithms based on sampling to compute various duplicate insensitive aggregates [6]. Their Count-Min sketch can be also easily adapted to compute duplicate insensitive quantiles.

Key Applications

Internet-scale network monitoring and database query optimization are two important applications that originally motivated the need for quantiles summaries over data streams. Gigascope [7] is a streaming database system that employs statistical summaries such as quantiles for monitoring network applications and systems. Quantile estimates are also widely used in query optimizers to estimate the size of intermediate results, and use those estimates to choose the best execution plan [5]. Distributed quantiles have been used to succinctly summarize the distribution of values occurring over a sensor network [8]. In a similar context, distributed quantiles are also used to summarize performance of websites and distributed applications [17].

Future Directions

The field of computing approximate quantiles over streams have led to a fertile research program and is expected to bring up new challenges in both theory and implementation. Although there is an obvious lower bound of $\Omega\left(\frac{1}{\varepsilon}\right)$ memory required to compute ε-approximate quantiles, there is no known non-trivial lower bound on memory. Since the current best algorithms require $O\left(\frac{1}{\varepsilon}\log(\epsilon n)\right)$ or $O\left(\frac{1}{\varepsilon}\log(U)\right)$ memory, it will be useful to either lower the memory usage or prove a better lower bound.

Another direction in which improvements are highly desirable is running time. The current deterministic algorithms require amortized running time of $O\left(\log\frac{1}{\epsilon} + \log\log(\epsilon n)\right)$ or $O\left(\log\frac{1}{\epsilon} + \log\log(U)\right)$ per item. In high data-rate streams, even such low processing times are not fast enough: what is desired is a $O(1)$ insert time, or even a sublinear time quantile algorithm. As of now, there is no memory efficient sublinear time quantile algorithm known except for random sampling.

Cross-References

▶ Adaptive Stream Processing
▶ Approximate Query Processing
▶ Continuous Query
▶ Data Aggregation in Sensor Networks
▶ Data Stream
▶ Distributed Data Streams
▶ Distributed Query Processing
▶ Geometric Stream Mining
▶ Hierarchical Heavy Hitter Mining on Streams
▶ In-Network Query Processing
▶ Stream Mining
▶ Stream Processing
▶ Streaming Applications

Recommended Reading

1. Arasu A, Manku GS. Approximate counts and quantiles over sliding windows. In: Proceedings of the 23rd ACM SIGACT-SIGMOD-SIGART Symposium on Principles of Database Systems; 2004. p. 286–96.

Foundations

Frequent Rules

To demonstrate the procedures of the frequent quantitative association rule discovery, take the toy database in Fig. 1a as an example [2], where attribute "Age" is numerical, and "Married" and "NumCars" are categorical attributes. Assuming the user specified *Support* and *Confidence* values are 40% and 50% respectively, it means that a prospective rule (the left-hand and the right-hand together) should cover at least two records ($5 \times 40\%$). For all data records satisfying the left-hand of the rule, 50% of them should also contains the right-hand sides of the rule ($X \rightarrow Y$). The major steps of discovering frequent quantitative association rules can then be summarized as follows [2]:

1. Determining the partition numbers and the region of partitioning for each numerical attribute. E.g., Fig. 1a lists four partitions for "Age" (denoted by [20,24], [25,29], [30,34],

and [35,39]), with each region mapping to one integer value {1,2,3,4}.

2. Applying each mapping table to all records of the corresponding numerical attribute, with numerical values replaced by the matching integer values. The example of the mapped database is shown in Fig. 1c.

3. Generating frequent itemsets based on the mapped database and the user specified *Support* value (the existing Apriori like association rule mining methods can be applied directly).

4. Using discovered frequent itemsets to generate quantitative association rules, with each frequent itemset decomposed into two (left- and right-hand) components. For example, if itemset "ABCD" is found frequent, a possible quantitative association rule can be made by decomposing "ABCD" as "AB" $\rightarrow$ "CD." As long as the validate check asserts that confidence value of this rule ("AB" $\rightarrow$ "CD") is greater than the user specified value (*Confidence*), the rule is taken as a valid rule.

a   People

RecordID	Age	Married	Numcars
100	23	No	1
200	25	Yes	1
300	29	No	0
400	34	Yes	2
500	38	Yes	2

b   Mapping age

Interval	Integer
20..24	1
25..29	2
30..34	3
35..39	4

c   After mapping attributes

RecordID	Age	Married	Numcars
100	1	2	0
200	2	1	1
300	2	2	1
400	3	1	2
500	4	1	2

d   Frequent itemsets: sample

Itemsets	Support
{ ⟨Age: 20..29⟩ }	3
{ ⟨Age: 30..39⟩ }	2
{ ⟨Married: Yes⟩ }	3
{ ⟨Married: No⟩ }	2
{ ⟨Numcars: 0..1⟩ }	3
{ ⟨Age: 30..39⟩, ⟨Married: Yes⟩ }	2

e   Rules: sample

Rule	Support	Confidence
⟨Age: 30..39⟩ and ⟨Married: Yes⟩ ⇒ ⟨Numcars: 2⟩	40%	100%
⟨Age: 20..29⟩ ⇒ ⟨Numcars: 0.1⟩	60%	66.6%

Quantitative Association Rules, Fig. 1 Example of problem decomposition for quantitative association rule mining (Revised from [3])

5. Collect all quantitative association rules generated from the above process and prune redundant rules. E.g., if "AB" → "CD" and "AB" → "CDE" are both valid rules, "AB" → "CD" can be pruned as it can be generalized (inferred) from "AB" → "CDE".

Distributional Rules

Different from the frequent quantitative association rules, where the main challenge is to determine the "optimum" number of partitions and the region of partitioning, the distributional rules represent a set of quantitative association rules, where the statistical features of the samples covered by the rule are different from the whole population. Similar to general association rules, the distributional rules also contain the left- and the right-hand. The left-hand side of the rule is a description of a subset of the database, while the right-hand side provides a description of outstanding behaviors of this subset. A rule is only interesting if the mean for the subset (specified by the left-hand side) is significantly different from the rest and is therefore unexpected. Consider the following distributional rule, the left-hand side consists of two categorical attributes (Non-smoker: {Yes or No}, and Wine-drinker: {Yes or No}), and the right-hand side is a numerical attribute (life expectancy). The rule is considered informative and meaningful as it indicates that individuals characterized by the left-hand side of the rule (Non-smoker and wine-drinker) have a longer average life expectancy (85) than the whole population (80).

$$\text{Non-smoker and wine-drinker} \rightarrow \text{expectancy} = 85 \,(\text{overall} = 80)$$

Following this definition, one can easily extend the framework to allow one or multiple numerical attributes to appear on the left- or right-hand sides of the rule (or both), or employ other statistical measures rather than the mean values to assess the rules.

In order to discover distributional rules, Aumann and Lindell [5] proposed two methods to discover the following two types of rules:

- $X \rightarrow Mean\ J\ (T\ X)$, where X and J denote a single numerical attribute, $T\ X$ denotes transactions confined by attribute X and $\text{Mean}_J(T\ X)$ represents the mean value of attribute J (for all samples in $T\ X$).
- $X \rightarrow M\ J\ (T\ X)$, where X denotes one or multiple categorical attributes, J consists of one or multiple quantitative attributes, and M means one particular statistical measure (there is no restriction on the number of attributes in X and J).

The solution to the first type of distributional rules is straightforward, since the rules only involve two numerical attributes (one on each side), an algorithm can afford to go through each pair of numerical attributes to discover meaningful rules. More specifically, for any two numerical attributes i and j, one can first sort all records in the database based on the attribute i, then any above or below average continuous region of values in j can form a prospective quantitative rule. For example, given the toy database in Fig. 2a which records the age and the size of the striped bass, the sorted database with respect to attribute i (age) is given in Fig. 2a. The average of attribute j for the top three records (001, 003, and 002) is 0.77, which is significantly lower than the mean of the whole population (1.83). Therefore, the below average region (001, 003, and 002) forms a meaningful quantitative association rule denoted by:

$$\text{Age} \leq 2 \rightarrow \text{Mean weight } 0.77\text{lbs (Overall } 1.83)$$

Because any above or below average continuous region of values in j can form a prospective quantitative rule, one can continuously span the region with respect to the attribute j, and discover the maximum region satisfying the user specified requirements (e.g., α times less/larger than the average).

In order to discover the second type of distributional quantitative association rules, one can employ a two-stage approach, which applies general association rule to the whole database by considering categorical attributes

Quantitative Association Rules, Fig. 2 Example of distributional association rule mining (age and weight of the striped bass)

a

ID	i (Age)	j weight (lbs)
001	1.0	0.5
002	2.0	0.8
003	1.5	1.0
004	2.5	1.7
005	4.0	3.8
006	3.5	3.2
Mean	2.42	1.83

Original table

b

ID	i (Sorted age)	j weight (lbs)	
001	1.0	0.5	⎫
003	1.5	1.0	⎬ 0.77
002	2.0	0.8	⎭
004	2.5	1.7	
006	3.5	3.2	
005	4.0	3.8	
Mean	2.42	1.83	

Sorted by attribute i

only, followed by a refining process to check each rule by considering the numerical attribute values [4].

1. *Discovering frequent itemsets:* Finding all frequent itemsets by considering categorical attributes of the database only (this can be easily achieved through existing Apriori-like algorithms [1].

2. *Calculating statistical distribution values:* For each numerical attribute, calculate the value of the distribution measures (mean/variance) over samples confined by each frequent itemset. For example, if "Non-smoker and Wine-drinker" are found frequent (i.e.,, a frequent itemset), all samples matching this itemset form a sample set P, from which the statistical distribution value of a numerical attribute can be calculated.

3. *Refining quantitative association rules:* For every frequent itemset (denoted by X) and one numerical attribute e, the algorithm continuously check if $X \rightarrow Mean\ e\ (T\ X)$ and $X \rightarrow Variance\ e\ (T\ X)$ are meaningful rule (comparing to the whole population). In addition, for any two rules $X \rightarrow Mean\ e\ (T\ X)$ and $Y \rightarrow Mean\ e\ (T\ Y)$, the algorithm will check whether the former is a sub-rule of the latter, or vice versa, such that the algorithm can output compact rules with minimum redundancy.

Structure Patterns

Structure patterns represent a special type of association rules, where items share interconnected relationships or linkages. For example, chemical compounds in biochemistry are often represented as graph structured data [6, 7], where each graph represents a chemical

compound with nodes denoting molecular atoms, such as a carbon, and linkages representing their bond relationships, such as a hydrogen bond. Finding structure patterns, such as frequent subgraphs, can help discover significant substructures commonly exist in the data in order to carry out Quantitative Structure-Activity Relationship (QSAR) modeling [6,7], or build graph classification models [8].

In bioinformatics, protein interaction networks can also be represented as a large network, with each node denoting a protein and edges denoting their interactions. Finding recurrent substructures commonly appearing in the network, i.e., a network motif [9], is a key step to search patterns potentially linked to significant biological functions.

In order to discover structure patterns, the mining process needs to consider both the items and their structure dependency relationships in order to generate patterns, and evaluate their significance of being a legitimate pattern. The structure pattern growth is much more complicated than frequent item sets, mainly because when growing a pattern, a new node can connect to any of the nodes in a candidate pattern, which is a combinatorial issue. In addition, two graphs may be identical but showing in different structures, i.e., a known graph isomorphism problem. Common practice is to introduce a specific coding mechanism, such as assigning a unique Depth-First-Search (DFS) code for each subgraph, according to the time the subgraph is discovered during the search process. Two isomorphism graphs will have the same DFS code, so graph enumeration and structure pattern growth for quantitative structure rule mining can be carried out efficiently.

Key Applications

Business intelligence, market basket analysis, fraud detection (fraud medical insurance claims), biochemical and bioinformatics

Future Directions

All of the above techniques intend to discover quantitative association rules in the forms of the conjunction of individual attributes. One interesting problem is to find quantitative association rules with (linearly or nonlinearly) combined attributes. For example, finding rules like "Age $\leq \alpha \to$ Length/Weight $\leq \beta$." Here the right-hand of the rule is a non-linear combination of the numerical attributes (Length and Height), and α and β are some discovered values. In [5], *Ruckert* et al. proposed a quantitative association rule mining approach which is able to discover similar rules with linearly weighted attributes like "Age $\leq \alpha \to \alpha 1 \cdot$ Length $+ \alpha 2$ Weight $\leq \beta$." Future research may emphasize the generalized quantitative association rule discovery, where rules consist of non-linearly combined attributes.

Another interesting problem concerning quantitative association rules is to discover relational patterns of the quantitative association rules across multiple databases. E.g., Finding patterns that are frequent with a support level of α in database A, but significantly infrequent with a support level of β in databases B or/and C. In [10], Zhu and Wu proposed a hybrid frequent pattern tree based solution to address this problem with a focus on the general association rules. Extending the problem of relational frequent pattern discovery to quantitative association rules is another interesting topic for future research.

Cross-References

▶ Association Rules

Recommended Reading

1. Agrawal R, Imielinski T, Swami A. Mining association rules between sets of items in large databases. In: Proceedings of the ACM SIGMOD International Conference on Management of Data; 1993. p. 207–216.
2. Srikant R, Agrawal R. Mining quantitative association rules in large relational tables. In: Proceedings of the ACM SIGMOD International Conference on Management of Data; 1996. p. 1–12.
3. Aumann Y, Lindell Y. A statistical theory for quantitative association rules. In: Proceedings of the 5th ACM SIGKDD International Conference on Knowledge Discovery and Data Mining; 1999. p. 261–70.
4. Webb GI. Discovering associations with numeric variables. In: Proceedings of the 7th ACM SIGKDD International Conference on Knowledge Discovery and Data Mining; 2001. p. 383–8.
5. Ruckert U, Richter L, Kramer S. Quantitative association rules based on half-spaces: an optimization approach. In: Proceedings of the 2004 IEEE International Conference on Data Mining; 2004. p. 507–10.
6. Frid A, Matthews E. Prediction of drug-related cardiac adverse effects in humans-B: use of QSAR programs for early detection of drug-induced cardiac toxicities. Regul Toxicol Pharmacol. 2010;56(3):276–89.
7. Harpaz R, DuMouchel W, Shah NH, Madigan D, Ryan P, Friedman C. Novel data-mining methodologies for adverse drug event discovery and analysis. Clin Pharmacol Therapeut. 2012;91(6):1010–21.
8. Pan S, Zhu X. Graph classification with imbalanced class distributions and noise. In: Proceedings of the 23rd International Joint Conference on Artificial Intelligence; 2013. p. 1585–92.
9. Alon U. Network motifs: theory and experimental approaches. Nat Rev Genet. 2007;8(6):450–61.
10. Zhu X, Wu X. Discovering relational patterns across multiple databases. In: Proceedings of the 23rd International Conference on Data Engineering; 2007. p. 726–35.

QUEL

Tore Risch
Department of Information Technology, Uppsala University, Uppsala, Sweden

Definition

QUEL was the query language used in the original Ingres system from Berkeley University.

Key Points

QUEL was one of the first relational database query languages. It can be seen as a syntactically sugared tuple relational calculus language. The Postgres extension of Ingres originally used an extention of QUEL called PostQUEL, but was later replaced with SQL.

Query by Humming

Yingyi Bu[1], Raymond Chi-Wing Wong[2], and Ada Wai-Chee Fu[1]
[1]Chinese University of Hong Kong, Hong Kong, China
[2]Department of Computer Science and Engineering, The Hong Kong University of Science and Technology, Clear Water Bay, Kowloon, Hong Kong

Synonyms

Music retrieval; Time series database querying

Definition

With the appearance of large scale audio and video databases in various application areas, novel information retrieval methods adapted to the specific characteristics of these data types are required. A natural way of searching in a musical audio database is by humming the tune of a song as a query, which is so-called "query by humming". In this entry, state-of-the-art techniques for effective and efficient querying by humming are described.

Historical Background

In 1995, Asif Ghias et al. [2] proposed the basic architecture for a system supporting query by humming. Three main components are introduced in the system: a pitch-tracking module, a melody database, and a query engine. Queries are hummed into a microphone, digitized, and fed into a pitch-tracking module. Then, a symbol sequence representation upon the relative pitch transitions of the hummed melody is sent to the query engine, which produces a ranked list of matching melodies.

Foundations

In recent researches, musical data are modeled as time series which are real valued sequences rather than symbol sequences. In speech comparisons, small fluctuation of the tempo of the speaker could be allowed in order to identify similar contents. There have been some works that match a melody more effectively by considering warping and scaling in humming queries. This generally gives better query results because it is free from the error-prone note segmentation. However, those works rely on distance measures such as universal scaling (US), dynamic time warping (DTW) [3, 5] and scaling and time warping (SWM) [1], the efficiency might be rather poor. Fortunately, tight lower bounds for DTW and SWM could greatly improve the efficiency by pruning large portions of non-candidate data at an early stage.

Comparisons of Distance Measures on Examples

Figure 1 demonstrates the effects of different distance measures with a typical piece of music, *Happy Birthday to You*, from top to bottom:

1. Since the query sequence is performed at a much faster tempo, direct application of DTW fails to produce an intuitive alignment;
2. Rescaling the shorter performance by a scaling factor of 1.54 seems to improve the alignment, but the higher pitched note produced on the third "*birth...*" of the candidate is forced to align with the lower note of the third "*happy...*" in the query;
3. Only the application of *both* uniform scaling and DTW produces the appropriate alignment.

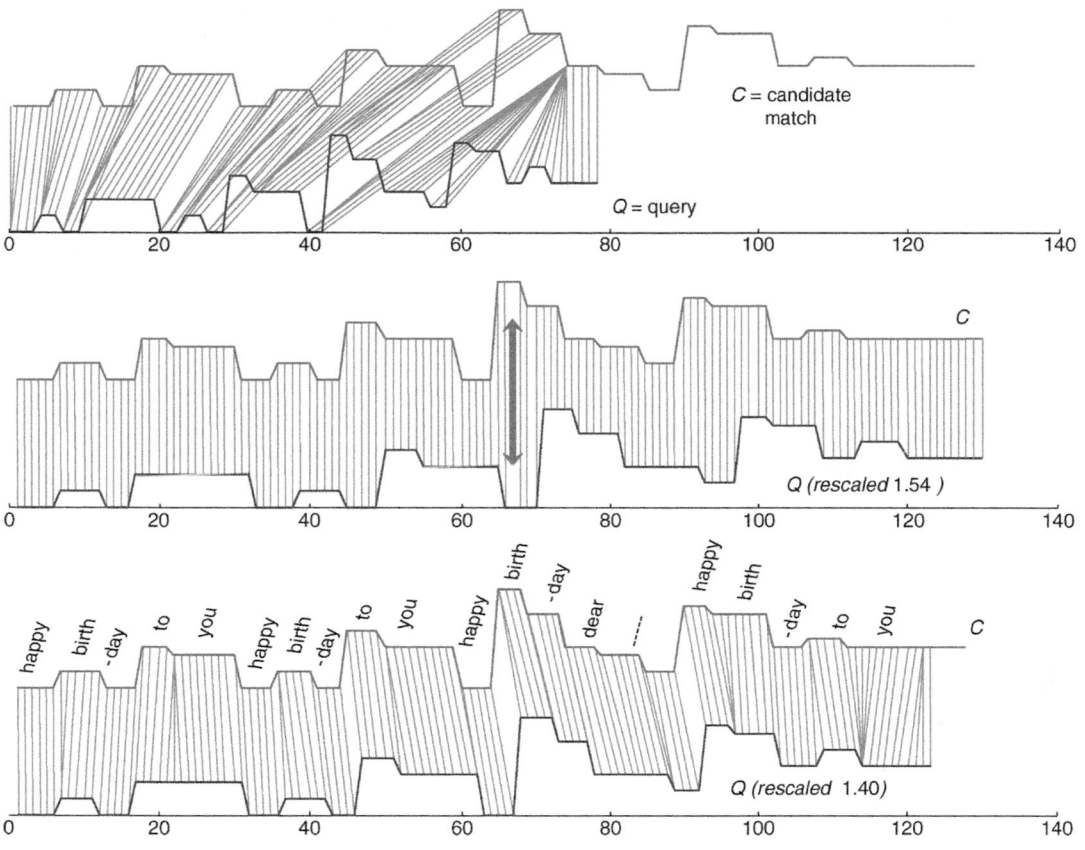

Query by Humming, Fig. 1 Motivating example

Dynamic Time Warping (DTW)

Intuitively, *dynamic time warping* is a distance measure that allows time series to be *locally* stretched or shrunk before the base distance measure is applied. Given two sequences $C = C_1, C_2, \ldots, C_n$ and $Q = Q_1, Q_2, \ldots, Q_m$, the time warping distance DTW is defined recursively as follows:

$$
\begin{aligned}
&\text{DTW}(\phi, \phi) = 0 \\
&\text{DTW}(C, \phi) = \text{DTW}(\phi, Q) = \infty \\
&\text{DTW}(C, Q) = D_{base}(\text{First}(C), \text{First}(Q)) + \\
&\quad \min \begin{cases} \text{DTW}(C, \text{Rest}(Q)) \\ \text{DTW}(\text{Rest}(C), Q) \\ \text{DTW}(\text{Rest}(C), \text{Rest}(Q)) \end{cases}
\end{aligned}
$$

where Φ is the empty sequence, $\text{First}(C) = C_1$, $\text{Rest}(C) = C_2, C_3, \ldots, C_n$, and D_{base} denotes the distance between two entries. Several L_p

measures were used as the D_{base} distance in previous literature, such as Manhattan Distance (L_1), squared Euclidean Distance (L_2) and maximum difference (L_∞). Typically *Squared Euclidean Distance* is used as the D_{base} measure. That is,

$$
D_{base}(C_i, Q_j) = (C_i - Q_j)^2.
$$

Thus in the following parts, without loss of generality, it is assumed that D_{base} is the squared Euclidean Distance and D is also used to denote it. However, the time complexity of DTW distance calculation is $O(mn)$, and intensive computations are employed for the corresponding dynamic programming. Thus, lower bounds on the distance are adopted to effectively prune the search space and support efficient search.

Constraints and Lower Bounds on Dynamic Time Warping

Keogh et al. [3] viewed a global constraint as a constraint on the warping path entry $w_k = (i, j)_k$ and gave a general form of global constraints in terms of inequalities on the indices to the elements in the warping matrix,

$$j - r \leq i \leq j + r$$

where r is a constant for the Sakoe-Chiba Band and r is a function of i for the Itakura Parallelogram. Incorporating the global constraint into the definition of dynamic time warping distance, DTW can be modified as follows.

Given two sequences $C = C_1, C_2, \ldots, C_n$ and $Q = Q_1, Q_2, \ldots, Q_m$, and the time warping constraint r, the constrained time warping distance cDTW is defined recursively as follows:

$$Dist_r (C_i, Q_j) = \begin{cases} D_{base} (C_i, Q_j) & if |i-j| \leq r \\ \infty & otherwise \end{cases}$$

$$cDTW (\phi, \phi, r) = 0$$

$$cDTW (C, \phi, r) = cDTW (\phi, Q, r) = \infty$$

$$cDTW (C, Q, r) = Dist_r (First(C), First(Q))$$

$$+ \min \begin{cases} cDTW \left((C, Rest(Q), r \right) \\ cDTW (Rest(C), Q, r) \\ cDTW \left((Rest(C), Rest(Q), r \right) \end{cases}$$

where Φ is the empty sequence, $First(C) = C_1$, $Rest(C) = C_2, C_3, \ldots, C_n$. The *upper bounding sequence UW* and the *lower bounding sequence LW* of a sequence C are defined using the time warping constraint r as follows.

Let $UW = UW_1, UW_2, \ldots, UW_n$ and $LW = LW_1, LW_2, \ldots, LW_n$,

$$UW_i = \max (C_{i-r}, \ldots, C_{i+r}) \text{ and}$$
$$LW_i = \max (C_{i-r}, \ldots, C_{i+r})$$

Considering the boundary cases, the above can be rewritten as

$$UW_i = \max (C_{\max(1,i-r)}, \ldots, C_{\min(i+r,n)}) \text{ and}$$
$$LW_i = \min (C_{\min(1,i-r)}, \ldots, C_{\min(i+r,n)})$$

$E(C) = <UW, LW>$ is called the envelope sequences of Keogh et al. [3] propose the lower bound distance LB_Keogh based on envelope sequences. The time warping distance between two sequences Q and C is lower bounded by the squared Euclidean distance between Q and the envelope sequences of C. Equation (1) below formally defines the lower bounding distance.

$$LB_{Keogh}(Q, C) = D (Q, E(C))$$

$$= \sum_{i=1}^{m} \begin{cases} (Q_i - UW_i)^2 & if \quad Q_i > UW_i \\ (Q_i - LW_i)^2 & if \quad Q_i < LW_i \\ 0 & otherwise \end{cases}$$

Zhu et al. [4] further improve on LB_Keogh. If a transformation T is a linear transform and lower-bounding, and $Env_r(C_i)$ is the envelope of C_i by global constraint r then

$$D(T(Q), T(Env_r (C_i))) \leq cDTW (Q, C_i, r) \tag{2}$$

Therefore transforms such as PAA, DWT, SVD and DFT on the envelope sequence of a candidate sequence could still lower bound DTW distance, since those transforms are both linear and lower bounding.

Uniform Scaling (US)

Given two sequences $Q = Q_1, \ldots, Q_m$ and $C = C_1, \ldots, C_n$ and a scaling factor bound $l, l \geq 1$. Let $C(q)$ be the prefix of C of length q, where $\lceil m/l \rceil \leq q \leq lm$ and $C(m, q)$ be a rescaled version of $C(q)$ of length m,

$$C(m, q)_i = C(q)_{\lceil i.q/m \rceil} \text{ where } 1 \leq i \leq m$$
$$US(C, Q, l) = \min_{q = \lceil m/l \rceil}^{\min(lm, n)} D (C(m, q), Q)$$

where D(X, Y) denotes the Euclidean distance between two sequences X and Y.

Lower Bounding Uniform Scaling

The two sequences $UC = UC_1, \ldots, UC_m$ and $LC = LC_1, \ldots, LC_m$, such that

$$UC_i = \max \left(C_{\lceil i/l \rceil}, \ldots, C_{\lceil il \rceil}\right)$$
$$LC_i = \min \left(C_{\lceil i/l \rceil}, \ldots, C_{\lceil il \rceil}\right)$$

bound the points of the time series C that can be matched with Q. The lower bounding function, which lower bounds the distance between Q and C for any scaling ρ, $1 \le \rho \le l$, can now be defined as:

$$LB_s(Q,C) = \sum_{i=1}^{m} \begin{cases} (Q_i - UC_i)^2 & \text{if } Q_i > UC_i \\ (Q_i - LC_i)^2 & \text{if } Q_i < UC_i \\ 0 & \text{otherwise} \end{cases} \tag{3}$$

Scaling and Time Warping (SWM)

Having reviewed time warping, uniform scaling, and lower bounding, this part introduces *scaling and time warping* (SWM) distance. Given two sequences $Q = Q_1, \ldots, Q_m$ and $C = C_1, \ldots, C_n$, a bound on the scaling factor $l, l \ge 1$ and the Sakoe-Chiba Band time warping constraint r which applies to sequence length m. Let $C(q)$ be the prefix of C of length q, where $\lceil m/l \rceil \le q \le \min(lm, n)$ and $C(m, q)$ be a rescaled version of $C(q)$ of length m,

$$C(m, q)i = C(q)_{\lceil i.q/m \rceil} \text{ where } 1 \le i \le m$$

$$\text{SWM}(C, Q, l, r) = \min_{q = \lceil m/l \rceil}^{\min(lm, n)} \text{cDTW}(C(m, q), Q, r)$$

If time warping is applied on top of scaling, i.e., the sequence is first scaled, and then measure the time warping distance of the scaled sequence with the query. Typically, time warping with Sakoe-Chiba Band constrains the warping path by a fraction of the data length, which is translated into a constant r. Hence, if the fraction is 10%, then $r = 0.1|C|$. If the length of C is changed according to the scaling fraction ρ, that is, C is changed to ρC, then the Sakoe-Chiba Band time warping constraint is $r = 0.1|\rho C|$. Hence, $r = r'\rho$, where r' is the Sakoe-Chiba Band time warping constraint on the unscaled sequence, and ρ is the scaling factor.

Lower Bounding SWM

The lower envelope L_i and upper envelope U_i on C can be deduced as follows: recall that the upper and lower bounds for uniform scaling between $1/l$ and l is given by the following:

$$UC_i = \max \left(C_{\lceil i/l \rceil}, \ldots, C_{\lceil il \rceil}\right)$$
$$LC_i = \min \left(C_{\lceil i/l \rceil}, \ldots, C_{\lceil il \rceil}\right)$$

and the upper and lower bounds for a Sakoe - Chiba Band time warping constraint factor of r for a point C_i is given by:

$$UW_i = \max \left(C_{\max(1, i-r)}, \ldots, C_{\min(i+r, n)}\right)$$
$$LW_i = \min \left(C_{\max(1, i-r)}, \ldots, C_{\min(i+r, n)}\right)$$

Therefore, when time warping is applied on top of scaling the upper and lower bounds will be:

$$\begin{aligned} U_i &= \max \left(UW_{\lceil i/l \rceil}, \ldots, UW_{\lceil il \rceil}\right) \\ &= \max \big(C_{\max}(1, \lceil i/l \rceil - r'), \ldots, \\ &\quad C_{\min}(\lceil i/l \rceil + r', n), \ldots, \\ &\quad C_{\max}(1, \lceil il \rceil - r'), \ldots, \\ &\quad C_{\min}(\lceil il \rceil + r', n) \big) \\ &= \max \big(C_{\max}(1, \lceil i/l \rceil - r'), \ldots, \\ &\quad C_{\min}(\lceil il \rceil + r', n) \big) \end{aligned} \tag{4}$$

$$\begin{aligned} L_i &= \min \left(LW_{\lceil i/l \rceil}, \ldots, LM_{\lceil il \rceil}\right) \\ &= \min \big(C_{\max}(1, \lceil i/l \rceil - r'), \ldots, \\ &\quad C_{\min}(\lceil i/l \rceil + r', n), \ldots, \\ &\quad C_{\max}(1, \lceil il \rceil - r'), \ldots, \\ &\quad C_{\min}(\lceil il \rceil + r', n) \big) \\ &= \min \big(C_{\max}(1, \lceil i/l \rceil - r'), \ldots, \\ &\quad C_{\min}(\lceil il \rceil + r', n) \big) \end{aligned} \tag{5}$$

In [5], the lower bound function which lower bounds the distance between Q and C for any scaling in the range of $\{1/l, l\}$ and time warping with the Sakoe-Chiba Band constraint factor of r'

on C is given by:

$$LB\left(Q,C\right) = \sum_{i-1}^{m} \begin{cases} (Q_i - U_i)^2 & \text{if } Q_i > U_i \\ (Q_i - L_i)^2 & \text{if } Q_i < L_i \\ 0 & \text{otherwise} \end{cases}$$

$$(6)$$

Efficient Pruning Algorithm by Lower Bounds

Algorithm 1 gives the pseudocode for the search algorithm, which utilizes the computational efficient lower bounds on computational intensive distance measures to prune candidate sequences at an early stage. "real_distance" could be DTW, US, or SWM distance, while "lower_bound_distance" denotes the corresponding lower bound.

Algorithm   1: Lower_Bounding_Sequential_Scan(Q)

Key Applications

Query by humming is essential for audio information retrieval in terms of both effectiveness and efficiency.

Future Directions

In [4], a unified framework is proposed to explain the existing lower-bound functions for dynamic time warping distance. A new lower-bound function that is shown by experiments to be superior is also proposed. For future studies, this function can be investigated for the effectiveness in query by humming.

Experimental Results

This section describes the experiments carried out to verify the effectiveness of the proposed lower bounding distance for the most effective distance measure: SWM. The *Pruning Power P* is defined in [3] as follows:

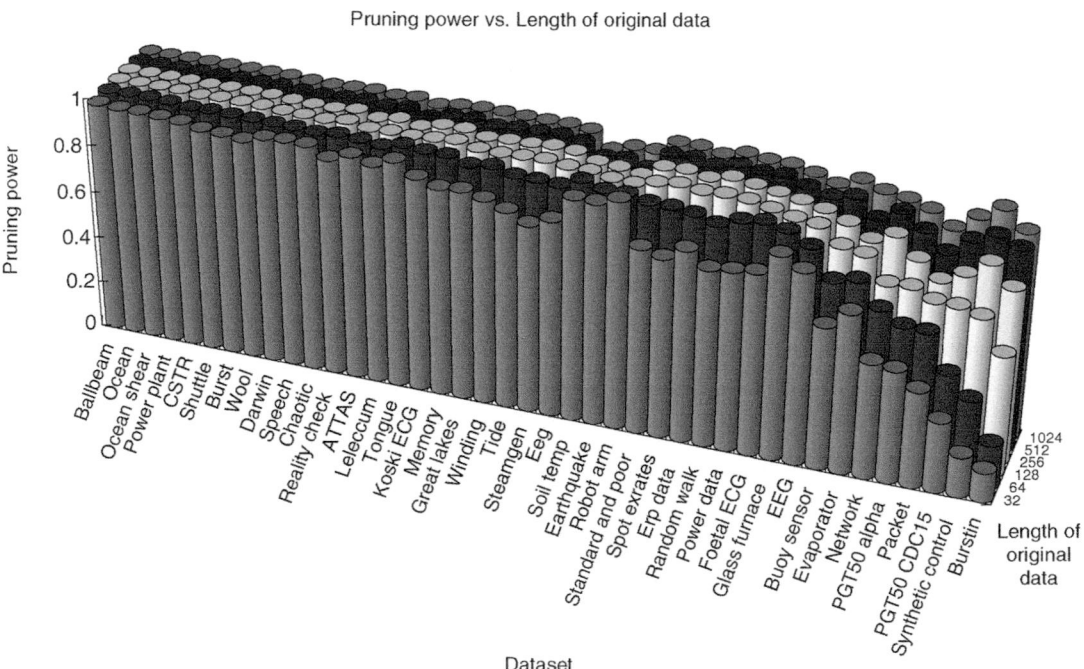

Query by Humming, Fig. 2 Pruning power vs. length of original data

$$p = \frac{\text{Number of objects that do not require full SWM}}{\text{Number of objects in database}}$$

The pruning power is an objective measure because it is free of implementation bias and choice of the underlying spatial index. This measure has become a common metric for evaluating the efficiency of lower bounding distances, therefore, it was adopted in evaluating the proposed lower bounding distance.

Figure 2 shows how the pruning power of the lower bounding measure varies as the length of data changes for different datasets. More than 78% (32 out of 41) of the datasets achieved a pruning power above 90%.

Data Sets

http://www.cs.ucr.edu/~eamonn/VLDB2005/

Cross-References

▶ Multimedia Information Retrieval Model
▶ Spatial Network Databases

Recommended Reading

1. Fu AW-C, Keogh EJ, Lau LYH, Ratanamahatana CA. Scaling and time warping in time series querying. In: Proceedings of the 31st International Conference on Very Large Data Bases; 2005. p. 649–60.
2. Ghias A, Logan J, Chamberlin D, Smith BC. Query by humming: musical information retrieval in an audio database. In: Proceedings of the 3rd ACM International Conference on Multimedia; 1995. p. 231–6.
3. Keogh EJ. Exact indexing of dynamic time warping. In: Proceedings of the 28th International Conference on Very Large Data Bases; 2002. p. 406–17.
4. Zhou M, Wong MH. Boundary-based lower-bound functions for dynamic time warping and their indexing. In: Proceedings of the 23rd International Conference on Data Engineering; 2007. p. 1307–11.
5. Zhu Y, Shasha D. Warping indexes with envelope transforms for query by humming. In: Proceedings of the ACM SIGMOD International Conference on Management of Data; 2003. p. 181–92.

Query Containment

Rada Chirkova
North Carolina State University, Raleigh, NC, USA

Definition

One query is contained in another if, independent of the values of the "stored data" (that is, database), the set of answers to the first query on the database is a subset of the set of answers to the second query on the same database. A formal definition of containment is as follows: denote with $Q(D)$ the result of computing query Q over database D. A query Q_1 is said to be contained in a query Q_2, denoted by $Q_1 \sqsubseteq Q_2$, if for all databases D, the set of tuples $Q_1(D)$ is a subset of the set of tuples $Q_2(D)$, that is, $Q_1(D) \subseteq Q_2(D)$. This definition of containment, as well as the related definition of query equivalence, can be used to specify query containment and equivalence on databases conforming to both relational and nonrelational data models, including XML and object-oriented databases.

Historical Background

Testing for query containment on finite databases is, in general, co-recursively enumerable: The procedure is going through all possible databases and simultaneously checking for noncontainment via bottom-up evaluation. See [6] for the details and for a discussion of the relationship between regular and unrestricted (that is, on not necessarily finite databases) containment and equivalence of relational expressions.

Chandra and Merlin [3] have shown that the problems of containment, equivalence, and minimization of conjunctive queries are NP complete. (Conjunctive queries are a subset of Datalog that is equivalent in expressive power to SQL select-project-join queries with only equality comparisons permitted.) It is also shown [3] that there is a simple test for containment and thus for equivalence. While the question of whether one conjunctive query is contained in another is NP complete, all the complexity is caused by "repeated predicates," that is, predicates appearing three or more times in the body. In the very common case that no predicate appears more than twice in any query, containment can be tested in linear time [11]. Moreover, conjunctive queries tend to be short, so in practice containment testing is not likely to be too inefficient.

Klug [8] has shown that containment for conjunctive queries with arithmetic – that is, inequality or disequality – comparisons (CQAC) is in II_2^P while being in NP for some proper subclasses. In addition to reporting, some new results on the complexity of CQAC query containment, Afrati and colleagues [2] provide a survey, comprehensive as of 2006, on the complexity of query containment, for query classes including conjunctive queries, CQAC, and its subclasses, as well as recursive and nonrecursive Datalog. More original work and further references on query containment in relational and nonrelational databases can be found in [1, 5, 7, 9, 10, 12–15].

Scientific Fundamentals

This overview of key ideas concerning conjunctive queries, Datalog programs, and their containment is based on [9] as well as on [15], which includes the details and references to the original sources of the results discussed here. The first item on the agenda is a review of containment of conjunctive queries. A conjunctive query is a single Datalog rule with subgoals that are assumed to have predicates referring to stored relations. (The standard Datalog notation is reviewed in, e.g., Sect. 6.1 of [4].) A conjunctive query is applied to the stored relations in a given database by considering all possible substitutions of values for the variables in the body. If a substitution makes all the subgoals true, then the same substitution, applied to the head of the rule, is an element of the set of answers to the head's predicate on the given database.

For example, rule

$$p(X, Z) \; : - a(X, Y), \, a(Y, Z)$$

talks about predicate a, for stored relation that contains information about arcs in a directed graph: $a(X, Y)$ means that there is an arc from node X to node Y in the graph. The rule also talks about predicate p whose relation is constructed by the rule. The rule says "$p(X, Z)$ is true if there is an arc from node X to some node Y and also an arc from node Y to Z." That is, conjunctive query p represents paths of length 2, in the sense that $p(X, Z)$ will be inferred exactly when there is a path of length 2 from X to Z in the graph represented by stored relation a.

It was proved in [3] that $Q_1 \sqsubseteq Q_2$ if and only if there is a homomorphism $h : Q_2^D \to Q_1^D$, where Q^D is the canonical database associated with conjunctive query Q. The canonical database Q^D for query Q is defined as the result of "freezing" the body of Q, which turns each subgoal of Q into a fact in the database. That is, the "freezing" procedure replaces each variable in the body of Q by a distinct constant, and the resulting subgoals are the only tuples in the canonical database Q^D.

Consider now an example of checking conjunctive-query containment using canonical databases. Rule

$$r(W, W) \; : - a(W, U), \, a(U, W)$$

defines conjunctive query r, whose answer represents circular paths of length 2 in a graph specified by arcs stored in relation a. As each circular path of length 2 is also an (arbitrary) path of length 2, the containment $r \sqsubseteq p$ is expected to hold. Indeed, the canonical database r^D, for the query defining predicate r, is a set of tuples $a(w, u)$, $a(u, w)$, while the canonical database p^D, for the query defining predicate p, is a set of tuples $a(x, y)$, $a(y, z)$. There exists a homomorphism from p^D to r^D that maps x into w, y into u, and z into w (or, alternatively, maps $a(x, y)$ into $a(w, u)$ and $a(y, z)$ into $a(u, w)$). Thus, by [3], the set of tuples in the answer to r on every database D is a subset of the set of tuples in the answer to p on D.

As no homomorphism exists from r^D to p^D, the conclusion is that p is not contained in r, which is to be expected from an intuitive interpretation of the two queries. However, if conjunctive query t defined by the rule

$$t(L, N) : - a(L, M), a(M, N), a(M, S)$$

is also considered, it is possible to ascertain both $p \sqsubseteq t$ and $t \sqsubseteq p$, by constructing a homomorphism from p^D to t^D (for the query defining predicate t) and another homomorphism from t^D to p^D. Thus, conjunctive queries p and t are *equivalent*. Indeed, each of p and t represents arbitrary paths of length 2 in a graph. However, p is more efficient to evaluate than t, because computing the set of answers to p requires only one join on the stored relation a, while evaluating t requires two joins. As described in [3], one can *minimize* conjunctive queries for more efficient evaluation; in fact, query p in this example can be obtained as a result of minimizing query t.

Another test for containment of conjunctive query Q_1 in conjunctive query Q_2 consists in computing the set of answers to Q_2 on the canonical database Q_1^D for Q_1. The test succeeds if the frozen head of Q_1 is an element of $Q_2(Q_1^D)$; otherwise the database Q_1^D is a counterexample to the containment.

An important extension of the theory of containment of conjunctive queries is the inclusion of arithmetic comparisons as subgoals, with the so-called built-in, or interpreted, predicates (e.g., subgoal $X \leq Y$ with built-in predicate $\leq$). When testing two conjunctive queries with arithmetic comparisons (CQACs) for containment $Q_1 \sqsubseteq Q_2$ using canonical databases, one must consider the set of values in the database as belonging to a totally ordered set, e.g., the integers or reals.

The containment test using canonical databases is conducted as follows. Each basic canonical database is constructed from only those subgoals of Q_1 that have uninterpreted predicates. Each basic canonical database is the canonical database (Q^D for query Q) defined above, together with a partition of the variables of the query into a list of blocks, such that each block is associated with a distinct integer value for all its variables, in increasing order of the integer values on the list of the blocks. The containment test succeeds if the frozen head of Q_1 is an element of $Q_2(Q_1^{D_i})$ on *all* basic canon-

ical databases $Q_1^{D_i}$ for Q_1; otherwise any $Q_1^{D_i}$ on which the test fails is a counterexample to containment. Another more general containment test for conjunctive queries with interpreted (not necessarily arithmetic comparison) predicates uses homomorphisms on the uninterpreted predicates and logical implication on the built-in predicates; see [15] for the details.

The problem of testing CQACs for containment is complete for Π_2^p, at least in the case of a dense domain such as real numbers. (In fact, the problem is Π_2^p complete even for conjunctive queries with disequalities $\neq$ as the only comparison predicate.) A containment test for conjunctive queries with negation is outlined in [15]. The test in this case is also complete for Π_2^p, as it also involves exploring an exponential number of canonical databases. The query-containment problem is also Π_2^p complete for unions of conjunctive queries [1].

Containment questions involving Datalog programs are often harder than for conjunctive queries. It is known that containment of Datalog programs is undecidable, while containment of a Datalog program in a conjunctive query is doubly exponential. However, the important case for purposes of information integration is the containment of a conjunctive query in a Datalog program, and this question turns out to be no more complex than containment of conjunctive queries. To test whether a conjunctive query Q is contained in a Datalog program P, one would "freeze" the body of Q to make a canonical database D. Then a check is done to see if $P(D)$ contains the frozen head of Q. The only significant difference between containment in a conjunctive query and containment in a Datalog program is that in the latter case, one must keep applying the rules until either the head of Q is derived or no more facts can be inferred in evaluating $P(D)$.

Key Applications

Query containment was recognized fairly early as a fundamental problem in database query evaluation and optimization. The reason is, for conjunctive queries – a broad class of frequently used queries, whose expressive power is equivalent to that of select-project-join queries in relational algebra – query containment can be used as a tool in query optimization, since the problem of conjunctive-query equivalence is equivalent to the problem of conjunctive-query containment. Specifically, to find a more efficient *and* answer-preserving formulation of a given conjunctive query, it is enough to "try all ways" of arriving at a "shorter" query formulation, by removing a query subgoal, in a process called query minimization [3]. In this process, a subgoal-removal step succeeds only if a containment test ensures equivalence (via containment) of the "original" and "shorter" query formulations.

Note that the problems of query containment and query equivalence are equivalent for conjunctive queries under the common setting of *set semantics for query evaluation,* where both stored relations and query answers are interpreted as sets of tuples. Interestingly, the relationship between the problems of containment and equivalence is very different under *bag semantics for query evaluation,* where both stored relations and query answers are allowed to have duplicates. See Jayram and colleagues [5] for a discussion and references on containment and equivalence under bag semantics, for conjunctive queries as well as for more expressive classes of queries, including CQACs and queries with grouping and aggregation. Jayram and colleagues [5] also present original undecidability results for containment of conjunctive queries with inequalities under bag and bag-set semantics for query evaluation.

In recent years, there has been renewed interest in the study of query containment, because of its close relationship to the problem of answering queries using views [4]. Intuitively, the problem of answering queries using views is as follows: Given a query on a database schema and a set of views (i.e., named queries) over the same schema, can the query be answered (efficiently) using only the views? Alternatively, what is the maximum set of tuples in the answer to the query that can be obtained from the views? As another alternative, in case it is possible to access both the views and

the database relations, what is the cheapest query-execution plan for answering the query?

The problem of answering queries using views has emerged as a central problem in integrating information from heterogeneous sources, an area that has been the focus of concentrated research efforts for a number of years [4, 15]. An information-integration system can be described logically by views that specify what queries the various information sources can answer. These views might be conjunctive queries or Datalog programs, for example. The "database" of predicates over which these views are defined is not a concrete database but rather a collection of "global" predicates whose actual values are determined by the sources, via the views. Information-integration systems provide a uniform query interface to a multitude of autonomous data sources, which may reside within an enterprise or on the World Wide Web. Data-integration systems free the user from having to locate sources relevant to a query, interact with each one in isolation, and manually combine data from multiple sources.

Given a user query Q, typically a conjunctive query, on the global predicates, an information-integration system determines whether it is possible to answer Q by using the various views in some combination. In addressing the problem of answering queries using views in the information-integration setting, query containment appears to be more fundamental than query equivalence. In fact, answering a query using only the answers to the views is considered "good enough" even in cases where equivalence does not hold (or cannot be demonstrated), provided that the view-based query rewriting can be shown to be contained in the query and is a maximal (i.e., returning the maximal set of answers) rewriting of the query using the available views and a given rewriting language. (For the details and references on maximally contained rewritings, see, e.g., [2].)

Besides its applications in information integration, the problem of answering queries using views is of special significance in other data-management applications. (Please see [4] for the details and references.) For instance, in query optimization finding a rewriting of a query using a set of materialized views (i.e., the answers to the queries defining the views) can yield a more efficient query-execution plan, because part of the computation necessary to answer the query may have already been done while computing the materialized views. Such savings are especially significant in decision support applications when the views and queries contain grouping and aggregation.

In the context of database design, view definitions provide a mechanism for supporting the independence of the logical and physical views of data. This independence enables the developers to modify the storage schema of the data (i.e., the physical view) without changing its logical schema and to model more complex types of indices. Provided the storage schema is described as a set of views over the logical schema, the problem of computing a query-execution plan involves figuring out how to use the view answers (i.e., the physical storage) to answer the query posed on the logical schema.

In the area of data warehouse design, the desideratum is to choose a set of views (and indexes on the views) to materialize in the warehouse. Similarly, in web site design, the performance of a web site can be significantly improved by choosing a set of views to materialize. In both problems, the first step in determining the utility of a choice of views is to ensure that the views are sufficient for answering the queries expected to be posed over the data warehouse or the web site. The problem, again, translates into the view-rewriting problem.

The problem of query containment is also of special significance in artificial intelligence, where conjunctive queries, or similar formalisms such as description logic, are used in a number of applications. The design theory for such logics is reducible to containment and equivalence of conjunctive queries. Original results and a detailed discussion concerning an intimate connection between conjunctive-query containment in database theory and constraint satisfaction in artificial intelligence can be found in [9].

Cross-References

► Conjunctive Query
► Query Optimization
► Query Optimization (in Relational Databases)
► Rewriting Queries Using Views
► SQL

Recommended Reading

1. Abiteboul S, Hull R, Vianu V. Foundations of databases. Reading: Addison-Wesley; 1995.
2. Afrati FN, Li C, Mitra P. Rewriting queries using views in the presence of arithmetic comparisons. Theor Comput Sci. 2006;368(1–2):88–123.
3. Chandra AK, Merlin PM. Optimal implementation of conjunctive queries in relational data bases. In: Proceedings of the 9th Annual ACM Symposium on Theory of Computing; 1977. p. 77–90.
4. Halevy AY. Answering queries using views: a survey. VLDB J. 2001;10(4):270–94,
5. Jayram TS, Kolaitis PG, Vee E. The containment problem for REAL conjunctive queries with inequalities. In: Proceedings of the 25th ACM SIGACT-SIGMOD-SIGART Symposium on Principles of Database Systems; 2006. p. 80–89.
6. Kanellakis PC. Elements of relational database theory. In: Handbook of theoretical computer science. Volume B: formal models and sematics (B). New York/Cambridge: Elsevier/MIT Press; 1990. p. 1073–156.
7. Kimelfeld B, Sagiv Y. Revisiting redundancy and minimization in an XPath fragment. In: Advances in Database Technology, Proceedings of the 11th International Conference on Extending Database Technology; 2008. p. 61–72.
8. Klug AC. On conjunctive queries containing inequalities. J ACM. 1988;35(1):146–60.
9. Kolaitis PG, Vardi MY. Conjunctive-query containment and constraint satisfaction. J Comput Syst Sci. 2000;61(2):302–32.
10. Miklau G, Suciu D. Containment and equivalence for a fragment of XPath. J ACM. 2004;51(1):2–45.
11. Saraiya Y. Subtree elimination algorithms in deductive databases. Ph.D. thesis, Stanford University; 1991.
12. Ullman JD. CS345 lecture notes. http://infolab.stanford.edu/~ullman/cs345-notes.html.
13. Ullman JD. Principles of database and knowledge-base systems, vol. II. Rockville: Computer Science Press; 1989.
14. Ullman JD. The database approach to knowledge representation. In: Proceedings of 13th National Conference on Artificial Intelligence and 8th Innovative Applications of AI Conference; 1996. p. 1346–48.
15. Ullman JD. Information integration using logical views. Theor Comput Sci. 2000;239(2):189–210.

Q

Query Evaluation Techniques for Multidimensional Data

Amarnath Gupta
San Diego Supercomputer Center, University of
California San Diego, La Jolla, CA, USA

Synonyms

Spatial data

Definition

There are two senses of the term "multidimensional data." The first relates to the data warehousing and online analytical processing (OLAP). The second sense of the term, used mostly in the context of scientific data, refers to variants of array-representable data where the dimensionality refers to the dimensions of the array. Query processing for this class of data uses array algebras [3] and array-specific storage [1] indexing techniques [4].

Key Points

In many scientific applications, data can be represented as multidimensional arrays. For example, the current flow in oceans is a time-varying vector field and can be roughly viewed as 4-dimensional data. In many applications the array may not be uniform, but it can be nested, and even irregular. Query evaluation on this kind of data is greatly dependent on applications. Some applications need to perform operations like value-based clustering on the data, while other applications need fast computation of aggregates such as temporal trends in regions that are selected by user queries. It has been established that storing the multidimensional data as relational tables and developing special routines to handle such relational data is feasible but not optimal, especially when the data is large. ESRI, the GIS provider, has assembled a number of operations collec-

tively called the MapAlgebra for the case where the arrays are two dimensional and the applications are spatial. Recently [2] developed a generic algebra called the gridfield algebra for manipulating arbitrary gridded datasets (i.e., arrays), and present algebraic optimization techniques in these applications. This system is implemented in a system called CORIE (http://www.ccalmr.ogi.edu/CORIE/) for an environmental observation and forecasting system.

Cross-References

▶ Multidimensional Modeling
▶ Online Analytical Processing

Recommended Reading

1. Furtado P Baumann P. Storage of multidimensional arrays based on arbitrary tiling. In: Proceedings of the 15th International Conference on Data Engineering; 1999. p. 480–9.
2. Howe B, Maier D. Algebraic manipulation of scientific datasets. VLDB J. 2005;14(4):397–416.
3. Marathe AP, Salem K. Query processing techniques for arrays. VLDB J. 2002;11(1):68–91.
4. Sinha RR, Winslett M. Multi-resolution bitmap indexes for scientific data. ACM Trans Database Syst. 2007;32(3):16.

Query Expansion for Information Retrieval

Olga Vechtomova
University of Waterloo, Waterloo, ON, Canada

Synonyms

QE, Query enhancement; Term expansion

Definition

Query expansion (QE) is a process in Information Retrieval which consists of selecting and adding terms to the user's query with the goal of minimizing query-document mismatch and thereby improving retrieval performance.

Historical Background

The work on query expansion following relevance feedback dates back to 1965, when Rocchio [1] formalized relevance feedback in the vector-space model. Early work on using collection-based term co-occurrence statistics to select query expansion terms was done by Spärck Jones [2] and van Rijsbergen [3].

Foundations

The central task of information retrieval (IR) is to find documents that satisfy the user's information need. This is usually taken to mean finding documents or some parts of them, such as passages, which contain information that would help resolving the user's information need. Therefore, at least in a more traditional sense, IR does not involve providing the user directly with the information needed. The user usually expresses his/her information need in free-text using natural language words and phrases (terms). Sometimes, prior to retrieving the documents, the user's free-text query is translated into controlled vocabulary which is a subset of the words that comprise a natural language. A document is similarly indexed either by the natural language terms that constitute its content, or by a set of controlled index terms that map its contents to the concepts in a given domain. Both directly matching free-text query terms to free-text index terms and translating natural language words to controlled vocabularies are inherently imprecise. In the first case, the main problem is that the user and the author of a document may express the same idea by means of different terms. In the second case, the shades of meanings that natural language terms carry may be lost in the translation process. In addition to these problems, the user's query may be incomplete or inaccurate, i.e., the user may not specify his/her information need exactly or express it accurately.

The goal of query expansion is to enrich the user's query by finding additional search terms, either automatically, or semiautomatically that represent the user's information need more accurately and completely, thus avoiding, at least to an extent, the aforementioned problems, and increasing the chances of matching the user's query to the representations of relevant ideas in documents. Query expansion techniques may be categorized by the following criteria:

- Source of query expansion terms;
- Techniques used for weighting query expansion terms;
- Role and involvement of the user in the query expansion process.

Query expansion can be performed automatically or interactively. In automatic query expansion (AQE), the system selects and adds terms to the user's query, whereas in interactive query expansion (IQE), the system selects candidate terms for query expansion, shows them to the user, and asks the user to select (or deselect) terms that they want to include into (or exclude from) the query.

There are four main sources of QE terms: (i) hand-built knowledge resources such as dictionaries, thesauri, and ontologies; (ii) the documents used in the retrieval process; (iii) external text collections and resources (e.g., the WWW, Wikipedia); (iv) search engine query logs.

Hand-built knowledge resources have three main limitations: they are usually domain-specific, have to be kept up-to-date, and typically do not contain proper nouns. Experiments with QE using knowledge resources did not show consistent performance improvements. For example, in [4] QE with words manually selected from WordNet, a large domain-independent lexical resource with lexical-semantic relations between words, did not improve well-formulated queries, but significantly improved performance of poorly-constructed queries.

Search engine query logs have been investigated as a source of query expansion terms in a number of works. For example, Billerbeck

Q

et al. [5] proposed a method of extracting QE terms from past user queries associated with the retrieved documents. Their experiments showed statistically significant improvements over no expansion and an optimised conventional expansion method. Beeferman and Berger [6] generated a bipartite graph from clickthrough data, where the queries and documents (URLs) are nodes and user clicks are edges between them, and then applied a clustering algorithm to find related queries and URLs. Yin et al. [7] used a random walk (RW) method on a bipartite query-URL graph to find related queries. They compared the RW approach to a method for extracting QE terms from snippets of documents retrieved by a search engine, and the latter method was found to outperform the RW approach on almost all queries.

The most common source of QE terms is the text collection used in the retrieval process or its subset (e.g., retrieved documents). These QE techniques showed overall better performance than techniques using hand-built knowledge resources. They can be subdivided into the following categories:

- QE following relevance feedback. QE terms are extracted from the documents retrieved in response to the user's query and judged relevant by the user.
- QE following blind (pseudo-relevance) feedback. QE terms are extracted from the top-ranked documents retrieved in response to the user's query.
- QE using automatically built association thesauri and collection-wide word co-occurrences.

Query Expansion Following Relevance and Pseudo-relevance Feedback

Relevance feedback (RF) is a process by which the system, having retrieved some documents in response to the user's query, asks the user to assess their relevance to his/her information need. Documents are typically shown to the user in some surrogate form, for example, as document titles, abstracts, snippets of text, query-biased, or general summaries, keywords and key-phrases. The user may also have an option to see the whole document before making the relevance judgement. After the user has selected some documents as relevant, query expansion terms are extracted from them, weighted and the top-weighted terms are either added to the query automatically, or shown to the user for further selection.

Query expansion following relevance feedback has consistently yielded substantial gains in performance in experimental settings. Many term selection methods have been proposed for query expansion following RF. The general idea behind all such methods is to select terms that will be useful in retrieving previously unseen relevant documents. Below is a brief description of a query expansion method [8] which showed consistently high performance results on Text REtrieval Conference (TREC) test collections. The first step is to retrieve documents in response to the user's initial query, which is done by calculating document matching score (Eq. 1) using the Robertson/Spärck-Jones probabilistic model.

$$MS = \sum_{i \in Q} \frac{(k_1 + 1) \times tf_i}{k_1 \times NF + tf_i} \times w_i \qquad (1)$$

where i is a term in the query Q, tf_i is the frequency of i in the document, k_1 is term frequency normalization factor, NF is document length normalization factor and is calculated as $NF = (1 - b) + b \times DL/AVDL$, where DL is document length, $AVDL$ is average document length, b is a tuning constant, w_i is term collection weight, calculated as $w_i = \log(N/n_i)$, where N is the number of documents in the collection, n_i is the number of documents containing i.

After the user looks through the top-retrieved documents, and judges some of them as relevant, the system extracts all terms from these documents, ranks them according to the Offer Weight (OW) in Eq. 2, and either adds a fixed number of terms to the query (automatic query expansion), or asks the user to perform the term selection.

$$OW = r \times RW \qquad (2)$$

where r is the number of relevant documents containing the candidate query expansion term and RW is Relevance Weight calculated as shown in Eq. 3:

$$RW = \log \left(\frac{(r + 0.5)(N - n - R + r + 0.5)}{(R - r + 0.5)(n - r + 0.5)} \right) \qquad (3)$$

where R is the number of documents judged relevant; r is the same as above; N is the number of documents in the collection; n is the number of documents containing the term.

The subsequent document retrieval with the expanded query is performed using Eq. 1 for document ranking with RW used instead of w.

A related approach to RF is pseudo-relevance or blind feedback (BF), which uses a number of top-ranked documents in the initially retrieved set for query expansion without asking the user to assess their relevance. The query expansion method described above can be used in BF with R being the number of top documents *assumed* to be relevant. Many other QE methods following blind feedback have been proposed. Two of the methods that showed good performance on large test collections are briefly introduced below.

Local Context Analysis (LCA) [9] technique consists of extracting terms (nouns and noun phrases) from n top ranked passages retrieved in response to the user's query. The extracted terms are ranked by their similarity to the entire user's query, and top m terms are added to the query. Carpineto et al. [10] proposed a term ranking method for QE based on Kullback-Leibler divergence (KLD) measure. The method ranks candidate QE terms based on the difference between their distribution in pseudo-relevant documents and in the entire collection.

In general, blind feedback has been demonstrated to be less robust in performance than relevance feedback. The QE performance following BF depends greatly on the performance of the user's initial query: if many of the top-ranked documents retrieved in response to the initial query are relevant, then it is likely that QE terms from these documents will be useful in retrieving other relevant documents. However, if the initial query is poorly formulated or ambiguous, and many top-ranked documents are nonrelevant, then QE terms extracted from such documents may deteriorate performance. Billerbeck and Zobel [11] report that blind feedback in their experiments improves performance of less than a third of queries. They also conclude that the best values for such BF parameters as the number of documents and terms used for QE vary widely across topics.

Query Expansion Using Association Thesauri and Term Co-occurrence Measures

Unlike QE following relevance or pseudo-relevance feedback, where terms are selected from documents at search time, QE techniques in this category rely on lexical resources automatically constructed prior to the search process. Typically, statistical measures of term similarity are applied to identify terms in a large document collection that co-occur in the same contexts, and which, therefore, are likely to be conceptually related.

For example, Qiu and Frei [12] developed a query expansion method where query expansion terms are selected from an automatically constructed co-occurrence based term-term similarity thesaurus on the basis of the degree of their similarity to all terms in the query. Jing and Croft [13] developed a technique for automatic construction of a co-occurrence thesaurus. Each indexing unit, defined through a set of phrase rules, is recorded in the thesaurus with its most strongly associated terms. An evaluation of different similarity measures (Dice, Jaccard, Cosine, Average Conditional Probability, and Normalized Mutual Information) for selecting query expansion terms from a document collection is reported in [14]. The Dice, Jaccard, and Cosine led to better QE results than Average Conditional Probability and Normalized Mutual Information.

Interactive Query Expansion

In interactive query expansion the task of selecting and adding terms to the user's query is split between the user and the system in such a way that the system selects the candidate query expansion terms, but it is the user who makes the final decision which terms to include into the query. Similarly to automatic query expansion, the most common source of terms used in IQE is a set of documents resulting from either relevance, or pseudo-relevance feedback. Terms are extracted by the system, ranked, and the top-ranked terms are shown to the user for selection. The process of IQE is iterative, and may be triggered either by the system, or the user.

Several studies comparing the effectiveness of AQE and IQE have been conducted. Intuitively, since the user is the one who decides which document is relevant to his/her information need, the user should be able to make better decisions than the system with respect to which terms to add to the query. However, experiments do not offer conclusive results that IQE is more effective than AQE. For instance, Beaulieu [15] showed that AQE is more effective than IQE in operational settings. On the other hand, Koenemann and Belkin [16] reported higher subjective user satisfaction with an IQE system, as well as better performance. This suggests that the effectiveness of IQE is highly variable, depending on the specific interactive system, the users, and the task. Ruthven [17] did a simulation study of a potential benefit of IQE. He concludes that while IQE has potential to achieve higher performance than AQE, this potential may not be easy to realize, because it is difficult for the users to make decisions about which terms are better in differentiating between relevant and nonrelevant documents.

Key Applications

Query expansion is used in some web search engines and enterprise search systems. Although QE following relevance feedback has been experimentally shown as one of the most successful IR techniques, its use in Web search has been limited.

Cross-References

► BM25
► Information Retrieval
► Information Retrieval Models
► Probabilistic Retrieval Models and Binary Independence Retrieval (BIR) Model
► Query Expansion Models
► Relevance Feedback for Text Retrieval
► Web Search Query Rewriting

Recommended Reading

1. Salton G. The SMART retrieval system (Chapter 14). Englewood Cliffs: Prentice- Hall. (Reprinted from Rocchio JJ. (1965). Relevance feedback in information retrieval. In: Scientific Report ISR-9, Harvard University), 1971.
2. Spärck JK. Automatic keyword classification for information retrieval. London: Butterworths; 1971.
3. van Rijsbergen CJ. A theoretical basis for the use of co-occurrence data in information retrieval. J Doc. 1977;33(2):106–19.
4. Voorhees E. Query expansion using lexical-semantic relations. In: Proceedings of the 17th Annual International ACM SIGIR Conference on Research and Development in Information Retrieval; 1994. p. 61–9.
5. Billerbeck B, Scholer F, Williams HE, Zobel J. Query expansion using associated Queries. Proceedings of the 12th International Conference on Information and Knowledge Management; 2003.
6. Beeferman D, Berger A. Agglomerative clustering of a search engine query log. In: Proceedings of the 6th ACM SIGKDD International Conference on Knowledge Discovery and Data Mining; 2000. p. 407–16.
7. Yin Z, Shokouhi M, Craswell N. Query expansion using external evidence. In: Proceedings of the 31st European Conference on Information Retrieval; 2009. p. 362–74.
8. Spärck JK, Walker S, Robertson SE. A probabilistic model of information retrieval: development and comparative experiments. Inf Proc Manage. 36(6):779–808 (Part 1), 2000; 809–840 (Part 2), 2000.
9. Xu J, Croft B. Query expansion using local and global document analysis. In: Proceedings of the 19th Annual International ACM SIGIR Conference on Research and Development in Information Retrieval; 1996. p. 4–11.

10. Carpineto C, de Mori R, Romano G, Bigi B. An information-theoretic approach to automatic query expansion. ACM Trans Inf Syst. 2001;19(1):1–27.
11. Billerbeck B, Zobel J. Questioning query expansion: an examination of behaviour and parameters. In: Proceedings of the 15th Australasian Database Conference; 2004. p. 69–76.
12. Qiu Y, Frei HP. Concept based query expansion. In: Proceedings of the 16th Annual International ACM SIGIR Conference on Research and Development in Information Retrieval; 1993. p. 160–9.
13. Jing Y, Croft B. An association thesaurus for information retrieval. In: Proceedings of the 4th International Conference on Computer-Assist IR; p. 146–60.
14. Kim M-C, Choi K-S. A comparison of collocation-based similarity measures in query expansion. Inf Process Manag. 1999;35(1):19–30.
15. Beaulieu M. Experiments with interfaces to support query expansion. J Doc. 1997;53(1):8–19.
16. Koenemann J, Belkin NJ. A case for interaction: a study of interactive information retrieval behavior and effectiveness. In: Proceedings of the SIGCHI Conference on Human Factors in Computing Systems; 1996. p. 205–12.
17. Ruthven I. Re-examining the potential effectiveness of interactive query expansion. In: Proceedings of the 26th Annual International ACM SIGIR Conference on Research and Development in Information Retrieval; 2003. p. 213–20.
18. Carpineto C, Romano G. Automatic query expansion in information retrieval. ACM Comput Surv. 2012;44(1):Article 1.

Query Expansion Models

Ben He
University of Glasgow, Glasgow, UK

Synonyms

Term expansion models

Definition

In information retrieval, the query expansion models are the techniques, algorithms or methodologies that reformulate the original query by adding new terms into the query, in order to achieve a better retrieval effectiveness.

Historical Background

The idea of expanding a query to achieve better retrieval performance emerged around the early 1970's. A classical query expansion algorithm is Rocchio's relevance feedback technique proposed in 1971 [10] for the Smart retrieval system. Since then, many different query expansion techniques and algorithms have been proposed.

Foundations

Query expansion models can be classed into three categories: manual, automatic, and interactive. Manual query expansion relies on searcher's knowledge and experience in selecting appropriate terms to add to the query. Automatic query expansion weights candidate terms for expansion by processing the documents returned from the first-pass retrieval, and expands the original query accordingly. Interactive query expansion automates the term weighting process, but it is the user who decides which are the expanded query terms.

Manual query expansion inspires and motivates the user to refine the initial query through heuristics, for instance, by providing a list of candidate terms mined from query log, and allowing user the to choose appropriate terms to add.

Automatic query expansion, different from manual query expansion, expands search queries without any form of human interaction. According to [15], the reason for not involving human interaction could be that the searcher does not want to make an effort perhaps he/she simply does not understand and does not bother adding more terms to the query. Efthimiadis [4] categorized the automatic query expansion methods into three subgroups on which the expansion process is based: search results, collection dependent data structures (e.g., term co-occurrence, term frequency distribution etc.), and collection independent data structures (e.g., lexicon relation between terms, synonyms etc.).

The first category (automatic query expansion based on search results) uses the returned documents from the first-pass retrieval with or with-

Q

out relevance information, as feedback for query expansion. In [10], Rocchio proposed a classical query expansion algorithm based on the Vector Space model. His algorithm has the following steps:

1. The first-pass retrieval consists of ranking the documents for the given query.
2. A term weight w(t, d) is assigned to each term occurring in one of the top-ranked document set D_{psd}. Such a document set is usually called a *pseudo relevance set*. A term weight is first assigned to each term document pair, that is the same weight assigned in the first-pass retrieval. The weighting model used is tf·idf (see TF IDF model).
3. Add the most weighted terms in the pseudo relevance set to the query, and modify the query term weights by taking into account both the original query term weight (qtw) used in the first-pass retrieval, and the weight assigned by the term weighting model. The query used in the first-pass retrieval is called the *original query*. Moreover, the query with the modified query term weights is called the *reweighed query*. The reweighed query with the added query terms is called the *expanded query*. Using Rocchio's method, the new query term weight qtw_m is given by Rocchio's query expansion formula as follows.

$$qtw_m = \alpha \cdot qtw + \beta \cdot \sum_{d \in D_{psd}} \frac{w(t, d)}{|D_p sd|} \quad (1)$$

If an expanded query term is not in the original query, qtw is zero. α and β are free parameters. The new query term weight is given by an interpolation of the original query term weight and the average term weight in the pseudo relevance set with $\alpha + \beta = 1$.

Another popular and successful automatic query expansion algorithm was proposed by Robertson [7, 8] in the development of the Okapi system. Okapi's query expansion algorithm is similar to Rocchio's, while using a different term weighting function. It takes the top R documents returned from the first-pass retrieval as the

pseudo relevance set. Unique terms in this set are ranked in descending order of the Robertson Selection Value (RSV) weights [7]. A number of top-ranked terms, including a fixed number of non-original query terms, are then added to the query. A major difference between Rocchio's and Robertson's methods is that the former explicitly uses relevant documents for feedback, while the latter assumes a pseudo relevance set. Okapi's query expansion method has been shown to be very effective for ad-hoc information retrieval. For example, in TREC-3 ad-hoc retrieval task, a 15.43% improvement over the Okapi BM25 baseline brought by Okapi's query expansion was reported [9].

Recently, Amati proposed a query expansion algorithm in his Divergence from Randomness (DFR) framework [1, 2], which similarly follows the steps in Rocchio's algorithm. However, in Amati's method, term weights are assigned by a DFR term weighting model. Two DFR term weighting models have been proposed, namely Bo1 based on Bose-Einstein statistics, and KL based on Kullback-Leibler divergence. For example, using the Bo1 model, the weight of a term t in the pseudo relevance document set D(Rel) is given as:

$$w(t) = tf_x \cdot \log_2 \frac{1 + \lambda}{\lambda} + \log_2 (1 + \lambda) \quad (2)$$

where λ is the mean of the assumed Poisson distribution of the term in the pseudo relevant document set D(Rel). It is given by $\frac{tf_{rel}}{N_{rel}}$. tf_{rel} is the frequency of the term in the pseudo relevant documents, and N_{rel} is the number of pseudo relevant documents. tf_x is the frequency of the query term in the pseudo relevance document set.

Once the first-pass retrieval is finished, using a document weighting model, a weight is assigned to each term in the top-ranked documents returned from the first-pass retrieval. This corresponds to the first and second steps of the relevance feedback process as introduced above.

In the next step, the original query terms are reweighed. The modified query term weight qtw_m is given by the following parameter-free formula [1]:

$$qtw_m = qtw + \frac{w(t)}{M} \qquad (3)$$

where qtw is the original query term weight. $w(t)$ is the weight of the query term that is given by Bo1. M is the upper bound of the weight of a term in the top-ranked documents. For Bo1, it is computed by:

$$M = \lim_{tf_c \to tf_{c,max}} w(t)$$
$$= tf_{c,max} \log_2 \frac{1+P_{max}}{P_{max}} + \log_2(1 + P_{max}) \qquad (4)$$

where tf_c is the frequency of the query term in the whole collection, and $tf_{c,\,max}$ is the frequency of the query term with the highest $w(t)$ weight in the top-ranked documents. P_{max} is given by $tf_{c,\,max}/N$. N is the number of documents in the collection. An obvious advantage of Amati's approach is that the query term reweighting formula is parameter free. The parameter α in Rocchio's formula (see Eq. (1)) is omitted.

In addition to automatic query expansion based on search results, various query expansion methods have been proposed based on collection dependent data structures (e.g., [6, 11, 12]), or collection independent data structures (e.g., [3, 5, 13]).

In interactive query expansion, on one hand, the search system offers the user a list of terms for expansion. On the other hand, it relies on the user to choose the appropriate expanded query terms. Similarly to automatic query expansion, interactive query expansion methods can also be categorized into three subgroups by on which the expansion process is based [4]: search results, collection dependent data structures, and collection independent data structures.

Key Applications

Query expansion models are employed in many information search systems such as library search systems and Web search engines for boosting their retrieval effectiveness.

Experimental Results

Query expansion models are usually helpful in improving retrieval effectiveness for general search tasks such as ad-hoc retrieval [14]. For example, a 15.43% improvement over the BM25 baseline was reported in the TREC-3 adhoc task using Okapi's query expansion method [9].

Table 1 demonstrates the effectiveness of query expansion for ad-hoc retrieval on the TREC (http://trec.nist.gov/) test data. The weighting model used for retrieval is DFRee proposed by G. Amati and implemented in the

Query Expansion Models, Table 1 The mean average precision (MAP) obtained with and without query expansion

Task	MAP, No QE	MAP QE	Difference (%)	p-value
TREC-1 ad-hoc	0.2148	0.2478	15.36	4.665e-06
TREC-2 ad-hoc	0.1821	0.2370	30.15	5.306e-08
TREC-3 ad-hoc	0.2557	0.3070	20.06	2.818e-07
TERC-8 small-web	0.2829	0.3164	11.84	9.261e-05
TREC2004 robust	0.2485	0.2920	17.50	2.41e-20
TREC-9 Web	0.2034	0.2180	7.18	0.01561
TREC-10 Web	0.2027	0.2526	24.62	1.715e-06
TREC-2004 Terabyte	0.2646	0.3027	14.40	4.537e-05
TREC-2005 Terabyte	0.3293	0.3828	16.25	2.818e-07
TREC-2006 Terabyte	0.2859	0.3348	17.10	1.582e-05
TREC-2004 Genomics	0.2921	0.3398	16.33	0.0005686
TREC-2005 Genomics	0.2172	0.2500	15.10	0.001759

Terrier Platform (http://ir.dcs.gla.ac.uk/terrier/). Terrier's default query expansion setting is applied, which expands the original query with the 10 most informative terms from the top-3 returned documents. Table 1 shows markable improvement in the retrieval performance, measured by mean average precision, over the baseline. In all cases, the improvement is statistically significant according to the Wilcoxon matched-pairs signed-ranks test.

However, the effectiveness of query expansion becomes unreliable when there are only very few relevant documents, or when the information needed is very specific, in which cases expanding the query does not bring up more relevant documents. There are also other factors that may affect the effectiveness of query expansion, such as the quality of the top-ranked documents, how much noise the document collection contains etc. For example, query expansion does not improve retrieval performance on the Blog06 test collection (http://ir.dcs.gla.ac.uk/test_collections/blog 06info.html), possibly due to the large amount of spam.

Cross-References

▶ Probabilistic Retrieval Models and Binary Independence Retrieval (BIR) Model
▶ Relevance Feedback for Content-Based Information Retrieval
▶ Rocchio's Formula
▶ Web search Relevance Feedback

Recommended Reading

1. Amati G. Probabilistic models for information retrieval based on divergence from randomness. Ph.D thesis, Department of Computing Science, University of Glasgow, Glasgow, UK. 2003.
2. Carpineto C, de Mori R, Romano G, Bigi B. An information-theoretic approach to automatic query expansion. ACM Trans Inf Syst. 2001;19(1):1–27.
3. Croft B, Das R. Experiments with query acquisition and use in document retrieval systems. In: Proceedings of the 12th Annual International ACM SIGIR Conference on Research and Development in Information Retrieval; 1989. p. 349–68.
4. Efthimiadis NE. Query expansion. Annu Rev Inf Syst Technol. 1996;31:121–83.
5. Jarvelin K, Kristensen J, Niemi T, Sormunen E, Keskustalo H. A deductive data model for query expansion. Technical report. Department of Information Studies. 1995.
6. Minker J, Wilson G, Zimmerman B. An evaluation of query expansion by the addition of clustered terms for a document retrieval system. Inf. Storage Retrieval. 1972;8(6):329–48.
7. Robertson SE. On term selection for query expansion. J Doc. 1990;46(4):359–64.
8. Robertson SE, Walker S, Beaulieu MM, Gatford M, Payne A.Okapi at TREC-4. In: Proceedings of the 4th Text Retrieval Conference; 1995.
9. Robertson SE, Walker S, Jones S, Hancock-Beaulieu MM, Gatford M.. Okapi at TREC-3. In: Proceedings of the 3rd Text Retrieval Conference; 1994.
10. Rocchio J. Relevance feedback in information retrieval. USA: Prentice-Hall; 1971. p. 313–23.
11. Sparck JK. Automatic keyword classification for information retrieval. London: Butterworths; 1971.
12. Sparck JK. Collection properties influencing automatic term classification. Inf Storage Retrieval. 1973;9(9):499–513.
13. Voorhees E. Query expansion using lexical-semantic relations. In: Proceedings of the 17th Annual International ACM SIGIR Conference on Research and Development in Information Retrieval; 1994. p. 61–9.
14. Voorhees E. TREC: experiment and evaluation in information retrieval. Cambridge, MA: MIT; 2005.
15. Walker S. The Okapi Online Catalogue Research Projects. London: Library Association; 1989.

Query Language

Tore Risch
Department of Information Technology, Uppsala University, Uppsala, Sweden

Synonyms

Data manipulation language

Definition

A query language is a specialized programming language for searching and changing the contents of a database. Even though the term originally refers to a sublanguage for only searching (querying) the contents of a database, modern query languages such as SQL are general languages for interacting with the DBMS, including statements for defining and changing the database schema, populating the contents of the database, searching the contents of the database, updating the contents of the database, defining integrity constraints over the database, defining stored procedures, defining authorization rules, defining triggers, etc. The data definition statements of a query language provide primitives for defining and changing the database schema, while data manipulation statements allow populating, querying, as well as updating the database. Queries are usually expressed declaratively without side effects using logical conditions. However, modern query languages also provide general programming language capabilities through the definition of stored procedures. Most query languages are textual, meaning that the queries are expressed as text string processed by the DBMS. There are also graphical query languages, such as Query-By-Example (QBE), where the queries are expressed graphically and then translated into textual queries interpreted by the DBMS.

Key Points

Since data is represented in terms of the data model used by the DBMS, the syntax and semantics of a query language depends on the data model of the DBMS. For example, the relational data model is based on representing data as tables which allows expressing queries over the tables using expressions expressed in variant of predicate logic called relational calculus. Such queries are non-procedural since the user need not specify the details how a query is executed, but rather only how to select and match data from the tables. It is up to the query optimizer to translate the non-procedural logical queries into an efficient program for searching the database. Queries can also be expressed as relational algebra expressions. A query language is said to be relationally complete if its power is equivalent to the power of the relational algebra or the relational calculus.

The most widespread query language is SQL the standard language used for interacting with relational DBMSs. Queries in SQL are mainly expressed as syntactically sugared relational calculus expressions. However SQL also has query constructs based on the relation algebra and other paradigms. The logical query language Datalog, as well as SQL's predecessor QUEL, is purely based on predicate calculus. Other data models use other query languages. For example, OQL is used for searching object-oriented databases and Daplex for functional databases.

Cross-References

► Daplex
► Datalog
► OQL
► Relational Calculus
► SQL
► Stored Procedure

Query Languages and Evaluation Techniques for Biological Sequence Data

Sandeep Tata[1] and Jignesh M. Patel[2]
[1]IBM Almaden Research Center, San Jose, CA, USA
[2]University of Wisconsin-Madison, Madison, WI, USA

Synonyms

Querying DNA sequences; Querying protein sequences

Definition

A common type of data that is used in life science applications is biological sequence data. Data such as DNA sequence and protein sequence data are growing at a very fast rate. For example, the data at GenBank[GB07] has been growing exponentially, doubling roughly every 18 months. These sequence datasets are often queried in complex ways and the methods required to query these sequences go far beyond the simple string matching methods that have been used in more traditional string applications. In order to enable users to easily pose sophisticated queries on these biological sequences, different languages have been designed to support a rich library of functions. In addition, some database systems have been extended to support a rich set of operators on the sequence data type. Compared to the stand-alone approach, the database method brings the power of algebraic query optimization and the use of indexes making it possible to find efficient execution plans for sophisticated queries. Furthermore, biological sequence processing can be integrated with traditional database processing, such as selecting a subset of data for analysis, or combining data from multiple sources, to produce a powerful sequence analysis and mining system.

Historical Background

Several research efforts have tackled the problem of enabling complex and efficient querying on biological data. Early efforts such as [4] investigated the idea of a Genomics Algebra, which would abstract several biological processes and enable users to construct complex expressions, thereby providing a sophisticated platform for processing biological sequences. As part of another effort, the Periscope project, an algebra for biological sequences called PiQA (Protein Query Algebra) was proposed in [12]. PiQA can be used to construct complex expressions that allow sophisticated manipulation of both genetic and protein sequence data. The Periscope/SQ [13] system and the PQL query language are based on this algebra.

One of the primary operations in biological sequence processing is local alignment. A dynamic programming algorithm was proposed in [9] for this problem. The popular BLAST algorithm [1]

is a heuristic version proposed in 1990 to support local alignment search on massive datasets.

Foundations

The amount of biological sequence data available to scientists for analysis has grown rapidly. While small sequences can be analyzed and queried fairly quickly on computers, large datasets require careful design of algorithms and data structures. Database systems have dealt with the problem of managing and processing large datasets for several years, and the area of query processing for biological sequences aims to bring the benefits of this research to the domain of biological sequences.

One of the most common computations in the context of sequence data is that of finding "similar" sequences. This is often the first step biologists perform to begin investigation of a particular biological sequence they have discovered. For instance, finding a known protein with a sequence similar to the one under investigation may yield some clues to the function of the protein. Sequence similarity is defined in many ways and the type of similarity used may depend on the sequence and the application for which the similarity is being computed. Some distance measures are based on the evolutionary distance between the sequences, while in some cases they simply count the number of mismatches between an alignment of two sequences of the same length. Examples of evolutionary distance matrices include the PAM [3] and BLOSUM [5] matrices for comparing protein sequences. Simple models for mismatches include simply finding sequences that have up to a specified number of mismatches, or models in which mismatches are only tolerated in specific positions (such at the middle portion of the sequence).

The problem of finding a local alignment for two sequences requires finding a region of maximum similarity in a longer sequence that will match with a shorter query sequence. This version of sequence similarity is extremely popular, especially for large DNA sequence repositories. While the Smith-Waterman algorithm [9] is an elegant dynamic programming technique to compute an optimal alignment with $O(n^2)$ cost, for large sequence databases this cost can be prohibitive. In practice, methods that approximate Smith-Waterman but are much faster are used. Popular methods in this category include BLAST [1] and FASTA [8]. While these methods allow the fast evaluation of simple sequence queries, for sophisticated queries, they do not address sophisticated queries, such as a complex string pattern query. Additional methods are required that include a language to express such quires and algorithms to efficiently evaluate the operations in these queries.

Query languages provide a convenient way to express common computational tasks. The class of query languages can be divided into two categories: (i) procedural and (ii) declarative. Several procedural query languages are often general purpose programming languages or scripting languages with a library of functions that support operations on biological sequences. BioPerl and BioPython are two such examples. Figure 1 shows an example of a BioPerl code snippet that uses library function to translate the DNA sequence into the corresponding protein sequence, and another function to compute the reverse complement. These libraries also provide functions to read and translate between several popular sequence file formats. Although such languages make it somewhat easier to develop biological sequence processing applications, they suffer from two major drawbacks: (i) inability to rapidly express sophisticated queries (ii) lack of an indexing and optimization infrastructure to choose efficient query plans. Applications written in procedural languages often need to solve several database problems in order to scale to larger data sets. As a result, researchers have recently focused on declarative query processing for biological data.

In addition to an algebra, a declarative query infrastructure needs efficient physical operators corresponding to the operators in the algebra in order to produce query evaluation plans. The research prototype system Periscope [14] uses the PiQA algebra and the PiQL query language. Figure 2 shows an example PQL query that finds

```
use Bio::PrimarySeq;
my $str = "GATTACATACAT";
my $pseq = Bio::PrimarySeq->new(-seq => $str,
    -display_id=>"testseq");
print $pseq->seq, "\n";
print $pseq->translate->seq,"\n";
print $pseq->revcom->seq;
```

Query Languages and Evaluation Techniques for Biological Sequence Data, Fig. 1 Sample BioPerl code

```
SELECT * FROM AUGMENT(
MATCH(T1, seqid, sequence, 0, "ACAC"),
MATCH(T1, seqid, sequence, 1, "TTACAGGG"), 0, 100)
```

Query Languages and Evaluation Techniques for Biological Sequence Data, Fig. 2 Sample PiQL query

all instances of the sequence "ACAC" followed by "TTACAGGG" within 100 symbols in the given sequence database. The Periscope system is built as an extension to an object-relational system, which allows the user to mix sequence queries with traditional relation data manipulation queries such as selections and joins. Relational database systems are now commonly used to manage large biological datasets (for example the GMOD project). Arguably, the declarative query processing with the relational frameworks provides a better framework (compared to the procedural framework) for complex biological data analysis.

Index Structures

To speed up the evaluation of complex biological sequence operations, a number of index-based methods have been designed. Exploiting these indexes is crucial in having efficient query execution plans, especially to cope with the increasing data volumes and query complexity. The suffix tree is one of the useful data structures in the world of string processing [15]. A Suffix tree is a tree that is built using an input string such that every path from the root of the tree to a leaf node corresponds to a suffix in the string. The edges are labeled with substrings from the string. A suffix tree can be used to evaluate exact substring queries in time proportional to the length of the

Q

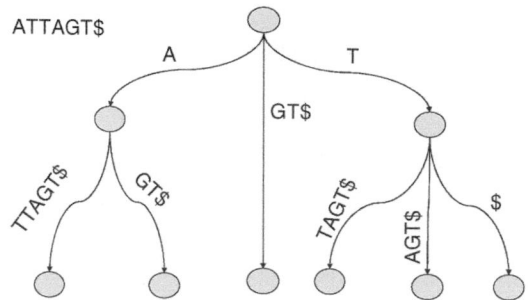

ATTAGT$

Query Languages and Evaluation Techniques for Biological Sequence Data, Fig. 3 Sample suffix tree built on the string "ATTAGT"

query. Figure 3 shows an example suffix tree on the string ATTAGT$. In order to check if the string TAG is a substring of the database string (on which the suffix tree is built), one simply follows the labeled edges from the root of the tree while trying to consume the symbols (T,A,G) from the query string.

Suffix trees can be constructed in time proportional to the length of the input string, i.e., in time O(n). Several algorithms have been designed to accomodate very large input datasets and construct disk based suffix trees. Suffix trees can also be used to efficiently answer approximate matching queries and even compute the local alignment of a query string with the database string.

Another interesting approach to indexing sequence data in the context of similarity search is the use of metric space indexing [7]. While extremely efficient for distance measures that are metrics, such indexing techniques tend to be less flexible than suffix trees.

Given the popularity of BLAST for sequence query processing, a number of approaches have added support for invoking BLAST from a database engine [2, 6, 11], and regular expression engines in SQL engines have also been adapted for more sophisticated biological pattern matching [10]. Their support for scanning sequences with complex patterns (including arbitrary position-specific weight matrices containing probabilistic models for matching a nucleic or protein sequence) is limited. However,

such extensions have recently been made for the declarative framework using suffix trees [12].

Key Applications

Any application that requires complex sequence matching, which include combinity sequence homology matching, and TFBS prediction.

Cross-References

► Biological Sequences
► Data Types in Scientific Data Management
► Index Structures for Biological Sequences
► Query Languages for the Life Sciences
► Scientific Databases
► Suffix Tree

Recommended Reading

1. Altschul SF, Gish W, Miller W, Myers EW, Lipman DJ. Basic local alignment search tool. J Mol Biol. 1990;215(3):403–10.
2. Barbara A, Eckman AK. Querying BLAST within a data federation. Q Bull IEEE TC Data Eng. 2004;27(3):12–9.
3. Dayhoff MO, Schwartz RM, Orcutt BC. A model of evolutionary change in proteins. Atlas Protein Seq Struct. 1978;5(3):345–52.
4. Hammer J, Schneider M. Genomics algebra: a new, integrating data model, language, and tool for processing and querying genomic information. In: Proceedings of the 1st Biennial Conference on Innovative Data Systems Research; 2003. p. 176–87.
5. Henikoff S, Henikoff J. Amino acid substitution matrices from protein blocks. In Proc Natl Acad Sci. 1992;89(22):10915–9.
6. Hsiao R-L, Stott Parker Jr D, Yang H-C. Support for BioIndexing in BLASTgres. In: In Data Integration in the Life Sciences (DILS), LNCS, vol. 3615. Berlin: Springer; 2005. p. 284–7.
7. Mao R, Weijia X, Neha S, Miranker DP. An assessment of a metric space database index to support sequence homology. In: Proceedings of the IEEE 3rd International Symposium on Bioinformatics and Bioengineering; 2003. p. 375–82.
8. Pearson WR, Lipman DJ. Improved tools for biological sequence comparison. Proc Natl Acad Sci. 1988;85(8):2444–8.

9. Smith TF, Waterman MS. Identification of common molecular subsequences. J Mol Biol. 1981;147(1): 195–7.
10. Stephens S, Chen JY, Davidson MG, Thomas S, Trute BM. Oracle database 10 g: a platform for BLAST search and regular expression pattern matching in life sciences. Nucleic Acids Res. 2005;33(Database-Issue):675–9.
11. Stephens S, Chen JY, Thomas S. ODM BLAST: sequence homology search in the RDBMS. Q Bull IEEE TC Data Eng. 2004;27(3):20–3.
12. Tata S, Lang W, Patel JM. Periscope/SQ: interactive exploration of biological sequence databases. In: Proceedings of the 33rd International Conference on Very Large Data Bases; 2007. p. 1406–9.
13. Tata S, Patel JM. PiQA: an algebra for querying protein data sets. In: Proceedings of the 15th International Conference on Scientific and Statistical Database Management; 2003. p. 141–50.
14. Tata S, Patel JM, Friedman JS, Swaroop A. Declarative querying for biological sequences. In: Proceedings of the 22nd International Conference on Data Engineering; 2006. p. 87.
15. Weiner P. Linear pattern matching algorithm. In: Proceedings of the 14th Annual IEEE Symposium on Switching and Automata Theory; 1973. p. 1–11.

Query Languages for the Life Sciences

Zoé Lacroix
Arizona State University, Tempe, AZ, USA

Synonyms

Biological data retrieval, integration, and transformation; Biological query Languages; Scientific query Languages

Definition

A *scientific query language* is a query language that expresses the data retrieval, analysis, and transformation tasks involved in the dataflow pertaining to a scientific protocol (or equivalently workflow, dataflow, pipeline). Scientific query languages typically extend traditional database query languages and offer a variety of operators expressing scientific tasks such as ranking, clustering, and comparing in addition to operators specific to a category of scientific objects (e.g., biological sequences).

Historical Background

A scientific query may involve data retrieval tasks from multiple heterogeneous resources and perform a variety of analysis, transformation, and publication tasks. Existing approaches used by scientists include hard coded scripts, data ware houses, link-based federations, database mediation systems, and workflow systems. Hard coded scripts written in Perl and Python are widely used in the biological community. Unlike an approach based on a query language, scripts are very limited in terms of scalability and flexibility. Extending or altering a protocol over time requires writing new scripts. Further, scripts are limited in the degree to which they capture biological expertise so they can be re-used for future related queries. Finally, unless explicitly written to do so, scripts do not assist the user in filtering retrieved data, resolving inconsistencies and sorting and ranking the results, or optimizing the overall execution. Data warehousing consists in collecting data from many possible data sources, curating the data, and creating a new database. Data warehouses typically are relational databases designed to provide users an integrated platform to answer a predefined set of scientific tasks.

A federation, e.g., NCBI Entrez and the Sequence Retrieval System (SRS) [4], links semiautonomous distributed databases with powerful full text indexing and keyword based search techniques for cross database retrieval. The approach expresses navigational queries over linked entries retrieved from multiple databases. In addition, useful tools such as BLAST are often made available. However, the approach is limited in that a federation relies on a materialized index that is difficult to keep up-to-date. Moreover, it focuses on data retrieval and does not support complex queries. Most existing data integration approaches rely on an internal query language that captures data management operations and a user query interface that aims at expressing queries meaningful to the scientists. Existing mediation approaches rely on traditional database query languages such as SQL adapted to handle biological data. K2 follows the Object Data Management Group (ODMG 1999) and its internal query language, called K2 Mediator Definition Language (K2MDL), is a combination of the Object Query Language (OQL) and the Object Definition Language (ODL) both specified by the ODMG. Kleisli (also known as Discovery Hub) provides the collection programming language (CPL) [2], a nested version of SQL, and Java and Perl access programming interfaces (API). Users may express their queries in SQL or through a graphical interface that limits access to query capabilities. The system that supports the Object Protocol Model (OPM) [3] provides a query language similar to OQL (ODMG 1999) and also exploits SQL when integrated data sources are retrofitted from a relational data model. Additional query capabilities may be integrated through CORBA classes. Tools such as BLAST are wrapped through an Application Specific Data Type (ASDT). Users may express their queries in a simplified version of OQL and generate query forms that do not offer access to all query capabilities. Discovery Link (also known as Information Integrator) offers a rich subset of SQL3 (SQL 1993) including user-defined functions, stored procedures, recursion, row types, object views, etc. P/FDM provides support to access specific capabilities of sources. P/FDM uses Daplex that offers a richer syntax than SQL [10] and Prolog. In particular it allows uses of function calls within queries. In addition to these query languages, P/FDM offers users a visual interface that enables users to build their queries by browsing and clicking through the database schema. The P/FDM Web interface uses HTML forms and access the mediator through CGI programs. Both interfaces restrict the query capabilities of the mediator. TAMBIS, which uses CPL internally, is primarily concerned with overcoming semantic heterogeneity through the use of ontologies. It provides users an ontology-driven browsing interface. Thus it too restricts the ex-

tent to which sources can be exploited for scientific discovery. To summarize, these systems have made many inroads into the task of data integration from diverse data sources. However they all rely on significant programming effort to adjust to specific scientific tasks, are difficult to maintain and provide user's query language that require programming ability such as SQL, OQL, Daplex, etc. or user's friendly interfaces that significantly limit the query capabilities. A comparative study of these systems and the many causes of failure of traditional database approaches to support the process of scientific discovery are detailed in [6].

Workflows are used in business applications to assess, analyze, model, define, and implement business processes. A workflow automates the business procedures where documents, information or tasks are passed between participants according to a defined set of rules to support an overall goal. In the context of scientific applications, a workflow approach may address the collaboration among scientists, as well as the integration of scientific data and tools. The procedural support of a workflow resembles the query-driven design of scientific problems and facilitates the expression of scientific pipelines (as opposed to a database query). However, because workflows are designed to orchestrate various applications (e.g., Web services) into a combined dataflow they do not provide a query language and do not express the specific queries against biological datasets. Peer-to-Peer (e.g., Chinook) and grid approaches (e.g., myGRID) may offer the support for resource and data sharing without providing a query language.

Foundations

Systems that focus on specific datasets such as sequences, protein structure, phylogenetic trees, metabolic pathways, rely on query languages designed to handle the characteristics of the biological datasets. Life sciences data are exceptionally diverse. Biological data include string data (e.g., sequences over various alphabets), 3D geometric data (e.g., protein structure), tree data (e.g., phylogenetic tree), and graph data (e.g., pathways). When their description and annotations are mostly textual and deeply nested in one or several documents, the intrinsic structure of the datasets are often poorly represented by traditional data models, thus poorly accessed and transformed by the corresponding query languages [7]. Each of these specific biological datasets has motivated the design of suitable query languages. For example PiQA, an algebra designed to query sequences, offers a match operator that expresses sequence alignment.

Query languages that express search queries (using regular expressions, wildcards, and Boolean operators) are often developed for life scientific applications. This is because scientific data are mostly textual (annotation on scientific instances) and scientists seem to mimic the manual query process they follow when exploring data sources on the Web. Querying systems that are compatible with scripting languages such as Perl are also favored because most scientific protocols are implemented with scripts. Additional features expected by scientists include the ability to compare and rank results. The ability to express cross-database queries and to integrate scientific analysis tools such as BLAST has a significant value to the scientist.

Scientific queries may exploit metadata as well as data. Querying scientific data through their meaning expressed in a terminology or an ontology is scientist-friendly because it hides the complex structure of data as they are organized and stored in the various repositories. A domain ontology may also provide a global schema for multiple integrated scientific resources. It facilitates scientific tasks such as *profiling* which aims at combining all information known about a scientific object. TAMBIS is an example of a system that provides a ontology-driven query interface.

The challenge for the design of a biological query language is to express the data management tasks involved in the scientific dataflow while offering meaningful and scientist-friendly query functionality. A scientific query language typically handles complex datatypes including text, string, tree, graph, list, variant, etc. for data and metadata. Scientific queries may invoke an increadibly diverse range of functions including extracting information in a textual document, identifying a feature common to two biological sequences, exploring a biological pathway, profiling a gene, etc.

Key Applications

The Acedb Query Language (AQL) is the language developed for the AceDB genome database. It is inspired by the Object Query Language (OQL), and Lorel the query language developed for semi-structured and XML data (also based on OQL). AQL expressions use the Select, From, Where syntax. Queries in AQL express traditional database retrieval and transformation against a data warehouse mostly populated with annotated scientific data (textual). It provides AcePerl and object-oriented Perl module access local or remote AceDB databases transparently. Many biological resources use the AceDB system that is still updated and maintained by the Sanger institute. The former version of AQL consisting of search statements of the form "find Gene TP*" is currently used to query WormBase.

SRS has its own query language that express retrieval queries using string comparison, wildcards, regular expressions, numeric comparison, Boolean operations, and the *link* operator specific to SRS that allows the navigation through resources exploiting the interal cross-references (hyperlinks, indices, and composite structures) made available by the providers. SRS uses Icarus as its internal interpreted object-oriented programming language. SRS queries return sets of entries or lists of entry identifiers that can be sorted with respect to various criteria. Because SRS integrates scientific analysis tools (e.g., BLAST) as tool-specific databanks, SRS queries may exploit the expressive power of the scientific analysis tools integrated in the federation. However, the SRS query language mostly expresses retrieval queries over cross databases. For a detailed description of the

Q

system and its query language see Chapter 5 in [6]. SRS is currently updated and mainted by Biowisdom Ltd.

The Life Sciences community has developed unique approaches to handle complex retrieval and integration tasks. They express the query "retrieve everything known about [a specific scientific instance]." Unlike a query language, the resources where the data are to be retrieved and the criteria for retrieval cannot be selected by the user. These approaches include sequence profiling tools that express complex data retrieval and integration queries over multiple heterogeneous resources to generate summaries of all information related to a particular biological sequence. For example, the Karlsruhe Bioinformatik Harvester (KIT) crawls and crosslinks 25 biological resources. Data integration is limited as queries against KIT are species-specific boolean expressions of keywords and the result is a list of summaries (each summary corresponds to a biological sequences), where each summary is the concatenation of relevant retrieved pages. Other approaches focus on the textual material contained in the biological data sources and use biological language processing to achieve information retrieval, extraction, classification, and integration [5]. These approaches retrieve documents from various resources and generate a document summary as input. Textpresso uses the Gene Ontology (GO) to markup documents related to a particular organism retrieved from PubMed. It provides a query language that expresses Boolean expressions on keyword, category, and attribute. Because the approach is document-driven, the query language expresses advanced search queries. Patent databases are other documented sources of biological information. In addition to the traditional features of querying documented data, Patent Lens offers the useful feature of querying the biological sequences contained in the patent documents with BLAST [9].

The use of a domain ontology as an interface (or view) for biological data is usually well received by the scientists. Such approaches typically support exploration, navigation, and search queries. TAMBIS uses an ontology to

provide an homogeneous layer over the integrated resources (data sources and applications) that acts as a global schema, to hide the heterogeneities of the integrated resources, and to be used as a query interface. WormBase is a warehouse derived from ACeDB and BioMart databases and uses biomaRt, a bioconductor package that provides a platform for data mining. BioMart databases are annotated with the Gene Ontology (GO) and the eVOC ontologies. These annotations provide support for querying data. EnsMart organizes data with respect to *foci*, that are central scientific objects (e.g., gene). All data are linked to the instances of those central objects. Queries consist of the selection of the species and focus of interest and the specification of filters and outputs. AmiGO provides a query interface to GO, genes, and gene products.

The Pathway Tools developed and maintained at SRI International to support pathway databases produces a query form for basic or advanced search against BioCyc databases. Each form consists of the selection of the BioCyc database (queries are limited to the scope of a single resource), the field(s) to be searched or the ontology term(s), and the format of the output. Most of the databases developed with BioCyc allow the use of BLAST. The BioVelo query language is designed to retrieve data from BioCyc databases. It is based on the object-oriented BioCyc data model composed of object classes that are identified to ontology classes and express set comprehension statements used in functional programming languages such as Python.

Future Directions

The design of query languages for the Life Sciences is a challenging problem. The desired expressive power and features of the languages are not well understood. The expectations of the users (life scientists) and the developers (computer scientists) are often dissimilar. Life scientists expect support for all their scientific questions which often go beyond the scope of traditional database query languages handling data retrieval and transformation. Scientific questions are often complex protocols that invoke not only data retrieval and transformation tasks but ranking, comparison, clustering and classification as well as sophisticated analysis tasks. The implementation of such a "query" would require accessing multiple resources and coordinating the dataflow as currently achieved by scientific workflow systems such as Taverna. However, workflow systems lack the elegance of a query language and its benefits such as query planning and optimization. Because many of the scientific tasks invoked in a scientific protocol could be expressed by operators of a query language, the design of a generic query language to handle data retrieval, transformation, comparison, clustering, and integration operations would be valuable to scientific data management.

Data Sets

AceDB databases http://www.acedb.org/Databases/
Public SRS servers http://downloads.biowisdomsrs.com/publicsrs.html
Biopathways Graph Data Manager (BGDM) http://hpcrd.lbl.gov/staff/olken/graphdm/graphdm.htm
BioCyc database collection http://www.biocyc.org/
Textpresso for C. elegans http://www.textpresso.org/
Patent Lens http://www.patentlens.net/daisy/patentlens/patentlens.html
WormBase http://www.wormbase.org/ and http://www.wormbase.org/db/searches/wb/_query
BioMart http://www.biomart.org/
EnsMart http://www.ensembl.org/EnsMart

URL to Code

Chinook http://www.bcgsc.bc.ca/chinook/
myGRID http://www.mygrid.org.uk/
AceDB Query Language http://www.acedb.org/Cornell/aboutacedbquery.html

AQL – Acedb Query Language http://www.acedb.org/Software/whelp/AQL/
Lorel – Lore Query Language http://infolab.stanford.edu/lore/
BioVelo Query Language http://www.biocyc.org/bioveloLanguage.html
BioCyc Pathway Tools http://bioinformatics.ai.sri.com/ptools/ptools-overview.html
Karlsruhe Bioinformatik Harvester (KIT) http://harvester.fzk.de/harvester/
biomaRt and Bioconductor http://www.bioconductor.org/
AmiGO http://amigo.geneontology.org/cgi-bin/amigo/go.cgi

Cross-References

▶ Graph Management in the Life Sciences
▶ Mediation
▶ Scientific Workflows

Recommended Reading

1. Bartlett JC, Toms EG. Developing a protocol for bioinformatics analysis: an integrated information behaviors and task analysis approach. J Am Soc for Inf Sci Technol. 2005;56(5):469–82.
2. Buneman P, Naqvi SA, Tannen V, Wong L. Principles of Programming with Complex Objects and Collection Types. Theor Comput Sci. 1995;149(1):3–48.
3. I-Min Chen A, Markowitz Victor M. An overview of the object-protocol model (OPM) and OPM data management tools. Inf Syst. 1995;20(5):393–418.
4. Etzold T, Harris H, Beaulah S. SRS-5 Chapter: an integration platform for databanks and analysis tools in bioinformatics,, In: Lacroix, Critchlow. 2003. p. 109–46. [6].
5. Hunter L, Cohen KB. Biomedical language processing: perspective what's beyond PubMed? Mol Cell. 2006;21(5):589–94.
6. Lacroix Z, Critchlow T. Bioinformatics: managing scientific data. San Francisco: Morgan Kaufmann; 2003.
7. Lacroix Z, Ludaescher B, Stevens R. Integrating biological databases. chapter 42. 2007. p. 1525–72. Vol. 3 of Lengauer [8].
8. Lengauer T. Bioinformatics – from genomes to therapies. Weinheim: Wiley-VCH Publishers; 2007.

Q

9. Seeber I. Patent searches as a complement to literature searches in the life sciences-a 'how-to' tutorial. Nat Protoc. 2007;2(10):2418–28.
10. Shipman DW. The functional data model and the data language DAPLEX. ACM Trans Database Syst. 1981;6(1):140–73.
11. Stevens R, Goble C, Baker P, Brass A. A classification of tasks in bioinformatics. Bioinformatics. 2001;17(2):180–8.

Query Load Balancing in Parallel Database Systems

Luc Bouganim
INRIA Saclay and UVSQ, Le Chesnay, France

Synonyms

Resource scheduling

Definition

The goal of parallel query execution is minimizing query response time using inter- and intraoperator parallelism. Interoperator parallelism assigns different operators of a query execution plan to distinct (sets of) processors, while intraoperator parallelism uses several processors for the execution of a single operator, thanks to data partitioning. Conceptually, parallelizing a query amounts to divide the query work in small pieces or tasks assigned to different processors. The response time of a set of parallel tasks being that of the longest one, the main difficulty is to produce and execute these tasks such that the query load is evenly balanced within the processors. This is made more complex by the existence of dependencies between tasks (e.g., pipeline parallelism) and synchronizations points. Query load balancing relates to static and/or dynamic techniques and algorithms to balance the query load within the processors so that the response time is minimized.

Historical Background

Parallel database processing appeared very early in the context of database machines in the 1970s. Parallel algorithms (e.g., hash joins) were later proposed in the early 1980s, where tuples are uniformly distributed at every stage of the query execution. However several works (e.g., [8]) gave considerable evidence that data skew, i.e., nonuniform distribution of tuples, exists, and its negative impact on query execution was shown in, e.g., [7]. This motivated numerous studies [10] on intra- and interoperator load balancing in the 1990s.

Foundations

Load-balancing problems can appear with intraoperator parallelism (variation in partition size, namely, *data skew*) and interoperator parallelism (variation in the complexity of operators, synchronization problems). Intra- and interoperator load-balancing problems are first detailed on a simplified query execution plan example considering a static allocation of processors to the query operators. The main load-balancing techniques proposed to address these problems are described next.

Load-Balancing Problems

Figure 1a shows a simplified query execution plan for the following query: "Select T.b from R, S, T where R.Rid = S.Rid and S.Sid = T.Sid and R.a = value" (the scan and project operators were omitted to simplify the drawing). The following assumptions are made: (i) the degree of parallelism (i.e., number of processors allocated) for the selection on R (called σR), the join with S (called $\triangleright\triangleleft S\triangleright\triangleleft S$) and the join with T (called $\triangleright\triangleleft\triangleright\triangleleft T$), has been statically fixed, using a cost model, to, respectively, 2, 3, and 2, and (ii) these operators are processed in pipeline, thus leading to a total degree of parallelism of 7.

Intraoperator load-balancing issues are first illustrated using the classification proposed in [13]. As shown in Fig. 1c, R and S are poorly partitioned because of *attribute value skew (AVS)* inherent in the data set and/or *tuple placement skew (TPS)*. The processing time of the two instances σR1 and σR2 are thus not equal. The case of ▷◁▷◁S is likely to be worse (see Fig. 1b). First, the number of tuples received can be different from one instance to another because of poor redistribution of the partitions of R (*redistribution skew, RS*) or variable selectivity according to the partition of R processed (*selectivity skew, SS*). Finally, the uneven size of S partitions (*AVS/TPS*) yields different processing times for tuples sent by the σR operator, and the result size is different from one partition to the other due to join selectivity (*join product skew, JPS*). The skew effects are therefore propagated toward the query tree, and even with a perfect partitioning of T, the processing time of ▷◁▷◁T1 and ▷◁▷◁T2 can be highly different (uneven size of their left input resulting from ▷◁▷◁S). Intraoperator load balancing is thus difficult to achieve statically, given the combined effects of different types of data skew.

In order to obtain good load balancing at the interoperator level, it is necessary to choose how many and which processors to assign to the execution of each operator. This should be done while taking into account pipeline parallelism, which requires interoperator communication and introduces precedence constraints between operators (i.e., an operator must be terminated before the next one begins). In [15], three main problems are described:

(i) The degree of parallelism and the allocation of processors to operators, when decided in the parallel optimization phase, are based on a possibly inaccurate cost model. Indeed, it is difficult, if not impossible, to take into account highly dynamic parameters like interference between processors, memory contentions, and, obviously, the impacts of data skew.

(ii) The choice of the degree of parallelism is subject to errors because both processors and operators are discrete entities. For instance, considering Fig. 1b, the number of processors for σR, ▷◁▷◁S, and ▷◁▷◁T may have been computed by the cost model as, respectively, 1.5, 3.8, and 2.4 and have been rounded to 2, 3, and 2 processors. But the good distribution, taking into account data skew on S partitions, should have been 1, 4, and 2.

(iii) The processors associated with the latest operators in a pipeline chain may remain idle a significant time. This is called the pipeline delay problem. For instance, while tuples do not match the selection on R or the

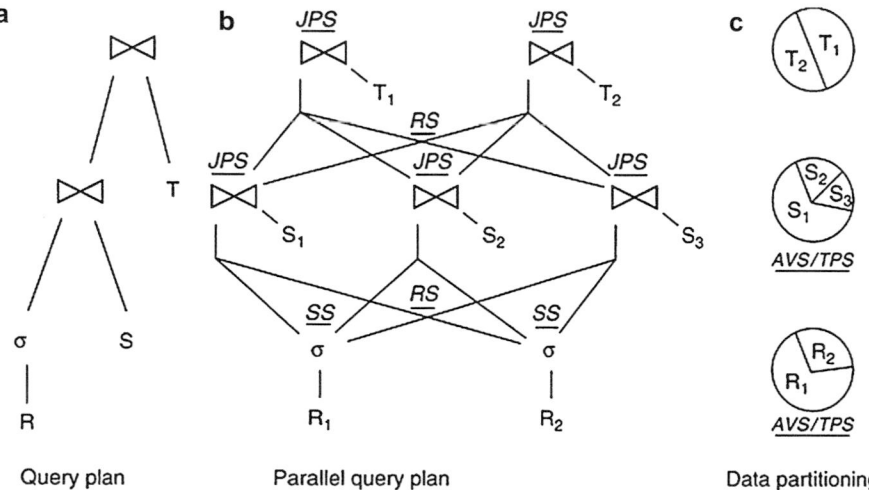

Query Load Balancing in Parallel Database Systems, Fig. 1 Intra- and interoperator load-balancing problems on a simple example

join with S, processors assigned to ▷◁▷◁T remain idle.

In a shared-nothing architecture, the inter-operator load-balancing problem is even more complex, since the degree of parallelism and the set of processors assigned for some operators are constrained by the physical placement of the manipulated data. For instance, if R is partitioned on two nodes, σR must be executed on these nodes.

This simple example thus shows that static allocation of processors to operators is usually far from optimal, thus advocating for more dynamic strategies. In the following section, existing proposals at the intra- and interoperator level are detailed.

Intraoperator Load Balancing

Good intraoperator load balancing depends on the degree of parallelism and on the allocation of processors for the operator. For some algorithms, e.g., the parallel hash join algorithm, these parameters are not constrained by the placement of the data. Thus, the home of the operator (the set of processor where it is executed) must be carefully decided. The skew problem makes it hard for a parallel query optimizer to make this decision statically (at compile-time) as it would require a very accurate and detailed cost model. Therefore, the main solutions rely on adaptive techniques or specialized algorithms which can be incorporated in the query optimizer/processor. These techniques are described below in the context of parallel joins, which has received much attention. For simplicity, each operator is given a home either statically or just before execution, as decided by the query optimizer/ processor.

Adaptive techniques: The main idea is to statically decide on an initial allocation of the processors to the operator (using a cost model) and, at execution time, adapt this decision to skew using load reallocation. A simple approach to load reallocation is detecting the oversized partitions and partition them again onto several processors (among those already allocated to the operator) to increase parallelism [6]. This approach is generalized in [2] to allow for more dynamic adjustment of the degree of parallelism. It uses specific *control operators* in the execution plan to detect whether the static estimates for intermediate result sizes differ from the runtime values. During execution, if the difference between estimate and real value is high enough, the control operator performs data redistribution in order to prevent join product skew and redistribution skew. Adaptive techniques are useful to improve intraoperator load balancing in all kinds of parallel architectures. However, most of the work has been done in the context of shared-nothing where the effects of load unbalance are more severe on performance.

Specialized algorithms: Parallel join algorithms can be specialized to deal with skew. The approach proposed in [3] is to use multiple join algorithms, each specialized for a different degree of skew, and to determine the best at execution time. It relies on two main techniques: range partitioning and sampling. Range partitioning is used instead of hash partitioning (in the parallel hash join algorithm) to minimize redistribution skew of the building relation. Thus, processors can get partitions of equal number of tuples, corresponding to different ranges of join attribute values. To determine the values that delineate the range values, sampling of the building relation is used to produce a histogram of the join attribute values, i.e., the numbers of tuples for each attribute value. Sampling is also useful in determining which algorithm to use and which relation to use for building or probing. The parallel hash join algorithm can then be adapted to deal with skew as follows: (i) Sample the building relation to determine the partitioning ranges. (ii) Redistribute the building relation to the processors using the ranges. Each processor builds a hash table containing the incoming tuples. (iii) Redistribute the probing relation using the same ranges to the processors. For each tuple received, each processor probes the hash table to perform the join. This algorithm can be further improved to deal with high skew using additional techniques and different processor allocation strategies [3]. A similar approach is to modify the join algorithms by inserting a scheduling step which is in charge of redistributing the load at runtime [14].

Interoperator Load Balancing

The interoperator load-balancing problem was extensively addressed during the 1990s. Since then many processor allocation algorithms have been proposed for different target parallel architectures and considering CPU, I/Os, or other resources, such as available memory.

The main approach in shared-nothing is to determine dynamically (just before the execution) the degree of parallelism and the localization of the processors for each operator. For instance, the rate match algorithm [9] uses a cost model in order to define the degree of parallelism of operators having a producer-consumer dependency such that the producing rate matches the consuming rate. It is the basis for choosing the set of processors which will be used for query execution (based on available memory, CPU, and disk utilization). Many other algorithms are possible for the choice of the number and localization of processors, for instance, by a dynamic monitoring and adjustment of the use of several resources (e.g., CPU, memory, and disks) [11].

Shared-disk and shared-memory architectures provide more flexibility since all processors have equal access to the disks. Hence there is no need for physical relation partitioning and any processor can be allocated to any operator [12].

Considering the shared-disk architecture, Hsiao et al. [5] propose to assign processors recursively from the root up to the leaves of a so-called allocation tree. This tree is derived from the query tree, each pipeline chain (i.e., set of operators having pipeline dependencies) being represented as a node. The edges of the allocation tree represent precedence constraints. All available processors are assigned to the root node of the allocation tree (the last pipeline chain to be executed). Then, a cost model is used to divide the CPU power between each child of the root in order to ensure that all the data necessary for the execution of the root pipeline chain will be produced synchronously.

The approach proposed in [4] for shared memory allows the parallel execution of independent pipeline chains called tasks. The main idea is combining IO-bound and CPU-bound tasks to increase system resource utilization. Before execution, a task is classified as IO-bound or CPU-bound using cost model information. CPU-bound and IO-bound tasks can then be run in parallel at their optimal IO-CPU balance, by dynamically adjusting the degree of intraoperator parallelism of the tasks.

Intra-query Load Balancing

Intra-query load balancing combines intra- and interoperator parallelism. To some extent, given a parallel architecture, the load-balancing techniques presented above can be extended or combined. For instance, the control operators used a priori for intraoperator load balancing can modify the degree of parallelism of an operator, thus impacting interoperator load balancing [2].

A general load-balancing solution in the context of hierarchical parallel architectures (a shared-nothing system whose nodes are shared-memory multiprocessors) is the execution model called dynamic processing (DP) [1]. In such systems, the load-balancing problem is exacerbated because it must be addressed both locally (among the processors of each shared-memory node) and globally (among all nodes). The basic idea of DP is decomposing the query into self-contained units of sequential processing, each of which can be carried out by any processor. Intuitively, a processor can migrate horizontally (intraoperator parallelism) and vertically (interoperator parallelism) along the query operators. This minimizes the communication overhead of internode load balancing by maximizing intra and interoperator load balancing within shared-memory nodes.

Key Applications

Load-balancing techniques are essential in applications dealing with very large databases and complex queries, e.g., data warehousing, data mining, business intelligence, and more generally all OLAP (online analytical processing) applications.

Data Sets

DBGen, a synthetic data generator, can be used to generate biased data distribution, for studying intraoperator load-balancing issues. It allows generating data with nonuniform distribution (Zipfian, Poisson, Gaussian, etc.). See http://research. microsoft.com/~Gray/DBGen/

Cross-References

▶ Parallel Database Management
▶ Parallel Data Placement
▶ Parallel Query Processing
▶ Storage Resource Management

Recommended Reading

1. Bouganim L, Florescu D, Valduriez P. Dynamic load balancing in hierarchical parallel database systems. In: Proceedings of the 22th International Conference on Very Large Data Bases; 1996. p. 436–47.
2. Brunie L, Kosch H. Control strategies for complex relational query processing in shared nothing systems. ACM SIGMOD Rec. 1996;25(3):34–9.
3. De Witt DJ, Naughton JF, Schneider DA, Seshadri S. Practical skew handling in parallel joins. In: Proceedings of the18th International Conference on Very Large Data Bases; 1992. p. 27–40.
4. Hong W. Exploiting inter-operation parallelism in XPRS. In: Proceedings of the ACM SIGMOD International Conference on Management of Data; 1992. p. 19–28.
5. Hsiao H, Chen MS, Yu PS. On parallel execution of multiple pipelined hash joins. In: Proceedings of the ACM SIGMOD International Conference on Management of Data; 1994. p. 185–96.
6. Kitsuregawa M, Ogawa Y. Bucket spreading parallel hash: a new, robust, parallel hash join method for data skew in the super database computer. In: Proceedings of the 16th International Conference on Very Large Data Bases; 1990. p. 210–21.
7. Lakshmi MS, Yu PS. Effect of skew on join performance in parallel architectures. In: Proceedings of the International Symposium on Databases in Parallel and Distributed Systems; 1988. p. 107–20.
8. Lynch C. Selectivity estimation and query optimization in large databases with highly skewed distributions of column values. In: Proceedings of the 14th International Conference on Very Large Data Bases; 1988. p. 240–51.
9. Metha M, De Witt D. Managing intra-operator parallelism in parallel database systems. In: Proceedings of the 21th International Conference on Very Large Data Bases; 1995. p. 382–94.
10. Özsu T, Valduriez P. Principles of Distributed Database Systems (2nd edn.). Prentice Hall; 1999 (3rd edn., forthcoming).
11. Rahm E, Marek R. Dynamic multi-resource load balancing in parallel database systems. In: Proceedings of the 21th International Conference on Very Large Data Bases; 1995.
12. Shekita EJ, Young HC. Multi-join optimization for symmetric multiprocessor. In: Proceedings of the 19th International Conference on Very Large Data Bases; 1993. p. 479–92.
13. Walton CB, Dale AG, Jenevin RM. A taxonomy and performance model of data skew effects in parallel joins. In: Proceedings of the 17th International Conference on Very Large Data Bases; 1991. p. 537–48.
14. Wolf JL, Dias DM, Yu PS, Turek J. New Algorithms for parallelizing relational database joins in the presence of data skew. IEEE Trans Knowl Data Eng. 1994;6(6):990–7.
15. Wilshut N, Flokstra J, Apers PG. Parallel evaluation of multi-join queries. In: Proceedings of the ACM SIGMOD International Conference on Management of Data; 1995. p. 115–26.

Query Optimization

Evaggelia Pitoura
Department of Computer Science and Engineering, University of Ioannina, Ioannina, Greece

Synonyms

Query compilation

Definition

A query optimizer translates a query into a sequence of physical operators that can be directly carried out by the query execution engine. The output of the optimizer is called a *query execution plan*. The execution plan may be thought of as a dataflow datagram that pipes data through a graph of query operators. The goal of query optimization is to derive an efficient execution plan in terms of relevant performance measures, such as memory usage and query response time. To achieve this, the optimizer needs to provide: (i) a space of execution plans (search space), (ii) cost estimation techniques to assign a relevant cost to each plan in the search space, and (iii) an enumeration algorithm to search through the space of plans.

Key Points

The query optimizer takes as input a parsed query and produces as output an efficient execution plan for the query. The task of the optimizer is nontrivial, since given a query (i) there are many logically equivalent algebraic expressions (for instance, resulting from the commutativity property among the logical operators), and (ii) for each expression, there are many physical operators supported by the query execution engine for implementing each logical operator (for example, there are several join algorithms, e.g., nested-loop and sort-merge join, for implementing join). The task of finding an equivalent algebraic expression is often called *query rewriting*.

The optimizer needs to enumerate all possible execution plans, estimate their cost and select the one with the smallest cost. The cost assigned to each plan is based on a *cost model* that provides an estimation of the resources needed for its execution, where the resources include CPU time, I/O cost, memory, and communication bandwidth. The cost model relies on statistics maintained on relations and indexes, and uses cost formulas for estimating the selectivity of the various operators and their expected recourse usage. Often, dynamic programming techniques are used to enumerate different plans. These techniques, exploit the fact that to obtain an optimal plan for an expression, it suffices to consider only the optimal plans for its sub-expressions [1–3].

Cross-References

▸ Query Optimization (in Relational Databases)
▸ Query Optimization in Sensor Networks
▸ Query Plan
▸ Query Processing
▸ Query Rewriting

Recommended Reading

1. Chaudhuri S. An overview of query optimization in relational systems. In: Proceedings of the 17th ACM SIGACT-SIGMOD-SIGART Symposium on Principles of Database Systems; 1998. p. 34–43.
2. Ioannidis Y. Query optimization. In: Tucker AB, editors. Handbook of computer science. CRC Press; 1996.
3. Jarke M, Koch J. Query optimization in database systems. ACM Comput Surv. 1984;16(2):111–52.

Query Optimization (in Relational Databases)

Thomas Neumann
Max-Planck Institute for Informatics, Saarbrücken, Germany

Synonyms

Query compilation

Definition

Database queries are given in declarative languages, typically SQL. The goal of query optimization is to choose the best execution strategy for a given query under the given resource constraints. While the query specifies the user intent (i.e., the desired output), it does not specify how the output should be produced. This allows for optimization decisions, and for many queries there is a wide range of possible execution strategies, which can differ greatly in their resulting performance. This renders query optimization an important step during query processing.

Historical Background

One of the first papers to discuss query optimization in relational database systems was the seminal System R paper [2]. It introduced a dynamic programming algorithm for optimizing the join order, and coined the concept of interesting orders for exploiting available orderings. Later approaches increased the set of optimized operators, and included a rule based description of optimization techniques (whereas the optimizations in System R were hard wired). One prominent example is the Starburst [3] optimizer. It introduced a different internal representation (query graph model), that could express complex queries suitably for optimization, and proposed using grammar-like rules to combine low level physical operators (LOLEPOPs) in execution plans. Besides being rule-based, the optimization itself used a bottom-up approach similar to System R. Another family of optimizers was introduced by Volcano [4] (which itself evolved from Exodus) and Cascades [5]. Instead of bottom-up constructive optimization they used transformative top-down optimization with memorization. Besides these more fundamental approaches, a rich literature of optimization techniques exist, ranging from support for specific operators like outer joins or expensive predicates to fundamental data reduction like magic sets.

Foundations

The key idea behind query optimization is the observation that the same query can be formulated in different ways. When a user gives a query in SQL, the query is first parsed and analyzed, and then brought into an internal representation, for example in relational algebra. This translation first creates a canonical representation, as shown in Fig. 1. As a first translation step, the *select-from-where* queries in a) can be answered by combining all relations in the *from* clause with a cross product and then checking the *where* condition on each resulting tuple (shown in b)). But cross products are expensive operations, therefore it is preferable to move a part of

the *where* condition into the cross product to form a join. The remaining part of the condition can be checked before the join, which further reduces the effort for the join (shown in c)). Both operator's trees produce the same tuples when executed, and thus are different representations of the same query. But executing the tree in (i) is most likely cheaper than executing the tree in (ii), which means that (iii) is preferable. The query optimizer therefore starts by translating the query into a canonical representation, which is easy to construct but inefficient to execute, and therefore finds better representations of the same query.

The base for finding better alternatives is the concept of *algebraic equivalences*. Two algebra expressions are equivalent if they produce the same result when executed. As this is difficult to decide in general, query optimizers instead rely on known equivalences. For example the join operator is commutative and associative:

$$A \bowtie B \equiv B \bowtie A$$
$$A \bowtie (B \bowtie C) \equiv (A \bowtie B) \bowtie C$$

Many equivalences are known from the literature [7]. The equivalences form the search space that is explored by the query optimizer: When two (sub-)expressions are equivalent, the optimizer is free to choose any of the two. Optimizers that are directly based upon this principle are called *transformative* optimizers, as they transform algebra expressions into other algebra expressions by applying algebraic equivalences. Transformative optimizers are relatively easy to build and can potentially make use of arbitrary equivalences, but an efficient exploration of the search space is very difficult. Most transformative optimizers are therefore only heuristical. The family of *constructive* optimizers does not apply the equivalences directly, but builds expressions bottom-up from smaller expressions such that the resulting expression is still equivalent to the original expression. This allows for a much more efficient exploration of the search space, but is difficult to organize for more complex equivalences. Therefore most constructive optimizers are at least partially transformative, applying transformative rewrite heuristics for complex

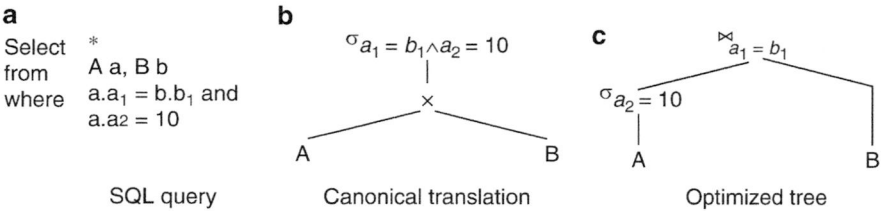

a

Select   *
from   A a, B b
where $a.a_1 = b.b_1$ and
          $a.a2 = 10$

SQL query

b

$\sigma_{a_1 = b_1 \wedge a_2 = 10}$

$\times$

A   B

Canonical translation

c

$\bowtie_{a_1 = b_1}$

$\sigma_{a_2 = 10}$

A   B

Optimized tree

Query Optimization (in Relational Databases), Fig. 1   Translating a SQL query

equivalences before (and after) the constructive optimization step.

The goal of query optimization is improving query processing, which means that the query optimizer needs to take into account the runtime effect of different alternatives. This is done by estimating the *costs* of executing an alternative. A primitive way to estimate the costs is to estimate the number of tuples processed. The intuition here is that a larger number of processed tuples implies more effort spent on executing the query. This estimation requires statistical information, in particular the cardinalities of the relations and the selectivities of the operators involved, but given these it can be computed directly from the algebra expression. Unfortunately this is much too inaccurate for practical purposes. A proper cost function should model the expected execution time (as this is the most common optimization goal), which implies taking into account access patterns on disk, costs for evaluating expensive predicates, etc. The cost function is therefore usually a linear combination of expected I/O costs and expected CPU costs. But this information cannot be derived from the algebra expression, as it is not detailed enough.

Optimizers therefore distinguish between *logical algebra* and *physical algebra*. The logical algebra consists of all operator concepts known to the optimizer, while the physical algebra consists of the operator implementations supported by the execution engine. For example the logical algebra contains one inner join operator $\bowtie$, while the physical algebra contains one join operator for each supported implementation, like nested loop join $\bowtie^{NL}$ or sort-merge join $\bowtie^{SM}$. While the initial query is represented in logical algebra (or an equivalent calculus), the final result must be in

physical algebra, e.g., the optimizer must decide which operator implementations should be used. As these physical algebra expressions (including some annotations) could be executed by the runtime system, they are often called *query execution plans*. The logical algebra is more abstract, which can be useful for optimization, but ultimately the optimizer must construct physical algebra expressions. In particular, cost-based optimization requires physical algebra, as then only costs can be estimated.

Optimizing Simple Queries

The Select-Project-Join queries (SPJ) are relatively simple yet commonly used. They correspond to SQL queries of the form SELECT ... FROM ... WHERE ... without any nested queries. They can be answered by using only selections (σ), joins/cross-products ($\bowtie$/$\times$), and projections (Π). Nevertheless, it has been shown that finding the optimal execution plan is NP hard in general, even for these simple queries. Several simplification are commonly used to reduce the optimization time. As a first step, the optimization concentrates on the join operators. The projections can be added as needed, i.e., whenever an operator materializes its input (and thus breaks the processing pipeline), all attributes that are no longer needed are projected away. Selections are added greedily, i.e., a selection is applied as early as possible. The rationale is that most selection predicates are cheap to evaluate and the selections reduce the work required by the following operators. Thus the optimizer only has to order the join/cross-product operators, and the other operators are added greedily. Unfortunately the problem remains NP hard even with this simplification.

Q

The problem of finding the optimal join order can be seen as finding the optimal binary tree whose leaves correspond to the relations in the *from* clause. The inner nodes are joins or cross products, depending on available predicates suitable for a join, and are determined implicitly by the relations involved in their subtrees. Therefore, only the structure of the binary tree and the labeling of the leaf nodes has to be specified by the query optimizer. However the number of binary trees with n leaf nodes is $C(n-1)$, where $C(n)$ are the Catalan Numbers, which grow in the order of $\Theta\left(4^n/n^{\frac{3}{2}}\right)$. As this grows very fast, some approaches reduce the search space by considering only a limited set of binary trees. A popular restriction is the limitation to *left-deep* join trees (or more general to *linear* join trees). A linear join tree is a join tree where only one of the two subtrees of a join operator may contain other join operators. If only the left subtree may contain other join operators, the trees are called left-deep. Figure 2 shows an example. General join trees without restrictions are called *bushy* join trees. Left-deep join trees are attractive (e.g., System R used them), as they are potentially easier to execute and the number of left-deep join trees is much smaller than the number of bushy join trees. As there are $n!$ ways to label the leaf nodes of the join tree for n relations, there are $n!C(n-1)$ bushy trees, but "only" $n!$ left-deep trees. Unfortunately the optimal join tree can be a bushy tree, and these cases are not uncommon, which means that generating only left-deep trees can hurt query execution performance. Most modern query optimizers therefore construct bushy join trees.

The huge factor of $n!$ is caused by the fact that all combinations of relations are considered valid. The search space can be significantly reduced by avoiding the creation of cross products. When combining two relations, the optimizer can either use a cross-product, or use a join if there is a suitable join predicate in the *where* condition. Joins are much more efficient than cross-products, and although it is sometimes beneficial to use cross-products between separate relations, these cases are rare. When allowing cross-products, any relations can be combined,

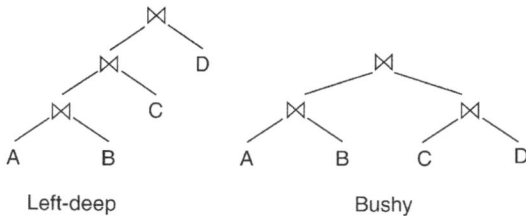

Query Optimization (in Relational Databases), Fig. 2
Left-deep Versus bushy join trees

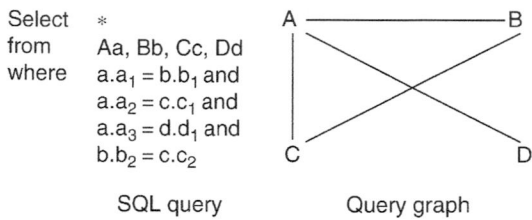

Query Optimization (in Relational Databases), Fig. 3
A query and its query graph

otherwise combinations are only possible if a suitable join predicate exists. The join possibilities implied by the query are captured in the query graph, as shown in Fig. 3. The relations from the *from* clause form the nodes, while the potential join conditions form the edges. Now the optimizer only has to consider sub-problems, i.e., sets of relations, that are connected in the query graph (this assumes that the query graph itself is connected, which can be guaranteed by adding additional edges). For example the relations A,B,C can be joined and thus could be part of a join tree, while no join tree will consist only of B,C,D, as this would require a cross-product. The structure of the query graph greatly affects the size of the search space. If the query graph forms a chain, the join ordering problem is no longer NP hard and can be solved in $O(n^3)$. If the query graph forms a clique, the problem is still NP hard (and just as difficult as when including cross products). Most queries are between those two extremes, and more like a chain than like a clique, and can thus be optimized much more efficiently by avoiding cross products.

Putting these observations together, SPJ queries can be optimized by the following strategy:

1. The only optimization decision is the join order, selections and projections are added greedily
2. The relations in a join tree must be connected in the query graph
3. When constructing left-deep trees, the right hand side of a join must contain only one relation.

The seminal System R paper on query optimization [9] introduced a dynamic programming (DP) strategy to optimize the join order. A slightly generalized version that generates bushy trees is shown in Fig. 4. It computes the optimal join order of the relations $R_1, \ldots, R_n$ (ignoring selections and projections for a moment). The basic strategy is to construct solutions for larger (sub-)problems (i.e., problems involving more relations) from optimal solutions of smaller problems. For example, consider the top-most join in a join tree with four relations. It will either combine two join trees with two relations each or a single relation with a join tree with three relations. The problem of joining four relations can thus be expressed as combining smaller problems with one to three relations each. Accordingly, the DP table is first organized by the size (i.e., number of relations) of a problem. For a given size, the table stores the optimal execution plan for a given set of relations. In lines 1–2 the algorithm initializes the DP table by adding table scans as the optimal solution for problems involving a single relation. The loop starting in line 3 now creates solutions of size s by combining smaller solutions. Lines 4–5 find all pairs of problems already solved (S_1, S_2) that have a combined size of s. If S_1 and S_2 overlap (i.e., they have relations in common) the pair is ignored, as no valid join tree can be constructed (line 6). Similarly (S_1, S_2) is ignored if S_1 and S_2 are not connected in the query graph to avoid the creation of cross-products (line 7). Otherwise it is possible to construct a new execution plan p that joins the known solutions for S_1 and S_2 (line 8). If no solution for $S_1 \cup S_2$ is

known, (or the estimated costs of the new plan are less than for the currently known solution), p is added as a solution of size s for $S_1 \cup S_2$ in the DP table (lines 9–10). At the end of the algorithm the DP table entry for size n contains the optimal solution (line 11).

The algorithm in Fig. 4 is simplified, as it ignores selections and projections. They can be added greedily in lines 2 and 8 and do not affect the algorithm otherwise. A more complex part that is missing is the selection of the physical join operator. Line 8 simply states ⋈, but in a real system, there are multiple join operators available, typically at least nested-loop join, hash join, and sort-merge join. Lines 8–10 should therefore loop over the different join implementations and try all of them. What complicates this choice is that the different implementations behave differently, in particular the sort-merge join. When the input has to be sorted, the sort-merge join is relatively expensive, but when it is already sorted it is very cheap. And the output of a sort-merge join is itself sorted, which can render a following sort-merge join cheap if the order can be reused. Tuple orders that could be used by other operators are called *interesting orderings*, and the set of interesting orderings for the current query is computed before starting the *plan generation* (i.e., the DP algorithm, which constructs execution plans). During plan generation, plans that are more expensive than others have to be preserved, but provide an interesting order the others do not. This can be generalized by the concept of *physical properties*. A physical property is a characteristic of a plan that affects its runtime behavior (i.e., the costs of subsequent operators), but not its logical equivalence. The physical properties define a partial order between plans, describing which plans satisfy "more" properties. For example a plan with sorted output dominates another plan with unsorted output (concerning the physical property "ordered"), while two plans with differently sorted output are not directly comparable. During plan generation, a plan only dominates another plan if both the physical properties are dominating and the estimated costs are lower. As a consequence, the DP table entries no longer consist of single optimal plans, but of sets of

Query Optimization (in Relational Databases), Fig. 4 Dynamic programming strategy for SPJ queries

$\text{DPsize}(R = \{R_1, ..., R_n\})$
for each $R_i \in R$
    $\text{dpTable}[1][\{R_i\}] = R_i$
for each $1 < s \le n$ **ascending** // size of plan
    for each $1 \le s_1 < s$ // size of left subplan
        for each $S_1 \in \text{dpTable}[s_1], S_2 \in \text{dpTable}[s - s_1]$
            if $S_1 \cap S_2 \ne \emptyset$ **continue**
            if $\neg(S_1$ connected to $S_2)$ **continue**
            $p = \text{dpTable}[s_1][S_1] \bowtie \text{dpTable}[s - s_1][S_2]$
            if $\text{dpTable}[s][S_1 \cup S_2] = \emptyset$ V $\text{cost}(p) < \text{cost}(\text{dpTable}[s][S_1 \cup S_2])$
                $\text{dpTable}[s][S_1 \cup S_2] = p$
return $\text{dpTable}[n][\{R_1, ..., R_n\}]$

plans, in which no plan dominates the other. Note that physical properties are an example for re-establishing the principle of optimality required for dynamic programming, which is also required in other contexts.

More Complex Queries

While Selection-Projection-Join queries are a important class of queries, queries can be much more complex. When only optimizing (inner) joins, the joins are freely reorderable, which means that any join order is valid as long as syntax constraints are satisfied (i.e., the relations required for the join predicates are available). This is no longer the case for other operators like outer joins and aggregations. A simple approach to handle these is to split the query into blocks that are freely reorderable (e.g., above and below an outer join) and to optimize the blocks individually. But this is too restrictive, and outer joins still allow for some reorderings, which have been described as algebraic equivalences. The challenge is to incorporate these equivalences into a cost-based (and potentially constructive) query optimizer. For outer joins, it has been shown how the possible reorderings can be expressed as dependencies on input relations [8]. The algorithm analyzes the original query and computes the set of relations that have to be part of a join tree before a specific outer join is applicable. Using this information the outer joins can be integrated easily, the plan generator just has to check the additional requirements before inserting an outer join. The main difficulty is computing this dependency information, but once it is available, the optimization is relatively

simple. Other operators like aggregations are more difficult to integrate. Aggregations can be moved down a join if the join itself does not affect the aggregation results (e.g., a 1:1 join involving the grouping attribute where the join behaves like a selection). Pushing the aggregation down can reduce the effort for the join itself and might thus be beneficial. But in cases where the aggregation can simply be moved are relatively limited. Here, a more general movement of aggregations is possible by allowing for compensation actions. These are computations added to the plan that compensate the fact that the aggregation is performed at a different position. Adding these kinds of optimizations into a cost-based, constructive query optimizer is very challenging, which is why they are usually implemented as heuristical rewrite operations.

Another important aspect of optimizing complex queries is unnesting nested queries and the related problem of view resolution. The SQL query language allows for nesting queries inside other queries, either explicitly by including a nested *select* block, or implicitly by accessing a view. The nested query could be optimized independently from the outer query, and then during the optimization of the outer query treated like a base relation with specific costs. But this will often lead to inferior plans, for example if selection predicates from the outer query could be pushed down into a view. Instead, the optimizer tries to merge the nested query with the outer query into one flat query, which is then optimized in one step. Perhaps even more important than the unified optimization is a decoupling between the nested query and the outer query. The evaluation of the nested query can depend on the attributes of the outer query, which implies a very expensive nested-loop evaluation. In many cases it is possible to unnest these queries such that the nested (and apparently dependent) part can be evaluated independently, and then joined appropriately with the outer part of the query [9]. These optimizations greatly improve the evaluation of nested queries.

Key Applications

Query optimization techniques can improve the query execution time by orders of magnitude. All modern relational database systems therefore implement at least some optimization techniques.

Cross-References

► Parallel data placement
► Parallel database management
► Parallel query execution algorithms
► Parallel query processing

Recommended Reading

1. Chaudhuri S. An overview of query optimization in relational systems. In: Proceedings of the 17th ACM SIGACT-SIGMOD-SIGART Symposium on Principles of Database Systems; 1998. p. 34–43.
2. Garcia-Molina H., Ullman J.D., and Widom J. Database system implementation. Prentice-Hall; 2000.
3. Graefe G. The cascades framework for query optimization. Q Bull IEEE TC Data Engineering. 1995;18(3):19–29.
4. Graefe G, McKenna WJ. The volcano optimizer generator: extensibility and efficient search. In: Proceedings of the 9th International Conference on Data Engineering; 1993. p. 209–218.
5. Haas LM, Freytag JC, Lohman GM, Pirahesh H. Extensible query processing in starburst. In: Proceedings of the ACM SIGMOD International Conference on Management of Data; 1989. p. 377–88.
6. Moerkotte G. Building query compilers, available at http://db.informatik.uni-mannheim.de/moerkotte.html. en. 2006.
7. Muralikrishna M. Improved unnesting algorithms for join aggregate SQL Queries. In: Proceedings of the 18th International Conference on Very Large Data Bases; 1992. p. 91–102.
8. Rao J, Lindsay BG, Lohman GM, Pirahesh H, Simmen DE. Using EELS, a practical approach to outerjoin and antijoin reordering. In: Proceedings of the 17th International Conference on Data Engineering; 2001. p. 585–94.
9. Selinger PG, Astrahan MM, Chamberlin DD, Lorie RA, Price TG Access path selection in a relational database management System. In: Proceedings of the ACM SIGMOD International Conference on Management of Data; 1979. p. 23–34.

Query Optimization in Sensor Networks

Kian-Lee Tan
Department of Computer Science, National University of Singapore, Singapore, Singapore

Synonyms

In-network aggregation; Query optimization; Query processing

Definition

Query optimization is the process of producing a query evaluation plan (QEP) for a query that minimizes or maximizes certain objective functions. A query in a sensor network has additional clauses that specify the life time of a query, the frequency in which the sensor data should be monitored, and even the rate in which query answers should be returned. As such, the query plan must reflect these. In addition, a typical query plan comprises two main components: a communication component that sets up the communication structure for data delivery and a computation component that performs the operation in the sensor network and/or the root node. Because sensor nodes are low-powered, besides minimizing computation cost, optimization criterion includes minimizing energy consumption (e.g., by minimizing transmission cost) or maximizing the life span of the entire sensor network. As such, the cost model must consider these various factors.

Historical Background

Query optimization has always been an important area of research in database query processing. This is because a poorly chosen query execution plan can result in significant waste of resources and, more importantly, user dissatisfaction. In sensor network, the problem is more critical because of the resource constrains of the sensor nodes which are limited in computation power, bandwidth, memory, and energy.

Most of the existing works focused on in-network aggregation [3, 5, 8, 10, 14]. While work in [5, 14] aim at precise answers, work in [3, 10] assume errors (approximations) can be tolerated. In particular, the work in [10] exploits the trade-off between data quality and energy consumption to extend the lifetime of the sensor network. Both the Cougar project [14] and the TinyDB project [5], in their prototype design of a sensor network management system, offer broader insights into query processing and optimization issues.

Scientific Fundamentals

To support applications in sensor networks, the database community has viewed the sensor network as a database [5, 14]. This provides a good logical abstraction for sensor data management. Users can issue declarative queries without having to worry about how the data are generated, processed, and transferred within the network and how sensor nodes are (re)programmed. As such, query optimization techniques can also be applied to optimize the network operations.

Queries

Queries are typically expressed using an extended SQL that include additional clauses that specifies the duration and sampling rates. As an example, the following query counts the number occupied nests in each loud region of a certain island [5, 14]:

        SELECT region, CNT(occupied), AVG(sound)
        FROM sensors
        GROUP BY region
        HAVING AVG(sound) > 200
        SAMPLE PERIOD 10 FOR 3600

Here, the clause "SAMPLE PERIOD ...FOR" indicates that the sensors will sample the environment every 10 s for a duration of 3600 s.

Query Plan

To evaluate a query, a query plan has to be generated for the sensors. A query plan specifies the role of each sensor (the computation to be performed, the rate at which it should sample the data) and the communication structure between sensors. For the sample query above, three alternative plans are considered here: (1) Each sensor samples its data every 10 s and then transmits the data back to the base station; at the base station, it will perform the grouping and aggregate computation as in a centralized system. (2) A sensor within a region is selected as a cluster head (CH); all sensors within the region send their sampled data to the corresponding CH; the CH performs the aggregate and sends it back to the base station if it satisfies the HAVING clause (i.e., >200). (3) This is similar to Plan 2, except that sensors within a region can be further hierarchically partitioned so that each partition has a leader that performs partial aggregation of the sampled data from its child nodes.

For Plan 1, each sensor node takes on at most two roles: (a) sample the data and transmit the sampled value back to the base station and (b) relay the data it receives from its child nodes if it is an internal node along the path to the base station. For Plan 2, the CH has the additional computation task of performing the aggregate and determining whether it should be routed back to the base station. For Plan 3, leaders have the responsibility to perform partial aggregates.

Now all the plans may also specify the sensor nodes in which one should be transmitting data to and/or receiving data from, i.e., the communication structure.

Clearly, there are many other possible plans that can be generated, and it is the optimizer's task to pick the one that best suits the objectives.

Metadata

The optimizer makes its decision based on certain metadata. For example, the metadata for each sensor include the static information about its location (region), the sensor types, the amount of memory, the energy consumption per operation type, the energy consumption per sampling per sensor type, and so on. It may also be necessary to maintain dynamic statistics on a sensor's estimated (remaining) battery life and the selectivity of query predicates. These metadata are periodically collected (via the routing tree) by the optimizer and used in several ways. For example, for sensor nodes with short battery life, the sampling rate may be adjusted so that they can operate in a doze mode longer in order to extend the life span of the entire network. As another example, nodes with long battery life may be selected as the cluster heads or leaders, nodes with moderate amount of energy can relay messages, and nodes with low battery power will simply transmit their sampled data. As such, it is possible to balance the energy across all nodes. As another example, knowing the energy consumption per sampling per sensor type may allow the optimizer to reorder predicates on different sensor types.

Cost Model

For each query plan, the optimizer estimates the cost of the plan. The cost model depends on the objective function. In a sensor network environment, there are two key metrics: energy consumption and transmission cost (as communication consumes most energy).

As an example, consider an arbitrary sensor node along the path of the routing tree. Let N be the number of sensor types and k be the number of predicates. Let E_i be the cost to sample sensor type i, E_{trans} and E_{recv} be the energy to transmit and receive a message, respectively, E^i_{pred} be the energy consumed to evaluate predicate i, and E_{agg} be the energy to compute a (partial) aggregate, and C be the number of child nodes routing through this node. The energy at a node s to collect a sample, and transmit its partial aggregate, including the costs to forward data, can be estimated as follows:

$$e_s = \sum_{i=0}^{N} E_i + \sum_{i=0}^{k} E^i_{pred} + E_{recv} \times C$$
$$+ E_{agg} + E_{trans}$$

The energy consumed by the node is the cost to read all the sensors at the node (first component), plus the cost to evaluate the predicates (second component), plus the cost to receive partial aggregates from its child nodes (third component), plus the cost to compute a partial aggregate (fourth component), plus the cost to transmit the partial aggregate (fifth component).

Suppose sensor s has a remaining battery capacity of B_s Joules. Then, s has enough power to last B_s/e_s sample collections. To extend life span of the system, a plan that maximizes this value has to be determined. Let there be P plans, then a plan that maximizes the minimum number of sample collections is the one that should be selected (to maximize the life span of the system), i.e.,

$$\max_{i=1}^{P} (\min_{j=1}^{N} B_j / e_j)$$

Depending on different objectives and model, similar cost models can be derived.

Centralized Optimization

Most of the existing work on optimizing sensor network queries are based on a centralized optimizer, i.e., the optimizer at the base station determines the query plan [5, 14]. There are two dimensions in which query processing can be optimized. The first dimension deals with the grouping of the sensor nodes to support in-network computation (e.g., aggregates). Essentially, the optimizer enumerates the possible alternative plans based on different ways in which nodes are grouped. Some heuristics are:

(a) On one extreme, all nodes belong to a single group;
(b) On another extreme, each node forms a single group;
(c) Between the two extremes, different group-based heuristics can be adopted – proximity-based schemes cluster nodes that are close-by together and semantic-based schemes cluster nodes that are semantically related, e.g., same set of sensor types, same resources, same metadata, etc.

The second dimension covers heuristics that are used to reorder the operations of a query. Some of these are:

(a) When a sensor node is required to sample multiple attributes (temperature, humidity, light) and the query involves a number of predicates, different orderings of the sampling and predicate evaluations may result in different energy consumption. This is the case because sampling consumes more power than evaluating a predicate. As such, to conserve energy, it makes sense to order selective predicates first before the non-selective predicates; in addition, data should be sampled only when necessary. Intuitively, once a sampled value is discarded (because it does not satisfy a predicate), there is no need to expend energy to sample other attributes. Thus, the optimizer enumerates different sequences of sampling and predicate evaluations to find the one that is most energy efficient. As an example, consider a query that finds sensor nodes whose accelerometer and magnetometer readings exceed thresholds a1 and m1, respectively. As the magnetometer consumes 50 times more energy than the accelerometer to sample a reading, if the selectivity of the predicate on the accelerometer reading is low, then it makes sense to sample the accelerometer first; moreover, the magnetometer should only be sampled when the accelerometer reading is larger than a1.

(b) For certain event monitoring queries, they can be rewritten into a window join queries that can be more efficiently processed in the network.

Distributed Optimization

As centralized optimization schemes require certain metadata that have to be periodically obtained from the sensor nodes, it may be costly (in terms of power) to collect these information. An alternative approach is to perform distributed optimization [7]. Here, the basic idea is to iden-tify clusters of nodes with its associated cluster head (CH) so that these CHs optimize the queries for processing within the cluster. Thus, the base stations disseminate the query to the CHs, each of which optimizes the query based on the local metadata and controls the processing within the cluster.

With distributed optimization, the metadata is collected at the CH, rather than the base station. This means that the transmission overhead for metadata is reduced. Moreover, "local" metadata are more accurate as some globally collected metadata may be too coarse (e.g., distribution of data). However, the CHs incur the overhead of optimizing the query.

Here, the challenge is to determine the clusters (either spatially or semantically). In spatial-based clustering scheme, the number of groups are based on the radius of a group, which is the number of hops between a non-CH node and its CH. In the semantic-based clustering scheme, nodes with similar metadata are formed into groups.

Complex Event Detection: Join Processing

In many applications such as surveillance and environmental monitoring, it is not uncommon to track the correlations among sensor data within a time window to detect events of interest. This calls for data from multiple sensor nodes to be joined. In [1], static join queries are considered, and in [13], continuous join queries are examined. In both cases, join processing is pushed into the sensor network. In the former, sensor nodes are grouped so that a static table can be partitioned across them; new sampled records are broadcast within a group, and only answers need to be sent back to the base stations. In the latter, the focus is on minimizing the number of subqueries that need to be injected into the sensor network. Stern et al. [9], on the other hand, adopted a filtering approach. The idea is to continuously compute optimal filters that minimize communication overhead, push them to the nodes, and update these filters at runtime.

Key Applications

Sensor databases can be applied in many applications, e.g., environmental monitoring and military surveillance. See Sensor Databases for details. To process queries in these applications, it is necessary to optimize the queries in order to ensure that resources are well utilized, and the system life span can be sufficiently large to minimize any need to replace the batteries regularly.

Future Directions

Most of the existing work in the literature have focused on optimizing aggregate and join queries. As sensor nodes become more powerful, there is much opportunity for optimizing complex queries that have not been previously studied. A promising direction that has recently received attention is the multi-query optimization problem [11, 12]. Here, multiple queries submitted by users are optimized collectively to exploit commonality among them so that any redundant data accesses from the sensors can be eliminated. Current focus has been on non-join queries. Extending these schemes to handle more complex queries is an interesting direction to explore. Yet another direction is to consider a large-scale sensor network deployment that involves multiple base stations. Here, the challenge is to optimize the network life span as well as the load across the base stations. Another direction that has gained interests in recent years is the computation of outliers [4] and top-k queries [2]. Finally, real-time analytics is becoming a promising area of active research [6].

Cross-References

▶ Continuous Queries in Sensor Networks
▶ Data Aggregation in Sensor Networks
▶ In-Network Query Processing
▶ Sensor Networks

Recommended Reading

1. Abadi DJ, Madden S, Lindner W. REED: robust, efficient filtering and event detection in sensor networks. In: Proceedings of the 31st International Conference on Very Large Data Bases; 2005. p. 769–80.
2. Chen B, Liang W, Yu JX. Energy-efficient top-k query evaluation and maintenance in wireless sensor networks. Wirel Netw. 2014;20(4):591–610.
3. Deligiannakis A, Kotidis Y, Roussopoulos N. Hierarchical in-network data aggregation with quality guarantees. In: Advances in Database Technology, Proceedings of the 9th International Conference on Extending Database Technology; 2004. p. 658–75.
4. Giatrakos N, Kotidis Y, Deligiannakis A, Vassalos V, Theodoridis Y. In-network approximate computation of outliers with quality guarantees. Inf Syst. 2013;38(8):1285–308.
5. Madden S, Franklin MJ, Hellerstein JM, Hong W. TINYDB: an acquisitional query processing system for sensor networks. ACM Trans Database Syst. 2005;30(1):122–73.
6. Palpanas T. Real-time data analytics in sensor networks. In: Managing and mining sensor data. Boston: Springer; 2013. p. 173–210.
7. Rosemark R, Lee W-C. Decentralizing query processing in sensor networks. In: Proceedings of the 2nd Annual International Conference on Mobile and Ubiquitous Systems; 2005. p. 270–80.
8. Silberstein A, Yang J. Many-to-many aggregation for sensor networks. In: Proceedings of the 23rd International Conference on Data Engineering; 2007. p. 986–95.
9. Stern M, Bohm K, Buchmann E. Processing continuous join queries in sensor networks: a filtering approach. In: Proceedings of the ACM SIGMOD International Conference on Management of Data; 2010. p. 267–78.
10. Tang X, Xu J. Extending network lifetime for precision-constrained data aggregation in wireless sensor networks. In: Proceedings of the 25th Annual Joint Conference of the IEEE Computer and Communication Societies; 2006.
11. Trigoni N, Yao Y, Demers AJ, Gehrke J, Rajaraman R. Multi-query optimization for sensor networks. In: Proceedings of the 1st IEEE International Conference on Distributed Computing in Sensor Systems; 2005. p. 307–21.
12. Xiang S, Lim HB, Tan KL, Zhou Y. Two-tier multiple query optimization for sensor networks. In: Proceedings of the 27th IEEE International Conference on Distributed Computing Systems; 2007.
13. Yang X, Lim HB, Ozsu T, Tan KL. In-network execution of monitoring queries in sensor networks. In: Proceedings of the ACM SIGMOD International Conference on Management of Data; 2007. p. 521–32.
14. Yao Y, Gehrke J. Query processing in sensor networks. In: Proceedings of the 1st Biennial Conference on Innovative Data Systems Research; 2003.

Q

Query Plan

Evaggelia Pitoura
Department of Computer Science and
Engineering, University of Ioannina, Ioannina,
Greece

Synonyms

Operator tree; Qucry evaluation plan; Query execution plan; Query plan; Query tree

Definition

A query or execution plan for a query specifies precisely how the query is to be executed. Most often, the plan for a query is represented as a tree whose internal nodes are operators and its leaves correspond to the input relations of the query. The edges of the tree indicate the data flow among the operators. The query plan is executed by the query execution engine.

Key Points

During query processing, an input query is transformed into an internal representation, most often, into a relational algebra expression. To specify how to evaluate a query precisely, each relational operator is then mapped to one or more physical operators, which provide several alternative implementations of relational operators.

A query plan for each query specifies the physical operators to be used for its execution and the order of their invocation. Most commonly, a query plan is represented as a tree. The leaf nodes of the query tree correspond to the input (or base) database relations of the query and its internal nodes to physical operators. The edges of the tree specify the data flow among the operators. Each operator receives input from its child nodes in the tree and, in turn, its output is used as input to its parent node.

The query optimizer enumerates alternative query plans for each query and selects the most efficient among them using a cost estimation model. The selected query plan is then passed for execution to the query execution engine that results in generating answers to the query.

Query plans can be divided into prototypical shapes and query execution engines can be divided into groups based on which shapes of plans they can evaluate. Such shapes include left-deep, right-deep and bushy plans. Deep plans are plans in which each join involves at least one base relation, whereas bushy plans are more general in that a join could involve one or two base relations or the results of one or two other join operations. Finally, for queries with common sub-expressions, the query evaluation plan may be an acyclic directed graph (DAG) instead of a tree.

Cross-References

▶ Evaluation of Relational Operators
▶ Query Optimization
▶ Query Processing

Recommended Reading

1. Graefe G. Query evaluation techniques for large databases. ACM Comput Surv. 1993;25(2):73–170.
2. Ramakrishnan R, Gehrke J. Database management systems. New York: McGraw-Hill; 2003.

Query Point Movement Techniques for Content-Based Image Retrieval

Kien A. Hua and Danzhou Liu
University of Central Florida, Orlando, FL, USA

Definition

Target search in content-based image retrieval (CBIR) systems refers to finding a specific (target) image such as a particular registered logo or a specific historical photograph. To search for such an image, query point movement techniques iteratively move the query point closer to the target image for each round of the user's relevance feedback until the target image is found. The goals of query point movement techniques include avoiding local maximum traps, achieving fast convergence, reducing computation overhead, and guaranteeing to find the target.

Historical Background

Images in a database are characterized by their visual features, and represented as points in a multidimensional feature space. A *query point* is one of these image points, selected to find similar images represented by image points nearest to the query point in the feature space. This cluster of nearby or relevant image points has a shape (see Figs. 1 and 2) referred to as the *query shape*.

For each iteration, a query point movement technique attempts to move the query point closer to the target image by refining the query based on user's relevance feedback. Existing query point movement techniques can be divided into two categories: single-point and multiple-point movement techniques. A technique is classified as a single-point movement technique if the refined query Q_r at each iteration consists of only one query point; otherwise, it is a multiple-point movement technique. In the latter category, the query result is the set of images nearest to the set of query points. Typical query shapes of single-point movement and multiple-point movement techniques are illustrated in Figs. 1 and 2, respectively, where the contours represent equisimilarity surfaces. Single-point movement techniques, such as MARS [9] and MindReader [5], construct a single query point close to relevant images and away from irrelevant ones. MARS uses a weighted distance (producing shapes as shown in Fig. 1b), where each dimension weight is inversely proportional to the standard deviation of the relevant images' feature values in that dimension. The rationale is that a small variation among the values is more likely to express

restrictions on the feature, and thereby should carry a higher weight. On the other hand, a large variation indicates that this dimension is not significant in the query, and should thus assume a lower weight. MindReader achieves better results by using a generalized weighted distance, see Fig. 1c for its shape.

In multiple-point movement techniques such as Query Expansion [1], Qcluster [6], and Query Decomposition [4], multiple query points are used to define the ideal space that is most likely to contain relevant results. Query Expansion groups query points into clusters and choose their centroids as the representatives of the query Q_r (see Fig. 2a). The distance of a point to Q_r is defined as a weighted sum of individual distances to those representatives. The weights are proportional to the number of relevant objects in the clusters. Thus, Query Expansion treats local clusters differently, compared to the equal treatment in single-point movement techniques. In some queries, clusters are too far apart for a unified, all-encompassing contour to be effective; separate contours can yield more selective retrieval. This observation motivated Qcluster to employ an adaptive classification and cluster-merging method to determine optimal contour shapes for complex queries. Qcluster supports disjunctive queries, where similarity to any of the query points is considered relevant (see Fig. 2b). To bridge the semantic gap more effectively, A Query Decomposition technique was presented in [4]. Based on user's relevance feedback, this scheme automatically decomposes a given query into localized subqueries, which more accurately capture images with similar semantics but with very different appearance (see Fig. 2c).

Standard query point movement techniques, explained above, allow re-retrieval of previously determined relevant images when they fall in the search range again. This leads to two major disadvantages:

1. *Local maximum traps.* Since query points in relevance feedback systems have to move through many regions before reaching a target, it is possible that they get trapped in one of these regions. Figure 3 illustrates a possible

Query Point Movement Techniques for Content-Based Image Retrieval, Fig. 1 Single-point movement query shapes

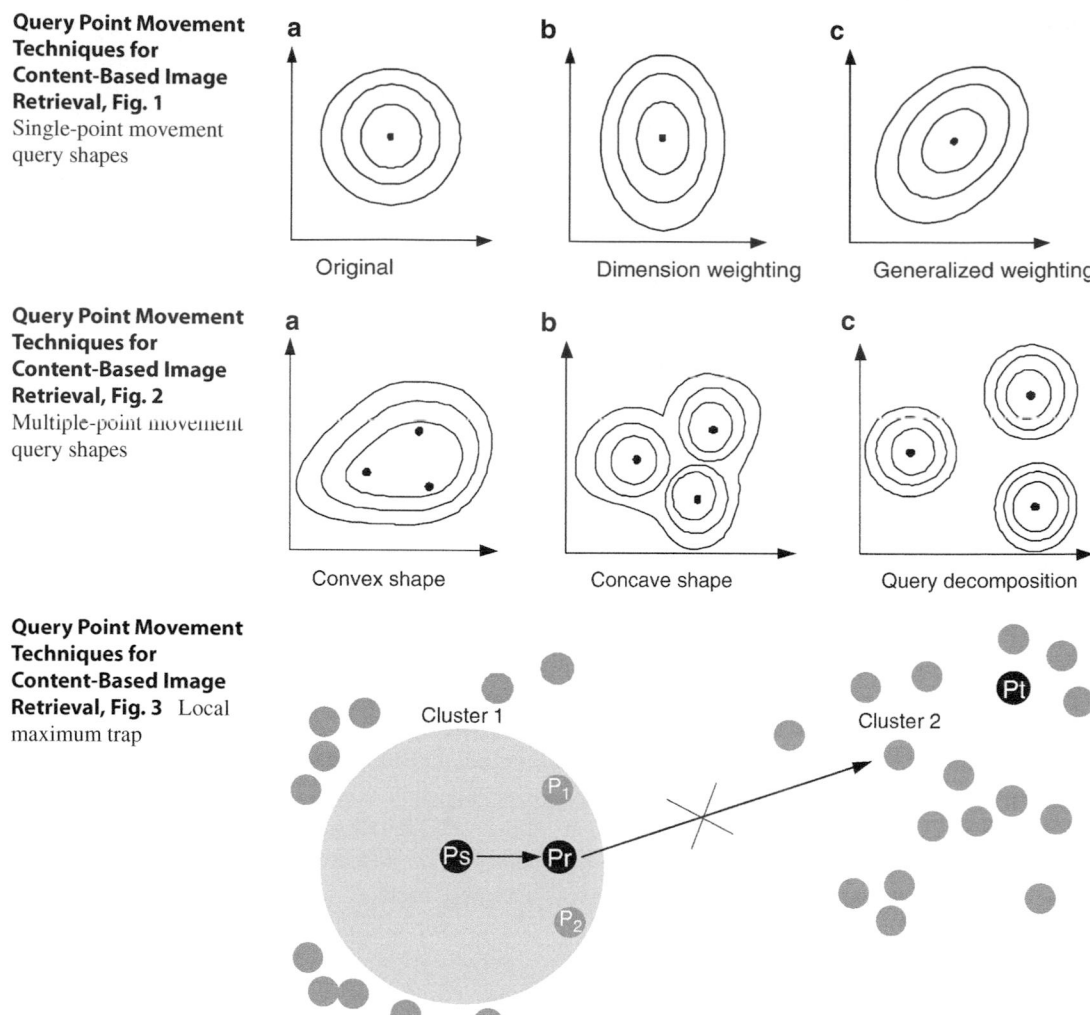

Original   Dimension weighting   Generalized weighting

Query Point Movement Techniques for Content-Based Image Retrieval, Fig. 2 Multiple-point movement query shapes

Convex shape   Concave shape   Query decomposition

Query Point Movement Techniques for Content-Based Image Retrieval, Fig. 3 Local maximum trap

scenario where p_s and p_t denote the starting query point and the target point, respectively. As a result of a 3-NN search at p_s, the system returns points p_1 and p_2, in addition to query point p_s. Since both p_1 and p_2 are relevant, the refined query point p_r is their centroid and the anchor of the next 3-NN search. In this situation, the system will retrieve exactly the same set; from which, points p_1 and p_2 are again selected. In other words, the system can never get out of the subspace because the retrieval set is saturated with the k selected images. Although the system can escape with a larger k, it is difficult to guess a proper threshold. Consequently, the user might not even know a local maximum trap is occurring, and there is no guarantee to find the target image.

2. *Slow convergence*. The centroid of the relevant points is typically selected as the anchor of refined queries. This, coupling with possible retrieval of already visited images, prevents aggressive movement of the search process (see Fig. 4, where $k = 3$). Slow convergence incurs longer search time, and significant computation and disk access overhead.

Foundations

To address the limitations of standard query point movement techniques, four target search methods have been proposed [8]: Naïve Random Scan (NRS), Local Neighboring Movement (LNM), Neighboring Divide and Conquer (NDC), and

Query Point Movement Techniques for Content-Based Image Retrieval, Fig. 4 Slow convergence

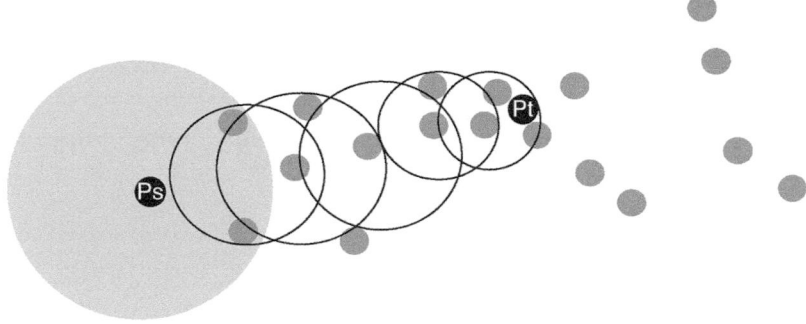

Global Divide and Conquer (GDC) methods. All these methods are designed around a common strategy: they do not retrieve previously selected images (i.e., shrink the search space). Furthermore, NDC and GDC exploit Voronoi diagrams to aggressively prune the search space and move towards the target image faster.

More formally, a query in target search is defined as $Q = \langle n_Q, P_Q, W_Q, D_Q, \mathbb{S}, k \rangle$, where n_Q denotes the number of query points in Q, P_Q the set of n_Q query points in the search space $\mathbb{S}$, W_Q the set of weights associated with P_Q, D_Q the distance function, and k the number of points to be retrieved in each iteration. Using these notations, the four target search techniques can be described as follows:

Naïve Random Scan Method (NRS)

This method randomly retrieves k different images at a time until the user finds the target image or the remaining set is exhausted. Specifically, at each iteration, a set of k not previously selected images are randomly retrieved from the candidate set $\mathbb{S}'$ for relevance feedback, and $\mathbb{S}'$ is then reduced by k for the next iteration. Clearly, this strategy does not suffer local maximum traps and is able to locate the target image after some finite number of iterations. In the best case, NRS takes one iteration, while the worst case requires $\left\lceil \frac{|\mathbb{S}|}{k} \right\rceil$. On average, NRS can find the target in $\left\lceil \sum_{i=1}^{\left\lceil \frac{|\mathbb{S}|}{k} \right\rceil} i / \left\lceil \frac{|\mathbb{S}|}{k} \right\rceil \right\rceil = \left\lceil \left(\left\lceil \frac{|\mathbb{S}|}{k} \right\rceil + 1 \right) / 2 \right\rceil$ iterations. In other words, NRS takes $\mathcal{O}(|\mathbb{S}|)$ to reach the target point. Therefore, NRS is only suitable for a small database set.

Local Neighboring Movement Method (LNM)

This method applies the non-re-retrieval strategy to MindReader [5]. Specifically, Q_r is constructed so that it moves towards neighboring relevant points and away from irrelevant ones, and k-NN query is now evaluated against S' instead of S. When LNM detects a local maximum trap, it requests that the user selects the most relevant image. This way, LNM can overcome local maximum traps, although it could take many iterations to do so. Again, one iteration is required in the best case. If data is uniformly distributed in the n-dimensional hypercube, the worst and average cases are $\left\lceil \sqrt{n} \sqrt[n]{|S|} / \lceil \log_{2^n} k \rceil \right\rceil$ and $\left\lceil \left(\frac{\sqrt{n} \sqrt[n]{|S|}}{\lceil \log_{2^n} k \rceil} + 1 \right) \middle/ 2 \right\rceil$, respectively. If the data are arbitrarily distributed, then the worst case could be as high as that of NRS, i.e., $\left\lceil \frac{|S|}{k} \right\rceil$ iterations (e.g., when all points are on a line). In summary, in the worst case LNM could take anywhere from $O\left(\sqrt[n]{|S|} \right)$ to $O\left(|S|\right)$.

Neighboring Divide and Conquer Method (NDC)

Although LNM can overcome local maximum traps, it does so inefficiently, taking many iterations and in the process returning numerous false hits. To speed up convergence, Voronoi diagrams are used in NDC to reduce the search space after each round of relevance feedback. That is, the k-NN search in each iteration is performed only within the Voronoi cell containing the query point. It can be proved that the target image must reside in this Voronoi cell [8]. Since this strategy

Query Point Movement Techniques for Content-Based Image Retrieval, Fig. 5 Example of NDC

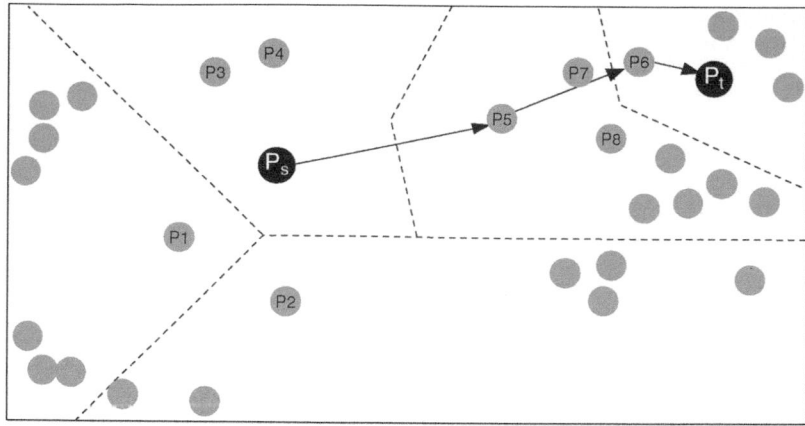

aggressively prunes the search space and moves rapidly towards the target image, it can overcome local maximum traps and achieve fast convergence. Figure 5 illustrates how NDC approaches the target after pruning the search space three times. In the first iteration, points p_1, p_2, and p_s are randomly chosen by the system, assuming $k = 3$; and they are used to construct a Voronoi diagram partitioning the search space into three regions. The user identifies p_s as the most relevant point (i.e., most similar to the target image). Since the target image must reside in the Voronoi cell containing p_s, the computation of the k-NN query anchored at p_s can be confined to this cell while the other two Voronoi regions can be safely ignored. This step retrieves three new nearest neighbors p_3, p_4, and p_5. Their Voronoi diagram further partitions the current search space into three regions. The user again correctly selects p_5 as the most relevant point, and therefore the query point for the third iteration. This refined query results in another set of relevant points p_6, p_7, and p_8; and another Vonronoi diagram is constructed. This time, the user selects p_6 as the most relevant image. Using it as the query point for the fourth iteration, the system returns three relevant points and the user identifies p_t as the target image. If the data points are uniformly distributed, NDC reaches the target point in no more than $O(\log_k |S|)$ iterations. When S is arbitrarily distributed, the worst case could take up to $\lceil \frac{S}{k} \rceil$ iterations (e.g., all points are on a line), the same as that of NRS. In other words, NDC could still

require $O(|S|)$ iterations to reach the target point in the worst case.

Global Divide and Conquer Method (GDC)

To reduce the number of iterations in NDC under the worst case scenario, GDC constructs the Voronoi diagram based on points randomly selected from the current search space, instead of using points from the query result. An example is given in Fig. 6. In the first iteration, a Voronoi diagram is constructed based on three randomly sampled points p_1, p_2, and p_s, assuming $k = 3$. The user selects p_3 as the most relevant point, and this results in p_4, p_5, and p_6 as the query result as computed in NDC. The user now selects p_5 as most relevant in the second iteration. The system randomly selects three points in the Vonoroi cell associated with p_5, and uses them to further partition this cell into three smaller Vonoroi cells. The computation of the k-NN query anchored at p_5 over the smaller Vonoroi cell containing p_5 returns three new points, and the user identifies p_t as the target image in the third round. As proved in [8], the worst case for GDC is bounded by $O(\log_k |S|)$. This implies that for arbitrarily distributed datasets, GDC converges faster than NDC in general, although NDC might be as fast as GDC for certain queries, e.g., the starting query point is close to the target point. In the previous example (Fig. 5), NDC could also take three iterations, instead of four, to reach the target point if the initial k points were the same as in Fig. 6, as opposed to Fig. 5.

Query Point Movement Techniques for Content-Based Image Retrieval, Fig. 6 Example of GDC

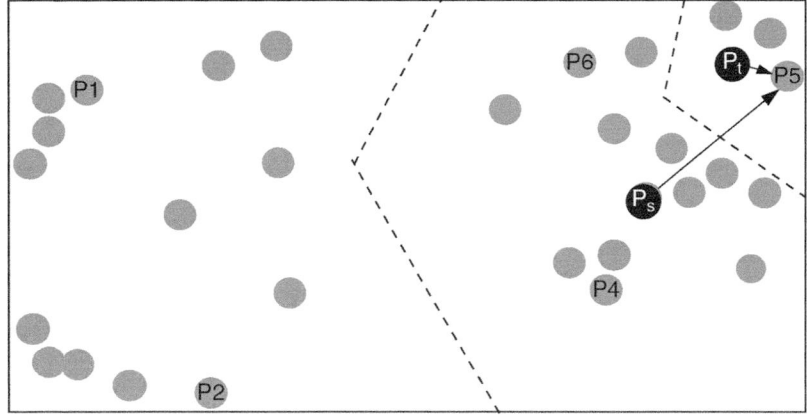

Key Applications

Multimedia Search Engine, Crime Prevention, Graphic Design.

Future Directions

1. Incorporate information from the log file on user relevance feedback to determine the query results in each feedback iteration, instead of performing the traditional k-NN computation. This strategy can minimize the effect of the semantic gap between the low-level visual features and the high-level concepts in the images.
2. Multiple query points, as in standard query point movement techniques such as Query Expansion [1] and Qcluster [6], can be used in each iteration to better convey user's relevance feedback.
3. With the growing interest in Internet-scale image search applications, it is desirable to extend the target search techniques because it will enable concurrent users to share computation [7].

Experimental Results

Figure 7 shows that standard techniques MARS [9], MindReader [5], and Qcluster [6] have poor false hit ratio when k is small. This is due to the

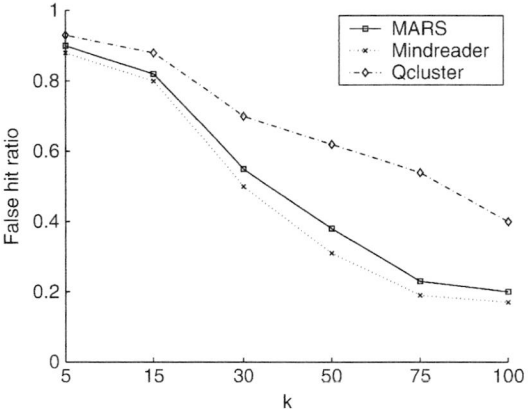

Query Point Movement Techniques for Content-Based Image Retrieval, Fig. 7 False hit ratio

effect of local maximum traps. Even for fairly large k, their false hit ratios remain very high. As a result, users of these techniques have to examine a large number of returned images, but might still not find their intended targets. Figure 8 shows that NDC and GDC perform more efficiently when k is small, with GDC being slightly better than NDC. Specifically, when $k = 5$, the average numbers of iterations for LNM, NDC, and GDC are roughly 21, 10, and 7, respectively. Experimental studies based on a prototype [8] showed that only seven iterations on average were needed to locate a given target image. Additional performance results can be found in [8].

declarative query language. Query processing for CODASYL databases is mainly about efficient access to data, where queries are implemented as procedural programs by software developers. There was no query optimization in the CODASYL model.

In 1970, Edgar Codd invented a relational algebra over tables. This relational data model was much more flexible than navigational data models. Relational algebra allowed the specification of arbitrary queries over relations. A combination of five primitive operations (selection, projection, union, filter, Cartesian product) over tables define a query in the relational algebra. Initially, the relational model was considered to be impractical, as it required data to be normalized into relations, and queries were very expensive operations. Queries often have to use the join operation (a combination of a Cartesian product with a selection) in order to reconstruct normalized relations. However, over the years, many inventions (database indexes, join methods like nested-loop join, merge join or hash-join, and techniques for query optimization) have together made relational database management systems feasible.

The prime example of a relational query language is SQL, which is based on a logical calculus called *tuple calculus*. An SQL query specifies *what* data to access and process in order to compute a query result, but not *how* to access and process the data. An SQL query over a relational database can be implemented in many different ways. For complex queries on databases with many indexes and tables, there may be millions of different ways to implement an SQL query. Each implementation may use different orders for combining the normalized tables, or use different join methods like hash join or nested-loop join, or use different types of indexes like B-Trees or bitmap indexes, or using indexes on different columns to efficiently process a selection over a table. Choosing the right method for processing a query over a large database can produce a query result in seconds (or faster), whereas choosing of the wrong method can result in queries running for hours, or even days.

Query optimization aims at selecting the most efficient access path (often called *query execution plan*, or *plan*) for any given query. The task of a query optimizer is to find the most efficient overall implementation of the query. Query optimization has been an active area of research since the 1970s, with advances still being made today. Some standard optimizations are based on simple heuristics. A typical objective may be to avoid large intermediate results during query processing by applying selections as early as possible. However, determining the most efficient access method for each table as well as the best join methods and join orders to combine tables cannot be carried out by simple heuristics alone. In order to find the best plan, the query processor actually needs to know some characteristics of the data that the query will process. For example, a nested-loop join method works best if one of the two tables to be joined is relatively small. A hash join has higher overhead for small tables, but will produce the query result much faster if both tables are large. Similarly, it is good to use an index if a selection results in very few rows of a table. If many rows qualify, a table scan will be much faster, as it can use sequential I/O, avoiding the consecutive repositioning of the read/write head of a hard disk as required when processing an index.

In 1979, Pat Selinger proposed a cost-based query optimizer for System R which determines the optimal query plan based on a mathematical model of the execution cost of each operator in a query execution plan. The System R optimizer, which has become the basis for many commercial database systems (e.g., IBM's DB2 and Oracle), enumerates all possible physical query execution plans for a query in a stratified bottom-up fashion. This is done by first determining the cheapest table accesses, then processing all two-table joins, three-table joins, and so on, and uses dynamic programming to prune the search space. In 1994, volcano introduced an alternative approach of a top-down optimizer based on goal-directed search and branch-and-bound pruning, which has found its way into commercial DBMS like Microsoft's SQL Server.

During the 2000s, advanced techniques for query processing have been proposed and implemented into commercial systems. One of these developments relies on feedback loops to improve query execution. Feedback obtained by monitoring some parameters of the mathematical cost model during query execution is used to either alter the query execution plan while the query is running, or to adjust the mathematical model to increase its accuracy for subsequent queries. In addition, integration of semi-structured data (XML) and the Xquery language are active areas of research, with new requirements imposed on a query processor due to the navigational operators like XPath needed for processing hierarchical data.

Foundations

The figure below gives an architectural overview of a query processor. The query compiler of a relational database translates a declarative SQL query into a procedural program. Initially, a parser carries out tokenization and creates a parse tree of the query based on the grammar of the query language. Semantic analysis then tests semantical correctness of the query. Those tests usually validate if the table names or column names in the query exist and have the right types. These parsing steps are similar to the steps a programming language parser would carry out, except that the query compiler has to generate a program for a data flow engine, not for a microprocessor. The query is therefore usually represented as a data flow graph (also called *query graph*) in a query compiler. A query graph is a graph whose edges represent the data flow and whose nodes represent operations on the data. Typical data flow operations are table access and join, group, and filter. Many query optimizers, like the optimizer of IBM's DB2 or Oracle's rule based optimizer, utilize a rule-based query rewrite phase before carrying out cost-based optimization. Query rewrite translates the query graph into a semantically equivalent query graph, which is preferable to execute.

Rewrite rules include, among others, translations from subqueries into joins, or rules to generate transitive predicates to involve indexes in query processing, which would otherwise not be applicable.

The System R approach to cost-based query optimization enumerates all possible physical query execution plans for a the query graph model, associates a cost with them, and selects the cheapest plan to be executed. The cost of a query plan is expressed as a linear combination of intermediate result sizes (cardinalities) weighed by factors for CPU cost, I/O cost. Cost models for distributed systems also factor in a the communication cost for transferring data between the processing components. The optimizer uses dynamic programming and prunes the search space as early as possible by only retaining the plan with the lowest cost whenever possible. The search space can be pruned whenever some (intermediate) plans are comparable. Plans are comparable, when they are semantically equivalent, i.e., when their execution produces the same intermediate result. However, in practice, there are cases where semantically equivalent plans are not yet comparable. For example, a semantically equivalent plan may not be comparable to an intermediate plan that has an interesting property, like a particular sort order for a an intermediate result, which could be exploited at a later stage during query processing to lower the overall cost.

Plans are enumerated bottom-up in a stratified way. This stratification starts with enumerating table access plans for each table then enumerates all (intermediate) plans for combining two of these table access plans, then all (intermediate) plans for three tables, and so on, until the all plans have been enumerated that produce the overall query result. As soon as several semantically equivalent (intermediate) plans have been computed, the optimizer retains only the plan with the lowest cost for further consideration. Dynamic programming has a complexity, which is exponential in the number of tables joined in a query. If the exponential memory requirement of dynamic programming is too high due to too many joins in a query, the optimizer uses a greedy algorithm as a fallback. While dynamic programming is guaranteed to determine the optimal plan with respect to the cost model, greedy algorithms usually do not return the optimal plan. Moreover, greedy algorithms tend to have a bias towards bushy plans (i.e., balanced join trees as opposed to left-deep join trees), which usually has negative impact on pipelining and transactional queries with small intermediate results. Many commercial database systems like DB2 and Oracle use a System R style bottom-up optimizer.

Volcano's alternative approach refines the query graph model by replacing logical operators like join with physical implementations like hash-join or merge-join and uses branch and bound to limit recursion. For instance, the top-down approach is used by the query optimizer of Microsoft's SQL Server.

Both bottom-up and top-down optimizers achieve the goal of determining the optimal plan with respect to a cost model. The quality of the plan does not depend on which of these search methods is used, but rather on the repertoire of rules available for generating or expanding plans. Top-down optimizers have an advantage, though, in that they always maintain a feasible execution plan, so it is possible to stop optimization at any time and execute the currently cheapest plan. Bottom-up optimizers have the risk of running out of memory without having produced a feasible plan, thus the necessary fallback to greedy or other heuristics.

Both bottom-up and top-down optimizers need a good model for the cost of producing the intermediate results that occur during query processing. This execution cost is largely dependent upon the number of rows - often called *cardinality* - that will be processed by each operator in the plan. Typically, an estimate for the cardinality of some intermediate result relies on statistics of database characteristics. Many database systems use simple statistics to approximate the size of an intermediate result, like the number of rows for each table and as well as the number of distinct values for each column. For a simple selections with an equality predicate "$C = value$" on column C of a table with n rows

and c distinct values in column C, many cost models use the simple formula $1/c * n$ to estimate the cardinality of the selection predicate. This simple formula assumes a uniform distribution of all values in column C. The cardinality estimate may be vastly incorrect if some values in column C occur more frequently than others.

Estimating the number of rows (after one or more predicates have been applied) has been the subject of much research since the 1980s. The percentage of the number of rows in a table or intermediate result satisfying a predicate P is often called *selectivity* of P. The selectivity of P effectively represents the probability that any row in the database will satisfy P. Database statistics are expensive to compute and cannot always easily be maintained incrementally, when the database changes. In general, there is a trade-off between accuracy of statistics and their storage and maintenance cost. The goal of database statistics is to be good enough to enable the optimizer to produce a robust and efficient plan. The cost model of a database systems therefore uses simplifying assumptions to compute selectivities and cardinalities from the available database statistics. Examples of these assumptions include:

1. *Currency of information*: The statistics are assumed to reflect the current state of the database, i.e., that the database characteristics are relatively stable. This may not be true, if a table is changing rapidly. In this case, statistics need to be collected frequently, using the database statistics collection tool. Outdated statistics are a major source for performance problems in database queries. Many modern database management systems have an infrastructure that tries to automatically detect when statistics are outdated by monitoring update, insert, and delete operations on a table. Those systems can be configured to (re-)collect statistics automatically when a certain amount of changes have occurred.

2. *Uniformity*: As describe above, without detailed statistics on a column, the data values within a column are assumed to be uniformly distributed. If that is not the case, the database administrator can create histograms for par-

ticular columns to deal with skew in values. This will improve the accuracy of selectivity estimation for selection predicates on a single table. Only recently have researchers begun to explore ways to improve the estimation of selectivities for join predicates which combine multiple tables.

3. *Independence of predicates*: Without any knowledge about interactions of predicates, selectivities for each predicate are calculated individually and multiplied together, even though the underlying columns may be related, e.g., by a functional dependency. This independence assumption usually results in severe underestimation of the selectivity of predicates on correlated columns. For a table containing cars with two columns, *make* and *model*, the independence assumption will result in severe estimation errors for a selection predicate like "make='VW' and model = 'Jetta'." If 10% of the cars in the table are VW, and 1% of the cars are Jetta, the independence assumption would result in a selectivity:

$$^{s}\text{VW Jetta} = {^{s}\text{VW}} \times {^{s}\text{Jetta}} = 10\%$$

$$\times 1\% = 0.1\% = 0.001$$

Since only VW makes Jettas, the real number is 1%, with an order of magnitude error in the estimation. This means that the intermediate result size for that part of a query will be ten times larger than what the optimizer assumes. This can result in not allocating enough memory for processing the query, or choice of a suboptimal plan. Overall, this estimation error progressively worsens, as more predicates are present in a query.

Collecting multivariate statistics across multiple columns can overcome the independence assumption when data is correlated. A simple correlation parameter is the number of distinct values over a set of attributes. This statistic is employed for instance by IBM's DB2 to correct the independence assumption for correlated local predicates, or to better estimate the selectivity of a join predicate with correlation between the two tables. While the number of distinct values over a set of columns addresses the correlation between these columns, it assumes all combinations of values in these columns to be uniformly distributed. Correlations inflate errors more than the uniformity assumption. Thus distinct values have proved worthwhile in practice as correlation statistics. In case of non-uniform correlations, multidimensional histograms have been proposed. These histograms store and maintain multivariate distribution statistics. However, multidimensional histograms do not work well for equality predicates or for maintaining correlations over a large set of columns. For that reason, they have not been widely adopted in commercial databases.

4. *Principle of inclusion:* The selectivity for a join predicate X . a = Y . b is typically defined to be $1/\max\{|a|, |b|\}$, where $|b|$ denotes the number of distinct values of column b. This implicitly assumes the "principle of inclusion," i.e., that each value of the smaller domain has a match in the larger domain (which is frequently true for joins between foreign keys and primary keys). Again, correlation statistics can help overcome this assumption. Products like IBM's DB2 or Microsoft's SQL Server offer statistics on views, which can address incorrect assumptions about correlations of columns between tables or within tables.

Applications commonly used today have hundreds of columns in each table and thousands of tables, making it very hard for a database administrator to know on which columns to collect and maintain multivariate statistics or statistics on views. Tooling, either based on query feedback or proactive sampling, is addressing that problem by determining the most important correlated columns that joint statistics need to be collected on, employing statistics methods like χ^2-testing for correlation detection. Methods such as entropy maximization have been proposed to generalize the independence assumption and allow for dealing with arbitrary correlation between columns, either within or across tables.

Once the best overall query plan according to the optimizer's model has been determined, it

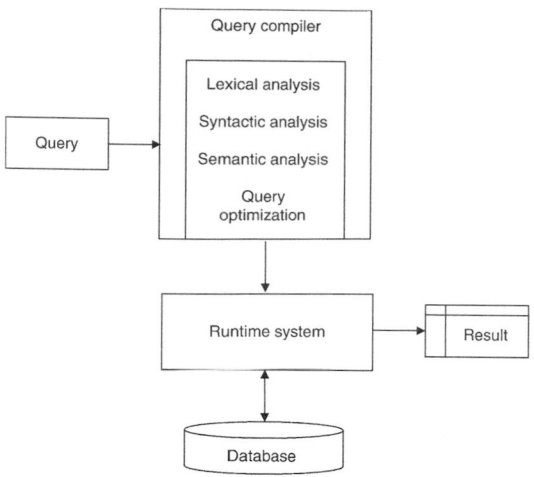

Query Processing (In Relational Databases),
Fig. 1 An overview of query processing

is handed to the runtime system for execution. In some architectures, e.g., Informix, the query plan is directly interpreted by the runtime system. Other architectures follow the System R design and employ a code generator to produce code for a database machine, which is then executed by the runtime system. This additional code generation step allows for optimizations typically found in programming languages, to reduce path length and copying of data between CPU and memory (Fig. 1).

Key Applications

All major database systems (DB2, SQL Server, Oracle, Sybase, Informix, MySQL, PostGres) implement a query processor that largely follows the previously described architecture and concepts. DB2 and Oracle follow the System R style optimizer as described above, Microsoft SQL server uses a top-down optimizer.

Recommended Reading

1. Codd EF. A relational model of data for large shared data banks. Commun ACM. 1970;13(6):377–87.
2. Freytag JC, Maier D, Vossen G. 1994. Query processing for advanced database systems. Morgan Kaufmann.
3. Graefe G. Volcano – an extensible and parallel query evaluation system. IEEE Trans Knowl Data Eng. 1994;6(1):120–35.
4. Graefe G. Query evaluation techniques for large databases. ACM Comput Surv. 1993;25(2):73–170.
5. Lorie RA, Fischer NJ. An access specification language for a relational data base system. IBM J Res Dev. 1979;23(3):286–98.
6. Markl V, Haas PJ, Kutsch M, Megiddo N, Srivastava U, Tran TM. Consistent selectivity estimation via maximum entropy. VLDB J. 2007;16(1):55–76.
7. Selinger PG, Astrahan MM, Chamberlin DD, Lorie RA, Price TG. Access path selection in a relational database management system. In: Proceedings of the ACM SIGMOD International Conference on Management of Data; 1979. p. 23–34.
8. Yu CT, Meng W. Principles of database query processing for advanced applications. Morgan Kaufmann; 1997.

Query Processing and Optimization in Object Relational Databases

Johann-Christoph Freytag
Humboldt University of Berlin, Berlin, Germany

Synonyms

Query evaluation; Query planning and execution

Definition

In an (object) relational database management system (DBMS) query processing comprises all steps of processing a user submitted query including its execution to compute the requested result. Usually, a user query – for example a SQL query – declaratively describes *what* should be computed. Then, it is the responsibility of the DBMS to determine *how* to compute the result by generating a (procedural) query execution plan (QEP) that is semantically equivalent to the original query. Query processing also includes the execution of this generated QEP.

While generating a QEP for a user submitted query the DBMS explores a large number of potential execution alternatives. To choose the best ones among those alternatives requires one or more (query) optimization phases.

Historical Background

In the late 1960s and early 1970s it became clear that existing DBMSs were too complex and cumbersome to use. The programmer was forced to know the physical layout of the data on disc for efficient access. The answer to those problems was the Relational Model developed by E.F. Codd in the early 1970s.

With the design and implementation of relational DBMSs in the mid-1970s it became clear very quickly that query processing and optimization are the key for implementing DMBSs with acceptable performance for the end user. In 1979, the first relational DBMS, IBM's System-R laid the foundation for today's query processing and optimization approaches: In System-R, query processing consists of three major phases:

1. Syntactic and semantic checking
2. (physical) query optimization
3. Query execution

The INGRES DBMS (Stonebreaker, UC Berkeley) took a similar approach to implement query processing.

Over the last 25 years this phase-based model has been extended by

1. adding more query processing phases such as logical query optimization or query rewrite
2. considering more alternatives during query optimization to generate better QEPs

Furthermore, the complexity of query processing has increased due to more complex query languages (such as SQL-2 or SQL-3), more complex query execution environments such as parallel processors, distributed data collections, and more complex data types, richer data structuring capabilities including object orientation.

As a response to the development of object oriented database management systems (OODBMS) the Relational Model was extended with object oriented concepts and features to embrace the major concepts of those languages. The extension of the language forced the database vendors to extend query optimization and query extension to handle those new concepts correctly efficiently such as objects, classes, path expressions, inheritance, methods, and polymorphism [10]. Most of those language features are reflected in the SQL-3 Standard.

Foundations

The Four Phases of Query Processing

Today's DBMSs usually implement query processing (including query optimization) in three different phases before executing the query to generate the requested result [2, 8].

During the first phase, the query is checked for syntactic and semantic correctness. Then, during the second phase the query is rewritten into a semantically equivalent one using additional (logical/conceptual) information, such as schema information (candidate keys/Primary key, uniqueness of values, foreign keys) or integrity constraints. Rewriting might also strive for a normalized form of the query to simplify the processing during the following phases [7].

Example 1:

The SQL query

```
SELECT *
FROM CUSTOMER
WHERE EXIST (
SELECT *
FROM ORDER
WHERE CUSTOMER.ID = ORDER.CUSTOMER_ID
AND ORDER.VALUE > 10.000)
could be rewritten into the query
SELECT *
FROM CUSTOMER, ORDER
WHERE CUSTOMER.ID = ORDER.CUSTOMER_ID
AND ORDER.VALUE > 10.000
```

If the inner query block returns zero or one tuple.

The third phase – commonly known as the (physical) query optimization – translates a user query into a query execution plan (QEP). While the initial query declaratively expresses which properties the result should have, the QEP determines the execution steps to evaluate the query generating the requested result.

In general, there are many different ways (QEPs) to evaluate a user submitted query based on the physical properties of the tables accessed, such as available indexes, available materialized views, sorting order, or clustering. Thus, the query optimizers must consider and evaluate many different alternatives to execute a query using different access paths and different operators.

During the QEP generation the optimizer determines

- The best (optimal) way to access each individual relation referenced in the query using indexes (including how and in which order to evaluate local predicates)
- The best (optimal) order to join two or more tables
- The best algorithm to perform the join between two tables

Example 2:

The rewritten query of Example 1

```
SELECT *
FROM CUSTOMER, ORDER
WHERE CUSTOMER.ID = ORDER.CUSTOMER_ID
AND ORDER.VALUE > 10.000
```

Might be translated into the following QEP. For the sake of clarity the QEP uses a LISP-like notation to represent the relational algebra-like expression [4].

```
(OUTPUT
  (PROJECT
  (ID CUSTOMER_NAME ADDRESS)
  (NESTED_LOOP_JOIN
    (CUSTOMER.ID = ORDER.CUSTOMER_ID)
    (SCAN CUSTOMER)
    (FILTER
      (ORDER.VALUE > 10.000)
    (SCAN ORDER)))))) □
```

Finally, during the last phase the DBMS executes the generated QEP to compute the re-quested result, i.e., the set of tuples that match the submitted query.

The Fundamentals of a (Physical) Query Optimizer

To generate the best (optimal) QEP any query optimizer embodies the following three aspects [4]:

1. A search strategy
2. (A set of) Cost functions
3. A QEP generation strategy

The search strategy determines which alternative (partial) QEPs should be generated while looking for the best (optimal) QEP among possibilities. Strategies such as exhaustive search strategies (together with dynamic programming) or heuristics (Greedy heuristic) are approaches used in current DBMSs. However it seems that exhaustive strategies – with some restrictions – are still the preferred approach to generate the best QEP despite a large search space whose size is determined by various parameters, such as the number of tables in the query, the available indexes, or the available join methods (algorithms).

While generating different alternatives for query execution, the optimizer must determine which of these alternatives is better. Therefore, cost functions assign cost values to (partially generated) QEPs by determining the resource consumption for that plan. The resources used might be CPU time, space (amount of memory), communication time (distributed query processing), and – most importantly – the number of disk IOs since disk IOs are the dominating cost in almost all QEPs. Cost functions therefore estimate ("foresee") the amount of resources that a plan will consume when being executed.

To feed the cost functions with input, most DBMSs maintain so-called *histograms* which record value distributions for various (sets of) attributes in different tables [5]. Those value distributions allow the optimizer to make informed decisions regarding the different operators of a QEP with respect to their efficiency and effectiveness.

Once the optimizer settled for one alternative during partial QEP generation it must determine how to advance the QEP generation in its search for the best (complete) QEP. This progress might be implemented by extending the current partial QEP with new operators possibly on additional tables not yet considered, or by generating alternatives for the existing QEP by transforming it into a syntactically different QEP. The latter might include considering alternative access paths (using different indexes) or replacing an existing join algorithm with a different one.

The query optimization phase terminates once the optimizers found the best QEP based on the given (searched) alternatives and based on the given cost functions. However, since the search might take time, the optimizer might also terminate after having reached a pre-determined tie limit or after having evaluated a certain number of alternatives.

Formalisms and Approaches for Query Processing and Query Optimization

The (research) literature reports on many different approaches to query processing and query optimization. However, there does not seem to be one formalism to describe those different approaches. Some of them use a relational algebra like notation (with extensions) to show operational approaches. Others use logic based notation such domain/tuple relational calculus or the tableaux notation. Some presentations use proprietary notations based on innovative data structures to present sophisticated algorithms.

Of course, the approach to query processing and query optimization is heavily influenced by the data model and the expressiveness of the query language. The continuous extension of SQL to SQL-2 or SQL-3 has lead to an extended portfolio of query processing and query optimization techniques. The same is true object oriented and object relational DBMSs where introduced. Similarly, the extension of SQL to express Data Warehouse queries leads to new techniques (algorithms and "tricks") for query processing and query optimization.

Implementing Query Optimizers

The first optimizers were implemented for the database prototypes IBM System-R [8] and Berkley's INGRES. The former system laid the foundation conceptually and architecture wise for many optimizers to come including those that are used in today's DBMS products.

In many cases, DBMS products only provide access to the architecture, the optimization strategies, or the cost functions used. Most of the dominant products allow the user to view the QEP as generated by the optimizer together with cost values and cardinality estimates.

However, for several DBMS products, there exist well known prototypes that provide hints how some of the optimizers in existing products might work. The newly researched ideas and novelties in those prototypes are often described in research papers published at well-know conferences like the ACM Sigmod conference, the VLB conference, the IEEE ICDE conference, or the European EDBT conference.

Key Applications

The query optimizer is an integral part of any DBMS (product). The degree and extent of optimization is determined by the vendors or implementers. The repertoire of the optimization step is continuously extended, depending on the requirements coming from OLTP queries, data warehousing/OLAP queries (star schema, aggregate queries), data mining queries, or queries coming from systems such as geographical information systems booking systems, and others.

Future Directions

Over the last several decades, query processing and optimization has adapted to the changes in computing hardware. In the 1990s, parallel hardware became readily available at a reasonable cost, DBMSs and query processing had to be extended and adapted to parallel QEP execution.

In addition, many database experts doubt that the general two-phase query processing approach (optimize the query, then execute) is the right one for future DBMs that should run on the next generation of hardware. Especially when the underlying system continuously changes query optimization and query execution must be closely intertwined to react to those changes in (almost) real time. Furthermore, when executing queries on large tables, it might be desirable to change execution strategies "on the fly." Currently, once the query processing phase is done (determining the best plan) the QEP is fixed. Recent research in the area of query processing has developed approaches to change QEPs either "on the fly" when necessary.

Now, multi-core CPUs seem to be the new development direction for hardware. At the same time, new storage technology comes to life (flash disks) with different operating characteristics that will change the architecture and therefore the processing models and processing capabilities of existing DBMSs. In addition, large CPU farms, large storage farms, and increasing main memory sizes already have a dramatic impact on existing approaches and techniques in query processing and query optimization. Understanding the Web as one large database could completely change today's DBMS architecture standards and assumptions on how to build future DBMSs.

Experimental Results

Unfortunately, there are no benchmarks that consider the database query optimizer as a separate component for benchmarking. However, there are efforts to ensure the stability of the optimizers despite estimation errors during query optimization and despite changes in the expected resources available during query execution.

Cross-References

▶ Histogram
▶ Logical Database Design: From Conceptual to Logical Schema
▶ Parallel Database Management
▶ Query Language

▶ Relational Algebra
▶ Relational Model
▶ System R (R*) Optimizer

Recommended Reading

1. Deshpande A, Ives Z, Raman V. Adaptive query processing. Found Trends Databases. 2007;1(1):1–140.
2. Freytag JC. The basic principles of query optimization in relational database management systems. In: Proceedings of the IFIP 11th World Computer Congress; 1989. p. 801–7.
3. Freytag JC, Maier D, Vossen G, editors. Query processing for advanced database systems. San Mateo: Morgan Kaufmann; 1994.
4. Graefe G. Query evaluation techniques for large databases. ACM Comput Surv. 1993;25(2):73–170.
5. Ioannidis YE. The history of histograms (abridged). In: Proceedings of the 29th International Conference on Very Large Data Bases; 2003. p. 19–30.
6. Jarke M, Koch J. Query optimization in database systems. ACM Comput Surv. 1984;16(2):111–52.
7. Pirahesh H, Hellerstein JM, Hasan W. Extensible/rule based query rewrite optimization in starburst. In: Proceedings of the ACM SIGMOD International Conference on Management of Data; 1992. p. 39–48.
8. Selinger PG, Astrahan MM, Chamberlin DD, Lorie RA, Price TG. Access path selection in a relational database management system. In: Proceedings of the ACM SIGMOD International Conference on Management of Data; 1979. p. 23–34.
9. Yu CT, Meng W. Principles of database query processing for advanced applications. San Francisco: Morgan Kaufmann; 1998.

Query Processing in Data Integration Systems

Zachary G. Ives
Computer and Information Science Department, University of Pennsylvania, Philadelphia, PA, USA

Synonyms

Adaptive query processing; Distributed query processing; Query processing for mediators

Definition

In (virtual) data integration, also known as enterprise information integration, queries are posed over a virtual mediated schema and answered on-the-fly using data from remote sources, which may themselves be DBMSs, Web sites, or applications. This requires two main stages that of *query reformulation* where the user's query is composed with schema mappings to produce a combined (distributed) query and *query optimization and execution* where the query is executed efficiently across the sources.

The query optimization and execution problem for data integration is, in principle, quite similar to that for distributed databases. However, it is actually significantly more complex because (1) remote data sources may have different data models and their own query capabilities; (2) statistics on the data at each source may be unavailable; (3) remote data sources may require the requestor to supply search terms or input values, in order to retrieve results; and (4) most data integration scenarios emphasize rapid response times and thus need fast initial answers.

Historical Background

Early work on query processing for data integration arose out of both the database community's work on federated querying across databases [1] and the Web [2], as well as the artificial intelligence community's work on general-purpose planning (search) for integrating Web sources [3]. Eventually the two branches of work unified around common abstractions such as specialized search and pruning for query operators and input-binding restrictions. Later, database researchers focused on introducing *pipelined* and adaptive query execution techniques [4] to improve responsiveness.

Scientific Fundamentals

Query processing techniques for data integration build heavily upon those for distributed database management systems. However, query optimization and execution are adjusted to emphasize *flexibility and parallelism* under unpredictable circumstances. Flexibility and parallelism are required because query processing for data integration occurs at uncertain speeds; data source statistics may not be available, making all cost estimates uncertain; parts of a data integration query may be "pushed" to a data source which may choose its own execution strategy, and this may not match the data integration system's predictions; and network connectivity may be bursty or unpredictable, delaying certain parts of the query. We briefly sketch how this changes query optimization and execution [4], and how it motivates adaptive query processing techniques [5].

Optimizing a Data Integration Query

As with a traditional database, a data integration system typically takes a query and performs a *search* over the space of possible query plans, estimating the cost of each plan and choosing the lowest-estimated-cost plan. The plan enumeration "engine" may use standard bottom-up (dynamic programming) or top-down search strategies, but it typically enumerates a slightly different plan space.

Bushy plans. Even today, relational database systems often restrict their search to "left linear" query plans, where every join is between a query subexpression and a base relation. This reduces the space of plans that must be considered but may result in a suboptimal execution plan. However, every query subexpression has at least one input with indices and statistics available, meaning that the error introduced during query optimizer cost and cardinality estimation is likely to be limited. In contrast, data integration query optimizers must search a broader space of plans (the full set of "bushy" query evaluation trees). This broader search space is often necessary to satisfy the constraints on the query (see below) and also to maximize the number of query operations that can run in parallel (see discussion of query execution below).

Query subexpression pushdown. In a data integration system, the remote sources often have the ability to perform database query operations internally. The data integration system's query optimizer must take an overall query over the mediated schema and determine which portions to "push" into the remote data source. To determine the most efficient "push-down" subexpression, it must often guess how the remote data source will execute the subexpression – i.e., it will simulate the optimizer of the remote source!

Input bindings. Sometimes, a data source *requires* that part of the query be pushed down, e.g., by "binding" certain fields within the remote source to particular values. An example is an e-commerce site where a list of items will only be returned if certain values (such as product name or manufacturer) are provided. Many sources have *binding restrictions* on inputs, and the query optimizer must choose plans that fetch values from a table with no restrictions, then "feed" the values to the source with binding restrictions in a *dependent join* operator.

Uncertain cost parameters. The final key challenge faced by a data integration query optimizer is limited availability of source statistics, cost parameters within the remote source, and parameters related to network burstiness. Sometimes simple statistics like input cardinalities are not even known! Most data integration query optimizers are thus extremely conservative in their estimates.

Executing a Data Integration Query

Due to its need to fetch data from remote sources with unpredictable latencies, a data integration query engine optimizes parallel and pipelined execution. It is typically "dataflow"-driven, using either event-based scheduling (each time a tuple arrives, it is propagated through the query plan and combined with existing state) or multithreaded execution (each input has a separate control thread). Most of the physical query operators are the same as those in a distributed database management system, with three exceptions.

The "leaf" operators in a data integration query plan are typically *wrappers*, adapter modules that interface with the external remote sources in their protocol of choice (e.g., REST, JDBC, HTTP POST, etc.), tell the remote source what operations to push down and fetch and

translate the data retrieved by the remote source. Note that remote sources often have semi-structured or even unstructured data, so the data integration query engine must be able to accommodate this (or the wrapper must translate the data into a form compatible with the engine).

A key algorithm in the data integration setting is the *pipelined hash join*, which has an in-memory hash table for each input. When a tuple is received, it is probed against the hash table from the opposite relation and the results are output. Simultaneously, the tuple is added to the hash table for its corresponding relation. The pipelined hash join always produces output at the earliest moment possible. Several variations on the basic ideas have been proposed and studied, with some of the key differences being in how state is overflowed to disk if there is insufficient memory [4].

Finally, the *dependent join* operator takes data from one input, projects out certain fields, and "feeds" them to the other input (binding the fields to particular values) via the wrapper operator. When it gets the corresponding results from the second source, it joins those results with the tuples from the other input. The dependent join in many ways takes on the role of the index-based join in a relational DBMS.

Adaptive Query Processing

In many data integration scenarios, the query optimizer has inadequate information to produce the best query plan possible. Moreover, across long streams of data from the inputs, there may be points at which it would be beneficial to switch query plans. These factors have motivated work on *adaptive* query processing [5], where the running query plan can be modified in mid-execution. Adaptive query processing typically requires iteration among *monitoring* and information gathering, *cost estimation* and *planning* or determining the optimal plan given the current information, and *execution plan modification* to change to the new plan. These correspond closely to the so-called OODA loop steps (*observe, orient, decide, act*) used to describe human decision-making and to the *measure, analyze, plan, actuate* steps in the main loop for control theory.

Broadly, we can divide adaptive query processing into *event-driven* adaptivity, where the system responds to an event (such as out-of-memory, remote source failed, network delay, or cardinality deviates from estimate) with a change in strategy, versus *performance-driven* adaptivity, where the system monitors and continuously reassesses performance at runtime and potentially changes its strategy towards a more optimal one. Event-driven adaptivity is typically low-overhead and can be done at periodic "checkpoints" (e.g., at a point where a relation expression is materialized to desk). It fits quite naturally into a conventional query optimization/execution architecture. Performance-driven adaptivity typically requires closer coupling between monitoring, cost re-estimation, plan switching, and plan execution – necessitating a more uniform treatment of the different aspects of query processing. Many different strategies [4, 5] have been considered for monitoring, which must be done in a way that adds minimal overhead to common-case query execution; for cost re-estimation, which needs to handle correlations among predicates; for replanning, which needs to be done very rapidly; and for changing plans in mid-execution. Each such option excels under certain conditions.

Key Applications

Virtual data integration has seen industry-wide adoption under the term *enterprise information integration*, where it is widely used to combine structured and unstructured data sources in real-time. IBM, Oracle, and Microsoft, as well as many other smaller players, offer middleware tools and technologies to support enterprise information integration, as part of a broader portfolio alongside *enterprise application integration* (EAI) and data warehousing *extract-transform-load* (ETL) tools. Virtual data integration is used to enable cross-data-platform applications such as customer relationship management (CRM), ad hoc discovery, and report generation combining structured and unstructured data. Additionally, virtual data integration is often used in the scientific community to provide "meta search" across existing data repositories. An emerging area of study is in data integration query systems that combine certain aspects of structured and unstructured data, e.g., using keyword search across uncertain linked data.

Cross-References

▶ Adaptive Query Processing
▶ Distributed Query Processing
▶ Query Optimization

Recommended Reading

1. Smith JM, Bernstein PA, Dayal U, Goodman N, Landers TA, Lin KWT, Wong E. Multibase: integrating heterogeneous database systems. In: Proceedings of the AFIPS National Computer Conference; 1981. p. 487–99.
2. Levy AY, Rajaraman A, Ordille JJ. Querying heterogeneous information sources using source descriptions. In: Proceedings of the 22th International Conference on Very Large Data Bases; 1996. p. 251–62.
3. Arens Y, Knoblock CA. SIMS: retrieving and integrating information from multiple sources. In: Proceedings of the ACM SIGMOD International Conference on Management of Data; 1993. p. 562–3.
4. Doan A, Halevy A, Ives Z. Query processing. In: Principles of data integration. Waltham: Morgan Kaufmann; 2012. p. 209–41.
5. Deshpande A, Ives Z, Raman V. Adaptive query processing. Found Trends Database Syst. 2007;1(1):1.

Query Processing in Data Warehouses

Wolfgang Lehner
Dresden University of Technology, Dresden, Germany

Synonyms

Data warehouse query processing; Query execution in star/snowflake schemas; Query optimization for multidimensional systems

Definition

Data warehouses usually store a tremendous amount of current and historical data, which is advantageous and yet challenging at the same time, since the particular querying/updating/-modeling characteristics make query processing rather difficult due to the high number of degrees of freedom.

Typical data warehouse queries are usually generated by *online analytical processing* (OLAP), *data mining* software components, or in an ad hoc manner using toolkits for data scientists in the form of statistical packages and homegrown analytical tools. They show an extremely complex structure and usually address a large number of rows of the underlying database. For example, consider the following query: "Compute the monthly variation in the behavior of seasonal sales for all European countries but restrict the calculations to stores with >1 million turnover in the same period of the last year and incorporate the top ten products with more than 8% market share." In a first step, this query has to identify certain stores based on the previous year's sales statistics, and it needs to define the top-selling products on a monthly basis. In a second step, the system is able to compute the differing behavior based on individual countries.

The structural complexity as well as the huge data volume addressed by data warehouse queries makes it a true challenge to optimize and execute these queries. Therefore, query processing in data warehouse systems encompasses the definition, the logical and physical optimization, as well as the efficient execution of analytical queries. The specific techniques range from classic rewrite rules to rules considering the special structure of a query. For example, data warehouse queries usually target *star schemas*. Such star queries with references from a fact table and foreign-key joins to multiple dimension tables can be beneficially exploited during query optimization and query processing. Moreover, the embedding of specialized operators (like *CUBE* or ROLLUP) or approximate query processing methods using samples or general-purpose synopses has to be addressed in the context of query processing in data warehouse environments. Finally, query processing in data warehouse systems has to address the support of non-*SQL* query languages (e.g., specialized analytical query languages like MDX [1]), alternative storage structures (e.g., column-wise organization), or alternative processing models based purely on main-memory structures.

Historical Background

Query processing has a long history with research and commercial products. The generation of efficient plans for the smooth execution of database queries thus represents one of the most challenging research directions in database history. With the advent of descriptive query languages, this has turned into a crucial problem, since systems now have to figure out the optimal way to execute queries.

Traditional *transactional*-style (online transaction processing, OLTP) interaction patterns between an application and the database system consist of short read and write queries with correlated sub-queries, which typically compete with a large number of concurrently running transactions. DWH-style query patterns follow a significantly different style. Since the update/load of data warehouse databases is under the control of data warehouse job scheduling components, user queries are typically read queries that touch a large number of rows with star joins and group-by operations as their functional core. Around 1995, the concept of data warehousing emerged and showed that OLTP-style query processing and optimization techniques were insufficient in dealing with this specific class of requirements. This deficit triggered a tremendous amount of research activities to push the envelope in multiple directions. After more than 10 years of research and development, a fairly large number of extensions/modifications have turned it into many commercial systems. However, due to the increasing need for analytical tasks performed within (and not on top of) database systems, the improvement of query processing techniques for analytical applications still represents an active research area with significant potential for innovation. In addition, different efforts are taken to cope with transactional as well as analytical workloads within a single system by providing different low-level query engines (e.g., row and column store) and adaptive scheduling algorithms [2, 3].

Foundations

Query processing in data warehouse systems does not address individual methods or techniques. In contrast, query processing in the analytical domain is characterized by a wide range of individual techniques ([4, 5]; see 3.13). The major challenge in this context is to orchestrate the different and individually deployed methods. The following introduction to important query processing aspects, including a brief discussion of the requirements and specifics of query processing in data warehouse systems and possible solutions, provides a comprehensive overview of the topic.

Requirements and Specifics of the Analytical Context

While the specific character of data warehouse systems poses challenges for query processing from multiple perspectives, it also has constraints that can be beneficially exploited, for example, for star joins, placement of group-by operators as well as special implementations of aggregation operators [6]. Although the processing and optimization of *SQL* queries attracts most of the attention, query processing in data warehouse environments addresses other query languages like MDX [1], DMX [7], or SQL/XML [8] with analytical functionality specified within the *XPath/XQuery* fragments. The complexity of handling domain-specific query languages and of mapping these expressions either to highly specialized storage structures or to SQL is further increased by the need for support of domain-specific operators. In this context, the integration of data mining algorithms for cluster searches or for the computation of association rules into database systems can be seen as a prominent example.

To counterbalance these requirements, query processing in data warehouse systems shows a number of limitations or specializations that facilitate adequate solutions. For example, a high ratio of analytical queries is issued to compute standardized reports (e.g., legal reporting in the financial sector, cockpit solutions, etc.). These canned queries show huge potential to pre-compile the execution plans, to apply multiple-query optimization, or to precompute partial results using materialized views. Furthermore, the explicitly controlled update transactions in data warehouse systems typically show append-only characteristics, which enable the system to apply specifically tailored optimization strategies ([9]). For example, adding incoming rows to a single partition and periodically attaching it to the analytical database in an atomic way affects the concurrently running analytical queries only to a minimal extent. Finally, although analytical queries usually apply aggregation techniques over a large number of rows, the analytical queries may exhibit quite selective predicates. For example, consider the following query: "Return the weighted sales distribution of all Californian stores selling Mac and Linux machines with more than 24 GB RAM in 2016 and divide it by month." It shows a highly selective predicate but may still incorporate a huge number of rows within the computation of the measure. Query processing may tackle this phenomenon by introducing special index structures to support selective predicates of multiple columns from different tables or by applying specialized processing techniques such as online aggregation or approximate query answering in general.

Potential and Solutions for Efficient Data Warehouse Query Processing

The specific characteristics as well as the specific circumstances create a wide range of options to perform query processing in data warehouse scenarios in an efficient way. The following list provides the most prominent classes of methods with an outline of their specific role in this context:

Part I: Query Planning and Execution

Optimizing the Optimization Process

Since analytical queries usually show a complex structure, the optimization of a query may require a considerable fraction of the overall query execution cost. Therefore, modern techniques propose a variety of solutions ranging from the a priori limitation of the search space (usually by applying certain heuristics, e.g., the optimization pattern of star queries) to the pruning of alternative plans as early as possible [10].

Optimization Goal

In contrast to OLTP-style query optimization, the general goal of OLAP-style query optimization is to generate only a good (i.e., not an optimal) but robust plan. Due to the complex structure of a query, potential errors in the cost estimation usually have their origin at the leaf nodes of an operator tree, and – after having been propagated along an operator path – they may result in imprecise estimates. A robust plan is required to expose well-performing behavior even if the data show different characteristics than those used in the planning phase ([11]). The notion of robust plans and the design of more adequate statistics (e.g., sample or wavelet synopses) are subject of intense discussions in the research community.

Specific Rewrite Rules

As already mentioned, data warehouse environments imply a specific characteristic of queries – usually following the notion of star queries – which consist of a join of the fact table and multiple dimension tables followed by a complex selection predicate and a grouping condition with aggregation. Query processing has to detect these situations and has to apply specific optimization patterns. For example, group-by operations can be pushed down in certain cases to reduce the number of rows for join operations [12]. Another example addresses star joins: the star-join technique may decide to compute the Cartesian product of the selected parts of all dimension tables in a first step and then join the (large) fact table in a last step [13, 14].

HW-Assisted Query Execution

In order to boost the performance of query execution within data warehouse environments, hardware-assisted execution models have been devised. The basic idea is to exploit graphic cards (GPUs) to execute scans, filters, joins, as well as aggregation operations more efficiently than traditional CPUs or a combination of both [15]. Furthermore, specifically tuned hardware components like FPGAs are used to improve query – and especially join – operations [16].

Multi-query Optimization

A final class of techniques – which have not been considered in OLTP-style query processing due to restrictions of *ACID properties* – addresses the issue of optimizing a set of queries simultaneously, thus resulting in shared use of different resources. Key applications for these techniques can be found on the application level (in the computation of multiple similarly structured statistical reports) or within a system (in the propagation of changes of a base table to multiple dependent materialized views to share update efforts) [17].

Part II: Considering Logical Access Paths

Partition Management

Partitions reflect a concept that is useful for the administration of a database system as well as for the optimized execution of partition-aware queries. Within a data warehouse environment, incoming data items are typically stored within staging tables. After transformation and cleaning steps, tables are converted into partitions and attached to the global data warehouse database. Whenever data items in a partition are obsolete, a partition can be detached and moved (as a table) to the archive. From a query optimization perspective, partition pruning represents a powerful mechanism to potentially reduce the number of rows accessed within a query. Whenever a selection predicate refers to a partitioning criterion, the system may restrict the query execution to only those partitions that are actually referenced by the query. The main partitioning criterion usually addresses the time dimension, e.g., by month, but may also be used to partition other dimensions, e.g., by product category or geographical entities. Usually, large data warehouse databases deploy a two-level partitioning scheme: a primary partitioning by time followed by sub-partitioning by some other dimensional attributes, e.g., product group and regional information.

Materialized Views

A second component of the logical access path consideration is the transparent use and the implicit maintenance of materialized views. Similar to classic index structures, a query is matched against a view description and internally rewritten to exploit an existing materialized view by producing the same result as the original query. Especially in the context of aggregation queries, the concept of materialized views represents an extremely powerful mechanism to speed up analytical queries. For example, a materialized view may hold summary data grouped on the family level within a product dimension, with an additional grouping based on city and month. Every query with a compatible aggregation function and a grouping condition that is "coarser" than the grouping combination of the materialized view, e.g., sum per quarter (with sales on a daily basis), product category, and state, may benefit from the pre-aggregated data stored within a materialized view. In addition (and similar to physical index structures), materialized views are transparently maintained in the case of changing base tables to provide a consistent view of the data.

Part III: Considering Physical Access Paths

Data Organization

While classic relational database engines favor the concept of row-based storage, a variety of specialized systems exists that follow the concept of column-based storage layout (also known as decomposition storage model, proposed in 1985 [18]). Query processing in column-based systems benefits from reading only those columns that are actually required to answer an incoming query. This feature is extremely beneficial in the presence of wide dimension tables and queries referring to just a few attributes. In addition, hardware prefetchers are supported by sequential access behavior over a column and soften the problem of cache misses in case of random memory accesses.

Compression and Main-Memory Techniques

Although (persistent) storage costs have been decreasing significantly, a compressed version of data may speed up the query processing because of the reduced number of I/O operations. While block- and table-based compression schemes have been well understood, the static characteristics of data warehouse data have provided the motivation to push compression techniques into commercial systems. For column-based systems with the primary goal to be main-memory centric, compression is essential; a large variety of techniques (e.g., Huffman coding schemes) is applied in different systems [19].

Additional Index Structures

The history of database research has shown that different application areas can be supported with specialized index structures. Particularly aimed at the support of queries within the analytical domain, *bitmap index* structures have experienced a rejuvenation and are now part of the prominent commercially available systems [20, 21]. Another approach taking advantage of the static characteristic of a data warehouse database is the concept of *star index*es (also called join indexes or foreign-table indexes) that follow the basic idea to index the result of a join operation [22]. Join indexes are particularly well suited to exploit the relationship of dimension tables with the corresponding fact table.

Multidimensional Clustering Schemes

Clustering in general tries to preserve the topological relationship of entries in a database. Since data warehouse datasets are typically multidimensional by nature (e.g., sales by shop, date, and product), multidimensional clustering schemes represent a valuable solution to store logically related facts within the same block with the overall goal to reduce the number of I/O operations.

Part IV: Alternative Query Answering Models

Online Aggregation
Analytical queries typically show a response time in the range of multiple seconds, minutes, and sometimes hours. Online aggregation [23] is a promising technique to return intermediate results of the queries while the query is still running; the intermediate results are refined step by step until the final result is computed or until the application is satisfied with the precision of the answer. However, online aggregation has not been implemented in commercial systems since it requires slow, random disk accesses and would lead to major changes of the query processor. Instead, approximate query processing based on synopses is far more widespread.

Approximate Query Answering/Approximate Query Processing
Many application areas that perform statistical analyses prefer to yield an approximate but quick answer instead of delaying an exact answer. The concept of approximate query answering addresses this idea and proposes techniques for the online or offline design of database synopses, which are used to answer incoming queries [24]. The design of synopses ranges from simple uniform samples over specifically tailored (stratified) samples to wavelet synopses. The main challenge of approximate query answering consists of the creation and exploitation of synopses on the fly (i.e., within the context of the execution of a single query) and of the provision of error bounds to derive a quality measure to be returned to the user.

This list of methods and techniques outlines the most influential factors for query processing in data warehouse environments. Behind every perspective touched in this description, there are huge numbers of specific research issues – solved and unsolved ones.

Key Applications

Over the last 10 years, database systems have become the foundation of any large analytical infrastructure (usually embedded within a data warehouse system). Therefore, all analytically flavored applications heavily exploit specialized database support. This obviously large spectrum of applications ranges from mass reporting for the computation of thousands of parameterized reports to the support of interactive data cube exploration (OLAP). As data mining methods are becoming more and more standardized techniques, the computation of association rules, classification trees, and (hierarchical) clusters represents key applications.

Future Directions

In the future, query processing in analytical domains will be driven by two factors. On the one hand, data volumes will continue to grow significantly, mainly because of advances in data integration efforts (to ease the pain of manual integration of additional data sources) and the growing presence of sensors to track individual items. More data will go hand in hand with the presence of increased main-memory capacity and the integration of storage class memory (SCM)-based phase change memory, spin-transfer torque RAM, magnetic RAM, or memristor technology into the memory hierarchy. Physical database design and the corresponding exploitation of specialized structures will be major challenges. Additionally, data warehouse query processing is stretched beyond traditional relational systems to big data platforms like Hadoop infrastructures. While interactive queries will be processed by a relational (in memory) system, the detailed and historical data will be situated in Hadoop environments; research questions will center around optimal distribution and replication schemes between different systems in large data warehouse environments.

On the other hand, the topic of query processing will face more sophisticated statistical methods, which have to be supported natively by the database engine to yield reasonable query performance. While for many analytical application infrastructures, the "bring the data to the statistical package" method still holds, this approach will be inverted at some point. Statistical packages will work more closely with database systems, implying that computational components are taken closer to the data (as an integral part of a database engine) to be integrated within the overall optimization process. Also, the scope of analytical applications will widen and incorporate comprehensive *data visualization* techniques.

Both future developments on the application level will have significant consequences for the design of query processing techniques ranging from low-level data organization up to the support of a specialized query interface.

Cross-References

▶ Approximate Query Processing
▶ Bitmap Index
▶ Cube
▶ Data Sampling
▶ Data Warehousing in Cloud Environments
▶ Optimization and Tuning in Data Warehouses
▶ Main Memory DBMS
▶ Multidimensional Modeling
▶ Parallel and Distributed Data Warehouses
▶ Parallel Query Processing
▶ Query Optimization
▶ Query Processing
▶ Rewriting Queries Using Views
▶ Snowflake Schema
▶ Star Schema
▶ SQL

Recommended Reading

1. N.N. Multidimensional Expressions (MDX) Reference. Available at: http://msdn2.microsoft.com/en-us/library/ms145506.aspx
2. Plattner H. The impact of columnar in-memory databases on enterprise systems. Proc VLDB Endow. 2014;7(13):1722–9.
3. Raman V, Attaluri GK, Barber R, Chainani N, Kalmuk D, Samy VK, Leenstra J, Lightstone S, et al. DB2 with BLU acceleration: so much more than just a column store. Proc VLDB Endow. 2013;6(11):1080–91.
4. Chaudhuri S, Dayal U. An overview of data warehousing and OLAP technology. ACM SIGMOD Rec. 1997;26(1):65–74.
5. Gray J, et al. The Lowell database research self assessment. 2003. Available at: http://research.microsoft.com/~gray/lowell/
6. Müller I, Sanders P, Lacurie A, Lehner W, Färber F. Cache-efficient aggregation: hashing is sorting. Proceedings of the ACM SIGMOD International Conference on Management of Data; 2015. p. 1123–36.
7. Data Mining Extensions (DMX) reference. Available at: http://msdn2.microsoft.com/en-us/library/ms132058.aspx
8. N.N. ISO/IEC 9075–14. Information technology – database languages – SQL – part 14: XML-related specifications (SQL/XML). 2003. Available at: http://www.iso.org/iso/iso_catalogue/catalogue_tc/catalogue_detail.htm?csnumber=35341
9. Faerber F, May N, Lehner W, Grosse P, Mueller I, Rauhe H, Dees J. The SAP HANA database – an architecture overview. IEEE Data Eng Bull. 2012;35(1):28–33.
10. Tao Y, Zhu Q, Zuzarte C, Lau W. Optimizing large star-schema queries with snowflakes via heuristic-based query rewriting. In: Proceedings of the Conference of the IBM Centre for Advanced Studies on Collaborative Research; 2003. p. 279–93.
11. Graefe G, Guy W, Kuno HA, Paulley G. Robust query processing (Dagstuhl seminar 12321). Dagstuhl Rep. 2012;2(8):1–15. https://doi.org/10.4230/DagRep.2.8.1
12. Weipeng PY, Larson P. Eager aggregation and lazy aggregation. In: Proceedings of the 12th International Conference on Very Large Data Bases; 1995. p. 345–57.
13. Star Schema processing for complex queries. White Paper, Red Brick Systems, Inc., 1997. http://www.redbrick.com/products/white/whitebtm.html.
14. O'Neil B, Schrader M, Dakin J, Hardy K, Townsend M, Whitmer M. Oracle data warehousing unleashed. Indianapolis: SAMS Publishing; 1997.
15. He J, Lu M, He B. Revisiting co-processing for hash joins on the coupled CPU-GPU architecture. Proc VLDB Endow. 2013;6(10):889–900.
16. Teubner J, Woods L. Data processing on FPGAs. In: Morgan Claypool Publishers. Data processing on FPGAs. 2013. p. 1–118.
17. Sellis TK. Multiple-query optimization. ACM Trans Database Syst. 1988; 13(1):23–52.
18. Copeland GP, Khoshafian SN. A decomposition storage mode. SIGMOD Rec. 1985;14(4):268–79.
19. Abadi D, Madden S, Ferreira M. Integrating compression and execution in column-oriented database systems. In: Proceedings of the ACM SIGMOD International Conference on Management of Data; 2006. p. 671–82.
20. Chan C-Y. Bitmap index design and evaluation. In: Proceedings of the ACM SIGMOD International Conference on Management of Data; 1998. p. 355–66.
21. Valduriez P. Join indices. ACM Trans Database Syst. 1987;12(2):218–46.

Q

22. Weininger A. Efficient execution of joins in a star schema. In: Proceedings of the ACM SIGMOD International Conference on Management of Data; 2002. p. 542–45.
23. Hellerstein JM, Haas PJ, Wang HJ. Online aggregation. In: Proceedings of the ACM SIGMOD International Conference on Management of Data; 1997. p. 171–82.
24. Garofalakis M, Gibbon B. Approximate query processing: taming the TeraBytes. In: Proceedings of the 27th International Conference on Very Large Data Bases; 2001.
25. Celko J. Joe Celko's data warehouse and analytic queries in SQL. Morgan Kaufmann; 2006.
26. Clement TY, Meng W. Principles of database query processing for advanced applications. Morgan Kaufmann; 1997.
27. Graefe G. Query evaluation techniques for large Databases. ACM Comput Surv. 1993;25(2):73–170.
28. Gupta A, Mumick I. Materialized views: techniques, implementations and applications. Cambridge, MA: MIT Press; 1999.
29. Inmon WH. Building the data warehouse. 2nd ed. New York: Wiley.
30. Niemiec R. Oracle database 10g performance tuning tips & techniques; 2007.
31. Roussopoulos N. The logical access path schema of a database. IEEE Trans Softw Eng. 1982;8(6):563–73.

Query Processing in Deductive Databases

Letizia Tanca
Computer Science, Politecnico di Milano, Milan, Italy

Synonyms

Datalog query processing and optimization; Logical query processing and optimization; Recursive query evaluation

Definition

Most of the research work on deductive databases has concerned the *Datalog language*, a query language based on the logic programming paradigm which was designed and intensively studied for about a decade. Its origins date back to the beginning of logic programming, but it became prominent as a separate area around 1978, when Hervé Gallaire and Jack Minker organized a workshop on logic and databases. In this entry, the definition of the typical computation styles of Datalog will be given, the most important optimization types will be summarized, and some developments will be outlined.

Historical Background

The research on deductive databases was concentrated mostly between the mid-1980s and the mid-1990s. In those years, substantial efforts were made to merge Artificial Intelligence technologies with those of the database area, with the aim of building *large and persistent knowledge bases*. An important contribution toward this goal came from database theory, which concentrated on the formalization of Datalog – specifically designed for the logic-based interaction with large knowledge bases – and on the definition of computation and optimization methods for Datalog rules [1–4]. In parallel, various experimental projects showed the feasibility of Datalog as a data-oriented logic programming environment [5, 6].

The reaction of the database community to Datalog has often been marked by skepticism. In particular, the immediate practical use of research on sophisticated rule-based interaction has often been questioned. However, the research experience on Datalog, properly filtered, has taught important lessons to database researchers, setting the basis for the theoretical systematization of several related issues [7].

Foundations

Datalog is in many respects a simplification of the more general *logic programming* paradigm [8], where a program is a finite set of *facts* and *rules*. Facts are assertions about the reality of interest, like *John is a child of Harry*, while rules are sentences which allow deducing of facts from other facts. For instance, a rule might say *If X is a child of Y, and Y is a child of Z, then X is a grandchild of Z.* Facts and rules may contain variables; facts that only contain constants are called *ground facts.*

In the formalism of Datalog, both facts and rules are represented as *Horn clauses*:

$$L_0 : -L_1, \ldots, L_n$$

where each L_i is a *literal* of the form $p(t_1...tn)$, such that p is a *predicate symbol* and the t_i are *terms*. A term is either a constant or a variable, while functional symbols are not allowed, at least in the basic syntax of Datalog. The left-hand side (LHS) of a Datalog clause is also called its *head*, while the right-hand side (RHS) is its *body*. Clauses with an empty body are the facts: indeed, the body contains the clause premises; thus, if there are no premises, this means that the head is an assertion. Clauses with a nonempty body are the rules. A set of ground facts can easily be thought of as a relational database, since each fact *parent (John, Harry)* (between John and Harry there is a child-parent relationship) can be written as a tuple ⟨*John, Harry*⟩ stored into a relation PARENT. Both facts and rules are a form of *knowledge*; indeed, the knowledge stored in the database (alternatively represented as the set of ground facts) is enriched by the knowledge which can be deduced from the rules.

In the context of general Logic Programming, it is usually assumed that all application-relevant knowledge (facts and rules) are contained within a single logic program. On the other hand, Datalog has been developed for applications which use large, relational databases; therefore, two sets of clauses will be considered: the set of ground facts, called *Extensional DataBase (EDB)*, and the set of rules, i.e., the Datalog program, called the *Intensional DataBase (IDB)*. The *Herbrand Base* is the set of all ground facts that can be expressed in the language of Datalog, by using all the constants present in the database and all the predicates of *EDB* ∪ *IDB*.

A *Datalog program* is a set of ground facts (possibly stored into a relational database) and

rules, satisfying the safety condition that *all the variables contained in the rule head must also be present in its body.*

Note that a Datalog program can be considered as the specification of a *query against the EDB*, producing as answer the (set of) relation(s) of the IDB. It often happens that a user is interested in a subset of the (large) relation(s) that can be defined from a Datalog program: a *goal* is a single literal, preceded by a question mark and a dash, used to express constraints on the relations specified by the Datalog program. For example, $? - parent(John,X)$. specifies all the X such that *John* is a child of X.

Consider, as an example, the EDB constituted by a unary relation PERSON and a binary relation PARENT (containing all the pairs ⟨*child, parent*⟩) and the following program:

$$r_1 : sgc\,(X, X) : -person(X).$$
$$r_2 : sgc(X, Y) : -parent(X, X1), sgc\,(X1, Y1),$$
$$parent\,(Y, Y1)\,.$$

Rule r_1 simply states that a person is a cousin at the same generation of himself/herself, while rule r_2 says that two persons are same-generation cousins if they have parents who, in turn, are same-generation cousins. Note that r_2 is *recursive* and that rule r_1 constitutes the recursion base. A typical goal against the set $\{r_1, r_2\}$ is $? - sgc\,(John, Y)$, asking for all the same-generation cousins of John.

Evaluation of Datalog Programs

Consider a Datalog rule $R = L_0 : -L_1, \ldots, L_n$ and a set of ground facts $F = \{F_1, \ldots F_n\}$. If a substitution θ exists, such that $\forall 1 \leq i \leq n$, $L_i\theta = F_i$, then, from the rule R and from the facts F, the fact $L_0\theta$ can be inferred in one step. This fact might be new, or already known.

The just-described inference rule – which is actually a metarule – is called *EPP (Elementary Production Principle)*. Consider now a whole Datalog program S. New knowledge can be obtained from the Datalog program by applying it to the set of all ground facts of the database, to obtain new ground facts. Informally, it can be said that a ground fact F is *inferred* from S ($S \vdash F$) if either (a) $F \in S$ or (b) F can be obtained by applying the EPP *a finite number of times*. More precisely:

- $S \vdash F$ if $F \in S$.
- $S \vdash F$ if a rule R and ground facts $F_1, \ldots, F_n$ exist such that $\forall 1 \leq i \leq n, S \vdash F_i$, and F_i can be inferred in one step by the application of EPP to R and $F_1, \ldots, F_n$.

The sequence of applications of *EPP* which is used to infer a ground fact F from S is called a *proof* of F from S. The *proof-theoretical framework* thus established allows to infer new ground facts from an original set of Datalog clauses; on the other hand, there is a model-theoretic approach, which provides a definition of *logical consequence* ($\models$). It is possible to prove that:

Theorem 1. (Soundness and completeness of Datalog) Let S be a set of Datalog clauses, and let F be a ground fact. Then $S \vdash F$ if and only if $S \models F$.

A proof of this theorem can be found, for example, in [2].

In order to check whether EPP applies to a rule $R : L_0 : -L_1, \ldots, L_n$ and to an ordered list of ground facts $F_1, \ldots, F_n$, an appropriate substitution θ for the variables of R must be found, such that $\forall 1 \leq i \leq n, L_i \theta = F_i$.

Given a finite set S of Datalog clauses, i.e., a Datalog program, according to the soundness and completeness theorem, the set of all facts which are *derivable* from S is the set *cons(S)* of the *logical consequences* of S, which can be computed by the following algorithm:

- FUNCTION INFER(S)
- INPUT: a finite set S of Datalog clauses
- OUTPUT: cons(S)
- Begin
- W:= S
- While EPP produces a new ground fact F $\notin$ W
- Do W: = W $\cup$ F
- Return (facts(W))

- End

The *INFER* algorithm always terminates and produces as output a *finite* set of ground facts, *cons(S)*, since the number of constants and predicates symbols, as well as the number of arguments of these predicates, is finite. The order in which *INFER* generates new facts corresponds to the *bottom-up* order of a proof tree; thus, the principle underlying *INFER* is called *bottom-up evaluation*, or, as in Artificial Intelligence, *forward chaining* (forward in the sense of the logical implication contained in the Datalog rules).

The set *cons(S)* can also be characterized as *the least fixpoint* of the transformation T_S, a mapping from 2 *HB* to itself defined as follows:

$$\forall W \in HB, T_S(W) = W \cup FACTS(S) \cup$$
$$INFER1(RULES(S) \cup W)$$

where *INFER1(S)* denotes the set of *all* ground facts that can be inferred in one step from S via *EPP*.

Accordingly, *cons(S)* can be computed by fixpoint iteration, i.e., by computing, in order, $T_S(\varnothing), T_S(T_S(\varnothing), T_S(T_S T_S(\varnothing)))$,...,until a term which is equal to its predecessor is reached. This final term is *cons(S)*.

The *top-down* evaluation of a Datalog program is based on a radically different approach, where proof trees are built from the top to the bottom, by applying *EPP* "backward," which is much more appropriate when a goal is specified together with S. The general principle of *backward chaining* corresponds, in Prolog, to the *SLD resolution (SL resolution with Definite clauses)* inference rule, introduced by Robert Kowalski [9]. Its name is derived from SL resolution, which is both sound and refutation complete for the unrestricted clausal form of logic.

The logic programming formalism is now related to a database query language, in order to show how easy the integration between the two realms is. Each clause of a Datalog program can be translated into an inclusion relationship of Relational Algebra; then, the set of relationships which refer to the same predicate are interpreted as Relational Algebra *equations*, whose constants are the EDB relations and whose variables are the IDB predicates, defining virtual relations. Determining a solution of the thus composed system corresponds to determining the values of the variable relations, i.e., to finding the ground facts in the IDB predicates. Consider a Datalog clause:

$$C : p\left(\alpha_1 \ldots \alpha_n\right) : -q_1\left(\beta_{k_1} \ldots \beta_{k_h}\right), \ldots$$
$$q_m\left(\beta_{k_j} \ldots \beta_{k_m}\right)$$

the translation associates to C an inclusion relationship $Expr(Q_1, \ldots, Q_m) \subseteq P$ among the relations $Q_1, \ldots, Q_m, P$ that correspond to the predicates $q_1, \ldots, q_m, p$, adopting the convention that relation attributes are named according to the number of the corresponding argument of the related predicate. For example, the Datalog rules of the *same-generation cousins* program are translated into the inclusion relationships:

$$\pi_{1,1}PERSON \subseteq SGC$$

$$\pi_{1,5}\left((PARENT^{\bowtie}{}_{2=1}SGC)^{\bowtie}{}_{4=2}PARENT\right) \subseteq SGC$$

The rationale behind this translation is that literals with common variables correspond to equijoins over those variables, while the variables exported to the rule head correspond to projections. The new (virtual) relation SGC (actually, a *view*) is defined as the extension of the predicate *sgc*. For each IDB predicate, all the related inclusion relationships are collected, generating an algebraic equation that obtains the predicate by performing the union:

$$SGC = \pi_{1,1}PERSON$$

$$\cup \pi_{1,5}\left(\left(PARENT^{\triangleright\triangleleft}{}_{2=1}SGC\right)^{\triangleright\triangleleft}{}_{4=2}PARENT\right)$$

Logical goals are also translated into algebraic queries, over the EDB or over the just defined views:? $- sgc\ (John,Y)$. corresponds to $\sigma_{1="John"}$ SGC.

Note that this translation is based on the use of all classical algebraic operators, except *the difference*. In fact, it can be shown that Datalog

without recursion is equivalent to Relational Algebra deprived of the difference operator.

Optimization of Datalog Programs

The evaluation of Datalog programs according to various forms of fixpoint computation, similar to the INFER algorithm, is called *naive*, as opposed to better, more performant techniques whose mutual relationship is not always obvious. Optimization methods can be observed with regard to different orthogonal dimensions: the *formalism* (logical vs. algebraic), the *search strategy* (bottom-up vs. top-down), the *technique* (rewriting programs into more efficient ones vs. directly applying an efficient evaluation method), and the *type of information exploited by the optimization process* (semantic vs. syntactic).

Since Datalog programs can equivalently be written as sets of algebraic equations, a Datalog program can actually be evaluated in the same way as any algebraic query, provided that a way to process recursion is available. Actually, algebraic evaluation methods that mimic the naive method have been introduced, and the classical results of algebraic query optimization, like common subexpression analysis and equivalence transformations, have been profitably transformed to be applied to recursive queries [2].

As far as the search strategy is concerned, observe that bottom-up methods actually consider rules as *productions*, generating all possible consequences of $EDB \cup IDB$ until no new facts can be deduced; thus, these methods are applied in a set-oriented fashion, which is a desirable feature in the database context, where large amounts of data are stored in mass memory and must be retrieved in the buffer in a "set-at-a-time" way. On the other hand, also observe that bottom-up methods do not take in immediate advantage the selectivity due to the existence of bound arguments in the goal predicate.

By contrast, top-down methods [2, 9] use rules as *subproblem generators*, since each goal is considered as a problem to be solved. The initial goal ? $- p(\alpha_1...\alpha_n)$. is matched against some rule $C : p\left(\alpha_1 \ldots \alpha_n\right) :$

$-q_1 \left(\beta_{k_1} \ldots \beta_{k_h} \right), \ldots q_m \left(\beta_{k_j} \ldots \beta_{k_m} \right)$ and generates *subgoals*? $- q_i \left(\beta_1 \ldots \beta_i \right)$ that represent new subproblems to be solved. In this case, if the goal contains some bound (i.e., constant) argument, then only facts that are related to the goal constants are involved in the computation. Suppose, for instance, that the goal? $- sgc(John, Y)$. be given. Then, when applying top-down rule r_2, the subgoal $parent(X, X1)$ is only further analyzed with regard to the parents of John, that is, *only the subrelation* $\sigma_{1\,=\,\text{``John''}}$ *PARENT* is involved in the computation of the first literal. However, although the algorithms based on this evaluation method already produce some optimization, they may be inappropriate for the database context, because many of them work "one-tuple-at-a-time."

Another analysis dimension for optimization methods is whether they directly interpret the program or first rewrite it into an equivalent, more efficient form and then evaluate it in a naive way. To this category belong, for instance, the *Magic Sets* and *Counting* methods [1], where the authors "simulate" the binding propagation achieved by top-down evaluation by applying a simple program transformation to a certain class of programs, in a similar way as algebraic database optimization techniques (e.g., push of the selections) are applied. A similar approach, directly introduced by means of the algebraic formalism, is taken in the *Reduction of Variables* and *reduction of Constants* methods [2].

Finally, also *semantic information* can be used to optimize programs: for instance, [10] base the optimization on the additional semantic knowledge provided by *database constraints*. For example, a constraint might state that *all sailing vessels in Ischia are sheltered in the main harbour*; thus, the query asking for *the harbour in Ischia where the sailing boat "Roxanne" is located* can be answered without even accessing the DB.

Negation in Datalog

In pure Datalog, the negation sign $\neg$ is not allowed; however, negative facts can be inferred from Datalog programs by adopting the *Closed-*

World Assumption (CWA), which, in this context, reads as follows:

> *If a fact is not derivable from a set of Datalog clauses, then the negation of that fact is true.*

For example, if, after computing the SGC relation, the tuple $\langle LUCY, JOHN \rangle$ is not found, the fact $\neg sgc\,(lucy, john)$ – that is, Lucy is not a child of John, is inferred.

The CWA applied to Datalog clauses allows the deduction of negative facts, but not their use within the Datalog rules in order to deduce some new facts. For instance, in Relational Algebra, all the pairs of persons who are not same-generation cousins are specified as $NONSGC = (PERSON \times PERSON) - SGC$. Accordingly, one would like to write

$$r_3 : nonsgc\,(X, Y) : -person(X), person(Y), \\ \neg sgc\,(X, Y).$$

The language $Datalog^\neg$ has the same syntax as Datalog, but here negated literals are allowed in the rule bodies. For safety reasons, all the variables occurring within a negated literal must also occur in a positive literal of the body. Without delving into semantic details, note that unfortunately, because of recursion, the computation of $Datalog^\neg$ programs is not as straightforward as that of a pure Datalog program; indeed, by simply applying the *EPP* and the INFER algorithm to a recursive $Datalog^\neg$ program, one may incur into a contradiction, since some negative facts that are inferred at step n of the computation might be derived as positive at some step $n + k$. Intuitively, consider the program composed by rules r_1, r_2, r_3, and think of the set of same-generation cousins derived at the first computation step. Suppose that $sgc(lucy, john)$ has not been derived; then, at the second step, $\neg sgc(lucy, john)$ may be derived by applying r_3; yet, it might be that $sgc(lucy, john)$ is derived from r_1, r_2 at some later step $k(k > 2)$.

The most common policy used to avoid such problems is only allowing $Datalog^\neg$ programs which are *stratified* [3], according to the following intuition: when evaluating a rule with one or more negative literals in the body, *evaluate first the predicates corresponding to these negative literals*. Then, the CWA is "locally" applied to these predicates. Consider the example above: since the *sgc* predicate appears in negative form in rule r_3, it is first evaluated by applying only rules r_1, r_2 (the "first stratum" of the program); then, once the whole extension of *sgc* has been computed, all the pairs of persons who are not in the SGC relation can be derived (actually by difference).

However, one can intuitively guess that not necessarily all the Datalog programs can be stratified in this way, that is, it may be possible that there is another rule r_4 which in turn contains the predicate *nonsgc* in negative form, and so on and so forth. More formally, define the *Dependency Graph DG(P)* = $\langle N(P),E(P)\rangle$ of a Datalog⁻ program P as follows: $N(P)$ is the set of predicates p occurring in the rule heads of P, while an edge $\langle p,q\rangle$ belongs to $E(P)$ if the predicate symbol q occurs positively or negatively in some rule whose head predicate is p. Moreover, $\langle p,q\rangle$ is labeled by ¬ if there is at least one rule in P with head predicate p, whose body contains a negative occurrence of q.

A Datalog⁻ program P is *stratified* iff DG(P) does not contain any cycle involving an edge labeled by¬. If P is stratified, it is quite easy to construct a stratification of P [2, 4, 7], that is, a sequence of subprograms $P_0 = EDB, P_1, \ldots, P_n$ of P such that $P_0 \cup P_1 \cup \ldots \cup P_n = P$ and, by evaluating them separately and in order from P_0 to P_n, and by applying the CWA to P_k when computing P_{k+1}, the result does not contain any contradiction. Note that stratifications are not unique, that is, if the condition on the $DG(P)$ is satisfied, P can be stratified in several different ways. It is easy to see that the sample program $P = \{r_1, r_2, r_3\}$ is stratified and that a stratification is $P_0 = EDB, P_1 = \{r_1, r_2\}, P_2 = \{r_3\}$.

Much further work has been done on the semantics of negation and of non-monotonic programs [7]. Semantics based on various kinds of partial models, like the *stable models* [11] and the *well-founded models* [12], are often studied. An example of more recent work on the subject is [13], where the new concept of soft stratification, based on a new bottom-up query evaluation

method based on the Magic Set approach, is proposed.

Key Applications

The research on deductive databases was concentrated in the decade between the mid-1980s and the mid-1990s. This work, and all logic-based approaches to database problems, constitutes the foundational experience for speculation that have led to a number of results in different fields. The field of active databases is one interesting example of the application of the theoretical foundations of Datalog. An active database system is a DBMS endowed with active rules, i.e., stored procedures activated by the system when specific events occur. The processing of active rules is characterized by two important properties: termination and confluence. In [14], a set of active rules is translated into logical clauses, taking into account the system's execution semantics, and simple results about termination and determinism available in the literature for deductive rules are transferred to the active evaluation process. Another, more recent application is the integration of Databases with the Semantic Web, an interesting example of which is the *SWRL (Semantic Web Rule Language)* [15], a proposal combining sublanguages of the OWL Web Ontology Language with the rule-based paradigm. SWRL is (roughly) the union of Horn logic and OWL, where rules are of the form of an implication between an antecedent (body) and consequent (head), with a semantics very similar to that of Datalog.

A very recent revival of Datalog concerns its use to formalize *declarative networking*, that is, the attempt to make *programming on the cloud* declarative. Cloud programming refers to maximizing efficiency by sharing computation and storage resources: computation is distributed over a network of nodes to increase parallelism, and declarative networking means allowing the programmer to express only the desired result and leave the implementation details to the responsibility of the run-time system. See [16] for an interesting and comprehensive survey.

Cross-References

▶ Active Database, Active Database (Management) System
▶ Query Language
▶ Views

Recommended Reading

1. Bancilhon F, Maier D, Sagiv Y, Ullman JD. Magic sets and other strange ways to implement logic programs. In: Proceedings of the 5th ACM SIGACT SIGMOD Symposium on Principles of Database Systems; 1986. p. 1–15. Available at: https://pdfs.semanticscholar.org/1289/85b85556c30ad405863f2a34340049957616.pdf
2. Ceri S, Gottlob G, Tanca L. Logic programming and databases. Berlin: Springer; 1990.
3. Chandra AK, Harel D. Horn clauses queries and generalizations. J Log Program. 1985;2(1):1–15.
4. Ullman JD. Principles of database and Knowledge-Base systems. IRockville: Computer Science Press; 1988.
5. Bocca JB. EDUCE: a marriage of convenience: Prolog and a Relational DBMS. In: Proceedings of the Symposium in Logic Programming; 1986. p. 36–45.
6. Tsur S, Zaniolo C. LDL: a logic-based data language. In: Proceedings of the 12th International Conference on Very Large Data Bases; 1986. p. 33–41. Available at: https://dl.acm.org/citation.cfm?id=671478.
7. Abiteboul S, Hull R, Vianu V. Foundations of databases. Reading: Addison-Wesley; 1995.
8. Lloyd JW. Foundations of logic programming. 2nd ed. Berlin: Springer; 1987. ISBN:3-540-18199-7.
9. Kowalski RA, Kuehner D. Linear resolution with selection function. Artif Intell. 1971;2(3–4):227–60.
10. Chakravarthy US, Minker J, Grant J. Semantic query optimization: additional constraints and control strategies. In: Proceedings of the Expert Database Conference; 1986. p. 345–79.
11. Sacca D, Zaniolo C. Stable models and non-determinism in logic programs with negation. In: Proceedings of the 9th ACM SIGACT-SIGMOD-SIGART Symposium on Principles of Database Systems; 1990. p. 205–217. Available at: https://dl.acm.org/citation.cfm?id=298572.
12. Laenens E, Vermeir D. Assumption-free semantics for ordered logic programs: on the relationship between well-founded and stable partial models. J Log Comput. 1992;2(2):133–72.
13. Behrend A. Soft stratification for magic set based query evaluation in deductive databases. In: Proceedings of the 22nd ACM SIGACT-SIGMOD-SIGART Symposium on Principles of Database Systems; 2003. p. 102–10. Available at: https://dl.acm.org/citation.cfm?id=773164.
14. Comai S, Tanca L. Termination and confluence by rule prioritization. IEEE Trans Knowl Data Eng. 2003;15(2):257–70.
15. W3C Member Submission. SWRL: a semantic Web rule language combining OWL and RuleML. 2004. Available at: http://www.w3.org/Submission/SWRL/
16. Ameloot TJ. Declarative networking: recent theoretical work on coordination, correctness, and declarative semantics. SIGMOD Rec. 2014;43(2):5–16. Available at: http://www.sigmod.org/publications/sigmod-record/1406/pdfs/03.principles.Ameloot.pdf.

Query Processing over Uncertain Data

Nilesh Dalvi[1] and Dan Olteanu[2]
[1]Airbnb, San Francisco, CA, USA
[2]Department of Computer Science, University of Oxford, Oxford, UK

Synonyms

Query processing over probabilistic data

Definition

An uncertain or probabilistic database is defined as a probability distribution over a set of deterministic database instances called *possible worlds*.

In the classical deterministic setting, the query processing problem is to compute the set of tuples representing the answer of a given query on a given database. In the probabilistic setting, this problem becomes the computation of all pairs (t, p), where the tuple t is in the query answer in some random world of the input probabilistic database with probability p.

Scientific Fundamentals

Representation of Uncertain Data

All aspects of query processing over uncertain data, and in particular its complexity and existing techniques, highly depend on data representation. Since it is prohibitively expensive to explicitly represent the extremely large set of all possible worlds of a probabilistic database, one has to settle for succinct data representations. Three such representations of increasing expressiveness are discussed next [34].

In *tuple-independent (TI) databases*, the tuples are independent probabilistic events. A TI database of n tuples represents the probabilistic database that is the powerset of the input set of n tuples and thus has 2^n possible worlds. In *block-independent-disjoint (BID) databases*, the tuples are partitioned into blocks such that tuples within the same block are disjoint events and tuples from different blocks are independent. A BID database represents all possible worlds with at most one tuple from each block. In *probabilistic conditional (PC) databases*, the tuples are associated with propositional formulas over independent random variables. Each total assignment ψ of these random variables defines a world consisting of those tuples whose formulas are satisfied by ψ. The probability of the world is the product of the probabilities of the assignments of the random variables in ψ. A TI databases is a PC database where each tuple is associated with a distinct Boolean random variable. Also, a BID database is a PC database where each relation is made up of blocks of tuples such that the formulas of the tuples within a block are mutually exclusive and the formulas of the tuples from different blocks are over disjoint sets of variables.

PC databases can represent the answers of any relational query over PC databases. The formula associated with an answer tuple t is called the lineage of t and explains which tuples must be present in the input database in order for t to be in the query answer. Given two tuples with formulas ϕ_1 and ϕ_2, their join is a tuple with lineage $\phi_1 \wedge \phi_2$, their union is a tuple with lineage $\phi_1 \vee \phi_2$, and their difference is a tuple with lineage $\phi_1 \wedge \neg \phi_2$.

In contrast to BID and TI databases, PC databases are complete in the sense that they can represent any probabilistic database. However, the formalism defined by conjunctive queries over BIDs is complete. Despite their restricted

nature, TI databases are practical; for instance, inference in Markov Logic Networks can be reduced to relational query processing over TI databases.

Prime examples of TI databases are the Google Knowledge Vault [8] and NELL (Never-Ending Language Learner, http://rtw.ml.cmu.edu/rtw/) knowledge bases, of BID databases are the Google Squared tables [10], and of PC databases are the answers and their lineage for relational queries on probabilistic databases in any of the above formalisms.

Complexity of Query Processing

The *data complexity* of queries over probabilistic databases represented as TI, BID, or PC databases is #P-hard. This high computational complexity is already witnessed for simple join queries on TI databases, since the computation of marginal probabilities may require to enumerate all possible worlds. Several classes of relational queries, e.g., non-repeating conjunctive queries [5], their ranking version [27] and extension with negation [12], unions of conjunctive queries [7], and aggregate queries [30], exhibit an interesting dichotomy property: the data complexity of every query in any of these classes over TI databases is either polynomial time or #P-hard. Beyond relational queries and TI databases, hardness and tractability results have been shown, e.g., for XML queries over Recursive Markov Chains and their restrictions [3, 24], for graph queries over probabilistic RDF graphs [25], for event queries over Markovian streams [29], and for conjunctive queries over graphical models [31].

Query Processing Techniques

To overcome the high complexity of probabilistic query processing, two avenues of research have been pursued [34]. The first avenue is to identify tractable queries, i.e., queries computable in polynomial time, and develop efficient processing techniques for this subset. Techniques for tractable queries developed in the context of the MystiQ [5] and SPROUT [11] probabilistic database systems show performance for tractable queries over probabilistic databases close to that for queries over classical deterministic databases.

The second avenue is to develop techniques that approximate the probabilities of the query answers. In many applications, probabilities are only needed to rank the query answers and approximate probabilities may suffice for ranking. Even when probabilities are returned to the user, it is often desirable to improve performance by sacrificing precision. In addition to MystiQ and SPROUT, Trio [36], Orion [32], and MayBMS [17] also employ approximate techniques for hard queries.

An orthogonal classification considers whether the probabilistic inference task is performed inside or outside the database engine. In-database techniques cast the inference task as query processing in deterministic databases. Out-database techniques rely on specialized inference methods beyond the capabilities of a database engine, e.g., knowledge compilation techniques. Prime examples of systems using out-database techniques are Trio, MayBMS, and PrDB [31], whereas MystiQ, Orion, and MCDB use in-database techniques; SPROUT uses in-database techniques for classes of tractable queries and out-database techniques for hard queries. The two types of techniques can also be combined: In-database approaches are first applied to tractable subproblems, and then out-database approaches are used for the remaining hard subproblems.

We next highlight several techniques for relational query processing over probabilistic databases. We leave aside processing techniques for complex queries such as nearest neighbors and skyline queries as they represent a topic on their own [4]. We also only refer to a few approaches to probabilistic ranking and refer the reader to a recent monograph for in-depth treatment [18].

In-Database Techniques

In-database techniques are used for exact evaluation of tractable queries in tuple-independent probabilistic databases, e.g., using safe plans [6] as discussed below, and also for approximate evaluation of hard queries, e.g., computing lower and upper bounds on answer probabilities via dissociation of input probabilistic events [14] or running Monte Carlo simulations that aggregate

the query answers over several possible worlds sampled from complex probabilistic models [20].

Such techniques extend standard operators in relational plans to compute both tuples and their probabilities. Examples of extended operators are the independent join operator, which multiplies the probabilities of the tuples it joins under the assumption that they are independent, and the independent project operator, which computes the probability of an output tuple t as $1 - (1 - p_1) \cdots (1 - p_n)$ where $p_1, \ldots, p_n$ are the probabilities of all tuples that are input to the operator and project into t, again assuming that these tuples are independent. The selection operator retains the tuples that satisfy its condition, along with their probabilities.

For a given relational query, not all of its query plans with extended operators compute the probabilities correctly since the independence assumption for the input to these operators may not hold. If the plan does compute the probabilities correctly for any input database, then it is called a *safe query plan*. Safe plans are easily added to a relational database engine, either by small modifications to the relational operators or even without any change in the engine by simply rewriting the SQL query to add aggregates that manipulate the probabilities explicitly. If a query admits a safe plan, then its data complexity is polynomial time because any safe plan can be computed in polynomial time in the size of the input database by simply evaluating its operators bottom-up. Consequently, #P-hard queries cannot admit a safe plan. For the class of non-repeating conjunctive (select-project-join) queries, the selection, independent join, and independent project operators are complete since the tractable queries are precisely those that admit safe plans [5]. For TI databases under functional dependencies, hard queries may become tractable and computable using safe plans [26].

A variant of this technique is to decouple the computation of answer tuples from the computation of their probabilities and to use different query plans for the two computation tasks. This is motivated by the observation that safe plans, while necessary for correct probability computation, can be suboptimal for computing the answer tuples [26]. This approach first computes the answer tuples and a relational encoding of their lineage using an optimized query plan, and then it uses the safe plan over the lineage to compute the probabilities of the answer tuples.

Safe plans can be extended to cope with richer classes of queries. Plans for tractable unions of conjunctive query terms use an inclusion-exclusion (IE) operator to compute their probabilities as a function of the probabilities of conjunctions of subsets of the query terms [7]. Safe plans with an additional disjoint project operator can compute non-repeating conjunctive queries over BID databases [28]. This operator sums up the probabilities of all the input tuples that project into the same output tuple.

Out-Database Techniques

Out-database query processing techniques are more general than in-database ones. They work for arbitrary relational queries and probabilistic data representations beyond TI databases.

An important class of out-database techniques draws on connections between probabilistic query processing and knowledge compilation. They compile the lineage of answer tuples into decision diagrams, e.g., ordered binary decision diagrams (OBDDs) and deterministic decomposable negation normal forms (d-DNNFs), that admit linear-time probability computation. While in general the compilation can take time exponential in the number of random variables in the lineage, it takes polynomial time for several known classes of tractable queries on tuple-independent databases, e.g., the tractable non-repeating queries with negation [12], the non-repeating inversion-free unions of conjunctive queries [34], and queries with inequalities [21].

Recall that lineage is a propositional formula Φ over Boolean random variables x_1 to x_n (The extension to multi-valued variables is straightforward.). The idea behind lineage compilation is to recursively apply the following two steps in the given order:

1. If $\Phi = \psi_1 \vee \cdots \vee \psi_m$ or $\Phi = \psi_1 \wedge \cdots \wedge \psi_m$ such that ψ_1 to ψ_m are pairwise independent

or mutually exclusive, then the probability of Φ can be computed from the probabilities of ψ_1 to ψ_m in linear time. This step precisely captures the operators independent/disjoint project/join operators used by safe plans.

2. If the previous step is not applicable (such as for hard queries), then we apply *Shannon expansion* (DPLL) on any variable x_i in Φ: Φ is equivalent to a disjunction of two mutually-exclusive expressions $x_i \wedge \Phi_{x_i} \vee \neg x_i \wedge \Phi_{\neg x_i}$, where Φ_{x_i} and $\Phi_{\neg x_i}$ are Φ where x_i is set to true and false, respectively. The probability of Φ is the sum of the probabilities of the two mutually exclusive formulas, where Φ_{x_i} and $\Phi_{\neg x_i}$ have at least one variable (x_i) less than Φ.

We exhaustively repeat these two steps until we reach the Boolean constants true or false. Since at each compilation step we can compute in linear time the probability of the formula using the probabilities of the child subformulas, the algorithm runs in time linear in the number of steps. The order of variables x_i in Shannon expansion steps drastically influences the number of compilation steps, which can be at most exponential in the number of variables [34].

If the compilation is stopped at any time before reaching a Boolean constant, we obtain lower and upper bounds on the true probability; the more compilation steps we run, the tighter the probability interval becomes and thus the smaller the approximation error, with the new probability interval included in the previous one. In this sense, this is an anytime approximation algorithm [11].

This lineage compilation approach has been applied to slightly different settings, e.g., top-k query evaluation in the presence of non-materialized views [9], sensitivity analysis and explanation for queries [22], and lifted to lineage of queries with aggregates expressible in the formalism of provenance semimodules [1]. ProApproX evaluates queries on probabilistic XML, where the lineage of the query answer is compiled using the first step only [33]. When this step cannot be applied anymore, ProApproX resorts to Monte Carlo approaches on the subformulas. Top-k queries can be evaluated by incrementally approximating the probabilities of the results using the above anytime algorithm until the lower bounds of k tuples are greater than the upper bounds of the remaining tuples [34].

A further out-database approach that has been coupled with the above compilation approach is model-based approximation [11]: Given a lineage formula Φ, we can derive two formulas Φ_L and Φ_U such that the satisfying assignments of Φ_L are also of Φ and those of Φ are also of Φ_U. This implies that the probability of Φ_L is less than or equal to that of Φ, which is less than or equal to that of Φ_U. The benefit of this approach is immediate if the bounds Φ_L and Φ_U admit probability computation in polynomial time and can be derived in polynomial time from Φ.

The first step from the above lineage compilation is also used for computing tractable non-repeating queries with negation [12] and so-called inversion-free unions of conjunctive queries [21]. The processing of such a query critically draws on the observation that the traces of the compilation for subqueries of the input query are OBDDs, whose variable orders are compatible, depths are linear in the database size, and widths only depend on the number of relations in the query. Boolean operations (conjunction, disjunction, negation) on these OBDDs yield OBDDs with the same properties.

In contrast to the lifted approach using inclusion-exclusion that reasons at the query level, the above lineage compilation variants, which rely on grounding the query to its lineage, are limited in that they cannot compute in polynomial time all tractable unions of conjunctive queries [2].

A common out-database probability approximation technique exploits decades-old seminal work on Monte Carlo algorithms including fully polynomial-time randomized approximation schemes (FPRAS) for model counting of propositional formulas in disjunctive normal form [23, 35]. MystiQ and MayBMS use adaptations of such approximations to probabilistic inference: They repeatedly choose at random a possible world and computes the truth value of query lineage. The probability is then approximated by the frequency with which

the lineage was true. A common expectation that users have from database management systems is that simple queries run fast and complex queries run slower. Monte Carlo algorithms do not fulfill this expectation because they make no distinction between simple and complex queries. An approach matching the expected behavior would identify simple (tractable) queries and process them at peak performance, while for more complex (hard) queries, its performance should degrade smoothly.

Historical Background

The quest for understanding the complexity of probabilistic query processing began two decades ago as a study on query reliability [15], where the Boolean query $R(x), S(x, y), R(y)$ was shown to be hard for #P via a reduction from model counting for positive bipartite formulas in disjunctive normal form. A solid body of work starting a decade later showed complexity dichotomies for various classes of queries.

Query lineage and PC databases draw on c-tables, a formalism for incomplete information put forward three decades ago [19]. It has been formally investigated more recently in the context of provenance semirings [16] and semimodules [1]. Intensional query semantics was first used in probabilistic databases by Fuhr and Rölleke [13].

MystiQ pioneered in-database techniques for exact query processing on TI databases. SPROUT pioneered in- and out-database techniques based on lineage compilation [11]. Reminiscent of lifted inference in AI, follow-up work lifts compilation to first-order lineage [9].

A detailed account to query processing on uncertain relational and XML data is given in several recent research monographs [4,18,24,34].

Key Applications

Probabilistic databases have a very diverse set of applications [34]. They arise naturally in data integration settings owing to approximate schema and data mappings. Automated information extraction and classification using machine learned models also generate probabilistic facts that can be naturally represented using a probabilistic database. Data cleaning, which involves resolving inconsistencies in data as well as inferring missing values, is another source of uncertainty; likewise, data generated by physical devices that succumb to faults and measurement errors, which is common in scientific databases and sensor networks. In all of these settings, probabilistic databases allow to represent the underlying uncertainties in a unified, consistent way and enable ad-hoc querying over the uncertain data.

Cross-References

▶ Uncertain Data Models

Recommended Reading

1. Amsterdamer Y, Deutch D, Tannen V. Provenance for aggregate queries. In: Proceedings of the 30th ACM SIGACT-SIGMOD-SIGART Symposium on Principles of Database Systems; 2011. p. 153–64.
2. Beame P, Li J, Roy S, Suciu D. Counting of query expressions: limitations of propositional methods. In: Proceedings of the 17th International Conference on Database Theory; 2014. p. 177–88.
3. Benedikt M, Kharlamov E, Olteanu D, Senellart P. Probabilistic XML via Markov Chains. Proc VLDB Endow. 2010;3(1-2):770–81.
4. Chen L, Lian X. Query processing over uncertain databases. Synthesis lectures on data management. San Rafael: Morgan & Claypool Publishers; 2012.
5. Dalvi N, Suciu D. Efficient query evaluation on probabilistic databases. VLDB J. 2007;16(4):523–44.
6. Dalvi NN, Suciu D. Efficient query evaluation on probabilistic databases. In: Proceedings of the 30th International Conference on Very Large Data Bases; 2004. p. 864–75.
7. Dalvi NN, Suciu D. The dichotomy of probabilistic inference for unions of onjunctive queries. J ACM. 2012;59(6):30:1–30:87. https://doi.org/10.1145/2395116.2395119.
8. Dong X, Gabrilovich E, Heitz G, Horn W, Lao N, Murphy K, Strohmann T, Sun S, Zhang W. Knowledge vault: a web-scale approach to probabilistic knowledge fusion. In: Proceedings of the 20th ACM SIGKDD International Conference on Knowledge Discovery and Data Mining; 2014. p. 601–10.
9. Dylla M, Miliaraki I, Theobald M. Top-k query processing in probabilistic databases with non-materialized views. In: Proceedings of the 29th International Conference on Data Engineering; 2013. p. 122–33.
10. Fink R, Hogue A, Olteanu D, Rath S. SPROUT^2: a squared query engine for uncertain web data. In: Proceedings of the ACM SIGMOD International Conference on Management of Data; 2011. p. 1299–302.
11. Fink R, Huang J, Olteanu D. Anytime approximation in probabilistic databases. VLDB J. 2013;22(6): 823–48.
12. Fink R, Olteanu D. Dichotomies for queries with negation in probabilistic databases. ACM Trans Database Syst. 2016;41(1):4.
13. Fuhr N, Rölleke T. A probabilistic relational algebra for the integration of information retrieval and database systems. ACM Trans Information Syst. 1997;15(1):32–66.
14. Gatterbauer W, Suciu D. Oblivious bounds on the probability of boolean functions. ACM Trans Database Syst. 2014;39(1):5.
15. Grädel E, Gurevich Y, Hirsch C. The complexity of query reliability. In: Proceedings of the 17th ACM SIGACT-SIGMOD-SIGART Symposium on Principles of Database Systems; 1998. p. 227–34.
16. Green TJ, Karvounarakis G, Tannen V. Provenance semirings. In: Proceedings of the 26th ACM SIGACT-SIGMOD-SIGART Symposium on Principles of Database Systems; 2007. p. 31–40.
17. Huang J, Antova L, Koch C, Olteanu D. MayBMS: a probabilistic database management system. In: Proceedings of the ACM SIGMOD International Conference on Management of Data; 2009. p. 1071–74.
18. Ilyas IF, Soliman MA. Probabilistic ranking techniques in relational databases. Synthesis lectures on data management. San Rafael: Morgan & Claypool Publishers; 2011.
19. Imieliński T, Lipski W Jr. Incomplete information in relational databases. J ACM. 1984;31(4):761–91.
20. Jampani R, Xu F, Wu M, Perez LL, Jermaine C, Haas PJ. The Monte Carlo database system: stochastic analysis close to the data. ACM Trans Database Syst. 2011;36(3):18.
21. Jha AK, Suciu D. Knowledge compilation meets database theory: compiling queries to decision diagrams. Theory Comput Syst. 2013;52(3): 403–40.
22. Kanagal B, Li J, Deshpande A. Sensitivity analysis and explanations for robust query evaluation in probabilistic databases. In: Proceedings of the ACM SIGMOD International Conference on Management of Data; 2011. p. 841–52.
23. Karp RM, Luby M, Madras N. Monte-Carlo approximation algorithms for enumeration problems. J Algorithms. 1989;10(3):429–48.
24. Kimelfeld B, Senellart P. Probabilistic XML: models and complexity. In: Advances in Probability Database for Uncertain Information Management; 2013. p. 39–66.
25. Lian X, Chen L. Efficient query answering in probabilistic RDF graphs. In: Proceedings of the ACM SIGMOD International Conference on Management of Data; 2011. p. 157–68.

Q

26. Olteanu D, Huang J, Koch C. SPROUT: lazy vs. eager query plans for tuple-independent probabilistic databases. In: Proceedings of the 25th International Conference on Data Engineering; 2009. p. 640–51.
27. Olteanu D, Wen H. Ranking query answers in probabilistic databases: complexity and efficient algorithms. In: Proceedings of the 28th International Conference on Data Engineering; 2012. p. 282–93.
28. Ré C, Dalvi NN, Suciu D. Query evaluation on probabilistic databases. IEEE Data Eng Bull. 2006;29(1):25–31.
29. Ré C, Letchner J, Balazinksa M, Suciu D. Event queries on correlated probabilistic streams. In: Proceedings of the ACM SIGMOD International Conference on Management of Data; 2008. p. 715–28.
30. Ré C, Suciu D. The trichotomy of having queries on a probabilistic database. VLDB J. 2009;18(5):1091–116.
31. Sen P, Deshpande A, Getoor L. PrDB: managing and exploiting rich correlations in probabilistic databases. VLDB J. 2009;18(5):1065–90.
32. Singh S, Mayfield C, Mittal S, Prabhakar S, Hambrusch SE, Shah R. Orion 2.0: native support for uncertain data. In: Proceedings of the ACM SIGMOD International Conference on Management of Data; 2008. p. 1239–42.
33. Souihli A, Senellart P. Optimizing approximations of DNF query lineage in probabilistic XML. In: Proceedings of the 29th International Conference on Data Engineering; 2013. p. 721–32.
34. Suciu D, Olteanu D, Ré C, Koch C. Probabilistic databases. Synthesis lectures on data management. San Rafael: Morgan & Claypool Publishers; 2011.
35. Vazirani VV. Approximation algorithms. Springer; 2001. ISBN:978-3-540-65367-7.
36. Widom J. Trio: a system for integrated management of data, accuracy, and lineage. In: Proceedings of the 2nd Biennial Conference on Innovative Data Systems Research; 2005. p. 262–76.

Query Processor

Anastasia Ailamaki[1] and Ippokratis Pandis[2,3]
[1]Informatique et Communications, Ecole Polytechnique Fédérale de Lausanne, Lausanne, Switzerland
[2]Carnegie Mellon University, Pittsburgh, PA, USA
[3]Amazon Web Services, Seattle, WA, USA

Synonyms

Query execution engine; Query engine; Relational query processor

Definition

The query processor in a database management system receives as input a query request in the form of SQL text, parses it, generates an execution plan, and completes the processing by executing the plan and returning the results to the client.

Key Points

In a relational database system the query processor is the module responsible for executing database queries. The query processor receives as input queries in the form of SQL text, parses and optimizes them, and completes their execution by employing specific data access methods and database operator implementations. The query processor communicates with the storage engine, which reads and writes data from the disk, manages records, controls concurrency, and maintains log files.

Typically, a query processor consists of four sub-components; each of them corresponds to a different stage in the lifecycle of a query. The sub-components are the query parser, the query rewriter, the query optimizer and the query executor [3].

The parser initially reads the SQL text, renames the table references to the *schema.table* template, and validates the structure. As a second step, it uses the database catalog to check the existence of the referenced tables, as well as ensuring that the user who submitted the specific query has the appropriate privileges for the particular operation on the particular data. If everything succeeds, the output from the parser is a data structure understood internally by both the rewriter and the optimizer. This data structure is handed over to the query rewriter.

The query rewriter modifies the query without changing its semantics. The rewriter replaces references to views as references to base tables, simplifies arithmetic expressions, and applies logical transformations to predicates. After the query is parsed and rewritten (and before it is passed on to the optimizer), the system checks a cache of execution plans of recently optimized queries for an execution plan for the specific query in order to avoid the (usually expensive) optimization phase.

The optimizer generates an efficient execution plan for answering a specific query. The decision on which specific access method or database operator implementation will be used relies heavily on the statistics kept by the system and the selectivity estimation. The output of the optimizer is the query execution plan. In the common case, the query plan is an interpretable dataflow directed acyclic graph, where each node is a specific implementation of a database operation. There are some systems where the optimizer generates directly executable machine code, such as in Daytona [2]. The optimizer can choose from many techniques that can speed up the execution of a query. For example, it may decide to generate a plan where multiple threads or processes work in parallel to answer a specific query. Such an execution strategy works well especially if the machine where the system is running contains multiple processors. Query optimization is the responsibility of a fairly sophisticated software module [1, 3, 4].

Although the query plan describes in detail the various operations needed for the execution of the query, it is the query executor that contains the algorithms for accessing base tables and indexes, as well as various database operator execution algorithms. The objective of the query executor is to execute the plan as fast as possible and return the answer to the client. The query optimizer and query executor are tightly coupled together. The query executor determines which algorithms implement the plans generated by the optimizer. For instance, if a query executor supports only Hash and Sort-Merge joins, then the optimizer is restricted to producing plans that use only those two join implementations.

Typically, query executors employ the *iterator* model, a simple and intuitive way to filter data. Each operator is implemented as a subclass of the iterator class, using as interface functions such as *init()*?????, *get_next()*?????, and *close()*?????. An iterator can be used as input to any other iterator, thereby enabling universal handling of iterators or iterator combinations by the system, regardless of the particular function they implement. However, recently researchers argue that the overhead of processing data in a tuple-at-a-time, iterator-based fashion leads to inefficient execution of queries when running on modern hardware and deep memory hierarchies [5].

There are many interpretations of what constitutes a query processor. The query executor is the sub-component that does the "real" job of answering a query. It pulls the data out of the database and employs the various data manipulation algorithms towards them. Thus, a frequent misinterpretation is to consider the query processor and query executor as being synonyms.

Cross-References

▶ Hash Join
▶ Iterator
▶ Parallel Query Processing
▶ Query Optimization
▶ Query Plan
▶ Query Rewriting
▶ Sort-Merge Join

Recommended Reading

1. Graefe G. The cascades framework for query optimization. Q Bull IEEE TC Data Eng. 1995;18(3): 19–29.
2. Greer R. Daytona and the fourth-generation language Cymbal. In: Proceedings of the ACM SIGMOD International Conference on Management of Data; 1999. p. 525–6.
3. Hellerstein JM, Stonebraker M, Hamilton J. Architecture of a database system. Found Trends Databases. 2007;1(2):141–259.
4. Selinger PG, Astrahan M, Chamberlin D, Lorie R, Price T. Access path selection in a relational database management system. In: Proceedings of the ACM SIGMOD International Conference on Management of Data; 1979. p. 23–34.
5. Zukowski M, Boncz P, Nes N, Heman S. MonetDB/X100 – a DBMS in the CPU cache. Q Bull IEEE TC Data Eng. 2005;28(2):17–22.

Q

Query Rewriting

Evaggelia Pitoura
Department of Computer Science and
Engineering, University of Ioannina, Ioannina,
Greece

Synonyms

Query transformations

Definition

Query rewriting is one of the initial phases of query processing. After the original query is parsed and translated into an internal representation, query rewrite transforms it to an equivalent one by carrying out a number of optimizations that are independent of the physical state of the system. Typical transformations include unnesting of subqueries, views expansions, elimination of redundant joins and predicates and various other simplifications.

Key Points

Query rewriting is one of the phases of query processing. It refers to the application of a number of transformations to the original query in order to produce an equivalent optimized one. Such transformations do not depend on the physical state of the system (such as the size of the relations, the system workload, etc). They are usually based on well-defined rules that specify how to transform a query expression into a logically equivalent one.

The goal of query rewriting is threefold: (i) the construction of a standardized starting point for query optimization (standardization), (ii) the elimination of redundancy (simplification), and (iii) the construction of expressions that are improved with respect to evaluation performance (amelioration).

To satisfy this goal, common responsibilities of the query rewriter include:

- View expansion
- Logical rewriting of predicates. For example, improving the match between expressions and the capabilities of index-based access methods
- Various semantic optimizations such as elimination of redundant joins and predicates
- Sub-query flattening

Typically, query rewriting is performed after parsing the original query. It can be thought of as either being the first part of query optimization or as an independent component preceding the query optimizer and the generation of the alternative execution plans. Rewriting is particularly important for complex queries, including queries with many sub-queries or many joins.

Cross-References

▶ Query Optimization
▶ Query Processing

Recommended Reading

1. Hellerstein JM, Stonebraker M, Hamilton J. Architecture of a database system. Found Trends Databases. 2007;1(2):141–259.
2. Jarke M, Koch J. Query optimization in database systems. ACM Comput Surv. 1984;16(2):111–52.
3. Pirahesh H, Hellerstein JM, Hasan W. Extensible/rule based query rewrite optimization in starburst. In: Proceedings of the ACM SIGMOD International Conference on Management of Data; 1992. p. 39–48.

Query Translation

Zhen Zhang
University of Illinois at Urbana-Champaign,
Urbana, IL, USA

Synonyms

Query mapping; Query translation

Definition

Given a source query Q_s over a source schema and a target query template over a target schema, query translation generates a query that is *semantically closest* to the source query and *syntactically valid* to the target schema. The semantically closest is measured by a closeness metrics, typically defined by precision and/or recall of a translated query Versus a source query over a database content. Syntax validness indicates the answerability of a translated query over the target schema. Therefore, the goal of query translation is to find a query that is answerable over the target schema and meanwhile retrieves the closest set of results as the source query would retrieve over a database content.

Historical Background

Query translation is an essential problem in any data integration system and has been studied extensively in the database area. Since a data integration system needs to integrate many different sources, query translation is thus needed to mediate heterogeneous query capabilities presented by those sources. A source typically only accepts and processes queries of certain formats. Such restrictions on acceptable queries form the query capability of the source. For instance, a Web database may only accept queries through their Web query interfaces, a relational database may accept SQL queries, and a legacy system may only accept selection queries over certain attributes through their wrappers.

Foundations

To represent query capabilities, different description languages have been proposed. Vassalos [13] proposes p-Datalog, a datalog-like language for describing query capabilities. Halevy [8] uses capability records to describe, for accepted queries, their binding requirements of input parameters and the attribute name of output parameters. Rajaraman [12] proposes query templates in the for-

mat of parameterized queries to specify binding patterns of acceptable queries. Similarly, Zhang [15] uses predicate templates and form templates to represent query capabilities of Web databases through Web query interfaces. As an example, Fig. 1 shows two query forms with different query capabilities. The source query form S accepts conjunctive queries over four predicates: such as s_1 : [*author*; *contain*; Tom Clancy], s_2 : [*title*; *contain*; red storm], s_3 : [*age*; >; 12], and s_4 : [*price*; $\leq$; 35], i.e., $Q_s = s_1 \wedge s_2 \wedge s_3 \wedge s_4$. A target query form T supports predicate templates on *author, title, subject, ISBN* one at a time with an optional template on predicate *price*.

More specifically, the heterogeneity of query capabilities can be categorized into three levels:

Attribute Heterogeneity

Two sources may query a same concept using different attribute names. For instance, the source schema S in Fig. 1 supports querying the concept of reader's age, while the target schema T does not. Also, S denotes book price using *price range*, while T using *price*.

Predicate Heterogeneity

Two sources may use different predicates for the same concept. For instance, the *price* predicate in T has a different set of value ranges from those of S. As a result, a translated target predicate can only be as "close" to the source predicate as possible. Therefore, a closeness metrics needs to be introduced to set up a goal of translation. For instance, a *minimal subsumption* translation requires that a translated target query subsume the source query with fewest extra answers.

Query Structure Heterogeneity

Two sources may support different sets of valid combinations of predicates. In the above example, the target schema T only supports queries on one of the four attributes *author*, *title*, *subject* and *ISBN* at a time with an optional attribute *price*. Therefore, T cannot query *author* and *title* together, while S can.

The goal of query translation is to generate an appropriate query expressed upon the target schema T. Such a query, as Fig. 1 shows, in

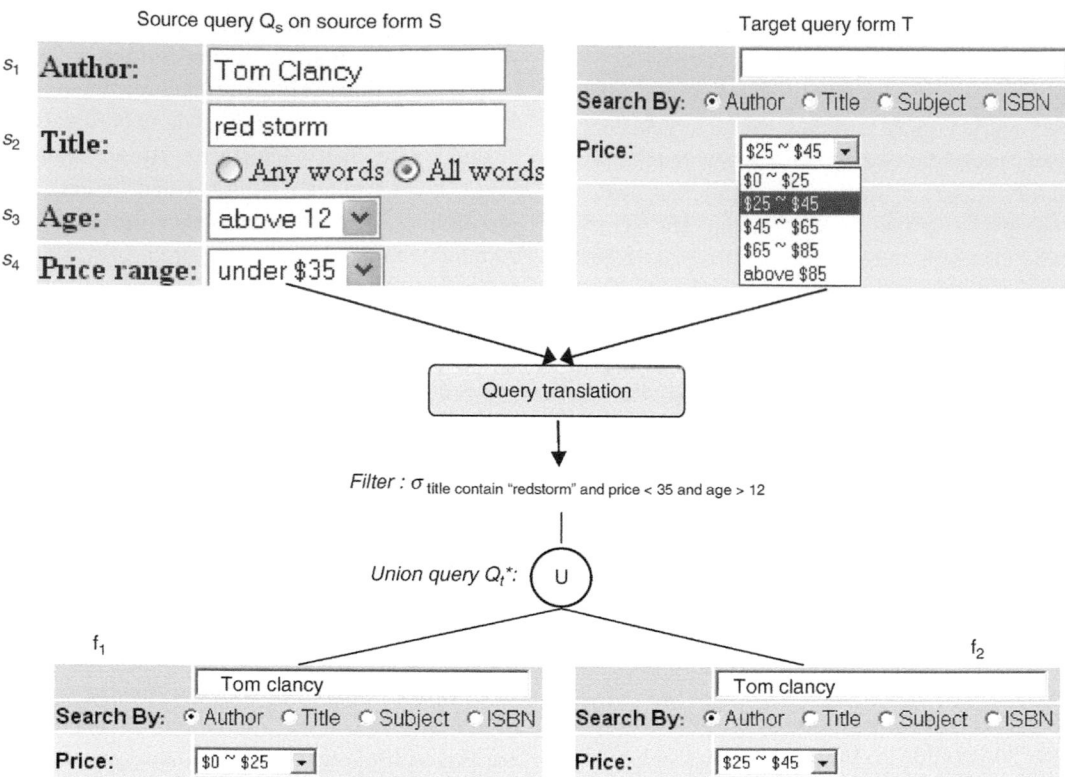

Query Translation, Fig. 1 Form assistant: a translation example

general, consists of two parts: a *union query* Q_t^* which is a union of queries upon the target schema to retrieve relevant answers from a target database, and a *filter* σ which is a selection condition to filter out false positives retrieved by Q_t^*. To minimize the cost of post processing, i.e., filtering, translation aims at finding a union query Q_t^* that is as "close" to the source query Q_s as possible so that it retrieves fewest extra answers. Q_t^* in Fig. 1 is such a query.

To realize the translation, query translation needs to reconcile the heterogeneities at the three levels – attribute, predicate and query. Techniques have been studied extensively for addressing the heterogeneity at each level.

approaches follow two different forms - pairwise matching and holistic matching. The pairwise matching approaches (e.g., [3, 7]) take two schemas as input, and find best attribute matchings between the two. The holistic matching approaches, pursued by, e.g., [5, 6, 14], take a collection of schemas as input and generate a set of matchings over all these schemas. Different approaches suit different application settings. The holistic matching approaches are suitable for applications that would dynamically query a large scale of sources simultaneously, while the pairwise matching approaches are suitable for applications that query a small set of pre-configured sources.

Schema Matching for Attribute Heterogeneity

Schema matching (e.g., see the survey of [11]) focuses on mediating the heterogeneity at the attribute level. Recent schema matching

Predicate Mapping for Predicate Heterogeneity

Predicate mapping focuses on addressing the heterogeneity at the predicate level. Existing solutions can be categorized into two categories:

static predicate mapping mechanism and dynamic predicate mapping mechanism. The static predicate mapping mechanism works with a set of pre-configured data sources. It pre-defines mapping knowledge between a source schema and a target schema. In such scenarios, it is common, e.g., as [1] studies, to use pairwise rules to specify the mapping. Figure 2 gives some example rules that encode the mapping knowledge required for translation in the example of Fig. 1.

In contrast, the dynamic predicate mapping mechanism works with dynamically discovered sources in a domain. It does not encode the mapping knowledge for specific sources, but instead defines common domain-based translation knowledge that handles most sources in a domain. Such a system may use rules to encode domain knowledge or alternatively may use a search-driven mechanism to dynamically search for the best mapping. Such a search-driven mechanism "materializes" the semantics of a query as results over a database. For instance, to realize rule r_3 in Fig. 2 in the search mechanism, the dynamic mechanism projects both the source and target predicates onto an axis of real numbers, and thus compares their semantics based on their coverage. Finding the closest mapping thus naturally becomes a search problem - to search for the ranges expressible in the target form that minimally cover the source predicate.

Query Rewriting for Query Structure Heterogeneity

Capability-based query rewriting focuses on mediating the heterogeneity at query structure level. Most query rewriting works [4, 8, 9, 10, 12] are studied for data integration systems following a *mediator-wrapper* architecture, where a global mediator integrates local data sources through their wrappers. Query rewriting specifically studies the problem of how to mediate a "global" query (from the mediator) into "local" subqueries

(for individual sources) based on their query capabilities. There are two basic approaches for addressing query rewriting in data integration system – *global as view* (GAV) and *local as view* (LAV). In global as view, each relation in the mediated (global) schema is defined as a view over schemas of local data source. Query rewriting in GAV is straightforward - simply replacing relation names in a query (over global schema) with their view definitions will yield a valid rewriting of query (over local schemas). In contrast, in local as view, each relation of a local schema is defined as a view over the mediated global schema. Query rewriting in LAV is thus to find a query plan or query expression which uses only views (i.e., local schemas) to answer queries over global schema. In particular, this problem is often abstracted as *answering queries using views*. For a thorough survey of related techniques for answering query using views, please refer to [4].

Key Applications

Query translation is a key component in any data integration system. The broad range of applications for data integration gives rise to the diverse applications of query translation. Two examples are:

Vertical Integration Systems

A vertical integration system integrates information from multiple pre-configured sources (usually in the same domain of data), and thus requires translating queries from a unified query interface to individual data sources. As data sources are usually pre-configured, such a system usually replies on static query translation mechanism such as [1] to handle translation with pre-defined source knowledge.

Query Translation, Fig. 2
Example mapping rules of source S and target T

r_1 [author; *contain*; \$t] → *emit:* [author; *contain*; \$t]
r_2 [title; *contain*; \$t] → *emit:* [title; *contain*; \$t]
r_3 [price; *under*; \$t] → **if** \$t ≤ 25, *emit:* [price; *between*; 0,25]
       elif \$t ≤ 45, *emit:* [price; *between*; 0,25] ∨ [price; *between*; 25,45]
       ......

Meta Querying Systems

A meta querying system, e.g., [2], integrates dynamically selected sources relevant to user's queries, and on-the-fly translates user's queries to these sources. As sources are dynamically discovered without pre-defined source knowledge, such a system needs a dynamic query translation mechanism such as [15] which handles translation without relying on source-specific knowledge.

Cross-References

▶ Information Integration
▶ Query Rewriting
▶ Rewriting Queries Using Views
▶ Schema Matching
▶ View-based Data Integration

Recommended Reading

1. Chen-Chuan CK, Garcia-Molina H. Approximate query mapping: accounting for translation closeness. VLDB J. 2001;10(2–3):155–81.
2. Chen-Chuan CK, He B, Zhang Z. Toward large scale integration: building a metaquerier over databases on the web. In: Proceedings of the 2nd Biennial Conference on Innovative Data Systems Research; 2005. p. 44–55.
3. Doan A, Domingos P, Halevy AY. Reconciling schemas of disparate data sources: a machine-learning approach. In: Proceedings of the ACM SIGMOD International Conference on Management of Data; 2001. p. 509–20.
4. Halevy AY. Answering queries using views: a survey. VLDB J. 2001;10(4):270–94.
5. He B, Cheng-Chuan CK. Statistical schema matching across web query interfaces. In: Proceedings of the ACM SIGMOD International Conference on Management of Data; 2003. p. 217–28.
6. He B, Cheng-Chuan CK, Han J. Discovering complex matchings across web query interfaces: a correlation mining approach. In: Proceedings of the 10th ACM SIGKDD International Conference on Knowledge Discovery and Data Mining; 2004. p. 148–57.
7. Kang J, Naughton JF. On schema matching with opaque column names and data values. In: Proceedings of the ACM SIGMOD International Conference on Management of Data; 2003. p. 205–16.
8. Levy AY, Rajaraman A, Ordille JJ. Querying heterogeneous information sources using source descrip-

tions. In: Proceedings of the 22th International Conference on Very Large Data Bases; 1996. p. 251–62.
9. Papakonstantinou Y, Gupta A, Garcia-Molina H, Ullman JD. A query translation scheme for rapid implementation of wrappers. In: Proceedings of the 4th International Conference on Deductive and Object-Oriented Databases; 1995. p. 161–86.
10. Papakonstantinou Y, Gupta A, Haas L. Capabilities-based query rewriting in mediator systems. Proceedings of the 4th international conference on Parallel and distributed information systems; 1996. p. 170–81.
11. Rahm R, Bernstein PA. A survey of approaches to automatic schema matching. VLDB J. 2001;10(4): 334–50.
12. Rajaraman A, Sagiv Y, Ullman JD. Answering queries using templates with binding patterns. In: Proceedings of the 14th ACM SIGACT-SIGMOD-SIGART Symposium on Principles of Database Systems; 1995. p. 105–12.
13. Vassalos V, Papakonstantinou Y. Expressive capabilities description languages and query rewriting algorithms. J Logic Program. 2000;43(1):75–122.
14. Wu W, Yu CT, Doan A, Meng W. An interactive clustering-based approach to integrating source query interfaces on the deep web. In: Proceedings of the ACM SIGMOD International Conference on Management of Data; 2004. p. 95–106.
15. Zhang Z, He B, Chen-Chuan Chang K. Light-weight domain-based form assistant: querying web databases on the fly. In: Proceedings of the 31st International Conference on Very Large Data Bases; 2005. p. 97–108.

Quorum Systems

Marta Patiño-Martínez[1,2] and Bettina Kemme[3]
[1]Distributed Systems Lab, Universidad Politecnica de Madrid, Madrid, Spain
[2]ETSI Informáticos, Universidad Politécnica de Madrid (UPM), Madrid, Spain
[3]School of Computer Science, McGill University, Montreal, QC, Canada

Synonyms

Continuous availability; Tolerance to network partitions

Definition

Data replication is a technique to provide high availability and scalability by introducing redundancy. The data remains available as long as some replicas are accessible and, as the load can be distributed across replicas, adding more replicas potentially allows for increased throughput. Challenges arise when the data has to be updated as the replicas must be kept consistent. The most intuitive approach is to always execute all write operations at all replicas. Then, all replicas always have the same state and a read operation can read any replica. The main problem with this Read-One-Write-All (ROWA) approach is that as soon as one replica is no more available write operations cannot be performed anymore. A further problem is that executing all updates always on all replicas makes write operations very expensive.

Quorum systems address both these issues. They allow write operations to succeed if they execute on a subset of replicas called a quorum. Reducing the number of copies to be accessed implies increased availability, tolerance to network partitions, reduced communication costs and the possibility of balancing the load among replicas. In order to preserve data consistency, two write operations on the same data item must overlap at least at one replica, that is, their quorums must intersect. In order to guarantee that read operations read the latest state, reads also have to read a quorum of replicas.

The literature distinguish between approaches where each write operation is only executed on a quorum of replicas, not even sending the requests to other replicas, and approaches where a write operation is sent to all replicas but the operation is considered successful once it has been executed on a quorum of replicas.

The first approach is formally referred to as a *quorum system* that consists of a collection S of subsets $S_i \subseteq \mathcal{N}$ over a set of nodes $\mathcal{N} = \{N_1, \ldots N_n\}$, each of them having replicas of all data items, such that each two subsets $S_i, S_j \in S$ have a non-null intersection, that is, $S_i \cap S_j \neq \emptyset$. Each subset $S_i \subseteq \mathcal{N}$ is called a *write quorum*. A write operation asks all sites of a quorum for permission. If all of them grant permission, the write operation is executed at all nodes in the quorum and the updated replica at each of the nodes is tagged with the same version value, namely a value higher than any current version among the nodes in the quorum. The non-empty intersection property guarantees that only one quorum can make decisions at a time. That is, if two concurrent write operations on the same data item ask for permission to two write quorums, at least one node is member of both write quorums and gives permission to only one of the write operations. Only the members of a single quorum need to be contacted. If this does not succeed because of some failures, the write operation chooses another quorum. In order to guarantee that a read operation reads the latest version, a read operation must be executed on a *read quorum*, a set of sites $R_j \in \mathcal{N}$ such that $\forall S_i \in S, S_i \cap R_j \neq \emptyset$. The read operation compares the versions attached to the replicas it has read, and selects the value corresponding to the highest version. Since a read quorum intersects with all write quorums, it is guaranteed that at least one of the versions read is the latest one. In failure free cases, any write operation is only executed on a quorum of replicas, and thus, the write load is reduced compared to a ROWA approach. The read load, however, is higher.

In the second approach, a write operation is submitted to all replicas but is considered successful once a quorum of replicas has confirmed that the write executed successfully. The most common implementation is based on Paxos protocols. In this case, the overall write load is not reduced compared to ROWA as in a failure free environment all replicas will execute the write. But availability is still better as only a quorum needs to be available for success. Also, the write operation potentially completes earlier because success is achieved once a quorum has applied the update. The updates on the other replicas will still occur but they can happen asynchronously in the background. This entry does not further discuss Paxos based quorum replication but refers to the entry *Paxos for Replication*.

Historical Background

Quorum systems for data replication were proposed concurrently by [18] and [7] in order to provide availability despite individual node failures and network partitions. *Majority* quorum (also known as *quorum consensus*) [18] exploits the concept of majority to guarantee the intersection property. A quorum can be any majority of nodes. *Weighted voting* [7] generalizes majority by assigning votes to each site and defining a quorum to be a majority over the total number of votes in the system. An overview of early quorum systems is given in [4].

Maekawa [12] initiated a research line for increasing the scalability of quorum systems by reducing their size (in the order of $O(\sqrt{n})$). After this seminal work a large number of quorum systems exploiting different schemes have been proposed, mainly in the 1990s. Extensive surveys on quorum systems can be found in [8, 15].

Many of the proposed quorum systems exploit geometrical properties to satisfy the intersection property. *Grid quorums* arrange sites as a grid and then define read and write quorums as rows and columns to enforce their intersection. Grids quorums can be rectangular [5] or can have other shapes, such as triangles [15]. Grid quorums were generalized into hierarchical grid quorums by [9]. Another popular way to arrange sites are trees. Tree quorums were introduced by [1].

Quorums have been extended to tolerate Byzantine (arbitrary) failures requiring the intersection of two quorums to be bigger than one. Such quorums have been termed Byzantine quorums [14].

The properties of quorum systems have been studied extensively, especially availability and scalability. Optimal availability for sites with homogeneous failure probability were studied in [2, 3, 16]. Scalability (also known as load) has been first studied for symmetric update processing (all the sites in the write quorum fully execute the update transaction or operation) in [15], and then under asymmetric update processing (only one site executes the operations, the others only apply the changes) in [8].

Despite this large body of literature on quorum systems, they have barely been used in real systems. Only recently, with the advent of cloud computing and the use of commodity hardware to store and manage extremely large databases, quorums have become attractive. Cassandra [10] offers, among others, an implementation of the traditional majority quorum consensus [18].

Others [6, 13, 17] focus mainly on fault-tolerance and availability by relying on Paxos-based protocols [11], which are used to reach consensus among a set of processes. In these cases each write operation is propagated to all replicas but only a majority has to confirm for the write operation to complete. More details can be found in the entry *Paxos for Replication*.

Scientific Fundamentals

There are two basic flavors to define quorum systems, plain (or exclusive) quorums and read/write quorums. Read/write quorums can reduce the cost of read operations by exploiting the knowledge of whether an operation is a read or write. They are more adequate for data replication in which this knowledge is typically exploited. Plain quorums, or just quorums, are typically defined for mutual exclusion purposes, but can also be used for data replication by using exclusive quorums for both reads and writes.

A set system S is a collection of subsets $S_i \subseteq \mathcal{N}$ of a finite universe $\mathcal{N}$. A quorum system defined over a set of sites $\mathcal{N}$ is a set system S that fulfils the following property: $\forall S_i, S_j \in S$, $S_i \cap S_j \neq \emptyset$. Given a quorum system S, each $S_i \in S$ is a quorum. A read-write quorum system, over the set of sites $\mathcal{N}$, is a pair (R,W) where W is a quorum system (write quorums), and R a set system (read quorums) with the following property: $\forall W_i \in W$, $\forall R_j \in R$, $W_i \cap R_j \neq \emptyset$.

Quorum Types and Their Sizes

For scalability purposes it is beneficial to keep both read and write quorums as small as possible. Furthermore, in order to distribute the load fairly, each node should ideally participate in the same number of quorums. In the following n denotes the number of nodes in $\mathcal{N}$.

Majority. In the majority quorum system [18] read and write quorums must fulfill the following constraints (wq and rq stand for the write and read quorum sizes): $2 \cdot wq > n$ and $rq + wq > n$. The minimum quorum sizes satisfying these constraints are: $2 \cdot wq = n + 1$ and $rq + wq = n + 1$ and therefore, $wq = \lfloor \frac{n}{2} \rfloor + 1$ and $rq = \lceil \frac{n}{2} \rceil = \lfloor \frac{n+1}{2} \rfloor$. The ROWA approach can be seen as an extreme case of majority, in which $rq = 1$ (a single replica is read) and $wq = n$ (all replicas are written). An example of a majority quorum system for three sites $(1, 2, 3)$ with $rq = wq = 2$ is: $\{\{1, 2\}, \{2, 3\}, \{1, 3\}\}$. The majority quorum system is fair since each node has the same probability to be part of a quorum. In weighted majority [7], each site has a non-negative weight (votes). A write quorum consists of nodes such that the sum of their votes is more than half of the total number of votes. A read quorum consists of nodes that have at least half of the total number of votes. Assigning votes allows to adjust to heterogeneous environments where nodes have different processing power and availabilities.

Grids. Another family of quorums is grid quorums. The simplest form of grid quorum is the rectangular grid [5]. A rectangular grid quorum organizes n sites in a grid of r rows and c columns (i.e., $n = r \cdot c$). Figure 1a depicts a 3×4 rectangular grid. A read quorum consists of accessing an element of each column of the grid ($rq = c$). A write quorum consists of a full column and one element from each of the remaining columns ($wq = r + c - 1$). In the quorum system of Fig. 1a $\{1,10,7,12\}$ would be a read quorum and $\{2,6,10,5,3,8\}$ a write quorum. The rectangular grid with the optimal (smallest) quorum size is the square. In this case, $rq = \sqrt{n}$ and $wq = 2 \cdot \sqrt{n} - 1$.

A variation of rectangular grids are *hierarchical grids* [9]. For instance, 16 sites can be configured into a 2-level-grid with 2×2 grids at each level (Fig. 1c).

A hierarchical grid organizes sites into a multi-level hierarchy, such that they reside on the leaves of this hierarchy, while other levels are represented by logical nodes. Each node at level i of the hierarchy (beside leaves) is defined by a rectangular $m \times n$ grid of nodes at level $i + 1$. A quorum consists of the union of a full row and a row cover obtained recursively on the hierarchy. A full row at level i is defined as the set of $(i + 1)$-level nodes all pertaining to a single row of the grid, while a row cover consists in a set of $(i + 1)$-level nodes where each node pertains to a different row of the grid. In Fig. 1c, a read quorum would be $\{1, 6, 7, 8\}$ and a write quorum would be $\{1, 5, 10, 14\}$. For rectangular grids, this results in a read quorum size of $\sqrt{n}$ and a write quorum size of $2 \cdot \sqrt{n} - 1$, i.e., identical to its non-hierarchical counterpart.

Another way to arrange a grid quorum is a triangular grid. Sites in a *triangular grid* [15] are arranged in d rows such that row i ($i \geq 1$) has i elements (Fig. 1b). A write quorum is defined as the union of one complete row and one element from every row below the full row. Therefore, the quorum size is always d. Read quorums are either a write quorum or an element from each row. In Fig. 1b a sample quorum would be $\{2,3,5\}$.

Triangle quorums are not fair since nodes that are higher in the triangle are more likely to be part of a quorum. In contrast, both rectangular and hierarchical grid quorums are fair.

Trees. Another important family of quorum systems is tree quorums. Tree quorums were introduced in [1]. Similar to grid quorums, tree quorums are arranged as a logical structure over the nodes in order to reduce quorum sizes. The nodes are organized into a tree of height h and degree d, i.e., each inner node in the tree has d children. Figure 1d shows a tree with $d = h = 3$. A *tree quorum* $q = \langle l, b \rangle$ over a tree with height h and degree b is a tree of height l and degree b constructed as follows. Read and write quorums have the same structure. The quorum contains the root of the tree and b children of the root. Then, recursively for each selected child, b of its children have to be selected, and so on, until a depth l is reached. In the case all nodes are

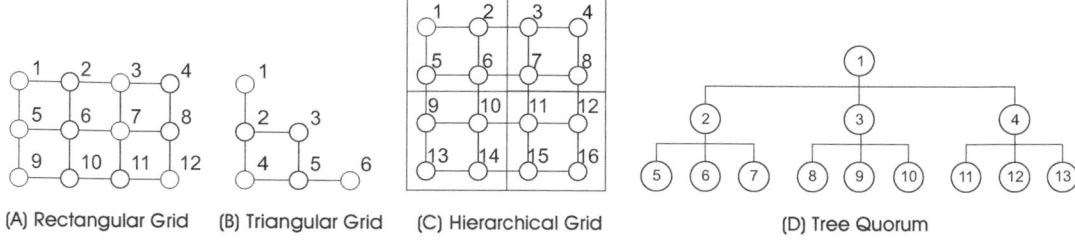

Quorum Systems, Fig. 1 Different quorum systems

accessible the quorum forms a tree of height l and degree b. If some node is inaccessible at depth h' from the root, the node is replaced by b tree quorums of height $l - h'$ starting from the children of the inaccessible node. In order to guarantee intersecting quorums, quorums must overlap both in height and degree. A read quorum $rq = \langle l_r, b_r \rangle$ and write quorum $wq = \langle l_w, b_w \rangle$ overlap if $l_r + l_w > h$ and $b_r + b_w > d$. Two write quorums overlap if $2 \cdot l_w > h$ and $2 \cdot b_w > d$. Tree quorums are generally not fair since nodes that are closer to the root take part in more quorums.

Depending on the values of l and b different tree quorum systems can be defined. The most well-known is the *ReadRoot* in which a write quorum $wq = \langle h, d/2 + 1 \rangle$, i.e., at each level in the tree a majority of nodes needs to be accessed while a read quorum is $rq = \langle 1, (d+1)/2 \rangle$. That is, all reads go to the root. If the root fails, reads go to the next level while writes cannot be performed anymore. More availability is provided by *MajorityTree* where a majority approach is used both for the degree and height parameters. That is, a read quorum is $rq = \langle (h+1)/2, (d+1)/2 \rangle$ and a write quorum is $wq = \langle h/2 + 1, d/2 + 1 \rangle$. This increases the availability of write operations (since write quorums can be built without the root) but also the access costs of read operations. For the tree depicted in Fig. 1d, $\{1, 2, 3\}$ is a read and a write quorum, or, if the root is down, $\{2, 5, 6, 3, 8, 9\}$.

Availability

Availability of quorums has also been studied extensively. Barbara and Garcia-Molina [3] demonstrated that majority is the most available quorum

system for homogeneous failure probabilities (all sites have the same failure probability), if the failure probability p is higher than 0.5. Later, it was shown in [16] that for $p < 0.5$ monarchy (all access goes to a single node) was the most available quorum system. When the failure probabilities are heterogeneous, the most available quorum system is weighted majority. The computation of the optimal weights has been provided for the different cases: all sites with $p > 0.5$ [19] and the general case in which $0 < p < 1$ [2]. An extensive comparative study of the availability of different quorum systems is presented in [16].

Experimental

A comparison of the performance of quorum systems has been performed by [8, 15]. In [8], two different forms of update processing were considered. Using symmetric update processing all the nodes in the write quorum fully execute update transactions (i.e., update operations). In contrast, asymmetric processing lets one node execute the operation while the others only apply the changes. As applying changes typically requires less resources than fully executing the operation, asymmetric processing imposes less load per write operation on the system than symmetric processing. Figure 2 compares the scalability of ROWA, majority and rectangular grids using the analytical model from [8]. The x-axis shows the fraction of writes ($w = 1.0$ means 100 % write operations, $w = 0.0$ means 100 % reads). The y-axis shows the total number of nodes (replicas). The z-axis shows the scalability (how many times the throughput of a single-node system is

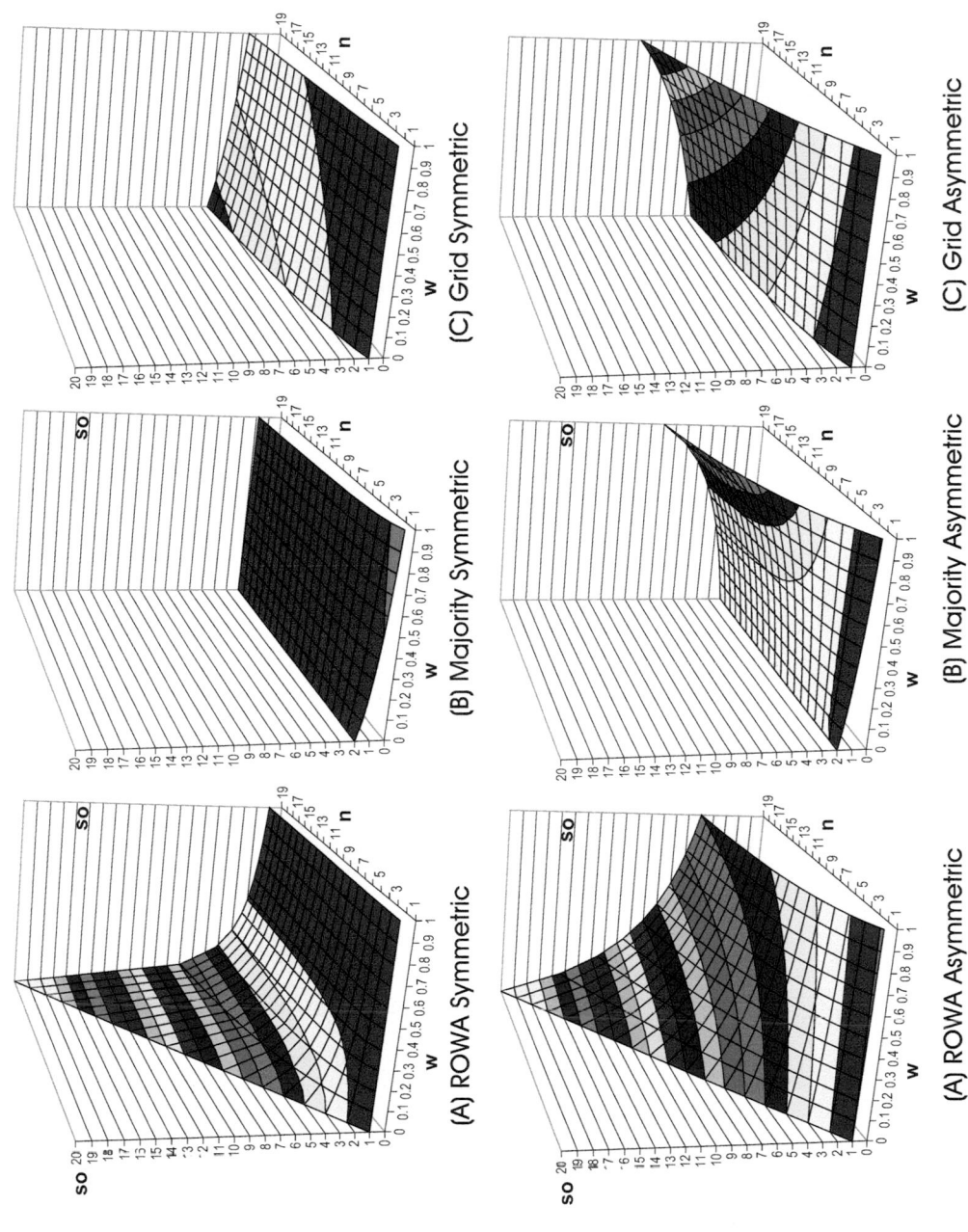

Quorum Systems, Fig. 2 Scalability of majority

multiplied by the replicated system). The first row of figures shows the performance for symmetric update processing. The scalability is generally very poor, especially for majority and grid. The second row shows results for asymmetric update processing when applying writes has 15 % of the costs of fully executing the operation. Scalability improves substantially. Majority and grid quorums can improve over ROWA for write-intensive workloads, but at the cost of performing worse in read-intensive environments.

Key Applications

One of the main applications of quorum systems is data replication. However, they are also used for other decentralized control protocols such as distributed mutual exclusion, distributed consensus, Byzantine replication, and group membership.

Quorum systems are currently offered in large-scale advanced storage systems such as Cassandra [10]. Apart of tradiitional, often majority-based read and write quorums, these systems also often offer variations that provide better performance but less consistency. For instance, Cassandra offers for both write and reads to be executed on one, two, a majority, or all replicas. If, for instance, writes are executed on a majority and reads only on a single node, reads might read stale data. If writes are executed only on a single replica, the update might get lost if the replica fails.

Future Directions

One of the main open issues with data replication based on quorum systems is how to manage efficiently collections of objects such as tables. Quorum systems might work reasonably well for accesses to individual objects. However, the access of collections of objects has not been addressed adequately so far. When accessing collections of objects the system is forced to collect all the instances of the collection from a read quorum to obtain the latest version of every item

and only then, it becomes possible to select the subset of objects from the collection (typically by means of a predicate as in a SELECT statement). This compilation of the full collection from all the sites in the quorum at a single site ruins the performance and scalability of the quorum approach.

Recommended Reading

1. Agrawal D, El Abbadi A. The generalized tree quorum protocol: an efficient approach for managing replicated data. ACM Trans Database Syst. 1992;17(4):689–717.
2. Amir Y, Wool A. Optimal availability quorums systems: theory and practice. Inf Process Lett. 1998;65(5):223–28.
3. Barbara D, Garcia-Molina H. The reliability of vote mechanisms. IEEE Trans Comput Syst. 1987;36(10):1197–1208.
4. Bernstein PA, Hadzilacos V, Goodman N. Concurrency control and recovery in database systems. Reading: Addison Wesley; 1987.
5. Cheung SY, Ahamad M, Ammar MH. The grid protocol: a high performance scheme for maintaining replicated data. In: Proceedings of the 6th International Conference on Data Engineering; 1990. p. 438–45.
6. Corbett JC, Dean J, Epstein M, Fikes A, Frost C, Furman JJ, Ghemawat S, Gubarev A, Heiser C, Hochschild P, Hsieh WC, Kanthak S, Kogan E, Li H, Lloyd A, Melnik S, Mwaura D, Nagle D, Quinlan S, Rao R, Rolig L, Saito Y, Szymaniak M, Taylor C, Wang R, Woodford D. Spanner: Google's globally distributed database. ACM Trans Comput Syst. 2013;31(3):8.
7. Gifford DK. Weighted voting for replicated data. In: Proceedings of the seventh ACM symposium on Operating systems principles; 1979. p. 150–62.
8. Jiménez-Peris R, Patiño-Martínez M, Alonso G, Kemme B. Are quorums an alternative for data replication. ACM Trans Database Syst. 2003;28(3):257–294.
9. Kumar A. Hierarchical quorum consensus: a new algorithm for managing replicated data. IEEE Trans Comput. 1991;40(9):996–1004.
10. Lakshman A, Malik P. Cassandra: a decentralized structured storage system. Op Syst Rev. 2010;44(2):35–40.
11. Lamport L. The part-time parliament. ACM Trans Comput Syst. 1998;16(2):133–69.
12. Maekawa M. A $\sqrt{N}$ algorithm for mutual exclusion in decentralized systems. ACM Trans Comput Syst. 1985;3(2):145–59.
13. Mahmoud HA, Nawab F, Pucher A, Agrawal D, El Abbadi A. Low-latency multi-datacenter databases

using replicated commit. Proc VLDB Endow. 2013;6(9):661–72.

14. Malkhi D, Reiter MK, Wool A. The load and availability of Byzantine quorum systems. SIAM J Comput. 2000;29(6):1889–1906.

15. Naor M, Wool A. The load, capacity, and availability of quorum systems. SIAM J Comput. 1998;27(2):423–47.

16. Peleg D, Wool A. The availability of quorum systems. Inf Comput. 1995;123(2):210–23

17. Rao J, Shekita EJ, Tata S. Using paxos to build a scalable, consistent, and highly available datastore. Proc VLDB Endow. 2011;4(4):243–54.

18. Thomas RH. A majority consensus approach to concurrency control for multiple copy databases. ACM Trans Database Syst. 1979;4(9):180–209.

19. Tong Z, Kain RY. Vote assignments in weighted voting mechanisms. In: IEEE international symposium on reliable distributed systems (SRDS). West Lafayette: IEEE Computer Society Press; 1988.

Q

R

Randomization Methods to Ensure Data Privacy

Ashwin Machanavajjhala and Johannes Gehrke
Cornell University, Ithaca, NY, USA

Synonyms

Perturbation techniques

Definition

Many organizations, e.g., government statistical offices and search engine companies, collect potentially sensitive information regarding individuals either to publish this data for research, or in return for useful services. While some data collection organizations, like the census, are legally required not to breach the privacy of the individuals, other data collection organizations may not be trusted to uphold privacy. Hence, if U denotes the original data containing sensitive information about a set of individuals, then an untrusted data collector or researcher should only have access to an *anonymized* version of the data, U^*, that does not disclose the sensitive information about the individuals. A randomized anonymization algorithm R is said to be a *privacy preserving randomization method* if for every table T, and for every output $T^* = R(T)$, the privacy of all the sensitive information of each individual in the original data is provably guaranteed.

Historical Background

This is a brief survey of state of the art randomization methods. The reader is referred to the classical survey by Adam and Wortman [1] for a more comprehensive description of older randomization techniques.

Existing literature can be classified based on whether the individuals sharing the data trust the data collector or not. The untrusted data collector scenario is discussed first. Randomization methods have been historically used to elicit accurate answers to surveys of sensitive yes/no questions. Respondents may be reluctant to answer such questions truthfully when the data collector (surveyor) is untrusted. Warner's classical paper on *randomized response* [17] proposed a simple technique, where each individual i independently randomized the answer as follows: i answers truthfully with probability p_i, and lies with probability $(1 - p_i)$. Randomized response intuitively ensures privacy since no individual reports the true value. However, Warner did not formalize this intuition.

Subsequent works [7, 2] generalized the above randomized response technique to other domains. Evfimievski et al. [7] studied the problem where individuals share itemsets (e.g., a set of movies rented) with an untrusted server (e.g., an online

© Springer Science+Business Media, LLC, part of Springer Nature 2018
L. Liu, M. T. Özsu (eds.), *Encyclopedia of Database Systems*,
https://doi.org/10.1007/978-1-4614-8265-9

movie rental company) in return for services (e.g., movie recommendations), and they proposed a formal definition of privacy breaches. They invented a provably private randomization technique where users submit independently randomized itemsets to the server. They also proposed data reconstruction algorithms to help the server mine association rules from these randomized itemsets, and experimentally illustrated the accuracy of the reconstruction techniques. The above methods are called *local randomization techniques* since every individual perturbs his/her data locally before sharing it with the data collector.

In the trusted data collector scenarios, while the individuals trust the data collector, they may not trust any third party with whom their data is shared. Hence, randomization methods have been proposed to help privately share the collected data. These techniques can be broadly categorized into *input randomization* and *output randomization* techniques. Input randomization techniques publish a perturbed version of the table; queries are answered using the perturbed data. Output randomization techniques, on the other hand, execute queries on the real data, and return perturbed answers.

One thread of work on input randomization techniques was initiated by Agrawal and Srikant [3]. They proposed an input randomization technique wherein 0-mean random noise is added to the numeric attributes of each individual in the table. The algorithm was experimentally shown to be utility preserving. Nevertheless, Kargupta et al. [10] and Huang et al. [9] showed that adding noise independently to each record in the table does not guarantee privacy.

Yet another thread of work involves publishing synthetic data that has the same properties as the original data, but preserves privacy. Synthetic data generation is a very popular technique in the statistics community, and real applications (like OnTheMap [13]) publish sensitive information using this technique. First proposed by Rubin [16], these techniques build a statistical model using a noise infused version of the data, and then generate synthetic data by randomly sampling from this model. While much research

has focused on deriving variance and confidence estimators from synthetic data, only recently has the privacy of these techniques been formally analyzed [12, 15].

Among output randomization techniques, the SULQ framework proposed by Blum et al. [5] stands out since it has provable guarantees of privacy. Here, numeric query answers are perturbed by adding Laplace noise. Unlike in input perturbation techniques, privacy is guaranteed if and only if the number of queries that are answered is sub-linear in the number of entities in the table. Nevertheless, Blum et al. show that a large number of useful data mining tasks can be performed using this framework. However, exploratory research could be hindered in this framework, since the researchers need to formulate their queries before seeing the data.

Foundations

Local Randomization Techniques

Let U be the original data and let D_U be its (potentially multi-dimensional) domain. Each record $u \in U$ corresponds to the sensitive information of a distinct individual. Each u is independently randomized using a perturbation matrix $\mathbf{A}$. The entry $\mathbf{A}[u, v]$ describes the transition probability $\Pr[u \to v]$ of perturbing a record $u \in D_U$ to a value v in the perturbed domain D_V. This random process maps to a Markov process, and the perturbation matrix $\mathbf{A}$ should therefore satisfy the following properties:

$$A \geq 0, \quad \sum_{v \in D_V} A\,[u, v] = 1 \forall_u \in D_U \quad (1)$$

Privacy

Since each record $u \in U$ is perturbed independent of the rest of the records, it is sufficient to reason about the privacy of each record separately. A *privacy breach* [7] is said to occur if for some predicate φ of an individual's private information, the prior belief in the truth of φ is very different from the posterior belief in its truth after seeing the randomized record $R(u)$. More precisely, an *upward* (ρ_1, ρ_2) *privacy breach* with respect to a predicate φ occurs if

$$\exists u \in U, \exists v \in Dv, \text{s.t.},$$
$$\Pr[\phi(u)] \leq \rho_1 \text{ and } \Pr[\phi(u)|R(u) = v] \geq \rho_2 \quad (2)$$

Similarly, a *downward* (ρ_1, ρ_2) *privacy breach* with respect to a predicate φ occurs if

$$\exists u \in U, \exists v \in Dv, \text{s.t.},$$
$$\Pr[\phi(u)] \leq \rho_1 \text{ and } \Pr[\phi(u)|R(u) = v] \leq \rho_2 \quad (3)$$

A randomization method R is defined to be γ-*amplifying* if

$$\forall v \in D_V, \forall u_1, u_2 \in D_U, \frac{\Pr[u_1 \to v]}{\Pr[u_2 \to v]} \leq \gamma \quad (4)$$

Evfimievski et al. [8] showed that a local randomization method that is γ-amplifying permits a (ρ_1, ρ_2) privacy breach if and only if

$$\frac{\rho_2}{\rho_1} \cdot \frac{1-\rho_1}{1-\rho_2} \geq \gamma \quad (5)$$

Algorithms

Randomized Response: Warner's randomized response technique [17] can be instantiated in this model as follows. Each entry $u \in U$ is a *yes/no* answer given by a distinct individual to a sensitive question Q (e.g., *"Have you ever used illegal drugs?"*). Hence, $D_U = \{0, 1\}$. In order to preserve privacy, each individual flips a coin with bias p, and answers honestly if the coin lands heads and lies otherwise. The perturbation matrix is the 2×2 matrix with $\mathbf{A}[0,0] = \mathbf{A}[1,1] = p$ and $\mathbf{A}[0,1] = \mathbf{A}[1,0] = (1-p)$.

Given n such perturbed answers, the aggregate answer can be estimated as follows. Let π be the fraction of the population for which the true response to Q is *yes*. Then the expected proportion of *yes* responses is

$$\Pr[yes] = \pi \cdot p + (1-\pi) \cdot (1-p) \quad (6)$$

$$\pi = \frac{\Pr[yes] - (1-p)}{2p-1} \quad (7)$$

If m out of the n individuals answered *yes*, then the following $\hat{\pi}$ is an unbiased estimator for π.

$$\hat{\pi} = \frac{\frac{m}{n} - (1-p)}{2p-1} \quad (8)$$

Warner also proposed a second randomization technique wherein, instead of lying with probability $(1-p)$, the respondent answers the question Q honestly with probability p and answers a different innocuous question Q_I with probability $(1-p)$. For instance, with probability p, the respondent truly answers if she had used illegal drugs, and with probability $(1-p)$, the respondent flips a coin with bias α and answers *yes* if the respondent got a head. In this case, the probability that the answer to Q_I is *yes* is α. Hence, if m out of the n individuals answered *yes*, then the following $\overline{\pi}$ is an estimator for π.

$$\Pr[yes] = \pi \cdot p + \alpha \cdot (1-p) \quad (9)$$

$$\pi = \frac{\Pr[yes] - (1-p) \cdot \alpha}{p} \quad (10)$$

$$\overline{\pi} = \frac{\frac{m}{n} - (1-p) \cdot \alpha}{p} \quad (11)$$

Typically, the innocuous question method is better than the former method, since the estimator $\overline{\pi}$ has a smaller variance than $\hat{\pi}$ when the probability of answering the correct question p is not too small.

Itemset Randomization

Itemset randomization is a useful tool that allows users to privately share their (e.g., shopping) histories with a centralized server in return for recommendation services. Let $\mathcal{I}$ represents the set of all items (e.g., products bought by at least one of the users). Then each entry $u \in U$ corresponds to a set of items $\mathcal{I}_u \subseteq \mathcal{I}$. Suppose, for simplicity, all the itemsets $\mathcal{I}_u$ are assumed to have the same number of items, say m. The server wants to learn the *frequent itemsets*, i.e., itemsets

$A \subseteq \mathcal{I}$ whose support $\left(\sup(A) := \dfrac{\left| \{u \in U | A \subseteq \mathcal{I}_u\} \right|}{|U|} \right)$ is $\geq s_{min}$. The following *Select-a-Size* algorithm, with parameters ρ and $\{p[j]\}_j^m = 0$, is used to randomize itemsets.

- Select an integer $j \in [1, m]$, with probability $p[j]$.
- Select a simple random sample of size j of $\mathcal{I}_u$, called $\mathcal{I}'_u$.
- For every $a \in \mathcal{I} - \mathcal{I}_u$, add a to $\mathcal{I}'_u$ with probability ρ.

Evfimievski et al. [8] proved sufficient conditions on the parameters, ρ and $\{p[j]\}_j^m = 0$, in order for this algorithms to be γ-amplifying while simultaneously maximizing the utility of the randomization method (e.g., maximizing the number of original items retained in the randomized itemset, $|\mathcal{I}_u \cap \mathcal{I}'_u|$). Algorithms for recovering the original data from the randomized itemsets and unbiased estimators for the mean and the covariance of these estimates are provided in [8].

Output Perturbation Techniques

Again let U be the original data. Output perturbation techniques execute queries on the real data, and then return perturbed answers. More precisely, if Q is a query on the data U, and R is the perturbation algorithm, $R(Q(U))$ is returned as the answer. In any such technique, there needs to be a limit on the number and the type of queries that can be posed to the database; for instance, answering the same query Q a large number of times discloses the exact answer to Q.

Privacy

Recall that each record in U corresponds to a distinct individual, and that the value of every record in U is independent of the other records. If each query accesses only one record in the table, then output perturbation essentially reduces to local randomization and the privacy breach analysis can be used.

When a query Q accesses multiple records ($U_Q \subseteq U$), however, one cannot reason about the privacy of a single record $u \in U$ (and hence, the privacy of an individual i) in isolation from the rest of the records $U_Q - \{u\}$. Moreover, depending on the amount of prior information known about U_Q, the extent of u's disclosure varies. For instance, if u is, say, the salary of Betty, Q is the query returning the total salary of women in the table, and the adversary knows that Betty is the only woman in the department ($|U_Q| = 1$), then disclosing $Q(U)$ discloses the value of u completely.

Differential privacy [6] can be used to quantify privacy in this case. Answering a query $Q(.)$ using $R(Q(.))$ is ε- *differential private* if for every pair of original tables U_1 and U_2 that differ in the value of a single record u, and for every possible answer A,

$$\left| \log \left(\frac{\Pr[R(Q(U_1)) = A]}{\Pr[R(Q(U_2)) = A]} \right) \right| \leq \in \qquad (12)$$

Intuitively, the above definition preserves privacy as follows. Consider a worst-case adversary who knows the exact values of all the records in $U - \{u\}$ and who is attempting to discover the value of record u_i. Now, if α_S denote the adversary's prior belief that $u \in S(S \subseteq D_U)$, then after seeing the answer $R(Q(U))$, the adversary's posterior belief β_S conditional on his knowledge of the rest of the records in the table is bounded by,

$$\alpha_S / e^\in \leq \beta_S \leq e^\in \times \alpha_S \qquad (13)$$

Algorithms

The SULQ framework, introduced by Blum et al. [6], answers aggregate queries by adding random noise. Let Q be a function $D_U{}^n \to \mathfrak{R}$ The *sensitivity* of query Q, is the smallest number $S(Q)$, such that

$$\begin{aligned} \forall U_1, U_2 \text{ that differ in one record,} \\ |Q(U_1) - Q(U_2)| \leq S(Q) \end{aligned} \qquad (14)$$

Let $\mathrm{Lap}(\lambda)$ denote the *Laplace* distribution which has a density function $h(y) \propto \exp(-|y|/\lambda)$. Suppose a query $Q(U)$ posed to a database U is answered using $Q(U) + Y$, where $Y \sim \mathrm{Lap}(S(Q)/\varepsilon)$.

This perturbation scheme satisfies ε-differential privacy. For every U_1, U_2 that differ in only one record u,

$$\frac{\Pr[Q(U_1) + Y = x]}{\Pr[Q(U_2) + Y = x]} = \frac{h(x - Q(U_1))}{h(x - Q(U_2))}$$

$$= \frac{\exp(-|x - Q(U_1)| \times \in /S(Q))}{\exp(-|x - Q(U_2)| \times \in /S(Q))} \tag{15}$$

$$\leq \exp(\in \times |Q(U_1) - Q(U_2)| /S(Q)) = \exp(\in) \tag{16}$$

This technique is useful when the amount of noise added is small; i.e., when the Q has low sensitivity. Examples of queries with low sensitivity are histograms, linear aggregation queries.

Input Perturbation Techniques

Most research is exploratory; the output perturbation techniques can be inconvenient, since researchers must specify queries before seeing the data. Input perturbation techniques on the other hand publish a perturbed version of the data that the researchers can then directly query. Though the two techniques seem different, technically they are the same; input perturbation uses a single perturbed query over the data, namely the sanitization algorithm. Dwork et al. [6] and Kifer et al. [11] show that in many cases publishing perturbed answers to multiple queries gives more utility than publishing a single perturbed dataset like in input perturbation.

Privacy

A simple technique to perturb the data is to independently add 0-mean noise to each attribute of each record. Let V be a noise matrix, then the perturbed data is $U_p = U + V$. The random noise added to each cell ($v \in V$) is usually either a uniform random variable in $[-\alpha, \alpha]$ or distributed as a Gaussian with 0 mean and a known variance. The privacy of such a scheme is unclear; in fact, Kargupta et al. [10] and Huang et al. [9] showed that the very accurate estimates of the original data can be recovered from such additively perturbed data due to dependencies inherent in U.

For instance, suppose an adversary knows that all the records in U have the same value, say z. Then, additive randomization does not guarantee any privacy; the mean of the perturbed data accurately estimates z if there are enough records in U.

Additive randomization can be broken using Principal Components Analysis (PCA). Suppose the data has m dimensions and is perturbed by adding noise independently to each dimension. Usually, different attributes in the data are correlated; hence, it can projected onto a smaller number, $p < m$, of dimensions. The first principle component (PC) of the data is the direction, e_1, along which the data has the highest variance. The ith PC, e_i, is a vector orthogonal to the first $(i - 1)$ PC's with the largest variance. These vectors are the eigenvectors of the covariance matrix of the data. In correlated data, only the variances along p directions are large. However, for the random data, the variances are the same along all directions. The variances of the perturbed data are roughly the sum of the variances of the original data and the random noise. Hence, by dropping $(m - p)$ directions along which the perturbed data has the least variance, while much information is not lost about the original data, a $(1 - p/m)$ fraction of the noise added is removed; this might lead to privacy breaches.

Algorithms

In order to guarantee privacy, the noise added should be correlated to the data. This is ensured by *synthetic data generation*. Here, a statistical model is generated from a noise infused version of the existing data, and synthetic data points are sampled from this model. Noise is introduced into the synthetic data on two counts; the noise infused prior to building the model, and the noise due to random sampling.

Different algorithms for generating synthetic data can be created by varying the synthetic model that is built using the data. One simple technique is based on Dirichlet resampling [12]. Let H denote the histogram of U, i.e., $H = \{f(v) \mid v \in D_U, f(v) = \text{multiplicity of } v \text{ in } U\}$, and let R denote the noise histogram. Then the statistical model is $D(H + R)$, where D denotes the Dirichlet distribution. Synthetic data is gener-

R

ated as follows. Draw a vector of probabilities, X, from $D(H + R)$, and generate m points according to the probabilities in X. The above process is mathematically equivalent to the following re-sampling technique. Consider an urn with balls marked with values $v \in D_U$ such that the number of balls marked with v equals the sum of the frequency of v in U and the frequency of v in the noise histogram. Synthetic data is generated in m sampling steps as follows. In each sampling step, a ball, say marked v, is drawn at random and two balls marked v are added back to the urn. In this step, the synthetic data point is v.

Machanavajjhala et al. [12] characterized the privacy guaranteed by this algorithm in terms of noise distribution. Specifically, they showed that in order to guarantee ε-differential privacy, the frequency of every $v \in D_U$ in the noise histogram should be at least $m/(e^\varepsilon - 1)$. For large m and small ε the noise required for privacy overwhelms all of the signal in the data and renders the synthetic data completely useless. Such large requirements of noise is due to the following worst case requirement of differential privacy. Consider a scenario where an adversary knows that U contains exactly one record u_i that takes either the value v_1 or v_2. Now suppose that in the output sample, every record takes the value v_1. If m is large, then the adversary's belief that $r_u = v_1$ is close to 1. In order to guard against such adversaries, differential privacy requires a large amount of noise. However, the probability that such synthetic data is output is negligibly small. This can be remedied using a weaker (ε, δ)-probabilistic differential privacy definition, where an algorithm is private if it satisfies ε-differential privacy for all outputs that are generated with a cumulative probability of at least$(1 - \delta)$. Under this weaker definition, the Dirichlet resampling technique is private with much smaller noise requirements.

Barak et al. [4] propose a solution to publish marginals of a contingency table (i.e., a histogram) using the SULQ framework. Publishing a set of noise infused marginals is not satisfactory; such marginals may not be consistent, i.e., there may not exist a contingency table that satisfies all these marginal contingency tables. Barak et al. solve this problem by adding noise to a small number of Fourier coefficients; any set of Fourier coefficients correspond to a (fractional and possibly negative) contingency table. They show that only a "small" number of Fourier coefficients are required to generate the required marginals, and hence only a small amount of noise (proportional to the size of the marginal domain) is required. The authors employ a linear program solution (in time polynomial in the size of multidimensional domain) to generate the final non-negative integral set of noise infused marginals.

Rastogi et al. [14] propose the $\alpha\beta$ algorithm for publishing itemsets. It is similar to the select-a-size randomization operator. Given an itemset I that is a subset of the domain of all items D, the $\alpha\beta$ algorithm creates a randomized itemset V by retaining items in I with probability $\alpha + \beta$ and adding items in $D - I$ with probability β. This algorithm satisfies a variant of ε-differential privacy. Moreover, the authors show that for queries $Q : 2^D \to \mathbb{R}$, $Q(I)$ can be estimated as follows:

$$\widehat{Q}(I) = (Q(V) - \beta Q(D))/\alpha \qquad (17)$$

where $Q(V)$ and $Q(D)$ are the answers to the query Q on the randomized itemset V and the full domain D, respectively. $\widehat{Q}(I)$ is shown to provably approximate $Q(I)$ with high probability.

Key Applications

Privacy preserving techniques have been used in Census applications and various web-applications. The Dirichlet resampling based synthetic data generation technique is used in the web-based OnTheMap Census application that plots worker commute patterns on the U.S. map to study workforce indicators [13]. Warner's randomized response has been used in eliciting responses to sensitive survey questions.

Future Directions

One common property of all provably private randomization methods is that the probability of perturbing a value $u \in D_U$ to every value $v \in D_V$ (the perturbed domain, which is usually the same as D_U) should be positive. As shown by Machanavajjhala et al. [12], this causes a problem when the size of the domain is very large. For instance, in the On The Map [13] application, the domain D_U is the set of census blocks on the U.S. map and there are about 8 million such blocks. However, given a destination, there is only a few hundred workers commuting to it. Hence, even if a small amount of noise is added to each block on the map, spurious commute patterns will arise in the synthetic data. All of the techniques discussed in this article should be revisited in the context of real scenarios with sparse data.

Cross-References

▶ Association Rule Mining on Streams
▶ Principal Component Analysis
▶ Statistical Disclosure Limitation for Data Access
▶ Synthetic Microdata

Recommended Reading

1. Adam NR, Wortmann JC. Security-control methods for statistical databases: a comparative study. ACM Comput Surv. 1989;21(4):515–56.
2. Agrawal R, Srikant R. Privacy preserving data mining. In: Proceedings of the ACM SIGMOD International Conference on Management of Data; 2000. p. 439–50.
3. Agrawal S, Haritsa JR. A framework for high-accuracy privacy-preserving mining. In: Proceedings of the 21st International Conference on Data Engineering; 2005. p. 193-204.
4. Barak B, Chaudhuri K, Dwork C, Kale S, McSherry F, Talwar K. Privacy, accuracy and consistency too: a holistic solution to contingency table release. In: Proceedings of the 26th ACM SIGACT-SIGMOD-SIGART Symposium on Principles of Database Systems; 2007.
5. Blum A, Dwork C, McSherry F, Nissim K. Practical privacy: the SuLQ framework. In: Proceedings of the 24th ACM SIGACT-SIGMOD-SIGART Symposium on Principles of Database Systems; 2005. p. 128–38.
6. Dwork C, McSherry F, Nissim K, Smith A. Calibrating noise to sensitivity in private data analysis. In: Proceedings of the 3rd Theory of Cryptography Conference; 2006. p. 265–84.
7. Evfimievski A, Gehrke J, Srikant R. Limiting privacy breaches in privacy preserving data mining. In: Proceedings of the 22nd ACM SIGACT-SIGMOD-SIGART Symposium on Principles of Database Systems; 2003. p. 211–22.
8. Evfimievsky A, Srikant R, Gehrke J, Agrawal R. Privacy preserving data mining of association rules. In: Proceedings of the 8th ACM SIGKDD International Conference on Knowledge Discovery and Data Mining; 2002. p. 217–28.
9. Huang Z, Du W, Chen B. Deriving private information from randomized data. In: Proceedings of the 23th ACM SIGMOD Conference on Management of Data; 2004.
10. Kargupta H, Datta S, Wang Q, Sivakumar K. On the privacy preserving properties of random data perturbation techniques. In: Proceedings of the 2003 IEEE International Conference on Data Mining; 2003. p. 99–106.
11. Kifer D, Gehrke J. Injecting utility into anonymized datasets. In: Proceedings of the ACM SIGMOD International Conference on Management of Data; 2006.
12. Machanavajjhala A, Kifer D, Abowd J, Gehrke J, Vihuber L. Privacy: from theory to practice on the map. In: Proceedings of the 24th International Conference on Data Engineering; 2008.
13. On The Map (Version 2) http://lehdmap2.dsd.census.gov/.
14. Rastogi V, Suciu D, Hong S. The boundary between privacy and utility in data publishing. Tech. rep., University of Washington; 2007.
15. Reiter J. Estimating risks of identification disclosure for microdata. J Am Stat Assoc. 2005;100(472):1103–13.
16. Rubin DB. Discussion statistical disclosure limitation. J Off Stat. 1993;9(2):461–8.
17. Warner SL. Randomized response: a survey technique for eliminating evasive answer bias. J Am Stat Assoc. 1965;60(309):63–9.

R

Range Query

Mirella M. Moro
Departamento de Ciencia da Computaçao,
Universidade Federal de Minas Gerais – UFMG,
Belo Horizonte, MG, Brazil

Synonyms

Range search; Range selection

Definition

Consider a relation R with some numeric attribute A taking values over an (ordered) domain D. A *range query* retrieves all tuples in R whose attribute A has values in the interval *[low, high]*. That is, *low$\leq$R.A$\leq$high*. The range interval may be closed as aforementioned, open (e.g., *low<R.A<high*), or half-open in either side (e.g., *low<R.A$\leq$high*). A range query can also be one-sided (e.g., low $\leq$ R.A retrieves all tuples with R.A value greater or equal to *low*). When *low=high*, the range query becomes an *equality* (or *membership*) query.

Key Points

Range queries involve *numeric* (or numerical) attributes. These are attributes whose domain is totally ordered and thus a query interval (e.g., *[low, high]*) can be formed. In contrast, attributes whose domain is not naturally ordered are called *categorical* (or *nominal*). Range queries correspond to selections and are thus amenable to indexing. The standard access method for a range query on some attribute A is a B+-tree built on the values of attribute A. Since the B+-tree maintains the order of the indexed values in its leaf pages, a range query is implemented as a search for the leaf page with the lower value of the range interval, followed by the accessing of sibling pages until a page that contains the higher value of the range interval is reached. So far, the discussion has covered the one-dimensional range search. Multidimensional range queries are also important. A typical example is the *spatial range query* that retrieves all objects which fall within (or intersect, overlap, etc.) a region (a rectangle specified by ranges in each dimension). Such multidimensional range queries are typically indexed by R-trees.

Cross-References

▸ B+-Tree
▸ R-Tree (and Family)

Rank-Aware Query Processing

Ihab F. Ilyas
Cheriton School of Computer Science,
University of Waterloo, Waterloo, ON, Canada

Synonyms

Top-k Query Processing

Definition

Rank-aware query processing refers to the efficient processing of a *top-k query* taking into account the ranking requirements on output results. A naïve way to process a *top-k query* is to calculate the full set of results and then sort them based on the ranking function; the top-k results are presented as the final query answers. Such a naïve *materialize-then-sort* scheme can be prohibitively expensive. Integrating top-k queries in SQL query engines requires addressing the challenge of making an RDBMS *rank-aware*. This requires introducing new constructs in the whole system including the data model, algebra, query operators, and query optimization techniques.

Historical Background

The need for rank-aware query processing arose since the introduction of *top-k queries* with *score aggregation* and *rank joins* to the database community. Fagin et al. [1] first introduced the problem of ranking a database of objects, given several rankings of the objects, by aggregating their scores from all the different rankings. Efficient algorithms such as TA and NRA introduced to answer *score aggregation* queries were incorporated in some prototype systems such as the IBM Garlic Middleware [9]. Ilyas et al. [4, 5] introduced pipelined query operators, known as *rank-join operators*, to compute the output of joining several ranked relational tables.

Scientific Fundamentals

The main theme of the techniques addressing rank-aware query processing is their tight coupling with the RDBMS query engine. This tight coupling has been realized through multiple approaches. Some approaches focus on the design of efficient rank-join query operators [2, 4, 5]. Other approaches introduce an algebra to formalize the interaction between ranking and other relational operators (e.g., joins and selections) [6]. A third category addresses modifying query optimizers, for example, plan enumeration and cost estimation procedures, to recognize the ranking requirements of top-k queries [7]. All of these provide significant support for efficient execution of *top-k queries*.

Rank-Relational Algebra

Rank-relational query processing introduced by Li et al. [6] supports ranking as a first-class database construct. This includes viewing ranking as another logical property of the data and subsequently extending the relational algebra to what is called *rank-relational algebra*. Consider the following example to illustrate rank-relational algebra.

Example 1 Consider a user who wants to plan her trip to Chicago. She wants to stay in a hotel, have lunch in an Italian restaurant (condition c_1 : $r.cuisine = Italian$), and walk to a museum after lunch; the hotel and the restaurant together should cost less than $100 ($c_2$: $h.price + r.price < 100$); the museum and the restaurant should be in the same area (c_3 : $r.area = m.area$). Further, to rank the qualified results, she specifies several ranking criteria, such as low hotel price (p_1 : $cheap(h.price)$), for close distance between the hotel and the restaurant (p_2 : $close(h.addr, r.addr)$) and for matching her interests with the museum's collections (p_3 :

$related(m.collection,$ 'dinosaur')). Here p_1 and p_2 are *rank-selection* predicates, and p_2 is a *rank-join* predicate. These ranking predicates return numeric scores, and the overall scoring function sums up their values. The query is shown below in PostgreSQL syntax.

```
Q: SELECT *
      FROM Hotel h, Restaurant r,
Museum m
      WHERE c₁ AND c₂ AND c₃
      ORDER BY p₁ + p₂ + p₃ LIMIT k
```

The ranking query in Example 1 can be expressed in rank-relational algebra as $\pi_* \lambda_k \tau_{F(p_1,p_2,p_3)} \sigma_{B(c_1,c_2,c_3)}(h \times r \times m)$ which shows two logical properties of the ranking queries: (1) **Filtering**, performed by the boolean function B and by the top-k restriction λ_k, induces the *membership* logical property to the ranking queries and (2) **Ranking**, performed by the ranking function F, induces the *order* logical property. Boolean filtering is modeled as a first-class construct in relational algebra, and rank-relational algebra does the same to ranking. This enables the notions of *splitting* and *interleaving* which are traditionally associated with processing Boolean predicates to also be applied with ranking operators. Splitting helps evaluate ranking in stages, predicate by predicate instead of the monolithic way in *materialize-then-sort* approach discussed at the beginning. Interleaving enables ranking to be interleaved with other operators, instead of doing it toward the end. A ranking algebra with rank-aware operators and algebraic equivalence laws is proposed [6] to enable query optimizers to transform the canonical form of a ranking query into efficient query plans.

Rank-Relational Operators

To begin, *rank relation* needs to be defined. Given a scoring function $F(p_1, \ldots, p_n)$, we need to know how the ranking predicates p_i's are processed and how the results of intermediate relations must be ranked. At any given point, not all predicates are processed. Hence, if P denotes the processed predicates, then the results of a relation can be ranked based on the upper bound

Rank-Aware Query
Processing, Fig. 1
Rank-relational operators

Rank: μ, with a ranking predicate p
- $t \in \mu_p(R_{\mathcal{P}})$ iff $t \in R_{\mathcal{P}}$
- $t_1 <_{\mu_p(R_{\mathcal{P}})} t_2$ iff $\overline{\mathcal{F}}_{\mathcal{P} \cup \{p\}}[t_1] < \overline{\mathcal{F}}_{\mathcal{P} \cup \{p\}}[t_2]$

Selection: σ, with a boolean condition c
- $t \in \sigma_c(R_{\mathcal{P}})$ iff $t \in R_{\mathcal{P}}$ and t satisfies c
- $t_1 <_{\sigma_c(R_{\mathcal{P}})} t_2$ iff $t_1 <_{R_{\mathcal{P}}} t_2$, i.e., $\overline{\mathcal{F}}_{\mathcal{P}}[t_1] < \overline{\mathcal{F}}_{\mathcal{P}}[t_2]$

Union: $\cup$
- $t \in R_{\mathcal{P}_1} \cup S_{\mathcal{P}_2}$ iff $t \in R_{\mathcal{P}_1}$ or $t \in S_{\mathcal{P}_2}$
- $t_1 <_{R_{\mathcal{P}_1} \cup S_{\mathcal{P}_2}} t_2$ iff $\overline{\mathcal{F}}_{\mathcal{P}_1 \cup \mathcal{P}_2}[t_1] < \overline{\mathcal{F}}_{\mathcal{P}_1 \cup \mathcal{P}_2}[t_2]$

Intersection: $\cap$
- $t \in R_{\mathcal{P}_1} \cap S_{\mathcal{P}_2}$ iff $t \in R_{\mathcal{P}_1}$ and $t \in S_{\mathcal{P}_2}$
- $t_1 <_{R_{\mathcal{P}_1} \cap S_{\mathcal{P}_2}} t_2$ iff $\overline{\mathcal{F}}_{\mathcal{P}_1 \cup \mathcal{P}_2}[t_1] < \overline{\mathcal{F}}_{\mathcal{P}_1 \cup \mathcal{P}_2}[t_2]$

Difference: $-$
- $t \in R_{\mathcal{P}_1} - S_{\mathcal{P}_2}$ iff $t \in R_{\mathcal{P}_1}$ and $t \notin S_{\mathcal{P}_2}$
- $t_1 <_{R_{\mathcal{P}_1} - S_{\mathcal{P}_2}} t_2$ iff $t_1 <_{R_{\mathcal{P}_1}} t_2$, i.e., $\overline{\mathcal{F}}_{\mathcal{P}_1}[t_1] < \overline{\mathcal{F}}_{\mathcal{P}_1}[t_2]$

Join: $\bowtie$, with a join condition c
- $t \in R_{\mathcal{P}_1} \bowtie_c S_{\mathcal{P}_2}$ iff $t \in R_{\mathcal{P}_1} \times S_{\mathcal{P}_2}$ and satisfies c
- $t_1 <_{R_{\mathcal{P}_1} \bowtie_c S_{\mathcal{P}_2}} t_2$ iff $\overline{\mathcal{F}}_{\mathcal{P}_1 \cup \mathcal{P}_2}[t_1] < \overline{\mathcal{F}}_{\mathcal{P}_1 \cup \mathcal{P}_2}[t_2]$

---

function $\bar{F}_{\mathcal{P}}$ given by :

$$\bar{F}_{\mathcal{P}}(p_1, \ldots, p_n)[t]$$
$$= F \left(\begin{array}{l} p_i = p_i[t] \text{ if } p_i \in \mathcal{P} \\ p_i = 1 \qquad\qquad \text{otherwise} \end{array} \right) \quad (1)$$

The rank relation $R_{\mathcal{P}}$ of a relation R w.r.t a monotonic scoring function F has its tuples augmented by a score computed using the above function and the order induced by a ranking w.r.t those scores. To accommodate ranking, a new rank operator μ is defined. The complete set of operators defined in the rank-relational algebra is given in Fig. 1. This list also extends the semantics of existing operators with rank awareness. Here, selection (σ) and projection (π) process tuples in their input rank relation similar to their original semantics, while maintaining the order. Binary operators such as $\cup, \cap, \bowtie$ perform their normal Boolean operations and at the same time output tuples in the aggregate order of their

operands. And, \ orders tuples in the order of its outer input operand.

Ranking Algebraic Laws

The algebraic laws governing the operators introduced in previous section are given in Fig. 2. Specifically, Proposition 1 allows us to *split* a scoring function with several predicates $(p_1, \ldots, p_n)$ into a series of rank operations $(\mu_1, \ldots, \mu_n)$. And, Propositions 4 and 5 together specify that rank operations can swap with other operators, thus achieving the interleaving requirement.

RankSQL Framework [7] fully integrates the rank-aware operators into a relational query engine. Based on the ranking algebra, the framework generates rank-aware query plans using two different ways: (1) by extending System R dynamic programming algorithm in both enumeration and pruning and estimating input cardinality of rank-join operators using a probabilistic model and (2) by treating ranking predicates as

**Rank-Aware Query
Processing, Fig. 2**
Algebraic laws

Proposition 1: Splitting law for μ

- $R_{\{p_1,p_2,\ldots,p_n\}} \equiv \mu_{p_1}(\mu_{p_2}(\ldots(\mu_{p_n}(R))\ldots))$

Proposition 2: Commutative law for binary operator

- $R_{\mathcal{P}_1} \Theta S_{\mathcal{P}_2} \equiv S_{\mathcal{P}_2} \Theta R_{\mathcal{P}_1}, \forall \Theta \in \{\cap, \cup, \bowtie_c\}$

Proposition 3: Associative law

- $(R_{\mathcal{P}_1} \Theta S_{\mathcal{P}_2}) \Theta T_{\mathcal{P}_3} \equiv R_{\mathcal{P}_1} \Theta (S_{\mathcal{P}_2} \Theta T_{\mathcal{P}_3}), \forall \Theta \in \{\cap, \cup, \bowtie_c{}^a\}$

Proposition 4: Commutative laws for μ

- $\mu_{p_1}(\mu_{p_2}(R_{\mathcal{P}})) \equiv \mu_{p_2}(\mu_{p_1}(R_{\mathcal{P}}))$
- $\sigma_c(\mu_p(R_{\mathcal{P}})) \equiv \mu_p(\sigma_c(R_{\mathcal{P}}))$

Proposition 5: Pushing μ over binary operators

- $\mu_p(R_{\mathcal{P}_1} \bowtie_c S_{\mathcal{P}_2})$
    $\equiv \mu_p(R_{\mathcal{P}_1}) \bowtie_c S_{\mathcal{P}_2}, \text{ if only } R \text{ has attributes in } p$
    $\equiv \mu_p(R_{\mathcal{P}_1}) \bowtie_c \mu_p(S_{\mathcal{P}_2}), \text{ if both } R \text{ and } S \text{ have}$
- $\mu_p(R_{\mathcal{P}_1} \cup S_{\mathcal{P}_2}) \equiv \mu_p(R_{\mathcal{P}_1}) \cup \mu_p(S_{\mathcal{P}_2}) \equiv \mu_p(R_{\mathcal{P}_1}) \cup S_{\mathcal{P}_2}$
- $\mu_p(R_{\mathcal{P}_1} \cap S_{\mathcal{P}_2}) \equiv \mu_p(R_{\mathcal{P}_1}) \cap \mu_p(S_{\mathcal{P}_2}) \equiv \mu_p(R_{\mathcal{P}_1}) \cap S_{\mathcal{P}_2}$
- $\mu_p(R_{\mathcal{P}_1} - S_{\mathcal{P}_2}) \equiv \mu_p(R_{\mathcal{P}_1}) - S_{\mathcal{P}_2} \equiv \mu_p(R_{\mathcal{P}_1}) - \mu_p(S_{\mathcal{P}_2})$

Proposition 6: Multiple-scan of μ

- $\mu_{p_1}(\mu_{p_2}(R_\phi)) \equiv \mu_{p_1}(R_\phi) \cap_r \mu_{p_2}(R_\phi)$

**Rank-Aware Query
Processing, Fig. 2**
Algebraic laws

another dimension in plan enumeration and using sampling-based statistics [6].

Physical Operators for Ranking

Rank-join algorithms are implemented as pipelined query operators, enabling greater flexibility to shuffle the evaluation plan operators, generate candidate execution plans, and seek the best plan. The HRJN and HRJN* [5], J* [8], and an optimal class of operators called PBRJ [3, 10] are some known implementations of *rank-join operators*.

Key Applications

The key applications of rank-aware query processing involve efficient computation of rank-join queries, which in turn is normally used for multi-criteria decision analysis, retrieving and ranking database objects such as hotels and house rentals based on user preferences. An application of rank join involves ranking combined packages involv-ing hotels and flights or ranking house rentals with schools.

Cross-References

▶ Score Aggregation
▶ Top-K Queries

Recommended Reading

1. Fagin R. Combining fuzzy information: an overview. ACM SIGMOD Rec. 2002;31(2):109–18.
2. Fagin R, Lotem A, Naor M. Optimal aggregation algorithms for middleware. J Comput Syst Sci. 2003;66(4):614–56.
3. Finger J, Polyzotis N. Robust and efficient algorithms for rank join evaluation. In: Proceedings of the 2009 ACM SIGMOD International Conference on Management of Data; 2009. p. 415–28.
4. Ilyas IF, Aref WG, Elmagarmid AK. Joining ranked inputs in practice. In: Proceedings of the 28th International Conference on Very Large Data Bases; 2002. p. 950–61.
5. Ilyas IF, Aref WG, Elmagarmid AK. Supporting top-k join queries in relational databases. VLDB J – Int J Very Large Data Bases. 2004;13(3):207–21.

R

6. Li C, Chang KC-C, Ilyas IF, Song S. Ranksql: query algebra and optimization for relational top-k queries. In: Proceedings of the 2005 ACM SIGMOD International Conference on Management of Data; 2005. p. 131–42.
7. Li C, Soliman MA, Chang KC-C, Ilyas IF. Ranksql: supporting ranking queries in relational database management systems. In: Proceedings of the 31st International Conference on Very Large Data Bases; 2005. p. 1342–45.
8. Natsev A, chi Chang Y, Smith JR, Li C-S, Vitter JS. Supporting incremental join queries on ranked inputs. In: Proceedings of the 27th International Conference on Very Large Data Bases; 2001.
9. Roth MT, Arya M, Haas LM, Carey MJ, Cody W, Fagin R, Schwarz PM, Thomas J, Wimmers EL. The garlic project. Sigmod Rec. 1996;25(2):557.
10. Schnaitter K, Polyzotis N. Evaluating rank joins with optimal cost. In: Proceedings of the 27th ACM SIGACT-SIGMOD-SIGART Symposium on Principles of Database Systems; 2008. p. 43–52.

Ranked XML Processing

Amélie Marian[1], Ralf Schenkel[2], and
Martin Theobald[3,4]
[1]Computer Science Department, Rutgers
University, New Brunswick, NJ, USA
[2]Campus II Department IV – Computer Science,
Professorship for databases and information
systems, University of Trier, Trier, Germany
[3]Institute of Databases and Information Systems
(DBIS), Ulm University, Ulm, Germany
[4]Stanford University, Stanford, CA, USA

Synonyms

Aggregation and threshold algorithms for XML;
Approximate XML querying; Top-k XML query
processing

Definition

When querying collections of XML documents
with heterogeneous or complex schemas, existing
query languages like XPath or XQuery with their
exact-match semantics are often not the perfect
choice. Such exact querying languages will typically miss many relevant results that do not
conform to the strict formulation of the query.

Top-k query processing for XML data, which
focuses on finding the k top-ranked XML elements to an XPath (or XQuery) query with full-
text search predicates, is a particularly appropriate query model for querying semi-structured
data when the actual content or structure of the
underlying data is not fully known. Challenges
in processing top-k queries over XML data include scoring individual answers based on how
closely they match the query, supporting IR-style
vague search over both content and structure,
and ranking the k best answers in an efficient
manner.

Historical Background

Non-schematic XML data that comes from
many different sources and inevitably exhibits
heterogeneous structure and annotations in the
form of hierarchical tags and deeply nested XML
elements often cannot be adequately searched
using pure database-style query languages like
XPath or XQuery. Typically, queries either return
too many or too few results using only Boolean
search predicates. Rather, the ranked-retrieval
paradigm needs to be called for, with relaxable
search conditions, various forms of similarity
predicates on tags and contents, and quantitative
relevance scoring. The information retrieval (IR)
community has historically focused on scoring
documents based on how closely they match
a user's keyword query. Intense research on
applying IR techniques to XML data has started
in the early 2000's and has meanwhile gained
considerable attention. Recent IR extensions
to XML query languages such as XPath 1.0
Full-Text or the NEXI query language used
in the INEX benchmark series [5] reflect this
emerging interest in IR-style ranked retrieval
over semi-structured data. So far, various work
on scoring answers to XML queries have focused
on adapting IR-style scoring techniques from
unstructured text to the semi-structured world,
see, e.g., [2, 5, 11]. On the IR side, some foray

into adding structure to standard IR search has taken place before the advent of XML [13]. The popularity of XML provides an opportunity to combine efforts led separately by both the DB and IR communities and provide robust techniques to query semi-structured data.

Threshold Algorithms

The method of choice for efficient processing of top-k similarity queries is the family of threshold algorithms (TA), most notably presented by Fagin et al. [8] and originally developed for multi-media databases and structured records stored in relational database systems (RDBMS). These algorithms rely on making dynamic choices for scheduling index lookups during query execution in order to prune low-scoring candidate items as early as possible. They typically scan pre-computed index lists for text terms or attribute values of structured records in descending order of local (i.e., per-term) scores and aggregate these scores for the same data item into a global score, using a monotonic score aggregation function such as (weighted) summation. Based on clever bookkeeping of score intervals and thresholds for the top-k matches, these index scans can often terminate early, namely as soon as the final top-k results can be safely determined, and thus the algorithm often only has to scan short prefixes of the inverted lists. In contrast to the heuristics adopted by many Web search engines, these threshold algorithms compute exact results and are provably optimal in terms of asymptotic costs.

XML and IR

Efficient evaluation and ranking of XML path conditions is a very fruitful research area. Solutions include various forms of structural joins, multi-predicate merge joins, the staircase join based on index structures with pre- and postorder encodings of elements within document trees [10], and holistic twig joins [6]. The latter, also known as path stack algorithm, is probably the most efficient method for twig queries using a combination of sequential scans over index lists stored on disk and linked stacks in memory. However, these approaches are not dealing with

uncertain structure and do not support top-k-style threshold-based early termination.

IR on XML data has become popular in recent years. Some approaches extend traditional keyword-style querying to XML data [7, 11], introduced full-fledged XML query languages with rich IR models for ranked retrieval [9, 19], or developed extensions of the vector space model for keyword search on XML documents. FleXPath [4] was among the first approaches to combine this theme with full-text conditions over search predicates. Meanwhile, various groups have started adding IR-style keyword conditions to existing XML query languages. TeXQuery is the foundation for the W3C's official full-text extensions to XPath 2.0 and XQuery 1.0. TIX and TAX are query algebras for XML that integrate IR-style query processing into a pipelined query evaluation engine. TAX furthermore comes with an efficient algorithm for computing structural joins. Here, the results of a query are scored subtrees of the data, and TAX already provides a threshold operator that drops candidate results with low scores from the result set. TOSS is an extension of TAX that integrates ontological similarities into the TAX algebra.

XIRQL [9], a pioneer in the field of ranked XML retrieval, presents a path algebra based on XQL, an early ancestor of W3C's XQuery, for processing and optimizing structured queries. It combines Boolean query operators with probabilistically derived weights for ranked result output, thus carrying the probabilistic IR paradigm over to the XML case. Finally, XXL [19], specifies a full-fledged, SQL-oriented query language for ranked XML-IR with a high semantic expressiveness that made it stand apart from the Boolean XQL and XPath language standards being predominant at that time. For ranked result output, XXL leverages both a standard IR vector space model and an ontology-oriented similarity search for the dynamic relaxation of structure and term conditions. TopX [17], the actual successor of XXL, on the other hand, focuses on a smaller, XPath-like, subset of the XXL query language which allows for a radically different query processing architecture that outperforms XXL in terms of efficiency by a large margin.

R

Foundations

Applying the TA paradigm for inverted index lists to XML ranked retrieval is not straightforward. In a data-centric XML setting, a ranking of query results to a query is typically induced by defining some form of structural similarity, whereas in a more text-centric view (with a rich mixture of XML tags and text contents), ranking is derived from IR-style text relevance measures, or a combination of structural similarity and text relevance. More precisely, the XML-specific difficulties arise from the following challenges:

Query Processing and Index Structures: Relevant intermediate results to a mixture of structural and content-related search conditions must be tested as to whether they satisfy the path conditions of the query, and this may incur repetitive and expensive *random access* to large, disk-resident index structures. Furthermore, instead of enforcing conjunctive query processing, it is desirable to relax path conditions and rather rank documents by a combination of content scores and an additional degree to which the structural query conditions are satisfied. *Incremental path evaluations* are required when the index structures are accessed mostly using efficient *sequential disk access* in order to limit or entirely avoid the more expensive random accesses. Yet all incremental updates to candidate score bounds during the query processing need to stay monotonic, in order to guarantee a correct algorithmic basis for top-k query evaluation with early candidate pruning.

IR Scoring Models and Vague Search: Existing IR scoring models for text documents cannot be directly carried over to the XML case, because they would not consider the specificity of content terms in combination with hierarchical elements or attribute tags. For example, the term "transactions" in a bibliographic data set should be viewed as specific (and lead to a high score) when occurring within elements of type section or caption but be considered less informative within tags like journalname. Fur-

thermore, it should be possible to relax search terms and, in particular, tag names, using tree editing operations, or ontology- and thesaurus-based similarities. For example, a query for a book element about "XML" should also consider a monograph element on "semi-structured data" as a relevant result candidate.

Result Granularity: Scores and index lists refer to individual XML elements and their content terms, but from an IR point-of-view it is desirable to aggregate scores at the document level and return the most relevant XML subtrees, up to the entire XML document, as results. Thus, the query evaluation has to weigh *different result granularities* against each other dynamically in the top-k query processing, and the relevance scoring model should consider XML-specific ranking aspects such as *exhaustiveness* and *specificity* when choosing the most suitable result granularity to address the user's information need.

Scoring Structure
Structural similarity is considered in the sense that documents can qualify even if they do not satisfy all path conditions, i.e., if there were too few results otherwise. For dynamically relaxing tag names and structural relationships of tags in path queries, various tree editing operations can be employed, such that only the most similar matches in the collection are returned.

XML data can typically be represented as forests of node-labeled trees. Figure 1 shows a database instance containing fragments of heterogeneous news documents. Figure 2 gives examples of several queries drawn as trees: the root nodes represent the returned answers, single and double edges represent the descendant and child axes, respectively, and node labels stand for names of elements or keywords to be matched. Hence, different queries match the different news documents in Fig. 1. For example, query (a) in Fig. 2 matches document (a) exactly, but would neither match document (b) (since link is not a child of item) nor document (c) (since item is entirely missing). Query (b) matches document (a), also since the only difference between this query and query (a) is the descendant axis be-

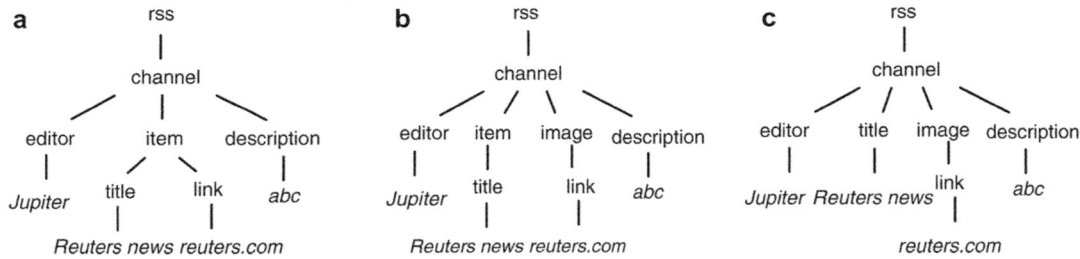

Ranked XML Processing, Fig. 1 Heterogeneous XML database example

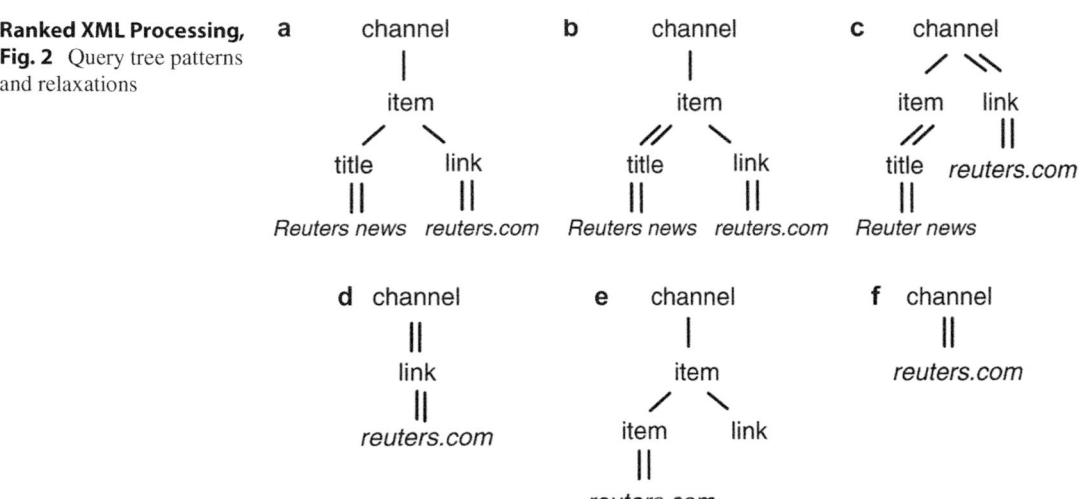

Ranked XML Processing, Fig. 2 Query tree patterns and relaxations

tween item and title. Query (c) matches both documents (a) and (b) since link is not required to be a child of item while query (d) matches all documents in Fig. 1. Intuitively, it makes sense to return all three news documents as candidate matches, suitably ranked based on their similarity to query (a) in Fig. 2. Queries (b), (c), and (d) in Fig. 2 correspond to *structural relaxations* of the initial query (a) as defined in [4]. In the same manner, none of the three documents in Fig. 1 matches query (e) because none of their title elements contains reuters.com. Query (f), on the other hand, is matched by all documents because the scope of reuters.com is broader than in query (e). It is thus desirable to return these documents suitably ranked according to their similarity to query (e).

In order to achieve the above goals, [4] defines query relaxations, including *edge generalization* (replacing a child axis with a descendant axis),

leaf deletion (making a leaf node optional), and *subtree promotion* (moving a subtree from its parent node to its grand-parent). These relaxations capture approximate answers but still guarantee that exact matches to the original query continue to be matches to the relaxed query. For example, query (b) can be obtained from query (a) by applying edge relaxation to the axis between *item* and title and still guarantees that documents where title is a child of item are matched. Query (c) is obtained from query (a) by composing edge generalization between item and title and subtree promotion (applied to the subtree rooted at link). Finally, query (d) is obtained from query (c) by applying leaf deletion to the nodes *ReutersNews*, title and item. Query (d) is a relaxation of query (c) which is a relaxation of query (b) which is in turn a relaxation of query (a). Similarly, query (f) in Fig. 2 can be obtained from query (e) by a combination of subtree promotion and leaf deletion. Other works have considered additional query relaxation such as *node renaming*, *node generalization*, *node insertion*, and *node deletion* [1, 9, 15, 16]. The use of schema knowledge can reduce the number of possible relaxed queries by ignoring relaxations that are guaranteed not to lead to additional matches [16].

Amer-Yahia et al. [3] presented strategies to assign scores to query relaxations. These strategies are based on the traditional *tf · idf* measure derived from IR-style ranking of keyword queries against an unstructured document collection. The *twig scoring* method introduced in [3] computes the score of an answer taking occurrences of *all* structural and content-related (i.e., keyword) predicates in the query. For example, a match to query (c) would be assigned an *inverse document frequency* score, *idf*, based on the fraction of the number of channel nodes that have a child item with a descendant title containing the keyword *ReutersNews* and a descendant link that contains the keyword reuters.com. Such a match would then be assigned a *term frequency* score, *tf*, based on the number of query matches for the specific channel answer.

Scoring Text

A variety of IR-style scoring functions has been proposed and adopted for XML retrieval, ranging from the classic vector space model with its *tf · idf* family of scoring approaches, typically using *Cosine* measure for score aggregations, over to the theoretically more sound *probabilistic scoring models*, with *Robertson & Sparck-Jones* and *Okapi BM25* being the most widely used ranking approaches in current IR benchmark settings such as TREC or INEX, up to even more elaborated *statistical language models*. An important lesson from text IR is that the influence of the term and document frequency values, *tf* and *df* – in the following referred to as their element-specific counterparts *ftf* and *ef*, should be sub-linearly dampened to avoid a bias for short elements with a high term frequency of a few rare terms. To address these considerations, the TopX engine, presented by Theobald et al. [17, 18], adopts the empirically very successful Okapi BM25 proba-

bilistic scoring model to a generic XML setting by computing individual relevance models for each element type occurring in the collection.

For a typical NEXI query pattern of the form $q = //A[about(.//,t_1,...,t_m)]$, the following *relevance score* is computed for an element e with tag name A:

$$score\,(e, q) = \sum_{i=1}^{m} \frac{(k_1+1)\,ftf\,(t_i,e)}{k+ftf\,(t_i,e)}$$

$$\cdot \log\left(\frac{N_A - ef_A(t_i) + 0.5}{ef_A(t_i) + 0.5}\right)$$

$$\text{with } K = k_1\left((1-b) + b\frac{length(e)}{avg\_length_A}\right)$$

Here, $ftf(t_i, e)$ models the *relevance* of a term t_i for an element's *full content*, i.e., the frequency of t_i in all the descending text nodes of element e; while $ef_A(t_i)$ models the *specificity* of t_i for a particular element with tag name A by capturing how many times t_i occurs under a tag A across the whole collection having N_A elements with this tag name.

That is, this extended BM25 model computes a separate relevance model for each term t_i with respect to its enclosing tag name A, thus maintaining detailed element frequency statistics $ef_A(t)$ of each individual tag-term pair that occurs in the collection. It provides a smoothed (i.e., dampened) influence of the *ftf* and *ef* components, as well as a compactness-based normalization that takes the average length of each element type into account. Note that the above function also includes the tunable parameters k_1 and b just like the original BM25 model that now even allows for fine-tuning the influence of the *ftf* components and the length normalization for each element type individually – if desired. For an about operator with multiple keyword conditions as used in the NEXI query language of the INEX benchmark series (or similarly for ftcontains in the XPath 2.0 Full-Text specification), that is attached to an element e with tag name A, the aggregated score of e is simply computed as the sum of the element's scores over the individual tag-term conditions. For path queries with more than one structural tag condition or with multiple full-text operators, the content scores of each element can be combined with the structural scores described above.

Various extensions for more specific full-text predicates such as keyword proximity and phrase matching, as well as incorporating ontological concept similarities for query expansion, have been proposed in the literature – some of which are leaving some interesting research questions for a top-k-style query processor, since most of the more sophisticated proximity- or graph-based compactness measures inherently lead to non-monotonic score aggregation functions.

XML Top-k Query Evaluation Techniques

Combining Structure Indexes and Inverted Lists: Kaushik et al. [12] proposed one of the first, exact-match, top-k algorithms for XML by employing various path index operations as basic steps for non-relaxed evaluations of branching path queries. Their strategy combines two forms of auxiliary indexes, a *DataGuide*-like path index for the structure whose extent identifiers are linked to an inverted index for processing relevance-ranked keyword conditions. The index processing steps are then invoked within a TA-style top-k algorithm, involving eager random access to these inverted index structures.

XRank: Among the most prominent IR-related approaches for ranked retrieval of XML data is XRank [11]. It generalizes traditional link analysis algorithms such as *PageRank* for authority ranking in linked Web collections and conceptually treats each XML element as an interlinked node in a large element graph. Then the *element rank* of an XML element corresponds to the authority weight computed over a mixture of containment edges, obtained from the XML tree structure, and hyperlink edges, obtained from the inter-document XLink structure, similar to the HTML case. XRank may indeed return deeply nested elements but merely supports conjunctive keyword search; it does not yet support structured and/or path query languages such as XPath. For efficient retrieval of multi-keyword queries,

it also uses inverted lists sorted in descending order of element ranks and sketches the usage of standard threshold algorithms for pruning the search space.

FlexPath: FlexPath [4] integrates structure and keyword queries and regards the query structure as templates for the context of a full-text keyword search. The query structure (as well as the content conditions) can be dynamically relaxed for ranked result output according to predefined tree editing operations when matched against the structure of the XML input documents. The Flex-Path query processor already comprises the usage of top-k-style query evaluations for a slightly modified, XPath-like, query language that later evolved as part of the official W3C Full-Text extensions to XPath 2.0 and XQuery 1.0. Like [12], it uses separate index structures for storing and retrieving the structural and content-related conditions of an XPath 2.0 Full-Text query; it may thus require a substantial amount of random access to disk-resident index structures for resolving the final structure of a result candidate.

Whirlpool: The Whirlpool system introduced by Marian et al. [14] provides a flexible architecture for processing top-k queries on XML documents *adaptively*. Whirlpool allows partial matches to the same query to follow different execution plans, and takes advantage of the top-k query model to make dynamic choices during query processing. The key features of Whirlpool are: (i) a partial match that is highly likely to end up in the top-k set is processed in a prioritized manner, and (ii) a partial match unlikely to be in the top-k set follows the cheapest plan that enables its early pruning. Whirlpool provides several adaptivity policies and supports parallel evaluation; details on experimental results can be found in [14].

TopX: The biggest challenge in further accelerating full-text query evaluations over large, semi-structured data collections lies in finding appropriate encodings of the XML data, for indexes that can be read *sequentially* in big

R

chunks directly from disk when the collection (or index) no longer fits into the main memory of current machines. Thus, the TopX engine [17, 18] operates over a *combined inverted index* for content- and structure-related query conditions by precomputing and materializing joins over tag-term pairs, the most common query patterns in full-text search. This simple precomputation step makes the query processing more scalable, with an encoding of the index structure that is easily serializable and can directly be stored sequentially on disk just like any inverted index, for example using conventional B^+-tree indexes or inverted files.

At query processing time, TopX scans the inverted lists for each tag-term pair in the query in an interleaved manner, thus fetching large element blocks into memory using only sorted access to these lists and then iteratively joining these blocks with element blocks previously seen at different query dimensions for the same document. Using pre-/postorder tree encodings [10] for the structure, TopX only needs a few final random accesses for the potential top-k items to resolve their complete structural similarity to a path query. An extended hybrid indexing approach using a combination of DataGuide-like path indexes and pre-/postorder-based range indexes can even fully eliminate the need for these random accesses – however at the cost of more disk space.

TopX further introduces pluggable extensions for *probabilistic candidate pruning*, as well as a *probabilistic cost-model* for adaptively scheduling the sorted and random accesses, that help to significantly accelerate query evaluations in the presence of additional, precomputed index list statistics such as index list selectivities, score distribution histograms or parameterized score estimators, and even index list (i.e., keyword) correlations. For *dynamic query expansions* of tag and term conditions, TopX can incrementally merge the inverted lists for similar conditions obtained from an exchangeable background thesaurus such as *WordNet* or *OpenCyc*. Thus, TopX provides a whole toolkit of specialized top-k operators for efficient full-text search, including

incremental merge operators for dynamic query expansion and *nested top-k operators* for high-dimensional phrase expansions.

Key Applications

Scalable, Web-Style Search over Semi-structured Collections: Efficient IR over large Web collections will remain one of the most challenging applications for XML-top-k query processing with full-text search, with an ever-increasing demand for scalability, interactive runtimes, and vague search involving dynamic query relaxation and/or expansion over heterogeneous collections or unknown schemata.

INEX Benchmark Series: INEX provides a comprehensive forum for IR research on semi-structured data, that goes beyond using the formerly prevalent synthetic data collections such as XMark or XBench for evaluating retrieval quality in true IR-style settings, with a variety of subtasks, XML-IR-specific evaluation metrics, and peer assessments of retrieval results [5].

Future Directions

Graph Top-k: Current XML-top-k algorithms are restricted to XML data trees. Future work could focus on further generalizing the scoring approach, in order to handle cycles arising from inter- or intra-document XLinks, with the need to still derive tight and accurate bounds for early candidate pruning. This may involve efficient index structures for arbitrary graphs and potentially non-monotonic score aggregation functions to incorporate graph compactness measures such as *Steiner trees*.

More XQuery: Similarly, current work has only focused on implementing various subsets of the XPath query language. Providing top-k-style bounds and pruning thresholds for more complex

XQuery constructs such as *loops* and *if-cases* would be an intriguing issue for future work.

Experimental Results

Extensive experiments can be gleaned from the various approaches presented in the literature, see, e.g., [3, 4, 12, 14, 17].

Data Sets

Links to the INEX IEEE and Wikipedia collections can be obtained from the INEX homepage: http://inex.is.informatik.uni-duisburg.de

URL to Code

http://topx.sourceforge.net

Cross-References

▶ Text Indexing and Retrieval
▶ XML Information Integration
▶ XQuery Full-Text

Recommended Reading

1. Amer-Yahia S, Cho S, Srivastava D. Tree pattern relaxation. In: Advances in Database Technology, Proceedings of the 8th International Conference on Extending Database Technology; 2002. p. 496–513.
2. Amer-Yahia S, Curtmola E, Deutsch A. Flexible and efficient XML search with complex full-text predicates. In: Proceedings of the ACM SIGMOD International Conference on Management of Data; 2006. p. 575–86.
3. Amer-Yahia S, Koudas N, Marian A, Srivastava D, Toman D. Structure and content scoring for XML. In: Proceedings of the 31st International Conference on Very Large Data Bases; 2005.
4. Amer-Yahia S, Lakshmanan LVS, Pandit S. FleX-Path: flexible structure and full-text querying for XML. In: Proceedings of the ACM SIGMOD International Conference on Management of Data; 2004. p. 83–94.
5. Amer-Yahia S, Lalmas M. XML search: languages, INEX and scoring. ACM SIGMOD Rec. 2006;35(4):16–23.
6. Bruno N, Koudas N, Srivastava D. Holistic twig joins: optimal XML pattern matching. In: Proceedings of the ACM SIGMOD International Conference on Management of Data; 2002. p. 310–21.
7. Cohen S, Mamou J, Kanza Y, Sagiv Y. XSEarch: a semantic search engine for XML. In: Proceedings of the 29th International Conference on Very Large Data Bases; 2003. p. 45–56.
8. Fagin R, Lotem A, Naor M. Optimal aggregation algorithms for middleware. J Comput Syst Sci. 2003;66(4):614–56.
9. Fuhr N, Großjohann K. XIRQL: a query language for information retrieval in XML documents. In: Proceedings of the 24th Annual International ACM SIGIR Conference on Research and Development in Information Retrieval; 2001. p. 172–80
10. Grust T, van Keulen M, Teubner J. Staircase join: teach a relational DBMS to watch its (axis) steps. In: Proceedings of the 29th International Conference on Very Large Data Bases; 2003. p. 524–5.
11. Guo L, Shao F, Botev C, Shanmugasundaram J. XRank: ranked keyword search over XML documents. In: Proceedings of the ACM SIGMOD International Conference on Management of Data; 2003.
12. Kaushik R, Krishnamurthy R, Naughton JF, Ramakrishnan R. On the integration of structure indexes and inverted lists. In: Proceedings of the ACM SIGMOD International Conference on Management of Data; 2004.
13. Kilpeläinen P, Mannila H. Retrieval from hierarchical texts by partial patterns. In: Proceedings of the 16th Annual International ACM SIGIR Conference on Research and Development in Information Retrieval; 1993. p. 214–22.
14. Marian A, Amer-Yahia S, Koudas N, Srivastava D. Adaptive processing of top-k queries in XML. In: Proceedings of the 21st International Conference on Data Engineering; 2005. p. 162–73.
15. Schenkel R, Theobald A, Weikum G. Semantic similarity search on semistructured data with the XXL search engine. Inf Retr. 2005;8(4):521–45.
16. Schlieder T. Schema-driven evaluation of approximate tree-pattern queries. In: Advances in database technology, proceedings of the 8th international conference on extending database technology. 2002. p. 514–32.
17. Theobald M, Schenkel R, Weikum G. An efficient and versatile query engine for TopX search. In: Proceedings of the 31st International Conference on Very Large Data Bases; 2005.
18. Theobald M, Schenkel R, Weikum G. The TopX DB&IR engine. In: Proceedings of the ACM SIGMOD International Conference on Management of Data; 2007. p. 1141–3.
19. Theobald A, Weikum G. Adding relevance to XML. In: Proceedings of the 3rd International Workshop on the World Wide Web and Databases; 2000. p. 105–24.

R

Ranking Views

Vagelis Hristidis
Department of Computer Science and
Engineering, University of California, Riverside,
Riverside, CA, USA

Synonyms

Ranked materialized views; Ranked views

Definition

Let R be a relation with n attributes $(A_1, \ldots, A_n)$, and let f(t) be a ranking function that assigns a score to each tuple t in R. Then, ranking view R_f is a view on R where tuples are ranked by their (most commonly decreasing) f(t) scores. For example, R may store restaurants with attributes price and rating, and f may be 0.4*price+0.6*rating.

A key problem is how to efficiently maintain one or more materialized ranking views (corresponding to ranking functions) over a relation R. Another key problem is how to use a set $V = \{R_{f1}, R_{f2}, \ldots, R_{fs}\}$ of ranking views over R to efficiently compute a new ranking view R_g not in V.

Historical Background

Ranking views were intensely studied from 2000 to 2006, as the popularity of the Web created large multi-attribute datasets, which users may wish to rank according to their personal preferences. For instance, different users may assign different weight (importance) to different attributes (price, rating, distance) of a restaurant.

Scientific Fundamentals

Almost all work on ranking views has focused on the top-k prefix of the views, that is, the k tuples with the highest score. For instance, a user interested in restaurants would likely only view the few top-ranked restaurants.

We start with the problem of *maintaining a top-k materialized ranking views*. The main difficulty of this problem is that a top-k view R_f is not self-maintainable with respect to deletions and updates on the base relation R. That is, sometimes we must query R in order to maintain the top-k view properly. For example, consider a top-k materialized ranking view containing ten stocks with the highest price/earning ratios currently on the market. If one of these stocks plummets and its price/earning ratio drops below the current top 10, the view still contains the top 9 stocks. But in order to find the stock with the 10th ranked price/earning ratio, we need to query the base table of all stocks. This query can be expensive. Another extreme solution is to store all tuples in R_f. The main disadvantage of this approach is the large space overhead, as users rarely need lower ranked tuples.

Yi et al. [1] propose an approach that balances these two extreme approaches. In particular, they maintain the top-k' tuples, where k'≥k. k' is dynamically changing between k and kmax, where kmax is computed based on a probabilistic cost model that tries to minimize the amortized update cost.

We now turn to the problem of computing a top-k ranking view, given a user-specified ranking function. Almost all works have focused on monotone ranking functions on the attributes, as monotonicity provides interesting optimization opportunities. Further, all works rely on smart indexing techniques to minimize the computation time when the user selects a ranking function.

Chang et al. [2] were the first to study this problem. They proposed an indexing technique called the Onion technique. The key observation behind Onion is that if we plot the points on the n-dimensional space (n is the number of attributes), then the point (tuple) with the highest score lies on the convex hull of the tuple space. Thus, the Onion technique in a preprocessing step computes the convex hull of the tuple space, storing all points of the first hull in a file, and proceeds iteratively computing the convex hulls of the remaining points; it stops when all points in the tuple space have been placed in one of the convex hull files. Then, at query time, it is guaranteed that the top 1 tuple with respect to any linear ranking function will be at the first convex hull, the top 2 tuples will be in the first two convex hulls, and so on.

PREFER [3] takes a different approach, where instead of storing the points on convex hulls, it stores a set V of materialized ranking views, corresponding to a carefully selected set of ranking functions. Then, when a user wants to rank the tuples by a new ranking function g, the "closest" view R_{fi} in V is selected. Then, a prefix of R_{fi} is read to compute the top-k ranking view R_g. For that, they define the "watermark" T of R_{fi} with respect to function g to be the minimum fi score that guarantees that the top tuple in R_g has score greater or equal to T in R_{fi}. PREFER also involves heuristic algorithms to define the materialized views that should constitute V, such that no more than one tuple should be read from the selected view to compute any user-defined top-k ranking view. Hristidis and Papakonstantinou [4] extends PREFER to only store a prefix of each materialized view in the set V of views to save space.

Das et al. [5] present linear programming-based techniques to compute a top-k ranking view given a set of available materialized top-m views, where each view may have a different m. This is in contrast to PREFER, where a single materialized view is selected to compute a ranking view.

Key Applications

Answering personalized user preference queries. For instance, different users may assign different weight (importance) to different attributes (price, rating, distance) of a restaurant or attributes (bedrooms, bathrooms, price, square feet) of a home.

Cross-References

▶ Preference Queries
▶ Rank-Aware Query Processing

Recommended Reading

1. Yi K, Yu H, Yang J, Xia G, Chen Y. Efficient maintenance of materialized top-k views. In: Proceedings of the 19th International Conference on Data Engineering; 2003. p. 189–200.
2. Chang Y, Bergman L, CastelliV, Li C, Lo ML, Smith J. The Onion technique: indexing for linear optimization queries. In: Proceedings of the ACM Special Interest Group on Management of Data Conference; 2000.
3. Hristidis V, Koudas N, Papakonstantinou Y. PREFER: a system for the efficient execution of multi-parametric ranked queries. In: Proceedings of the ACM SIGMOD International Conference on Management of Data; 2001.
4. Hristidis V, Papakonstantinou Y. Algorithms and applications for answering ranked queries using ranked views. VLDB J. 2004;13(1):49–70.
5. Das G, Gunopulos D, Koudas N, Tsirogiannis D. Answering top-k queries using views. In: Proceedings of the 32nd International Conference on Very large Data Bases; 2006. p. 451–62.
6. Ilyas IF, Beskales G, Soliman MA. A survey of top-k query processing techniques in relational database systems. ACM Comput Surv. 2008;40(4):11.

Rank-Join

Ihab F. Ilyas
Cheriton School of Computer Science,
University of Waterloo, Waterloo, ON, Canada

Definition

A rank-join operator $RJ(R, J, F, k)$ joins a set of relations $R = \{R_1, \ldots, R_n\}$ on a set of join predicates J, and returns the k join results with the largest combined scores. The combined score of each join result is computed according to a scoring function $F(p_1, \ldots, p_n)$, where $p_1, \ldots, p_n$ are scoring predicates defined over the input relations $\{R_1, \ldots, R_n\}$ respectively. A naïve implantation of a rank-join operator involves computing the full set of join results and then sorting the results on F to report the top k answers. However, the order of the inputs and the properties of the scoring function can be leveraged to avoid materializing the full join results and sorting. For instance, when the inputs are ordered on the scoring predicates $p_1, \ldots, p_n$, and the scoring function F is monotone, early termination can be achieved by computing tight bounds on the scores of join results that are not yet produced. Most rank-join algorithms are generalization of the Threshold algorithms [2] for top-k *score aggregation*.

Historical Background

Multimedia retrieval, such as the Querying Images By Content (QBIC) project [8] studied searching for images by various visual characteristics such as color, shape, and texture of image objects and regions. This required combining the score from different attributes of every image and globally ranking them based on their similarity of multiple features. This problem was developed in [1] and [2] into aggregating ordered results from multiple systems, also known as the problem of *score aggregation*. Rank join was introduced in [4,5,7] as a generalization of *score aggregation* to allow scores to be associated with a combination of (joinable) objects from different relations instead of a single object.

Scientific Fundamentals

There have been multiple different implementations of the *RJ* operation to achieve scalability with different optimizations and allowance for seamless integration in query plans. Most implementations require that the inputs are ordered on the scoring predicates and that the combing function, F, is monotone. The implementations mainly differ on how to compute bounds on the scores of unseen join results.

J*
The J^* algorithm [7] is based on the A^* search algorithm. The idea is to maintain a priority queue of partial and complete join combinations, ordered on the upper bounds of their scores. At each step, the algorithm tries to complete the

join combination at queue top by selecting the next input stream to join, and retrieving the next object from that stream. The algorithm reports the next join result as soon as the join result at queue top includes an object from every input. The score of a complete join combination is computed by aggregating its objects' scores. The score upper bound of a partial join combination is computed by exploiting the monotonicity of the score aggregation function, by aggregating the scores of objects from inputs already joined, and score upper bounds (last seen score) of inputs not yet joined with the combination. More details and other variants of the algorithm can be found in [7].

HRJN
The HRJN operator is based on symmetric hash join [6] of two input relations. It maintains a hash table for each of the relations to store objects read. The HRJN operator implements the traditional iterator interface of query operators, and a $getNext()$ method reads the left or right input alternatively and probes the hash table of the other to generate join results. A priority queue is used to buffer the valid join results in the order of their scores. It computes a threshold T based on the top scores and the last seen scores of each input, which is given by the following:

$$max(F(p_1^{top}, p_2^{last}), F(p_1^{last}, p_2^{top})) \quad (1)$$

where p_1^{last} and p_2^{last} are the last seen scores of the two inputs, p_1^{top} and p_2^{top} are the top scores in the two inputs, and F is the scoring function. A join result is reported as the next answer to $getNext()$ if the result has a combined score greater than or equal the threshold T.

The HRJN* extends the HRJN operator by adaptively switching between relations to read based on the threshold values, instead of reading relations alternatively at each step as HRJN does. Recall that the threshold is computed as the maximum between two virtual scores $T_1 = F(p_1^{top}, p_2^{last})$ and $T_2 = F(p_1^{last}, p_2^{top})$. If $T_1 > T_2$ more inputs should be retrieved from the right input to reduce the value of T_1 and hence the value of the threshold, leading to possible faster reporting of ranked join results.

Optimal Rank-Joins

Evaluation studies of rank-join algorithms [3, 9] introduced techniques to analyze and to benchmark known algorithms based on a notion of a *cost* metric and computational efficiency. In [9], a *cost* metric is a function $cost(A, I)$ that yields the cost of solving a rank join problem instance I with algorithm A. This metric is defined using the idea of depth. The depth on an input relation R_i of a rank join problem instance is the number of tuples read sequentially from R_i before returning a solution. For a rank join algorithm A and an instance I with input relations $R_1, \ldots, R_n$, $sumDepths(A, I)$ is defined as the sum of depths on all input relations. Clearly, $sumDepths$ is an interesting cost metric as it indicates the amount of I/O performed by an algorithm. A notion of optimality known as *instance optimality* is also defined in [9]. Given a class of algorithms $\mathcal{B}$, a class of problem instances $\mathcal{J}$, and a cost metric, we say that a rank-join algorithm $A \in \mathcal{B}$ is instance-optimal if there exist constants c_1 and c_2 such that $cost(A, I) \leq c_1 \cdot cost(A', I) + c_2 \; \forall A' \in \mathcal{B}$ and $I \in \mathcal{J}$. The constant c_1 is called the optimality ratio.

To provide a template to the analysis of various rank join algorithms, an algorithm, Pull/Bound Rank Join (PBRJ), which generalizes previous rank-join algorithms is introduced. The idea of PBRJ is to alternate between pulling tuples from input relations and upper bounding the score of join results that use the unread part of the input. There are many possible ways that an algorithm may pull from its input and compute an upper bound on unseen join results. The PBRJ provides a foundation for analyzing a wide variety rank join algorithms for their instance optimality and computational efficiency. For instance, HRJN is a special case of PBRJ which is proved to be instance optimal but its performance can be arbitrarily bad as it uses a loose upper bound on the scores of remaining join results.

Key Applications

The rank join operation can be used in any application that requires the processing of ranking queries based on multiple attributes. For example in information retrieval, the search queries different sources to find documents containing the search topics. While each source provides documents sorted by relevance, the returned result collection must be sorted in a combined relevance order. Other applications include ranking multimedia objects (e.g., images) based on different features that are separately scored, ranking travel packages or products in online shopping according to different features, rank-aware query optimization, and joining in uncertain databases.

Cross-References

▶ Score Aggregation
▶ Top-K Queries

Recommended Reading

1. Fagin R. Combining fuzzy information from multiple systems. In: Proceedings of the ACM SIGACT-SIGMOD Symposium on Principles of Database Systems; 1996.
2. Fagin R, Lotem A, Naor M. Optimal aggregation algorithms for middleware. In: Proceedings of the ACM SIGACT-SIGMOD Symposium on Principles of Database Systems; 2001.
3. Finger J, Polyzotis N. Robust and efficient algorithms for rank join evaluation. In: Proceedings of the ACM SIGMOD International Conference on Management of Data; 2009. p. 415–28.
4. Hristidis V, Koudas N, Papakonstantinou Y. Prefer: a system for the efficient execution of multi-parametric ranked queries. In: Proceedings of the ACM SIGMOD International Conference on Management of Data; 2001. p. 259–70.
5. Ilyas IF, Aref WG, Elmagarmid AK. Joining ranked inputs in practice. In: Proceedings of the 28th International Conference on Very Large Data Bases; 2002. p. 950–61.
6. Ilyas IF, Aref WG, Elmagarmid AK. Supporting top-k join queries in relational databases. VLDB J Int J Very Large Data Bases. 2004;13(3):207–21.
7. Natsev A, chi Chang Y, Smith JR, Li C-S, Vitter JS. Supporting incremental join queries on ranked inputs. In: Proceedings of the 27th International Conference on Very Large Data Base; 2001.

R

Cross-References

▶ Top-K Selection Queries on Multimedia Datasets

Recommended Reading

1. Agrawal R, Wimmers E. A framework for expressing and combining preferences. In: Proceedings of ACM SIGMOD International Conference on Management of Data; 2000. p. 297–306.
2. Bruno N, Gravano L, Marian A. Evaluating top-k queries over web accessible databases. In: Proceedings of ICDE; 2002.
3. Chang K, Huang S-W. Minimal probing: supporting expensive predicates for top-k queries. In: Proceedings of the ACM SIGMOD International Conference on Management of Data; 2002.
4. Chang YC, Bergman L, Castelli V, Li CS, Lo ML, Smith J. The onion technique: indexing for linear optimization queries. In: Proceedings of the ACM SIGMOD International Conference on Management of Data; 2000. p. 391–402.
5. Donjerkovic D, Ramakrishnan R. Probabilistic optimization of top-N queries. In: Proceedings of the 25th International Conference on Very Large Data Bases; 1999.
6. Fagin R. Combining fuzzy information from multiple systems. In: Proceedings of the ACM SIGACT-SIGMOD Symposium on Principles of Database Systems; 1996. p. 216–26.
7. Fagin R. Fuzzy queries in multimedia database systems. In: Proceedings of the ACM SIGACT-SIGMOD Symposium on Principles of Database Systems; 1998. p. 1–10.
8. Fagin R, Wimmers E. Incorporating user preferences in multimedia queries. In: Proceedings of the 6th International Conference on Database Theory; 1997. p. 247–61.
9. Gravano L, Chaudhuri S. Evaluating top-k selection queries. In: Proceedings of the 23th International Conference on Very Large Data Bases; 1999.
10. Hristidis V, Koudas N, Papakonstantinou Y. Efficient execution of multiparametric ranked queries. In: Proceedings of the ACM SIGMOD International Conference on Management of Data; 2001.
11. Ilyas IF, Aref WG, Elmagarmid AK. Joining ranked inputs in practice, Hong Kong. 2003. p. 950–61.
12. Natsev A, Chang Y, Smith J, Li C-S, Vitter JS. Supporting incremental join queries on ranked inputs. In: Proceedings of the 27th International Conference on Very Large Data Bases; 2001.
13. Singh S. Ranked selection indexes for linear preference queries. MSc thesis, Wichita State University. 2011.
14. Tsaparas P, Palpanas T, Kotidis Y, Koudas N, Srivastava D. Ranked join indices. In: Proceedings of the 19th International Conference on Data Engineering; 2003. p. 277–88.

RDF Stores

Katja Hose[1] and Ralf Schenkel[2]
[1]Department of Computer Science, Aalborg University, Aalborg, Denmark
[2]Campus II Department IV – Computer Science, Professorship for databases and information systems, University of Trier, Trier, Germany

Synonyms

Knowledge base; RDF database system; Triple store

Definition

An RDF store is a system that is optimized to manage data given in RDF (Resource Description Framework). Similar to other database systems, an RDF store uses an appropriate internal storage layout (native RDF, relational, graph, etc.,) and enables efficient query processing over the data. While there are alternative query languages to query RDF data, SPARQL has become the de facto standard that is supported by RDF stores. Inferencing, i.e., logically deducing information that is not explicitly contained in the stored data based on RDFS and OWL ontologies, is a special extension that is not supported by all RDF stores.

Historical Background

RDF has been developed as a general-purpose format to describe metadata on the Web. But RDF has gained particular popularity as a data format for the Semantic Web and as a very flexible data format to store information obtained by information extraction, e.g., from Web sources such as Wikipedia.

The first frameworks for RDF storage and querying became available around the year 2000. The main obstacles were the lack of a standardized and commonly used query language and the efficiency of the underlying data management systems. Hence, early approaches [3, 17] represent frameworks that initially made use of third-party storage systems (mostly relational database systems) to store the data. After the W3C published the first SPARQL working draft being published in 2004, SPARQL quickly became the de facto standard query language and is now supported by basically all RDF stores.

RDF store frameworks can use existing relational database systems as underlying back-end and map query languages developed for RDF such as SPARQL to SQL. As storing a set of triples is trivially possible by using a relational table with three columns, basically no transformation is necessary in the straightforward layout. Mapping all constructs that SPARQL – and especially the recent SPARQL 1.1 standard – defines and supporting inferencing, however, is difficult and not always possible. Moreover, the high number of self-joins imposes restrictions on the applicability of relational back ends.

Therefore, native RDF stores [23] have been developed to address the particular problems and challenges of RDF data and the SPARQL query language. Despite great progress in recent years regarding the support of large amounts of data [4], facilitating inferencing based on RDFS and OWL at large scale efficiently is still a challenge.

Foundations

The main advantage of the RDF format and one of the reasons for its growing popularity is its flexibility. The flexibility originates from the triple format of the data with subject, predicate, and object. For instance, the information that Berlin is located in Germany can be expressed as: (Berlin, locatedIn, Germany). We can easily extend this information by the fact that Frankfurt also lies in Germany: (Frankfurt, locatedIn, Germany). By using well-defined ontologies and vocabularies, we can define restrictions and describe the meaning of classes (e.g., Germany can be defined as an instance of a class Country) and the usage of properties (e.g., subjects or objects of triples with locatedIn as predicate can be restricted to instances of specific classes). Ontologies themselves can conveniently be expressed in RDF as well and stored along with the data it describes.

The increasing availability and use of RDF-based information in the last decade has led to an increasing need for systems that can store RDF and, more importantly, efficiently evaluate complex queries over large bodies of RDF data. Traditionally, the problem of storing and querying RDF data has been considered by various research communities. As a consequence, architectures and solutions range from relational database systems over column stores to graph stores and cloud-based solutions. Extensive surveys can be found in [13, 16, 22].

Centralized RDF Stores

Various systems for storing and querying RDF data have been proposed, partly reusing and adapting well-established techniques from relational databases. The majority of these systems can be grouped into one of the following three classes:

1. *Triple stores* where RDF triples are stored in a single relational table, usually with additional indexes and statistics,
2. *Vertically partitioned tables* that maintain one table for each property,
3. Solutions with a custom schema that represent frequently co-occurring properties together in *property tables*.

Triple Stores

In triple stores RDF triples are stored in a simple relational table with three or four attributes. A large number of such systems have been developed; its success is mainly due to its low implementation overhead and its very generic nature. Important systems from this category are RDF-3X [18] and HexaStore [24] from the database community and 3store [9] and Virtuoso [7] from the Semantic Web community.

RDF facts are represented in a so-called *triple table*, a generic three-attribute table of the form (subject, property, object). Most systems do not store resource identifiers and constants in this table, but first convert them to numeric ids, for example, by hashing; this saves spaces and allows for more efficient access structures. If a system stores data from more than one source (or more than one RDF graph), the relation is often extended by a fourth numeric attribute, the graph id, that uniquely identifies

R

the source of a triple. In this case, the relation is also called a *quadruple table*, and such a system is called a *quadruple store.*

For efficient query processing, systems maintain indexes on (a subset of) all combinations of subject, property, object, and graph id. With such indexes, any single triple pattern of a query can be processed by a single index lookup. The idea of such a combination of covering indexes was first implemented in the YARS system [10]. Since each index is about as large as the relation and needs to be maintained whenever a triple is modified, added, or deleted, most systems use only a subset of all possible indexes and usually apply compression methods. Query processing in a triple or quadruple store is conceptually done in two steps: it first *converts* the SPARQL query into an equivalent SQL query or an equivalent logical operator tree and then creates, optimizes, and executes a *query plan* for this SQL query.

Converting a SPARQL query to an equivalent SQL query in step 1 is done in a very systematic process. For each triple pattern in the SPARQL query, a copy of the triple relation is added to the query. For each join between two patterns (expressed by a variable common to both patterns), a join of the corresponding relation instances is created. Constants in the triple patterns are mapped to constraints on the corresponding relation's attribute. Similarly, filter conditions are mapped to equivalent constraints. The SPARQL 1.1 features' grouping and aggregation are directly translated to SQL grouping and aggregation.

The resulting SQL query can be directly evaluated on a relational system, but generic relational query optimizers often have problems to find good execution plans, caused by the usually large number of self-joins. Triple stores therefore often provide their own relational back end with custom operator implementations and a specialized query optimizer. As in standard relational systems, the query is first translated into an abstract operator tree, which the query optimizer then translates into an efficient physical execution plan. In the latter step, the optimizer needs to decide how the abstract operators (joins, projection, selection) should be mapped to physi-

cal implementations (such as merge joins, hash joins, or nested loop joins) and in which order these operations should be executed, such that the plan can be efficiently executed. Since index accesses are efficient, many systems generate an index scan for each triple pattern in the query. Finding an efficient execution plan is difficult since existing methods for estimating the cost of a plan from relational databases often fail to deliver precise estimates; these techniques usually ignore correlation of attributes, since statistics are available only separately for each attribute. Multidimensional histograms which could capture this correlation can easily grow too large for large-scale RDF data. Triple stores like RDF-3X therefore maintain specialized statistics and reuse precomputed aggregate indexes initially built to improve query processing performance.

Vertically Partitioned Tables

Queries in real applications have usually triple patterns with fixed properties. To optimize storage for such queries, a separate table is created for each property that has two attributes, one for storing subjects and one for storing objects of triples with that property; for quadruples, the graphid is stored in a third attribute. String literals are usually encoded as numeric ids. For further improvements, these tables can be stored in a *column store* that stores tables as a collection of columns. Since all entries within the same column have the same data type, they can be compressed very efficiently. The idea of using column stores for RDF was initially proposed by Abadi et al. [1].

Triple patterns with a fixed property can be evaluated by scanning the (columns of the) table for this property, which can be done very efficiently. For query optimization, per-table statistics are used that can represent correlation of subjects and objects. In contrast, triple patterns with a variable at the property position are very expensive to evaluate since they need to access all two-column tables.

Property Tables

It is very common in RDF data collections that many subjects have the same or at least are highly overlapping set of properties, and queries will often access many of these properties at the same time. It is now a very natural approach to represent all properties of such a subject in the same table. A query accessing some or all of these properties will be executed very efficiently since it does not require to join information from different tables. In the property table approach, subjects that have similar properties are clustered and represented by a single table where each attribute corresponds to a property. A set of facts for one of these subjects is then stored as one row in that table; properties that do not exist with a subject are represented by NULL values. The most prominent example for this storage structure is Jena [17]. Property tables perform best for RDF data collections with regular structure, where they can perform better than other solutions.

Property tables cannot easily handle subjects with multivalued properties; a standard solution for this is storing the corresponding triples in a standard triple table.

Graph Databases

It is very natural to consider a set of RDF triples as a graph: subjects and objects of triples form the nodes, and there is an edge between the subject and object of each triple labeled by the triple's property. A triple store can therefore be seen as a graph database for specific type of graphs; similarly, graph databases can often be utilized for storing and querying RDF data as well, but they may not support querying in SPARQL.

Distributed RDF Stores

Centralized systems are limited regarding the amount of data they can handle efficiently. Hence, distributed architectures have been proposed that make use of multiple machines and exploit parallel processing.

Data Storage

Strategies similar to those originally developed as scale-out architectures for distributed and parallel database systems, which build upon centralized systems and commodity hardware, can be applied to distributed RDF stores as well.

The main principle is to use multiple machines each running an instance of a centralized RDF store and adding an additional conceptual layer with a coordinator that assigns the data to the instances. There are several alternatives on how to implement this general architecture. The main difference between them is how the data is distributed between the instances; some approaches use hash functions on different components of an RDF triple [11], whereas others use graph partitioning [14] or query workloads [8, 12].

An alternative way to store triples and share them between multiple machines is using a distributed file system that all participating machines have efficient access to. Conceptually, the data is stored in large files that are divided into chunks and assigned to machines in the system. The crucial step is the order in which the data is stored in the files, how big these files are, how they are partitioned into chunks, and which machines these chunks are assigned to. Whereas some systems [21] maintain the triple structure, others [15] apply the principles of vertical partitioning and property tables that are also used by centralized systems.

Another alternative is to store the RDF triples in a key-value store, i.e., an id (key) is determined for an RDF triple (value) and used for storage and retrieval. The key is constructed based on the triples' components (subject, predicate, and object). Hence, the components and the order in which they are used to construct the key determine in which ways a triple can efficiently be found during query processing. As query processing might once prefer lookups based on subjects and another time based on predicates, systems usually store a triple multiple times, for instance, based on subject-predicate-object, predicate-object-subject, and object-subject-predicate. Implementations of RDF stores [2, 19, 20] then additionally exploit properties of the particular key-value stores they are built upon.

Instead of focusing on the triple as basic building block, sets of RDF triples can also be considered as a graph. Hence, distributed RDF stores [25] can also be built on top of distributed graph stores.

Query Processing

Efficient techniques for query processing strongly depend on the way the data is stored. If the system is built upon a layer of centralized RDF stores, then query processing can follow the principles of traditional distributed and parallel database systems. The coordinator has information about which machine is storing which partition of the data. Based on this information, the coordinator can optimize queries and create subqueries for each instance. Then, the coordinator collects the partial results from the instances and computes the final result.

To facilitate efficient join processing, some systems make use of MapReduce [6]. In principle, each join can be computed in a separate MapReduce job by first partitioning the data according to the components the join is defined on and then performing the join over the partitions. The task of the query optimization component is then to try to exploit these phases in a way that additional joins can benefit and the overall number of necessary MapReduce jobs can be reduced [16].

If the data is stored in a distributed graph-based RDF store [25], then query processing can benefit from graph exploration, i.e., instead of trying to find matching triples to compute a join, the graph is traversed via edges matching the join condition. As multiple subgraphs might represent answers to a particular query, an important cost factor is the number of nodes in the data graph that the traversal starts from.

Key Applications

Semantic Web: With RDF being a standard data format on the Semantic Web, huge datasets have become available. Publicly available SPARQL endpoints represent RDF stores with a Web-accessible interface and an underlying RDF store to manage the data and evaluate queries.

Enhancing search results: In addition to displaying a list of websites matching the user's list of keywords, search engines have begun to enhance their search results based on semantic information. This information is not only used to interpret that user query and finding the results best matching the user's intention but also to display concise information about entities that represent answers to the query, e.g., information about a particular city, such as inhabitants, location, and country.

Future Directions

Given the increasing volume of RDF information within a single collection, there has been a clear trend toward distributed and cloud-based triple stores. A key for success of these systems is an effective distribution of the RDF triples across the nodes in a distributed cluster. While first solutions exist and have been sketched above, finding good partitions that allow for fast and scalable query processing is still an open research issue.

Once data distribution is fixed, finding efficient execution plans to process a query, i.e., the problem of query optimization, is another difficult problem that will need some further work; this is also true for centralized systems where some progress has been already made. In the distributed case, a core handle for an optimized execution is avoiding to generate and send large sets of intermediate results over the network. Given the early state of most systems, it can be very beneficial to exploit, adapt, and extend the large body of work on query processing in distributed relational databases; here, extensions are required for example regarding statistics and selectivity estimations.

In centralized but a lot more in distributed triple stores, the support for transactional guarantees is still at its beginning, and some work needs to be done especially regarding synchronization of large numbers of concurrent clients accessing and updating the same triple store.

Some existing triple stores already implement and support reasoners that enable inference in combination with RDFS and OWL. Due to the complexity, however, usually only a subset of all possibilities are actually supported. Developing efficient solutions for the full language on large amounts of data is a challenge to be solved in future research.

Experimental Results

A large number of experimental results on the efficiency and scalability of single RDF stores exist, but there are only few recent comparative results. The home page of the BSBM benchmark (http://wifo5-03.informatik.uni-mannheim.de/bizer/berlinsparqlbenchmark/, see below) provides some recent results for a number of RDF stores. In addition, the W3C maintains a list of scalability results for large triple stores at http://www.w3.org/wiki/LargeTripleStores. Cudré-Mauroux et al. [5] compare four NoSQL-based distributed RDF stores with regard to their performance under the BSBM and DBPedia SPARQL benchmarks.

Datasets

Many RDF datasets spanning across a wide range of topics are available free of charge on the Web in the so-called Linked Open Data cloud (http://linkeddata.org/), registered at CKAN (http://ckan.org/), and accessible via datahub.io (http://datahub.io/group/lodcloud). The most popular RDF datasets are DBpedia (http://dbpedia.org/), YAGO (http://mpii.de/yago), and Freebase (http://www.freebase.com/) – all representing information that is not restricted to any particular domain and similar in coverage and spirit to Wikipedia (http://www.wikipedia.org/).

Another series of real-world datasets originates from the Semantic Web Challenge (http://challenge.semanticweb.org/) – a yearly challenge since 2010 with the goal of providing a platform for Semantic Web applications. One of these tracks is the Big Data track (formerly known as the billion triple challenge) that traditionally provides a big dataset originating from real-world RDF data available on the Web.

A number of benchmarks for RDF stores have been proposed that focus on the efficiency of processing complex SPARQL queries, sometimes including inference. Important examples include the Berlin SPARQL Benchmark (BSBM, http://wifo5-03.informatik.uni-mannheim.de/bizer/berlinsparqlbenchmark/), the DBPedia SPARQL Benchmark (DPBSB, http://aksw.org/Projects/DBPSB.html), the Linked Data Benchmark Council (LDBC, http://www.ldbc.eu/), the SP^2 Bench SPARQL Performance Benchmark (http://dbis.informatik.uni-freiburg.de/forschung/projekte/SP2B/), the Lehigh University Benchmark (LUBM, http://swat.cse.lehigh.edu/projects/lubm/), and WatDiv (http://db.uwaterloo.ca/watdiv/). The W3C maintains a (not necessarily complete) list of such benchmarks at http://www.w3.org/wiki/RdfStoreBenchmarking.

URL to Code

In addition to RDF and SPARQL support by commercial relational database systems, most triple stores and related tools are community efforts or research prototypes that are available on the Web. Incomplete and tentative lists of triple stores are provided by the W3C (http://www.w3.org/2001/sw/wiki/Category:Triple_Store) and maintained on Wikipedia (http://en.wikipedia.org/wiki/Triplestore).

The majority of RDF stores is available in an open-source version, including Big Data (http://www.systap.com/download), RDF-3X (https://code.google.com/p/rdf3x/), Apache Jena (https://jena.apache.org/), and Virtuoso (http://virtuoso.openlinksw.com/download/).

Cross-References

▶ Graph Database
▶ Linked Open Data
▶ RDF Technology
▶ Resource Description Framework (RDF) Schema (RDFS)
▶ Semantic Web
▶ Semantic Web Query Languages
▶ SPARQL

Recommended Reading

1. Abadi DJ, Marcus A, Madden S, Hollenbach K. SW-Store: a vertically partitioned DBMS for semantic Web data management. VLDB J. 2009;18(2):385–406.

2. Aranda-Andújar A, Bugiotti F, Camacho-Rodríguez J, Colazzo D, Goasdoué F, Kaoudi Z, Manolescu I. Amada: Web data repositories in the Amazon cloud. In: Proceedings of the 21st ACM International Conference on Information and Knowledge Management; 2012. p. 2749–51.
3. Broekstra J, Kampman A, van Harmelen F. Sesame: a generic architecture for storing and querying RDF and RDF schema. In: Proceedings of the 1st International Semantic Web Conference; 2002. p. 54–68.
4. Bugiotti F, Camacho-Rodríguez J, Goasdoué F, Kaoudi Z, Manolescu I, Zampetakis S. SPARQL query processing in the cloud. In: Harth A, Hose K, Schenkel R, editors. Linked data management. Boca Raton: CRC Press; 2014. p. 165–92.
5. Cudré-Mauroux P, Enchev I, Fundatureanu S, Groth PT, Haque A, Harth A, Leif Keppmann F, Miranker DP, Sequeda J, Wylot M. NoSQL databases for RDF: an empirical evaluation. In: Proceedings of the 12th International Semantic Web Conference; 2013. p. 310–25.
6. Dean J, Ghemawat S. Mapreduce: simplified data processing on large clusters. In: Proceedings of the 6th USENIX Symposium on Operating System Design and Implementation; 2004. p. 137–50.
7. Erling O, Mikhailov I. RDF support in the virtuoso DBMS. In: Pellegrini T, Auer S, Tochtermann K, Schaffert S, editors. Networked knowledge – networked media. Studies in computational intelligence. Berlin/Heidelberg: Springer; 2009. vol. 221, p. 7–24.
8. Galarraga L, Hose K, Schenkel R. Partout: a distributed engine for efficient RDF processing. In: Proceedings of the 23rd International World Wide Web Conference; 2014. p. 267–68.
9. Harris S, Gibbins N. 3store: efficient bulk RDF storage. In: Proceedings of 1st International Workshop on Practical and Scalable Semantic Systems; 2003.
10. Harth A, Decker S. Optimized index structures for querying RDF from the web. In: Proceedings of the Third Latin American Web Congress; 2005. p. 71–80.
11. Harth A, Umbrich J, Hogan A, Decker S. YARS2: a federated repository for querying graph structured data from the Web. In: Proceedings of the 6th International Semantic Web Conference and 2nd Asian Semantic Web Conference; 2007. p. 211–24.
12. Hose K, Schenkel R. WARP: workload-aware replication and partitioning for RDF. In: Proceedings of the 4th International Workshop on Data Engineering meets the Semantic Web (In conjunction with ICDE 2013); 2013. p. 1–6.
13. Hose K, Schenkel R, Theobald M, Weikum G. Database foundations for scalable RDF processing. In: Proceedings of the 7th International Conference on Reasoning Web: Semantic Technologies for the Web of Data; 2011. p. 202–49.
14. Huang J, Abadi DJ, Ren K. Scalable SPARQL querying of large RDF graphs. Proc. VLDB Endow. 2011;4(11):1123–34.
15. Husain MF, McGlothlin JP, Masud MM, Khan LR, Thuraisingham BM. Heuristics-based query processing for large rdf graphs using cloud computing. IEEE Trans Knowl Data Eng. 2011;23(9):1312–27.
16. Kaoudi Z, Manolescu I. RDF in the clouds: a survey. VLDB J. 2014;24(1):1–25.
17. McBride B. Jena: a semantic web toolkit. IEEE Internet comput. 2002;6(6):55–59.
18. Neumann T, Weikum G. The RDF-3X engine for scalable management of RDF data. VLDB J. 2010;19(1):91–113.
19. Papailiou N, Konstantinou I, Tsoumakos D, Koziris N. H2RDF: adaptive query processing on RDF data in the cloud. In: Proceedings of the 21st International Conference on World Wide Web (Companion Volume); 2012. p. 397–400.
20. Punnoose R, Crainiceanu A, Rapp D. Rya: a scalable RDF triple store for the clouds. In: Proceedings of the 1st International Workshop on Cloud Intelligence; 2012. p. 4.
21. Rohloff K, Schantz RE. High-performance, massively scalable distributed systems using the mapReduce software framework: the SHARD triple-store. In: Programming Support Innovations for Emerging Distributed Applications; 2010. p. 4:1–4:5.
22. Sakr S, Al-Naymat G. Relational processing of RDF queries: a survey. SIGMOD Rec. 2009;38(4):23–28.
23. Thompson B, Personick M, Cutcher M. The bigdata RDF graph database. In: Harth A, Hose K, Schenkel R, editors. Linked data management. Boca Raton: CRC Press; 2014. p. 193–237.
24. Weiss C, Karras P, Bernstein A. Hexastore: sextuple indexing for Semantic Web data management. Proc. VLDB Endow. 2008;1(1):1008–19.
25. Zeng K, Yang J, Wang H, Shao B, Wang Z. A distributed graph engine for web scale RDF data. Proc. VLDB Endow. 2013;6(4):265–76.

RDF Technology

Christian Bizer[1], Maria-Esther Vidal[2], and Michael Weiss[3]
[1]Web-based Systems Group, University of Mannheim, Mannheim, Germany
[2]Computer Science, Universidad Simon Bolivar, Caracas, Venezuela
[3]Carleton University, Ottawa, ON, Canada

Definition

Family of technologies that are standardized by the World Wide Web Consortium (W3C) and built on the Resource Description Framework (RDF). RDF technologies encompass data exchange formats, query languages, and various vocabularies and ontologies [1]. RDF technologies provide the basis for data integration of heterogeneous data sources as well as for the semantic description of resources in terms of assertions on the properties of these resources and relationships among them.

Historical Background

RDF technologies integrate ideas for knowledge representation commonly used by Database and Artificial Intelligence communities, e.g., semantic data models, deductive databases, semantic networks, frames, and relational databases. In 1998, Tim Berners-Lee presented the RDF data model as a framework to describe the metadata of resources in the Semantic Web [2]. Since then, several extensions have been defined to provide more precise descriptions of resources as well as more powerful inference systems to deduce properties of the RDF-based described resources [1].

Scientific Fundamentals

RDF technologies are defined on top of the RDF data model [3] which relies on triples to represent assertions about the properties and relationships of a resource. A triple consists of a subject, a predicate, and an object. Universal Resource Identifiers (URIs) are used to uniquely identify resources that play the role of subject, predi-

cate, or object in a triple of an RDF document; additionally, objects can correspond to literals. RDF technologies all meet a fundamental data exchange assumption: An RDF document is a set of triples, i.e., triples are not duplicated, and the order of the triples in the document is irrelevant. Detailed information about the RDF data model is provided in the article about the Resource Description Framework. RDF technologies can be grouped as follows.

Serialization Formats

The W3C has standardized various formats for encoding and exchanging RDF graphs [3]: The RDF/XML and the JSON-LD syntaxes provide for encoding RDF data as XML and JSON documents. The RDFa syntax allows RDF data to be embedded into HTML documents. N-Triples is a simple plain-text serialization of RDF, while the Turtle and TriG syntaxes focus on the compact, human-readable serialization of RDF. Figure 1 shows a RDF/XML serialization of the example RDF triples introduced in the entry about the Resource Description Framework (RDF). The serialization starts with several namespace declarations, one for the RDF syntax and the others for the FOAF vocabulary and the DBpedia ontology. The `rdf:Description` element introduces the following lines to describe William Shakespeare. The `rdf:datatype` attribute of the `dbo:birthdate` element specifies the birth date to have the XML datatype `date`.

```
<?xml version="1.0"?>
<rdf:RDF
    xmlns:rdf="http://www.w3.org/1999/02/22-rdf-syntax-ns#"
    xmlns:foaf="http://xmlns.com/foaf/0.1/"
    xmlns:dbo-"http://dbpedia.org/ontology/">
    <rdf:Description rdf:about="http://dbpedia.org/resource/William_Shakespeare">
        <rdf:type rdf:resource=""http://xmlns.com/foaf/0.1/Person" />
        <dbo:birthdate rdf:datatype="http://www.w3.org/2001/XMLSchema#date">
            1564-04-26
        </dbo:birthdate>
    </rdf:Description>
</rdf:RDF>
```

RDF Technology, Fig. 1 Example RDF/XML serialization of RDF

Query Language and RDF Stores

The standard query language for RDF is SPARQL [4]. SPARQL queries use a select-from-where syntax that resembles the SQL query language used in the relational model. Figure 2 shows an example of a SPARQL query. The query returns a table containing persons together with their birthdates which have been born before the year 1600. The query starts with namespace definitions (lines 1–3). The SELECT clause specifies that the result table should contain bindings for the variables ?person and ?birthdate. The WHERE clause consists of two triple patterns and a FILTER condition. The triple patterns are matched against the RDF graph http://dbpedia.org, and a set of variable bindings is generated for each match (result row). The sets of variable bindings are then filtered to contain only bindings which fulfill the FILTER condition on the ?birthdate variable.

Most current RDF stores implement SPARQL as the language to manage RDF data in both centralized and federated environments. An overview of different benchmarks comparing these triple stores is given by the website RDF Store Benchmarking (http://www.w3.org/wiki/RdfStoreBenchmarking). The SPARQL query language is accompanied by the SPARQL Protocol [4] which provides for querying remote SPARQL endpoints using HTTP requests.

Vocabularies and Ontology Languages

RDF has provided the basis for the definition of a wide variety of controlled vocabularies and ontology languages, allowing for the precise definition of the semantics of the represented data. RDF schema (RDFS) [5] and OWL 2 [6] correspond to the extensions of RDF that provide a set of logical operators and abstractions that facilitate the description of resources in terms of their properties, hierarchical relationships, and complex relationships. Additionally, SKOS [7], Dublin Core [8], and schema.org are popular vocabularies to describe general properties of resources. SKOS enables the publication of thesauri, taxonomies, and classification schemas. The Dublin Core vocabulary composes a set of terms to describe metadata of any type of record, application, or document. Finally, the schema.org vocabulary defines terms for describing things that are of special interest to search engines such as product offers, local businesses, events, and reviews.

Key Applications

RDF technologies have been used to define ontologies that encode knowledge of different do-

```
PREFIX rdf: <http://www.w3.org/1999/02/22 rdf syntax ns#>
PREFIX dbo: <http://dbpedia.org/ontology/>
PREFIX foaf: <http://xmlns.com/foaf/0.1/>
SELECT ?person ?birthdate
FROM <http://dbpedia.org>
WHERE {
    ?person rdf:type foaf:Person .
    ?person dbo:birthDate ?birthdate .
     FILTER (?birthdate < "1600-01-01"^^xsd:date)
}
```

RDF Technology, Fig. 2 Example SPARQL query

mains. Different techniques have been developed to exploit knowledge represented in ontologies and facilitate thus the integration and exchange of heterogeneous data.

Life Sciences and Chemistry [9, 10]: RDF technologies such as RDFS, OWL, and SKOS have been extensively used to define domain-specific ontologies. The Gene Ontology, the Human Phenotype Ontology, and ChEBI are examples of ontologies that have been collaboratively built. Abstractions and logical operators provided by RDF vocabularies and ontology languages are exploited to precisely define functional information of proteins, genes, human phenotypes, and chemical concepts. These ontologies have been used to semantically describe scientific concepts by replacing textual descriptions with ontological terms, e.g., the Gene Ontology Annotation (UniProt-GOA) database. Furthermore, general frameworks such as ChemAxiom make use of domain-specific ontologies and ontological annotations to facilitate integration and exchange of scientific data. Finally, RDF technologies are used in pattern and link prediction techniques; some applications include discovery and repurposing of drugs and the semantic annotation of genes.

Bibliographic Metadata: RDF technologies such as the Dublin Core vocabulary [8] have been used to establish best practices to describe and documenting bibliographic metadata. Case studies of such application profiles include The Europeana Data Model, the public Library of America, the German Digital Library, and the DINI AG KIM RDF-Representation of Bibliographic Data.

Semantic Annotation of HTML Pages: The schema.org vocabulary has been used to describe text in HTML pages as structured data [11]. Data annotated using schema.org terms is embedded in HTML documents using different RDF serialization formats, e.g., RDFa or JSON-LD. Knowledge encoded in schema.org data annotations is exploited by the search engines to enhance the quality of Web search results.

Cross-References

▶ Dublin Core
▶ Linked Open Data
▶ OWL: Web Ontology Language
▶ RDF Stores
▶ Resource Description Framework
▶ Resource Description Framework (RDF) Schema (RDFS)
▶ Semantic Web
▶ Semantic Web Query Languages

Recommended Reading

1. Hitzler P., Krötzsch M., Rudolph S. Foundations of semantic web technologies. CRC Press; 6 Aug 2009.
2. Berners-Lee T. What the semantic web can represent. Available online at: http://www.w3.org/DesignIssues/RDFnot.html, 1998.
3. Schreiber G., Raimond Y.: RDF 1.1 primer, W3C working group note. 2014. Available online at: http://www.w3.org/TR/2014/NOTE-rdf11-primer-20140624/
4. W3C SPARQL Working Group: SPARQL 1.1 Overview. W3C Recommendation. 2013. Available online at: http://www.w3.org/TR/sparql11-overview/
5. Brickley D., Guha RV. RDF schema 1.1. W3C recommendation. 2014. Available online at: http://www.w3.org/TR/2014/REC-rdf-schema-20140225/
6. Hitzler, P., et al: OWL 2 web ontology language primer (second edition). W3C recommendation. 2012. Available online at: http://www.w3.org/TR/2012/REC-owl2-primer-20121211/
7. Summers E, Isaac A: SKOS simple knowledge organization system primer. W3C note, W3C. 2009. Available online at: http://www.w3.org/TR/2009/NOTE-skos-primer-20090818/
8. DCMI Usage Board: Dublin Core metadata element set, version 1.1. DCMI recommendation. 2012. Available online at: http://dublincore.org/documents/2012/06/14/dces/
9. Frey J., and Colin B. Cheminformatics and the semantic web: adding value with linked data and enhanced provenance. Wiley Interdiscip Rev Comput Mol Sci 2013;35(5):465–481. PMC. Web. 18 Sept 2015.
10. Jupp S, Malone J, Bolleman J, Brandizi M, Davies M, Garcia L, Gaulton A, Gehant S, Laibe C, Redaschi N, et al. The EBI RDF platform: linked open data for the life sciences. Bioinformatics. 2014;30(9):1338–9.
11. Meusel R, Petrovski P, Bizer C. The WebDataCommons microdata, RDFa and microformat dataset series. In: Proceedings of the 13th International Semantic Web Conference; 2014. p. 277–92.

R

Real and Synthetic Test Datasets

Thomas Brinkhoff
Institute for Applied Photogrammetry and
Geoinformatics (IAPG), Oldenburg, Germany

Synonyms

Spatio-temporal benchmarking; Spatio-temporal
data generator

Definition

In the area of mobile and ubiquitous data
management, real and synthetic test datasets
are used for experimental investigations of
performance and robustness. Typical applications
are the examination of access methods for
spatio-temporal databases and the simulation
of mobility for location-based services. Besides
synthetic and real datasets, combinations of these
two types of data are often used that integrate a
predefined infrastructure.

Historical Background

Beginning in the mid 1990s, the development
of algorithms and data structures for spatio-
temporal data took place of the research in the
field of pure (geo-)spatial applications. Previous
spatial test datasets were not sufficient for
investigating the performance and robustness
of those algorithms and data structures.
Consequently, first data generators for spatio-
temporal test datasets were published in the end
of the 1990s.

Foundations

Comprehensible performance evaluations are an
important requirement in the field of mobile and
ubiquitous data management. This demand covers the preparation and use of well-defined test
datasets and benchmarks enabling the systematic and comprehensible evaluation and comparison of algorithms and data structures in this
area.

In experimental investigations, synthetic data
following some statistical distributions as well
as data from real-world applications are used as
test datasets. The use of *synthetic datasets* allows
testing the behavior of an algorithm or of a data
structure under exactly specified conditions or
in extreme situations. In addition, for testing the
scalability, synthetic data sets are often suitable.
However, it is difficult to assess the performance
of real applications by employing synthetic data.
The use of *real datasets* tries to solve this problem. In this case, the selection of the data is crucial. For non-experts it is often difficult to decide
whether a dataset reflects a "realistic" situation
or not. Furthermore, real datasets are typically
connected with a special type of application. For
example, a dataset recording cars driving within
a city may have completely different properties
than the traffic in rural areas or the movement of
vehicles in battlefields.

In the area of mobile and ubiquitous data management, *infrastructure-based dataset generators*
have become popular. These generators compute
moving objects that are restricted by some infrastructure like a network or prohibited areas.
The infrastructure may be a real-world dataset
or completely artificial. The number, distribution,
speed and other properties of the moving objects
are influenced by the infrastructure as well as
by parameters specified by the user of the generator. Dataset produced by infrastructure-based
dataset generators are a compromise between real
and synthetic datasets in respect of realism on
the one hand and of controllability on the other
hand.

Synthetic Datasets

A prominent example for a program providing
synthetic spatio-temporal test datasets is
the *GSTD (Generate SpatioTemporal Data)*
algorithm by Theodorides, Silva and Nascimento
[8]. The basic idea of this algorithm is to start
with a distribution of point or rectangular objects,
e.g., a uniform, a Gaussian or a skewed data

Real and Synthetic Test Datasets, Fig. 1 Raster regions generated by G-TERD

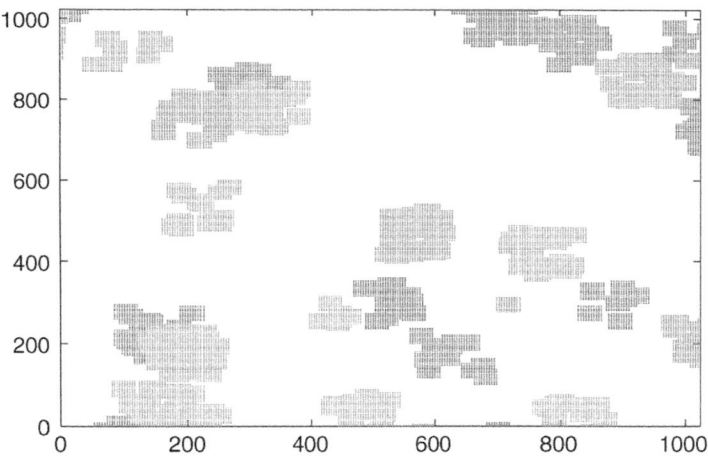

distribution. These objects are modified by computing positional and shape changes using parameterized random functions. If a maximum time stamp is exceeded, an object will become invalid. If an object leaves the predefined spatial data space, different approaches can be applied: the position remains unchanged, the position is adjusted to fit into the data space, or the object re-enters the data space at the opposite edge of the data space. In order to create more realistic movements, a later version of the GSTD algorithm considers rectangles for simulating an infrastructure: each moving object has to be outside of these rectangles.

The *Oporto generator* by Saglio and Moreira [7] is designed for a specific scenario: fishing at sea. The generator supports different object types, which allow modeling different behavior. In order to generate smooth movements, objects of one type may be attracted (e.g., fish by plankton) or repulsed by objects of other types (e.g., ships by stormy areas).

G-TERD (Generator for Time-evolving Regional Data) by Tzouramanis, Vassilakopoulos and Manolopoulos [9] generates two-dimensional raster data (see Fig. 1). The user can control the behavior of the generator by defining parameters and statistical models. The parameters influence the color, the maximum speed, size and rotation of the moving regions. Other moving or static objects may have impact on the speed and on the direction of a moving object.

Real Datasets

There exist many spatio-temporal datasets in the WWW [6]. Very often GPS positions of cars or other vehicles like buses, trains or bikes are recorded. An example is the INFATI dataset from the Aalborg University [3]. It represents a collection of spatio-temporal data that was collected for a project about intelligent speed adaptation. The dataset contains the GPS tracks of about two dozen cars equipped with GPS receivers and logging equipment.

Other popular objects are animals equipped with active satellite tags, e.g., whales, sharks and turtles. The recordings of weather phenomena like hurricanes and the orbits of satellites can also be used as spatio-temporal test datasets.

The section "Data Sets" of this entry gives a small overview on real datasets that are provided for download.

Datasets from Infrastructure-Based Generators

The *Network-based Generator* by Brinkhoff [1] is based on the observation that the motion of objects is often restricted by a network. Examples are streets, railways, air corridors or waterways. Therefore, the generator computes moving objects according to a network that is provided by a file or by a spatial database (see Fig. 2). For each edge of the network, a speed limit and a maximum capacity can be defined. If the number of objects traversing an edge at the same time exceeds the specified capacity, the speed limit on

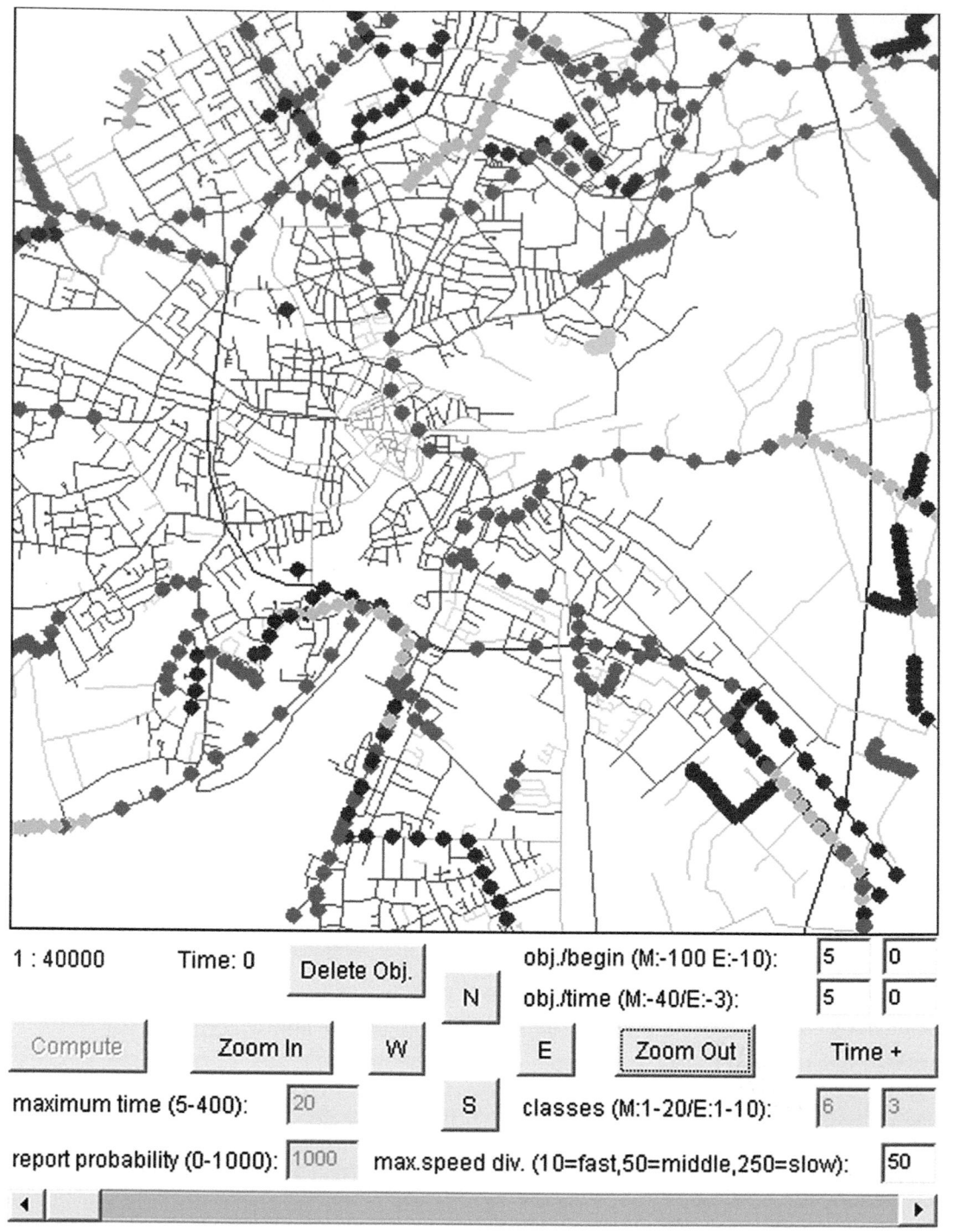

(c) Th. Brinkhoff, 1999-2001, tbrinkhoff@acm.org

Real and Synthetic Test Datasets, Fig. 2 Demo of the network-based generator

this edge may decrease. In addition, each moving object has a (maximum) speed. The computations of the number of new objects per time stamp, of the start location, of the length of a new route and of the location of the destination are done by time-dependent Java functions that can be overloaded by the user of the generator. This concept allows modeling daily commuting and rush hours. The route of a moving object is computed at the time of its creation. However, the fastest path may change over the time by the motion of other objects and of other influences. Therefore, the re-computation of a route is triggered by events depending on the travel time and on the deviation between the current speed and the expected speed on an edge.

The *City Simulator* by Kaufman, Myllymaki, and Jackson [4] is a scalable, three-dimensional model city that enables the creation of dynamic spatial data simulating the motion of up to one million moving objects. The data space of the city is divided into different types of places that influence the motion of the moving objects: roads, intersections, lawns and buildings are such places that define together a city plan (see Fig. 3).

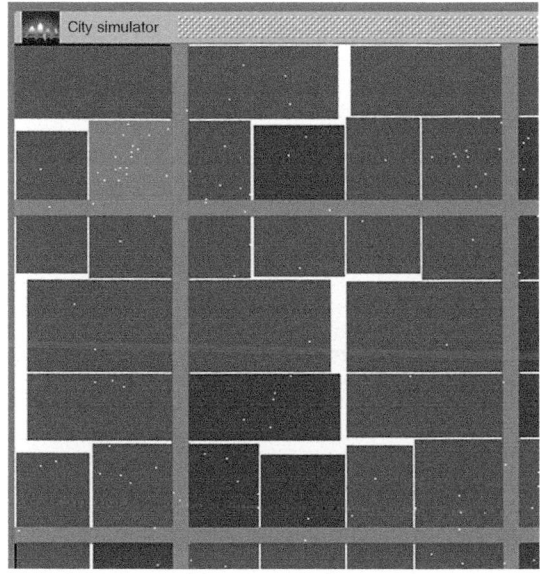

Real and Synthetic Test Datasets, Fig. 3 Visualization of a city plan by the city simulator

Each building consists of an individual number of floors for modeling the third dimension.

SUMO (Simulation of Urban Mobility) [5] developed by Institute of Transport Research at the German Aerospace Center and the Center for Applied Informatics (ZAIK) is open-source software for traffic simulation. Its objective is to support the traffic research community with a common platform for testing and comparing models of vehicle behavior, traffic light optimization, routing etc. Therefore, the car movement model of SUMO is much more evaluated than the models of the other generators.

Key Applications

The primary field of application of the presented test datasets is the evaluation of spatio-temporal databases, but also in other fields the datasets are used. According to Citeseer and the ACM Portal, applications cover (among others) the evaluation of spatio-temporal data structures, the analysis of spatio-temporal queries and of mobility patterns, the simulation of wireless environments by mobile agents, the design of (web) server architectures for moving objects, and the test of car-to-car communication.

Future Directions

It can be expected that spatio-temporal test datasets will be more often used for evaluations of sensor networks, peer-to-peer communication and positioning techniques.

Data Sets

INFATI Dataset: http://arxiv.org/abs/cs.DB/0410001

Pfoser, D. Where can I get spatio-temporal data? http://dke.cti.gr/people/pfoser/data.html

Sea Turtle Migration-Tracking: http://www.cccturtle.org/satellitetracking.php

Unisys Weather – Hurricane/Tropical Data: http://weather.unisys.com/hurricane/index.html

WhaleNet: http://whale.wheelock.edu/whalenet-stuff/stop_cover.html

URL to Code

GSTD: see http://www.cs.ualberta.ca/~mn/ for further notices.

G-TERD: http://delab.csd.auth.gr/stdbs/g-terd.html

Network-based generator: http://www.fh-oow.de/institute/iapg/personen/brinkhoff/generator/

Oporto: http://www.inf.enst.fr/~saglio/etudes/oporto/

SUMO: http://sumo.sourceforge.net/

Cross-References

▶ Geographic Information System
▶ Indexing Historical Spatiotemporal Data
▶ Indexing of the Current and Near-Future Positions of Moving Objects
▶ Location-Based Services
▶ Moving Objects Databases and Tracking
▶ Road Networks
▶ Spatial and Spatiotemporal Data Models and Languages
▶ Spatial Network Databases
▶ Spatiotemporal Trajectories

Recommended Reading

1. Brinkhoff T. A framework for generating network-based moving objects. GeoInformatica. 2002;6(2):153–80.
2. Jensen CS. Special issue on infrastructure for research in spatio-temporal query processing. Bull Tech Comm Data Eng. 2003;26(2):51–5.
3. Jensen CS, Lahrmann H, Pakalnis S, Runge J. The INFATI Data. 2004. Available at: http://oldwww.cs.aau.dk/research/DP/tdb/TimeCenter/TimeCenter Publications/TR-79.pdf.
4. Kaufman J, Myllymaki J, Jackson J. City simulator. IBM alphaworks emerging technologies. 2001. Available at: https://secure.alphaworks.ibm.com/aw.nsf/techs/citysimulator.
5. Krajzewicz D, Hertkorn G, Rössel C, Wagner P. SUMO (Simulation of Urban MObility): an open-source traffic simulation. In: Proceedings of the 4th Middle East Symposium on Simulation and Modelling; 2002. p. 183–7.
6. Nascimento MA, Pfoser D, Theodoridis Y. Synthetic and real spatiotemporal datasets. Q Bull IEEE TC Data Eng. 2003;26(2):26–32.
7. Saglio JM, Oporto MJ. A realistic scenario generator for moving objects. GeoInformatica. 2001;5(1):71–93.
8. Theodoridis Y, Silva JRO. Nascimento MA. On the generation of spatiotemporal datasets. In: Proceedings of the International Symposium on Large Spatial Databases; 1999. p. 147–64.
9. Tzouramanis T, Vassilakopoulos M, Manolopoulos Y. On the generation of time-evolving regional data GeoInformatica. 2002;6(3):207–31.

Real-Time Transaction Processing

Jörgen Hansson[1] and Ming Xiong[2,3]
[1]University of Skövde, Skövde, Sweden
[2]Bell Labs, Murray Hill, NJ, USA
[3]Google, Inc., New York, NY, USA

Synonyms

Time-constrained transaction management

Definition

Real-time transaction processing focuses on (i) enforcing time constraints of transactions, i.e., meet time constraints on invocation and completion, and (ii) ensuring temporal consistency of data, i.e., data should be valid/fresh at the time of usage.

The successful integration of time-cognizant behavior and transaction processing into a database system is generally referred to as a real-time database system (RTDB).

Historical Background

The area of real-time transaction processing has emerged from the need for real-time systems, which often are safety-critical, to handle large amounts of data in a systematic fashion, and the increasing expectation of non-critical applications that have used conventional databases but are now needed to deal with "soft" real-time data applications, e.g., multimedia. Real-time systems have traditionally managed data in an ad hoc manner, i.e., system developers have stored and manipulated data in regular data structures resident in the application code. This approach does not scale well as applications increase in complexity and in their needs for managing large amounts of data. Conventional databases, however, are considered inadequate for handling real-time requirements. Conventional databases impose a throughput-centric design, e.g., maximizing average throughput, and is thus not concerned about the specific outcome of a transaction (in fact, transactions are considered equally important). Furthermore, conventional databases are general-purpose transaction processing system design to be satisfy the needs of a multitude of non-real-time applications. In contrast, a predominant part of the real-time systems are very resource limited, normally these systems have a couple of orders of magnitude less resources, e.g., primary memory, and are not using hard disks. Thus, system functionality need to be tailored to application-specific needs to accommodate a minimum footprint and overcome different architectural assumptions and data requirements.

Note that the term real-time has many connotations in industry and in the general literature it is often used interchangeably to denote that something is fast. This differs from the notion used here, where real-time is synonymous to the notion of predictability. The notion of real-time indicates that the correctness of a result depends not only on the logical result of its computation but also on the time at which the result is derived. Thus, the system correctness depends both on the functional and temporal behavior of the system execution. The temporal behavior is tightly connected to the time constraints associated with transactions and data, e.g., deadlines. This requires that a transaction management system needs to enforce that the timeliness of transactions and data, in turn requiring that

scheduling and concurrency control algorithms are time-cognizant.

Examples of real-time applications are control systems, multimedia, to air-traffic control, and as evident, these applications differ in their real-time requirements, complexity, and the type of data they manage. In order to maintain external/internal correctness, real-time systems have to respond to input stimuli and by an actuator, i.e., produce an output result/action within a finite and sufficiently small time bound. These systems must feature predictability, i.e., the ability to show that the system meets the specified requirements under various conditions the system is expected work under [11].

Several research platforms have developed, e.g., ARTS-RTDB & BeeHive (both from University of Virginia, USA), COMET (Linköping University, Sweden), DeeDS (University of Skövde, Sweden), REACH (Technical University of Darmstadt, Germany), Rodain (University of Helsinki, Finland), and STRIP (Stanford University, USA).

Foundations

Real-transaction processing systems operate under and are benchmarked against different performance goals, correctness criteria, and assumptions applications than throughput-centric systems. The performance metrics adopted for real-time transaction processing system reflect that real-time transactions have time constraints on the execution, and the validity of the data. These requirements impose constraints on the system, which is reflected in the performance metrics used for measuring the timeliness of the system, in addition to conventional metrics for measuring consistency. Typical performance metrics, out of which several originate from the area of real-time systems, include deadline miss ratio of transactions, tardiness/lateness of transactions, cost/effects due transactions missing their deadlines, data temporal consistency (external and logical). The requirements of timeliness can exceed that of consistency and isolation, i.e., it might be more important to deliver a result of adequate precision/quality in time than delivering an exact result late. This implies that correctness of a result can be traded for timeliness by relaxing consistency (referring to ACID properties).

Transaction Model

Real-time transactions are typically characterized along the dimensions of the temporal scope, criticality, and transaction type.

The temporal scope of a transaction is typically described by an arrival time a_i, a release time r_i, a (worst case) execution time e_i, and a deadline d_i, where the following conditions hold (the times are expressed as absolute variables): $a_i \leq r_i < d_i$, $r_i + e_i \leq d_i$. A system is, thus, considered schedulable if it can be shown that all transactions can meet their deadlines.

Real-time transactions can be categorized given the criticality and the stringency of meeting their deadlines and the notions of hard, firm, and soft deadlines are generally used. A hard deadline is critical and, thus, the transaction must always complete within the deadline. Missing a hard deadline will have severe or even cataclysmal consequences. In contrast, firm and soft deadlines are used to for transactions when it system can tolerate occasional time constraint violations. The distinction between them is in the utility of completing a task after its deadline; a soft deadline indicates there is often value of completing the task albeit late. A firm deadline indicates that there is no value for late completion and tardy transactions, i.e., tasks running beyond their deadline, can thus be aborted to minimize resource usage.

Transactions may have precedence constraints that put constraints on the relative order of execution among transactions.

The temporal consistency of a data object has two parts: *absolute consistency*, which reflects the state of an external environment and how that state is reflected in the database, and *relative consistency,* which concerns the consistency among data elements used to derive other data. Thus, absolute consistency (a.k.a. external consistency) is necessary to ensure that a system's view of the external environment (e.g., the controlled system) is consistent with the actual state

of the environment. Relative consistency (a.k.a. logical consistency) ensures that only valid data is used to derive new data. To express temporal consistency, a real-time data object o_j is annotated with an absolute validity interval avi_j denoting the time length and a timestamp ts_j when the data object was last updated/sampled. The data object is considered valid in the time interval $[ts_j, ts_j + avi_j]$.

To define the notion of a relative validity interval rvi, a *relative consistency set R* is introduced for each derived data object, which contains the set of data objects used for its derivation. Furthermore, each such set is has a relative validity interval denoted R_{rvi}. A data object o_j is temporally consistent if the following two conditions hold: (i) $(t + ts_j) \leq avi_j$ (absolute consistency) and (ii) (assume R is the set for o_j) for all $o_k \in R| ts_j - ts_k | \leq R_{rvi}$.

Concurrency Control

A conventional non-real-time two-phase-locking scheme (2PL) is prone to priority inversion, unknown blocking times, and deadlocks, making it infeasible in a real-time context. Priority inversion occurs when a high-priority transaction is blocked due to a resource locked by a low-prioritized transaction. Thus, the high-priority transaction experiences blocking delays that can jeopardize this timeliness. Several 2PL variants have been developed to overcome these deficiencies, and the most well-known scheme is two-phase-locking with high priority resolution (2PL-HP). Consider the case a transaction T_R is requesting a lock, which is held by another transaction T_H. In 2PL-HP, T_R will abort T_H if the priority of T_R exceeds that of T_H; otherwise T_R will wait until T_H completes. There are additional schemes that deploy more elaborate conditions to avoid that a process is aborted unnecessarily, e.g., considering the remaining execution time of T_R and the needed execution time of T_R.

In real-time optimistic concurrency control (OCC), transactions execute in three stages: read, validation, and write. In the read stage, transactions read and update data items freely, storing their updates into private workspaces. Note that the updates of a transaction stored in the private workspaces are installed as global copies in the write stage (i.e., after the transaction is validated). In the validation stage, a validating transaction may conflict with ongoing transactions. Depending on the transaction characteristics, there are several real-time conflict resolution mechanisms for the validation [5], i.e., a validating transaction may commit, abort, or be "put on the shelf" to wait for the conflicting transactions.

In priority-wait conflict resolution mechanism [5], which is demonstrated to have superior performance for OCC, a transaction that reaches validation and finds higher priority transactions in its conflict set is "put on the shelf", that is, it is made to wait and not allowed to commit immediately. This gives the higher priority transactions a chance to make their deadlines first. After all conflicting higher priority transactions leave the conflict set, either due to committing or due to aborting, the on-the-shelf waiter is allowed to commit. Note that a waiting transaction might be restarted due to the commit of one of the conflicting higher priority transactions.

Distributed Real-Time Transaction Processing

The distributed transaction execution model for a real-time two-phase commit protocol is presented next. A commonly adopted sub-transaction model has been presented [4] in which there is one process, called the master, which is executed at the site where the transaction is submitted, and a set of other processes, called cohorts, which execute on behalf of the transaction at the various sites that are accessed by the transaction. Cohorts are created by the master sending a STARTWORK message to the local transaction manager at that site. This message includes the work to be done at that site and is passed on to the cohort. Each cohort sends a WORKDONE message to the master after it has completed its assigned data processing work. The master initiates the commit protocol (only) after it has received this message from all its cohorts. Within the above framework, a transaction may execute in either sequential or parallel fashion. The distinction is that cohorts in a sequential transaction execute one after another,

whereas cohorts in a parallel transaction execute concurrently.

Real-Time Two-Phase Commit

The master implements the classical two-phase commit protocol [3] to maintain transaction atomicity. In this protocol, the master, after receiving the WORKDONE message from all its cohorts, initiates the first phase of the commit protocol by sending PREPARE (to commit) messages in parallel to all its cohorts. Each cohort that is ready to commit first force-writes a prepare log record to its local stable storage and then sends a YES vote to the master. At this stage, the cohort has entered a prepared state wherein it cannot unilaterally commit or abort the transaction, but has to wait for the final decision from the master. On the other hand, each cohort that decides to abort force writes an abort log record and sends a NO vote to the master. Since a NO vote acts like a veto, the cohort is permitted to unilaterally abort the transaction without waiting for the decision from the master.

After the master receives votes from all its cohorts, the second phase of the protocol is initiated. If all the votes are YES, the master moves to a committing state by force-writing a commit log record and sending COMMIT messages to all its cohorts. Each cohort, upon receiving the COMMIT message, moves to the committing state, force-writes a COMMIT log record, and sends an ACK message to the master. On the other hand, if the master receives one or more NO votes, it moves to the aborting state by force-writing an abort log record and sends ABORT messages to those cohorts that are in the prepared state. These cohorts, after receiving the ABORT message, move to the aborting state, force-write an abort log record and send an ACK message to the master. Finally, the master, after receiving ACKs from all the prepared cohorts, writes an end log record and then "forgets" the transaction (by removing from virtual memory all information associated with the transaction).

A real-time two-phase commit protocol such as PROMPT (Permits Reading Of Modified Prepared data for Timeliness) [6], works differently from traditional two-phase commit

R

protocol in that transactions requesting data items held by other transactions in the prepared state are allowed to access this data. That is, prepared cohorts lend their uncommitted data to concurrently executing transactions (without releasing the update locks) in the optimistic belief that this data will be committed. If the lender is aborted later, the borrower is also aborted since it has utilized dirty data. On the other hand, if the borrowing cohort completes its local data processing before the lending cohort has received its global decision, the borrower is "put on the shelf", that is, it is made to wait until either the lender receives its global decision or its own deadline expires, whichever happens earlier. In this case, the borrower can only commit if the lender commits.

In contrast to centralized databases where transactions that validate successfully *always* commit, a distributed transaction that is successfully locally validated might be aborted later because it fails during global validation. This can lead to "wasteful" aborts of transactions- a transaction that is locally validated may abort other transactions in this process. If this transaction is itself later aborted during global validation, it means that all the aborts it caused during local validation were unnecessary.

Key Applications

The number of real-time applications that handle real-time data is large. Thus, a range of applications that manage data with temporal constraints is given in the following.

- *Air-traffic control systems.* This system is used to monitor air-traffic around an airport, which requires continuous monitoring of aircraft positions and weather data, as well as (non-real-time) data of different aircraft types.
- *Control system.* A specific case is an engine control system, which approximately uses readings of 20 sensors, and these sensor values are fundamental in the computation of 400 other variables, where the validity of the derived data is a function of the sensed

value. Derived data is used control the mixture of fuel and air in the ignition, as well as diagnostics.
- Wireless networking systems for retrieval of subscriber information such as service subscription and device location stored in Home Location Register (HLR) as well as billing information in the case of prepaid phone calls. For example, the network communication protocols have various timers for call delivery that may cause connection failure if information is not retrieved before a transaction deadline.

Future Directions

The increasing number of applications that handle large amounts of real-time data calls for a strong support from the underlying transaction processing system to satisfy timeliness and consistency requirements. The development of time-cognizant concurrency control and scheduling algorithms has provided a foundation for real-time transaction processing systems. There are several challenging areas where progress would be conducive to enforce the predictability, and below a few examples are provided

- The underlying assumption which most real-time scheduling approaches build upon is the knowledge of the worst-case execution times of the transactions, either a priori to the system starts its execution or they are made available upon their arrival to the system. The dynamic nature of transaction workloads can cause the system to experience transient overloads. Techniques that are shown to effectively manage the uncertainty of execution times or can measure tight and not overly conservative worst case execution times would increase schedulability and resource utilization, as well as techniques for resolving and minimizing the effects of transient overloads.
- *Extended transaction models supporting relaxed ACID properties.* Enforcement of ACID in real-time applications has shown to be costly, jeopardize timeliness, and does not utilize the potential parallelism

that can be exploited by relaxation of the ACID properties. This has resulted in the development of, e.g., epsilon-serializability and the notion of data similarity. There is a need for additional application-centric consistency that would capture the tolerance of the applications and the presence of multiple models coexisting in parallel.

- *Transaction scheduling for guaranteeing data temporal consistency.* There is need to maintain coherency between the state of the environment and data used in applications such as control systems. For example, in order to react to abnormal situations in time, it is necessary to monitor the environment continuously. Such requirements pose a great challenge for maintaining the freshness of data while scheduling transactions to meet their deadlines.

- The notion of data precision and data confidence are becoming increasingly important in real-time applications, i.e., data is annotated with additional attributes to represent the confidence in the data value, (i.e., the level of trust in how the data was derived) and the precision/accuracy of the data. It is unclear how these two notions can be effectively used as decision parameters for concurrency control and schedulability.

Cross-References

▶ ACID Properties

Recommended Reading

1. Abbott R, Garcia-Molina H. Scheduling real-time transactions: a performance evaluation. In: Proceedings of the 14th International Conference on Very Large Data Bases; 1988.
2. Bestavros A, Fay-Wolfe V. Real-time database and information systems – research advances. Bostan: Kluwer Academic Publishers; 1997.
3. Carey M, Livny M. Conflict detection tradeoffs for replicated data. ACM Trans Database Syst. 1991;16(4):703–46.
4. Gray J, Reuter A. Transaction processing: concepts and techniques. Burlington: Morgan Kaufmann; 1992.
5. Haritsa J, Carey M, Livny M. Data access scheduling in firm real-time database systems. J Real Time Syst. 1992;4(3):203–41.
6. Haritsa J, Ramamritham K, Gupta R. The PROMPT real-time commit protocol. IEEE Trans Parall Distr Syst. 2000;11(2):160–81.
7. Ramamritham K. Real-time databases. Int J Distrib Parallel Databases. 1993;1(2):199–226.
8. Ramamritham K, Son SH, Cingiser DL. Real-time databases and data services. Real Time Syst J. 2004;28(2/3):179–215.
9. Son SH. Advances in real-time systems. Englewood Cliffs: Prentice-Hall; 1995.
10. Soparkar N, Korth HF, Silberschatz A. Time-constrained transaction management – real-time constraints in database transaction systems. Bostan: Kluwer Academic Publishers; 1996.
11. Stankovic J, Ramamritham K. What is predictability for real-time systems? Real Time Syst. 1990;2(4):247–54.
12. Xiong M., Han S., and Lam K-Y. A deferrable scheduling algorithm for real-time transactions maintaining data freshness. In: Proceedings of the 26th IEEE Real-Time Systems Symposium; 2005. p. 27–37.

Recall

Ethan Zhang[1] and Yi Zhang[2]
[1]University of California, Santa Cruz, CA, USA
[2]Yahoo! Inc., Santa Clara, CA, USA

Definition

Recall measures the coverage of the relevant documents of an information retrieval (IR) system. It is the fraction of all relevant documents that are retrieved. Consider a test document collection and an information need Q. Let R be the set of documents in the collection that are relevant to Q. Assume an IR system processes the information need Q and retrieves a document set A. Let $|R|$ and $|A|$ be the numbers of documents in R and A, respectively. Let $|R \cap A|$ denote the number of documents that are in both R and A. The *recall* of the IR system for Q is defined as $R = |R \cap A|/|R|$.

Key Points

Precision and recall are the most frequently used and basic retrieval performance measures. Many other standard performance metrics are based on the two concepts.

Cross-References

▶ Eleven Point Precision-Recall Curve
▶ F-Measure
▶ Precision
▶ Standard Effectiveness Measures

Receiver Operating Characteristic

Pang-Ning Tan
Michigan State University, East Lansing, MI, USA

Synonyms

Operating characteristic; Relative operating characteristic; ROC

Definition

Receiver operating characteristic (ROC) analysis is a graphical approach for analyzing the performance of a classifier. It uses a pair of statistics – true positive rate and false positive rate – to characterize a classifier's performance. The statistics are plotted on a two-dimensional graph, with false positive rate on the x-axis and true positive rate on the y-axis. The resulting plot can be used to compare the relative performance of different classifiers and to determine whether a classifier performs better than random guessing.

Historical Background

ROC analysis was originally developed in signal detection theory to deal with the problem of discriminating known signals from a random noise background [11]. It was first applied to the radar detection problem to quantify how effective targets such as enemy aircrafts can be identified according to their radar signatures. In the 1960s, ROC analysis was applied to experimental psychology and psychophysics [6]. The approach has subsequently found its application in a variety of areas including radiology, epidemiology, finance, weather forecasting, and social sciences. In machine learning, the benefits of using ROC analysis for evaluating and comparing the performance of classifiers was first demonstrated by Spackman [14]. The approach was brought to the attention of the data mining community by Provost and Fawcett [12], who proposed the idea of using a convex hull of ROC curves to compare multiple classifiers in imprecise and changing environments.

Foundations

Predictive accuracy has traditionally been used as the primary evaluation measure for classifiers. However, its limitation is well-documented, particularly for data sets with skewed class distributions [12]. ROC analysis provides an alternative way for measuring performance by examining the trade-off between the successful detection of positive examples and the misclassification of negative examples. The approach was originally developed for binary classification problems, where each example is assigned to either a positive or a negative class. When applying the classifier to a given example, four possible outcomes may arise: (i) true positive (TP), when a positive example is classified correctly, (ii) true negative (TN), when a negative example is classified correctly, (iii) false positive (FP), when a negative example is misclassified as positive, and (iv) false negative (FN), when a positive example is misclassified as negative. These outcomes can be tabulated in a

Receiver Operating Characteristic, Table 1 Confusion
matrix for a 2-class problem

		Predicted class	
		+	−
Actual class	+	TP	FN
	−	FP	TN

2×2 table known as the confusion matrix, as
shown in Table 1.

An ROC graph is constructed by examining
the true positive rate and false positive rate of
a classifier. The true positive rate (TPR), also
known as hit rate or sensitivity, corresponds to
the proportion of positive examples that are correctly labeled by the classifier, whereas the false
positive rate (FPR), also known as false alarm
rate, corresponds to the proportion of negative
examples that are incorrectly labeled. Mathematically, these statistics are computed from a given
confusion matrix as follows:

$$TPR = \frac{TP}{TP + FN} \qquad (1)$$

$$FPR = \frac{FP}{FP + TN} \qquad (2)$$

The number of points in an ROC graph depends on the type of output produced by the
classifier. A classifier that produces a discrete-
valued output is mapped to a single point in the
ROC graph because there is only one confusion
matrix. Other classifiers such as naïve Bayes and
neural networks can produce numeric-valued outputs to indicate the degree to which an example
belongs to the positive class. A threshold must be
specified to determine the class membership – if
the classifier's output exceeds the threshold, the
example is assigned to the positive class. Each
threshold setting leads to a different point in the
ROC graph. By varying the threshold, a piecewise
linear curve, known as the ROC curve, is formed.
For example, the solid line in Fig. 1c shows the
ROC curve obtained by varying the threshold
on classification outputs produced by a neural
network classifier.

There are several critical points in the diagram
that are of practical significance. The critical
point (FPR = 0, TPR = 0) corresponds to an
extremely conservative classifier, i.e., one that
assigns every example to the negative class. In
contrast, the critical point (FPR = 1, TPR = 1)
corresponds to a classifier that liberally declares
every example to be positive. Since the ideal classifier corresponds to the critical point (FPR = 0,
TPR = 1), points closer to the upper-left corner
of the ROC graph are generally better classifiers.
A random classifier, on the other hand, produces
points that reside along the diagonal line connecting the bottom-left to the upper-right corner of
the ROC graph, as shown by the dashed line in
Fig. 1c. For example, a classifier that randomly
assigns one-fourth of the examples to the positive
class has TPR = 0.25 and FPR = 0.25.

An ROC curve X dominates another ROC
curve Y if X lies above and to the left of Y. The
more dominant the ROC curve, the better the
classifier is. For example, Fig. 2 shows that classifier A is better than classifiers B and C because
it has a more dominant ROC curve. In practice,
however, one seldom finds an ROC curve that
completely dominates other ROC curves. Instead,
one would find different ranges of FPR values in
which one classifier is better than another. For
example, in Fig. 2, classifier C is better than B
when FPR below 0.2 or above 0.95. Therefore,
Provost and Fawcett [12] introduced the ROC
convex hull (ROCCH) method to combine the
ROC curves from a set of classifiers to obtain the
most dominant ROC curve. The convex hull can
be used to identify ranges of FPR values in which
a classifier is potentially optimal.

Although an ROC curve provides a visual
display of a classifier's performance, it is often useful to summarize the curve into a single
metric to estimate the overall classifier's performance. By analyzing the statistical properties of
the metric [10], this provides a quantitative way
to determine whether the observed difference
in performance of two classifiers is statistically
significant. Examples of ROC-derived metrics include area under ROC curve (AUC), slope inter-
cept index, and the ROC breakeven point. AUC is
also equivalent to the Wilcoxon-Mann-Whitney

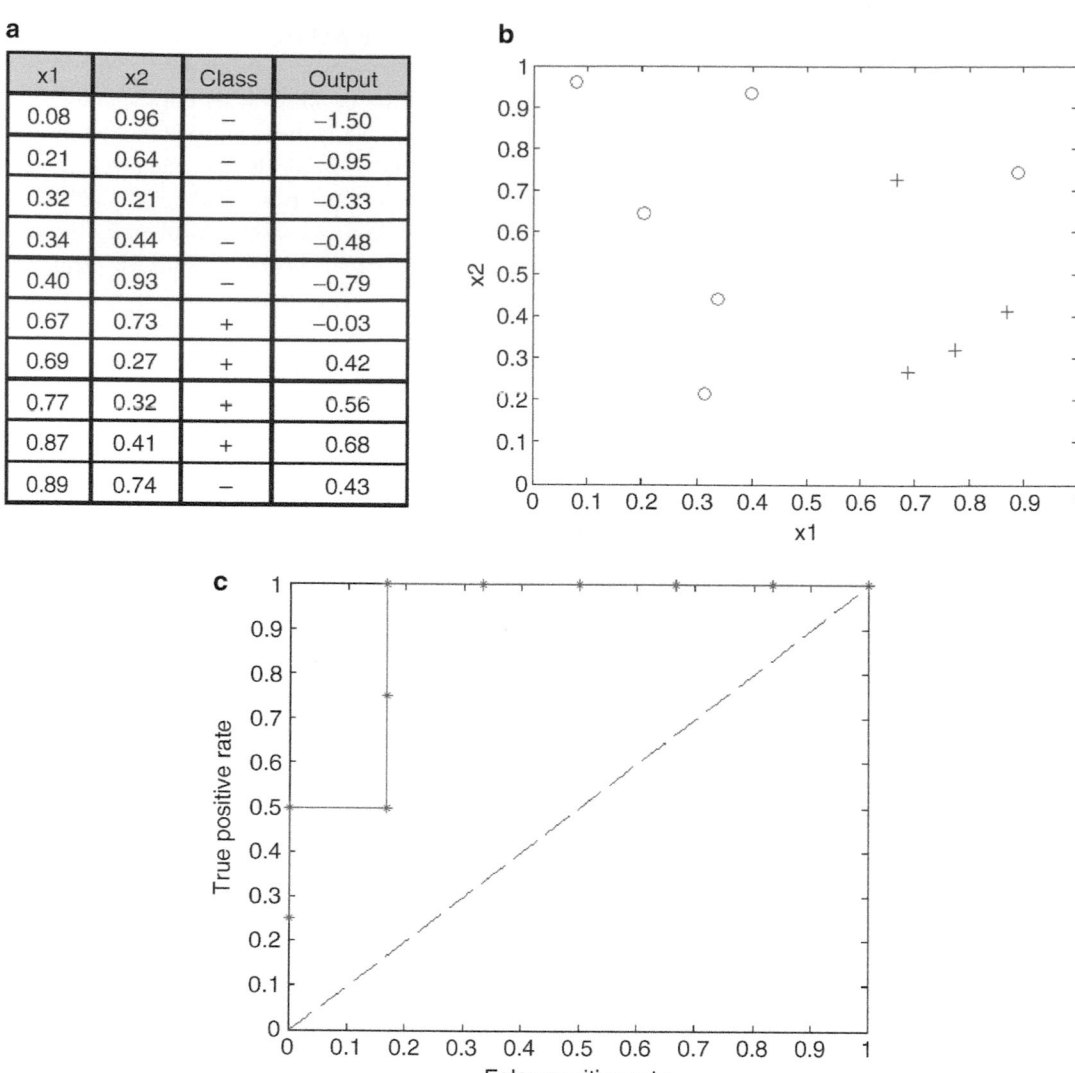

Receiver Operating Characteristic, Fig. 1 ROC curve for a two-dimensional data set classified using a single layer neural network

statistic [7], thus allowing us to compute its statistical properties such as standard error and confidence interval.

In addition to ROC curves, there are alternative ways to visualize the performance of a classifier such as precision-recall curves [2] and cost curves [3]. A precision-recall curve plots the tradeoff between precision, which is the fraction of examples classified as positive that are actually positive, against recall, which is equivalent to true positive rate. Davis and Goadrich [2] has shown that a curve that

dominates in the ROC space also dominates in the precision-recall space. Nevertheless, an ROC curve may not effectively capture important differences between classifiers when applied to data sets with skewed class distributions. For example, the confusion matrices shown in Table 2 have very similar TPR and FPR values even though their precision values are very different. The ROC representation also does not commit to any particular cost function or class distribution. As a result, it does not convey information such as the misclassification costs

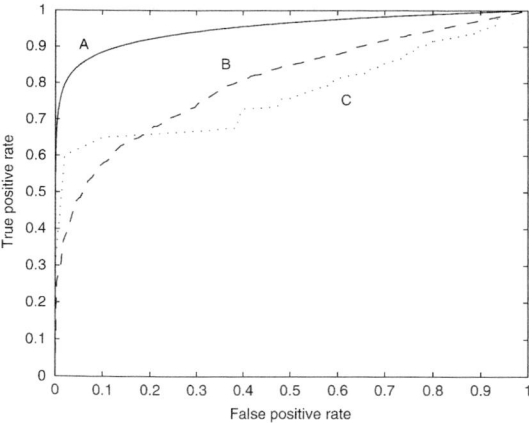

Receiver Operating Characteristic, Fig. 2
Performance comparison of classifiers using ROC curve

Receiver Operating Characteristic, Table 2
Confusion matrices to illustrate the difference between ROC and precision-recall curves

(a)

		Predicted class	
		+	−
Actual class	+	35	5
	−	25	435

(b)

		Predicted class	
		+	−
Actual class	+	35	5
	−	5	455

and class probabilities for which a classifier performs better than another. To overcome this limitation, Drummond and Holte [3] proposed the idea of using cost curves to explicitly represent the cost information. A cost curve plots the expected cost of a classifier against a probability cost function which is defined as follows:

$$\text{Probability Cost Function}$$
$$= \frac{P(+)\,C(-|+)}{P(+)\,C(-|+) + P(-)\,C(+|-)} \quad (3)$$

where $P(+)$ and $P(-)$ are the prior probabilities of each class, $C(-|+)$ is the cost of misclassifying a positive example as negative, while $C(+|-)$ is the cost of misclassifying a negative example as positive.

Key Applications

ROC analysis has been successfully applied to many application domains including psychology (in the studies of perception to resolve the issue of sensory threshold) [15], radiology (to distinguish between subjective judgment and objective detectability in imaging systems) [9], and epidemiology [13]. In addition to classification problems, ROC analysis is also applicable to other ranking problems such as recommender systems, information retrieval, and anomaly detection.

Future Directions

Most of the previous work on ROC analysis are limited to binary class problems. For multi-class problems, the analysis is more complicated as the confusion matrix is no longer a simple 2×2 table. An obvious way to extend the approach is to generate a different ROC graph for each class. However, with this approach, the ROC analysis is no longer insensitive to the class distribution [4]. Furthermore, combining the AUC statistics from multiple ROC graphs remains an open problem.

Another research direction that has attracted considerable interests in recent years is designing classification algorithms that directly optimize the area under ROC curve, or equivalently, the Wilcoxon-Mann-Whitney statistic. For instance, Cortes and Mohri [1] showed that, under certain conditions, the objective function optimized by the RankBoost algorithm is identical to AUC. New algorithms have also been developed to incorporate AUC into decision tree induction [5] and support vector machines [8].

URL to Code

The WEKA data mining software provides codes for plotting ROC curves and cost curves (http://www.cs.waikato.ac.nz/~ml/weka/.) The software for plotting ROC convex hull is available at http://home.comcast.net/~tom.fawcett/public_html/ROCCH/index.html.

R

Cross-References

► Classification

Recommended Reading

1. Cortes C, Mohri M. Auc optimization vs. error rate minimization. In: Advances in Neural Information Proceedings of the Systems 16, Proceedings of the Neural Information Proceedings of the Systems; 2003.
2. Davis J, Goadrich M. The relationship between precision-recall and roc curves. In: Proceedings of the 23rd International Conference on Machine Learning; 2006.
3. Drummond C, Holte RC. Explicitly representing expected cost: an alternative to roc representation. In: Proceedings of the 6th ACM SIGKDD International Conference on Knowledge Discovery and Data Mining; 2000. p. 198–207.
4. Fawcett T. An introduction to roc analysis. Pattern Recogn Lett. 2006;27(8):861–74.
5. Ferri C, Flach PA, Hernandez-Orallo J. Learning decision trees using the area under the roc curve. In: Proceedings of the 19th International Conference on Machine Learning; 2002.
6. Green DM, Swets JA, editors. Swets signal detection theory and psychophysics. New York: Wiley; 1966.
7. Hanley JA, McNeil BJ. The meaning and use of the area under a receiver operating characteristic (roc) curve. Radiology. 1982;143(1):29–36.
8. Joachims T. A support vector method for multivariate performance measures. In: Proceedings of the 22nd International Conference on Machine Learning; 2005.
9. Lusted LB. Signal detectability and medical decision making. Science. 1971;171(3977):1217.
10. McNeil BJ, Hanley JA. Statistical approaches to the analysis of the receiver operating characteristic (roc) curves. Med Decis Mak. 1984;4(2):137–50.
11. Peterson WW, Birdsall TG, Fox WC. The theory of signal detectability. IRE Trans. 1954;PGIT-4(4):171.
12. Provost FJ, Fawcett T. Analysis and visualization of classifier performance: comparison under imprecise class and cost distributions. In: Proceedings of the 3rd International Conference on Knowledge Discovery and Data Mining; 1997. p. 43–8.
13. Sackett DL. Clinical diagnosis and the clinical laboratory. Clin Invest Med. 1978;1:37.
14. Spackman KA. Signal detection theory: valuable tools for evaluating inductive learning. In: Proceedings of the 6th International Workshop on Machine Learning; 1989. p. 160–3.
15. Swets JA. The relative operating characteristics in psychology. Science. 1973;182(4116):990.

Recommender Systems

Mohamed Sarwat[1] and Mohamed F. Mokbel[2]
[1]School of Computing, Informatics, and Decision Systems Engineering, Arizona State University, Tempe, AZ, USA
[2]Department of Computer Science and Engineering, University of Minnesota-Twin Cities, Minneapolis, MN, USA

Synonyms

Recommendation engine

Definition

A recommender system (abbrv. *RecSys*) is a software artifact that suggests interesting items to users from a large pool of items. Let $\mathcal{U}$ be the set of users registered in the system and $\mathcal{I}$ denote the set of items. *RecSys* filters items based upon a utility function $\mathcal{F}(u, i)$ that predicts how much a user u (such that $u \in \mathcal{U}$) would like an item i (such that $i \in \mathcal{I}$). The system then recommends a set of items $\mathcal{I}'$ such that $\mathcal{I}' \subset \mathcal{I}$ and $\forall i \in \mathcal{I}'$: $\mathcal{F}(u, i) \geq \mathrm{argmax}_{j \in \mathcal{I} - \mathcal{I}'} \mathcal{F}(u, j)$. In other words, the system selects a set items $\mathcal{I}'$ among $\mathcal{I}$ that maximize the recommendation utility function $\mathcal{F}(u, i)$. The cardinality of $\mathcal{I}'$ is usually much less than $\mathcal{I}(|\mathcal{I}'| << |\mathcal{I}|)$. In many applications where the user $u(u \in \mathcal{U})$ already explored a set of items $\mathcal{I}_u \subset \mathcal{I}$, the system recommends only items that belong to the set of items $\mathcal{I}'_u$ (such that $\mathcal{I}'_u = \mathcal{I} - \mathcal{I}_u$) that was not explored by the user before.

Historical Background

Recommendations, as a general concept, existed and exist everywhere. In the early 1990s, the concept of recommender (software) systems has emerged in the computer science world [9]. Computer scientists have come up with several algorithms to generate relevant recommendations to users. In the early 2000s, recommender systems have become increasingly popular. Many Web users have experienced recommender systems in some way since then. Recommender systems helped internet users consume less time and effort in finding products (e.g., movies, books, music), information (e.g., news articles), or even other people (i.e., friends) of their interest. For instance, an online movie streaming website (like Netflix) recommends movies based on users' past preferences or interests. Recommender systems help users in two main ways. Firstly, they help users find what they are looking for. Secondly, they help with information overload as there is a lot of information on the Web, and recommender systems can automatically filter it out so that users only receive the most relevant information.

A recommender system takes as input a set of users $\mathcal{U}$, items $\mathcal{I}$, and ratings (history of user opinions over items) $\mathcal{R}$. It then estimates a utility function $\mathcal{F}(u, i)$ that predicts how much a certain user $u \in \mathcal{U}$ will like an item $i \in \mathcal{I}$ such that i has not been already seen by u [1]. To estimate such utility function, many recommendation algorithms have been proposed in the literature that can be classified as follows:

(1) Non-personalized: this class of algorithms leverages statistics and/or summary information to recommend the same interesting (e.g., the most highly rated) items to all users.
(2) Content-Based Filtering: this family of algorithms analyzes the items' content information and recommends to a user a set of items similar (in content) to those she liked before.
(3) Collaborative Filtering: these algorithms harness the historical preferences (tastes) of many users to predict how much a specific user would like a certain item.

Collaborative filtering recommenders falls into two main categories: (a) neighborhood based [1, 9, 10] that leverages the similarity between system users or items to estimate how much a user like an item and (b) matrix factorization [2, 5] that leverages linear algebra techniques to analyze the user/item ratings data

and hence predict how much a user would like an unseen item. In this entry, we focus more on collaborative recommendation algorithms.

Scientific Fundamentals

Collaborative Recommendation

Collaborative recommendation systems suggest items based on a collection of similar users and items. For example, if two users have shared similar interest in the past, this technique predicts that they will have similar interests in the future as well. Collaborative filtering assumes a set of n users $\mathcal{U} = \{u_1, \ldots, u_n\}$ and a set of m items $\mathcal{I} = \{i_1, \ldots, i_m\}$. Each user u_j expresses opinions about a set of items $\mathcal{I}_{u_j} \subseteq \mathcal{I}$. Many applications assume opinions are expressed through an explicit numeric rating (e.g., one through five stars), but other methods are possible (e.g., hyperlink clicks, Facebook "likes"). An active user u_a is given a set of recommendations $\mathcal{I}_r$ such that $\mathcal{I}_{u_a} \cap \mathcal{I}_r = \emptyset$, i.e., the user has not rated the recommended items. Collaborative recommendation techniques can be further divided into user-based nearest neighbor recommendation and item-based nearest neighbor recommendation. The user-based approach predicts a rating for a user based on the ratings that similar users have given to that item. The Pearson correlation coefficient is used as a primary measure to find out a set of similar users. The item-based recommendation is preferred as an alternative to the user-based approach when large data is involved because the user-based approach is too slow for real-time applications since it has to measure large number of possible neighbors. The item-based approach calculates the similarity between items instead of users. A cosine similarity measure could be employed to calculate the similarity between pairs of items. Probabilistic recommendation approach falls under collaborative recommendation. This approach uses conditional probability to calculate probabilities for every rating and then selects the rating that has the highest probability. It uses Bayesian classifier to estimate the

R

recommendation utility function $\mathcal{F}(u, i)$. Matrix factorization is another collaborative filtering approach which uses projections of a matrix in a 2D space to predict where the user exists in the 2D space.

Content-Based Recommendation

The family of content-based recommendation techniques [7] leverages items' content and the user profile to estimate the recommendation utility function $\mathcal{F}(u, i)$. This approach recommends only items for which the content is similar to whatever the user's preferences are. Unlike collaborative recommendation, this approach does not require a lot of similar users to give an accurate rating and does not have to wait for users to rate a new item as it can immediately predict rating for a new item once its basic attributes are known. However, the content-based approach requires updating the item descriptions and adding details at a regular basis which is costly and time consuming. Also, it is challenging to judge qualitative features of an item. A main component in the content-based recommendation system is content representation in which Dice coefficient may be used to measure similarity between two items. A recommender system can then match that item to a user whose requirements are a close fit. According to the importance of an items' elements, weights are assigned and stored in a vector using measures like TF-IDF (term frequency-inverse document frequency). The elements with higher weights are then recommended to end users. Similarity-based retrieval is another method that falls into the content-based category. The goal of this method is to recommend items similar to the ones that the user has liked in the past. Nearest neighbors approach is used to find out if an item will be of interest to a user. The recommender system finds out if the user has liked similar items in the past. To this end, the system analyzes the history of user preferences (like/dislike) of previous items. We also need to measure similarity between items which can be done using the cosine similarity function. Vector-space models can be used in content-based recommendation approaches. Relevance feedback approaches allow users to give feedback whether the retrieved items are relevant. This feedback allows the algorithm to separate the relevant document from the nonrelevant document which enables more refined results in the future. There is an increased trend in the leading search engines to show personalized results from a query that are specific to a user based on the user feedback in the past.

Hybrid Recommendation

Hybrid recommendation [1, 3] is obtained by combining multiple recommendation techniques into one system. Types of hybrid recommendation are as follows: (1) Monolithic hybridization consists of only one recommender which combines various approaches by integrating multiple knowledge sources. This method uses feature combination as well as feature augmentation. The feature combination hybrid combines item characteristics from content-based systems with user preferences from collaborative systems which enables it to have the best of both worlds. Feature augmentation combines multiple algorithms to predict a rating for the user. This approach is more complex than feature combination as it uses rating matrix to calculate the appropriate algorithm. (2) Parallelized hybridization has several recommenders and one hybridization mechanism to combine their outputs. This parallelized hybridization can employ mixed, weighted, or switching strategies. In the mixed hybrid approach, each recommendation component will generate a list of independent recommendations ranked according to their recommendation utility function $\mathcal{F}(u, i)$. Therefore, the recommendation generation algorithm combines the ranks of each item and makes a newly ranked list of recommendations. The weighted hybrid approach combines the recommendation scores calculated using all recommendation approaches. This approach requires that all involved recommenders provide a recommendation score $\mathcal{F}(u, i)$, which can be put in a linear formula to calculate the final recommendation scores. Finally, the switching hybrid recommender system switches between different recommendation components depending on situation, external criteria, and confidence value.

(3) Pipelined hybridization follows a sequential process where different recommendation techniques build on each other to produce one final improved recommendation for the user. Pipelined hybridization systems can employ cascade or meta-level hybrids. Cascade hybrids represent a chain that keeps on improving as every recommender improves or refines the recommendation of its previous recommender. Meta-level hybrids employ a single recommender to build a model that is examined by other recommenders to generate the final recommendation.

Key Applications

Movie Recommendation
A popular application for recommender systems is movie recommendation. For instance, Netflix uses both content-based and collaborative recommendation techniques to recommend movies to their users. MovieLens is another movie recommendation website that emerged in academia [8].

Personalized Search Engines and News Feed
Recommender system has been leveraged by major search engines to filter Web information based on users' personal preferences. Recommender systems have been also used to filter news feed [4] (and social media) to generate relevant news feed to users of social media websites (e.g., Facebook, Twitter).

Location-Based Services
Recommender systems have also been used to boost the performance of classical location-based services. For instance, recommender systems have been employed to recommend point of interests (POI) that are relevant to end users [12].

Data Sets

MovieLens
Three data sets (of different sizes) extracted from the MovieLens System (a real movie recommendation website). The largest MovieLens data set consists of 10 million ratings (scale from 1 to 5) assigned by 72,000 users to 10,000 movies. The MovieLens data sets also consist of users' demographic information like age and gender as well as movies' attributes like genre and release year.

Netflix Prize
A data set provided by Netflix to participants in the Netflix prize. The data set consists of 100 million ratings (scale from 1 to 5) assigned by 480,000 anonymous users to over 17,000 movies.

Yahoo! Music
This data set collected by Yahoo! Music services between 2002 and 2006. It contains over 717 million ratings of 136,000 songs given by 1.8 million users of Yahoo! Music services. The data set also provides attributes for each song like the artist, album, and genre.

URL to Code

LensKit
LensKit (http://lenskit.org) is a java-based open-source software that provides a generic implementation of various recommendation algorithms (e.g., collaborative filtering).

Apache Mahout
Mahout (https://mahout.apache.org) is an open-source software that implements machine learning and data mining algorithms on top of scalable cluster computing platforms (currently implemented on top of Hadoop). Mahout provides support for major recommendation algorithms that include user-based collaborative filtering, item-based collaborative filtering, and matrix factorization.

RecDB
RecDB (http://www-users.cs.umn.edu/~sarwat/RecDB/) [6,11] is a recommendation engine built entirely inside PostgreSQL (relational DBMS) to answer recommendation-based queries. RecDB provides support for major collaborative filtering

algorithms that includes user-/item-based and singular value decomposition.

Cross-References

▶ Collaborative Filtering

Recommended Reading

1. Adomavicius G, Tuzhilin A. Toward the next generation of recommender systems: a survey of the state-of-the-art and possible extensions. IEEE Trans Knowl Data Eng TKDE. 2005;17(6):734–49.
2. Breese JS, Heckerman D, Kadie C. Empirical analysis of predictive algorithms for collaborative filtering. In: Proceedings of the 14th Conference on Uncertainty in Artificial Intelligence; 1998.
3. Burke R. Hybrid recommender systems: survey and experiments. User Model User-Adap Inter. 2002;12(4):331–70.
4. Das A, Datar M, Garg A, Rajaram S. Google news personalization: scalable online collaborative filtering. In: Proceedings of the 16th International World Wide Web Conference; 2007.
5. Koren Y, Bell RM, Volinsky C. Matrix factorization techniques for recommender systems. IEEE Comput. 2009;42(8):30–7.
6. Levandoski JJ, Sarwat M, Mokbel MF, Ekstrand MD. RecStore: an extensible and adaptive framework for online recommender queries inside the database engine. In: Proceedings of the 15th International Conference on Extending Database Technology; 2012.
7. Lops P, de Gemmis M, Semeraro G. Content-based recommender systems: state of the art and trends. In: Recommender systems handbook. Springer; 2011. p. 73–105. https://link.springer.com/book/10.1007/978-0-387-85820-3
8. Miller BN, Alber I, Lam SK, Konstan JA, Riedl J. MovieLens unplugged: experiences with an occasionally connected recommender system. In: Proceedings of the International Conference on Intelligent User Interfaces; 2002.
9. Resnick P, Iacovou N, Suchak M, Bergstrom P, Riedl J. GroupLens: an open architecture for collaborative filtering of netnews. In: Proceedings of the ACM Conference on Computer Supported Cooperative Work; 1994.
10. Sarwar B, Karypis G, Konstan J, Riedl J. Item-based collaborative filtering recommendation algorithms. In: Proceedings of the 10th International World Wide Web Conference; 2001.
11. Sarwat M, Avery J, Mokbel MF. RecDB in action: recommendation made easy in relational databases. Proc. VLDB Endow. 2013;6(12):1242–5.
12. Sarwat M, Levandoski JJ, Eldawy A, Mokbel MF. LARS*: an efficient and scalable location-aware recommender system. IEEE Trans Knowl Data Eng. 2014;26(6):1384–99.

Record Linkage

Josep Domingo-Ferrer
Universitat Rovira i Virgili, Tarragona, Catalonia, Spain

Synonyms

Record matching; Re-identification

Definition

Record linkage is a computational procedure for linking each record a in file A (e.g., a file masked for disclosure protection) to a record b in file B (original file). The pair (a, b) is a match if b turns out to be the original record corresponding to a.

Key Points

Record linkage techniques were created for data fusion and to increase data quality. However, they have also found an application in measuring the risk of identity disclosure in statistical disclosure control. In the SDC context, it is assumed that an intruder has an external dataset sharing some (key or outcome) attributes with the released protected dataset and containing additionally some identifier attributes (e.g., passport number, full name, etc.). The intruder is assumed to attempt to link the protected dataset with the external dataset using the shared attributes. The number of matches gives an estimation of the number of protected records whose respondent can be re-identified by the intruder. Accordingly, disclosure risk is defined as the proportion of matches among the total number of records in A.

There are two main types of record linkage used to measure identity disclosure in SDC: distance-based record linkage and probabilistic record linkage.

Distance-based record linkage consists of linking each record a in file A to its nearest record b in file B. Therefore, this method requires a definition of a distance function for expressing *nearness* between records. This record-level distance can be constructed from distance functions defined at the level of attributes. Construction of record-level distances requires standardizing attributes to avoid scaling problems and assigning each attribute a weight on the record-level distance. A straightforward choice is to use the Euclidean distance, but other distances can be used.

The main advantages of using distances for record linkage are simplicity for the implementer and intuitiveness for the user. Another strong point is that subjective information (about individuals or attributes) can be included in the re-identification process by properly modifying distances.

The main difficulty of distance-based record linkage consists of coming up with appropriate distances for the attributes under consideration. For one thing, the weight of each attribute must be decided and this decision is often not obvious. Choosing a suitable distance is also especially difficult in the cases of categorical attributes and of masking methods such as local recoding where the masked file contains new labels with respect to the original dataset.

Like distance-based record linkage, probabilistic record linkage aims at linking pairs of records (a, b) in datasets A and B, respectively. For each pair, an index is computed. Then, two thresholds LT and NLT in the index range are used to label the pair as linked, clerical or non-linked pair: if the index is above LT, the pair is linked; if it is below NLT, the pair is non-linked; a clerical pair is one that cannot be automatically classified as linked or non-linked and requires human inspection. When independence between attributes is assumed, the index can be computed from the following conditional probabilities for each attribute: the probability $P(1|M)$ of coincidence between the values of the attribute in two records a and b given that these records are a real match, and the probability $P(0|U)$ of non-coincidence between the values of the attribute given that a and b are a real unmatch.

To use probabilistic record linkage in an effective way, one needs to set the thresholds LT and NLT and estimate the conditional probabilities $P(1|M)$ and $P(0|U)$ used in the computation of the indices. In plain words, thresholds are computed from: (i) the probability $P(LP|U)$ of linking a pair that is an unmatched pair (a *false positive* or *false linkage*) and (ii) the probability $P(NP|M)$ of not linking a pair that is a match (a *false negative* or *false unlinkage*). Conditional probabilities $P(1|M)$ and $P(0|U)$ are usually estimated using the EM algorithm.

Cross-References

▶ Disclosure Risk
▶ Inference Control in Statistical Databases
▶ Microdata
▶ Record Matching

Recommended Reading

1. Fellegi IP, Sunter AB. A theory for record linkage. J Am Stat Assoc. 1969;64(328):1183–210.
2. Torra V, Domingo-Ferrer J. Record linkage methods for multidatabase data mining. In: Torra V, editor. Information fusion in data mining. Berlin: Springer; 2003. p. 101–32.

R

Record Matching

Arvind Arasu[1] and Josep Domingo-Ferrer[2]
[1]Microsoft Research, Redmond, WA, USA
[2]Universitat Rovira i Virgili, Tarragona, Catalonia, Spain

Synonyms

Deduplication in Data Cleaning ; Duplicate detection; Entity resolution; Instance identification; Merge-purge; Name matching; Record linkage

Definition

Record matching is the problem of identifying whether two records in a database refer to the same real-world entity. For example, in Fig. 1, the customer record $A1$ in Table A and record $B1$ in Table B probably refer to the same customer, and should therefore be matched. (The example in Fig. 1 was adapted from an example in [21].) As Fig. 1 suggests, the same entity can be encoded in different ways in a database; this phenomenon is fairly common and occurs due to a variety of natural reasons such as different formatting conventions, abbreviations, and typographic errors. Record matching is often studied in the following setting: Given two relations A and B, identify all pairs of matching records, one from each relation. For the two tables in Fig. 1, a reasonable output might be the pairs $(A1, B1)$ and $(A2, B2)$. In some settings of the record matching problem, there is a constraint that each record of table A be matched with at most one record of table B. This asymmetric setting is typically used when records in table B are "clean" and those of table A, "dirty." The record matching problem is closely related to the *deduplication* problem. The focus of record matching is to identify pairs of matching records while that of deduplication is to partition records so that records in the same partition refer to the same entity. In practice, the output produced by record matching is inaccurate; it is not an equivalence relation and

does not correspond to a natural partitioning. The above distinction between record matching and deduplication is often ignored and the two terms are used synonymously.

Historical Background

Record matching has a rich history dating back to the work by Newcombe and others [16] in 1959. Fellegi and Sunter [10] formalize the intuition of Newcombe and others. Specifically, they cast the record matching problem as a classification problem that can be stated as follows: Given a vector of *similarity scores* between attribute values for a pair of records, classify the pair as a *match* or a *nonmatch*. For a given attribute, a similarity score indicates how similar the two records are on that attribute. A simple similarity measure assigns a score 1 if the records agree on the attribute and 0 otherwise. More sophisticated similarity measures are discussed subsequently. Fellegi and Sunter [10] use a naive Bayes classifier, but subsequent work has considered other kinds of classifiers such as decision trees [7] and SVMs [3]. The classifiers are typically trained using learning examples comprising of a set of record pairs, each labeled as a match or a nonmatch [3, 7, 10]. One of the problems with this approach is that learning examples required to train an accurate classifier are hard to generate, since they should not be either obvious

Record Matching, Fig. 1 Record matching example

a

Id	Name	Address	Age
A1	J H Smith	16 Main St	17
A2	Javeir Marteenez	49 Apple cross Road	33
A3	Gillian Jones	645 Reading Ave	24

b

Id	Name	Address	Age
B1	John H Smith	16 Main Street	17
B2	Javier Martinez	49 E Apple cross Road	33
B3	Jilliam Brown	123 Norcross Blvd	43

matches or obvious non-matches. Sarawagi and Bhamidipaty [17] address this problem and propose an approach based on active learning to interactively identify useful learning examples. Jaro [13] and Winkler [20] propose an alternate approach where the classifier is learned in an unsupervised setting without learning examples using a variant of EM (expectation maximization) algorithm.

A large class of work on record matching has focused on more sophisticated measures of similarity between attribute values of records. As mentioned above, these form the basis for classifying record pairs as a match or a non-match. A variety of classic similarity functions such as edit distance and variants [5] and cosine similarity with tf-idf (term frequency-inverse document frequency) weights [8] have been used for record matching. Jaro [12] proposes a more domain specific similarity measure designed for people names. Sarawagi and Bhamidipaty [17] propose using a weighted linear combination of simple similarity functions such as those mentioned above, and present techniques for learning the weights using training examples. Bilenko and Mooney [3] present a generalization of edit distance whose parameters can be learnt from a large text corpus. Arasu and others [1] present a framework for *programmable similarity*, where a similarity function can be programmed to be sensitive to synonymous words or phrases such as *Robert* and *Bob* or *US* and *United States*.

Another body of work on record matching has focused on efficiency issues. For large inputs, it is impractical to exhaustively consider all pairs of records and check if they are a match or not. Hernandez and Stolfo [11] present the *sorted neighborhood approach*, which linearly orders all the records based on a carefully selected key and considers only pairs of records that are close to each other in the linear ordering. McCallum and others [15] present an approach based on *canopies*, which are overlapping clusters of the input records. Only pairs of records within a cluster are checked for a match. Chaudhuri and others [6] identify *set-similarity join* as a useful primitive for large scale record matching.

Foundations

String Similarity

Most record matching approaches are based on the observation that two matching records have similar values for their attributes. For example, the records A2 and B2 have similar names and similar addresses and the same age, and are therefore likely to be matches. The similarity between two values is typically determined using a *similarity function* that takes two values and produces as output a number that quantifies the similarity of the values. String similarity functions that quantify the similarity between two strings are particularly relevant for record matching since many attributes in record matching are textual in nature. One of the earliest and well-known string similarity measure is *edit distance* or *Levenshtein distance*. The edit distance between two strings $s1$ and $s2$ is defined as the smallest number of edit operations required to produce $s2$ from $s1$, where an edit operation is an insertion, deletion, or substitution at the character level. For example, the edit distance between *Martinez* and *Marteenez* is 2 since one can derive the second string from the first using one substitution and one insertion. Edit distance is often a poor fit for record matching since two strings representing different entities can have small edit distance (e.g., *148th Ave NE* and *147th Ave NE*) and two strings representing the same entity, a large edit distance (e.g., *148th Ave NE* and *148th Ave Northeast*).

An alternate approach that works well for some domains is to treat strings as a bag of words (or *tokens*) and use a token-based similarity function such as *cosine* or *jaccard* (defined below). For a string s, let $Tokens(s)$ denote the set of tokens in s. For a token t, let $w(t)$ denote the *weight* of token t. The weight of a token represents its "importance" for the purposes of computing similarity. The jaccard similarity of two strings s_1 and s_2 is defined as:

$$\frac{|\ Tokens\ (s_1) \cap Tokens\ (s_2)\ |}{|\ Tokens\ (s_1) \cup Tokens\ (s_2)\ |}$$

For a set S, $|S|$ denotes the weighted cardinality of S, i.e., the sum of weights of the tokens in B.

The cosine similarity between two strings s_1 and s_2 is defined as:

$$\frac{\| Tokens\,(s_1) \cap Tokens\,(s_2) \|_2}{\| Tokens\,(s_1) \|_2 \cdot \| Tokens\,(s_2) \|_2}$$

For a set S, $\|S\|_2$ is defined as $\sqrt{\sum_{t \in S} w(t)^2}$. Token-based similarity functions differ from edit distance in two aspects: First, they ignore the ordering of tokens; for example, *148th Ave NE* represents the same bag of tokens as *NE 148th Ave*. This feature is desirable in some domains. For example, the list of authors appears before the title string in some citations and after the title in other citations. An order-sensitive similarity function would produce a low similarity score for two citation strings referring to the same publication but with different author-title orderings. Second, token-based similarity functions are not sensitive to intra-token edits. Two tokens are considered different even if they are textually very similar. As mentioned earlier, this feature is useful in domains such as addresses where, for example, two street names can differ by a single character. Token-based similarity functions enable weighting of tokens to capture their relative importance. A commonly used weighting scheme is the idf-based one, where the weight $w(t)$ of a token t is defined to be $log(N/N_w)$, where N is the number of records in a reference table and N_w denotes the number of records in the reference table containing the token t. With idf-based weighting, the similarity of the pair (*Applecross Rd*, *Applecross Road*) would be higher than the similarity of the pair (*Applecross Rd*, *Maltby Rd*) although both pairs have one token that is common to the two strings in the pair. This happens since the idf-weights of the rarer words *Applecross* and *Maltby* are higher than the weights for the common words *Road* and *Rd*. A related class of similarity functions is obtained by tokenizing strings to their character-level n-grams instead of words. For example, the set of 2-grams of the string *Applecross* is: {*Ap, pp, pl, le, ec, cr, ro, os, ss*}. These similarity functions share some characteristics with edit distance since they can

capture intra-word edits, but they are not as order sensitive as edit distance.

A variety of other domain specific similarity functions have been used for record matching. These are covered in depth in the survey by Elmagarmid and others [9] and the tutorial by Koudas and others [14].

Record Matching

Let R and S denote the input tables of record matching. The record matching problem can be viewed as a binary classification problem, where a given pair of records (r, s), $r \in R$ and $s \in S$, has to be classified either as a *match* or a *nonmatch*. A slight variant, not discussed here, is to consider a third category *possible match* consisting of hard-to-classify pairs that require manual inspection. One common approach for record matching is to train a standard binary classifier using learning examples consisting of a set of record pairs prelabeled as a match or nonmatch. The original Fellegi and Sunter approach uses a naive Bayes classifier and is discussed below. Other kinds of classifiers such as SVMs [3] and decision trees [7] have also been used.

Fix a record pair $(r, s) \in R \times S$. Define a *similarity vector* $\bar{x} = [x_1, ..., x_n]$, where each x_i denotes the similarity between r and s on some attribute A. The similarity can be computed using one of the functions discussed earlier (for string attributes). It can also be a simple binary value based on equality, i.e., $x_i = 1$ if r and s agree on the attribute and 0, otherwise. Let M denote the event that the pair (r, s) is a match and U the event that it is a nonmatch. Using Bayes rule, one gets

$$\Pr(M \,|\, \bar{x}) = \frac{\Pr(\bar{x}|M)\Pr(M)}{\Pr(\bar{x})}$$
$$\Pr(U \,|\, \bar{x}) = \frac{\Pr(\bar{x}|U)\Pr(U)}{\Pr(\bar{x})}$$

Therefore, $Pr\,(M\,|\,\bar{x}) \geq Pr\,(U\,|\,\bar{x})$ whenever,

$$\frac{Pr\,(\bar{x}|M)}{Pr\,(\bar{x}|U)} \geq \frac{Pr\,(U)}{Pr\,(M)}$$

The expression $\ell\,(\bar{x}) = Pr\,(\bar{x}|M)\,/\,Pr\,(\bar{x}|U)$ is called the *likelihood function*. The construction of the naive Bayes classifier assumes that x_i and x_j, $i \neq j$, are conditionally independent given M or U, implying:

$$Pr\left(\overline{x}|M\right) = \prod_{i=1}^{n} p\left(x_i|M\right)$$

$$\Pr\left(\overline{x}|U\right) = \prod_{i=1}^{n} p\left(x_i|U\right)$$

The probabilities $p(x_i|M)$ and $p(x_i|U)$ can be estimated using the learning examples, which can be used to estimate $Pr\left(\overline{x}|M\right)$ and $Pr\left(\overline{x}|U\right)$ using the expressions above. It is harder to estimate $Pr(U)$ and $Pr(M)$. A simple approach is to empirically pick a threshold T and classify the pair (r, s) as a match if $\ell(x) \geq T$ and a nonmatch otherwise.

A simple classifier that has performance advantages over more sophisticated machine-learning (ML) classifiers is the *threshold-based* classifier. Here, the record pair (r, s) is classified as a match if the similarity vector $\overline{x}$ satisfies:

$$(x_1 \geq T_1) \wedge (x_2 \geq T_2) \wedge \cdots \wedge (x_n \geq T_n)$$

where $T_1,...,T_n$ are thresholds that can be learned using labeled examples or set manually based on domain expertise. Record matching based on a threshold-based classifier can be performed efficiently using the string similarity join primitive discussed below. Chaudhuri and others [4] show that the record matching accuracy of union of several threshold-based classifiers is comparable to that of state-of-art ML classifiers.

A related approach, sometimes called *distance-based* [19] record matching, is to define a distance measure between two records by suitably combining the similarity (distance) scores of the attributes of the two records. A pair (r, s) is considered a match if the distance between r and s is smaller than a given threshold value. This approach is also easily extended to the asymmetric version of record matching where each record in R is constrained to match at most one record in S: For each record r in R, the record in S with the smallest distance to r is selected as a match.

More details of the other approaches used for record matching can be found in the survey by Elmagarmid et al. [9].

Performance

For large input tables R and S, it is impractical to consider every pair of records $(r, s) \in R \times S$ and classify them as a match or a nonmatch. Therefore, an important challenge in record matching is to identify the matching pairs without exhaustively considering every pair in $R \times S$. An important primitive for efficient record matching is *string similarity join*: Given two tables R and S, identify all pairs of records $(r, s) \in R \times S$ such that the string similarity of $r.A$ and $s.B$ is above some specified threshold, where the string similarity is measured using one of the similarity functions mentioned earlier. A string similarity join can be used to heuristically identify promising matching candidates, which can then be subjected to more complex classification. The string similarity join primitive can also be used to perform record matching based on the threshold-based classifier discussed above [4].

Since many of the string similarity measures such as jaccard and cosine are set-based, a more foundational primitive is *set-similarity join* [6]. A set-similarity join conceptually takes two collections of sets U and V as input and outputs all pairs of sets $(u, v) \in U \times V$ having high set-similarity. String-similarity joins can be evaluated using set-similarity joins as a primitive even for many non-set based similarity functions such as edit distance. A simple algorithm for set-similarity join involves building an inverted index over the collection V. The inverted index efficiently retrieves for any given element e all sets in V containing e. For each set u in U, the inverted index is used to retrieve all sets v in V that share one or more elements with u; the pair (u, v) is then produced as output if its satisfies the set-similarity join condition. More efficient algorithms for set-similarity joins are presented in [2, 18].

R

Key Applications

The primary application of record matching is data cleaning. The presence of duplicate records in a database adversely affects the quality of the database and its utility for analysis and mining. Another related application of record matching is data integration. The same real-world entity is often represented differently in other databases being integrated, and record matching is necessary to evolve a common representation for the entity. Another application of record matching is to measure the risk of identity disclosure in statistical disclosure control (SDC) [19]. In the SDC context, a dataset A (for public release) is produced from a source dataset B by masking values for identity protection. The goal of this step is to make it hard for an adversary to link a record in A to the record in B from which it was generated. The number of records in A that can be matched to records in B using the record matching techniques discussed above gives an estimate for disclosure risk of the released dataset A. Distance-based record matching is often used for this purpose.

Cross-References

▶ Column Segmentation
▶ Constraint-Driven Database Repair
▶ Data Cleaning
▶ Deduplication in Data Cleaning
▶ Record Linkage

Recommended Reading

1. Arasu A, Chaudhuri S, Kaushik R Transformation-based framework for record matching. In: Proceedings of the 24th International Conference on Data Engineering; 2008. p. 40–9.
2. Arasu A, Ganti V, Kaushik R. Efficient exact set-similarity joins. In: Proceedings of the 32nd International Conference on Very Large Data Bases; 2006. p. 918–29.
3. Bilenko M, Mooney, RJ. Adaptive duplicate detection using learnable string similarity measures. In: Proceedings of the 10th ACM SIGKDD International Conference on Knowledge Discovery and Data Mining; 2004. p. 39–48.
4. Chaudhuri S, Chen B.C, Ganti V, Kaushik R. Example-driven design of efficient record matching queries. In: Proceedings of the 33rd International Conference on Very Large Data Bases; 2007. p. 327–38.
5. Chaudhuri S, Ganjam K, Ganti V, Motwani R. Robust and efficient fuzzy match for online data cleaning. In: Proceedings of the ACM SIGMOD International Conference on Management of Data; 2003. p. 313–24.
6. Chaudhuri S, Ganti V, Kaushik R. A primitive operator for similarity joins in data cleaning. In: Proceedings of the 22nd International Conference on Data Engineering; 2006.
7. Cochinwala M, Kurien V, Lalk G, Shasha D. Efficient data reconciliation. Inf Sci. 2001;137(1–4):1–15.
8. Cohen WW. Data integration using similarity joins and a word-based information representation language. ACM Trans Inf Syst. 2000;18(3):288–321.
9. Elmagarmid AK, Ipeirotis PG, Verykios VS. Duplicate record detection: a survey. IEEE Trans Knowl Data Eng. 2007;19(1):1–16.
10. Felligi IP, Sunter AB. A theory for record linkage. J Am Stat Soc. 1969;64(328):1183–210.
11. Hernandez M, Stolfo S. The merge/purge problem for large databases. In: Proceedings of the ACM SIGMOD International Conference on Management of Data; 1995. p. 127–38.
12. Jaro MA. Unimatch: a record linkage system: user's manual. Technical Report. Washington, DC: US Bureau of the Census; 1976.
13. Jaro MA. Advances in record-linkage methodology as applied to matching the 1985 census of Tampa. Florida J Am Stat Assoc. 1989;84(406):414–20.
14. Koudas N, Sarawagi S, Srivastava D. Record linkage: similarity measures and algorithms. In: Proceedings of the ACM SIGMOD International Conference on Management of Data; 2006. p. 802–3.
15. McCallum A, Nigam K, Ungar LH. Efficient clustering of high-dimensional data sets with application to reference matching. In: Proceedings of the 6th ACM SIGKDD International Conference on Knowledge Discovery and Data Mining; 2000. p. 169–78.
16. Newcombe HB, Kennedy JM, Axford SJ, James AP. Automatic linkage of vital records. Science. 1959;130(3381):954–9.
17. Sarawagi S, Bhamidipaty A. Interactive deduplication using active learning. In: Proceedings of the 8th ACM SIGKDD International Conference on Knowledge Discovery and Data Mining; 2002. p. 269–78.
18. Sarawagi S, Kirpal A. Efficient set joins on similarity predicates. In: Proceedings of the ACM SIGMOD International Conference on Management of Data; 2004. p. 743–54.
19. Torra V, Domingo-Ferrer J. Record linkage methods for multidatabase data mining. In: Torra V, editor. Information fusion in data mining. Springer; 2003. p. 101–32.

20. Winkler W. Improved decision rules in the felligi-sunter model of record linkage. Technical Report. Washington, DC: Statistical Research Division/US Bureau of the Census; 1993.
21. Winkler W. The state of record linkage and current research problems. Technical Report. Washington, DC: Statistical Research Division/US Bureau of the Census; 1999.

Redundant Arrays of Independent Disks

Kazuhisa Fujimoto
Hitachi Ltd., Tokyo, Japan

Synonyms

Array; Disk array; RAID; Storage array

Definition

A set of disks from one or more commonly accessible disk subsystems is combined with a body of control software in which part of the physical storage capacity is used to store redundant information about user data stored on the remainder of the storage capacity. The redundant information enables regeneration of user data in a storage emergency in which a disk in the array or an access path fails.

Key Points

The term *RAID* was adopted from the 1988 SIG-MOD paper "A Case for Redundant Arrays of Inexpensive Disks (RAID)." In the paper, RAID refers to a group of storage schemes that divide and replicate data among multiple disks to provide better performance, cost and power consumption rate with reasonable availability comparing with conventional Single Large Expensive Disk (SLED) used for Mainframe. Currently, the term of "independent" is usually used rather than "inexpensive" because SLED becomes obsolete.

A number of standard schemes have evolved which are referred to as *levels*. Originally, five RAID levels were conceived, but many more variations have evolved. Currently, there are several sublevels as well as many non-standard levels.

The two key concepts in RAID are striping, and error correction using parity. The striping of RAID means dividing user data into fixed length blocks (striping units). The error correction using parity conceals one disk failure in an array of disks from the host computer and user data lost by one disk failure is regenerated using the parity.

Original five RAID levels are as follows:

1. Level 1: Mirrored disk. Two or more identical copies of data are maintained on separate disks.
2. Level 2: Using Hamming Code and striping with small striping unit (each read or write spread across all disks in an array). During read or write, on the fly ECC is adopted using Hamming Code in order to conceal some number of disk failures in an array.
3. Level 3: Using parity and striping with small striping unit (each read or write spread across all disks in an array). If one disk fails in an array, data on the failed disk are regenerated using corresponding parity and data on other disks in the array.
4. Level 4: Using parity and striping with large striping unit (small read or write can fit in one disk). One disk in an array is allocated to store the parity and user data is striped across remaining disks.
5. Level 5: Using parity and striping with large striping unit (small read or write can fit in one disk). Parity and user data is striped across all disks in an array.

In addition to these 5 levels, Level 0 is often used to represent arrays using striping without data redundancy in which fixed-length sequences of virtual disk data addresses are mapped to sequences of member disk addresses in a regular rotating pattern.

Cross-References

► Disk

Recommended Reading

1. Patterson D, Gibson G, Katz R. A case for redundant arrays of inexpensive disks (RAID). In: Proceedings of the ACM SIGMOD International Conference on Management of Data; 1988.

Reference Knowledge

Chintan Patel and Chunhua Weng
Columbia University, New York, NY, USA

Definition

Reference knowledge is the knowledge about a particular part of the world in a way that is independent from specific objectives, through a theory of the domain [2]. Different knowledge bases reuse or extend subsets of the reference knowledge for application specific tasks.

Key Points

Developing knowledge bases is a laborious process involving domain experts, knowledge engineers and computer scientists. To minimize this effort, often the core theory and concepts of a given domain are represented in a reference knowledgebase that are reused or extended by other application specific knowledge bases. Consider for example, in the biomedical domain, the Foundational Model of Anatomy [3] is considered a reference knowledge base for representing anatomy terms that can be used for representing the mouse anatomy.

One of the important advantages of using reference knowledge bases is that they provide interoperability [1] across the applications using the knowledge. The applications that heavily reuse or extend the reference knowledge achieve higher levels of interoperability.

The reference knowledge bases are generally developed without any application requirements or objectives. The reference knowledge imposes a rigid set of constraints on representing the application specific or local knowledge base. Such constraints sometimes conflict with the application or turn out to be difficult to implement or maintain, hence impeding the use of reference knowledge. Secondly, the reference knowledge bases tend to be very large [1] and creating subsets appropriate for small applications is challenging.

Recommended Reading

1. Brinkley J, Suciu D, Detwiler L, Gennari J, Rosse C. A framework for using reference ontologies as a foundation for the semantic web. AMIA Ann Symp Proc. 2006;2006:96–100.
2. Burgun A. Desiderata for domain reference ontologies in biomedicine. J Biomed Inform. 2003;39(3): 307–13.
3. Rosse C, Mejino J. A reference ontology for biomedical informatics: the foundational model of anatomy. J Biomed Inform. 2003;36(6):478–500.

Region Algebra

Matthew Young-Lai
Sybase iAnywhere, Waterloo, ON, Canada
Google, Inc., Mountain View, CA, USA

Definition

A region algebra is a collection of operators, each of which returns a set of regions as a result and takes as arguments one or more sets of regions. A *region* of a string is a pair of natural number positions (s, e) that correspond to the substring starting at s and ending at e. A position is a count of bytes, characters, or words from the beginning of the string.

Choosing a set of operators defines a particular region algebra. Operators are chosen for efficiency as well as utility. For example, if regions correspond to structure elements such as chapters and sections in a document, then many operators for querying structure conditions are useful and can be implemented efficiently. One example of such an operator is *containedIn(X,Y)* which takes two sets of regions X and Y and returns the subset of regions in X that are contained in some region of Y, i.e., $\{(s_x, e_x) \in X | \exists (s_y, e_y) \in Y (s_x \geq s_y) \wedge (e_x \leq e_y)\}$. A similar operator is *contains(X,Y)* which returns the subset of regions in X that contain some region of Y, i.e., $\{(s_x, e_x) \in X | \exists (s_y, e_y) \in Y (s_x \leq s_y) \wedge (e_x \geq e_y)\}$.

Another defining characteristic of a particular region algebra is the restrictions that apply to the sets of regions. If sets are unrestricted, then a string of length n has $\binom{n}{2}$ regions. This means that the worst case cost of evaluating a single operator cannot be linear in the length of the string. Thus, region algebras are often defined with restrictions on the nesting or overlap of regions in a set.

Historical Background

The first use of a region algebra for text search was in the PAT system [10]. The operators in PAT assume the use of a PAT array (also known as a suffix array) as the underlying implementation. This data structure can provide sets of matches for various lexical patterns (e.g., find all positions corresponding to a given substring of the text). It can also filter matches based on frequency or length conditions (e.g., return only the most frequent words matching a substring). The query language includes operators for dynamically combining matches into regions - a very flexible capability. It also provides set operators such as *union* and *difference* and structure operators such as *including*. Thus, it is a good example of the idea of using a region algebra to mix structure and content operations in text search. The system makes a distinction between sets of regions and sets of points which means that it does not completely fit the definition of

a region algebra. Its operators are typed in that their arguments and return values must be either sets of regions or sets of points. This means that the language is not fully compositional. Also, region sets are not allowed to include nesting or overlapping regions which causes semantic problems. For example, when an operator such as *union* results in overlapping regions, only the start points are returned.

Burkowski describes a region algebra for combined content and structure search in text [2, 1]. It treats both single words and structure elements spanning many words uniformly as regions. Thus it avoids the problems that result from distinguishing points and regions. Like PAT, region sets do not include nesting or overlapping regions. Later extensions to this work allow regions to overlap which is just as efficient [3]. An advantage of overlapping regions is that it makes the ability to dynamically define regions outside of a fixed hierarchical structure even more flexible and useful. The earlier papers by Burkowski describe simple structure operations such as *containing*, simple content operators for selecting words, and ranking operators based on inverse document frequency from traditional information retrieval. Later work explores ranking for information retrieval in more depth [6].

Jaakkola continues the pattern of imposing fewer restrictions on region sets [7, 8]. Sets are allowed to nest and overlap arbitrarily, abandoning the guarantee of linear size relative to the length of the string. However, the maximum depth of nesting tends to be constant in most data, and independent of the length of the data. This is true, at least, when talking about structure in text documents where region algebras are usually applied. Therefore, allowing arbitrary sets gives even more flexibility while not abandoning efficiency in practice.

Compared to relational algebra, there has been relatively little work exploring expressiveness or optimization issues with region algebras. Many systems make an informal effort to balance expressiveness with efficiency. For many useful algebras, the operators are restricted enough that the efficiency of evaluating arbitrary compositions is obvious without need of formal proof.

R

Some work does explore more general issues. For example, Consens and Milo examine equivalence testing and whether operators like direct inclusion can be efficiently supported [4, 5]. Young-Lai and Tompa look at characterizing operators that allow the possibility of efficient evaluation [12].

Foundations

The operators of a region algebra can combine and select regions. An important question, however, is where the regions originate. The most flexible approach is to scan the string for matches at every query. These matches can then be treated as regions and further manipulated using the algebra. In principle, any type of pattern language can be used for scanning. For example, it is possible to search for simple substrings, for regular expressions, or for words (possibly with linguistic processing such as stemming or lemmatization). It is also possible to parse the string with a grammar and return structures as regions. This is useful for data such as programming language source code where a well-defined grammar exists. For a scanning operation such as parsing with a grammar, the functionality of the parser starts to overlap with what can be accomplished with the region algebra. In fact, it is possible to use appropriately defined region algebra operators to simulate the process of parsing a string of tokens with a grammar.

If a string is too long to efficiently scan at every query, then regions can be pre-computed and stored persistently. This means making choices before-hand about what queries to support, and involves a tradeoff between flexibility and space. For example, consider storing a list of regions for every unique word. It is possible to apply a stoplist of common words to save space, but then it is impossible to search for phrases involving these stop words. There is also a tradeoff between efficiency and flexibility. For example, consider stemming the input by converting all forms of a word such as "stemming," "stemmed," and "stemmer" to the root word "stem." This does not reduce the number of stored regions (ignoring for the moment the fact that it can reduce the

total space requirement in some compressed representations). However, it reduces the flexibility since it is no longer possible to search for just one of the forms. On the other hand, it increases the efficiency of searching for all forms together since there is no need to combine separate lists. Overall, choosing what regions to pre-compute means predicting the types of queries that need to be supported.

Various representations are possible for persistently stored regions. The simplest is a collection of unordered lists. Given that region lists generated by scanning are ordered by position in the string, however, it makes sense to store them in that order. This can be done in a flat file which allows sequential access or binary searching. Alternatively, it can be in an index structure such as a B-tree which allows more efficient searches and updates. Ordered representations allow the possibility of particularly efficient compression using various techniques.

For a non-nesting set of regions, there is a single, unique sort order. For a set of regions with nesting there are two possible sort orders. That is, regions may ordered primarily by either s or by e. The choice has some effect on evaluation strategies for expressions that compose multiple operators. If regions are stored persistently, then a single sort order must be chosen. Of course, a physical operator can be provided to convert between the two orderings, although this is not possible in linear time and constant memory. Note that considering such an operator part of the algebra itself implies modifying the definition to use lists of regions with physical order properties rather than sets of unordered regions. Alternatively, one can make a distinction between the logical algebra that works with un-ordered sets, and the physical operators that are used for optimization and evaluation.

One way to avoid choosing between the two possible sort orders in persistent storage is to only index non-nested regions. This is less of a limitation than it might appear since it is easy to generate nested regions dynamically given two lists of endpoints. For example, it is possible to index the start and end tags for nested sections in a text in separate lists. The tags themselves have a unique order since they are not nested. A physical operator can scan and merge the two lists of tag regions, pushing the start tags onto a stack and popping the stack at every end tag. This results in a region list ordered primarily by e. A very similar operator can be used to dynamically generate regions that contain two words of interest. In this case, there is no natural pairing of word occurrences and the result may be as big as the cross product of the two lists. A useful solution is to define the operator so that it discards all but the shortest regions in the cross product, or equivalently, those that do not contain another nested region.

Given operands that are stored in an ordered list representation, there are many useful operators such as *contains* and *containedIn* that can be evaluated by scanning both lists and performing a type of merge join. This produces an ordered result that can then be used as an input to another operator. If an algebra consists exclusively of operators that can be evaluated with such merge strategies, then it is possible to evaluate arbitrarily composed expressions with very little query optimization effort and relatively good efficiency. A simple strategy is to evaluate the operator with the two smallest available inputs at each step until all operators in the tree are finished. Intermediate results can be buffered.

There are many ways to improve on this basic strategy. Some of them are implied by viewing the problem as an instance of relational query optimization and have not been explicitly described in the context of region algebras. For example, it is possible to consider additional physical operators such as index nested loop join. This can be used in the above strategy when joining a small operand with a larger one. Rather than scan the entire large operand, the join operator performs a binary search or index lookup within the large operand for each region in the small operand.

It is also possible to consider optimizing an entire query rather than making a local, greedy decision at each step about which operator to evaluate next. This implies the need to determine equivalences between expressions, to consider different physical evaluation plans based on these

equivalences, and to choose between the plans. The choice must involve estimating the costs of physical plans which is a difficult problem in general. However, it may admit heuristic or approximate solutions that do well in practice as is the case with relational query optimization.

There are some inherent limitations of the region model. One has to do with the limited information contained in a region. Given just numeric endpoints, it is easy to tell whether two given regions overlap or nest. However, nothing is known about their relationship with other regions that may exist. For example, given two regions a and b such that a contains b, there is no way to tell if a directly contains b meaning that there is no region c in the system such that a contains c and c contains b. Ignoring the possibility of constraints that may allow this to be inferred, the only way to know if this is the case is to search every other region list. This is likely to be inefficient.

By definition, operators in a region algebra are also excluded from referring to the content of a region. Thus, it is not possible to define an operator that takes the region endpoints and goes back to the original string to check some condition. Note that this is related to the issue of which regions have been indexed in a string. If there is a set of regions for "stem," an operator cannot check the string to see which of those regions were really the word "stemmed." However, even if there are separate region lists for "stem" and "stemmed," it is not possible to look up the value of words that follow "stem" in the text.

Of course, both of these limitations can be overcome by extending from regions to arbitrary tuples of information. To solve the direct containment problem, it is possible to store a third number *depth* with each region that indicates its nesting level. Then b is directly contained in a if it is contained in a and its depth is higher by one. Similarly, it is possible to add a text column to a region to store the word that follows the region and carry it with subsequent intermediate results. Essentially then, it is possible to consider extending to a full relational model. However, the region columns in such an extended model might admit the possibility of useful physical operators not generally provided in relational systems. For example, many region algebra operators can be executed with list merges but cannot be formulated as joins with equality conditions.

Another general limitation of region algebras has to do with the use of regions as a structure model. Consider a collection of region sets where each set has an associated type or label. This can represent arbitrarily nested structures in a strictly non-overlapping hierarchy. It can also represent multiple independent hierarchies over the same string. It can represent overlapping structures which goes beyond what is possible with a strict hierarchy. One thing that it cannot do is represent an arbitrary graph structure. For example, if regions are mapped to nodes in a graph, and there is an edge between any two nodes that overlap, then not all graphs are possible. For example, this can represent a chain of nodes with edges between them, but cannot add an edge from the first node in the chain to the last node without adding other edges at the same time. This limitation applies even if using regions only to encode a structure without requiring that the endpoints correspond to physical locations in a contiguous string.

Key Applications

A region algebra can be used as a query language for searching structured or semi-structured text, for searching structured or semi-structured data encoded in a text format such as SGML or XML, or for information retrieval. Alternatively, it can be used as an underlying model for query optimization or execution for some other query language.

As a query language, a region algebra can also serve as a component of a larger text processing task. One example is structure recognition where features such as composability and a loose structure model give advantages over alternatives such as grammars [11]. Other examples include constraint definition, outlier finding, and editing by example [9].

Cross-References

▶ Information Retrieval
▶ Relational Calculus
▶ Semi-Structured Query Languages
▶ XML
▶ XML Indexing

Recommended Reading

1. Burkowski FJ. Retrieval activities in a database consisting of heterogeneous collections of structured text. In: Proceedings of the 15th Annual International ACM SIGIR Conference on Research and Development in Information Retrieval; 1992. p. 112–125.
2. Burkowski FJ. An algebra for hierarchical organized text-dominated databases. Inf Process Manag. 1994;28(3):313–24.
3. Clarke CLA, Cormack GV, Burkowski FJ. An algebra for structured text search and a framework for its implementation. Comput J. 1995;38(1): 43–56.
4. Consens M.P. and Milo T. Algebras for querying text regions. In: Proceedings of the 14th ACM SIGACT-SIGMOD-SIGART Symposium on Principles of Database Systems; 1995. p. 11–22.
5. Consens MP, Milo T. Algebras for querying text regions: expressive power and optimization. J Comput Syst Sci. 1998;57(3):272–88.
6. Cormack GV, Clarke CLA, Palmer CR, Good RC. The multitext retrieval system (demonstration abstract). In: Proceedings of the 22nd Annual International ACM SIGIR Conference on Research and Development in Information Retrieval; 1999. p. 334.
7. Jaakkola J, Kilpelinen P. Using sgrep for querying structured text files. In: Proceedings of Standard Generalized Markup Language Finland 1996, Saarela J, editor. 1996. p. 56–67. Available as Report C-1996–83, Department of Computer Science, University of Helsinki, Nov 1996.
8. Jaakkola J, Kilpeläinen P. Nested text-region algebra. Technical Report C-1999-2, Department of Computer Science, University of Helsinki, Jan 1999.
9. Miller RC. Lightweight structure in text. Ph.D thesis, School of Computer Science, Carnegie Mellon University, 2002.
10. Salminen A, Tompa F. PAT expressions: an algebra for text search. Acta Linguistica Hungarica. 1992;41(1–4):277–306.
11. Young-Lai M. Text structure recognition using a region algebra. Ph.D thesis, Department of Computer Science, University of Waterloo, 2000.
12. Young-Lai M, Tompa FW. One-pass evaluation of region algebra expressions. Inf Syst. 2003;28(3):159–68.

Regulatory Compliance in Data Management

Radu Sion and Sumeet Bajaj
Stony Brook University, Stony Brook, NY, USA

Definition

Regulatory compliance in data management refers to information access, processing, and storage mechanisms designed in accordance to regulations. For example, in the United States, health-related data falls under the purview of the Health Insurance Portability and Accountability Act (HIPAA). Any associated healthcare data management systems need to be compliant with HIPAA requirements, including provision of data confidentiality and retention assurances. Such compliance has potential for far-reaching impact in the design of data processing systems.

Historical Background

In recent times, the increasing collection and processing of data have raised several concerns regarding data confidentiality, access, and retention. Driven by the concerns, regulators have enacted laws that govern all facets of data management. In the United States alone, over 10,000 regulations can be found in financial, life sciences, healthcare, and government sectors, including the Gramm-Leach-Bliley Act, the Health Insurance Portability and Accountability Act, and the Sarbanes-Oxley Act. A recurrent theme in data management regulations is the need for regulatory-compliant data management to ensure data confidentiality, data integrity, audit trail maintenance, data retention, and guaranteed deletion.

Regulations are enacted in the form of a legal document referred to as an *act*, a *directive*, a *program*, or an *agreement*. While each regulation has its own unique characteristics, certain assurance features are commonly found in most regulations. The common features include data confidentiality, audits, and data retention.

Data Confidentiality

Many entities collect personal information from individuals in order to provide services. Since personal information can be linked to specific individuals, illegitimate access to personal information violates individuals' privacy. Hence, regulations demand service providers to ensure that personal information is not accessed for any purpose other than the intended provision of services. Specifically, regulations require service providers to employ data confidentiality, de-identify personal information before distribution, seek explicit permission from individuals before disclosing personal information, and notify individuals in case of unauthorized access to individuals' personal information.

Audits

The purpose of audits is to collect evidence for noncompliance. Typically, regulations require periodic audits to verify operating practices of regulated entities. For example, the Corporate Law Economic Reform Program Act (CLERP9) in Australia requires audit of semiyearly and yearly financial statements of companies by certified auditors. CLERP9 mandates auditors to report within 28 days any circumstances that give the auditor reasonable grounds to suspect noncompliance. Similar laws exist in other countries, such as the German Corporate Governance Code [24], the Financial Instruments and Exchange Act (J-SOX) in Japan, and Keeping the Promise for a Strong Economy Act in Canada.

Data Retention

To ensure the availability of data for future audits, regulations mandate minimum retention periods for certain data. For example, the EU Data Retention Directive requires certain communications data to be retained for a minimum period of 6 months. Additionally, the directive also lists storage and protection requirements for the retained data. Section 103 of the Sarbanes-Oxley Act (SOX) requires corporations to maintain audit reports for a period of 7 years.

Regulations also stipulate the maximum retention periods for certain data. Once maximum retention period ends, a thorough and safe disposal of data is required. The goal of limiting retention periods is to prevent data retention past the intended use, thereby reducing the risks of data misuse. Examples of retention regulations that mandate maximum retention periods include the EU Data Retention Directive and the UK Data Protection Act.

Fundamentals

Regulatory-Compliant Commercial Systems

Regulatory-compliant systems have been brought to the market by several storage vendors, including IBM, HP, EMC, Hitachi, Oracle, Network Appliance, and Quantum Inc. A set of representative instances are discussed in the following.

Write-Once Read-Many (WORM) Storage

Under WORM semantics, once written, data cannot be undetectably altered or deleted before the end of regulation-mandated life span, even with physical access to the hosting server (Fig. 1). Therefore, WORM storage plays an important role in regulatory-compliant data retention. Following are some of the commercially available WORM storage systems.

Quantum Inc DLTSage: DLTSage [16] provides tape-based WORM assurances. WORM assurances are provided under the assumption that only quantum tape readers are used. In DLTSage, WORM integrity is ensured by placing an electronic key on each tape. It is claimed that the unique identifier cannot be altered ensuring a tamper-proof WORM storage system.

GreenTech WORMDisk: WORMDisk [10] was initially developed for a government agency but us now being provided to other organizations. WORMDisk protects data at the physical disk level. WORMDisk is compatible with existing server and laptop hardware and provides petabyte-level storage capacities.

Trusted Hardware

Achieving a secure, cost-effective, and efficient design in the presence of insider adversaries is extremely challenging. To defend against insiders, processing components that are tamper-resistant, such as general-purpose trusted hardware, are needed. By offering the ability to run logic within a secured enclosure, trusted hardware devices allow fundamentally new paradigms of trust. Trust chains spanning untrusted and possibly hostile environments can now be built by deploying secure tamper-resistant hardware at the storage components' site.

Regulatory Compliance in Data Management, Fig. 1 WORM prevents history "re-writing"

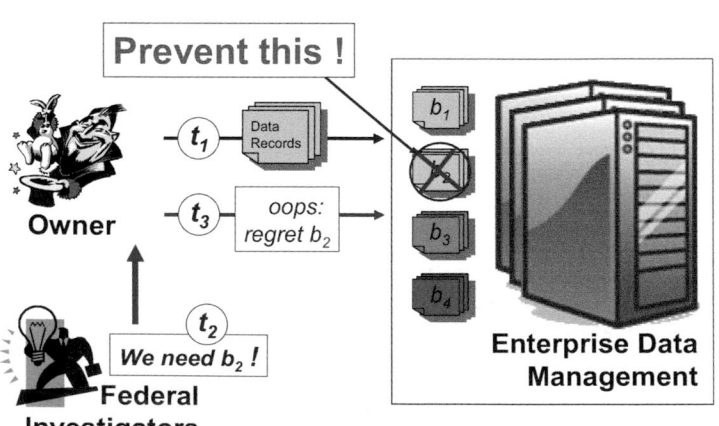

However, trusted hardware devices are not a panacea. Their practical limitations pose significant challenges in achieving efficient regulatory-compliance assurances. Specifically, heat dissipation concerns under tamper-resistant requirements limit the maximum allowable spatial gate density. As a result, general-purpose trusted hardware, such as the IBM 4764 [12], are significantly constrained in both computational ability and memory capacity, being up to one order of magnitude slower than host CPUs. Such constraints mandate careful consideration for the use of trusted hardware in data management. Direct implementations of full processing logic inside trusted hardware are bound to fail in practice due to lack of performance. Instead, efficient protocols are needed that access the trusted hardware sparsely and asynchronously from the main dataflow.

Software-Based Compliance Storage

Magnetic disk recording currently offers better overall cost and performance than optical or tape recording. Moreover, while immutability is often specified as a requirement for records, what is required in practice is that they be term "immutable," that is, immutable for a specified retention period. Thus, almost all recently introduced WORM storage devices are built atop conventional rewritable magnetic disks with write-once semantics enforced through software ("soft WORM"). Following are some examples of soft-WORM storage.

EMC Centera: The EMC Centera Compliance Edition [7] is a content-addressed storage (CAS) solution that also offers regulatory compliance. Each data record "has two components: the content and its associated content descriptor file (CDF) that is directly linked to the stored object (business record, e-mail, etc.). A digital fingerprint derived from the content itself is the content's locator (content address). The CDF contains metadata record attributes (e.g., creation date, time, format) and the object's content address. The CDF is used for access to and management of the record. Within this CDF, the application will assign a retention period for each individual business record. Centera will permit deletion of a pointer to a record upon expiration of the retention period. Once the last pointer to a record has been so deleted, the object will be eliminated" [7].

Network appliance snaplock compliance and enterprise software: The NetApp SnapLock software suite [14] is designed to work on top of NetApp NearStore and FAS storage systems. It provides soft-WORM assurances, "preventing critical files from being altered or deleted until a specified retention date." As opposed to other vendors, NetApp SnapLock supports open industry standard protocols such as NSF and CIFS.

Oracle StorageTek compliance archiving software: Oracle offers soft-WORM assurances through its StorageTek compliance archiving software [15]. The software runs on top of the Sun StorageTek 5320 NAS Appliance to "provide compliance-enabling features for authenticity, integrity, ready access, and security."

IBM LockVault compliance software: IBM offers multiple soft-WORM solutions. The compliance software is a layer that operates on top of IBM System Storage N series to provide "cost-effective, long-term retention of rapidly restorable disk-based backup" [11].

Regulatory-Compliant Data Management Systems Research

Over the last two decades, a large body of research has focused on regulatory compliance in data management. Also, with increasing awareness of and emphasis on regulatory research areas that were not intended to serve regulatory compliance at inception have now been identified as a solution toward regulatory compliance. Following is a broad overview of several research areas targeting regulatory compliance in data management.

Privacy-Preserving Data Publishing

The sharing of personal information is permitted by certain regulations as long as appropriate measures are taken to render data not individually identifiable. To be practical, data usefulness must be preserved when data is de-identified. Research in the area of Privacy-Preserving Data Publishing (PPDP) [4] has strived to achieve de-identification while preserving data useful-ness. The objective under PPDP is transforma-tion of the original data to render inferences about individually identifiable information un-likely.

Data Confidentiality

Encryption is commonly used to protect data residing with untrusted cloud services. The hope is that encryption will help data owners to stay compliant and benefit from the use of cloud services. However, encryption limits the type of operations that can be performed on data reduc-ing the functionality that cloud services can offer.

Both theoretical and systems research have focused to overcome the limitations imposed by encryption on computation. On the theoretical front, new mathematical constructs, such as ho-momorphic encryption [9], have been devised to enable computation over encrypted data without the need for decryption. However, implementa-tions of these constructs are not yet practical. For example, even for primitive operations, such as addition of two integers, the cost associated with homomorphic encryption are orders of mag-nitude higher that processing of plaintext data [1].

In order to be efficient, systems research has focused on specific scenarios, such as range and aggregation query processing over encrypted data. However, limiting functionality to a small subset of query operations reduces practicality.

Verification of Computation

Outsourcing data to potentially untrusted envi-ronments, such as a third-party cloud, raises con-cern over the correct operation of the remote services. The concern is especially serious when the liability for compliance is on cloud users. Using techniques for verifiable computation [8], a remote client can verify the correct execution of an outsourced computation. However, current techniques for verifiable computation are im-practical, consuming several orders of magnitude more resources as compared to unverified com-putation. To lower the costs of verifiable com-putation, research has focused on solutions for specific scenarios, such as range query verifica-tion in outsourced relational databases. Although the overheads of verification are relatively lower in specific targeted scenarios, the overheads are high enough to be a significant deterrent for widespread use [2].

Audit Logs

Regulations require maintenance of audit logs in various applications, such as drug approval data, medical information disclosure, financial records, and electronic voting. The collection and logging of system activity are not a particularly difficult task. The real challenges lie in protecting the integrity of recorded data and in analyzing the recorded data to determine compliance. To ad-dress the challenges of data integrity and analysis in audit logs, researches have designed tamper-proof audit logs [17], audit frameworks [5], and forensic tools. Audit frameworks enable applica-tion designers to specify policies in high-level languages, which can then be auto-enforced in data management systems. Forensic tools aid in analysis of audit logs.

Data Retention

Retention regulations stipulate both minimum and maximum data retention periods for the pur-pose of audit and privacy, respectively. Minimum retention period indicates the duration for which data must be retained. Maximum retention period mandates the time when data must be disposed and made irrecoverable.

Compliance storage [13] and trustworthy in-dexes [18] have been designed for secure reten-tion of data records. Compliance storage facili-tates the storage and protection of audit-related data. Data stored with compliance storage is protected from both tampering and deletion. In addition to data storage and protection, trust-worthy indexes also permit verified searches on stored data. Currently, trustworthy indexes have been designed for key-based lookups and range queries.

Secure deletion [6] mechanisms are proposed to render data irrecoverable on deletion. Secure deletion is achieved by either overwriting the data to be deleted or by using encryption. Overwriting deleted data or using encryption as in secure deletion does not ensure irrecoverability of deleted data [3]. Since the past existence of delete data affects the current system state implicitly at all layers, even after secure deletion, evidence of past existence of deleted data can be recovered by analyzing data side effects and current system state. Data side effects can be eliminated using history independence and untraceable deletion.

Key Applications

Recent compliance regulations are intended to foster human trust in digital information records and, more broadly, in our businesses, hospitals, and educational enterprises. As increasing amounts of information are created and stored digitally, compliance data management is a vital tool in ensuring trust and ferreting out corruption and data abuse.

Future Directions

In spite of recent efforts, many challenges yet lie on the path toward regulatory-compliant data management. The challenges are faced by both regulators and regulated entities. Challenges are in the form of scale and complexity of modern systems; advent of new computing environments, such as cloud services; complexity of regulatory framework; and high costs of compliance. Future research will need to explore novel data management solutions for low-cost, efficient, and increasingly automated regulatory compliance.

Cross References

▶ Privacy-Preserving Data Mining
▶ Trusted Hardware

Recommended Reading

1. Bajaj S, Sion R. Trusteddb: a trusted hardware based database with privacy and data confidentiality. In: Proceedings of the ACM SIGMOD International Conference on Management of Data; 2011. p. 205–16.
2. Bajaj S, Sion R. CorrectDB: SQL engine with practical query authentication. In: Proceedings of the 39th International Conference on Very Large Data Bases; 2013.
3. Bajaj S, Sion R. HIFS: history independence for file systems. In: Proceedings of the 20th ACM Conference on Computer and Communications Security; 2013.
4. Benjamin CM, Fung KW, Chen R, Yu PS. Privacy-preserving data publishing: a survey of recent developments. ACM Comput Surv. 2010;42(4): 14:1–53.
5. Cederquist JG, Corin R, Dekker MAC, Etalle S, den Hartog JI, Lenzini G. Audit-based compliance control. Int J Inf Secur. 2007;6(2):133–51.
6. Diesburg SM, Andy Wang An-I. A survey of confidential data storage and deletion methods. ACM Comput Surv. 2010;43(1):2:1–37.
7. EMC. Centera compliance edition plus. http://www.emc.com/data-protection/centera/compliance-edition-plus.htm.
8. Gennaro R, Gentry C, Parno B. Non-interactive verifiable computing: outsourcing computation to untrusted workers. In: Proceedings of the 30th Annual Conference on Advances in Cryptology; 2010. p. 465–82.
9. Gentry C. Fully homomorphic encryption using ideal lattices. In: Proceedings of the Annual ACM Symposium on Theory of Computing; 2009. p. 169–78.
10. GreenTec. Wormdisk. http://greentec-usa.com/wp-content/uploads/2012/05/GreenTec-WORM-Flyer-12-15-20131.pdf.
11. IBM. IBM system storage n series with open system snapvault. http://www-03.ibm.com/systems/storage/network/software/.
12. IBM 4764 PCI-X Cryptographic Coprocessor. Online at http://www-03.ibm.com/security/cryptocards/pcixcc/order4764.shtml.
13. Li T, Ma X, Li N. Worm-seal: trustworthy data retention and verification for regulatory compliance. In: Proceedings of the 14th European Conference on Research in Computer Security; 2009. p. 472–88.
14. Network Appliance Inc. Snaplock compliance and snaplock enterprise software. http://www.netapp.com/us/products/protection-software/snaplock.aspx.
15. Oracle. Storagetek 5320 nas appliance. http://docs.oracle.com/cd/E19783-01/index.html.
16. Quantum Inc. Dltsage: Write once read many solution. http://www.quantum.com/products/tapedrives/dlt/dltsageworm/index.aspx.
17. Schneier B, Kelsey J. Secure audit logs to support computer forensics. ACM Trans Inf Syst Secur. 1999;2(2):159–76.
18. Zhu Q, Hsu WW. Fossilized index: the linchpin of trustworthy non-alterable electronic records. In: Proceedings of the ACM SIGMOD International Conference on Management of Data; 2005. p. 395–406.

R

Relational Algebra

Val Tannen
Department of Computer and Information
Science, University of Pennsylvania,
Philadelphia, PA, USA

Definition

The operators of the relational algebra were already described in Codd's pioneering paper [2]. In [3] he introduced the term *relational algebra* and showed its equivalence with the tuple relational calculus.

This entry details the definition of the relational algebra in the *unnamed perspective* [1], with selection, projection, cartesian product, union and difference operators. It also describes some operators of the *named perspective* [1] such as join.

The flagship property of the relational algebra is that it is equivalent to the (undecidable!) set of domain independent relational calculus queries thus providing a standard for *relational completeness*.

Key Points

Fix a countably infinite set $\mathbb{D}$ of constants over which Σ-*instances* are defined for a relational schema Σ.

The relational algebra is a *many-sorted* algebra, where the sorts are the natural numbers. The idea is that the elements of sort n are finite n-ary relations. The *carrier* of sort n of the algebra is the set of finite n-ary relations on $\mathbb{D}$ If f is a many-sorted k-ary operation symbol that takes arguments of sorts $n_1, \ldots, n_k$ (in this order) and returns a result of sort n then its type is written as follows: $f: n_1 \times \ldots \times n_k \to n_0$, and this is simplified to n for nullary ($k = 0$) operations. Bold letters $\mathbf{x}$, $\mathbf{y}$ used for tuples and x_i for the i'th component of $\mathbf{x}$. The operations of the algebra, with their types and their interpretation over the relational carriers are the following:

constant-singletons $\{c\}$: $1(c \in \mathbb{D})$.

selection1 $\sigma_{ij}{}^n$: $n \to n$ ($1 \le i < j \le n$) interpreted as $\sigma_{ij}{}^n (R) = \{\mathbf{x} \in R \mid x_i = x_j\}$

selection2 $\sigma_{ic}{}^n$: $n \to n$ ($1 \le i \le n, c \in \mathbb{D}$) interpreted as $\sigma_{ic}{}^n (R) = \{\mathbf{x} \in R \mid x_i = c\}$

projection $\pi i_1{}^n \ldots i\ k$: $n \to k$ ($1 \le i_1, \ldots, i_k \le n$, not necessarily distinct) interpreted as $\pi_{i_1 \ldots i_k}^n (R) = \{x_{i_1}, \ldots, x_{i_k} \mid \mathbf{x} \in R\}$

cartesian(cross-) product $\times^{mn}$: $m \times n \to m + n$ interpreted as $\times^{mn} (R, S) = \{x_1, \ldots, x_m, y_1, \ldots, y_n \mid \mathbf{x} \in R \wedge \mathbf{y} \in S\}$

union$\cup^n$: $n \times n \to n$ interpreted as $\cup^n (R, S) = \{\mathbf{x} \mid \mathbf{x} \in R \vee \mathbf{x} \in S\}$

difference $-^n$: $n \times n \to n$ interpreted as $-^n(R, S) = \{\mathbf{x} \mid \mathbf{x} \in R \wedge \mathbf{x} \notin S\}$

Relational algebra *expressions* are built, respecting the sorting, from these operation symbols, using the relational schema symbols *as variables*.

Note that an obvious operation, intersection, is missing. Of course, intersection can be defined from union and difference, by De Morgan's laws. Interestingly, intersection is also definable just from cartesian product, selection, and projection. Other useful operations are definable also.

Given a relational schema Σ, a relational algebra *query* is an algebraic expression constructed from the symbols in Σ and the relational algebra operation symbols. Given a database instance I as input, such a query e returns a relation I as output. For example if R, S are binary, the expression $\pi_{2414}(\sigma_{13}(R \times S)) - (R \times R)$ defines a query that returns a 4-ary relation (omit the operation's superscripts because they can usually be reconstructed and use infix notation for the binary operations). Clearly, each of the operations of the relational algebra maps finite instances to finite relations, and more importantly, it is easily seen that *relational algebra queries are domain independent*. Moreover, there exists an (easily) computable translation that takes any relational algebra query into an equivalent domain independent FO (first-order) query.

The converse of this last fact is the main "raison d'être" for the relational algebra. However, because the set of domain independent FO queries is not decidable, it is not possible to define an effective translation just for these queries.

Instead, define a "translation" for *all* FO queries, such that domain independent FO queries are indeed translated to equivalent relational algebra queries. Recalling the notation $q(I/D)$ gives

Theorem

There exists a total computable translation that takes any FO query q into a relational algebra query e such that for any instance I, gives $e(I) = q(I/adom(I) \cup adom(q))$ (and for domain independent queries the right-hand side further equals $q(I)$.

The key to the proof is the observation that active domains can be computed in the relational algebra.

This result justifies Codd calling a query language *relationally complete* whenever it has the expressive power of the relational algebra. However, it is necessary to note that the relational algebra also inherits the negative results about first-order logic: it is undecidable whether there exists some instance on which a given query returns a non-empty answer (satisfiability) and it is undecidable whether two queries are equivalent.

The presentation above assumes that the relational symbols have just arity. This is the so-called *unnamed perspective* [1] and it is convenient for theoretical investigations. The practical descriptions of the relational model, e.g., [4], use the *named perspective* [1] in which a set of *attributes* is fixed and each relation is organized vertically by a finite set of them. For such as relation, a tuple is function from its attributes to constants. The relational algebra operators in the named perspective are similar to the ones above except that attributes are used instead of the integers identifying components of tuples. Complications arise with cartesian product when the two relations have attributes in common. Thus, in the named perspective an additional operator is needed for the *renaming* of a relation's attributes. On the positive side, using attributes leads to an nice generic definition of *natural join*, and operation that generalizes both cartesian product (when attribute sets are disjoint) and intersection (when attribute sets are identical) and which has many elegant properties.

Cross-References

▸ Cartesian Product
▸ Computationally Complete Relational Query Languages
▸ Difference
▸ First-Order Logic: Semantics
▸ First-Order Logic: Syntax
▸ Join
▸ Projection
▸ Relational Algebra
▸ Selection
▸ Union

Recommended Reading

1. Abiteboul S, Hull R, Vianu V. Foundations of databases: the logical level. Reading: Addison Wesley; 1994.
2. Codd EF. A relational model of data for large shared data banks. Commun ACM. 1970;13(6):377–87.
3. Codd EF. Relational completeness of database sublanguages. In: Rustin R, editor. Courant computer science symposium 6: data base systems. Englewood Cliffs: Prentice-Hall; 1972. p. 65–98.
4. Ramakrishnan R, Gehrke J. Database management systems. 3rd ed. New York: McGraw-Hill; 2003.

Relational Calculus

Val Tannen
Department of Computer and Information Science, University of Pennsylvania, Philadelphia, PA, USA

Synonyms

Domain relational calculus; First-order query; Tuple relational calculus

Definition

The relational database model was proposed by Codd in [2] where he assumed that its "data sublanguage" would be based on the predicate calculus (FOL) and where he introduced various algebraic operations on relations. Only in [3] did he introduced the terms *relational algebra* and *relational calculus*.

Later, it became customary to talk about the *domain relational calculus* (detailed below), which is closely related to the syntax of first-order logic and has quantified variables ranging over individual constants, and about the *tuple relational calculus* which is in fact the one given by Codd in [3] and whose variables range over tuples of constants. The two calculi are equivalent, via easy back and forth translations. However, both calculi allow the formulation of *domain dependent* queries which are inappropriate for database languages. While domain independence is undecidable, it is possible to define decidable sublanguages of *safe* queries which are themselves domain independent and such that any domain independent query is equivalent to a safe one.

Key Points

A *relational (database) schema* is a finite first-order vocabulary consisting only of relation symbols. In fact, relational formalisms also permit constants so it is necessary to fix a countably infinite set $\mathbb{D}$ of constants and work with formulae over the first-order vocabulary $\Sigma \cup \mathbb{D}$ where Σ is a relational schema. Also, the set $\mathbb{D}$ is taken as the sole *universe of discourse* for the interpretation of formulae. A *relational (database) instance* for a given schema Σ (a Σ-*instance*) is a first-order structure whose domain, or universe, is $\mathbb{D}$ and in which the relation symbols are interpreted by *finite* relations, while the constants are interpreted as themselves.

For a given schema Σ, a *domain relational calculus query* (a.k.a. first-order query) has the form $\{\langle e_1, \ldots, e_n \rangle \mid \phi\}$ where $e_1, \ldots, e_n$ are (not necessarily distinct) variables or constants and ϕ is a first-order formula over the vocabulary $\Sigma \cup \mathbb{D}$ with equality such that the all free variables of ϕ occur among $e_1, \ldots, e_n$. The *inputs* of the query are the

Σ-instances. For each input I, the *output* of the query $q \equiv \{\langle e_1, \ldots, e_n \rangle \mid \phi\}$ is the n-ary relation

$$q(\mathcal{I}) = \left\{ \langle \bar{\mu}(e_1), \ldots, \bar{\mu}(e_n) \rangle \mid \text{assignment } \mu \text{ such that } \mathcal{I}, \mu \models^2 \varphi \right\} \quad <?pag?>$$

Let $\mathcal{I}$ be an instance. The *active domain* of I, notation adom(I), is the set of all elements of $\mathbb{D}$ that actually appear in the relations that interpret Σ in I. While $\mathbb{D}$ is infinite, adom(I) is always finite. Moreover, given a query $q \equiv \{\langle e_1, \ldots, e_n \rangle \mid \phi\}$ adom(q) is denoted by the (finite) set of constants that occur in ϕ or among $e_1, \ldots, e_n$. One expects that the instance I together with the query q completely determines the output q(I). In particular, only the elements in adom(I) $\cup$ adom(q) can appear in the output. However, this is not the case for all first-order queries. For example, the outputs of $\{x \mid \neg R(x)\}$ or $\{\langle x, y \rangle \mid R(x) \vee S(y)\}$ are in fact infinite! More subtly, the following query is also problematic: $\{x \mid \forall y R(x, y)\}$. Here the output contains only elements from adom(I) but whether a tuple is in the output or not depends on the set of elements that y ranges over. These queries are "dependent on the domain." More precisely, for any instance I and any D such that adom(I) $\cup$ adom(q) $\subseteq$ D $\subseteq$ D, denote by q(I/D) the output of the query q on the input structure obtained by restricting the domain to D. A query q is *domain independent* if for any I and any D_1, D_2 where adom(I) $\cup$ adom(q) $\subseteq D_i \subseteq D$, $i = 1, 2$ one has $q(I/D_1) = q(I/D_2)$. It is generally agreed that in a reasonable *query language*, all the queries should be domain independent. Therefore, general first-order queries do not make a good query language. Worse, it is undecidable whether a first-order query is domain independent [1]. So how does one get a reasonable query language? It is possible (in several ways) to define decidable *safety* restrictions on general first-order formulae such that the safe queries are domain independent and moreover for any domain independent query there exists an equivalent safe query [4, 1]. The safety restrictions tend to be complicated and have little practical value. A better idea is the *relational algebra*.

Clearly, not all functions that map instances to relations are meanings of first-order queries. However, the meanings of first-order queries are all *generic*, i.e., invariant under bijective renamings of the constants in D\adom(q). Conversely, a function f that maps instances to relations and is generic is said to be *first-order definable* (a.k.a. definable in the relational calculus) if there exists a first-order query q whose semantics is f (contrast this with the definability within a given structure described in FIRST-ORDER LOGIC: Semantics). A well-known example of function that is not first-order definable is taking the *transitive closure* of a binary relation.

Codd's *tuple relational calculus* [4, 3] differs from the domain relational calculus in that its variables range over tuples and its terms are either constants or of the form $t.k$ where t is a variable and k is a positive integer selecting one of the components of the tuple (for example if t is assigned to (a,b,c) then the meaning of $t.2$ is b).

Cross-References

▶ Computationally Complete Relational Query Languages
▶ First-Order Logic: Semantics
▶ First-Order Logic: Syntax
▶ Relational Algebra

Recommended Reading

1. Abiteboul S, Hull R, Vianu V. Foundations of databases: the logical level. Reading: Addison Wesley; 1994.
2. Codd EF. A relational model of data for large shared data banks. Commun ACM. 1970;13(6):377–387.
3. Codd EF. Relational Completeness of Database Sublanguages. In: Rustin R, editor. Courant computer science symposium 6: data base systems. Englewood Cliffs: Prentice-Hall; 1972. p. 65–98.
4. Ullman JD. Principles of database and knowledge-base systems volume, I. Rockville: Computer Science Press; 1988.

Relational Model

David W. Embley
Brigham Young University, Provo, UT, USA

Synonyms

Relational database

Definition

The *Relational Model* describes data as named relations of labeled values. For example, customer ID's can relate with customer names and addresses in the relational model as *Customer*: {<(*CustomerID*, *11111*), (*Name*, *Pat*), (*Address*, *12 Maple*)>, <(*CustomerID*, *22222*), (*Name*, *Tracy*), (*Address*, *44 Elm*)>}. In this example, there is a name for the relation - *Customer*; label-value pairs - e.g., (*CustomerID*, *11111*), which provide the labeled values; and tuples - e.g., <(*CustomerID*, *11111*), (*Name*, *Pat*), (*Address*, *12 Maple*)>, which are the tuples of the named relation.

Usually, the relations of the relational model are viewed as tables. Fig. 1 shows an example of several relations viewed as tables. Together, they constitute a relational database. The first table in Fig. 1 is the table view of the relation described in the previous paragraph.

Besides their structures, tables in the relational model also have constraints. Typical constraints include key constraints (e.g., *CustomerID* values must be unique), type constraints (e.g., *Date-Ordered* values must be of type *Date*), and referential integrity constraints (e.g., the *CustomerID* values in table *Order* must refer to existing *CustomerID* values in table *Customer*).

Historical Background

Codd's seminal paper [2] introduced the relational model as a model based on n-ary relations. The seminal paper also introduces normal forms

R

```
create table Customer (
    CustomerID numeric(5) primary key,
    Name varchar(20),
    Address varchar(25),
    unique (Name, Address)
);

create table Item (
    ItemNr char(7) primary key,
    Description varchar(50)
);

create table Order (
    OrderNr varchar(10),
    CustomerID numeric(5) references Customer,
    ItemNr char(7) references Item,
    DateOrdered date,
    NrOrdered smallint,
    primary key (OrderNr, CustomerID, ItemNr ),
);
```

Relational Model, Fig. 3 SQL implementation schemas

must declare each of the table's attributes. *Customer* is the name for the first table declared in Fig. 3.

- A developer declares an attribute by giving its name and then listing the constraints that apply to the attribute. The attributes in the *Customer* table declared in Fig. 3 are *CustomerID*, *Name*, and *Address*.

- A developer declares domain constraints for an attribute by giving a type declaration. In Fig. 3 the type declaration for *CustomerID* is *numeric* (5), declaring that customer IDs are 5-digit numbers. SQL provides various types such as numbers, strings, dates, time, and money. Type declarations are unfortunately not uniform across all database systems.

- A developer declares key constraints for an attribute either by stating that it is the *primary key* or that it is *unique* – a key but not the primary key. In Fig. 3 *CustomerID* in the first table schema and *ItemNr* in the second table schema are primary keys.

- A developer declares foreign-key constraints for an attribute with a *references* clause. The *references* clause designates the table in which the referenced attribute is found. When the attribute in the referenced table has the same

name as the attribute in the referencing table, this simple declaration is sufficient. If the name is different, then the *references* clause must also include the name of the attribute being referenced. In Fig. 3, *CustomerID* in the *Order* table references *CustomerID* in the *Customer* table, and *ItemNr* references *ItemNr* in the *Item* table. These foreign-key constraints ensure that the *CustomerID* and *ItemNr* values in the *Order* table refer to existing values in the *Customer* and *Order* tables.

- When constraints involve multiple attributes, SQL provides syntax that allows a developer to declare these constraints in a separate entry in a table declaration. Thus, as Fig. 3 shows, a developer can declare that the attribute combination consisting of *Name* and *Address* constitutes a key for the *Customer* table, and that the attribute combination consisting of *OrderNr*, *CustomerID*, and *ItemNr* constitutes the primary key for the *Order* table. The SQL syntax also provides for multiple-attribute foreign keys. Thus, although neither necessary nor even desirable in the example in Fig. 3, the attributes *Name* and *Address* could be added to the *Order* table and a foreign-key constraint could then be declared as *foreign key (Name, Address) referencesCustomer (Name, Address)*.

- Typical additional constraints declarable with SQL include null constraints and check constraints. Null constraints let developers decide whether null values can or cannot appear as values for attributes. Values for primary-key attributes may never be null; other attributes require a *not null* designation (otherwise they can have null values). Check constraints let developers add conditions that must hold. For example, a developer can declare that *NrOrdered* can be neither negative nor zero by adding the constraint *check(NrOrdered > 0)* to the *NrOrdered* attribute in the *Order* table in Fig. 3.

Formal View

The formal view of the relational model captures the essence of a relational schema in terms of

Relational Model, Fig. 4
Sample table for
illustrating the formal
definition

$$r = \begin{array}{cc} A & B \\ a & 1 \\ b & 1 \\ b & 2 \end{array}$$

mathematical concepts. The definition is based on the concepts of sets, relations, and functions.

A *relational schema R* is a non-empty set of attribute names $R = \{A_1, A_2,...,A_n\}$. Usually an "attribute name" as just called an "attribute." Further, as a notational convenience, when an attribute name is a single letter, the set notation is reduced by dropping the braces, commas, and spaces. Thus, a *relational schema R* is a non-empty set of attributes $R = A_1 A_2... A_n$. Each attribute A has a domain, denoted $dom(A)$, which is a set of values.

A relation is always defined with respect to a relational schema. The notation $r(R)$ denotes that relation r is defined with respect to a relational schema R and is read "r is a relation on schema R" or just "r on R" when the context is clear. A *relation* is a set of n-tuples, $\{t_1,...,t_k\}$, where n is $|R|$, the cardinality of R. Let schema R be $A_1 A_2... A_n$. Then, an *n-tuplet* for a relation $r(R)$ is a function, from R to the union of domains $D = dom(A_1) \cup dom(A_2) \cup ... \cup dom(A_n)$, with the restriction that $t(A_i) \in dom(A_i)$, $1<= i <= n$.

A relation is usually written as a table - the table in Fig. 4, for example. In Fig. 4 the relation r has relational schema $R = AB$. To declare the attribute domains, assume $dom(A)$ is $\{a, b, c\}$ and $dom(B)$ is $\{0,...,9\}$. The set of n-tuples for r is the set of discrete functions $\{\{(A, a), (B, 1)\}, \{(A, b), (B, 1)\}, \{(A, b), (B, 2)\}\}$.

Another way to view the relational model formally is to view it as an interpretation in first-order logic. In this view of the relational model each relation r, whose schema has n attributes, is an n-place predicate. An *interpretation* for a first-order language consists of (1) a non-empty domain D of values, which under the unique name assumption each represent themselves, and (2) for each n-place predicate, an assignment of *True* or *False* for each possible substitution of n values from D.

As an example, consider the relation in Fig. 4. To see this relation as an interpretation, let $r(x, y)$ be a two-place predicate. The domain D is {a, b, c, 0, 1, 2, 3, 4, 5, 6, 7, 8, 9}. *True* is assigned to the substitutions $(x = a, y = 1)$, $(x = b, y = 1)$, and $(x = b, y = 2)$ and *False* is assigned to all other substitutions.

Most often in this first-order-logic view, the "closed world assumption" is used to conveniently assign *True* and *False* to each substitution. The "closed world assumption" states that whatever is not *True* is *False*, and thus only the *True* facts need to be recorded. This, then, reduces to the equivalent of just giving a table such as the one in Fig. 4 – the rows in the table represents the substitutions for which the predicate is *True* and the only substitutions for which the predicate is *True*.

In a first-order-logic view of the relational model, it is also possible to express constraints. For the table in Fig. 4, for example, the constraint that A should be a key can be expressed by the closed formula $\forall x_1 \forall y_1 \forall x_2 \forall y_2 (r(x_1, y_1) \wedge r(x_2, y_2) \wedge x_1 = x_2 \Rightarrow y_1 = y_2)$. This does not hold for the table in Fig. 4, however, since the second and third tuples have $x_1 = x_2$, but $y_1 \neq y_2$. A check constraint stating that the values in the B column must all be less than 5 can be expressed as $\forall x \forall y(r(x, y) \Rightarrow y < 5)$. This constraint does hold in the table in Fig. 4.

For an interpretation for a first-order language, when all the closed formulas hold, the interpretation is said to be a *model*. Database instances in which all constraints hold are therefore models – models of the world they represent.

Key Applications

Relations, stored in relational databases, are widely used in industry. Indeed, their combined usage constitutes a mega-billion-dollar industry. Relational databases range from relatively small databases used as backends to web applications such as "items for sale" to relatively large databases used to store corporate data.

Relevance Feedback for Content-Based Information Retrieval, Fig. 1 CBIR with RF

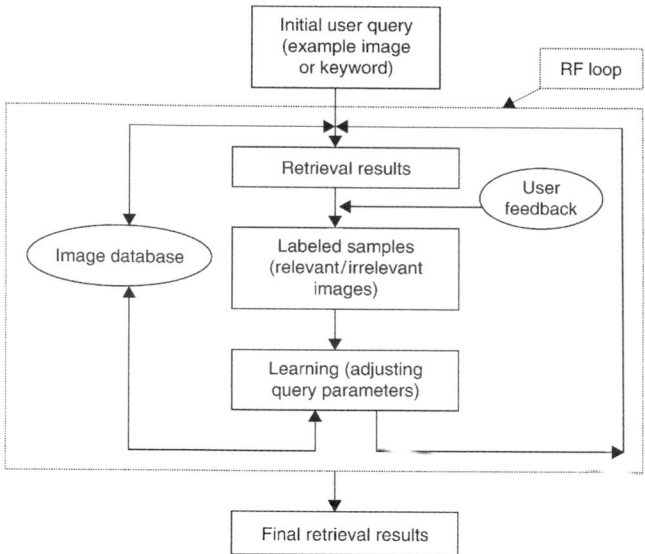

2. There are some relatively simple relations between the topology of the description space and the characteristics shared by the images that the user is searching for.
3. Relevant images are a small part of the entire image database.
4. While some of the early work on RF assumed that the user could (and would be willing to) provide a rather rich feedback, including relevance notes for many images, currently the assumption is that this feedback information is scarce. The user will only mark a few relevant images as positive and some very different images as negative.

Figure 1 shows the flowchart of a typical Content-based image retrieval process with relevance feedback [3]. Based on the above general assumptions, a typical scenario for relevance feedback in content-based image retrieval is as below [3, 9]:

1. The system provides initial retrieval results given query examples;
2. User judges the above results as to what degree, they are relevant (positive examples) or irrelevant (negative examples) to the query.
3. Machine learning algorithm is applied to learn a new ranking model based on the user's feedback. Then go back to (2).

Steps (2)-(3) are repeated till the user is satisfied with the results.

Step (3) is comparably the most important step and different approaches can be used to learn the new query. A few generally adopted approached are introduced in the following.

Re-weighting Approaches

A typical approach in step (3) is to automatically adjust the weights of low-level features to accommodate the user's need, rather than asking the user to specify the weights as adopted in earlier content-based image retrieval systems. This re-weighting step dynamically updates the weights embedded in the query (not only the weights to different types of low-level features such as color, texture, shape, but also the weights to different components in the same feature vector) to model the high-level concepts and perception subjectivity [4].

Query Point Movement Approaches

Another method is called query-point-movement (QPM) [3]. It improves the estimation of the query point by moving it towards the positive examples and away from the negative examples. A widely adopted query point removing technique is called the Rocchio's formula [1] (see (Eq. 1) below):

$$Q' = \circledR Q + {}^{-} \left(\frac{1}{N_{R'}} \Sigma_{i \in D'_R} D_i \right)$$

$$- \left(\frac{1}{N_{N'}} \Sigma_{i \in D'_N} D_i \right) \quad (1)$$

In (Eq. 1), Q and Q' are the original query and updated query, respectively, and D'_R and D'_N are sets of the positive and negative images returned by the user, and $N_{R'}$ and $N_{N'}$ are the set sizes. α, β and γ are weights.

Machine Learning Approaches

Machine learning techniques are also widely used. As mentioned previously, support vector machine (SVM) is used to capture the query concept by firstly applying the kernel trick which projects images onto a hyperspace and then separates the relevant images from irrelevant ones using maximum margin strategy. The advantages of adopting SVM are that (i) it has high generalization ability, and (ii) it works for small training sets.

Another step-forward approach is proposed by Tong and Chang [8] called SVM active learning, which was reported to be able to effectively use negative and non-labeled samples, and learn the query concept faster and with better accuracy.

Since manually labeling images is tedious and expensive, training image set is usually very small. This is called the small sample problem. To handle this problem, some researchers proposed boosting methods, e.g., Discriminant-EM (D-EM) [7], which boosts the classifier learnt from the limited labeled training data.

Decision-tree learning methods such as C4.5, ID3 were also used in RF loop to classify the database images into relevant and irrelevant.

Key Applications

Relevance feedback approach can help not only content-based image retrieval, but also applications like image annotation, segmentation, etc.

Future Directions

There are still many research work on adopting relevance feedback to content-based image retrieval [2]:

1. It is better to exploit prior information, such as domain-specific similarity, clustering, context of session, etc., in the RF mechanisms.
2. The impact of the data and of the policy of the user on both the learner and the selector must be addressed.
3. How to scale up RF to handle very large image databases is an important issue which was not extensively studied.

Cross-References

▸ Feature Extraction for Content-Based Image Retrieval

Recommended Reading

1. Chen Z, Zhu B. Some formal analysis of Rocchio's similarity-based relevance feedback algorithm. Inf Retr. 2002;5(1):61–86.
2. Crucianu M, Ferecatu M, Boujemaa N. Relevance feedback for image retrieval: a short survey. In: State of the art in audiovisual content-based retrieval, Information Universal Access and Interaction, Including Datamodels and Languages. Report of the DELOS2 European Network of Excellence (FP6). (2004).
3. Liu Y, Zhang D, Lu G, Ma W-Y. A survey of content-based image retrieval with high-level semantics. Pattern Recogn. 2007;40(1):262–82.
4. Rui Y, Huang TS, Ortega M, Mehrotra S. Relevance feedback: a power tool for interactive content-based image retrieval. IEEE Transactions on Circuits and Systems for Video Technology. 1998;8(5):644–55.
5. Ruthven I, Lalmas M. A survey on the use of relevance feedback for information access systems, The knowledge engineering review. London: Cambridge University Press; 2003.
6. Schölkopf B, Smola A. Learning with kernels. Cambridge, MA: MIT Press; 2002.
7. Tian Q, Yu Y, Huang TS. Incorporate discriminant analysis with EM algorithm in image retrieval. In: Proceedings of the IEEE International Conference on Multimedia and Expo; 2000. p. 299–302.

R

the performance. Both kernel based replication solutions (e.g., [12, 16]) as well as middleware-based replication tools (e.g., [5,14,18]) have been developed.

In the last years, the emergence of cloud data stores has led to a further wave of research. This time, the main focus was again on high availability and handling of catastrophic failures such as the shutdown of a complete data centre. Thus both replication within a data center [9, 19] as well as across centers [7, 9, 22] (often referred to as geo-replication) is important. In this context some systems chose weak consistency levels in order to guarantee availability despite network partitions [9]. Others offer strong levels up to 1-copy-serializability in order to be able to serve applications with demanding consistency requirements [7, 19].

All these efforts have led to a wide range of replica control protocols. Which one to choose for a specific scenario depends on the environment and the application requirements. A trade-off between correctness, performance, generality, and potential of scalability is nearly always unavoidable.

Scientific Fundamentals

Replica control algorithms and their implementations can be categorized by a wide range of parameters.

Architecture. One aspect of replica control is *where* the protocol is implemented. It can be implemented within the database what is known as *kernel-based* or *white box* approach (e.g., Postgres-R [12] and the database state machine [16]). A client connects to any database replica which then coordinates with the other replicas. Typically, replica control is tightly coupled with the concurrency control mechanism of the database system.

Alternatively, replica control can be implemented outside the database as a *middleware* layer (e.g., [1, 14, 15, 18, 20]). Clients connect to the middleware that appears as a database system. The middleware then controls the exe-

cution and directs the read and write operations to the individual database replicas. This can be instrumented in two ways: (1) A *black-box approach* uses standard database systems to store the database replicas [1, 18, 20]; (2) A *gray-box* approach expects the database system to export some minimal functionality that can be used by the middleware for a more efficient implementation of replica control, e.g., providing the tuples that a transaction has updated so far in form of a writeset [15]. A middleware-based approach typically has its own concurrency control mechanism which might partially depend on the concurrency control of the underlying database systems. There might be a single middleware component (centralized approach), or the middleware might be replicated itself. For example, the middleware could have a backup replica for fault-tolerance. Other approaches have one middleware instance per database replica, and both together build a replication unit.

Replica Control Phases. A replica control protocol has to control the execution of transactions during regular operation (no failure), while failures occur, and during recovery (failed sites with stale data replicas rejoin the system or new sites are added to the system). As shown in Fig. 1, the execution of a transaction during regular operation may have a number of phases (adjusted from [17]):

1. Client connection. The client interacts with the database through a client proxy (such as a JDBC or an ODBC driver, or the database client API). The client connects to the replicated database by invoking the client proxy connection method. In the middleware approach, the client connects to the middleware (or one of the middleware instances if they are replicated). In the kernel-based approach, it connects to one of the database replicas. Connecting to a replicated database may require some mechanism to enable *replica discovery* in a transparent way (such as a well-known registry or IP multicast). With this, the client can connect to the database independently of which replicas are currently available. If trans-

Replica Control, Fig. 1
Replica control phases

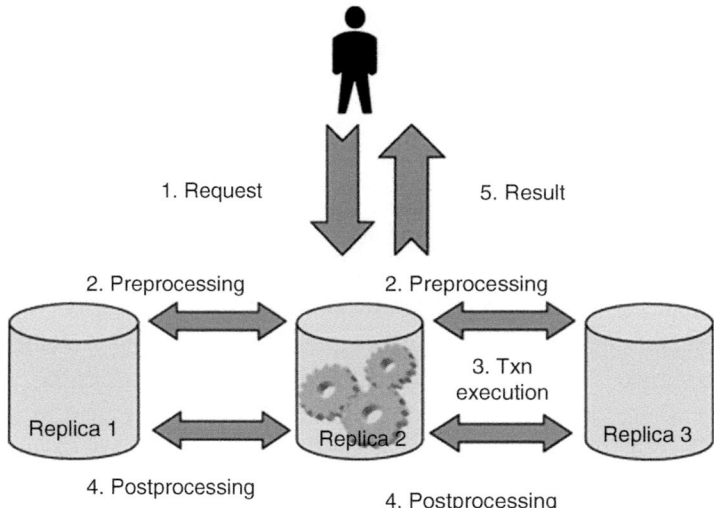

parency is required, i.e., the client application should not be aware of replication, then replica discovery can be hidden in the client proxy.

2. Request submission. The client submits a request via the client proxy that forwards it to the middleware or database replica to which it is connected.

3. Pre-processing coordination. Some protocols require sending requests to all replicas. As a result, the replica to which the client is connected might forward the request to one, some or all replicas. This phase may also be used for other purposes such as distributed concurrency control or load-balancing.

4. Request processing. The request is processed by one or more replicas.

5. Post-processing coordination. Once a request is processed some protocols perform concurrency control tasks, propagate changes, aggregate results from a quorum, or guarantee atomicity by executing a two-phase commit protocol or another termination protocol.

6. Result return. The result of the request is sent back to the client proxy that returns it to the client.

A request can be either an entire transaction (consisting of one or multiple read and write operations), or an operation within a transaction. In the latter case, phases 2–6 may be repeated for each operation and for the final commit of a trans-

action (as in distributed locking [3]). Moreover, the order of the last two phases could be reversed, i.e., first a result is sent to the client and after that, some coordination takes place. Replica control protocols are implemented in the pre-processing phase, in the post-processing phase or in both of them.

Mapping approaches. One of the main tasks of replica control is to map read and write operations on logical data items to operations on the physical data copies such that replicas eventually converge to the same value and reads see consistent data. With ROWA (read-one-write-all) protocols, read-only operations are processed at a single replica, while write operations are executed at all replicas. The extension to ROWAA (read-one-write-all-available) protocols aims at handling site failures. However, in case of network partitions ROWAA might result in two partitions executing transactions. In contrast, quorum-based replica control can handle both site and network failures. Write operations need to access a quorum of replicas before they can complete. Any two write quorums have to overlap in at least one replica in order to guarantee data consistency. If read operations want to the latest and consistent data, they have to access a read quorum of replicas that overlaps with any write quorum. If reads can handle stale data and do not require to read a consistent snapshot, then read operations might be allowed to read any replica. Quorums are handled in more detail in the entry *Quorum systems*.

Correctness Criteria. One of the main aspects of replica control is the correctness criterion to be supported. Several, relatively strong correctness criteria extend the notion of isolation from a non-replicated database to a replicated setting. *1-copy correctness* states that the replicated data should appear as one logical non-replicated database. Depending on the isolation level used in the non-replicated execution different correctness criteria can be defined.In a non-replicated database, *serializability* guarantees that the concurrent execution of transactions is equivalent to a serial execution. *1-copy-serializability* [3] (1CS) extends this notion and guarantees that the concurrent execution of transactions over the replicated database is equivalent to a serial execution over a non-replicated database. Non-replicated database systems usually offer more relaxed forms of isolation, such as the ANSI isolation levels or snapshot isolation. 1-copy-snapshot-isolation has been defined as a correctness criterion for a replicated database (e.g., [14], and several replica control protocols consider snapshot isolation (e.g., [8, 14, 18]).

Weak consistency models do not require 1-copy correctness. Depending on the consistency level, clients might read replicas that are stale, that is, have not yet applied the latest updates, or copies might even diverge if different replicas are allowed to concurrently apply conflicting updates [21]. The weakest consistency criteria that appears acceptable is *Eventual Consistency* that guarantees that all copies of a given data item will eventually converge. The degree of staleness or divergence might be bound. Correctness criteria are discussed in more detail in the entry *Consistency Models for Replicated Data*.

Concurrency Control. In cases where *1-copy* correctness is required, distributed concurrency control is needed to enforce the correctness criteria by restricting the execution of concurrent conflicting transactions. That is, many replica

control protocols extend concurrency control to a replicated system.

Concurrency control protocols can be either optimistic or pessimistic. A *pessimistic* approach restricts concurrency to enforce consistency across replicas. The easiest way consists in executing all update transactions sequentially in the same order at all sites, what would provide 1CS trivially. Other protocols increase concurrency by exploiting knowledge about the data that will be accessed. With this information, transactions accessing disjoint data sets can be executed in parallel while those that potentially access common data are executed sequentially. The granularity of data can be at different levels, such as tables, tuples or conflict classes (i.e., data partitions). Pessimistic replica control protocols are implemented in the pre-processing phase. *Optimistic* approaches, in contrast, execute potentially conflicting transactions concurrently. Only when the transaction has completed execution, a validation phase (also known as certification) takes place. It checks whether the transaction being validated conflicts with concurrent transactions. If there is a conflict, some transaction must be aborted. A standard mechanism to guarantee serializability is to abort the validating transaction if the set of data items it read during execution overlaps with the set of data items written by a concurrent transaction that already validated. With snapshot isolation, a transaction fails validation if it wrote some data item that a concurrent, already validated transaction also wrote. The entries *Traditional Concurrency Control for Replicated Data* and *Replication based on Group Communication* present some concrete optimistic and pessimistic protocols.

Processing Update Transactions. Another important feature of replica control is how update transactions are processed. With *symmetric update processing* each update transaction is fully executed at all replicas. In contrast, *asymmetric update processing* executes update transactions at one site (or subset of sites) and then, the resulting changes (known as *writeset*) are propagated to the rest of the replicas. Some protocols [1] lie in between, being symmetric at the statement level, but asymmetric at the transaction level. That is, if an update transaction contains both write and read statements, the read statements are executed at one site while write statements are executed at all sites. Asymmetric update processing has typically much less overhead than symmetric update processing and thus, allows for better scalability in update-intensive environments [12].

Timepoint of Synchronization. As described earlier, a client typically connects to one replica and submits its transactions to this replica. An important question is *when* this replica coordinates with other replicas to guarantee data consistency. In *eager* (aka synchronous) protocols, coordination takes place before the transaction commits locally. Typically, this means that the replica sends the changes (asymmetric processing) or the operation request (symmetric processing) and the concurrency control component decides on a serialization order for this transaction before it commits. However, this does not necessarily mean that all replicas have applied the changes before commit; that might be done in the background. In contrast, with *lazy replication* (aka asynchronous), no coordination takes place before commit. Lazy replication usually always applies asymmetric processing. The propagation of changes can be done immediately after the commit, periodically, or be triggered by some weak consistency criteria such as freshness. With lazy replication, transaction execution has usually no pre-processing coordination phase, and phases five and six are switched. This means, the results are first returned to the client, and then, the post-processing phase is run. Eager replication usually results in longer client response times since communication is involved but can more easily provide strong consistency.

Who executes transactions. *Primary copy replication* requires that all update transactions are executed at a given site (the primary) (e.g., [8, 18, 20]). The primary propagates the changes to the other replicas (secondary replicas). Secondaries are only allowed to execute read-only transactions themselves. In order to be able to forward read-only transactions to secondaries and update transactions to the primary, the system must be aware at the start time of a transaction whether it is read-only or not (e.g., through a tagging mechanism). If update transactions can be executed at any site, the replica control protocol follows an *update everywhere* approach (also referred to as *update anywhere*). An alternative to primary copy replication and update everywhere is based on partitioning the database items such that each partition has a primary, but different partitions may have different primaries. This avoids that the primary becomes a bottleneck under write-intensive workloads. However, if a transaction has to access data of more than one partition, additionally to the coordination between replicas, there needs to be coordination among the participating partitions to agree on the outcome of the transaction. applications where the database can be partitioned such that each transaction only accesses data of one partition.

Degree of Replication. A further dimension is the *degree of replication*. In *full replication* every data item is replicated at each site. At the other extreme, in a *distributed database* each data item is stored at only one site and there is no replication. In partial replication each data item is replicated at a subset of nodes. The replication degree is further discussed in the entry *partial replication*.

Coordination Steps. Another aspect considers the number of coordination steps, and thus, the number of message rounds, per transactions. Some protocols use a constant number of message rounds, while others require a linear number of message rounds, depending on the number of (write) operations within the transaction (e.g., in distributed locking). In the former case, the pre-processing and/or the post-processing phases are executed once for each transaction. In the latter case, these phases are executed per (write) operation.

Furthermore, some replica control protocols require a coordination protocol among the replicas in order to decide the outcome of

R

a transaction (*voting termination*), similar to a distributed commit protocol. In others, each replica decides by itself deterministically about the outcome of a transaction (*non-voting termination*).

In early eager approaches, atomicity was achieved by running a commit protocol, such as 2-phase-commit, at the end of transaction. This was not only time-consuming by itself but also required that all sites had completely executed the transaction before the transaction was committed at any site. In more recent approaches, atomicity is often achieved by other means such as reliable multicast that provides the required failure atomicity (see the entry *Replication Based on Group Communication*), or by running the Paxos protocol [13], which requires a majority of replicas to have applied the write operation before it is considered successful. Even in eager approaches where the participating sites agree on a serialization order before the transaction commits at any site, this allows the user to receive the commit outcome before the transaction is actually executed at all replicas.

If the database is partitioned for scalability, and each partition is replicated for scalability, then transactions that access more than one partition have to coordinate all the participating partitions as well as all replicas within each of these partitions. For instance, a two-phase commit protocol across the primary copies of all participating partitions can guarantee that all partitions decide on the same outcome, while a Paxos protocol within each partition guarantees that all updates are applied in the same order at all replicas of the partition [7].

Restrictions. Another point to consider when analyzing replica control protocols are the possible constraints they set on the kind of transactions that are supported. Some protocols only allow single statement transactions (known as auto-commit mode in JDBC). Other protocols allow several statements within a transaction, but they have to be known at the beginning of the transaction. This is typically implemented using stored procedures (or prepared statements in JDBC)

[15, 23]. The more general protocols do not have any restriction on the number of statements a transaction contains [12, 14, 18].

Fault-Tolerance and Availbility. One of the main reasons to deploy replication is to handle failures. Thus, replica control also needs to work correctly in the case of failures and when new or repaired replicas are added to the system. These issues are explored in more detail in the entries *Replication for Availbility* and *Online Recovery in Parallel Database Systems*.

Other Aspects Replica control can also be affected by the implementation of self-* properties or autonomic behavior. The entry *autonomous replication* explores these self-* properties for data replication.

Another crucial feature of replica control is the environment for which it is designed. Many approaches target local area networks (LAN) where the bandwidth is high and partitions are rare. With the advent of cloud computing and the access of data across the globe, geo-replication, that is replicating data across wide area networks (WANs), e.g., across two cloud locations, has become equally importnt. Other kinds of networks have also been explored such as mobile networks and peer-to-peer networks. There are specific entries for all them.

Key Applications

Replica control is always needed if replicated data is updated. In particular, in database replication, where updates occur in the context of transactions, replica control is also needed to guarantee transactional correctness.

Future Directions

An important future direction is to combine replica control with modern multi-tier and service-oriented architectures. Providing high availability and scalability for a multi-tier architecture is a non-trivial task. It implies replicating all the tiers to avoid single points of failure and performance bottlenecks. In this setting, guaranteeing consistency and scalability at the same time is very challenging.

Cross-References

▶ Autonomous Replication
▶ Consistency Models for Replicated Data
▶ Online Recovery in Parallel Database Systems
▶ Optimistic Replication and Resolution
▶ Replication Based on Group Communication
▶ Replication for Scalability
▶ Replicated Database Concurrency Control
▶ WAN Data Replication

Recommended Reading

1. Amza C, Cox AL, Zwaenepoel W. Distributed versioning: consistent replication for scaling back-end databases of dynamic content web sites. In: Proceedings of the ACM/IFIP/USENIX International Middleware Conference; 2003. p. 282–304.
2. Bernstein PA, Fekete A, Guo H, Ramakrishnan R, Tamma P. Relaxed-currency serializability for middle-tier caching and replication. In: Proceedings of the ACM SIGMOD International Conference on Management of Data; 2006. p. 599–610.
3. Bernstein PA, Hadzilacos V, Goodman N. Concurrency control and recovery in database systems. Addison-Wesley: Reading; 1987.
4. Bornea MA, Hodson O, Elnikety S, Fekete A. One-copy serializability with snapshot isolation under the hood. In: Proceedings of the 27th International Conference on Data Engineering; 2011. p. 625–36.
5. Cecchet E, Candea G, Ailamaki A. Middleware-based database replication: the gaps between theory and practice. In: Proceedings of the ACM SIGMOD International Conference on Management of Data; 2008. p. 739–52.
6. Chairunnanda P, Daudjee K, Tamer Özsu M. Confluxdb: multi-master replication for partitioned snapshot isolation databases. Proc. VLDB Endow. 2014;7(11):947–58.
7. Corbett JC, Dean J, Epstein M, Fikes A, Frost C, Furman JJ, et al. Spanner: Google's globally distributed database. ACM Trans Comput Syst. 2013;31(3):8.
8. Daudjee K, Salem K. Lazy database replication with snapshot isolation. In: Proceedings of the 32nd International Conference on Very Large Data Bases; 2006. p. 715–26.
9. DeCandia G, Hastorun D, Jampani M, Kakulapati G, Lakshman A, Pilchin A, et al. Dynamo: Amazon's highly available key-value store. In: Proceedings of the 21st ACM Symposium on Operating System Principles; 2007. p. 205–20.
10. Gray J, Helland P, O'Neil P, Shasha D. The dangers of replication and a solution. In: Proceedings of the ACM SIGMOD International Conference on Management of Data; 1996. p. 173–82.
11. Jung H, Han H, Fekete A, Röhm U. Serializable snapshot isolation for replicated databases in high-update scenarios. Proc. VLDB Endow. 2011;4(11):783–94.
12. Kemme B, Alonso G. Don't be lazy, be consistent: Postgres-r, a new way to implement database replication. In: Proceedings of the 26th International Conference on Very Large Data Bases; 2000. p. 134–43.
13. Lamport L. The part-time parliament. ACM Trans Comput Syst. 1998;16(2):133–69.
14. Lin Y, Kemme B, Patiño-Martínez M, Jiménez-Peris R. Middleware based data replication providing snapshot isolation. In: Proceedings of the ACM SIGMOD International Conference on Management of Data; 2005. p. 419–30.
15. Patiño-Martínez M, Jiménez-Peris R, Kemme B, Alonso G. Middle-r: consistent database replication at the middleware level. ACM Trans Comput Syst. 2005;23(4):375–23.
16. Pedone F, Guerraoui R, Schiper A. The database state machine approach. Distrib Parallel Databases. 2003;14(1).
17. Pedone F, Wiesmann M, Schiper A, Kemme B, Alonso G. Understanding replication in databases and distributed systems. In: Proceedings of the 20th IEEE International Conference on Distributed Computing Systems; 2000. p. 464–74.
18. Plattner Ch, Alonso G. Ganymed: scalable replication for transactional web applications. In: Proceedings of the ACM/IFIP/USENIX 5th International Middleware Conference; 2004. p. 155–74.
19. Rao J, Shekita EJ, Tata S. Using paxos to build a scalable, consistent, and highly available datastore. PVLDB. 2011;4(4):243–54.
20. Röhm U, Böhm K, Schek H-J, Schuldt H. FAS – a freshness-sensitive coordination middleware for a cluster of OLAP components. In: Proceedings of the 28th International Conference on Very Large Data Bases; 2002. p. 754–65.
21. Saito Y, Shapiro M. Optimistic replication. ACM Comput Surv. 2005;37(1):42–81.
22. Sovran Y, Power R, Aguilera MK, Li J. Transactional storage for geo-replicated systems. In: Proceedings of the 23rd ACM Symposium on Operating System Principles; 2011. p. 385–400.
23. Thomson A, Diamond T, Weng S-C, Ren K, Shao P, Abadi DJ. Calvin: fast distributed transactions for partitioned database systems. In: Proceedings of the ACM SIGMOD International Conference on Management of Data; 2012. p. 1–12.

R

Replica Freshness

Alan Fekete
University of Sydney, Sydney, NSW, Australia

Synonyms

Divergence control; Freshness control; Incoherency bounds

Definition

In a distributed system, information is often *replicated* with copies of the same data stored on several sites. Ideally, all copies would be kept identical, but doing this imposes a performance penalty. Many system designs allow replicas to lag behind the latest value. For some applications, it is acceptable to use out-of-date copies, provided they are not too far from the true, current value. *Freshness* refers to a measure of the difference between a replica and the current value.

Historical Background

The tradeoff between consistency and performance or availability is an old theme in distributed computing. In the database community, many researchers worked on ideas connected with explicitly allowing some discrepancy between replicas during the late 1980s and early 1990s. Early papers identified many of the diverse freshness measures discussed here, from groups at Princeton, Bellcore and Stanford [1, 10, 11]. A mixed model that integrated freshness limits of several kinds was Pu's epsilon-serializability [6, 7]. TACT is a more recent mixed model, introduced by Yu and Vahdat [12]. Much of the research since the 1990s focused on optimization decisions to improve the performance of a system with slightly stale replicas. Many different system assumptions and metrics to optimize have been considered. Among influential papers are [4, 5, 9]. Research continues on defining different models which bound the staleness of replicas. Unlike earlier work, the focus has recently been on setting bounds which relate to the clients' view of divergence, rather than to the underlying state of the replicas. Röhm et al. designed a system with transaction-level client-defined staleness limits for read-only transactions, within the framework of 1-copy serializability [8]. Guo et al. suggested SQL extensions to express query-specific limits on the perceived staleness and inter-object drift [3], and later developed a theory to express these constraints even when stale reads are allowed within update transactions [2].

Foundations

Consider a distributed system, where information is stored at multiple sites, connected by a communications network. If several sites all store values that are intended to represent the same information in the real world, one can call these copies or *replicas*; it is usual to write x^A to represent the replica at site A of the logical data item x. At any instant, the true or current value of the logical item is the value assigned to x in the most recent non-aborted update of x. The ideal, of course, has this value stored in every replica, always (or at least, at any instant when a read operation can occur). This is possible using traditional eager replication (see entry on Traditional Concurrency Control for Replicated Databases). However, many systems prefer to propagate updates lazily, and to allow replicas or cached copies that are not completely up-to-date, as described in the entry on Optimistic Replication and Resolution. While there are many applications that can tolerate data that is not current, there are usually limits on the application's tolerance for inaccuracy. Thus many system designs allow for a bound to be placed on how far a replica can diverge from the true value; such a bound expresses a constraint on the *freshness* of the replica. This entry describes some of the main ideas that have been proposed to quantify the freshness of replicas.

The focus here is on the properties that define what is allowed, rather than on implementation details that control how these bounds are enforced. For example, some system designs have a single master replica, which always has the true value, while in others the most up-to-date information is sometimes found in one copy and sometimes in another. However, the definitions can be stated without concern for this issue. Another axis of variation is whether the bounds have to apply always, or only when a replica is used in a query. Some papers give a numeric value for freshness or precision (which should be high), others measure staleness or imprecision. This paper uses measures where a low value is better, so zero means that the replica is completely up-to-date.

Value-Based Divergence

Suppose the logical data item comes from a numeric domain such as real numbers. In that case, one can just use the metric on the values themselves. For example, if the true current value is 10.5 and a replica has the value 9.2, then the divergence of that replica is measured as 1.3. In some proposals, the interest is in the value-based divergence between two replicas, even if neither has the current true value. If each replica is within δ of the true value, the difference between any pair of replicas is bounded above by 2δ.

Delay-Based Staleness

For some applications, a useful measure of tolerance for imprecise data is to quantify how recently the data was correct. For example, when a person moves house, the post-office will often redirect mail that had the former address, for a short period. Thus staleness can be measured by how long the replica has to wait before learning of an update that has occurred. For example, suppose the true value was 9.2 until time 100, and which time the value was updated to 10.5; if at time 103 a replica still contains 9.2, one says it is stale by 3 time units. Where the values are representing a real-world quantity, and the quantity can't change by more than v units each time interval, then a bound of δ on the delay implies a bound of $v\delta$ on the value divergence.

Many system designs can't determine, at a replica, how stale it is. Instead they keep a timestamp with each value, indicating when that update instruction was first applied. If they bound the difference between the timestamp and the current time, they also bound the staleness. Suppose one wants to keep delay below 5 time units; one can say that this holds at time 103, if the replica contains a value whose timestamp is 98 or greater. However, the argument does not work in reverse: even at time 103, a replica with timestamp 93 might be stale by less than $103 - 93 = 10$, depending on when the next update occurred after time 93. One says that a value has a valid period, which is a half-closed interval from the time at which the value first appeared in an update, until just before the next update occurred which changed the value.

Measures of Missed Updates

In many system designs, only a subset of updates are propagated to the replica (in order to save bandwidth). This motivates a definition where one measures how many updates occurred on the logical item, without being recorded (yet) at the replica. For example, suppose the logical item is updated from 9.2 to 10.5 at time 100, and then to 10.1 at time 102, and then to 9.6 at time 104. A replica with value 9.2 at time 103 has missed 2 updates that have occurred by that time. This definition can be related to a delay-based measure if the updates come periodically, as is common for sensor readings.

Inter-Object Consistency Drift

When a query reads several logical objects, one might bound the drift between the versions seen in the two reads. For example, if a query examines copies of the temperature and humidity from a sensor, the user might care that the two measurements were taken at almost the same time, because the humidity affects the accuracy of the thermometer. Suppose a replica x^A contains a value whose valid period is the interval $[T,U)$ and y^B contains a value valid in the interval $[V,W)$. Measure the drift between the replicas, as the smallest separation between the valid intervals; this is zero if $[T, U) \cap [V, W) \neq \emptyset$, and otherwise it is $\min(|V - U|, |T - W|)$. A drift bound of zero is called a "snapshot" condition, as all the data examined must come from a common state of the database.

Transaction Semantics

Often, each update and each query is a separate operation. However, several updates or several queries can be placed in a single transaction. In 1-copy serializability (q.v.), all the operations in a transaction have to appear to take place together, but a transaction with reads may be serialized in the past, and so one might desire a common maximum delay for all the reads done in that transaction. A different approach is taken in epsilon-serializability. Here one ascribes a numeric measure to each of the various situations of divergence between the value read and the correct value, and adds up the divergence measure seen in each read, there is then a bound on the total divergence accumulated during the reads in a given query transaction.

Mixed Measures

Each of the definitions of freshness seems to work for some applications but not for others. Thus the TACT model has separate bounds of each measure which must be applied to all the logical data items within a user-specified grouping of items. For example, a particular logical item might require that whenever it is read, it is separated from the true value by no more than 1.5 in numeric value, it is stale by no more than 10 in delay, and it must be missing no more than 3 updates done at the master.

Key Applications

The commercial database platforms generally offer best-effort update propagation, rather than giving applications guaranteed bounds on freshness. Thus the properties described here are usually found in research prototypes or special-purpose data management, for example in cache management for web content or sensor data.

Future Directions

The ideas of replica freshness reappear in new domains, where applications might be able to tolerate some imprecision and where there is a high cost to keeping data up-to-date. For example, somewhat stale data might be used in a sensor network, or delivered in dynamic web page content. Many of these ideas are being generalized, in a broad research agenda that deals with uncertain or imprecise data.

Cross-References

▶ Data Replication
▶ Real-Time Transaction Processing
▶ Sensor Networks
▶ Uncertainty Management in Scientific Database Systems

Recommended Reading

1. Alonso R, Barbará D, Garcia-Molina H. Data caching issues in an information retrieval system. ACM Trans Database Syst. 1990;15(3):359–84.
2. Bernstein PA, Fekete A, Guo H, Ramakrishnan R, Tamma P. Relaxed-currency serializability for middle-tier caching and replication. In: Proceedings of the ACM SIGMOD International Conference on Management of Data; 2006. p. 599–610.
3. Guo H, Larson PÅ, Ramakrishnan R, Goldstein J. Relaxed currency and consistency: how to say "good enough" in SQL. In: Proceedings of the ACM SIGMOD International Conference on Management of Data; 2004. p. 815–26.
4. Olston C, Loo BT, Widom J. Adaptive precision setting for cached approximate values. In: Proceedings of the ACM SIGMOD International Conference on Management of Data; 2001. p. 355–366.
5. Pacitti E, Coulon C, Valduriez P, Özsu MT. Preventive replication in a database cluster. Distrib Parall Databases. 2005;18(3):223–51.
6. Pu C and Leff A Replica control in distributed systems: as asynchronous approach. In: Proceedings of the ACM SIGMOD International Conference on Management of Data; 1991. p. 377–86.
7. Ramamritham K, Pu C. A formal characterization of epsilon serializability. IEEE Trans Knowl Data Eng. 1995;7(6):997–1007.
8. Röhm U, Böhm K, Schek HJ, Schuldt H. FAS – a freshness-sensitive coordination middleware for a cluster of OLAP components. In: Proceedings of the

28th International Conference on Very Large Data Bases; 2002. p. 754–65.

9. Shah S, Ramamritham K, Shenoy PJ. Resilient and coherence preserving dissemination of dynamic data using cooperating peers. IEEE Trans Knowl Data Eng. 2004;16(7):799–812.

10. Sheth AP, Rusinkiewicz M. Management of interdependent data: specifying dependency and consistency requirements. In: Proceedings of the Workshop on the Management of Replicated Data; 1990. p. 133–6.

11. Wiederhold G, Qian X. Consistency control of replicated data in federated databases. In: Proceedings of the Workshop on the management of replicated data. Houston. 1990. p. 130–2.

12. Yu H, Vahdat A. Design and evaluation of a conit-based continuous consistency model for replicated services. ACM Trans Comput Syst. 2002;20(3):239–82.

Replicated Data Types

Marc Shapiro
Inria Paris, Paris, France
Sorbonne-Universités-UPMC-LIP6, Paris, France

Synonyms

Commutative replicated data types (CmRDTs); Conflict-free replicated data types (CRDTs); Convergent replicated data types (CvRDTs); Replicated abstract data types (RADTs); Replicated data types (RDTs)

Definition

Conflict-free replicated data types (CRDTs) were invented to encapsulate and hide the complexity of managing ▶ Eventual Consistency. A CRDT is an abstract data type that implements some familiar object, such as a counter, a set, or a sequence. Internally, a CRDT is replicated, to provide reliability, availability, and responsiveness. Encapsulation hides the details of replication and conflict resolution.

In a sequential execution, the CRDT behaves like its sequential counterpart. Thus, a CRDT is reusable by programmers without detailed knowledge of its implementation. Furthermore, a CRDT supports concurrent updates and encapsulates some strategy that provably ensures that replicas of the CRDT will converge despite this concurrency. Concurrent updates are never conflicting.

Historical Background

There is precursor work on specific CRDTs, before the concept was formally identified as an independent abstraction. Johnson and Thomas [12] proposed the so-called last-writer-wins (LWW) or greatest-timestamp-wins approach for a replicated register, i.e., an untyped memory that an update completely overwrites. Wuu and Bernstein [21] studied more complex data types, the log and dictionary (a.k.a. map or key-value store). The whole area of operational transformation (OT) studied replicated strings or sequences, intended for concurrent editing applications [19]. Baquero and Moura [4] identified some convergence conditions for data types used in mobile computing. The Dynamo system is based on a Multi-Value Register construct [10]. Related topics include replicated file systems and version control systems.

The concept of CRDTs was identified by Preguiça et al. [14], formalized by Shapiro et al. [17] and [18], and studied systematically in Shapiro et al. [16]. A similar concept, called replicated abstract data types (RADTs), was proposed independently by Roh et al. [15]. These works consider symmetric replicas, in which concurrent updates must be commutative and associative. In related work, Burckhardt et al. [8] consider so-called cloud types with asymmetric main and secondary branches, thus relaxing the commutativity requirement.

The distinction between *operation*- and *state-based* CRDTs was established by Shapiro et al. [17]. Burckhardt et al. [9] established lower-bound and optimality results for some representative state-based CRDTs. *Delta CRDTs*

R

were proposed to decrease the footprint of state-based CRDTs while keeping most of their advantages [1]. *Pure operation-based* CRDTs leverage causal-order delivery to streamline the design and implementation of operation-based CRDTs [5].

Foundations

Encapsulating Replication and Concurrency

In a distributed system, shared data is often replicated to improve the availability and latency of *reads*. However, requiring strong consistency between replicas will actually degrade the availability and latency of *writes*. According to the ► CAP Theorem, to improve write availability and latency requires to relax the consistency requirement: a replica should accept updates without synchronizing with other replicas and propagate them in the background.

If multiple replicas accept updates (a so-called "multi-master" system), inevitably, there will be concurrent updates to separate replicas. Managing and reconciling conflicting concurrent updates, in order to ensure ► Eventual Consistency, is a major issue of such systems.

CRDTs were invented to resolve this issue, by encapsulating a familiar object abstraction with a mathematically sound conflict resolution protocol.

CRDT Behavior

A CRDT supports the interface of the corresponding abstraction. Thus, a *register* CRDT will support mutation methods such as *read* and *write* methods; a *counter* supports *increment*, *decrement*, and *value* methods; a set methods to *add*, *remove*, and *query* elements; and so on.

A number of CRDT types have been proposed in the literature. The most basic ones are the LWW Register [12] and the Multi-Value Register [10]. A widely studied CRDT is the sequence or list, used in particular for cooperative editing [14, 15]. Other common CRDTs include counters [2, 16], sets [16], and maps [15].

Consider, for instance, a set data type, supporting operations to add and remove elements, ignoring duplicates. An archetypical CRDT set is the so-called *observed-remove set* (OR-Set). In any sequential execution, it behaves exactly like a sequential set. Concurrently adding and removing different elements *e* and *f*, or adding the same element *e* twice, or removing the same element *e* twice, commute per the sequential specification. However, to ensure commutativity of two updates that concurrently add and remove the same element *e*, the OR-Set makes the "add win," i.e., any replica that observes both operations concludes that *e* is a member of the set. To do this, the implementation of the remove operation effectively cancels out those add operations that it previously observed and only those. We return to this example later in this entry.

Implementation Approaches and Requirements

A CRDT is typically designed to behave like its sequential counterpart in any sequential execution. Furthermore, a CRDT is replicated, supports *concurrent updates* for availability, and encapsulates some strategy to *merge* concurrent updates and ensure that its replicas eventually converge. One such strategy is the "last-writer-wins" approach [12] that uses timestamps to totally order updates and discard all but the highest-timestamped one. Another is to record concurrent updates side by side, so that the application can deal with them later, as in the Multi-Value Register of Dynamo [10] and in many file systems or version control systems.

The literature distinguishes two implementation strategies for CRDTs. In the *state-based* approach, a mutation method changes only the state of the origin replica. Periodically, a replica sends its full state to some other. The receiver *merges* the received state into its own. A state-based CRDT manages its state space as a join-semilattice, where every mutator method is an inflation, and the *merge* method computes the join (a.k.a. least upper bound) of the states to be merged [4, 17]. Join-semilattice is associative, commutative, and idempotent. The first two properties ensure that all replicas converge deterministically to the same outcome. The latter ensures that the system tolerates duplicated merges. As long as replicas communicate their state sufficiently often, and the communication graph is connected, replicas eventually converge, and each object's history is causally consistent.

The *operation-based* approach consists of sending updates rather than states. A mutator method consists of two steps. The *generator* step reads the state of the origin replica and generates an *effector*, a state transformation that is sent and eventually applied to all replicas in the second step [11, 13, 17]. (This vocabulary is not standardized. Other names for the generator phase are upstream, prepare, or prepare-update. Alternative names for the effector phase are downstream, effect, effect-update, or shadow operation.) Concurrent effectors must commute with one another since they may be received in any order. Associativity is not required but, if available, enables batching multiple effectors into a single one. The operation-based approach requires that the underlying communication layers deliver updates to the object in causal order and never deliver the same (non-idempotent) update twice.

The state-based approach is generally considered less efficient (state may be very large) but more elegant and simpler to understand. It makes very few assumptions about the underlying network; for instance, the number and identity of replicas may be unknown and variable. Conversely, the operation-based approach appears more efficient but is more complex to implement and requires a more elaborate communication layer.

While CRDTs replicas are guaranteed to eventually converge, this may be insufficient for application correctness. Many applications also require ▸ Causal Consistency to avoid ordering anomalies across objects. Furthermore, maintaining the integrity of structural invariants may require synchronization to disallow certain concurrent updates [3, 11] (see also ▸ Multi Datacenter Consistency).

Example: OR-Set

The following pseudocode, in the style of Shapiro et al. [17], illustrates a state-based implementation of an OR-Set [16]. The local variables of a replica are a set E of added elements and a set T of removed elements or tombstones. Adding an element e puts it into E along with a unique tag. The tag remains internal to the implementation and is not visible through the interface. Removing an element e moves all pairs of the form $(e, \_)$ from S into T. An element e is contained in the set if there exists a pair of the form $(e, \_)$ in E. Merging two states retains the element pairs that are contained in both states and makes tombstones of element pairs that are tombstones in either state.

R

---

variables set E, set T

initial ∅, ∅
query *contains* (element e) : boolean b
   let $b = (\exists n : (e, n) \in E)$
update *add* (element e)
   let $n = unique()$
   $E := E \cup \{(e, n)\}$
update *remove* (element e)
   let $R = \{(e, n) | \exists n : (e, n) \in E\}$
   $E := E \setminus R$
   $T := T \cup R$
merge (B)
   $E := (E \setminus B.T) \cup (B.E \setminus T)$
   $T := T \cup B.T$

-- *State-based OR-Set specification, with tombstones*
-- *E : elements; T : tombstones*
-- *sets of pairs { (element e, unique-tag n), . . . }*

-- *unique() returns a unique tag*
-- *e + unique tag*

-- *Collect all unique pairs containing e*

-- *Make pairs observed at origin into tombstone*

Unfortunately the memory usage of this specification grows, without bound, with every *add* and *remove* operation. However, observe that adding an element pair necessarily happens before removing the same pair. Leveraging this observation, Bieniusa et al. [7] propose an implementation whose size is bounded by the number of currently contained elements; since the state-based approach does not assume any

particular delivery order, it is somewhat complex. As the operation-based approach already assumes causal-order delivery, avoiding tombstones is straightforward, as shown next [18]. A replica maintains a set of contained element pairs E. Adding an element e creates the corresponding pair, and removing an element e simply removes all pairs of the form $(e, \_)$ observed at the origin replica.

```
                                                 -- Operation-based Observed-Remove Set, without tombstones
  variables set E                                -- set of pairs { (element e, unique-tag u), ...}
    initial ∅
  query contains (element e) : boolean b
    let b = (∃u : (e, u) ∈ E)
  update add (element e)
    generator (e)
      let u = unique()                           -- unique() returns a unique value
    effector (e, u)
      E := E ∪ {(e, u)}                          -- e + unique tag
  update remove (element e)
    generator (e)
      let R = {(e, u)|∃u : (e, u) ∈ E}

                                                 -- Generator: Collect all unique pairs containing e

    effector (R)
      E := E \ R
                                                 -- Effector: remove pairs observed at source
```

Usage

Several implementations of CRDTs have been reported, in languages such as C++, Clojure, Erlang, Go, Java, Python, Ruby, and Scala.

The Riak NoSQL database, as of Version 2.0, implements a number of replicated data types, including flags, registers, counters, sets, and maps [6]. Bet365, a large online betting company, manages 2.5 million simultaneous users with Riak OR-Sets. League of Legends, an online multiplayer game, implements online chat for 70 million users with Riak sets. TomTom extend Riak's CRDTs to share navigation data between a user's different devices. SoundCloud uses a Go implementation on top of the Redis database to store time-series information.

Antidote [20] is a geo-distributed, open-source cloud database, designed to support CRDTs. In order to help applications maintain their struc-

tural invariants, Antidote ensures a transactional causal consistency model.

Cross-References

▶ CAP Theorem
▶ Causal Consistency
▶ Eventual Consistency
▶ Multi-datacenter Consistency Properties
▶ Optimistic Replication and Resolution
▶ Weak Consistency Models for Replicated Data

Recommended Readings

1. Almeida PS, Shoker A, Baquero C. Efficient state-based CRDTs by delta-mutation. In: Proceedings of the Third International Conference on Networked Systems; 2015. p. 62–76.

2. Attiya H, Burckhardt S, Gotsman A, Morrison A, Yang H, Zawirski M. Specification and complexity of collaborative text editing. In: Proceedings of the 2016 ACM Symposium on Principles of Distributed Computing; 2016. p. 259–68.
3. Balegas V, Preguiça N, Rodrigues R, et al. Putting consistency back into eventual consistency. In: Proceedings of the 10th ACM SIGOPS/EuroSys European Conference on Computer Systems; 2015. p. 6:1–6:16.
4. Baquero C, Moura F. Using structural characteristics for autonomous operation. Oper Syst Rev. 1999;33(4):90–96. ISSN:0163-5980.
5. Baquero C, Almeida PS, Shoker A. Making operation-based CRDTs operation-based. In: Proceedings of the 14th IFIP International Conference on Distributed Applications and Interoperable Systems; 2014. p. 126–40.
6. Basho Inc. Data types, version 2.1.1. https://docs.basho.com/riak/kv/2.1.1/developing/data-types/, Viewed May 2016.
7. Bieniusa A, Zawirski M, Preguiça N, et al. An optimized conflict-free replicated set. Rapport de Recherche RR-8083, Institut National de la Recherche en Informatique et Automatique (Inria), Rocquencourt. Oct 2012.
8. Burckhardt S, Fahndrich M, Leijen D, et al. Cloud types for eventual consistency. In: Proceedings of the European Conference on Object-Oriented Programming; 2012. p. 283–307.
9. Burckhardt S, Gotsman A, Yang H, et al. Replicated data types: specification, verification, optimality. In: Proceedings of the 41st ACM SIGPLAN-SIGACT Symposium on Principles of Programming Languages; 2014. p. 271–84.
10. DeCandia G, Hastorun D, Jampani M, et al. Dynamo: Amazon's highly available key-value store. In: Proceedings of the Symposium on Operating Systems Principles (SOSP). Operating systems review; 2007. p. 205–20.
11. Gotsman A, Yang H, Ferreira C, et al. 'Cause I'm strong enough: reasoning about consistency choices in distributed systems. In: Proceedings of the 43th ACM SIGACT-SIGPLAN Symposium on Principles of Programming Languages; 2016. p. 371–84.
12. Johnson PR, Thomas RH. The maintenance of duplicate databases. Internet Request for Comments RFC 677, Information Sciences Institute. Jan 1976.
13. Li C, Porto D, Clement A, et al. Making geo-replicated systems fast as possible, consistent when necessary. In: Proceedings of the 10th USENIX Symposium on Operating System Design and Implementation; 2012. p. 265–78.
14. Preguiça N, Marquès JM, Shapiro M, et al. A commutative replicated data type for cooperative editing. In: Proceedings of the 29th IEEE International Conference on Distributed Computing Systems; 2009. p. 395–403.
15. Roh H-G, Jeon M, Kim J-S, et al. Replicated abstract data types: building blocks for collaborative applications. J Parallel Distrib Comput. 2011;71(3):354–68.
16. Shapiro M, Preguiça N, Baquero C, et al. A comprehensive study of convergent and commutative replicated data types. Rapport de Recherche 7506, Institut National de la Recherche en Informatique et Automatique (Inria), Rocquencourt. Jan 2011.
17. Shapiro M, Preguiça N, Baquero C, et al. Conflict-free replicated data types. In: Proceedings of the 13th International Symposium on Stabilization, Safety, and Security of Distributed Systems; 2011. p. 386–400. Oct 2011.
18. Shapiro M, Preguiça N, Baquero C, et al. Convergent and commutative replicated data types. Bull EATCS. 2011;104:67–88.
19. Sun C, Ellis C. Operational transformation in real-time group editors: issues, algorithms, and achievements. In: Proceedings of the 1998 Conference on Computer Supported Cooperative Work; 1998. p. 59.
20. The SyncFree Consortium. Antidote DB: a planet-scale, available, transactional database with strong semantics. http://antidoteDB.eu/.
21. Wuu GTJ, Bernstein AJ. Efficient solutions to the replicated log and dictionary problems. In: Proceedings of the ACM SIGACT-SIGOPS 3rd Symposium on the Principles of Distributed Computing; 1984. p. 233–242.

Replicated Database Concurrency Control

Bettina Kemme
School of Computer Science, McGill University, Montreal, QC, Canada

Synonyms

Replica and concurrency control

Definition

In a replicated database, there exist several database servers each of them maintaining a (partial) copy of the database. Thus, each logical data item of the database has several physical copies or replicas. As transactions submit their read and write operations on the logical data items, the *replica control* component of the replicated system translates them into operations on the physical data copies. The *concurrency control component* of the replicated system controls the execution order of these operations such that the global execution obeys the desired correctness criteria. For a tightly coupled system with strong consistency requirements, 1-copy-serializability is the standard correctness criterion, requiring that the concurrent execution of transactions on the replicated data has to be equivalent to the serial execution of these transactions over a logical copy of the database. Thus, a global concurrency control strategy is needed. For loosely coupled systems with weak consistency requirements, each database server uses its local concurrency control mechanism. Inconsistencies are only resolved later via reconciliation techniques.

Key Points

Global concurrency control strategies for replicated databases can be centralized or distributed. In a centralized solution, there is a single or main concurrency control module which could be installed on one of the database servers or in an independent component (e.g., a middleware layer between clients and the individual database servers). This single scheduler decides on the global execution order. Using this architecture, standard concurrency control methods found in non-replicated database systems can be easily extended to the replicated environment. However, the single module becomes a single point of failure, and might lead to considerable message overhead. In a distributed solution, each database server has its own concurrency control component, and the different components have to work in a coordinated fashion in order to guarantee the correct global execution order.

The standard strict two-phase-locking (2PL) protocol can be extended very easily to a replicated system. In a centralized solution, there is no real difference to a non-replicated database. Before an operation on a data item is issued, a lock for this data item has to be acquired and all locks are released at commit time. In a distributed solution, each local lock manager performs strict 2PL. The difference is that locks are now set on the physical copies and not on the logical data item. As many replica control algorithms follow the read-one-copy/write-all-copies strategy it means that a write operation on a data item will lead to write locks on all database servers holding a copy of the data item. With this, a distributed deadlock might occur that involves only a single data item.

Optimistic concurrency control mechanisms are attractive in a replicated database system since they allow a transaction to be executed first locally at only one database server. Only at commit time, the modified data items are propagated to the other replicas and a validation phase checks whether a proper global execution order can be found. This keeps the communication overhead low.

In general, communication is an important issue in a replicated database. An option is to take advantage of advanced communication primitives, e.g., multicast protocols that provide delivery guarantees and a global total order of all messages. As all available database servers receive all messages in the same total order, this order can be used as a guideline to determine at each site locally the very same global serialization order.

Cross-References

▶ Concurrency Control: Traditional Approaches
▶ Concurrency Control for Replicated Databases
▶ Data Replication
▶ One-Copy-Serializability
▶ Replica Control
▶ Replication Based on Group Communication
▶ Serializability

Recommended Reading

1. Bernstein PA, Hadzilacos V, Goodman N. Concurrency control and recovery in database systems. Reading: Addison Wesley; 1987.
2. Carey MJ, Livny M. Conflict detection tradeoffs for replicated data. ACM Trans Database Syst. 1991;16(4):703–46.
3. Wiesmann M, Schiper A. Comparison of database replication techniques based on total order broadcast. IEEE Trans Knowl Data Eng. 2005;17(4):551–66.

Replication

Kazuhisa Fujimoto
Hitachi Ltd., Tokyo, Japan

Synonyms

Data replication; Duplication

Definition

Replication (or data replication) is the process to make copy of a collection of data. Copied data made by replication is referred as replica (or data replica).

Replication can be categorized by data type to be copied. Most DBMSs support database replication for high availability or parallel transaction processing. Some filesystems support filesystem replication for high availability or workload distribution. Filesystem replication is commonly used in distributed filesystems. Volume replication can be processed by Logical Volume Manager, Filesystem or Storage system.

Most replication techniques try to keep replicas consistent and updated so that in case of system failures, the replica can be used to recover from the failure with minimal data loss.

Key Points

During data replication, original data is usually copied on an ongoing basis. The goal of this data replication is to provide high availability. There are two types of data replication in terms of data copy method: synchronous and asynchronous. Synchronous replication guarantees, "zero data loss" by means of atomic write operations (i.e., a write either completes on both sides or not at all). A write is not complete until there is acknowledgment for both original and replica. Most applications wait for a write transaction to complete before proceeding with further work,

so synchronous replication may lower overall performance considerably.

In asynchronous replication, a write is complete as soon as the original is updated. The replica is usually updated after a small lag. This decreases write costs, but "zero data loss" is not guaranteed.

Data replication can be categorized by the layer of doing it. In case of Server-based replication, some entity (e.g., DBMS, Filesystem, Logical Volume Manager, and Application) in a server performs replication. Most Database replication and filesystem replication falls into this category.

Appliance-based replication is another category. Some appliance in a network connecting servers and storage systems makes replication. Most storage virtualization appliances have volume replication functions. File virtualization appliances may provide filesystem replication.

Storage-based replication is done by storage system. Block storage can provide volume replication and file storage such as NAS can provide filesystem replication.

Cross-References

▶ Backup and Restore
▶ Data Replication

Replication Based on Group Communication

Fernando Pedone
Università della Svizzera Italiana (USI), Lugano, Switzerland

Definition

Database replication based on group communication encompasses protocols that implement database replication using the primitives available in group communication systems. Most to join, a new view $v_{i+1}(g)$ is installed, which reflects the membership change.

Group communication primitives for dynamic groups can guarantee properties relating views to multicast messages. One such a property is *sending view delivery*, according to which a message can only be delivered in the context of the view in which it was multicast. This means that if a server multicasts message m while in view $v_i(g)$, then m can only be delivered before view $v_{i+1}(g)$ is installed.

In the following sections we will use two multicast primitives with different properties: abcast(m, g) ensures message agreement and total order in the context of static groups; vscast(m, g) assumes dynamic groups and guarantees message agreement, FIFO ordering, and sending view delivery properties. In both cases, messages are delivered to the application using deliver(m).

A Functional Model for Database Replication

Replication protocols can be generically decomposed into five phases, as shown in Fig. 2 [19]. Some protocols may skip some phases, while others may apply some of the phases several times during the execution of transactions. In Phase 1 the client contacts one or more database replicas to submit its requests. Phase 2 involves coordination among servers before executing requests. (Notice that message exchanges in Fig. 2 are

illustrative only; different communication protocols may present different message patterns.) The actual execution of a request takes place in Phase 3; one or more servers may be involved in the execution of the operation. In Phase 4 servers agree on the result of the execution, usually to ensure atomicity, and in Phase 5 one or more servers send the result to the client. There are basically two parts in which group communication primitives can help: during *server coordinaton* (Phase 2) and *agreement coordination* (Phase 4).

Database replication protocols can be *eager* or *lazy*, according to when update propagation takes place; and *update everywhere* or *primary copy*, according to who performs the updates (see the entry *Replica Control* for more details on this categorization). Group communication has traditionally been used by eager protocols, both in the case of update everywhere and primary copy. We briefly review some eager approaches next; a more detailed account can be found in [19].

Distributed two-phase locking (D2PL) is a traditional mechanism to coordinate servers during transaction execution. In replicated scenarios, implementing D2PL with point-to-point communication may lead to many distributed deadlocks [9]. If lock requests are propagated to servers using abcast, however, total order is ensured, and the probability of deadlocks is reduced [3]. In such a protocol, the client submits its requests to one database server (Phase 1) which abcasts them to all servers (Phase 2). Upon delivering

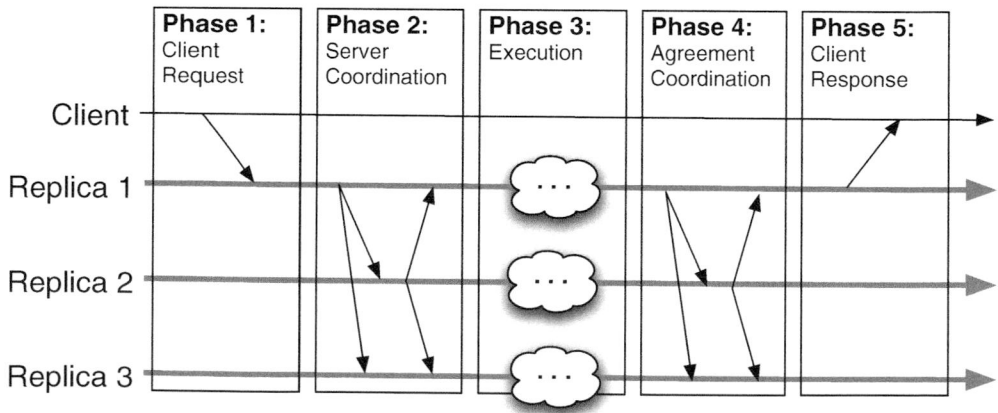

Replication Based on Group Communication, Fig. 2 A functional model for database replication

the request, servers should make sure that conflicting locks are obtained in a consistent manner (Phase 3). Once the servers have executed the operation they send the response to the client (Phase 5). Commit and abort requests are processed like any other operation and so there is no need for agreement coordination (Phase 4). Notice that Phases 1–3, 5 may be executed multiple times, once per request. If transactions can be predefined (e.g., storage procedures), then they can be abcast to all servers as a single request, and then Phases 1–3, 5 are executed only once per transaction [4, 13].

The deferred update approach is another form of eager update-everywhere replication mechanism based on group communication. In this case, the client selects one database server and submits all its requests to this server, without communicating with other servers (Phase 1). During the execution of the transaction, there is no synchronization among servers (i.e., no Phase 2), and only the selected server executes the transaction (Phase 3). Read-only transactions are committed locally by the selected server. Update transactions are propagated to all servers at commit time using abcast. The delivery of a terminating update transaction triggers a certification procedure used to ensure consistency. If the transaction passes the certification test its updates are committed against the database (Phase 4). Only then the reply is sent to the client (Phase 5). Several group communication-based replication protocols follow this approach, which we discuss in more detail in the next section.

Eager primary copy replication can also take advantage of group communication primitives. In this case clients interact only with the primary copy during the execution of the transactions (Phase 1). Therefore, there is no Phase 2. After the primary executes a request it sends the result to the client (Phase 3 executed by the primary only). Several requests from the client may be executed against the primary copy until commit is requested, at which point the primary communicates the new database state (e.g., redo logs) to the secondary copies (Phase 4). The secondaries apply the modifications to their local database. Communication between the primary and the sec-

ondaries is through vscast. FIFO ordering ensures that updates from the primary are received in the order they are sent. Sending view delivery guarantees correct execution in case of failure of the primary. For example, if the primary fails before all secondaries receive the updates for a certain request and another replica takes over as new primary, vscast ensures that updates sent by the new primary will be properly ordered with regard to the updates sent by the faulty primary.

Deferred Update Database Replication Protocols

Database replication protocols based on the deferred update technique, sometimes called certification-based protocols, are optimistic in that transactions execute on a single server without synchronization among servers. Abcast ensures that at termination transactions are delivered in the same order by all servers. The total order property together with a deterministic certification test guarantee that all servers agree on which transactions should be committed and which ones should be aborted.

The precise way in which the certification test is implemented and the information needed to implement it depend on the consistency criterion. Two typical consistency criteria are serializability and snapshot isolation. Serializability specifies that a concurrent execution of transactions in a replicated setting should be equivalent to a serial execution of the same transactions using a single replica. With snapshot isolation, transactions obtain at the beginning of their execution a "snapshot" of the database reflecting previously committed transactions; a transaction can commit as long as its writes do not intersect with the writes of the transactions that committed since the snapshot was taken.

The certification test may require transaction readsets and writesets. Transaction readsets and writesets refer to the data items the transaction reads and writes during its execution (e.g., the primary keys of the rows read or written). Let $RS(T)$ and $WS(T)$ denote, respectively, the readset and writeset of transaction T, and let $CC(T)$ be the set of transactions that executed

R

Replication Based on Paxos

Fernando Pedone
Università della Svizzera Italiana (USI), Lugano,
Switzerland

Synonyms

Consensus; State machine replication

Definition

Paxos is a consensus protocol designed for state machine replication in asynchronous environments subject to crash failures. State machine replication is a technique to increase the availability of a service by replicating the service in multiple replicas and regulating how client commands are propagated to and executed by the replicas: every non-faulty replica must receive and execute every command in the same order. State machine replication provides strong consistency, that is, from the perspective of the clients, the behavior of a service implemented by replicated servers is no different than the behavior of the service when implemented by a single server. Paxos ensures that commands submitted by the clients are delivered to the replicas in the same total order, despite the crash of some replicas.

Historical Background

State machine replication, sometimes called active replication, was introduced in [6] for environments in which failures could not occur and later extended to the context of failures in [20]. State machine replication requires to solve a sequence of consensus instances, where the i-th instance decides on the i-th command to be executed by the replicas (this can be trivially extended so that multiple commands are decided in a single instance of consensus). This scheme guarantees that replicas can receive and execute

commands in the same order, a requirement of state machine replication. A fundamental result in distributed computing states that consensus cannot be solved in a completely asynchronous system, where there exists no bound on transmission delays and no bound on process relative speeds, if even one process can experience a crash failure [4]. Paxos can be implemented in a partially synchronous system, that is, a system that is initially asynchronous and eventually becomes synchronous. However, Paxos is safe (i.e., no two processes decide different values in the same consensus instance) in both asynchronous and synchronous periods; it ensures progress (i.e., processes decide some value) when the system becomes synchronous and a majority quorum of processes is operational. Paxos was originally described in [7] and later published in [8].

Scientific Fundamentals

Paxos solves consensus, a fundamental problem in distributed computing. Paxos distinguishes three roles: *proposers*, *acceptors*, and *learners*. Proposers propose a value, acceptors choose a value, and learners learn the decided value. In state machine replication, clients are typically proposers and replicas are learners. Consensus requires that (i) If a learner learns a value, then the value was previously proposed by a proposer; (ii) No two learners learn different values; (iii) Every non-faulty learner eventually learns the decided value. A process can execute any of these roles and multiple roles simultaneously. One process, typically among the proposers or acceptors, plays the role of *coordinator*. To propose a value, proposers send the value to the coordinator.

In Paxos, the execution of one consensus instance proceeds in two phases: In the *first phase*, the coordinator picks a unique *round number* and requests the acceptors to "promise" not to accept any future message with a smaller round number (this is message "Phase 1A" in Fig. 1). The acceptors reply to the coordinator with the highest-numbered round in which they have cast a vote, if any, and the value they voted for (Phase

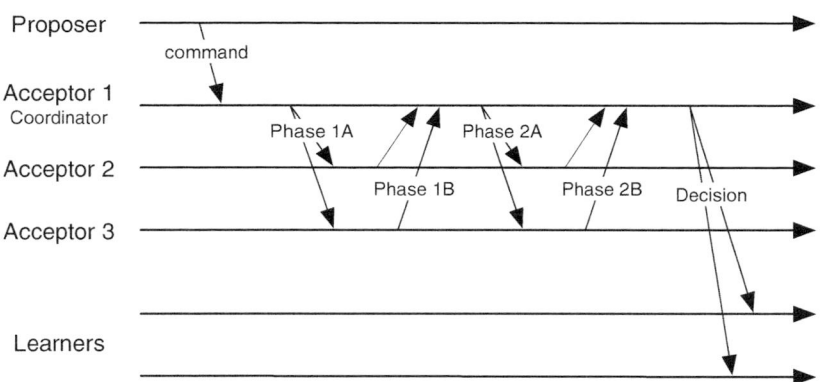

Replication Based on Paxos, Fig. 1 Illustrative execution of Paxos

1B). The coordinator starts the *second phase* after it receives a majority quorum of replies from the acceptors, possibly including its own reply if the coordinator is also an acceptor. If the coordinator receives at least one vote from the acceptors, it chooses the value that corresponds to the highest-round number and proposes this value to the acceptors in the second phase; if the coordinator does not receive any votes in the first phase, it can propose any value to the acceptors (Phase 2A). If the coordinator can choose any value in the second phase, it proposes the command it received from the proposer. An acceptor accepts the value proposed by the coordinator in the second phase and replies positively to the coordinator if it has not received a message with a higher round number (Phase 2B). When the coordinator receives positive replies from a majority quorum of acceptors, it knows that its proposed value is accepted and can communicate this value to the learners (Decision).

In the execution depicted in Fig. 1, it takes six communication steps from the instant in which the proposer proposes a command and the learners learn the decision. Two optimizations can reduce this procedure to three steps. First, since the coordinator only sends a proposal in the second phase of Paxos, the first phase can be executed before the coordinator receives the proposal from a proposer. Typically, the coordinator can execute the first phase for multiple instances of Paxos [8]. Second, the acceptors can send the reply message (Phase 2B) in the second phase of the protocol

to the coordinator and to the learners. When a learner receives replies from a majority quorum of acceptors, it can learn the decision.

If two or more coordinators coexist, it is possible that neither one can have its proposal chosen by the acceptors, which violates the properties of consensus. This can happen in the unlucky situation in which one coordinator executes the first phase of the protocol with round number r_1 and before it executes the second phase, the second coordinator executes the first phase with a round number $r_2 > r_1$. Upon failing to execute the second phase for r_1, the first coordinator re-executes the first phase with round $r_3 > r_2$, which prevents the proposal of the second coordinator from being chosen, and so on. Although this situation may seem unlikely, it shows that choosing a value is not always possible. Paxos prevents such cases by assuming that eventually a single non-faulty coordinator exists (something that can be implemented in a partially synchronous system, but not in an asynchronous system).

Paxos has been extended in many ways. Fast Paxos [10] allows multiple proposers, not only the coordinator, to have their proposal learned in two message steps under certain conditions, which is latency optimum [11]. Disk Paxos [5] is a variation of Paxos in which acceptors are not processes but disks, capable of executing read and write operations on blocks of data. Cheap Paxos [12] exploits the fact that only a quorum of acceptors is needed for progress. Thus, the set of acceptors can be divided into

Redundancy is one of the major mechanisms to achieve the desired goal.

In the database community, looking for high-availability solutions was the main reason to start research into database replication [2], while replication for scalability and performance was only explored later. The research community has mainly focused on the maintenance of transactional properties despite various kinds of failures. Apart of mere correctness, there exist many implementation alternatives that can have a large impact on the performance of the system, and current major database systems have developed a wide range of high-availability solutions with different trade-offs.

Data availability has also been explored early on in many other data-centric domains such as file systems [15, 16], mobile systems [8], or fault-tolerant processes [3].

A recent rise in high-availability solutions came with the onset of cloud computing and cloud storage. Given that an individual cloud center maintains thousands of machines, it is quite likely that some of the machines are down and unavailable; furthermore, network connectivity, especially in wide-area networks, is not always given. And albeit much less likely, the outage of a complete data center is possible. Thus, fault-tolerant and high-availability solutions have quickly been implemented in cloud storage management systems, both at the storage level [6, 12] and at the data management layer [4, 5, 9, 14], covering both intra-cloud and inter-cloud outages and differing in the level of consistency they provide.

Scientific Fundamentals

This entry focuses solely on availability for database management systems. However, many of the issues and solutions discussed here are also valid outside the database domain, such as file systems and object management systems.

Basic Fault-Tolerance Architecture

Let's first have a look at the case when a non-replicated database system crashes. In this case,

it does not perform any further actions, and the data is no more available. Only after restart and recovery it becomes again operational (see entry "▸ Logging and Recovery"). As the recovery procedure can take considerable time, a fault-tolerance solution maintains more than one copy of the database. In the following, the terms replica and site refer to a database management instance controlling a copy of the complete database.

A fault-tolerance replication architecture has to tackle several issues. Firstly, when data is updated, the data copies have to be kept consistent, resulting in additional measures and thus *overhead during normal processing when no failures occur*. Secondly, when a replica fails, a *failover procedure* has to reconfigure the system. Thirdly, after a failure, the failed replica or a new replica has to be started and added to the system in order to handle future failures. For this, the joining replica has to receive the current state of the database from the available replicas. We refer this third step as *replica recovery*.

As the most common fault-tolerance architecture uses primary-backup replication (Fig. 1), the following discussion focuses on this architecture. There is one primary replica and one or more backup replicas in the system. All clients are connected to the primary replica which executes all transactions. In case the primary fails, the clients have to be reconnected to one of the backups which takes over the tasks of the primary.

Execution While No Failures Occur The primary executes all transactions and decides on a serialization order (via the concurrency control module). At specific timepoints, the replication module of the primary replica propagates updates to the backup replica(s). Each backup replica applies the changes in an order that conforms to the serialization order determined by the primary.

Propagation can be done in several ways. In a *2-safe* approach, the changes performed by a transaction are propagated before the transaction commits, and an agreement protocol, such as a two-phase commit protocol (see entry *Commit protocols*) or Paxos [13], guarantees atomicity (the transaction either commits or aborts at all participating replicas). In a *1-safe* approach, the

Replication for Availability and Fault Tolerance, Fig. 1 Primary-backup architecture

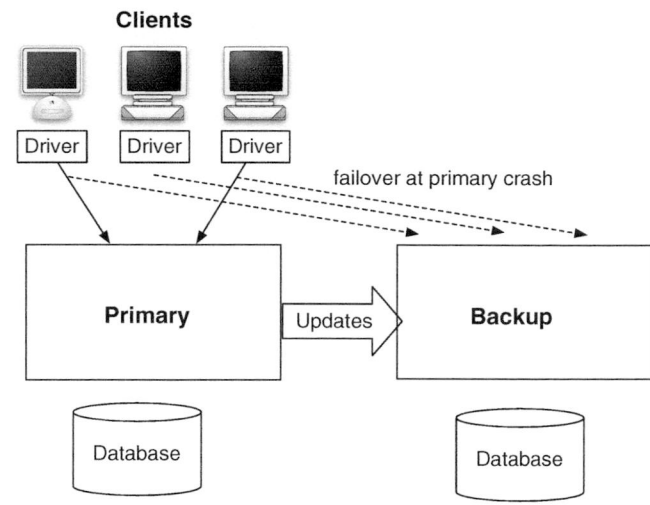

primary can commit a transaction before update propagation is completed.

1-safe propagation is faster than 2-safe propagation. However, if the primary fails after committing a transaction but before propagating its changes, then the transaction is lost because the backup that becomes the new primary has no information about it. Neither atomicity nor durability of the transaction is given. Recovery of this transaction can be complex. In other context, 1-safe propagation is also often referred to as lazy, asynchronous, or optimistic replication, while 2-safe propagation is often referred to as eager or synchronous replication.

There are several other design choices in regard to propagation. For instance, a message could be sent per update operation, per transaction, or even per set of transactions. Furthermore, the update message could contain the physically changed records (e.g., taken from the log used for local recovery) or the update statement. These issues are relevant for replica control in general and are discussed in more detail in the entry "▶ Replica Control".

Changes can be applied at the backup at different timepoints. A backup is called *hot standby* if it immediately applies any changes it receives to its own local database. Therefore, at the time of the failover, the backup is immediately operational and can accept client requests. However, it requires a powerful backup because of the

high overhead during normal processing. A *cold standby* defers applying the changes, e.g., to a timepoint when it is idle or when it actually becomes the new primary at failover time. This results in low overhead during normal processing. Thus, a cold standby could run on a less-expensive machine. However, failover will likely take longer, and performance will be compromised until the powerful machine has completed recovery and can become again the primary.

Failover If there is more than one backup, an agreement protocol can be used to decide which of the backups take over as new primary.

As part of the failover procedure, the clients have to be reconnected to the new primary. This reconnection is typically implemented within the connectivity software (e.g., JDBC driver) running on the client side (see Fig. 1). This *driver* component is connected to the primary but knows the backup(s). When it looses connection to the primary, it connects to the backups to see who is the new primary.

Furthermore, the backups have to agree on the set of transactions that were committed before the primary failure and apply their changes. A difficult issue is the handling of transactions that were active at the time of the failure. If the client had not yet submitted the commit request, the driver can simply throw an abort exception for this transaction before it connects to the new primary. This is correct since independently of 1 safe or 2 safe, the new primary neither has committed nor is involved in the commit protocol for this transaction. It is up to the client to resubmit the transaction (as would be the case for other types of transaction aborts). A different situation arises if the application program has already submitted the commit request, and the driver is waiting for the confirmation from the primary when the failure occurs. In this case, the transaction might or might not be committed at the backups. The driver has to ask the new primary for the outcome of the transaction and inform the client accordingly with a commit confirmation or an abort exception.

Replica Recovery Replica recovery is the task of integrating a new or failed replica into the replicated system. The recovery procedure can be performed *online* or *offline*. In offline recovery, transaction processing is halted in the system, the joining replica receives the accurate state of the data, and then transaction processing can resume. The problem is that availability is compromised. Online recovery, in contrast, performs recovery without stopping transaction processing in the rest of the system.

Data Transfer Strategies There are several possibilities to provide a recovering replica with the latest state of the database [11]. One possibility is to transfer the entire database state to the recovering replica. This approach is simple but leads to unnecessary data transfer if large parts of the data did not change during the downtime of the recovering replica. Thus, an alternative solution first determines for each data item, whether this data item was actually changed during the downtime. Only if this is the case, the current version is transferred to the recovering replica. While this might decrease the size of data to be transferred, it needs additional measures during normal processing and recovery to determine the changed objects. Instead of the data items, one can also transfer the log with all updates that the recovering replica has missed. Then, the recovering replica applies these missing updates to its database. The latter two approaches are only possible for failed replicas

that later recover, while completely new replicas need a complete state transfer.

Recovery Procedure A failed replica or a new replica typically joins as a backup. It can become the primary once recovery has completed (e.g., if it runs on the most powerful machine). The joining replica can receive the state from the primary or from another backup.

Using online recovery, the state transfer takes place concurrently to transaction processing at the primary. Thus, one has to make sure that the recovering replica does not miss any transactions. That is, for a given transaction, the updates performed by this transaction at the primary either are reflected by the state transferred through recovery or are propagated by the primary to the new backup after recovery has completed. More information can be found in the entry "▶ Online Recovery in Parallel Database Systems".

Replication for Performance and Fault Tolerance

The research literature has proposed many approaches where replication serves both performance and fault tolerance by allowing transaction execution to be distributed across all available replicas. Scalability solutions often create several copies of the database in the same LAN cluster and evenly distribute the load among replicas to handle ever increasing workloads, while geo-replication creates replicas at strategic locations so that all clients have a database copy close by, achieving low response times for clients.

Many protocols follow a read-one-write-all-available (ROWAA) replication strategy [1], where read operations are executed at one replica (which allows load-balancing) and write operations are executed at all replicas that are currently available (to keep copies consistent). Some of them execute all update transaction on the primary and only allow read-only transactions to execute on the backups, also referred to as secondaries. Others allow update transactions to be submitted everywhere. Depending on whether update propagation follows a 2-safe or 1-safe approach, reads might access stale data because

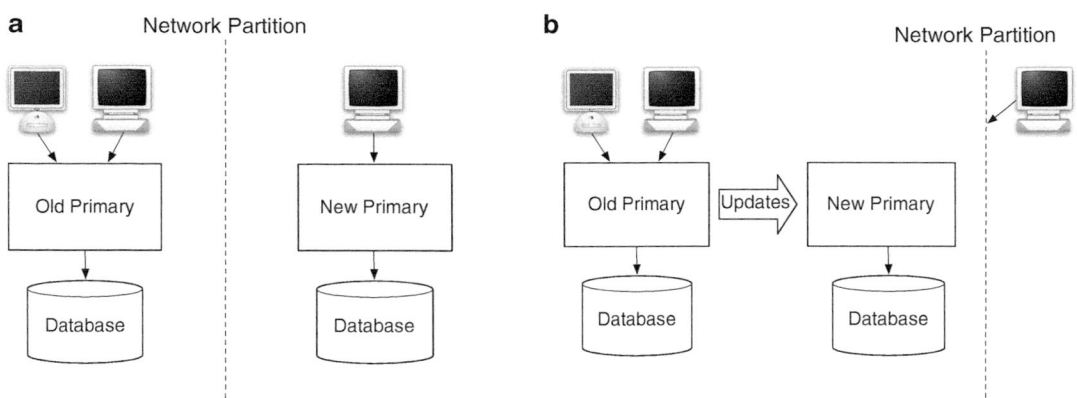

Replication for Availability and Fault Tolerance, Fig. 2 Network partitions. (**a**) Two primaries. (**b**) Unavailability for some clients

the most current update is not yet applied at the copy that is read.

When a replica fails, it is excluded from the system. Clients served by the replica are transferred to available replicas, and writes are only performed on the remaining available replicas. Failover is complicated by the fact that transaction execution is ongoing during the failover period.

Network Partitions

As mentioned before, site crash is not the only failure type that can occur. A network problem might (temporarily) partition the replicas, allowing each replica to only communicate with a subset of other replicas. Network partitions are not frequent in LANs but are common in WANs or wireless networks. They are often transient and only hold for short periods of time. In general, it is impossible for a replica to determine whether another replica that is not reachable has failed, is currently not connected, or is simply slow and overloaded making it temporarily nonresponsive. The CAP theorem [7] formally states that in a distributed system prone to network partitions, it is impossible to achieve consistency and availability at the same time. Thus, a fault-tolerant distributed system that appears as a single system that never fails is not possible once network partitions occur. It is easy to see why standard primary-backup replication does not work correctly if network partitions occur.

Figure 2 depicts two cases. When a network partition occurs between primary and backup (Fig. 2a), the backup and all clients within the partition of the backup will suspect the primary to have failed leading to both partitions having one primary serving local clients. If only clients but not the backup are disconnected from the primary (Fig. 2b), the system continues to run correctly, but from the disconnected clients' point of view, the system is unavailable.

Quorums One major approach to handle network partitions is to use quorums [10, 13, 17]. In order to maintain consistency, each write operation has to access a quorum of replicas, which could be, e.g., the majority of replicas. Write quorums are defined such that any two write operations on the same data item access at least one common replica. This is needed to serialize write operations and assure that any given write quorum has at least one replica with the latest version of the data item. Only having to write to a quorum of replicas and not all allows writes to make progress despite network partitions and node failures, but it avoids that two disjoint sets of nodes (disconnected due to a partition) make concurrently progress and diverge. Availability of writes is compromised in that part of the network that cannot access a quorum of replicas. In order to guarantee that a read operation reads the latest version of the data, read operations also have to read a quorum of nodes, where every read quorum must overlap with every write quorum. Again, the system is not available to clients who cannot connect to a read quorum.

Many current cloud storage systems use Paxos [13] to replicate updates. In Paxos, a leader is elected (either for every single update or more commonly for a longer period of time in order to be more efficient), and then the update is propagated and coordinated by the leader. An update is successful once a majority of replicas has agreed to it (while the other replicas might be updated in the background). In many systems, to guarantee consistent reads, reads either have to be performed by the current leader (who is guaranteed to be up to date) or a majority of replicas. But many systems allow reads to go to any replica in which case the client has to be aware that it might not read the latest version but stale data. Thus, availability for reads is achieved at the price of a lower level of consistency.

The entry "▶ Quorum Systems" discusses quorums in detail.

Data Consistency and Performance Aspects with 1-Safe (lazy) Replication Lazy replication provides generally better performance and provides availability, at least for some transactions, even if there are network partitions. Given that no communication is needed before commit, transactions are typically much faster than in 2-safe solutions, in particular, in wide-area settings. Read operations usually only access one copy of the data item, typically the one that is the closest. Thus, queries are usually also fast, and high availability is given since only one replica needs to be reachable.

For update transactions, one approach is to still have a primary replica and execute all update transactions at the primary. The primary then propagates the updates to all available secondary replicas sometime after commit. Having a single primary for update transactions allows for a globally correct serialization order. Propagating lazily allows the primary to commit a transaction without caring whether the other replicas are available. However, the secondaries can have outdated data, and thus local read operations might read stale data. Furthermore, if a client cannot reach the primary, its update transactions cannot be executed.

A second approach allows an update transaction to execute and commit at any replica, typically the closest. This provides high availability for update transactions. However, data copies might diverge, requiring some form of conflict detection and reconciliation, and read operations can read inconsistent data. Still, for many WAN applications, this might be the only feasible option to provide full availability. Also, in mobile environments with planned periods of disconnection, this is the only way to provide availability to the mobile units. The entry "▶ Optimistic Replication and Resolution" discusses the issues of data consistency in more detail, in particular, how these systems can achieve *eventual consistency*, meaning that the data eventually converges.

In these lazy strategies, recovery is done incrementally. When a site wants to propagate an update but cannot reach a replica, it stores the update locally on persistent storage. When the replica is again available, the update will be propagated. That is, propagation is typically done via a persistent queue guaranteeing that an update is propagated and applied exactly once despite transient failures.

Other Failure Types

There exist other failures that are not considered here. For instance, message loss can be handled by the underlying communication system. Byzantine or malicious behavior of the database system is addressed, e.g., by the entry "▶ Secure Data Outsourcing".

Key Applications

Basically all commercial database systems provide a high-availability solution based on primary-backup replication. Furthermore, other data replication strategies are also often supported and can be deployed for high availability and performance reasons. Most businesses that have update-intensive workloads and require 24/7 availability deploy one or more of these solutions.

With the rise of cloud computing, high availability has become a major key point for cloud storage systems, cloud data management systems, cloud application servers, and cloud coordination systems. This affects anything and everything that is stored and maintained in the cloud.

Future Directions

As there is no universal solution that can achieve full availability and consistency of data in current network infrastructures, trade-offs between availability, consistency, and performance have to be taken into account to determine the level of fault tolerance and availability that is needed for the particular application. Understanding all options and their implications in terms of degrees of these performance parameters (e.g., exact level of consistency, cases in which unavailability can occur) will be of interest to choose the best solution for any given application.

Cross-References

▶ Concurrency Control for Replicated Databases
▶ Data Replication
▶ Logging and Recovery
▶ Online Recovery in Parallel Database Systems
▶ Optimistic Replication and Resolution
▶ Quorum Systems
▶ Replica Control
▶ Secure Data Outsourcing
▶ WAN Data Replication

Recommended Reading

1. Bernstein PA, Goodman N. An algorithm for concurrency control and recovery in replicated distributed databases. ACM Trans Database Syst. 1984;9(4):596–615.
2. Bernstein PA, Hadzilacos V, Goodman N. Concurrency control and recovery in database systems. Reading: Addison Wesley; 1987.
3. Budhiraja N, Marzullo K, Schneider FB, Toueg S. The primary-backup approach. In: Mullender S, editor. Distributed systems. 2nd ed. Harlow/Munich: Addison Wesley; 1993. p. 199–216.
4. Corbett JC, Dean J, Epstein M, Fikes A, Frost C, Furman JJ, Ghemawat S, Gubarev A, Heiser C, Hochschild P, Hsieh WC, Kanthak S, Kogan E, Li H, Lloyd A, Melnik S, Mwaura D, Nagle D, Quinlan S, Rao R, Rolig L, Saito Y, Szymaniak M, Taylor C, Wang R, Woodford D. Spanner: Google's globally distributed database. ACM Trans Comput Syst. 2013;31(3):8
5. DeCandia G, Hastorun D, Jampani M, Kakulapati G, Lakshman A, Pilchin A, Sivasubramanian S, Vosshall P, Vogels W. Dynamo: Amazon's highly available key-value store. In: Proceedings of the 21st ACM Symposium on Operating System Principles; 2007. p. 205–20
6. Ghemawat S, Gobioff H, Leung S. The google file system. In: Proceedings of the 19th ACM Symposium on Operating System Principles; 2003. p. 29–43
7. Gilbert S, Lynch NA. Brewer's conjecture and the feasibility of consistent, available, partition-tolerant web services. SIGACT News. 2002;33(2):51–9.
8. Gray J, Helland P, O'Neil P, Shasha D. The dangers of replication and a solution. In: Proceedings of the ACM SIGMOD International Conference on Management of Data; 1996. p. 173–82.
9. Hunt P, Konar M, Junqueira FP, Reed B. Zookeeper: wait-free coordination for internet-scale systems. In: Proceedings of the USENIX 2010 Annual Technical Conference; 2010.
10. Jiménez-Peris R, Patiño-Martínez M, Alonso G, Kemme B. Are quorums an alternative for data replication? ACM Trans Database Syst. 2003;28(3):257–94.
11. Kemme B, Bartoli A, Babaoglu Ö. Online reconfiguration in replicated databases based on group communication. In: Proceedings of the International Conference on Dependable Systems and Networks; 2001. p. 117–30.
12. Lakshman A, Malik P. Cassandra: a decentralized structured storage system. Oper Syst Rev. 2010;44(2):35–40.
13. Lamport L. The part-time parliament. ACM Trans Comput Syst. 1998;16(2):133–69.
14. Rao J, Shekita EJ, Tata S. Using paxos to build a scalable, consistent, and highly available datastore. Proc. VLDB Endow. 2011;4(4):243–54.
15. Satyanarayanan M, Kistler JJ, Kumar P, Okasaki ME, Siegel EH, Steere DC. Coda: a highly available file system for a distributed workstation environment. IEEE Trans Comput. 1990;39(4):447–59.
16. Terry DB, Theimer M, Petersen K, Demers AJ, Spreitzer M, Hauser C. Managing update conflicts in Bayou, a weakly connected replicated storage system. In: Proceedings of the 15th ACM Symposium on Operating System Principles; 1995. p. 172–83.
17. Thomas RH. A majority consensus approach to concurrency control for multiple copy databases. ACM Trans Database Syst. 1979;4(2):180–209.

R

Replication for Scalability

Ricardo Jiménez-Peris[1] and Marta
Patiño-Martínez[1,2]
[1]Distributed Systems Lab, Universidad
Politecnica de Madrid, Madrid, Spain
[2]ETSI Informáticos, Universidad Politécnica de
Madrid (UPM), Madrid, Spain

Synonyms

Cluster replication; Scale out; Scalable database
replication

Definition

One of the main uses of data replication is to
increase the scalability of databases. The idea is
to have a cluster (of possibly inexpensive) nodes,
to replicate the data across the nodes, and then
distribute the load among them. In order to be
scalable, the more nodes are added to the system,
the higher the achievable throughput should be.
The scale reached today is on tens of nodes
(i.e., below 100 nodes). Communication is not an
issue since CPU and IO overheads are dominant.
The approach in the last years has been to learn
from the traditional approaches but change some
fundamentals so that the limitations of these traditional approaches are avoided.

In order to attain scalability each transaction
should not be fully processed by every replica.
This depends on how transactions are mapped to
replicas. For read only transactions, it is easy to
avoid redundant processing since they can be executed at any single replica. Update transactions
are more challenging, since updates should be
reflected at all replicas where there are copies of
the updated data items. This should be achieved
in an atomic way across the replicated system
what makes scalability challenging. Under the
traditional approaches, atomicity was attained by
means of distributed locking (for isolation) and
two-phase-commit (for failure atomicity), what
resulted in solutions lacking scalability.

A second aspect introduced by updates is their
overhead. With symmetric update processing no
scalability is achieved for updates, since all replicas pay the cost of fully executing the transaction. However, for asymmetric processing some
scalability is possible, since only one replica
pays the cost of fully executing the transaction,
whilst the others just pay the cost of propagating and installing the resulting updated tuples.
Another of the major factors that impact the
scalability is the degree of replication. Under full
replication, all replicas keep a full copy of the
database. This means that an update transaction
executed at one of the replicas should propagate
the resulting updated tuples to all other replicas.
This update propagation overhead increases its
relative weight with an increasing number of
replicas, what inherently limits the scalability.
Some research efforts pursue partial replication to
overcome the scalability limit of full replication.

There are a number of additional factors
that influence scalability. One of such factors
is the consistency being provided. The traditional
consistency criterion for replicated databases,
1-copy-serializability, constrains the potential
concurrency therefore limiting the scalability. A
lot of research has targeted to relax consistency
in order to increase the scalability.

Historical Background

Traditional textbook approaches for data replication [3] were concerned with consistency and
did not provide any scalability. The paper from
Gray et al. [9] provided analytical evidence of
this lack of scalability triggering a large body
of research performed during the last decade on
scalable database replication.

One set of approaches that aim at attaining
scalability were based on lazy replication [4, 8].
In order to provide consistency, some of these
lazy approaches constrained update propagation
to attain 1-copy-serializability [4].

A second batch of research explored scalable
eager (synchronous) replication. The seminal
approaches in this direction were Postgres-R
[12] and the database state machine [17]. These

approaches explored optimistic concurrency control under a white box (or kernel-based) replication approach and triggered the research performed in the last decade around scalable data replication. Due to the complexity of white box approaches, middleware-based replication was proposed to simplify the engineering of replication. The seminal approach to middleware-based replication was Middle-R [16, 22] that explored scalable pessimistic replica control. The middleware approach from Middle-R became very popular and today it is one of the main approaches to engineer database replication [1, 5, 15, 20]. Middleware approaches took two different flavors, those based on group communication [15, 16, 22] and those based on schedulers [1, 5, 20].

Another batch of research has looked at trading off some consistency in favor of scalability. There are two main families within this batch: those stemming from lazy approaches and those coming from eager replication. On the lazy replication side, consistency was relaxed through the concept of freshness. Freshness quantifies consistency, e.g., by the potential number of missed updates [8]. On the eager replication side, consistency was relaxed by resorting to isolation levels lower than serializability. The most popular one has been snapshot isolation [13] provides full 1-copy correctness (1-copy-snapshot-isolation, 1CSI) for snapshot isolation replicated databases. 1CSI is provided by a number of replication protocols both under update-everywhere [13] and primary-copy [20]. Another consistency criterion proposed based on snapshot isolation is Generalized Snapshot Isolation (GSI) [7]. It enables queries to see older snapshots that are prefix consistent. More recently, snapshot isolation has also been exploited in the context of lazy approaches in [6] and also for multi-tier architectures [18].

The fourth batch of research has tried to overcome the lack of scalability of full replication. The paper [11] demonstrated analytically the scalability limits of full replication. Two ways to overcome this inherent scalability limitation are quorum-based replication and partial replication. Quorum-based replication can improve the scalability for extreme update workloads, but introduces a number of technical problems, the most important ones related to the scalability of predicate reads [11]. Partial replication approaches have aimed at improving scalability by limiting the number of copies of each data item [19, 21].

Foundations

One of the uses of database replication is to increase the scalability of the system. The use of data replication to scale is also known as *scale-out* approach, in which the scalability increases by adding new nodes. This contrasts with the *scale-up* approach in which a system scales by substituting the current node with a more powerful computer. This section discusses the major factors that influence scalability of data replication, namely: how to attain atomicity, the mapping of transactions to replicas, and the consistency criterion.

Attaining Atomicity and Isolation

The traditional way to attain transactional properties in data replication [3] was based on using distributed locking to serialize transactions in the same order (isolation at replicated level) and using two-phase-commit (2PC) to guarantee that either all replicas commit the transaction or none (failure atomicity). Gray et al. [9] showed analytically the lack of scalability of traditional approaches. Basically, distributed locking has a probability of deadlock that increases exponentially with the number of transactions and 2PC does not scale and produces high response times. Two trends were followed to escape from the bottlenecks of traditional approaches. Lazy protocols commit transactions without waiting for transaction update propagation, that is, updates are propagated lazily or asynchronously with respect to transaction commit at the replica that processed the transaction. Alternatively, eager protocols propagate transaction updates atomically using alternative ways to enforce atomicity more scalable than 2PC.

R

Lazy protocols have mainly resorted to two approaches to attain some levels of consistency: primary-copy (aka single-master) and multi-master data replication. The *primary-copy* approach simplifies the consistency problem by enabling updates only at a single replica known as primary. All the other replicas, known as secondaries, only allow the execution of read-only queries. In this way, the consistency problem boils down to installing the updates from the primary in FIFO order or a more relaxed order that guarantees the same relative order of conflictive transactions. A popular replication protocol providing primary-copy is [20].

Primary-copy replication has its bottleneck in the primary that has to fully process all update transactions. As an alternative, *multi-master* replication has been proposed in which update transactions can be processed by any replica. Different approaches have been adopted for multi-master replication. In some approaches the way update transactions are propagated is constrained to guarantee high levels of consistency [4]. In others consistency is relaxed by providing quantitative levels of consistency such as freshness. A more detailed discussion on freshness is provided later in this text. Leganet is one of the systems with a lazy replication protocol providing freshness [8].

Eager protocols aim at high levels of qualitative consistency. Most eager protocols are either based on group communication or on a scheduler. *Approaches based on group communication* use atomic multicast to enforce atomicity in a more scalable way than 2PC. Atomic multicast provides failure atomicity by guaranteeing that all or none of the replicas receive messages and also guarantees that all replicas receive the messages in the same relative order. In this way, each replica is able to schedule the messages containing updates in an order consistent with respect to the common total order. There are a number of data replication protocols based on group communication such as Postgres-R [12], database state machine [17], and Middle-R [13, 16].

The *scheduler-based approach* has also been proposed for attaining consistency in eager protocols. The idea is to use a node as transaction scheduler. This scheduler is in charge of enforcing consistency. It labels transactions with a sequence number and forwards them to the replicas. In this way, replicas are able to order conflictive transactions in the same way and atomicity can also be enforced by preventing gaps in the transaction processing. Some of the database replication systems based on scheduler are conflict-aware data replication [1] and C-JDBC [5].

Transaction Mapping

The mapping of transactions to replicas is crucial in attaining scalability. There are several issues related to the transaction mapping: which replicas processes read-only transactions (queries), how update transactions are processed, and how many copies of each data item are kept.

The first issue is *how read-only transactions (queries) are managed.* Since queries only read data, for fully replicated systems it becomes possible to execute queries at any single replica, whilst update transactions are fully executed at all replicas. This mapping is known as read-one write-all-available approach (ROWAA) [3]. It enables to scale under read workloads, since for reads the load is shared among replicas. However, this approach fails to scale even with a small percentage of update transactions [11]. ROWAA has been used by the early replication protocols that are surveyed in [3].

The second issue is *how update transactions are processed.* Update transactions can be executed at only one of the replicas, as far as the other replicas get the resulting updates and install them. This mapping is known as *asymmetric update processing*, in contrast with *symmetric update processing* in which update transactions are fully executed by all replicas. This mapping also enables sharing the load introduced by update transactions, enabling scalability with update workloads [11].

Figure 1 compares the scalability of symmetric and asymmetric processing under ROWAA. The x-axis shows the fraction of writes (1.0 = 100% writes), the y-axis shows the number of nodes. The z-axis shows the

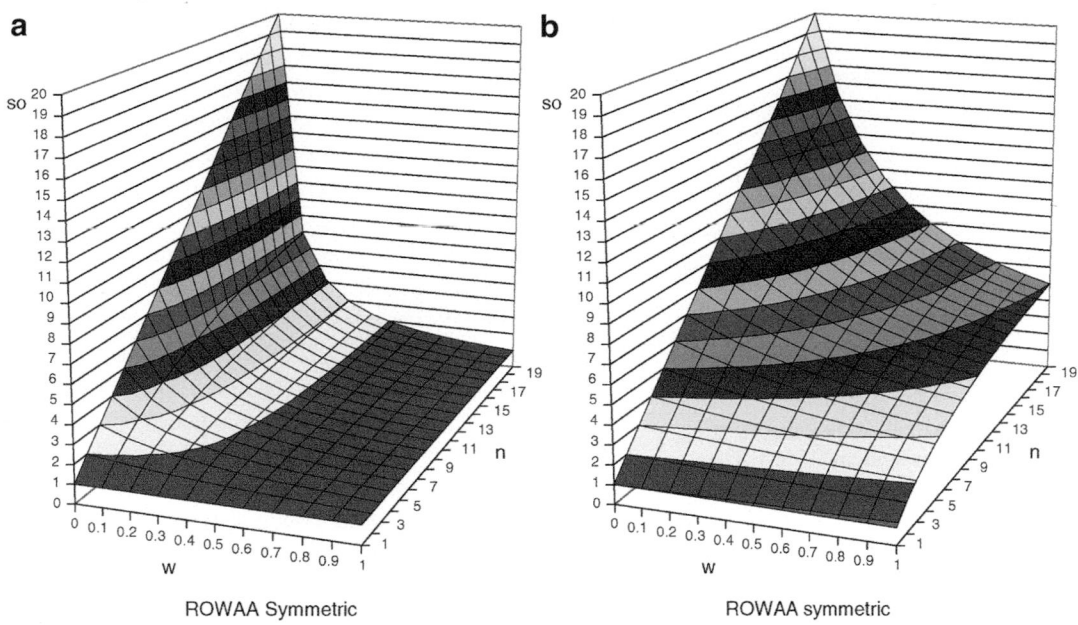

Replication for Scalability, Fig. 1 Scalability of symmetric and asymmetric processing (wo = 0.15 for asymmetric)

scale-out, i.e., how many times the throughput of a single-node system is multiplied by the replicated system. Figure 1 shows that symmetric processing only achieves a high scale-out for very low values of w. With asymmetric processing, assuming that installing updates has around 15% of the costs of executing the update transaction itself ($wo = 0.15$), the scale-out is relatively high even for higher update rates.

A different mapping that has been proposed to further increase the scalability of update workloads is based on quorum systems. Quorum systems enable to write just in a subset of the replicas (write quorum). This has the associated tradeoff that reads should also be performed on a subset of replicas (read quorum). Quorums can compete with ROWAA when the percentage of updates in the workload is very high (80–100%) [11]. However, they only work for transactions accessing individual objects. When combined with collections of objects such as tables, queries become too expensive for being practical.

The third issue is *how many copies of each data item are kept (aka degree of replication).* The paper [11] demonstrated analytically the scalability limits of full replication as a function

of the ratio of the cost of fully executing the transaction and installing the resulting updates termed write overhead, *wo*.

With typical write overheads of 0.15 and below, the scalability becomes very reasonable for a few tens of nodes. However, full replication does not scale beyond that. The reason is that the overhead introduced by the update propagation and installation consumes higher and higher fractions of the capacity of the full system what finally is translated in consuming any extra additional capacity to process updates from the existing replicas.

The alternative to scale beyond the limits of full replication is to use partial replication. Partial replication, however, also introduces other challenges. If there is no single replica with the full database, what is called pure partial replication, transactions become inherently distributed. As an alternative, if there is application knowledge available, it can be exploited to guarantee that there will be at least a replica that can execute a transaction fully locally. A potential solution is to have hybrid partial replication in which there are some replicas containing the full database and some replicas storing a fraction of the database.

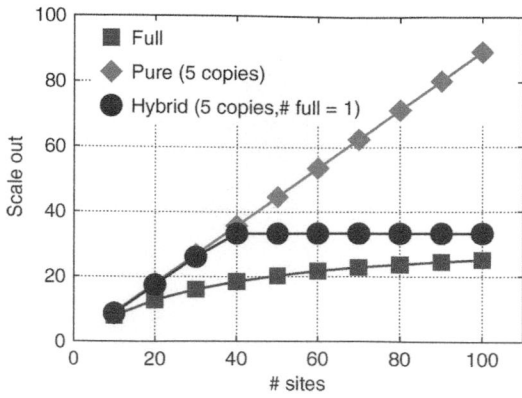

Replication for Scalability, Fig. 2 Scalability of full versus partial data replication

The former enables to avoid distributed transactions, and the latter enables to scale. Unfortunately, this approach, although it scales significantly better than full replication, reaches a limit of scalability. This limit is reached when the full replicas are saturated with update installation from partial replicas (even if they do not process any local transaction). The way to scale beyond hybrid partial replication is to use pure partial replication, what implies to be able to process replicated distributed transactions in an efficient way. The scalability of partial replication, both hybrid and pure, has been studied recently both analytically and empirically [21].

Figure 2 compares the scale-out of full replication, pure partial replication where each data item has five copies equally distributed among the nodes, and hybrid replication where one node has a copy of the entire database, and each data item has a total of five copies. The update workload in this case is 20% one can clearly see, that only pure replication has no scalability limit while the others are limited by the fact that all updates have to be performed on all copies, and there is at least one server that has copies of all data items.

Consistency Criterion
Another of the major factors influencing the scalability of replication is the consistency criterion. Some replication protocols provide a qualitative consistency. These protocols are usually *eager* protocols in which replica control is tightly in-

tegrated with update propagation to guarantee the qualitative consistency criterion. Other protocols have aimed at *quantitative consistency*, typically *lazy* protocols. In lazy protocols replica control is more relaxed and typically offers quantitative consistency, and in some cases simply eventual consistency.

Qualitative consistency for data replication provides a formal definition of the attained consistency. The consistency criterion can be seen as the extension of the isolation concept of centralized systems to a replicated setting. Isolation formalizes the consistency for concurrent executions in a centralized system. The replication consistency criterion extends this formal notion to the consistency of the concurrent transaction execution in a replicated system. The traditional criterion for replicated databases has been 1-copy-serializability (1CS) [3]. This criterion states that the concurrent execution of a set of transactions in a replicated system should be equivalent to the execution in a serializable centralized system. 1CS has been the only criterion used till very recently. 1CS, as its centralized counterpart, serializability, constrains the potential concurrency in the system by making reads and writes conflicting. Constraining the potential concurrency is harmful for scalability (even for performance in a centralized system) since it might prevent to fully utilize the available capacity in the replicated system. In order to overcome the scalability limitation of 1CS, some researchers have explored snapshot isolation (SI) as isolation notion on which to build a replication consistency criterion. 1-copy snapshot-isolation [13] (1CSI) provides the notion of 1-copy correctness underlying in 1CS for a SI database. That is, the concurrent execution of a set of transactions in a replicated system should be equivalent to a centralized (1-copy) execution in a centralized SI database. The main advantage of SI is that reads and writes do not conflict. The only conflicts are between writes on the same tuples that are rare in most applications. 1CSI enables replica control protocols that are based on SI databases. This means that they do not have contention problems due to read-write conflicts, what results in higher scalability. Snapshot isolation is currently being used for different replica control protocols, lazy primary-copy replication [20], eager update-every-where [13], lazy replication [6], and application server replication [18].

Quantitative consistency aims at increasing the scalability by relaxing the consistency enabling queries to see outdated data bounded quantitatively. It is typically used in lazy propagation schemes. Freshness criteria bound how much the value a query reads differs from the actual value of the data item. Freshness, e.g., could bound the number of updates a query has missed or the time since the copy read was last updated. The concept of freshness has been exploited by many different systems, and has been used in the context of primary-copy [20], multi-master [8], or application server replication [2]. The entry *Consistency Models for Replicated Data* provides a more detailed discussion on the different qualitative and quantitative consistency levels that exist.

for providing transparent scalability by means of self-provisioning. In this way, the hosted application can increase the number of replicas as needed depending on the received load and paying only for the resources it needs. SaaS platforms require an underlying scalable storage system such as a replicated database. Another area in which database replication is becoming a competitive solution is edge computing. *Edge computing* aims at moving web contents closer to clients located at the edge of Internet. Centralized solutions to edge computing are not adequate since distant clients observe high latencies. With edge computing, the contents is cached or replicated at data centers geographically close to the client to mask the latency. Early approaches only managed to reduce the latency of static contents. Recently, it has been shown that by using database replication in wide area networks it becomes possible also to mask the latency for dynamic contents [23].

Key Applications

Database replication has a wide number of applications. In principle any database that reaches saturation can be substituted by a replicated database to scale out. *Enterprise data centers* are certainly one of the main targets of database replication in order to scale for high loads typical of these systems. *Web farms* hosting dynamic content require the use of a database for storing it. In this kind of systems the main bottleneck is precisely the database and therefore, it can benefit from replicated databases. Grid systems are able to scale for applications based on the paradigm of bag of tasks, problems that can be split in a myriad of small subproblems that can be solved independently. In some cases, some grid applications require database access that is what finally becomes the bottleneck of the system. By employing a replicated database as a *data grid*, these grid applications can increase their scalability. A recent and increasingly important trend is *Software as a Service (SaaS)*. In this new paradigm, applications are hosted at a remote data center such as Google. SaaS aims

Cross-References

▶ Data Replication

Recommended Reading

1. Amza C, Cox AL, Zwaenepoel W. Distributed versioning: consistent replication for scaling back-end databases of dynamic content web sites. In: Proceedings of the ACM/IFIP/USENIX International Middleware Conference; 2003.
2. Bernstein PA, Fekete A, Guo H, Ramakrishnan R, Tamma P. Relaxed-currency serializability for middle-tier caching and replication. In: Proceedings of the ACM SIGMOD International Conference on Management of Data; 2006. p. 599–610.
3. Bernstein PA, Hadzilacos V, Goodman N. Concurrency control and recovery in database systems. Reading: Addison Wesley; 1987.
4. Breitbart Y, Komondoor R, Rastogi R, Seshadri S, Silberschatz A. Update propagation protocols for replicated databases. In: Proceedings of the ACM SIGMOD International Conference on Management of Data; 1999.
5. Cecchet E, Marguerite J, Zwaenepoel W. C-JDBC: flexible database clustering middleware. In: Proceedings of the USENIX 2004 Annual Technical Conference; 2004.

6. Daudjee K, Salem K. Lazy database replication with snapshot isolation. In: Proceedings of the 32nd International Conference on Very Large Data Bases; 2006. p. 715–26.
7. Elnikety S, Zwaenepoel W, Pedone F. Database replication using generalized snapshot isolation. In: Proceedings of the 24th Symposium on Reliable Distributed Systems; 2005. p. 73–84.
8. Gançarski S, Naacke H, Pacitti E, Valduriez P. The leganet system: freshness-aware transaction routing in a database cluster. Inf Syst. 2007;32(2):320–43.
9. Gray J, Helland P, O'Neil P, Shasha D. The dangers of replication and a solution. In: Proceedings of the ACM SIGMOD International Conference on Management of Data; 1996.
10. Jiménez-Peris R, Patiño-Martínez M, Alonso G, Kemme B. Scalable database replication middleware. In: Proceedings of the 22nd IEEE International Conference on Distributed Computing Systems; 2002.
11. Jiménez-Peris R, Patiño-Martínez M, Alonso G, Kemme B. Are quorums an alternative for data replication. ACM Trans Database Syst. 2003;28(3):257–94.
12. Kemme B, Alonso G. Don't be lazy, be consistent: Postgres-R, a new way to implement database replication. In: Proceedings of the 26th International Conference on Very Large Data Bases; 2000.
13. Lin Y, Kemme B, Patiño-Martínez M, Jiménez-Peris R. Middleware based data replication providing snapshot isolation. In: Proceedings of the ACM SIGMOD International Conference on Management of Data; 2005.
14. Lin Y, Kemme B, Patiño-Martínez M, Jiménez-Peris R. Enhancing edge computing with database replication. In: Proceedings of the 26th Symposium on Reliable Distributed Systems; 2007.
15. Muñoz-Escoí FD, Pla-Civera J, Ruiz-Fuertes MI, Irún-Briz L, Decker H, Armendáriz-Iñigo JE, de Mendívil JRG. Managing transaction conflicts in middleware-based database replication architectures. In: Proceedings of the 25th Symposium on Reliable Distributed Systems; 2006. p. 401–20.
16. Patiño-Martínez M, Jiménez-Peris R, Kemme B, Alonso G. Middle-R: consistent database replication at the middleware level. ACM Trans Comput Syst. 2005;23(4):375–423.
17. Pedone F, Guerraoui R, Schiper A. The database state machine approach. Distributed Parallel Databases. 2003;14(1):71–98.
18. Perez-Sorrosal F, Patiño-Martínez M, Jiménez-Peris R, Kemme B. Consistent and scalable cache replication for multi-tier J2EE applications. In: Proceedings of the ACM/IFIP/USENIX 8th International Middleware Conference; 2007. p. 328–47.
19. Pinto AL, Oliveira R, Moura F, Pedone F. Partial replication in the database state machine. In: Proceedings of the IEEE International Symposium on Networking Computing and Applications; 2001. p. 298–309.
20. Plattner C, Alonso G. Ganymed: scalable replication for transactional web applications. In: Proceedings of the ACM/IFIP/USENIX 5th International Middleware Conference; 2004.
21. Serrano D, Patiño-Martínez M, Jiménez-Peris R, Kemme B. Boosting database replication scalability through partial replication and 1-copy-snapshot-isolation. In: Proceedings of the IEEE Pacific Rim Dependable Computing Conference; 2007. p. 328–47.
22. Serrano D, Patiño-Martínez M, Jiménez-Peris R, Kemme B. An Autonomic Approach for Replication of Internet-based services. In: Proceedings of the 27th Symposium on Reliable Distributed Systems; 2008.

Replication in Multitier Architectures

Ricardo Jiménez-Peris[1] and Marta Patiño-Martínez[1,2]
[1]Distributed Systems Lab, Universidad Politecnica de Madrid, Madrid, Spain
[2]ETSI Informáticos, Universidad Politécnica de Madrid (UPM), Madrid, Spain

Synonyms

Application server clustering; Cloud computing; Cluster replication; Scalable replication; Scale out; SOA replication

Definition

Modern middleware systems are commonly used in multi-tier architectures to enable separation of concerns. For each tier, a specific component container is provided, tailored to its mission, web interface, business logic, or persistent storage. Data consistency across tiers is guaranteed by means of transactions. This entry focuses on the main three tiers: web, application server, and database tiers.

Middleware systems are at the core of enterprise information systems. For this reason they require high levels of availability and scalability. Replication is the main technique to achieve these two properties. First middleware replication approaches addressed the replication of individual tiers and focused initially on providing availability and later, on scalability. However, an integral approach is needed to provide availability and scalability of multi-tier systems. Replicating an individual tier increases its availability and might increase its scalability as well. However, replicating a single tier is not sufficient for providing data availability and scaling for the whole system; the non-replicated tiers eventually become a single point of failure and/or a potential bottleneck. A potential solution is to combine independently replicated tiers. However, due to their independent replication, they are not aware of the replication of each other; this results in inconsistency problems in the event of failures.

Recent research has introduced new approaches to replication of multi-tier architectures. The two main replication architectures are horizontal and vertical replication. In horizontal replication, the tiers are replicated independently, then measures are taken to make at least one of them aware of the other to deal in a consistent way with failures and failovers. In vertical replication, a set of tiers (typically the application and database tiers) are replicated as a unit, and the replication logic is encapsulated in the upper tier (typically the application server). Another important aspect of replication of multi-tier systems is related to the definition of the correctness criterion, i.e., the conditions a replicated multi-tier architecture should fulfill in order to provide consistency. There are two important criteria in this area, namely, exactly-once semantics and 1-copy correctness.

Historical Background

The first wave of research on the replication of multi-tier architectures focused solely on the availability of individual tiers, mainly the application server and database tiers. In the first wave, the theoretical basis for process replication was set [22]. The state machine approach was proposed to guarantee the consistency of replicated servers. This approach stated that a server could be consistently replicated by guaranteeing that all server replicas receive the requests in the same order and the server deterministically processed them. This determinism requirement was interpreted as the server had to be sequential, which was too restrictive for real servers that are multi-threaded in nature.

The efforts on the database tier concentrated on how to attain data consistency in the advent of failures and how to characterize data consistency. These initial efforts led to lock-based data replication approaches. These approaches were not scalable but they enabled the development of correctness criterion. The criterion characterizing correctness was 1-copy-serializability, i.e., replication should be semantically transparent, the replicated database should allow only those executions that had an equivalent in a non-replicated (1-copy) system.

A second wave of research on middleware replication addressed the fault-tolerance of real systems in the context of CORBA, which led to the (fault-tolerant) FT-CORBA standard. These approaches were based on the theoretical foundations of the state machine and tried to solve the difficulties of engineering replication in real middleware systems [2, 7, 16]. A new line of research opened in this wave enabled the use of the state machine approach for multi-threaded servers. A seminal paper [11] showed that it was possible to devise a scheduler that can guarantee deterministic behavior of multi-threaded servers. Later approaches studied how to increase the potential concurrency [3].

The efforts on the database tier focused on how to attain scalability. Some approaches relaxed consistency in the quest for scalability relying on lazy replication approaches (see the entry "► Replication for Scalability" for a deeper view) [5, 17]. Other approaches aimed at attaining scalability while preserving full consistency [1, 12, 18, 19]. The three later approaches leveraged group communication to provide consistency in any failure scenario.

The third wave of research on middleware replication focused on dealing with specific

R

aspects of multi-tier architectures, more concretely, on how to deal with transactional processing in a consistent way, the consistency criteria, and how to attain scalability. The issue of the lack of transactional consistency in replication approaches for multi-tier architectures was then raised [8]. In the context of FT-CORBA, it was studied how to connect a replicated CORBA application server with a shared database guaranteeing exactly once semantics [25], therefore providing the same semantics as the non-replicated system. Also in the J2EE context transactional consistency has been studied. First approaches replicated session state in the application server replicas sharing a common database [24] and enforced transactional consistency in this context.

More recently, replication of multiple tiers has also been studied, e.g., in a vertical replication approach [20]. That is, an application server and database server pair is the replication unit. Database servers are centralized and are not aware of the replication. The application server encapsulates the replication logic transparently to the database. The application server replicas interact among them to enforce consistency (i.e., 1-copy-serializability). Additionally, in this approach transactions are not aborted in the advent of failures from the client perspective, that is, it provides high transaction availability. Another recent approach has considered the issue of replicating application servers accessing multiple databases [14]. In this approach a primary-backup model is adopted. Clients interact with the primary application server. The most general case in which the application server accesses multiple databases is considered. This case is more complex since it involves dealing with two-phase-commit (2PC) and the failover of the 2PC coordinator (the transaction manager of the application server). The primary application server accesses multiple databases to process client requests. When the client transaction ends, 2PC is started and coordinated by the primary application server. The primary checkpoints session state as well as transaction management information to the backups. The approach is able to handle the primary failure and enforce transaction consistency for distributed transactions, providing transaction manager (2PC coordinator) failover.

On the correctness side, it was studied how to guarantee exactly once semantics in multi-tier architectures in which the application server tier is stateless [9]. That is, how to guarantee transaction atomicity and exactly once semantics in the advent of a failover in replicas of stateless application servers. A thorough study of the different approaches for multi-tier replication that guarantee transactional consistency was presented in [13].

On the database tier, the third wave of research has focused on boosting the scalability beyond a few tens of replicas attained in the second wave. Two aspects were mainly addressed: How to overcome the scalability limits of 1-copy serializability and full replication. 1-copy-serializability limits significantly the potential concurrency of the system which had an impact on the attainable scalability. This resulted in approaches based on a more relaxed consistency criterion, 1-copy-snapshot-isolation [15]. The scalability limits of full replication were studied analytically in [10] which led to the exploration of partial replication. Partial replication has resulted in boosting the scalability of full data replication [23], although it has some complexities such as how to deal with data partitioning and distributed transactions.

Nowadays, integral approaches for the replication of multi-tier architectures are becoming common. Research in the area is aiming at enhancing the scalability of the seminal approaches to replication of multi-tier systems. More concretely [4] studies how to replicate the application server tier with a shared database and boost the scalability by relaxing data currency through freshness constraints. Another recent approach is looking at the scalable replication of application server and database tiers [21]. In order to overcome the scalability bottleneck of serializability, this approach is based on 1-copy-snapshot isolation. Additionally, it deals for the first time with the issue of the cache consistency at the application server for relaxed isolation levels, in particular, for snapshot isolation.

Foundations

This section reviews the main replication approaches for the different tiers. Both approaches (replicating a single tier and replicating multiple tiers) will be considered.

The web tier is simple to replicate due to its stateless nature. On the other hand, it has to deal with a very specific aspect that lies in the availability and scalability of the connectivity to the Internet. This has resulted in multiple approaches to virtualize URLs and IP addresses. IP virtualization enables providing transparent failover since the client sees a single logical IP despite the fact that two physical IPs might get involved in the processing of its requests upon a failover. IP virtualization also enables to provide transparent load balancing since a network switch can transparently route requests to a logical IP to different physical IPs to balance the load across different sites. A thorough survey on the replication of the web tier can be found in [6].

The replication of the application server tier is more challenging due to its stateful nature and also due to its support for transactional processing. It is possible to distinguish between replication for availability and replication for scalability. Replication for availability aims at providing fault-tolerance but typically without taking care of scalability, even providing negative scalability (a performance below the one of the non-replicated server). Replication for scalability, in addition to providing availability, it is able to increase the system capacity by increasing the number of replicas.

Replication for availability approaches rely on the state machine approach [22]. The state machine replication is based on two basic principles: (i) All replicas receive requests in the same global order; (ii) Each replica behaves deterministically and then, given the same input request sequence produces the same output sequence. Therefore, replication protocols for availability rely on a method for ordering client requests in the same order at all replicas to attain the required global order, typically, using group communication and total order multicast. Then, they remove the non-determinism from the application. In this class,

it is possible to find all the efforts around FT-CORBA such as Eternal [16].

One of the main sources of non-determinism is multi-threading. Most of the approaches simply tackle with single-threaded (sequential) servers to guarantee determinism. However, multi-threading is a necessary feature for real-world servers. Some research has been conducted in how to guarantee determinism of replicas in the presence of multi-threading [3, 11]. This research has produced different multi-threading schedulers that enforce deterministic scheduling enabling the replication of multi-threaded servers. The seminal approach [11] showed that by using a deterministic scheduler and setting a deterministic method to schedule requests integrated in the deterministic scheduler it became possible to guarantee the determinism of multi-threaded servers. Later on some other deterministic schedulers have been proposed aiming at increasing the real concurrency [3].

An important issue that needs to be dealt with in the replication of application servers is how to preserve transactional consistency [8]. The paper [9] proposes e-transactions to preserve consistency and identifies exactly once semantics as a key consistency criterion for replication of transactional multi-tier architectures. e-Transactions also provide a protocol for satisfying exactly-once semantics for replicated stateless application server tiers with a shared database (see Fig. 1). Exactly-once semantics states that each transaction should be executed exactly once despite failures and re-executions due to failovers.

A more evolved approach [24] deals with the replication of stateful application servers also with a shared database. In this approach, the relationship between request and transaction is not constrained to be 1 to −1 as in e-transactions, instead, it can be arbitrary. That is, it can be 1:1, N:1, and 1:N. In [24] it is proposed a protocol providing exactly-once semantics [9] despite failures. The protocol intercepts database accesses labeling them with a global sequence identifier and the associated client and request identifier. Each database requests is multicast to the other replicas with the associated information. In the

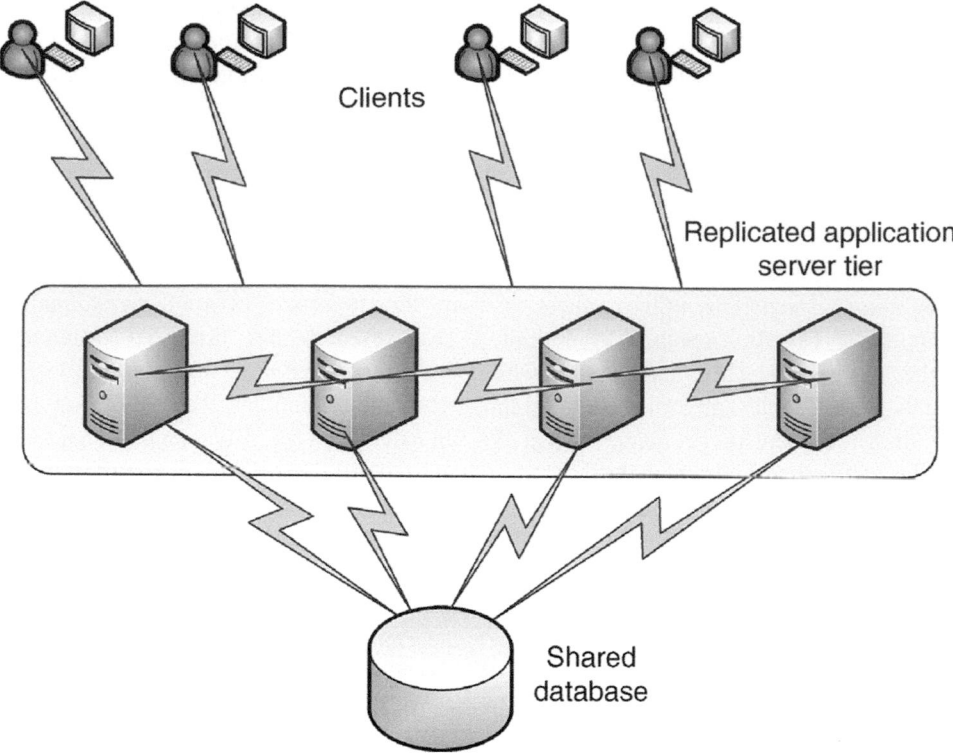

Replication in Multitier Architectures, Fig. 1 *Horizontal* replication with a replicated application server tier and a shared database

advent of a failure of the primary replica of the application server, a backup replica will take over. Those clients that have not received a reply to their last request will resubmit it to the new primary. The new primary will execute resubmitted requests intercepting the requests submitted to the database. This interception takes care of executing the database requests in the same order as in the old primary in order to guarantee a deterministic execution and exactly once semantics. New requests will be delayed till the requests received from the former primary are replayed.

All the aforementioned approaches deal exclusively with the replication of the application server tier, which results in the database becoming a single point of failure and performance bottleneck. More recently, research has been conducted to deal with the replication of both the application server and database tiers guaranteeing consistency despite failures. Two alternative architectures have been considered: horizontal and vertical replication [13].

In horizontal replication each tier is replicated independently (see Fig. 2). However, the integration of two independently replicated tiers results in inconsistency problems in the advent of failures [13]. In order to avoid inconsistencies the replicated tiers should become aware of each other to handle failovers consistently. However, this is typically unrealistic since it implies the cooperation of two different vendors very often competitors in the market. Kemme et al. [13] proposes two solutions to attain consistent failover by incorporating awareness of replication in only one of the two tiers, either the application server or the database tier.

Vertical replication lies in taking as replication unit an application server – database server pair [20] (see Fig. 3). The database server is not aware of the replication. The application server encapsulates all the replication logic. In the ver-

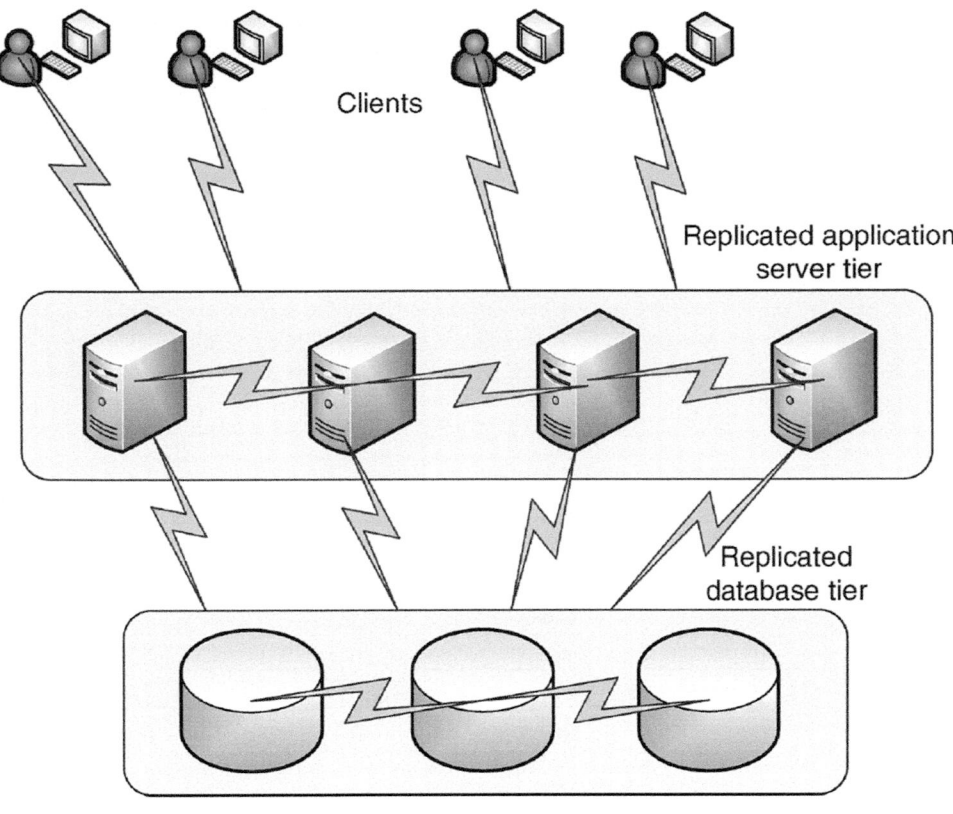

Replication in Multitier Architectures, Fig. 2 *Horizontal* replication with two replicated tiers

tical replication approach [20] both changes to session (session beans in the context of J2EE) and persistent state (entity beans in the context of J2EE) are captured. The two kinds of changes are propagated to the backup replicas at the end of each client request to provide *highly available transactions*, that is, transactions that do not abort despite failovers.

Both horizontal and vertical replication architectures are also used to attain scalability. Horizontal replication has mainly focused on replicating the middleware tier to replicate sessions while sharing a single database. In these approaches, consistency is attained through the shared database that serializes concurrent database accesses from different replicas. The replication of sessions enables to increase the scalability of applications in which the bottleneck is located in the application server, they are CPU intensive (e.g., image processing). However, these approaches fail to scale when the bottleneck

is on the database, which is the most common case in practice.

Vertical approaches have aimed at increasing the scalability independently of whether they are application server or database intensive while providing strong consistency [4, 21]. To date there are two different approaches to scalable vertical replication: [21] that provides strong consistency and [4] that provides relaxed consistency.

Bernstein et al. [4] uses a master-slave approach and aims at scaling read-intensive workloads. The master processes all update transactions that get reflected in the underlying database and takes care of propagating a sequence of committed updates to the slaves. Each slave updates the cache and also commits the updates to the database in commit order. Read-only queries can be executed at any replica and can read stale data. The system can be configured to set the staleness or freshness upon which each transaction should be executed.

**Replication in Multitier
Architectures, Fig. 3**
Vertical replication.
Replicated application
server and database tiers

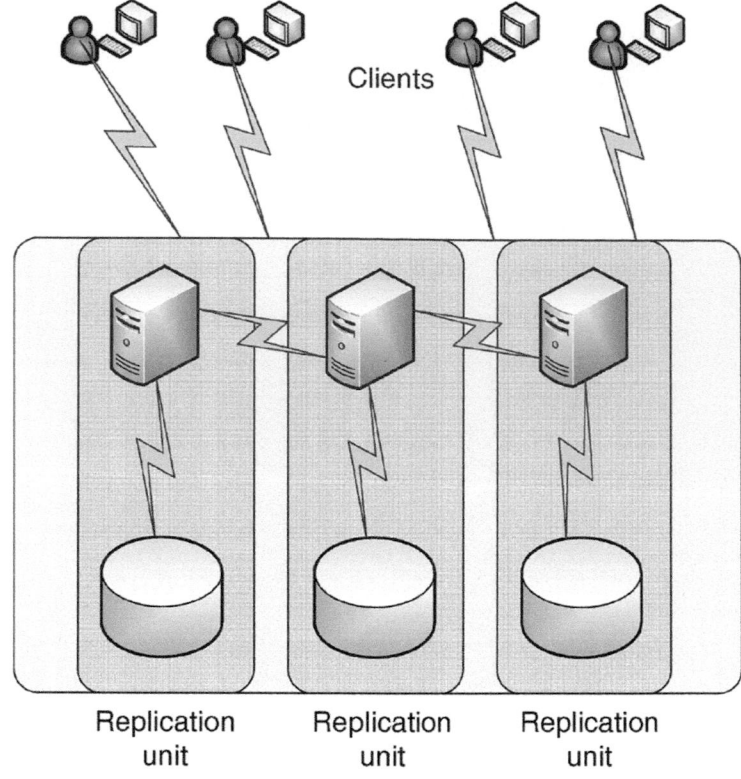

On the other hand, [21] is an update-everywhere approach (all replicas can process update transactions) and aims at scaling both read and write workloads. Unlike [4, 21], provides strong consistency, namely, 1-copy-snapshot-isolation. Perez-Sorrosal et al. [21] firstly provides a consistency criterion for application server caching, named cache-transparency. An application server is cache-transparent if it allows the same executions as the application server with the cache disabled. Cache transparency extends the isolation notion of databases to a multi-tier architecture. This is crucial since current application servers may work incorrectly with databases running on some relaxed isolation levels, such as, snapshot isolation. Secondly, it guarantees 1-copy correctness for the multi-tier server which provides replication transparency. Perez-Sorrosal et al. [21] builds on 1-copy-snapshot isolation to obtain higher scalability levels since 1-copy-serializability has some strict scalability limits as discussed earlier. Basically, [21] uses multi-versioning in the cache of the application server replicas that is synchronized with the snapshots of the underlying database and among replicas in order to provide cache and replication transparency, respectively.

Cross-References

▶ Data Replication

Recommended Reading

1. Amza C, Cox AL, Zwaenepoel W. Distributed versioning: consistent replication for scaling back-end databases of dynamic content web sites. In: Proceedings of the ACM/IFIP/USENIX International Middleware Conference; 2003.
2. Baldoni R, Marchetti C. Three-tier replication for FT-CORBA infrastructures. Softw Pract Exper. 2003;33(8):767–97.
3. Basile C, Kalbarczyk Z, Iyer RK. Active replication of multithreaded applications. IEEE Trans Parallel Dist Syst. 2006;17(5):448–65.

4. Bernstein PA, Fekete A, Guo H, Ramakrishnan R, Tamma P. Relaxed-currency serializability for middle-tier caching and replication. In: Proceedings of the ACM SIGMOD International Conference on Management of Data; 2006. p. 599–610.
5. Breitbart Y, Komondoor R, Rastogi R, Seshadri S, Silberschatz A. Update propagation protocols for replicated databases. In: Proceedings of the ACM SIGMOD International Conference on Management of Data; 1999.
6. Cardellini V, Casalicchio E, Colajanni M, Yu PS. The state of the art in locally distributed web-server systems. ACM Comput Surv. 2002;34(2): 263–311.
7. Felber P, Guerraoui R, Schiper A. The implementation of a CORBA object group service. Theory Pract Object Syst. 1998;4(2):93–105.
8. Felber P, Narasimhan P. Reconciling replication and transactions for the end-to-end reliability of CORBA applications. In: Proceedings of the International Symposium on Distributed Objects and Applications; 2002.
9. Frølund S, Guerraoui R. E-transactions: end-to-end reliability for three-tier architectures. IEEE Trans Softw Eng. 2002;28(4):378–95.
10. Jiménez-Peris R, Patiño-Martínez M, Alonso G, Kemme B. Are quorums an alternative for data replication. ACM Trans Database Syst. 2003;28(3): 257–94.
11. Jiménez-Peris R, Patiño-Martínez M, Arevalo S. Deterministic scheduling for transactional multi-threaded replicas. In: Proceedings of the 19th Symposium on Reliable Distributed Systems; 2000. p. 164–73.
12. Kemme B, Alonso G. Don't be lazy, be consistent: postgres-R, a new way to implement database replication. In: Proceedings of the 26th International Conference on Very Large Data Bases; 2000.
13. Kemme B, Jiménez-Peris RR, Patiño-Martínez MM, Salas J. Exactly once interaction in a multi-tier architecture. In: Proceedings of the VLDB Workshop on Design, Implementation and Deployment of Database Replication; 2005.
14. Kistijantoro AI, Morgan G, Shrivastava SK. Enhancing an application server to support available components. IEEE Trans Softw Eng. 2008;SE-34(4): 531–45.
15. Lin Y, Kemme B, Patiño-Martínez M, Jiménez-Peris R. Middleware based data replication providing snapshot isolation. In: Proceedings of the ACM SIGMOD International Conference on Management of Data; 2005.
16. Moser LE, Melliar-Smith PM, Narasimhan P, Tewksbury L, Kalogeraki V. The eternal system: an architecture for enterprise applications. In: Proceedings of the International Enterprise Distributed Object Computing Conference; 1999. p. 214–22.
17. Pacitti E, Simon E. Update propagation strategies to improve freshness in lazy master replicated databases. VLDB J. 2000;8(3):305–18.
18. Patiño-Martínez M, Jiménez-Peris R, Kemme B, Alonso G. Middle-R: consistent database replication at the middleware level. ACM Trans Comput Syst. 2005;23(4):375–423.
19. Pedone F, Guerraoui R, Schiper A. The database state machine approach. Distrib Parallel Database. 2003;14(1):71–98.
20. Perez-Sorrosal F, Patiño-Martínez M, Jiménez-Peris R, Vuckovic J. Highly available long running transactions and activities for J2EE applications. In: Proceedings of the 23rd IEEE International Conference on Distributed Computing Systems; 2006.
21. Perez-Sorrosal F, Patiño-Martínez M, Jiménez-Peris R, Kemme B. Consistent and scalable cache replication for multi-tier J2EE applications. In: Proceedings of the ACM/IFIP/USENIX 8th International Middleware Conference; 2007. p. 328–47.
22. Schneider FB. Implementing fault-tolerant services using the state machine approach: a tutorial. ACM Comput Surv. 1990;22(4):299–319.
23. Serrano D, Patiño-Martínez M, Jiménez-Peris R, Kemme B. Boosting database replication scalability through partial replication and 1-copy-snapshot-isolation. In: Proceedings of the IEEE Pacific Rim Dependable Computing Conference; 2007. p. 328–47.
24. Wu H, Kemme B. Fault-tolerance for stateful application servers in the presence of advanced transactions patterns. In: Proceedings of the 24th Symposium on Reliable Distributed Systems; 2005. p. 95–108.
25. Zhao W, Moser LE, Melliar-Smith PM. Unification of transactions and replication in three-tier architectures based on CORBA. IEEE Trans Depend Secure Comput. 2005;2(1).

R

Replication with Snapshot Isolation

Carsten Binnig
Computer Science-Database Systems, Brown University, Providence, RI, USA

Synonyms

Replication with Snapshot Isolation; Snapshot Replication

Definition

Database replication is concerned with the management of data copies residing on different database nodes (also called sites). The main propose of replication is to achieve scalability and availability of database systems which are both very important design goals in modern cloud deployments. A main challenge of database replication is replica control, which keeps the copies at different sites consistent. Replica control needs to be combined with mechanisms for concurrency control in replicated databases to guarantee a globally correct execution of transactions under replication.

The ideal goal for a replicated database is to provide the same consistency level as in a centralized database system without any replication. Early work has concentrated on providing a consistency level that is called *one-copy serializability*; that is, the replicated database should appear just like an unreplicated database with serializable transactions. However, as pointed out in the seminal paper by Gray et al. [1] one-copy serializability typically scales poorly with the number of replicas. Therefore, other replica control techniques (e.g., based on lazy replication) have been developed that allow replicas to be temporarily inconsistent. These techniques have shown to increase scalability of replicated databases but they make it much harder for programmers to be used correctly since the application needs to deal with (potential) inconsistencies when accessing data copies at different sites.

More recent work has focused on providing other consistency levels that provide scalability and avoid temporary inconsistencies. A promising direction are replicated database systems such as Postgres-RSI that implement *one-copy snapshot isolation*. Snapshot isolation (SI) is a popular consistency level in many centralized database systems such as Oracle and Postgres since read-only transactions never cause update transactions to block or to abort. In one-copy SI the replicated database appears like an unreplicated one using snapshot isolation (SI) as consistency level. One-copy SI has shown to achieve much better scalability as one-copy serializability [2] while providing a consistency level that programmers can better understand than other techniques which result in temporarily inconsistent databases.

More recent SI-based replication techniques [3, 4] have focused on relaxing some properties of one-copy SI to further increase the scalability or to be applicable in geo-replicated scenarios which typically suffer from high latencies to keep data copies up to date.

Overview

Historical Background

Snapshot Isolation (SI) was originally defined in [5]. The key properties of SI are that a transaction reads data from a snapshot, which captures a consistent state of the database before the transaction started. In strong SI (or only SI for short), the snapshot that is read by a transaction must represent the most recent consistent state that has been committed before the transaction started. There exist also weaker forms of SI, such as generalized SI (GSI), that allows a transaction to also read any older committed snapshot, which is an interesting alternative in replicated databases where each replica might not have the most recent snapshot. In SI (as well as in GSI) a transaction never sees updates of a concurrent transaction. Moreover, lost updates are prevented by the so-called "First-Committer-Wins" rule that prevents two concurrent transactions from both committing if they modify the same data item; i.e., two concurrent transactions are only allowed to commit if their write-sets are disjoint.

Foundation: Snapshot Replication

The main challenge of replication is replication control; i.e., keeping copies consistent. All recent solutions for replication control perform reads on one replica and writes on all (available) replicas which is called read-once-write-all(-available) or ROWA(A) for short. In the following, we only consider ROWAA protocols. Moreover, replication control strategies can be further classified by two parameters. The first parameter determines on which replicas updates can take place: In a primary copy approach, each database object has a primary replica which performs all updates and forwards the updates to all other replicas. In

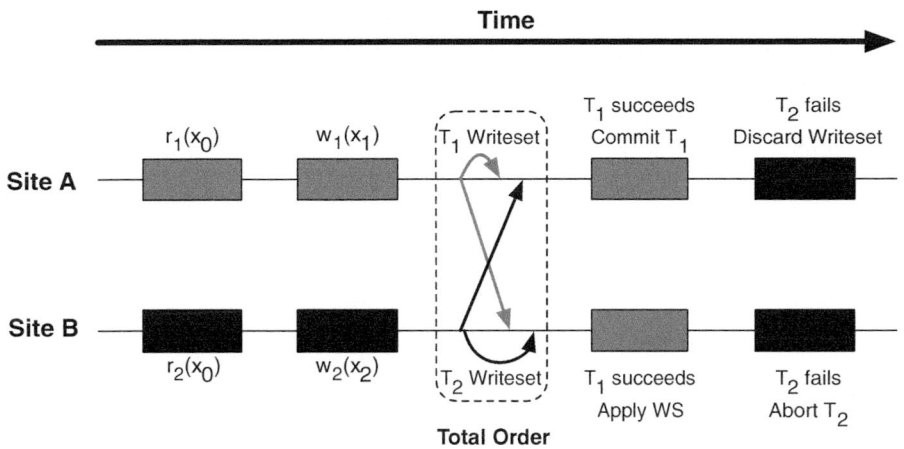

an update everywhere approach each replica accepts updates. The second parameter determines when replicas coordinate: In an eager approach, updates must be applied to all replicas before the transaction commits. In a lazy approach coordination between replicas takes place after a transaction commits.

In the following, we describe a simple protocol of how we can achieve one-copy snapshot isolation in a replicated database when using a primary copy for updates and eager coordination to propagate updates to replicas. Other approaches to allow update everywhere or lazy coordination can be found in the recommended reading section. The main idea of the simple protocol for one-copy SI is that the update transactions first execute locally on the primary copy without committing. Afterwards, the writeset of each update transaction is multicast to all remote replicas before the transaction commits globally. This ensures that clients do not see different snapshots, when accessing copies on different sites when using read-only transactions. Moreover, to ensure that concurrent update transactions do not execute in different orders on remote replicas, writesets of all update transactions have to be executed serially in the same order at all remote sites. One way to achieve transaction ordering is to use a group communication system as outlined in [6] to distribute the writesets. Finally, read transactions in our scheme can execute at any replica without interfering with write transactions just as in a centralized SI scheme.

In the following, we see an example where transactions T_1 on site A and transaction T_2 on site B update the same item x. Since the writeset is distributed to all nodes in the same order (T_1 before T_2 in the example), only the writeset of T_1 is applied on all nodes while the writeset of T_2 is discarded.

Replica control for one-copy SI that follows the simple protocol outlined before has been implemented directly in the database kernel (e.g., in Postgres-RSI [2]) or in a separate middleware component outside the database kernel (e.g., in [7]).

Key Applications

SI is used as a concurrency control mechanism in a wide range of commercial and open source database systems (e.g., Microsoft SQL Server, Oracle, SAP HANA, as well as Postgres). Many research prototypes have shown that single-node SI databases can be extended using a replication protocol (implemented in the database nodes or in a middleware) such that a cluster of single-node SI databases acts transparently as a global one-copy SI database. However, these solutions are not widely used in practice yet. Recently, driven by cloud databases, a variant of snapshot isolation called parallel snapshot isolation (PS) was applied to replicate data to different geo-distributed data centers [4] in a commercial setup.

R

Future Directions

Generalized snapshot isolation (GSI) [3] extends snapshot isolation in a manner that it is better suited for replicated databases. While strong snapshot isolation requires that transactions observe the most recent committed snapshot, GSI allows transactions to use of older snapshots thus enabling lazy replication. In replicated GSI, many of the properties of one-copy SI remain. In particular, read-only transactions can be executed on any site and never block update transactions.

Another approach for SI-based replication over multiple geo-replicated data centers is called Parallel Snapshot Isolation (PSI). Geo-replication suffers from high latencies between data centers. With PSI, all copies within a data center observe data according to a consistent snapshot and a common ordering of transactions. Across data centers, PSI enforces no global ordering of transactions, allowing the system to replicate transactions asynchronously across sites. To that end, if a user session of an application only accesses data at same data center (which is often the case), the user will see a consistent snapshot.

Cross-References

▶ Distributed Concurrency Control
▶ One-Copy-Serializability
▶ Replica Control
▶ Replicated Database Concurrency Control
▶ Replication Based on Group Communication
▶ Replication for Availability and Fault Tolerance
▶ Replication for Scalability
▶ Snapshot Isolation

Recommended Reading

1. Gray J, Helland P, O'Neil PE, Shasha D. The dangers of replication and a solution. In: Proceedings of the ACM SIGMOD International Conference on Management of Data; 1996. p. 173–82.
2. Wu S, Kemme B. Postgres-R(SI): combining replica control with concurrency control based on snapshot isolation. In: Proceedings of the 21st International Conference on Data Engineering; 2005. p. 422–33.
3. Elnikety S, Zwaenepoel W, Pedone F. Database replication using generalized snapshot isolation. In: Proceedings of the 24th IEEE Symposium on Reliable Distributed Systems; 2005. p. 73–84.
4. Sovran Y, Power R, Aguilera MK, Li J. Transactional storage for geo-replicated systems. In: Proceedings of the 23rd ACM Symposium on Operating System Principles; 2011. p. 385–400.
5. Berenson H, Bernstein PA, Gray J, Melton J, O'Neil EJ, O'Neil PE. A critique of ANSI SQL isolation levels. In: Proceedings of the ACM SIGMOD International Conference on Management of Data; 1995. p. 1–10.
6. Kemme B, Alonso G. Database replication: a tale of research across communities. Proc. VLDB Endow. 2010;3(1): 5–12.
7. Lin Y, Kemme B, Patiño-Martínez M, Jiménez-Peris R. Middleware based data replication providing snapshot isolation. In: Proceedings of the ACM SIGMOD International Conference on Management of Data; 2005. p. 419–30.
8. Daudjee K, Salem K. Lazy database replication with snapshot isolation. In: Proceedings of the 32nd International Conference on Very Large Data Bases; 2006. p. 715–26.

Reputation and Trust

Zoran Despotovic
NTT DoCoMo Communications Laboratories
Europe, Munich, Germany

Synonyms

Feedback systems; Word of mouth

Definition

Trust means reliance on something or someone's action. As such, it necessarily involves risks on the side of the subject of trust, i.e., trustor. The main goal of a trust management system is to reduce the involved risks. Reputation systems present a possible solution to do that. They use relevant information about the participants' past behavior (feedback) to encourage trustworthy behavior in the community in question. The key presumptions of a reputation system are that the participants of the considered online community engage in repeated interactions and that the information about their past doings is informative of their future performance and as such will influence it. Thus, collecting, processing, and disseminating the feedback about the participants' past behavior is expected to boost their trustworthiness.

Key Points

The goal of a reputation system is to encourage trustworthy behavior. It is up to the system designer to define what trustworthy means in her specific setting. There are two possible ways to do this: *signaling and sanctioning reputation systems*. In a signaling reputation system, the interacting entities are presented with signals of what can go wrong in the interactions if they behave in specific ways. Having appropriate signals, the entities should decide what behavior is most appropriate for them. An important assumption of the signaling reputation systems is that the involved entities do not change their behavior in response to a change of their reputation. As an example, the system may just provide a prospective buyer with indications of the probability that the seller will fail to deliver a purchased item. This probability is the main property of the seller. It can change with time, but independently of the seller's reputation.

The other possibility is *sanctioning reputation systems*. The main assumption they make is that the involved entities are aware of the effect the reputation has on their benefits and thus adjust their behavior dynamically as their reputation changes. The main task of a reputation system in this case is to *sanction* misbehavior through providing correlation between the feedback the agent receives and the long-run profit she makes.

Cross-References

▸ Distributed Hash Table
▸ Peer-to-Peer System
▸ Similarity and Ranking Operations
▸ Social Networks
▸ Trust in Blogosphere

Request Broker

Aniruddha Gokhale
Vanderbilt University, Nashville, TN, USA

Synonyms

Event broker; Object request broker; Storage broker

Definition

A Request Broker is a software manifestation of the Broker architectural pattern [3] that deals primarily with coordinating requests and responses, and managing resources among communicating entities in a distributed system. A Request Broker is usually found as part of middleware, which are layers of software that sit between applications, and the underlying operating systems, hardware and networks.

Historical Background

The mid to late 1980s established the TCP/IP protocol suite as the de facto standard suite of protocols for building networked applications. Contemporary operating systems provided a number of application programming interfaces (APIs) for network programming. The famous among these were the *socket* API that were tailored towards building TCP/IP-based applications. This era also saw the widespread use of killer TCP/IP-based applications, such as the File Transfer Protocol (FTP).

Despite the success of the socket API to build distributed applications, there were a number of challenges involved in developing these applications. First, the socket API was tedious to use and incurred a number of accidental complexities stemming from the use of type-unsafe C-language data types. Second, application programmers were responsible for handling the marshaling and demarshaling

R

of the data types transferred between the communication entities, which involved a number of challenges. Notable among these were the need to address the byte ordering, and word size and padding issues arising from the heterogeneity in the hardware architectures. Third, server applications were able to provide only one primary functionality due to the lack of an object-oriented service abstraction. Finally, client applications had to explicitly bind to a location-dependent service, which made the applications inflexible and brittle.

Every distributed application that was developed had to reinvent the wheel and address these complexities. There was a compelling need to overcome these challenges by factoring out the commonly occurring patterns of network programming and provide them within reusable frameworks. This led to the notion of middleware [1, 2], which provide reusable capabilities in one or more layers of software that sit between the application logic, and the operating systems, hardware and networks. A Request Broker is at the heart of such a middleware and performs a number of functions on behalf of the applications.

Request Brokering capabilities at the middleware layer started appearing with the advent of the Remote Procedure Call (RPC). Sun RPC was among the earlier middleware platforms that illustrated Request Brokers. Others that emerged thereafter included the Distributed Computing Environment (DCE), the Common Object Request Broker Architecture (CORBA), Java Remote Method Invocation (RMI), Distributed Component Object Model (DCOM) and .NET Remoting.

Foundations

A Request Broker implements the Broker architectural pattern [3]. The primary objectives of a request broker are to decouple clients and servers of a distributed application and provide reusable services, such as concurrency management, connection management, seamless transport-level networking support, data marshaling, and location transparency. Figure 1 illustrates the commonly found functional blocks [5] in a Request Broker discussed below.

- *Proxy* - A Request Broker allows the separation of interface from implementation thereby decoupling the client of a service from the implementation of the service. A service is often described using interface definition languages to define the interfaces, the operations they support and data types that can be exchanged. Versioning of interfaces can also be provided within these descriptions. A proxy gives an illusion of the real implementation to the client.

- *Discovery services* - A Request Broker manages service discovery on behalf of clients. Some form of a service description, such as a URL or service name, is used by the broker to lookup a potentially remote implementation that offers the service. The broker will return to the client application a handle to the external service in the address space of the client so that the client can seamlessly invoke services on the handle via the Proxy. Depending on whether *pass by reference* or *pass by value* semantics are used, the execution of client requests may occur remotely on the server machine or locally in the client's address space, respectively.

- *Marshaling engine* - A Request Broker often specifies an encoding scheme or a serialization format for representing application data. Often there exist tools that read the interface descriptions of services and synthesize code that can marshal and unmarshal all the data types that are defined as serializable in the interface descriptions. This tool-generated code, which is usually called a *stub*, is linked into the fabric of the Request Broker, which manages data marshaling and unmarshaling on behalf of the application.

- *Concurrency control* - A server application may require finer-grained control on concurrency for scalability and performance. Request Brokers often provide sophisticated concurrency control mechanisms, such as *thread pools*, to handle application requests in a scalable and concurrent manner.

Request Broker, Fig. 1
Request broker functional
architecture

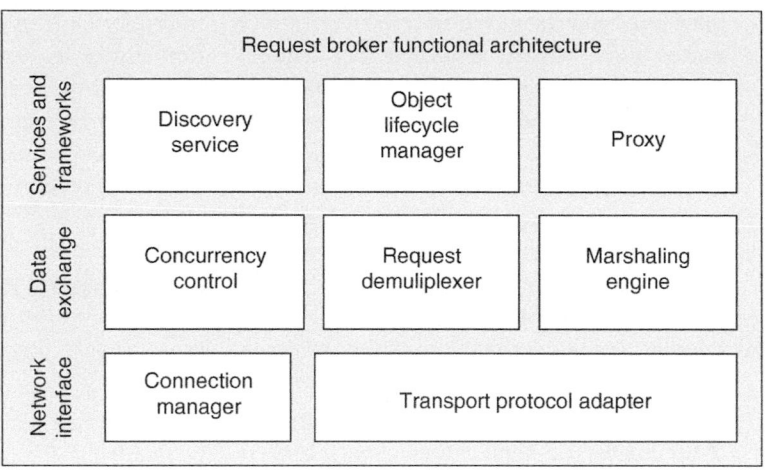

• *Object lifetime manager* - A server application may host multiple different services to optimally utilize resources. Often these services are provided in the form of objects implemented in a programming language. The Request Broker must manage several objects simultaneously in the system according to the policies dictating their lifetimes, e.g., transient or persistent. Other policies that tradeoff between memory footprint and performance include the *activate-on-demand* policy, which activates a service only on demand and for the duration of the request. This conserves resources but impacts performance.

• *Request demultiplexing and dispatching* - A broker may manage several hundreds of objects at a time. When requests arrive at a server, they must be efficiently demultiplexed and dispatched to the right object implementation.

• *Connection management and Transport Adapter* - Since applications are distributed they must communicate over different networking protocols. Request Brokers often enable applications to configure the choice of protocol to use for communication. For example, the unpredictable behavior of TCP/IP will not suffice for real-time applications in which case special transport protocols must be used. But the application must be shielded from these differences. Transport adapters [4] can provide these capabilities. For the transport

protocol used, appropriate connection management capabilities are required. For example, connections may need to be purged periodically to conserve resources.

Apart from the functional architecture, a taxonomy of the capabilities provided by a Request Broker can be developed along the following orthogonal dimensions.

1. *Communication models* - This dimension includes different classifications, such as (i) Remote Procedure Calls (RPC) versus Message Passing; (ii) Synchronous versus Asynchronous communications; (iii) Request/Response versus Anonymous Publish/Subscribe semantics; (iv) Client/Server versus Peer-to-Peer.
2. *Location transparency* - This dimension includes mechanisms, such as object references or URLs, used by the Request Broker to hide the details of the service.
3. *Type system support* - This dimension includes the richness of the data types that can be exchanged between the communicating entities, and the semantics of data exchange. For example, Request Brokers can support passing objects by value or by reference or both.
4. *Interoperability and portability* - This dimension includes the degree of heterogeneity supported by the Request Brokers. For example, technologies such as CORBA are both

R

platform- and language-independent, which makes them widely applicable but requires complex mapping between the platform- and language-independent representations to platform- and language-specific artifacts. Often this impacts the richness of data types that can be exchanged. On the other hand some technologies are language-dependent, e.g., Java RMI, or platform-dependent, e.g., DCOM and. NET Remoting.

5. *Quality of service (QoS) support* - This dimension includes the capabilities provided by the Request Broker to support different QoS requirements of applications. Some brokers may be tailored to support real-time support while others are customized for persistence and transaction support.

Key Applications

Request Brokers are at the heart of distributed computing, and span a wide range of application domains including telecommunications, finance, healthcare, industrial automation, retail, grid computing, among others. Request Brokers with enabling technologies to support real-time applications have also been used to build distributed real-time and embedded systems found in domains, such as automotive control, avionics mission computing, shipboard computing, space mission computing, among others. Request Brokers are also found in systems that require management and scheduling of storage resources, or in large, event-based or content-management systems.

Future Directions

As applications become more complex, heterogeneous, and require multiple different and simultaneous quality of service properties, such as real-time, fault tolerance and security, the responsibilities of the Request Broker increase substantially. Brokering capabilities themselves will need to be distributed requiring coordination among the distributed brokers. Supporting multiple QoS properties will require design-time tradeoffs due to the mutually conflicting objectives of each QoS property. Run-time resource management will be a key in supporting the QoS properties since applications are increasingly demanding autonomic capabilities.

Experimental Results

Experimental research on Request Brokers is proceeding along the discussion articulated in the future trends. Service oriented computing is requiring Request Brokers to move beyond simple client-server or peer-to-peer computing to more advanced scenarios where the brokering capabilities are required across entire application workflows.

URL to Code

The following URLs provide more information on different Request Broker technologies and sample code.

CORBA is standardized by the Object Management group (http://www.omg.org).

Java RMI is a technology of Sun Microsystems (http://java.sun.com/javase/technologies/core/basic/rmi/index.jsp).

DCOM and .NET Remoting are technologies from Microsoft (http://msdn2.microsoft.com/en-us/library/ms809340.aspx and http://msdn2.microsoft.com/en-us/library/2e7z38xb.aspx, respectively).

DCE is a technology standardized by the Open Group (http://www.opengroup.org/dce/).

Cross-References

▶ CORBA
▶ DCE
▶ DCOM
▶ .NET Remoting
▶ RMI
▶ Service-Oriented Architecture

Recommended Reading

1. Bakken DE. Middleware. In: Urban J, Dasgupta P, editors. Encyclopedia of distributed computing. Dordrecht: Kluwer; 2001.
2. Bernstein PA. Middleware: a model for distributed system services. Commun ACM. 1996;39(2):86–98.
3. Buschmann F, Meunier R, Rohnert H, Sommerlad P, Stal M. Pattern-oriented software architecture – a system of patterns. New York: Wiley; 1996.
4. Gamma E, Helm R, Johnson R, Vlissides J. Design patterns: elements of reusable object-oriented software. Reading: Addison-Wesley; 1995.
5. Schmidt DC. Evaluating architectures for multithreaded CORBA object request brokers. Commun ACM. 1998; Special Issue on CORBA.41(10):54–60.

Residuated Lattice

Vilém Novák
Institute for Research and Applications of Fuzzy Modeling, University of Ostrava, Ostrava, Czech Republic

Synonyms

Structure of truth values

Definition

The residuated lattice is a basic algebraic structure accepted as a structure of *truth values* for fuzzy logic and fuzzy set theory. In general, it is an algebra

$$\langle L, \vee, \wedge, \otimes, \rightarrow, 0, 1 \rangle$$

where L is a support, $\vee$ and $\wedge$ are binary lattice operations of join and meet, 0 is the smallest and 1 is the greatest element. The $\otimes$ is additional binary operation of *product* that is associative and commutative, and $a \otimes 1 = a$ holds for every $a \in$ L. The $\rightarrow$ is a binary *residuation* operation that is adjoined with $\otimes$ as follows:

$$a \otimes b \leq c \text{ if and only if } a \leq b \rightarrow c$$

for arbitrary elements $a, b, c \in L$. The residuation operation is a generalization of the classical implication.

Key Points

The residuated lattice is naturally ordered by the classical lattice ordering relation defined by

$$a \leq b \quad \text{if and only if} \quad a \wedge b = a.$$

In general, of course, there can exist incomparable elements in L. A typical property of the residuation operation is $a \rightarrow b = 1$ iff $a \leq b$. In words: $a \leq b$ iff the degree of implication $a \rightarrow b$ is equal to 1.

A typical example of residuated lattice is the *standard Łukasiewicz MV-algebra*

$$\langle [0, 1], \max, \min, \otimes, \rightarrow, 0, 1 \rangle$$

where $a \otimes b = \max\{0, a + b - 1\}$ is Łukasiewicz product (conjunction) and $a \rightarrow b = \min\{1, 1 - a + b\}$ is Łukasiewicz implication.

Another widely used residuated lattice is an MTL-algebra, where $L = [0, 1]$, $\otimes$ as some left continuous *t-norm* (see "▶ Triangular Norms") and $\rightarrow$ is the corresponding residuation.

Every boolean algebra is a residuated lattice. Thus, a special case of residuated lattice is also the boolean algebra for classical logic

$$\langle \{0, 1\}, \vee, \wedge, \otimes, \rightarrow, 0, 1 \rangle$$

where $\otimes = \wedge$ (i.e., $\otimes$ coincides with minimum) and $\rightarrow$ is the classical boolean (material) implication.

Both $\wedge$ as well as $\otimes$ are natural interpretations of logical conjunction. This means that there are two, in general different, conjunctions in fuzzy logic. In classical logic they coincide, though.

R

The residuation $\rightarrow$ is natural interpretation of implication.

Negation is defined by

$$\neg a = a \rightarrow 0.$$

This operation coincides with classical negation in boolean algebra for classical logic, i.e., $\neg 0 = 1$ and $\neg 1 = 0$. In the standard Łukasiewicz MV-algebra (based on [0, 1]) this operation reduces to $\neg a = 1 - a$.

The *biresiduation* operation is defined by

$$a \leftrightarrow b = (a \rightarrow b) \wedge (b \rightarrow a).$$

This operation coincides with with classical equivalence in boolean algebra for classical logic. In fuzzy logic, it is used as a natural interpretation of logical equivalence. In standard Łukasiewicz MV-algebra it gives $a \leftrightarrow b = 1 - |a - b|$, $a, b \in [0, 1]$.

There are many more special kinds of residuated lattices. Except for boolean algebra, the most important are BL-algebra, Gödel algebra, product algebra and MV-algebra.

Cross-References

► Fuzzy Relation
► Fuzzy Set
► Triangular Norms

Recommended Reading

1. Esteva F, Godo L. Monoidal t-norm based logic: towards a logic for left-continuous t-norms. Fuzzy Set Syst. 2001;124(3):271–88.
2. Hájek P. Metamathematics of fuzzy logic. Dordrecht: Kluwer; 1998.
3. Klement EP, Mesiar R, Pap E. Triangular norms. Dordrecht: Kluwer; 2000.
4. Novák V, Perfilieva I, Močkoř J. Mathematical principles of fuzzy logic. Boston/Dordrecht: Kluwer; 1999.
5. Gottwald S. A treatise on many-valued logics. Baldock, Herfordshire: Research Studies; 2001.

Resource Allocation Problems in Spatial Databases

Donghui Zhang[1] and Yang Du[2]
[1]Paradigm4, Inc., Waltham, MA, USA
[2]Northeastern University, Boston, MA, USA

Synonyms

Facility-location problem

Definition

Assume that a franchise plans to open one or more branches in a state. How shall the locations of the new branches be allocated to maximally benefit the customers? Depending on whether some branches already exist and how to quantify the benefits to the customers, there are multiple forms of such resource allocation problems. In spatial databases, distance plays an important role. A customer is assumed to always visit the closest branch. Therefore it is beneficial to a customer if a new branch is opened at a location closer than her closest existing branch. The *max-inf optimal-location query* assumes the existence of a set of sites (already opened franchise branches), and aims to find a new location within a given area which benefits the largest number of customers. The *min-dist optimal-location query* also assumes the existence of a set of sites and aims to find a location for a new site which is optimal; but here the optimality is defined as minimization of the average distance from each customer to the nearest site. Compared with its max-inf counterpart, the min-dist optimal-location query takes into account the saved distance for each customer. The *k-medoid query* assumes that there does not exist any p as residential blocks that minimize the average distance from every customer to the nearest picked location.

Historical Background

There exists an extensive literature in operations research on resource allocation problems, named the *facility location problems*. The most widely studied version is the *uncapacitated facility location (UFL)* problem.

Given a set of customers, a set of potential sites, a non-negative cost for opening each site, and a non-negative service cost between each site and a customer, the *UFL* problem is to find a subset of the potential sites so that the total cost (opening cost plus service cost) is minimum. Here, the term "uncapacitated" refers to the assumption that there is no limit on the number of customers that each site can serve.

The k-medoid problem is a variation of the UFL problem, where the set of customers and sites are identical, the opening cost of each site is zero, the service cost between a site and a customer is the distance between them, and there is an additional constraint that exactly k sites will be opened. The k-means problem is a related problem, where the potential sites can be anywhere.

Facility location problems are (typically) NP-hard. Therefore research has focused on approximate algorithms with low computational complexity and small approximation errors. A recent survey of approximate algorithms for facility location problems appeared in [5]. Unlike spatial database research, the operations research assumes that all objects fit in memory and existing approaches typically scan through the dataset multiple times.

Foundations

When considering resource allocation problems in a large spatial database, the data usually reside in secondary storage and are indexed by spatial index structures so that scanning through all objects is avoided. The research goal is not to derive good asymptotic complexities, but to design efficient algorithms which, when applied to real datasets, incur a small number of I/O operations.

For both the max-inf and min-dist optimal-location queries under L_1 distance, there exist straightforward $O(n^2)$ solutions which find *exact* answers. Note that this differentiates the optimal-location queries from the NP-hard facility-location problem. However, such $O(n^2)$ algorithms are prohibitively slow in large spatial databases. Exploiting spatial index structures, [1, 6], proposed solutions which are much faster (sub-linear in practice).

Similarly, the k-medoid query was examined in the context of spatial databases by utilizing spatial index structures. In particular, [2], relies on the clustering of higher-level index entries in an R-tree to avoid scanning through all objects.

Max-Inf Optimal-Location Query
The max-inf optimal location (OL) query aims to find the location of a new site which benefits (or influences) the largest number of customers. Here the *influence* of a location is the number of customers which are closer to the location than to any existing site. To answer the max-inf OL query under L_1 metric, Du et al. [1] introduced the concept of *nn_buffer* and reduced the problem into finding a location with maximum overlap among *nn_buffers* of objects. Formally, the *nn_buffer* of an object is defined as follows.

Definition *Given an object o and its closest site s, the nn_buffer of o is a contour such that $\forall l$ on the contour, $d(l, o) = d(o, s)$.*

In other words, a location l is inside *o.nn_buffer* if and only if o is closer to l than to any site. As shown in Fig. 1, the *nn_buffer* of

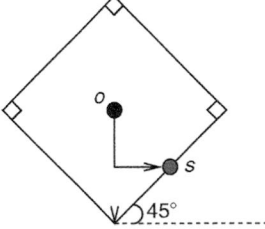

Resource Allocation Problems in Spatial Databases, Fig. 1 The *nn_buffer* of an object (or customer) under L_1 metric is a diamond (*rotated square*)

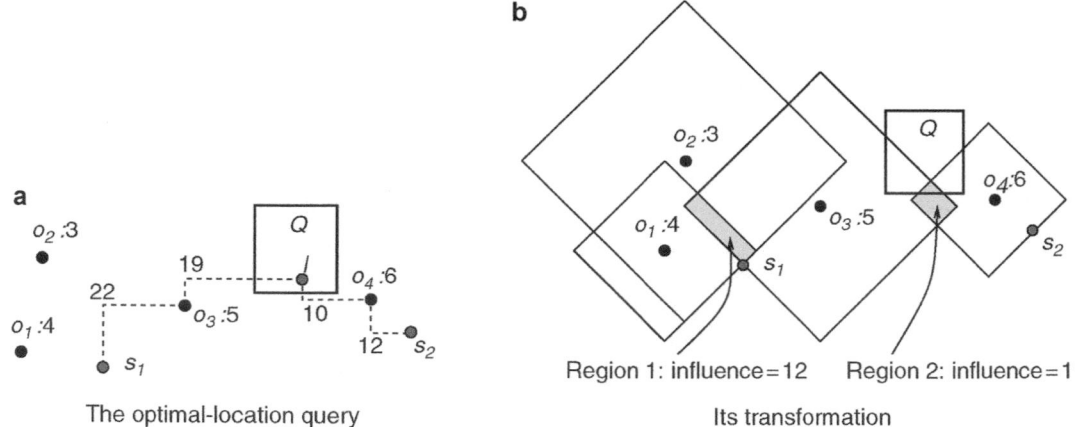

Resource Allocation Problems in Spatial Databases, Fig. 2 In (**a**), l is an optimal location, with influence 11. The transformation in (**b**) shows that any location in the intersection between Q and region 2 is an optimal location

an object o under L_1 metric is a diamond. The object o contributes to the influence of a location l, if and only if l is inside the *nn_buffer* of o. Furthermore, if the coordinates are rotated by 45° counter-clockwise, the *nn_buffers* become axis-parallel squares. Therefore, given a query region Q, an optimal location is a location l inside Q which maximizes the total weight of overlapped *nn_buffers*. Here the weight of an object tells how important an object is. For instance, if the object is an apartment complex, its weight can be the number of residents living in it. Figure 2 gives an example of the query with four objects and two sites and its corresponding transformation. Based on this observation, three solutions are proposed to answer the max-inf OL query under L_1 metric.

The first one is an R-tree based solution. It assumes that an R-tree is used to index the set O of objects. Furthermore, the R-tree is augmented with some extra information. Every object stores the L_1 distance to its closest existing site, and every index entry stores the maximum such distance of objects in the sub-tree. The solution follows two steps. The first step is to retrieve those objects from the R-tree whose *nn_buffers* intersect with Q. The objects are identified in increasing order of X coordinate in the rotated coordinate, even though the R-tree was built in the original coordinate. The second step is a plane-sweep process, which consumes the objects streamed in from the first step and computes

the weight of overlapped *nn_buffers*. A naive plane-sweep solution has $O(n^2)$ cost, where n is the number of retrieved objects. However, it can be improved to $O(n \log n)$ by using a data structure called *aggregation SB-tree*. After processing all retrieved objects, the algorithm reports the location with maximum overlap as the answer.

The second solution is based on a specialized aggregation index called the *OL-tree*. The *OL-tree* is a balanced, disk-based, dynamically updateable index structure extended from the k-d-B-tree. This index is built in the rotated coordinate. Objects to be inserted into the OL-tree are *nn_buffers* versus points in the original k-d-B-tree. Local optimal location information in each subtree is stored along with the index entry referencing the sub-tree. Such information enables efficient processing algorithms that can answer the max-inf optimal-location query without examining all *nn_buffers* intersecting the query region Q. In the best case, the algorithm only needs to examine the root node of the tree.

These two solutions have interesting tradeoffs between storage cost and query efficiency. The R-tree based solution utilizes the existing R-tree structure. The storage cost is linear to the number of objects. However, as the objects are not pre-aggregated, a query needs to examine all objects whose *nn_buffers* intersect with the query range Q. If Q has large size, the query performance is poor. On the other hand, the OL-tree is a specialized aggregation index, whose space over-head is much higher, but provides faster query processing.

The third solution combines the benefits of the previous two approaches. As in the R-tree based solution, it uses an R-tree to store the objects. But to guide the search, it uses a small, in-memory OL-tree-like index structure. This index is named the *Virtual OL-tree (VOL-tree)*. It resembles the top levels of an OL-tree and it does not phys-ically store any *nn_buffer*. A leaf entry plays a similar role to an index entry. It corresponds to a spatial range, and it logically references a node that stores (pieces of) *nn_buffers* in that range. If necessary, these *nn_buffers* can be retrieved from the R-tree dynamically. To prune the search space, the VOL-tree stores some (but not all) aggregated information of the *nn_buffers* in each level. This solution provides the best trade-off between space overhead and computational costs, as shown experimentally in [1].

Min-Dist Optimal-Location Query

The min-dist OL query aims to find the loca-tion of a new site which minimizes the average distance from each customer to the nearest site. Zhang et al. [6] proposed a progressive algorithm to find the exact answer of the Min-Dist OL query under L_1 metric. The algorithm works in two steps. In the first step, it limits the candidate locations to finite locations, which are the inter-sections of certain horizontal and vertical lines. The second step is a recursive partition-and-refine step. In this step, it partitions the query range Q into a few cells (by using some of the vertical and horizontal lines), and calculates $AD(\cdot)$ for the corners of these cells. Here $AD(l)$ is denoted as the average distance from each customer to the nearest site after opening a new site l. The smaller $AD(l)$ is, the better the new location l is. For each cell, the algorithm estimates the lower-bound of $AD(\cdot)$ among all locations in the cell and prunes the cell if its lower bound is larger than the minimum $AD(\cdot)$ already found. It repeatedly partitions the unpruned cells into smaller cells to further refine the result, until the actual optimal location is found.

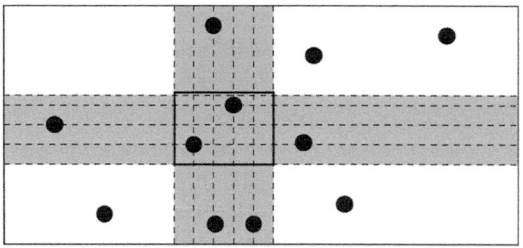

Resource Allocation Problems in Spatial Databases, Fig. 3 The candidate locations are limited to the intersec-tions of the *dashed lines*

To understand how the candidate locations are limited to finite locations, consider the example in Fig. 3. The black dots are the objects and the thick-bordered rectangle is the query region Q. The shadowed region is composed of the horizontal extension of Q and the vertical ex-tension of Q. The following theorem guarantees that the optimal location can be limited to finite candidates.

Theorem *Consider the set of horizontal (and vertical) lines that go through some object in the horizontal (and vertical) extension of Q or go through some corner of Q. The min-dist optimal location under L_1 metric can be found among the intersection points of these lines.*

In order to prune some cells, the second step of the algorithm utilizes a novel lower-bound estimator for $AD(\cdot)$ of all locations in a cell. The following theorem describes how the estimator works.

Theorem *Let the corners of a cell C be c_1, c_2, c_3, and c_4, where $\overline{c_1 c_4}$ is a diagonal. Let the perimeter of C be p.*

$$\max \left\{ \frac{AD(c_1) + AD(c_4)}{2}, \frac{AD(c_2) + AD(c_3)}{2} \right\} - \frac{p}{4}$$

is a lower bound of $AD(l)$ for any location $l \in C$. The progressive algorithm, which starts with one cell (the query region Q) and keeps partitioning it into smaller cells and trying to prune cells from the processing queue, has two advantages. First, it avoids computing $AD(\cdot)$ for all candi-

R

date locations, as all candidate locations inside a pruned cell are ignored. Second, it responds fast. An approximate answer is reported at the very beginning (after computing $AD(\cdot)$ for the four corners of Q), within a guaranteed error bound. As the algorithms runs, the candidate optimal location and the associated error rate are improved progressively, until the exact optimal location is found.

Disk-Based k-Medoid Query

Mouratidis et al. [3] studied the k-medoid problem in large spatial databases. They assume that the data objects are spatial points indexed by an R-tree, and the service cost between two points is defined by the distance between them. They propose the TPAQ (Tree-based PArtition Querying) algorithm that achieves low CPU and I/O cost. The TPAQ algorithm avoids reading the entire dataset by exploiting the grouping properties of the existing R-tree index.

Initially, TPAQ traverses the R-tree in a top-down manner, stopping at the topmost level that provides enough information for answering the given query. In the case of the k-medoid problem, TPAQ finds the topmost level with more than (or equal to) k entries. For instance, if $k = 3$ in the tree of Fig. 4, TPAQ stops at level 1, which contains five entries, n_1 through n_5.

Next, TPAQ groups the entries of the partitioning level into k slots (i.e., groups). To utilize the grouping properties of the R-tree index, TPAQ augments each retrieved entry n_i with a weight w and a center c. Here, the weight w is the

number of points in the subtree referenced by n_i and the center c is the geometric centroid of the entry, assuming that the points in the sub-tree are uniformly distributed. To merge the initial entries into exactly k groups, TPAQ utilizes space-filling curves to select k seed entries which capture the distribution of points in the dataset. Then, each remaining entry is inserted into the slot whose weighted center is the closest to its center.

The final step of TPAQ is to pick k medoid objects, one from each group; TPAQ reports the weighted center of each group as the corresponding medoid.

Since the k-medoid problem is NP-hard, like any other practical algorithm TPAQ can only provide an approximate answer to the query. Experiments show that, compared to previous approaches, TPAQ achieves comparable or better quality, at a small fraction of the cost (seconds as opposed to hours).

In [3], the authors also extended the above method to solve the medoid-aggregate query, where k is not known in advance. Given a user-specified parameter T, the medoid-aggregate query determines the smallest value of k and computes k-medoids such that the average distance from each object to the nearest medoid is smaller than T. The solution extends from TPAQ in two ways. First, without knowing k in advance, the R-tree top-down traversal stops at the level decided by the spatial extents and the expected cardinality of the entries. Second, multiple passes over the initial entries might be required to find the proper way of grouping the entries into slots.

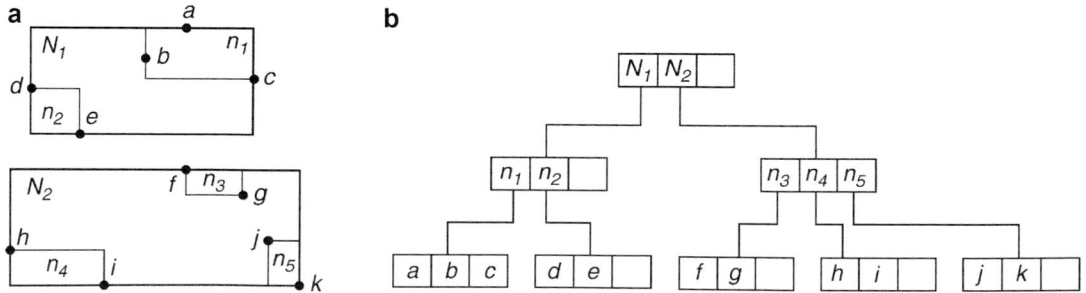

Resource Allocation Problems in Spatial Databases, Fig. 4 R-tree example

Key Applications

Location-Based Services

Research on resource allocation problems is expected to help enhance the performance of location-based services based on the geographic proximity of clients to potential facilities of interest.

Spatial Decision Making

The efficient solutions to the resource allocation problems can provide valuable information to decision making. For example, to help provide candidate locations of a new branch.

Future Directions

One future research direction is to extend the solutions to other practical distance metrics. The existing solutions to both the max-inf and min-dist OL queries assume L_1 distance. Extending to L_2 (i.e., Euclidean) distance and road network distance is desirable. Similarly, extending the existing k-medoid solution to handle road network distance is interesting. Another future direction is to extend the queries to allow both the pre-existence of certain sites and the ability to find multiple locations. Existing solutions to the OL queries are limited to only one optimal location, and existing solutions to the k-medoid query are limited to zero existing site. Relaxing/avoiding these limitations will lead to more practical problems and solutions. The third future direction is to consider moving objects instead of static ones. In this case, the optimal locations and k-medoids should be continuously monitored over a set of moving objects [3, 4].

Cross-References

▶ Nearest Neighbor Query
▶ Reverse Nearest Neighbor Query

Recommended Reading

1. Du Y, Zhang D, Xia T. The optimal-location query. In: Proceedings of the 9th International Symposium on Advances in Spatial and Temporal Databases; 2005. p. 163–80.
2. Mouratidis K, Papadias D, Papadimitriou S. Medoid queries in large spatial databases. In: Proceedings of the 9th International Symposium on Advances in Spatial and Temporal Databases; 2005. p. 55–72.
3. Papadopoulos S, Sacharidis D, Mouratidis K. Continuous medoid queries over moving objects. In: Proceedings of the 10th International Symposium on Advances in Spatial and Temporal Databases; 2007. p. 38–56.
4. U LH, Mamoulis N, Yiu ML. Continuous monitoring of exclusive closest pairs. In: Proceedings of the 10th International Symposium on Advances in Spatial and Temporal Databases; 2007. p. 1–19.
5. Vygen J. Approximation algorithms for facility location problems (lecture notes). Technical Report, University of Bonn; 2005. p. 1–59.
6. Zhang D, Du Y, Xia T, Tao Y. Progressive computation of the min-dist optimal-location query. In: Proceedings of the 32nd International Conference on Very Large Data Bases; 2006. p. 643–54.

Resource Description Framework

Christian Bizer[1], Maria-Esther Vidal[2], and Michael Weiss[3]
[1]Web-based Systems Group, University of Mannheim, Mannheim, Germany
[2]Computer Science, Universidad Simon Bolivar, Caracas, Venezuela
[3]Carleton University, Ottawa, ON, Canada

R

Synonyms

RDF

Definition

The Resource Description Framework (RDF) is a graph-based data model for sharing information on the Web [1, 2]. RDF has been standardized by the World Wide Web Consortium (W3C). The data model puts special emphasis on the globally unique identification of all model elements with Uniform Resource Identifiers (URIs). The globally unique identification of the described objects (called resources in the RDF context) enables data from different sources to be represented in a single RDF graph and lays the foundation for the integration of the data in a pay-as-you-go fashion. The globally unique identification of schema elements (called vocabulary terms in the RDF context) provides for mixing terms from different schemata in a single RDF graph and makes RDF a schema-less data model. The RDF data model forms the foundation for a wide range of RDF-based technologies including the SPARQL query language, various data exchange formats such as RDF/XML, RDFa, JSON-LD, and Turtle, as well as RDF vocabularies for data interchange (FOAF, Dublin Core, Schema.org, Open Graph Protocol, Data Cube, Provenance Vocabulary) and knowledge representation (SKOS, RDF Schema, OWL).

Historical Background

In 1998, Tim Berners-Lee described RDF in his vision for how data should be represented on the Semantic Web [3]. The first W3C Recommendation of RDF as a standard was published in 1999. The original specification was refined in 2004 with a more explicit description of the abstract RDF data model as well as revised version of the RDF/XML data exchange format. In the following years, the family of W3C standards that build on RDF was extended by the OWL Web Ontology Language (2004 and 2009), the RDFa syntax for embedding RDF data into HTML pages (2007 and 2012), the SPARQL query language for RDF

(2008 and 2013), the SKOS knowledge organization system (2009), JSON-LD data exchange format (2011), and the Provenance Vocabulary (2013). In 2014, W3C published version 1.1 of the RDF Recommendation. This new version extends the original RDF data model with the concept of RDF datasets which group RDF statements into multiple named graphs and thus provide a simple means for representing original data and meta-information about this data in a single model.

Scientific Fundamentals

The RDF data model has its basis in the theory of semantic networks. RDF uses triples to represent assertions about the properties and relationships of entities (called resources in the context of RDF). Each RDF triple can be understood as a simple sentence consisting of a subject (the entity that is described), a predicate (which aspect of the entity is described), and an object (the actual value of the aspect). A set of RDF triples forms an RDF graph. Figure 1 shows three RDF triples that describe William Shakespeare. The first two triples state his birthplace and birthdate. The last triple asserts that he is of type person. RDF uses URI references to identity the subjects and predicates of RDF triples (e.g., dbr:William_Shakespeare is a qName shorthand form of http://dbpedia. org/resource/William_Shakespeare). The object of a triple can either be another URI (such as dbr:Stratford-upon-Avon) or a literal (such as 1564-04-26). The meaning of literals can further be refined using language tags (such as @en or @de) or XML datatypes. As an

Subject	Predicate	Object
dbr:William_Shakespeare	dbo:birthPlace	dbr:Stratford-upon-Avon
dbr:William_Shakespeare	dbo:birthDate	"1564-04-26"^^xsd:date
dbr:William_Shakespeare	rdf:type	foaf:Person

Resource Description Framework, Fig. 1 Example RDF triples

Resource Description Framework, Fig. 2
Example of an RDF graph set

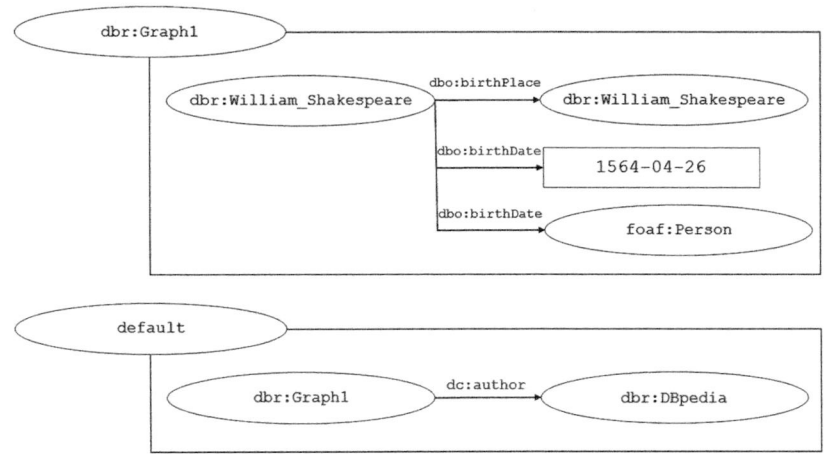

alternative to URI references, RDF resources may also be identified using blank nodes.

The RDF 1.1 specification [1] extends the RDF data model with the concept of RDF datasets. RDF datasets consist of a set of named graphs (each named with a URI reference) and a default graph. Figure 2 shows an example of a RDF dataset consisting of the named graph dbr:Graph1 and the default graph. Oval nodes represent resources, and arrows between them represent predicates. Objects of RDF triples that are themselves resources are shown as ovals, while literal values are shown as small boxes. RDF graphs are shown as large boxes with the graph name in the upper left corner. As we can see in Fig. 2, the graph name dbr:Graph1 is picked up by the RDF triple in the default graph in order to express provenance meta-information (dc:author) about the named graph.

The RDF data model aims to ease the interpretation of RDF data on the Web by providing for mixing terms from proprietary vocabularies (dbo:birthDate) with terms from commonly used vocabularies (rdf:type, foaf:Person) in a single RDF graph as well as by allowing vocabulary definitions to refer to each other using vocabulary links (e.g., rdfs:subClassOf or owl:owl:equivalentClass).

The RDF data model forms the foundation for a wide range of RDF-based technologies including the SPARQL query language, various data exchange formats, as well as RDF vocabularies for data interchange and knowledge representa-

tion. Further details about these technologies are provided in the entry on "RDF Technologies."

Key Applications

The RDF data model is applied in a variety of use cases:

1. **Semantic annotation of web documents**: RDF together with the RDFa syntax is used to represent meta-information in web documents and to annotate data in HTML documents that describes entities such as products, people, events, local businesses, reviews, and ratings. It is estimated that in 2013 at least 470,000 websites use RDFa for including semantic annotations into their HTML documents [6]. The annotations are used major search engines such as Google, Bing, Yahoo!, and Yandex to display rich snippets in search results (for instance, average rating and price range for a restaurant) and are partly included into the knowledge graphs built by these companies. RDFa annotations that follow the Open Graph protocol specification are used by Facebook to display entities from external websites inside Facebook.

2. **Publication of structured data on the web**: The RDF data model is also used to directly publish structured data on the web and to interconnect data from different sources into a single global logical data graph according

R

to the Linked Data principles (see entry on "▸ Linked Open Data"). The main application domains in which data is published according to the Linked Data principles are e-government (with the UK and US governments being forerunners), libraries (including different national libraries), and scientific data sharing in life sciences, geography, and linguistics [4, 5]. It is estimated that in 2014, around 1000 datasets were available as Linked Data on the web [5].

3. **Knowledge representation**: The OWL Web Ontology Language is increasingly adopted as a main standard for exchanging ontologies between knowledge-based systems [8]. OWL ontologies are further used for knowledge representation in life science applications. The simple knowledge organization system SKOS has been widely adopted in the libraries and cultural heritage domains for representing and exchanging topic taxonomies between systems. Large-scale cross-domain knowledge bases that employ RDF to represent knowledge that has been extracted from Wikipedia are YAGO and DBpedia.

4. **Application integration**: The RDF data model is also used in application integration scenarios. For instance, IBM Rational uses RDF for enabling data interoperability between different software development and application lifecycle management (ALM) tools [7]. The British Broadcasting Company (BBC) uses RDF to integrate the content management systems of different TV and radio stations. A main goal in these use cases is to decouple the stored data from the specific application and to allow cross-application data access.

Data Sets

A variety of RDF datasets is available for public download. The Web Data Commons project regularly extracts all RDFa, microdata, and microformat data from large web crawls and provides the extracted data for download. Their 2014 dataset (http://www.webdatacommons.org/

structureddata/) consist of 20 billion RDF triples that were extracted from 600 million HTML pages. The web of Linked Data is regularly crawled in order to provide evaluation data for the Billion Triples Challenge. The resulting datasets (http://km.aifb.kit.edu/projects/btc-2014/) consist of several billion RDF triples originating from different Linked Data sources. The DBpedia and YAGO knowledge bases are available for download from the respective project websites.

Cross-References

▸ Dublin Core
▸ Linked Open Data
▸ OWL: Web Ontology Language
▸ Resource Description Framework (RDF) Schema (RDFS)
▸ RDF Stores
▸ Semantic Web
▸ SPARQL

Recommended Reading

1. Cyganiak R, Wood D, Lanthaler M. RDF 1.1 Concepts and abstract syntax. W3C recommendation. 2014. Available online at: http://www.w3.org/TR/2014/REC-rdf11-concepts-20140225/
2. Schreiber G, Raimond Y. RDF 1.1 Primer, W3C working group note. 2014. Available online at: http://www.w3.org/TR/2014/NOTE-rdf11-primer-20140624/
3. Berners-Lee T. Semantic web road map. Available online at: http://www.w3.org/DesignIssues/Semantic.html
4. Heath T, Bizer. C. Linked data: evolving the web into a global data space. San Rafael: Morgan & Claypool; 2011.
5. Schmachtenberg M, Bizer C, Paulheim H. Adoption of the linked data best practices in different topical domains. In: Proceedings of the 13th International Semantic Web Conference; 2014. p. 245–60.
6. Meusel R, Petrovski P, Bizer C. The WebDataCommons Microdata, RDFa and Microformat Dataset Series. In: Proceedings of the 13th International Semantic Web Conference; 2014. p. 277–92.
7. Le Hors A, Speicher S. Case Study: Open Services Lifecycle Collaboration framework based on Linked Data. 2012. Available online at: http://www.w3.org/2001/sw/sweo/public/UseCases/IBM/
8. Allemang D, Hendler J. Semantic web for the working ontologist: modeling in RDF, RDFS and OWL. 2nd ed. San Francisco: Morgan Kaufmann; 2011.

Resource Description Framework (RDF) Schema (RDFS)

Vassilis Christophides
INRIA Paris-Roquencourt, Paris, France

Synonyms

Conceptual schemas

Definition

An RDF schema (RDFS) is represented in the basic RDF model and provides (i) *abstraction* mechanisms, such (multiple) class or property subsumption and (multiple) classification of resources; (ii) *domain and range* class specifications to which properties can apply; (iii) *documentation facilities* for names defined in a schema.

RDF/S follow the W3C design principles of interoperability, evolution and decentralization. In particular, it is possible to interconnect in an extensible way resource descriptions (by superimposing different statements using the same resource URIs) or schema namespaces (by reusing or refining existing class and property definitions) regardless of their physical location on the Web.

Key Points

Over the last decade, RDF and its accompanying RDFS specifications has been the subject of an extensive collaborative design effort. ([1]http://www.w3.org/RDF) RDF/S was originally developed as an application-neutral model to represent various kinds of descriptive information about Web resources, i.e., metadata. The W3C published the RDF Schema as Candidate Recommendation in 2000. Under the boost of the Semantic Web for transforming the Web into a universal medium for data, information, and knowledge exchange, refined versions of the RDF/S family of specifications have been published

in 2004 [1–5]. As a matter of fact, RDF/S serves today as general purpose languages for consistent encoding, exchange and processing of available Web content, through a variety of syntax formats (XML or others). Their current design has been strongly influenced by recent advances in knowledge representation and aims to provide a simple *semantic layer* to the Web (no negation), as a base for more advanced query answering and reasoning services.

To present the core RDF/S modeling primitives, the example of a catalog of a cultural Portal (see Fig. 1) is used. To build this catalog, various cultural resources (e.g., Web pages of cultural sites) must be described from both a Portal and a Museum curator perspective. Figure 1 relies on an almost standard graphical notation, where nodes represent RDF/S resources (circles) or literals (rectangulars) while edges represent either user-defined (single arrows) or build-in (double and dashed arrows) properties.

The lower part of Fig. 1 depicts the description of an image (e.g., *museoreinasofia.mcu.es/guernica.jpg*) available on the Web. Hereforth the prefix "&" is used to abbreviate the involved resource URIs (e.g., &**r1**). In a first place, &**r1** is described from a Portal perspective as instance of the class named adm:*ExtResource* (uniqueness of names is ensured by using as prefix the corresponding schema URIs, like adm or *cult* in the example). More precisely, the statement (a triple in the RDF jargon) <&*r1,rdf:type,adm:ExtResource*> asserts that the resource &**r1** (subject) is of type (predicate) adm:*ExtResource* (object). Additionally &**r1** is stated to have two properties: one with name title and value the string "Guernica" (triple <&*r1,adm:title,Guernica*>) and, the other, with name file_size and value the integer 200 (triple <&*r1,adm:file_size,200*>). Resource &**r1** is also asserted as an instance of the class named Painting (triple <&*r1,rdf:type,cult:Painting*>) having a property technique with the literal value (string) "*oil on canvas*" (triple <&*r1,cult:technique,oil on canvas*>). &**r1** is further described by considering additional resources as for instance, &**r2** whose identifier (*#artist132*) is local to the Portal, i.e., not universally accessible in

R

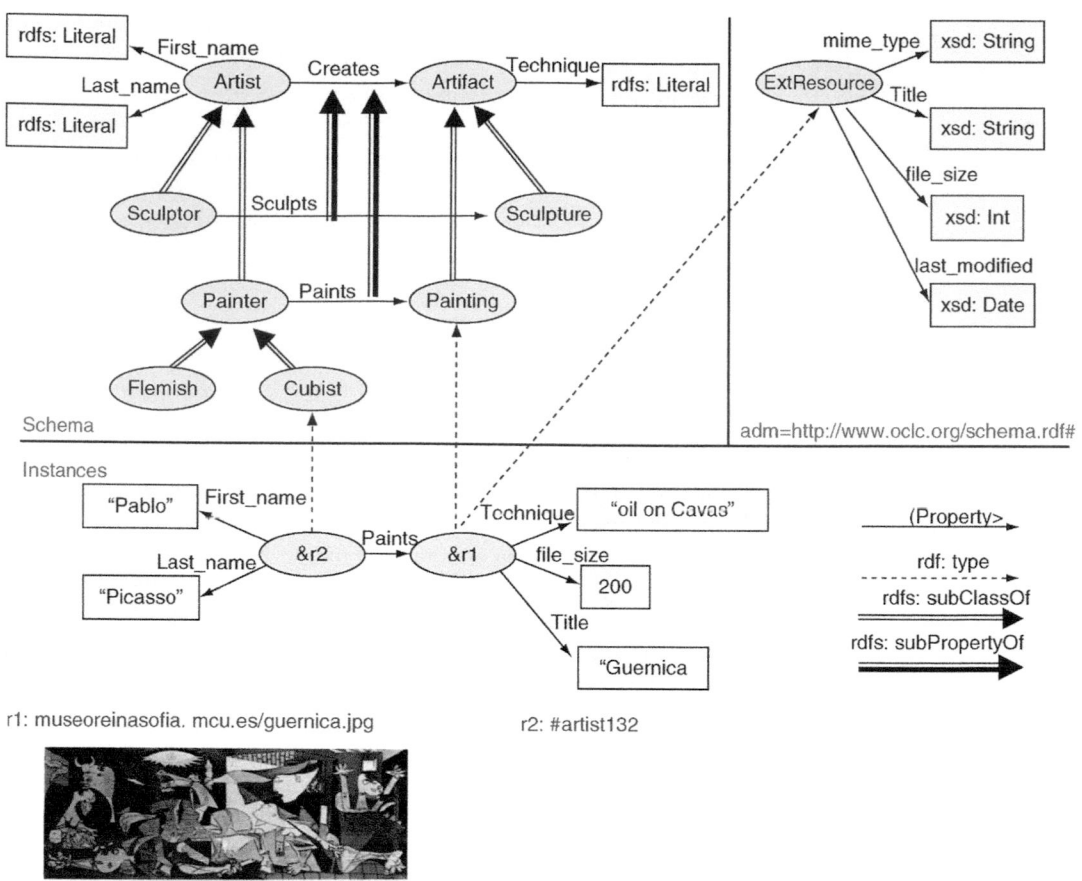

Resource Description Framework (RDF) Schema (RDFS), Fig. 1 A cultural portal catalog in RDF/S

the Web. Resource &r2 is then classified under *Cubist* and has a property first_name with value "*Pablo*" and a property last_name with value "*Picasso*." Finally, it is asserted that &r2 and &r1 are related thought the property paints (triple *<&r2,cult:paints,&r1>*).

The upper part of Fig. 1 depicts two RDF/S schemas (i.e., the namespace URIs *cult* = http://www.icom.com/schema.rdf and *adm* = http://www.oclc.org/schema.rdf#) which define the classes and properties (i.e., the names) employed by the previous resource descriptions. In schema *cult*, the property creates (triple *<cult:creates,rdf: type,rdf:Property>*), is defined with domain the class Artist (triple *<cult:creates,rdf:domain,cult: Artist>*) and range the class Artifact (triple *<cult:createsrdf:range,cult:Artifact>*). Note that properties are binary relations used to represent

attributes of resources (e.g., *technique* with range a literal type (XML Schema datatypes could be used in this respect.) as well as *relationships* between resources (e.g., *creates* with range a class of resources). Furthermore, both classes and properties can be organized into *taxonomies* carrying inclusion semantics (multiple subsumption is also supported). For example, the class *Painter* subsumes *Cubist* (*<cult:Cubist,   rdfs:subClassof,cult:Painter>*) while the property paints is subsumed by creates (*<cult:paints,rdfs:subPropertyof,cult:creates>*).

According to the RDFS semantics [3], rdfs:sub Classof (or *rdfs:subPropertyof*) relations are transitive (and reflexive), thus enabling one to infer more triples (not depicted in Fig. 1.) than those explicitly stated (e.g., *<cult:Cubist,rdfs: subClassof,cult:Artist>*

<&r2,rdf:type,cult: Painter> <&r2,rdf:type,cult: Artist> etc.).

To summarize, RDF/S properties are by default are *unordered* (e.g., there is no order between the properties *first_name* and *last_name*), *optional* (e.g., the property *mime-type* is not used) or with *multi-occurrences* (e.g., **&r2** may have two properties paints). Cardinality constraints for property domains/ranges (as well as inverse properties and Boolean class expressions) can be captured in more expressive ontology languages such as OWL. Furthermore, utility properties like *rdfs:label, rdfs:comment, rdfs:isDefinedBy and rdfs:seeAlso* are also available for documenting the development of a schema. Although not illustrated in Fig. 1, RDF/S also support structured values called *containers* for grouping statements, namely *rdf:Bag* (i.e., multi-sets) and *rdf:Sequence* (i.e., tuples), as well as, higher-order statements (i.e., *reified statements* whose subject or object can be another RDF statement). Figure 2 summarizes RDF/S axiomatic triples.

The RDF/S modeling primitives are reminiscent of knowledge representation languages (like Telos). Compared to traditional object or relational database models, RDF/S blurs the distinction between schema and instances. RDF/S schemas are *descriptive* (and not prescriptive designed by DB experts), *interleaved with the instances* (i.e., may cross abstraction layers when a resource is related through a property with a class) while may be *large* (compared to the size of instances). In particular, unlike objects (or tuples) RDF/S resources *are not strongly typed*:

- *RDF/S Classes do not define object or relation types*: an instance of a class is just a resource URI without any value/state (e.g., **&r1** is an instance of *Painting* regardless of any property associated to it);
- *RDF Resources may be instances of different classes* not necessarily pair wise related by subsumption: the instances of the same class may have associated quite different properties (e.g., see the properties of **&r1** which is multiply classified under the classes *ExtResource* and *Painting*);
- *RDF/S Properties are self-existent individuals* (i.e., decoupled from class definitions) which *may also be related through subsumption* (e.g., the property creates).

```
rdf:type rdfs:domain rdfs:Resource .
rdfs:domain rdfs:domain rdf:Property .
rdfs:range rdfs:domain rdf:Property .
rdfs:subPropertyOf rdfs:domain rdf:Property .
rdfs:subClassOf rdfs:domain rdfs:Class .
rdf:subject rdfs:domain rdf:Statement .
rdf:predicate rdfs:domain rdf:Statement .
rdf:object rdfs:domain rdf:Statement .
rdfs:member rdfs:domain rdfs:Resource .
rdf:first rdfs:domain rdf:List .
rdf:rest rdfs:domain rdf:List .
rdfs:seeAlso rdfs:domain rdfs:Resource .
rdfs:isDefinedBy rdfs:domain rdfs:Resource .
rdfs:comment rdfs:domain rdfs:Resource .
rdfs:label rdfs:domain rdfs:Resource .
rdf:value rdfs:domain rdfs:Resource .

rdf:type rdfs:range rdfs:Class .
rdfs:domain rdfs:range rdfs:Class .
rdfs:range rdfs:range rdfs:Class .
rdfs:subPropertyOf rdfs:range rdf:Property .
rdfs:subClassOf rdfs:range rdfs:Class .
rdf:subject rdfs:range rdfs:Resource .
rdf:predicate rdfs:range rdfs:Resource .
rdf:object rdfs:range rdfs:Resource .
rdfs:member rdfs:range rdfs:Resource .
rdf:first rdfs:range rdfs:Resource .
rdf:rest rdfs:range rdf:List .
```

```
rdfs:seeAlso rdfs:range rdfs:Resource .
rdfs:isDefinedBy rdfs:range rdfs:Resource .
rdfs:comment rdfs:range rdfs:Literal .
rdfs:label rdfs:range rdfs:Literal .
rdf:value rdfs:range rdfs:Resource .

rdf:Alt rdfs:subClassOf rdfs:Container .
rdf:Bag rdfs:subClassOf rdfs:Container .
rdf:Seq rdfs:subClassOf rdfs:Container .
rdfs:ContainerMembershipProperty rdfs:subClassOf
rdf:Property .

rdfs:isDefinedBy rdfs:subPropertyOf rdfs:seeAlso .

rdf:XMLLiteral rdf:type rdfs:Datatype .
rdf:XMLLiteral rdfs:subClassOf rdfs:Literal .
rdfs:Datatype rdfs:subClassOf rdfs:Class .

rdf:_1 rdf:type rdfs:ContainerMembershipProperty .
rdf:_1 rdfs:domain rdfs:Resource .
rdf:_1 rdfs:range rdfs:Resource .
rdf:_2 rdf:type rdfs:ContainerMembershipProperty .
rdf:_2 rdfs:domain rdfs:Resource .
rdf:_2 rdfs:range rdfs:Resource .
...
```

Resource Description Framework (RDF) Schema (RDFS), Fig. 2 RDF/S axiomatic triples

In addition, less rigid data models, such as those proposed for semi-structured databases, when they are not totally schemaless (such as OEM, UnQL), they cannot certainly exploit the RDF class (or property) subsumption taxonomies (as in the case of YAT). Finally, XML DTDs and Schemas have substantial differences from RDF schemas: (i) they cannot represent directed label *graphs* (Formally speaking, *RDF graphs* are not quite classical directed labeled graphs. First, a resource (e.g., paints) may occur both as a predicate (e.g., <&r2,cult:paints,&r1>) and a subject (e.g., <cult:paints,rdf:domain, cult:Painter>) of a triple. This compromises one of the more important aspects of graph theory: the intersection between the nodes and arcs labels must be empty. Second, in an RDF graph a predicate (e.g., *rdfs:subPropertyof*) may relate other predicates (<cult:paints,rdfs: subPropertyof,cult: creates>). Thus, the resulting structure is not a graph in the strict mathematical sense, because the set of arcs must be a subset of the Cartesian product of the set of nodes. There is an ongoing research on formalizing RDF using adequate graph models (e.g., bipartite graphs, directed hypergraphs).) (vs. rooted labeled *trees*); (ii) they cannot distinguish between *entity labels* (e.g., *Artist*) and *relationship labels* (e.g., *creates*); and (iii) they *constrain the structure* of XML documents, whereas an RDF/S schema simply *defines the vocabulary of class and property names* employed in RDF descriptions.

Cross-References

▶ Ontology
▶ Resource Description Framework
▶ Semantic Web

Recommended Reading

1. Beckett D. RDF/XML syntax specification (revised). W3C recommendation. Available online at: http://www.w3.org/TR/rdf-syntax-grammar/. 2004.
2. Brickley D, Guha RV. RDF Vocabulary Description Language 1.0: RDF Schema. W3C Recommendation. Available online at: http://www.w3.org/TR/rdf-schema/. 2004.
3. Hayes P. RDF Semantics. W3C Recommendation. Available online at: http://www.w3.org/TR/rdf-mt/. 2004.
4. Klyne G, Carroll J. Resource Description Framework (RDF): Concepts and Abstract Syntax. W3C Recommendation. Available online at: http://www.w3.org/TR/rdf-concepts/. 2004.
5. Manola F, Miller E. RDF Primer W3C Recommendation. Available online at: http://www.w3.org/TR/rdf-primer/. 2004.

Resource Identifier

Greg Janée
University of California-Santa Barbara, Santa Barbara, CA, USA

Synonyms

Document identifier; GUID; Uniform resource identifier; URI; UUID

Definition

In a networked information system, a *resource identifier* is a compact surrogate for a resource that can be used to identify, retrieve, and otherwise operate on the resource. An identifier typically takes the form of a short textual string. An identifier must be *resolved* to yield the associated resource.

Key Points

Resource identifiers can be broadly characterized as either *locations*, which identify resources by where they reside, or *names*, which identify resources by properties intrinsic to the resources [2]. This distinction is not absolute, and identifiers can exhibit characteristics of both classes. Nevertheless, the distinction is useful in defining the relationship between identifiers and resources. Consider:

> Can two distinct, yet identical resources have the same identifier?
>
> If a resource changes, must its identifier change?

If the answer to these questions is yes, then the identifiers should be considered names; if no, locations. To take two well-known examples, International Standard Book Numbers (ISBNs) are names, while HTTP URLs on the World Wide Web are locations.

Uniqueness

Uniqueness is the property that an identifier resolves to a single resource. The converse property - that every resource is identified by a single identifier, i.e., that identifier "aliasing" is avoided - is generally desirable, but is often not enforceable in systems that allow free generation of identifiers.

Broadly speaking, two approaches have been employed to guarantee uniqueness. The first is to incorporate into each identifier unique characteristics of the identified resource, for example a content-based signature, or characteristics of the context in which the resource and/or identifier system reside, for example a network address and timestamp. UUIDs incorporate both types of characteristics. The second approach is to acquire identifiers from an "authority" that maintains a centralized store of previously generated identifiers (identifier-resource associations are often stored as well). For scalability such systems are often arranged hierarchically so that a root authority, located at a well-known address, may delegate identifier generation and resolution requests to distributed sub-authorities. DNS and the Handle system are two well-known examples of this approach.

Persistence

Persistence is the property that an identifier continues to reference the associated resource over time. Strictly speaking, persistence is not a property of an identifier, or even a property at all; it's an outcome of the commitment of the operator of the identifier resolution system. A persistent identifier system is one that attempts to address known risks to persistence.

The risk of identifier breakage due to resource movement is universally mitigated by employing indirection: identifiers identify intermediate quantities which are maintained by resource owners to track current resource locations. In principle the indirection may be hidden from users, but for scalability reasons it is typically exposed. For example, the persistent uniform resource locator (PURL) system employs HTTP's;s redirection mechanism. The risk of breakage due to resource renaming has been mitigated in some systems by issuing so-called "semantics-free" identifiers; for example, DOIs are strings of digits with no external referent. However, the benefit of this approach must be balanced by the inscrutability of such identifiers to humans. Other notable persistent identifier systems include Open URLs, which identify objects by metadata constraints, i.e., by intrinic resource properties; "robust hyperlinks," which append content-based signatures to locations, specifically URLs; and archival resource keys (ARKs), which incorporate a protocol for obtaining resource persistence guarantees and policies.

Other Properties

Additional desirable properties of resource identifiers include global scope, global uniqueness, extensibility, machine readability, recognizability in text, and human transcribability [3]. Identifiers that are subject to transcription errors may benefit from having error-correcting codes incorporated into them.

Cross-References

▶ Citation
▶ Digital Signatures

► Distributed Architecture
► Object Identity

Recommended Reading

1. Hilse HW, Kothe J. Implementing persistent identifiers: overview of concepts, guidelines and recommendations. London/Amsterdam: Consortium of
   European Libraries and European Commission on
   Preservation and Access; 2006. http://nbn-resolving.
   de/urn:nbn:de:gbv:7-isbn-90-6984-508-3-8.
2. Jacobs I, Walsh N, editors. Architecture of the World
   Wide Web, Volume One. 2004. http://www.w3.org/TR/
   webarch/.
3. Sollins K, Masinter L. Functional requirements for
   uniform resource names. IETF RFC. 1737, 1994. http:/
   /www.ietf.org/rfc/rfc1737.txt.

Result Display

Catherine Plaisant
University of Maryland, College Park, MD, USA

Synonyms

Preview; Result display;Result overview

Definition

After formulating and initiating a search in a
database, users review the results. The complexity of this task varies greatly depending on the
users' needs, from selecting one or more top
ranked items to conducting a complex analysis of the results in the hope of discovering
an unknown phenomena. Displaying results includes providing and overview of the results and
previews of items, manipulating visualizations,
changing the sequencing of the results, adjusting
the size of the results, clustering results by topic
or attribute values, providing relevance feedback,
examining individual items, and presenting ex-
planatory messages such as restating the initial
query.

Key Points

Displaying results is part of a dynamic and
iterative decision-making process in which users
initiate queries, review results and refine their
queries. Users scan objects rapidly to determine
whether to examine them more closely, or
move on in the dataset. This process continues
until the information need is satisfied, or the
search is abandoned [3]. A visual result display
typically includes interactive widgets to further
filter the results, blurring the boundary between
search and result display, e.g., dynamic queries
entirely blend searching and result display
[1].

The visual display of results relies on previews and overviews of the items returned by the
search [2]. Graphical overviews indicate scope,
size or structure and help gauge the relevance
of items retrieved. Those overviews can vary
from simple bar charts displaying the distribution
of results over important attributes such as size
or type, or consist of specialized visualizations
such as interactive geographical maps, timelines,
node link diagrams, conceptual topic maps etc.
When no natural representation exist, more abstract overviews of the results can be used e.g., a
treemap (Fig. 1). Multiple overviews are tightly
coupled to facilitate synchronized browsing in
multiple representations.

Previews consist of samples or summaries
and help users select a subset of the results
for detail review. Multimedia database require
specialized preview mechanisms such as
thumbnail browsers or video summaries allowing
zooming and scene skipping. Both previews and
overviews help users define more productive
queries as they learn about the content of the
database.

Users should be given control over what the
size of the result set is, which fields are displayed, how results are sequenced (alphabeti-
cal, chronological, relevance ranked, and how results are clustered (by attribute value, by topics).
One strategy involves automatic clustering and
naming of the clusters for example in Vivisimo.
Studies show that clustering according to more
established and meaningful hierarchies such as

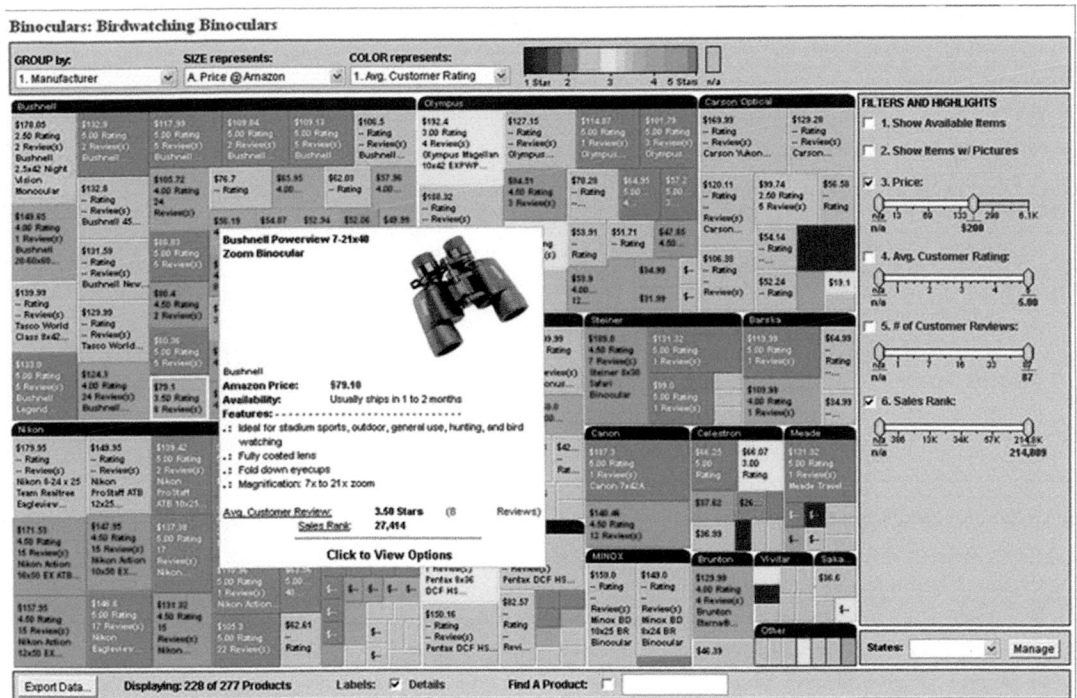

Result Display, Fig. 1 After querying for birdwatching binoculars, users can review the results of their query using the Hive Group's treemap. Each box corresponds to a pair of binoculars and the size of the box is proportional to its price. *Green boxes* are best sellers, *gray* indicates unavailability. Results are grouped by manufacturer. Using the sliders on the right, users can filter results e.g., showing only items under $200

the open directory might be effective. Translations may be proposed. Finally users need to gather information for decision making, therefore results need to be saved, annotated, sent by email or used as input to other programs such as visualization and statistical tools.

Cross-References

▶ Information Retrieval
▶ Video Querying

Recommended Reading

1. Ahlberg C, Shneiderman B. Visual information seeking: tight coupling of dynamic query filters with starfield displays. In: Proceedings of the SIGCHI Conference on Human Factors in Computing Systems; 1994. p. 313–7.
2. Greene S, Marchionini G, Plaisant C, Shneiderman B. Previews and overviews in digital libraries: designing surrogates to support visual information-seeking. J Am Soc Inf Sci. 2000;51(3):380–93.
3. Shneiderman B, Plaisant C. Designing the user interface. 4th ed. New York: Addison-Wesley; 2005.

Retrospective Event Processing

Opher Etzion
IBM Software Group, IBM Haifa Labs, Haifa University Campus, Haifa, Israel

Definition

Retrospective event processing is the detection of patterns on past events i.e., not done when the event occur, this can be done as part of existing event processing.

Historical Background

While the concept of "event pattern" typically refers to the events on-the-move, there are cases, in which it is required to use these patterns on past events, this typically happens in one of the following cases:

1. *Situation Reinforcement*: An event pattern designates the possibility that a business situation has occurred; in order to provide positive or negative reinforcement, as part of the on-line pattern detection, there is a need to find complementary pattern in order to assert or refute the occurrence of the situation.
2. *Retrospective Context*. In regular cases, contexts to start and look for patterns start with a certain event, and go forward until either some event occurs, or the context expires. There are cases in which the arrival of an event indicates the end of the context, however the context has not been monitored in the forward-looking way, and need to be monitored backwards.
3. *Patterns as queries*. Patterns are higher-level abstractions relative to SQL queries; in some cases it is more convenient to use patterns, as higher-level languages on top of queries.

The issue of querying past information have been introduced in the area of *temporal databases* in which dealt with maintaining, querying and even updating past (and future) information. This included regular database queries or updates, however provided the infrastructure for storing past events.

The emerging of the event processing area have lead to the development of *event processing patterns* that initially was applied on the current events, retrospective event processing is the next logical step in getting the patterns on the past. It should be noted that there are three approaches to implementation of retrospective pattern language:

1. Extending pattern language to look at retrospective contexts
2. Adding patterns as an extension to SQL
3. Providing a hybrid language

Foundations

Taking the approach of extending event processing patterns for past information, this materializes in the following areas:

- *Retention policies*: in order to enable retrospective processing, event should be available beyond the original context of their processing, this leads to issues such as retention policies and vacuuming policies.
- *Grid storage*: one of the emerging areas in storing events and states for pre-determined term is storing events on in-memory stores on the grid.
- *Extending the notion of context*: the notion of temporal context should be extended to include past time intervals.
- *Automatic translation to SQL*: Assuming that the events are stored on a database, the pattern language should be translated to SQL. Again, it should be noted that an alternative approach is to include pattern extensions to SQL.

Key Applications

Use Cases for Situation Reinforcement

Anti-money Laundering
A person that has deposited (in aggregate) more than $20,000 within a single working day is a SUSPECT in money laundering. To reinforce the suspicion the following retrospective patterns are sought:

- There has been a period of week within the last year in which the same person has deposited (in aggregate) $50,000 or more and has withdrawn (in aggregate) at least $50,000 within the same week.
- The same person has already been a "suspect" according to this definition within the last 30 business days.

If any of these patterns are satisfied – the event "confirmed suspect" is derived.

The Greedy Seller Alert

An electronic trade site provides the opportunity to customers to offer items for sale, but letting them conduct a bid, and provide bid management system (using a CEP system, of course). One of the services it provides to the customer is "alert on expensive sales":

If there have been at least two bidders, however, and none of them have matched the minimum price of the seller then this may be an indication of "too expensive bid."

To reinforce it, if at least two-thirds of the past bids of the same sellers have also resulted in a "too expensive bid" situation, then reinforce and send the seller a notification "you are too greedy."

Monitored Patient Alert

A patient is hooked up to multiple monitors, the monitors are uncorrelated and each of them issues an alert when a certain threshold is passed (either up or down), results in most cases being false alarms. The physician can set up a "global monitoring system" which checks a pattern over recent time, e.g.,

- An alert has been given from the blood pressure monitor
- and Reinforcement condition:
- If fever is more than 103F despite medication taken less than 2 h ago, and blood pressure is strictly increasing in the last five measures then alert nurse.

Use Cases for Retrospective Contexts

Smart Retail

The detected pattern is – "no item of a certain product reached the checkout in the last hour." This is a pattern that is detected hourly. If it is detected there is a retrospective context opened for this hour to check two retrospective patterns:

- If during that hour more than five customers took an item of that product from the shelf, but returned it after they have taken a competitive product (the information can be obtained by

RFID tag on each item, an RFID reader for putting and removing items from the shopping cart).
- If no customer has taken an item of the product from the shelf, and did not take a competitor's product either.

Luggage Handling

The reported event is – Luggage did not arrive.

Retrospective context – start at luggage check-in time.

- Collect all events related to this luggage (using the tag reading at various points).
- If no events found – notify the source airport to trace in their video tracking system.

Utilities Billing System

Identified situation – customer has not paid 30 days after due date.

- Find out over the last billing cycle – has the customer addressed the customer center around issues with this bill, and obtain status.
- Look at the customer billings over the last year – determine maximal and average days of late in paying.
- Look at customers with the same Zip-code over the last billing cycle and determine the percentage of non-payment, late-payment (to determine if they are mail has significant delays in that area).

Use Cases for Patterns as Queries

Stock Trends

Find all stocks that during the last month have satisfied the following conditions:

- The stock closing values at the end of the day were strictly increasing over a period of five consecutive working days, anywhere during this month.
- The stock value in the beginning at the end of the 5 days value was at least 30% more than its value at the beginning of the 5 days period.

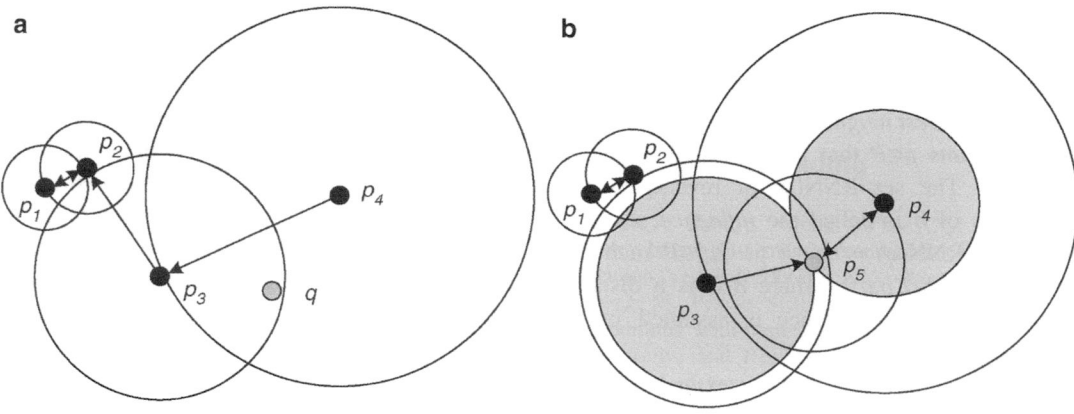

Reverse Nearest Neighbor Query, Fig. 2 Illustration of *KM*

In order to avoid the maintenance of two separate structures, *YL* [13] combines the two indexes in the RdNN-tree. Similar to the RNN-tree, a leaf node of the RdNN-tree contains vicinity circles of data points. On the other hand, an intermediate node contains the MBR of the underlying points (not their vicinity circles), together with the maximum distance from every point in the sub-tree to its nearest neighbor. As shown in the experiments of [13], the RdNN-tree is efficient for both RNN and NN queries because, intuitively, it contains the same information as the RNN-tree and has the same structure (for node MBRs) as a conventional R-tree. *MVZ* [7] is also based on precomputation. The methodology, however, is applicable only to 2D spaces and focuses on asymptotical worst-case bounds (rather than experimental comparison with other approaches).

The problem of *KM*, *YL*, *MVZ*, and all techniques that rely on preprocessing is that they cannot deal efficiently with updates. This is because each insertion or deletion may affect the vicinity circles of several points. Consider Fig. 2b, where a new point p_5 needs to be inserted in the database. First, a RNN query is performed to find all objects (in this case p_3 and p_4) that have p_5 as their new nearest neighbors. Then, the vicinity circles of these objects are updated in the index. Finally, the update algorithm computes the NN of p_5 (i.e., p_4) and inserts the corresponding circle. Similarly, each deletion must update the vicinity circles of the affected objects. In order to alleviate

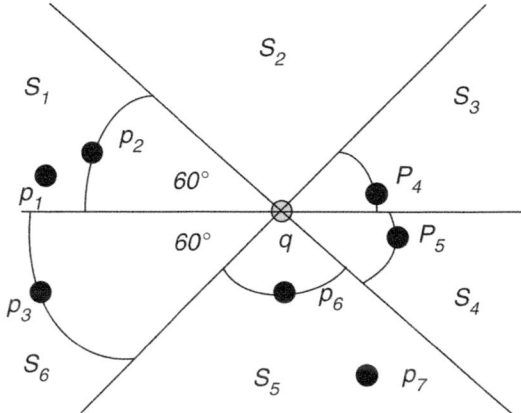

Reverse Nearest Neighbor Query, Fig. 3 Illustration of *SAA*

the problem, Lin et al. [6] propose a method for bulk insertions in the RdNN-tree.

SAA [9] eliminates the need for precomputing all NNs by utilizing some interesting properties of RNN retrieval. Consider Fig. 3, which divides the space around a query q into six equal regions S_1 to S_6. Let p be the NN of q in some region S_i; it can be proved that (i) either $p \in$ RNN(q) or (ii) there is no RNN of q in S_i. For instance, in Fig. 3 the NN of q in S_1 is point p_2. However, the NN of p_2 is p_1. Consequently, there is no RNN of q in S_1 and it is not necessary to search further in this region. The same is true for S_2 (no data points), S_3, S_4 (p_4, p_5 are NNs of each other) and S_6 (the NN of p_3 is p_1). The actual result is RNN(q) = {p_6}. Based on the above property, *SAA* adopts a two-step processing method. First, six constrained NN queries [2] retrieve the nearest neighbors of q in regions S_1 to S_6. These points constitute the *candidate* result. Then, at a second step, a *nearest neighbor query* is applied to find the NN p' of each candidate p. If $dist(p, q) < dist(p, p')$, p belongs to the actual result; otherwise, it is a false hit and discarded.

The number of regions to be searched for candidate results increases exponentially with the dimensionality, rendering *SAA* inefficient even for three dimensions. *SFT* [8] follows a different approach that (i) finds (using an R-tree) the K NNs of the query q, which constitute the initial candidates; (ii) eliminates the points that are closer to some other candidate than q; and (iii) applies *Boolean range queries* on the remaining candidates to determine the actual RNNs. Consider, for instance, the query of Fig. 4 assuming that K (a system parameter) is 4. *SFT* first retrieves the four NNs of q: p_6, p_4, p_5, and p_2. The second step discards p_4 and p_5 since they are closer to each other than q. The third step uses the circles $(p_2, dist(p_2,q))$ and $(p_6, dist(p_6,q))$ to perform two Boolean ranges on the data R-tree. The difference with respect to conventional range queries is that a Boolean range terminates immediately when (i) the first data point is found

or (ii) the entire side of a node MBR lies within the circle. For instance, N_1 contains at least a point within the range. Thus, p_2 is a false hit and *SFT* returns p_6 as the only RNN of q. The major shortcoming of the method is that it may incur false misses. In Fig. 4, although p_3 is a RNN of q, it does not belong to the four NNs of the query and will not be retrieved.

Similar to *SAA* and *SFT*, *TPL* [11] follows a filter-refinement framework. As opposed to *SAA* and *SFT* that require multiple queries for each step, the filtering and refinement processes are combined into a single traversal of the R-tree. In particular, *TPL* traverses the data R-tree and retrieves potential candidates in ascending order of their distance to the query point q because the RNNs are likely to be near q. Each candidate is used to prune node MBRs (data points) that cannot contain (be) candidates. For instance, consider the perpendicular bisector $\perp(p,q)$ between the query q and an arbitrary data point p as shown in Fig. 5a. The bisector divides the data space into two half-planes: PL_q (p,q) that contains q and PL_p (p,q) that contains p. Any point (e.g., p') in PL_p (p,q) cannot be a RNN of q because it is closer to p than q. Similarly, a node MBR (e.g., N_1) that falls completely in PL_p (p,q) cannot contain any candidate.

In some cases, the pruning of an MBR requires multiple half-planes. For example, in Fig. 5b, although N_2 does not fall completely in $\mathrm{PL}_{p1}(p_1,q)$ or $\mathrm{PL}_{p2}(p_2,q)$, it can still be pruned since it lies entirely in the *union* of the two half-planes. In general, if p_1, p_2, ..., p_{n_c} are candidate results, then any node whose MBR falls inside $\cup_{i=1\sim n_c} \mathrm{PL}_{p_i}(p_i, q)$ cannot contain any RNN result. The filter step terminates, when there are no more candidates inside the remaining (i.e., non-pruned data space). Each pruned entry is inserted in a *refinement set* S_{rfn}. In the refinement step, the entries of S_{rfn} are used to eliminate false hits.

Table 1 summarizes the properties of each algorithm. Precomputation methods cannot efficiently handle updates. *MVZ* is suitable only to 2D spaces, while *SAA* is practically inapplicable for three or more dimensions. *SFT* incurs false misses, the number of which depends on

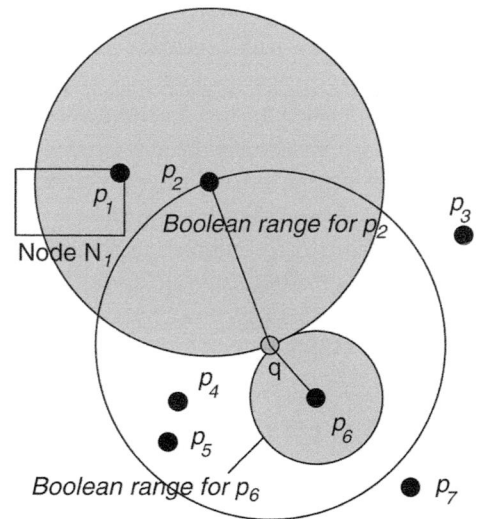

Reverse Nearest Neighbor Query, Fig. 4 Illustration of *SFT*

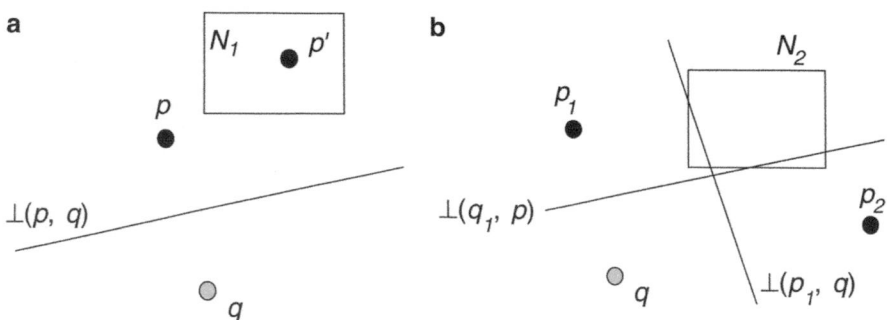

Reverse Nearest Neighbor Query, Fig. 5 Illustration of half-plane pruning in *TPL*

Reverse Nearest Neighbor Query, Table 1 Summary of algorithm properties

	Dynamic data	Arbitrary dimensionality	Exact result	Arbitrary k
KM,YL	No	Yes	Yes	No
MVZ	No	No	Yes	No
SAA	Yes	No	Yes	Yes
SFT	Yes	Yes	No	Yes
TPL	Yes	Yes	Yes	Yes

the parameter K: a large value of K decreases the false misses but increases significantly the processing cost. Regarding the applicability of the existing algorithms to arbitrary values of k, precomputation methods only support a specific value (typically equal to 1), used to determine the vicinity circles. *SFT* can be adapted for retrieval of RkNN by setting a large value of K ($\gg k$) and replacing the Boolean with *count* queries (that return the number of objects in the query range instead of their actual ids). *SAA* can be extended to arbitrary k as discussed in [11]. *TPL* can handle dynamic data for arbitrary values of k and dimensionality.

Key Applications

A number of applications for RNN can be found in [4, 14]. Examples include:

Profile-Based Marketing
Assume that a real estate company keeps profiles of its customer set P based on their goals, i.e., each customer is a point in a vector space defined

by the features of interest (e.g., house area, neighborhood, etc.). When a new estate q enters the market, a RNN query could retrieve the clients for which q constitutes the closest match to their interests.

Decision Support Systems
Consider that a franchise wants to open a new branch at location q so that it attracts a large number of customers from competitors based on proximity. This can be modeled as a bichromatic RNN query where P_1 corresponds to the competitor set and P_2 to the customer dataset. The result for a potential location q is the set of customers that are closer to q than any competitor.

Peer-to-Peer Systems
Assume that a new user q enters a P2P system. A RNN query retrieves among the existing users, the ones for which q will become their new NN based on the network latency. In a collaborative environment, q would inform such users about its arrival, so that they could address future requests directly to q, minimizing the network cost. Furthermore, the set RNN(q) reflects the potential workload of q; thus, by knowing this set, each peer could manage/control its available resources.

Future Directions

Stanoi et al. [10] solve bichromatic RNN queries using *R-trees* to prune the search space.

Benetis et al. [1] extend the *SAA* algorithm for continuous RNN queries. Yiu et al. [14] deal with reverse nearest neighbors in large graphs. Tao et al. [12] focus on RNN processing in *metric spaces*. Kang et al. [3] discuss the continuous evaluation of RNN queries in highly dynamic environments.

Experimental Results

Tao et al. [11] contain a comprehensive comparison of *SAA*, *SFT*, and *TPL*. Each of the following references, except for [7], also contains an experimental evaluation of the proposed algorithm.

Datasets

A common benchmark for RNN queries in the Euclidean space is the Tiger dataset: http://www.census.gov/geo/www/tiger/

In addition, the DBLP graph has been used for RNN queries in large graphs [14], while road networks have been applied in [11, 12].

Cross-References

▶ Metric Space
▶ Nearest Neighbor Query
▶ R-Tree (and Family)

Recommended Reading

1. Benetis R, Jensen C, Karciauskas G, Saltenis S. Nearest neighbor and reverse nearest neighbor queries for moving objects. VLDB J. 2006;15(3):229–50.
2. Ferhatosmanoglu H, Stanoi I, Agrawal D, Abbadi A. Constrained nearest neighbor queries. In: Proceedings of the 7th International Symposium on Advances in Spatial and Temporal Databases; 2001.
3. Kang J, Mokbel M, Shekhar S, Xia T, Zhang D. Continuous evaluation of monochromatic and bichromatic reverse nearest neighbors. In: Proceedings of the 23rd International Conference on Data Engineering; 2007. p. 806–15.
4. Korn F, Muthukrishnan S. Influence sets based on reverse nearest neighbor queries. In: Proceedings of the ACM SIGMOD International Conference on Management of Data; 2000. p. 201–12.
5. Korn F, Muthukrishnan S, Srivastava D. Reverse nearest neighbor aggregates over data streams. In: Proceedings of the 28th International Conference on Very Large Data Bases; 2002. p. 814–25.
6. Lin K, Nolen M, Yang C. Applying bulk insertion techniques for dynamic reverse nearest neighbor problems. In: Proceedings of the International Conference on Database Engineering and Applications; 2003. p. 290–7.
7. Maheshwari A, Vahrenhold J, Zeh N. On reverse nearest neighbor queries. In: Proceedings of the Canadian Conference Computational Geometry; 2002. p. 128–32.
8. Singh A, Ferhatosmanoglu H, Tosun A. High dimensional reverse nearest neighbor queries. In: Proceedings of the 12th International Conference on Information and Knowledge Management; 2003.
9. Stanoi I, Agrawal D, Abbadi A. Reverse nearest neighbor queries for dynamic databases. In: Proceedings of the ACM SIGMOD Workshop on Research Issues in Data Mining and Knowledge Discovery; 2000. p. 44–53.
10. Stanoi I, Riedewald M, Agrawal D, Abbadi A. Discovery of influence sets in frequently updated databases. In: Proceedings of the 27th International Conference on Very Large Data Bases; 2001. p. 99–108.
11. Tao Y, Papadias D, Lian X. Reverse kNN search in arbitrary dimensionality. In: Proceedings of the 30th International Conference on Very Large Data Bases; 2004. p. 744–55.
12. Tao Y, Yiu M, Mamoulis N. Reverse nearest neighbor search in metric spaces. IEEE Trans Knowl Data Eng. 2006;18(9):1239–52.
13. Yang C, Lin K. An index structure for efficient reverse nearest neighbor queries. In: Proceedings of the 17th International Conference on Data Engineering; 2001. p. 482–95.
14. Yiu M, Papadias D, Mamoulis N, Tao Y. Reverse nearest neighbors in large graphs. IEEE Trans Knowl Data Eng. 2006;18(4):540–53.

Reverse Top-k Queries

Akrivi Vlachou
Athena Research and Innovation Center,
Institute for the Management of Information
Systems, Athens, Greece

Synonyms

R top-k queries

Definition

Given a query point q and a d-dimensional dataset S, a reverse top-k query returns the scoring functions for which q belongs to the top-k result set. The most important and commonly used case of scoring functions is the weighted sum function $f_w()$ that is defined by a weighting vector w and computes the score of any data point $p \in S$, i.e., $f_w(p) = \sum_{i=1}^{d} w[i] \cdot p[i]$, where $w[i] \geq 0$. For a point q and a positive number k, as well as two datasets S and W, where S represents data points and W is a dataset containing different weighting vectors, a weighting vector $w_i \in W$ belongs to the reverse top-k result set of q, if and only if there exists a point p that belongs to the top-k result set of w_i such that $f_{w_i}(q) \leq f_{w_i}(p)$ (assuming that minimum values are preferable).

Main Text

Reverse top-k queries [1, 2] are useful in various real-life applications that exploit user preferences for data analysis tasks. For example, manufacturers need to assess the potential market and impact of their products based on customer preferences and the competitor products. Given a database of products, an important analysis task is to identify the most influential products to customers [3], where the influence of a product is defined as the cardinality of its reverse top-k result set. Intuitively, the influence of a product relates to its visibility to the customers who express their preferences using top-k queries. Therefore, reverse top-k queries are a nice fit for such applications as their result set is directly related to the number of customers that value a particular product.

A naive algorithm to compute the reverse top-k result set requires processing a top-k query for each weighting vector $w_i \in W$, in order to determine whether q belongs to the top-k result of w_i or not. Several algorithms have been proposed that improve the performance of query processing by taking into account the properties of the reverse top-k queries as well as the benefits of spatial indexing. Also, techniques that exploit pre-computation of top-k or reverse top-k results have also been shown to improve performance significantly.

The above definition corresponds to the bichromatic version of the reverse top-k query. An alternative type of reverse top-k queries is the monochromatic reverse top-k query, where no explicit knowledge of user preferences is assumed. Given a query point q, a positive number k and a dataset S, the result set of the monochromatic reverse top-k query of point q is the locus containing d-dimensional points representing the weighting vectors, for which the query point belongs to the top-k result set. Processing a monochromatic reverse top-k query leads to geometric problem and has been studied mainly in the two-dimensional space. Notice that it is not possible to enumerate all possible weighting vectors, since the number of possible vectors is infinite, but the solution space can be split into a finite set of partitions such that the query point q has the same ranking position for all the weighting vectors in the partition. Then, it is sufficient to examine each partition to find those partitions that belong to the result set. Furthermore, the reverse top-k query may be defined by other families of scoring functions, and the above definition can be adapted.

Cross-References

▶ Top-k Queries

Recommended Reading

1. Vlachou A, Doulkeridis C, Kotidis Y, Nørvåg K. Reverse top-k queries. In: Proceedings of the 26th International Conference on Data Engineering; 2010; p. 365–76.
2. Vlachou A, Doulkeridis C, Kotidis Y, Nørvåg K. Monochromatic and bichromatic reverse top-k queries. IEEE Trans. Knowl. Data Eng. 2011;23(8): 1215–29.
3. Vlachou A, Doulkeridis C, Nørvåg K, Kotidis Y. Identifying the most influential data objects with reverse top-k queries. Proc. VLDB Endow. 2010;3(1):364–72.

Rewriting Queries Using Views

Chen Li
University of California – Irvine, School of
Information and Computer Sciences, Irvine, CA,
USA

Definition

Given a query on a database schema and a set
of views over the same schema, the problem
of query rewriting is to find a way to answer
the query using only the answers to the views.
Rewriting algorithms aim at finding such rewritings efficiently, dealing with possible limited
query-answering capabilities on the views, and
producing rewritings that are efficient to execute.

Historical Background

Query rewriting is one of the oldest problems
in data management. Earlier studies focused on
improving performance of query evaluation [9],
since using materialized views can save the execution cost of a query. In 1995, Levy et al. [10]
formally studied the problem and developed complexity results. The problem became increasingly
more important due to new applications such as
data integration, in which views are used widely
to describe the semantics of the data at different
sources and queries posed on the global schema.
Many algorithms have been developed, including
the bucket algorithm [11] and the inverse-rules
algorithm [7, 15]. See [8] for an excellent survey.

Foundations

Formally, a query Q_1 is *contained* in a query
Q_2 if for each instance of their database, the
answer to Q_1 is always a subset of that to Q_2.
The queries are *equivalent* if they are contained
in each other. Let T be a database schema, and V
be a set of views on T. The *expansion* of a query P
using the views in V, denoted by P^{exp}, is obtained
from P by replacing all the views in P with their
corresponding base relations. Given a query Q on

T, a query P is called a *contained rewriting* of
query Q using V if P uses only the views inV, and
P^{exp} is contained in Q as queries. P is called an
equivalent rewriting of Q using V if P^{exp} and Q
are equivalent as queries.

Examples Consider a database with the following three relations about students, courses, and
course enrollments:

Student(sid, name, dept);
Course(cid, title, quarter);
Take(sid, cid, grade).

Consider the following query on the database:

Query Q1: SELECT C.title, T.grade
FROM Student S, Take T, Course C
WHERE S.dept = 'ee' AND S.sid = T.sid AND
    T.cid = C.cid;

The query asks for the titles of the courses
taken by EE students and their grades. Queries
and views are often written as conjunctive queries
[4]. For instance, the above query can be rewritten
as:

Q1(T, G) :- Student(S, N, ee), Take(S, C, G),
    Course(C, T, Q).

Lower-case arguments (such as "ee") are used
for constants, upper-case arguments (such as "T")
for variables. The right-hand side of the symbol
":-" is the *body* of the query. It has three *subgoals*,
each of which is an occurrence of a relation in
the body. The constant "*ee*" in the first subgoal
represents the selection condition. The variable
S shared by the first two subgoals represents the
join between the relations *Student* and *Take* on
the student-id attribute. The variables T and G in
the head of the query, which is the left-hand side
of the symbol ":-", represent the final projected
attributes.

Consider the following materialized views defined on the base tables:

Views: V1(S, N, D, C, G) :- Student(S, N, D),
    Take(S, C, G);
V2(S, C, T) :- Take(S, C, G), Course(C, T, Q).

The SQL statement for the view definition of *V* 1 is the following:

CREATE VIEW V1 AS
SELECT S.sid, S.name, S.dept, T.cid, T.grade
FROM Student S, Take T
WHERE S.sid = T.sid;

This view is the natural join of the relations *Student* and *Take*. Similarly, view *V2* is the natural join of the relations *Take* and *Course*, except that the attributes about grades and quarters are dropped in the final results. The following is a rewriting of the query *Q*1 using the two views.

answer(T, G) :- V1(S, N, ee, C, G), V2(S, C, T).

This rewriting takes a natural join of the two views on the attributes of student ids and course ids, then does a projection on the title and grade attributes. This rewriting can always compute the answer to the query on every instance of the base tables. In particular, after replacing each view in the rewriting with the body of its definition, the rewriting becomes the following expansion:

answer(T, G) :- Student(S, N, ee), Take(S, C, G), Take(S, C, G'), Course(C, T, Q').

G' and *Q'* are fresh variables introduced during the replacements. This expansion is equivalent to the query, thus the rewriting is an equivalent rewriting of the query.

Now, assume in the definition of *V2*, there is another selection condition on the quarter attribute. The following is the view definition:

V2'(S, C, T) :- Take(S, C, G), Course(C, T, fall2006).

That is, it only includes the information about the courses offered in the fall quarter of 2006. If only views *V1* and *V2'* are given, then the following is a rewriting of the query *Q*1:

answer(T, G) :- V1(S, N, ee, C, G), V2'(S, C, T).

In particular, its expansion, which is obtained by replacing each view with the body of its definition, is the following:

answer(T, G) :- Student(S, N, ee), Take(S, C, G), Take(S, C, G'), Course(C, T, fall2006).

This expansion is contained in the original query, thus this rewriting is a contained rewriting of the query *Q*1. It is not an equivalent rewriting, since it does not include information about courses offered in other quarters. On the other hand, each fact in the answer to this rewriting is in the answer to the original query.

Suppose the view definition of *V2* does not have the attribute about course ids. Then using this modified view and *V1*, there is no rewriting of the query, since the modified view does not have the course id to join with view *V1*. As another example, if the view definition of *V1* does not keep the grade information, the following is the new view:

V1'(S, N, D, C) :- Student(S, N, D), Take(S, C, G).

Using this new view and the original view *V2*, there is no rewriting to answer the query, since the views do not provide any information about grades, which is requested by the query. All these examples show that, when deciding how to answer a query using views, it is important to consider the conditions in the query and the views, including their selections, joins, and projections.

Algorithms There are two classes of algorithms for rewriting queries using views: the first one includes the bucket algorithm [11] and its variants, and the second one includes the inverse-rules algorithm [7, 15]. Notice that the number of possible rewritings of a query using views is exponential in the size of the query. Here the main idea of the bucket algorithm is explained using the running example, in which the query *Q*1 needs to be answered using the views *V1* and *V2*. Its main idea is to reduce the search space of rewritings by considering each subgoal in the query separately, and deciding which views could be relevant to the query subgoal.

The bucket algorithm has two steps. In step 1, for each subgoal in the query, the algorithm considers each view definition, and checks if the body (definition) of the view also includes a subgoal that can be used to answer this query subgoal. For each view, if it includes a subgoal that can be unified with the query subgoal, and the query and the view are compatible after the unification, the corresponding head of the view definition is added to the bucket of this query subgoal. The following shows the buckets for the three query subgoals.]

Student(S, N, ee): {V1(S, N, ee, C', G')};
Take(S, C, G): {V1(S, N', D', C, G)};
Course(C, T, Q): {V2(S', C, T)}.

Each primed variable is a fresh variable introduced in the corresponding unification process. The bucket of the second query subgoal does not include the view $V2$ because the query subgoal requires the grade information be included in the answer, while the corresponding grade information in the view subgoal is not exported in the head of $V2$.

In step 2, the algorithm selects one view from each bucket, and combines the views from these buckets to construct a contained rewriting. The following is a contained rewriting:

Q1(T, G) :- V1(S, N, ee, C', G'), V1(S, N', D', C, G), V2(S', C, T).

The final output of the algorithm is the union of contained rewritings in order to maximize the set of answers to the query using the views, since these rewritings could produce different pieces of information.

One main advantage of the bucket algorithm is that it can prune those views that do not contribute to a condition in the query, thus it can reduce the number of candidate rewritings to be considered. One limitation of the algorithm is that each query subgoal introduces a view in a rewriting. For instance, in the example above, view $V1$ could be used to answer the first two query subgoals. But the algorithm needs to use three view instances in each candidate rewriting, which requires more postprocessing steps to simplify this rewriting. In addition, the algorithm does not use the fact that if a view can be used to cover a query subgoal using a view variable that is not exported in the head of the view, then the view has to cover all the query subgoals that use

the corresponding query variable. Based on these observations, a new algorithm, called MiniCon, was developed to make the rewriting process significantly more efficient [14]. A similar idea was used in the shared-bucket-variable (SVB) algorithm [13].

In some cases, especially in the context of data integration, where a view is a description of the content at a data source, the views could have limited query capabilities. For instance, imagine the case where the view $V1$ above is a materialized table, such that it can be accessed only if a student id is provided to the table, and the table can return its information about that student id. The table does not accept arbitrary queries such as "return all records, " or "retrieve all information about students from the CS department. " These limitations on the views present new challenges for the development of query-rewriting algorithms. The problem in this setting was studied in [16]. It is shown that the the inverse-rules algorithm [7] can handle such restrictions with minor modifications.

Other algorithms have been developed to study variants of the query-rewriting problem. The CoreCover algorithm [2] was developed for the problem of generating an *efficient* equivalent rewriting efficiently. There was also a study [1] for the case where the query and the views can have comparison conditions such as *salary > 30 K* and *year <= 2004*. The work in [6] studied how to compute a set of views with a minimal size to compute the answers to a set of queries. In some settings, applications need to find a rewriting called "maximally contained rewriting, " which can compute the maximal set of answers to the query using the views. The problem is also different depending on whether the closed-world assumption is taken (as in data warehousing, in which each materialized view is assumed to include all the facts satisfying the view definition) or the open-word assumption is taken (as in data integration, in which each view includes a subset of the facts satisfying the view definition). In the literature there is another related problem called "query answering." See [3] for a comparison between "query rewriting" and "query answering."

R

Key Applications

The problem of rewriting queries using views is related to many data-management applications, including information integration [12, 18], data warehousing [17], and query optimization [5].

Cross-references

▶ Answering Queries Using Views
▶ Data Warehouse
▶ Query Containment
▶ Query Optimization

Recommended Reading

1. Afrati FN, Li C, Mitra P. Answering Queries using views with arithmetic comparisons. In: Proceedings of the 21st ACM SIGACT-SIGMOD-SIGART Symposium on Principles of Database Systems; 2002. p. 209–20.
2. Afrati F, Li C, Ullman JD. Generating efficient plans using views. In: Proceedings of the ACM SIGMOD International Conference on Management of Data; 2001. p. 319–30.
3. Calvanese D, Giacomo GD, Lenzerini M, Vardi MY. View-based query processing: on the relationship between rewriting, answering and losslessness. In: Proceedings of the 10th International Conference on Database Theory; 2005. p. 321–36.
4. Chandra AK, Merlin PM. Optimal implementation of conjunctive queries in relational data bases. In: Proceedings of the 9th Annual ACM Symposium on Theory of Computing; 1977. p. 77–90.
5. Chaudhuri S, Krishnamurthy R, Potamianos S, Shim K. Optimizing queries with materialized views. In: Proceedings of the 11th International Conference on Data Engineering; 1995. p. 190–200.
6. Chirkova R, Li C. Materializing views with minimal size to answer queries. In: Proceedings of the 22nd ACM SIGACT-SIGMOD-SIGART Symposium on Principles of Database Systems; 2003. p. 38–48.
7. Duschka OM, Genesereth MR. Answering recursive queries using views. In: Proceedings of the ACM SIGACT-SIGOPS 16th Symposium on the Principles of Distributed Computing; 1997. p. 109–16.
8. Halevy AY. Answering queries using views: a survey. VLDB J. 2001;10(4):270–94.
9. Larson PÅ, Yang HZ. Computing queries from derived relations. In: Proceedings of the 11th International Conference on Very Large Data Bases; 1985. p. 259–69.
10. Levy A, Mendelzon AO, Sagiv Y, Srivastava D. Answering queries using views. In: Proceedings of the 14th ACM SIGACT-SIGMOD-SIGART Symposium on Principles of Database Systems; 1995. p. 95–104.
11. Levy A, Rajaraman A, Ordille JJ. Querying heterogeneous information sources using source descriptions. In: Proceedings of the 22th International Conference on Very Large Data Bases; 1996. p. 251–62.
12. Li C. Query processing and optimization in information-integration systems. Ph.D. Thesis, computer science department. Stanford University. 2001.
13. Mitra P. An algorithm for answering queries efficiently using views. In: Proceedings of the 12th Australasian Database Conference; 2001. p. 99–106.
14. Pottinger R, Levy A. A scalable algorithm for answering queries using views. In: Proceedings of the 26th International Conference on Very Large Data Bases; 2000.
15. Qian X. Query folding. In: Proceedings of the 12th International Conference on Data Engineering; 1996. p. 48–55.
16. Rajaraman A, Sagiv Y, Ullman JD. Answering queries using templates with binding patterns. In: Proceedings of the 14th ACM SIGACT-SIGMOD-SIGART Symposium on Principles of Database Systems; 1995. p. 105–12.
17. Theodoratos D, Sellis T. Data warehouse configuration. In: Proceedings of the 23th International Conference on Very Large Data Bases; 1997. p. 126–35.
18. Ullman JD. Information integration using logical views. In: Proceedings of the 6th International Conference on Database Theory; 1997. p. 19–40.

RMI

Aniruddha Gokhale
Vanderbilt University, Nashville, TN, USA

Synonyms

Remote method invocation

Definition

Java Remote Method Invocation (RMI) [1, 2] is a Java language-based technology to achieve distributed computing among distributed Java virtual machines.

Key Points

Java RMI is a Java language-dependent technology for distributed computing. It provides seamless distributed communication between Java virtual machines. RMI uses object serialization and offers true object-oriented polymorphism even across distributed address spaces. Since RMI is based on Java, it brings the power of safety, concurrency and portability to distributed applications. In developing their applications, programmers must explicitly indicate which interfaces will be available as a remote service by extending the java.rmi. Remote interface.

A special quality of RMI is its ability to dynamically load new objects into an address space. For example, if a remote service has undergone change and extended its capabilities, it is feasible for RMI to dynamically load the new class in the client's address space.

Another attractive feature of RMI is its ability to allow entire behaviors of objects to be sent to remote entities. At the remote end, it is then feasible to activate a local copy of the passed object with its behavior. These techniques are useful in load balancing and faster response times.

RMI can allow clients behind firewalls to contact remote servers. This capability enables clients to reside within applets. RMI also provides interoperability with other broker technologies, such as CORBA, by supporting RMI over CORBA IIOP.

Cross-References

▶ .NET Remoting
▶ Client-Server Architecture
▶ CORBA
▶ DCE
▶ DCOM
▶ Request Broker
▶ SOAP

Recommended Reading

1. Sun Microsystems. Java remote method invocation. 1996.
2. Sun Developer Network. Remote method invocation. Available at: http://java.sun.com/javase/technologies/core/basic/rmi/index.jsp.

Road Networks

Cyrus Shahabi
University of Southern California, Los Angeles, CA, USA

Synonyms

Road network databases; Road vector data; Spatial network databases

Definition

In *vector* space, the distance between two objects can be computed as a function of the components of the vectors representing the objects. A typical distance function for multidimensional vector spaces is the well-known Minkowski metric, with the Euclidean metric as a popular case for two-dimensional space. Therefore, the distance computation in multidimensional vector spaces is fast because its complexity depends on the number of dimensions which is limited to two or three in geospatial applications. However, with *road-networks*, the distance between two objects is measured by their *network distance*, i.e., the length of the shortest path through the network edges that connects the two locations. This computationally expensive network-based metric mainly depends on the connectivity of the network. For example, representing a road network as a graph with e weighted edges and v vertices, the complexity of the Dijkstra algorithm to find the minimum weighted path between two vertices is $O(e + vLogv)$. Therefore, several distance-based queries, such as nearest-neighbor queries, cannot be performed efficiently in road-networks. One option is to estimate the distance between two objects by their Euclidean distance. Unfortunately, as shown in [13], the Euclidean distance is not a good approximation of the network distance with real-world road-networks. Therefore, in the past several years, many studies have been investigating new techniques to index road-networks and/or pre-compute distances in order

R

to expedite distance-based query processing on road-networks.

Historical Background

The main challenge with query processing in road-networks is the high complexity of network distance computation. This is important since many applications, especially those dealing with moving objects, require frequent, fast and on-the-fly computation of distances between a query point and several points of interests. The earliest work that studied this challenge of fast distance computation in road-network databases is by Shahabi et al. [13]. In this entry, the authors proposed an embedding technique to transform the road network to a high dimensional space in which fast Minkowski metrics can be utilized for distance measurement. The results are still approximation of the actual distances but much more accurate than using Euclidean distances on the points' geographical coordinates.

Besides this transformation-based approach to perform fast network distance computation, three other main approaches are based on either indexing the network or pre-computing the network distances or a hybrid of the two. A prominent work in the indexing category is by Papadias et al. [10], which proposes an architecture for road-networks that uses a disk-based network representation. Their approach is based on the fact that the current algorithms (e.g., Dijkstra) for computing the distance between a query object q and an object O in a network will automatically result in the computation of the distance between q and the objects that are (relatively) closer to q than O. This approach applies an optimized network expansion algorithm with the advantage that the network expansion only explores the objects that are closer to q and computes their distances to q during expansion. The advantages of this approach are: (i) it offers a method that finds the exact distance in networks, and (ii) the architecture can support other spatial queries like range search and closest pairs. There are other studies that similar to this work try to combine traditional spatial access methods with some sort

of network representation and expansion method such as [3, 5, 7]. The main disadvantage of these network-expansion approaches is that they perform poorly when the objects are not densely distributed in the network because then they require to retrieve a large portion of the network for distance computation. A rather different indexing approach is proposed in [4], termed distance signature. This approach categorizes the distances between objects and network nodes into groups and then encodes these groups.

A representative work for pre-computation approaches is the work by Sankaranarayanan et al. [12], which proposes a framework called SILC for computing the shortest distance between vertices on a spatial network. The proposed framework pre-computes the shortest path between all pairs of vertices, which in turn results in fast distance computation. Examples of other studies that used some sort of pre-computation to expedite network distance computation are [2, 6].

A hybrid approach that combines an indexing technique with pre-computation is proposed in [8]. This approach is based on partitioning and then indexing a large network to small *network Voronoi* regions, and then pre-computing distances both within and across the regions.

In reality the travel-time (weight) to traverse a street (edge) of the road network (graph) is subject to variations such as traffic jams, weather conditions, accidents and much more. The key factor to compute the fastest path on road networks is thus the accuracy of the travel times on the edges. Recent advances in sensor networks and location based services have enabled collecting traffic data from roadside sensors and smartphones, and hence it is now feasible to accurately capture the travel-time on the edges in real-time. Consequently, some of the car navigation systems and smartphone apps (e.g., Waze) use the real-time traffic data to find the shortest travel time path from a source to a destination.

Some studies (e.g., [16, 17]) suggest that relying on the real-time edge travel time is still not sufficient for an accurate computation of the shortest path. This is because traffic is a dynamic phenomenon and thus the edge travel times obviously keep changing while the user who issued the query is traveling. To cope with this issue, a number of algorithms for time-varying traffic networks have been proposed, including some classics [18, 19] and some newer studies [20–22]. In practice, the application of these algorithms is complicated by the fact that the future travel times of edges are not known by the time the query is issued but have to be predicted. Therefore, several studies [23–25] focused on using the historical traffic data to predict future traffic patterns, even in the presence of accidents [26].

There are also many variations in the applications and query types on road-networks that require special-purpose index structures and/or query processing algorithms for efficient network distance computation. Some of these applications and query-types are reviewed below.

As mentioned, the fast computation of network distance becomes more critical when objects are moving. Most of the aforementioned techniques can indeed be used for querying points of interests from a moving object. However, some optimization can be performed if one wants to update the query results as the query point moves. One main representative query type here is the Continuous k-nearest-neighbor (C-kNN) queries. C-kNN on road-networks maintains the k nearest neighbors in network distance as the query object moves on the road network, which has its own unique challenges as opposed to C-kNN in Euclidean spaces. Sample studies focusing on moving objects in road-networks are [1, 11].

Other novel applications try to address NN queries the time-dependent road network that integrates both spatial networks and real-time traffic information [9, 27]. In [15], the authors propose and solve Aggregate nearest neighbor queries (ANN) in the context of large road networks. In [14], the authors study the novel problem of optimal sequenced route (OSR) query in both vector and road-network spaces. The OSR query tries to find a route of minimum length starting from a given source location and passing through a number of typed locations in a specific sequence imposed on the types of the locations. The paper proposes a pre-computation approach to OSR query by exploiting the geometric properties of the solution space and relating it to additively weighted Voronoi diagrams.

Foundations

A popular class of queries in geospatial applications is the class of distance-based queries. To answer these queries, the distance between one or more query points or regions with some points or areas of interests must be computed. A frequently used member of this class is the k nearest neighbor (kNN) query where the k closest points to a query point are requested. For example, many car navigation systems provide the feature to ask for the k closest gas stations to the vehicle's current location. One way to answer this query is to estimate the distance by computing the Euclidean distance between the vehicle's geographical coordinates (i.e., latitude and longitude) and all the gas stations' coordinates in the vicinity. The problem with this approach is that the Euclidean distance corresponds to the *air-distance* between the vehicle and the gas station. Basically, if the car could have flown, then the Euclidean distance would have been the accurate distance between the car and the gas station. Unfortunately, most cars are restricted to the underlying road network. Hence, the actual distance between the car and the gas station depends on the connectivity of the underlying network and usually is very different than the Euclidean distance. In [13], it is shown that in real-world road-networks, Euclidean distance does not yield a good approximation of the network distance.

Formal Definition of the Problem
The challenge with computing network-distance is that the complexity of its computation depends on the number of vertices and edges of the underlying network, which is normally very large for real road-networks. Now imagine computing this distance continuously from a moving vehicle to several gas stations and one would appreciate the efforts in developing new techniques to compute this distance as fast as possible. To formally define this problem, first a formal model to represent road networks is provided and then using this

R

Road Networks, Fig. 1
(**a**) Graph model of a road
network, (**b**) ten points of
interest on the edges of the
graph

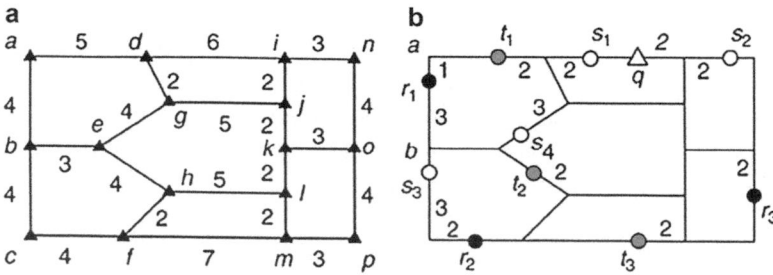

model, the concept of network-distance is defined more accurately.

A road-network can be modeled as a weighted graph. Consider the weighted undirected (Many spatial networks consist of directed edges and hence must be modeled as directed graphs. Throughout this entry, undirected graphs are used for simplicity.) graph $G = (V, E)$ as the two sets V of vertices, and $E \subseteq V \times V$ of edges. Each edge of E, directly connecting vertices u and v, is represented as the pair $[u, v]$. Each vertex v represents a 2-d point $(v.x, v.y)$ in a geometric space (e.g., an intersection in a road network). Hence, each edge is also a line segment in that space (e.g., a road segment). A numeric weight (cost) w_{uv} is associated with the edge $[u, v]$. In road-networks, this is the distance or the travel time between intersections u and v. N refers to the space of points located on the edges/vertices of graph G. For a point $p \in N$ located on the edge $[u, v]$, $w_{up} = \frac{|up|}{|uv|} w_{uv}$ where $|uv|$ is the Euclidean distance between u and v. Figure 1a shows the graph model of a road network including the vertex set $V = \{a, \dots, p\}$. Each edge of the graph is labeled by its weight. Figure 1b shows points $s_1, \dots, s_4$, $r_1, \dots, r_3$, and $t_1, \dots, t_3$ on the edges of the same graph. As shown in the figure, point $r_1 \in N$ corresponds to the weights $w_{r_1 a} = 1$ and $w_{r_1 b} = 3$.

The main challenge with query processing in road-networks is the expensive cost of computing the network distance between two points. Hence, *network distance* for road-networks is formally defined below.

Definition 1 Given a graph G, a *path P* from $p_1 \in N$ to $p_2 \in N$ is an ordered set $P = \{p_1, v_1, \dots, v_n, p_2\}$ consisting of a sequence of connected edges

from p_1 to p_2. Here, p_1 and p_2 are located on the edges $[u, v_1]$ and $[v_n, w]$, respectively. Also, v_i is connected to v_{i+1} by the edge $[v_i, v_{i+1}]$ for $1 \leq i < n$. As shown in Fig. 1b, $P = \{t_1, a, b, s_3\}$ is a path from t_1 to s_3.

Definition 2 Given a path $P = \{p_1, v_1, \dots, v_n, p_2\}$, *Path Cost* of P, $pcost(P)$, is defined as the sum of the costs of all edges in P. Formally, for the path P,

$$pcost(P) = w_{p_1 v_1} + \sum_{i=1}^{n-1} w_{v_i v_{i+1}} + w_{v_n p_2}$$

In Fig. 1b, the cost of path $P = \{t_1, a, b, s_3\}$ is calculated as $pcost(P) = 3 + 4 + 1 = 8$. For the points $p_1, p_2 \in N$, $P_{p_1 p_2}$ is used to denote the *shortest path* from p_1 to p_2 in G; the path $P = \{p_1, \dots, p_2\}$ with minimum cost $pcost(P)$.

Definition 3 Given the two points p_1 and p_2 in N, the *network distance* between p_1 and p_2, $D_n(p_1, p_2)$, is the cost of the shortest path between p_1 and p_2 (i.e., $D_n(p_1, p_2) = pcost(P_{p_1 p_2})$). For instance, $D_n(t_1, s_3) = 8$. The network distance $D_n(.,.)$ is non-negative and obeys identity, symmetry and the triangular inequality. Hence, together with N, it forms a metric space.

As discussed in the background section several techniques have been proposed to expedite the computation of this network-distance for efficient processing of distance-based queries. Here, one hybrid approach is reviewed which combines indexing the space (using voronoi diagrams) with a distance pre-computation approach and has shown to be superior in performance to its competitors [8].

A Voronoi-Based Solution for Road-Networks

A comprehensive solution for spatial queries in road-networks must fulfill the following real-world requirements: (i) be able to incorporate the network connectivity to provide exact distances between objects, (ii) efficiently answer the queries in real-time in order to support distance-based queries (e.g., kNN) for moving objects, (iii) be scalable in order to be applicable to usually very large networks, (iv) be independent of the density and distribution of the points of interest, (v) be adaptive to efficiently cope with database updates where nodes, links, and points of interest are added/deleted, and (vi) be extendible to consider query constraints such as direction or range.

In [8], the authors proposed a novel approach that fulfills the above requirements by reducing the problem of distance computation in a very large network, in to the problem of distance computation in a number of much smaller networks plus some additional table lookups.

The main idea behind this approach, termed Voronoi-based Network Nearest Neighbor (VN3), is to first partition a large network in to smaller/more manageable regions. This is achieved by generating a first-order *network* Voronoi diagram over the points of interest. Each cell of this Voronoi diagram is centered by one point of interest (e.g., a restaurant) and contains the nodes that are closest to that object in *network* distance (and not the Euclidean distance). Next, the intra and inter distances for each cell are pre-computed. That is, for each cell, the distances between all the edges (or border points) of the cell to its center are pre-computed. In addition, the distances across the border points of the *adjacent* cells are also pre-computed. This will reduce the pre-computation time and space by localizing the computation to cells and handful of neighbor-cell node-pairs.

Now, to find the k nearest-neighbors of a query object q, first the first nearest neighbor is found by simply locating the Voronoi cell that contains q. This can be easily achieved by utilizing a spatial index (e.g., R-tree) that is generated for the Voronoi cells. In [8], it is shown that the next nearest neighbors of q are within the adjacent cells of the previously explored ones, which can be efficiently retrieved from a lookup table. Next, the intra-cell pre-computed distances are utilized to find the distance from q to the borders of the Voronoi cell of each candidate, and finally the inter-cell pre-computed distances are used to compute the actual network distance from q to each candidate. The local pre-computation nature of VN3 also results in low complexity of updates when the network is modified.

Key Applications

The applications of distance-based queries on road-networks are numerous. In emergency response, the first responders may want to find k closest hospitals to a crisis area (kNN query). In urban planning, one may need to find the set of parks that are closest to a set of houses (known as *spatial skyline queries*). In location-based services, a group of mobile users want to find a meeting location where traveling towards which minimizes their total travel distance (Aggregate-NN query). In Location-based Services (LBS), a driver wants to find all the gas stations within its 4-mile distance (spatial range queries). In On-Line map services such as Yahoo! Maps, Google Earth or Microsoft Virtual Earth, one may want to minimize distance when planning a day trip to a shopping center, a restaurant and a movie theater (OSR query). In all these applications and queries, the accurate and fast computation of network distance given the underlying road network is critical.

Future Directions

Even though, as reviewed in this entry, several techniques for fast computation of network distances have been proposed, there are still no real-world deployment of these or any other techniques to enable accurate and fast network distance computation for spatial queries. This may be attributed to the fact that many users can tolerate or are used to the inaccuracy in distance computations. However, as the collected geospatial data becomes more accurate and the users become more sophisticated, the competition between different geospatial services would lead into the adaptation of a simple but effective technique to provide accurate network distance for query processing. Therefore, a technique that can easily be integrated into the current information infrastructure used by geospatial applications is of substantial importance.

On the research front, new queries and applications for multidimensional spaces are often proposed in academic conferences and journals. Most of these queries are applicable to geospatial applications and road-networks. Therefore, a rather effortless way of finding a new research topic is to take any of these new queries and extend it to work in the road-network space.

Finally, adding other attributes that would affect the distance or time-to-travel in road-networks renders most if not all of the current solutions insufficient. For example, one can consider efficient on-the-fly computation of network distance in the presence of traffic flow, road direction, road closure, road elevation, or navigational data such as costs of left-turns, right turns, U-turns, and stops.

Cross-References

► R-Tree (and Family)
► Spatial Data Analysis
► Spatial Data Mining
► Spatial Data Types
► Spatial Indexing Techniques
► Spatial Network Databases
► Spatial Operations and Map Operations
► Spatiotemporal Trajectories
► Voronoi Diagrams

Recommended Reading

1. Almeida VTD, Güting RH. Indexing the trajectories of moving objects in networks. GeoInformatica. 2005;9(1):33–60.
2. Cho H.-J, Chung C.-W. An efficient and scalable approach to cnn queries in a road network. In: Proceedings of the 31st International Conference on Very

Large Data Bases; 2005. p. 865–876.
3. Güting H, de Almeida T, Ding Z. Modeling and querying moving objects in networks. VLDB J. 2006;15(2):165–90.
4. Hu H, Lee DL, Lee VCS. Distance indexing on road networks. In: Proceedings of the 32nd International Conference on Very Large Data Bases; 2006. pp. 894–905.
5. Hu H, Lee DL, Xu J. Fast nearest neighbor search on road networks. In: Advances in Database Technology, Proceedings of the 10th International Conference on Extending Database Technology; 2006. p. 186–203.
6. Huang X, Jensen CS, Saltenis S. The islands approach to nearest neighbor querying in spatial networks. In: Proceedings of the 9th International Symposium on Advances in Spatial and Temporal Databases; 2005. p. 73–90.
7. Jensen CS, Kolářvr J, Pedersen TB, Timko I. Nearest neighbor queries in road networks. In: Proceedings of the 11th ACM International Symposium on Advances in Geographic Information Systems; 2003. p. 1–8.
8. Kolahdouzan MR, Shahabi C. Voronoi-based k nearest neighbor search for spatial network databases. In: Proceedings of the 30th International Conference on Very Large Data Bases; 2004. p. 840–51.
9. Ku W-S, Zimmermann R, Wang H, Wan C-N. Adaptive nearest neighbor queries in travel time networks. In: Proceedings of the 13th ACM International Symposium on Advances in Geographic Information Systems; 2005. p. 210–19.
10. Papadias D, Zhang J, Mamoulis N, Tao Y. Query processing in spatial network databases. In: Proceedings of the 29th International Conference on Very Large Data Bases; 2003. p. 790–801.
11. Pfoser D, Jensen CS. Indexing of network constrained moving objects. In: Proceedings of the 11th ACM International Symposium on Advances in Geographic Information Systems; 2003. p. 25–32.
12. Sankaranarayanan J, Alborzi H, Samet H. Efficient query processing on spatial networks. In: Proceedings of the 13th ACM International Symposium on Advances in Geographic Information Systems; 2005. p. 200–09.
13. Shahabi C, Kolahdouzan MR, Sharifzadeh M. A road network embedding technique for k-nearest neighbor search in moving object databases. In: Proceedings of the 10th ACM International Symposium on Advances in Geographic Information Systems; 2002. p. 94–100.
14. Sharifzadeh M, Shahabi C. Processing optimal sequenced route queries using voronoi diagrams. GeoInformatica. 2008;12(4):411–33.
15. Yiu ML, Mamoulis N, Papadias D. Aggregate nearest neighbor queries in road networks. IEEE Trans Knowl Data Eng. 2005;17(6):820–3.
16. Fleischmann B, Gietz M, Gnutzmann S. Time-varying travel times in vehicle routing. Transp Sci. 2004;38(2):160–73.
17. Demiryurek U, Banaei-Kashani F, Shahabi C. A case for time-dependent shortest path computation in spatial networks. In: Proceedings of the 18th SIGSPA-

TIAL ACM International Symposium on Advances in Geographic Information Systems; 2010. p. 474–77.

18. Cooke L, Halsey E. The shortest route through a network with time-dependent internodal transit times. J Math Anal Appl. 1966.

19. Dreyfus SE. An appraisal of some shortest-path algorithms. Oper Res. 1969;17(3).

20. Ding B, Yu JX, Qin L. Finding time-dependent shortest paths over large graphs. EDBT In: Advances in Database Technology, Proceedings of the 11th International Conference on Extending Database Technology; 2008. p. 205–16.

21. Delling D, Wagner D. Time-dependent route planning. Robust and Online Large-Scale Optimization 2009; p. 207–30.

22. Demiryurek U, Kashani FB, Shahabi C, Ranganathan A. Online computation of fastest path in time-dependent spatial networks, SSTD, 2011.

23. Smith B, Demetsky M. Traffic flow forecasting: comparison of modeling approaches. J Transp Eng. 1997;123(4):261–6.

24. Pan B, Demiryurek U, Shahabi C. Utilizing real-world transportation data for accurate traffic prediction. In: Proceedings of the 12th IEEE International Conference on Data Mining; 2012. p. 595–604.

25. Clark S. Traffic prediction using multivariate nonparametric regression. J Transp Eng. 2003;129(2):161–7.

26. Pan B, Demiryurek U, Shahabi C, Gupta C. Forecasting spatiotemporal impact of traffic incidents on road networks. In: Proceedings of the 13th IEEE International Conference on Data Mining; 2013. p. 587–96.

27. Demiryurek U, Kashani FB, Shahabi C. Efficient K-nearest neighbor search in time-dependent spatial networks. In: Proceedings of the 21th International Conference on Database and Expert Systems Applications; 2010; p. 432–49.

Rocchio's Formula

Ben He
University of Glasgow, Glasgow, UK

Definition

Rocchio's formula is used to determine the query term weights of the terms in the new query when Rocchio's relevance feedback algorithm is applied.

Key Points

In 1971, Rocchio proposed a classical query expansion algorithm based on the Vector Space model [1]. The basic algorithm assumes that the user identifies a set R of relevant documents and a set N of non relevant documents and the improved query is the result of a linear combination of the mean frequencies tf of the terms in the original query and in these two sets (*the centroids of R and N*), that is the weight of each term in the new query is:

$$qt\,f_m = \alpha.qtf + \beta.\sum_{d \in R} tf - \gamma.\sum_{d \in N} tf$$

Feedback variations assume that positive feedback exhibits a much clear impact on the reformulation of the query. Also positive feedback can be applied to expand query without the explicit feedback from the user (*blind relevance feedback*). Blind relevance feedback can be obtained in four steps:

1. All documents are ranked for the given query using a particular Information Retrieval model, for example the TF-IDF term weighting of the vector space model. This step is called *first-pass retrieval*. The user identifies a set R of relevant documents and a set N of non relevant documents.

2. A weight qtf_{exq} is assigned to each term appearing in the set of the k highest ranked documents. In general, qtf_{exq} is the mean of the weights provided by the Information Retrieval model, for example the TF-IDF weights, computed over the set of the k highest ranked documents.

3. The vector of query terms weight is finally modified by taking a linear combination of the initial query term weights qtf used for the first-pass retrieval and the new weight qtf_{exq}, that is:

$$qt\omega_m = qtf + \beta.qtf_{exq} \qquad (1)$$

4. To remove noisy terms from the expanded query, automatic query expansion techniques

usually selects only the highest informative terms from the set of top-ranked documents. The informativeness of a term is determined by the terms with highest weights assigned in step 2.

Cross-References

▶ Query Expansion Models

Recommended Reading

1. Rocchio J. Relevance feedback in information retrieval. Englewood Cliffs: Prentice-Hall; 1971. p. 313–23.

Role-Based Access Control

Yue Zhang and James B. D. Joshi
University of Pittsburgh, Pittsburgh, PA, USA

Synonyms

RBAC; Role based security

Definition

Access control is a security service responsible for defining which subjects can perform what type of operations on which objects. A subject is typically an active entity such as a user or a process, and an object is an entity, such as a file, database table or a field, on which the subject can perform some authorized operations. A permission indicates the mode of operation on a particular object.

Role based access control (RBAC) involves controlling access to computer resources and information by (i) defining *users*, *roles*, and *permissions*, and (ii) assigning users and permissions to roles. A user can create a *session* in which he/she can activate a subset of the roles he/she has been assigned to and use the permissions associated with the activated roles. RBAC approach is based on the understanding that a user's access needs are defined by the roles that he/she plays within his/her organization. In general, a role is considered as a group of permissions. RBAC approach also uses *role-role relation*, known as *role-hierarchy*, to provide permission inheritance semantics, and *constraints* on the assignment relations and the activation of roles to capture various access control requirements.

Historical Background

The origin of the concept of roles can be traced back to organizational theory much earlier than the advent of computerized information systems. However, it was primarily in the early 1990s that the security researchers and practitioners became interested in adopting the notion of a role to address the access control issues for information systems. In particular, in 1992, Ferraiolo and Kuhn showed that the existing mandatory and discretionary access control (MAC and DAC) approaches were inadequate in addressing the complex and diverse access control needs of various organizations and proposed the use of a Role based approach. While the MAC approach uses the predefined set of system rules to control accesses to resources, thus not giving users discretionary power to grant the rights they have on objects (e.g., files, databases) to other users/subjects, the DAC approach allowed the users to grant the permissions that they have on objects to others freely. Later, the initial work by Ferraiolo and Kuhn was followed by Nyanchama and Osborn's role graph model in 1995 [9], and later by the seminal paper by Sandhu, Coyne, Feinstein and Youman in 1996 [12], where they defined a family of RBAC models each with different sets of capabilities, known as the RBAC96 model. This model later evolved into the NIST RBAC model. The NIST model was later modified into the ANSI/INCITS (ANSI/INCITS stands for American National Standards. Institute and International Committee for Information Technology Standards) RBAC standard in 2004 [2]. The significant interest in this area can be seen by the establishment of the ACM Workshop on RBAC in 1996 which later evolved into the current ACM Symposium in Access Control Method and Technologies (SACMAT). Several extensions of the RBAC model have been proposed and several newer RBAC issues have been identified over the last one decade. A key contribution is also related to the demonstration by Osborn et al. that the RBAC96 model can also be configured to represent the MAC and DAC policies [10], establishing its usefulness as a uniform model for addressing very diverse set of access control needs.

Foundations

Role is a prevalent organizational concept and it identifies the various job functions and responsibilities within an organization. As an information system has become a crucial component for carrying out an organization's job functions, it has motivated the use of roles that users within the organization play to define what accesses should be authorized to them so that they can carry out their job functions and responsibilities efficiently. Based on the premise that a role represents the set of permissions that are needed for carrying out the job functions, various RBAC approaches have been proposed. RBAC96 model is the most widely recognized initial model which has evolved into the ANSI/INCITS RBAC standard. The RBAC96 family of models is briefly de-

scribed next followed by a discussion on other extensions made to it and its standardized version.

Figure 1 depicts the RBAC96 family of models which include $RBAC_0$, $RBAC_1$, $RBAC_2$, and $RBAC_3$. $RBAC_0$ is the base model containing only basic elements; $RBAC_1$ augments $RBAC_0$ with role hierarchy; $RBAC_2$ augments $RBAC_0$ with constraints; and $RBAC_3$ combines $RBAC_1$ and $RBAC_2$, and provides the most complete set of features.

RBAC₀ Base Model

In the base model $RBAC_0$, the key elements are the sets of users (U), roles (R), permissions (P) and sessions (S). Various relations are defined among them, as depicted in Fig. 1. A *user* in this model is a human being but can also be generalized to include intelligent autonomous agents such as robots, immobile computers, or even networks of computers. A *role* typically represents a job function within the organization with some authority and responsibility. The objects are data objects as well as other computer resources within a computer system. RBAC96 model only supports "positive permissions" that grants accesses to objects, and does not support "negative permissions" that deny accesses. Constraints, as defined in $RBAC_2$, are used as a mechanism to achieve the denial of accesses.

The *user assignment* (*UA*) and *permission assignment* (*PA*) relations shown in Fig. 1 are many-to-many relations. A user can be a member of many roles, and a role can have many users. Similarly, a role can have many permissions and the same permission can be assigned to many

R

Role-Based Access Control, Fig. 1 RBAC96 model

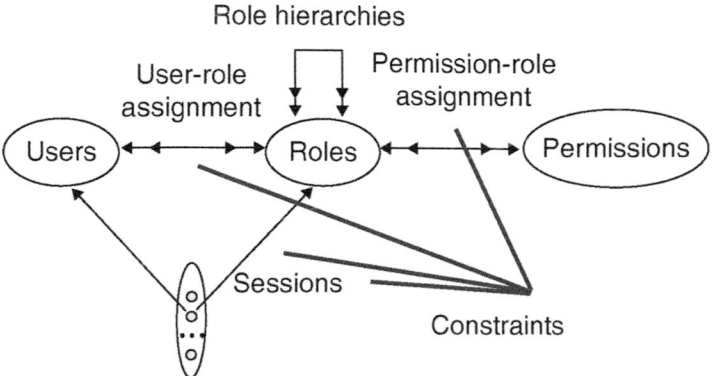

Role hierarchies

User-role assignment

Permission-role assignment

Users ◀▶ Roles ◀▶ Permissions

Sessions

Constraints

Role-Based Access Control, Fig. 2 Sample role hierarchies

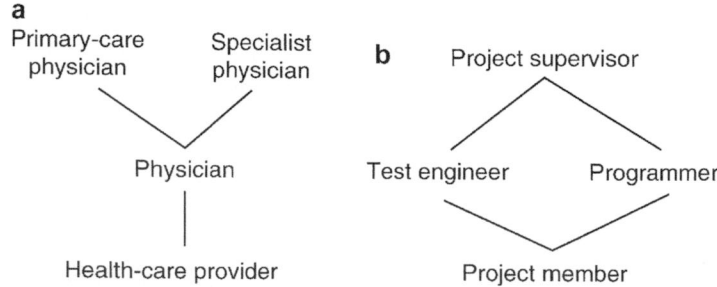

roles. The placement of a role as an intermediary to enable a user to exercise a permission provides much greater control over access configuration and review than does directly relating users to permissions.

A *session* is essentially a mapping of one user to possibly many roles, i.e., a user can establish a session and activate within it some subset of roles that he/she has been assigned to. The set of permissions that can be used by a user is the union of the permissions assigned to the roles that the user activates in one session. The association between a user and the activated set of roles within a session remains constant for the life of the session. A user may have multiple sessions open at the same time and each session may have a different combination of assigned roles activated in it. This feature of RBAC$_0$ supports the principle of least privilege that involves use of the minimal set of permissions needed for a particular task. A user who is a member of several roles can invoke any subset of these within a session to complete a task. The concept of a session is equivalent to that of the traditional notion of a *subject* in the access control literature. A subject (or session) is a unit of access control, and a user may have multiple subjects (or sessions) with different sets of permissions [12].

RBAC$_1$: RBAC with Role Hierarchy

RBAC$_1$ introduces role hierarchies (*RH*) on top of RBAC$_0$ to indicate which roles can inherit which permissions from which other roles. Role hierarchies are a natural way of structuring roles to reflect an organization's lines of authority and responsibility. Figure 2 depicts some examples of role hierarchies. By convention senior roles are shown toward the top of these diagrams,

and junior roles toward the bottom. For example, in Fig. 2a, the junior-most role is *health-care provider*. The *physician* role is senior to *health-care provider* and hence it inherits all the permissions from *health-care provider*. In addition to the permissions inherited from the *health-care provider* role, the *physician* role can have other permissions assigned directly to it. Inheritance of permissions is transitive. For example, in Fig. 2a, the *primary-care physician* role inherits permissions from the *physician* and *health-care provider* roles. *Primary-care physician* and *specialist physician* both inherit permissions from the *physician* role. Figure 2b illustrates multiple inheritances of permissions, where the *project supervisor* role inherits from both *test engineer* and *programmer* roles. Mathematically, these hierarchies are partial orders. A partial order is a reflexive, transitive and anti-symmetric relation.

RBAC$_2$: RBAC with Constraints

RBAC$_2$ introduces the concept of constraints. Constraints are an important aspect of RBAC and are sometimes argued to be a principal motivation for using RBAC. Constraints are a powerful mechanism for capturing higher-level organizational security policies. Constraints can be applied to the *UA* and *PA* relations, as well as to the association between a user and its activated set of roles within a session.

The most frequently mentioned constraint in the context of RBAC is the *mutually exclusive roles* (MER) constraint that is used to enforce Separation of Duty (SoD) requirements. For instance, a user can be restricted to assume at the most one role in a mutually exclusive role set. This supports separation of duty requirements where a person should not be able to assume two different roles to carry out critical steps of a task. Consider two mutually exclusive roles, *accounts-manager* and *purchasing-manager*. Mutual exclusion in terms of *UA* specifies that one individual cannot be a member of both these roles. Mutual exclusion in terms of *PA* specifies that the same permission cannot be assigned to both these roles. For example, the permission to issue checks should not be assigned to both these roles. Normally such a permission would be assigned to the *accounts-manager* role only. The mutual exclusion constraint on *PA* helps prevent the permission from being unintentionally or intentionally assigned to the *purchasing-manager* role.

Another example of a user assignment constraint is that a role can have a maximum number of members. For instance, the number of roles to which and individual user can belong to could also be limited. These are called *cardinality constraints*. Similarly, the number of roles to which a permission can be assigned can have cardinality constraints to control the distribution of powerful permissions.

RBAC$_3$: The Consolidated Model

RBAC$_3$ combines RBAC$_1$ and RBAC$_2$ to provide both role hierarchies and constraints. There are several issues that arise by bringing these two concepts together.

Constraints can be applied to the role hierarchy itself. For example, constraints on *RH* can limit the number of senior (or junior) roles that a given role may have. Two or more roles can also be constrained to have no common senior (or junior) role. These kinds of constraints are useful in situations where the authority to change the role hierarchy has been decentralized, but the chief security administrator desires to restrict the cases in which such changes can be made.

Benefits of the RBAC Approach

The RBAC approach has been recognized for several beneficial features. First, it supports the principle of least privilege which has been considered very important for better security of systems. When RBAC is used, a user's access

requirements can be easily changed when his role is changed within his organization. All that should be done is removing him as a member of his current role and assigning him to the new role. Also, the total number of assignment relationships that needs to be maintained in RBAC is $n_r (n_u + n_p)$ where n_r, n_u, and n_p represent the number of roles, users and permissions, respectively, in a system. In general subject-object based authorization system will need to maintain $n_u . n_p$ subject-to-permission associations. Role hierarchy makes security administration significantly easy as it eliminates the need for explicitly assigning the permissions to multiple roles. Constraints can be used to capture various types of policies including very important SoD requirements. It has been shown that by configuring constraints and relationships in RBAC, one can configure an RBAC system to express the traditional DAC and MAC policies [10]. Because of its capability to capture very diverse sets of requirements, it has been considered as a very promising approach to address emerging multidomain security problems where various domains with different access control policies need to securely interact with each other.

RBAC Standards

The NIST RBAC model was the first attempt towards establishing an RBAC standard. Compared to the RBAC96 models, the most distinct feature in NIST RBAC is the clear specification of incremented 4-level RBAC models. The first level, called the core RBAC is similar to the RBAC$_0$ model with the explicit specification of user-role view where the roles assigned to a specific user and the users assigned to a specific role can be determined. The second level, called the hierarchical RBAC model is based on the RBAC$_1$ model and includes the separation of general hierarchy (the same as in RBAC$_1$) and the limited hierarchy, where the structure of the hierarchy is restricted to a certain type, such as, a tree or an inverted tree structure. The third level is the same as the constrained RBAC. The fourth level further adds the permission-role view where

R

the roles assigned to a specific permission and the permissions assigned to a specific role can be determined. Later, the NIST model was further refined to establish the ANSI/INCITS RBAC standard.

Administration Models for RBAC

In large organizational systems the number of roles can be in the hundreds or thousands. Managing these roles and their relationships is a daunting task that is often highly centralized. A key issue is how the benefits of the RBAC model can be used to construct an administration model for managing the RBAC policies. Several researchers have addressed this issue, including the Administrative RBAC (ARBAC) family of models by Sandhu et al. [13] and Scoped AR-BAC (SARBAC) family of models by Crampton et al. [6].

RBAC Extensions

The RBAC96 model and its standardized versions also have been extended in several ways to address emerging applications. One notable among the emerging requirements is the need to capture context based access control requirements. Several extensions of RBAC models have been made to address such a need. Notable among these are extensions of RBAC to capture temporal and/or location context. Temporal RBAC (TRBAC) model [3] and its generalized version Generalized TRBAC (GTRBAC) [7] are main work related to temporal extensions of the RBAC model. While these capture the time context, several works have tried to address the need of capturing location context within an RBAC framework. Covington et al. propose a Generalized RBAC (GRBAC) model, where they introduce the concept of environment roles, which are roles that can be activated based on the value of conditions in the environment where the request has been made. Bertino et al. [4] has proposed the GEO-RBAC model that integrates RBAC with a spatial model based on the OpenGIS system. A significant aspect of the GEO-RBAC model includes (i) its specification of role schema that are location-based, and (ii) the separation of role schema and role instances

to provide different authorizations for different logical and physical locations. Location and time-based RBAC (LoT-RBAC) model was proposed by Chandran et al. [5] to address the access control requirements of highly mobile, dynamic environments to provide both location and time based control. Significant work has also been done to try to develop specification languages for RBAC models. This includes the inclusion of RBAC profile in OASIS's XACML. Joshi et al. propose X-RBAC that generalizes X-GTRBAC, which is the XML-based language for specifying the GTRBAC policies. Other extensions to the RBAC work include developing better constraint frameworks and analyzing their complexity issues.

Support for SoD constraints is a major issue for the RBAC model. Various SoD constraints proposed in the literature include: Static SoD, Dynamic SoD, History Based SoD, Object Based SoD, Operational SoD, Order Dependent/Independent SoD. The ANSI RBAC standard, however, only supports the basic SoD constraints. Several researchers have developed RBAC extensions to support these advanced SoD constraints. Ahn et al. have proposed the RCL2000 language for specifying several role-based SoD constraints [1]. Joshi et al. have proposed time-based SoD constraints in [8].

RBAC approach has also been shown to be very beneficial for multi-domain security which refers to the need to facilitate secure interactions among multiple security domains with very diverse set of access control requirements. Several researchers have recently focused on using RBAC to address the multi-domain security challenge [11, 14]. One approach uses mapping roles from multiple domains to ensure secure cross-domain accesses. Such inter-domain role mapping problem has been recently formalized and solutions for both tightly coupled environments as well as lightly coupled environments are being sought [11, 14]. Some researchers have employed RBAC delegation models to address such multi-domain sharing issues [15]. Several RBAC delegation models have been proposed that use roles as the central entity in the process of delegating rights from one user to another.

Key Applications

RBAC has been used in operating systems, databases and applications. In particular, RBAC is very promising for large scale application environments and enterprises. Emerging systems and applications have more context and content based access control requirements as can be seen in mobile and peer to peer applications and several extensions of RBAC has been motivated by such requirements.

Future Directions

Future applications will require efficient access control techniques to protect large number of objects and manage huge number of privileges that may need to be given to unknown users. Furthermore, interactions among multiple domains with different access control policy requirements are crucial in emerging applications. RBAC has been found to be promising for these emerging access control requirements. Several researchers have advocated use of RBAC to address large scale enterprise security as well as multidomain security. Work related to multidomain security using RBAC approach is in its initial stages of development. Policy verification and evolution management issues are crucial for access control in large and dynamic systems but little work currently have addressed these.

URL to Code

NIST provides an implementation of some RBAC versions http://csrc.nist.gov/rbac/.

ANSI/INCITS website: http://csrc.nist.gov/groups/SNS/rbac/standards.html.

Cross-References

▶ Access Control Policy Languages
▶ ANSI/INCITS RBAC Standard
▶ Discretionary Access Control
▶ Mandatory Access Control
▶ Temporal Access Control

Recommended Reading

1. Ahn G, Sandhu R. Role-based authorization constraints specification. ACM Trans Inf Syst Secur. 2000;3(4):207–26.
2. American National Standard for Information Technology (ANSI). Role based access control. ANSI INCITS 359-2004, February 2004.
3. Bertino E, Bonatti PA, Ferrari E. TRBAC: a temporal role-based access control model. ACM Trans Inf Syst Secur. 2001;4(3):191–233.
4. Bertino E, Catania B, Damiani ML, Perlasca P. GEO-RBAC: a spatially aware RBAC. In: Proceedings of the 10th ACM Symposium on Access Control Models and Technologies; 2005. p. 29–37.
5. Chandran SM, Joshi JBD. LoT RBAC: a location and time-based RBAC model. In: Proceedings of the 6th International Conference on Web Information Systems Engineering; 2005. p. 361–75.
6. Crampton J, Loizou G. Administrative scope: a foundation for role-based administrative models. ACM Trans Inf Syst Secur. 2003;6(2):201–31.
7. Joshi JBD, Bertino E, Latif U, Ghafoor A. A generalized temporal role-based access control model. IEEE Trans Knowl Data Eng. 2005;17(1):4–23.
8. Joshi JBD, Shafiq B, Ghafoor A, Bertino E. Dependencies and separation of duty constraints in GTR-BAC. In: Proceedings of the 8th ACM Symposium on Access Control Models and Technologies; 2003. p. 51–64.
9. Nyanchama M, Osborn SL. The role graph model. In: Proceedings of the 1st ACM Workshop on Role-Based Access Control; 1995.
10. Osborn S, Sandhu R, Munawer Q. Configuring role-based access control to enforce mandatory and discretionary access control policies. ACM Trans Inf Syst Secur. 2000;3(2):85–106.
11. Piromruen S, Joshi JBD. An RBAC framework for time constrained secure interoperation in multi-domain environment. In: Proceedings of the IEEE Workshop on Object-oriented Real-time Dependable Systems; 2005. p. 36–45.
12. Sandhu RS, Coyne EJ, Feinstein HL, Youman CE. Role-based access control models. IEEE Comput. 1996;29(2):38–47.
13. Sandhu R, Bhamidipati V, Munawer Q. The AR-BAC97 model for role-based administration of roles. ACM Trans Inf Syst Secur. 1999;2(1):105–31.
14. Shafiq B, Joshi JBD, Bertino E, Ghafoor A. Secure interoperation in a multi-domain environment

employing RBAC policies. IEEE Trans Knowl Data Eng. 2005;17(11):1557–77.

15. Zhang L, Ahn G, Chu B. A role-based delegation framework for healthcare information systems. In: Proceedings of the 7th ACM Symposium on Access Control Models and Technologies; 2002. p. 125–34.

R-Precision

Nick Craswell
Microsoft Research Cambridge, Cambridge, UK

Definition

For a given query topic Q, R-precision is the precision at R, where R is the number of relevant documents for Q. In other words, if there are r relevant documents among the top-R retrieved documents, then R-precision is $\frac{r}{R}$.

Key Points

R-precision is defined as the proportion of the top-R retrieved documents that are relevant, where R is the number of relevant documents for the current query. This requires full knowledge of a query's relevant set, and will be a shallow evaluation for a query with few relevant documents and a deep evaluation for a query with many relevant documents. Rank cutoff R is the point at which precision and recall are equal, since at that point both are $\frac{r}{R}$.

Cross-References

► Average R-Precision
► Precision
► Precision at n
► Precision-Oriented Effectiveness Measures

R-Tree (and Family)

Apostolos N. Papadopoulos[1], Antonio Corral[2], Alexandros Nanopoulos[3], and Yannis Theodoridis[4]
[1]Aristotle University of Thessaloniki, Thessaloniki, Greece
[2]University of Almeria, Almeria, Spain
[3]Aristotle University, Thessaloniki, Greece
[4]University of Piraeus, Piraeus, Greece

Definition

The R-tree is an indexing scheme that has been originally proposed towards organizing spatial objects such as points, rectangles and polygons. It is a hierarchical data structure suitable to index objects in secondary storage (disk) as well as in main memory. The R-tree has been extensively used by researchers to offer efficient processing of queries in multi-dimensional data sets. Queries such as *range*, *nearest-neighbor* and *spatial joins* are supported efficiently leading to considerable decrease in computational and I/O time in comparison to previous approaches. The R-tree is capable of handling diverse types of objects, by using approximations. This means that an object is approximated by its minimum bounding rectangle (MBR) towards providing an efficient filtering step. Objects that survive the filtering step are inspected further for relevance in the refinement step. The advantages of the structure, its simplicity as well as its resemblance to the B^+-tree "persuaded" the database industry to implement it in commercially available systems in addition to research prototypes.

Historical Background

The R-tree index was proposed by Guttman [6] in 1984 in order to solve an organization problem regarding rectangular objects in VLSI design. Later on, the structure was revised to become even more efficient and to adapt to the particular problem. The set of variants comprise the so called R-tree family of access methods.

The first variant, the R^+-tree, was proposed in 1987 [15]. The main difference between the two schemes is that while in the R-tree the MBR of an object is placed to one leaf node only, in the R^+-tree the MBR may break to several smaller MBRs and stored in different leaf nodes. The motivation behind this new index is that large MBRs may cause performance degradation. However, storage utilization decreases because of the MBR decomposition.

The next most important variation, the R^*-tree [1], retains the properties of the original R-tree and provides better strategies for inserting MBRs. New heuristics are offered towards improving the shape of the tree during insertions and node splits. Performance evaluation results have shown that the R^*-tree offers significantly better performance in *range query* processing in comparison to R-trees and R^+-trees. The trade-off is that more time is spent during insertions and deletions of objects, since the heuristics used require more computational overhead.

Another important variation of the R-tree is the Hilbert R-tree [8]. It is a hybrid structure based on the R-tree and the B^+-tree. Actually, it is a B^+-tree with geometrical objects being characterized by the Hilbert value of their centroid. The structure is based on the Hilbert space-filling curve. According to the authors' experimentation in [8], Hilbert R-trees were proven to be the best dynamic version of R-trees as of the time of publication. The term *dynamic* is used to denote that insertions and deletions are allowed in the underlying data set.

The aforementioned contributions focused on the use of heuristics during tree construction with the aim to provide a well-formed structure towards better query processing. However, by inserting objects one-by-one two problems may arise: (i) the shape of the structure may not be compact and (ii) the storage utilization will never reach 100% due to the rules applied in node splits. In cases where objects are known in advance, a bottom-up building process (also called bulk-loading) may be applied in contrast to the usual top-down tree construction. The first contribution

in this area is due to Roussopoulos and Leifker in [13] who proposed the use of packed R-trees. Along the same lines, Kamel and Faloutsos [7] proposed a packed version of the R-tree where packing is performed by the use of the Hilbert space filling curve. Other packed variants have been proposed in [3, 4, 9].

The R-tree inspired subsequent work on handling high-dimensional data. It has been observed by many researchers that the performance of R-trees degrades rapidly when the number of dimensions increases above a threshold (10 or 15). To tackle this dimensionality curse problem a number of R-tree variations appeared that show better scaling capabilities for high-dimensional spaces. Some of these contributions are the TV-tree [10], the X-tree [2] and the A-tree [14].

Due to space limitations, the description of more variations (actually these are more than 70) is not feasible. A more detailed presentation of the R-tree family can be found in [11].

Foundations

Let O be a set of objects in the two-dimensional space. Each object $o_i \in O$ is represented by its MBR, which is the minimum rectangle that completely encloses the object. The MBR of object o_i is denoted by o_i .*mbr*. Since MBRs are forced to be orthogonal with respect to the axes, each MBR is completely defined by its lower-left and upper-right corners. The R-tree organizes data in a hierarchical manner. It is a height-balanced tree where object MBRs are hosted in leaf nodes, whereas internal nodes contain auxiliary entries to guide the search process. More formally, each leaf node contains pairs of the form (*objMBR*, *objPtr*) where *objMBR* is the MBR of an object, and *objPtr* is the pointer to the object's detailed information (geometry and/or other attributes). Each internal node contains entries of the form (*entryMBR*, *childPtr*) where *entryMBR* is the MBR enclosing all descendants in the corresponding subtree and *childPtr* is the pointer to the subtree. The format of leaf and internal nodes is illustrated in Fig. 1.

a

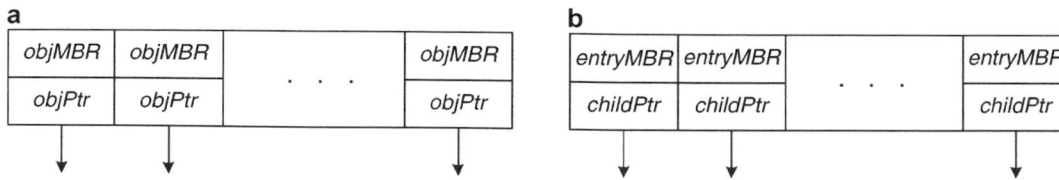

b

R-Tree (and Family), Fig. 1 Format of leaf and internal nodes. (**a**) Format of a leaf node. (**b**) Format of an internal node

Each R-tree node corresponds to one disk page which has a limited capacity. The number of entries in each internal node determines the tree *fanout*, which is the maximum number of subtrees that can emanate from an internal node. Let M_{int} and M_{leaf} denote the maximum number of entries that can be hosted in an internal node and a leaf node, respectively. To guarantee acceptable storage utilization, a minimum number of entries per node must also be defined. Therefore, the tree construction procedure forces that each internal node (leaf) must contain no less than m_{int} (m_{leaf}) entries. In the case where the R-tree stores MBR of objects, then evidently $M_{int} = M_{leaf}$ and $m_{int} = m_{leaf}$. However, there are cases where the above equalities do not hold. For example, when the R-tree stores point objects, then the number of objects in a leaf node increases, since a point requires less data for representation than a rectangle. For ease of illustration, for the rest of the article it is assumed that the R-tree stores rectangles, which means that the minimum and the maximum number of entries is the same for all tree nodes (the symbols m and M are used, respectively). Therefore, each tree node contains at least $m \geq M/2$ entries and at most M entries. A violation of this rule is only allowed for the root of the tree, which may contain less than m entries.

Since MBR is an approximation of the original object, intersection tests must be performed in two steps to guarantee correctness of results. Figure 2 depicts three polygonal objects with their corresponding MBRs. Searching for objects intersected by o_2, it is evident that both o_1 and o_3 may be contained in the final answer, because $o_2.mbr$ intersects both $o_1.mbr$ and $o_3.mbr$. This information is easily obtained by performing efficient intersection tests for rectangular objects. However, to produce the final result the answer

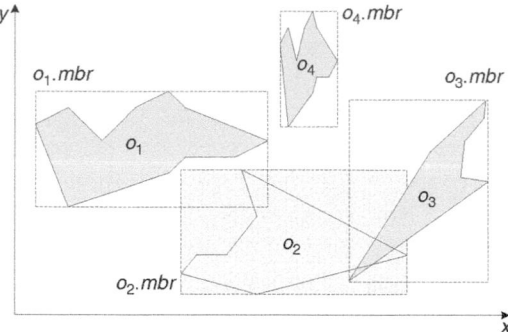

R-Tree (and Family), Fig. 2 Intersection of objects and MBRs

requires further refinement by considering the detailed geometric characteristics of objects o_1, o_2 and o_3. This refinement step involves more complex geometric computations and it is applied only for candidate objects. Evidently, if an MBR does not intersect $o_2.mbr$ (e.g., $o_4.mbr$) there is no need to investigate it further.

Figure 3 shows a set of objects in the 2-d space (left) and the corresponding R-tree index (right). For simplicity only the MBRs of objects are shown using the symbols r_1 through r_{12}. The R-tree is composed of seven nodes. There are four leaf nodes containing the object MBRs, and three internal nodes containing MBRs that enclose all the descendants in the corresponding subtree. For example, the MBR R_1 which is hosted at the root, encloses MBRs R_3 and R_4. Moreover, R_3 encloses the object MBRs r_1, r_2 and r_3. In this example, it is assumed that each node can host up to three entries. However, in real implementations this number is significantly higher and depends on the page size and the number of dimensions.

It is evident that the R-tree for a collection of object is not unique. The shape of the tree depends significantly on the insertion order. As

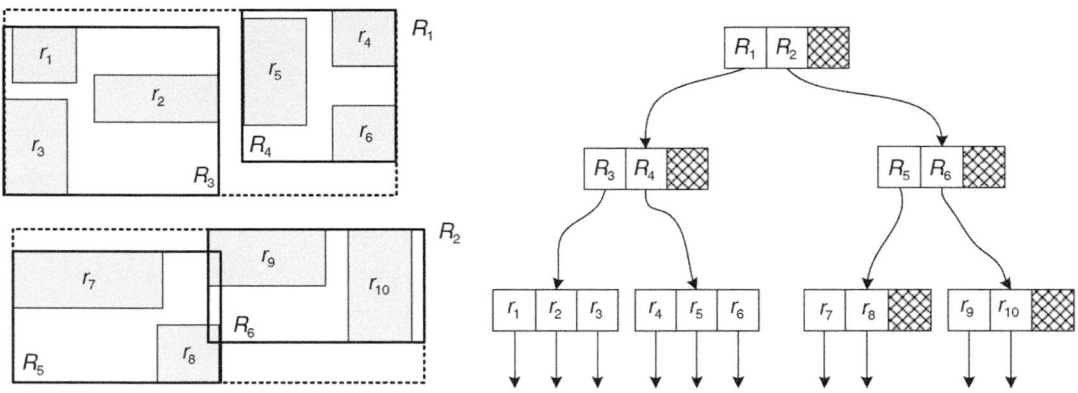

R-Tree (and Family), Fig. 3   R-tree example

it has been pointed out previously, there are two ways to build an R-tree: (i) by individual insertion of objects and (ii) by bulk loading. In dynamic data sets, where objects may be inserted and deleted in an ad hoc manner, the first method is applied. In the sequel, the insertion process is discussed briefly.

Insertions in an R-tree are handled similarly to insertions in a B^+-tree. In particular, the R-tree is traversed to locate an appropriate leaf node L to accommodate the new entry. The selection of L involves a number of internal node selections towards determining an appropriate path from the root to the most convenient leaf. It is important to guarantee that after the insertion, the tree will be in a good shape. Towards this goal, the insertion process tries to select the next node in the insertion path, aiming at low MBR enlargement, because the size of MBRs is directly connected to search efficiency. The new entry is inserted in L and then all nodes within the path from the root to L are updated accordingly (adjustment of MBRs). In case the found leaf cannot accommodate the new entry because it is full (it already contains M entries), then it is split into two nodes. Splitting in R-trees is different from that of the B^+-tree, because it considers different criteria. Guttman in his original paper [6] proposed three different split policies:

Linear Split. Choose two objects as seeds for the two nodes, where these objects are as far apart as possible. Then consider each remaining object in a random order and assign it to the node requiring the smallest enlargement of its respective MBR.

Quadratic Split. Choose two objects as seeds for the two nodes, where these objects if put together create as much dead space as possible (*dead space* is the space that remains from the MBR if the areas of the two objects are ignored). Then, until there are no remaining objects, insert the object for which the difference of dead space if assigned to each of the two nodes is maximized in the node that requires less enlargement of its respective MBR.

Exponential Split. All possible groupings are exhaustively tested and the best is chosen with respect to the minimization of the MBR enlargement.

Guttman suggested using the quadratic algorithm as a good compromise between insertion speed and retrieval performance. The linear split policy is the most efficient but the resulting R-tree does not keep its nice characteristics, whereas the exponential split policy although it takes the best split decision it requires significant computational overhead. Albeit the split policy being used, splits may propagate upwards up to the root node. If there is a split in the root, a new root node is created and the height of the tree increases. To demonstrate the importance of the splitting policy an example is given in Fig. 4. Figure 4a depicts an overflowing node, assuming

R

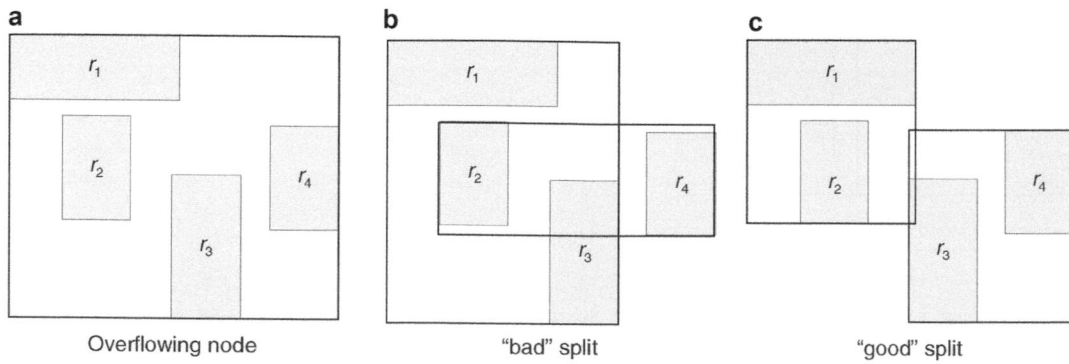

R-Tree (and Family), Fig. 4 Splitting example

that at most three entries can be stored. A "bad" split is shown in Fig. 4b whereas Fig. 4c depicts a "good" split choice. The first split should be avoided since the quality of the resulting nodes degrades (large MBRs with a lot of overlap), whereas the second is more preferable (small MBRs with small overlap).

Deletions are performed by first searching the tree to locate the corresponding leaf L which contains the deleted object. After the removal of the entry from L the node may contain fewer than m entries (node underflow). The handling of an underflowing node is different in the R-tree, compared with the case of B$^+$-tree. In the latter, an underflowing case is handled by merging two sibling nodes. Since B$^+$-trees index one-dimensional data, two sibling nodes will contain consecutive entries. However, for multi-dimensional data, this property does not hold. Although one still may consider the merging of two R-tree nodes that are stored at the same level, reinsertion is more appealing for the following reasons:

1. Reinsertion achieves the same result as merging. Additionally, the algorithm for insertion is used. Moreover, the pages required during reinsertion are likely to be available in the buffer memory, because they have been retrieved during the search of the deleted entry.
2. As described, the insertion process tries to maintain the good quality of the tree during the query operations. Therefore, it sounds rea-

sonable to use reinsertion, because the quality of the tree may degrade after several deletions.

In all R-tree variants that have appeared in the literature, tree traversals for any kind of operations are executed in a way similar to the one applied in the original R-tree. An exception is the R$^+$-tree which decomposes an object to smaller parts and therefore, the search process requires some modifications in comparison to other variants. Basically, the dynamic variations of R-trees differ in how they perform splits and how they handle insertions in general.

To demonstrate the way queries are executed, a simple range query example is given. Figure 5 shows the region of interest Q and the query asks for all objects that intersect Q. The nodes of the R-tree are labeled using the symbols A through G. The first accessed node is A (the root). The entries of A are tested for intersection with Q. Evidently, Q intersects both R_1 and R_2, therefore both subtrees require further investigation. The next accessed node is B which contains the MBRs R_3 and R_4. Among them, only R_3 intersects the region of interest, which means that node D should be accessed next. Node D contains the object MBRs r_1, r_2 and r_3 and none of them intersects Q. At this point, the search process backtracks to node B and since no eligible entries are found it backtracks to node A. Next, node C is accessed which contains the MBRs R_5 and R_6. Only R_5 intersects the region of interest, and therefore node F is accessed next. By inspecting the entries of F it is evident that both object

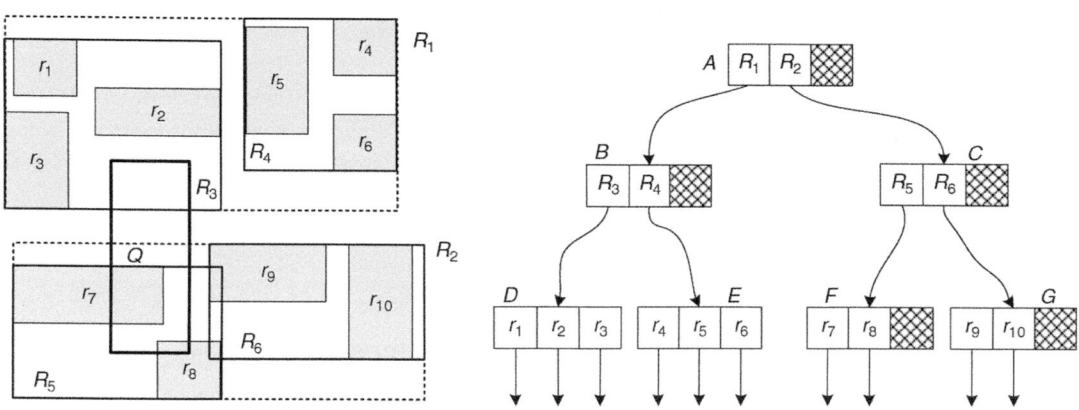

R-Tree (and Family), Fig. 5 Range query example using an R-tree

MBRs r_7 and r_8 intersect Q, and both are included in the final answer. The search process backtracks to node C and since no more promising branches can be followed, it backtracks to node A. At this point, all promising branches have been examined and the range query execution terminates. The object MBRs that intersect Q are r_7 and r_8. Note that if r_7 and r_8 correspond to the real objects then no further actions should be taken. Otherwise, the detailed geometry of these objects must be tested for intersection with Q (refinement).

In the previous lines, the main issues related to R-tree construction and search have been briefly discussed. The interested reader is directed to [11] for an exhaustive list of algorithmic techniques regarding R-trees and related structures.

Key Applications

Geographic Information Systems
R-tree is an excellent choice for indexing spatial data sets. Algorithms have been proposed to answer fundamental query types like *range queries*, *nearest-neighbor queries* and more complex queries like *spatial joins* and *closest pairs* using R-trees to index the underlying data. Therefore, the structure is equipped with all necessary tools to organize and index geographic information.

Location-Based Services
Many location-aware algorithmic techniques are based on the R-tree index or its enhancements. Queries involving the current location of moving objects are handled efficiently by R-tree variants and therefore these schemes are a convenient tool for organizing objects that potentially change their location.

Multimedia Database Systems
Since the R-tree index is capable of organizing multi-dimensional data, it can be utilized as an indexing scheme for multimedia data (e.g., images, audio). A common approach to organize multimedia data is to represent the complex multimedia information by using feature vectors. These feature vectors can be organized by means of R-trees, to facilitate similarity search towards multimedia retrieval by content [5].

Future Directions

R-trees have been successfully applied to offer efficient indexing support in diverse disciplines such as query processing in spatial and spatio-temporal databases, multi-attribute data indexing, preference query processing to name a few. An important research direction towards more efficient query processing in modern systems is to provide efficient indexing schemes towards distributed processing. A significant effort towards this direction has been reported in [12], which

R

proposes a distributed R-tree indexing scheme. Taking into account that huge volumes of multi-attribute data are scattered across different systems, such distributed schemes are expected to offer enormous help towards efficient query processing. The challenge is to provide efficient implementations of the corresponding centralized algorithms developed so far, since the methods used for centralized structures are likely to fail when applied to distributed data.

Experimental Results

The interested reader will find a plethora of performance evaluation results in the corresponding literature. Usually, when a new indexing scheme is proposed it is experimentally compared to other methods. Although these comparisons are not based on a common framework, they reveal the advantages and disadvantages of the indexing schemes under study. Usually, the comparison is based on the number of disk accesses and the computational time required to process queries.

Data Sets

Performance evaluation results regarding R-trees and related indexing schemes are produced based on real-life as well as on synthetically generated data sets with diverse distributions. The interested reader can browse the data sets hosted in http://www.rtreeportal.org to view some representative real-life data sets that are consistently being used by researchers for comparison purposes. Synthetically generated data sets follow different distributions (e.g., uniform, normal, Zipf) and their use provide additional hints for the index performance. Moreover, the use of synthetically generated data sets offers the flexibility to choose the cardinality of the data set, the distribution and the size of the objects (e.g., extent of MBRs). Hence, the comparison among indexing schemes is more reliable.

URL to Code

R-tree portal (http://www.rtreeportal.org) contains the code for most common spatial access methods (mainly R-tree and variations), as well as data generators and several useful links for researchers and practitioners interested in spatial database issues.

Cross-References

- ► Closest-Pair Query
- ► Nearest Neighbor Query
- ► Range Query
- ► Spatial Join

Recommended Reading

1. Beckmann N, Kriegel HP, Seeger B. The R.*-tree: an efficient and robust method for points and rectangles. In: Proceedings of the ACM SIGMOD International Conference on Management of Data; 1990. p. 322–31.
2. Berchtold S, Keim DA, Kriegel HP. The X-tree: an index structure for high-dimensional data. In: Proceedings of 22nd International Conference on Very Large Data Bases; 1996. p. 28–39.
3. Chen L, Choubey R, Rundensteiner EA. Bulk-insertions into R-trees using the small-tree-large-tree approach. In: Proceedings of 6th International Symposium on Advances in Geographic Information Systems; 1998. p. 161–2.
4. Choubey R, Chen L, Rundensteiner EA. GBI – a generalized R-tree bulk-insertion strategy. In: Proceedings of 6th International Symposium on Advances in Spatial Databases; 1999. p. 91–108.
5. Faloutsos C. Searching multimedia databases by content. Dordecht: Kluwer; 1996.
6. Guttman A. R-trees: a dynamic index structure for spatial searching. In: Proceedings of the ACM SIGMOD International Conference on Management of Data; 1984. p. 47–57.
7. Kamel I, Faloutsos C. On packing R-trees. In: Proceedings of the 2nd International Conference on Information and Knowledge Management; 1993. p. 490–9.
8. Kamel I, Faloutsos C. Hilbert R-tree – an improved R-tree using fractals. In: Proceedings of the 20th International Conference on Very Large Data Bases; 1994. p. 500–9.
9. Leutenegger S, Edgington JM, Lopez MA. STR – a simple and efficient algorithm for R-tree packing. In: Proceedings of the 13th International Conference on Data Engineering; 1997. p. 497–506.

10. Lin K, Jagadish HV, Faloutsos C. The TV-tree: an index structure for high-dimensional data. VLDB J. 1994;3(4):517–42.
11. Manolopoulos Y, Nanopoulos A, Papadopoulos AN, Theodoridis Y. R-trees: theory and applications. Berlin: Springer; 2006.
12. du Mouza C, Litwin W, Rigaux P. SD-Rtree: a scalable distributed R-tree. In: Proceedings of the 23rd International Conference on Data Engineering; 2007. p. 296–305.
13. Roussopoulos N, Leifker D. Direct spatial search on pictorial databases using packed R-trees. ACM SIGMOD Rec. 1985;14(4):17–31.
14. Sakurai Y, Yoshikawa M, Uemura S, Kojima H. Spatial indexing of high-dimensional data based on relative approximation. VLDB J. 2002;11(2):93–108.
15. Sellis T, Roussopoulos N, Faloutsos C. The R+-tree: a dynamic index for multidimensional objects. In: Proceedings of the 13th International Conference on Very Large Data Bases; 1987. p. 507–18.

Rule-Based Classification

Anthony K. H. Tung
National University of Singapore, Singapore, Singapore

Definition

The term rule-based classification can be used to refer to any classification scheme that make use of IF-THEN rules for class prediction. Rule-based classification schemes typically consist of the following components:

- *Rule Induction Algorithm* This refers to the process of extracting relevant IF-THEN rules from the data which can be done directly using sequential covering algorithms [1, 2, 5–7, 9, 12, 14–16] or indirectly from other data mining methods like decision tree building [11, 13] or association rule mining [3, 4, 8, 10].
- *Rule Ranking Measures* This refers to some values that are used to measure the usefulness of a rule in providing accurate prediction. Rule ranking measures are often used in the rule induction algorithm to prune off unnecessary rules and improve efficiency. They are also

used in the class prediction algorithm to give a ranking to the rules which will be then be utilized to predict the class of new cases.

- *Class Prediction Algorithm* Given a new record with unknown class, the class prediction algorithm will predict the class of the new record based on the IF-THEN rules that are output by the rule induction algorithm. In many cases where multiple rules could be matched by the new case, the rule ranking measures will be used to either select the best best matching rule based on the ranking or to compute an aggregate from the multiple matching rules in order to arrive at a final prediction.

Historical Background

Earlier rule-based classification methods includes AQ [7], CN2 [1] and more recently RIPPER [2]. These methods induce rules using the sequential covering algorithm where. Rules are learned one at a time. Decision tree classification methods like C4.5 [*13*] can also be considered as a form of rule-based classification. However, decision tree induction involved parallel rule induction, where rules are induced at the same time. Even more recently, advances in association rule mining had made it possible to mine association rules efficiently in order to build a classifier [3, 4, 8, 10]. Such an approach can also be considered as rule-based classification.

Foundations

The discussion first looks at how IF-THEN rules can be used for classification before proceeding to look at their ranking measures and induction algorithms.

Using IF-THEN Rules for Classification

An IF-THEN rule is typically an expression of the form $LHS \Rightarrow RHS$ where LHS is a set of conditions that much be meet in order to derive a conclusion represented by RHS. In much literature, LHS is called the *antecedent* of the rule and

RHS is called the *consequent* of the rule. For rule-based classification, the rule antecedent typically consists of a conjunction (AND) of attribute tests while the rule consequent is a class value. An example of a IF-THEN rule will be:

(No. of years ≥ 6) and (Rank =

Associate Professor) ⇒ (Tenured = YES)

A rule is say to *cover* a record if the record matches all the antecedent conditions in the rule. Given a new record with unknown class value, rules which cover the record will be used to determine the class value of the record. In the case where only one rule matches the tuple, the class value at the consequent of such a rule will be assigned as the predicted class for the tuple. On the other hand, it is also possible for none of the rules to match, in which case a *default class value* will be assigned.

For more complex situation in which multiple rules are matched, there are usually two approaches:

(i) **Top Rule Approach**

In this approach, all the rules that matched the new record are ranked based on the rule ranking measures. The consequent of the rule that is rank top based on this approach will be the predicted class value of the record.

(ii) **Aggregation Approach**

In the aggregation approach, rules that match the new record are separated into groups based on their consequent. For each of the rule group with the same consequent, an aggregated measure will be computed based on the rule ranking measure for each rule in the group. Each group of rules are then ranked based on their aggregated measure and the consequent of the rule group that are ranked highest will be the predicted class value for the new record.

Note that the two approaches described above are not mutually exclusive. For example it is possible to pick the top-k rules based on the first approach and then apply the aggregation approach on the top-k rules so as to determine the final predicted class value.

Rule Ranking Measures

Rule ranking measures are important components in a rule-based classification scheme because of two reasons. First, they can improve the efficiency of constructing and using the classifier. Rules that are not deemed to be useful based on these measures can be pruned off during rule induction making the process more efficient and also reducing the number of rules that must be processed during classification. Second, they can enhance the effectiveness of a rule-based classifier by removing rules which have weak prediction power.

Among the various rule ranking measure, the most basic ones are *coverage, accuracy* and *length of the rule*. Let n be the number of record in the training database, n_r be the number of records that match both the antecedent and consequent of a rule r and n_{lr} be the number of records that are covered by r. The coverage of a rule r, denotes as $cov(r)$ is defined to be n_{lr}/n while the accuracy of rule r is defined to be n_r/n_{lr} and denote as $acc(r)$. A rule should have high coverage and accuracy in order to be useful for classification. Besides coverage and accuracy, the length of a rule, $len(r)$, which is the number of terms at the antecedent of r, is also important in determining the usefulness of a rule in a rule-based classification scheme.

More complex rule ranking measures try to integrate both coverage and accuracy. They include *information gain* and *likelihood ratio*. Information gain was proposed in FOIL (First Order Inductive Learner) for comparing rules whose antecedents are subset/superset of each other. Let r' be a rule which antecedent is a superset of r. FOIL assess the information gain of r' over r to be

$$FOIL\_Gain = p' \times \left(\log_2 \frac{p'}{p' + n'} - \log_2 \frac{p}{p+n} \right)$$

$$(1)$$

Algorithm 1 Generic Sequential Covering Algorithm

Algorithm: Sequential covering. Learn a set of IF-THEN rules for classification.
Input: D, a database of class-labelled records; Attvals, the set of all attributes and
their possible values.
Output: A set of IF-THEN rules. Method:

```
        Rule set = { };
    for each class c do
        repeat
            Rule = Learn OneRule (D, Att vals, c);
            remove tuples covered by Rule from D;
        until terminating condition;
        Rule set = Rule set + Rule; // add new rule to rule set
    return Rule Set;
```

where p', n', p' and n are the number of records that are of positive class and negative class and which are covered by r' and r respectively. FOIL_gain favors rules that have high accuracy and cover many positive tuples.

Besides, information gain, one can also use a statistical test of significance to determine if the apparent effect of a rule is not attributed to chance but instead indicates a genuine correlation between attribute values and classes. The test compares the observed distribution among classes of tuples covered by a rule with the expected distribution that would result if the rule made predictions at random. Given m classes, let p_i be the probability distribution of class i within the database and $p_i(r)$ be the probability distribution of class i within the set of records that are covered by r. The likelihood ratio is computed as:

$$Likelihood\_Ratio = 2 \sum_{1}^{m} p_i(r) \log \left(\frac{p_i(r)}{p_i} \right)$$

(2)

The likelihood ratio is then used to perform a significant test against a χ^2 distribution with m-1 degrees of freedom. The higher the likelihood ratio is, the more likely that there is a significant difference in the number of correct predictions made by the rule in comparison with a random guess that following the class probability distribution of the database.

Rule Induction

Various forms of rule induction can be performed for rule-based classification. Here, the *sequential covering* algorithm will be described.

Algorithm 1 presents a generic algorithm for sequential covering rule induction. The algorithm iteratively learn one rule from the database and then remove all records that are covered by the rule before learning the next rule. This is done repeatedly until a terminating condition is met. The terminating condition can varies across different algorithms but is typically linked to the fact no more interesting rules can be induced once many of the records are removed from the database. All rules that are induced during the process are then output.

Each single invocation of LearnOneRule typically involve a greedy search for interesting rule based on the rule ranking measures. This is done by searching for attribute conditions which will improve the rule ranking measure when they are appended to the antecedent of the rule. A single attribute condition that best improve the measure is appended each time and this is done repeatedly until the measure cannot be improved. The rule will then be return as the output for LearnOneRule.

Key Applications

Rule-based classification has been very popularly used in machine learning for classification of data. It is applicable whenever other classification scheme is applicable.

Future Directions

Despite efforts to reduce the number of rules in a rule-based classifier, the number of rules being used are still substantially higher than what a human can handle. More studies on new interestingness measure and visualization techniques are needed to further enhance the interpretability of a rule-based classifier.

Cross-References

▶ Association Rule Mining on Streams
▶ Decision Tree Classification

Recommended Reading

1. Clark P, Niblett T. The CN2 induction algorithm. Mach Learn. 1989;3(4):261–83.
2. Cohen W. Fast effective rule induction. In: Proceedings of the 12th International Conference on Machine Learning; 1995. p. 115–23.
3. Cong G, Tung AKH, Xu X, Pan F, Yang J. FARMER: finding interesting rule groups in microarray datasets. In: Proceedings of the ACM SIGMOD International Conference on Management of Data; 2004. p. 143–54.
4. Cong G, Tan K, Tung A, Xu X. Mining top-K covering rule groups for gene expression data. In: Proceedings of the ACM SIGMOD International Conference on Management of Data; 2005. p. 670–81.
5. Domingos P. The RISE system: conquering without separating. Tools with artificial intelligence, 1994. In: Proceedings of the 6th IEEE International Conference on Tools with Artificial Intelligence; 1994. p. 704–7.
6. Furnkranz J, Widmer G. Incremental reduced error pruning. In: Proceedings of the 11th International Conference on Machine Learning; 1994. p. 70–7.
7. Hong J, Mozetic I, Michalski R. AQ15: incremental learning of attribute-based descriptions from examples: the method and user's guide. Reports of the Intelligent Systems Group, ISG, p. 86–5.
8. Liu B, Hsu W, Ma Y. Integrating classification and association rule mining. In: Proceedings of the 4th International Conference on Knowledge Discovery and Data Mining; 1998.
9. Major J, Mangano J. Selecting among rules induced from a hurricane database. J Intell Inf Syst. 1995;4(1):39–52.
10. Pan F, Cong G, Tung AKH. CARPENTER: finding closed patterns in long biological datasets. In: Proceedings of the 9th ACM SIGKDD International Conference on Knowledge Discovery and Data Mining; 2003.
11. Quinlan J. Simplifying decision trees. Int J Man Mach Stud. 1987;27(3):221–34.
12. Quinlan J. Learning logical definitions from relations. Mach Learn. 1990;5(3):239–66.
13. Quinlan J. C4.5: programs for machine learning. San Mateo: Morgan Kaufmann; 1993.
14. Quinlan J, Cameron-Jones R. FOIL: a midterm report. In: Proceedings of the European Conference on Machine Learning; 1993.
15. Smyth P, Goodman R. An information theoretic approach to rule induction from databases. IEEE Trans Knowl Data Eng. 1992;4(4):301–16.
16. Weiss S, Indurkhya N. Predictive data mining: a practical guide. Los Altos: Morgan Kaufmann; 1998.

S

Safety and Domain Independence

Rodney Topor
Griffith University, Nathan, Australia

Synonyms

Finiteness

Definition

The values in the relations of a relational database are elements of one or more underlying sets called domains. In practical applications, a domain may be infinite, e.g., the set of natural numbers. In this case, the value of a relational calculus query when applied to such a database may be infinite, e.g., $\{n \mid n \geq 10\}$. A query Q is called *finite* if the value of Q when applied to any database is finite.

Even when the database domains are finite, all that is normally known about them is that they are some finite superset of the values that occur in the database. In this case, the value of a relational calculus query may depend on such an unknown domain, e.g., $\{x \mid \forall y R(x, y)\}$. A query Q is called *domain independent* if the value of Q when applied to any database is the same for any two domains containing the database values or, equivalently, if the value of Q when applied

to a database contains only values that occur in the database.

The term *safe* query has been used ambiguously in the literature. Often safe queries have been identified with finite queries. Sometimes safe queries have been members of a large, simple, decidable class of queries that are guaranteed to be finite, or, in other cases, domain independent. The use of word *safe* is preferred to denote a large, simple, decidable class of queries that are guaranteed to be domain independent and hence, normally finite.

Obviously, it is desirable that queries be finite and domain independent. Unfortunately, the classes of finite queries and domain independent queries are undecidable, which leads to a search for decidable classes of queries that can represent all, or as many as possible, finite (resp., domain independent) queries.

Historical Background

DiPaola [3] and independently Vardi [11] recognized the desirability that queries be domain independent and proved that the class of domain independent queries was undecidable.

Many researchers then attempted to define decidable classes of queries that were guaranteed to be domain independent. This work was summarized by Topor [9], Kifer [6], Ullman [10], and Abiteboul et al. [1]. Many different names such as range-restricted, allowed, safe, with subtle differences, were used in these definitions. Ullman

© Springer Science+Business Media, LLC, part of Springer Nature 2018
L. Liu, M. T. Özsu (eds.), *Encyclopedia of Database Systems*,
https://doi.org/10.1007/978-1-4614-8265-9

[10], Van Gelder and Topor [12], and Abiteboul et al. [1] gave algorithms for translating queries in these classes into relational algebra for efficient evaluation.

Other researchers such as Escobar-Molano et al. [4], Hull and Su [5] and Suciu [8] attempted to define decidable classes of queries, *safe* queries, that were guaranteed to be finite in the presence of functions (e.g., arithmetic functions) over infinite domains (e.g., the natural numbers).

Whether or not there is a decidable class of queries that can express every finite (resp., domain independent) query depends critically on the particular set of functions on the domains. Stolboushkin and Taitslin [7] showed that, for many common domains, there is a decidable class of queries that can express every finite (resp. Domain independent) query, but that there do exist domains for which there is no such decidable class. Benedikt and Libkin [2] extended and generalized these results to a wider class of domains.

Foundations

Following standard practice in the literature, assume that every database is defined over a single domain. The use of multiple domains complicates the presentation without introducing any substantially new concepts.

A *domain* $D = (U, O)$ consists of an underlying set U and a set of operations O on the set. The set may be infinite or finite. The operations may be represented as (infinite) relations over the set. Technically, each operation must be decidable. Often, one also wants the first-order theory of the domain to also be decidable. These conditions are satisfied in most common cases.

Examples of such domains include (a) any finite set of symbols possibly with an equality operator, (b) the set of natural numbers with addition and linear order operators, and (c) the set of finite strings over some finite alphabet with concatenation and lexicographic order operators.

A *database scheme* S is a finite set of pairs $\{(S_i, p_i) \mid 1 \leq i \leq k\}$, where each S_i is a relation name with arity $p_i \geq 1$.

Given a domain $D = (U, O)$, *a database (instance)* I of a scheme $\{(S_i, p_i) \mid 1 \leq i \leq k\}$ over D is a family of finite sets $\{R_i \mid 1 \leq i \leq k\}$, where each $R_i \subseteq U^{p_i}$.

A *query* over a database scheme is a first-order formula constructed from the relations in the scheme and the operators (and constants) in its domain. That is, the domain relational calculus is used as our query language.

A query Q over a database scheme S and domain $D = (U, O)$ is called *finite* if, for every database instance I of S over D, the value of Q when applied to I is a finite relation over U.

For example, over the domain of natural numbers, the query $Q_1 = \{n \mid n + 1 \leq 10\}$ is finite, but the query $Q_2 = \{n \mid 10 \leq n + 1\}$ is infinite.

It is desirable that queries be finite so that they may be composed and so that their results may be displayed.

The *active domain* of a database I is the finite set of domain elements that occur in the relations of I.

A query Q over a database scheme S and domain $D = (U, O)$ is called *domain independent (d.i.)* if, for every database instance I of S over D, the value of Q when applied to I contains only elements in the active domain of I. Equivalently, a query Q is domain independent if and only if, for every database instance I, and for every two extensions U_1 and U_2 of the active domain of D, the value of Q when applied to I over U_1 equals the value of Q when applied to I over U_2.

For example, over a finite domain of symbols without equality, the query $Q_3 = \{x \mid \exists y (P(x) \lor R(y))\}$ is *not* domain independent because, when applied to any database instance over this domain, the value consists all elements x that occur in P if R is empty, and *all* elements of the domain otherwise.

It is desirable that queries be domain independent so that their values are predictable despite the possibly unknown underlying domain.

It is natural to ask about the difference between finiteness and independence.

Clearly, if the domain is finite, all queries are finite, but the query Q_3 above is still not domain independent.

However, if the domain is infinite and the only operation on the domain is equality, then a query is finite if and only if it is domain independent [7].

Further, if the domain is infinite, even if there are operations other than equality, every domain independent query is also finite (as the active domain of every database instance is finite).

More interestingly, if the domain is the set of natural numbers and the only operation on the domain is linear order, then the query.

$$Q_4 = \{x \mid \forall_y (\Delta(y) \to x > y) \\ \land \forall_y (y < x \to \exists z (\Delta(z) \land z \geq y)),$$

where $\Delta(y)$ is true if and only if y is in the active domain of the database, defines the smallest integer greater than all the active domain elements, and is hence finite but not domain independent [7].

Next, it is natural to ask whether it is possible to effectively recognize finite or domain independent queries. The answer is no. By reduction from standard undecidable problems in first-order logic, DiPaola [3] and, independently, Vardi [11] showed that, over any infinite domain, if the database scheme contains at least one relation of arity 2 or more, the classes of finite and domain independent queries are both undecidable. This undecidability result extends to domains such as the natural numbers with addition and linear order operations.

However, Benedikt and Libkin [2] show that finiteness (resp., d.i.) *is* decidable for Boolean combinations of conjunctive queries for a large class of domains over the real numbers.

Given that it is not possible to recognize whether or not a given query is finite (resp., d.i.) over a given domain D, two further questions arise naturally. First, is there is a *decidable* class of queries over D that can express all finite (resp., d.i.) queries over D? That is, is there a decidable class C of queries over D such that, for every finite (resp., d.i.) query Q there exists an equivalent query Q' in C? Second, can a large, simple, decidable class C' of queries be defined such that every query in C' is finite (resp., d.i.)? Historically, the second question was considered

first, but they were considered in the order given.

For many domains, there is a decidable class of queries that can express every finite (resp., d.i.) query. Stolboushkin et al. [7] say that the finite (resp. d.i.) queries hence have an "effective syntax" for such domains. Examples of domains for which the finite (resp., d.i.) queries have an effective syntax include (a) an infinite domain of symbols in which the only operation is equality, (b) the domain of natural numbers with only the linear order operation, (c) the domain of natural numbers with the addition and linear order operations (Pressburger arithmetic), and (d) the domain of finite strings over a finite alphabet with the lexicographic order operation [7].

However, Stolboushkin and Taitslin [7] show that it is possible to construct an (artificial) domain with decidable operations and decidable first-order theory that does *not* have an effective syntax for the finite (resp., d.i.) queries. They also give an example of a domain with *undecidable* first-order theory (arithmetic) that has an effective syntax for the finite (resp., d.i.) queries.

Given the undecidability results above, many researchers, e.g., [1, 6, 9, 10] have defined specific, decidable classes of queries that are guaranteed to be finite (resp. d.i.). Researchers have attempted to make these classes both simple and as large as possible. In many cases, they showed that queries in their class could express *all* finite (resp. d.i.) queries. These classes were given names such as safe, range-restricted, allowed, and many others. Most researchers restricted attention to domains with equality as the only operation, but some extended their work to the domain of the natural numbers with arithmetic and linear order operations, and some, e.g., [2, 4, 5] considered more arbitrary domains. Others, e.g., [9] applied these ideas to the domain independence of deductive databases.

The basic idea of these definitions is to ensure that every free variable in the query is somehow bound to an element in the active domain of the database or, in the presence of nontrivial operations, to one of a finite number of domain elements. In the absence of operations, this is typically done by ensuring that every free or existentially quantified variable in a query occurs positively in its scope, every universally quantified variable occurs negatively in its scope, and that the same free variables occur in every component of a disjunction. For example, the query $\{x \mid P(x) \wedge \forall y(Q(x, y) \rightarrow R(x, y))\}$ is safe according to these ideas. The equality operation is used to propagate a positive variable (or constant) from one side of the equality to the other. For example, the query $\{x \mid P(y) \wedge x = y\}$ is safe according to this idea. Finiteness dependencies are used with arithmetic operations over the natural numbers of concatenation operations over strings to propagate a positive variable (or constant) from one or more positions in the operation to one or more other positions. For example, the query $\{x \mid P(z) \wedge x + y = z \wedge \neg Q(y)\}$ is safe according to this idea. With deductive databases or, equivalently, with Datalog queries, it is necessary to require that every variable in the head of a rule occurs positively in the body of a rule and that the body of a rule is itself safe. In every case, the details of the definitions are too complicated to present here.

Finally, as relational calculus queries are evaluated in real database systems by translation into relational algebra, many researchers have studied techniques for translating safe queries (as defined in the previous two paragraphs) into equivalent relational algebra expressions. (A basic result that many researchers proved is that the class of safe queries is equivalent to the class of relational algebra queries.) These translations typically involve a sequence of transformations into increasingly restricted forms, until the translation into relational algebra is direct. Again, the details of the transformations too complicated to present here. See [1, 10, 12] for more information.

Key Applications

The concepts of finiteness, domain independence and safety are fundamental to our understanding of database queries over different domains. Tools that generate queries over specific domains should only generate safe queries. Query processors should check that input queries over specific domains are safe and should report warnings otherwise.

Future Directions

These questions of finiteness, domain independence and safety are well-understood and largely resolved for relational databases. However, additional work extending these ideas and methods to other data models and query languages may still be required.

Cross-References

▸ Computationally Complete Relational Query Languages
▸ Conjunctive Query
▸ Constraint Query Languages
▸ Query Language
▸ Relational Algebra
▸ Relational Calculus
▸ Relational Model

Recommended Reading

1. Abiteboul R, Hull R, Vianu V. Foundations of databases, chapter 5. Reading: Addison-Wesley; 1995. p. 70–104.
2. Benedikt M, Libkin L. Safe constraint queries. SIAM J Comput. 2000;29(5):1652–82.
3. DiPaola RA. The recursive unsolvability of the decision problem for the class of definite formulas. J ACM. 1969;16(2):324–7.
4. Escobar-Molano M, Hull R, Jacobs D. Safety and translation of calculus queries with scalar functions. In: Proceedings of the 12th ACM SIGACT-SIGMOD-SIGART Symposium on Principles of Database Systems; 1993. p. 253–64.
5. Hull R, Su J. Domain independence and the relational calculus. Acta Inform. 1994;31(6):513–24.
6. Kifer M. On Safety, Domain Independence, and Capturability of Database Queries (Preliminary Report). In: Proceedings of the 3rd International Conferences on Data and Knowledge Bases; 1988. p. 405–15.

7. Stolboushkin AP, Taitslin MA. Finite queries do not have effective syntax. Inf Comput. 1996;153(1): 99–116.
8. Suciu D. Domain-independent queries on databases with external functions. Theor Comput Sci. 1998;190(2):279–315.
9. Topor RW. Domain independent formulas and databases. Theor Comput Sci. 1987;52(3):281–306.
10. Ullman JD. Principles of database and knowledge-base systems Volume I, Sections 3.2 and 3.8. Rockville: Computer Science Press; 1988. p. 100–6.
11. Vardi MY. The decision problem for database dependencies. Inf Process Lett. 1981;13(5):251–4.
12. Van Gelder A, Topor RW. Safety and translation of relational calculus queries. ACM Trans Database Syst. 1981;16(2):235–78.

Sagas

Kenneth Salem
University of Waterloo, Waterloo, ON, Canada

Definition

A saga [3] is a sequence of atomic transactions $T_1, \ldots, T_n$ for which the following execution guarantee is made. Either the component transactions T_i will all commit in the order:

$$T_1 T_2, \ldots, T_n$$

in which case that saga is said to have committed or one of the transaction sequences

$$T_1, \ldots, T_j C_j, \ldots, C$$

will be executed (for some $0 \leq j < n$), in which case the saga is said to have aborted. The transactions C_i are *compensating transactions* for the corresponding saga transactions T_i. Each transaction T_i in a saga must have a corresponding compensating transaction, which is responsible for undoing the T_i's effects.

Key Points

A saga is a type of extended transaction model [1]. Each component transaction in a saga is executed atomically, but the saga itself is not atomic. Effects of the component transactions are visible to other operations as soon as those transactions commit, which may be well before the saga has finished.

The saga model guarantees execution of a compensating transaction for each component transaction that has already committed at the time that a saga aborts. This guarantee is known as *semantic atomicity* [2]. The saga model does not specify the nature of these compensations. Rather, the definition or identification of appropriate compensations is an application-specific task.

Sagas were originally proposed as a weaker substitute for the traditional atomic transaction model in situations for which atomicity would be expensive to enforce, e.g., when the traditional transaction would be long running. Sagas and other extended transaction models have since found a variety of applications, such as workflow systems.

Cross-References

▶ Compensating Transactions
▶ Extended Transaction Models and the ACTA Framework
▶ Open Nested Transaction Models
▶ Semantic Atomicity
▶ Workflow Management

Recommended Reading

1. Chrysanthis PK, Ramamritham K. Synthesis of extended transaction models using ACTA. ACM Trans Database Syst. 1994;19(3):450–91.
2. Garcia-Molina H. Using semantic knowledge for transaction processing in a distributed database. ACM Trans Database Syst. 1983;8(2):186–213.
3. Garcia-Molina H, Salem K. Sagas. In: Proceedings of the ACM SIGMOD International Conference on Management of Data; 1987. p. 249–59.

Sampling Techniques for Statistical Databases

Amarnath Gupta
San Diego Supercomputer Center, University of
California San Diego, La Jolla, CA, USA

Definition

A sampling technique is a method by which
one inspects only a small portion of data from
a database to reduce the time to compute an
aggregate query, but simultaneously ensuring that
result computed on the sample faithfully represents the true results of the query for the entire
data population.

Example Acceptance-Rejection   sampling
(AR sampling) is sampling technique.

Key Points

Sampling is used in a database for different
reasons such as (i) to estimate the results of
aggregate queries (e.g., SUM, COUNT, or
AVERAGE), (ii) to retrieve a sample of records
from a database query for subsequent processing,
(iii) for internal use by the query optimizer
for selectivity estimation, (iv) to provide
privacy protection for records on individuals
contained in statistical databases. It has been
determined that fixed size random sampling
of data does not yield a true representation
of the population. *Acceptance/rejection (A/R)
sampling* is used to construct weighted samples in
which the inclusion probabilities of a record are
proportional to some arbitrary weight. *Reservoir
sampling* is a form of sequential scan sampling
algorithms which are used on files of unknown
size to perform on-the-fly sampling from the
results of a query. Methods have been developed
to perform sampling not only from raw records,
but also from B+−trees, hash structures, spatial
data structures and so on.

Cross-References

▶ Online Analytical Processing
▶ Privacy
▶ Secure Database Development
▶ Summarizability

Recommended Reading

1. Olken F, Rotem D. Random sampling from databases:
   a survey. Stat Comput. 1995;5:25–42.

SAN File System

Kazuo Goda
The University of Tokyo, Tokyo, Japan

Synonyms

Shared-disk file system

Definition

The term SAN file system refers to a file system
which transfers file data directly to/from a storage
device through a SAN. A SAN file system often
has the capability of coherency control such that
multiple servers may share the file system volume
and simultaneously access files stored in the volume. The term shared disk file system is also used
to refer to a SAN file system.

Key Points

In contrast to local file systems, SAN file systems allow multiple servers to share file system
volumes directly and to access the same file
simultaneously. To achieve this, the SAN file
system has the capability of coherency control

between servers. File system software running on each server may cache file data in a main memory buffer. Suppose that two servers, A and B, have cached the same file X. If A and B were to update X independently at the same time, two versions of X might be generated. The SAN file system needs to let each server be aware of the other servers. When server A updates a fragment of the file X, server B must be informed of a message of cache invalidation or changed information regarding the file X. Some SAN file systems exchange mutual exclusion messages to synchronize write accesses between servers.

The beneficial property of SAN file systems is high performance in comparison with network file systems such as NFS and CIFS. SAN file systems can directly transfer file data between servers and storage devices e.g., on a high-speed Fibre Channel network. In addition, SAN file systems do not need a central file server, which often becomes a bottleneck.

A variety of SAN file systems have been proposed and some of them have been deployed mainly into cluster computer systems often used to run scientific calculations. Recently, SAN file systems are also used as back-end file systems for large-scale enterprise NAS systems.

Cross-References

▶ Storage Network Architectures

Recommended Reading

1. Barrios M, Jones T, Kinnane S, Landzettel M, Al-Safran S, Stevens J, Stone C, Thomas C, Troppens U. Sizing and tuning GPFS. IBM Redbook. SG24-5610-00, 1999.
2. Burns RC, Rees RM, Long DDE. Semi-preemptible locks for a distributed file system. In: Proceedings of the 19th IEEE International Performance, Computing and Communications Conference; 2000. p. 397–404.
3. Soltis SR, Ruwart TM, O'Keefe MT. The global file system. In: Proceedings of the 15th NASA Goddard Conference on Mass Storage Systems; 1996. p. 319–42.

Scalable Decision Tree Construction

Johannes Gehrke
Cornell University, Ithaca, NY, USA

Synonyms

Scalable classification tree construction; Scalable top-down decision tree construction; Tree-structured classifier

Definition

Decision trees are popular classification models. Decision trees are usually contructed greedily top-down from a training dataset. In many modern applications, the training dataset is very large and thus decision tree construction algorithms that scale with the size of the training dataset are needed.

Historical Background

Decision trees, in particular classification trees, have a long history both in the statistics [4] and the machine learning communities [12, 13]. Scalability was not much a concern until the advent of data mining brought training datasets that were orders of magnitude larger than in traditional applications in machine learning and statistics.

Scalability concerns in classification started with the work by Agrawal et al. who presented an interval classfier that generated classification functions that distinguishes the different groups of training records based on their class label [1]. A follow-up paper introduces scalable construction of classification models as one of the three important classes of database mining problems [2], the other two being associations and sequences. The first scalable classification tree construction algorithm in the literature was SLIQ

[10], which was then quickly followed by more algorithms that improved performance and allowed scaling up a more general class of algorithms from the machine learning and statistics literature [5, 6, 14–17].

Foundations

The input to a classification or regression problem is a dataset of *training records* (also called the *training database*). Each record has several attributes. Attributes whose domain is numerical are called *numerical attributes*, whereas attributes whose domain is not numerical are called *categorical attributes*. A *categorical* attribute takes values from a set of categories. Some authors distinguish between categorical attributes that take values in an unordered set (*nominal* attributes) and categorical attributes having ordered domains (*ordinal* attributes).

There is one distinguished attribute called the *dependent attribute*. The remaining attributes are called *predictor attributes*; they are either numerical or categorical. If the dependent attribute is categorical, the problem is referred to as a *classification problem* and the dependent attribute is called the *class label*. The elements of the domain of the class label attribute will also be denoted as *class labels*; the meaning of the term class label will be clear from the context. If the dependent attribute is numerical, the problem is called a *regression problem*. This entry concentrates on classification problems.

The goal of classification is to build a concise model of the distribution of the dependent attribute in terms of the predictor attributes. The resulting model is used to assign values to a database where the values of the predictor attributes are known but the value of the dependent attribute is unknown. This entry surveys research on scalable classification tree construction from the database literature. An excellent survey of other aspects of decision tree construction can be found in Murthy [11].

Problem Definition

Let $X_1, \ldots, X_m$, C be random variables where X_i has domain $dom(X_i)$; assume without loss of generality that $dom(C) = \{1, 2, \ldots, J\}$. A *classifier* is a function

$$d : \dom(X_1) \times \cdots \times dom(X_m) \mapsto \dom(C).$$

Let $P(X', C')$ be a probability distribution on $dom(X_1) \times \ldots \times dom(X_m) \times dom(C)$ and let $t = \langle t.X_1, \ldots, t.X_m, t.C \rangle$ be a record randomly drawn from P, i.e., t has probability $P(X', C')$ that $\langle t.X_1, \ldots, t.X_m \rangle \in X'$ and $t.C \in C'$. Define the *misclassification rate* R_d of classifier d to be $P(d(\langle t.X_1, \ldots, t.X_m \rangle) \neq t.C)$. The training database D is a random sample from P, the X_i correspond to the predictor attributes and C is the class label attribute.

A *decision tree* is a special type of classifier. It is a directed, acyclic graph T in the form of a tree. The focus here is on binary decision trees although these techniques can be generalized to non-binary decision trees. If a node has no outgoing edges it is called a *leaf node*, otherwise it is called an *internal node*. Each leaf node is labeled with one class label; each internal node n is labeled with one predictor attribute X_n called the *splitting attribute*. Each internal node n has a predicate q_n, called the *splitting predicate* associated with it. If X_n is a numerical attribute, q_n is of the form $X_n \leq x_n$, where $x_n \in dom(X_n)$; x_n is called the *split point* at node n. If X_n is a categorical attribute, q_n is of the form $X_n \in Y_n$ where $Y_n \subset dom(X_n)$; Y_n is called the *splitting subset* at node n. The combined information of splitting attribute and splitting predicates at node n is called the *splitting criterion* of n. An example training database is shown in Fig. 1, and a sample classification tree is shown in Fig. 2.

With each node $n \in T$ there is associated a predicate $f_n : \dom(X_1) \times \cdots \times \dom(X_m) \mapsto \{\text{true, false}\}$, called its *node predicate* as follows: For the root node n, $f_n \overset{\text{def}}{=} true$. Let n be a non-root node with parent p whose splitting predicate is q_p. If n is the right child of p, define $f_n \overset{\text{def}}{=} f_p \wedge q_p$; if n is the right child of p, define f_n

$\stackrel{\text{def}}{=} f_p \wedge \neg q_p$. Informally, f_n is the conjunction of all splitting predicates on the internal nodes on the path from the root node to n. Since each leaf node $n \in T$ is labeled with a class label, n encodes the classification rule $f_n \to c$, where c is the label of n. Thus the tree T encodes a function $T : dom(X_1) \times \ldots \times dom(X_m) \mapsto dom(C)$ and is therefore a classifier, called a *decision tree classifier*. (Both the tree as well as the induced classifier will be denoted by T; the semantics will be clear from the context.) For a node $n \in T$ with parent p, the *family of tuples* F_n is the set of records in D that follows the path from the root to n when being processed by the tree, formally

$$F_n \stackrel{\text{def}}{=} \{t \in D : f_n(t)\}.$$

Also define F_n^i for $i \in \{1, \ldots, J\}$ as the set of records in F_n with class label i, formally

$$F_n^i \stackrel{\text{def}}{=} \{t \in D : f_n(t) \wedge t.C = i\}.$$

The problem of classification tree construction can now be stated formally: Given a dataset $D = \{t_1, \ldots, t_n\}$ where the t_i are independent random samples from an unknown probability distribution P, find a decision tree classifier T that minimizes the misclassification rate $R_T(P)$.

A classification tree is usually constructed in two phases. In phase one, the *growth phase*, an overly large decision tree is constructed from the training data. In phase two, the *pruning phase*, the final size of the tree T is determined with the goal to minimize R_T. It is possible to interleave growth and pruning phase for performance reasons as in the PUBLIC Pruning Method described later in this entry. Nearly all decision tree construction algorithms grow the tree top-down in the following greedy way: At the root node n, the training database is examined and a splitting criterion for n is selected. Recursively, at a non-root node n, the family of n is examined and from it a splitting criterion is selected. This schema is depicted in Fig. 3.

During the tree growth phase, two different algorithmic issues need to be addressed. The first issue is to devise an algorithm such that the resulting tree T minimizes R_T; this part of

Record Id	Car	Age	Children	Subscription
1	sedan	23	0	yes
2	sports	31	1	no
3	sedan	36	1	no
4	truck	25	2	no
5	sports	30	0	no
6	sedan	36	0	no
7	sedan	25	0	yes
8	truck	36	1	no
9	sedan	30	2	yes
10	sedan	31	1	yes
11	sports	25	0	no
12	sedan	45	1	yes
13	sports	23	2	no
14	truck	45	0	yes

Scalable Decision Tree Construction, Fig. 1 Example training database

Scalable Decision Tree Construction, Fig. 2 Magazine subscription example classification tree

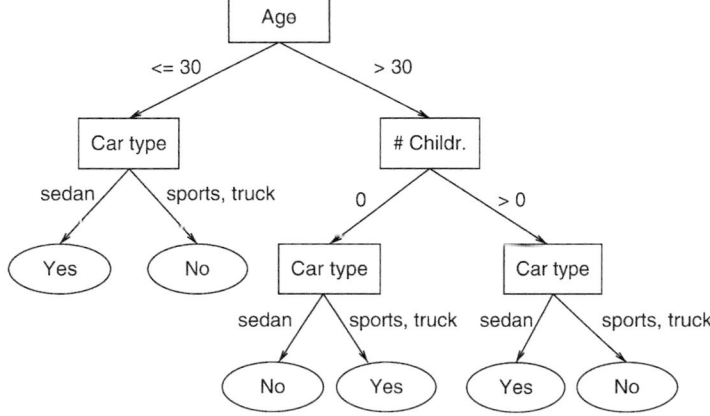

the overall decision tree construction algorithm is called the *split selection method*. The second issue is to devise a *data access method* for data management in the case that the training database is very large. During the pruning phase a third issue arises, namely how to find a good estimator $\widehat{\mathcal{R}}_T$ of R_T and how to efficiently calculate $\widehat{\mathcal{R}}_T$.

A popular class of split selection methods are *impurity-based* split selection methods [4, 12]. Impurity-based split selection methods find the splitting criterion by minimizing a concave *impurity function* imp_θ such as the entropy [12] or the *gini*-index [4]. (Arguments for the concavity of the impurity function can be found in Breiman et al. [4].) The most popular split selection methods such as CART [4] and C4.5 [12] fall into this group. At each node, all predictor attributes X are examined and the impurity imp_θ of the best split on X is calculated. The final split is chosen such that the combination of splitting attribute and splitting predicates minimizes the value of imp_θ.

Data Access

There exist many scalable data access methods for classification tree construction. Some of the issues in scalable decision tree construction are

Input: Node n, partition D, split selection method SS

Output: Decision tree for D rooted at node n

Top-down decision tree induction schema:

BuildTree (Node n, dataset D, split selection method SS)
(1)   Apply SS to D to find the splitting criterion
(2)   **if** n splits
(3)   Use best split to partition D into D_1 and D_2
(4)   BuildTree(n_1, D_1, SS)
(5)   BuildTree (n_2, D_2, SS)
(6)   **endif**

Scalable Decision Tree Construction, Fig. 3
Classification tree construction

Scalable Decision Tree Construction, Fig. 4
RainForest refinement

introduced by briefly discussing one method, RainForest [6].

An examination of the split selection methods in the literature reveals that the greedy schema can be refined to the generic *RainForest Tree Induction Schema* shown in Fig. 4. A broad class of split selection methods, namely those that generate splitting criteria involving a single splitting attribute, proceed according to this generic schema. Split selection methods that generate linear combination splits cannot be captured by RainForest. Consider a node n of the decision tree. The split selection method has to make two decisions while examining the family of n: (i) It has to select the splitting attribute X, and (ii) it has to select the splitting predicates on X. Once decided on the splitting criterion, the algorithm is recursively applied to each of the children of n. Denote by SS a representative split selection method.

Note that at a node n, the utility of a predictor attribute X as a possible splitting attribute is examined independent of the other predictor attributes: The *sufficient statistics* are the class label distributions for each distinct attribute value of X. Define the *AVC-set* of a predictor attribute X at node n to be the projection of F_n onto X and the class label where counts of the individual class labels are aggregated. Denote the AVC-set of predictor attribute X at node n by $AVC_n(X)$. (The acronym AVC stands for **A**ttribute-**V**alue, **C**lasslabel.) To give a formal definition, let $a_{n,X,x,i}$ be the number of records t in F_n with attribute value $t.X = x$ and class label $t.C = i$. Formally,

$$a_{n,X,x,i} \overset{\text{def}}{=} |\{t \in F_n : t.X = x \ \wedge \ t.C = i\}|$$

For a predictor attribute X, let $S \overset{\text{def}}{=} dom(X) \times \mathbf{N}^J$ where $\mathbf{N}$ denotes the set of natural numbers. Then

RainForest refinement to the schema in Fig 3:

(1a)   **for** each predictor attribute X
(1b)   Construct the AVC-set of X
(1c)   Call SS.find_best_partitioning(AVC-set of X)
(1d)   **endfor**
(2a) SS.decide_splitting_criterion();
(2b)   **if** n splits...

Scalable Decision Tree Construction, Fig. 5
Rainforest AVC-sets

Car	Subscription	
	Yes	No
sedan	5	2
sports	0	4
truck	1	2

Age	Subscription	
	Yes	No
23	1	1
25	1	2
30	1	1
31	1	1
36	0	3
45	2	0

$$\mathrm{AVC}_n(X) \overset{\text{def}}{=} \{(x, a_1, .., a_J) \in S : \exists t \in F_n :$$

$$(t.X = x \wedge \forall i \in \{1, .., J\} : a_i = a_n, X, x, i)\}.$$

Car	Subscription	
	Yes	No
sedan	3	0
sports	0	3
truck	0	1

Age	Subscription	
	Yes	No
23	1	1
25	1	2
30	1	1

Scalable Decision Tree Construction, Fig. 6
Rainforest AVC-sets of the left child of the root node

Define the *AVC-group* of a node n to be the set of the AVC-sets of all predictor attributes at node n. Note that the size of the AVC-set of a predictor attribute X at node n depends only on the number of distinct attribute values of X and the number of class labels in F_n.

As an example, consider the training database shown in Fig. 1. The AVC group of the root node is depicted in Fig. 5. Assume that the root node splits as shown in Fig. 2. The AVC-group of the left child node of the root node is shown in Fig. 5 and the AVC-group of the left child node of the root node is shown in Fig. 6.

If the training database is stored inside a database system, the AVC-set of a node n for predictor attribute X can be retrieved through a simple SQL-query:

```
SELECT      D.X, D.C, COUNT (*)
FROM        D
WHERE       f_n
GROUP BY    D, X, D.C
```

In order to construct the AVC-sets of all predictor attributes at a node n, a *UNION*-query would be necessary. (In this case, the *SELECT* clause needs to retrieve also some identifier of the attribute in order to distinguish individual AVC-sets.) Graefe et al. observe that most database systems evaluate the *UNION*-query through several scans and introduce a new operator that allows gathering of sufficient statistics in one database scan [7].

Based on this observation, there exist several algorithms that construct as many AVC-sets as possible in main memory while minimizing the number of scans over the training database. As an example of the simplest such algorithm, assume that the complete AVC-group of the root node fits into main memory. Then the tree can be constructed according to the following simple schema: Read the training database D and construct the AVC-group of the root node n in-memory. Then determine the splitting criterion from the AVC-sets through an in-memory computation. Then make a second pass over D and partition D into children partitions D_1 and D_2. This simple algorithm reads the complete training database twice and writes the training database once per level of the tree; more sophisticated algorithms are possible [6]. Experiments show that RainForest outperforms SPRINT on the average by a factor of three. Note that RainForest has a large memory requirement: RainForest is only applicable if the AVC-group of the root node fits in-memory (this requirement can be relaxed through more sophisticated memory management [6]).

Tree Pruning

The pruning phase of classification tree construction decides on the tree of the right size in order to prevent overfitting to minimize the misclassification error R_T (P). In bottom-up pruning, in the tree growth phase the tree is grown until the size of the family of each leaf node n falls below a user-defined threshold c; the pruning phase follows the growth phase. Examples of bottom-up pruning strategies are cost-complexity pruning, pruning with an additional set of records called a test set [4], and pruning based on the MDL-principle [9]. In top-down pruning, during the growth phase a statistic s_n is computed at each node n, and based on the value of s_n, tree growth at node n is continued or stopped [13]. Bottom-up pruning results usually in trees of higher quality [4, 8, 13], but top-down pruning is computationally more efficient since no parts of the tree are first constructed and later discarded.

The section describes the PUBLIC pruning algorithm [14], an algorithm that integrates bottom-up pruning into the tree growth phase; thus PUBLIC preserves the computational advantages of top-down pruning while preserving the good properties of top-down pruning. PUBLIC uses pruning based on the MDL principle [9], which sees the classification tree as a means to encode the values of the class label attribute given the predictor attributes $X_1, \ldots, X_m$. The MDL principle states that the "best" classification tree is the tree can be encoded with the least number of bits. Thus there needs to be an encoding schema that allows encoding of any binary decision tree. Given an encoding schema, a classification tree can be pruned by selecting the subtree with minimum code length.

In the MDL encoding schema for binary splitting predicates from Mehta et al. [9], each node requires one bit to encode its type (leaf or intermediate node). An intermediate node n needs to encode its splitting criterion, consisting of the splitting attribute X (log m bits since there are m predictor attributes) and splitting predicate. Let X be the splitting attribute at node n and assume that X has v different attribute values. If X is a numerical attribute, the split will be of the form $X \leq c$. Since c can take $v - 1$ different values, encoding of the split point c requires $\log(v - 1)$ bits. If X is a categorical attribute, the split will be of the form

$X \in Y$. Since Y can take $2^v - 2$ different values, the encoding requires $\log(2^v - 2)$ bits. Denote the cost of a split at a node n by C_{Split} (n). For a leaf node n, Metha et al. show that the cost of encoding the leaf is [9]:

$$C_{\text{leaf}}(n) = \sum_i n_i \log \frac{|F_n|}{|F_n^i|} + \frac{k-1}{2} \log \frac{|F_n|}{2}$$

$$+ \log \frac{\pi^{k/2}}{\Gamma(k/2)}.$$

Given this encoding schema, a fully grown tree can be pruned bottom-up by deciding for each node whether it should be pruned or whether it should remain [10].

The PUBLIC algorithm integrates the building and pruning phase by computing a lower bound $L(n)$ on the MDL-cost of any subtree rooted at a node n. A trivial lower bound is $L(n) = 1$ (for the encoding of n). Rastogi and Shim give in their PUBLIC algorithms more sophisticated lower bounds including the following:

PUBLIC Lower Bound

Consider a (sub-)tree T with $s > 1$ nodes rooted at node n. Then the cost $C(n)$ of encoding T has the following lower bound [14]:

$$C(n) \geq 2 \cdot (s-1) + 1 + (s-1) \cdot \log m$$

$$+ \sum_{i=s+1}^{k} |F_n^i|$$

With this lower bound the MDL Pruning Schema can be used even during top-down tree construction. PUBLIC distinguishes two different types of leaf nodes: "True" leaf nodes that are the result of pruning or that cannot be expanded any further, and "intermediate" leaf nodes n, where the subtree rooted at n might be grown further. During the growth phase, the PUBLIC Pruning Schema is executed from the root node of the tree. Rastogi and Shim show experimentally that integration of pruning with the growth phase of the tree results in significant savings in overall tree construction. More sophisticated ways of integrating tree growth with pruning in addition to tighter lower bounds are possible [14].

Key Applications

Classification has a wide range of applications, including scientific experiments, medical diagnosis, fraud detection, credit approval, and target marketing.

Recommended Reading

1. Agrawal R, Ghosh SP, Imielinski T, Iyer BR, Swami AN. An interval classifier for database mining applications. In: Proceedings of the 18th International Conference on Very Large Data Bases; 1992. p. 560–73.
2. Agrawal R, Imielinski T, Swami AN. Database mining: a performance perspective. IEEE Trans Knowl Data Eng. 1993;5(6):914–25.
3. Alsabti K, Ranka S, Singh V. Clouds: a decision tree classifier for large datasets. In: Proceeding of the 4th International Conference on Knowledge Discovery and Data Mining. 1998. p. 2–8.
4. Breiman L, Friedman JH, Olshen RA, Stone CJ. Classification and regression trees. Wadsworth: Belmont; 1984.
5. Gehrke J, Ganti V, Ramakrishnan R, Loh W-Y. BOAT – optimistic decision tree construction. In: Proceedings of the ACM SIGMOD International Conference on Management of Data; 1999. p. 169–80.
6. Gehrke J, Ramakrishnan R, Ganti V. Rainforest – a framework for fast decision tree construction of large datasets. Data Min Knowl Dis. 2000;4(2/3):127–62.
7. Graefe G, Fayyad U, Chaudhuri S. On the efficient gathering of sufficient statistics for classification from large SQL databases. In: Proceedings of the 4th International Conference on Knowledge Discovery and Data Mining; 1998. p. 204–8.
8. Lim T-S, Loh W-Y, Shih Y-S. A comparison of prediction accuracy, complexity, and training time of 33 old and new classification algorithms. Mach Learn. 2000;40(3):203–28.
9. Mehta M, Rissanen J, Agrawal R. MDL-based decision tree pruning. In: Proceedings of the 1st International Conference on Knowledge Discovery and Data Mining; 1995.
10. Mehta M, Agrawal R, Rissanen J. SLIQ: a fast scalable classifier for data mining. In: Advances in Database Technology, Proceedings of the 5th International Conference on Extending Database Technology; 1996.
11. Murthy SK. Automatic construction of decision trees from data: a multi-disciplinary survey. Data Min Knowl Dis. 1998;2(4):345–89.
12. Quinlan JR. Induction of decision trees. Mach Learn. 1986;1(1):81–106.
13. Quinlan JR. C4.5: programs for machine learning. San Mateo: Morgan Kaufman; 1993.
14. Rastogi R, Shim K. PUBLIC: a decision tree classifier that integrates building and pruning. In: Proceedings of the 24th International Conference on Very Large Data Bases; 1998. p. 404–15.
15. Shafer J, Agrawal R, Mehta M. SPRINT: a scalable parallel classifier for data mining. In: Proceedings of the 22th International Conference on Very Large Data Bases; 1996.
16. Sreenivas MK, AlSabti K, Ranka S. Parallel out-of-core decision tree classiers. In: Kargupta H, Chan P, editors. Advances in distributed and parallel knowledge discovery. Cambridge, MA: AAAI; 2000. p. 317–36.
17. Srivastava A, Han E, Kumar V, Singh V. Parallel formulations of decision-tree classication algorithms. Data Min Knowl Disc. 1999;3(3):237–261.

Scheduler

Nathaniel Palmer
Workflow Management Coalition, Hingham, MA, USA

Synonyms

Queuing mechanism; Workflow scheduler

Definition

The mechanism that identifies and initiates the sequence for activities and work items are executed.

Key Points

A Scheduler initiates work assignments based on precedence relationships and the state of a workflow instance. A Scheduler is not a "queue" which is typically a set of work items or activities waiting to be scheduled, however, a queue of items may be managed by the Scheduler. The role of the Scheduler is to minimize queue time and optimize executive efficiency.

Cross-References

▶ Activity

Scheduling Strategies for Data Stream Processing

Mohamed Sharaf[1] and Alexandros Labrinidis[2]
[1]Electrical and Computer Engineering, University of Toronto, Toronto, ON, Canada
[2]Department of Computer Science, University of Pittsburgh, Pittsburgh, PA, USA

Synonyms

Continuous query scheduling; Operator scheduling; Scheduling policies

Definition

In a Data Stream Management System (DSMS), data arrives in the form of continuous streams from different data sources, where the arrival of new data triggers the execution of multiple continuous queries (CQs). The order in which CQs are executed in response to the arrival of new data is determined by the CQ scheduler. Thus, one of the main goals in the design of a DSMS is the development of scheduling policies that leverage CQ characteristics to optimize the DSMS performance.

Historical Background

The growing need for *monitoring applications* [8] has forced an evolution on data processing paradigms, moving from Database Management Systems (DBMSs) to Data Stream Management Systems (DSMSs) [4, 11]. Traditional DBMSs employ a store-and-then-query data processing paradigm, where data are stored in the database and queries are submitted by the users to be answered in full, based on the current snapshot of the database. In contrast, in DSMSs, monitoring applications register continuous queries which continuously process unbounded data streams looking for data that represent events of interest to the end-user.

The data stream concept permeated the data management research community in the mid- to late 90's, with general-purpose research prototypes of data stream management systems materializing shortly afterwards, for example Aurora [8], TelegraphCQ [10] and STREAMS [5].

Scheduling is one of the fundamental research challenges for effective data stream management systems; as such, it has received a lot of attention, with early works on scheduling in 2003 [2, 9].

Foundations

System Model

A continuous query evaluation plan can be conceptualized as a data flow tree [2, 8], where the nodes are operators that process tuples and edges represent the flow of tuples from one operator to another (Fig. 1). An edge from operator O_x to operator O_y means that the output of O_x is an input to O_y. Each operator is associated with a *queue* where input tuples are buffered until they are processed.

Multiple queries with common sub-expressions are usually merged together to eliminate the repetition of similar operations. For example, Fig. 1 shows the global plan for two queries Q_1 and Q_2. Both queries operate on data streams M_1 and M_2 and they share the common sub-expression represented by operators O_1, O_2 and O_3, as illustrated by the half-shaded pattern for these operators.

A *single-stream query* Q_k has a single *leaf* operator Q_l^k and a single *root* operator Q_r^k, whereas a *multi-stream* query has a single root operator and more than one leaf operators. In a query plan

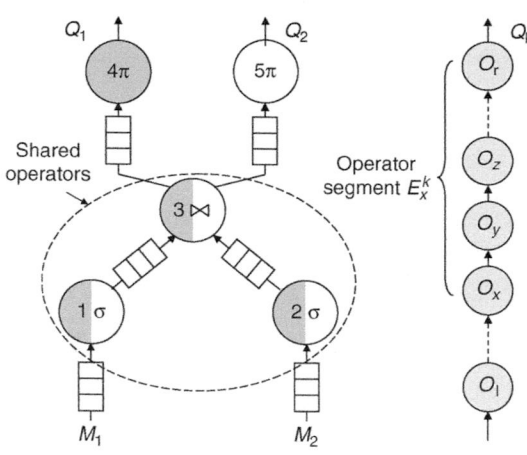

Scheduling Strategies for Data Stream Processing, Fig. 1 Continuous queries plans

Q_k, an *operator segment* $E^k_{x,y}$ is the sequence of operators that starts at O^k_x and ends at O^k_x. If the last operator on $E^k_{x,y}$ is the root operator, then that operator segment is simply denoted as E^k_x. For example, in Fig. 1, $E^1_1 = < O_1, O_3, O_4 >$, whereas $E^2_1 = < O_1, O_3, O_5 >$.

In a query, each operator $O_x^{\ k}$ (or simply O_x) is associated with two parameters:

1. *Processing cost* or *Processing time* (c_x) is the amount of time needed to process an input tuple.
2. *Selectivity* or *Productivity* (s_x) is the number of tuples produced after processing one tuple for c_x time units. s_x is less than or equal to 1 for a filter operator and it could be greater than 1 for a join operator.

Multiple CQ Scheduling

At the arrival of new data, the MCQ scheduler decides the execution order of CQs, or more precisely, the execution order of operators within CQs. The execution order is decided with the objective of optimizing the DSMS performance under certain metrics. Towards this, the scheduler assigns a priority to each operator and operators are executed according to these priorities.

For a single-stream query Q_k which consists of operators $O^k_l, \ldots, O^k_x, O^k_y, \ldots, O^k_r >$ (Fig. 1), the function for computing the priority of operator

O^k_x typically involves one or more of the following parameters:

- *Operator Global Selectivity* (S^k_x) *is the number of tuples produced at the root* O^k_r *after processing one tuple along operator segment* E^k_x.

$$S^k_x = s^k_x \times s^k_y \times \cdots \times s^k_r$$

- *Operator Global Average Cost* $(\overline{C}^k_x)$ *is the expected time required to process a tuple along an operator segment* C^k_x.

$$\overline{C}^k_x = \left(c^k_x\right) + \left(c^k_y \times s^k_x\right) + \cdots \\ + \left(c^k_r \times s^k_{r-1} \times \cdots \times s^k_x\right)$$

If O^k_x is a leaf operator $(x = l)$, when a processed tuple actually satisfies all the filters in E^k_l, then $\overline{C}^k_l$ represents the ideal total processing cost or time incurred by any tuple *produced* or *emitted* by query Q_k. In this case, $\overline{C}^k_l$ is denoted as T_k:

- *Tuple Processing Time* (T_k) is the ideal total processing cost required to produce a tuple by query Q_k.

$$T_k = c^k_l + \cdots + c^k_x + c^k_y + + c^k_r$$

The exact priority function depends on the performance metric to optimize, and in turn on the employed scheduling strategy.

Metrics and Strategies

Response Time: Processing a tuple by a CQ might lead to discarding it (if it does not satisfy some filter predicate) or it might lead to producing one or more tuples at the output, which means that the input tuple represents an event of interest to the user who registered the CQ. Clearly, in DSMSs, it is more appropriate to define response time from a data/event perspective rather than from a query perspective as in traditional DBMSs. Hence, the *tuple response time* or *tuple latency* is defined as follows:

Definition 1

Tuple response time, R_i, for tuple t_i is $R_i = D_i - A_i$, where A_i is t_i's arrival time and D_i is t_i's output time. Accordingly, the average response time for N tuples is: $\frac{1}{N}\sum_{i=1}^{N} R_i$.

For a single CQ over multiple data streams, the *Rate-based* policy (*RB*) has been shown to improve the average response time of tuples processed by that CQ [17].

For multiple CQs, the Aurora DSMS [9], uses a two-level scheduling strategy where *Round Robin (RR)* is used to schedule queries and *RB* is used to schedule operators within the query. The work in [14] proposes the *Highest Rate* policy (*HR*) which extends the *RB* to schedule both queries and operators. Basically, *HR* views the network of multiple queries as a set of operators and at each scheduling point it selects for execution the operator with the highest priority (i.e., output rate).

Specifically, under *HR*, each operator O_x^k is assigned a value called *global output rate* (GR_x^k). The output rate of an operator is basically the expected number of tuples produced per time unit due to processing one tuple by the operators along the operator segment starting at O_x^k all the way to the root O_r^k. Formally, the output rate of operator O_x^k is defined as follows:

$$GR_x^k = \frac{S_x^k}{\overline{C}_x^k} \quad (1)$$

where S_x^k and $\overline{C}_x^k$ are the operator's global selectivity and global average cost as defined above. The intuition underlying *HR* is to give higher priority to operator paths that are both productive and inexpensive. In other words, the highest priority is given to the operator paths with the minimum latency for producing one tuple.

Slowdown: Under a heterogeneous workload, the processing requirements for different tuples may vary significantly and average response time is not an appropriate metric, since it cannot relate the time spent by a tuple in the system to its processing requirements. Given this realization, other on-line systems with heterogeneous workloads such as DBMSs, OSs, and Web servers have adopted *average slowdown* or *stretch* [13] as another metric. This motivated considering the stretch metric in [14].

The definition of slowdown was initiated by the database community in [12] for measuring the performance of a DBMS executing multi-class workloads. Formally, the slowdown of a job is the ratio between the time a job spends in the system to its processing demands [13]. In a DSMS, the slowdown of a tuple is defined as follows [14]:

Definition 2

The slowdown, H_i, for tuple t_i produced by query Q_k is $H_i = \frac{R_i}{T_k}$, where R_i is t_i's response time and T_k is its ideal processing time. Accordingly, the average slowdown for N tuples is: $\frac{1}{N}\sum_{i=1}^{N} H_i$.

Intuitively, in a general purpose DSMS where all events are of equal importance, a simple event (i.e., an event detected by a low-cost CQ) should be detected faster than a complex event (i.e., an event detected by a high-cost CQ) since the latter contributes more to the load on the DSMS.

The *HR* policy schedules jobs in descending order of output rate which might result in a high average slowdown because a low-cost query can be assigned a low priority since it is not productive enough. Those few tuples produced by this query will all experience a high slowdown, with a corresponding increase in the average slowdown of the DSMS.

The work in [14] proposes the *Highest Normalized Rate (HNR)* policy for minimizing the slowdown in a DSMS. Under HNR, each operator O_x^k is assigned a priority V_x^k which is the *weighted rate* or *normalized rate* of the operator segment E_x^k that starts at operator O_x^k and it is defined as:

$$V_x^k = \frac{1}{T_k} \times \frac{S_x^k}{\overline{C}_x^k} \quad (2)$$

The *HNR* policy, like *HR*, is based on output rate, however, it also emphasizes the ideal tuple processing time in assigning priorities. As such, an inexpensive operator segment with low productivity will get a higher priority under *HNR* than under *HR*.

Worst-Case Performance: It is expected that a scheduling policy that strives to minimize the average-case performance might lead to a poor worst-case performance under a relatively high load. That is, some queries (or tuples) might starve under such a policy. The worst-case performance is typically measured using *maximum response time* or *maximum slowdown* [7].

Intuitively, a policy that optimizes for the worst-case performance should be pessimistic. That is, it assumes the worst-case scenario where each processed tuple will satisfy all the filters in the corresponding query.

The work in [14] shows that the traditional *First-Come-First-Serve (FCFS)* minimizes the maximum response time. Similarly, it shows that the traditional *Longest Stretch First (LSF)* [1] optimizes the maximum slowdown.

Average- vs. Worst-Case Performance: On one hand, the average value for a QoS metric provided by the system represents the expected QoS experienced by any tuple in the system (i.e., the average-case performance). On the other hand, the maximum value measures the worst QoS experienced by some tuple in the system (i.e., the worst-case performance). It is known that each of these metrics by itself is not enough to fully characterize system performance.

The most common way to capture the trade-off between the average-case and the worst-case performance is to measure the ℓ_2 norm [6]. For instance, the ℓ_2 norm of response times, R_i, is defined as:

Definition 3
The ℓ_2 norm of response times for N tuples is equal to $\sqrt{\sum_N^1 R_i^2}$.

The definition shows that the ℓ_2 norm considers the average in the sense that it takes into account all values, yet, by considering the second norm of each value instead of the first norm, it penalizes more severely outliers compared to the average metrics.

In order to balance the trade-off between the average- and worst-case performance, the *Balance Slowdown (BSD)* and the *Balance Response Time (BRT)* policies have been proposed in [14]. To avoid starvation, the two policies consider the amount of time an operator O_x^k has been waiting for scheduling (i.e., W_x^k). Specifically, under *BSD*, each operator O_x^k is assigned a priority value V_x^k which is the product of the operator's normalized rate and the current highest slowdown of its pending tuples. That is:

$$V_x^k = \left(\frac{S_x^k}{\overline{C}_x^k T_k} \right) \left(\frac{W_x^k}{T_k} \right) \tag{3}$$

As such, under *BSD*, an operator is selected either because it has a high weighted rate or because its pending tuples have acquired a high slowdown.

Application-Specific QoS: Aurora also proposes a QoS-aware scheduler which attempts to satisfy application-specified QoS requirements [9]. Specifically, under that QoS-aware scheduler, each query is associated with a QoS graph which defines the utility of stale output.

Given, a QoS graph, the scheduler computes for each operator a *utility* value which is basically the slope of the QoS graph at the tuple's output time. The scheduler also computes for each operator its *urgency* value which is an estimation of how close is an operator to a critical point on the QoS graph where the QoS changes sharply. Then, at each scheduling point, the scheduler chooses for execution the operators with the highest utility value and among those that have the same utility, it chooses the one that has the highest urgency.

Memory Usage: Multi-query scheduling has also been exploited to optimize metrics beyond QoS. For example, *Chain* is a multi-query scheduling policy that optimizes memory usage in order to minimize space requirements for buffering tuples [2]. Towards this, for each query plan, *Chain* constructs what is called a *progress chart*. A progress chart is basically a set of segments where the slope of each segment represents the rate of change in the size of a tuple being processed by a set of consecutive operators along the query plan. Given that progress chart, at each scheduling point, *Chain* schedules for execution the tuple that lies on the segment with the steepest slope. The intuition is to give higher priority to segments of operators with higher tuple consumption rate which will lead to quickly freeing more memory.

Quality of Data (QoD): Another metric to optimize is Quality of Data (*QoD*). For instance, the work in [15] proposes the *freshness-aware* scheduling policy for improving the QoD of data streams, when QoD is defined in terms of freshness. The proposed scheduler exploits the variability in query costs, divergence in arrival patterns, and the probabilistic impact of selectivity in order to maximize the freshness of output data streams.

Multiple-Objective Scheduling: In DSMSs, and in computer systems in general, it is often desirable to optimize for multiple metrics at the same time. However, those metrics might be in conflict most of the time. This motivated the proposals of schedulers that are able to balance the trade-off between certain conflicting metrics.

For instance, the work in [3] attempts to balance the trade-off between memory usage and latency by formalizing latency requirements as a constraint to the *Chain* scheduler. This formulation lead to the *Mixed* policy which can be viewed as a heuristic strategy that is intermediate between Chain and FIFO. Specifically, *Mixed* is tuned via a parameter where a high value of that parameter causes *Mixed* to behave more like *FIFO*, whereas a lower value makes it behave more like *Chain*.

In another attempt towards multiple-objective scheduling, the work in [16] proposes *AMoS* which is an Adaptive Multi-objective Scheduling selection framework. Given several scheduling algorithms, AMoS employs a learning mechanism to learn the behavior of the scheduling algorithms over time. It then uses the learned knowledge to continuously select the algorithm that has statistically performed the best.

Scheduler Implementation: To ensure the applicability of scheduling policies in DSMSs, a low-overhead implementation is needed in order to reduce the amount of computation involved in computing priorities. For static policies (i.e., policies where an operator priority is constant over time), priorities are computed only once when a query is registered in the DSMS which naturally leads to a low-overhead implementation. Examples of such static policies include *HR*, *HNR*, and *Chain*. On the other hand, for dynamic policies where priority is a function of time, the priority of each operator should be re-computed at each instant of time. Such a naive implementation renders that class of policies very impractical. This motivated several approximation methods for efficient implementation of dynamic policies to balance the trade-off between scheduling overhead and accuracy. For instance the work in [9] proposes using bucketing as well as pre-computation for an efficient implementation of the QoS-aware scheduling in Aurora. Similarly, [14] proposes using search space reduction and pruning methods in addition to clustered processing of continuous queries.

Key Applications

There is a plethora of applications that require data stream management systems and, as such, proper scheduling strategies. The most well-known class of applications is that of *monitoring applications* [8], be it environmental monitoring (e.g., via sensor networks), network monitoring (e.g., by collecting router data), or even financial monitoring (e.g., by observing stock-market data). In all such cases, the sheer amount of input data precipitates the use of the data stream processing paradigm and proper scheduling strategies.

Cross-References

▶ Adaptive Query Processing
▶ Adaptive Stream Processing
▶ Data Stream
▶ Event Stream
▶ Stream Processing
▶ Streaming Applications
▶ Stream-Oriented Query Languages and Operators

Recommended Reading

1. Acharya S, Muthukrishnan S. Scheduling on-demand broadcasts: new metrics and algorithms. In: Proceedings of the 4th Annual International Conference on Mobile Computing and Networking; 1998.
2. Babcock B, Babu S, Datar M, Motwani R. Chain: operator scheduling for memory minimization in data stream systems. In: Proceedings of the ACM SIGMOD International Conference on Management of Data; 2003.
3. Babcock B, Babu S, Datar M, Motwani R, Thomas D. Operator scheduling in data stream systems. VLDB J. 2004;13(4):333–53.
4. Babcock B, Babu S, Datar M, Motwani R, Widom J. Models and issues in data stream systems. In: Proceedings of the ACM SIGMOD International Conference on Management of Data; 2002.
5. Babu S, Widom J. Continuous queries over data streams. ACM SIGMOD Rec. 2001;30(3):109–120.
6. Bansal N, Pruhs K. Server scheduling in the Lp norm: a rising tide lifts all boats. In: Proceedings of the 35th Annual ACM Symposium on Theory of Computing; 2003.
7. Bender MA, Chakrabarti S, Muthukrishnan S. Flow and stretch metrics for scheduling continuous job streams. In: Proceedings of the 9th Annual ACM-SIAM Symposium on Discrete Algorithms; 1998.
8. Carney D, Cetintemel U, Cherniack M, Convey C, Lee S, Seidman G, Stonebraker M, Tatbul N, Zdonik S. Monitoring streams: a new class of data management applications. In: Proceedings of the 28th International Conference on Very Large Data Bases; 2002.
9. Carney D, Cetintemel U, Rasin A, Zdonik S, Cherniack M, Stonebraker M. Operator scheduling in a data stream manager. In: Proceedings of the 29th International Conference on Very Large Data Bases; 2003.
10. Chandrasekaran S, Cooper O, Deshpande A, Franklin MJ, Hellerstein JM, Hong W, Krishnamurthy S, Madden S, Raman V, Reiss F, Shah MA. TelegraphCQ: continuous dataflow processing for an uncertain world. In: Proceedings of the 1st Biennial Conference on Innovative Data Systems Research; 2003.
11. Golab L, Özsu MT. Issues in data stream management. ACM SIGMOD Rec. 2003;32(2):5–14.
12. Mehta M, DeWitt DJ. Dynamic memory allocation for multiple-query workloads. In: Proceedings of the 19th International Conference on Very Large Data Bases; 1993.
13. Muthukrishnan S, Rajaraman R, Shaheen A, Gehrke J.E. Online scheduling to minimize average stretch. In: Proceedings of the 40th Annual Symposium on Foundations of Computer Science; 1999.
14. Sharaf MA, Chrysanthis PK, Labrinidis A, Pruhs K. Efficient scheduling of heterogeneous continuous queries. In: Proceedings of the 32nd International Conference on Very Large Data Bases; 2006.
15. Sharaf MA, Labrinidis A, Chrysanthis PK, Pruhs K. Freshness-aware scheduling of continuous queries in the Dynamic Web. In: Proceedings of the 8th International Workshop on the World Wide Web and Database; 2005.
16. Sutherland T, Pielech B, Zhu Y, Ding L, Rundensteiner EA. An adaptive multi-objective scheduling selection framework for continuous query processing. In: Proceedings of the International Database Engineering and Applications Symposium; 2005.
17. Urhan T, Franklin M.J. Dynamic pipeline scheduling for improving interactive query performance. In: Proceedings of the 27th International Conference on Very Large Data Bases; 2001.

Schema Evolution

John F. Roddick
Flinders University, Adelaide, SA, Australia

Definition

Schema evolution deals with the need to retain current data when database schema changes are performed. Formally, *Schema Evolution* is accommodated when a database system facilitates database schema modification without the loss of existing data, (q.v. the stronger concept of *Schema Versioning*) (Schema evolution and schema versioning has been conflated in the literature with the two terms occasionally being used interchangeably. Readers are thus also encouraged to read also the entry for *Schema Versioning*.).

Historical Background

Since schemata change and/or multiple schemata are often required, there is a need to ensure that extant data either stays consistent with the revised schema or is explicitly deleted as part of the change process. A database that supports schema evolution supports this transformation process.

The first schema evolutioning proposals discussed database conversion primarily in terms of a set of transformations from one schema to another [10]. These transformations focused on the relational structure of the database and included name changing, changing the membership of keys, composing and decomposing relations both vertically and horizontally and so on. In all cases only one schema remained and all data (that still remained) was coerced (ie. copied from one type to another) to the new structure.

Schema evolution has also been covered in the proposals to manage issues such as data coercion [5, 12], authority control [2] and query language support [9].

Foundations

Schema evolution is related to the view-update problem, discussed in-depth when the relational model was introduced [1], and is strongly linked to the notion of information capacity [4, 7]. Specifically, non-loss evolution can only be guaranteed when the information capacity of the new schema exceeds that of the existing schema. Formally, if $I(S)$ is the set of all valid instances of S, then for non-loss evolution $I(S_{new}) \supseteq I(S_{old})$. One novel solution is the integration of schema evolution with the database view facilities. When new requirements demand schema updates for a particular user, then the user specifies schema changes to a personal view, rather than to the shared base schema [8].

Once a schema change is accepted, the common procedure is for the underlying instances to be coerced to the new structure. Since the old schema is obsolete, this presents few problems and is conceptually simple. However, results in an inability to reverse schema amendments. Schema versioning support provides two other options (q.v.).

Four classes of schema evolution can be envisaged. Each type brings different problems.

1. *Attribute Evolution* occurs when attributes are added to, deleted from, or renamed in a relation. Issues here include the values

to be ascribed to attributes in tuples stored under a new version that does not possess the attribute.

2. *Domain Evolution* occurs when the domain over which an attribute is defined is altered. Issues here include implying accuracy that does not exist in existing data when, for example, attributes defined as integers are converted to reals, and in truncation when character fields are shortened.

3. *Relation Evolution* occurs when the relational structure is altered through the definition, deletion, decomposition or merging of a relation. Such changes are almost always irreversible.

4. *Key Evolution* occurs when the structure of a primary key is altered or when foreign keys are added or removed. The issues here can be quite complex. For example, removing an attribute from a primary key may not violate the primary key uniqueness constraint for current data (the amendment can be rejected if it does) but in a temporal database may still do so for historical information.

Note that one change may involve more than one type of evolution, such as changing the domain of a key attribute.

These changes may also be reflected in the conceptual model of the system. For example, the addition of an entity in an EER diagram would result in the addition of a relation in the underlying relational model; deleting a 1-to-many relationship would remove a foreign key constraint, and so on.

Key Applications

Schema changes are linked to either error correction or design change. It is therefore useful if the design decisions can be consulted and the users can interact with schema changes at a high level. One way is to propagate requirements changes to database schemas [3] or provide better support for metadata management by providing a higher level view in which models can be mapped to each other [6].

In order to quantify the types of schema evolution, Sjøberg [11] investigated change to a database system over 18 months, covering 6 months of development and 12 months of field trials. A more recent study complements this by following the changes in an established database system over many years [13].

Future Directions

The major directions for schema versioning research have moved from low-level handling of syntactic elemental changes (such as adding an attribute or demoting an index attribute) to more model-directed semantic handling of change (such as propagating changes in a conceptual model to a database schema) [3]. Research has also moved from schema evolution to the more complex problem of providing versions of schema.

Cross-References

► Schema Versioning
► Temporal Algebras
► Temporal Conceptual Models
► Temporal Query Languages

Recommended Reading

1. Bancilhon F, Spyratos N. Update semantics of relational views. ACM Trans Database Syst. 1981;6(4):557–75.
2. Bretl R, Maier D, Otis A, Penney J, Schuchardt B, Stein J, Williams EH, Williams M. The GemStone data management system. In: Kim W, Lochovsky F, editors. Object-oriented concepts, databases and applications. New York: ACM; 1989. p. 283–308.
3. Hick JM, Hainaut JL. Database application evolution: a transformational approach. Data Knowl Eng. 2006;59(3):534–58.
4. Hull R. Relative information capacity of simple relational database schemata. Soc Ind Appl Math. 1986;15(3):856–86.
5. Kim W, Chou H.T. Versions of schema for object-oriented databases. In: Proceedings of the 24th International Conference on Very Large Data Bases; 1988. p. 148–59.
6. Melnik S, Rahm E, Bernstein PA. Rondo: a programming platform for generic model management. In: Proceedings of the ACM SIGMOD International Conference on Management of Data; 2003. p. 193–204.
7. Miller R, Ioannidis Y, Ramakrishnan R. The use of information capacity in schema integration and translation. In: Proceedings of the 19th International Conference on Very Large Data Bases; 1993. p. 120–33.
8. Ra YG, Rundensteiner EA. A transparent schema-evolution system based on object-oriented view technology. IEEE Trans Knowl Data Eng. 1997;9(4):600–24.
9. Roddick JF. SQL/SE – a query language extension for databases supporting schema evolution. ACM SIGMOD Rec. 1992;21(3):10–6.
10. Shneiderman B, Thomas G. An architecture for automatic relational database system conversion. ACM Trans Database Syst. 1982;7(2):235–57.
11. Sjøberg D. Quantifying schema evolution. Inf Softw Technol. 1993;35(1):35–44.
12. Tan L, Katayama T. Meta operations for type management in object-oriented databases - a lazy mechanism for schema evolution. In: Proceedings of the 1st International Conference on Deductive and Object-Oriented Databases; 1989. p. 241–58.
13. de Vries D, Roddick JF. The case for mesodata: an empirical investigation of an evolving database system. Inf Softw Technol. 2007;49(9–10):1061–72.

Schema Mapping

Ariel Fuxman[1] and Renée J. Miller[2]
[1]Microsoft Research, Mountain View, CA, USA
[2]Department of Computer Science, University of Toronto, Toronto, ON, Canada

Synonyms

Mapping

Definition

The problem of establishing associations between data structured under different schemas is at the core of many data integration and data sharing tasks. *Schema mappings* establish semantic connections between schemas. Given a source schema $\mathbf{S}$ and a target schema $\mathbf{T}$, a *schema mapping* $\mathcal{M}$ is a specification of a relation between instances of $\mathbf{S}$ and instances of $\mathbf{T}$. Given an instance of the source I and an instance of the target J that satisfy the mapping, say that $(I, J)| = \mathcal{M}$. Research on schema mapping has focused on the formal specification of schema mappings, the semantics of mappings, along with techniques for creating schema mappings.

Historical Background

Schema mappings have been developed primarily to solve two different problems, each of which has led to a substantial body of research: *data integration* [11] and *data exchange* [4]. In both problems, one is given a source schema $\mathbf{S}$ (or a set of source schemas) and an instance I of $\mathbf{S}$, along with a target schema $\mathbf{T}$, which is sometimes called a global schema. A user knows the schema of the target and would like to retrieve source data by posing queries on the target. In data integration, queries posed on the global schema are translated, at query time, to the schema(s) of the local data source(s) and answered using source data. Data integration is sometimes called *virtual data integration* to emphasize that the translated source data are not materialized in the target. In contrast, in data exchange the goal is to translate the source data into a target instance that conforms to the target schema and reflects the source data as accurately as possible. Target queries are then answered using the materialized target instance.

For both problems, schema mappings are used to describe the semantic relationship between the schemas and their instances, and to determine how queries are translated (in data integration) and what is the best target instance to materialize (in data exchange).

Foundations

Semantics of Schema Mappings

Schema mappings establish semantic connections between schemas. These connections can be represented formally using logical formulas. Assume the existence of two schemas, called the source schema $\mathbf{S}$, and the target schema $\mathbf{T}$. The specifications of $\mathbf{S}$ and $\mathbf{T}$ may include a set of constraints that instances of the schema must satisfy. A *schema mapping* $\mathcal{M}$ is a specification of a relation between instances of $\mathbf{S}$ and instances of $\mathbf{T}$. Given an instance of the source I and an instance of the target J that satisfy the mapping, one could say that $(I, J) \models \mathcal{M}$. A *mapping setting* is a triple $\mathcal{S} = \langle \mathbf{S}, \mathbf{T}, \mathcal{M} \rangle$. In order to give semantics to a mapping setting, it is assumed that an instance of the source is given, and the goal is to reason about the instances of the target that satisfy the constraints imposed by the schema mapping. That is, given an I, are would like to reason about all J such that $(I, J) \models \mathcal{M}$. These target instances are called *solutions*. (This terminology comes from the data exchange literature, and was introduced by Fagin et al. [4]. However, the same concepts are equally applicable to data integration, where the target schema may be referred to as the global or mediated schema.).

Definition

[solution] Let $\mathcal{S} = \langle \mathbf{S}, \mathbf{T}, \mathcal{M} \rangle$ be a mapping setting. Let I be an instance over the source schema $\mathbf{S}$. One could say that an instance J over the target schema $\mathbf{T}$ is a *solution* for I in $\mathcal{S}$ if $\langle I, J \rangle \Big| = \mathcal{M}$.

A mapping may be a function which defines a single target instance J for each source instance I. If $\mathbf{T}$ consists of a single relation, then any view over $\mathbf{S}$ defines such a function. Alternatively, in many industrial mapping tools, a mapping is a program which, given an instance of $\mathbf{S}$, outputs an instance of $\mathbf{T}$. But in general, a mapping does not need to be a function and there are clear advantages in having a declarative specification language for mappings. Declarative mapping specifications permit easier reasoning about the relationship between mappings, something that is essential in designing, maintaining, and evolving mappings.

The most commonly used specification for schema mappings are source-to-target tuple-generating-dependencies (TGDs) which have the form

$$\forall x \, (\phi_S \, (x) \rightarrow \exists y \, \psi_T \, (x, y)),$$

where $\varphi_S \, (x)$ is a conjunction of atomic formulas over **S** and $\psi_T \, (x, y)$ is a conjunction of atomic formulas over **T**. TGDs have been generalized in some approaches to permit more general source (φ_S) and target (ψ_T) queries (formulas) in the mapping.

Since a solution is any target instance that satisfies the mapping, there may be more than one solution for a given source instance. This fact must be accounted for in the semantics of query answering. The prevalent semantics adopted in the literature is based on the notion of *certain answers*. This semantics takes the conservative approach of returning only the answers that are valid in *every* solution.

Definition

[certain answer] Let $\mathcal{S}$ be a mapping setting. Let I be a source instance such that there exists some solution for I in $\mathcal{S}$. Let q be a query. We one could that a tuple **t** is a *certain answer* to q in $\mathcal{S}$, denoted $t \in \text{certain} \, (q, I, \mathcal{S})$, if for every solution J for I in $\mathcal{S}$, it is the case that $\mathbf{t} \in q(J)$.

In addition to certain answers, other semantics have been explored in the literature, such as epistemic interpretations and probabilistic notions.

To illustrate the notions introduced so far, let **S** be a schema with relation symbol *Country* (*person, country*). Let **T** be a schema with relation symbols *Home*(*person, city*) and *Loc*(*city, country*). As an example of a mapping setting, let $\mathcal{S} = \langle \mathbf{S}, \mathbf{T}, \mathcal{M} \rangle$, where $\mathcal{M}$ consists of the following source-to-target TGD:

$$\forall p, \text{cou} \, (\mathbf{S}. \, \text{Country} \, (p, \text{cou}) \rightarrow$$
$$\exists \text{cit} \, \mathbf{T}. \, \text{Home} \, (p, \text{cit}) \wedge \mathbf{T}. \, \text{Loc} \, (\text{cit}, \text{cou}))$$

There may be more than one solution for a given instance I. For example, let $I = \{Country(john, canada)\}$. Consider $J_1 = \{Home \, (john, \, toronto), \, Loc \, (toronto, \, canada)\}$ and $J_2 = \{Home(john, \, montreal), \, Loc(montreal, \, canada)\}$. It is easy to see that both J_1 and J_2

are solutions for I in $\mathcal{S}$. The reason for this is that the mapping states that the people and countries of the source must be in the target, but the city is left unspecified.

Now, consider a query q_1 that retrieves all people from the database. Let $q_1(p) = \exists cit$: $\mathbf{T}.Home(p, cit)$. Since $\mathcal{M}$ must be satisfied by all solutions, there are tuples *Home* (*john, c*) and *Loc* (*c, canada*) for some city c, in every solution for I. Thus, (*john*) $\in q_1(J)$, for every solution J. Say that (*john*) is a certain answer to q_1, and denote this by (john) $\in$ certain $(q_1, I, \mathcal{S})$. Next, consider a query q_2 that returns all cities. Let $q_2 \, (\text{cit}) = \exists p : \mathbf{T}. \, \text{Home} \, (p, \text{cit})$. In this case, there are no certain answers to q_2. To see why, notice that there is no tuple **t** such that $\mathbf{t} \in q_2 \, (J_1) \cap q_2 \, (J_2)$. The intuition is that John is the only person in the database, but different solutions may assign him a different city (as long as it is within Canada).

Types of Schema Mappings

Sound, Complete, and Exact Mappings

A common assumption in the data integration and data exchange literature is that the mappings consist of implications, where each side of the implication contains relation symbols coming from the same schema. This results in the following three types of mappings [6, 11]. Let φ_s be a formula over the source schema, and ψ_t be a formula over the target schema. *Sound mappings* are rules of the form $\varphi_s \, (x) \rightarrow \psi_t \, (x)$; *complete mappings* are of the form $\psi_t \, (x) \rightarrow \varphi_s \, (x)$; and *exact mappings* are of the form $\varphi_s \, (x) \leftrightarrow \psi_t \, (x)$. (Variable quantifiers are omitted for generality).

There is a substantial body of work on sound mappings. Such systems are sometimes called *open* because the mapping specifies what source data *must* be in the solutions, but it does not give negative information (i.e., it does not specify what *must not* be in any solution). A setting $\mathcal{S}$ containing only sound mappings is referred to as an *open mapping setting*. If J is a solution for an instance I in an open setting $\mathcal{S}$, and J' is such that $J \subseteq J'$, then J' is a solution for I in $\mathcal{S}$. As an example, consider an open setting with the following sound mapping.

$$\forall p, \text{cou. } \mathbf{S}. \text{ Country } (p, \text{cou}) \rightarrow$$
$$\exists \text{cit. } \mathbf{T}. \text{ Home } (p, \text{cit}) \wedge \mathbf{T}. \text{ Loc } (\text{cit}, \text{cou})$$

Let $I = \{Country(john, canada)\}$. Let $J = \{Loc(calgary, canada), Home(john, ottawa), Loc(ottawa, canada)\}$, which is a solution for I in S. The tuple $Loc(calgary, canada)$ does not seem to be related to the sources, and the mapping does not *force* its addition to the solution. However, it does not forbid its inclusion in the solution either. In fact, one could add any arbitrary tuple to J, and still have a solution for I.

Research on data integration and data exchange has focused primarily on open settings. One reason for this is that, given a source instance, open settings always have a solution. From a practical standpoint, open settings are better suited than settings containing exact or complete mappings in dynamic or autonomous environments, where new sources may be added independently of other sources. With open settings, sources can be described without requiring any knowledge of the other sources, or their relationship to the target. In particular, consider the problem of adding a new data source to an existing data integration (or exchange) system. With open settings, it is not necessary to change any of the existing mappings. A mapping for a new data source cannot conflict with existing mappings.

In contrast, in mapping settings containing complete or exact rules, the addition of a new source may lead to conflicts resulting in a setting for which there may be no solutions. In fact, even for a single setting the use of complete or exact mappings may sometimes preclude the existence of a solution. The problem of deciding the existence of a solution for mapping settings has been studied by Fuxman et al. [5]. As an example of a case in which there may be no solution, consider a setting $S = \langle \mathbf{S}, \mathbf{T}, \mathcal{M} \rangle$, where $\mathcal{M}$ consists of the following rules:

$$\forall p, \text{cou. } \mathbf{S}. \text{ Country } (p, \text{cou}) \rightarrow$$
$$\exists \text{cit. } \mathbf{T}. \text{ Home } (p, \text{cit}) \wedge \mathbf{T}. \text{ Loc } (\text{cit}, \text{cou}) \quad (1)$$

$$\forall p, \text{cit}, \text{cou.} \mathbf{T}. \text{ Home } (p, \text{cit}) \wedge \mathbf{T}. \text{ Loc } (\text{cit}, \text{cou})$$
$$\rightarrow \mathbf{S}. \text{ Capital } (\text{cit}, \text{cou}) \quad (2)$$

For the source instance $I = \{Country(john, canada), Capital(washington, us)\}$, there is no solution in S. To see this, assume that there is a solution J for I in S. By rule (1) of $\mathcal{M}$, J has tuples $Home$ (john, c) and $Loc(c, canada)$, for some city c. By rule (2) of $\mathcal{M}$, I is required to have a tuple $Capital$ (canada, c) which it does not. Intuitively, there is no solution for I in S since the source has no information on what city is the capital of Canada. Notice the effect of rules (1) and (2) on the solutions. The former is a sound mapping and specifies what *must* be in the solutions; the latter is a complete mapping and constrains what can be in the solutions.

Global-as-View and Local-as-View

In many data integration systems, each relation (or element) of the target schema is defined in terms of the source schemas. In many early systems, there was also an implicit assumption that the schema mapping was a function (for example a view). This approach is known as *global-as-view*. An alternative approach, known as *local-as-view*, was later proposed in which each relation of the source schema is defined in terms of the target schema. For relational systems, these notions can be defined formally as follows [11].

- In *global-as-view* systems (GAV), mappings are of the form $\forall \mathbf{x}.\varphi_s (\mathbf{x}) \leftrightarrow R_t (\mathbf{x})$, where R_t is a relation symbol from $\mathbf{T}$ and $\varphi_s (\mathbf{x})$ is a formula over $\mathbf{S}$ (i.e., rather than a formula, there is just one atom on the right-hand-side with no repeated variables). (Note that GAV mappings as originally defined, were typically exact mappings, though more recently sound GAV mappings have also been studied [11].)
- In *local-as-view* systems (LAV), mappings are of the form $\forall \mathbf{x}.R_s (\mathbf{x}) \rightarrow \psi_t (\mathbf{x})$, where R_s is a relation symbol from $\mathbf{S}$ and $\psi_t (\mathbf{x})$ is a formula over $\mathbf{T}$ (i.e., rather than a formula, there is just one atom on the left-hand-side with no repeated variables).

Recall that queries are posed in terms of the target (global) schema. For this reason, query answering in GAV is easy: the mapping indicates explicitly how to retrieve data from the sources. In particular, given a query q over the target schema, it suffices to *unfold* q using the mapping in order to obtain a rewriting q' of q (i.e., a query that computes the certain answers to q). As an example, consider the following GAV setting $\mathcal{S} = \langle \mathbf{S}, \mathbf{T}, \mathcal{M} \rangle$, where $\mathcal{M}$ consists of the following rule.

$$\forall \text{cit}, \text{cou} \exists p. \mathbf{S}. \text{Capital (cit, cou)} \wedge$$
$$\mathbf{S}. \text{Country } (p, \text{cou}) \leftrightarrow \mathbf{T}. \text{Loc (cit, cou)}$$

Suppose that a user wants to retrieve the cities and countries from relation *Loc*. Thus, she issues the following query over the global schema:

$$q \text{ (cit, cou)} = \mathbf{T}. \text{Loc (cit, cou)}$$

In order to obtain a rewriting of q, it suffices to replace *Loc(cit, cou)* by its definition in $\mathcal{M}$

$$q' \text{ (cit, cou)} = \exists p. \mathbf{S}. \text{Capital (cit, cou)} \wedge$$
$$\mathbf{S}. \text{Country } (p, \text{cou})$$

Query processing in LAV is more involved than in GAV. The reason is that queries are written over the target schema, but a LAV mapping associates views to relations of the source schemas. Thus, the unfolding strategy no longer works in this case; and it is not immediate how to rewrite queries over the source schema. In general, the complexity of query answering in LAV is higher than in GAV. The intuitive explanation is that in GAV, given a source instance I, it suffices to concentrate on a single solution J, whereas in LAV there may be many such solutions.

Rather than unfolding, the problem of reformulating a target query using a LAV mapping boils down to the problem of *answering queries using views* [12].

For many classes of open schema mappings (including source-to-target TGDs) query answering has tractable *data complexity*. For example, query answering is tractable when the schema mappings and queries are conjunctive. In contrast, if complete or exact rules are allowed, the same problem becomes coNP-complete. To show the jump in complexity, consider Figs. 1 and 2 (due to Abiteboul and Duschka [1]). The former gives results for open LAV mappings.

mapping	query				
	CQ	$CQ^{\neq}$	UCQ	Datalog	FO
CQ	PTIME	coNP	PTIME	PTIME	undec.
$CQ^{\neq}$	PTIME	coNP	PTIME	PTIME	undec.
UCQ	coNP	coNP	coNP	coNP	undec.
Datalog	coNP	undec.	coNP	undec.	undec.
FO	undec.	undec.	undec.	undec.	undec.

Schema Mapping, Fig. 1 Complexity of query answering using open LAV mappings

mapping	query				
	CQ	$CQ^{\neq}$	UCQ	Datalog	FO
CQ	coNP	coNP	coNP	coNP	undec.
$CQ^{\neq}$	coNP	coNP	coNP	coNP	undec.
UCQ	coNP	coNP	coNP	coNP	undec.
Datalog	undec.	undec.	undec.	undec.	undec.
FO	undec.	undec.	undec.	undec.	undec.

Schema Mapping, Fig. 2 Complexity of query answering using exact LAV mappings

The latter gives results under the "closed world assumption," where mappings consist of exact LAV mappings of the form $\forall x : R_s(x) \leftrightarrow \psi_t(x)$, where $R_s(x)$ is a relation from the source, and $\psi_t(x)$ is a formula over the target. The results are given for different logical languages used for the mappings and queries: conjunctive queries (CQ), conjunctive queries with inequalities ($CQ^{\neq}$), union of conjunctive queries (UCQ), Datalog, and first-order logic (FO). In addition to these results, the problem of query answering in open mapping settings has also been studied for other query languages (e.g., description logics) and on other data models (e.g., semi-structured models).

Schema Mappings in Peer Data Sharing

The success of peer-to-peer (P2P) technology in the domain of file exchange motivated the research community to consider peer-to-peer architectures for data sharing. In a P2P system, participants (peers) rely on one another for service, blurring the distinction between clients and servers, source and target. P2P systems are founded on the principles of *peer autonomy* and *decentralized coordination*. As a result, peers do not have a global view of the system. Rather, global behavior emerges from local interactions.

In a Peer Data Management System (PDMS), each peer has a schema that describes the structure of its data, and can establish connections (typically specified as schema mappings) with other peers in order to exchange data [2, 7]. A PDMS is expected to satisfy the desirable properties of P2P systems. For example, the requirement of decentralized coordination precludes the existence of a central catalog. Rather, knowledge about schemas and mappings should be distributed among the peers.

A PDMS should support an arbitrary network of mappings among peers. More importantly, it should be able to exploit the transitive relationships of the network during query answering. A PDMS is essentially a directed graph whose nodes are individual mapping settings, and whose arcs correspond to mappings that relate the schemas in the network. More precisely, there is an arc from $\mathbf{P}_1$ to $\mathbf{P}_2$ if there is a sound rule in some peer setting $\langle \mathbf{P}_1, \mathbf{P}_2, \mathcal{M} \rangle$ or a complete rule in some peer setting $\langle \mathbf{P}_2, \mathbf{P}_1, \mathcal{M} \rangle$. It turns out that the topology of this graph has a direct impact on query answering. In particular, Halevy et al. [7] showed the undecidability of the problem of obtaining the certain answers for a PDMS of arbitrary topology, where conjunctions are used for the mapping rules and the queries. Contrast this to the case of a single mapping setting with exact mappings, which is not undecidable, but coNP-complete and tractable for open settings. Fuxman et al. [5] studied this problem for a special case of PDMS, called Peer Data Exchange, and gave a class of mappings and queries for which the problem is decidable under the existence of cycles. Calvanese et al. [3] proposed an alternative semantics for query answering, based on epistemic interpretations, for which obtaining the certain answers in a PDMS of arbitrary topology is decidable and, in some cases, tractable. An alternative approach involves specifying schema mappings using *mapping tables* which specify how data values are mapped between peers [10]. Research is on-going on what are good mapping formalisms to support PDMS.

Creating Schema Mappings

Creating a schema mapping between independently designed schemas can be a tremendous challenge. Schemas that are designed independently, even if they represent the same or similar information may use different names and structures to describe the same or similar data. Designing schema mappings by hand is known to be a very difficult task requiring expert users familiar with both source and target schemas. Even experts can often make errors leading to specifications that omit information or produce incorrect answers to target queries.

To help automate this task, Milo and Zohar proposed the use of *schema matchings* (or matchings) which indicate potential associations between elements within different schemas [14]. A matching is most often represented as a set of pairs of schema attributes from two different schemas. Schema matchings can be semi-automatically inferred by using a variety of matching tools. These tools use schema and data characteristics such as lexical similarities, structural proximity, data values, etc. to infer potential matches between attributes of different schemas. Matchings, however, do not represent the full semantic relationship between schemas and their instances.

Figure 3 shows two schemas that represent information about companies and grants. The left-most schema (the source **S**) is a relational schema (with three tables *companies*, *grants*, and *contacts*), presented in a nested relational representation that is used as a common platform for modeling relational and XML schemas. The curved lines *f1*, *f2* and *f3* in the figure represent either foreign keys or simple inclusion dependencies, specified as part of the schema (or discovered using a dependency or constraint miner).

The schema on the right (the target **T**), records the funding (*fundings*) that an organization (*organizations*) receives, nested within the respective organization element. The amount of each funding is recorded in the *finances* record along with a contact phone number (*phone*).

Figure 3 indicates, using the dotted lines v_1 through v_4, a matching entered by a schema expert or discovered by a matching tool. Due to the heterogeneity and the different requirements under which the two databases were developed, the same real world entity (for example, a company 'IBM') may be represented in very different ways

Schema Mapping, Fig. 3
A matching between
source and target schemas

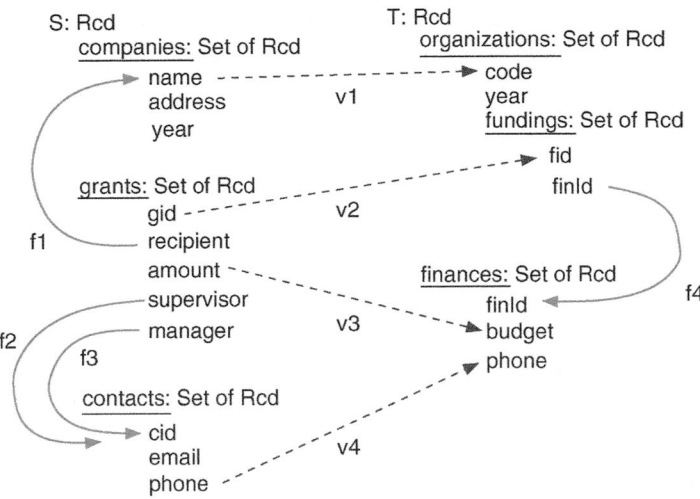

in the two databases, and structures that appear to be the same may actually model different concepts. In our example, the matching v_1 indicates that what is called a company *name* in the first schema, is referred to as an organization *code* in the second. On the other hand, both schemas have an element *year*, but there is no match between these attributes indicating that they likely do not represent the same concept. For instance, element *year* in the source schema may represent the time the company was founded, while in the target it may represent the time the company had its initial public offer.

Note that matchings are far from sufficient to tell us how the data instances of **S** and **T** are related. While the matching may indicate that *companies.name* data should appear in the *organizations.code* attribute of the target and that *grants.gid* data should appear in *fundings.fid*, it does not tell us which grant should be associated with which company. Similarly, if one relies solely on the matching, one could map *grants.gid* to *fundings.fid* and leave the *finId* attribute value empty (since there is no matching for *finId*). If one does this, in the target data, there will be no way to associate a funding record with a finance record. However, the foreign key f_4 indicates there is a real world relationship between these concepts.

The generation of schema mappings have been considered in a number of research projects and industrial tools. The Clio project [13] was the first mapping system to exploit logical reasoning about the semantics embedded in the schemas and their instances to help automate mapping creation. In this work, the mapping discovery process has been referred to as *query discovery* in that the goal is to discovery a query over the source (φ_S), a query over the target (ψ_T), and their relationship, in order to create a set of possible source-to-target TGDs: $\forall x\, (\varphi_S\, (x) \rightarrow \exists y\, \psi_T\, (x, y))$ [15]. These mappings can then be shown to a mapping designer (visually or using data examples) who can decide if they correctly represent the relationship between source and target instances.

To illustrate Clio's approach, consider our example schemas. In the target, the nesting structure within *organizations* indicates that there is a real-world relationship between organizations and their fundings- that is, the association of a specific funding record with a specific organization has some natural semantics in the domain. For example, the nesting may represent fundings given to (or alternatively given by) an organization. Hence, in creating a mapping, Clio will consider related associations in the source. In the example, the data that matches organizations and fundings comes from companies and grants (via the matches v_1 and v_2). There is a relationship between these two tables represented by the inclusion dependency on *grants.recipient*.

Hence, by ignoring v_3 and v_4 for the moment, Clio will suggest the following source-to-target TGD to a user. (Notation is slightly abused and let $\underline{F}$ represent identifiers for sets nested inside of *organizations*.)

$$\forall n, \forall d, \forall y, \forall a, \forall s, \forall m$$
$$\text{companies}\,(n, d, y)\,, \text{grants}\,(g, n, a, s, m) \rightarrow$$
$$\exists y';, \underline{F}, f, \text{organizations}\,(n, y';, \underline{F})\,, \underline{F}\,(g, f)$$
$$(3)$$

Notice that this mapping retains the semantic association between companies and their grants, and uses this to associate organizations with a set of related fundings. This mapping is correct if these two associations represent the same real world association, something a user must verify. By extending this example to include v_3, Clio will see that there is an association between *gid* and *amount* in the source (because these values are paired in the same record) and consider possible ways of associating the matched attributes (in this case *fid* and *budget*) in the target. These attributes are not in the same record in the target, but are associated through a foreign key on *finId*. To maintain the source association in the target, Clio creates a mapping containing a target join on *finId*. Finally, if one considers how to create finances tuples in the target, notice that there are two possible ways of associated the related source data (*grants.amount* and *contacts.phone*) – using a join on *supervisor* and *cid*, or using a join on *manager* and *cid*. These joins represent different semantic associations and Clio will create mappings corresponding to each and let the user decide which (if any) of these associations should be preserved in the target data. One of the mappings created by Clio, which uses the source association represented by f_2, is illustrated below.

$$\forall n, \forall d, \forall y, \forall g, \forall a, \forall s, \forall m, \forall e, \forall p \text{ companies}$$

$$(n, d, y) \wedge \text{grants}\,(g, n, a, s, m)$$

$$\wedge \text{ contacts}\,(s, e, p) \rightarrow \exists y',$$

$$\underline{F}, f, \text{organizations}\,(n, y', \underline{F}) \wedge \underline{F}\,(g, f) \wedge$$

$$\text{finances}\,(f, a, p)$$
$$(4)$$

Clio's mapping discovery algorithm is based on an extension of standard relational dependency inference (based on the chase) to nested relational schemas. The schemas may be relational or nested relational containing source and target TGDs (e.g., inclusion dependencies) and ends (e.g., functional dependencies).

Data Translation

Mapping tools that create declarative mappings provide a way of translating these specifications into programs (transformation code) that given a source instance produce a single target instance for data exchange [15, 16], Industrial mapping systems, such as Altova Mapforce (http://www.altova.com/products/mapforce/data_mapping.html), Stylus Studio (http://www.stylusstudio.com), or Aqualogic (http://www.bea.com/aqualogic) are often visual programming systems which compile visual specifications of mappings into executable code including SQL, XSLT, Java or C.

Within this area, there has been a great deal of work on producing data transformation code that is modular and efficient [9, 16]. For schema mappings that permit many solutions, a decision must be made as to what is the "best" solution to materialize. Notice that there may be target data (for example, *organizations.year*, or *fundings.finId* attributes from Fig. 3) that do not correspond to any source data. It may not be sufficient to simply fill in null values for this information. Consider the *fundings.finId* and *finances.finId* from Fig. 3. If one fills both with null values, it is possible to join *fundings* with *finances* to find the budget of a specific funding. As an alternative, to maintain the association between the source values *grants.gid* and *grants.amount* as this data are translated into the target, one option is to create identifiers (using Skolem functions) that represent the desired association [8]. The Clio project was the first to consider systematically how to create Skolem functions that fill in missing target data specifically for data exchange [15].

Key Applications

Schema mappings are foundational to enabling data integration, data exchange, schema evolution, and data translation (between data models). Applications of schema mappings include Enterprise Information Integration (EII), e-commerce, object-to-relational wrappers, XML-to-relational mapping, data warehousing, and portal design tools.

Cross-References

▶ Answering Queries Using Views
▶ Certain (and Possible) Answers
▶ Data Exchange
▶ Peer Data Management System
▶ Peer-to-Peer Data Integration
▶ Schema Mapping Composition
▶ Schema Matching

Recommended Reading

1. Abiteboul S, Duschka OM. Complexity of answering queries using materialized views. In: Proceedings of the 17th ACM SIGACT-SIGMOD-SIGART Symposium on Principles of Database Systems; 1998. p. 254–63.
2. Bernstein PA, Giunchiglia F, Kementsietsidis A, Mylopoulos J, Serafini L, Zaihrayeu I. Data management for peer-to-peer computing: a vision. In: Proceedings of the 5th International Workshop on the World Wide Web and Databases; 2002.
3. Calvanese D, De Giacomo G, Lenzerini M, Rosati R. Logical foundations of peer-to-peer data integration. In: Proceedings of the 23rd ACM SIGACT-SIGMOD-SIGART Symposium on Principles of Database Systems; 2004. p. 241–51.
4. Fagin R, Kolaitis PG, Miller RJ, Popa L. Data exchange: semantics and query answering. Theor Comput Sci. 2005;336(1):89–124.
5. Fuxman A, Kolaitis PG, Miller RJ, Tan W-C. Peer data exchange. ACM Trans Database Syst. 2006;31(4):1454–98.
6. Grahne G, Mendelzon AO. Tableau techniques for querying information sources through global schemas. In: Proceedings of the 7th International Conference on Database Theory; 1999. p. 332–47.
7. Halevy A, Ives Z, Suciu D, Tatarinov I. Schema mediation in peer data management systems. In: Proceedings of the 9th International Conference on Data Engineering; 2003. p. 505–18.
8. Hull R, Yoshikawa M. ILOG: declarative creation and manipulation of object identifiers. In: Proceedings of the 16th International Conference on Very Large Data Bases; 1990. p. 455–68.
9. Jiang H, Ho H, Popa L, Han WS. Mapping-driven xml transformation. In: Proceedings of the 16th International World Wide Web Conference; 2007. p. 1063–72.
10. Kementsietsidis A, Arenas M, Miller RJ. Mapping data in peer-to-peer systems: semantics and Algorithmic Issues. In: Proceedings of the ACM SIGMOD International Conference on Management of Data; 2003. p. 325–36.
11. Lenzerini M. Data integration: a theoretical perspective. In: Proceedings of the 21st ACM SIGACT-SIGMOD-SIGART Symposium on Principles of Database Systems; 2002. p. 233–46.
12. Levy AY, Mendelzon AO, Sagiv Y, Srivastava D. Answering queries using views. In: Proceedings of the 14th ACM SIGACT-SIGMOD-SIGART Symposium on Principles of Database Systems; 1995. p. 95–104.
13. Miller RJ, Haas LM, Hernández M. Schema mapping as query discovery. In: Proceedings of the 26th International Conference on Very Large Data Bases; 2000. p. 77–88.
14. Milo T, Zohar S. Using schema matching to simplify heterogeneous data translation. In: Proceedings of the 24th International Conference on Very Large Data Bases; 1998. p. 122–33.
15. Popa L, Velegrakis Y, Miller RJ, Hernández MA, Fagin R. Translating web data. In: Proceedings of the 28th International Conference on Very Large Data Bases; 2002. p. 598–609.
16. Shu NC, Housel BC, Taylor RW, Ghosh SP, Lum VY. EXPRESS: a data eXtraction, processing, amd restructuring system ACM Trans. Database Syst. 1997;2(2):134–74.

Schema Mapping Composition

Wang-Chiew Tan
University of California-Santa Cruz, Santa Cruz, CA, USA

Synonyms

Mapping composition; Semantic mapping composition

Definition

A *schema mapping* (or *mapping*) is a triple $\mathcal{M}$ = (S_1, S_2, Σ), where S_1 and S_2 are relational schemas with no relation symbols in common and Σ is a set of formulas of some logical formalism over (S_1, S_2). An *instance* of $\mathcal{M}$ is a pair (I, J) where I is an instance of S_1 and J is an instance of S_2 such that (I, J) satisfies every formula in the set Σ. The set of all instances of $\mathcal{M}$ is denoted as Inst $\mathcal{M}$.

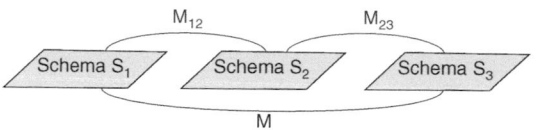

Let $\mathcal{M}_{12} = (S_1, S_2, \Sigma_{12})$ and $\mathcal{M}_{23} = (S_1, S_2, \Sigma_{23})$ be two consecutive mappings such that there are no relation symbols in common between any two schemas of S_1, S_2 and S_3. A mapping $\mathcal{M} = (S_1, S_3, \Sigma)$ is a *composition* of $\mathcal{M}_{12}$ and $\mathcal{M}_{23}$ if Inst $(\mathcal{M})$ = Inst $(\mathcal{M}_{12})$ ∘ Inst $(\mathcal{M}_{23})$ Inst $(\mathcal{M})$ = Inst $(\mathcal{M}_{12})$ ∘ Inst $(\mathcal{M}_{23})$. In other words, Inst($\mathcal{M}$) is the set of all pairs (I, J) such that I is an instance of S_1, J is an instance of S_3 and there exists an instance K of S_2 such that (I, K) ∈Inst$\mathcal{M}_{12}$ and (K, J) ∈Inst$\mathcal{M}_{23}$.

Historical Background

Mappings are widely used in the specification of relationships between data sources in applications such as data integration, data exchange, and peer data management systems. The model management framework introduced by Bernstein et al. [2], where the primary abstractions are models and mappings between models, can be used to model these applications. Several operators for manipulating mappings between models were introduced in this framework. Among them, the *composition operator* is one of the most fundamental operators for manipulating mappings between models.

Madhavan and Halevy [10] first studied the problem of composing mappings between relational schemas (mappings are also called

semantic mappings in [10]). They gave a definition of the semantics of the composition operator in the context where mappings are specified by sound *Global–local-As-View (GLAV)* formulas [9]. Sound GLAV formulas are the most widely used and studied form of mappings in data integration systems and they are equivalent to *source-to-target tuple generating dependencies (s-t tgds)* [13]. (See "Foundations" section for a definition of s-t tgds.) The Madhavan and Halevy semantics of composition is different from the one stated above; The set of formulas that specifies the composition $\mathcal{M}$ of two successive mappings, $\mathcal{M}_{12}$ and $\mathcal{M}_{23}$, is relative to a class of queries Q that is defined over the schema S_3. This means that for every query q in Q, the certain answers of q according to $\mathcal{M}$ coincide with the certain answers of q that would be obtained by applying the consecutive mappings $\mathcal{M}_{12}$ and $\mathcal{M}_{23}$.

The Madhavan and Halevy notion of composition is termed a composition that is *certain answer adequate for Q* in [8]. Fagin et al. [8] showed that while certain answer adequacy may be sufficient for obtaining the certain answers of any query in Q, whether via Madhavan and Halevy's notion of composition $\mathcal{M}$ or through the successive mappings $\mathcal{M}_{12}$ and $\mathcal{M}_{23}$, there may be formulas that are logically inequivalent to Σ of $\mathcal{M}$ that are also certain answer adequate for Q. In other words, there can be two mappings $\mathcal{M}$ and $\mathcal{M}'$ that are both certain answer adequate for the consecutive mappings $\mathcal{M}_{12}$ and $\mathcal{M}_{23}$ with respect to the class of queries Q, but the sets of formulas Σ and Σ' of $\mathcal{M}$ and $\mathcal{M}'$, respectively, are logically inequivalent. It was also shown in [8] that certain answer adequacy is a rather fragile notion. A mapping $\mathcal{M}$ that is a composition of $\mathcal{M}_{12}$ and $\mathcal{M}_{23}$ with respect to a class Q of conjunctive queries may no longer be a composition of the same two mappings when Q is extended to the class of conjunctive queries with inequalities.

Fagin et al. [8] introduced a definition of composition that is based entirely on the set-theoretic composition of instances of the two successive mappings $\mathcal{M}_{12}$ and $\mathcal{M}_{23}$. (See the "Definition" section for the set-theoretic definition given in [8].) Unlike the definition of composition given by Madhavan and Halevy, Fagin et al.'s [8] definition is not relative to a class of queries. Furthermore, the set of formulas in $\mathcal{M}$ that defines the composition of $\mathcal{M}_{12}$ and $\mathcal{M}_{23}$ is unique up to logical equivalence. Hence, with Fagin et al.'s [8] notion of composition, $\mathcal{M}$ is referred to as *the* composition of $\mathcal{M}_{12}$ and $\mathcal{M}_{23}$. It was also shown in [8] that the composition $\mathcal{M}$ is always certain answer adequate for $\mathcal{M}_{12}$ and $\mathcal{M}_{23}$ for *every* class of queries.

The results established in [8] were based on mappings specified by s-t tgds. In other words, Fagin et al. [8] assumes that the sets of formulas Σ_{12} and Σ_{23} in the successive mappings $\mathcal{M}_{12}$ and $\mathcal{M}_{23}$, respectively, are finite sets of s-t tgds. A subsequent paper by Nash et al. [12] also studied the composition operator where mappings are specified by *embedded dependencies*. Embedded dependencies are more general than s-t tgds and can model constraints such as keys. Among the results established in [12] is an algorithm that computes the composition of two successive mappings specified by embedded dependencies. Their algorithm may not terminate in general and the authors characterized sufficient conditions on the input mappings for which their composition algorithm is guaranteed to produce a composition of the input mappings. An implementation of the composition operator that extends the composition algorithm of [12] is described in [3].

Foundations

In [8], mappings are specified by finite sets of s-t tgds which are equivalent to sound GLAV assertions used in [10]. A *s-t tgd* is a first-order formula of the form:

$$\forall \mathbf{X} \left(\phi_S (\mathbf{X}) \to \exists \mathbf{y} \psi_T (\mathbf{x}, \mathbf{y}) \right),$$

where $\phi_S (\mathbf{x})$ is a conjunction of atomic formulas over the schema $\mathbf{S}$ and where $\psi_T (\mathbf{x}, \mathbf{y})$ is a conjunction of atomic formulas over the schema $\mathbf{T}$. Every variable in $\mathbf{x}$ and $\mathbf{y}$ must appear in φ_S and ψ_T respectively. However, some variables in $\mathbf{x}$ need not appear in ψ_T. A *full source-to-target tuple generating dependency (full s-t tgd)* is a special s-t tgd of the form

$$\forall \mathbf{x} \left(\phi_S (\mathbf{x}) \to \psi_T (\mathbf{x}) \right),$$

where no existentially-quantified variables appears on the right-hand-side of the s-t tgd. As before, $\phi_S (\mathbf{x})$ is a conjunction of atomic formulas over $\mathbf{S}$ and $\psi_T (\mathbf{y})$ is a conjunction of atomic formulas over $\mathbf{T}$. Every variable in $\mathbf{x}$ must appear in ϕ_S.

Example 1 Let $\mathcal{M}_{12} = (\mathbf{S}_1, \mathbf{S}_2, \Sigma_{12})$ and $\mathcal{M}_{23} = (\mathbf{S}_2, \mathbf{S}_3, \Sigma_{12})$ be two successive mappings. The schema $\mathbf{S}_1$ consists of a binary relation symbol *Takes*, the second schema $\mathbf{S}_2$ consists of two binary relation symbols *Takes*$_1$ and *Student*, and the third schema $\mathbf{S}_3$ consists of a single binary relation symbol *Enrollment*. The formulas in Σ_{12} and Σ_{23} are s-t tgds stated below:

$$\sum 12 = \{\forall_n \forall_c \, (\text{Takes} \, (n, c) \to \text{Takes}_1 \, (n, c)),$$
$$\forall_n \forall_c \, (\text{Takes} \, (n, c) \to \exists s$$
$$\text{Student} \, (n, s))\}$$
$$\sum 23 = \{\forall_n \forall_s \forall_c \, (\textbf{Student} \, (n, s)$$
$$\wedge \textbf{Takes}_1 \, (\mathbf{n}, \mathbf{c})) \, , \textbf{Enrollment} \, (s, c)))\}$$

Observe that the first s-t tgd in Σ_{12} and the s-t tgd in Σ_{23} are full s-t tgds. The s-t tgds in Σ_{12} state that the *Takes*$_1$ relation in $\mathbf{S}_2$ contains the *Takes* relation in $\mathbf{S}_1$ and that every student with name n who takes a course c (in *Takes*) has a associated tuple in *Student* with name n and some student id s. The s-t tgd in Σ_{23} states that every student with name n and student id s (in *Student*) who is also taking a course c (in *Takes*$_1$) must have a corresponding tuple (s, c) that associates the student id s with the course c in the *Enrollment* relation of $\mathbf{S}_3$.

Recall from the definition that the composition $\mathcal{M}_{12}$ and $\mathcal{M}_{23}$ is a mapping $\mathcal{M}$ that captures exactly the set of instances $\text{Inst}(\mathcal{M}_{12})^\circ \, \text{Inst}(\mathcal{M}_{23})$. The set of instances $\{(I_1, I_3) \mid (I_1, I_3) \in \text{Inst}\,(\mathcal{M}_{12}) \,^\circ \text{Inst}\,(\mathcal{M}_{23})\}$ is called the *composition query* of $\mathcal{M}_{12}$ and $\mathcal{M}_{23}$. Among the issues investigated by Fagin et al. [8] in composing mappings under this semantics are:

- Is the language of s-t tgds always sufficient to define the composition of two successive mappings?
- What is the complexity of the instances associated with the composition query of two successive mappings?
- What is a right language for composing mappings?
- How does data exchange and query answering behave in the chosen language for composing mappings?

Composing s-t TGDs: Definability and Complexity. The answer to the first question above is no. It was shown in [8] that if $\mathcal{M}_{12}$ and $\mathcal{M}_{23}$ are specified by finite sets of full s-t tgds and s-t tgds respectively, then the composition of $\mathcal{M}_{12}$ with $\mathcal{M}_{23}$ is always definable by a finite set of s-t tgds. Furthermore, the associated composition query is in PTIME. However, if $\mathcal{M}_{12}$ is specified by a set of s-t tgds, not necessarily full, then the composition of $\mathcal{M}_{12}$ with $\mathcal{M}_{23}$ may not always be definable by a finite set of s-t tgds. For instance, the composition of the successive mappings in Example 1 is not definable by any finite set of s-t tgds. However, the composition of $\mathcal{M}_{12}$ and $\mathcal{M}_{23}$ is definable by an infinite set of s-t tgds and, in fact, definable by a first-order formula. As a consequence, the composition query of $\mathcal{M}_{12}$ and $\mathcal{M}_{23}$ is a PTIME query.

Fagin et al. [8] further showed that there exists successive mappings $\mathcal{M}_{12}$ and $\mathcal{M}_{23}$ where the composition query of $\mathcal{M}_{12}$ and $\mathcal{M}_{23}$ is NP-complete and the composition of the two mappings is not definable in least fixed-point logic LFP. The mappings used are such that $\mathcal{M}_{12}$ is specified by a finite set of s-t tgds each having at most one existentially-quantified variable and $\mathcal{M}_{23}$ consists of only one full s-t tgd. Essentially, the NP-hardness result is obtained by a reduction from 3-Colorability to the composition query of two fixed mappings with the above properties. They showed that the composition query of mappings specified by finite sets of s-t tgds is always in NP.

Second-Order TGDs. Since the composition query of mappings specified by finite sets of s-t tgds is always in NP, it follows from

Fagin's theorem [5] that the composition of the two mappings is always definable by an existential second-order formula, where the existential second-order variables are interpreted over relations on the union of the set of values in I_1 with the set of values in I_3. Here, $(I_1, I_3) \in \text{Inst}(\mathcal{M}_{12}) \circ \text{Inst}(\mathcal{M}_{23})$ Fagin et al. [8] showed that, in fact, the composition of mappings specified by finite sets of s-t tgds is always definable by a restricted form of existential second-order formula, called *second-order tgds* (SO tgds).

SO tgds are s-t tgds that are extended with existentially quantified functions and with equalities. SO tgds are the "right" language for composing mappings because they form the smallest well-behaved extension to the class of s-t tgds that is closed under conjunction and composition. In addition, as explained later, SO tgds also possess good properties for data exchange and query answering. The precise definition of SO tgds is given next, after the definition of *terms*.

Given a collection **x** of variables and a collection **f** of function symbols, a *term (based on* **x** *and* **f**) is defined recursively as follows: (i) Every variable in **x** is a term. (ii) If f is a k-ary function symbol in **f** and $t_1, \ldots, t_k$ are terms, then $f(t_1, \ldots, t_k)$ is a term. Let **S** be a source schema and **T** a target schema. A *second-order tgd (SO tgd)* is a formula of the form:

$$\exists \mathbf{f} \left((\forall \mathbf{x_1} (\phi_1 \rightarrow \psi_1)) \wedge \cdots \wedge (\forall \mathbf{x_n} (\phi_n \rightarrow \psi_n)) \right),$$

where

1. Each member of **f** is a function symbol.
2. Each ϕ_i is a conjunction of
   - atomic formulas of the form $S(y_1, \ldots, y_k)$, where S is a k-ary relation symbol of schema **S** and $y_1, \ldots, y_k$ are variables in $\mathbf{x_i}$, not necessarily distinct, and
   - equalities of the form $t = t'$ where t and t' are terms based on $\mathbf{x_i}$ and **f**.
3. Each ψ_i is a conjunction of atomic formulas $T(t_1, \ldots, t_l)$, where T is an l-ary relation symbol of schema **T** and $t_1, \ldots, t_l$ are terms based on $\mathbf{x_i}$ and **f**.
4. Each variable in $\mathbf{x_i}$ appears in some atomic formula of ϕ_i.

Each subformula $\forall \mathbf{x}_i(\phi_i \rightarrow \psi_i)$ is a *conjunct* of the SO tgd. The last condition is a safety condition similar to that made for s-t tgds. As an example, the formula $\exists f \forall x \forall y (S(x) \wedge (y = f(x)) \rightarrow T(x, y))$ is not a SO tgd because the safety condition is violated. In particular, the variable y does not appear in an atomic formula on the left-hand-side of the formula. As another example, the formula

$$\exists f \forall_n \forall_c \Big(\text{Takes}\,(n, c) \rightarrow \text{Enrollment}\,(f(n), c)$$

is a SO tgd. Recall that the composition of the successive mappings in Example 1 is not definable by any finite set of s-t tgds. Fagin et al. [8] showed that the SO tgd above defines the composition of the two mappings given in Example 1. Hence, existentially quantified function symbols are a necessary extension to the language of s-t tgds for composing mappings. Intuitively, the SO tgd states that if a student with name n takes a course c in *Takes*, then the student id of n, which is denoted by the function $f(n)$, is associated with c in *Enrollment*. An example of SO tgds where equalities are involved is described next.

Example 2 Let $\mathcal{M}_{12} = (\mathbf{S}_1, \mathbf{S}_2, \Sigma_{12})$ and $\mathcal{M}_{23} = (\mathbf{S}_2, \mathbf{S}_3, \Sigma_{23})$ be two successive mappings. The first schema $\mathbf{S}_1$ consists of a unary relation symbol *Emp*, the second schema $\mathbf{S}_2$ consists of a binary relation symbol Mgr_1, and the third schema $\mathbf{S}_3$ consists of a binary relation symbol *Mgr* and a unary relation symbol *SelfMgr*. The s-t tgds in Σ_{12} and Σ_{23} are:

$$\sum 12 = \{\forall e\,(\mathbf{Emp}(e) \rightarrow \exists m\ \mathbf{Mgr}_1(e, m))\}$$

$$\sum 23 = \{\forall e \forall m\,(\mathbf{Mgr}_1(e, m) \rightarrow \mathbf{Mgr}(e, m))$$
$$\forall e\,(\mathbf{Mgr}_1(e, e) \rightarrow \mathbf{SelfMgr}(e))\}$$

Intuitively, the s-t tgd in Σ_{12} asserts that every employee e in **Emp** must have a manager m and this association can be found in $\mathbf{Mgr}_1$. The first s-t tgd in Σ_{23} asserts that the *Mgr* relation contains the $\mathbf{Mgr}_1$ relation and the second s-t tgd in Σ_{23} asserts that **SelfMgr** contains employees

who are their own managers according to the $\mathbf{Mgr}_1$ relation.

Fagin et al. [8] showed that the following SO tgd defines the composition of $\mathcal{M}_{12}$ and $\mathcal{M}_{23}$.

$$\exists f\,(\forall e\,(\mathbf{Emp}(e) \rightarrow \mathbf{Mgr}\,(e,\ f(e)) \wedge \forall e$$
$$(\mathbf{Emp}(e) \wedge (e = f(e)) \rightarrow \mathbf{SelfMgr}(e)))$$

They also showed that the above SO tgd is not logically equivalent to any finite or infinite sets of SO tgds without equalities. In other words, the composition of $\mathcal{M}_{12}$ and $\mathcal{M}_{23}$ is not definable by SO tgds without equality. Hence, equalities are a necessary extension to the language of s-t tgds for composing mappings.

The extensions of s-t tgds with function symbols and equalities in SO tgds are necessary to compose mappings specified by s-t tgds. Fagin et al. [8] also showed that SO tgds are closed under composition. This means that the composition of two mappings, each specified by an SO tgd, is another SO tgd.

Example 3 An illustration of the algorithm described in [8] for composing two mappings, based on Example 2, is described next. The algorithm takes as input two mappings, $\mathcal{M}_{12}$ and $\mathcal{M}_{23}$, specified by s-t tgds and returns as output a mapping $\mathcal{M}_{13}$ that defines the composition of the two input mappings.

The first step of the algorithm is to transform the s-t tgds in Σ_{12} and Σ_{23} into SO tgds by introducing Skolem functions to replace existentially quantified variables. For example, Σ_{12} and Σ_{23} will now become Σ'_{12} and Σ'_{23} respectively.

$$\sum\nolimits'_{12} = \{\exists f \forall e\,(\mathbf{Emp}\,(e) \rightarrow \mathbf{Mgr}_1\,(e, f(e)))\}$$

$$\sum\nolimits'_{23} = \{\forall e \forall m\,(\mathbf{Mgr}_1\,(e, m) \rightarrow \mathbf{Mgr}\,(e, m)),$$
$$\forall e\,(\mathbf{Mgr}_1\,(e, e) \rightarrow \mathbf{SelfMgr}(e))\}$$

In particular, observe that Σ'_{12} now consists of an SO tgd with function $f(e)$ that denotes the manager of employee e. The arguments of f consist of all universally quantified variables in the s-t tgd. The next step of the algorithm combines Σ'_{12} with Σ'_{23} to obtain Σ_{13}' by replacing atomic formulas on the left-hand-side of conjuncts of SO tgds in Σ'_{23} with atomic formulas from $\mathbf{S}_1$ through conjuncts of SO tgds in Σ'_{12}. In the running example, $Emp(e_0) \rightarrow Mgr_1(e_0, \mathrm{f}(e_0))$ is combined with $Mgr_1(e,m) \rightarrow Mgr(e,m)$ to obtain

$$\mathbf{Emp}(e) \wedge (e = e) \wedge (m = f(e)) \rightarrow \mathbf{Mgr}\,(e,m)\,\big)$$

and $\mathbf{Emp}(e_1) \rightarrow \mathbf{Mgr}_1(e_1, f(e_1))$ is combined with $\mathbf{Mgr}_1(e, e) \rightarrow \mathbf{SelfMgr}(e)$ to obtain

$$\mathbf{Emp}\,(e_1) \wedge (e = e_1) \wedge (e = f(e_1)) \rightarrow \mathbf{SelfMgr}(e)$$

Observe that the equalities generated by this step has the form $y = t$ where y is a variable in Σ'_{23} and t is a term based on the variables and functions of Σ'_{12}. The next step of the algorithm removes variables from Σ'_{23} according to the equalities. For example, e is replaced with e_0 in the first formula and e is replaced with e_1 in the second formula to obtain the following SO tgds:

$$\mathbf{Emp}(e) \wedge (m = f(e)) \rightarrow \mathbf{Mgr}\,(e_0, m)$$

$$\mathbf{Emp}\,(e_1) \wedge (e_1 = f(e_1)) \rightarrow \mathbf{SelfMgr}\,(e_1)$$

At this point, the variable m is replaced with $f(e_0)$ to obtain the following SO tgds:

$$\mathbf{Emp}(e) \rightarrow \mathbf{Mrg}\,(e, f(e))$$

$$\mathbf{Emp}\,(e_1) \wedge (e_1 = f(e_1)) \rightarrow \mathbf{SelfMgr}\,(e_1)$$

Finally, when no more variables of Σ'_{23} can be replaced, the following SO tgd is returned.

$$\exists f\,(\forall e\,(\mathbf{Emp}(e) \rightarrow \mathbf{Mgr}\,(e, f(e))) \wedge \forall e$$

$$(\mathbf{Emp}(e) \wedge (e = f(e)) \rightarrow \mathbf{SelfMgr}(e)))$$

Data Exchange and Query Answering. Let $\mathcal{M} = (\mathbf{S}_1, \mathbf{S}_2, \Sigma)$ be a mapping. Given a finite instance *I* over the schema $\mathbf{S}_1$, the *data exchange problem* [7] is to construct a finite instance J over the schema $\mathbf{S}_2$ such that (I, J) satisfies all the formulas specified in Σ. Such an instance J is

called a *solution* of *I* under the mapping $\mathcal{M}$. If Σ is specified by a finite set of s-t tgds, many solutions for *I* under $\mathcal{M}$ may exist in general because s-t tgds underspecify the data exchange process in general. In [7], the classical *chase* procedure [1, 11] has been used to construct *universal* solutions of *I* under a mapping $\mathcal{M}$. Universal solutions are the most general type of solutions in the following sense: If J is a universal solution for *I* under $\mathcal{M}$, this means that J is a solution for *I* under $\mathcal{M}$ with the additional property that J has a *homomorphism* into every solution for *I* under $\mathcal{M}$. Intuitively, an instance K has a homomorphism into an instance K' if K can be embedded in K' (modulo the renaming of nulls that occur in K). It was shown in [7] that universal solutions can be computed in polynomial time when the mapping is fixed. Universal solutions are desirable not only because they are the most general, but also because they can be used to compute the certain answers of unions of conjunctive queries that are posed against the schema $\mathbf{S}_2$ in polynomial time. If q is a k-ary query over $\mathbf{S}_2$, then the certain answers of q with respect to an instance *I* over $\mathbf{S}_1$, denoted as $\mathrm{certain}_{\mathcal{M}}(q, I)$, is the set of all k-tuples t of constants from *I* such that $t \in q(J)$ for every solution J of *I* under $\mathcal{M}$. It was shown in [7] that if J is a universal solution for *I* under $\mathcal{M}$ and q is a union of conjunctive queries, then $\mathrm{certain}_{\mathcal{M}}(q, I)$ can be computed as follows: (i) Evaluate $q(J)$ and then (ii) discard tuples from $q(J)$ that contain nulls. The remaining tuples obtained from this process, denoted as $q(J)_\downarrow$, form the certain answers of q with respect to I.

In the case where there are two or more successive mappings specified by s-t tgds and only the target instance over the last schema is of interest, one approach to obtain a universal solution of the last schema when given an instance *I* over the first schema is to perform a series of data exchanges (using the chase procedure [7]) starting from *I* according to the sequence of mappings. The final target instance that is arrived through this process is a universal solution for *I* for the sequence of mappings. (The series of data exchanges produces a universal solution. This result can be found in Proposition 7.2 of

[6].) Obviously, one drawback of this approach is the unnecessary construction of potentially many intermediate instances which are not of interest. Another approach that avoids the construction of intermediate instances altogether is to first compose the sequence of mappings to obtain a composed mapping over the first source schema and the final target schema. After this, data are exchanged (by using the chase procedure) according to the composed mapping. However, as described earlier, the language of s-t tgds may no longer be sufficient for defining the composition of two or more successive mappings. Instead, SO tgds are needed to describe the composition of successive mappings in general. In [8], the classical chase technique is extended to SO tgds. They showed that chasing with SO tgds is again a polynomial time procedure and that the chase with SO tgds produces a universal solution as in s-t tgds. Hence, a universal solution for the final target schema can be computed by simply chasing I over the composed mapping. As a consequence, the certain answers of unions of conjunctive queries that are posed over the target schema of a mapping specified by SO tgds can also be computed in polynomial time: First, a universal solution J of I is computed by chasing I with the mapping. After this, compute $q(J)_{\downarrow}$, as described earlier.

Key Applications

One important application of composing mappings is schema evolution. Consider the figure shown in the "Definition" section. If S_3 is an evolved schema of S_2 and the mappings $\mathcal{M}_{12}$ and $\mathcal{M}_{23}$ are given, then it is possible to derive the direct relationships between S_1 and S_3 by composing $\mathcal{M}_{12}$ with $\mathcal{M}_{23}$. See [14] for an application of composition to schema evolution.

Another important application of composition is to optimize the migration of data through a sequence of mappings. An end-to-end mapping is first assembled from a sequence of two or more mappings by composition before data are migrated through the assembled mapping. The benefit of using an end-to-end mapping for migrating

data from the first schema to the last schema in the sequence of mappings is the potential savings from the sequence of unnecessary data migration steps through the intermediate schemas along the sequence of mappings. For a similar reason, the end-to-end mapping could also be used to optimize query rewriting. Referring back to the figure in the "Definition" section, if a query that is posed against the schema S_3 needs to be rewritten into a query against the schema S_1, it is potentially rewarding to first compose the sequence of mappings $\mathcal{M}_{12}$ and $\mathcal{M}_{23}$ and then reason about the rewriting through the composition rather than through the sequence of mappings $\mathcal{M}_{23}$ and $\mathcal{M}_{12}$.

Composed mappings can also be used as an abstraction for a sequence of data migration steps. A recent work [4] on Extract-Transform-Load (ETL) systems illustrates this point. An ETL script can be modeled as a network of mappings describing the flow of data from a source to a target. By composing various sequences of mappings in the network of mappings, an abstraction of the overall ETL transformation can be achieved.

Cross-References

▶ Data Exchange
▶ Schema Mapping

Recommended Reading

1. Beeri C, Vardi MY. A proof procedure for data dependencies. J ACM. 1984;31(4):718–41.
2. Bernstein PA, Halevy AY, Pottinger R. A vision of management of complex models. ACM SIGMOD Rec. 2000;29(4):55–63.
3. Bernstein PA, Green TJ, Melnik S, Nash A. Implementing mapping composition. In: Proceedings of the 32nd International Conference on Very Large Data Bases; 2006. p. 55–66.
4. Dessloch S, Hernández M, Wisnesky R, Radwan A, Zhou J. Orchid: integrating schema mapping and ETL. In: Proceedings of the 24th international conference on data engineering; 2008. p. 1307–16.
5. Fagin R. Generalized first-order spectra and polynomial-time recognizable sets. In: Karp RM,

editor. Complexity of computation, SIAM-AMS Proceedings, vol. 7. 1974. p. 43–73.

6. Fagin R. Inverting schema mappings. ACM Trans Database Syst. 2007;32(4):24.

7. Fagin R, Kolaitis PG, Miller RJ, Popa L. Data exchange: semantics and query answering. Theor Comput Sci. 2005a;336(1):89–124.

8. Fagin R, Kolaitis PG, Popa L, Tan WC. Composing schema mappings: second-order dependencies to the rescue. ACM Trans Database Syst. 2005b;30(4): 994–1055.

9. Lenzerini M. Data integration: a theoretical perspective. In: Proceedings of the 21st ACM SIGACT-SIGMOD-SIGART Symposium on Principles of Database Systems; 2002. p. 233–46.

10. Madhavan J, Halevy AY. Composing mappings among data sources. In: Proceedings of the 29th International Conference on Very Large Data Bases; 2003. p. 572–83.

11. Maier D, Mendelzon AO, Sagiv Y. Testing implications of data dependencies. ACM Trans Database Syst. 1979;4(4):455–69.

12. Nash A, Bernstein PA, Melnik S. Composition of mappings given by embedded dependencies. ACM Trans Database Syst. 2007;32(1):4.

13. Popa L, Velegrakis Y, Miller RJ, Hernández MA, Fagin R. Translating web data. In: Proceedings of the 28th International Conference on Very Large Data Bases; 2002. p. 598–609.

14. Yu C, Popa L. Semantic adaptation of schema mappings when schemas evolve. In: Proceedings of the 31st International Conference on Very Large Data Bases; 2005. p. 1006–17.

Schema Matching

Anastasios Kementsietsidis
IBM T.J. Watson Research Center, Hawthorne, NY, USA

Synonyms

Attribute or value correspondence

Definition

Schema matching is the problem of finding potential associations between elements (most often attributes or relations) of two schemas. Given two schemas S_1 and S_2, a solution to the schema matching problem, called a *schema matching* (or

more often a matching), is a set of *matches*. A match associates a schema element (or a set of schema elements) in S_1 to (a set of) schema elements in S_2. Research in this area focuses primarily on the development of algorithms for the discovery of matchings. Existing algorithms are often distinguished by the information they use during this discovery. Common types of information used include the schema dictionaries and structures, the corresponding schema instances (if available), external tools like thesauri or ontologies, or combinations of these techniques. Matchings can be used as input to *schema mappings* algorithms, which discover the semantic relationship between two schemas.

Historical Background

A *schema matching* is most often a binary relation between the elements of two schemas, but may, in a few approaches, be a relation between sets of elements in different schemas. In general, a matching represents a potential semantic relationship but does not specify the semantics. For example, a matching between attributes $S_1.A$ and $S_2.B$ indicates that there may be some semantic relationship between these attributes. Examples of possible semantic relationships include subset relationships (e.g., all values of $S_1.A$ are also values of $S_2.B$) or has-a relationships (e.g., each value of $S_2.B$ has-a $S_1.A$ value). Consider the matches (v_1 through v_4) in Fig. 1. The matching helps in understanding the possible relationship between the schemas but is not sufficient to determine how to transform data or queries from one schema to another. In contrast, a *schema mapping* (or mapping) is a specification of the semantic relationship between schemas [8]. The discovery of matchings between elements of different schemas has been studied for decades, most notably in the context of the *schema integration* problem [1]. A solution to the schema integration problem presumes the ability to discover elements in the various schemas that are potentially *semantically related*, including those that may represent the same real-world concepts. Schema matching algorithms attempt to find candidate

Schema Matching, Fig. 1
A matching between
source and target schemas

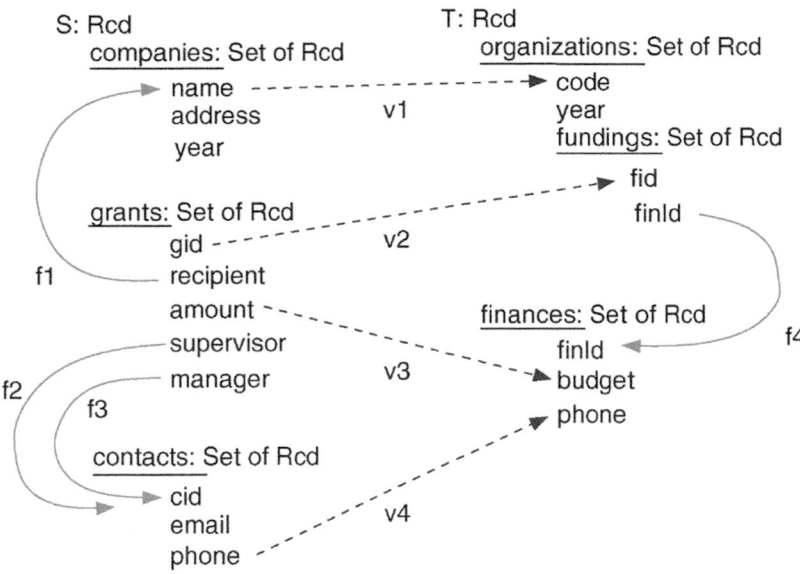

elements that may have a semantic relationship, though notably, they do not attempt to specify (or differentiate) the semantics of the relationship. These algorithms have been motivated by the presence of naming or structural differences (referred to as *conflicts*) among schemas that have been developed independently. Such differences are due to the fact that a real-world concept might have a different name or representation in different schemas. Schema integration deals with the development of methodologies to *discover* matchings in the presence of conflicts, but the main focus is on how each methodology *resolves* such conflicts (e.g., through schema transformations) so that the real-world concept is *uniquely* represented in the global schema. Indeed, from the five generic schema integration steps in each methodology, identified in [1], three of them deal with the resolution of conflicts. On the other hand, schema matching deals exclusively with the development of algorithms to *discover* matchings (see [9] for a survey of matching approaches).

Foundations

A matching between a pair of schemas S_1 and S_2 is typically a binary relation between the elements of the two schemas. In such cases, the

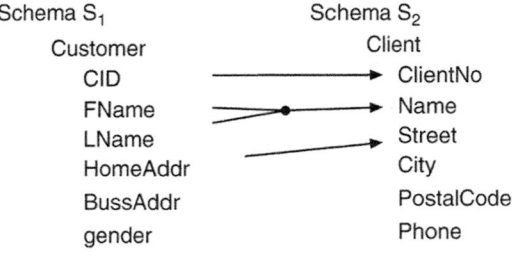

Schema Matching, Fig. 2 A matching with n:1 local cardinality

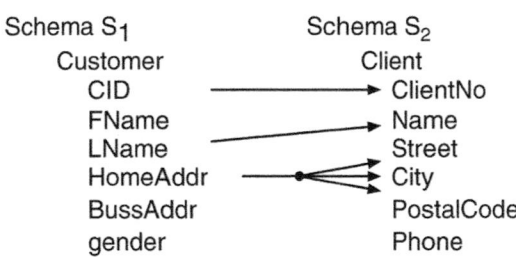

Schema Matching, Fig. 3 A matching with 1:n local cardinality

so-called *local cardinality* of the matching is said to be one-to-one (1:1). Some algorithms consider matchings between sets of elements and, in the terminology of [9], are said to have a many-to-many local cardinality (see Figs. 2 and 3). In most approaches, the matching is between individual attributes of the schemas (Fig. 1). Matching algorithms compute a matching between S_1 and S_2 through a process which can abstractly be described by the following steps:

1. Consider (possibly all) pairs of elements s_1 and s_2 with $s_1 \in S_1$ and $s_2 \in S_2$.
2. Compute a *score* indicating the confidence in the validity of each match between s_1 and s_2.
3. Compute a matching by filtering and selecting a subset of the matching elements of the previous step.

Existing matching algorithms differ on how they implement each of these steps. Each implementation needs to make some key decisions in each step, hence the substantial diversity of existing solutions. During the first step, an important consideration is whether every possible pair of schema elements will be considered as a candidate for a match [6] or whether there is a more sophisticated mechanism in place to prune the potentially unrelated element pairs considered [3]. Many approaches assume that an element can be associated with at most one element in another schema (a restriction referred to as (1:1) global cardinality in [9]). For matching algorithms that only associate elements (not sets of elements), this means the resulting matching is a simple 1:1 relation over the schema elements.

The second step is probably the most important, and there is a huge space of alternatives for the computation of scores. Score computation may take into account the name and types of schema elements, the structure of schemas and the corresponding nesting depths of elements, and instance-level information (if instances of the schemas are available) like value ranges and patterns (e.g., the frequency or position of substrings appearing in attribute values), or it may combine various types of information to compute a score. Matching algorithms are commonly classified by the type of information used during scoring computation (see the classification in [9]). The term *individual* matcher is commonly used to describe algorithms that consider only a single (or a limited) type of information during score computation. Individual matchers are further classified

into *schema based* and *instance based* depending on the type of information (i.e., schema versus instance) used during this phase. In contrast to individual matchers, *hybrid* or *composite* matchers rely on several types of information during score computation, where each type essentially corresponds to a different individual matcher. While hybrid matchers combine the results of multiple individual matchers in a *prespecified* manner, composite matchers are more flexible and allow for a dynamic composition of individual matchers which can be customized for the specific schemas being matched.

In the final step, there are two key considerations which influence the selection of matching elements that will comprise the result matching. First, the selection of matching elements is influenced by the supported *cardinality* which determines whether, or not, sets of elements (e.g., street, city, and postal code as a set as in Fig. 3) are considered in the matching.

The second consideration in this final step relates to how matching elements are selected based on their score. A common approach is to select matching elements whose scores are above a certain threshold and then select the matching elements with the maximum score, among the alternatives [6]. While such an approach results in matchings that are locally optimal, a more sophisticated approach considers maximizing the cumulative score of the matching elements in a matching [7].

State of the Art

There are many approaches to schema matching, so we offer a non-comprehensive overview of some of the representative approaches.

Cupid [6] is one of the first hybrid schema matching algorithms proposed in the context of model management. The algorithm considers initially every possible pair of elements in the two input schemas and thus its local cardinality is 1:1. It computes a linguistic and structural similarity score between these elements from which a weighted mean is computed using these two scores. The selection of matching elements in the resulting matching is performed by using a threshold over the computed scores, and the supported global cardinality is 1:1, although it is suggested that matchings with global cardinality 1:n can also be supported (Fig. 4).

Coma [4] is a composite matcher with an extensible library of single and hybrid matchers. For example, Cupid might become one of the matchers used by Coma. Both the local and global cardinality is 1:1, and each *component* matcher of Coma computes a score between every pair of elements in the input schemas. Being a composite matcher, emphasis in Coma is given on how the results of component matchers are combined, and four alternative strategies are proposed to this end. The four strategies compute the score of a matching element by taking the max, min, average, or weighted sum of scores, computed for this element, of the component matchers. Experiments with these four strategies show that average gives the best results on the schemas tested.

The majority of matchers discover matchings whose local cardinality is 1:1. The iMap [3] matcher in contrast emphasizes the discovery of complex matching elements between two schemas, i.e., matchings with a local cardinality of n:1. For each element in the target schema, iMap employs a set of *specialized searchers* to discover candidate sets of elements in the source schema that together can be associated to the target schema element. Examples of the matchers supported are a numerical matcher, which discovers matches between elements containing numerical values; a categorical matcher, for categorical attributes; a unit conversion matcher; and a date matcher. In terms of global cardinality, iMap supports n:1 matchings (Fig. 5), since it

allows an element to participate in more than one complex matching. An interesting feature of iMap is that apart from discovering (candidate) matchings, it also provides a module which traces the key decisions made by the system during matching discovery, and it can therefore present the reasoning behind a suggested matching, in a human understandable format.

Kang and Naughton [5] make the interesting observation that matching algorithms often rely on *interpreting* the element names and values in the two input schemas, that is, they assume that the names used to described the same real-world concept or entity are syntactically and semantically related (e.g., a relational column named COLOR in S_1 versus one called PAINT in S_2). Therefore, when different element names are used, for the same elements in the two schemas (e.g., the former column is called CID in S_1 and PMS (PMS stands for the Pantone Color Matching System used in various industries) in S_2) or different data encodings are used for the same real-world domain (e.g., different encodings for colors), then existing matching algorithms fail to discover appropriate matching elements. To this end, Kang and Naughton propose an instance-based matching algorithm that does not interpret values. Their matcher relies on the well-known notions of entropy and mutual information, from information theory, to discover a matching between two input schemas. In a nutshell, their approach consists of two main steps. First, for each input schema, it computes the mutual information between each pair of attributes within the schema. The second step considers each possible matching with local and global cardinality

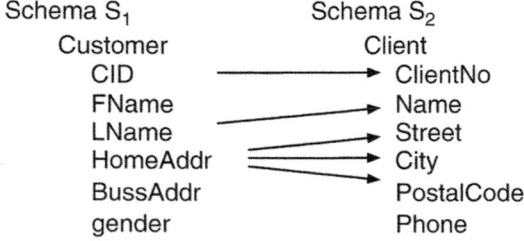

Schema Matching, Fig. 4 A 1:1 local and 1:n global cardinality matching

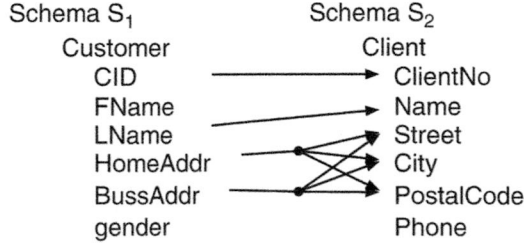

Schema Matching, Fig. 5 A 1:n local and n:1 global cardinality matching

of 1:1 and computes a score for this matching, a computation that takes into account the mutual information and entropy of the matched elements.

Key Applications

Schema mapping, schema integration

Future Directions

Schema matchings represent potential associations between a pair of schema elements (or between two sets of schema elements). Matching algorithms do not discover the meaning of the association. A match between elements $S_1.A$ and $S_2.B$ may be discovered because these attributes contain similar values, their names have a small edit distance, their names are related in a domain ontology, or of numerous other reasons. Hence, matchings by themselves are not directly useful until they have been interpreted, either by a human or a system designed to infer the semantic relationship between the elements. The most common example of the latter is schema mapping algorithms, which are designed to infer the semantic relationship between two schemas.

Cross-References

▶ Information Integration
▶ Metadata
▶ Schema Mapping

Recommended Reading

1. Batini C, Lenzerini M, Navathe SB. A comparative analysis of methodologies for database schema integration. ACM Comput Surv. 1986;18(4):323–64.
2. Bernstein PA, Halevy AY, Pottinger RA. A vision for management of complex models. ACM SIGMOD Rec. 2000;29(4):55–63.
3. Dhamankar R, Lee Y, Doan A, Halevy A, Domingos P. iMap: discovering complex semantic matches between database schemas. In: Proceedings of the ACM SIGMOD International Conference on Management of Data; 2004.
4. Do H, Rahm E. Coma – a system for flexible combination of schema matching approaches. In: Proceedings of the 28th International Conference on Very Large Data Bases; 2002. p. 610–21.
5. Kang J, Naughton JF. On schema matching with opaque column names and data values. In: Proceedings of the ACM SIGMOD International Conference on Management of Data; 2003. p. 205–16.
6. Madhavan J, Bernstein PA, Rahm E. Generic schema matching with cupid. In: Proceedings of the 27th International Conference on Very Large Data Bases; 2001. p. 49–58.
7. Melnik S, Garcia-Molina H, Rahm E. Similarity flooding: a versatile graph matching algorithm. In: Proceedings of the 18th International Conference on Data Engineering; 2002. p. 117–28.
8. Miller RJ, Haas LM, Hernández MA. Schema matching as query discovery. In: Proceedings of the 26th International Conference on Very Large Data Bases; 2000. p. 77–88.
9. Rahm E, Bernstein PA. A survey of approaches to automatic schema matching. VLDB J. 1994;10(4): 334–50.

Schema Tuning

Philippe Bonnet[1] and Dennis Shasha[2]
[1]Department of Computer Science, IT University of Copenhagen, Copenhagen, Denmark
[2]Department of Computer Science, New York University, New York, NY, USA

Definition

Schema tuning is the activity of organizing a set of table designs in order to improve overall query and update performance.

Historical Background

Table design entails deciding which tables to implement and which attributes to put in those tables. Other sections of this encyclopedia (design theory, normalization theory) discuss a mathematical model of table design to eliminate redundancy. Sometimes however redundancy can be good for performance, so database tuners must consider the possibility of a principled incorporation of redundancy.

Foundations

Normalization tends to break up the attributes of an application into separate tables. Consider the normalized schema consisting of two tables:

Blog(blog_id, author_id, title, numreaders) and *Author*(author_id, author_city).

If one frequently wants to associate blogs with the city of their authors, then this table design requires a join on author_id for each of these queries. A denormalized alternative is to add author_location to *Blog*, yielding *Blog*(blog_id, author_id, product, numreaders, author_city) and *Author*(author_id, author_city).

The *Author* table avoids anomalies such as the inability to store the location of an author whose blogs are perhaps temporarily offline.

Comparing these two schemas, one can see that the denormalized schema requires more space and more work on insertion of a blog. On the other hand, the denormalized schema is much better for finding the authors in a particular city.

The tradeoff of space plus insertion cost vs. improved speeds for certain queries is the characteristic one in deciding when to use a denormalized schema. Good practice suggests starting with a normalized schema and then denormalizing sparingly.

Redundant Tables
The previous example showed that redundancy can be helpful. The form of redundancy there was to repeat an association between two fields (in this case between authors and their address) for every blog.

One may also consider a completely redundant table.

For example:

Blog(blog_id, author_id, product, numreaders, author_city)
Author(author_id, author_city).
City_Agg(city, totalreaders)

This improves performance if one frequently wants to know the total readers per author city, but imposes an update time as well as a small space overhead. The trade-off is worthwhile in situations where many aggregate queries are issued and an exact answer is required.

Key Applications

Schema tuning is relevant for all applications, but it is especially important for complex multi-table queries, particularly involving aggregates. Data warehousing applications typically include denormalized, redundant schemas, because data warehouses can be engineered to be updated at off hours and then intensively queried during the work day.

Experimental Results

Denormalization
This experiment compares the performance impact of denormalization in the example presented above. Consider a query that finds the author in a given city. That query requires a join in the normalized schema whereas it requires a simple lookup in the denormalized schema.

Figure 1 presents the performance figures running this example on MySQL 6.0. The author table is populated with 100,000 tuples, and the blog table with 50,000. Note that the denormzalized schema provides a significant speed-off whether the cache is cold (in which case IOs are issued) or warm (the data already resides in the database cache).

Materialized Views
This experiment illustrates the trade-off between query speed-up and insert slow-down when using a materialized view. Consider the schema from

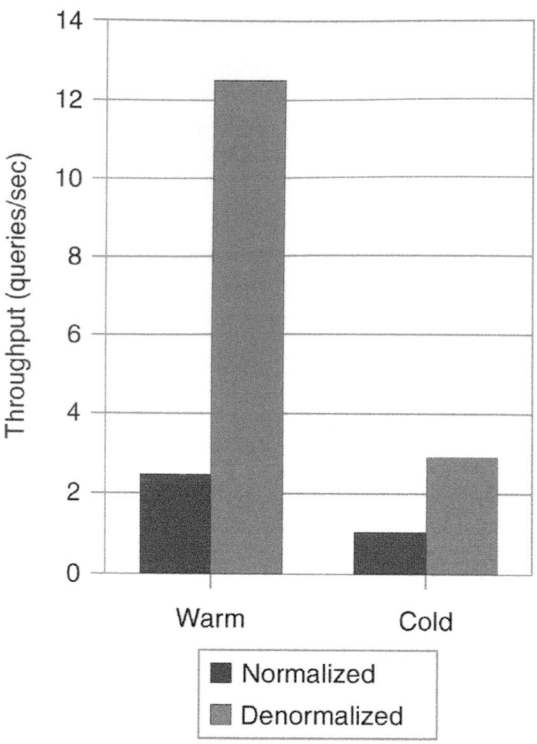

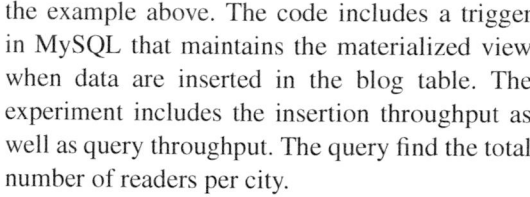

Schema Tuning, Fig. 1 Denormalization experiment on MySQL 6.0

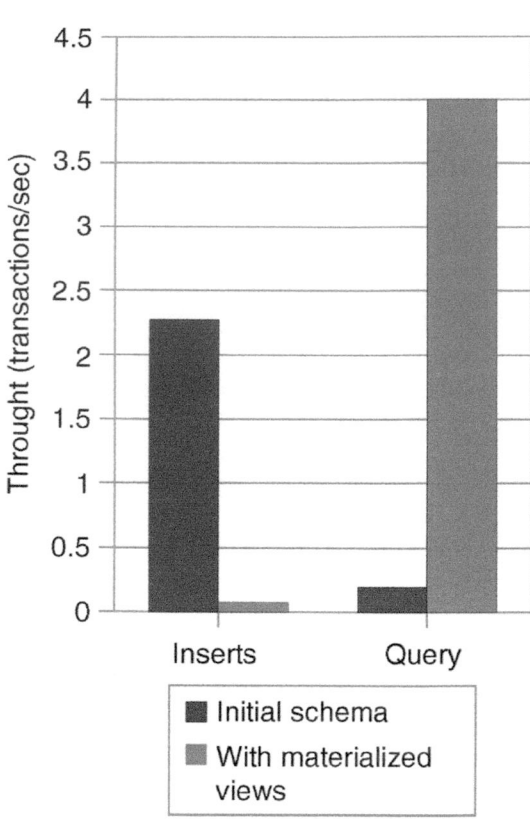

Schema Tuning, Fig. 2 Materialized view experiment on MySQL 6.0

the example above. The code includes a trigger in MySQL that maintains the materialized view when data are inserted in the blog table. The experiment includes the insertion throughput as well as query throughput. The query find the total number of readers per city.

Figure 2 shows the expected trade-off. Insertions are much slower with the materialized view due to trigger execution. Queries are much faster, however, because the query requires a join and an aggregate computation under the initial schema, whereas the query requires only a simple lookup when using the materialized view.

URL to Code and Data Sets

Denormalization experiment: http://www.datab asetuning.org/sec=denormalization

Materialized view experiment: http://www. databasetuning.org/sec = materalized_views

Cross-References

▶ Application-Level Tuning
▶ Performance Monitoring Tools

Recommended Reading

1. Celko J. Joe Celko's SQL for smarties: advanced SQL programming. 3rd ed. San Fransisco: Morgan Kaufmann; 2005.
2. Kimball R, Ross M. The data warehouse toolkit: the complete guide to dimensional modeling. 2nd ed. New York: Wiley; 2002.
3. Shasha D, Bonnet P. Database tuning: principles, experiments and troubleshooting techniques. San Fransisco: Morgan Kaufmann; 2002.
4. Tow D. SQL tuning. North Sebastopol: OReilly; 2003.

Schema Versioning

John F. Roddick
Flinders University, Adelaide, SA, Australia

Definition

Schema versioning deals with the need to retain current data, and the ability to query and update it, through alternate database structures. (The structure of a database is held in a *schema* (pl. *schemata* or *schemas*). Commonly, particularly in temporal databases, these schemata represent the historical structure of a database but this may not always be the case.) *Schema Versioning* requires not only that data are not lost in schema transformation but also requires that all data are able to be queried, both retrospectively and prospectively, through user-definable version interfaces. *Partial schema versioning* is supported when data stored under any historical schema may be viewed through any other schema but may only be updated through one specified schema version – normally the current or active schema. (Schema evolution and schema versioning has been conflated in the literature with the two terms occasionally being used interchangeably.)

Historical Background

Multiple versions of a database schema may exist for a number of reasons. First, as a result of changes in system functionality and the external environment, the structure of a database system might change over time but the historical shape of the database might need to be retained. Second, future versions might be created to develop and test later versions of a system. Third, more than one schema may be required in parallel to access the same data in a number of ways. Temporal databases, because of their requirement to maintain the context of historical information, are particularly affected by schema change.

The idea of schema versioning was introduced in the context of OODBs with a number of systems implementing techniques to handle multiple schema (such as *Encore* [12], *Gemstone* [7] and *Orion* [1]), including those that might be required for reasons other than simple historical succession. For example, parallel, alternate schema might be required to conceptualize an idea from a number of semantically consistent but different perspectives. In particular, polymorphism was suggested as a mechanism for providing some stability when faced with changing schema [6].

In order to maintain long-established concepts such as soundness and completeness, algebraic extensions have also been discussed [3]. More recently, schema versioning has also been considered in the context of spatio-temporal databases [10] and meta-data management [2, 4].

Foundations

Schema versioning is closely related to the concepts of schema integration and data integration – all deal with the problems of accessing data through schema that were not used when the data were originally stored. However, the idea of maintaining multiple schemata, and allowing data to be accessed through them, raises a number of issues.

- What is the significance of a difference between two schema (or two databases) and therefore what is the informational cost of the change?
- What are the atomic operations of schema translation or transformation and what happens to the data during these operations?
- Are there any modelling techniques that can be used?
- Are there any other side-effects or opportunities (for instance in query language support)?

Types of Schema Evolution

As outlined elsewher, four forms of schema evolution can be envisaged – attribute, domain, relation and key evolution. Moreover, one change may involve more than one type of evolution, such as changing the domain of a key attribute and may also be reflected in the conceptual model of the system. Importantly for schema versioning, the inverse function for each of these must be considered. For example, when a schema (merely) evolves by vertically splitting a relation in two with data being suitably transformed, for schema versioning to be allowed, active transformation functions must be provided if the old schema is still to be utilized.

Practical and Theoretical Limits of Schema Versioning

It has been shown that in order to update data stored under two different schemata using the opposite schemata, they must have equivalent information capacity – all valid instances of some schema S_1 must be able to be stored under S_2 and vice-versa [5]. Specifically, $S_1 \equiv S_2$ if $I(S_1) \rightarrow I(S_2)$ is bijective where $I(S)$ is the set of all valid instances of S. This means that, in theory, *full* schema versioning across nonequivalent versions of a schema is unattainable and much research in the area adopts the weaker concept of *partial schema versioning* in which data stored under any historical schema may be viewed through any other schema but may only be updated through one specified schema version – normally the current or active schema.

However, in practice, many schema changes that expand or reduce the information capacity of a schema can be done without loss of information. This is the case, for example, for domains defined too large for any of the data, or for the creation of subclass relations from a single relation where the subclass type attribute already exists. It is a common practice, where there is some ambiguity in the requirements definition of a system, to allow for a larger schema capacity – some of which may never materialize and, as a result, changes to schemata to adhere to the data actually collected are not uncommon. For example, allowing for time and date when only date is recorded in practice.

Thus the limits for *practical schema versioning* in a database $\mathcal{D}$ are that $S_1 \overset{p}{\equiv} S_2$ (S_1 and S_2 have practical equivalent information capacity)

if $I'(\mathcal{D} \mid S_1) \rightarrow I'(\mathcal{D} \mid S_2)$ is bijective where $I'(\mathcal{D}|S_n)$ is the set of all instances of S_n inferrable from $\mathcal{D}$ given the constraints of S_n. This means that whether the integration of two schema is possible is dependent on the data held as well as the schema definition and while this makes the ability to undertake wholesale change less predictable, it may provide an acceptable level of support in many practical situations.

Completed Schemas

In order to make all data for a relation available without the need to issue multiple queries, each targeting different time periods, the concept of a *completed schema*, C, can be employed that includes all attributes that have ever been defined over the life of a relation. The domain of each attribute in C is considered syntactically general enough to hold all data stored under every version of the relation and the implicit primary key of C is defined as the maximal set of key attributes for the relation over time. Depending on the mechanism used to implement schema versioning, the completed schema can then be used by a series of view functions. For example, in Fig. 1, V_{t_5} maps the completed schema C to a subset of the attributes in a schema S_{t_5} active during t_5. A converse view function W_{t_2} maps from S_{t_2} to C.

Thus the data stored during t_2 may be mapped to the format specified during t_5 through invocation of $V_{t_5}(W_{t_2}(S_{t_2}))$.

Query Language Support

Support for schema versioning does not yet exist in commercially available query languages. However, the TSQL2 proposal [9] and an earlier SQL/SE proposal [8] outlined some parts of the solution. As examples of such extensions:

- Reference to the *completed schema* can be included to provide access to all data;
- The specification of the schema could be done either through the specification of a global *schema-time* as in TSQL2, which would be useful for SQL embedded in a program with

**Schema Versioning,
Fig. 1** Versions of
Schemata over time

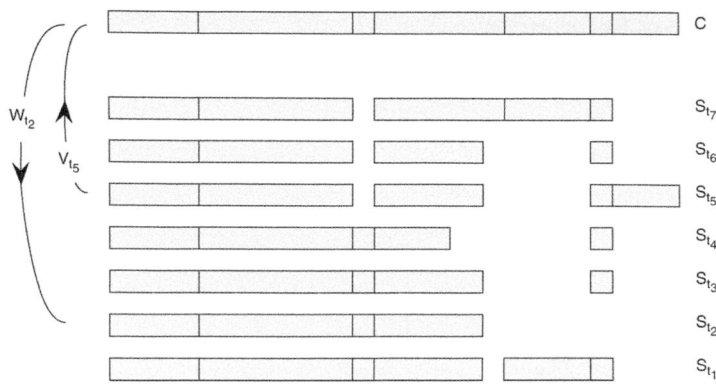

the schema-time set to compile time, or explicitly as part of the query;

- Attribute definition might be able to be tested by adding a test to see if a value was missing because it was *not defined* rather than being merely *null*;
- The language may also include meta-data queries such as the ability to ask what version of the schema a given piece of data adheres to.

Instance Amendment

For schema versioning in which the old schemata are still considered valuable, once a schema change is accepted, there are three options regarding the change to existing data. First, the underlying instances may be coerced to the new structure. While conceptually simple, this may result in lost information and an inability to reverse schema amendments. Secondly, data are retained in the format in which it was originally stored. This retains information content at the expense of more complex (and slower) translation of data when needed. Third, data are initially retained in the format in which it was originally stored but is converted when amended. While the most complex option, it has the advantage of identifying data that has not been amended since the schema change.

Future Directions

Schema versioning research has moved from low-level handling of syntactic elemental changes to more model-directed semantic handling of change. There are a number of other issues that make schema versioning non-trivial. Some of these represent future issues to be investigated.

- Many schema change requirements involve composite operations and thus a mechanism for schema level commit and rollback functions could be envisaged which could operate at a higher level to the data level commit and rollback operations.
- Access rights considerations are particularly a problem in object-oriented database systems. Consider, for example, a change to a class (e.g., Employees) from which attributes are inherited to a sub-class (e.g., Engineers) for which the modifying user has no legitimate access. Any change to the definition of attributes inherited from the superclass can be considered to violate the access rights of the subclass. Moreover, in some systems ownership of a class does not imply ownership of all instances of that class.
- In temporal databases the concept of vacuuming (q.v.) allows for the physical deletion of temporal data in cases where the utility of holding the data are outweighed by the cost of doing so [11]. Similar consideration must be given to the deletion of obsolete schema definitions, especially in cases where no data exists adhering to either that version (physically) or referring, through its transaction-time values, to the period in which the definition was active.

S

Cross-References

▶ Conceptual Modeling Foundations: The Notion of a Model in Conceptual Modeling
▶ Schema Evolution
▶ Temporal Algebras
▶ Temporal Query Languages

Recommended Reading

1. Kim W, Ballou N, Chou HT, Garza JF, Woelk D. Features of the orion object-oriented database system. In: Kim W, Lochovsky F, editors. Object-oriented concepts, databases and applications. New York: ACM Press; 1989. p. 251–82.
2. Madhavan J, Halevy AY. Composing mappings among data sources. In: Proceedings of the 29th International Conference on Very Large Data Bases; 2003; p. 572–83.
3. McKenzie L, Snodgrass R. Schema evolution and the relational algebra. Inf Syst. 1990;15(2):207–32.
4. Melnik S, Rahm E, Bernstein PA. Rondo: a programming platform for generic model management. In: Proceedings of the ACM SIGMOD International Conference on Management of Data; 2003. p. 193–204.
5. Miller R, Ioannidis Y, Ramakrishnan R. The use of information capacity in schema integration and translation. In: Proceedings of the 29th International Conference on Very Large Data Bases; 1993. p. 120–33.
6. Osborn S. The role of polymorphism in schema evolution in an object-oriented database. IEEE Trans Knowl Data Eng. 1989;1(3):310–7.
7. Penney D, Stein J. Class modification in the gemstone object-oriented DBMS. In: Proceedings of the 1987 Conference on Object-Oriented Programming Systems, Languages, and Applications; 1987. p. 111–17.
8. Roddick JF. SQL/SE – a query language extension for databases supporting schema evolution. ACM SIGMOD Rec. 1992;21(3):10–6.
9. Roddick JF, Snodgrass R. Schema versioning support, chapter 22. In: Snodgrass R, editor. The TSQL2 temporal query language. Boston: Kluwer; 1995. p. 427–49.
10. Roddick JF, Grandi F, Mandreoli F, Scalas MR. Beyond schema versioning: a flexible model for spatio-temporal schema selection. Geoinformatica. 2001;5(1):33–50.
11. Skyt J, Jensen CS, Mark L. A foundation for vacuuming temporal databases. Data Knowl Eng. 2003;44(1):1–29.
12. Zdonik S. Version management in an object-oriented database. In: Proceedings of the International Workshop on Advanced Programming Environments; 1986. p. 405–22.

Scientific Databases

Amarnath Gupta
San Diego Supercomputer Center, University of California San Diego, La Jolla, CA, USA

Definition

Scientific data refers to data that arise from scientific experiments, instruments, analytical tools, and computations. A chemistry experiment, for example, can yield data about the experimental setup, the pressure and temperature conditions under which the experiment was set up, measured variable like the heat released, initial and final masses the ingredients and products of the experiment, and so forth. The output of an instrument like a radio telescope, after running signal processing algorithms, will produce "images" of the radio-frequency sources in a part of the sky that the telescope was looking at. A biologist, after obtaining the image of a dye-filled nerve cell, uses image analysis software to produce a set of measurements that reflect the structure of the cell and its subparts. Recently, environmental sensors are cast in oceans and send real-time data on ocean temperature, salinity, oxygen content, and other parameters. A scientific database refers to an information management framework need to store, organize, index, query, analyze, maintain, and mine such heterogeneous scientific data.

Historical Background

Investigation in data management techniques for scientific data started with studying file organization principles that are tuned toward specific kinds of scientific data [1], where the problem explored was to develop a multi-query indexing scheme for scientific records. Discipline-specific systems with data retrieval capabilities were also developed. Coughran [2] describes a systems called Hydrosearch for worldwide hydrographic data from oceans that supported range queries like:

OCEAN = PACIFIC
OUTPUT = REPORT
LATHEM = N
LONHEM = W
LATDEG GEQ 31 AND LATDEG LEQ 33
LONDEG GEQ 121 AND LONDEG LEQ 123
MONTH GEQ 3 AND YEAR = 63 OR MONTH
    LEQ 2
MXDPTH GEQ.98 DBOT

Around the same time, the National Laboratories dealing with a wide category of data related to energy research recognized that "such diverse data applications as material compatibility, laser fusion, magnetic fusion, test, equation of state, weather, environmental and demographic data, has an acute need for a Scientific Data Base Management System (SDBMS)... The large volume of data, the numeric values within an epsilon of accuracy, the unknown data relationships, the changing requirements, coupled with the overall goal of extracting new intelligence from the raw data, dictate a data base system tailored toward scientific applications. Such an SDBMS should support scientific data types, a relational end user view, an interactive user language, interfaces to graphical and statistical packages, a programming language interface, interfaces to existing facilities, extensibility, portability, and use in a distributed environment" [3]. The system in [3] was developed on the CODASYL model [4].

It was in the 1980s that these different efforts started to take a coherent form where some general characteristics of scientific data management were identified. It was recognized through papers like [5] that unlike business data that is usually defined by a data schema and values conforming to the schema, scientific data can come with a measurement framework, a metadata specification, and often a summarization framework. In [6] the database requirements for scientific data were characterized, and a new conference started in 1986 devoted to the issues managing scientific and statistical data.

Foundations

As alluded to in the previous paragraph, scientific data are usually more than the values of attributes – they are often accompanied by additional descriptors which together specify the semantics of the data and therefore determines how the data can be interpreted for query and analytical operations. Some broad categories of these additional components that specify the context of the raw scientific data are described below.

Measurement Framework

A measurement framework is a specification of the setting of the experimental data. For time series data, it may be the sampling frequency. For spatial data represented as a raster, it includes the resolution of the grid and how the measurement is obtained per grid (e.g., once at the center of the grid, average of n measurements within the grid...). For finite element data such as data from a fluid mechanics model, it may be the nature and regularity of the mesh over which data are recorded. The measurement framework is important in understanding the semantics of the data. If there are two raster data sets containing the measured temperature of two overlapping regions such that (i) one has a finer resolution than another and (ii) one has the average-of-the-grid semantics, while the other has a single-sample-at-center semantics, how can one define a join operation to combine the two data sets? One cannot simply define a band join without considering a way to homogenize the data sets before they can be joined. Another aspect of the measurement framework is an assessment of the uncertainty associated with the measured, estimated, or computed data. When data are associated with uncertainty, the traditional data models do not suffice – probabilistic or uncertainty – aware data models and query evaluation techniques are needed.

Metadata Framework

Metadata refers to descriptors that provide additional semantics beyond the value of an attribute.

These include the unit of measurement, the precision and accuracy of the data value, the uncertainty associated with the temporal or spatial position at which the data are taken, and the experimental setting including whether the data are absolute or relative to any other reference and what if any computational corrections should be made on the data before it can be delivered to an end user. Metadata also covers constraint statements that limit the allowable domain of data values or additional conditions that must be satisfied for the data to be interpreted. These may range from simple encoding schemes that specify "out of range" or "unknown" values to multi-attribute constraints like "data are valid only if the cloud cover coefficient at the location is less than 0.2." For data that are produced by computational algorithms (such as simulations of natural phenomena), the metadata also consists of the parameter settings of the algorithms, which must be taken into account to interpret, compare, and analyze the data.

Summarization Framework

Scientific data are often voluminous due to high degree of sampling or the total time or space over which data are acquired. In many applications, the total amount of data are too much and noninformative, and the scientists maintain only summarized versions of the data. For example, one may keep only weekly average temperature obtained from satellites; alternatively, one might keep the information only when there is a significant local change in the data. Since this is a common practice in many scientific disciplines, specification of how the data was summarized is a form of information that needs to go together with the data itself.

Heterogeneity of Types

An important characteristic of scientific data is the wide range of complexity and heterogeneity of the data types that are needed to model the applications.

Complexity and Heterogeneity of Formats

Distinct from the issue of data types, scientific data demonstrates a wide variety of formats for the same kind of data. In some domains, there is a lack of a single standard, and vendors of instruments that provide data, or vendors of software that manipulate the data define their own formats to facilitate their respective needs for data generation, analysis, and visualization. For example, biological pathways, which are essentially graphs with node and edge attributes, are represented differently by software such as Cytoscape [7], PATIKA [8], Pathway Studio [9], and the standardization effort called BioPAX (www.biopax. org) In other domains, there is more standardization. But the formats are very complex. HDF (hierarchical data format) is a complex scientific data format for storing multidimensional data, raster data, and tables. It is also designed to be self-describing and contains additional metadata. The multiplicity of supported models and the embedded metadata requires special data management tools [10] to be developed for indexing and querying HDF data. A consequence of this format heterogeneity and complexity is that interoperability of scientific data remains a research challenge.

Data Management Issues in Scientific Databases

Traditional Issues
In 1985, [6] identified a number of data management issues that pertain to scientific data management. Many of these issues that hold equally well today are:

- *Data volume and compression*: Much of scientific data are multidimensional. While the data sets can be very large, the fraction of the multidimensional space that is occupied by the data is smaller. This brings up the need to compress the data as well as to choose a data organization that will exploit the sparsity of data. Further, data manipulation and query evaluation techniques that utilize the compressed data or a new data organization are needed. Managing large-scale scientific data is now considered to be an important challenge. A notable project in this area is led by the Stanford Linear Accelerator (SLAC – http://www.slac.stanford.edu/) that is attempting to build a data management system for a petabyte of data.

- *Data structures*: With new scientific data types, there is a need for new data structures and access methods. In recent years, a number of index structures have been proposed for multidimensional data. For example, Zhang et al. [11] proposed a data structure for sampling multidimensional data; Rotem et al. [12] have developed bitmap indexes for very large-scale multidimensional data.

- *New operations*: Data manipulation and search in scientific databases need operators that go beyond traditional relational or tree manipulation algebras. It requires new operations such as sampling, neighborhood searching in metric space, estimation and interpolation operators for sampled data over a dense data space, and novel join methods for complex data types. New generation databases like SciDBfor arrays (http://www.paradigm4.com/) and Neo4j for graphs (http://www.neo4j.org/) allow type-specific operations as part of their data manipulation and query languages.

- *Analysis support*: The ultimate goal of scientific data acquisition and storage is some form of analysis and derivation of scientific truth. Scientists, the primary users of the database, are not often willing to learn complex query languages – instead, they want to query the data as part of their analytical tasks. The management of the entire analysis process require analysis-friendly user interfaces that are sufficiently expressive but not overly complex, a way to facilitate repeated use of the same query with differing parameters, management of long-running queries, handling large volumes of intermediate data, and optimal execution of an entire analytical workflow.

- *Quality management*: Quality awareness of data is important when the data collected in a database comes from any error prone process. For scientific data, errors and approximations arise often due to factors like resolution limits of instruments, malfunctioning of devices, unforeseen environmental or experimental confounding factors, biases introduced by sampling, approximations used by preprocessing computations and, of course, human error. The problem gets compounded when a data product is derived from an existing data product. In many cases, data for a given application may come from data sources with different quality and "believability." Query languages, evaluation techniques, and analysis for scientific data need to be quality-aware and give a user the ability to filter data based on its quality and integrity.

Recent Trends

More recently, the scientific data management community has identified newer challenges over and above the issues above. Some of them are:

- *Annotation management*: Annotating data is a common practice in science. An annotation is a piece of user-imposed data that references an arbitrary data element in an existing data store. One can annotate a block of data with a statement about its quality; one may annotate a fragment of data with information of its provenance (i.e., where the data was obtained from and how it was transformed before it appeared in its present form); one may annotate data by tagging it with keywords or terms from an ontology so that it can be easily related to other data. Annotation management attempts to create a uniform way to store the annotation and their referent data so that both the primary data and the annotations can be queried together.

- *Semantics*: In many domain sciences, the semantics of data is not adequately represented in data repositories. This makes it very difficult to one user to interpret data from another user and even harder to combine multiple kinds of data together. This recognition has led to a renewed interest in developing semantic data models for scientific data. As part of this effort ontologies are being created to standardize and define the terms and the interterm relationships in a discipline using standards like the Web Ontology Language, OWL, so that data producers can either use the ontological terms to represent their data or map their existing data to the ontologies. At the same time, efforts are underway to develop query, integration, and data mining techniques that make use of the semantic framework.

• *In-database analytics*: While the need for combining data with analytics is well known, the traditional ways to accomplish this were less than optimal. The most common practice was to decouple querying and analysis; the user would query a DBMS to get a subset of data, put the results in a file, postprocess the file to fit an input format accepted by an analysis engine, and perform the analysis in the engine. Over time, the analysis engines (e.g., SAS, R) provide connectivity to a DBMS; the user queries the DBMS from inside the analysis engine, the data is transformed into a convenient format inside the engine, and the analysis is performed. When the data to analyze is very large, this leads to a bottleneck because the data transfer happens in memory, and the available memory of the analysis engine may be smaller than the data volume. The recent trend is to take the computation to the data, i.e., perform the analysis inside the DBMS environment. This takes two forms. In the first, the DBMS provides more primitive operations (e.g., statistical computations, time-window operations) so that the scientist can program with DB functions (e.g., Vertica, see https://my.vertica.com/). In the second, an external analysis engine is customized to perform smooth data transfer (e.g., to take advantage of the paging offered by the DBMS) between the DBMS and the engine (e.g., Fuzzy Logix, see http://www.fuzzl.com). The users would write their own analysis functions in a familiar language (e.g., R or C++). In one variant, these analyses will be treated as user-defined

functions and will be executed in the same process space as the DBMS. In another variant, the DBMS will invoke the external analysis engine (e.g., R) but manage the data transfer efficiently.

• *Model databases*: Statistical modeling of data lies on the border of data mining applications as well as scientific data management. There is now a recent upsurge in combining statistical modeling with data operations. The widespread use of sensor networks encourages researchers to not only acquire and store the data but also to construct and store statistical models of the data along with the data itself. These systems [13, 14] can be viewed as data sources with an infinite number of values – only a small portion of the data are actually observed, while the rest is computed by the model during query processing time. For this purpose, some systems like Splash [14] have defined a model specification language in addition to the system's query and data manipulation languages.

Cross-References

▶ Data Types in Scientific Data Management

Recommended Reading

1. Ikeda H, Naito M. Evaluation of a combinatorial file organization scheme of order one. In: Proceedings of the Study on Scientific Database Management Systems; 1979. p. 195–199 (in Japanese).
2. Coughran E. HYDROSEARCH, an easy-to-use retrieval system for hydrographic station data. In: Proceedings of the First Combined IEEE Conference on Engineering in the Ocean Environment and the Annual Meeting of the Marine Technology Society; 1975. p. 418–21.
3. Birss EW, Jones SE, Ries DR, Yeh JW. Scientific data base management at Lawrence Livermore Laboratory: needs and a prototype system. Technical Report UCRL-80146;CONF-771062–1, Lawrence Livermore Lab, California University. 1977.
4. Olle TW. The Codasyl approach to data base management. Chichester: Wiley; 1978.
5. Shoshani A, Olken F, Wong HKT. Characteristics of scientific databases. In: Proceedings of the 10th

International Conference on Very Large Data Bases; 1984. p. 147–60.

6. Shoshani A, Wong HKT. Statistical and scientific database issues. IEEE Trans Softw Eng. 1985;11(10):1040–7.

7. Shannon P, Markiel A, Ozier O, Baliga NS, Wang JT, Ramage D, Amin N, Schwikowski B, Ideker T. Cytoscape: a software environment for integrated models of biomolecular interaction networks. Genome Res. 2003;13(11):2498–504.

8. Demir E, Babur O, Dogrusoz U, Gursoy A, Nisanci G, Cetin-Atalay R, Ozturk M. PATIKA: an integrated visual environment for collaborative construction and analysis of cellular pathways. Bioinformatics. 2002;18(7):996–1003.

9. Nikitin A, Egorov S, Daraselia N, Mazo I. Pathway studio – the analysis and navigation of molecular networks. Bioinformatics. 2003;19(16):2155–7.

10. Gosink L, Shalf J, Stockinger K, Wu K, Bethel W. HDF5-FastQuery: accelerating complex queries on HDF datasets using fast bitmap indices. In: Proceedings of the 18th International Conference on Scientific and Statistical Database Management; 2006. p. 149–58.

11. Zhang X, Kurc T, Saltz J, Parthasarathy S. Design and analysis of a multi-dimensional data sampling service for large scale data analysis applications. In: Proceedings of the 20th International Parallel and Distributed Processing Symposium; 2006.

12. Rotem D, Stockinger K, Wu K. In: Proceedings of the 18th International Conference on Scientific and Statistical Database Management; 2006. p. 33–44.

13. Deshpande A, Samuel M. MauveDB: supporting model-based user views in database systems. In: Proceedings of the ACM SIGMOD International Conference on Management of Data; 2006. p. 73–84.

14. Fang L, Kristen L. Splash: ad-hoc querying of data and statistical models. In: Proceedings of the 13th International Conference on Extending Database Technology; 2010.

Scientific Visualization

Ronald Peikert
ETH Zurich, Zurich, Switzerland

Definition

Scientific visualization [1] provides graphical representations of numerical data for their qualitative and quantitative analysis. In contrast to a fully automatic analysis (e.g., with statistical methods), the final analytic step is left to the user, thus utilizing the power of the human visual system. Scientific visualization differs from the related field of information visualization in that it focuses on data that represent samples of continuous functions of space and time, as opposed to data that are inherently discrete.

The challenge in scientific visualization is to cope with massive data, which cannot be presented to the user in an unprocessed way for several reasons:

1. Volumetric data, i.e., data given on a three-dimensional domain, occlude each other. This problem becomes even more challenging if data are not scalars, but vectors or even tensors.
2. Visualization should provide a global picture of the spatial and temporal behavior of the data, but also allow for interactive exploration of details.
3. There can be multiple data (different physical quantities, multiple data channels, etc.) at each point in the domain.
4. Visualization of scientific data should also include visualization of their uncertainty.
5. The amount of raw data often exceeds limitations of processor speed, transfer rates, memory size, and display resolution.

Applications of scientific visualization cover a wide spectrum of science and engineering disciplines. Currently, some of the most active fields are medical and biomedical image data, simulation and measurement data from fluid or solid mechanics, molecular data, data from geology and geophysics, astronomy, weather and climate.

Key Points

Scientific visualization evolved in the 1980s from earlier graphing techniques when 3D computer graphics opened new ways of displaying numerical data. The abstraction from the application domain led to interdisciplinary visualization software systems with a modular dataflow architecture. This approach is still successful [2], since for many basic visualization tasks, the semantics of the data is less relevant than mathematical properties such as the discretization type or the categorization into scalars, vectors, and tensors.

For volumetric scalar data, important visualization techniques are isosurfaces and direct volume rendering. For vector fields, examples are arrow glyphs, integral lines (streamlines, streaklines) and texture advection. Tensor fields are visualized with glyphs (ellipsoids, superquadrics) or tensor lines. In the special case of diffusion tensor MRI data, fiber tracking techniques are applied. Scalar, vector, and tensor fields all are amenable to topology-based visualization, which provides both the singularities and a segmentation of the domain into regions of "similar data behavior". More generally, feature extraction and feature tracking techniques aim at reducing the data complexity and providing the viewer with only the most salient information. Features (e.g., edges, ridges, flow structures) are typically defined in terms of data and their spatial derivatives. User-defined feature definitions are possible in visualization systems built on the linked views paradigm, where simultaneous views of both physical and data space are available all of which allow for interactions such as data coloring and subsetting. The visualization of very large data requires optimization techniques including multi-resolution, parallel and out-of-core algorithms, as well as view-dependent visualization.

Cross-References

▶ Data Visualization
▶ Visualization Pipeline

Recommended Reading

1. Hansen CD, Johnson CR, editors. Visualization handbook. San Diego: Academic; 2004.
2. Schroeder W, Martin K, Lorensen B. The visualization toolkit: an object-oriented approach to 3D graphics. 4th ed. New York: Kitware; 2006.

Scientific Workflows

Bertram Ludäscher, Shawn Bowers, and Timothy McPhillips
University of California-Davis, Davis, CA, USA

Synonyms

Grid workflow; In silico experiment

Definition

A *scientific workflow* is the description of a process for accomplishing a scientific objective, usually expressed in terms of *tasks* and their *dependencies*. Typically, scientific workflow tasks are computational steps for scientific simulations or data analysis steps. Common elements or stages in scientific workflows are acquisition, integration, reduction, visualization, and publication (e.g., in a shared database) of scientific data. The tasks of a scientific workflow are organized (at design time) and orchestrated (at runtime) according to dataflow and possibly other dependencies as specified by the workflow designer. Workflows can be designed visually, e.g., using block diagrams, or textually using a domain-specific language.

Historical Background

Workflows have a long history in the database community and in business process modeling, in which case they are sometimes called *business workflows* to distinguish them from scientific workflows. The database community realized early [10] that scientific data management has different characteristics from more traditional business data management. Early work on scientific workflows within the database community took a database-centric view by defining data models and query languages suitable for scientific experiment management systems. The MOOSE data model and FOX query language have their roots in the late 1980s [5] and early 1990s [13] and gave rise to the ZOO experiment management environment [6], an early system based on an underlying object-oriented database. Another pioneering work that emphasized the importance of workflow concepts in scientific data management is WASA, a *Workflow-based Architecture for Scientific Applications* [8]; the related publication [12] introduced the term "scientific workflow" and contrasted such workflows with office automation and business workflows. An early benchmark comparing different database architectures for scientific workflow applications is LabFlow-1 [1].

Other roots of scientific workflow systems include *problem solving environments*, which emerged in the 1990s in the computational sciences community as intuitive tools to "solve a target class of problems for scientific computing" [4], and *laboratory information management systems* (LIMS) [9], which can be seen as special scientific workflow systems that are used in a laboratory environment for the management of samples, instrument-based measurements, and other functions, including data analysis and workflow automation. Similar to many scientific workflow systems, problem solving environments and LIMS sometimes employ a visual programming paradigm to link together components. An early, if not the first, visual language that allowed simple interfacing with lab instruments was G in LabVIEW1.0, released in 1986 for the Apple Macintosh. Modern incarnations of LIMS can include functions of enterprise resource planning (ERP) systems and thus go beyond the scope of current scientific workflow systems.

With the advent of *e-Science* as a paradigm, scientific workflow research and development has seen a major resurgence. Similar to the related term *cyberinfrastructure*, e-Science brings together computational techniques and tools from the computational sciences, distributed and high-performance computing, databases, data analysis, visualization, sharing, and collaboration. There are now a number of new open source as well as commercial scientific workflow systems available

and under active development. For example, a special journal issue of *Concurrency and Computation: Practice and Experience* covers a number of systems, including Kepler, Taverna, and Triana among others [2]. For a high-level overview and attempt at a classification of current scientific workflow systems see [14], which includes also references to many other systems, such as Askalon, Pegasus/DAGMan, Karajan, etc.

Foundations

Science is an exploratory process involving cycles of observation, hypothesis formation, experiment design and execution. Today, scientific knowledge discovery is increasingly driven by data analysis and computational methods, e.g., due to ever more powerful instruments for observation and the use of commodity clusters for high-performance scientific computing and simulations in the computational sciences. Scientific workflows can be applied during various phases of the larger science process, specifically modeling and automation of computational experiments, data analysis, and data management. The results from workflow runs can yield new data and insights and thus may lead to affirmation, modification, or refutation of a given hypothesis or experiment outcome.

Scientific workflow systems automate the execution of scientific workflows, and may additionally assist in workflow design, composition, and the management and sharing of workflow descriptions. Other important functions include support for workflow execution monitoring, for recording and querying provenance information, for workflow optimization (e.g., exploiting dataflow and concurrency information for parallel execution), and for fault-tolerant execution. These additional features also distinguish a scientific workflow systems approach from more traditional script-based solutions in which such functionality is usually not provided. Workflow provenance information can be used, e.g., to facilitate the interpretation, debugging, and reproducibility of scientific analyses. An increasing number of scientific workflow systems now offer support for various forms of provenance. One can distinguish *data provenance*, i.e., the processing history of data, and provenance information describing the *workflow evolution*, i.e., the history of changes of a workflow definition and the parameter settings used for a particular workflow instance.

Scientific workflows are often visually represented as directed graphs (Figs. 1 and 2) linking atomic tasks or composite components, so-called *subworkflows*. Tasks can include native functions of the workflow system, but often correspond to invocations of localapplications, remote (web) services, or subworkflows. Scientific workflows differ from conventional programming in that the workflows are often more coarse-grained and involve wiring together of pre-existing components

and specialized algorithms. Figure 1 shows a simple bioinformatics workflow in the Taverna system, consisting of multiple (soaplab) services.

There is currently no standard *scientific workflow language*, and standards from related communities (e.g., BPEL4WS) have not found widespread adoption in the scientific workflow community. For example, job-based *grid workflows* are often represented as directed acyclic graphs (DAGs), which are then scheduled on a computational grid or cluster computer according to the implied task dependencies. In this *model of computation*, each task is executed only once per workflow run and task scheduling amounts to finding a *topological sort* for the partial order implied by the DAG. Other more sophisticated models of computation consider tasks as independent and continuously executing processes which can receive and

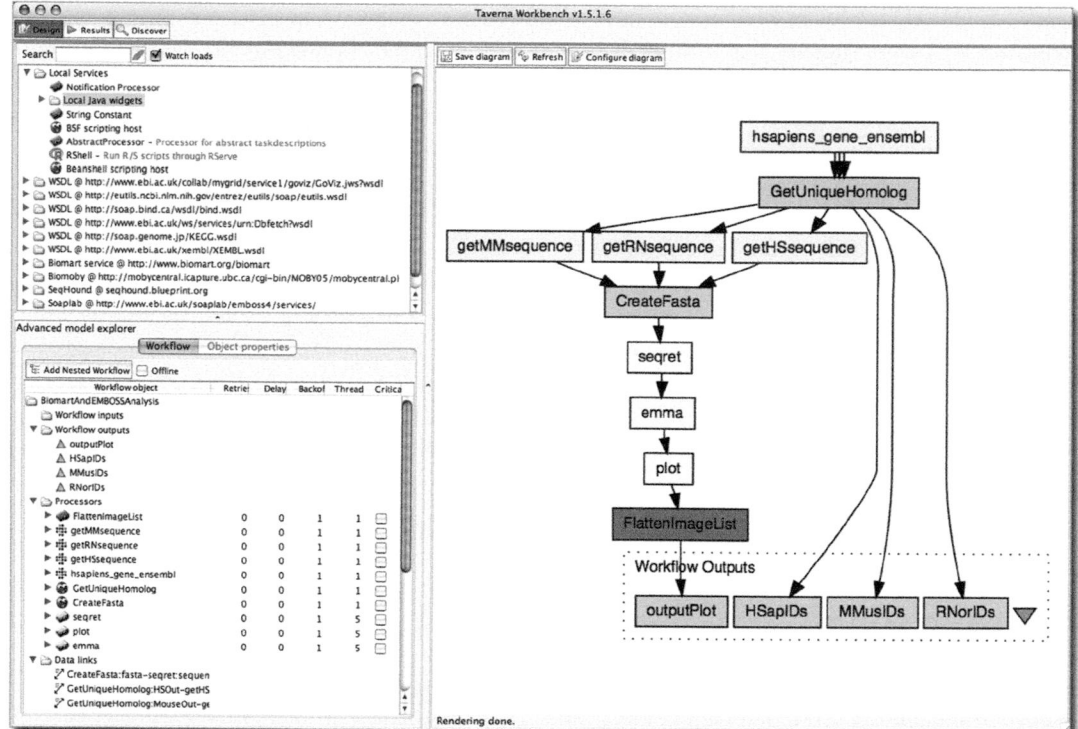

Scientific Workflows, Fig. 1 Example workflow represented in the Taverna workflow system. This workflow extracts gene IDs from human chromosome 22 with mappings to disease functions and homologues in mouse and rat; fetches base pairs of the associated DNA sequences; combines the sequences into a FASTA file; performs a multiple sequence alignment; and renders the result. The workflow uses three soaplab-based analysis operations (seqret, emma, plot) that run on the EBI compute cluster

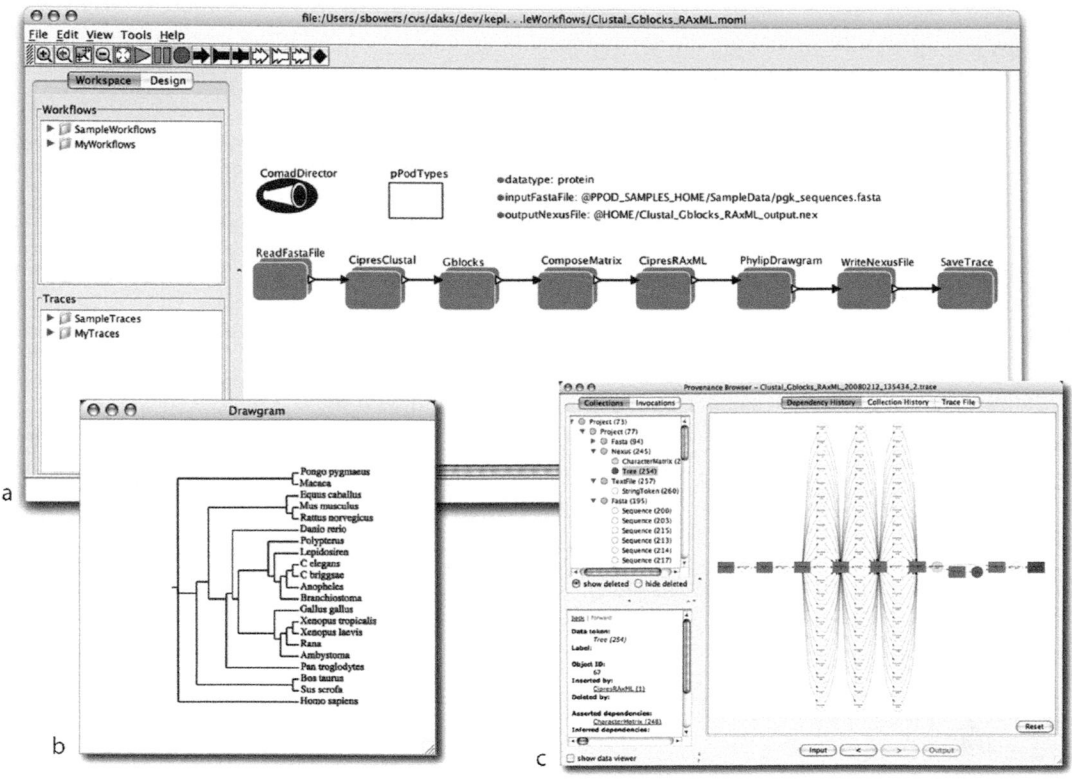

Scientific Workflows, Fig. 2 Example scientific workflow in the Kepler system: (**a**) user interface for creating, editing, and executing scientific workflows; (**b**) a visual representation of the data product (a phylogenetic tree) computed by a workflow run; and (**c**) a viewer for navigating the data provenance (lineage) captured in an execution trace. This workflow uses a combination of local and remote (web) services to perform multiple sequence alignment and phylogenetic tree inference on input DNA sequences

send many different data items per workflow run. Scientific workflow systems that support such models of computation may thus be used for *data stream processing* and *continuous queries*. Similar to business workflows, formal approaches such as *Petri nets* can be used to describe scientific workflow execution semantics. However, the dataflow models of computation of many scientific workflow systems can exhibit both task- and pipeline-parallelism where token order is important. A standard computation model for such dataflow systems is the *Kahn Process Network* model. The structurally simple linear Kepler workflow in Fig. 2 is achieved via a special model of execution, implemented by a so-called *director*. (Kepler inherits from the underlying Ptolemy II system the capability to use distinct directors at different workflow

modeling levels and thus to combine different models of computation in a single workflow.) The COMAD (*Collection-Oriented Modeling And Design*) director in Fig. 2 specifies that workflow components work on a continuous, XML-like data stream which passes through all components eventually. Each component is configurable to compute only on certain (tagged) data collections. Results are injected back into the stream. The resulting more linear workflows are easy to comprehend and evolve over time, another important advantage over script-based solutions.

Key Applications

Scientific workflows now span virtually all areas of the natural sciences. Bioinformatics is a particularly active application area (cf. Figs. 1 and 2), but the spectrum of disciplines employing scientific workflow systems is much wider and includes particle physics, chemistry, neurosciences, ecology, geosciences, oceanography, atmospheric sciences, astronomy and cosmology, among others.

URL to Code

A number of open source scientific workflow systems are available, among them:

Kepler: http://www.kepler-project.org
Taverna: http://taverna.sourceforge.net
Triana: http://www.trianacode.org
For a list including many other systems, see http://www.extreme.indiana.edu/swf-survey/.

Cross-References

▶ Business Process Modeling

Recommended Reading

1. Bonner AJ, Shrufi A, Rozen S. LabFlow-1: a database benchmark for high-throughput workflow management. In: Advances in Database Technology, Proceedings of the 5th International Conference on Extending Database Technology; 1996. p. 463–78.
2. Fox GC, Gannon D, editors. Concurrency and computation: practice and experience. Spec Issue Workflow Grid Systems. 2006;18(10):1009–1019.
3. Gil Y, Deelman W, Ellisman W, Fahringer T, Fox G, Gannon D, Goble C, Livny M, Moreau L, Myers J. Examining the challenges of scientific workflows. Computer. 2007;40(12):24–32.
4. Houstis E, Gallopoulos E, Bramley R, Rice J. Problem-solving environments for computational science. IEEE Comput Sci Eng. 1997;4(3):18–21.
5. Ioannidis YE, Livny M. MOOSE: modeling objects in a simulation environment. In: IFIP Congress, Ritter GX, editors. North-Holland; 1989. p. 821–6.
6. Ioannidis YE, Livny M, Gupta S, Ponnekanti N. ZOO: a desktop experiment management environment. In: Proceedings of the 22th International Conference on Very Large Data Bases; 1996. p. 274–85.
7. Ludäscher B, Goble C, editors. ACM SIGMOD Rec. 2005;34(3):44–9. Special Issue on Scientific Workflows.
8. Medeiros CB, Vossen G, Weske M. WASA: a workflow-based architecture to support scientific database applications. In: Proceedings of the 6th International Conference on Database and Expert Systems Applications; 1995. p. 574–83.
9. Nakagawa AS. LIMS: implementation and management. Cambridge: The Royal Society of Chemistry, Thomas Graham House, The Science Park, CB4 4WF; 1994.
10. Shoshani A, Olken F, Wong HKT. Characteristics of scientific databases. In: Proceedings of the 10th International Conference on Very Large Data Bases; 1984. p. 147–60.
11. Taylor I, Deelman E, Gannon D, Shields M, editors. Workflows for e-Science: scientific workflows for grids. Berlin: Springer; 2007.
12. Wainer J, Weske M, Vossen G, Medeiros CB. Scientific workflow systems. In: Proceedings of the NSF Workshop on Workflow and Process Automation in Information Systems: State of the Art and Future Directions; 1996.
13. Wiener JL, Ioannidis YE. A moose and a fox can aid scientists with data management problems. In: Proceedings of the 4th International Workshop on Database Programming Languages; 1993. p. 376–98.
14. Yu J, Buyya R. A taxonomy of scientific workflow systems for grid computing. ACM SIGMOD Rec. 2005;34(3):44–9. Special Issue on Scientific Workflows.

Score Aggregation

Ronald Fagin
IBM Almaden Research Center, San Jose, CA, USA

Assume that there is a fixed collection **O** of objects and that there are m attributes of the objects. Assume that for attribute i (with $1 \leq i \leq m$), there is a function f_i that assigns a score $f_i(x)$ to each object x in **O**. Typically we have $0 \leq f_i(x) \leq 1$. Intuitively, $f_i(x)$ tells the extent to which object x has attribute i. For example, if attribute i represents "redness" (telling how red an object is), then a redness score $f_i(x)$ near 1 means that object x is very red and a redness score $f_i(x)$ near 0 means that object x is far from being red.

We assume that there is a scoring function (or aggregation function) F with m arguments, so that $F(f_1(x), \ldots, f_m(x))$ gives the overall score of object i (the result of aggregating the scores of object x over all of the attributes). It is natural to assume that F is monotone, in the sense that if $y_i \leq z_i$, for $1 \leq i \leq m$, then $F(y_1, \ldots, y_m) \leq F(z_1, \ldots, z_m)$. Typical scoring functions are the min, which is used in fuzzy logic [2] to represent the conjunction, and various types of averages (such as the arithmetic mean). For a discussion of various scoring functions and their properties, see Zimmermann's textbook [3].

In a number of applications, we wish to obtain the "top k" objects under some scoring function F, that is, k objects with the highest overall scores. The naive algorithm for finding these top k objects would be to compute the overall score of every object and then take k objects with the highest scores. But in practice, this is too inefficient, since it involves examining every object in the database (i.e., every member of **O**). Under certain assumptions about what type of database accesses are allowed (e.g., sorted access and random access), an optimally efficient algorithm, which is widely used, is the Threshold Algorithm of [1].

Recommended Reading

1. Fagin R, Lotem A, Naor M. Optimal aggregation algorithms for middleware. J Comput Syst Sci. 2003;66(4):614–56.
2. Zadeh LA. Fuzzy sets. Inf Control. 1969;8:338–53.
3. Zimmermann HJ. Fuzzy set theory. 3rd ed. Boston: Kluwer Academic Publishers; 1996.

Screen Scraper

Harald Naumann
Vienna University of Technology, Vienna, Austria

Synonyms

Data extraction; Screen scraping; Screen wrapper

Definition

A screen scraper is a program which extracts relevant data from the visual user interface of an application. Input data are commonly represented using text-only or graphically enhanced tables, lists, and forms, tailored to a human audience. Scraping is the task of collecting data from its presentation, not directly from its source for lack of access. The scraper output has a structured and machine-readable format, where extracted data are usually annotated with its semantics (metadata), suitable for automatic post-processing. The process can be thought of as reverse-engineering a data store from its presentation, abstracting content from layout. Using this approach, application data are taken from the human-oriented screen output rather than the application's hidden proprietary data structures.

Key Points

Traditionally, screen scrapers have been used to interface legacy systems residing on old mainframes, which often host critical data processing applications. Although both hardware and software are obsolete, they cannot be replaced for various reasons. Screen scraping offers a cost-effective alternative to access and leverage underlying data stores. Typical applications include capturing emulated IBM 3270 screens (a widely used text-based protocol for dumb terminals). Combined with macros to enable navigation throughout different screens, scrapers can be used to integrate with modern architectures.

Common text scraping methods make heavy use of syntactic tools such as regular expressions to identify relevant data. Recently, more semantic approaches have been researched that furthermore allow scraping from unstructured documents such as PDF using generic document understanding techniques supplemented by domain-specific knowledge modeled with ontologies. Layout and table recognition can be performed on a visual level using top-down segmentation (recursive X-Y cut), bottom-up clustering, as well as probabilistic graph-matching algorithms [1]. Identified document segments can then be classified using semantically designed rules in order to annotate the original document with its implicit structure.

Another key area is web scraping, which locates data by exploiting the explicit underlying layout markup (HTML) of its presentation. As a means to build interfaces (APIs) for web sites not available otherwise, scrapers also serve as the basis for state-of-the-art semantic web applications called web mash-ups, such as MIT's SIMILE project [2]. Web scrapers filter relevant content, serializing it in annotated XML format. The main complexity issue arising with all scraper types is coping with change: scrapers are said to "break," when data presentation changes substantially. Visual IDEs can assist scraper design, offering lower maintenance effort compared to purely programmatical solutions.

Cross-References

▶ Information Extraction
▶ Web Data Extraction System
▶ Wrapper Maintenance

Recommended Reading

1. Hassan T, Baumgartner R. Intelligent text extraction from PDF documents. In: Proceedings of the International Conference on Intelligent Agents, Web Technologies and Internet Commerce; 2005. p. 2–6.
2. Huynh D, Mazzocchi S, Karger D. Piggy bank: experience the semantic web inside your web browser. In: Proceedings of the 4th International Semantic Web Conference; 2005.

SCSI Target

Kaladhar Voruganti
Advanced Development Group, Network Appliance, Sunnyvale, CA, USA

Definition

In SCSI protocol, the server which provides the storage is known as target. There can be multiple targets in a storage controller. Each target can offer access to either a single volume or multiple volumes. The volumes being offered by a storage target are mapped into LUNs by the host operating system.

Key Points

Storage controllers, JBOD (just a bunch of disks in an enclosure), direct attached disks, and storage virtualization boxes can all act as SCSI targets. Other types of storage media that can support the SCSI protocol can also act as SCSI targets. The transport protocol encapsulating the SCSI commands dictate the uniqueness of the SCSI target and initiator identifiers.

Cross-References

▶ Logical Unit Number
▶ Logical Unit Number Mapping
▶ Storage Protocols
▶ Volume

SDC Score

Josep Domingo-Ferrer
Universitat Rovira i Virgili, Tarragona, Catalonia, Spain

Definition

Statistical disclosure control (SDC) methods for microdata can be ranked based on information loss, disclosure risk or a combination of both. An SDC score is a combination of information loss and disclosure risk measures used to rank methods.

Key Points

The construction of an SDC score combining information loss and disclosure risk was first proposed in [1, 2]. For each method M and parameterization P, the following score is computed:

$$Score\left(\mathbf{V},\mathbf{V}'\right) = \frac{IL\left(\mathbf{V},\mathbf{V}'\right) + DR\left(\mathbf{V},\mathbf{V}'\right)}{2}$$

$$Score = \frac{IL + \frac{(0.5DLD+0.5PLD)+ID}{2}}{2}$$

where IL is an information loss measure, DR is a disclosure risk measure and $\mathbf{V}'$ is the protected dataset obtained after applying method M with parameterization P to an original dataset $\mathbf{V}$.

In the above references, IL and DR were computed using a weighted combination of several information loss and disclosure risk measures. With the resulting score, a ranking of a set of masking methods (and their parameterizations) was obtained. Yancey et al. [3] later followed the same approach to rank a different set of methods using a slightly different score.

To illustrate how a score can be constructed, the particular score used by [2] is next described. Let X and X' be matrices representing original and protected datasets, respectively, where all attributes are numerical. Let V and R be the covariance matrix and the correlation matrix of X, respectively; let $\overline{X}$ be the vector of attribute averages for X and let S be the diagonal of V. Define V', R', $\overline{X}'$, and S' analogously from X'. The Information Loss (IL) is computed by averaging the mean variations of $X - X'$, $\overline{X} - \overline{X}'$, $V - V'$, $S - S'$, and the mean absolute error of $R - R'$ and multiplying the resulting average by 100. Thus, the following expression is obtained for information loss:

$$IL = \frac{100}{5}\left(\frac{\sum_{j=1}^{p}\sum_{i=1}^{n}\frac{|x_{ij}-x'{ij}|}{|x_{ij}|}}{np}\right.$$

$$+\frac{\sum_{j=1}^{p}\frac{|\overline{x}_j - \overline{x}'_j|}{|\overline{x}_j|}}{p} + \frac{\sum_{j=1}^{p}\sum_{1\le i\le j}\frac{|v_{ij}-v'_{ij}|}{|v_{ij}|}}{\frac{p(p+1)}{2}}$$

$$\left.+\frac{\sum_{j=1}^{p}\frac{|v_{jj}-v'_{jj}|}{|v_{jj}|}}{p} + \frac{\sum_{j=1}^{p}\sum_{1\le i\le j}|r_{ij}-r'_{ij}|}{\frac{p(p-1)}{2}}\right)$$

The expression of the overall score is obtained by combining information loss and information risk as follows:

Here, DLD (Distance Linkage Disclosure risk) is the percentage of correctly linked records using distance-based record linkage, PLD (Probabilistic Linkage Record Disclosure risk) is the percentage of correctly linked records using probabilistic linkage, ID (Interval Disclosure) is the percentage of original records falling in the intervals around their corresponding masked values and IL is the information loss measure defined above.

Based on the above score, it turned out that, for the benchmark datasets and the intruder's external information they used in [2], two good performers among the set of methods and parameterizations they tried were: (i) rankswapping with parameter p around 15; (ii) multivariate microaggregation on unprojected data taking groups of three attributes at a time.

Cross-References

▶ Data Rank/Swapping
▶ Disclosure Risk
▶ Inference Control in Statistical Databases
▶ Information Loss Measures
▶ Microaggregation
▶ Microdata
▶ Record Matching

Recommended Reading

1. Domingo-Ferrer J, Mateo-Sanz JM, Torra V. Comparing SDC methods for microdata on the basis of information loss and disclosure risk. In: Proceedings of the Joint Conferences on New Techniques and Technologies for Statistics and Exchange of Technology and Know-How; 2001. p. 807–26.
2. Domingo-Ferrer J, Torra V. A quantitative comparison of disclosure control methods for microdata. In: Doyle P, Lane JI, Theeuwes JJM, Zayatz L, editors. Confidentiality, disclosure and data access: theory and practical applications for statistical agencies. Amsterdam: North-Holland; 2001. p. 111–34.
3. Yancey WE, Winkler WE, Creecy RH. Disclosure risk assessment in perturbative microdata protection. In: Domingo-Ferrer J, editor. Inference control in statistical databases, LNCS, vol. 2316. Berlin: Springer; 2002. p. 135–52.

Search Engine Metrics

Ben Carterette
University of Massachusetts Amherst, Amherst, MA, USA

Synonyms

Evaluation measures; Performance measures

Definition

Search engine metrics measure the ability of an information retrieval system (such as a web search engine) to retrieve and rank relevant material in response to a user's query. In contrast to database retrieval, relevance in information retrieval depends on the natural language semantics of the query and document, and search engines can and do retrieve results that are not relevant. The two fundamental metrics are *recall*, measuring the ability of a search engine to find the relevant material in the index, and *precision*, measuring its ability to place that relevant material high in the ranking. Precision and recall have been extended and adapted to many different types of evaluation and task, but remain the core of performance measurement.

Historical Background

Performance measurement of information retrieval systems began with Cleverdon and Mills in the early 1960s with the Cranfield tests of language indexing devices [3, 4]. Prior to that, retrieval systems had been measured primarily by their efficiency; as with databases, it was implicitly assumed that any document matching the query was relevant. Cleverdon and Mills recognized that information retrieval is not like database retrieval. Queries can be under- or over-specified, polysemy can confound the relationship between query and document, the wrong word can be chosen for a concept with

many names, and so on. Results that are not relevant to the user's request will be returned and results that are relevant will not be returned; there is a need to measure how often this can be expected to happen in general.

Cleverdon and Mills identified two primary dimensions on which to evaluate performance: the proportion of relevant material retrieved (the *recall ratio*) and the proportion of retrieved material that is relevant (originally the *relevance ratio*, later *precision*). Part of the goal of the Cranfield tests was to measure how different indexing strategies affected recall and precision. To this end, Cleverdon and Mills assembled a collection of 1,100 papers in high speed aerodynamics and asked the authors to list the research questions that inspired the paper. Each of the cited references was then judged for relevance to each of the questions. The resulting set of data – a collection of documents, a set of questions or queries, and judgments of the relevance of each document to each query – is called a *test collection*, and the use of test collections for information retrieval evaluation is now referred to as the Cranfield methodology.

Through his extensive evaluations of the SMART retrieval system in the 1960s and 1970s using the Cranfield collection and methodology, Gerald Salton cemented precision and recall as the primary evaluation metrics for search engine performance [8]. He additionally offered extensions and refinements to these basic measures: normalized precision and recall, precision-recall curves to demonstrate the tradeoff between the two, interpolated precision at standard levels of recall, average precision over different levels of recall or different queries.

The early 1990s saw the formation of the Text REtrieval Conference (TREC) and the first evaluations over hundreds of thousands of full-text documents rather than the tens of thousands of abstracts that had previously been the standard in research [12]. The TREC collections are large and heterogeneous, and thus are a prime proving ground for any automatic retrieval technique. However, with an order-of-magnitudes increase in the number of documents, it became impossible to know every relevant document, and thus to know recall with certainty. TREC also motivated the birth of a field of research on "meta-evaluation," the evaluation of performance measures themselves, the evaluation of test collections, and the estimation of retrieval measures.

TREC also led the way in defining and providing models for the evaluation of new retrieval tasks. Some of the tasks studied and evaluated at TREC over the years include routing, multimedia retrieval, cross-language retrieval, and passage retrieval. Closely related to TREC are conferences on machine translation, summarization, and document understanding. Precision- and recall-based metrics such as the BLEU score [5] have become standard in these fields as well.

The 1990s also saw the explosive growth of the web, which over the past 15 years has grown into a collection of billions of documents, and within which search is a multi-million dollar industry. Accurate measures of performance are more important than ever, as millions of dollars are at stake when decisions are made based on those measures.

Foundations

Automatic text retrieval systems such as web search engines return results in the form of a ranked list, with the documents most likely to be relevant to the user"s request at the top. Bad results can be returned if a query is over- or under-specified, a word with multiple meanings included in the query, or a word chosen to represent a concept that is in the index but not under that word, the ranked list will be "polluted" with nonrelevant results. To understand the extent to which this happens and how to fix it, it is necessary to evaluate the ability of the system to retrieve and rank relevant material independent of other factors affecting the utility of the search engine such as interface design or response time.

The two primary dimensions on which to evaluate a ranked list are its ability to find the indexed relevant material (recall) and its ability to rank that relevant material highly (precision). Formally, precision and recall are defined for a given rank cut-off in terms of binary relevance – each document is either relevant or not. Considering everything above the cut-off to be "retrieved" and everything below it to be "not retrieved" and comparing to the relevance of each document produces a 2×2 contingency table, as shown in Fig. 1.

Precision at rank n is defined as the proportion of relevant documents in the top n retrieved:

$$precision@n = \frac{number\ of\ documents\ relevant\ \&\ retrieved\ in\ the\ top\ n}{number\ retrieved}$$

Recall at rank n is the proportion of all relevant documents in the index retrieved in the top n:

$$recall@n = \frac{number\ of\ documents\ relevant\ \&\ retrieved\ in\ the\ top\ n}{total\ number\ relevant}$$

this means that the engine was able to find every relevant document without ever confusing a nonrelevant document for relevant.

As Fig. 2 shows, precision-recall curves appear jagged, as each new relevant document increases both precision and recall. The curve can be smoothed into a non-increasing curve by *interpolating* precision: k equally-spaced points of recall are chosen, and the interpolated precision at the ith point is defined as the highest precision at any point of recall greater than or equal to that point [7]. The interpolated curve demonstrates the trade-off between recall and precision: retrieving more documents increases

Search Engine Metrics, Fig. 1 A ranking of eight documents, four of which have been judged relevant (**R**) and four nonrelevant (**N**). Cut-offs at ranks one, three, five, and seven produce the 2 × 2 tables shown

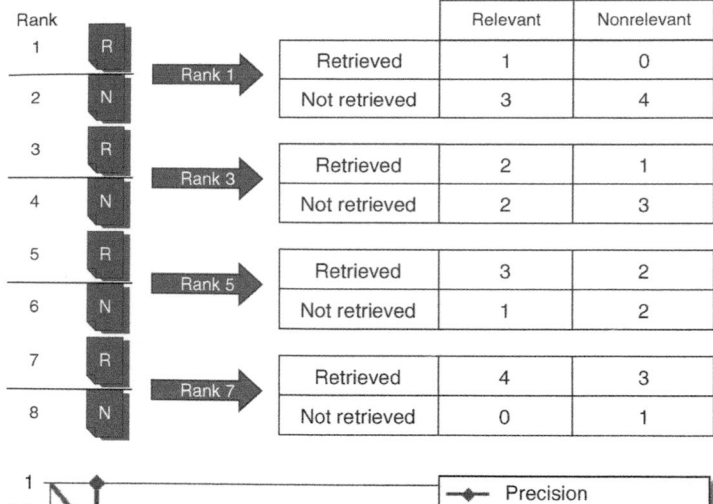

Search Engine Metrics, Fig. 2 An example precision-recall curve, along with its 11-point interpolated curve

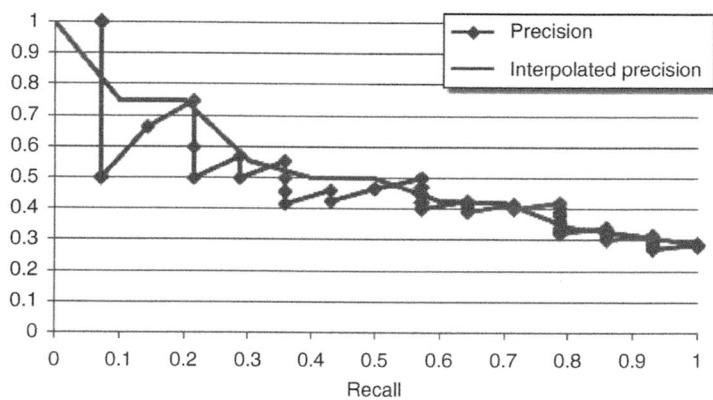

recall, but it also brings more nonrelevant material into the ranking, decreasing precision. Figure 2 shows an example of an 11-point (recall $= 0,0.1,\ldots,1$) interpolated curve.

Besides precision and recall (which readers may also know as "positive predictive value" and "sensitivity" respectively) there are many other statistics that can be calculated on a 2 × 2 contingency table as in Fig. 1: specificity, χ^2, mutual information, and accuracy, among others; besides the precision-recall curve, there are other curves that can be plotted over varying cut-off values, the ROC curve being the most famous. The utility of these to information retrieval is limited: they depend on counts of "true negatives," i.e., nonrelevant documents that were not retrieved. When retrieving documents over a large heterogeneous collection such as the web, nonretrieved nonrelevant documents make up the vast majority of indexed pages – to a close approximation, 100% of the index. Thus the difference in one

of these statistics for any two rankings is negligible, and certainly not distinguishable from chance.

A number of statistics that summarize the precision-recall curve have been invented over the years. The most common is *average precision*, the area under the precision-recall curve (originally the interpolated curve, now more commonly the non-interpolated curve). Another is *R-precision*, the point at which recall and precision are equal. These are both measures that reward systems for having both high recall and high precision, but in a nonlinear fashion. The F-measure is more linear: it is the weighted harmonic mean of precision and recall (weights are chosen depending on the relative importance of precision vs. recall); max-F is the highest of all such values.

Additionally, there are a number of other metrics in the literature that are based on the fundamental ideas behind precision and recall. One that has found widespread use in web measurement is *discounted cumulative gain* (DCG) [5]. Since it can handle graded relevance judgments, it is more flexible than precision as traditionally defined. *Normalized DCG* (NDCG) incorporates a recall component into DCG by dividing it by the best possible DCG for the query.

Relevance Judgments

As described above, calculating metrics requires judgments of the relevance of each document to each query. Precision requires a judgment on every retrieved document to the query. Recall requires that every document that is in the index and relevant to the query has been identified; thus until every document in the collection has been judged, the possibility remains that recall is being overestimated.

In Cleverdon and Mills' original experiments, judgments were made on how relevant a cited reference was to the research questions that inspired the citing paper [4]. They were made by the authors of the citing papers, the ones who came up with the research questions to begin with. To fill out the set needed for precise recall computation, additional judgments were made by students working for Cleverdon.

With the shift to much larger, much more heterogeneous document collections that was inaugurated by the National Institute for Standards and Technology (NIST) at the TREC conferences, it became impossible to judge every document to every query. Instead, the *pooling* method [10] was adopted: a set of queries is sent to participating sites without relevance judgments; sites run the queries through their retrieval systems and return the resulting ranked lists to NIST. The top N documents retrieved by each system are pooled, and the entire pool judged for relevance. Although this results in a small fraction of the total collection being judged, it is a biased sample that ensures that most of the documents that are likely to be retrieved by any system will be judged. Zobel has shown that although relevant documents are missed using this method (and thus recall overestimated), it is more than satisfactory for evaluation when the goal is comparing two or more different systems [13]. Additional

work has shown that reliable comparisons can be made with very few judgments; even when judgments needed to calculate precision are missing, system comparisons can often be made with high confidence, and even when confidence is not high, the degree of confidence can be reliably estimated [1, 2].

The judgments acquired for TREC are typically binary – relevant or not – or trinary – highly relevant, relevant, or not relevant – but binarized for evaluation. This is appropriate for the tasks studied at TREC, which tend to emphasize recall. For many types of web searches, recall is significantly less important than precision. For example, the query "microsoft" may return Microsoft's corporate web page, pages about Microsoft software, pages about court cases Microsoft is involved in, and pages about Microsoft's stock activity. All of these are relevant to some user's need, but it is unlikely that all of them are relevant to the same need. Since the query is so broad, the best ranking would probably put Microsoft's home page at rank 1. But if relevance is binary, any of those pages would be considered equally relevant, and a page about a small drop in Microsoft stock on a certain date could appear at rank 1 without affecting precision or recall, even though the user's utility is clearly negatively affected.

To resolve this, web judgments are often made on a graded scale. Examples of graded scales include the "highly relevant," "relevant," and "nonrelevant" sometimes used at TREC; "highly relevant," "relevant," "maybe relevant," "nonrelevant" to allow for some uncertainty on the part of the judge, or the five-point scale originally used by Cleverdon and Mills. There is a trade-off between finer performance distinctions and judgment quality, however: as more categories are added, it becomes harder to define what exactly distinguishes one category from another, and as a result the judgments become less reliable. Even with the binary judgments and highly-specified information needs used at TREC, there is a fair amount of disagreement about what is relevant [11]; when moving to finer scales and trying to infer user's needs on the basis of a 1–3 word query, disagreement may skyrocket.

Hypothesis Testing and Relative Performance

Measures like average precision and NDCG defy easy interpretation. What does it mean for a system to have an NDCG of 0.69? Thus the goal of performance measurement is often to compare the performance of two engines, one of which may be a minor modification of the other. But a small difference in performance can occur simply by chance. A decision based on such a difference should take into account the probability that it is "real," i.e., whether it is unlikely to have occurred only due to random factors.

Estimating this probability involves taking a random sample of queries likely to be input to the system. For the web, the sample can be obtained from search logs. The measure of interest is computed for each query, and some test statistic computed over the set. The ideal test statistic should have high power to detect the "real" differences when they exist.

There has been some debate over which test statistic (and therefore which hypothesis test) is applicable to information retrieval. If the same sample of queries can be treated as a random sample to either engine and both engines index the same documents, paired tests provide more powerful analysis [7]. The sign test makes no assumptions about the distribution of metrics like NDCG over queries, but is not very powerful. The Wilcoxon sign rank test, which has been popular, also makes no distributional assumptions, but as a test for difference in median has limited power to detect differences in mean performance. The t-test is a powerful test for detecting differences in means. Although it requires some distributional assumptions that may not hold in practice, it is robust to violations of those assumptions, and therefore is probably the best test to use when at least 25 queries can be sampled [13].

Key Applications

Measuring Search Engine Performance

Precision, recall, and DCG measure how well the engine ranks documents independent of other factors that can influence users' opinions, such as interface, extra tools, and so on. Each metric measures a different aspect of performance with varying degrees of fineness.

Comparing Search Engines

Metrics allow the comparison of two different search engines or two variations on a baseline ranking algorithm. The statistical significance of differences can be evaluated and used to make decisions about development and deployment.

Optimizing Search Engine Performance

Search engine algorithms can be optimized to maximize performance on one or more of these metrics.

Future Directions

There are many open problems in search performance measurement: how to evaluate personalized search (in which results are tailored to the user), how to evaluate novelty (ensuring that the same information is not duplicated in results), how to use context in evaluation, and so on.

A challenge of web evaluation is the everchanging nature of the query stream and the indexed documents [9]. The distribution of queries changes frequently, and there is always a long tail of queries that only appear in the logs once. As a result, queries need to be resampled and reevaluated constantly. Web pages disappear or fall out-of-date frequently, and judgments should be kept accordingly up-to-date. Finally, changes in the search engine's interface or its underlying algorithms can affect the way users interact with it, making comparisons between engines separated by long time periods difficult if not impossible.

Finally, there is still more work to be done on understanding how missing relevance judgments affect conclusions that can be drawn from evaluations.

Data Sets

The TREC test collections described in this article are available from NIST at http://trec.nist.gov/.

Cross-References

▶ Average Precision
▶ Discounted Cumulated Gain
▶ F-Measure
▶ Information Retrieval
▶ Mean Reciprocal Rank
▶ Relevance
▶ R-Precision
▶ Web Page Quality Metrics
▶ Web Search Relevance Ranking

Recommended Reading

1. Aslam JA, Pavlu V, Yilmaz E. A statistical method for system evaluation using incomplete judgments. In: Proceedings of the 32nd Annual International ACM SIGIR Conference on Research and Development in Information Retrieval; 2006. p. 541–8.
2. Carterette B, Allan J, Sitaraman RK. Minimal test collections for retrieval evaluation. In: Proceedings of the 32nd Annual International ACM SIGIR Conference on Research and Development in Information Retrieval; 2006. p. 268–75.
3. Cleverdon CW. The cranfield tests on index language devices. In: Jones KS, Willett P, editors. Readings in information retrieval. Morgan Kaufmann; 1967. p. 47–59.
4. Cleverdon CW, Mills J. The testing of index language devices. In: Jones KS, Willett P, editors. Readings in information retrieval. Morgan Kaufmann; 1963. p. 98–110.
5. Kekalainen J, Jarvelin K. Using graded relevance assessments in IR evaluation. JASIST. 2002;53(13): 1120–9.
6. Papineni K, Roukos S, Ward T, Zhu WJ. BLEU: a method for automatic evaluation of machine translation. In: Proceedings of the 40th Annual Meeting of the Association for Computational Linguistics; 2002. p. 311–8.
7. van Rijsbergen CJ. Information retrieval. London: Butterworths; 1979.
8. Salton G, Lesk ME. Computer evaluation of indexing and text processing. In: Jones KS, Willett P, editors. Readings in information retrieval. Morgan Kaufmann; 1967. p. 60–84.
9. Soboroff I. Dynamic test collections: measuring search effectiveness on the live web. In: Proceedings of the 32nd Annual International ACM SIGIR Conference on Research and Development in Information Retrieval; 2006. p. 276–83.
10. Sparck JK, van Rijsbergen CJ. Information retrieval test collections. J Doc. 1976;32(1):59–75.
11. Voorhees E. Variations in relevance judgments and the measurement of retrieval effectiveness. In: Proceedings of the 21st Annual International ACM SIGIR Conference on Research and Development in Information Retrieval; 1998. p. 315–23.
12. Voorhees EM, Harman DK, editors. TREC: experiment and evaluation in information retrieval. Cambridge, MA: MIT; 2005.
13. Zobel J. How reliable are the results of large-scale information retrieval experiments? In: Proceedings of the 21st Annual International ACM SIGIR Conference on Research and Development in Information Retrieval; 1998. p. 307–14.

Searching Digital Libraries

Panagiotis G. Ipeirotis
New York University, New York, NY, USA

Synonyms

Federated search

Definition

Searching digital libraries refers to searching and retrieving information from remote databases of digitized or digital objects. These databases may hold either the metadata for an object of interest (e.g., author and title), or a complete object such as a book or a video.

Historical Background

The initial efforts to standardize and facilitate searching of digital libraries date back to the 1970s, when the development of the Z39.50 protocol started. The Z39.50 protocol is an ANSI standard and defines how to search and retrieve items from a remote database catalog. The Z39.50 protocol was widely deployed within library environments, allowing users to perform searches to remote libraries.

With the advent of the Web, libraries started digitizing and making contents available on the

Web, and the Z39.50 protocol started losing its importance. Many libraries made their content "searchable" through standard Web forms, allowing users to search and retrieve content using simply a Web browser. However, due to the lack of a link structure, the contents of the libraries remained "hidden" from the modern search engine crawlers, forming part of the "Hidden-Web" (also known as Deep Web, or Invisible Web). Searching across multiple Hidden Web databases, despite the tremendous progress since 2000, is still an open research problem.

However, achieving interoperability across all Web databases is inherently harder than achieving interoperability across library databases, which are relatively more homogeneous. Therefore, a set of efforts focused on introducing protocols to facilitate integrating and searching digital libraries. The Open Archives Initiative focused on defining a protocol for exporting metadata about the objects in the collections hosted by each library. The SRU protocol aims to modernize the Z39.50 by making it similar to modern Web services. Such efforts allow programmers to leverage their existing skills and develop easier tools for the library market.

Foundations

Digital libraries host a variety of digital objects, including, but not limited to, textual documents, images, sounds, videos, or even multimodal objects that combine the above. The concept of searching digital libraries may refer either to the action of searching a *single* digital library or to the action of searching across *multiple* digital libraries.

Searching a *single* digital library typically refers to the action of searching and browsing the contents of the underlying relational, textual, or multimedia database.

Searching across *multiple* digital libraries is a concept that evolved significantly over the years. The development of these efforts is broadly divided in three periods:

- The pre-Web period (late 1970s-mid 1990s): Development of the Z39.50 standard.
- The early-Web period (mid 1990s-early 2000s): Emergence of the Web, and increased accessibility of libraries over the Web.
- The Web-services period (early 2000s-now): Definition of protocols for Web services, and development of library-focused search and discovery protocols.

The Pre-Web Period
The first attempts to define a standardized, common protocol for searching library databases date back to the 1970s. Then, the "Linked Systems Project" examined how to provide support for standardized access method to a small set of homogeneous, bibliographic databases. This effort led to the formation of a NISO committee in 1979, which after years of efforts defined the "American National Standard Z39.50, Information Retrieval Service Definition and Protocol Specifications for Library Applications" in 1987. The protocol was later revised in 1992, in 1995, and in 2003 (See [11] for a detailed history and timeline of the development of Z39.50).

The Z39.50 protocol was designed as a client-server protocol, defining how the client can search and retrieve information from a remote database. The protocol supports a significant number of actions, including searching across individual fields, such as author, abstract, title, and so on. Unfortunately, the protocol did not mandate the implementation of several aspects of the specifications, allowing the developers to choose the aspects of the protocol to implement. This led to unexpected behavior of some systems, as the same query, executed over the same underlying content, could return very different results, depending on the implementation. Furthermore, the extremely heavy specification made it difficult for vendors to develop systems that were fully compatible with each other.

The Early-Web Period
The emergence of the Web changed significantly the way that digital libraries make their content available. Many libraries, perhaps encouraged by

the *Digital Libraries Initiative* in 1994, started digitizing and making their content available over the Web. This meant that user could simply visit the Web site of a library and then, using simply Web forms, could query and browse the holdings of the library.

A significant fraction of these new digital libraries are only accessible via a search interface and the ability to browse through a static hyperlink structure is often missing. This means that the contents of these libraries are "hidden" from search engines, since traditional crawlers, which discover new pages by following links, cannot discover the contents of the library. Such libraries are part of the *hidden-Web* [2]. On the other hand, libraries that provide a link structure for accessing their holdings, are part of the *surface Web*, which is accessible by using general search engines, such as Google.

For libraries with content available as part of the *surface Web*, the common model for searching is through vertical search engines. The vertical search engines create topically-focused indexes of the material available on the Web by using *focused crawlers* [4] to identify and index the pages about a given topic. Under this model, the distributed digital libraries become searchable through a centralized search interface that indexes the remotely stored content. When a user issues a query, the vertical search engine identifies the most relevant pages in the index and returns to the user the URLs of the pages, which are stored remotely.

For libraries with *hidden Web* content, the typical way of searching their contents is through *metasearchers*. A complete metasearcher has to perform the following tasks:

- Discover the available digital libraries. This involves crawling the Web to identify pages with Web forms that are search interfaces for underlying databases [5].
- Understand the capabilities of the available query interface [1, 13, 16].
- Characterize the contents of the underlying database, typically by extracting a small sample of the stored contents through query-based sampling. The characterization may involve

classifying the database into a topic hierarchy [6], extracting a statistical summary of the content [3, 8], or it may involve keeping the actual sample as a surrogate for the contents of the database [7, 15].

- Use the database characterization to select the most promising databases for evaluating a given query [9, 15].
- Evaluate the queries in the selected databases, retrieve, and merge the results from multiple databases into a single list [14].

An alternative approach to the distributed search technique adopted by metasearchers is to try to download *all* contents of a hidden Web database [12]. Once all the contents of the remote digital libraries are retrieved and stored locally, the problem of searching multiple digital libraries is reduced to the problem of searching a single, centralized database. One of the issues in this case is the need to periodically refresh the local copy with the most recent contents of the remote database [10].

The Web-Services Period

During the early-Web period, the problem of integrating and searching across digital libraries was similar to the problem of integrating Web databases at large. The vision of the *semantic Web* promised a solution for this problem, and the implementation of a *Web services* framework was a first step towards this direction.

Inherently, though, the library integration problem is much easier than the problems involved in the full implementation of the semantic Web. Therefore, a set of niche solutions were developed for the library integration problem, focusing on the one hand on library-specific needs, but building on top of the existing tools for general Web services that are being developed and rapidly improved.

One of the first attempts to make effortless the discovery of the contents of a library database was the development of the *Open Archives Initiative Protocol for Metadata Harvesting (OAI-PMH)*. This protocol defines how a library can export metadata descriptions of its holdings. Then, *metadata harvesters* can easily collect the contents of the database and make these contents searchable through a centralized search interface. The OAI-PMH protocol is now widely adopted by many libraries and a set of OAI registries facilitate even further the discovery of libraries that support this protocol. Notably, major search engines, such as Google and Yahoo! also support the protocol, as an alternative of the *sitemaps protocol*. This support allows libraries to be an integral part of the general Web and at the same time use a protocol developed and customized for their own needs.

Beyond OAI, there are also attempts to modernize the Z39.50 protocol and make it part of the larger family of Web protocols. First, the *Bath profile* specifies the exact query syntax that Z39.50 clients should use, so that clients can interpret the results returned by Bath-compliant Z39.50 servers. A more significant development is the agreement for the *Search/Retrieval via URL (SRU) protocol*. SRU is a standard XML-focused search protocol for Internet search queries that uses *Contextual Query Language (CQL)* for representing queries. The SRU uses the REST protocol and introduces a standard method for querying library databases, by simply submitting URL-based queries. For example, consider the following URL-encoded query:

http://z3950.loc.gov:7090/voyager?version= 1.1&operation=searchRetrieve&query=dinosaur &maximumRecords=10

This example is a search for the term "dinosaur," requesting that at most ten records to be returned. The SRU protocol is easy to support and implement, and is familiar to programmers that also use such syntax to interact with other popular Web services.

Key Applications

Digital libraries are increasingly becoming part of everyday life. The book digitization projects undertaken by corporations (e.g., Google, Microsoft) and by many universities will generate enormous digital archives accessible over the Web. Similarly, the high-quality holdings of the existing libraries are becoming increasingly accessible over the Web, allowing users to reach easier authoritative sources of information.

Cross-References

▶ Digital Libraries
▶ Multimedia Databases
▶ Multimedia Information Retrieval Model
▶ Scientific Databases

Recommended Reading

1. Bergholz A, Chidlovskii B. Using query probing to identify query language features on the web. In: Proceedings of the Distributed Multimedia Information Retrieval, SIGIR 2003 Workshop on Distributed Information Retrieval; 2004. p. 21–30.
2. Bergman MK. The deep web: surfacing hidden value. J Electron Pub. 2001;7(1).
3. Callan JP, Connell M. Query-based sampling of text databases. ACM Trans Inf Syst. 2001;19(2):97–30.
4. Chakrabarti S, van den Berg M, Dom B. Focused crawling: a new approach to topic-specific web resource discovery. Comput Netw. 1999;31(11–16):1623–40.
5. Cope J, Craswell N, Hawking D. Automated discovery of search interfaces on the web. In: Proceedings of the 14th Australasian Database Conference; 2003. p. 181–189.
6. Gravano L, Ipeirotis PG, Sahami M. QProber: a system for automatic classification of hidden-web databases. ACM Trans Inf Syst. 2003;21(1):1–41.
7. Hawking D, Thomas P. Server selection methods in hybrid portal search. In: Proceedings of the 31st Annual International ACM SIGIR Conference on Research and Development in Information Retrieval; 2005. p. 75–82.
8. Ipeirotis PG, Gravano L. Distributed search over the hidden web: hierarchical database sampling and selection. In: Proceedings of the 28th International Conference on Very Large Data Bases; 2002. p. 394–405.
9. Ipeirotis PG, Gravano L. When one sample is not enough: improving text database selection using shrinkage. In: Proceedings of the ACM SIGMOD International Conference on Management of Data; 2004. p. 767–778.
10. Ipeirotis PG, Ntoulas A, Cho J, Gravano L. Modeling and managing content changes in text databases. In: Proceedings of the 21st International Conference on Data Engineering; 2005. p. 606–617.
11. Lynch CA. The Z39.50 information retrieval standard. D-Lib Mag. 1997;3(4).
12. Ntoulas A, Zerfos P, Cho J. Downloading textual hidden web content by keyword queries. In:

Proceedings of the 5th ACM/IEEE Joint Conference on Digital Libraries; 2005.

13. Raghavan S, García-Molina H. Crawling the hidden web. In: Proceedings of the 27th International Conference on Very Large Data Bases; 2001. p. 129–138.

14. Si L, Callan J. A semisupervised learning method to merge search engine results. ACM Trans Inf Syst. 2003;21(4):457–91.

15. Si L, Callano J. Modeling search engine effectiveness for federated search. In: Proceedings of the 31st Annual International ACM SIGIR Conference on Research and Development in Information Retrieval; 2005. p. 83–90.

16. Zhang Z, He B, Chang KC-C. Understanding web query interfaces: best-effort parsing with hidden syntax. In: Proceedings of the ACM SIGMOD International Conference on Management of Data; 2004. p. 107–118.

Second Normal Form (2NF)

Marcelo Arenas
Pontifical Catholic University of Chile, Santiago, Chile

Synonyms

2NF

Definition

Let $R(A_1,..., A_n)$ be a relation schema and Σ a set of functional dependencies over $R(A_1,..., A_n)$. An attribute A_i ($i \in \{1,...,n\}$) is a *prime* attribute if A_i is an element of some key of $R(A_1,..., A_n)$. Then specification (R, Σ) is said to be in second normal form (2NF) if for every nontrivial functional dependency $X \to A$ implied by Σ, it holds that A is a prime attribute or X is not a proper subset of any (candidate) key for R [1].

Key Points

In order to avoid update anomalies in database schemas containing functional dependencies, 2NF was introduced by Codd in [1]. This normal form is defined in terms of the notions of prime attribute and key as shown above. For example, given a relation schema $R(A, B, C)$ and a set of functional dependencies $\Sigma = \{A \to B\}$, it does not hold that $(R(A, B, C), \Sigma)$ is in 2NF since B is not a prime attribute and A is a proper subset of the key AC. On the other hand, $(S(A, B, C), \Gamma)$ is in 2NF if $\Gamma = \{A \to B, B \to C\}$, since A is a key (and thus it is not a proper subset of any candidate key) and B is not contained in any (candidate) key for S.

It should be noticed that relation schema $S(A, B, C)$ above is in 2NF if $\Gamma = \{A \to B, B \to C\}$, although this schema is not in 3NF. In fact, 3NF is strictly stronger than 2NF; every schema in 3NF is in 2NF, but there exist schemas (as the one shown above) that are in 2NF but not in 3NF.

Cross-References

▶ Boyce-Codd Normal Form
▶ Fourth Normal Form
▶ Normal Forms and Normalization
▶ Third Normal Form

Recommended Reading

1. Further CEF. Normalization of the data base relational model. In: Data base systems. Englewood Cliffs: Prentice-Hall; 1972. p. 33–64.

Secondary Index

Yannis Manolopoulos[1], Yannis Theodoridis[2], and Vassilis J. Tsotras[3]
[1]Aristotle University of Thessaloniki, Thessaloniki, Greece
[2]University of Piraeus, Piraeus, Greece
[3]University of California-Riverside, Riverside, CA, USA

Synonyms

Non-clustering index

Definition

A tree-based index is called a *secondary index* if the order which it maintains on the search-key values is *not* the same as the order of the file which it indexes. For example, consider a relation R with some numeric attribute A taking values over an (ordered) domain D. Assume that relation R is *not* physically stored on the values of attribute A (i.e., relation R is either stored as a heap – an unordered file, or is ordered on another attribute). Furthermore, assume that a tree-based index (e.g., B+ -tree) has been created on attribute A. Then this index is secondary.

Key Points

Tree-based indices are built on numeric attributes and maintain an order among the indexed search-key values. They are further categorized by whether their search-key ordering is the same with the file's physical order (if any). Note that a file may or may not be ordered. Ordered is a file whose records are stored in pages according to the order of the values of an attribute. Obviously, a file can have at most a single such order since it is physically stored once. For example, if the *Employee* relation is ordered according to the *name* attribute, the values in the other attributes will not be in order. A file stored without any order is called an unordered file or heap. An index built on any non-ordering attribute of a file is called *secondary* (or non-clustering) while an index built on the ordering attribute of a file is called *primary* (clustering).

Since the actual data record can be anywhere in the file, the secondary index needs an extra level of indirection, namely, a pointer to the actual position of a record with a given value in the relation file. In other words, a secondary index only clusters references to records (in the form of <value, pointer>fields), but *not* the records themselves. This extra indirection from a leaf page of a secondary index to the actual position of a record in a file has important subsequences on optimization. Consider, for example, a secondary index (B+ -tree) on the *ssn* attribute of the *Employee* relation (which assume is ordered by the *name* attribute). A query that asks for the salaries of employees with *ssn* in the range *[x, y]* can facilitate the B+ -tree on *ssn* to retrieve references to all records in the query range. Assume there are 1,000 such *ssn* values in the *Employee* file. Since the actual *Employee* records must be retrieved (so as to report their salaries), each such reference needs to be materialized by possibly a separate page I/O (since the actual records can be in different pages of the *Employee* file).

A relation can have several indices, on different search-keys; among them, at most one is primary (clustering) index while the rest are secondary ones.

Cross-References

▸ B+-Tree
▸ Indexed Sequential Access Method

Recommended Reading

1. Elmasri RA, Shamkant NB. Fundamentals of database systems. 5th ed. Reading: Addisson-Wesley; 2007.
2. Manolopoulos TY, Tsotras Y, Vassilis J. Advanced database indexing. Dordecht: Kluwer; 1999.
3. Ramakrishnan R, Gehrke J. Database management systems. 3rd ed. New York: McGraw-Hill; 2003.

Secure Data Outsourcing

Barbara Carminati
Department of Theoretical and Applied Science, University of Insubria, Varese, Italy

Synonyms

Secure third-party data management

Definition

Data outsourcing is an emerging data management paradigm in which the owner of data is no longer totally responsible for its management. Rather, a portion of data is outsourced to external providers, potentially hosted in cloud platforms, who offer data management functionalities. Secure data outsourcing is a discipline that investigates security issues associated with data outsourcing.

Historical Background

Service outsourcing is a paradigm widely used by many companies and organizations to achieve better service. The key idea is to delegate some of their business functions to external specialized service providers. A natural evolution of this paradigm is the recent emergence of *data outsourcing*. With this strategy, a company is no longer completely responsible for its own data management. Rather, it outsources some of its data functionalities to one or more external data management service providers (e.g., offering efficient query processing or large storage capability). The success of data outsourcing as well as, more in general, of service outsourcing is also due to the spread of cloud computing technologies that greatly increase the availability of service providers.

Regardless the framework where data outsourcing services are placed, this clearly leads to a range of security issues, because data owners have the potential to lose control over the outsourced data. Thus, the challenge is to ensure the highest level of security when data are managed by external service providers. With this aim, several research groups have started to investigate and propose mechanisms to achieve *secure data outsourcing*.

Foundations

In general, the enforcement of data outsourcing requires examination of new, challenging issues. With the traditional, well-known client server architecture (see Fig. 1a), data owners manage the DBMS and directly answer user queries. Data outsourcing relies on third-party architecture (see Fig. 1b), in which data owners outsource their data (or portions of it) to one or more data service providers. In real world environments, it cannot be assumed that third parties always operate according to the data owner's security policies. By contrast, to achieve secure data outsourcing, one needs to define techniques that satisfy

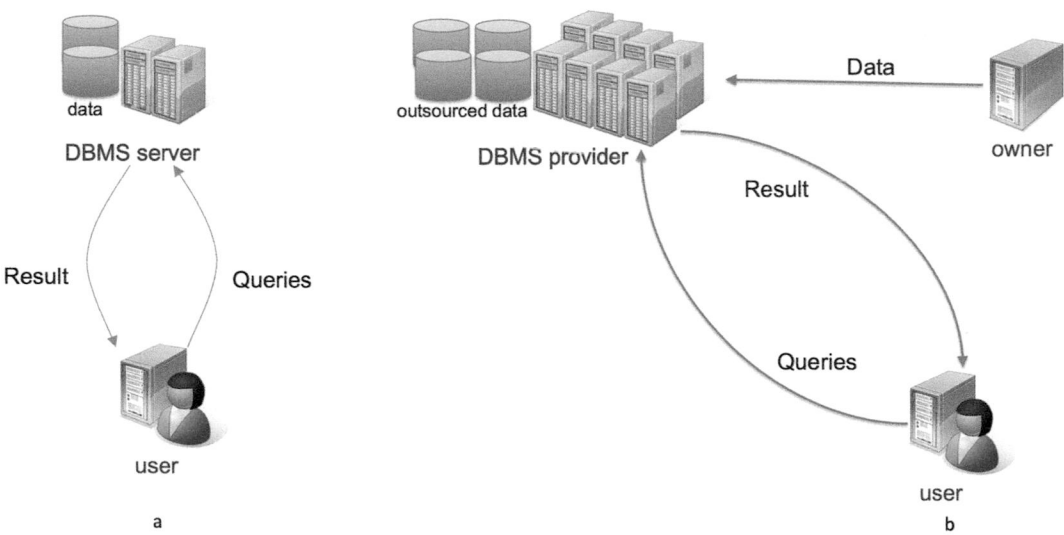

Secure Data Outsourcing, Fig. 1   (**a**) Two-party; (**b**) third-party architecture

the main security properties even in the presence of an *untrusted* third party – that is, a provider that could maliciously modify or delete the data it manages by, for instance, inserting fake records or sending data to unauthorized users. Several researchers have focused on this problem and have developed different proposals. However, before illustrating the techniques proposed to date, it is necessary first to identify the main security requirements in secure data outsourcing.

Security Requirements in Secure Data Outsourcing

According to a third-party architecture, data is managed by potentially untrusted providers. For this reason, secure data outsourcing must examine novel security issues as well as re-examine the traditional ones. The following presents the main security issues studied so far in secure data outsourcing.

Privacy: If a third-party architecture is adopted, a user could be concerned about his/her privacy for any query processing performed by the provider. This is due to the fact that by simply tracking a user's queries, an untrusted provider could infer sensitive information about the user (for instance, the user's preferences). For this reason, the privacy of the submitted queries needs to be protected. To ensure access privacy, a provider should not be able to know the details of the query, by at the same time being able to process it.

Authenticity and Integrity: Ensuring authenticity and integrity in third-party architectures implies to make a user able to verify that the data received as query result have been indeed generated by the data owner and not modified by the provider. It is relevant to note that *integrity* also has an additional meaning, that is, ensuring that unauthorized users have not modified the data; however, since data outsourcing is mainly conceived for read-only data access, this definition of *integrity* is not considered in this entry. In traditional architectures, both authenticity and integrity are ensured by means of digital signatures. When a user submits a query, the data owner computes and digitally signs the query

result. This, together with its digital signature, is sent to the user, thus enabling him/her to verify the query's authenticity and integrity. However, in third-party architectures, traditional signature techniques cannot be used. Indeed, a provider may return to the user only selected portions of the signed data in answer to the query evaluation. Thus, a user that is provided with only these portions is not able to validate the owner's digital signature, which has been generated on the whole data. To cope with these requirements, alternative ways must be found to digitally sign outsourced data so that a user is able to validate the digital signature even if he/she has only received selected portions of the signed data.

Completeness: Third-party architectures introduce a further novel security requirement, called completeness. If satisfied, this property ensures a user that the answer received by the service providers genuinely contains all of the data answering to the submitted query. In the literature, some works refer to this property as *query correctness*, by also implying authenticity and integrity requirements.

Confidentiality: Data confidentiality means ensuring that data are disclosed only to authorized users. However, it is obvious that when data are outsourced, confidentiality requirements are not limited to users, but extended also to providers. Thus, confidentiality in data outsourcing acquires a twofold meaning. The first deals with protecting the owner's data from access by a malicious or untrusted provider, we refer to it as *confidentiality w.r.t. the provider*. A further confidentiality requirement, hereafter called *confidentiality w.r.t. users*, refers to the protection of data from unauthorized users' access on the basis of the access control policies stated by the data owners. In traditional client-server architectures, this requirement is enforced by access control mechanisms, called *reference monitors*, which mediate each user request by authorizing only those in accordance with the owner's access control policies. This type of solution can hardly be applied in data outsourcing, since it implies the delegation of the reference monitor tasks to a potentially untrusted publisher. For this reason, alternative solutions should be devised for access control enforcement when data services are outsourced.

Techniques for Secure Data Outsourcing

Many research groups have investigated security in data outsourcing, resulting in several proposals for different security requirements. The following presents the main results proposed so far, grouped according to the addressed security properties.

Access Privacy: Private Information Retrieval protocols (PIR, for short), first introduced in [1], are one of the most relevant results of investigations into the problem of query privacy protection. The underlying idea of PIR protocols is that several replications of the same database are available in different servers (i.e., providers). To preserve access privacy, the user submits different queries to each different server. These are defined so that the user is able to obtain the information desired by combing the servers' answers, whereas each server is unable to infer the actual interest of the user, by analyzing the submitted query.

Another line of solutions in this field is motivated by the fact that access patterns can be inferred even by observing accesses on encrypted data. As an example, it has been shown in [2] that by analyzing accesses to an encrypted email repository, an attacker can infer up to 80% of these queries. To cope with this issue the Oblivious RAM (ORAM) algorithms, first introduced in [3], have been deeply investigated. The key idea of these solutions is to hide access patterns to the outsourced data by continuously shuffling and re-encrypting data as they are accessed as well as by mixing the real accesses with a sequence of random dummy queries.

Authenticity and Integrity: These are the first properties that have been investigated in secure data outsourcing. The aim is to devise alternative digital signature schemes for signing the data to be outsourced, enabling users to validate the signature even if users are only provided with selected portions of the signed data. Several schemes have been proposed so far, exploiting different strategies for achieving this result. In particular, two of the most widely used techniques are Merkle trees and aggregate signatures.

The following presents both of these concepts by introducing some of the related proposals.

Merkle Trees. Merkle proposed a method to authenticate, with a unique signature, a set of messages $\{m_1, \ldots, m_n\}$, by at the same time enabling an intended verifier to authenticate a single message without the disclosure of the other messages. The proposed solution exploits a binary tree, where each leaf contains the hash values of a message in $\{m_1, \ldots, m_n\}$, whereas internal nodes enclose the concatenation of the hash values corresponding to its left and right children (see Merkle tree entry for more details). The root node of the resulting binary hash tree can be considered the digest of all messages, and thus it can be digitally signed by using a standard signature technique. For this reason the Merkle tree plays the role of an *authenticated data structure*. The main benefit of this data structure is that a user is able to validate the signature by having a subset of messages, provided that he/she receives a *proof*, that is, a set of additional hash values of the missing messages. By using the proof, a user is able to build up locally the binary hash tree and thus to validate the signature. Merkle trees have been used in several computer areas. However, the first work to exploit them in relational data was the one by Devanbu et al. [4], which adapts these trees to relational data to prove the completeness, authenticity, and integrity of query answers. According to this approach, for each relation R, a different Merkle tree is generated in such a way that leaves contain hash values of tuples in R. Then, when a user submits a query on R, the provider replies to him/her with tuples answering the submitted query together with the signature generated on the root node of the corresponding Merkle tree. Moreover, the provider also sends the user the hash values of tuples of R not included in the result set. These additional hash values enable the user to locally generate the Merkle tree of R and to validate the signature. Since [4], several proposals based on Merkle tree for relational data outsourcing have been presented so as to deal with new scenarios (e.g., big data domain [5], edge computing [6]). Merkle trees have also been investigated for secure data outsourcing of other

each resulting encrypted word is XORed with a different pseudorandom number. Since different occurrences of the same encrypted word are XORed with a different pseudorandom number, information about the word distribution cannot be inferred by analyzing the distribution of the encrypted words. Each user is provided with the secret key k, and the used pseudorandom numbers. The scheme proposed in [21] is defined in such a way that users are able to compute pseudorandom numbers locally without any interaction with the data owner. By having this information, therefore, whenever a user intends to ask the third party for a keyword W, he/she first generates the encrypted word using the secret key k, and then computes the XOR of the result with the corresponding pseudorandom number. The user then submits the obtained ciphered word to the third party, which sequentially scans all ciphered words to search for the one matching the one submitted. Thus, this scheme allows the third party to search for a keyword W directly on the ciphered data without gaining any information on the clear text or on the required keyword W. In 2003, Eu-Jin Gon proposed an alternative solution, based on indexes, for searching keywords in an encrypted document [22]. According to this approach, a different index is associated with each document to be encrypted. These indexes, called security indexes, exploit Bloom filters and have the property to store hidden information about the keywords contained within the corresponding document. According to this scheme, the owner outsources the encrypted documents and the corresponding security indexes to service providers, which are then able to search for a keyword by simply accessing the indexes. A similar approach has been devised in [23], which proposes dictionary-based keyword indexes.

However, all of these work have the limitation that third parties are able to identify only documents matching with a given keyword, but are not able to support more expressive searches, such as Boolean combinations of keywords. A first step to overcome this limitation has been done by Golle et al. in [24], which proposes a public key scheme to support conjunctive keywords searches. More recently, in [25] a general frame-

work for searchable public-key systems supporting various families of predicates has been proposed. Here, the scheme supports comparison (such as greater-than) and general subset queries, defined in arbitrary conjunctions.

Key Applications

Data outsourcing offers several benefits. One of the most relevant is related to cost reduction. Indeed, the company pays only for services that it uses from providers, which are generally significantly less than the cost implied by deployment, installation, maintenance, and upgrades of DBMSs. Moreover, the data management services offered by specialized providers are more competitive than the ones provided by the company itself. A further benefit of data outsourcing is its scalability, since a company can outsource its data to as many providers as it needs according to the amount of data and the number of managed users, avoiding that provider might become a bottleneck for the system. All these benefits make secure data outsourcing suitable for a wide range of applications in different data domains. For geographical data, for instance, the secure data outsourcing paradigm can be adopted to support geomarketing services. A data owner can outsource some of its geographical data (for instance, maps at various levels of details) to a publisher that provides them to customers based upon different registration fees or different confidentiality requirements (for instance, maps of some regions that cannot be distributed to everyone because they show sensible objectives).

Future Directions

Given the attention that the data outsourcing paradigm is receiving, it is expected that secure data outsourcing will be intensely investigated in the future. Besides proposing more efficient strategies for the security requirements considered so far, it is necessary to consider further challenging security issues, like those related to user privacy and ownership protection. Moreover, more consideration must be given to complex data outsourcing scenarios to enable users to manipulate the outsourced data rather than just simply reading it.

Cross-References

► Access Control
► Data Encryption
► Digital Signatures
► Merkle Trees

Recommended Reading

1. Chor B, Goldreich O, Kushilevitz E, Sudan M. Private information retrieval. In: Proceedings of the 36th Annual Symposium on Foundations of Computer Science; 1995.
2. Islam M, Kuzu M, Kantarcioglu M. Access pattern disclosure on searchable encryption: ramification, attack and mitigation. In: Proceedings of the Network and Distributed Systems Security Symposium; 2012.
3. Goldreich O, Ostrovsky R. Software protection and simulation on oblivious rams. J ACM. 1996; 43(3):431–473.
4. Devanbu P, Gertz M, Martel C, Stubblebine SG. Authentic third-party data publication. In: Proceedings of the IFIP TC11/ WG11.3 14th Annual Working Conference on Database Security; 2000.
5. Tang T, Liu L, Wang T, Hu X, Sailer R, Pietzuch P. Outsourcing multi-version key-value stores with verifiable data freshness. In: Proceedings of the IEEE 30th International Conference on Data Engineering; 2014.
6. Pang H, Tan K. Authenticating query results in edge computing. In: Proceedings of the 20th International Conference on Data Engineering; 2004.
7. Bertino E, Carminati B, Ferrari E, Thuraisingham B, Gupta A. Selective and authentic third-party distribution of XML documents. IEEE Trans Knowl Data Eng. 2004;16(10):1263–78.
8. Carminati B, Ferrari E, Bertino E. Securing XML data in third-party distribution systems. In: Proceedings of the 14th ACM International Conference on Information and Knowledge Management; 2005.
9. Devanbu P, Gertz M, Kwong A, Martel C, Nuckolls G, Stubblebine SG. Flexible authentication of XML documents. In: Proceedings of the 8th ACM Conference on Computer and Communication Security; 2001.
10. Boneh D, Gentry C, Lynn B, Shacham H. Aggregate and verifiably encrypted signatures from bilinear maps. In: Proceedings of the Advances in Cryptology; 2003.
11. Mykletun E, Narasimha M, Tsudik G. Authentication and integrity in outsourced databases. In: Proceedings of the 11th Annual Symposium on Network and Distributed System Security; 2004.
12. Narasimha M, Tsudik G. Authentication of outsourced databases using signature aggregation and chaining. In: Proceedings of the 11th International Conference on Database Systems for Advanced Applications; 2006.
13. Pang H, Jain A, Ramamritham K, Tan K. Verifying completeness of relational query results in data publishing. In: Proceedings of the ACM SIGMOD International Conference on Management of Data; 2005.
14. Sahai A, Waters B. Fuzzy identity-based encryption. In: Advances in Cryptology - EUROCRYPT 2005: 24th Annual International Conference on the Theory and Applications of Cryptographic Techniques; 2005.
15. Goyal V, Pandey O, Sahai A, Waters B. Attribute-based encryption for fine- grained access control of encrypted data. In: Proceedings of the 13th ACM Conference on Computer and Communications Security; 2006.
16. Bethencourt J, Sahai A, Waters B. Ciphertext-policy attribute-based encryption. In: Proceedings of the IEEE Symposium of Security and Privacy; 2007.
17. Hacigümüş H, Iyer B, Li C, Mehrotra S. Executing SQL over encrypted data in the database-service-provider model. In: Proceedings of the ACM SIGMOD International Conference on Management of Data; 2002.
18. Hacigumus H, Iyer B, Li C, Mehrotra S. Efficient execution of aggregation queries over encrypted relational databases. In: Proceedings of the 11th International Conference on Database Systems for Advanced Applications; 2004.
19. Rivest R, Adleman L, Dertouzos M. On data banks and privacy homomorphisms. In: Lipton RJ, Dobkin DP, Jones AK, editors. Foundations of secure computation. Academic; 1978. p. 169–78.
20. Agrawal R, Kiernan J, Srikant R, Xu Y. Order-preserving encryption for numeric data. In: Proceedings of the ACM SIGMOD International Conference on Management of Data; 2004.
21. Song DX, Wagner D, Perrig A. Practical techniques for searches on encrypted data. In: Proceedings of the IEEE Symposium on Security and Privacy; 2000.
22. Goh E. Secure indexes, cryptology ePrint archive, Report 2003/216; 2003.
23. Chang Y, Mitzenmacher M. Privacy preserving keyword searches on remote encrypted data, Cryptology ePrint Archive, Report; 2004.
24. Golle P, Staddon J, Waters B. Secure conjunctive keyword search over encrypted data. In: Proceedings of the Applied Cryptography and Network Security Conference; 2004.
25. Boneh D, Waters B. Conjunctive, subset, and range queries on encrypted data. In: Proceedings of the 4th Conference on Theory of Cryptography; 2007.

S

Secure Database Development

Jan Jurjens[1] and Eduardo B. Fernandez[2]
[1]The Open University, Buckinghamshire, UK
[2]Florida Atlantic University, Boca Raton, FL, USA

Synonyms

Secure database design; Secure DBMS development

Definition

This entry considers how to build secure database system software. In particular, it describes how to build a general-purpose database management system where security is an important design parameter. For the database community, the words secure database design may refer to the schema design to produce a database for a specific application with some level of security properties. There is a large amount of literature on this latter subject and a related entry in this encyclopedia (Database security). This entry concentrates mostly on how to build the software of a DBMS such that it exhibits security properties, which is called secure database development. Both approaches are contrasted so that the reader can decide which one of these problems applies to their specific case but more space is dedicated to the general secure database development problem.

Historical Background

While there is a large number of papers on security models including authorization and other security aspects of databases [1–3], there is little work on how to implement a secure Database Management System (DBMS). It is true that many proposals for secure multilevel databases include details of implementation but most of them are ad hoc architectures that cannot be generalized to databases using different models or even to other multilevel databases with different requirements. Of the books on database security, [4] had several chapters on how to build secure relational database systems, and later [2] included also multilevel models. Those books do a good job of indicating the architectural units of such systems and their general requirements. However, software development aspects are not discussed in detail. It appears that [5] is the only work discussing these aspects explicitly.

Foundations

There are two aspects to the problem of developing software for secure databases: building a general (application-independent) secure DBMS and building a database system which is part of a secure application. These two problems are first briefly defined and then discussed in more detail. Other approaches and possible system architectures are also considered.

In the first approach the DBMS is just a complex software application in itself and a general secure software methodology can be applied without or with little change. Object-oriented applications typically start from a set of use cases, which define the user interactions with the system under development. In this particular case, use cases would define the typical functions of a DBMS, e.g. search, query, and update, and security would be included as part of its development life cycle. The DBMS would follow an appropriate model, e.g. Role-Based Access Control (RBAC), selectable in the design stage, which defines security constraints for the functions defined by the use cases. In some cases, it may be possible to support more than one security model. This would result in a secure DBMS where security would be a general nonfunctional requirement. The approach results in a general-purpose secure DBMS, where nothing is known about the specific applications that will be executed by its users. The DBMS itself is the application. The secure development methodologies of [6, 7] and others are applicable here.

Another view is the one from a designer who needs to build a specific user application (or type of application) that includes a DBMS as part of its architecture, e.g. a financial system (most applications require a database but the degree of security needed may vary). This is discussed in [2, 5, 8, 9]. In this case, the DBMS is rather ad hoc and tailored to the level of security desired for the specific type of application. For example, [9] separates the requirements into three types: functional, security, and database. Typically, these approaches emphasize how to define and enforce a set of application-specific rules that follow some security model and how to reflect them in the schema and other parts of the DBMS. Most of these studies emphasize the security of the database schema or some specific sections without much concern for the rest of the application. A methodology such as [6] or [7] can also be applied here, the DBMS being one of the architectural levels of a system that implements a specific application, although these methodologies have little to say about the contents of the specific rules that are needed in the schema (only their safe storage but not their consistency or security).

An interesting problem that applies to both approaches is the mapping from the conceptual security model (that may apply to a collection of DBMSs) to the authorization system of a specific database; for example, security constraints defined in a conceptual UML model defining authorizations in terms of classes must be mapped to an SQL-based authorization system which defines authorizations in terms of relations. Clearly, whatever is defined in the common conceptual model must be respected in the DBMS authorization system, although this latter may add further constraints related to implementation aspects.

General Secure Database Systems

In this case, as indicated earlier, the DBMS is a complex application requiring a general high level of security. There are several methodologies for this purpose and two of them are described below. A methodology for secure software development should include appropriate tools and provide a unified and consistent approach through all the life cycle stages. Ideally, a methodology should use a Model-Driven Development approach, where transformations between development stages are based on corresponding meta-models. Since the resulting software is independent of the access control model adopted, it does not provide for special requirements of the model; for example, multilevel models typically require data labeling. This means that the resulting software would be less secure than an ad hoc design (unless the multilevel model was the target in the example). Because of the generality of the resultant DBMS it may be difficult to prove formally security properties. An early approach in this direction was based on adding security functions to a general-purpose DBMS, e.g. INGRES or System/R.

Secure Database Development Using Patterns

A methodology to build secure systems is presented in [6]. A main idea in the proposed methodology is that security principles should be applied at every stage of the software lifecycle and that each stage can be tested for compliance with those principles. Another basic idea is the use of patterns at each stage. A pattern is an encapsulated solution to a recurrent problem and their use can improve the reusability and quality of software.

Domain analysis stage: A business model is defined. Legacy systems are identified and their security implications analyzed. Domain and regulatory constraints are identified and use as global policies. The suitability of the development team is assessed, possibly leading to added training. This phase may be performed only once for each new domain or team. The need for specialized database architectures should be determined at this point. The approach (general DBMS or application-oriented system) should also de defined at this stage.

Requirements stage: Use cases define the required interactions with the system. Each activity within a use case is analyzed to see which threats are possible. Activity diagrams indicate created objects and are a good way to determine which data should be protected. Since many possible threats may be identified, risk analysis helps to prune them according to their impact and probability of occurrence. Any requirements for degree of security should be expressed as part of the use cases.

Analysis stage: Analysis patterns can be used to build the conceptual model in a more reliable and efficient way. The policies defined in the requirements can now be expressed as abstract security models, e.g. access matrix. The model selected must correspond to the type of application; for example, multilevel models have not been successful for medical applications. One can build a conceptual model where repeated applications of a security model pattern realize the rights determined from use cases. In fact, analysis patterns can be built with predefined authorizations according to the roles in their use cases. Patterns for authentication, logging, and secure channels are also specified at this level. Note that the model and the security patterns should define precisely the requirements of the problem, not its software solution. UML is a good semi-formal approach for defining policies, avoiding the need for ad-hoc policy languages. The addition of OCL (Object Constraint Language) can make the approach more formal.

Design stage: When one has defined the policies needed, one can select mechanisms to stop attacks that would violate them. A specific security model, e.g. RBAC, is now implemented in terms of software units. User interfaces should correspond to use cases and may be used to enforce the authorizations defined in the analysis stage. Secure interfaces enforce authorizations when users interact with the system. Components can be secured by using authorization rules for Java or . NET components. Distribution provides another dimension where security restrictions can be applied. Deployment diagrams can define secure configurations to be used by security administrators. A multilayer architecture is needed to enforce the security constraints defined at the application level. In each level, one can use patterns to represent appropriate security mechanisms. Security constraints must be mapped between levels.

The persistent aspects of the conceptual model are typically mapped into relational databases. The design of the database architecture is done

according to the requirements from the uses cases for the level of security needed and the security model adopted in the analysis stage. Two basic choices for the enforcement mechanism include query modification as in INGRES and views as in System R. A tradeoff is using an existing DBMS as a Commercial Off-the-Shelf (COTS) component, although in this case security will depend on the security of that component.

Implementation stage: This stage requires reflecting in the code the security rules defined in the design stage. Because these rules are expressed as classes, associations, and constraints, they can be implemented as classes in object-oriented languages. In this stage one can also select specific security packages or COTS, e.g., a firewall product or a cryptographic package. Some of the patterns identified earlier in the cycle can be replaced by COTS (these can be tested to see if they include a similar pattern). Performance aspects become now important and may require iterations. As indicated, a whole DBMS could be such component.

An important aspect for the complete design is assurance. Experience shows that one can verify each pattern used but this does not in general verify their combination. One can however still argue that since one has used a careful and systematic methodology with verified and tested patterns, the design should provide a good level of security. The set of patterns can be shown to be able to stop or mitigate the identified threats.

Secure Database Development Using UMLsec

A general methodology for developing security-critical software which in particular can be used to develop secure DBMSs has been proposed in [7]. It makes use of an extension of the Unified Modeling Language (UML) to include security-relevant information, which is called UMLsec. The approach is supported by extensive automated tool-support for performing a security analysis of the UMLsec models against the security requirements that are included [10] and has been used in a variety of industrial projects [11]. The UMLsec extension is given in form of a UML profile using the standard UML extension mechanisms. Stereotypes are used together with tags to formulate the security requirements and assumptions. Constraints give criteria that determine whether the requirements are met by the system design, by referring to a precise semantics of the used fragment of UML. The security-relevant information added using stereotypes includes security assumptions on the physical level of the system, security requirements related to the secure handling and communication of data, and security policies that system parts are supposed to obey. The UMLsec tool-support can be used to check the constraints associated with UMLsec stereotypes mechanically, based on XMI output of the diagrams from the UML drawing tool in use. There is also a framework for implementing verification routines for the constraints associated with the UMLsec stereotypes. Thus advanced users of the UMLsec approach can use this framework to implement verification routines for the constraints of self-defined stereotypes. The semantics for the fragment of UML used for UMLsec is defined using so-called UML Machines, which is a kind of state machine which is equipped with UML-type communication mechanisms. On this basis, important security requirements such as secrecy, integrity, authenticity, and secure information flow are defined.

Applications Including Secure Databases

Since this approach is tailored to the application, one can add the required level of security using formal proofs when necessary. Specialized operating system and hardware are also possible and may be needed to reach the required level of security. High-security systems require faithful application of basic security principles; for example, multilevel databases apply complete mediation. Databases work through transactions and a concurrency control system serializes transactions to prevent inconsistencies. High-security multilevel databases also require that the concurrency control system preserves security. The methods described in the last section still apply here, except that additional requirements must be considered. Because of this, these approaches are discussed in less detail, describing only two recent papers that contain references to past work.

Designing Secure Databases Using OCL

An approach to designing the content of a security-critical data base uses the Object Constraint Language (OCL) which is an optional part of the Unified Modeling Language (UML). More specifically, [12] presents the Object Security Constraint Language V.2. (OSCL2), which is based in OCL. This OCL extension can be used to incorporate security information and constraints in a Platform Independent Model (PIM) given as a UML class model. The information from the PIM is then translated into a Platform Specific Model (PSM) given as a multilevel relational model. This can then be implemented in a particular Database Management System (DBMS), such as Oracle9*i* Label Security. These transformations can be done automatically or semi-automatically using OSCL2 compilers. Related to this, [8] presents a methodology that consists of four stages: requirements gathering; database analysis; multilevel relational logical design; and specific logical design. Here, the first three stages define activities to analyze and design a secure database. The last stage consists of activities that adapt the general secure data model to one of the most popular secure database management systems: Oracle9*i* Label Security. They later extended the approach to data warehouses, multidimensional databases, and on-line analytical processing applications.

In both cases, a particular multilevel database system, meaning a set of users organized in levels, compartments, and groups, is given access to specific items of a relational database, according to the characteristics of those items, which also include levels, compartments, and groups. A set of rules describes the allowed access of users to data items. The secure metamodel is stored in the labels of each row or user definition. As indicated, the extra requirements can be superimposed in a general secure software development methodology.

Other Approaches to Secure Software Development with Applicability to Databases

There are other approaches to developing security-critical software which can be applied to developing secure databases and database management systems [13] gives a recent overview.

[14] presents an approach for the predicative specification of user rights in the context of an object oriented use case driven development process. It extends the specification of methods by a permission section describing the right of some actor to call the method of an object. The syntactic and semantic framework is first-order logic with a built-in notion of objects and classes provided with an algebraic semantics. The approach can be realized in OCL.

[15] presents an approach to building secure systems where designers specify

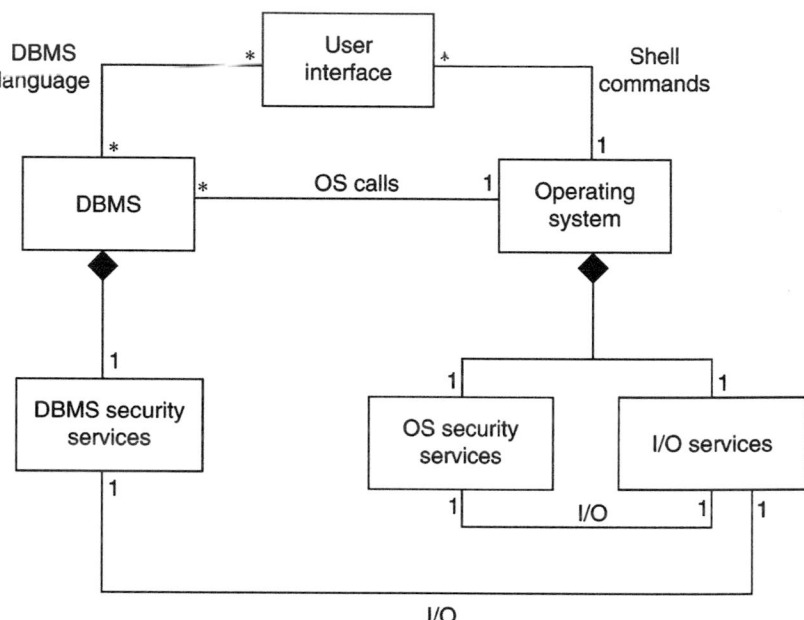

Secure Database Development, Fig. 1 Standard placement of security services

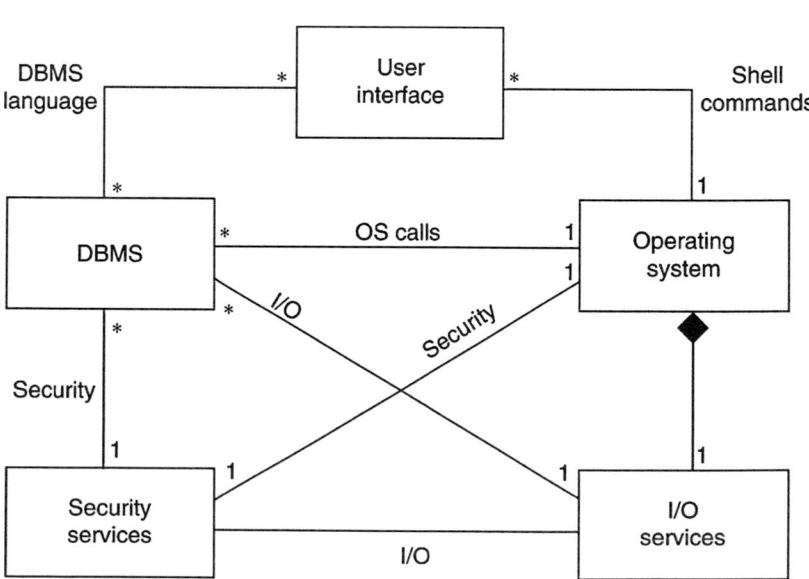

Secure Database Development, Fig. 2 Common security services

Secure Database Development, Fig. 3
Architecture using a web application server

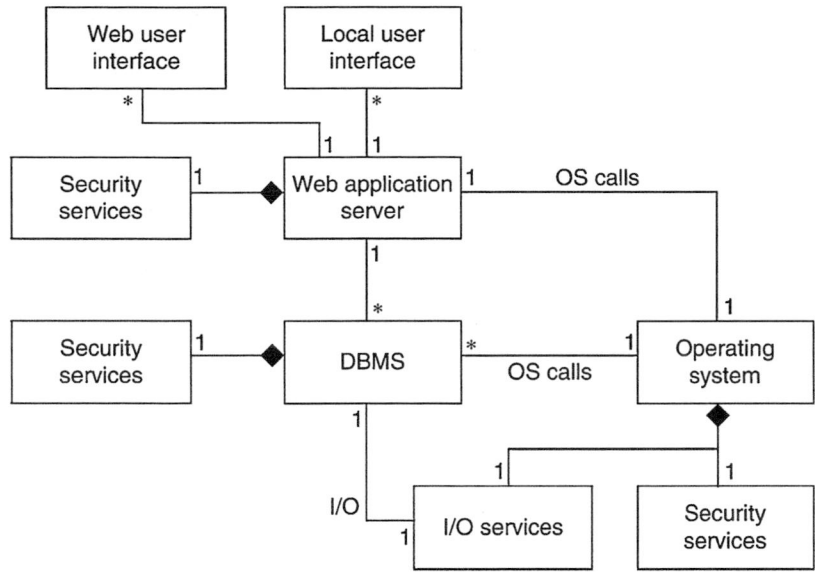

system models along with their security requirements and use tools to automatically generate system architectures from the models, including complete, configured access control infrastructures. It includes a combination of UML-based modeling languages with a security modeling language for formalizing access control requirements.

[16] presents an approach based on the high-level concepts and modeling activities of the secure Tropos methodology and enriched with low level security-engineering ontology and models derived from the UMLsec approach.

System Architecture for Security

Whichever approach is used, there are basically three general architectural configurations to include security functions:

1. Figure 1 shows the standard approach. Here the DBMS and the operating system have their own set of security services.
2. Figure 2 shows a way to unify the design of the DBMS with the design of the OS, using an I/O and file subsystem and a security subsystem to be used by both the DBMS and the OS.
3. Figure 3 is an extension of the standard approach where a Web Application Server

(WAS) unifies security for several databases. The WAS applies a common conceptual model to the information and can integrate different types of databases.

These configurations can be used in either of the approaches discussed earlier. Within each configuration it is possible to use security kernels and virtual machines.

Key Applications

Clearly, the first approach makes sense when the objective is a secure DBMS product, since it is not possible to know what user applications will be supported in the future. The only choice is then to build a system which is as secure as possible within these constraints and within a reasonable cost.

In the second case, the type of application to be supported is known. This gives the designers the flexibility of choosing an appropriate existing database system, as done in [8], or to build the DBMS to reach the required degree of security. If the complete DBMS is to be built, the first approach is appropriate, using as parameter the degree of security.

S

Cross-References

- ► Access Control Administration Policies
- ► Access Control Policy Languages
- ► Application Server
- ► Architecture-Conscious Database System
- ► Concurrency Control: Traditional Approaches
- ► Database Design
- ► Database Management System
- ► Database Middleware
- ► Database Security
- ► Data Stream Management Architectures and Prototypes
- ► Data Warehouse Life Cycle and Design
- ► Data Warehouse Security
- ► Discretionary Access Control
- ► Distributed Database Design
- ► Distributed Database Systems
- ► Distributed DBMS
- ► Mandatory Access Control
- ► Metamodel
- ► Object Constraint Language
- ► Object Data Models
- ► Object-Role Modeling
- ► Privacy
- ► Process Life Cycle
- ► Process Structure of a DBMS
- ► Role-Based Access Control

Recommended Reading

1. Bertino E, Sandhu R. Database security – concepts, approaches, and challenges. IEEE Trans Dependable Secur Comput. 2005;2(1):2–19.
2. Castano S, Fugini M, Martella G, Samarati P. Database security. Addison-Wesley; 1994.
3. Fernandez EB, Gudes E, Song H. A model for evaluation and administration of security in object-oriented databases. IEEE Trans Knowl Database Eng. 1994;6(2):275–92.
4. Fernandez EB, Summers RC, Wood C. Database security and integrity (Systems Programming Series). Addison-Wesley; 1981.
5. Fugini M. Secure database development methodologies. In: Landwehr CE editor. Database security: status and prospects. Elsevier; 1987. p. 103–29.
6. Fernandez EB, Larrondo-Petrie MM, Sorgente T, VanHilst M. A methodology to develop secure systems using patterns, Chapter V. In: Mouratidis H, Giorgini P, editors. Integrating security and software engineering: advances and future vision. IDEA Press; 2006. p. 107–26.
7. Jürjens J. Secure systems development with UML. New York: Springer; 2004.
8. Fernández-Medina E, Piattini M. Designing secure databases. Inf Softw Technol. 2005;47(7):463–77.
9. Ge X, Polack F, Laleau R. Secure databases: an analysis of Clark-Wilson model in a database environment. In: Proceedings of the 16th International Conference on Advanced Information Systems Engineering; 2004. p. 234–47.
10. Jürjens J, Wimmel G. Formally testing fail-safety of electronic purse protocols. In: Proceedings of the 16th IEEE International Conference on Automated Software Engineering; 2001. p. 408–11.
11. Jürjens J, Wimmel G. Security modelling for electronic commerce: the common electronic purse specifications. In: Proceedings of the 1st IFIP Conference on E-Commerce, E-Business, E-Government; 2001. p. 489–506.
12. Fernández-Medina E, Piattini M. Extending OCL for secure database development. In: Proceedings of the International Conference on the Unified Modeling Language; 2004. p. 380–94.
13. Fernández-Medina E, Jürjens J, Trujillo J, Jajodia S. Model-driven development for secure information systems E Fernández-Medina. Inf Softw Technol. 2009;51(5):809–14.
14. Hafner M, Breu R. Towards a MOF/QVT-based domain architecture for model driven security. In: Proceedings of the 9th International Conference Model Driven Engineering Language and Systems; 2006.
15. Basin DA, Doser J, Lodderstedt T. Model driven security: from UML models to access control infrastructures. ACM Trans Softw Eng Methodol. 2006;15(1):39–91.
16. Mouratidis H, Jürjens J, Fox J. Towards a comprehensive framework for secure systems development. In: Proceedings of the 18th International Conference on Advanced Information Systems Engineering; 2006. p. 48–62. CAiSE, Luxembourg. LNCS, (Eric Dubois, Klaus Pohl, eds.).

Secure Multiparty Computation Methods

Murat Kantarcıoğlu[1] and Jaideep Vaidya[2]
[1]University of Texas at Dallas, Richardson, TX, USA
[2]Rutgers University, Newark, NJ, USA

Definition

The problem of preserving privacy while allowing data analysis can be attacked in many ways.

One way is to avoid disclosing data beyond its source while still constructing data mining models equivalent to those that would have been learned on an integrated data set. This follows the approach of Secure Multiparty Computation (SMC). SMC refers to the general problem of computing a given function securely over private inputs while revealing nothing extra to any party except what can be inferred (in polynomial time) from its input and output. Since one can prove that data are not disclosed beyond its original source, the opportunity for misuse is not increased by the process of data mining.

The definition of privacy followed in this line of research is conceptually simple: no site should learn anything new from the *process* of data mining. Specifically, anything learned during the data mining process must be derivable given one's own data and the final result. In other words, nothing is learned about any other site's data that is not inherently obvious from the data mining result. In the context of data mining, the approach followed in this research has been to select a type of data mining model to be learned and develop a protocol to learn the model while meeting this definition of privacy.

Historical Background

Privacy-preserving data mining can be defined as the problem of how to mine data when it is not possible to see it. Two seminal papers [1, 9] first considered this problem and proposed different ways to attack it. Both looked at the problem of constructing decision trees from distributed data in a privacy-preserving manner. Agrawal and Srikant [1] proposed a randomization approach based on perturbing the input data and reconstructing the distribution. Lindell and Pinkas [9] proposed a cryptographic solution based on secure multiparty computation. This entry describes the second approach following the application of secure multiparty computation methods to data mining.

Secure Multiparty Computation (SMC) originated with Yao's Millionaires' problem [15]. The basic problem is that two millionaires would like to know who is richer, with neither revealing their net worth. Abstractly, the problem is to simply compare two numbers, each held by one party, without either party revealing its number to the other. Yao [15] presented a generic circuit evaluation based solution for this problem as well as generalizing it to any efficiently computable function restricted to two parties. Goldreich et al. [6] generalized this to multi-party computation and proved that there exists a secure solution for any functionality. There has been significant theoretical work in this area. The restriction of polynomially time bounded passive adversaries has been removed. Similarly, work has been extended to active adversaries, as well as mobile adversaries. While much effort has been due to efficiency reasons, it is completely infeasible to directly apply the theoretical work from SMC to form secure protocols for privacy-preserving data mining.

Thus, work in privacy-preserving data mining has focused on creating specialized efficient solutions in the context of data mining. Starting with the work of Lindell and Pinkas [9], secure methods have been proposed for various tasks such as association rule mining [7, 13], clustering [8], classification [2], and outlier detection [12] [14]. gives a good overview of much of this work.

Foundations

The basic ideas used in SMC based privacy-preserving data mining techniques are now illustrated using a commonly deployed public key encryption technique called homomorphic encryption [11]. More formally, let $E_{pk}(\cdot)$ denote the encryption function with public key pk and $D_{pr}(.)$ denote the decryption function with private key pr. A secure public key cryptosystem is called additively homomorphic if it satisfies the following requirements: (i) Given the encryption of m_1 and m_2, $E_{pk}(m_1)$ and $E_{pk}(m_2)$, there exists an efficient algorithm to compute the public key encryption of $m_1 + m_2$, denoted $E_{pk}(m_1 + m_2) := E_{pk}(m_1) +_h E_{pk}(m_2)$. (ii) Given a constant k and the encryption of m_1, $E_{pk}(m_1)$, there exists an efficient algorithm to compute the public key encryption of km_1, denoted $E_{pk}(km_1) := k \times_h E_{pk}(m_1)$.

Using the homomorphic encryption technique, one can easily develop many secure protocols. For example, consider the case where three sites S_1, S_2 and S_3 want to add their private values (resp.) v_1, v_2, and v_3 to learn $v_1 + v_2 + v_3$ securely. A simple protocol for the above task using homomorphic encryption can be given as follows: S_1 creates a homomorphic encryption public and private key pair, and sends the public key to S_2 and S_3. In addition, S_1 computes $e_1 = E_{pk}(v_1)$ and sends e_1 to S_2. Using the homomorphic encryption scheme, S_2 can calculate $e_2 = e_1 +_h E_{pk}(v_2)$ and can send e_2 to S_3. Similarly, S_3 can calculate $e_3 = e_2 +_h E_{pk}(v_3) = E_{pk}(v_1 + v_2) + E_{pk}(v_3)$. Finally, S_1 can decrypt the e_3 to compute $D_{pr}(e_3) = v_1 + v_2 + v_3$. If all the parties follow the protocol exactly, it can be shown that nobody learns anything other than the final result. The obvious question is what happens when the parties do not follow the protocol exactly. Clearly S_3 can collaborate with S_1 to learn the private value v_2 because if S_3 sends the message e_2 to S_1 then S_1 can compute $D_{pr}(e_2) - v_1 = (v_1 + v_2) - v_1 = v_2$ to learn v_2.

The above example indicates that when considering privacy, one must first model the different adversarial behaviors that an attacker can assume. The SMC literature defines two basic adversarial models:

- **Semi-Honest:** Semi-honest (or Honest but Curious) adversaries follow the protocol faithfully, but can try to infer the secret information of the other parties from the data they see during the execution of the protocol.
- **Malicious:** Malicious adversaries may do anything to infer secret information. They can abort the protocol at any time, send spurious messages, spoof messages, collude with other (malicious) parties, etc.

While the semi-honest model may seem questionable for privacy (if a party can be trusted to follow the protocol, why would they not be trusted with the data?), it does meet several practical needs for early adoption of the technology. Consider the case where credit card companies jointly build data mining models for credit card fraud detection. In many cases the parties involved already have authorization to see the data (e.g., the theft of credit card information from CardSystems involved data that CardSystems was expected to see during processing). The problem is that *storing* the data brings with it a responsibility (and cost) of protecting that data; CardSystems was supposed to delete the information once the processing was complete. If parties could develop the desired models without seeing the data, then they are saved the responsibility (and cost) of protecting it. Also the simplicity and efficiency possible with semi-honest protocols will help speed adoption so that trusted parties are saved the expense of protecting data other than their own. As the technology gains acceptance, malicious protocols will become viable for uses where the parties are not mutually trusted.

In either adversarial model, there exist formal definitions of privacy [5]. Informally, the definition of privacy is based on equivalence to having a trusted third party perform the computation. This is the gold standard of secure multiparty computation. Imagine that each of the data sources gives their input to a (hypothetical) trusted third party. This party, acting in complete isolation, computes the results and reveals them. After revealing the results, the trusted party forgets everything it has seen. A secure multiparty computation approximates this standard: no party learns more than it would in the trusted third party approach.

One fact is immediately obvious: no matter how secure the computation, some information about the inputs may be revealed. This is a result of the computed function itself. For example, if one party's net worth is $100,000, and the other party is richer, one has a lower bound on their net worth. This is captured in the formal SMC definitions: any information that can be inferred from one's own data and the result can be revealed by the protocol. Thus, there are two kinds of information leaks; the information leak from the function computed irrespective of the process used to compute the function and the information leak from the specific process of computing the function. Whatever is leaked from the function itself is unavoidable as long as the function has to be computed. In secure computation, the second kind of leak is provably prevented. There is *no* information leak whatsoever due to the process.

While the generic secure multi-party computation methods exist, they pose significant computational problems. The challenge of privacy-preserving distributed data mining is to develop algorithms that have reasonable computation and communication costs on real-world problems, and prove their security with respect to the SMC definition. The typical approach taken is to reduce the large domain problem to a series of smaller sub-tasks and to use secure cryptographic protocols to implement those smaller sub-tasks.

In the following, the common secure sub-protocols used in privacy-preserving distributed data mining are now described. As far as possible, for each sub-protocol, a version using only homomorphic encryption is described. Unless otherwise stated, all the sub-protocols are secure in the semi-honest model with no collusion, and all the arithmetic operations are defined in some large enough finite field.

Following the description of the subprotocols, it is shown how different algorithms could be implemented using these secure sub-protocols. Since these common building blocks are quite general, using the Composition theorem [5], they can be combined to create new privacy preserving algorithms in the future.

Secure Sum

Secure Sum securely calculates the sum of values from individual sites. As seen above, homomorphic encryption can easily be used to secure compute the sum local values. Assuming three or more parties and no collusion, a more efficient method can be found in [7].

Secure Comparison/Yao's Millionaire Problem

Assume that two sites, each having one value, want to compare the two values without revealing anything else other than the comparison result. Secure Comparison methods can be used to solve

the above problem. To the best of our knowledge, secure circuit evaluation based approaches still provide the best performance [15].

Dot Product Protocol

Securely computing the dot product of two vectors is another important sub-protocol required in many privacy-preserving data mining tasks. Many secure dot product protocols have been proposed in the past. Among those proposed techniques, the method of Goethals et al. [4] is quite simple and provably secure. It is now briefly described.

The problem is defined as follows: Alice has a n-dimensional vector $\overrightarrow{X} = (x_1, \ldots, x_n)$ while Bob has a n-dimensional vector $\overrightarrow{Y} = (y_1, \ldots, y_n)$. At the end of the protocol, Alice should get $r_a = \overrightarrow{X} \cdot \overrightarrow{Y} + r_b$ where r_b is a random number chosen from uniform distribution that is known only to Bob, and $\overrightarrow{X} \cdot \overrightarrow{Y} = \sum_{i=1}^{n} x_i \cdot y_i$. The key idea behind the protocol is to use a homomorphic encryption system that can be used to perform arithmetic operations over encrypted data. Using such a system, it is quite simple to build a dot product protocol. If Alice encrypts her vector and sends in encrypted form to Bob, using the additive homomorphic property, Bob can compute the dot product. The specific details can be found in [4].

Oblivious Evaluation of Polynomials

Another important sub-protocol required in privacy-preserving data mining is the secure polynomial evaluation protocol. Consider the case where Alice has a polynomial P of degree k over some finite field F. Bob has an element $x \in F$ and also knows k. Alice would like to let Bob compute the value $P(x)$ in such a way that Alice does not learn x and Bob does not gain any additional information about P (except $P(x)$). This problem was first investigated by [10]. Subsequently, there have been more protocols improving the communication and computation efficiency as well as extending the problem to floating point numbers.

S

Privately Computing In x

For entropy measures used in data mining, one must be able to privately compute $\ln x$, where $x = x_1 + x_2$ with x_1 known to Alice and x_2 known to Bob. Thus, Alice should get y_1 and Bob should get y_2 such that $y_1 + y_2 = \ln x = \ln(x_1 + x_2)$. One of the key results presented in [9] was a cryptographic protocol for this computation. One point to note is that $\ln x$ is *Real* while general cryptographic tools work over finite fields. Therefore, $\ln x$ is actually multiplied with a known constant to make it integral. The basic idea behind computing random shares of $\ln(x_1 + x_2)$ is to use the Taylor approximation for $\ln x$. Thus, shares for the Taylor approximation are actually computed. The actual details of the protocol, as well as the proof of security, can be found in [9].

Secure Intersection

Secure Intersection methods are useful in data mining to find common rules, frequent itemsets etc., without revealing the owner of the item. Many algorithms have been developed for calculating Secure Set Intersection. For example, [13] provides an efficient solution. However, a secure set intersection protocol that utilizes secure polynomial evaluation [3] is described below. Let us assume that Alice has set $X = \{x_1, \ldots, x_n\}$ and Bob has set $Y = \{y_1, \ldots, y_n\}$. Our goal is to securely calculate $X \cap Y$. By representing set X as a polynomial and using polynomial evaluation, Alice and Bob can calculate $X \cap Y$ securely.

Secure Set Union

Secure union methods are useful in data mining to allow each party to give its rules,decision trees etc. without revealing the owner of the item. Union of items can be easily evaluated using SMC methods if the domain of the items is small. Each party creates a binary vector (where the ith entry is 1 if the ith item is present locally). At this point, a simple circuit that *or's* the corresponding vectors can be built and securely evaluated using general secure multi-party circuit evaluation protocols. However, in data mining, the domain of the items are usually very large, potentially infinite. This problem can be overcome using approaches based on commutative encryption [7].

Key Applications

This section overviews how different sub-protocols described above could be used to create various privacy-preserving distributed data mining (PPDM) algorithms for different data models. In each of the discussed PPDM algorithms general data mining functionality is reduced to a computation of secure sub-protocols.

In the following discussion, horizontal partitioning of data implies that different sites collect the same set of information about different entities. Vertical data partitioning implies that different sites collect different features of information for the same set of entities. While this entry does not explicitly discuss arbitrary partitioning, the building blocks presented above are actually useful even in that case.

Classification

In the first work on privacy-preserving distributed data mining on horizontally partitioned data [9], the goal is to securely build an ID3 decision tree where the training set is horizontally distributed between two parties. The basic idea is that finding the attribute that maximizes information gain is equivalent to finding the attribute that minimizes the conditional entropy. The conditional entropy for an attribute for two parties can be written as a sum of the expression of the form $(v_1 + v_2) \times \log(v_1 + v_2)$. The authors use the secure log algorithm, secure polynomial evaluation, and secure comparison sub-protocols to securely calculate the expression $(v_1 + v_2) \times \log(v_1 + v_2)$ and show how to use this function for building the ID3 securely. Correspondingly, decision trees for vertically partitioned data can also be built if the attribute with maximum entropy gain can be found. If the class attribute is present with all parties, this can be easily done. But even when the class attribute is only present with one of the parties, the secure scalar product protocol can be used to compute counts of transactions having certain attribute values and class values. This can then be used to compute the information gain for the attribute, and thus to decide the best attribute. Naïve Bayes classifiers can also be built for both horizontally and vertically partitioned data using combinations of the secure sum, secure scalar product and secure comparison primitives. [14] provides more details.

Association Rule Mining

The essential problem in association rule mining is the problem of finding frequent itemsets (meeting some support threshold). Once frequent itemsets are found, it is easy to find association rules meeting certain confidence thresholds. For horizontally partitioned data, [7] showed that for every candidate itemset, the support at each site can be computed locally. Now, a secure sum followed by a secure comparison is sufficient to evaluate if a candidate itemset is indeed frequent. However, this still requires the knowledge of the candidate itemsets. [7] uses some additional techniques (such as Secure Union) to ensure that candidate itemsets contributed by each site are also kept secret along with their support values. Similarly, [13] shows that the problem of finding frequent itemsets in vertically partitioned data can be reduced to the problem of securely computing the scalar product of multiple vectors (or equivalently as the problem of finding the size of the intersection set). Once this is done, finding globally valid association rules is quite simple.

Clustering

Several solutions for privacy-preserving clustering have been proposed. Lin et al. [8] propose a privacy preserving EM algorithm for secure clustering of horizontally partitioned data. EM clustering is an iterative algorithm. Each iteration consists of an expectation (E) step followed by a maximization (M) step. In the E-step, the expected value of the cluster membership for each entity is determined. In the M-step, the each cluster distribution parameters are re-estimated to maximize the likelihood of the data, given the expected estimates of the membership. Lin et al. [8] show that computing the cluster parameters at each iteration can be easily done via secure summation once the total number of objects is known. Once computed, the cluster parameters are assumed to be public (i.e., known to all parties). Therefore each party can then locally assign its entities to the appropriate clusters.

This is repeated until the algorithm converges or until a sufficient number of iterations have been carried out. Similar solutions exist for vertically partitioned data as well.

Outlier Detection

The goal of outlier detection is to find anomalies or outliers in the data. This requires a definition/metric of outlyingness. Many such metrics (with varying degrees of sophistication) have been defined in the statistical literature. One of the simplest metrics is that of $DB(p, d)$ outliers. Under this definition, an entity e in the dataset DB is said to be an outlier if more than p percentage of the entities in the dataset DB are farther than distance d from e. Thus, to figure out if an entity is an outlier, several tasks need to be performed: first, the distance of this entity to other entities must be computed; next, one must check if the number of farther entities is more than the given threshold. Vaidya and Clifton [12] show how to do this for both horizontally and vertically partitioned data. The key primitives used are the secure sum, secure comparison and secure dot product primitives. More detail can be found in [12].

Future Directions

Now that the key concepts behind secure multiparty computation methods have been presented, it is necessary to discuss some of the problems and challenges still open in this area. Inherently, the primary challenge with secure multiparty computation techniques lies with efficiency. Even with cryptographic accelerators and faster machines, since data mining is typically done over millions of transactions, this cost significantly balloons up. Even for other application areas, more efficient protocols are clearly needed. One alternative that has not been well explored is that of approximation. Instead of computing the exact results, it may make a lot more sense to compute approximations of the final results, especially if it gives huge efficiency improvements. This will be critical for development of real solutions in this area.

Cross-References

▶ Horizontally Partitioned Data
▶ Privacy-Preserving Data Mining
▶ Vertically Partitioned Data

Recommended Reading

1. Agrawal R, Srikant R. Privacy-preserving data mining. In: Proceedings of the ACM SIGMOD International Conference on Management of Data; 2000. p. 439–50.
2. Du W, Zhan Z. Building decision tree classifier on private data. In: Clifton C, Estivill-Castro V, editors. Proceedings of the IEEE International Conference on Data Mining Workshop on Privacy, Security, and Data Mining; 2002. p. 1–8.
3. Freedman MJ, Nissim K, Pinkas B. Efficient private matching and set intersection. In: Proceedings of the International Conference on Theory and Application of Cryptographic Techniques; 2004.
4. Goethals B, Laur S, Lipmaa H, Mielikäinen T. On secure scalar product computation for privacy-preserving data mining. In: Proceedings of the 7th Annual International Conference in Information Security and Cryptology; 2004. p. 104–20.
5. Goldreich O. The foundations of cryptography, General cryptographic protocols, vol. 2. London: Cambridge University Press; 2004.
6. Goldreich O, Micali S, Wigderson A. How to play any mental game – a completeness theorem for protocols with honest majority. In: Proceedings of the 19th ACM Symposium on the Theory of Computing; 1987. p. 218–29.
7. Kantarcı oğlu M, Clifton C. Privacy-preserving distributed mining of association rules on horizontally partitioned data. IEEE Trans Knowl Data Eng. 2004;16(9):1026–37.
8. Lin X, Clifton C, Zhu M. Privacy preserving clustering with distributed EM mixture modeling. Knowl Inf Syst. 2005;8(1):68–81.
9. Lindell Y, Pinkas B. Privacy preserving data mining. J Cryptol. 2002;15(3):177–206.
10. Naor M, Pinkas B. Oblivious transfer and polynomial evaluation. In: Proceedings of the 31st Annual ACM Symposium on Theory of Computing; 1999. p. 245–54.
11. Paillier P. Public-key cryptosystems based on composite degree residuosity classes. In: Proceedings of the International Conference on Theory and Application of Cryptographic Techniques; 1999. p. 223–38.
12. Vaidya J, Clifton C. Privacy-preserving outlier detection. In: Proceedings of the 4th IEEE International Conference on Data Mining; 2004. p. 233–240.
13. Vaidya J, Clifton C. Secure set intersection cardinality with application to association rule mining. J Comput Security. 2005;13(4):593–622.
14. Vaidya J, Clifton C, Zhu M. Privacy-preserving data mining. In: Advances in information security, vol. 19. 1st ed. Berlin: Springer; 2005.
15. Yao AC. How to generate and exchange secrets. In: Proceedings of the 27th IEEE Symposium on Foundations of Computer Science; 1986. p. 162–7.

Secure Transaction Processing

Indrakshi Ray and Thilina Buddhika
Colorado State University, Fort Collins, CO, USA

Synonyms

Cloud Computing; Database Security; Data Confidentiality; Privacy; Multilevel Secure Database Management System; Transaction Processing

Definition

Secure transaction processing refers to execution of transactions that cannot be exploited to cause security breaches.

Historical Background

Research in making transaction processing secure has progressed along different directions. Early research in this area was geared toward military applications. Such applications are characterized by having a set of security levels which are partially ordered using the dominance relation. Information is transmitted through read and write operations on data items belonging to the various levels. Information is allowed to flow from a dominated level to a dominating level but all other flows are illegal. Traditional concurrency control and recovery algorithms cause illegal information flow. Most research in this area involved providing new architectures, concurrency control, and recovery mechanisms, which achieves the traditional database functionality and also prevents illegal information flow. Researchers have also looked into the problem of processing real-time multilevel secure transactions. These transactions must satisfy real-time requirements in addition to preventing illegal information flow.

In the commercial sector, subsequent research focused on how to deal with the effect of malicious transactions that may have compromised the integrity of the database. The malicious transactions can damage one or more data items. Other transactions reading from these committed data items help spread the damage. Traditional recovery mechanisms cannot undo the effects of committed transactions. The research in this area focused on developing efficient techniques that will remove the effects of malicious transactions while minimizing the impact on good transactions.

Subsequently, the concept of secure transaction processing is applied in the context of web services. Consequently, researchers have formalized web services atomic transactions which define a framework for applications to implement transactions in a heterogeneous setting. Security in these contexts can be incorporated with web services security related standards. In very recent years, organizations tend to rely on cloud computing platforms for their storage requirements. Such data is encrypted for reasons of privacy and security. Some ongoing research is focusing on how to provide transactional guarantees over encrypted data.

Scientific Fundamentals

A comprehensive survey on the work in multi-level secure (MLS) databases was performed by Atluri et al. [11]. We enumerate some of the important results discussed in this paper. An MLS database environment is characterized by a set of security levels L that are partially ordered by the relation $\prec$. $\prec$ is the dominance relation between classes and it is transitive, reflexive, and anti-symmetric. For any two levels, $l_i, l_j \in L$,

if level $l_i \prec l_j$, then l_j is said to dominate l_i. In this case, l_j and l_i are referred to as dominating and dominated level, respectively. If neither $l_i \prec l_j$, nor $l_j \prec l_i$, then l_i and l_j are said to be incomparable. Each data object o in an MLS environment is associated with a security classification, denoted by $l(o)$, where $l(o) \in L$. Each user u is also cleared to some security level $l(u)$, where $l(u) \in L$. A user u cleared to security level $l(u)$ can log in at any security level l', where l' is dominated by $l(u)$. All processes, including transactions, initiated during a session inherit the security level at which the user has logged in.

MLS systems allow information to flow from dominated to dominating levels but all other information flows are considered illegal. Direct information flow occurs by virtue of transactions reading and writing data items. When a transaction reads a data item, information flows from the data item to the transaction. Similarly, when a transaction writes a data item, information flows from the transaction to the data item. Such direct illegal information flow is prevented by using the simple security property and the $\star$-property of the Bell-LaPadula (BLP) model [12]. Simple security property states the condition under which a transaction can read a data item. A transaction T may read a data item O only if the security level of the data item, denoted by $L(O)$, is dominated by the security level of the transaction, denoted by $L(T)$, that is, only when $L(O) \prec L(T)$. During the read operation, information flows from the data item to the transaction. Thus, when $L(O) \prec L(T)$, information flows from dominated level to the dominating one. $\star$-property states the condition under which a transaction T can write a data item O. The write operation is allowed only when the security level of the data object, denoted by $L(O)$, is dominated by that of the transaction, denoted by $L(T)$. In other words, the write operation is allowed only when $L(T) \prec L(O)$. In this case the information flows from the dominated level to the dominating level. However, the $\star$-property does not prevent a transaction operating at the dominated level from corrupting data items at the dominating level. Thus, for reasons of integrity, a modified form of $\star$-property, known as restricted $\star$-property, is used in practice. The restricted $\star$-property allows a transaction T to write a data item O only if the security levels of the transaction are the same as that of the object, that is, $L(T) = L(O)$. The properties stated in the BLP model are not adequate in preventing illegal information flows that occur through indirect means. One such example is the covert channels. A covert channel is an information flow mechanism within a system that is based on the use of system resources; it is not intended for communication between the regular users of the system. Unfortunately, the traditional transaction processing mechanisms can be exploited to establish a covert channel.

First, we describe how the concurrency control protocols that are used in traditional transaction processing systems can be used to establish a covert channel. We consider two concurrency control mechanisms: two-phase locking (2PL) and timestamp ordering (TO). 2PL requires transactions to acquire read lock (write lock) before reading (writing) a data item. A read lock on a data item can be acquired if no other transaction has a write lock on the same data item. A write lock can be acquired if no other transaction have any lock on the data item. The locks acquired by a transaction must be eventually released. Moreover, once a transaction releases a lock, it can no longer lock any other data item. TO requires each transaction t to have a unique timestamp $ts(t)$. Each data item x is associated with a read timestamp $rts(x)$ and a write timestamp $wts(x)$ that denotes the latest transaction that have read and written, x respectively. The operations are executed on a first-come-first-serve basis. If the execution of an operation does not violate the serialization order specified by the timestamps of the transactions, it is executed. If not, the operation is not allowed and the transaction is aborted.

Suppose our MLS database system is associated with two security levels low and $high$ where the level $high$ dominates low. The transactions initiated by a $high$ user can read all data items and write high data items in accordance with the security and the restricted-$\star$ properties of the BLP model. The transactions initiated by a low user can read and write low data items only.

Suppose there are two transactions T_h and T_l that have decided to collude and there are no other transactions executing in the system. The security level of T_h and T_l are high and low, respectively. The database has a data item x whose security level is low – both transactions have decided to communicate by accessing this data item: T_h will read the data item x and T_l will write the data item. Let us assume that the concurrency control mechanism uses two-phase locking (2PL). When T_h wants to read data item x, a read lock is placed on x which prohibits T_l from acquiring a write lock on x. Thus, T_h can selectively issue lock request on x to transmit information. T_l can measure the delay in acquiring lock and interpret the information. Thus, a covert communication channel has been established between T_l and T_h. A similar problem occurs if a timestamp-based protocol is used. Suppose $ts(T_l) < ts(T_h)$. If T_l attempts to write x after T_h has read it, the write operation is rejected and T_l is aborted. Here again, a high transaction can selectively cause a low transaction to abort and communicate information. Thus, concurrency control mechanisms in an MLS database must not only ensure serializability but must also eliminate such illegal information flows.

Researchers have proposed several concurrency control algorithms for processing transactions in an MLS database. The algorithms are dependent on the underlying architecture of the DBMS. First, we focus on the algorithms developed for the kernelized architecture. Some of the early solutions proposed are by Schaefer [43], Lamport [25], and Reed and Kanodia [42]. In these solutions, the transactions are allowed to proceed. However, before a transaction can commit it is validated. Thus, if a transaction at the dominating level has read some data item which has been updated by a transaction at the dominated level, the dominating level transaction must abort. Such algorithms will cause starvation of transactions at the dominating levels. Keefe and Tsai [23] proposed a protocol based on multiversion timestamp ordering. Although this protocol ensures serializable histories without causing starvation of transactions at the dominating levels, it has several problems. First, it requires a large number of versions to be maintained. Second, transactions at the dominating levels read stale data. Third, performance is an issue. Subsequent research [2, 4, 34] focused on limiting the number of versions to two. The idea is that transactions reading at the dominated level read from the snapshot, while those writing data at their own level write it on the current state.

Researchers have also proposed solutions that are suitable for replicated architectures. In such architectures, an MLS DBMS is constructed from several single-level DBMSs. The DBMS at level l will contain a copy of every data item that a transaction at level l can access. The first protocol for this architecture was proposed by Jajodia and Kogan [20]. The protocol assumes that the set of security levels are totally ordered. Transactions are submitted to a global transaction manager (GTM) who is responsible for forwarding the transactions to the corresponding DBMSs. The updates made by the transactions must also be propagated to the DBMSs at the dominating levels. Two transactions that conflict at level l must be submitted at the dominating level l' in the order in which they commit. However, when transactions do not conflict, they are sent in an arbitrary order to the dominating levels. This may result in nonserializable histories as pointed out by Kang and Keefe [21]. Costich [15] improves upon the Jajodia-Kogan protocol in the following manner. First, it reduces the amount of trust required to implement GTM. Second, it does not require the security levels to form a total order. However, McDermott, Jajodia, and Sandhu [29] illustrate that certain security posets cause the protocol to deadlock and block update projections and produce non-serializable executions. Subsequently, researchers [3–5, 21] have characterized the posets that create this problem. An example will help illustrate the problem. Suppose we have the following security levels: A, B, C, AB, BC, AC, and ABC. The dominance relationship between the levels is as follows: $A \prec AB$, $A \prec AC$, $B \prec AB$, $B \prec BC$, $C \prec AC$, $C \prec BC$, $AB \prec ABC$, $BC \prec ABC$, and $AC \prec ABC$. Three update transactions T_1, T_2, and T_3 at levels A, B, and C are submitted. These transactions execute at the DBMS at these levels

and are then propagated to dominating levels. Thus, T_1, T_2 must execute in the DBMS at level AB, T_2, T_3 must execute at level BC, and T_3, T_1 must execute at level AC. Suppose T_1 is serialized before T_2 in level AB, T_2 is serialized before T_3 in level BC, and T_3 is serialized before T_1 in level AC. Now these updates must be propagated to level ABC. The DBMS at level ABC must respect the serialization orders of the dominated levels. Here we have a deadlock situation because the orders are conflicting. This problem is solved by ordering transactions according to the timestamps generated when the transactions commit for the first time and using a conservative TO protocol which ensures that update projections are never aborted.

The transactions that we have discussed so far have a single security level associated with them and are termed single-level transactions. These transactions can read at multiple security levels but can update data at one security level only. The problem with single-level transactions is that they cannot preserve integrity constraints spanning multiple security levels. Thus, for these databases serializability is an overly restrictive correctness criterion [19, 28]. Jajodia and Atluri [19] have proposed weaker notions of correctness, such as levelwise serializability, one-item read serializability, and pairwise serializability for MLS databases having single-level transactions. Other researchers have proposed the notion of multilevel transactions. A multilevel transaction is associated with a set of security levels. It allows the transaction to read and write data items at multiple security levels. A multilevel transaction is composed of a set of subtransactions. Each subtransaction is associated with a single security level and performs operations following the simple security and restricted-$\star$ property of the BLP model. Ideally, a multilevel transaction must have the properties of atomicity, consistency, isolation, and durability, and it also should not cause any illegal information flow by virtue of its execution. However, it is often not possible to guarantee atomicity without causing illegal information flow [14, 44]. Toward this end, Blaustein et al. [14] define varying degree of atomicity that can be achieved by multilevel

S

transactions. Ray, Ammann, and Jajodia [39] provide a notion of semantic atomicity which is suitable for multilevel transactions. The application containing the transactions is formally analyzed to give assurance of the satisfaction of this property. This work does not use serializability as the correctness criterion but uses the notion of semantic correctness.

Protocols for distributed transaction also may have to be modified for multilevel secure databases. Consider, for instance, the early prepare protocol that ensures atomicity for distributed transactions. When a transaction T_i is submitted, the coordinator decomposes it into subtransactions, say, T_{ij} and T_{ik}, which are distributed to the participants at sites j and k for processing. The participants execute the subtransactions and reply to the coordinator with a prepare/no vote. If all the sites have responded with a prepare vote, the transaction will be committed and the coordinator sends a commit message to all the participants. Now suppose that the transaction T_i is executing on a MLS database. Since T_i is a distributed transaction, it is possible that the subtransaction T_{ij} has finished its execution and entered the prepare state before T_{ik} completes. Some other subtransaction say T_{mj} at the dominated level may want to write a data item, say x, that has been read by T_{ij}. To prevent a covert channel, the lock on x must be released by T_{ij}. Since T_{ik} is executing at a different site, it is possible that T_{ik} will acquire a lock after T_{ij} releases the lock. This will violate the two-phase locking rule and may result in non-serializable executions.

This motivated Atluri, Bertino, and Jajodia [8] to propose a new protocol called secure early prepare (SEP). The coordinator decomposes the transaction T_i into subtransactions T_{i1}, T_{i2}, ..., T_{in} and sends them together with their security levels to the participants. The participants on completing their work successfully responds with a yes vote as before. However, if the participant has read a data item at the dominated level, it also sends a read-low indicator bit. If none of the participants have read low data items, the transaction proceeds like the early prepare protocol. However, if at least one participant

has read a data item at the dominated level, additional rounds of message are necessary. In such cases, the coordinator sends a confirm message to all participants who have read data items at dominated levels. If the participant has not released any locks so far, it responds with a confirmed message. Otherwise it responds with a non-confirmed message. When the coordinator receives confirmed messages from all the participants who have read data items at dominated levels, the coordinator sends a commit message to all the participants. Otherwise, an abort message is sent. In a subsequent work [8], the authors propose an optimization to SEP that avoids some unnecessary aborts caused by SEP and also reduces the number of messages. Ray et al. [40] also attempt to improve upon SEP – instead of aborting subtransactions that have read from low data items, it rolls back the subtransaction to an earlier savepoint and reexecutes it. On successful reexecution, the participant sends a yes message. When all subtransactions have responded with a yes message, the transaction is committed. Otherwise, it is aborted.

Some researchers [1, 22, 45] have also looked into secure real-time transaction processing. The goal in such systems is to prevent illegal information flow as well as maintain the timing constraints required for real-time applications. When both security constraints and real-time constraints cannot be satisfied, some works [1, 45] trade-off security in order to improve the performance. Others [17, 22] do not compromise security for the sake of performance. George and Haritsa [17] propose a concurrency control mechanism in which data conflicts are resolved in favor of the dominated level. Within a given level, data conflicts are resolved in favor of the earliest transaction deadline. Kang et al. [22] improve upon the work presented by George and Haritsa [17] by providing guarantees on average/transient miss ratios. A separate work [37] describes a new concurrency control protocol, known as, multiversion locking protocol with freezing, for processing secure real-time transactions.

Researchers have also investigated the impact of multilevel security on extended transaction models, such as workflows. A workflow is characterized as having a set of tasks and dependencies specified between the tasks. In an MLS workflow, each task is associated with a single security level and is allowed to read and write data items provided they obey the BLP rules. However, the dependencies between tasks at different levels may cause illegal information flow. Toward this end, Atluri et al. [10] have proposed an approach that redesigns the dependencies such that illegal information flow does not occur. A separate work [9] argues about how mandatory and discretionary access controls can be enforced in a workflow.

Secure transaction processing in non-MLS database systems focused on survivability. Attacks will occur in spite of sophisticated prevention mechanisms. The issue is how to identify the attack, confine it, assess the damage caused by it, and repair the damage in a timely manner. One of the early works in damage detection and recovery is by Ammann et al. [7]. After an attack occurs, the data items are marked with different colors to indicate the severity of the damaged. The authors define a notion of consistency for databases in which some data may have been damaged. Clean data must satisfy the integrity constraints defined over them. Damaged data must satisfy a set of relaxed integrity constraints. The authors classify transactions into three categories: *attack transactions*, *normal transactions*, and *countermeasure transactions*. Attack transactions damage data items. Normal transactions sometimes help spread the damage. The normal transaction access protocol determines how the damage is propagated and ensures that the database satisfy the consistency constraints. The countermeasure transactions detect and repair the effects of an attack. These transactions must execute as trusted processes. Detection transactions change the marking of data items whereas repair transactions alter the value of data items. When an attack has occurred, the state of the database before the execution of attack transactions can be retrieved using snapshots. Toward this end, the paper proposes a technique by which snapshots can be generated while the database is servicing normal transactions.

Survivability issues and repair from malicious attacks have received attention in the database context. Ammann et al. [6] propose repair algorithms for traditional database systems that help to recover from the damage caused by malicious transactions. A two-pass static algorithm is proposed where the first pass scans the log forward to locate all malicious and suspect tasks and the second pass goes backward from the end of the log to undo all malicious and suspect tasks. They also proposed a dynamic repair algorithm that continues to accept new transactions while repair is taking place. Panda et al. [24] have also proposed a number of algorithms on damage assessment and repair; some of these store the dependency information in separate structures so that the log does not have to be traversed for damage assessment and repair. Ray et al. [41] improve upon the time taken to assess the damage by using a dependency graph to store the dependencies and using depth-first search to retrieve the affected transactions. Liu and Jajodia [27] present a multiphase damage confinement model. In the initial confinement phase, an estimation is done with respect to the damaged items. The estimation may not be accurate. The authors propose several schemes about damage confinement. The first one maintains timestamps. The initial confinement confines all data items that were updated after the commitment of the bad transactions. A damage assessor unconfines data items that are written by transactions not dependent on the bad transactions. This simple scheme causes damage leakage because a data item unconfined by an unaffected transaction can be updated by an affected transaction. Thus, even temporarily releasing this data item can cause damage spreading. The second scheme takes care of this problem. Confining the data items and later unconfining them usually takes some time. To reduce the relaxation latency, the authors propose a third scheme which uses transaction access patterns to unconfine data items that were not affected by the bad transaction. Panda and Giordano [35] provide two techniques for performing damage assessment and recovery. The first algorithm does detection and recovery simultaneously and is not very efficient. Moreover, new transactions are blocked until recovery is complete. These two shortcomings are removed in the second algorithm. Most of the work on damage assessment are based on transaction dependency approach. In these approaches, the goal is to identify affected transactions which must be undone and then re-executed. However, this may involve unnecessary undoing and redoing of operations. This is because not all operations of an affected transaction are influenced by a bad transaction. Toward this end, Panda and Haque [36] propose a damage assessment technique based on data dependency approach – only the affected operations are undone and redone.

Damage assessment in distributed databases has also been studied by several researchers. Liu and Hao [26] propose a damage assessment technique for distributed database that incurs a high communication overhead. Zuo and Panda [51] propose two approaches for damage assessment in distributed databases. The first one is a peer-to-peer approach which does not require a coordinator to perform damage assessment. When a site knows that it has some global affected transactions, it sends a multicast message to other sites which were involved with these global transactions. The other sites on receiving this message may identify some more global affected transactions which are then broadcast. The process continues until no more new global affected transactions are detected. This approach incurs high communication overhead. The other approach requires a coordinator. There are three variants of this other approach. In the first one, known as receive and forward, the coordinator keeps information about all global transactions. Each site manager sends a list of global affected transaction identifiers to the coordinator. The coordinator informs the other sites of these affected transactions. The other sites check whether any new transactions are affected or not. If not, a clear message is sent to the coordinator. Otherwise, the identifiers of the affected transactions are sent. The process stops when all the other sites send a clear message. The second coordinator-based approach incurs less communication overhead. When a malicious transaction is identified, the coordinator requests other sites for their local dependency graphs. The coordinator builds a global dependency graph using this information. The global graph is used for identifying affected transactions. The third coordinator-based approach relies on sites sending their graphs periodically to the coordinator. In this approach, the local graphs are not merged.

Yu, Liu, and Zang [49] describe an algorithm for on-line attack recovery of workflows. The algorithm tries to build the list of redo and undo tasks, after an independent Intrusion Detection System reports malicious tasks. They also relax the restriction of executing order that exist in an attack recovery system; they introduced multi-version data objects to reduce unnecessary blocks in order to reduce degradation of performance in recovery. However, like in most existing papers, the authors only pay attention to restore consistency for data objects, while they do not analyze the correct actions needed in repair for different control-flow dependencies. In repair for advanced transactions, we need to ensure that constraints of all control-flow dependencies are satisfied, and we should not treat all types of control-flow dependencies in the same manner – need to distinguish different types of control-flow dependencies and adopt different treatment to enforce these dependencies during repair. This issue is addressed in a subsequent work [50].

With the growth of the world-wide web, web services transactions are on the rise. Researchers have taken the concept of transaction processing beyond traditional database systems. One example is Web Services Atomic Transactions specification [32] which defines a framework for applications to implement transactions in an interoperable and heterogeneous setting. The framework leverages the distributed two-phase commit protocol, and a coordinator will orchestrate the flow of the transaction. Security can be integrated with transaction processing by incorporating the web services security related standards such as WS-Security [30], WS-Trust [33], and WS-SecureConversation [31]. With all these modules in place, applications can achieve authentication, confidentiality, and integrity in transaction processing.

In recent years, organizations tend to rely more on cloud computing platforms for their storage requirements considering the reduced cost of acquiring and maintaining their own storage systems. With the popularity of "pay-as-you-go" models, organizations can benefit from reliable and cost effective cloud-based storage solutions. Unlike using cloud infrastructures for computational power, using cloud as a storage solution has its own drawbacks, more importantly data being stored outside the organizational perimeter. Organizations typically expect to preserve transactional capabilities over data while ensuring that data is securely stored in cloud servers.

The work done by Biswas [13] et al. proposes a model where transactional guarantees are provided over encrypted data. Data is encrypted mainly to ensure the confidentiality against the cloud provider. Apart from the data confidentiality and transactional guarantees, this model clearly distinguishes between data owners from transaction owners and supports to keep both parties anonymous from each other. For example, a client can initiate a transaction without being noticed by the data owners. The protocol is explained in the context of a distributed setup where multiple data owners and transaction owners participate and use a third-party cloud server as the storage. Data owners can restrict or grant read and update permissions for certain clients. Not only the data value itself, the data access permissions can also be kept confidentially. Data replication is also a possibility by allowing clients to maintain a cached copy in their local sites. These different possibilities lead to an array of combinations which are called "configurations." They demonstrate the capabilities of their model by using various configurations starting from a base model. Base model is comprised of multiple clients who maintain a copy of the database locally while the server is maintaining the transaction log. In this model, it is assumed that every client has access rights to update any data item. In an optimistic model, clients execute the transactions in their local copies and send their read and write sets (without the data values) to the server for conflict detection. If there are no conflicts, client creates a log of the update operations performed during the transaction and encrypts it with a symmetric key shared among clients in advance. This log is forwarded to the server to update its log and to propagate the update to the rest of the clients. The base model ensures that the data values modified during the transaction are not visible to the server. The base model is further extended to support more sophisticated models, for instance, the "single confidential owner" model. In this configuration, data items accessed by the transaction are owned by a single owner and its identity should remain anonymous from the transaction owner. Similarly this model is extended to support scenarios where data is stored in the server and the read operations are restricted.

Most of the works assume an "honest-but-curious" server in which the clients aim to protect data confidentiality. In one of the works [46], the authors assume an even stronger model where the server can launch *confidentiality attacks* and also *locking attacks*. Note that locking plays a vital role when achieving consistency and isolation in transactions. As a consequence of using a cloud server for storage, it becomes a control point of the locking protocol. For example, when there are multiple transactions active at the same time, a cloud server may try to reorganize them in order to reduce the locking overhead which in turn will aid them in meeting the turnaround times as per the service level agreements. CloudLock protocol proposed in [46] prevents untrusted cloud servers from maliciously exploiting the locking protocols for their benefit when executing user transactions. For example, a cloud server may deny issuing a lock by falsely claiming that the lock is busy. The authors propose extending the traditional locking protocol [48] to include a brief history about the previous transactions which can subsequently be used to detect any protocol violations by the cloud provider. This protocol is assisted by a data structure maintained at the data owner's end which tracks the details about locks (requested time, granted time, released time) as well as an independent time server for ordering the transactions. The data is stored as encrypted data which prevents possible confidentiality attacks.

Most of the outsourced storage solutions attempt to store encrypted data at service providers to provide privacy. However, the service providers may be untrusted and one or more clients may be malicious. The Blind Stone Tablet [47] proposes a novel approach in this regard. The data is stored in local databases maintained at clients and the transaction serializability and durability are achieved through the external storage provider. The external storage provider maintains a client-encrypted and authenticated log for this purpose. The authors propose two approaches. In the basic approach, a global lock is maintained and the clients will have to acquire this lock before executing and committing the transaction. This protocol ensures that the client's local database is brought to an up-to-date and consistent state with respect to the other client databases prior to transaction execution. Then it sends an encrypted description of the transaction updates back to the server to store it and distributes it among other clients. This update is digitally signed and encrypted using a symmetric key shared among clients to achieve privacy and authenticity. Further each client can validate the sequence of transactions applied at other clients by validating a signed hash chain maintained at another client which will be attached with each message sent by that particular client. In the second approach, they propose a wait-free protocol which relies on an optimistic conflict detection mechanism instead of a global lock. Each clients sends a pre-commit message to the server with transaction details and the server responds with the set of previous pre-commit messages sent by other clients. The client locally checks if there is a conflict between the current transaction and any of the transactions appearing in previous pre-commit messages. Depending on the result of this check, it will decide to commit or abort the transaction and finally it informs the decision to the server in a secure message. Similar to the previous step, the server will distribute this message among other clients.

Relational Cloud [16] is a "database-as-a-service" implementation where back-end database engines are powered by CryptDB [38].

Data tuples are stored in encrypted form to provide privacy. Clients use a modified JDBC driver to query the data, and the transactional guarantees are inherited from the CryptDB.

With the recent trends of preferring availability over consistency, the cloud databases are moving toward weaker consistency models such as eventual consistency. As Iskander et al. [18] point out, weaker consistency models may cause policy-based authorization systems associated with transaction handling to base their decisions on stale policies and credentials. Policies may be replicated across servers and the delay in propagating updates may cause policy inconsistencies. User credential inconsistencies are also a possibility during the lifetime of a transaction; for instance, a user credential may be revoked by the authorization server at a later stage of a transaction, but certain read and write operations may be allowed in the earlier stages based on that credential. In order to solve this problem, the authors propose the concept of *trusted transactions*, which are immune to credential and policy inconsistencies and redefine *safe* transactions as both trusted and satisfies the ACID properties. Two consistency models are defined as view consistency (every server has the same policy version) and global consistency (every server has the latest policy version). An extension to the two-phase commit protocol is proposed called *2 Phase Validation Commit protocol*. The basic underlying principle is to perform a policy and credential validation at individual servers upon receiving the *prepare-to-commit* message and include its results along with the policy versions in the reply to transaction manager. Depending on the consistency model used, the transaction manager will evaluate the responses from individual servers and proceed to the next phase. For example, in the view consistency model, the transaction manager will check if every server uses the same policy version. If not it will issue update requests to the appropriate servers and restart the protocol.

Although a lot of research appears in secure transaction processing, a lot more needs to be done. One research direction is how to process distributed transactions on the cloud, such that it satisfies privacy constraints. For such transactions, it will be useful to investigate how should logs be maintained so as to recover from failures and attacks. One other area of research that appears promising is how can we analyze log data in real time in order to predict attacks and take actions in a timely manner. In spite of preventive measures, attacks will occur. It will be useful to investigate how to confine the damage and take timely corrective actions in order to prevent the damage from spreading.

Key Applications

Critical information, such as health records and financial records, are stored in the databases of an organization. Organizations must also interact with each other and share critical information to accomplish a particular mission. Security and privacy breaches involving such critical information have disastrous consequences. In many of these applications for reasons of correctness and security, data items are updated through transactions that have the atomicity, consistency, isolation, and durability properties. Transaction processing mechanisms must be adapted to meet the security needs of the application. For example, the concept of secure information flow is important for military applications. Thus, transaction processing mechanisms must be adapted so as to ensure that there are no illegal information flows. For organizations storing data in an encrypted manner in the clouds, it is important to have mechanisms that support transaction processing over encrypted data. Since it is impossible to protect against all kinds of security and privacy breaches, it is also necessary to design systems that can automate to the extent possible the detection and repair from an attack.

Cross-References

▶ Access Control
▶ Cloud Computing
▶ Database Security
▶ Data Encryption
▶ Privacy
▶ Multilevel Secure Database Management System
▶ Transaction

Recommended Reading

1. Ahmed Q, Vrbsky S. Maintaining security in firm real-time database systems. In: Proceedings of the 14th Annual Computer Security Applications Conference; 1998.
2. Ammann P, Jaeckle F, Jajodia S. A two-snapshot algorithm for concurrency control in secure multi-level databases. In: Proceedings of the IEEE Symposium on Security and Privacy; 1992. p. 204–15.
3. Ammann P, Jajodia S. Distributed timestamp generation in planar lattice networks. ACM Trans Comput Syst. 1993;11(3):205–25.
4. Ammann P, Jajodia S. An efficient multiversion algorithm for secure servicing of transaction reads. In: Proceedings of the 1st ACM Conference on Computer and Communication Security; 1994. p. 118–25.
5. Ammann P, Jajodia S, Frankl P. Globally consistent event ordering in one-directional distributed environments. IEEE Trans Parallel Distrib Syst. 1996;7(6):665–70.
6. Ammann P, Jajodia S, Liu P. Recovery from malicious transactions. IEEE Trans Knowl Data Eng. 2002;14(5):1167–85.
7. Ammann P, Jajodia S, McCollum C, Blaustein B. Surviving information warfare attacks on databases. In: Proceedings of the 1997 IEEE Symposium on Security and Privacy; 1997.
8. Atluri V, Bertino E, Jajodia S. Degrees of isolation, concurrency control protocols, and commit protocols. In: Proceedings of the IFIP WG11.3 Working Conference on Database Security; 1995. p. 259–74.
9. Atluri V, Huang W-K. Enforcing mandatory and discretionary security in workflow management systems. J Comput Secur. 1997;5(4):303–40.
10. Atluri V, Huang W-K, Bertino E. A semantic-based execution model for multilevel secure workflows. J Comput Secur. 2000;8(1):3–42.
11. Atluri V, Jajodia S, Keefe TF, McCollum C, Mukkamala R. Multilevel secure transaction processing: status and prospects. In: Proceedings of the 10th IFIP WG11.3 Working Conference on Database Security. Como; 1996.
12. Bell DE, LaPadula LJ. Secure computer system: unified exposition and multics interpretation. Technical Report MTR-2997, MITRE Corporation, Bedford; 1975.
13. Biswas D, Vidyasankar K. Secure cloud transactions. Comput Syst Sci Eng. 2013;28(6):439–48.
14. Blaustein BT, Jajodia S, McCollum CD, Notargiacomo L. A model of atomicity for multilevel

S

transactions. In: Proceedings of the IEEE Symposium on Research in Security and Privacy; 1993.
p. 120–34.

15. Costich O. Transaction processing using an untrusted
scheduler in a multilevel database with replicated
architecture. In: Proceedings of the IFIP WG11.3
Working Conference on Database Security; 1992. p.
173–90.

16. Curino C, Jones EPC, Popa RA, Malviya N,
Wu E, Madden S, Balakrishnan H, Zeldovich
N. Relational cloud: a database service for the
cloud. In: Proceedings of the 5th Biennial Conference on Innovative Data Systems Research; 2011.
p. 235–40.

17. George B, Haritsa JR. Secure concurrency control
in firm real-time database systems. Distrib Parallel
Databases. 2000;8(1):41–83.

18. Iskander MK, Wilkinson DW, Lee AJ, Chrysanthis
PK. Enforcing policy and data consistency of cloud
transactions. In: Proceedings of the 31st International Conference on Distributed Computing Systems
Workshops; 2011. p. 253–62. IEEE.

19. Jajodia S, Atluri V. Alternative correctness criteria for concurrent execution of transactions in multilevel secure databases. In: Proceedings of the
IEEE Symposium on Security and Privacy; 1992.
p. 216–24.

20. Jajodia S, Kogan B. Integrating an object-oriented
data model with multilevel security. In: Proceedings
of the IEEE Symposium on Security and Privacy;
1990. p. 76–85.

21. Kang IE, Keefe TF. Transaction management for multilevel secure replicated databases. J Comput Secur.
1995;3(2/3):115–45.

22. Kang K, Son SH, Stankovic J. STAR: secure real-
time transaction processing with timeliness guarantees. In: Proceedings of the 23rd IEEE Real-time
Systems Symposium; 2002.

23. Keefe TF, Tsai WT. Multiversion concurrency control
for multilevel secure databases. In: Proceedings of
the IEEE Symposium on Security and Privacy; 1990.
p. 369–83

24. Lala C, Panda B. Evaluating damage from cyber
attacks: a model and analysis. IEEE Trans Syst Man
Cybern Part A. 2001;31(4):300–10.

25. Lamport L. Concurrent reading and writing. Commun
ACM. 1977;20(11):806–11.

26. Liu P, Hao X. Efficient damage assessment and repair
in resilient distributed database systems. In: Proceedings of the 15th IFIP WG11.3 Working Conference
on Data and Application Security; 2001. p. 75–89.

27. Liu P, Jajodia S. Multi-phase damage confinement in
database systems for intrusion tolerance. In: Proceedings of the 14th IEEE Computer Security Foundations Workshop; 2001.

28. Maimone WT, Greenberg IB. Single-level
multiversion schedulers for multilevel secure
database systems. In: Proceedings of the 6th Annual
Computer Security Applications Conference; 1990.
p. 137–47.

29. McDermott J, Jajodia S, Sandhu R. A single-level
scheduler for replicated architecture for multilevel
secure databases. In: Proceedings of the 7th Annual
Computer Security Applications Conference; 1991.
p. 2–11.

30. OASIS. Web services security: SOAP message security, 2; 2006.

31. OASIS.WS-SecureConversation, 3; 2007.

32. OASIS.Web services atomic transaction, 2; 2009.

33. OASIS. WS-Trust, 4; 2012.

34. Pal S. A locking protocol for multilevel secure
databases providing support for long transactions.
In: Proceedings of the 10th IFIP WG11.3 Working
Conference on Database Security; 1996. p. 183–98.

35. Panda B, Giordano J. Reconstructing the database
after electronic attacks. In: Proceedings of the 12th
IFIP WG11.3 International Working Conference on
Database Security; 1998.

36. Panda B, Haque KA. Extended data dependency
approach: a robust way of rebuilding database. In:
Proceedings of the 2002 ACM Symposium on Applied Computing; 2002.

37. Park C, Park S, Son SH. Multiversion locking protocol with freezing for secure real-time
database systems. IEEE Trans Knowl Data Eng.
2002;14(5):1141–54.

38. Popa RA, Redfield C, Zeldovich N, Balakrishnan
H. Cryptdb: protecting confidentiality with encrypted
query processing. In: Proceedings of the 12rd ACM
Symposium on Operating Systems Principles; 2011.
p. 85–100. ACM

39. Ray I, Ammann P, Jajodia S. A semantic-based transaction processing model for multi-level transactions.
J Comput Secur. 1998;6(3):181–217.

40. Ray I, Bertino E, Jajodia S, Mancini L. An advanced
commit protocol for MLS distributed database systems. In: Proceedings of the 3rd ACM Conference
on Computer and Communications Security; 1996. p.
119–28.

41. Ray I, McConnell RM, Lunacek M, Kumar V.
Reducing damage assessment latency in survivable
databases. In: Proceedings of the 21st British National Conference on Databases; 2004.

42. Reed DP, Kanodia RK. Synchronizations with
event counts and sequencers. Commun ACM.
1979;22(5):115–23.

43. Schaefer M. Quasi-synchronization of readers and
writers in a multi-level environment. Technical Report TM-5407/003, System Development Corporation; 1974.

44. Smith KP, Blaustein BT, Jajodia S, Notargiacomo
L. Correctness criteria for multilevel secure transactions. IEEE Trans Knowl Data Eng. 1996;8(1):32.

45. Son SH, Mukkamala R, David R. Integrating
security and real-time requirements using covert
channel capacity. IEEE Trans Knowl Data Eng.
2000;12(6):865–79.

46. Tan CC, Liu Q, Wu J. Secure locking for untrusted
clouds. In: Proceedings of the IEEE International
Conference on Cloud Computing; 2011. p. 131–8.

47. Williams P, Sion R, Shasha D. The blind stone tablet: outsourcing durability to untrusted parties. In: Proceedings of the Network Distributed System Security Symposium; 2009.
48. Wu J. Distributed system design. Boca Raton: CRC Press; 1998.
49. Yu M, Liu P, Zang W. Multi-version attack recovery for workflow systems. In: Proceedings of the 9th Annual Computer Security Applications Conference; 2003. p. 142–51
50. Zhu Y, Xin T, Ray I. Recovering from malicious attacks in workflow systems. In: Proceedings of the 16th International Conference on Database and Expert Systems; 2005.
51. Zuo Y, Panda B. Damage discovery in distributed database systems. In: Proceedings of the 18th IFIP WG11.3 Working Conference on Data and Applications Security; 2004.

Security Services

Athena Vakali
Aristotle University, Thessaloniki, Greece

Synonyms

Authentication; Data confidentiality; Data integrity services

Definition

Given a set of local or distributed resources to be protected, a security service is a task (or set of tasks) that coherently performs processing or communication on behalf of the underlying system infrastructure, in order to support and employ several security requirements of both the system and the data sources. Such requirements involve authentication, PKI accessing, etc. over the underlying resources. Security services typically implement portions of security policies and are implemented via particular processes which are called security mechanisms.

Key Points

Security services are well documented and described in X.800 documentation [1] for almost all the layers from the physical up to the application layer, whereas focus on security services applied on the Web is given in [2]. Physical and data layer security services focus on support confidentiality at various levels, namely, at the connection and the traffic flow (both full and limited), and they are designed for peer-to-peer or multi-peer communications. At the network and transport layers, security services involve authentication, access control, traffic flow, and confidentiality, and they are provided together or separately.

At the application layer, from X.800 the core security services are identified, and they may be supported either singly or in combination, to employ access control for enforcing authentication, data confidentiality, data integrity, and nonrepudiation. As highlighted in [2], these involve:

(a) Services for subjects/clients and resources identities: to verify the identity of the subject who requests a source access and prevent malicious client attempts. Therefore, such services may be either authentication services (to verify an identity claimed by/for an entity) or nonrepudiation services (to prevent either sender or receiver from denying a transmitted message).

(b) Services for subjects/clients authorizations over resources: to manage relationships among clients and protected resources. Therefore, such services involve either access control services (for protecting system resources against unauthorized access) or data privacy (for protecting data against unauthorized disclosure-data confidentiality and changes-data integrity).

Security services support is crucial and rather challenging in several domains such as in:

Current and future next-generation wireless sensor networks (WSN) due to their specific requirements and characteristics (e.g., scarcity

S

of energy and processing power). Security services supporting in WSN demand authentication, confidentiality, integrity, balancing the energy cost, nonreputation, and scalability services, and already two relevant protocols to address these main security services have appeared in the literature [3, 4]. The proposed protocols integrate public key and symmetric key algorithms to ensure optimal usage of sensors' energy and processing power to provide adequate security in next-generation WSN [5].

Intelligent power grids (also known as smart grids), where various objects and "things" access systems in several network environments, raising access control security critical requirements. Providing users with secure services in smart grids, access control security models are needed and dynamic access models for secure user services in the smart grid environment have been proposed with appropriate context types mapping to a context-based user security policy which allows user's access to services dynamically [6].

Mobile cloud environments based on a machine-to-machine service model is typically utilized at which a mobile device can use the cloud for searching, data mining, and multimedia processing. To protect the processed data, security services, i.e., encryption, decryption, authentications, etc., are performed in the cloud, and a general classification of cloud security services has identified two categories, namely, the critical security (CS) service and the normal security (NS) service [7–9].

Systems and methods for a security delegate module to select appropriate security services for web applications have been patented (http://www.google.com/patents/US8635671) with emphasis on a method which supports an application running a first instance in a network for a user; determining, by the security delegate module, a type of the application; and selecting, by the security delegate module, a security service based on the set of user authentication credentials and the type of application.

New trends in security services have placed emphasis on well-known security defenses as OpenFlow security services which are used to examine various performance and efficiency aspects [10].

Moreover relevant other patents for security services (such as methods, systems, and products for security services https://www.google.ch/patents/US9379915 and network-based security services for managed internet service http://www.google.st/patents/US20110145911 and http://www.google.com/patents/US8549610) indicate the interest in the field from different views and domains [11].

Cross-References

▶ Database Security
▶ Secure Database Development

Recommended Reading

1. Recommendation X.800 Security architecture for open systems, interconnection for CCITT applications. http://fag.grm.hia.no/IKT7000/litteratur/paper/x800.pdf
2. Stoupa K, Vakali A. Policies for web security services, chapter III. In: Ferrari E, Thuraisingham B, editors. Web and information security: Idea Group; 2006.
3. Faisal M, Al-Muhtadi J, Al-Dhelaan A. Integrated protocols to ensure security services in wireless sensor networks. Int J Distrib Sensor Networks. 2013;9(7):740392.
4. Kang J, Adibi S. A review of security protocols in mHealth wireless body area networks (WBAN). In: Proceedings of the International Conference on Future Network Systems and Security; 2015. p. 61–83.
5. Fouad MMM, Hassanien AE. Key pre-distribution techniques for WSN security services, vol. 70. 2014. p. 265–83.
6. Yeo S-S, Kim S-J, Cho D-E. Dynamic access control model for security client services in smart grid. Int J Distrib Sensor Networks. 2014;10(6):81760. 7 pages
7. Liang H, Huang D, Cai LX, Shen X, Peng D. Resource allocation for security services in mobile cloud computing. In: Proceedings of the IEEE Conference on Computer Communications Workshops; 2011.
8. Rosado DG, Fernández-Medina E, López J. Security services architecture for secure mobile grid systems. Spec Issue Secur Dependability Assur Software Archit. 2011;57(3):240–58.

9. Demchenko Y, De Laat C, Lopez DR, Garcia-Espin JA. Security services lifecycle management in on-demand infrastructure services provisioning. In: Proceedings of the 2nd IEEE International Conference on Cloud Computing Technology and Science; 30 Nov 2010–3 Dec 2010. p. 644, 650.
10. Shin S, Porras PA, Yegneswaran V, Fong MW, Gu G, Tyson M. FRESCO: modular composable security services for software-defined networks. In: Proceedings of the Network and Distributed Systems Security Symposium; 2013.
11. Sheng QZ, Qiao X, Vasilakos AV, Szabo C, Bourne S, Xu X. Web services composition: a decade's overview. Inf Sci. 2014;280(Oct):218–38.

Selection

Cristina Sirangelo
IRIF, Paris Diderot University, Paris, France

Synonyms

Selection (Relational Algebra)

Definition

Given a relation instance R over set of attributes U and a condition F, the selection $\sigma_F(R)$ returns a new relation over U consisting of the set of tuples of R which satisfy F. The condition F is an atom of the form $A = B$ or $A = c$, where A and B are attributes in U and c is a constant value.

The generalized selection allows more complex conditions: F can be an arbitrary Boolean combination of atoms of the form $A = B$ or $A \neq B$ or $A = c$ or $A \neq c$. Moreover, if a total order is defined on the domain of attributes, more general comparison atoms of the form $A \alpha B$ or $A \alpha c$ are allowed, where α ranges over $\{=, \neq, <, >, \leq, \geq\}$.

Key Points

The selection is one of the basic operators of the relational algebra. It operates by "selecting" rows of the input relation. A tuple t over U satisfies the condition $A = B$ if the values of attributes A and B in t are equal. Similarly t satisfies the condition $A = c$ if the value of attribute A in t is c. Satisfaction of generalized selection atoms is defined analogously.

As an example, consider a relation *Exams* over attributes (*course-number, student-number, grade*), containing tuples $\{(EH1, 1001, A), (EH1, 1002, A), (GH5, 1001, C)\}$. Then $\sigma_{grade=A \wedge course-number = EH1}(Exams)$ is a relation over attributes (*course-number, student-number, grade*) with tuples $\{(EH1, 1001, A), (EH1, 1002, A)\}$.

In the case that a relation schema is only specified by a relation name and arity, the result of the selection is a new relation having the same arity as the input one, containing the tuples which satisfy the selection condition. In this case the selection atoms are expressions of the form $j = k$ or $j = c$ (or $j \alpha k$ and $j \alpha c$ in the generalized selection). Here j and k are positive integers bounded by the arity of the input relation, identifying its j-th and k-th attributes, respectively.

Cross-References

▶ Relational Algebra

S

Selectivity Estimation

Evaggelia Pitoura
Department of Computer Science and Engineering, University of Ioannina, Ioannina, Greece

Synonyms

Cost estimation; Selectivity estimation

Definition

For each query, there are many equivalent execution plans. To choose the most efficient among these different query plans, the optimizer has to estimate their cost. Computing the precise cost of each plan is usually not possible without actually evaluating the plan. Thus, instead, optimizers use statistical information stored in the DBMS, such as the size of relations and the depth of the indexes, to estimate the cost of each plan. For large databases, this cost is dominated by the number of disk accesses.

Since, in general, the cost of each operator depends on the size of its input relations, it is important to provide good estimations of their selectivity, that is, of their result size.

Key Points

During query optimization, the optimizer enumerates potential execution plans for each query and evaluates their cost in order to choose the less expensive among them. In general, computing the exact cost of each query plan is not possible without actually evaluating the plan. Instead, optimizers make use of statistical information stored in the DBMS catalog to estimate the cost of each plan. Such statistics include the number of tuples in each relation, the size of each tuple and the number of distinct values that appear in each relation. The cost of a plan can be measured in terms of different resources, such as CPU, I/O, buffer utilization, and, in the case of parallel and distributed databases, communication costs. However, in large databases, the cost is usually dominated by disk accesses.

The cost of each plan is computed by combining the estimations of the cost of each of the operators appearing in the plan. Since the cost of an operator depends mainly on the sizes of its input relations, it is central to have good estimates of the result size of each operator that is going to be used as input to the next operator in the plan.

Take, for example, a selection condition consisting of a number of predicates. Each predicate

has a reduction factor, which is the relative reduction in the number of result tuples caused by this predicate. There are many heuristic formulas for the reduction factors of different predicates. In general, they depend on the assumption of uniform distribution of values and independence among the various relation fields. More accurate reduction factors can be achieved by maintaining more accurate statistics, for example in the form of histograms or multidimensional histograms. The accuracy of the estimates also depends on how frequently the available statistics are updated to reflect the current database state.

Cross-References

▶ Cost Estimation
▶ Evaluation of Relational Operators
▶ Query Optimization
▶ Query Plan

Recommended Reading

1. Chaudhuri S. An overview of query optimization in relational systems. In: Proceedings of the 17th ACM SIGACT-SIGMOD-SIGART Symposium on Principles of Database Systems; 1998. p. 34–43.
2. Ramakrishnan R, Gehrke J. Database management systems. New York: McGraw-Hill; 2003.
3. Selinger PG, Astrahan MM, Chamberlin DD, Lorie RA, Price TG. Access path selection in a relational database management system. In: Proceedings of the ACM SIGMOD International Conference on Management of Data; 1979. p. 23–34.

Self-Maintenance of Views

Himanshu Gupta
Stony Brook University, Stony Brook, NY, USA

Definition

A data warehouse is a collection of materialized views derived from base relations that may not reside at the warehouse. It is important to keep the views up to date in response to changes to the base relations. *Self-maintenance* of views involves maintaining the views, using information that is strictly local to the warehouse: the view definitions and the view contents. Such self-maintenance of views (whenever possible) is more efficient than incremental maintenance or recomputation of views.

Key Points

In a data warehouse, views are computed and stored in the database to allow efficient querying and analysis of the data. These views stored at the data warehouse are known as *materialized views* and are defined in terms of the base relations residing in data sources that may or may not be local to the warehouse. To keep the views consistent with the base data, any change reported by the data sources must be reflected in the views. In response to the changes at the base relations, the view can be either recomputed from scratch or incrementally maintained or self-maintained.

Incremental maintenance of a view involves propagating the changes at the source onto the view so that the view reflects the changes. While maintaining these views incrementally is often significantly more efficient than recomputing them from scratch (as done in most current data warehouses), it can still be expensive. For instance, in response to an update to a base relation, incremental maintenance of views defined as a join may involve looking up the non-updated base relations, which may reside in external sources. Some of these base relations may even be unavailable when the view needs to be maintained. Further, since base relations are independently updated, they may be read in an inconsistent state, often resulting in erroneous view updates. Thus, in data warehousing environments where maintenance is performed locally at the warehouse, an important incremental view-maintenance issue is how to minimize external base data access.

Motivated by the above, another approach of maintaining a view is called *self-maintenance*. In view self-maintenance, a view is maintained us-

ing information only local to the data warehouse, viz., the view definition and the view contents. In general, a view may not be self-maintainable. However, in certain cases, a view can still be maintained using only a specified subset of the base relations. With the above approach of self-maintenance, it is possible to minimize the cost to maintain the data warehouse, shorten the time window during which the warehouse is inconsistent with the updated data sources, and avoid view update anomalies due to asynchronous base data updates.

There can be two notions [2] of self-maintainability (SM): the *compile-time SM* where a view is self-maintainable independently of the view's contents and the contents of the base relations and under all updates of a certain type. The *runtime SM* is when a specific view is self-maintainable under a specific update and given the contents of a specified subset of base relations. Note that the run-time approach is more aggressive in the way that it may succeed in maintaining a view when the compile-time approach may fail. Below, works on run-time self-maintainability approach are described.

An important question is to determine whether a view is (run-time) self-maintainable. In other words, test for self-maintainability requires whether a unique new state is guaranteed, given an update to the base relations, an instance of the views, and an instance of a subset of the base relations. A second question is how to bring the view up to date using only the given information. Together, these two questions define the *view self-maintenance problem*.

The authors in [1, 5] give self-maintainability conditions for views that are select-project-join queries with no self-joins and for single-relation insertions or deletions. Huyn in [3] solved the problem more efficiently than [1] for select-project-join views with no self-joins for single insertions. In particular, Huyn generate SQL queries that test whether a view is self-maintainable and update the view if it is. More specifically, he shows that for insertion updates and conjunctive queries: (i) the self-maintainability test is extremely simple queries that look for certain tuples in the view to be maintained, (ii) these tests can be generated from just the view definition using a simple algorithm based on the concept of "minimum z-partition," and (iii) view self-maintenance can also be expressed as simple update query over the view.

In a follow-up work [4], Huyn considers the view self-maintenance problem in the presence of multiple views, under arbitrary mixes of insertions and deletions. In particular, for conjunctive queries, he gives an algorithm that generates, at view definition time, the query expressions required to maintain the view in response to base updates. He generalizes his techniques to the problem of generalized self-maintenance problem, where in addition to the warehouse views, one is also given access to some of the base relations. In general, he provides better insight into the problem by showing that view self-maintainability can be reduced to the problem of deciding query containment.

Cross-References

▸ Data Warehouse
▸ View Maintenance
▸ Views

Recommended Reading

1. Gupta A, Blakeley JA. Using partial information to update materialized views. Inf Syst. 1995;20(9):641–62.
2. Gupta A, Mumick IS. Maintenance of materialized views: problems, techniques, and applications. IEEE Data Eng Bull. 1995;18(2):3–18.
3. Huyn N. Efficient view self-maintenance. In: Proceedings of the Workshop on Materialized Views; 1996.
4. Huyn N. Multiple-view self-maintenance in data warehousing environments. In: Proceedings of the 23rd International Conference on Very Large Data Bases; 1997.
5. Tompa FW, Blakeley JA. Maintaining materialized views without accessing base data. Inf Syst. 1988;13(4):393–406.

Self-Management Technology in Databases

Surajit Chaudhuri[1] and Gerhard Weikum[2]
[1]Microsoft Research, Microsoft Corporation, Redmond, WA, USA
[2]Department 5: Databases and Information Systems, Max-Planck-Institut für Informatik, Saarbrücken, Germany

Synonyms

Auto-administration and auto-tuning of database systems; Autonomic database systems; Self-managing database systems; Self-tuning database systems

Definition

The total cost of ownership (TCO) for a database-centric information system is dominated by the expenses for highly skilled human staff in order to deploy, configure, administer, monitor, and tune the database system. Self-management technology for databases aims to automate these tasks to the largest possible extent and throughout the entire life cycle of the information system. This involves many dimensions that determine the system performance and availability such as workload analysis, capacity planning, physical database design, database statistics management for query optimization, load control, memory management, system-health monitoring, failure diagnosis and root-cause identification, configuration of backup procedures, and other self-healing capabilities. The self-managing capabilities can be incorporated in a system using a number of architectural options. For example, such capabilities can be either built into the system itself and integrated with its normal functionality or provided through external tools that for the database engine. The latter approach is often referred to as DB advisors, DB assistants, or DB wizards.

The notion of self-management for database systems comprises a wide spectrum of issues, and it is tempting to consider the partial automation of all database-related human activity as facets of self-management, for example, information integration. However, this entry defines self-management technology to be confined to system issues that arise with the operation of a database engine, thus excluding the tasks that do not directly affect the engine's operation.

self-management was revived by the AutoAdmin project, which then became the leading initiative in this area [7, 8, 9]. AutoAdmin initially focused on physical database design, but subsequently also considered many other issues such as adaptive statistics management, system monitoring, and online tuning techniques. Good overviews of the progress achieved in the past 10 years, in terms of both research contributions and product impact, are given by [8–10].

Historical Background

Needs for tuning tools and certain self-managing capabilities have been around for decades. For example, analytic models for capacity planning have a long tradition in the mainframe world [1], and methods for incremental online reorganization of storage and indexing systems can be seen as early forms of self-management. Selected issues of automatic tuning have been addressed already in the 1980s, most notably, on index selection [2].

In the late 1980s and throughout the 1990s, both database system functionality and workloads became much richer and increased the complexity of system management. Together with the proliferation of database systems across a wide spectrum of IT applications, this created a shortage of sufficiently skilled system administrators and tuning experts. At the same time, hardware- and software-licensing costs were rapidly decreasing, so that human staff for system management became the key factor in total cost of ownership (TCO). These trends alerted the database system industry in the mid-1990s and led to intensive research and development initiatives at all of the major system vendors.

Early work on more comprehensive strategies and principles of automatic tuning included the COMFORT project at ETH Zurich [3, 4] and the "DBMS autopilot" work at the University of Wisconsin [5]. These projects were in turn inspired by prior work toward adaptive and self-tuning operating systems [6] and particularly focused on resource control for dynamically evolving, mixed workloads. Starting in the mid-1990s, interest in

Foundations

General Framework

The overriding goal of self-management is to operate the database system at a satisfactory level of performance and service quality – at every point in time regardless of load peaks or shifts in workload characteristics. Specifically, the system should automatically adjust its configuration to evolving workloads. To this end, the key issue is to understand for each component of the system the dependencies that relate the system configuration and workload properties to the resulting performance measures:

- *configuration* × *workload* → *performance*

All three parameters of this relation need to be interpreted broadly:

- *System configuration* includes the *hardware setup* (number and speed of processor cores, memory size, number and characteristics of disks and access channels, etc.), the *software setup* at system-boot time (e.g., thresholds for lock escalation or memory-pressure handling), the *physical database design* (indexes, materialized views, etc.), the *backup or replication procedures* and their parameters (e.g., frequency and granularity of backups, consistency protocols for replication), and also the system's *run-time adaptation policies* to handle newly arising conditions (e.g., the scheduling policy and workload-class priorities, the memory allocation policy, etc.) as well as the system's *exception-handling policies* (e.g., load shedding under memory pressure).

- *Workload characteristics* include the types of queries, update operations, transactions, and workflows that access the database (i.e., the *workload classes*), their frequencies, their arrival rates during specific periods (e.g., main business hours vs. weekend vs. end-of-fiscal-year processing), their arrival patterns (e.g., typical sequences of different queries), their co-occurrence patterns (e.g., online transactions concurrently with certain batch jobs), the distribution of query parameter values, and so on.
- *Performance measures* include the *throughput* and *response times* (or properties of response-time distributions like quantiles) of different workload classes, but also metrics that capture the system's *dependability*, namely, *reliability* (e.g., probability of losing data by a permanent failure such as double or triple disk failures), *availability* (i.e., probability of being able to service requests at any timepoint in the presence of transient outages), *performability* (i.e., the performance level that can be sustained in degraded configurations, e.g., when servers or data replicas are temporarily unavailable), and, ultimately, even capabilities to react to the system environment such as *resilience to security attacks*, graceful handling of denial-of-service attacks, countering attempts to breach data privacy, and so on.

If one had a reasonably accurate and complete model of the configuration-workload-performance relation, a self-managing system could, in principle, solve the following "inverse problem":

Given specified goals for performance measures and the workload properties, find the lowest-cost configuration that satisfies the performance goals.

This would have to be solved dynamically whenever the workload exhibits major changes and necessitates adaptive reconfiguration. The goals in this setting should cover throughput, response time (e.g., a 95th quantile of at most 1 s), availability (e.g., expected downtime per year of

at most 10 min), and other measures, leading to a more general notion of *service level agreements (SLA)* or *quality-of-service (QoS) guarantees*. The specified SLA/QoS requirements would drive the self-management procedures.

All three elements of the configuration-workload-performance function can refer to an entire system or to individual *components*. The latter is highly preferable: the configuration parameters are then restricted to the ones that are relevant for the specific component at hand, and the same modularization holds for workload properties and performance measures. At the system level, the configuration is the union of the component-specific parameters, the workload at the system level is decomposed into workload properties for each component, and the performance observed at the component level is aggregated into system-level measures. As a consequence, predicting the system-wide performance requires *composability* of workload-characterization models and, most crucially, of performance-prediction models for the underlying components.

An inherent difficulty in achieving self-manageability is that many of the input parameters are not known a priori, may change rapidly thus posing difficulties to parameter estimation techniques, and are inherently difficult to model and thus bound to be incompletely captured. In particular, workload modeling is an extremely difficult task: queries, for example, can be modeled at different resolution levels from SQL statements down to internal operator trees (or actually, pipelined DAGs) or storage-level access operations; arrival rates can refer to steady state (long-term averages) or a specific look-ahead time horizon like the next hour or next minute; the dependencies in the arrival patterns of different query or transaction types can be determined by statistical correlations or by causal flow models via static program analysis of the application code.

Although the above discussion presents only a conceptual model, this framing does provide useful guidelines for approaching specific self-management issues. The salient aspects of the configuration-workload-performance framework and the general aim of solving for a suitable configuration suggest a methodology that is best characterized as an *observe-predict-react* cycle:

- *Observe:* Workload characteristics need to be observed as the system is running, so as to estimate parameters of the workload model or models. Observations may need to be collected at different resolution levels and time scales, and they must track drifts and anomalies in the workload evolution. This calls for a form of *introspection* or *self-monitoring*.
- *Predict:* As the current configuration is known and given the outcome of the observation step, one can aim to assess the performance in the near-term future, either with the current configuration or with various alternative configurations that could potentially improve performance. This calls for a notion of *what-if analysis*.
- *React:* Occasionally the system configuration needs to be reconsidered, in view of the observed workload changes and projected near-term performance. In the case that explicit performance goals are given, the search for and assessment of new configurations could be triggered whenever one of the goals is violated (or about to be violated) with the current configuration. In the case without explicit goals, one or more objective functions need to be specified, so that whenever there is a significant loss in the objective function(s), the reaction step is invoked. The react step needs a smart search strategy to identify promising new configurations.

The observe-predict-react approach can be applied to *different time scales* based on the specific self-management task at hand from long-term capacity planning, which may be done relatively infrequently, down to real-time decisions, for example, about memory management, which requires all three steps at the resolution of minutes or even seconds to handle sudden memory-pressure situations. The three steps may be implemented as tools outside the database engine, or could be incorporated into the

engine at *different integration levels*. Typically, short time scales mandate deeper integration, whereas long-term decisions could be made by external advisor tools.

Self-Management Paradigms

Ideally, the models and strategies for self-managing systems would follow a unified principle such as mathematical optimization theory, but such a "grand solution" is not in reach today. Instead, there are many diverse results on a variety of specific self-management issues; the state of the art is best characterized by a number of self-management paradigms: approaches that work well on paradigmatic example problems but bear the potential for being generalized into more broadly applicable principles. In the following, several such paradigms will be introduced, each with a general characterization and one or more exemplary use cases.

Trade-Off Elimination Tuning knobs exist because of trade-offs: there is no algorithm (or data structure) that performs near optimal under all possible workloads, and therefore, systems are equipped with different options. However, sometimes it is possible to design an algorithm or a strategy that performs very well across the full spectrum of workloads, and this algorithm should be knob-free or only have second-order parameters which are uncritical or easy to set once across (almost) all workloads [4]. Such an algorithm effectively eliminates the trade-off. Examples are modern cache-replacement algorithms (LRU-k, ARC, etc.) or $B + -$ tree indexes for both exact-match lookups and range scans. In the case of quantitative tuning parameters such as disk block sizes or striping units, similar considerations may lead to robust techniques that can effectively eliminate knobs.

Static Optimization Some self-tuning problems can be cast into mathematical optimizations with statically given input parameters. In principle, this opens up solutions based on combinatorial optimization methods such as branch and bound. Physical database design falls into this category; planning backup or replication procedures is another example. However, it is crucial to obtain input parameters and evaluate the objective function in a way that preserves the actual behavior of the database system. A major lesson from the research on physical design automation [8, 9] is that a hand-crafted cost model, for assessing the quality of a particular design configuration, is bound to be inappropriate no matter how detailed and seemingly accurate it may be. The crux is that a separate cost model does not consider the actual behavior of the engine's query optimizer in selecting indexes for particular queries. Instead, the practically viable solutions reference the query optimizer's cost model each time they need to assess the benefit of a design configuration for the given workload. To limit the computational overhead of these calls, thoroughly designed techniques for what-if assessment are needed (so as to estimate the benefit of a design configuration without actually building indexes). In addition, great care must be taken in the enumeration of candidate configurations. Although this is an offline optimization, the combinatorial explosion of the search space could easily lead to unacceptable run-time.

Stochastic Optimization In some situations, albeit dealing with an offline optimization problem, the input parameters can only be characterized as random variables or by means of stochastic processes. System capacity planning falls into this category, for example, deciding how many disks are needed or how big a shared database cache should be in order to satisfy throughput and response time goals. For such problems, the established methodology is stochastic modeling; most importantly, queuing theory as the nonlinear effects of resource contention under multiuser load is most critical [1, 11]. For tractability, stochastic models often need to make simplifying assumptions, and this entails the crucial issue of how accurate the model's predictions can still be. This concern can be addressed in two ways: (i) using more advanced mathematics to capture also nonstandard situations or combining analytic models with simulations (e.g., to capture realistic inter-arrival time distributions

rather than simply postulating an exponential distribution) and (ii) using stochastic models as a building block in a more comprehensive approach, where even somewhat inaccurate relative predictions are beneficial toward configuration decisions.

Online Optimization Many self-managing problems obtain their inputs – workload parameters – only dynamically (but then accurately, not stochastically) and need to optimize configuration parameters as the workload evolves. These situations typically entail periodic or even continuous re-optimization, possibly in an incremental manner, and face tight timing constraints for finding the solution [5, 12]. Memory governing for workspaces (e.g., for hash joins or sorting) and database statistics management (e.g., for multidimensional histograms) fall into this category, the former having a time horizon of minutes or seconds, the latter with a typical reconsideration cycle of days but then requiring expensive database accesses. Inspired by online optimization theory, a possible approach to tackle these problems is to identify fast approximation techniques to obtain a viable solution. However, the details of the specific approach heavily depend on the problem at hand.

Feedback Control When it is very difficult, if not impossible, to capture some aspect of the configuration-workload-performance relation in a causal model, the paradigm of feedback control loops offers a principled alternative even in the absence of causal understanding. Dynamic load control that adjusts the multiprogramming level (MPL, i.e., the maximum number of concurrent threads) of a server falls into this category. These methods consist of an admission control, to avoid overload effects that may result in performance thrashing, and a cancellation control to shed load when performance goals can no longer be met. This is crucially important especially for memory management, but potentially also for lock management and other resources. The main challenge that must be addressed here is that of sudden load surges: bursty arrivals of requests that require adjustment of MPL settings for smoother load (possibly even with workload class-specific MPL limits). These transient effects are inherently difficult to model analytically, even with stochastic models. Feedback loops treat the MPL settings as control variables that are adjusted based on simple differential equations over the measured performance values, the desired performance goals, and the control values. Control theory may be harnessed to ensure stability properties. Feedback control has been shown to work well for sufficiently simple functionality such as Web application servers [13].

Statistical Learning Machine-learning models that use statistics for regression of continuous functions or classification with discrete labels are becoming increasingly attractive for self-tuning tasks [14, 15]. Modern statistical learning methods can handle large numbers of input parameters (high-dimensional multivariate models) and can determine the most influential factors or can learn a predictor for an output variable. The only input is prior observations, sometimes along with manually assigned labels for training (if needed). In the context of self-managing database systems, the data are the system's event log: fine-grained information about configuration values (whenever they change), performance measurements, and also exceptions and potential problem situations. Statistical learning on this data can be used for diagnosis and root-cause analysis. A word of caution is in order, though: even if, in principle, all kinds of functions, labelings, and rankings can be learned, it is still an art to design the learning models and their feature spaces in the right way. So just like all the other self-management paradigms, statistical learning is not a panacea in itself.

Infrastructure

Self-management technology requires capabilities for adaptation, introspection, and self-healing. This in turn calls for a rich infrastructure in or around the database engine, so as to gather the necessary data and provide the mechanisms that will be enacted by the self-

managing strategies. In the last few years, all modern database systems have addressed these requirements and provide rich facilities in this regard:

- Many internal algorithms (e.g., hash joins or sorting) are *resource adaptive*: they can continue running even if their resources, like workspace memory, are reduced at run-time. (But this alone does not provide self-management; in addition, strategies are needed for deciding when and how to reduce or increase resource assignments.)
- For *continuous self-monitoring*, lightweight techniques have been developed to identify and trace relevant events (e.g., exceptions raised because of memory shortage) and aggregate large amounts of monitoring data. The difficulty that these techniques have successfully overcome is to do all this with very low run-time overhead so as not to adversely affect normal system operation [16].
- For self-healing, techniques have been added to isolate abnormal behavior and re-initialize potentially affected system components [17]. These approaches are eased by the fact that database systems generally follow the transactional paradigm for data accesses. By aborting ongoing transactions and sessions, resetting software components is greatly simplified. A similar argument holds for fail-over techniques among redundant servers (with data replication or shared disk storage), and this even works over large geographic distances to provide disaster recovery.

Future Directions

Self-management is an important topic across all kinds of computer systems. Ultimately, all IT systems should strive to become as easily usable as household appliances such as washing machines or TV sets. In some areas, the vision of administration-less and trouble-free solutions has almost been achieved, most notably, in storage systems [18]. However, database systems, with their very powerful languages like SQL and XQuery and their application-specific extensibility, have a much richer functionality than storage and thus face a much harder challenge.

In fact, database systems and their application workloads have become so complex that self-management is no longer just a desirable capability but will be a vital necessity in the long run. But despite the good progress on many issues of this theme, the quest for overriding principles has not yet achieved any breakthrough. This is partly due to the difficulty of the problems (and the quest is certainly ongoing); on the other hand, one could conjecture that the problem may, to some extent, be ill-defined given today's very sophisticated system architecture with their overwhelming richness of features and the limited set of abstractions.

This concern is reflected in considerations on componentizing database systems into building blocks with narrow interfaces and much fewer tuning choices [19, 20], in building specialized data-management engines for particular application domains, or drastically limiting the engine's options and functionality. Virtualization of resources is another trend toward simplifying system management, but its impact on database engine tuning is still unclear. A breakthrough may require radical departures from today's architectures as well as rethinking the functionality that is offered by the database system. Similar to the "design-for-recovery" position of [17], a completely new *design-for-manageability* approach may be needed.

Cross-References

▶ Database Tuning Using Combinatorial Search
▶ Database Tuning Using Online Algorithms
▶ Database Tuning Using Trade-Off Elimination

Recommended Reading

1. Lazowska ED, Zahorjan J, Scott GG, Sevcik KC. Quantitative system performance: computer analysis using queuing network models. Englewood Cliffs: Prentice-Hall; 1984.

2. Finkelstein SJ, Scholnick M, Tiberio P. Physical database design for relational databases. ACM Trans Database Syst. 1988;13(1):91–128.

3. Weikum G, Hasse C, Moenkeberg A, Zabback P. The COMFORT automatic tuning project. Inf Syst. 1994;19(5):381–432.

4. Weikum G, Moenkeberg A, Hasse C, Zabback P. Self-tuning database technology and information services: from Wishful thinking to viable engineering. In: Proceedings of the 28th International Conference on Very Large Data Bases; 2002.

5. Brown KP, Mehta M, Carey MJ, Livny M. Towards automated performance tuning for complex workloads. In: Proceedings of the 20th International Conference on Very Large Data Bases; 1994.

6. Reiner DS, Pinkerton TB. A method for adaptive performance improvement of operating systems. In: Proceedings of the 18th ACM Symposium on Operating System Principles; 1981.

7. Chaudhuri S, König AC, Narasayya VR. SQLCM: a continuous monitoring framework for relational database engines. In: Proceedings of the 20th International Conference on Data Engineering; 2004.

8. Chaudhuri S, Narasayya V, Syamala M. Bridging the application and DBMS profiling divide for database application developers. In: Proceedings of the 33rd International Conference on Very Large Data Bases; 2007.

9. Chaudhuri S, Weikum G. Rethinking database system architecture: towards a self-tuning RISC-style database system. In: Proceedings of the 26th International Conference on Very Large Data Bases; 2000.

10. Ailamaki A, editor. Special issue on self-managing database systems. IEEE Data Eng Bull 2006; 29(3):1–62.

11. Menasce DA, Almeida VAF. Capacity planning for web performance. Metrics, models and methods. Upper Saddle Rive: Prentice-Hall; 2001.

12. Bruno N, Chaudhuri S. To tune or not to tune? A lightweight physical design alerter. In: Proceedings of the 32nd International Conference on Very Large Data Bases; 2006. p. 499–510.

13. Diao Y, Hellerstein JL, Parekh SS, Griffith R, Kaiser GE, Phung DB. A control theory foundation for self-managing computing systems. IEEE J Select Areas Commun. 2005;23(12):2213–22.

14. Jiang N, Villafane R, Hua KA, Sawant A, Prabhakara K. ADMiRe: an algebraic data mining approach to system performance analysis. IEEE Trans Knowl Data Eng. 2005;17(7):888–901.

15. Stillger M, Lohman GM, Markl V, Kandil M. LEO – DB2's learning optimizer. In: Proceedings of the 27th International Conference on Very Large Data Bases; 2001.

16. Chaudhuri S, Narasayya V. Self-tuning database systems: a decade of progress. In: Proceedings of the 33rd International Conference on Very Large Data Bases; 2007.

17. Candea G, Brown AB, Fox A, Patterson DA. Recovery-oriented computing: building multitier dependability. IEEE Comput. 2004;37(11):60–7.
18. Wilkes J, Golding RA, Staelin C, Sullivan T. The HP AutoRAID hierarchical storage system. ACM Trans Comput Syst. 1996;14(1):108–36.
19. Chaudhuri S, Narasayya VR. An efficient cost-driven index selection tool for Microsoft SQL server. In: Proceedings of the 23th International Conference on Very Large Data Bases; 1997.
20. Lightstone S. Seven software engineering principles for autonomic computing development. Innov Syst Softw Eng. 2007;3(1):71–4.

Semantic Atomicity

Greg Speegle
Department of Computer Science, Baylor University, Waco, TX, USA

Synonyms

None

Definition

Let T be a transaction composed of subtransactions $S_0, S_1, \ldots S_{n-1}$. Let $C_0, C_1, \ldots C_{n-1}$ be a set of *compensating transactions*, such that C_i compensates for the corresponding S_i. T is *semantically atomic* iff all S_i have committed, or for all S_i that have committed, C_i has also committed. A schedule (or history) ensures semantic atomicity if all transactions are semantically atomic. If T requires compensating transactions, then the resulting database is semantically equivalent to one in which T did not execute at all, but it is not guaranteed to be identical. Typically, two database states are equivalent if they both satisfy all of the database constraints.

Historical Background

Semantic atomicity is first defined in [6], with the use of *countersteps* to remove parts of a failed transaction executing in a distributed database environment, without rolling back the entire transaction. The "step" grew in complexity to a subtransaction with the introduction of *sagas* [7]. This required a corresponding increase in the complexity of the countermeasure, now called *compensating transactions*. Within sagas, subtransactions are allowed to interleave with other transactions, and an execution is correct if every subtransaction commits or the corresponding compensating transaction is executed. In [10], transactions are extended to transaction programs with input and output constraints, thus allowing formal representations of the capabilities and requirements for compensating transactions. Semantic atomicity has been applied in multidatabases [3], multilevel secure databases [1], workflows [4], and real-time database systems [15].

Scientific Fundamentals

Semantic atomicity is appropriate for nontraditional database applications, such as long-duration design transactions [9], workflows [4], and distributed transactions [13]. For these applications, the traditional database criteria referred to as the ACID properties (atomicity, consistency, isolation, and durability) can significantly limit performance. For example, atomicity requires that all updates performed by a transaction are applied to the database or all effects are removed. Consider the case where transaction T_1 updates a data item and transaction T_2 reads that new value. If transaction T_1 aborts, the update must be removed, and transaction T_2 must not be allowed to commit (it read a value that never existed in the database). If T_2 is a long-duration transaction, then significant work can be lost. Alternatively, T_2 can be denied access to the data, which can cause a long-duration wait.

As a result, transactions in nontraditional applications are often structured to allow better performance while still maintaining correct execution. Most often, this structure is *nested* or *multilevel* transactions. See Fig. 1 as an example. One great, but not immediately obvious, benefit

S

of semantic atomicity is that subtransactions can commit as soon as possible, thus externalizing the effects of the subtransactions right away. This is because a later failure cannot cause a cascading abort. The appropriate compensating transaction is executed, and the transaction system continues forward. As a result, long-duration transaction systems do not have to impose long-duration waits to avoid cascading aborts. Thus, the two primary disadvantages for ensuring ACID with advanced transactions (long waits and large amounts of lost work) are prevented.

Consider the example transactions in Fig. 1. Specifically, assume U1 and U2 are two users reserving seats on an airline with the schedule as in Fig. 2. Under ACID properties, several undesirable actions occur:

- T2 may not commit, since it is dependent on uncommitted transaction T1
- S1 must abort, since T1 fails
- S4 must be aborted, since it read invalid data
- T2 (and therefore S5) must be aborted, since S4 fails

However, with airline reservation systems, none of these actions are required. In fact,

S1, S4, and S5 should all retain their state, and user 2 should be able to complete his transaction without any interference from user 1. Transactions are still required, since user 1 does not want a reservation that only gets to New York.

Semantic atomicity provides the transaction property of allowing all legs of the trip to be reserved or none of them while allowing greater flexibility for the completion of the subtransactions. In this case, a compensating transaction, C2, is executed. C2 returns the seat reserved by S2. A new transaction S2 reserves a different flight from Dallas to New York, and T1 can continue. Figure 3 shows the completed set of transactions. In this execution, all of the transactions have either committed or compensating transactions have been executed. Thus, this execution is semantically atomic.

Formally, semantic atomicity allows transactions to read dirty data if all failed transactions meet the following compensation requirement:

Definition 1 Let T be a transaction, let H be the set of all transactions concurrently executing with T (excluding T), and let C be the compensating transaction for T. Let D represent the database state resulting from executing THC on the database, and D' be the database state resulting from executing H alone. A transaction has been compensated if $D \equiv D'$.

In general, two database states are equivalent if they both satisfy all database consistency constraints. In our example, let the consistency constraint for our database be that only one user may reserve a seat. T_2 is allowed to see dirty data (the reserved seats on flight 1) because S_2 has been compensated. The database is not identical to the state if T_1 had not executed (the seat on

Semantic Atomicity, Fig. 1 User U1 starts transaction T1 to reserve a one way ticket from Waco to London. User U2 starts transaction T2 to reserve a round trip ticket from Waco to Dallas. Subtransactions S1 and S4 make reservations on flight 1 from Waco to Dallas. S2 is a reservation on flight 2 from Dallas to New York, S3 is a reservation on flight 3 from New York to London, and S5 is a reservation on flight 4, the return trip from Dallas to Waco

S1	R(flt 1)	W(flt 1)				
S4			R(flt 1)	W(flt 1)		
S5					R(flt4)	W(flt4)
S2						R(flt2) W(flt2) abort

Semantic Atomicity, Fig. 2 Schedule of Transactions in Fig. 1. S1 reserves a seat on flight 1. S4 reads the seats reserved, including the one reserved by S1. S5 reserves a seat on flight 2, and then S2 fails

Semantic Atomicity, Fig. 3 Starting from Fig. 1, user U1 aborts transaction S2. Compensating transaction C2 returns the seat reserved. U1 reserves a different flight from Dallas to New York, and then S3 reserves a seat from New York to London. Both T1 and T2 commit

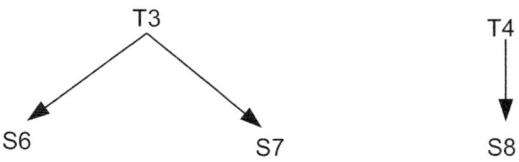

Semantic Atomicity, Fig. 4 User U3 starts transaction T3, which transfers $75 from account 1 (as transaction S6) and account 2 (as transaction S7) to account 3. User U4 starts transaction T4, which withdraws $100 from account 3. Initially, account 1 has $100, account 2 has $50, and account 3 has $50

S_1 is still reserved) but S_4 did not violate the database consistency constraint.

Likewise, it must be the case that the database state seen by H is consistent, otherwise the execution of some of the transactions in H is unpredictable. This is satisfied in the example, as T_2 (H for the example) executes on the consistent state.

Unfortunately, compensating transactions cannot always be applied automatically. Consider the banking example in Fig. 4. Once S_7 fails, transaction S_6 must be compensated. The obvious compensating transaction is to reverse the transfer to account 3. However, account 3 no longer has the funds available. Therefore, one of the accounts must have a negative balance, and the database consistency constraint is violated. Note that in this case, a compensating transaction of transferring $25 from account 1 to account 2 would satisfy the database constraint. Long-duration transactions with human interaction (such as design transactions) could generate such a compensating transaction.

One approach to address this issue is creating classes of transactions. If transactions are in compatible classes, they are able to execute as sagas

– the results of subtransactions are visible, and compensating transactions are available to correct errors. If transactions are not in compatible classes, traditional database correctness would apply. Thus, T_1 and T_2 would be classified as compatible, while T_3 and T_4 would not.

An alternative to creating transaction classes is to increase the semantics available. For example, in the *Nested Transactions with Predicates and Versions* (NT/PV) model in [8], transactions are allowed to be arbitrarily nested. Each transaction (and subtransaction) has an input predicate and an output predicate. An execution is considered correct if every transaction executes on a database state that satisfies its input predicate, executes its subtransactions according to a defined order, and terminates with a database state that satisfies its output predicate. Subtransactions (including compensating transactions) can be added dynamically as needed. The execution of Fig. 1 could be implemented by requiring an input predicate of at least one seat available on the flight and an output predicate of no dual booked seats. The subtransactions could arbitrarily interleave (although a race condition could exist such that only one transaction would be allowed to reserve the last seat on a plane), and the compensation of Fig. 3 would execute concurrently as before. For the execution in Fig. 4, transactions T_3 and T_4 could be added to the same parent transaction and ordered such that one must complete before the other. The partial order prevents the incorrect execution.

Nested transactions were originally designed for distributed databases [13], and semantic atomicity can be applied in that domain as well. Specifically, in the case where the local databases must maintain local autonomy (i.e., a remote site may not arbitrarily delay processing at the local site), semantic atomicity can replace standard atomic transactions by using compensating transactions where needed. In [11], semantic atomicity means either all subtransactions are locally committed (and thus the transaction is committed) or all locally committed subtransactions are compensated. Note this definition is equivalent to the definition of semantic atomicity given earlier. A protocol called *O2PC* for *Optimistic Two-Phase*

Commit achieves this result. See "▶ Two-Phase Commit" to compare with the traditional commit protocol.

In O2PC, a coordinator initiates the distributed commit process by requesting a vote by each participating site. If the participant can commit the transaction locally, it votes to commit. Otherwise, it votes to abort. After voting, the participant terminates the local transaction, performing either a local commit or a local abort. As a result of the local commit, all of the locks held by the transaction are released. This is the optimistic nature of the protocol, as another participant may vote to abort the transaction, which violates the traditional atomicity requirement. The coordinator collects all votes. If any participant votes to abort, the coordinator decides to abort the transaction. If all participants vote to commit, the coordinator decides to commit the transaction. This is the commit point for the transaction. The coordinator sends the message to all of the participants. If a participant locally committed the transaction, but it has globally aborted, the participant executes a compensating transaction to undo the locally committed transaction. As with Fig. 4, transactions which cannot be compensated are not suitable for the O2PC protocol.

Semantic atomicity is extended to a higher level of abstraction by ACTA [5]. Unlike the models presented earlier, ACTA, named for the Latin word for actions, is a formalism for defining extended correctness criteria (such as semantic atomicity). ACTA is based on five powerful building blocks:

- history – the series of actions performed by the transactions and the transaction manager; correct histories satisfy a set of invariants
- dependencies – constraints on the execution of concurrent transactions
- view of the transaction – the state of the objects visible to a transaction at a point in time
- conflict set – the set of operations in the history which could lead to an incorrect execution by a transaction

- delegation – allows another entity to control the result (typically commit or abort) of an operation

With these tools, ACTA is able to represent many formal models, including semantic atomicity. Fully defining semantic atomicity within the ACTA framework is beyond the scope of this article, but as an example, consider the definition of semantic atomic transactions. Within ACTA, this can be captured by the first-order logic predicate:

$$\forall S_i \in T, \text{commit}(S_i) \in H \rightarrow \text{commit}(C_i)$$
$$\in H \vee \forall S_j \in T, \text{commit}(S_j) \in H \quad (1)$$

where H is the history of all operations and C_i is the compensating transactions for S_i.

The complexity of using application semantics in building compensating transactions prevents the common deployment of semantic atomicity. In [10] these problems are studied in detail, with the creation of *operations* which are arbitrarily complex modifications to a single database entity. These operations are allowed local variables, thereby resembling functions in traditional programming languages. The operations are combined into *transaction programs*, which include conditional statements and statement blocks. Within this model, several aspects of compensating transactions are explored, such as compensation when the database states must be identical (not just equivalent) and compensation when some transaction program must follow the compensated-for transaction (called unsound).

Another key issue mentioned in [10] is the requirement that compensating transactions do not fail. Although this can be ensured during normal database operations (e.g., using a deadlock avoidance mechanism for compensating transactions), system failures cannot be prevented. The solution to this problem requires logging the internal state of the compensating transaction as well as any database modifications. During recovery, incomplete compensating transactions are not aborted but are continued from the saved internal state, similar to the notion of compensating log records in ARIES [12].

Note that semantic atomicity is distinctly different from the concept of a *savepoint*. A savepoint does not expose the updates of a transaction to outside processes (called externalized operations in [10]), while semantic atomicity supports this. Likewise, when a traditional database transaction performs a rollback to a savepoint, any other transaction which has read the aborted updates must also abort. Thus, savepoints do not prevent cascading aborts.

Key Applications

Semantic atomicity is appropriate for any application where the benefit of avoiding cascading aborts without long-duration waits is greater than the difficulty of creating the compensating transactions. Examples include:

- multidatabases [3] where each site can commit a subtransaction without the overhead of two-phase commit
- application services such as the Microsoft Phoenix project [2] which support applications surviving database failures in part by using semantic information
- workflows [4] where processes need to begin as soon as possible and often cannot be aborted
- web services [14] where actions are performed by loosely coupled systems
- real-time systems [15] where the ability to predict transaction length is greatly improved by avoiding cascading aborts and by allowing early commits
- secure databases [1] where cascading aborts can cause covert information exchange

Although not an application per se, support for compensating transactions, and thereby semantic atomicity, has been included in the Organization for the Advancement of Structured Information Standards (OASIS) Web Services Business Activity (WS-BusinessActivity) standard released in July 2007. Certainly, this will increase the number of commercial applications using semantic atomicity.

Cross-References

▶ ACID Properties
▶ Atomicity
▶ Compensating Transactions
▶ Extended Transaction Models and the ACTA Framework
▶ Nested Transaction Models
▶ Open Nested Transaction Models
▶ Sagas

Recommended Reading

1. Ammann P, Jajodia S, Ray I. Ensuring atomicity of multilevel transactions. In: Proceedings of the IEEE Symposium on Research in Security and Privacy; 1996. p. 74–84.
2. Barga R, Lomet D. Phoenix project: fault-tolerant applications. ACM SIGMOD Rec. 2002;31(2):94–100.
3. Breitbart Y, Garcia-Molina H, Silberscahtz A. Overview of multidatabase transaction management. VLDB J. 1992;1(2):181–240.
4. Breitbart Y, Deacon A, Schek H-J, Sheth A, Weikum G. Merging application-centric and data-centric approaches to support transaction-oriented multi-system workflows. SIGMOD Rec. 1993;22(3): 23–30.
5. Chrysanthis PK, Ramamritham K. Synthesis of extended transaction models using acta. ACM Trans Database Syst. 1994;19(3):450–91.
6. Garcia-Molina H. Using semantic knowledge for transaction processing in a distributed database. ACM Trans Database Syst. 1983;8(2):186–213.
7. Garcia-Molina H, Salem K. Sagas. In: Proceedings of the ACM SIGMOD International Conference on Management of Data; 1987. p. 249–59.
8. Korth HF, Speegle G. Formal aspects of concurrency control in long-duration transaction systems using the NT/PV model. ACM Trans Database Syst. 1994;19(3):492–535.
9. Korth HF, Kim W, Bancilhon F. On long duration CAD transactions. Inf Sci. 1988;46(1):73–107.
10. Korth HF, Levy E, Silberschatz A. A formal approach of recovery by compensating transactions. In: Proceedings of the 16th International Conference on Very Large Data Bases; 1990. p. 95–106.
11. Levy E, Korth HF, Silberschatz A. An optimistic commit protocol for distributed transaction management. In: Proceedings of the ACM SIGMOD International Conference on Management of Data; 1991. p. 88–97.
12. Mohan C, Haderle D, Lindsay B, Pirahesh H, Schwarz P. ARIES: a transaction recovery method supporting fine-granularity locking and partial roll-

S

backs using write-ahead logging. ACM Trans Database Syst. 1992;17(1):94–162.

13. Moss JEB. Nested transactions – an approach to reliable distributed computing. Cambridge: The MIT Press; 1985.

14. Puustjarvi J. Using advanced transaction and workflow models in composing web services. In: Advances in Computer Science and Technology – ACST 2007; 2007.

15. Soparkar N, Levy E, Korth HF, Silberschatz A. Adaptive commitment for distributed real-time transactions. In: Proceedings of the 3rd International Conference on Information and Knowledge Management; 1994. p. 187–94.

Semantic Crowdsourcing

Elena Simperl
Electronics and Computer Science, University of Southampton, Southampton, UK

Synonyms

Semantic crowdsourcing; Crowdsourcing for the Semantic Web

Definition

The term covers the bilateral relationship between two research areas, one associated with semantic technologies and related topics such as knowledge representation, linked data, and the Semantic Web, and the other one with approaches to problem solving involving crowd participation. One angle on the term refers to the use of different crowdsourcing paradigms, methods, and platforms to solve tasks that are related to semantic technologies. Complementarily, one could also consider the application of semantic technologies as a means to design and implement crowdsourcing tools.

How about Concepts and Applications?

Crowdsourcing is increasingly applied in various settings from using gamification techniques to motivate employees (employees as a crowd), to challenges and prizes rewarding ideas for product development and innovation (both enterprise internal and open crowds), to financially rewarded microtasks as a new form of outsourcing data management online. Scenarios with a semantic technologies component are in many ways naturally amenable to a crowd-participatory approach. When studying these scenarios, one can distinguish between two broad categories: those with an explicit grounding in a vertical domain, which use semantic technologies and standards such as RDF, SPARQL, linked data, ontologies, or reasoning to meet specific application requirements; and those that are formulated in a domain-independent fashion and focus rather on the technical aspects of semantic content management. In this context, the term "semantic content" refers to data encoded and processed using technologies and standards as those just mentioned. An example in the first category would be the Wikidata project (http://www.wikidata.org/), which is a free knowledge base available to everyone to edit and use. Wikidata relies on semantic technologies to represent structured knowledge and integrate it with external resources, and an open collaborative process to create and maintain the knowledge base. From a semantic crowdsourcing point of view, it offers an interesting testbed to study and experiment with social design principles, motivation theories, and incentive mechanisms in order to ensure that the contributions from the crowd are optimally fitted for the purpose of the project. The fact that the underlying implementation of Wikidata uses semantic technologies influences the type of tasks that are subject to crowdsourcing; most prominently, knowledge is collected in tuple form, including provenance information. In the second category of scenarios, we tend to find technical use cases such as crowdsourced ontology engineering, which, while relevant for many vertical domains, refers to the use of open participation to carry out specific aspects of the ontology engineering process, such as, for instance, the creation of a class hierarchy, or the definition of relationships between classes or entities [1]. In this context, researchers typically look into purposeful ways to apply specific crowdsourcing techniques to approach the task, including topics such as the choice of contributions to be solicited, the most appropriate type of crowd, the ways in which information produced by the crowd will lead to the resulting ontologies, means to validate crowd answers, and so forth. An overview of this area is provided in [1].

In the following, we will focus on this second category of scenarios because it covers those crowdsourcing use cases which are more indicative for the types of challenges that need to be addressed when engaging an open crowd in dealing with semantic content. There are various examples of crowdsourced semantic content management projects in the literature, which discuss these challenges. They use different forms of crowdsourcing (paid vs not paid crowdsourcing, games with a purpose, competitions etc.), sometimes in combination, or in combination with automatic tools. The tasks that are subject to these projects span the entire content management life cycle, from data collection and knowledge representation to curation, interlinking, or query formulation. For instance, Crowdmap [2] has used microtask platforms such as CrowdFlower to solve ontology alignment problems. In [3] the authors combine contests and microtasks to curate content from the DBpedia (http://dbpedia.org/) knowledge base, a structured version of Wikipedia. SemanticGames is a portal that collects games with a purpose related to Semantic Web tasks (http://semanticgames.org/). The Insemtives project (http://insemtives.org/) has looked into methodologies for the incentives-minded design of semantic applications [1].

Recommended Reading

1. Simperl E, Cuel R, Stein M. Incentive-centric Semantic Web application engineering. San Rafael: Morgan & Claypool; 2013.
2. Sarasua C, Simperl E, Noy N. CrowdMap: crowdsourcing ontology alignment with microtasks. In: Proceedings of the 11th International Semantic Web Conference; 2012. p. 525–41.
3. Acosta M, Zaveri A, Simperl E, Kontokostas D, Auer S, Lehmann J. Crowdsourcing linked data quality assessment. In: Proceedings of the 12th International Semantic Web Conference; 2013. p. 260–76.

Semantic Data Integration for Life Science Entities

Ulf Leser
Humboldt University of Berlin, Berlin, Germany

Synonyms

Data fusion; Duplicate detection; LSID; Object identification

Definition

An entity is the representation of a (not necessarily physical) real-world object, such as a gene, a protein, or a disease, within a database. To integrate information about the same entities from different databases, these representations must be analyzed to uncover the corresponding underlying objects. This process is called entity identification. A variation of entity identification is *duplicate detection*, which analyses two or more entities to determine whether they represent the same real-world object or not. Finally, *data fusion* is the process of generating a single, homogeneous representation from multiple, possibly inconsistent entities that represent the same real-world object.

When entities have globally unique *keys*, such as ISBN numbers in the case of books, entity identification and duplicate detection are simple. However, in life science databases, one usually has only descriptive information, such as the name or the sequence of a gene, which does not suffice to uniquely identify real-world objects. A *homonym* is a single name (or an ID) that is identifies multiple, different objects. For instance, the term "ACE" may reference many different proteins, such as "angiotensin converting enzyme" or "acetylcholinesterase." *Synonyms* are multiple names (or IDs) given to the same object.

Entity identification is particularly important in data curation, which is the (often manual) process of distilling a comprehensive description of a complex object from multiple data sources.

Historical Background

Duplicate detection has a long tradition in *census databases* where multiple representations of a person have to be identified to ensure reliable statistics. In the life sciences, the problem is particularly difficult, because many biological objects are much less stringently defined than, say, a human being. For instance, it is not clear whether two copies of a gene sequence on different sections of a chromosome should be considered the same gene or not, or if two highly similar and functionally identical genes in different species should be considered as different genes or not. At the other extreme, one often treats a gene and the protein it encodes as one object, especially in prokaryotes where differential splicing is almost non-existent, leading to a strict 1:1 relationship between genes and proteins. Thus, although a (DNA or protein) sequence is the fundamental identifying property of many biological objects, it does not always lead to unique entity identification.

First calls for establishing world-wide standards to identify biological objects appeared in the early 1990s, when the Human Genome Project (HGP) quickly increased the amount of available information on genes and other molecular objects [4]. The HGP from the very start was an internationally distributed effort with no central organization that could have enforced consistent naming of objects or assignment of IDs to objects.

Standardization efforts were initiated in the domains of genes, proteins, and clones. Other areas where identification of objects is an important issue are small molecules, diseases, and species. One way to achieve standardized names

was the installation of committees to define naming conventions for biological objects, such as the HUGO Gene Nomenclature Committee. On the other hand, some databases have become the de-facto standards for some types of biological objects, and their IDs are now commonly used as identifiers. An example is the use of UniProt-IDs as identifiers for proteins.

However, in many areas no standards exist or standards are not commonly used. Thus, many areas are facing the problem of identifying duplicates. The predominant approach is the comparison of biological sequences, i.e., DNA or protein sequences, using various variations of *edit distance* calculations. However, recent scientific discoveries, such as the importance of differential splicing (one gene forming several proteins) or the existence of paralogs (highly similar genes in the same genome with different function) render the pure usage of sequence similarity insufficient. Today, the method of choice for performing duplicate detection in a life sciences database therefore depends on the specific domain and the scientists' and databases' particular understanding of the biological object.

Foundations

Identity Versus Similarity

The entity identification problem exists for all types of objects in the life sciences. It is particularly pressing for genes and proteins. For most purposes one considers a gene to be a (not necessarily continuous) stretch of DNA, i.e., a sequence of the four nucleic acids, on a chromosome, which is – by some complex regulation mechanism – at times first transcribed into RNA and then translated into a protein. A protein is a molecule consisting of a linear chain of amino acids forming a complex 3D structure. The translation procedure moves through the RNA and appends for each triple of nucleic acids one amino acid to the growing protein chain; thus, the translation of a given gene sequence into a protein is unique. The function of a protein mostly depends on the topology of its 3D structure and the properties of atoms exhibited at the surface of the structure [3].

The revelation of the function(s) of a protein is a crucial task in Bioinformatics and especially important for drug development. Because similar structure may hint toward similar function (independent of the species), a standard procedure for revealing the function of a human protein is to experimentally analyze the structure of a similar protein in another well-studied species, such as mouse or fruit-fly. Because the sequence of a gene determines the sequence of its protein(s), which in turn determines the structure of the protein, the same principle holds for genes: Two genes with similar sequence quite likely code for proteins with similar function.

With the function being the most important attribute of a gene/protein, the predominant question for semantic integration of genes and proteins is one of *similarity of sequences* rather than identity. The similarity of two biological sequences usually is measured by their (weighted) *edit distance* [9]. Two genes are considered as identical if their sequences are very similar, with the exact definition of "very similar" being a subject of much debate. The fundamental tool for entity identification and duplicate detection of genes and sequences therefore still is sequence comparison, using tools such as BLAST.

Naming Standards

Once a group of genes or proteins have been identified as "one real-world" object, one has to select a common name or ID for this group. For human genes, the Human Genome Organization (HUGO) installed the HUGO Gene Nomenclature Committee (HGNC), which is responsible for assigning names to human genes. Gene names are often complex, multi-term phrases that might include parts of the name of the person who discovered the gene, the phenotype it produces when a defect occurs, an inherited disease the gene is associated to, the chromosome and species where it first was detected, etc. However, the work of the HGNC is still widely ignored (especially in scientific publications). For other species, similar bodies often were more successful in terms of

standardization. The reason for their success is (i) the number of researchers working on a particular other species is usually much smaller than for humans, (ii) the number of genes in other species is much smaller (e.g., Yeast ∼4000 genes, human ∼22,000 genes), and (iii) data collection was centralized early on; (see the Mouse Genome Database MGD for an example). Furthermore, certain database identifiers are commonly used to denote genes, especially those from the Entrez-Gene database.

No established naming convention exists for human proteins. The problem is worse than for genes, because on average every human gene codes for ∼8 proteins by means of differential splicing. Very often database identifiers are used instead of spoken names. De-facto standards are IDs from the Protein Data Bank (PDB) and the Uniprot Knowledge Base (formerly known as SwissProt).

For DNA sequencing, chromosomes are broken into pieces, which are cultivated, copied, and distributed inside living bacteria, called clones. Sets of clones are gathered in libraries that often contain hundreds of thousands of clones. Physically, a library is a set of so-called plates with 10–100 dwells, each containing a particular clone. To avoid duplication of work, clones must be identified uniquely. However, after generation nothing is known about a particular clone except the dwell where it is hosted. Researchers therefore identify clones by the plate number and column/row number inside the library. Since re-distribution of clones into different, specialized libraries is commonplace, there are clones with more than 20 different IDs of this form (see the Genome Database (GDB) for examples).

To overcome the problem of object names, the *Object Management Group* (OMG) recently has defined the LSID – Life Science Identifiers – standard for "*persistent, location-independent, resource identifiers for uniquely naming biologically significant resources including species names, concepts, occurrences, genes or proteins, or data objects that encode information about them.*" An LSID identifier consists of a network identifier (usually the fixed term

S

urn:lsid), an authority identifier (who defines this name), a namespace identifier, an object identifier, and a revision identifier (see discussion below on object versions). Whether or not this standard will be accepted by the community remains to be seen.

Evolution of Names

A particular problem with names and IDs is how to keep them stable and consistent. For instance, the question of which (of multiple) names of a particular gene is used in the literature is highly influenced by a kind of social-scientific "fashion" [10]. In biological databases, objects need to be merged and deleted, rendering existing IDs inconsistent. Furthermore, new findings may completely change what is known about a gene while keeping the ID unchanged, which makes previous studies based on the now outdated knowledge obsolete. This problem is usually solved by implementing a particular versioning model, which distinguishes major and versioned IDs; major ID always point to the most current version and a new version of the object is created with every update.

Key Applications

Semantic Integration of Entities

Duplicate detection is vital to ensure high data quality in integrated databases. It consists of two steps. First, multiple representations of same real-world entities have to be discovered. Second, for each group of duplicates, a uniform representation must be found [8]. Both steps are particularly difficult in the Life Sciences, because object definitions are vague (see above) and most biological data are obtained by complex experiments and are notoriously noisy [2].

Therefore, both tasks most often are performed manually, a process called *data curation*. It is common that large biological databases employ professional curators whose task it is to read new publications and to convert the most important information into some semi-structured database entry. There are also forms of community curation, where the correction of

errors in databases through a web interface is possible for registered users [5].

The general problem with curation is that it is very costly and highly subjective. At the same time, deciding on the correct value given two diverging experimental results is usually impossible without further experiments. Consequently, a typical approach to data integration in this field is to omit the second step. The resulting architecture has been called "entity-based" or "multidimensional." Different sources containing information about the same set of entities are considered as dimensions of the entities and are integrated into a schema similar to a *star schema* in *Data Warehouses* [11]. The advantage and disadvantage is that the different views on entities are reported to the user in a logically separated manner; thus, she has the ability but also the obligation to select the appropriate values herself. Alternatively, mixtures of manual and automatic data fusion are used (see [1] for an example).

Entity Identification in Text

A related problem is the identification of object names in scientific publications, i.e., in English sentences. Named Entity Recognition (NER) is the problem of judging for a given set of terms within a document whether they form a gene name. A standard technique for solving NER is the usage of *classification* based on machine learning algorithms. Named Entity Normalization (NEN) is the problem of assigning a unique name to an entity in text once it has been identified as such. NEN typically is tackled using large *dictionaries* of names and some kind of fuzzy string matching method. See [6, 7] for recent surveys on both tasks.

URL to Code

BLAST tool for search similar sequences in database: http://www.ncbi.nlm.nih.gov/BLAST/

EntrezGene database: http://www.ncbi.nlm.nih.gov/sites/entrez

HUGO Gene Nomenclature Committee: http://www.genenames.org/

Mouse Genome Database: http://www.informatics.jax.org/

OMG Life Science Research Task Force: http://www.omg.org/lsr/

Protein Data Bank (PDB): http://www.pdb.org/

Uniprot knowledge base: http://www.uniprot.org/

Cross-References

► Information Integration
► Information Integration Techniques for Scientific Data
► Object Identity
► Scientific Databases

Recommended Reading

1. Bhat TN, Bourne P, Feng Z, Gilliland G, Jain S, Ravichandran V, Schneider B, Schneider K, Thanki N, Weissig H, et al. The PDB data uniformity project. Nucleic Acids Res. 2001;29(1):214–8.
2. Brenner SE. Errors in genome annotation. Trends Genet. 1999;15(4):132–3.
3. Gibson G, Muse SV. A primer of genome science. Sunderland: Sinauer Associates; 2001.
4. Karp P.D. Models of identifiers. In: Proceedings of the 2nd Meeting on Interconnection of Molecular Biology Databases; 1995.
5. Kingsbury D. Consensus, common entry, and community curation. Nat Biotechnol. 1996;14(6):679.
6. Krauthammer M, Nenadic G. Term identification in the biomedical literature. J Biomed Inform. 2004;37(6):512–26.
7. Leser U, Hakenberg J. What makes a gene name? Named entity recognition in the biomedical literature. Brief Bioinform. 2005;6(4):357–69.
8. Müller H, Naumann F, Freytag J.-C. Data quality in genome databases. In: Proceedings of the 8th Conference on Information Quality; 2003.
9. Smith TF, Waterman MS. Identification of common molecular subsequences. J Mol Biol. 1981;147(1):195–7.
10. Tamames J, Valencia A. The success (or not) of HUGO nomenclature. Genome Biol. 2006;7(5):402.
11. Trissl S, Rother K, Müller H, Koch I, Steinke T, Preissner R, Frömmel C, Leser U. Columba: an integrated database of proteins, structures, and annotations. BMC Bioinformatics. 2005;6(1):81.

Semantic Data Model

David W. Embley
Brigham Young University, Provo, UT, USA

Synonyms

Conceptual data model; Conceptual model

Definition

A *semantic data model* represents data in terms of named sets of objects, named sets of values, named sets of relationships, and constraints over these object, value, and relationship sets. The *semantics* of a semantic data model are the intensional declarations: the names for object, value, and relationship sets that indicate intended membership in the various sets and the declared constraints that the data should satisfy. The *data* of a semantic data model is extensional and consists of instances of object identifiers and values for object and value sets and of m-tuples of instances for m-ary relationship sets. The *model* of a semantic-data-model instance describes intensionally a real-world domain of interest. The modeling components of the semantic data model specify the modeling elements from which a real-world model instances can be built.

For a general description of semantic data models, see [1]. This entry describes the generic properties of semantic data models and presents a representative collection of early semantic data models.

Key Points

Figure 1a gives a sample semantic data model. Semantic data models use graphical symbols to represent data semantics. Each semantic data model, however, has its own set of graphical symbols. The graphical symbols in Fig. 1b are meant to be generic-representative of the symbols used and

S

Semantic Data Model,
Fig. 1 Sample semantic
data model instance

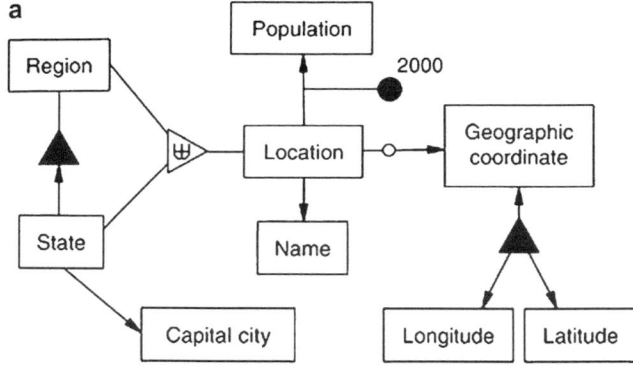

a

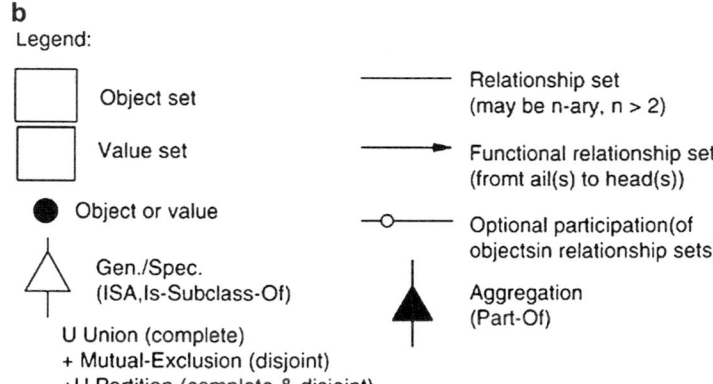

Region. Population = sum (Population); Region

b
Legend:

▢ Object set		——— Relationship set (may be n-ary, n > 2)
⬚ Value set		——▶ Functional relationship set (fromt ail(s) to head(s))
● Object or value		—o— Optional participation(of objectsin relationship sets)
△ Gen./Spec. (ISA,Is-Subclass-Of)		▲ Aggregation (Part-Of)

U Union (complete)
+ Mutual-Exclusion (disjoint)
+U Partition (complete & disjoint)

illustrative of the kinds of data sets and constraints included in typical semantic data models.

The legend in Fig. 1b tells what each symbol means.

- A box with a solid border designates a set of objects. In Fig. 1a *State* designates the set of US states (e.g., the state of California) and *Region* designates areas within the USA (e.g., the region of states in the northeastern part of the USA).
- A box with a dashed boarder represents a set of values. In Fig. 1a *Capital City* designates the names of the capital cities of the US states (e.g., "Sacramento" for California) and *Longitude* designates the set of longitude values for the geographic coordinates of locations (e.g., 120°4.9′W for the longitudinal part of the centerpoint of California).
- A large filled-in dot represents a single object or value. In Fig. 1a the object stands for the year 2000. Single objects or values can be

thought of as singleton object or value sets – an object set or value set with one object.

- Relationship-set names can appear explicitly or can be a composition of the names of connected object and value sets. In Fig. 1a the names are all compositions (e.g., *State-CapitalCity* names the binary relationship set between *State* and *Capital City*). In general, relationship sets are *m*-ary ($n \geq 2$) (e.g., the *Location-2000-Population* relationship set is ternary).
- Constraints on relationship sets include functional/ nonfunctional and mandatory/optional constraints. An arrowhead designates a functional relationship set from its tail(s) as domain space(s) to its head(s) as range space(s) (e.g., *Location-GeographicCoordinate* is functional from *Location* to *Geographic Coordinate*). A small "o" near a connection between an object (value) set and a relationship set allows for optional participation of the

objects in the object set (values in the value set) in relationships in the relationship set (e.g., the "o" near *Location* in the *Location-GeographicCoordinate* relationship set makes the participation of location objects in the relationship set optional – a geographic coordinate for a location such as northeast need not have a geographic coordinate).

- Generalization/specialization constraints, represented by a triangle, designate the specialization sets attached to the base of the triangle as subsets of the generalization set attached to the apex of the triangle (e.g., both the set of states and the set of regions are subset of the set of locations). If a union symbol ($\cup$) appears in the triangle, the generalization is a union of the specializations. If a mutual-exclusion symbol ($+$) appears, the specializations are pairwise nonintersecting. And if a partition symbol ($\uplus$) appears, both a union and mutual-exclusion constraint hold, making the constraint be a partition (e.g., in the semantic-data-model instance in Fig. 1, *Region* and *State* are mutually exclusive and the union constitutes *Location* – all the locations of interest for the semantic-data-model instance).

- Aggregation constraints, represented by a filled-in triangle, designate sub-part/super-part constraints. In Fig. 1a several states make up a region, and a longitude and latitude together constitute a geographic coordinate. The functional constraints associated with the aggregations allow a state to be part of at most one region and require that a longitude and latitude together correspond to one and only one geographic coordinate.

Formally, a populated semantic-data-model instance is an interpretation for a first-order language. Each object set and value set is a unary predicate (e.g., *State*(x) and *Latitude*(x)), and each m-ary relationship set is an m-ary predicate (e.g., *State-StateCapital*(x, y) and *Location-GeographicCoordinate*(x, y)). The domain for the interpretation is the set of object identifiers and values in the populated semantic-data-model instance (or more generally the set of

potential object identifiers and potential values for the semantic-data-model instance). The constraints are closed, well-formed formulas (e.g., $\forall x(State(x) \Rightarrow \exists! y State\text{-}StateCapital(x, y))$). A populated semantic-data-model instance whose data satisfies all the constraints is said to be a *model* – a valid semantic-data-model instance.

Cross-References

▸ Entity Relationship Model
▸ Extended Entity-Relationship Model
▸ Hierarchical Data Model
▸ Network Data Model
▸ Object Data Models
▸ Object-Role Modeling
▸ Ontology
▸ Unified Modeling Language

Recommended Reading

1. Peckham J, Maryanski F. Semantic data models. ACM Comput Surv. 1988;20(3):153–89.

Semantic Matching

Fausto Giunchiglia, Pavel Shvaiko, and Mikalai Yatskevich
University of Trento, Trento, Italy

Definition

Semantic matching: given two graph representations of ontologies G1 and G2, compute N1 × N2 *mapping elements* $\langle ID_{i,j}, n1_i, n2_j, R' \rangle$, with $n1_i \in$ G1, $i = 1, \ldots, $N1, $n2_j \in$ G2, $j = 1, \ldots, $N2 and R' the strongest *semantic relation* which is supposed to hold between the *concepts at nodes* $n1_i$ and $n2_j$.

A *mapping element* is a 4-tuple $\langle ID_{ij}, n1_i, n2_j, R \rangle$, $i = 1, \ldots, N1$; $j = 1, \ldots, N2$; where ID_{ij} is a unique identifier of the given mapping element; $n1_i$ is the i-th node of the first graph, N1 is the number of nodes in the first graph; $n2_j$ is the j-th node of the second graph, N2 is the number of nodes in the second graph; and R specifies a semantic relation which is supposed to hold between the concepts at nodes $n1_i$ and $n2_j$.

The *semantic relations* are within *equivalence* ($=$), *more general* ($\sqsupseteq$), *less general* ($\sqsubseteq$), *disjointness* ($\perp$) and *overlapping* ($\sqcap$). When none of the above mentioned relations can be explicitly computed, the special *idk* (I don't know) relation is returned. The relations are ordered according to decreasing binding strength, i.e., from the strongest ($=$) to the weakest (*idk*), with more general and less general relations having equal binding power. The semantics of the above relations are the obvious set-theoretic semantics.

Concept of a label is the logical formula which stands for the set of data instances or documents that one would classify under a label it encodes. *Concept at a node* is the logical formula which represents the set of data instances or documents which one would classify under a node, given that it has a certain label and that it is in a certain position in a graph.

Historical Background

An ontology typically provides a vocabulary that describes a domain of interest and a specification of the meaning of terms used in the vocabulary. Depending on the precision of this specification, the notion of ontology encompasses several data and conceptual models, for example, classifications, database schemas, or fully axiomatized theories. In open or evolving systems, such as the semantic web, different parties would, in general, adopt different ontologies. Thus, just using ontologies, just like using XML, does not reduce heterogeneity: it raises heterogeneity problems to a higher level. Ontology matching is a plausible solution to the semantic heterogeneity

problem faced by information management systems. Ontology matching aims at finding correspondences or mapping elements between semantically related entities of the input ontologies. These mapping elements can be used for various tasks, such as ontology merging, query answering, data translation, etc. Thus, matching ontologies enables the knowledge and data expressed in the matched ontologies to interoperate [6].

Many diverse solutions of matching have been proposed so far, see [7, 17, 18] for recent surveys which addressed the matching problem from different perspectives, including databases, artificial intelligence and information systems; while the major contributions of the last decades are provided in [2, 14, 19]. Some examples of individual approaches addressing the matching problem can be found in [4, 5, 8, 15, 16]. (See http://www.ontologymatching.org for a complete information on the topic.) Finally, ontology matching has been given a book account in [6]. This work provided a uniform view on the topic with the help of several classifications of the available methods, discussed these methods in detail, etc. In particular, the matching methods are primarily classified and further detailed according to (i) the input of the algorithms, (ii) the characteristics of the matching process and (iii) the output of the algorithms.

The work in [10] mixed the process dimension of matching together with the output dimension and classified matching approaches into *syntactic* and *semantic*. Syntactic are those approaches that rely on purely syntactic matching methods, e.g., edit distance between strings, tree edit distance. The semantic category, in turn, represents methods that work with concepts and compare their meanings in order to compute mapping elements. However, these have been also constrained by a second condition dealing with the output dimension: syntactic techniques return coefficients in the [0 1] range, while semantic techniques return logical relations, such as equivalence, subsumption (and justified by deductive techniques for instance). The work in [4] provided a first implementation of semantic matching.

Foundations

In order to motivate the matching problem two simple XML schemas are used. These are represented as trees in Fig. 1 and exemplify one of the possible situations which arise, for example, when resolving a schema integration task. Suppose an e-commerce company A1 needs to finalize a corporate acquisition of another company A2. To complete the acquisition, databases of the two companies have to be integrated. The documents of both companies are stored according to XML schemas A1 and A2, respectively. A first step in integrating the schemas is to identify candidates to be merged or to have taxonomic relationships under an integrated schema. This step refers to a process of ontology (schema) matching. For example, the elements with labels *Personal_Computers* in A1 and *PC* in A2 are the candidates to be merged, while the element with label *Digital_Cameras* in A2 should be subsumed by the element with label *Photo_and_Cameras* in A1.

Consider semantic matching as first motivated in [10] and implemented within the S-Match system [13]. Specifically, a schema-based solution is discussed, where only the schema information is exploited. It is assumed that all the data and conceptual models, e.g., classifications, database schemas, ontologies, can be generally represented as graphs. This allows for the statement and solution of a *generic (semantic) matching problem* independently of specific conceptual or data models, very much along the lines of what is done, for example, in Cupid [15].

The semantic matching takes as input two graph representations of ontologies and returns as output logical relations, e.g., equivalence, subsumption (instead of computing coefficients rating match quality in the [0 1] range, as it is the case with other approaches, e.g., [15, 16]), which are supposed to hold between the nodes in the graphs. The relations are determined by (i) expressing the entities of the ontologies as logical formulas, and (ii) reducing the matching problem to a logical validity problem. In particular, the entities are translated into logical formulas which explicitly express the concept descriptions as encoded in the ontology structure and in external resources, such as WordNet. (http://wordnet.princeton.edu/.) This allows for a

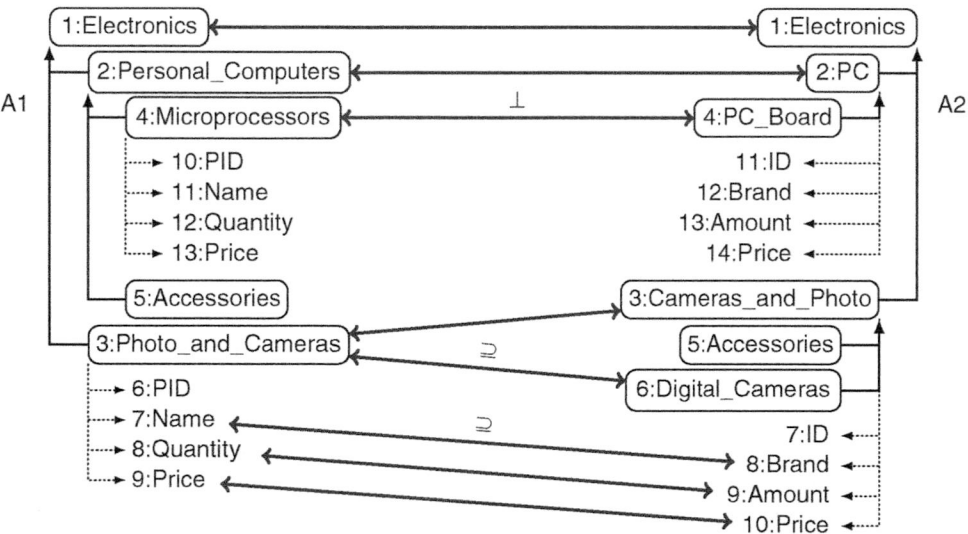

Semantic Matching, Fig. 1 Two simple XML schemas. The XML elements are shown in rectangles with rounded corners, while attributes are shown without them. Numbers before the labels of tree nodes are the unique iden-tifiers of the XML elements and attributes. In turn, the mapping elements are expressed by *arrows*. By default, their relation is =; otherwise, these are mentioned above the *arrows*

translation of the matching problem into a logical validity problem, which can then be efficiently resolved using (sound and complete) state of the art satisfiability solvers.

Consider tree-like structures, e.g., classifications, and XML schemas. Real-world ontologies are seldom trees, however, there are (optimized) techniques, transforming a graph representation of an ontology into a tree representation, e.g., the graph-to-tree operator of Protoplasm [3]. From now on it is assumed that a graph-to-tree transformation can be done by using existing systems, and therefore, the focus is on other issues instead.

Consider Fig. 1. "C" is used to denote concepts of labels and concepts at nodes. Also "$C1$" and "$C2$" are used to distinguish between concepts of labels and concepts at nodes in tree 1 and tree 2, respectively. Thus, in A1, $C1_{Photo\_and\_Cameras}$ and $C1_3$ are, respectively, the concept of the label $Photo\_and\_Cameras$ and the concept at node 3. Finally, in order to simplify the presentation whenever it is clear from the context, it is assumed that the formula encoding the concept of label is the label itself. Thus, for example in A2, $Cameras\_and\_Photo_2$ is a notational equivalent of $C2_{Cameras\_and\_Photo}$.

The algorithm inputs two ontologies and outputs a set of mapping elements in four macro steps. The first two steps represent the pre-processing phase. The third and the fourth steps are the element level and structure level matching, respectively. (Element level matching techniques compute mapping elements by analyzing entities in isolation, ignoring their relations with other entities. Structure level techniques compute mapping elements by analyzing how entities are related together.)

Step 1 For all labels L *in the two trees, compute* concepts of labels. The labels at nodes are viewed as concise descriptions of the data that is stored under the nodes. The meaning of a label at a node is computed by taking as input a *label*, analyzing its real-world semantics, and returning as output a *concept of the label*, C_L. Thus, for example, $C_{Cameras\_and\_Photo}$ indicates a shift from the natural language ambiguous label $Cameras\_and\_Photo$ to the concept $C_{Cameras\_and\_Photo}$, which codifies

explicitly its intended meaning, namely the data which is about cameras and photo. Technically, concepts of labels are codified as propositional logical formulas [9]. First, labels are chunked into *tokens*, e.g., $Photo\_and\_Cameras \rightarrow \langle photo, and, cameras\rangle$; and then, *lemmas* are extracted from the tokens, e.g., $cameras \rightarrow camera$. Atomic formulas are WordNet *senses* of lemmas obtained from single words (e.g., cameras) or multiwords (e.g., digital cameras). Complex formulas are built by combining atomic formulas using the connectives of set theory. For example, $C2_{Cameras\_and\_Photo} = \langle Cameras, senses_{WN\#2}\rangle \sqcup \langle Photo, senses_{WN\#1}\rangle$, where senses$_{WN\#2}$ is taken to be disjunction of the two senses that WordNet attaches to *Cameras*, and similarly for *Photo*. The natural language conjunction "and" has been translated into the logical disjunction "$\sqcup$."

Step 2 For all nodes N *in the two trees, compute* concepts at nodes. During this step the meaning of the positions that the labels at nodes have in a tree is analyzed. By doing this, concepts of labels are *extended* to *concepts at nodes*, C_N. This is required to capture the knowledge residing in the structure of a tree, namely the context in which the given concept at label occurs. For example, in A2, by writing C_6 it is meant the concept describing all the data instances of the electronic photography products which are digital cameras. Technically, concepts of nodes are written in the same propositional logical language as concepts of labels. XML schemas are hierarchical structures where the path from the root to a node uniquely identifies that node (and also its meaning). Thus, following an *access criterion* semantics, the logical formula for a concept at node is defined as a conjunction of concepts of labels located in the path from the given node to the root. For example, $C2_6 = Electronics_2 \sqcap Cameras\_and\_Photo_2 \sqcap Digital\_Cameras_2$.

Step 3

For all pairs of labels in the two trees, compute relations *among atomic* concepts of

Semantic Matching, Table 1 Element level semantic matchers

Matcher name	Execution order	Approximation level	Matcher type	Schema info
WordNet	1	1	Sense-based	WordNet senses
Prefix	2	2	String-based	Labels
Suffix	3	2	String-based	Labels
Edit distance	4	2	String-based	Labels
Ngram	5	2	String-based	Labels

labels. Relations between concepts of labels are computed with the help of a library of element level semantic matchers. These matchers take as input two atomic concepts of labels and produce as output a semantic relation between them. Some of them are re-implementations of the well-known matchers used, e.g., in Cupid. The most important difference is that these matchers return a semantic relation (e.g., $=$, $\sqsupseteq$, $\sqsubseteq$), rather than an affinity level in the [0 1] range, although sometimes using customizable thresholds.

The element level semantic matchers are briefly summarized in Table 1. The first column contains the names of the matchers. The second column lists the order in which they are executed. The third column introduces the matcher's approximation level. The relations produced by a matcher with the first approximation level are always correct. For example, *name* $\sqsupseteq$ *brand* returned by the *WordNet* matcher. In fact, according to WordNet *name* is a hypernym (superordinate word) of *brand*. In WordNet *name* has 15 senses and *brand* has 9 senses. Some sense filtering techniques are used to discard the irrelevant senses for the given context, see [13] for details. Notice that matchers are executed following the order of increasing approximation. The fourth column reports the matcher's type, while the fifth column describes the matcher's input. As from Table 1, there are two main categories of matchers. *String-based* matchers have two labels as input. These compute only equivalence relations (e.g., equivalence holds if the weighted distance between the input strings is lower than a threshold). *Sense-based* matchers have two WordNet senses as input. The *WordNet* matcher computes equivalence, more/less general, and disjointness relations. The

Semantic Matching, Table 2 The matrix of semantic relations holding between atomic concepts of labels

	Cameras$_2$	Photo$_2$	Digital_Cameras$_2$
Photo$_1$	*idk*	$=$	*idk*
Cameras$_1$	$=$	*idk*	$\sqsupseteq$

result of step 3 is a matrix of the relations holding between atomic concepts of labels. A part of this matrix for the example of Fig. 1 is shown in Table 2.

Step 4. For all pairs of nodes in the two trees, compute relations *among* concepts at nodes. During this step, initially the tree matching problem is reformulated into a set of node matching problems (one problem for each pair of nodes). Then, each node matching problem is translated into a propositional validity problem. Semantic relations are translated into propositional connectives in an obvious way, namely: equivalence ($=$) into equivalence ($\leftrightarrow$), more general ($\sqsupseteq$) and less general ($\sqsubseteq$) into implication ($\leftarrow$ and $\rightarrow$, respectively) and disjointness ($\perp$) into negation ($\neg$) of the conjunction ($\wedge$). The criterion for determining whether a relation holds between concepts at nodes is the fact that it is entailed by the premises. Thus, it is necessary to prove that the following formula:

$$axioms \rightarrow rel\,(context_1, context_2) \quad (1)$$

is valid, namely that it is *true* for all the truth assignments of all the propositional variables occurring in it. $context_1$ is the concept at node under consideration in tree 1, while $context_2$ is the concept at node under consideration in tree 2. *rel* (within $=$, $\sqsubseteq$, $\sqsupseteq$, $\perp$) is the semantic relation (suitably translated into a propositional connective) to be proved to hold

Semantic Matching, Table 3 The matrix of semantic relations holding between concepts at nodes (the matching result)

	$C2_1$	$C2_2$	$C2_3$	$C2_4$	$C2_5$	$C2_6$
$C1_3$	$\sqsubseteq$	idk	$=$	idk	$\sqsupseteq$	$\sqsupseteq$

between $context_1$ and $context_2$. The *axioms* part is the conjunction of all the relations (suitably translated) between atomic concepts of labels mentioned in $context_1$ and $context_2$. The validity of formula (1) is checked by proving that its negation is unsatisfiable. Specifically, it is done, depending on a matching task, either by using ad hoc reasoning techniques or standard propositional satisfiability solvers.

From the example in Fig. 1, trying to prove that $C2_6$ is less general than $C1_3$, requires constructing formula (2), which turns out to be unsatisfiable, and therefore, the less general relation holds.

$$((Electronics_1 \leftrightarrow Electronics_2) \wedge (Photo_1 \leftrightarrow Photo_2) \wedge$$
$$(Cameras_1 \leftrightarrow Cameras_2) \wedge$$
$$(DigitalCameras_2 \rightarrow Cameras_1)) \wedge (Electronics_2 \wedge$$
$$(Cameras_2 \vee Photo_2) \wedge DigitalCameras_2) \wedge$$
$$\neg(Electronics_1 \wedge (Photo_1 \vee Cameras_1)$$

$$(2)$$

A part of this matrix for the example of Fig. 1 is shown in Table 3.

Finally, notice that the algorithm returns $N1 \times N2$ correspondences, therefore the cardinality of mapping elements is *one-to-many*. Also, these, if necessary, can be decomposed straightforwardly into mapping elements with the *one-to-one* cardinality.

Key Applications

Semantic matching is an important operation in traditional metadata intensive applications, such as *ontology integration*, *schema integration*, or *data warehouses*. Typically, these applications are characterized by heterogeneous structural models that are analyzed and matched either manually or semi-automatically at design time.

In such applications matching is a prerequisite of running the actual system. A line of applications that can be characterized by their dynamics, e.g., *agent communication*, *peer-to-peer information sharing*, *web service composition*, is emerging. Such applications, contrary to traditional ones, require (ultimately) a run time matching operation and take advantage of more explicit conceptual models [18].

Future Directions

Future work includes development of a fully-fledged *iterative* and *interactive* semantic matching system. It will improve the quality of the mapping elements by iterating and by focusing user's attention on the critical points where his/her input is maximally useful. Initial steps have already been done in this direction by discovering automatically *missing background knowledge* in ontology matching tasks [11]. Also, an *evaluation methodology* is needed, capable of estimating quality of the mapping elements between ontologies with hundreds and thousands of nodes. Initial steps have already been done as well; see [1, 12] for details. Here, the key issue is that in these cases, specifying reference mapping elements manually is neither desirable nor feasible task, thus a semi-automatic approach is needed.

Experimental Results

In general, for the semantic matching approach, there is an accompanying experimental evaluation in the corresponding references. Also, there is the Ontology Alignment Evaluation Initiative (OAEI), (http://oaei.ontologymatching.org/.) which is a coordinated international initiative that organizes the evaluation of the increasing number of ontology matching systems. The main goal of the Ontology Alignment Evaluation Initiative is to be able to compare systems and algorithms on the same basis and to allow anyone for drawing conclusions about the best matching strategies. From such evaluations, matching system developers can learn and improve their systems.

Data Sets

A large collection of datasets commonly used for experiments can be found at: http://oaei. ontologymatching.org/.

URL to Code

The OntologyMatching. org contains links to a number of ontology matching projects which provide code for their implementations of the matching operation: http://www.ontologymatching.org/.

Cross-References

▶ Semantic Data Model

Recommended Reading

1. Avesani P, Giunchiglia F, Yatskevich M. A large scale taxonomy mapping evaluation. In: Proceedings of the 4th International Semantic Web Conference; 2005. p. 67–81.
2. Batini C, Lenzerini M, Navathe S. A comparative analysis of methodologies for database schema integration. ACM Comput Surv. 1986;18(4):323–64.
3. Bernstein P, Melnik S, Petropoulos M, Quix C. Industrial-strength schema matching. ACM SIGMOD Rec. 2004;33(4):38–43.
4. Bouquet P, Serafini L, Zanobini S. Semantic coordination: a new approach and an application. In: Proceedings of the 2nd International Semantic Web Conference; 2003. p. 130–45.
5. Doan A, Madhavan J, Dhamankar R, Domingos P, Halevy AY. Learning to match ontologies on the semantic web. VLDB J. 2003;12(4):303–19.
6. Euzenat J, Shvaiko P. Ontology matching. Berlin/New York: Springer; 2007.
7. Gal A. Why is schema matching tough and what can we do about it? ACM SIGMOD Rec. 2006;35(4):2–5.
8. Gal A, Anaby-Tavor A, Trombetta A, Montesi D. A framework for modeling and evaluating automatic semantic reconciliation. VLDB J. 2005;14(1):50–67.
9. Giunchiglia F, Marchese M, Zaihrayeu I. Encoding classifications into lightweight ontologies. J Data Semant. 2007;8:57–81.
10. Giunchiglia F, Shvaiko P. Semantic Matching. Knowl Eng Rev. 2003;18(3):265–80.
11. Giunchiglia F, Shvaiko P, Yatskevich M. Discovering missing background knowledge in ontology matching. In: Proceedings of the 17th European Conference on Artificial Intelligence; 2006. p. 382–86.
12. Giunchiglia F, Yatskevich M, Avesani P, Shvaiko P. A large scale dataset for the evaluation of ontology matching systems. Knowl Eng Rev. 2008;23:1–22.
13. Giunchiglia F, Yatskevich M, Shvaiko P. Semantic matching: algorithms and implementation. J Data Semant. 2007;9:1–38.
14. Larson J, Navathe S, Elmasri R. A theory of attributed equivalence in databases with application to schema integration. IEEE Trans Softw Eng. 1989;15(4): 449–63.
15. Madhavan J, Bernstein P, Rahm E. Generic schema matching with Cupid. In: Proceedings of the 27th International Conference on Very Large Data Bases; 2001. p. 48–58.
16. Noy N, Musen M. The PROMPT suite: interactive tools for ontology merging and mapping. Int J Hum Comput Stud. 2003;59(6):983–1024.
17. Rahm E, Bernstein P. A survey of approaches to automatic schema matching. VLDB J. 2001;10(4): 334–50.
18. Shvaiko P, Euzenat J. A survey of schema-based matching approaches. J Data Semant. 2005;4:146–71.
19. Spaccapietra S, Parent C. Conflicts and correspondence assertions in interoperable databases. ACM SIGMOD Rec. 1991;20(4):49–54.

Semantic Modeling and Knowledge Representation for Multimedia Data

Edward Y. Chang
Google Research, Mountain View, CA, USA

Synonyms

Image/Video/Music search; Multimedia information retrieval

Definition

Semantic modeling and knowledge representation is essential to a multimedia information retrieval system for supporting effective data organization and search. Semantic modeling and knowledge representation for multimedia data (e.g., imagery, video, and music) consists of three steps: *feature extraction, semantic labeling,* and *features-to-semantics mapping.* Feature extraction obtains perceptual characteristics such as color, shape, texture, salient-object, and motion features from multimedia data; semantic labeling associates multimedia data with cognitive concepts; and features-to-semantics mapping constructs correspondence between perceptual features and cognitive concepts. Analogically to data representation for text documents, improving semantic modeling and knowledge representation for multimedia data leads to enhanced data organization and query performance.

Historical Background

The principal design goal of a multimedia information retrieval system is to return data (images, video clips, or music) that accurately match users' queries (for example, a search for pictures of a deer). To achieve this design goal, the system must first comprehend a user's query concept thoroughly, and then find data in the low-level input space (formed by a set of perceptual features) that match the concept accurately. For traditional relational databases, a query concept is explicitly specified by a user using SQL. For multimedia information retrieval, however, articulating a query concept (e.g., a deer) using low level features (e.g., color, shape, texture, and salient-object features) is infeasible. Semantic modeling and knowledge representation thus plays a key role in query-concept formulation and query processing for a multimedia query.

The QBIC system [8] introduced in 1995 is the first query-by-example system. QBIC uses color histograms to represent an image/video clip; two images/clips containing similar color histograms are considered to be similar. Such knowledge representation for multimedia data is clearly inadequate. In the subsequent 5 years, many researchers in the *signal processing* and *computer vision* communities proposed techniques to extract perceptual features, such as textures, shapes, and segments of objects, for improve image rep-

resentation (see [13] for a survey). At the same time, the query-by-example paradigm was applied also to music retrieval.

Query by *just one example* was soon discovered insufficient to represent a query concept. Relevance feedback, a query refinement technique developed by the information retrieval community in the 1970's [12], was then borrowed to provide additional examples to augment the shortcoming of knowledge under-representation. In 2001, the work of [14] showed that relevance feedback could be much improved by using the *kernel methods* [1] with *active learning.* The kernel methods project data from their input space formed by perceptual features to a much higher (possibly infinite) dimensional space, where a linear classifier can be learned to separate desired data (with respect to the query) from the others. The kernel methods enjoy both rich semantic modeling (the linear class boundary in a high-dimensional space represents a non-linear boundary in the input space) and computational efficiency (computation is performed in the projected, linear space). Active learning is applied to select the most ambiguous and diversified training instances along the class boundary to query the user for labels. Once these training instances have been labeled, maximal information is gained for refining the class boundary. This process of active learning continues until the search result is satisfactory. In order to further improve the effectiveness of query-concept learning through active learning, keywords (tagged by users [9] or obtained from query logs [10]) were subsequently integrated into the semantic modeling and knowledge representation framework.

Over a decade of research since QBIC, though productive, has not yielded a large-scale real-world deployment of multimedia information retrieval system. The key reason is that semantic modeling and knowledge representation for multimedia data is intrinsically inter-disciplinary. Its success demands collaborative effort from researchers of *signal processing, computer vision, machine learning,* and *databases.* Recent works in addressing issues of perceptual similarity [11] and scalability in statistical learning [5] are inter-disciplinary approaches that hold promises to lead to a Web-scale deployment. The survey conducted by [6] provides a complementary view on the historical background.

Foundations

Semantic Modeling

There are two realistic ways for users to specify a multimedia query semantic: *query by keywords* and *query by examples*. In order to support query by keywords, *semantic annotation* provides data with semantic labels (for example, landscape, sunset, animals, and so forth). Several researchers (e.g., [2]) have proposed semi-automatic annotation methods to propagate keywords from a small set of annotated images to the other images. Although semantic annotation can provide some relevant query results, annotation is often subjective and narrowly construed. When it is, query performance may be compromised. To thoroughly understand a query concept, with all of its semantics and subjectivity, a system must obtain the target concept from the user directly via *query-concept learning*. Semantic annotation can assist, but not replace, query-concept learning.

Both semantic annotation and query-concept learning require mapping features to semantics. This semantic modeling consists of three steps. First, a set of perceptual features (e.g., color, texture and shape) is extracted from each training instance. Second, each training feature-vector x_i is assigned semantic labels g_i. Third, a classifier $f(.)$ is trained by a supervised learning (or semi-supervised learning) algorithm, based on the labeled instances, to predict the class labels of a query instance x_q. Given a query instance x_q represented by its low-level features, the semantic labels g_q of x_q can be predicted by $g_q = f(x_q)$. (About how multimedia data and knowledge can be represented is discussed in the Knowledge Representation section.)

At first it might seem that traditional supervised learning methods could be directly applied to perform *semantic annotation* and *query-concept learning*. Unfortunately, traditional learning algorithms are not adequate to deal with the technical challenges posed by these two tasks.

To illustrate, let D denote the number of low-level features. Let N denote the number of training instances, N^+ the number of positive training instances, and N^- the number of negative training instances ($N = N^+ + N^-$). And let U denote the number of unlabeled instances in the repository. Three major technical challenges arise:

1. *Scarcity of training data.* The features-to-semantics mapping problem often comes up against the $D > N$ challenge. For instance, in the query-concept learning scenario, the number of low-level features that characterize an image (D) is greater than the number of training instances that a user can provide (N) via her query history or relevance feedback. The theories underlying "classical" data analysis are based on the assumptions that $D < N$, and N approaches infinity. But when $D > N$, the basic methodology which was used in the classical situation is not similarly applicable [7].

2. *Imbalance of training classes.* The target class in the training pool is typically outnumbered by the non-target classes ($N^- >> N^+$). When the prior of the non-target class dominates the target class, a class prediction favors the non-target class. This skew can substantially reduce recall in search performance [16].

3. *Scalability.* A typical value of D can be in the order of hundreds, and U can be millions or even billions. Scalability challenges arise in at least two areas. First, searching data among U instances in a high-dimensional space is inefficient [3]. Second, when $U >> N$, training data may under-represent the knowledge required to model semantics.

Effective techniques for addressing the above challenges are inter-disciplinary. The *signal processing* and *computer vision* communities devise algorithms to extract useful features to represent multimedia data. The *machine learning* community develops models that can map features to semantics both effectively and efficiently. The *database* community improves indexing, meta-data fusion, and query processing techniques to deal with scalability issues. All these endeavors may consult experts in *neural processing* or *cognitive science* (e.g., [15]) to develop representations and models that fit human perception.

Knowledge Representation

As mentioned, a piece of multimedia data can be represented at two levels: low-level features and high-level semantics/concepts. A set of low-level features consists of perceptual features, and these features can be put in the form of a vector or a bag. High-level concepts are organized into an ontology structure, depicting relationship between concepts. In between, descriptors can be formulated either explicitly or implicitly to provide building blocks for low-level to high-level mapping and reasoning. For instance, a high-level ski concept can be formed by descriptors of snow, ski equipment, and people. Each of these descriptors is in turn composed of color, texture, shape, or salient-point features. Texts when available can be used to augment low-level perceptual features (e.g., using word "white" to depict the color of the mountain), to label descriptors (e.g., snow), or to directly annotate high-level semantics/concepts (e.g., ski). Statistical methods such as SVMs and Latent Semantic Analysis techniques (e.g., LDA [4]) can be employed to perform mapping between the three levels.

Efforts of standardizing knowledge representation have been embarked on for over a decade by academia and industry. For instance, digital cameras save JPEG files with EXIF (Exchangeable Image File) data. EXIF records camera settings, scene information, and time (and location where a photo is taken in the near future). DOLCE devises descriptive ontology for linguistic and cognitive engineering. MPEG-7 proposes different description granularity to depict multimedia data. Standard knowledge representation is essential for supporting metadata exchange and system interoperability.

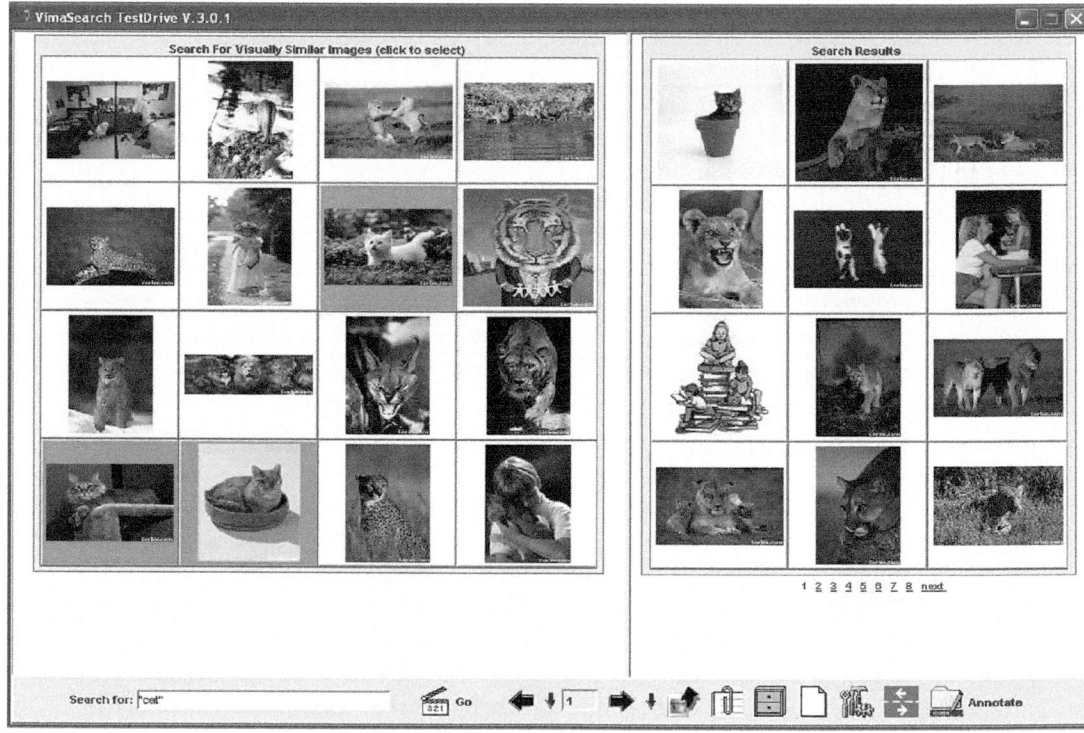

Semantic Modeling and Knowledge Representation for Multimedia Data, Fig. 1 Cat query initial screen

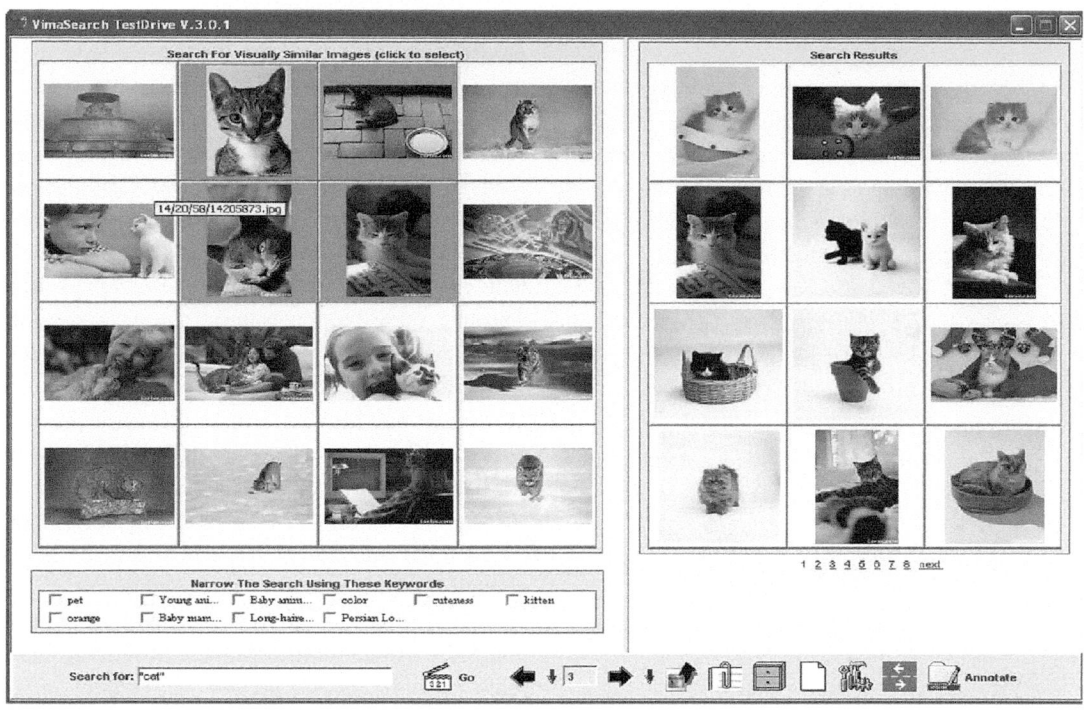

Semantic Modeling and Knowledge Representation for Multimedia Data, Fig. 2 Cat query after iteration #2

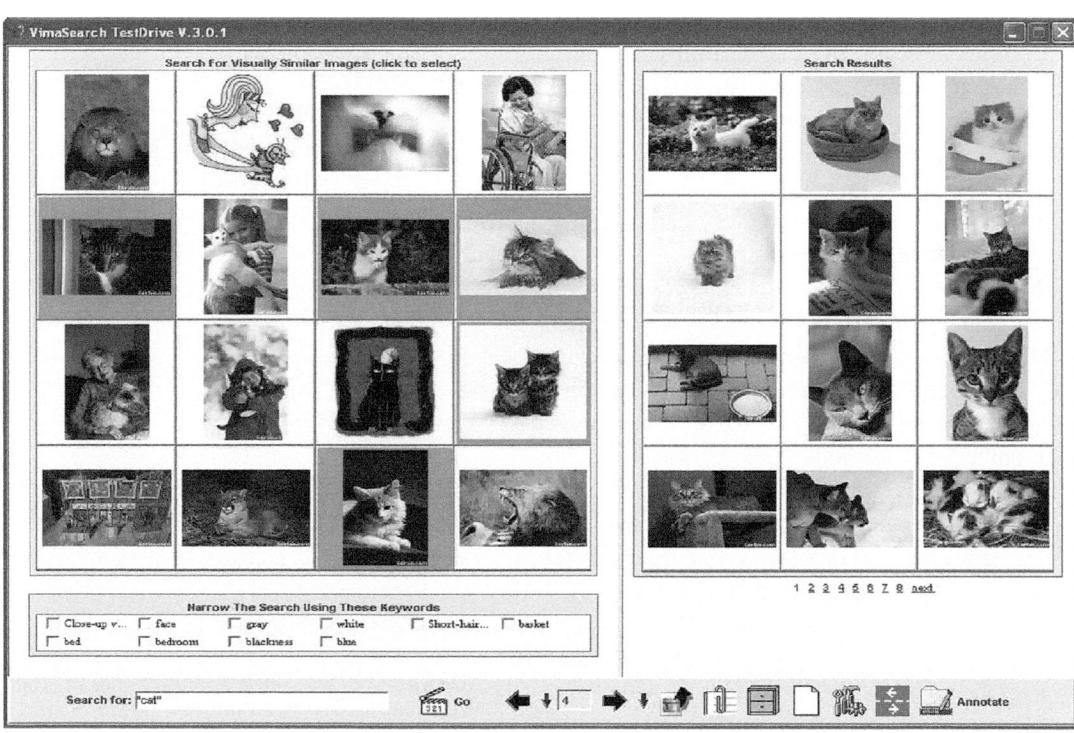

Semantic Modeling and Knowledge Representation for Multimedia Data, Fig. 3 Cat query after iteration #3

Key Applications

The launches of photo and video sharing sites such as Flickr, Google Photos, and YouTube between 2002 and 2008 renewed the interest on multimedia data management. The following applications are in high demand to manage large-scale multimedia data repositories:

1. Content-based Video, Image, Music Search Engines
2. Copy Right Infringement Detection
3. Multimedia Digital Libraries
4. Semi-automatic Photo/Video Annotation/-Classification

An application scenario is used to illustrate how aforementioned science fundamentals can improve multimedia information retrieval. Figures 1–3 show an example query using a *Perception-based Image Retrieval* (PBIR) prototype developed at UC Santa Barbara. The figures demonstrate how a query concept is learned in an iterative process by the PBIR search engine to improve search results. The user interface shows two frames. The frame on the left-hand side is the feedback frame, on which the user marks images relevant to his or her query concept. On the right-hand side, the search engine returns what it interprets as matching this far from the image database.

Most images were annotated by users. To query "cat," one first enters the keyword *cat* in the query box to get the first screen of results in Fig. 1. The right-side frame shows a couple of images containing domestic cats, but several images containing tigers or lions. This is because many tiger/lion images were annotated with "wild cat" or "cat." To disambiguate the concept, the user clicks on a couple of domestic cat images on the feedback frame (left side, in gray/green borders). The search engine refines the class boundary accordingly, and then returns the second screen in Fig. 2. In this figure, the images in the result frame (right side) have been much improved. All returned images contain a domestic cat or two.

After performing another couple of rounds of feedback to make some further refinements, more satisfactory results are shown in Fig. 3.

This example illustrates three critical points. First, keywords alone cannot retrieve images effectively because words may have varied meanings or senses. This is called the *word-aliasing* problem. Second, the number of labeled instances that can be collected from a user is limited. Through three feedback iterations, it is possible to gather just $16 \times 3 = 48$ training instances, whereas the feature dimension of this dataset is more than one hundred. Since most users would not be willing to give more than three iterations of feedback, the system encounters the problem of scarcity of training data. Third, the negatives outnumber the relevant or positive instances being clicked on. This is known as the problem of imbalanced training data. Besides, there are a large number of images in the repository. To achieve real-time performance in query refinement and in search, efficiently indexing schemes are needed to reduce search space.

Future Directions

Major advancements in three areas are necessary before large-scale multimedia systems can be realistic: *accurate and efficient object segmentation*, *scalable statistical learning*, and *high-dimensional indexing*. For details please consult the section of Foundations.

Cross-References

▶ Nearest Neighbor Query in Spatiotemporal Databases

Recommended Reading

1. Aizerman MA, Braverman EM, Rozonoer LI. Theoretical foundations of the potential function method in pattern recognition learning. Autom Remote Control. 1964;25:821–37.

2. Barnard K, Forsyth D. Learning the semantics of words and pictures. Int Conf Comput Vision. 2000;2:408–15.
3. Beyer K, Goldstein J, Ramakrishnan R, Shaft U. When is nearest neighbor meaningful. In: Proceedings of the 7th International Conference on Database Theory; 1999. p. 217–35.
4. Blei DM, Ng A, Jordan M. Latent Dirichlet allocation. J Machine Learning Res. 2003;3(4/5):993–1022.
5. Chang EY, et al. Parallelizing support vector machines on distributed computers. In: Advances in Neural Information Proceedings of the Systems 20, Proceedings of the 21st Annual Conference on Neural Information Proceedings of the Systems; 2007.
6. Datta R, Joshi D, Li J, Wang JZ. Image retrieval: ideas, influences, and trends of the new age. ACM Comput Surv. 2008;40(65).
7. Donoho DL. Aide-Memoire. High-dimensional data analysis: the curses and blessings of dimensionality (American Math. Society Lecture). In: Mathematical Challenges Of The 21st Century Explored at American Mathematical Society Conference; 2000.
8. Flickner M, et al. Query by image and video content: QBIC system. IEEE Comput. 1995;28(9).
9. Goh K, Chang EY, Lai W-C. Concept-dependent multimodal active learning for image retrieval. In: Proceedings of the 12th ACM International Conference on Multimedia; 2004. p. 564–71.
10. Hoi C-H, Lyu MR. A novel log-based relevance feedback technique in content-based image retrieval. In: Proceedings of the 12th ACM International Conference on Multimedia; 2004. p. 24–31.
11. Li B, Chang EY. Discovery of a perceptual distance function for measuring image similarity. ACM Multimedia Syst J. (Special Issue on Content-Based Image Retrieval). 2003;8(6):512–22.
12. Rocchio JJ. Relevance feedback in information retrieval. In: Salton G, editor. The SMART retrieval system – experiments in automatic document processing. Englewood Cliffs: Prentice-Hall; 1971. p. 313–23. Chapter 14.
13. Rui Y, Huang TS, Chang S-F. Image retrieval: current techniques, promising directions and open issues. J Visual Commn Image Represent. 1999;10(1):39–62.
14. Tong S, Chang EY. Support vector machine active learning for image retrieval. In: Proceedings of the 9th ACM International Conference on Multimedia; 2001. p. 107–18.
15. Tversky A. Features of similarity. Psychol Rev. 1997;84(4):327–52.
16. Wu G, Chang EY. KBA: Kernel Boundary Alignment considering imbalanced data distribution. IEEE Trans Knowl Data Eng. 2005;17(6):786–95.

Semantic Modeling for Geographic Information Systems

Esteban Zimányi[1], Christine Parent[2], and Stefano Spaccapietra[3]
[1]CoDE, Université Libre de Bruxelles, Brussels, Belgium
[2]University of Lausanne, Lausanne, Switzerland
[3]EPFL, Lausanne, Switzerland

Synonyms

Conceptual modeling; Conceptual modeling for Geographic Information System; Conceptual modeling for Spatio-temporal applications; Geographical databases; GIS.

Definition

Semantic modeling denotes the activity of designing and describing the structure of a data set using a *semantic data model*. Semantic data models (also known as *conceptual data models*) are data models whose aim is to provide designers with modeling constructs and rules that are well suited for representing the user's perception of data in the application world, abstracting from implementation concerns. They contrast with *logical* and *physical data models*, whose aim is to organize data in a way that is easily manageable by a computer. The most popular semantic data models are UML, a de facto standard, and ER (Entity-Relationship), still widely used in many design methodologies and favored by the academic community.

Semantic models were first created in the database community in the 1980s. They started to be developed for Geographical Information Systems (GIS) in the 1990s. Their aim is the same as for traditional databases, to free GIS users from the specificities of system-oriented data models and proprietary file formats (e.g., spaghetti and topological data models, triangulated irregular network (TIN) models, shape files, and raster models). While a number of semantic data models for geographic data have been developed, rapidly the focus has shifted from supporting spatial data to supporting data with both spatial and temporal features, leading to the development of several spatiotemporal semantic data models. Despite the fact that current GIS and Database Management Systems (DBMS) provide poor support for temporal features, semantic modeling advocates that space and time aspects are intrinsically correlated in the application world.

Historical Background

Most people consider the 1976 paper by Peter Chen [5], defining the basic ideas of Entity-Relationship modeling, as the foundational milestone for semantic modeling. The paper had indeed an enormous effect on the database design community, leading to considerable developments to further extend the semantic capabilities of the approach. It took more than 15 years to see the same idea spreading in the academic GIS community with, for example, the 1993 MODUL-R formalism [4], which extended with spatial data the ER approach used in the leading French design methodology, Merise. Further work on MODUL-R eventually resulted in the Perceptory UML-based approach and tool [3]. Semantic models for GIS bloomed in the 1990s, basically splitting into approaches stemming from the object-oriented paradigm (e.g., [6, 16]) and approaches following the ER or the UML paradigm (e.g., MADS [10], STER [15], GeoUML [2]). A survey of many spatial data models may be found in [12]. The industrial and application world has also developed GIS data modeling specifications to help promoting interoperability between different systems and different applications. The Open Geospatial Consortium (OGC) and the International Standards Organization (ISO) have produced specifications supporting conceptual modeling for data with spatial (and some temporal) features (http://www.opengeospatial.org/).

Thanks to the development of ubiquitous and mobile computing on the one hand, and of sensors and GPS technologies on the other hand, large-scale capture of the evolving position of mobile objects has become technically and economically feasible. This opened new perspectives for a large number of applications (e.g., from transportation and logistics to ecology and anthropology) built on the knowledge of objects' movements. Typical examples of moving objects include cars, persons, and planes equipped with a GPS device, animals bearing a transmitter whose signals are captured by satellites, and parcels tagged with RFIDs. This fostered the interest in spatiotemporal models, rather than purely spatial or purely temporal models at the logical and semantic levels. Güting's approach [8] defined a set of data types and associated operators for moving objects (points and surfaces), which allows one to record, for example, the changing geometry of pollution clouds and flooding waters. At the semantic level, examples of spatiotemporal models include MADS [10], Perceptory [3], STUML [15], STER [15], and ST USM [9]. Extending the limited capabilities of commercial data management systems, some research prototype systems [1, 11] do provide nowadays support for storing and querying the position of a moving object all along the lifespan of the object. The latest developments in this domain are the management of trajectories, which adds a semantic interpretation to the movement of objects of kind moving point [13]. Trajectory management is important in many application domains, e.g., for addressing traffic management issues, building social models of people's movements within a city, and optimizing the localization of resources (e.g., communication antennas, shops, advertisement panels) that have to be available to moving customers.

Foundations

Requirements for Semantic Modeling of Spatial Data
Semantic modeling of spatial data requires concepts for the description of both the discrete and

the continuous view of space, in a seamlessly integrated way. The *discrete view* (or *object-based view*) is the one that sees space as filled by objects with a defined location and shape. Parts of space where no object is located are considered as empty. This view typically serves application requests asking where certain objects are located, or which objects are located in a given surface. On the other hand, the *continuous view* (or *field-based view*) is the one that sees space as a continuum, holding properties whose values depend on the location in space but not on any specific object (i.e., the value for the property is given by a function whose domain is a spatial extent). Typical examples where this view applies are the recording of continuous phenomena such as temperature, altitude, soil coverage, etc. Both views are important for applications, which may use one or the other, or both simultaneously.

Assuming the discrete view, any traditional database schema can be enriched to become a spatiotemporal database schema by including the description of the spatial and/or temporal properties of the real-world phenomena represented in the schema. Consider, for instance, a Building object type, with properties name, address, usage, architect, and owner. Adding positional information on the geographic location of the building (e.g., its coordinates in some spatial reference system) turns Building into a spatial object type. If one adds information characterizing the existence of the building in time (e.g., when construction was first decided, when construction started, when it was completed, when it was abandoned, and when it was demolished), Building becomes a temporal object type. Space and time are independent dimensions. Some data may have spatial features, some may have temporal features, some may have both, and some may have none.

Objects, be they spatial or not, can have spatial properties, i.e., properties whose value domain is composed of spatial values rather than alphanumeric values. Spatial values conform to spatial data types (see the entry in this encyclopedia), e.g., point, line, polyline, surface. For example, a Building object type can have a property near-

estFireStation whose value for each building is the geographic location of the nearest fire station, e.g., a spatial value composed of two spatial coordinates defining a point.

Most basic types for space are Point, Line, and Surface (and volume for 3D databases). However, applications may require more than simple spatial data types. Some spatial objects have extents (the term "extent" denotes the set of points that an object occupies in space) that are made up of a set of elementary extents. For example, an archipelago is a set of surfaces; many coastal countries do have islands too; and facility networks may be represented by connected sets of lines. Moreover, some spatial objects have complex extents made up of a heterogeneous set of spatial values. For example, an avalanche zone is described by a surface and a set of oriented lines describing, respectively, its maximal extent and the usual avalanche paths. Similarly, a river may be described by lines when its bed is narrow and by surfaces when it is broad. Therefore, the set of spatial data types should include types for homogeneous or heterogeneous collections, like PointSet, LineSet, SurfaceSet, or SpatialHeterogeneousSet. The whole set of spatial data types is organized into a generalization hierarchy with generic data types, in order to support spatial object types whose extent may be of different types depending on the instance. For example, the object type City may contain larges cities represented by a surface and small ones represented by a point. The spatial extent of City could then be described by a generic spatial data type that would contain points and surfaces. The Open Geospatial Consortium (OGC) has defined such a hierarchy of spatial data types.

Geographical applications often need to enforce spatial or temporal constraints between spatial or temporal features. For example, harbors should be located along water bodies and bridges on roads or railways. Therefore, a spatial data model should support constructs allowing designers to specify constraints that will be automatically enforced by the system. A first kind of construct is the spatial (or temporal) relationship type. They link two spatial (and/or temporal) object types and bear a spatial (and/or temporal) condition that the linked objects must obey. Typically, conditions express topological relationships (e.g., inclusion, disjointedness, overlapping), metric relationships (e.g., based on distance), orientation relationships (e.g., North of), or the temporal predicates defined by Allen (e.g., during). Applications may need two different kinds of these spatial and temporal relationships:

- Spatially/temporally constraining relationship types: Users can link two spatial objects by a spatially constraining relationship only if their spatial/temporal extents abide by the condition.
- Derived spatial/temporal relationship types: The system automatically creates the instances of the relationship for all couples of objects that satisfy the condition.

Moving and deforming objects may also be linked by spatial relationships. For instance, an aeronautic database may need recording the trajectories of planes when they cross storms, the two being moving objects. In these cases, the condition of the relationship type is spatiotemporal: It bears on the location and the time.

Applications may also need constraints between composite and component elements. For example, a spatial aggregation relationship may enforce that the extent of the composite object is made up by the union of the extents of the component objects, as in a spatial aggregation linking the spatial object types Country and District. Another example is restricting the spatial (or temporal) values of attributes to be within the spatial (or temporal) extent of the object to which they belong. For example, the values of the spatial attribute major Cities (a multivalued attribute of type Point) of the object type Country should be within the spatial extent of the country. This kind of constraint may be frequent, but it is not always the case. Refer for example to the Building spatial object type with the spatial attribute nearestFireStation. Therefore, the data model should not automatically and implicitly enforce these constraints.

It should provide designers with a means for explicitly specifying which constraints should be enforced.

The modeling of the continuous view of space requires another construct for properties that are defined on a spatial extent and whose value depends on the exact location (point) of the spatial extent. The spatial extent may be the whole space covered by the database or a specific extent. For example, the water quality of a river exists only in the spatial extent of the river course. On the other hand, temperature, soil, and land coverage are information that exist and may be measured (if relevant) at any point of the geographical space covered by the database. *Field-based* models are well suited for applications that perceive the real world exclusively through varying properties. For the many applications that use both the discrete and continuous views, several spatial data models provide a predominant discrete view (i.e., based on spatial objects) in addition to a special construct for representing varying properties, the *space-varying attribute*, which is a function from a spatial extent to a range of values. Any object and relationship, be it spatial or not, should be able to bear space-varying attributes. Moreover, the range of space-varying attributes may be simple (e.g., elevation) or complex (e.g., weather composed of temperature, pressure, and rainfall), monovalued (e.g., altitude) or multivalued (e.g., insects in forests, assuming this information is captured using a space unit large enough to be the home of several kinds of insect, e.g., using cells of 1 m^2).

Another important requirement for space modeling is the ability to describe data at different granularity or resolution, for example to be able to support applications working with maps at different scales.

Finally, an essential requirement is the ability to model spatial features of a phenomenon irrespectively of the fact that the phenomenon has been modeled as an object, a relationship, or an attribute. This orthogonality of the space modeling dimension with the data structure modeling dimension is what avoids making the designs in the two dimensions dependent on each other.

Survey of Current Semantic Modeling Approaches

Semantic models are typically developed in the academic world. For example, MADS [10] has been purposely developed to match all the requirements discussed in the previous section. MADS belongs to the extended ER family of models. Its distinguishing feature is the full support for multiple perceptions and multiple representations of the same real-world objects. Another distinguishing feature of MADS is its support of explicit relationships equipped with topological and synchronization constraints. Multiple perceptions and representations are also supported, to a more limited extent, by Perceptory [3], an UML extension targeted to support spatiotemporal analysis in a data-warehousing framework. STUML [15] and GeoUML [2] are other UML-based approaches, although without multi-perception support. Other spatiotemporal approaches include ST USM [9], very similar to MADS but emphasizing support of multi-granularity, and STER [15], another extended ER formalism which supports both valid and transaction time but, compared to MADS, is weaker in data structures. Most of these academic proposals have been implemented in prototypes, but, with the exception of Perceptory, they have not yet turned in commercial products. These proposals deal with 2D data.

A very different approach, known as spatial constraint database modeling, relies on mathematical equations to define spatial extents. Some existing prototype systems (e.g., DEDALE [7]) use this approach.

Key Applications

Semantic modeling is an essential capability for organizations that need to develop a database that provides different applications and different categories of users with different sets of data, possibly organized in different ways. Designing a database in such a complex environment is a very challenging task, as has been extensively proven in traditional data management. Adding spatial features makes the design task even more complex, in particular since this inevitably leads to adding also temporal features. Indeed, what most applications in the geographical domain need to analyze is the temporal evolution of the spatial features of interest. Cartographic applications are the most traditional ones, but today the focus is rather on all kinds of planning and forecasting services to citizens and the society at the municipal, regional, and statewide levels. Examples of such services include environmental control management and global warming. Given the cost of developing databases for these applications and the need for people in charge (politicians and managers) to be successful, it is of the highest importance that the design of an operational database is carried out using the most suitable tools. Semantic modeling is the key to a successful design that determines what data are needed, to be complemented afterwards in the implementation phase by addressing performance aspects in order to guarantee that the data can be used effectively.

Semantic modeling is also the key to all data exchange, reuse, and integration efforts. Whether in database terms, as discussed here, or in ontological terms, semantic modeling is the kernel of the semantic web.

Future Directions

The economic trend towards worldwide enterprise operation and the technical trend towards web-based interoperability will significantly increase the complexity of GIS and the challenges designers will have to overcome. A key help in this context will come from ontologies about spatiotemporal application domains. These ontologies provide a common semantic basis to build repositories of domain knowledge that go beyond traditional enterprise boundaries. In this perspective, the current focus on ontology-assisted semantic modeling and ontology-assisted data integration is leading research into a fruitful direction [14].

In a complementary effort, ontologies and geographic markup languages facilitate the integration of geographical knowledge coming from multiple sources available through the Web. This will contribute to significantly enhance geographical knowledge, benefiting from geo-content actually hidden in Web pages.

Cross-References

▶ Database Design
▶ Field-Based Spatial Modeling
▶ Geographic Information System
▶ Multiple Representation Modeling
▶ Spatial Data Types
▶ Topological Data Models
▶ Topological Relationships

Recommended Reading

1. Almeida VT, Güting RH, Behr T. Querying moving objects in SECONDO. In: Proceedings of the 7th International Conference on Mobile Data Management; 2006. p. 47–51.
2. Belussi A, Negri M, Pelagatti G. GeoUML: a geographic conceptual model defined through specialization of ISO TC211 standards. In: Proceedings of the 10th EC GI & GIS Workshop, ESDI State of the Art; 2004.
3. Brodeur J, Bédard Y, Proulx M.J. Modelling geospatial application database using UML-based repositories aligned with international standards in geomatics. In: Proceedings of the 8th ACM Symposium on Adavances in Geographic Information System; 2000. p. 39–46.
4. Caron C, Bédard Y. Extending the individual formalism for a more complete modeling of urban spatially referenced data. Comput Environ Urban Syst. 1993;17(4):337–46.
5. Chen PP. The entity-relationship model: toward a unified view of data. ACM Trans Database Syst. 1976;1(1):9–36.
6. Egenhofer MJ, Frank AU. Object-oriented modeling for GIS. J Urban Reg Inf Syst Assoc. 1992;4(2):3–19.
7. Grumbach S, Rigaux P, Scholl M, Segoufin L. The DEDALE prototype. In: Constraint databases. Berlin: Springer; 2000. p. 365–82.
8. Güting RH, Böhlen MH, Erwig M, Jensen CS, Lorentzos NA, Schneider M, Vazirgiannis M. A foundation for representing and querying moving objects. ACM Trans Database Syst. 2000;25(1): 1–42.
9. Khatri V, Ram S, Snodgrass RT. On augmenting database design-support environments to capture the geo-spatio-temporal data semantics. Inf Syst. 2006;31(2):98–133.
10. Parent C, Spaccapietra S, Zimányi E. Conceptual modeling for traditional and spatio-temporal applications: the MADS approach. New York: Springer; 2006.
11. Pelekis N, Theodoridis Y, Vosinakis S, Panayiotopoulos T. Hermes – a framework for location-based data management. In: Advances in Database Technology, Proceedings of the 10th International Conference on Extending Database Technology; 2006. p. 1130–34.
12. Rios Viqueira JR, Lorentzos NA, Brisaboa NR. Survey on spatial data modelling approaches. In: Manolopoulos Y, Papadopoulos A, Vassilakopoulos M, editors. Spatial databases: technologies, techniques and trends. Hershey: Idea Group; 2005. p. 1–22.
13. Spaccapietra S, Parent C, Damiani ML, Macedo J, Porto F, Vangenot C. A conceptual view on trajectories. Data Knowl Eng. 2008;65(1):126–46.
14. Sugumaran V, Storey VC. The role of domain ontologies in database design: an ontology management and conceptual modeling environment. ACM Trans Database Syst. 2006;31(3):1064–94.
15. Tryfona N, Price R, Jensen CS. Spatiotemporal conceptual modeling. In: Spatiotemporal databases: the chorochronos approach (chapter 3), lecture notes in computer science, vol. 2520. Springer: Berlin; 2003. p. 79–116.
16. Worboys M, Hearnshaw H, Maguire D. Object-oriented data modelling for spatial databases. Int J Geogr Inf Syst. 1990;4(4):369–83.

Semantic Overlay Networks

George Anadiotis, Spyros Kotoulas, and Ronny Siebes
VU University Amsterdam, Amsterdam, The Netherlands

Synonyms

Semantic overlays; SONs

Definition

Semantic Overlay Networks are types of *Overlay Networks* where the topology is formed according to the resources (i.e., services or data provided) of the participants. This is done on the basis of similarity between participants, which is calculated by means of resource metadata exchange (e.g. keywords, term vectors, concepts from ontologies, histograms).

Historical Background

Semantic Overlay Networks were introduced in 2002, by H. Garcia-Molina and A. Crespo [3]. The motivation was to give an alternative to inefficient search methods for overlay networks, one that would provide more relevant results in less time and using less resources (mainly bandwidth).

Foundations

The original idea of Semantic Overlay Networks [3] was to organize the nodes that participate in a network into many different overlays, based on the content that the nodes are contributing and some classification scheme for this content. As one overlay is created for each class, the classification to be used is rather important for the operation of the system. Having classifications whose distribution is skewed would result in overlays that are not efficient (either too big or too small to be of real value). Classification as used in [3] entails hierarchy, so it is in fact a taxonomy.

After choosing an appropriate taxonomy, content needs to be classified (either manually or automatically). This is performed massively for all content belonging to incoming nodes, so then the results of content classification are used to assign nodes to overlays, by choosing the one(s) that is the best match according to the majority of the content's classification. Different techniques and policies can be applied here (for example the predecessors and the descendants of a class are also considered), and nodes can be assigned to more than one overlay.

Finally, in order to be able to answer queries, queries are also subjected to the same classification. When they have been classified, they are subsequently forwarded only to the appropriate overlay. Results can be incremental, meaning that the query can also be sent to overlays that correspond to predecessors of the querys assigned class (i.e., more general classes) in order to retrieve additional matches.

In Fig. 1, one can see an example of three semantic overlay networks, as described above. Each node may belong to more than one semantic overlay (e.g., node d belongs to all semantic overlays) and has to maintain a number of connections for each one of them.

Inter and intra overlay routing (i.e., finding the appropriate overlay and routing within the overlay) presents a series of challenges which, along with other optimizations, have been elaborated by recent research.

In [6], the metadata used to extract similarity are (schema-less) XML documents extracted from peer content, and supported queries are extended from keyword to path queries that exploit the structure of XML documents. Peers maintain specialized data structures that summarize large collections of documents (filters - specialized extensions of Bloom filters). Each node maintains a local filter that summarizes the documents stored locally. Nodes are organized in hierarchies (trees) based on similarity of local

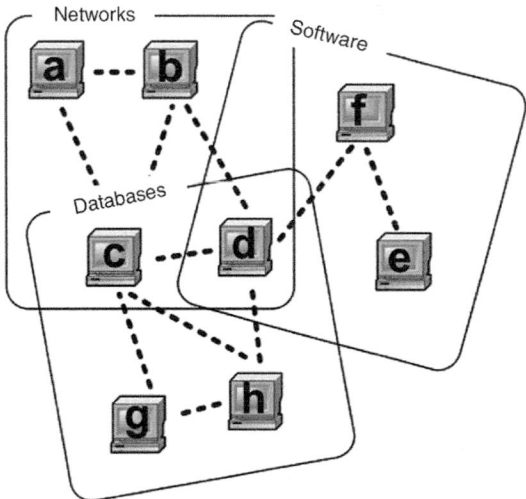

Semantic Overlay Networks, Fig. 1 Three overlay networks

filters; non-leaf nodes also contain merged filters summarizing the documents of its children, or in the case of root nodes, other root nodes as well.

With this organization, nodes belonging to the top levels receive more load and responsibilities, thus, the most stable and powerful nodes should be located to the top levels of the hierarchies. When receiving a query, nodes use their local filter to find results and then forward the query to their sub-tree.

In [12], the notion of *peer schemas* is defined. Peer schemas are virtually defined schemas that represent a peer's view of the world and are used for purposes of querying and mapping. Relations between peer schemas are called peer relations. Peers also contribute data to the system in the form of stored relationships, which correspond to the peer's local view of the data. Mappings are utilized in order to translate queries between different semantic networks. There are two types of mappings, namely mappings that relate two or more peer schemas (peer descriptions) as well as mappings that relate a stored schema to a peer schema (storage descriptions).

Sending queries only to peers that might provide answers is achieved using a two-fold approach: on the one hand, there is a query reformulation algorithm that works by combining global-as-view (*GAV*) and local-as-view (*LAV*) approaches and selectively applying unfolding and rewriting techniques to the original query. Since however the algorithm is only able to exploit information pertaining to schema mappings and not actual data stored at the peers, a (centralized) index structure that allows simple value lookup with partial match over structured attributes is also used. Participating peers upload data summaries as well as peer mappings to the index, thus enabling the index engine to correlate attributes from different peers and provide a simple type of schema mapping.

In [9], a variation on the approach presented in [6] is given. Instead of schema-less XML documents, metadata consist of RDF statements abiding to different RDFS schemas. There are normal peers (P) and super peers (SP) that normal peers attach to, as well as index structures (SP/P and SP/SP) that utilize frequency statistics used to define similarity measures responsible for clustering peers to super-peers, thus making them dynamic. Furthermore, an efficient topology is maintained for communication in the super peer network (HyperCup), which also offers mediation services between different peer schemas.

In [2], a layered architecture for SONs is proposed, comprising of a knowledge infrastructure layer and a communication infrastructure layer. Here fully blown ontologies are used instead of taxonomies or schemas. Each peer is able to store data as well as a peer ontology. When a peer sends a query Q over the H3 network, the request goes through a query processing module for rewriting in terms of the ontological description of target concept(s). The rewritten query is then forwarded to a semantic routing module that sends it only to peers that may provide results semantically related to the kind of concepts requested (semantic neighbors).

In order to choose the semantic neighbors, semantic routing exploits the services of a knowledge manager module to retrieve ontology location links to the peers whose contents are semantically related to the target concept(s) in the query. Location links are returned to the semantic routing module, which uses them for query routing. When a peer's routing module receives a query, it forwards it to the query processing manager, where it is analyzed and processed. If no matching concepts are found, the query is discarded and no reply is returned. Otherwise the query answer is composed and forwarded to the semantic routing module which sends back the reply to the requesting peer.

In some systems, SONs are not discretely clustered and are formed on per-peer basis. While in the original approach by [3] there is a separate SON for each concept, a category of systems use pair-wise peer interactions. They are used to construct a topology enriched with content information about every neighbor, and dynamically determine routing according to this information, instead of broadcasting on the entire SON.

Semantic Overlay Networks, Fig. 2 Peer neighborhood relations

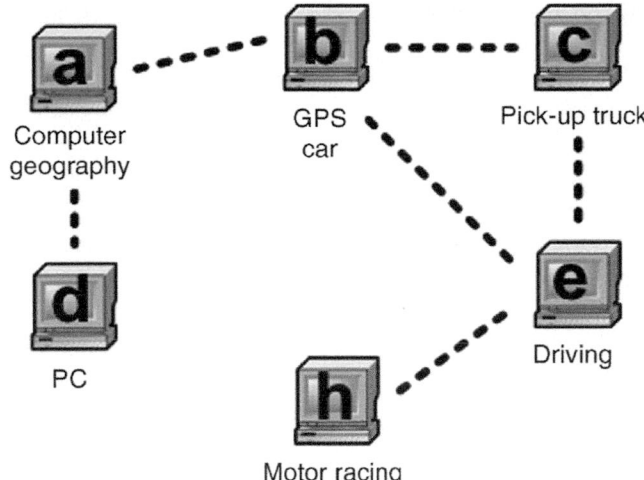

In [5, 10], the SON is built using *advertising*. Peers exchange descriptions of their content with their neighbors. Furthermore, they keep advertisements that are similar to theirs, and thus, a semantic topology is formed. An example of such a topology is given in Fig. 2. Peers keep neighborhood relations based on their content descriptions. Thus, locality is improved. Furthermore, each peer is aware of the description of the content of its neighbors. Thus, it can forward queries to the peers which are more likely to have relevant content. Research in the context of [10] indicates that either maintaining neighbor relations according to description similarity or forwarding queries to the peers with description most similar to the query perform much better than flooding approaches.

It is important to note a category of systems that use both a semantic and a structured overlay. P-Scarch [11] uses Latent Semantic Indexing *LSI* to extract vector representations of documents. These vectors are mapped into a multidimensional space maintained by a Content Addressable Network (CAN), a type of structured overlay. Nodes within this overlay participate in an an additional semantic overlay used for content-directed search.

Gridvine [1] uses semantic overlays to store, query and make mappings between RDFS schemas. They are split into triples and are indexed by subject, predicate and object using the P-Grid structured overlay. Each peer may define and use its own RDFS schema, so naturally incompatibilities in terms of resource description/query interpretation may appear. To cope with such incompatibilities, the notion of schema translations is introduced. A schema translation is a mapping between two different schemas. Since RDFS does not support schema mapping, these translations are encoded using OWL.

This is a very powerful feature, as it enables the gradual forwarding of requests from the originating peer to peers for which no direct schema translation exists. This is achieved through a procedure called Semantic Gossiping, in which each peer that receives a query expressed in a certain schema examines available translation links and evaluates (by performing syntactic and semantic analysis) if and where it should forward this query. In addition, schema inheritance is also supported, which enables peers to not only define their own schemas, but also either directly reuse or extend existing schema hierarchies. Sets of peers that share the same schema are called semantic neighborhoods.

Finally, other state-of-the-art features in SONs include observing past queries submitted/answers received in order to judge semantic proximity [13], proactively acquiring semantic links through gossiping [14], and applying ant-colony heuristics to improve semantic routing [8].

Key Applications

Semantic overlay networks use an order of magnitude less messages than flooding overlays. They enable Internet-scale systems, by providing the infrastructure for efficient search over large numbers of hosts. Despite keen interest in the area by the scientific community, none of the aforementioned systems has been commercially deployed yet.

The OpenKnowledge project (http://www.openk.org.) works on a P2P system where webservices and workflows, annotated by keywords, can be shared. The system uses a SON to store and retrieve them efficiently and to find peers that execute the webservices.

Bibster [5] is a P2P application for sharing bibliographic items. These items are annotated by concepts from an ontology. The SON is used to route queries, which are also concepts from the same ontology, to the peers that semantically match the query.

Another proposed application of SONs is using them to cluster content providers on the WWW in order to facilitate a comprehensive distributed search engine [4].

Future Directions

An interesting problem concerns optimizing multi-attribute search. Simply joining the results of single attributes is often not scalable, especially when the single attributes individually result in many answers.

Another topic is to do efficient information retrieval by using SONs. Currently, most resource discovery systems do not rank the results according to the relevance: either they match or they do not match.

Current peer-to-peer discovery systems using SONs fail to solve privacy infringement issues and are actually more vulnerable than centralized approaches. This is because, due to the nature of SONs, peers "know" about the content of other peers, which may be undesirable.

Distributed reasoning over a P2P network is another interesting topic where SONs may be of help, for example to cluster consistent parts of knowledge. Especially in the Semantic Web area, there is a desire to have an efficient (RDF) triple storage where reasoning is done via shared or local schema's. Some first solutions like Unistore (http://www.p-grid.org/publications/applications.html.) and Gridvine [1] are based on storing the triples in *DHTs*, which leads to many messages because each triple leads at least to three *DHT* storage- or lookup messages.

Cross-References

▶ Distributed Hash Table

Recommended Reading

1. Aberer K, Cudré-Mauroux P, Hauswirth M, Pelt TV. GridVine: building internet-scale semantic overlay etworks. In: McIlraith SA, Plexousakis D, Van Harmelen F, eds. The semantic web In: Proceedings of the 3rd International Semantic Web Conferences; 2004. p. 107–21.
2. Castano S, Ferrara A, Montanelli S, Pagani E, Rossi G. Ontology-addressable contents in p2p networks. In: Proceedings of the 1st Workshop on Semantics in Peer-to-Peer and Grid Computing; 2003.
3. Crespo A, Garcia-Molina H. Semantic Overlay Networks for P2P Systems, Technical Report 2003-75. Stanford University/InfoLab. 2003.
4. Doulkeridis C, Nørvåg K, Vazirgiannis M. DESENT: decentralized and distributed semantic overlay generation in P2P networks. IEEE Jour Selected Areas in Commun. 2007;25(1):25–34.
5. Haase P, Broekstra J, Ehrig M, Menken M, Mika P, Olko M, Plechawski M, Pyszlak P, Schnizler B, Siebes R, Staab S, Tempich C. Bibster – a semantics-based bibliographic peer-to-peer system. In: McIlraith SA, Plexousakis D, Van Harmelen F, eds. The semantic web In: Proceedings of the 3rd International Semantic Web Conferences; 2004. p. 122–36.
6. Koloniari G, Pitoura E. Content-Based Routing of Path Queries in Peer-to-Peer Systems. In: Advances in Database Technology, Proceedings of the 9th International Conference on Extending Database Technology; 2004. p. 29–47.
7. McIlraith SA, Plexousakis D, Van Harmelen F, eds. The semantic web In: Proceedings of the 3rd International Semantic Web Conferences; 2004.
8. Michlmayr E, Pany A, Kappel G. Using taxonomies for content-based routing with ants. In: Proceedings of the 2nd workshop on Innovations in Web Infras-

tructure (IWI2), 15th International Worldwide Web Conference; 2006.

9. Nejdl W, Wolpers M, Siberski W, Schmitz C, Schlosser M, Brunkhorst I, Löser A. Super-Peer-Based Routing and Clustering Strategies for RDF-Based Peer-To-Peer Networks. In: Proceedings of the 12th International World Wide Web Conferences; 2003.

10. Siebes R, Kotoulas S. pRoute: Peer selection using shared term similarity matrices. Web Intelligence Agent Systems. 2007;5(1):89–107.

11. Tang C, Xu Z, Dwarkadas S. Peer-to-Peer information retrieval using self-organizing semantic overlay networks. Technical reports, HP Labs. 2002.

12. Tatarinov I, Ives Z, Madhavan J, Halevy A, Suciu D, Dalvi N, Dong XL, Kadiyska Y, Miklau G, Mork P. The Piazza peer data management project. ACM SIGMOD Rec. 2003;32(3):47–52.

13. Tempich C, Staab S, Wranik A. Remindin': semantic query routing in peer-to-peer networks based on social metaphors. In: Proceeding of the 12th International World Wide Web Conferences; 2004. p. 640–9.

14. Voulgaris S, Kermarrec AM, Massoulie L, Van Steen M. Exploiting semantic proximity in Peer-to-peer content searching. In: Proceeding of the 10th International Workshop on Future Trends in Distributed Computing Systems; 2004.

Semantic Social Web

Maribel Acosta[1] and Fabian Flöck[2]
[1]Institute AIFB, Karlsruhe Institute of Technology, Karlsruhe, Germany
[2]GESIS – Leibniz Institute for the Social Sciences, Köln, Germany

Synonyms

Social Semantic Web

Definition

The Semantic Social Web refers to a global *data space* of semantically enriched and user-generated data. Interconnections between people comprise the Semantic Social Web as well as documents, activities, tags, and geo-markers. The Semantic Social Web combines technologies and methodologies from the Semantic Web and applies them in an effort to enhance the Social Web (also known as the Web 2.0).

The Semantic Social Web aims at interlinking heterogeneous data available from different social platforms by applying knowledge representation mechanisms of the Semantic Web. Ontologies and vocabularies are used to describe entities and relationships between those entities within the Social Web, facilitating interoperability among social websites. The integration of complementary segments of information dispersed among isolated websites and services allows for obtaining a more complete picture of certain topics and the development of novel applications [1, 2].

Although the Semantic Social Web is established on the foundations of the Semantic Web, the former differs from the latter in its social nature, i.e., the data is generated through – and mainly consumed by – human agents. Machine-readable data, encoding the semantics that characterize the Semantic Web, is still essential in the Semantic Social Web. Semantics is used to perform computational tasks, for example, executing queries over this data or applying reasoning techniques.

Historical Background

The first idea of a global web of knowledge produced by man was envisioned in 1945 by Vannevar Bush in the essay *As We May Think* (http://www.w3.org/History/1945/vbush/vbush.shtml). Bush explains his vision of a remote collective memory machine denominated *memex* defined as the *world's record* in which individuals store and consult all type of information. The memex includes the implementation of an index that would enable easy access to all the information. Bush also explains the idea of annotating and associating information, which conforms to the concept of the Semantic Social Web.

In 1962, Doug Engelbart publishes the report "Augmenting Human Intellect: A Conceptual Framework" (http://www.dougengelbart.org/pubs/augment-3906.html). Engelbart proposed a distributed collaborative system for augmented intelligence where people are able to gain certain understanding to devise solutions to complex problems by harvesting collective knowledge. The human-computer machine envisioned by Engelbart combined the strength of people and computers to support collective learning processes where the computers are conceived as artifacts to store and display information.

The concept of socio-semantic web (S2W) was coined by Manuel Zacklad and Jean-Pierre Cahier in 2004 in the article "Socio-Semantic Web" applications: Towards a methodology based on the Theory of Community of Actions" (http://cahier.tech-cico.fr/publi/JPC-MZ-COOP04.pdf). Zacklad and Cahier define S2W applications as systems designed to support collective knowledge management including the underlying semantic resources, denominated "local" semantics. S2W is founded on participative design principles in which users produce and consume knowledge and semantics, which may evolve over time and even become conflictual. The participative principle of S2W applications is also one of the principles of the Semantic Social Web.

Foundations

The Semantic Social Web is founded on the technologies of the Semantic Web. Similarly to the development of the Semantic Web, the Semantic Social Web structure is composed of consecu-

tive layers. Breslin et al. have denominated this structure the *vocabulary onion*, which enables a vertical creation of vocabularies by linking and combining vocabularies to describe social data on the Web [1]. Figure 1 depicts the current vision of the Semantic Social Web stack, an extended version of the vocabulary onion.

Core Vocabularies

This layer is comprised of the vocabularies that conform the pillars of the Semantic Web, i.e., Resource Description Framework (RDF), RDF Schema (RDFS), and Web Ontology Language (OWL). RDF(S) and OWL are fundamental languages to model objects on the Web and that can be used for writing other vocabularies on top of them.

Core Social Vocabularies

These vocabularies provide a formal representation of entities that coexist on the Social Web. They model knowledge about persons, organizations, activities, and other types of objects that are common for social interaction on the Web. This layer should be composed of lightweight semantic ontologies and vocabularies [1]. The vocabularies described in the following are characterized for being relatively simple facilitating their adoption for the development of the Semantic Social Web [1]:

- FOAF, Friend-of-a-friend (http://www.foaf-project.org/): FOAF is a vocabulary that

Semantic Social Web, Fig. 1 Semantic Social Web stack

Query language: SPARQL	Domain-specific vocabularies
	Interconnecting vocabularies
	Core social vocabularies
	Core vocabularies
	OWL
	RDF Schema

describes agents (person, organization, or group) and their actions.

- SIOC, Semantically Interlinked Online Communities (http://rdfs.org/sioc/spec/): SIOC is an ontology that describes information about online communities including blogs, forums, and wikis.
- SKOS, Simple Knowledge Organization System (http://www.w3.org/TR/skos-primer/): SKOS is a model to represent concept schemes including relationships between terms.
- DC, Dublin Core (http://dublincore.org/documents/dces/): DC is a vocabulary composed of terms to model resources that are published on the Web, including their creators.

Interconnecting Vocabularies

This layer supports interoperability among social platforms through nonsocial data, for example, geographical or temporal. Also, other types of metadata can be described in this layer to provide additional links among the data.

Domain-Specific Vocabularies

This layer contains all the other vocabularies and ontologies that are developed for specific applications and that could not be modeled in the previously described layers.

Query Language

SPARQL Protocol and RDF Query Language (SPARQL) is the W3C recommendation for querying RDF data. In the Semantic Social Web stack, SPARQL is the language used to access knowledge among the stack layers.

Key Applications

The potential of the Semantic Social Web becomes apparent in the following scenarios [1]: data sharing, knowledge creation and sharing, social tagging, social networking, and interlinking of online communities.

Data Sharing

Certain Social Web platforms allow for uploading and sharing different forms of data, including, for instance, documents, images, audio, video, or software. All this data can be annotated using Semantic Web ontologies. The semantically rich tags combined with other metadata can be used to enhance the results of search engines or recommendation systems.

The W3C Multimedia Semantics Incubator Group (http://www.w3.org/2005/Incubator/mmsem/) has compiled a set of relevant tools and resources for extracting the semantics encoded in the ontology-based annotations of multimedia data as well as for semantically annotating this data [4].

Regarding software sharing, the Description of a Project vocabulary (DOAP, https://github.com/edumbill/doap/wiki) describes metadata about software projects. DOAP can be used in combination with other vocabularies to properly annotate pieces of code, APIs, and other components of open source projects. The usage of vocabularies allows for searching software with a functionality that meets certain criteria.

Knowledge Creation and Sharing

On the Social Web, knowledge sources are characterized as being created collaboratively. One of the most prominent tools in this context are wikis, which are used for creating and sharing knowledge in a wide variety of domains including all-purpose encyclopedias such as Wikipedia. Wikis can be augmented with Semantic Web technologies, as showcased by Semantic MediaWiki (https://semantic-mediawiki.org/), to annotate and describe concepts in wiki articles with semantic metadata. The combination of semantics and wikis can be exploited to create (semi-)structured data sets like DBpedia (http://dbpedia.org/) thus supporting complex queries, beyond plain-text keywords, over wiki content.

Other mechanisms of knowledge sharing on the Social Web are blogs, microblogs, forums, message boards, and mailing lists. In particular, blogs have been the subject of study for the application of semantic technology. Blog posts are chronological entries that can be used to publish

S

any type of information, can be commented on by users, and frequently reference other blogs. "Semantic blogging" means annotating posts with ontologies that allow for flexible exchange of blog metadata as well as for post linking, discovery, and recommendation [5]. Applications of semantic blogging include named-entity extraction from post contents to enrich the blogging experience.

Social Tagging

Social tagging is one of the most common applications on the Social Web. Social or collaborative tagging refers to the annotation of content with keywords (or tags) by contributors, usually inside the boundaries of a themed online platform, e.g., Flickr.com for pictures. The resulting collective description of individual resources is an unordered, nonhierarchical collection of tags, often called "Folksonomy." Semantics can be added to tagging platforms to model user behavior or to extract the relationships (hierarchies) between concepts derived from the tags. The Social Semantic Cloud of Tags vocabulary (SCOT, http://rdfs.org/scot/spec/) can be used to describe the structure, relationships, and other information related to the creation of tags across different platforms in the Semantic Social Web. Some examples of semantic social tagging applications include Revyu (http://revyu.com/), a platform for providing reviews about any topic, e.g., books, movies, restaurants, etc., and Faviki (http://www.faviki.com/), a bookmarking tool that utilizes concepts available in Wikipedia which allows for attaching tags with meaning and performing disambiguation of terms.

Social Networks

In the past few years, the Web has seen a proliferation of online social networking services (OSNS) like Facebook, Google Plus, or LinkedIn. Social networks foster the generation of social graphs encoding the relationships among OSNS users.

Generally, OSNS precludes users from exporting or importing their personal social graph and hence forcing them to introduce duplicate data across different OSNS. There are currently several initiatives to support data portability among OSNS, for instance, WebID, the Open Graph protocol, and the OpenSocial standards, all employing semantic representations of the social graph data. WebID (http://www.w3.org/wiki/WebID) is a community group at the W3C which aims at providing a unique general identifier to agents on the Web, e.g., persons, groups, or organizations. WebID provides an identification mechanism that enables agents to control their identity on the Web. Agents can create WebID profiles using vocabularies like FOAF to represent personal data. The WebID protocol is designed to support secure authentication over several web sites. The Open Graph protocol (http://ogp.me/) provides a set of core concepts that allows for representing objects within a web page as entities in the social graph. The Open Graph protocol is currently consumed by Facebook and Google and published by IMDb, Microsoft, TIME, and others. OpenSocial (http://opensocial.org/) is a set of standards that enables portability of social applications, which are packaged in gadgets and embedded in other applications that can be deployed in several social platforms.

Interlinking of Online Communities

One of the main purposes of the Semantic Social Web is providing the means to interlink data that is available in different social platforms. Data within online communities on the Social Web can be linked in different ways, for example, unifying user accounts over different social websites or integrating shared content among platforms.

The Semantically-Interlinked Online Communities initiative (SIOC, http://www.sioc-project.org/) aims at connecting online content from social platforms. This project provides a lightweight ontology for modeling the structure and the participation of users within communities in the Social Web. The SIOC ontology can be used in combination with other vocabularies for [1] integration of distributed posts, interlinking related posts or online discussions, creation of unified communities, definition of content topic and social tagging, and unification of user accounts.

Cross-References

Recommended Reading

1. Breslin J, Passant A, Decker S. The semantic social web. Berlin/Heidelberg: Springer; 2009.
2. Breslin J, Passant A, Vrandečić D. Semantic social web. In: Domingue J, Fensel D, Hendler JA, editors. Handbook of semantic web technologies. Springer: Berlin/Heidelberg; 2011. p. 467–506.
3. Maccioni A. Towards an integrated semantic social web. In: Sheng QZ, Kjeldskov J, editors. Current trends in web engineering. Berlin: Springer; 2013. p. 207–14.
4. Obrenovic Z. Multimedia semantics: overview of relevant tools and resources. In: W3C Multimedia Semantics Incubator Group Wiki, Tools and Resources. 2005. http://www.w3.org/2005/Incubator/mmsem/wiki/Tools_and_Resources. Accessed 11 Aug 2014.
5. Cayzer S. Semantic blogging and decentralized knowledge management. In: Communications of the ACM – the blogosphere. New York: ACM; 2004. p. 47–52. https://doi.org/10.1145/1035134.1035164.
6. Weller K. Knowledge representation in the semantic social web. Berlin/New York: Walter de Gruyter; 2010.

Semantic Streams

Manfred Hauswirth[1,2], Danh Le Phuoc[1], and Josiane Xavier Parreira[3]
[1]Open Distributed Systems, Technical University of Berlin, Berlin, Germany
[2]Fraunhofer FOKUS, Galway, Germany
[3]Siemens AG, Galway, Austria

Synonyms

Linked streams; RDF streams

Definition

A semantic stream S is an unbounded partially ordered set of tuples $\langle G, \tau \rangle$ where G is a directed labeled graph that follows a semantic data model and the values of τ define a partial order among the tuples. In existing semantic stream models, stream elements are semantically annotated following the W3C Resource Description Framework (RDF) semantic data model, i.e., G is a set of RDF triples. Typical examples for τ would be integers (to define a simple ordering relationship, e.g., logical time), timestamps, coordinates, intervals, or combinations of those.

Historical Background

The heterogeneous nature of stream data sources makes accessing and managing their data a labor-intensive task, which currently requires a lot of manual programming and data integration. To remedy these issues, the RDF data model, based on its success for information integration on the Web, was investigated for its suitability for stream processing, as it enables the expression of knowledge in a generic manner, without requiring adherence to a specified schema a priori. Together with formal semantics, many data integration tasks can be automated. Pioneering works on semantic descriptions of stream data and stream data sources in different scientific communities [1–3] and the Open Geospatial Consortium (http://www.opengeospatial.org/) (OCG) have been integrated in the (extensible) ontology of the W3C Semantic Sensor Network Incubator Group (http://www.w3.org/2005/Incubator/ssn/) (SSN-XG) [4] which is the de facto standard at the moment (albeit not being a formal one). On the query language side, the W3C RDF Stream Processing Community Group (RSP-XG) [5] is in the process of defining a common model for producing, transmitting, and continuously querying RDF streams. This includes extensions to both RDF and SPARQL for representing and querying streaming data.

Initially, semantic stream query processors focused on data models and algebras, e.g.,

S

streaming SPARQL [6], and did not meet high standards in terms of throughput and scalability. Recent implementations, however, have significantly improved on throughput and scalability. Conceptually two classes of approaches were investigated based on, namely, logic programming (Datalog), e.g., EP-SPARQL [7], and databases (SPARQL), e.g., C-SPARQL [8].

Scientific Fundamentals

Traditionally, semantic data management has focused on classical data management areas such as relational database systems. As a typical example, consider the linked data cloud (http://lod-cloud.net/), in which RDF (**R**esource **D**escription **F**ramework) datasets can be queried through SPARQL (**SPARQL** **P**rotocol **a**nd **R**DF **Q**uery **L**anguage is a query language for data represented in RDF format.) endpoints (services that accept SPARQL queries and return results). Data points in these datasets may refer to other data points in the same or other datasets using URIs (**U**niform **R**esource **I**dentifiers), creating a labeled, directed multigraph of subject-predicate-object relationships (*triples*). By using RDFS (**R**esource **D**escription **F**ramework **S**chema is a set of classes with certain properties providing basic elements for the description of ontologies (RDF vocabularies)) and/or OWL (**W**eb **O**ntology **L**anguage (OWL) is a family of knowledge representation languages for authoring ontologies), formal semantics can be assigned to this graph.

Semantic streams extend this static data model with a time dimension in a similar way as relational stream models have extended the classical relational data and processing models. Semantic streams differ from relational streams in the following ways:

1. Graphs, more specifically RDF graphs, are the underlying data model instead of relations.
2. Graphs can change over time in relation to data values and structure, i.e., both the data and the schemas can change over time.

3. Inherently the data model uses an open-world assumption, i.e., the schema is not controlled.
4. The vast majority of semantic streams include both a static and a dynamic data component, calling for an integrated data management approach.

The main goals of semantic streams are:

1. integrated, "schema-less" management of data from heterogeneous stream data sources;
2. efficient, integrated management of stream data and static data;
3. making stream data available according to the linked data principles (http://www.w3.org/DesignIssues/LinkedData.html).

To illustrate the basic concepts of semantic streams, Fig. 1 provides an example of a semantic stream element that carries semantic metadata about the sensor from which the stream originates. The example uses the vocabulary provided by the Semantic Sensor Network (SSN) Ontology [4]. A semantic stream is divided into two layers: sensor metadata (upper level) and stream data (lower level). The sensor metadata corresponds to time-independent information, which is static (although in principle it could be dynamic too). It captures the context in which the sensor readings are obtained. In this example, the sensor metadata describes a "**weather station** that *observes* the **temperature** and **humidity** at *Dublin Airport*." The stream data contains the dynamic, graph-based stream data from the sensor readings over time. These readings have links to their meanings, e.g.*"tempValue* (**18 Celsius**) is the *temperature* of *Dublin Airport* at **21:32:52, 09/08/2011**."

To capture semantic data as shown in the above example, the semantic stream is realized as an RDF stream by extending the definitions of RDF nodes and RDF triples [5]. Stream elements of an RDF stream are represented as RDF graphs with temporal annotations. In some simple implementations [7–10], each RDF graph contains only one RDF triple. The temporal annotation defines the order among the semantic stream elements. A temporal annotation of an RDF triple can be an interval-based or point-based label.

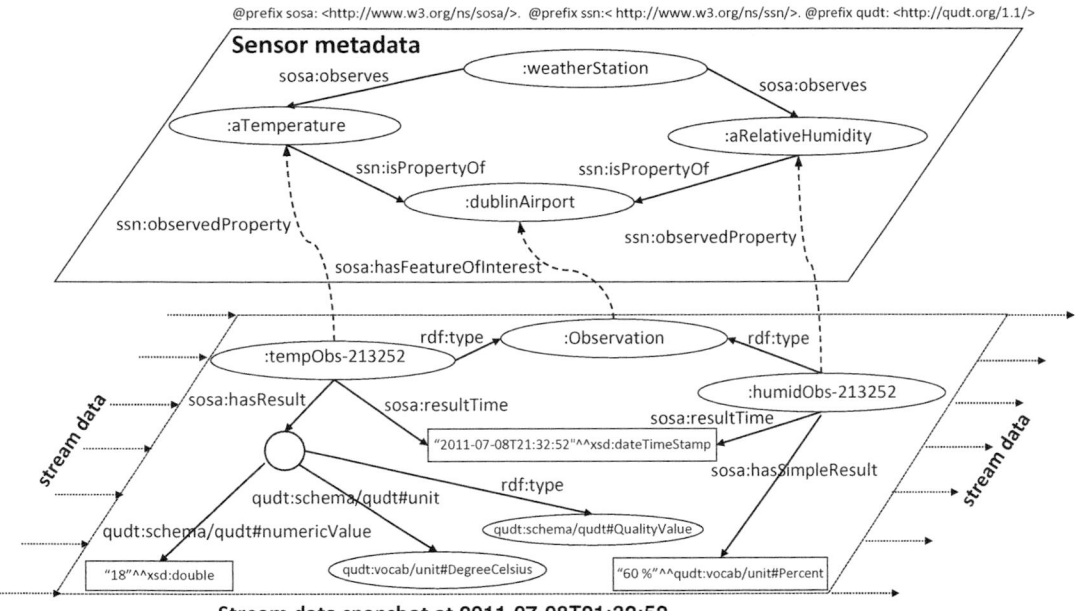

@prefix sosa: <http://www.w3.org/ns/sosa/>. @prefix ssn:< http://www.w3.org/ns/ssn/>. @prefix qudt: <http://qudt.org/1.1/>

Stream data snapshot at 2011-07-08T21:32:52

Semantic Streams, Fig. 1 Semantic stream example: the sensor metadata defines that the stream source is a weather station that observes temperature and humidity at Dublin Airport

An interval-based label is a pair of timestamps $[start, end]$ which are usually natural numbers representing logical time, specifying the interval in which the RDF triple is valid. For instance, $\langle :John \; :at \; :office, \; [7, \; 9] \rangle$ represents the fact that John was in the office from (logical times) 7 to 9.

A point-based label is a single natural number representing the point in (logical) time that the stream element was recorded or received. In the previous example, the triple $\langle :John \; :at \; :office \rangle$ might be continuously recorded by a tracking system, so three temporal triples are generated $\langle :John \; :at \; :office, \; 7 \rangle$, $\langle :John \; :at \; :office, \; 8 \rangle$, and $\langle :John \; :at \; :office, \; 9 \rangle$. Point-based labels can be considered a special case of interval-based labels where $start = end$.

The choice of the type of temporal annotation largely depends on the application and the type of data. Interval-based labels facilitate a more compact representation and fewer stream elements, whereas point-based labels enable the processing to immediately react to newly arriving stream elements.

Streaming SPARQL [6] uses interval-based labels for representing its physical data stream elements and EP-SPARQL [7] uses them for representing triple-based events. While point-based labels may look redundant and less efficient in comparison to interval-based ones, they are more practical for stream data sources because triples are generated unexpectedly and instantaneously, i.e., their validity may often not be known a priori. For example, a tracking system detecting people in an office can easily generate a triple with a timestamp whenever it receives a sensor reading and process it. If interval-based labels were to be used, the system would have to buffer the readings and do some further processing in order to infer the interval that the triple could be valid. Moreover, the instantaneity of processing is vital in some applications that need to process the data as soon as it arrives in the system. For instance, an application that notifies a supervisor where John is located at the moment should be triggered at time 7 in the above example and should not have to wait until time 9 to report that he was in the office from 7 to 9. Point-based labels are

S

supported in C-SPARQL [8], SPARQL$_{\text{stream}}$ [9], and CQELS [10].

Similar to Data Stream Management Systems (DSMS), query operators on semantic streams can be classified into three types: stream to graph, graph to graph, and graph to stream operators. A sliding window is a stream to graph operator which extracts a bounded multi-set of graphs from a semantic stream S by applying a function on the time annotations of the graph stream elements of S. Graph to stream operators take a directed labeled graph G as input and produce a semantic stream element (G, τ), by adding a temporal annotation. Graph to graph operators, such as pattern matching, correspond to the traditional query operators used in graph databases. For defining the semantics of continuous query languages, current state-of-the-art approaches extend the query operators of SPARQL such as join, union, and filter. As such operators consume and output mappings, these approaches also introduce operators on RDF streams to output mappings. For instance, in C-SPARQL, the stream operator is defined to access an RDF stream identified by its IRI (IRIs are a generalization of URIs [RFC3986] that permit a wider range of Unicode characters) and the window operator is defined to extract an RDF graph from an RDF stream based on a certain window. The definitions of window operators on RDF streams are adopted from window operators on relational streams as defined for CQL [11]. The semantics of a continuous query on RDF streams are defined as a composition of the query operators. In streaming SPARQL and C-SPARQL, a query is composed as an operator graph.

There are a number of query languages and processing engines for semantic streams in the literature. At the time of this writing, there exists no agreement on a common language and processing framework. Each language supports different expressivity, and the query processing depends on the internal implementation of semantic stream access methods and operators.

Architecturally, the current approaches are based on either blackbox architectures combining a query processing engine for static data with a stream query processor for data streams, e.g., C-SPARQL, or whitebox architectures building an integrated, optimized query processor from scratch. The advantage of the blackbox architecture is that existing engines can be reused, i.e., be combined and extended, but this comes at significant performance losses, as optimization can only be done within the engines but not across them, where most of the performance would be gained [12] and the architecture uses subcomponents which were initially not designed for semantic stream processing which causes additional overheads due to the need for data transformation, query rewriting, and orchestration among the subcomponents. This problem is avoided by the significantly more scalable and performant whitebox approaches at the cost of higher implementation efforts.

The above basic principles have been implemented in current semantic stream processing engines based on a variety of underlying concepts:

Streaming SPARQL [6] was one of the first systems proposed for semantic stream data processing. It extends the SPARQL algebra for stream processing with a temporal relational algebra based on multi-sets. The focus was on the conceptual work (feasibility), and the system does not meet processing efficiency and scalability criteria and is only of interest in terms of being one of the first approaches.

C-SPARQL [8] combines an existing SPARQL query engine (SESAME) to evaluate time-invariant query parts with a relational stream processor (STREAM) in a blackbox approach. The C-SPARQL language is an extension of SPARQL introducing concepts necessary for stream processing such as windows, similar to the relational CQL.

EP-SPARQL [7] also takes a blackbox approach which is grounded in logic programming. It uses an existing logic engine in the background and the query processing is translated into logic programs. The execution mechanism of EP-SPARQL – ETALIS – is based on event-driven backward chaining (EDBC) rules. EP-SPARQL queries are compiled into EDBC rules, which enable timely, event-driven, and incremental detection of complex events, i.e., answers to EP-SPARQL queries. EDBC rules are logic rules and hence can be mixed with time-invariant background knowledge. Since EP-SPARQL is routed in logic programming, it provides the biggest expressivity of all semantic stream processors and supports higher-complexity stream reasoning (at the cost of performance compared to other processors).

SPARQL$_{stream}$ [9] aims at supporting the integration of heterogeneous relational stream data sources. The SPARQL$_{stream}$ engine rewrites a query described in the SPARQL$_{stream}$ query language to a relational continuous query language, e.g., SNEEql. In order to transform the SPARQL$_{stream}$ query, expressed in terms of the ontology, into queries in terms of the data sources, a set of mappings must be specified in S2O, an extension of the R2O mapping language, which supports streaming queries and data, most notably window and stream operators.

CQELS [10] is an adaptive approach for processing semantic streams implementing a whitebox architecture from scratch to provide an optimized query processor. It is based on native support for the linked data model and tries to incorporate all the relevant latest results of database research combined with new data structures and optimization algorithms tailored to the specific requirements of semantic streams. CQELS's performance rivals the performance and scalability of relational stream processing and complex event approaches. A CQELS version (CQELS Cloud) for cloud platforms exists to achieve scalability in hosted environments.

Key Applications

Semantic streams address a number of requirements for open environments such as the Web and the Internet of things, supporting easier discovery and integration of data sources. They enable a higher degree of self-management due to the integrated use of ontologies and semantic descriptions for data, data sources, and the producing and consuming systems. Thus, they are being investigated in a number of areas where a large number of sources with no control of the data schemas must be used and be integrated

efficiently, for example, Web data management, data middleware, embedded systems, sensor networks, and data analytics. Typical application areas are analytics platforms for Smart Cities which integrate Web channels such as Twitter and social networks with embedded systems and sensing infrastructures in open platforms or building management where many heterogeneous (embedded) systems together with external participants (Web data, weather, building management systems) produce and consume information and act upon it. The potentially lower performance of semantic stream systems compared to solutions controlling the schemas and data production/consumption is compensated by the lower complexity of all forms of integration efforts.

Future Directions

The topic of semantic streams is a young research topic with a number of open research challenges. Generally, the scalability of systems must be increased. Currently only C-SPARQL and CQELS support distributed processing in cloud environments. Expressivity is another open challenge. Besides EP-SPARQL, no other system supports high-level temporal queries based on temporal logic. Also, support for spatial data management and query processing is in its infancy at the moment and will require significant research efforts. The vast body of existing approaches in the database world will have to be analyzed for applicability. Support for higher-level logical reasoning specific forms of analytics needs to be investigated (again, EP-SPARQL has some support for this) with specific focus on scalability of the reasoning tasks. Tailored rule-based languages and approaches, e.g., Answer Set Programming, are investigated for their applicability along with attempts to combine statistical with logical reasoning for higher scalability. Data quality, provenance, and correctness are major areas which are not well explored for semantic streams yet. A number of efforts are underway in the area of standardized performance tests (following the TPC example) to support better comparability of systems.

Experimental Results

Most of the semantic stream processing approaches found in the literature provide an experimental evaluation of their methods. The heterogeneity of their processing models as well as their evaluation setup (e.g., different input stream throughput, knowledge bases with different sizes, queries with varying complexity) prevents a direct performance comparison. The work in [13] was the first attempt to provide a detailed comparison of the different models of three semantic streams processing systems – C-SPARQL, CQELS, and JTALIS (a Java wrapper for ETALIS) – under a common evaluation framework. The differences in the operational semantics of the different systems are also subject of the work in [14], which suggests the need to include functional tests to verify correctness of evaluation results. There is a clear need for defining common semantic stream data models (which might be application specific) and the semantics of the different stream operators currently available. This is in fact the goal of the current W3C Community Group on RDF Stream Processing [5].

Datasets

The efforts in comparing and evaluating existing semantic stream processing systems have produced two benchmarks: SRBench [12] and LSBench [13]. SRBench provides a semantic stream dataset generated from real sensor measurements, as well as a number of time-invariant RDF datasets. The throughput rate of the stream is fixed and determined by the time annotation of the original sensor readings. LSBench provides the S2Gen Stream Social network data Generator, based on an ontology model derived in the context of social networks. S2Gen offers a number of parameters to control the dataset generation, from stream size to the skewness to the dataset.

URL to CODE

CQELS: http://cqels.org/

C-SPARQL: https://github.com/streamreasoning /CSPARQL-engine
EP-SPARQL/ETALIS: https://github.com/sspid er/etalis
SPARQLstream: https://github.com/jpcik/ morph-streams

Cross-References

▶ Ontology
▶ Resource Description Framework
▶ Semantic Web
▶ Stream Models
▶ W3C

Recommended Reading

See references below. References have been formatted according to the Vancouver style, and DOIs have been included in the BIBTEX source files but unfortunately are not shown by the Vancouver BIBTEX style used to generate the PDF.

1. Bouillet E, Feblowitz M, Liu Z, Ranganathan A, Riabov A, Ye F. A semantics-based middleware for utilizing heterogeneous sensor networks. In: Aspnes J, Scheideler C, Arora A, Madden S, editors. Proceedings of the 3rd IEEE International Conference on Distributed Computing in Sensor Systems; 2007. p. 174–188.
2. Sheth A, Henson C, Sahoo SS. Semantic sensor web. IEEE Internet Comput. 2008;12(4):78–83.
3. Whitehouse K, Zhao F, Liu J. Semantic streams: a framework for composable semantic interpretation of sensor data. In: Römer K, Karl H, Mattern F, editors. Proceedings of the 3rd European Workshop on Wireless Sensor Networks; 2006. p. 5–20.
4. Compton M, Barnaghi PM, Bermudez L, Garcia-Castro R, Corcho Ó, Cox S, et al. The SSN ontology of the W3C semantic sensor network incubator group. J Web Sem. 2012;17(Dec):25–32.
5. The W3C RDF Stream Processing Community Group; [cited May 2014]. Available from: http:// www.w3.org/community/rsp/.
6. Bolles A, Grawunder M, Jacobi J. Streaming SPARQL – extending SPARQL to process data streams. In: Bechhofer S, Hauswirth M, Hoffmann J, Koubarakis M, editors. Proceedings of the 5th European Semantic Web Conference on The Semantic Web: Research and Applications; 2008. p. 448–462.
7. Anicic D, Fodor P, Rudolph S, Stojanovic N. EP-SPARQL: a unified language for event processing and

stream reasoning. In: Srinivasan S, Ramamritham K, Kumar A, Ravindra MP, Bertino E, Kumar R, editors. Proceedings of the 20th International Conference on World Wide Web; 2011. p. 635–44.

8. Barbieri DF, Braga D, Ceri S, Grossniklaus M. An execution environment for C-SPARQL queries. In: Manolescu I, Spaccapietra S, Teubner J, Kitsuregawa M, Léger A, Naumann F, et al., editors. Proceedings of the 13th International Conference on Extending Database Technology; 2010. p. 441–52.

9. Calbimonte JP, Corcho Ó, Gray AJG. Enabling ontology-based access to streaming data sources. In: Patel-Schneider PF, Pan Y, Hitzler P, Mika P, Zhang L, Pan JZ, et al., editors. Proceedings of the 9th International Semantic Web Conference on the Semantic Web; 2010. p. 96–111.

10. Phuoc DL, Dao-Tran M, Parreira JX, Hauswirth M. A native and adaptive approach for unified processing of linked streams and linked data. In: Aroyo L, Welty C, Alani H, Taylor J, Bernstein A, Kagal L, et al., editors. Proceedings of the 10th International Semantic Web Conference; 2011. p. 370–88.

11. Arasu A, Babu S, Widom J. The CQL continuous query language: semantic foundations and query execution. The VLDB J. 2006;15(2):121–42.

12. Zhang Y, Duc PM, Corcho O, Calbimonte JP. SR-Bench: a streaming RDF/SPARQL benchmark. In: Cudré-Mauroux P, Heflin J, Sirin E, Tudorache T, Euzenat J, Hauswirth M, et al., editors. Proceedings of the 11th International Conference on The Semantic Web; 2012. p. 641–57. Available from: https://doi.org/10.1007/978-3-642-35176-1_40.

13. Phuoc DL, Dao-Tran M, Pham MD, Boncz PA, Eiter T, Fink M. Linked stream data processing engines: facts and figures. In: Cudré-Mauroux P, Heflin J, Sirin E, Tudorache T, Euzenat J, Hauswirth M, et al., editors. Proceedings of the 11th International Semantic Web Conference; 2012. p. 300–12.

14. Dell'Aglio D, Calbimonte JP, Balduini M, Corcho Ó, Valle ED. On correctness in RDF stream processor benchmarking. In: Alani H, Kagal L, Fokoue A, Groth PT, Biemann C, Parreira JX, et al., editors. Proceedings of the 12th International Semantic Web Conference; 2013. p. 326–42.

Semantic Web

Grigoris Antoniou and Dimitris Plexousakis
Foundation for Research and Technology-Hellas (FORTH), Heraklion, Greece

Definition

The central idea of the Semantic Web initiative is to enrich Web content by machine-processable semantics. The approach is based on the following ideas:

1. Use meta-data (data about data) as semantic annotations
2. Use ontologies to describe knowledge needed to understand collections of Web information. The semantic annotations are linked to such ontologies
3. Use logic-based techniques to process and query collections of meta-data and ontologies

In the current Semantic Web work, two main goals can be distinguished.

Interpretation 1: The Semantic Web as the Web of Data

In the first interpretation, the main aim of the Semantic Web is to enable the integration of structured and semi-structured data sources over the Web. The main recipe is to expose data-sets on the Web enriched with semantic annotations, to use ontologies to express the intended semantics of these data-sets, in order to enable the integration and unexpected re-use of these data.

A typical use case for this version of the Semantic Web is the combination of geo-data with a set of consumer ratings for restaurants in order to provide an enriched information source.

Interpretation 2: The Semantic Web as an Enrichment of the Current Web

In the second interpretation, the aim of the Semantic Web is to improve the current World Wide Web. Typical use cases here are improved search engines, dynamic personalization of Web sites, and semantic enrichment of existing Web pages.

The source of the required semantic meta-data in this version of the Semantic Web is mostly claimed to come from automatic sources: concept extraction, named-entity recognition, automatic classification, etc. More recently, the insight is gaining ground that the required semantic markup can also be produced by social mechanisms of communities that provide large-scale human-produced markup.

S

Historical Background

The Semantic Web sprang as a vision approximately 10 years after the birth of the World-Wide Web. Not surprisingly, it was the inventor of the WWW that shaped the vision of the Semantic Web, which in turn gave rise to the entire research field.

Up until that stage, the Web was (and still is to a great extent) purely about syntax, a specific syntax geared towards homogenizing the way in which information is presented to human users via a browser. As revolutionary as the concept may have been, it was making content available only for human consumption as the interpretation of the content relied on implicit semantics. In other words, meaningful representation of content was not possible in HTML or its precursor SGML. Information and its presentation were mixed in the form of HTML documents, many of which generated automatically by applications. The Web made it easy to fetch any Web page from any server, on any platform through a uniform interface. HTML has many benefits: it is simple, textual, portable, easily searchable by keyword-based search engines and connects pieces of information together through hypertext links. The browser is the universal application. If written properly, normal HTML markup may reflect document presentation, but it cannot adequately represent the semantics & structure of data. Newer applications require more than the publishing of HTML documents; data must be made available on the Web for use by Web-enabled applications.

XML was the incarnation of the paradigm shift on the Web: a new standard that could be easily generated and consumed by applications, facilitating data exchange across platforms and organizations, transforming the Web from a collection of documents to a collection of data published as documents. XML gained popularity very fast. It resembles HTML in that it is easy to read and learn, it is universal, portable and at the same time extensible and more flexible than HTML. However, XML cannot address all interoperability requirements as it only provides the means for solving syntactic heterogeneity problems. The challenge is to address the inherent structural but foremost semantic heterogeneities that are encountered on the Web. Modern applications need more than data on the Web; they need semantics on the Web. Applications themselves evolve into services on the Web that may exploit semantics.

The main motivation behind the Semantic Web (or Web of meaning) vision is to make vast amounts of information resources (data, documents, programs) available along with various kinds of descriptive information, i.e., metadata. Better knowledge about the meaning, usage, accessibility or quality of web resources considerably facilitates automated processing of available Web content/services especially when metadata are described in a form that is precise, human-readable and machine-interpretable. The Semantic Web enables syntactic and semantic/structural interoperability among independently-developed Web applications, allowing them to efficiently perform sophisticated tasks for humans. At the same time, it enables Web resources (data & applications) to be accessible by their meaning rather than by keywords and syntactic forms.

Foundations

The Semantic Web approach is based on the use of *semantic annotations* to describe the meaning of certain parts of Web information. For example, the Web site of a hotel could be suitably annotated to distinguish between hotel name, location, category, number of rooms, available services etc. Such metadata can facilitate the automated processing of the information on the Web site, thus making it accessible to machines.

However, the question arises as to how the semantic annotations of different Web sites can be combined, if everyone uses terminologies of their own. The solution lies in the organization of vocabularies in so-called *ontologies*. Recommended Reading to such shared vocabularies allow interoperability between different Web resources and applications. For example, an ontology of

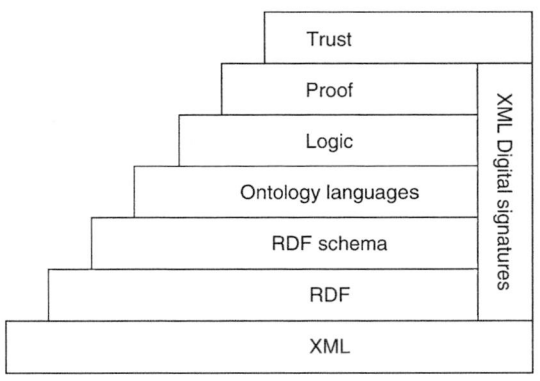

Semantic Web, Fig. 1 The semantic web tower

hotel classifications in a given country could be used to relate the rating of certain hotels. And a geographic ontology could be used to determine that Crete is a Greek island and Heraklion a city on Crete. Such information would be crucial to establish a connection between a requester looking for accommodation on a Greek island, and a hotel advertisement specifying Heraklion as the hotel location.

The development of the Semantic Web proceeds in steps, each step building a layer on top of another. The basic layered design is shown in Fig. 1, which is outlined below.

1. The bottom layer comprises *XML*, a language that lets one write structured Web documents with a user-defined vocabulary. XML is particularly suitable for sending documents across the Web, thus supporting syntactic interoperability

2. *RDF* (Resource Description Framework) is a basic data model, like the entity-relationship model, for writing simple statements about Web objects (resources). The RDF data model does not rely on XML, but RDF has an XML-based syntax. Therefore it is located on top of the XML layer

3. *RDF Schema* provides modeling primitives, for organizing Web objects into hierarchies. RDF Schema is based on RDF. RDF Schema can be viewed as a primitive language for writing ontologies

4. But there is a need for more powerful *ontology languages* that expand RDF Schema and allow the representations of more complex relationships between Web objects. Ontology languages, such as OWL (Ontology Web Language), are built on the top of RDF and RDF Schema

5. The *logic layer* is used to enhance the ontology language further and to allow writing application-specific declarative knowledge. Rule languages are the most popular logical languages used in Semantic Web applications

6. The *proof layer* involves the actual deductive process, as well as the representation and exchange of proofs in Web languages, for purposes such as explanation provision and proof validation

7. Finally, *trust* will emerge through the use of digital signatures, and other kind of knowledge, based on recommendations by agents that can be trusted, or rating and certification agencies and consumer bodies

RDF Basic Features

The language of RDF allows one to write statements. A statement consists of three parts (subject, predicate, object) and is often referred to as a triple. A triple of the form (x, P, y) corresponds to the logical formula $P(x, y)$, where the binary predicate P relates the object x to the object y; this representation is used for translating RDF statements into a logical language ready to be processed automatically in conjunction with rules.

There are other ways to describe an RDF document, using a graphical and an XML representation.

RDF Schema Basic Features

In RDF, Web resources are individual objects. In RDFS, objects sharing similar characteristics are put together to form *classes*. Examples for classes are hotels, airlines, employees, rooms, excursions etc. Individuals belonging to a class are often referred to as *instances* of that class. For example, John Smith could be an instance of the class of employees of a particular hotel.

Binary *properties* are used to establish connections between classes. For example, a property *works_for* establishes a connection between employees and companies. Properties apply to individual objects (instances of the classes involved) to form RDF statements, as seen above.

The application of predicates can be restricted through the use of *domain and range restrictions*. For example, the property *works_for* can be restricted to apply only to employees (domain restriction), and to have as value only companies (range restriction).

Classes can be put together in hierarchies through the *subclass relationship*: a class C is a subclass of a class D if every instance of C is also an instance of D. For example, the class of island destinations is a subclass of all destinations: every instance of an island destination (e.g., Crete) is also a destination.

The hierarchical organization of classes is important due to the notion of *inheritance*: once a class C has been declared a subclass of D, every known instance of C is *automatically* classified also as instance of D. This has far-reaching implications for matching customer preferences to service offerings. For example, a customer may wish to make holidays on an Indonesian island. On the other hand, the hotel Noosa Beach advertises its location to be Bali. It is not necessary (nor is it realistic) for the hotel to add information that it is located in Indonesia and on an island; instead, this information is inferred by the ontology automatically.

Key Applications

This section provides a bird's eye survey of key application areas. It should be noted that a healthy uptake of Semantic Web technologies is beginning to take shape in the following areas:

1. Knowledge management, mostly in intranets of large corporations
2. Data integration (Boeing, Verison and others)
3. e-Science, in particular the life-sciences
4. Convergence with Semantic Grid

If one considers the profiles of companies active in this area, they will see a distinct transition from small start-up companies such as Aduna, Ontoprise, Network Inference, Top Quadrant (to name but a few) to large vendors such as IBM (their Snobase ontology Management System), HP (with their popular Jena RDF platform), Adobe (with their RDF-based based XMP meta-data framework), and Oracle (now lending support for RDF storage and querying in their prime database product).

However, besides the application areas listed above, there is also a noticeable lack of uptake in some other areas. In particular, promises in the areas of

1. e-commerce
2. Personalization
3. Large-scale semantic search (on the scale of the World Wide Web, not limited to intranets),
4. Mobility and context-awareness

are largely unfulfilled, though there is significant ongoing activity in these directions.

A pattern that seems to emerge between the successful and unsuccessful application areas is that the successful areas are all aimed at *closed communities* (employees of large corporations, scientists in a particular area), while the applications aimed at the general public are still in the laboratory phase at best. The underlying reason for this could well be the difficulty of dealing with multiple ontologies and mappings among them.

Future Directions

At present, Semantic Web research focuses, among others, on:

1. Rule languages and their interaction or integration with ontology languages (RDF and OWL)
2. Scalable storage and retrieval systems
3. Knowledge and ontology evolution and change

4. Mapping mechanisms between different ontologies

A number of items on the research agenda are hardly tackled, but do have a crucial impact on the feasibility of the Semantic Web vision. In particular:

1. The mutual interaction between machine-processable representations and the dynamics of social networks of human users
2. Mechanisms to deal with trust, reputation, integrity and provenance in a (semi-) automated way
3. Inference and query facilities that are sufficiently robust to work in the face of limited resources (be it either computation time, network latency, memory or storage space), and that can make intelligent trade-off decisions between resource use and output-quality

Cross-References

▶ Interoperation of NLP-based Systems with Clinical Databases
▶ Ontology
▶ OWL: Web Ontology Language
▶ Resource Description Framework
▶ Resource Description Framework (RDF) Schema (RDFS)

Recommended Reading

1. Antoniou G, van Harmelen F. A semantic Web primer. 2nd ed. Cambridge: MIT Press; 2008.
2. Staab S, Studer R, editors. Handbook on ontologies. 2nd ed. New York: Springer; 2008.
3. Berners-Lee T, Hendler J, Lassila O. The semantic Web. Sci Am. 2001;284(5):34–43.
4. REASE. Available at: ubp.l3s.uni-hannover.de/ubp.
5. www.SemanticWeb.org.
6. www.w3.org/2001/sw/.
7. www.ontology.org.
8. The International Semantic Web Conference (http://iswc.semanticweb.org/).
9. Journal of Web Semantics (www.elsevier.com/locate/websem).

Semantic Web Query Languages

James Bailey[1], François Bry[2], Tim Furche[2], and Sebastian Schaffert[3]
[1]University of Melbourne, Melbourne, VIC, Australia
[2]University of Munich, Munich, Germany
[3]Salzburg Research, Salzburg, Austria

Synonyms

Ontology query languages; Web query languages

Definition

A number of formalisms have been proposed for representing data and meta data on the Semantic Web. In particular, RDF, Topic Maps and OWL allow one to describe relationships between data items, such as concept hierarchies and relations between the concepts. A key requirement for the Semantic Web is integrated access to data represented in any of these formalisms, as well the ability to also access data in the formalisms of the "standard Web," such as (X)HTML and XML. This data access is the objective of *Semantic Web query languages*. A wide range of query languages for the Semantic Web exist, ranging from (i) pure "selection languages" with only limited expressivity, to fully-fledged reasoning languages, and (ii) from query languages restricted to a certain data representation format, such as XML or RDF, to general purpose languages that support multiple data representation formats and allow simultaneous querying of data on both the standard and Semantic Web.

Historical Background

The importance of Semantic Web query languages can be traced back to the roots of the Semantic Web itself. In its original conception, Tim Berners-Lee viewed the Semantic Web as allowing Web-based systems to take advantage of "intelligent" reasoning capabilities [4]:

The Semantic Web will bring structure to the meaningful content of Web pages, creating an environment where software agents roaming from page to page can readily carry out sophisticated tasks for users. For the Semantic Web to function, computers must have access to structured collections of information and sets of inference rules that they can use to conduct automated reasoning.

As the representation format for the Semantic Web has grown to cover XML, RDF, Topic Maps and OWL, there has been a corresponding growth in query languages that support access to each of these kinds of data.

Foundations

A number of techniques have been developed to facilitate powerful data retrieval on the Semantic Web. This article follows the classification and taxonomy given in [1], which provides a comprehensive survey of the area. Several categories of query languages can be distinguished, according to the format of the Semantic Web data they can retrieve:

1. Query languages for XML
2. Query languages for Topic Maps
3. Query languages for RDF
4. Query languages for OWL

XML Query Languages: Although not a primary format, it is possible to specify information on the Semantic Web using XML. Hence query languages for XML are applicable to Semantic Web data. Most query and transformation languages for XML specify the structure of the data to retrieve using either of two approaches. In the navigational approach, path-based queries over the XML data are specified and the W3C standardized languages XPath, XSLT and XQuery are well known instances of this scheme. In the example based approach, query patterns are specified as "examples" of the XML data to be retrieved. Languages of this kind are mainly

research languages, with some well known representatives being XML-QL [7] and Xcerpt [3, 15].

Topic Maps Query Languages: Several different query languages for Topic Maps data exist, with representatives being tolog [9], AsTMA [2] and Toma [11]. tolog was selected as the initial straw man for the ISO Topic Maps Query Language and is inspired from logic programming, also having SQL style constructs. AsTMa is a functional query language, in the style of XQuery, whereas Toma combines both SQL syntax and path expressions for querying.

RDF Query Languages can be grouped into several families, that differ in aspects such as data model, expressivity, support for schema information, and type of queries. Principal among these families is the "SPARQL Family." This originated with the language SquishQL [12], which evolved into RDQL [12] and then was later extended to the language SPARQL [14]. These languages all "regard RDF as triple data without schema or ontology information unless explicitly included in the RDF source." SPARQL currently has W3C Candidate Recommendation status as being the "Query Language for RDF." In particular, SPARQL has facilities to:

1. Extract RDF subgraphs
2. Construct a new RDF graph using data from the input RDF graph queried
3. Return "descriptions" of the resources matching a query part
4. Specify optional triple or graph query patterns (i.e., data that should contribute to an answer if present in the data queried, but whose absence does not prevent an answer being returned 5. Test the absence, or non-existence, of tuples. The general format of a SPARQL query is:

Another family of languages for RDF, the "RQL family," consists of the language RQL [10], and its extensions such as SeRQL [5]. Common to this family is support for the combination of both data and schema querying. The RDF data model which is used slightly deviates from the standard data model for RDF and RDFS, disallowing cycles in the subsumption hierarchy and requiring both a domain and a range to be defined for each property. RQL itself has a large number of features and choices in syntactic constructs. This results in a complex, yet powerful language, which is far more expressive than other RDF query languages, especially those of the SPARQL family.

A number of other types of query languages for RDF also exist, using alternative paradigms. These include query languages using reactive rules, such as Algae [13] and deductive languages such as TRIPLE [6] and Xcerpt [15, 3]. The last of these is noteworthy, as it combines querying on both the Standard Web (HTML/XML), with querying on the Semantic Web (e.g. RDF, TopicMaps) and also allows pattern-based, incomplete specification of queries.

OWL Query Languages: Query languages for OWL are still in their infancy compared to those for RDF. OWL-QL [8] is a well known language for querying OWL data and is an updated version of the DAML Query language. Its design targets the assistance of query-answering dialogues between computational agents on the Semantic Web. Unlike the RDF query languages, it focuses on the querying of schema rather than instance data. An RDF language such as SPARQL may of course be used to query OWL data, but it is not well suited to the task, since it is not designed to be aware of OWL semantics.

Several themes emerge from considering the design of the various Semantic Web Query languages [1].

- *Choice of querying paradigm*: Semantic Web query languages express basic queries using either the path based (navigational) or logic based (positional) paradigm.
- *Choice of variable type*: When Semantic Web query languages have variables, they almost always are logical variables, as opposed to variables in imperative programming languages.
- *Provision of Referential Transparency and Answer-Closure*. Referential Transparency (i.e., within the same scope, an expression always means the same), a well known trait of declarative languages, is striven for by Semantic Web query languages. Answer

closedness is a property that allows answers to queries themselves to be used as input to queries and is a key design principle of the languages SPARQL and Xcerpt.
- *Degree of Incompleteness*: Many Semantic Web query languages offer a means for incomplete specifications of queries, a reflection of the semi-structured nature of data on the Semantic Web.
- *Reasoning Capabilities*. Interestingly, but not surprisingly, not all XML query languages have views, rules, or similar concepts allowing the specification of other forms of reasoning. Surprisingly, the same holds true of RDF query languages. Many authors of RDF query languages see deduction and reasoning to be a feature of an underlying RDF store offering materialization, i.e., completion of RDF data with derivable data prior to query evaluation. This is surprising, because one might expect many Semantic Web applications to access not only one RDF data store at one Web site, but instead many RDF data stores at different Web sites and to draw conclusions combining data from different stores.

Key Applications

Like classical query languages such as SQL, the first key application of Semantic Web query languages is the efficient and scalable access, classification, analysis and transformation of large collections of data in a Web format such as XML, RDF, OWL, or Topic Maps. Whereas classical query languages are most often used for accessing a single, centralized database, Semantic Web query languages need to be able to access also remote databases and data sources. This opens up new application scenarios, potentially utilizing any of the vast number of the data sources available on the Web.

For example, one might query researcher and publication information integrated over various sources, such as DBLP, Citeseer, IEEE and Cordis, combine that data with course and lecturer information from the Semantic Web School and then even further correlate it with the

S

US census data. All these resources would be far too large to download individually and query locally, but they provide interfaces known as *endpoints*, that can be used to select the relevant portions via a Semantic Web query interface. Another example application is the W3C Amaya browser, which can be used to enrich Web pages visited by a user, with annotations contained in remote data sources. The annotations relevant to a given Web page are accessed by querying an annotation server using Algae [13], an RDF query language similar to SPARQL. In such scenarios, the ability of RDF (and to some extent, XML) to define the names and concepts used in a database, reason about them and to map them to names and concepts used in another database, is essential. This clearly separates the use of Semantic Web query languages from the use of classical query languages for centralized databases.

Increasingly, current Web applications (often referred to as Web 2.0 applications) contain a Javascript-based user interface which is separate from the data processed by the application itself. Thus, the user interface can be loaded once and data then requested from the origin server or other data sources on the Web as required. Web query languages for XML, RDF, JSON and Topic Maps are now becoming recognized as the ideal interfaces between the client user interfaces of Web 2.0 applications and data sources, since they can target just the data that is needed in the current state of the application. Web query languages allow flexible, but fine-grained access to the required data, rather than the coarse-grained access provided by other solutions.

Future Directions

Most RDF query languages are RDF-specific, and even specifically designed for one RDF serialization, which of course limits their applicability. It is to be hoped that in the future, there will be an evolution towards data format "versatile" languages, capable of easily accommodating XML, RDF, Topic Maps and OWL, without requiring "serialization consciousness" from the programmer.

PREFIX	Specification of a name for a URI (like RDQL's USING)
SELECT	Returns all or some of the variables bound in the WHERE clause
CONSTRUCT	Returns a RDF graph with all or some of the variable bindings
DESCRIBE	Returns a "description" of the resources found
ASK	Returns whether a query pattern matches or not
WHERE	list, i.e., conjunction of query (triple or graph) patterns
OPTIONAL	list, i.e., conjunction of optional (triple or graph) patterns
AND	boolean expression (the filter to be applied to the result)

The method of query evaluation in current Semantic Web query languages is either backtracking-free logic programming (as used by positional languages) or set-oriented functional query evaluation. It seems likely these two paradigms may converge in future Semantic Web query languages. Language engineering issues, such as abstract data types and static type checking, modules, polymorphism, and abstract machines, have not yet made their way into Semantic Web query languages, as they did not in database query languages. This situation opens avenues for promising research of great practical, as well as theoretical relevance.

Data Sets

There are a number of SPARQL endpoints that can be browsed on the Web. These provide RDF data which can be viewed and then queried using a SPARQL client:

- The 2000 US Census Data endpoint: http://www.rdfabout.com/demo/census/
- The Semantic Web School endpoint: http://sparql.semantic-web.at/
- A compilation of endpoints including DBLP, Citeseer, IEEE and Cordis: http://www.rkbexplorer.com/

A collection of concrete query language use cases for accessing RDF data can be found in the W3C RDF Use Case document at http://www.w3.org/TR/rdf-dawg-uc/. A use case collection is also included in [2].

URL to Code

The D2R Server is a utility for publishing relational databases on the Semantic Web and can be found at: http://sites.wiwiss.fu-berlin.de/suhl/bizer/d2r-server/

Annotea is a project that aims to assist collaboration via shared semantic meta-data. The Annotea-Server with Amaya Browser and Algae QL can found at: http://www.w3.org/2001/Annotea/

Cross-References

▶ Ontology
▶ OWL: Web Ontology Language
▶ Resource Description Framework
▶ Resource Description Framework (RDF) Schema (RDFS)
▶ Semantic Web
▶ Topic Maps
▶ XML
▶ XPath/XQuery
▶ XSL/XSLT

Recommended Reading

1. Bailey J, Bry F, Furche T, Schaffert S. Web and semantic web query languages: a survey. In: Reasoning web, LNCS 3564. Springer; 2005. p. 35–133.
2. Barta R. AsTMa 1.3 language specification. Technical report, Bond University. 2003.
3. Berger S, Bry F, Furche T, Linse B, Schroeder A. Beyond XML and RDF: the versatile Web query language Xcerpt. In: Proceedings of the 15th International World Wide Web Conference; 2006. p. 1053–4.
4. Berners-Lee T, Hendler J, Lassila O. The semantic web-a new form of web content that is meaningful to computers will unleash a revolution of new possibilities. Sci Am. 2001; 29–37.
5. Broekstra J, Kampman A. SeRQL: a second generation RDF query language. In: Proceedings of the

SWAD-Europe Workshop on Semantic Web Storage and Retrieval; 2003.
6. Decker S, Sintek M, Billig A, Henze N, Dolog P, Nejdl W, Harth A, Leicher A, Busse S, Ambite JL, Weathers M, Neumann G, Zdun U. TRIPLE – an RDF rule language with context and use cases. In: Proceedings of the Rule Languages for Interoperability; 2005.
7. Deutsch A, Fernandez M, Florescu D, Levy A, Suciu D. A query language for XML. Comput Netw. 1999;31(11–16):1155–69.
8. Fikes R, Hayes P, Horrocks I. OWL-QL – a language for deductive query answering on the semantic Web. J Web Semant. 2004;2(1):19–29.
9. Garshol LM. Tolog – a topic maps query language. In: Proceedings of the 1st International Workshop on Topic Maps Research and Applications; 2005. p. 183–96.
10. Karvounarakis G, Magkanaraki A, Alexaki S, Christophides V, Plexousakis D, Scholl M, Tolle K. Querying the Semantic Web with RQL. Comput Netw ISDN Syst J. 2003;42(5):617–40.
11. Lacher M, Decker S. RDF, topic maps, and the semantic web. Markup Lang Theory Pract. 2001;3(3):313–31.
12. Miller L, Seaborne A, Reggiori A. Three implementations of SquishQL, a simple RDF query language. In: Proceedings of the International Semantic Web Conference; 2002. p. 423–35.
13. Prud'hommeaux E. Algae RDF Query Language. http://www.w3.org/2004/05/06-Algae/, 2004.
14. Prud'hommeaux E, Seaborne A. SPARQL query language for RDF. Candidate recommendation, W3C, June 2007, http://www.w3.org/TR/rdf-sparql-query/.
15. Schaffert S, Bry F. Querying the Web reconsidered: a practical introduction to Xcerpt. In: Proceedings of the Extreme Markup Languages; 2004.

Semantic Web Services

David Martin
Nuance Communications, Sunnyvale, CA, USA

Definition

Semantic Web Services (SWS) is a research area developing theory, technology, standards, tools, and infrastructure for working with distributed, networked services. As its name indicates, SWS has arisen from the cross-fertilization of challenges and approaches from the Web services and Semantic Web areas. The central theme of SWS is the enrichment of Web services technology with knowledge representation and reasoning technologies (including but not limited to those associated with the Semantic Web). The starting point for most SWS approaches is the use of expressive, declarative descriptions of the elements of dynamic distributed computation, with a particular focus on services, processes that are encapsulated by services, and message-based conversations between service providers and consumers. (Depending on the approach, other relevant concepts might include goals, transactions, roles, commitments, mediators of various kinds, etc.) These descriptions, in turn, are seen as the basis for fuller, more flexible automation of service provision and use, and the construction of more powerful components, architectures, tools and methodologies for working with services. In most cases, descriptions are expressed in a formal logical framework allowing for the use of well-understood reasoning procedures.

Many SWS researchers have articulated a broad and ambitious long-term vision of a Web where support for shared *activities* is as central as support for shared *information*. Many view SWS, developed to its full potential, as a technology foundation for distributed autonomous agents (and much SWS work draws on earlier work on agent-based systems). Another important theme in SWS is the development of a unified, comprehensive representation framework (often making use of ontologies) that can provide a foundation for a broad range of activities throughout the Web service lifecycle, including design and development, publication in registries, discovery and selection, negotiation and contracting, composition of services, monitoring and recovery from failure, and so forth.

long thereafter, including the Web Services Modeling Ontology (WSMO) [4], the Semantic Web Services Framework (SWSF) [2], WSDL-S [1], and the Internet Reasoning Service [3]. Many individual researchers and small teams have also done much valuable work, sometimes drawing on one of these larger efforts, sometimes not.

A fair amount of work in SWS has been focused on two central problems. Given a service request and a collection of service descriptions, *service discovery* is the problem of identifying those services that can satisfy the request, and possibly ranking them according to some measure of suitability. Given a goal to be satisfied and a collection of service descriptions, *service composition* is the problem of finding a procedure composed of service invocations that will achieve that goal. It should be emphasized, however, that SWS is a broad field with many challenging problems, of which these two are mentioned as illustrations.

Important application areas for SWS have included business (e.g., automated or partially automated discovery and use of needed services, enactment and composition of business processes and workflow, supply chain management, contracting, formation of virtual organizations, etc.), e-Government, and e-Science.

A few SWS standards activities have occurred. For example, the World Wide Web Consortium (W3C) has published a set of extensions to the Web Services Description Language (WSDL), known as Semantic Annotations for WSDL (SAWSDL), which makes it possible to associate elements of WSDL specifications with elements defined in a SWS framework (not defined by SAWSDL). W3C has also hosted workshops and study groups to consider the suitability of various aspects of SWS for standardization.

Key Points

SWS research, as a distinct field, began in earnest in 2001. In that year, the initial release of OWL for Services (OWL-S) [5] became available. Other major initiatives began not

Cross-References

▶ Semantic Web
▶ Web Services

Recommended Reading

1. Akkiraju R, Farrell J, Miller J, et al. Web service semantics – WSDL-S, vol. 1.0, tech. note, Apr. 2005. At http://www.w3.org/Submission/WSDL-S/.
2. Battle S, Bernstein A, Boley H, et al. Semantic Web Services Framework (SWSF) Overview, 2005. At http://www.w3.org/Submission/SWSF/.
3. Cabral L, Domingue J, Galizia S, et al. IRS-III: a broker for semantic web services based applications. In: Proceedings of the 5th International Semantic Web Conference; 2006. p. 201–14.
4. Fensel D, Lausen H, Polleres A, et al. Enabling semantic Web services: the Web service modeling ontology. New York: Springer; 2006.
5. Martin D, Burstein M, McDermott D, et al. Bringing semantics to web services with OWL-S. World Wide Web J. 2007;10(3):243–77.

Semantics-Based Concurrency Control

Krithi Ramamritham[1] and Panos K. Chrysanthis[2]
[1]Department of Computer Science and Engineering, Indian Institute of Technology Bombay, Mumbai, India
[2]Department of Computer Science, University of Pittsburgh, Pittsburgh, PA, USA

Definition

Specifications of data contain semantic information that can be exploited to increase concurrency. For example, two insert operations on a multiset object commute and hence, can be executed in parallel; further, regardless of whether one operation commits, the other can still commit. Applying the same rule, two push operations on a stack object do not commute and hence cannot be executed concurrently. Several schemes have been proposed for exploiting the semantics of operations have to provide more concurrency than obtained by the conventional classification of operations as *reads* or *writes*.

Key Points

In most semantics-based protocols, conflicts between operations is based on commutativity, an operation o_i which does not commute with other uncommitted operations will be made to wait until these conflicting operations abort or commit. Some protocols use operations' return value commutativity, wherein information about the results of executing an operation is used in determining commutativity, and some use the arguments of the operations in determining whether or not two operations commute. An example of the former, two increment operations on a counter object commute as long as they do not return the new or old value of the counter. An example of the latter, two insert operations on a set object commute as long as they do not insert the same item.

In the scheme reported in [1], non-commuting but *recoverable* operations are allowed to execute in parallel; but the order in which the transactions invoking the operations should commit is fixed to be the order in which they are invoked. If o_j is executed after o_i, and o_j is *recoverable relative to o_i*, then, if transactions T_i and T_j that invoked o_i and o_j respectively commit, T_i should commit before T_j. Thus, based on the recoverability relationship of an operation with other operations, a transaction invoking the operation sets up a dynamic commit dependency relation between itself and other transactions. If an invoked operation is not recoverable with respect to an uncommitted operation, then the invoking transaction is made to wait. For example, two pushes on a stack do not commute, but if the push operations are forced to commit in the order they were invoked, then the execution of the two push operations is serializable in commit order. Further, if either of the transactions aborts the other can still commit.

In [2] authors make an effort to discover, from first principles, the nature of concurrency semantics inherent in objects. Towards this end, they identify the dimensions along which object and operation semantics can be modeled. These dimensions are then used to classify and unify existing semantic-based concurrency control schemes. To formalize this classification, a graph representation for objects that can be derived from the abstract specification of an object is proposed. Based on this representation, which helps to identify the semantic information inherent in an object, a methodology is presented that shows how various semantic notions applicable to concurrency control can be effectively combined to improve concurrency. A new source of semantic information, namely, the ordering among component objects, is exploited to further enhance concurrency. Lastly, the authors present a scheme, based on this methodology, for *deriving* compatibility tables for operations on objects.

Cross-References

▶ ACID Properties
▶ Concurrency Control: Traditional Approaches

Recommended Reading

1. Badrinath BR, Ramamritham K. Semantics-based concurrency control: beyond commutativity. ACM Trans Database Syst. 1991;17(1):163–99.
2. Chrysanthis PK, Raghuram S, Ramamritham K Extracting concurrency from objects: a methodology. In: Proceedings of the ACM SIGMOD International Confernce on Management of Data; 1991.

Semijoin

Kai-Uwe Sattler
Technische Universität Ilmenau, Ilmenau, Germany

Synonyms

Bit vector join; Bloom filter join; Bloom join; Hash filter join; Semijoin filter

Definition

Semijoin is a technique for processing a join between two tables that are stored at different sites. The basic idea is to reduce the transfer cost by first sending only the projected join column(s) to the other site, where it is joined with the second relation. Then, all matching tuples from the second relation are sent back to the first site to compute the final join result.

Historical Background

The semijoin technique was originally developed by Bernstein et al. [3] as part of the SDD-1 project as a reduction operator for distributed query processing. The idea of applying hash filtering was proposed by Babb [1] as well as by Valduriez [9] particularly for specialized hardware (content addressed file stores and distributed database machines respectively). The theory of semijoin-based distributed query processing was presented in [2]. In [10] semijoins are also exploited for query processing on multiprocessor database machines. Results of detailed experimental work on semijoins in distributed databases were first reported by Lu and Carey [6] as well as by Mackert and Lohman [7].

Foundations

Semijoin is a join processing technique which was originally developed for distributed databases. A semijoin is the "half of a join" and is particularly useful as a reduction operator.

Relational Definition

Given two relations $R(A,B)$ and $S(C,D)$ with the join condition $R.A = S.C$ the semijoin $R \ltimes S$ is defined as follows:

$$R \bowtie_{A=C} S = \pi_{attr(R)}(R \bowtie_{A=C} S)$$

where $attr(R)$ denotes the set of attributes in R. The semijoin has two important characteristics:

1. It is a reducing operator, because $R \bowtie_{A=C} S \cup R$.
2. It is asymmetric, i.e., $R \bowtie_{A=C} S \neq S \bowtie_{A=C} R$.

Semijoin Filtering

The obvious approach of processing a join between a relation R stored at site 1 and S stored at site 2 is to ship the smaller relation to the other site and compute the join locally. This is also called "ship whole" approach. However, for computing the join one or both of relations can be replaced by a semijoin with the other relation, i.e.,:

$$R \bowtie_{A=C} S = (R \bowtie_{A=C} S) \bowtie_{A=C} S$$

$$= R \bowtie_{A=C} (S \bowtie_{A=C} R)$$

$$= (R \bowtie_{A=C} S) \bowtie_{A=C} (S \bowtie_{A=C} R)$$

In each case the semijoin acts as a reducer operation just like a selection operator. Which variant is chosen for the actual join processing has to decided by estimating the costs.

The principled approach of the semijoin filtering can be formulated in the following algorithm:

1. At site 1 compute $R' := \pi_A(R)$ and send it to site 2
2. At site 2 process the semijoin $S' = S \bowtie_{C=A} R'$ Note, relation S' contains only tuples matching the join condition and will appear in the final result. Furthermore, the result relation provides only the attributes from S
3. Send relation S' to site 1
4. At site 1, $R \bowtie_{A=C} S'$ is computed producing a result equivalent to $R \bowtie_{A=C} S$

In Fig. 1 the process is illustrated using an example.

In order to estimate the benefit of the semijoin compared to the "ship whole" approach it is sufficient to consider only the transfer costs. Let T denote the cost for transfer a data unit and $size(R) = |R| \cdot width(R)$ the size of the relation derived from the cardinality $|R|$ and the size $width(R)$ of a tuple in data units. Then, the cost for the ship whole strategy is

$$C \cdot size(S)$$

assuming S is the smaller relation ($size(S) < size(R)$). For the semijoin the main costs are in step 1 and step 3, i.e.,

Semijoin, Fig. 1 Example
of semijoin processing

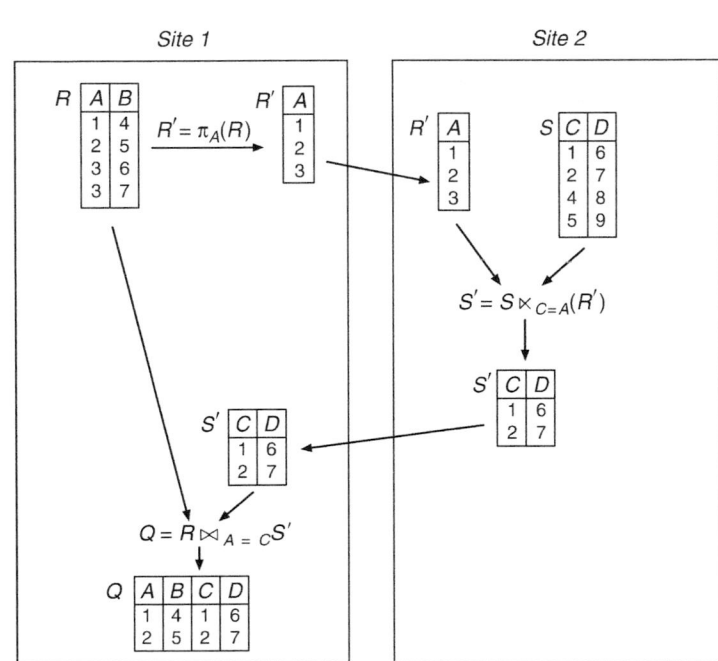

$$C \cdot size(\pi_A(R)) + C \cdot size(S \ltimes_{C=A} R)$$

Comparing these costs, one can observe that the semijoin approach is better if

$$size(\pi_A(R)) + size(S \bowtie_{C=A} R) < size(S)$$

More exactly due to $width(S \bowtie_{C=A} R) = width(S)$ the semijoin is the better choice if $|S \bowtie_{C=A} R| < |S|$, i.e., if the semijoin is really a reducer. At the other hand, the ship whole approach is better if nearly all tuples of S contribute to the join result. In this case, the semijoin has the disadvantage of the additional transfer of $\pi_A(R)$.

Thus, a decision for one of these join strategies requires an estimation of the join selectivity factor SF. For the semijoin the following approximation

$$SF_{R \bowtie_{C=A} S} = \frac{|\pi_C(S)|}{|distinct(C)|}$$

was proposed by [5], where $|distinct(C)|$ denotes the number of distinct values in attribute C.

Bit Vector Filtering

The effort for step 1 of the semijoin can be further reduced by sending only a compact bitmap representation of the column values instead of $\pi_A(R)$. This bitmap or bit vector is built using a hash function [4] and, thus, the approach is called bit vector filtering, hash filter join or bloom join.

For a hash function $h(v)$ returning values $0...n$ a bit vector $B[0...n]$ containing $n+1$ bits initially set to 0 is required. For each value $v \in \pi_A(R)$ the corresponding bit $B[h(v)]$ is set to 1. Instead of processing the semijoin $S \bowtie R$ at site 2, this bit vector and the hash function are used to probe the tuples of S for matching with the join values of R, i.e., if for a value v' of the join attribute B the corresponding bit is set: $B[h(v')] = 1$. The whole process is shown in the following algorithm:

1. At site 1: for each $v \in \pi_A(R)$ set $B[h(v)] = 1$ and sent B to site 2
2. At site 2: derive $S' = \{t \in S | B[h(t.C)] = 1\}$
3. Sent S' to site 1
4. At site 1: $R \bowtie_{A=C} S'$ is computed producing a result equivalent to $R \bowtie_{A=C} S$

This algorithm is illustrated by an example in Fig. 2 using a simple hash function $h(v) = v \bmod$

Semijoin, Fig. 2 Example of hash filter join

7. Applying h to column A of relation R produces the vector of seven bits ([0111000]) which is used to probe the S-tuples at site 2 by computing $h(C)$. Note, that $\checkmark$ indicates a match and $-$ a non-match.

Note that a hash function is usually not injective and therefore the problem of collision occurs, i.e., for different values $v_1 \neq v_2$ one can have $h(v_1) = h(v_2)$. Thus, useless tuples are sent to site 1 in step 3 which will not contribute to the final result, e.g. the tuple with $C = 8$ in this example. This problem can be mitigated by chosing a bit vector of an appropriate length. An alternate approach is to use multiple hash functions $h_1,...,h_k$ together with the associated bit vectors $B_1,...,B_k$ and to set the bits for a value v in each bit vector:

$$B_1 h_1(v)]=1, B_2[h_2(v)]=1, \ldots, B_k[h_k(v)]=1$$

All these bit vectors are sent to site 2 and used there for probing. A tuple $t \in S$ qualifies only to be a candidate tuple if all bits are set to 1, i.e., if the result of the bitwise AND is 1. It can be shown that with an increasing k the collision probability comes close to 0.

Key Applications

The main application of semijoin techniques is distributed join processing, where the semijoin acts as a reducer. Though, experimental work has shown that the computational overhead is typically higher than the savings in transfer cost, particularly the hash filter strategy is often an attractive alternative.

Variants of the semijoin are also used for processing queries in heterogeneous databases where a component database provides only limited query capabilities, e.g. selections with parameters (also called bindings). If a set of tuples is sent to the component database as binding parameter, this corresponds in fact to the semijoin strategy.

Finally, semijoins are also useful for processing star queries in data warehouses. Here, the semijoin technique is exploited for joining each dimension table with the fact table (or more exactly an index on the fact table) in order to collect the rowids of the fact tuples. Then, the intersection of all rowid sets is computed which is finally used to retrieve the tuples from the fact table.

Experimental Results

Mackert and Lohman [7] report results of an experimental analysis of the performance of distributed join strategies in the R* system. Though, the experiments were conducted on a hardware which was up-to-date in the 1980s (e.g., a high-speed network with 4 Mbit/s effective transfer rate), the general trend of the results is still valid.

Figure 3 shows the results of a comparison of several strategies for computing $R \bowtie S$ where

Semijoin, Fig. 3 Comparison of semijoin algorithms

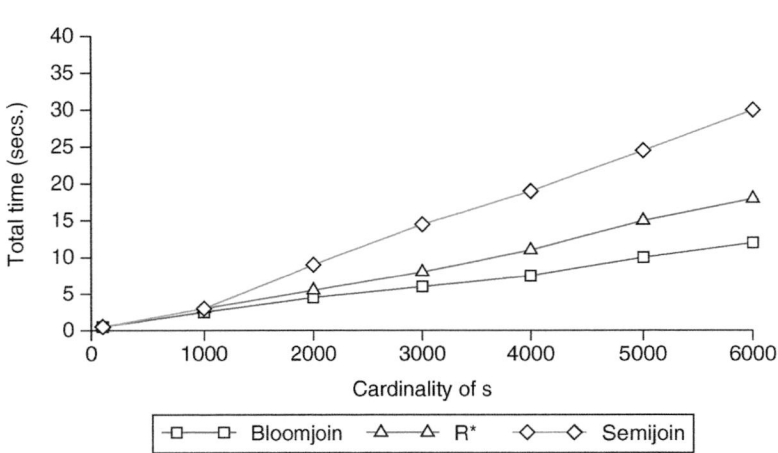

the cardinality of R was $|R| = 1000$ and the cardinality of S varied from 100 to 6,000.

In this experiment, the hash filter join clearly outperformed the other strategies. Only for small cardinalities where the inner relation S fits into the buffer, the semijoin has advantages. The third join variant was the R* strategy of shipping one relation to the other site and exploiting local indexes for join processing.

Cross-References

► Distributed Join
► Evaluation of Relational Operators
► Semijoin Program

Recommended Reading

1. Babb E. Implementing a relational database by means of specialized hardware. ACM Trans Database Syst. 1979;4(1):1–29.
2. Bernstein PA, Chiu D-MW. Using semi-joins to solve relational queries. J ACM. 1981;28(1):25–50.
3. Bernstein PA, Goodman N, Wong E, Reeve CL, Rothnie Jr. Query processing in a system for distributed databases (SDD-1). ACM Trans Database Syst. 1981;6(4):602–25.
4. Bloom BH. Space/time trade-offs in hash coding with allowable errors. Commun ACM. 1970;13(7):422–6.
5. Hevner AR, Yao SB. Query processing in distributed database systems. IEEE Trans Softw Eng. 1979;5(3):177–82.
6. Lu H, Carey M. Some experimental results on distributed join algorithms in a local network. In: Proceedings of the 11th International Conference on Very Large Data Bases; 1985. p. 229–304.
7. Mackert L.F., Lohman G. R* optimizer validation and performance evaluation for local queries. In: Proceedings of the ACM SIGMOD International Conference on Management on Data; 1986. p. 4–95.
8. Özsu MT, Valduriez P. Principles of distributed database systems. 2nd ed. Prentice-Hall; 1999.
9. Valduriez P. Semi-join algorithms for distributed database machines. In: Schneider J-J, editor. Distributed data bases. Amsterdam: North-Holland; 1982. p. 23–37.
10. Valduriez P, Gardarin G. Join and semi join algorithms for a multiprocessor database machine. ACM Trans Database Syst. 1984;9(1):133–61.

Semijoin Program

Stéphane Bressan
National University of Singapore, School of Computing, Department of Computer Science, Singapore, Singapore

Synonyms

Semijoin reducer

Definition

A semijoin program is a query execution plan for queries to distributed database systems that uses semijoins to reduce the size of relation instances before they are transmitted and further joined. Yet the reduction itself requires that a projection of the relation instances involved in the join onto the join attributes be transmitted. The maximum amount of reduction can be achieved by a semijoin program called a full reducer. Full reducers that do not require the computation of a fixpoint exist for acyclic queries. Fully reducing relation instances is rarely beneficial. However semijoin programs partially reducing selected relation instances may be an effective optimization when the dominant cost of query execution is communication. Considering semijoin programs considerably increases the distributed query optimization search space.

Historical Background

Semijoin programs were first introduced to improve the performance, input/output operations and communication of database applications running on database machines [7] and on parallel database machines [10]. These machine were possibly equipped with specialized hardware such as filters efficiently implementing semijoins. SDD-1 [3] is the first distributed database management system making use of semijoin programs for query optimization. Bernstein and Chiu, in [1], review the algorithms for these early applications. In these approaches, reduction of relation instances and the other steps of global and local optimization are conducted as successive phases of the distributed query optimization. Stoker et al. [8], revisit semijoin programs for modern applications and empirically evaluate their usefulness. The authors propose dynamic programming query optimization algorithms that integrate the selection of selected semijoin reducers with join ordering into a single phase.

The hyper-graph representation of relational queries and the notion of acyclic queries is from Fagin [5]. Bernstein et al. introduce comparable notions [2] and define the notion of full reducer. Several authors, see for instance [11] and, more recently [6], have discussed the relationship between reducers and constraint satisfaction problems in the context of database query optimization.

The textbook [4] gives a good overview of semijoin programs and their use in early distributed database systems. It describes the use of semijoin programs in SDD-1 while the presentation in [9] emphasizes the notions of reducers, full reducers, and reduction algorithms.

Foundations

Cyclic and Acyclic Query Hyper-graphs
For the sake of simplicity, queries are considered which contain natural joins only. The hyper-graph representing such queries is composed of vertices corresponding to attributes and hyper-edges corresponding to relations.

Given, for instance, the three relations $R(A, B)$, $S(A, C)$ and $T(C, B)$, the query $R \bowtie (S \bowtie T)$ is the natural join of R with the natural join of S and T. Its hyper-graph is represented on Fig. 1.

In addition, consider the relation $U(C, D)$. The query $R \bowtie (S \bowtie U)$ is the natural join of R with the natural join of S and U. Its hyper-graph is represented on Fig. 2.

An *ear* of a hyper-graph is a hyper-edge that contains a vertex that does not belong to any other hyper-edge. R and U in Fig. 2 are ears. Notice

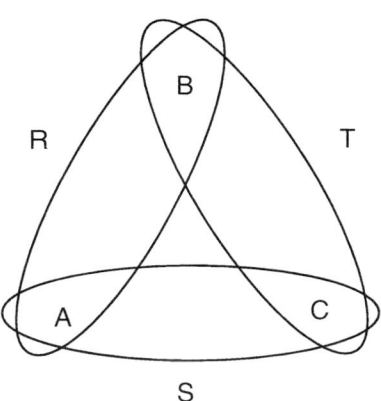

Semijoin Program, Fig. 1 Hyper-graph representing $R \bowtie (S \bowtie T)$

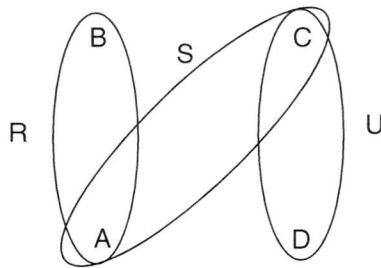

Semijoin Program, Fig. 2 Hyper-graph representing $R \bowtie (S \bowtie U)$

that removing one ear may create new ears. If all hyper-edges of a hyper-graph can be removed by iteratively removing ears, the hyper-graph is said to be acyclic. The hyper-graph of Fig. 2 is acyclic. The ears U, S and R can be removed in this order, for instance. The hyper-graph of Fig. 1 is cyclic: none of the hyper-edges is an ear.

Reducer and Full Reducer
A reducer for a relation R with respect to a query Q is a program of semijoins with other relations in Q applied to R, such that R can be replaced by the result of the program in Q without changes in the result of the query Q. In other words, the semijoins possibly remove some tuples that do not contribute to the query Q. The program $R \ltimes S$ is a reducer of R in both queries of Figs. 1 and 2, respectively. The reader can verify that with the instances of R and S given in Fig. 3, the

8. Stocker K, Kossmann D, Braumandl R, Kemper A. Integrating semi-join-reducers into state of the art query processors. In: Proceedings of the 17th International Conference on Data Engineering; 2001. p. 575–84.
9. Ullman JD. Principles of database and knowledge-base systems, vol. II. Rockville: Computer Science; 1989.
10. Valduriez P, Gardarin G. Join and semijoin algorithms for a multiprocessor database machine. ACM Trans Database Syst. 1984;9(1):133–61.
11. Wallace M, Bressan S, Provost TL. Magic checking: constraint checking for database query optimization. In: Proceedings of the ESPRIT WG CONTESSA Workshop on Constraint Databases and Applications; 1995. p. 148–66.

Semi-structured Data

Serge Abiteboul
Inria, Paris, France

Synonyms

XML (almost)

Definition

A semi-structured data model is based on an organization of data in labeled trees (possibly graphs) and on query languages for accessing and updating data. The labels capture the structural information. Since these models are considered in the context of data exchange, they typically propose some form of data serialization, i.e., a standard representation of data in files. Indeed, the most successful such model, namely XML (that is promoted by the W3C), is often confused with its serialization syntax. XML equipped with query/update language [10] is a semi-structured data model.

Semi-structured data models are meant to represent from very structured to very unstructured information, and in particular, irregular data. In a structured data model such as the relational model [9], one distinguishes between the type of the data (schema in relational terminology) and the data itself (instance in relational terminology). In semi-structured data models, this distinction is blurred. One sometimes speaks of schemaless data although it is more appropriate to speak of self-describing data. Semi-structured data may possibly be typed. For instance, tree automata have been considered for typing XML. However, semi-structured data applications typically use very flexible and tolerant typing or sometimes no typing at all.

Historical Background

Before the Web, publication of electronic data was limited to a few scientific and technical areas. With the Web and HTML, it rapidly became universal. HTML is a format meant for presenting documents to humans. However, a lot of the data published on the Web is produced by machines. Moreover, it is more and more the case that Web data are consumed by machines. Since HTML is not appropriate for machine processing, this lead in the 1990's to the development of *semi-structured data models* and most importantly of a new standard for the Web, namely XML. The use of a semi-structured data model as a standard for *data representation* and *data exchange* on the Web brought important improvement to the publication and reuse of electronic data by providing a simple syntax for data that is machine-readable and at the same time, human readable (with the help of the so-called "style-sheets").

Semi-structured data models may be viewed, in some sense, as bringing together two cultures that were for a long while seen as irreconcilable, document systems (with notably SGML [8]) and database systems (with notably relational systems [9]). From a model perspective, there are many similarities with the object database model [5]. Indeed, like XML, the object database model is also based on trees, provides an object API, comes equipped with a query language and offers some form of serialization. A main difference is that the very rigorous typing of object databases was abandoned in semi-structured data models.

The articulation of the notion of semi-structured data may be traced to two simultaneous origins, the OEM model at Stanford [3, 6] and the UnQL model at U. Penn [4].

Specific data formats had been previously proposed and even became sometimes popular in specific domains, e.g. ASN.1 [7]. The essential difference between data exchange formats and semi-structured data models is the presence of high level query languages in the latter. A query language for SGML is considered in [2]. Languages for semi-structured data models such as [3, 4] then paved the way for languages for XML [10].

Foundations

One can start with an idea familiar to Lisp programmers of association lists, which are nothing more than label-value pairs and are used to represent record-like or tuple-like structures:

{name: "Alan,??????" tel.: 2157786, email: "agb@abc.com"???????}

This is simply a set of pairs such as *name: "Alan"* consisting of a label and a value. The values may themselves be other structures as in:

{name: {first: "Alan,???????" last: "Black"???????},
   tel.: 2157786,
   email: "agb@abc.com"???????}

This data may be represented graphically with nodes denoting object, connected by edges to values, see Fig. 1. Departing from the usual assumption made about tuples or association lists that labels are unique, duplicate labels may be allowed as in:

{name: "alan," tel.: 2157786, tel: 2498762 }

The syntax makes it easy to describe sets of tuples as in:

{person: {name: "alan,???????" phone: 3127786, email: "agg@abc.com"???????},

   person: {name: "sara,???????" phone: 2136877, email: "sara@math.xyz.edu"???? ???},

   person: {name: "fred,???????" phone: 7786312, email: "fds@acme.co.uk"???????}?? ?????}

Furthermore, one of the main strengths of semi-structured data is its ability to accommodate variations in structure, e.g., all the *Person* tuples do not need to have the same type. The variations typically consist of missing data, duplicated fields or minor changes in representation, as in the following example:

{person: {name: "alan,???????" phone: 3127786, email: "agg@abc.com"???????},
   person: &314.
   {name: {first: "Sara,???????" last: "Green" ???????},
   phone: 2136877,
   email: "sara@math.xyz.edu,???????"
   spouse: &443
   person: &443
   {name: "fred,???????" Phone: 7786312, Height: 183,
   spouse: &314}}

Observe how identifiers (here &443 and &314) and references are used to represent graph data. It should be obvious by now that a wide range of data structures, including those of the relational and object database models, can be described with this format.

Semi-structured Data,
Fig. 1 Tree representation

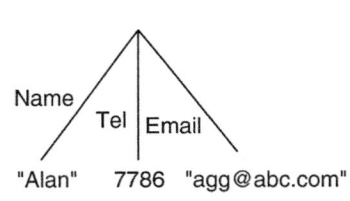

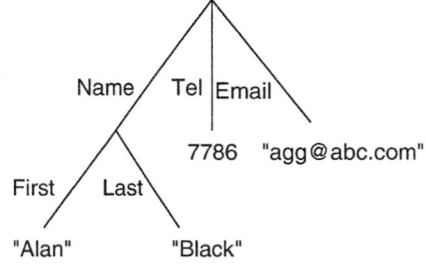

As already mentioned, in semi-structured data, the conscious decision is made of possibly not caring about the type the data might have, and serialize it by annotating each data item explicitly with its description (such as *name, phone*, etc). Such data are called *self-describing*. The term *serialization* means converting the data into a byte stream that can be easily transmitted and reconstructed at the receiver. Of course self-describing data wastes space, since these descriptions are repeated for each data item, but more interoperability is achieved, which is crucial in the Web context.

There have been different proposals for semi-structured data models. They differ in choices such as labels on nodes versus on edges, trees versus graphs, ordered trees versus unordered trees. Most importantly, they differ in the languages they offer.

Key Applications

The main applications of semi-structured data models are found on the Web.

First, semi-structured data models and XML are very useful for data *publication*. XML is also serving as a universal *data exchange* format in a wide variety of fields, from bioinformatics to e-commerce. It presents the advantage compared to previous formats that it comes equipped with an array of available software such as parsers or programming interfaces. Also, the flexibility of the typing in semi-structured data models turns out to be essential for *data integration*, and in particular in the integration of heterogeneous data in mediator systems.

Cross-References

► Document Representations (Inclusive Native and Relational)
► Semi-structured Data Model
► W3C
► XML
► XML Types

Recommended Reading

1. Abiteboul S, Buneman P, Suciu D. Data on the web: from relations to semistructured data and XML. San Francisco: Morgan Kaufmann; 1999.
2. Abiteboul S, Cluet S, Christophides V, Milo T, Moerkotte G, Simeon J. Querying documents in object databases. Int J Digit Libr. 1997;1(1):5–19.
3. Abiteboul S, Quass D, McHugh J, Widom J, Wiener J. The Lorel query language for semistructured data. Int J Digit Libr. 1997;1(1):68–88.
4. Buneman P, Davidson S, Suciu D. Programming constructs for unstructured data. In: Proceeding of the 5th International Workshop on Database Programming Languages; 1995.
5. Cattell RGG. The object database standard: ODMG-93. San Francisco: Morgan Kaufmann Publishers; 1994.
6. Papakonstantinou Y, Garcia-Molina H, Widom J. Object exchange across heterogeneous information sources. In: Proceeding of the 11th International Conferences on Data Engineering; 1995. p. 251–60.
7. Specification of Abstraction Syntax Notation One (ASN.1), ISO Standard 8824, Information Processing System. 1987.
8. Standard Generalized Markup Language (SGML), ISO 8879. 1986.
9. Ullman JD. Principles of database and knowledge-base systems, vol. I: classical database systems. Computer science. 1988.
10. XQuery. XQuery 1.0: An XML query language. http://www.w3.org/TR/Xquery.

Semi-structured Data Model

Dan Suciu
University of Washington, Seattle, WA, USA

Synonyms

Semi-structured data

Definition

The semi-structured data model is designed as an evolution of the relational data model that allows the representation of data with a flexible structure. Some items may have missing attributes, others may have extra attributes, some items may have two or more occurrences of the same attribute. The type of an attribute is also flexible: it may be an atomic value, or it may be another record or collection. Moreover, collections may be heterogeneous, i.e., they may contain items with different structures. The semi-structured data model is self-describing data model, in which the data values and the schema components co-exist. Formally:

Definition 1

A semi-structured data instance is a rooted, directed graph in which the edges carry labels representing schema components, and leaf nodes (i.e., nodes without any outgoing edges) are labeled with data values (integers, reals, strings, etc.).

There are two variations of semi-structured data, depending on how one interprets equality. In the object exchange model (OEM) introduced in Tsimmis [8], each node in the data has its own identity, and thus two data instances are "equal" if and only if they are isomorphic, and this corresponds to the bag semantics of collections. In the value-based model introduced in UnQL [2], two data graphs are equal if they are bisimilar; this corresponds to the set semantics of collections.

A variation of the semi-structured data model is one in which labels are placed on nodes rather than edges.

Historical Background

Theterm *semi-structured data* was introduced by Luniewski et al. in 1993 in a system called Rufus [6, 9]. In 1995, Papakonstantinou et al. introduced a data model for semi-structured data (called *object exchange model*, OEM) for a system for integrating heterogeneous databases called Tsimmis [5, 8]. In 1995, Buneman et al. introduced a data model for biological data where equality is based on bisimulation [1–3]. The connection between the semi-structured data model and XML was described in 1999 by Deutsch et al., who proposed a query language for XML called XML-QL [4].

r2	c	d
	c2	d2
	c3	d3
	c4	d4

Foundations

Semi-structured data are *schema-less* or *self-describing*. Both the data and its schema is described directly using a simple syntax for sets of label-value pairs, similar to association lists in Lisp. For example:

```
{name: "Alan",
    tel: 2157786, email: "agb@abc.com"}
```

This is simply a set of pairs such as *name: "Alan"* consisting of a label and a value. The values may themselves be other structures as in

```
{name: {first: "Alan", last: "Black"},
tel: 2157786,
email: "agb@abc.com"
}
```

One may represent this data graphically as a node that represents the object, connected by edges to values, see Fig. 1.

Unlike in traditional tuples or association lists the labels are not necessarily unique, and duplicate labels are allowed, as in:

```
{name: "alan, tel: 2157786,
    tel: 2498762}
```

The syntax makes it easy to describe sets of tuples as in

```
{person:
{name: "alan", phone: 3127786,
 email: "agg@abc.com"},
person:
{name: "sara", phone: 2136877,
email: "sara@math.xyz.edu"},
person:
{name: "fred", phone: 7786312,
 email: "fds@acme.co.uk"}
}
```

Person tuples do not necessarily have to be of the same type. One of the main strengths of semi-structured data is its ability to accommodate variations in structure. While in principle semi-structured data could become a completely random graph, data instances that are usually found

S

Semi-structured Data Model, Fig. 1 Graph representations of simple structures

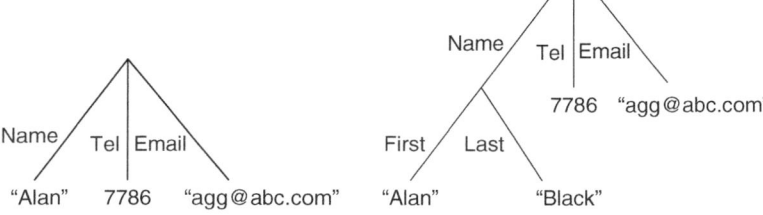

in practice are "close" to some type, and have only minor variations from that type. The variations typically consist of missing data, duplicated fields or minor changes in representation, as in the example below.

```
{person:
{name: "alan", phone: 3127786,
 email: "agg@abc.com"},
person:
{name: {first: "Sara", last: "Green"},
phone: 2136877,
email: "sara@math.xyz.edu"
},
person:
{name: "fred", Phone: 7786312
 Height: 183}
}
```

Representing relational databases. It is easy to represent every relational database as a semi-structured data, which happens to have a regular structure. For example the relational database instance:

r1	a	b	c
	a1	b1	c1
	a2	b2	c2

can be described as a set of rows:

```
{r1: {row: {a: a1, b: b1, c: c1},
row: {a: a2, b: b2, c: c2}
},
r2: {row: {c: c2, d: d2},
row: {c: c3, d: d3},
row: {c: c4, d: d4}
}
}
```

It is worth noting that this is not the only possible representation of a relational database. Figure 2 shows tree diagrams for the syntax given above and for two other representations of the same relational database.

Representing object databases. Consider for example the following collection of three persons, in which *Mary* has two children, *John* and *Jane*. Object identities may be used to construct structures with references to other objects.

```
{person: &o1{name: "Mary",
age: 45,
child: &o2,
child: &o3
},
person: &o2{name: "John",
age: 17,
relatives: {mother: &o1,
sister: &o3}
},
person: &o3{name: "Jane",
country: "Canada",
mother: &o1
}
}
```

The presence of a label such as *&o1* before a structure binds *&o1* to the identity of that structure. This makes it possible to use that label - as a value - to refer to that structure. In this graph representation it is allowed to build graphs with shared substructures and cycles, as shown in Fig. 3. The name *&o1*, *&o2*, *&o3* are called *object identities*, or oid's. In this figure, arrows are placed on the edges to indicate the direction, which is no longer implicit, like in the tree-like structure.

The object exchange model (OEM) was explicitly defined for the purpose of integrating heterogeneous data sources in Tsimmis. An OEM object is a quadruple *(label, oid, type, value)*, where *label* is a character string, *oid* it the object's identifier, *type* is either *complex*, or some identifier denoting an atomic type (like *integer*, *string*, *gif-image*, etc.). When *type* is *complex*,

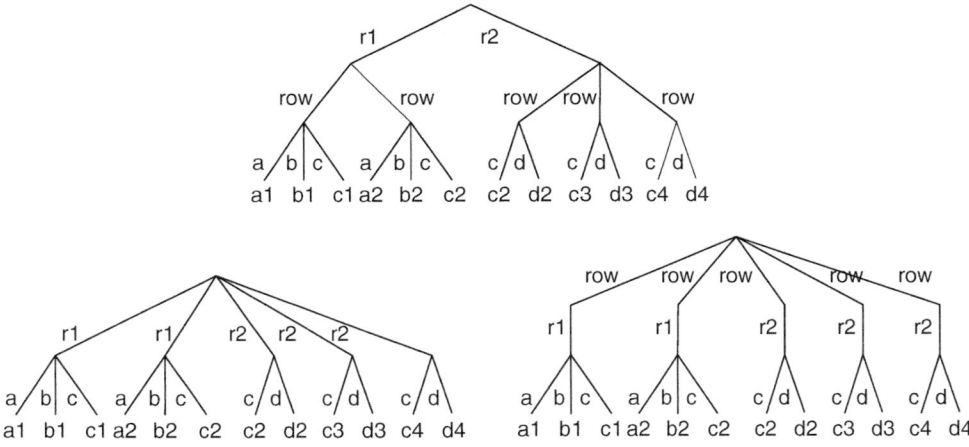

Semi-structured Data Model, Fig. 2 Three representations of a relational database

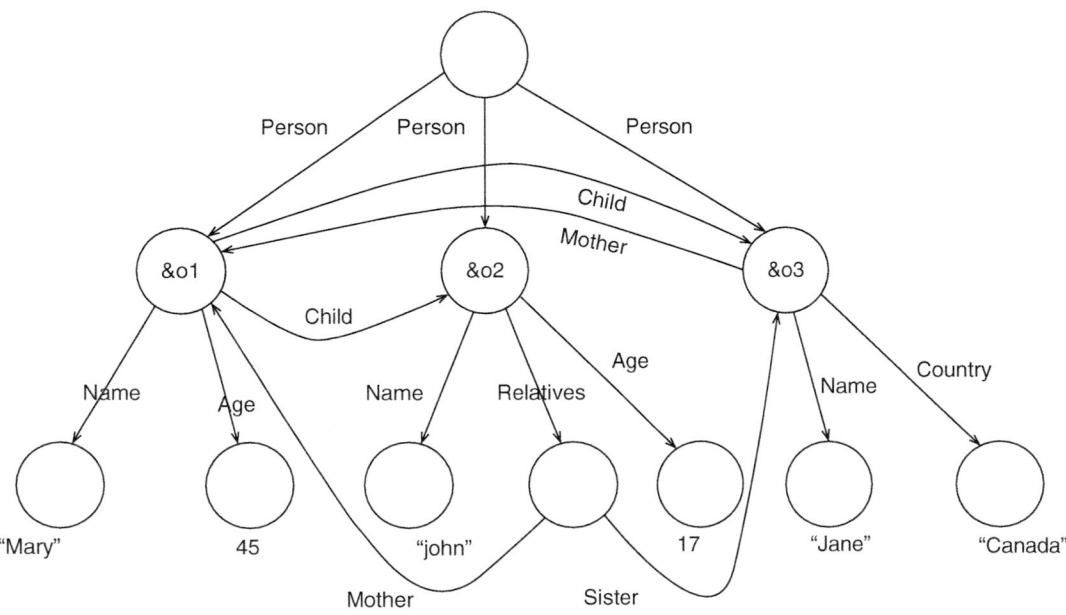

Semi-structured Data Model, Fig. 3 A cyclic structure

then the object is called a *complex object*, and *value* is a set (or list) of oid's. Otherwise the object is an *atomic object* and the *value* is an atomic value of that type. Thus OEM data are essentially a graph, but in which labels are attached to nodes rather than edges.

Equality in semi-structured data. A shallow notion of equality simply checks whether two object identifiers are the same, or two data values are equal. Beyond that a deep notion of equality is needed, which addresses the following question:

given two semi-structured data instances (i.e., two graphs), do they represent the same data? This question is fundamental in query optimization, since it allows replacement of one query expression with another if the instances they return are equal.

Two notions of deep equality have been considered. One is *graph isomorphism*: two data instances are equal if there exists an isomorphism that preserves the edge labels and the data values.

Definition 2

An isomorphism *between two semi-structured data instances D_1, D_2 is a function f mapping the nodes of D_1 to the nodes of D_2 such that:*

1. *f is a bijection.*
2. *f maps the root of D_1 to the root of D_2.*
3. *If there exists an edge from a node x to a node y in D_1 then there exists an edge from the node $f(x)$ to the node $f(y)$ in D_2 and both edges have the same label.*
4. *Conversely, if there exists an edge from $f(x)$ to $f(y)$ in D_2 then there exists an edge (It follows from the previous condition that the two edges have the same label.) from x to y in D_2.*
5. *If a leaf node x in D_1 is labeled with a data value v then the node $f(x)$ in D_2 is labeled with the same data value (It follows from the previous conditions that $f(x)$ is also a leaf node.)v.*

In the first interpretation two data instances are equal if there exists an isomorphism between them. For example the two instances $\{a:3, b:5, c:7, b:9\}$ and $\{b:5, b:9, a:3, c:7\}$ are equal, because they are isomorphic. On the other hand, the instances $\{a:3, a:3, b:5\}$ and $\{a:3, b:5\}$ are not equal. Thus, when restricted to collections this notion of equality corresponds to the bag semantics.

The second notion of equality is based on *bisimulation*: two data instances are equal if there exists a bisimulation:

Definition 3

A bisimulation between two semi-structured data instances D_1, D_2 is a relation $R(x, x')$ between the nodes in the two instances s.t.

1. *If r_1 is the root node in D_1 and r_2 is the root node in D_2, then $R(r_1, r_2)$.*
2. *If $R(x, x')$ holds, and D_1 contains an edge (x, y) with label a, then there exists an edge (x', y') labeled a in D_2, and $R(y, y')$ holds.*
3. *Symmetrically, if $R(x, x')$ holds and D_2 contains an edge (x', y') with label a then there exists an edge (x, y) in D_1 labeled a, and $R(y, y')$ holds.*

4. *If $R(x, x')$ holds and the node x in D_1 is a leaf node, and is labeled with the atomic value v then the node x' in D_2 is also a leaf node and labeled with the same atomic value v (the symmetric property follows automatically).*

In the second interpretation, two semi-structured data instances are said to be equal if there exists a bisimulation between them. For example $\{a:3, a:3,b:5\}$ is equal to $\{a:3, b:5\}$ because, denoting r, n_1, n_2, n_3 the nodes in the first graph and r', n_1', n_2' the nodes in the second graph, the relation $R(r, r')$, $R(n_1, n_1')$, $R(n_2, n_1')$, $R(n_3, n_2')$ is a bisimulation. Thus, the equality based on bisimulation corresponds to the set semantics on collections.

If two data instances are isomorphic, then there always exists a bisimulation between them; the converse does not always hold. Checking whether two data instances are isomorphic is a computationally hard problem. By contrast, checking if two data instances are bisimilar can be done efficiently [7].

Edge vs. node labeled graphs. The model described here is that of an *edge-labeled graph*. A minor variation is one in which nodes are labeled, and this has gained a lot of popularity since the introduction of XML.

Key Applications

The initial motivation for the introduction of semi-structured data was to support the integration of heterogeneous data, and to model nonstandard data formats, especially in the bioinformatics domain, s.a. ACEDB and ASN.1. After the introduction of XML, this became the main application of semi-structured data.

Cross-References

▶ Semi-structured Data
▶ Semi-structured Query Languages
▶ XML

Recommended Reading

1. Buneman P, Davidson S, Suciu D. Programming constructs for unstructured data. In: Proceedings of the 5th International Workshop on Database Programming Languages; 1995.
2. Buneman P, Davidson S, Hillebrand G, Suciu D. A query language and optimization techniques for unstructured data. In: Proceedings of the ACM SIGMOD International Conference on Management of Data; 1996. p. 505–16.
3. Buneman P, Fernandez M, Suciu D. UNQL: a query language and algebra for semistructured data based on structural recursion. VLDB J. 2000;9(1): 76–110.
4. Deutsch A, Fernandez M, Florescu D, Levy A, Suciu D. A query language for XML. In: Proceedings of the 8th International World Wide Web Conference; 1999. p. 77–91.
5. Garcia-Molina H, Papakonstantinou Y, Quass D, Rajaraman A, Sagiv Y, Ullman J, Widom J. The TSIM-MIS project: integration of heterogeneous information sources. J Intell Inf Syst. 1997;8(2):117–32.
6. Luniewski A, Schwarz P, Shoens K, Stamos J, Thomas J. Information organization using Rufus. In: Proceedings of the ACM SIGMOD International Conference on Management of Data; 1993. p. 560–1.
7. Paige R, Tarjan R. Three partition refinement algorithms. SIAM J. Comput. 1987;16(6):973–88.
8. Papakonstantinou Y, Garcia-Molina H, Widom J. Object exchange across heterogeneous information sources. In: Proceedings of the 11th International Conference on Data Engineering; 1995. p. 251–60.
9. Shoens K, Luniewski A, Schwarz P, Stamos J, Thomas II J. The Rufus system: information organization for semi-structured data. In: Proceedings of the 19th International Conference on Very Large Data Bases; 1993. p. 97–107.

Semi-structured Database Design

Gillian Dobbie[1] and Tok Wang Ling[2]
[1]University of Auckland, Auckland, New Zealand
[2]National University of Singapore, Singapore, Singapore

Synonyms

XML database design

Definition

From a requirements document, a database designer distills the real world constraints and designs a database schema. While the design process for structured data is well defined, the design process for semi-structured data is not as well understood. What is a "good" design for semi-structured databases that captures real world constraints, prevents data redundancy and update anomalies, and allows typical queries to execute quickly?

Historical Background

There was a lot of research into the design of relational databases in the 1970s, and it was found that the design of relational databases involves a trade off between the speed of execution of queries and the updating anomalies caused by maintaining redundant data when updates occur. During logical schema design normalization algorithms are used to reduce redundancy, and during physical design to improve performance some redundancy may be reintroduced, views can be created over the schema, and indexes may be introduced.

Semi-structured data differs from relational data in a number of ways: it is hierarchical, the queries that are posed are more complex, a database may or may not have a schema, and there is no generally agreed upon mathematical foundation. Because of these differences, there is a need for a different design process for semi-structured databases.

Foundations

Consider the XML document in Fig. 1a. It models the courses that students are taking within a department, and the grade of each student taking a course. What is the best way to organise this information? There are various possibilities, such as modeling *course* as a subelement of *student*, or modeling *student* as a subelement of *course*. While both of these options seem quite natural,

a
```
<department>
    <name>Computer Science</name>
    <student>
        <stuNo>123456</stuNo>
        <stuName>Bob Smith</stuName>
        <course>
            <code>CS101</code>
            <title>Introduction to Computer Science</title>
            <grade>A</grade>
        </course>
        <course>
            <code>CS105</code>
            <title>Data Structures</title>
            <grade>B</grade>
        </course>
    </student>
    <student>
        <stuNo>234567</stuNo>
        <stuName>Mary Brown</stuName>
        <course>
            <code>CS101</code>
            <title>Introduction to Computer Science</title>
            <grade>B</grade>
        </course>
    </student>
</department>
```
An XML Document

b
```
<department>
    <name>Computer Science</name>
    <student>
        <stuNo>123456</stuNo>
        <stuName>Bob Smith</stuName>
        <cGrade>
            <code>CS101</code>
            <grade>A</grade>
        </cGrade>
        <cGrade>
            <code>CS105</code>
            <grade>B</grade>
        </cGrade>
    </student>
    <student>
        <stuNo>234567</stuNo>
        <stuName>Mary Brown</stuName>
        <cGrade>
            <code>CS101</code>
            <grade>B</grade>
        </cGrade>
    </student>
    <course>
        <code>CS101</code>
        <title>Introduction to Computer Science</title>
    </course>
    <course>
        <code>CS105</code>
        <title>Data Structures</title>
    </course>
</department>
```
A normalized XML Document

Semi-structured Database Design, Fig. 1 (**a, b**) An original and normalized XML Document

Fig. 1a demonstrates that these options involve repeating information. The details of the course, i.e., title in this example, are repeated for each student that takes that course. Consequently, if the title of a course changes, it must be updated in every *title* element of the course. Brandin [2] describes a simple way to define XML documents, in which the modeler designs simple forms entailing the information to be modeled, and uses the form headings as tags and the entries on the form as data. While the simplicity of methods like this is appealing, they can lead to similar anomalies that arise with relational databases.

In order to ensure a "good" design for semi-structured databases, the following steps must be followed:

1. Choose a data model that is able to represent the semantics necessary for modeling semi-structured data.

2. Capture the semantics of the data that will be stored, either by:
   (a) Extracting the schema from a set of documents and discovering the semantics in a data model, or
   (b) Studying the constraints in the real world and capturing them in a data model.
3. Reorganize the schema into a normalized schema to avoid replication of data in the XML documents.
4. Consider the typical query set and reorganize the schema to improve the performance of typical queries, perhaps by introducing controlled replication of data.
5. Consider the users of the system and define views over the data for individual users or groups of users.

A data model for semi-structured database design needs to model the logical structure of the schema from a real world perspective, much like an ER diagram [4], while also modeling the physical aspects that are representative of XML documents, such as the hierarchical relationship between elements. This enables a designer to first model the real world constraints on the data, and using the same diagram, capture the extra constraints that are introduced with XML. For example, consider again the scenario captured in Fig. 1a. From a real world perspective the data model must model the relationship between students and courses, and from an XML perspective the data model must be able to capture whether student should be modeled as a subelement of course or vice-versa. The data model must also capture the participation constraints between elements, because they also change the way the data are modeled. For example, there is a many-to-many relationship between courses and students and this will be modeled differently than a one-to-many relationship say between department and employees. In order to capture these requirements, the data model must be able to model n-ary relationship sets, cardinality, participation and uniqueness constraints, ordering, irregular and heterogeneous structures, for both data- and document-centric data. One such data model is ORA-SS (Object Relationship Attribute Data Model for Semi-Structured Data) [6].

One of the advantages of semi-structured data is that it is self-describing, and so it does not require a schema. However, after XML was introduced, it was soon realized that schemas offer many benefits, and as a consequence schema languages have been defined for XML recently [3, 8]. Some semantics of a document can be extracted from the schema if one exists or extracted from the document. In [7, 9] and Chap. 4 of [6], the authors have described how an approximate schema can be extracted from semi-structured data. The algorithms generally follow two steps. The first step extracts the structural information, such as elements and their subelements. The second step infers semantics from the original document, such as key attributes, attributes of object classes vs attributes of relationship types, and participation constraints. Normalization is used to identify and reduce

anomalies, and a key ingredient in normalization algorithms is functional dependencies. Consider the real world constraints that should be modeled in the document in Fig. 1a. There is a many-to-many relationship between elements *student* and *course*, attributes *stuNo* and *stuName* belong to *student*, attributes *code* and *title* belong to *course*, and attribute *grade* belongs to the relationship type between *student* and *course*. These relationships may be modeled as functional dependencies, such as $stuNo \rightarrow stuName$, $code \rightarrow title$, and $\{stuNo, code\} \rightarrow grade$. Yu and Jagadish [10] describe a system, DiscoverXFD, which given an XML document, discovers XML functional dependencies and data redundancies. However, multi-valued dependencies (MVDs) are also very important in relational database design, such as in the definition of 4NF. Like with relational databases, functional dependencies cannot describe all redundancy in XML databases, so MVDs must also be considered.

The motivation for normalization algorithms for semi-structured databases is similar to that for relational databases, namely the identification of redundant data that lead to update anomalies. The normalization algorithms must recognize the hierarchical structure of semi-structured data, the participation constraints of both parents and children in hierarchical relationships, and whether an attribute is an attribute of an object class or the attribute of a relationship type. The algorithm described in [1] converts an arbitrary DTD into a well-designed one, using path functional dependencies. They show that the normal form that they define, XNF, generalizes BCNF for XML documents. This algorithm works with the semantics available in the DTD along with additional functional dependencies, however it does not consider MVDs, so documents in XNF may still contain redundancy. Using the semantics expressed in ORA-SS, it is possible to capture the multi-valued attributes of object classes and relationship types, avoiding the redundancy because of the existence of multi-valued attributes or MVDs. Note each multi-valued attribute of an object class is a MVD, e.g. if a student can have many hobbies, $studNo \rightarrow hobby$. As hobby is not involved in any functional dependencies, XNF

Note that a query language consisting of path expressions is not compositional, since queries return sets of nodes, not pieces of semi-structured data.

Patterns. A convenient way to combine multiple path expressions is through patters. Consider the following query, returning all titles that were published before 1979:

% Query q2:
select X
from biblio.paper T, T.title X, T.year Y
where Y < 1979

Note that here the expression *T.title* starts from the node *T* rather than from the root. The query can be written more concisely as:

% Query q3:
select X
from {biblio: {paper: {title: X, year: Y}}}
where Y < 1979

Here *biblio: paper: title: X, year:Y* is a pattern.

Constructors. To return semi-structured data rather than nodes a query language needs to have *constructors*. For example the expressions *authorResult:X* and *result:{title X, author Z}* are constructors below:

% Query q4:
select authorResult:X
from biblio.book.author X
% Query q5:
select result:{title: X, author: Z}
from biblio.paper T, T.title X, T.year Y,
    T.author Z
where Y < 1979

More complex constructors can built by nesting subqueries inside constructors, as in:

% Query q6:
select row:(select author: Y from X.author Y)
from biblio.book X
% Query q7:
select row: (select author: Y, title: T
from X.author Y X.title T)

from biblio.paper X
where "Roux" in X.author

Declarative semantics. The formal semantics of a query is given in three steps. In the first step a set of bindings to all variables is computed, which results in a relation that has one column for every variable in the query, and one row for every binding of these variables. In the second step the predicates in the *where* clause are applied to select a subset of the rows. In the third step, the constructor is applied to each remaining row. For example in query *q6* above, the effect of the first step is to construct a relation of tuples of the form (t, x, y, z), where t, x, y, z are nodes or atomic values in the input databases. The second step selects only those tuples for which $z < 1979$. The third step constructs a partial graph for each remaining tuple (t, x, y, z) consisting of a root (common among all these graphs), an edge labeled *result*, and from there two more edges labeled *title* and *author*.

Skolem functions. A Skolem function takes as input some arguments and constructs as output a fresh new node. The essential property of a Skolem function is that if called again at a later time, on the same arguments, then it returns the same node for those argument, not a new one. This allows for duplicate elimination, grouping, and the construction of cyclic outputs. The query below is a standard grouping query that groups publications by their years:

% Query q8:
select resultYear f(Y): {paper: {title: X,
    author: Z}
from biblio.paper T, T.title X, T.year Y,
    T.author Z

Here $f(Y)$ is the Skolem function. Without it, the query would construct a separate *resultYear* node for every binding of the variables T, X, Y, and Z. The Skolem function determines the query to construct only one node for every year, thus performing a duplicate elimination on years.

The combination of Skolem functions and regular path expressions leads to transformation languages that are not compositional. Consider a semi-structured data instance that represents a binary table R:

R: { row: {a:4, b:8},
    row: {a:3, b:4},
    row: {a:3, b:9},
    . . .
}

It is known that the relational calculus cannot express the transitive closure of R. The same holds for the language described so far, since its only addition to the relational calculus consists of regular path expressions, but on this simple data instance these expressions can only be applied to very short paths, namely $R.row.a$ and $R.row.b$, and therefore do not give extra power. However, the transitive closure can be expressed by first transforming the relation R into a graph, i.e., making all atomic values into nodes, then using a regular expression on this graph to compute the transitive closure. The two queries are

% Query q9:
select node f(X): {value: X, next: f(Y)}
from row T, T.a X, T.b Y
% Query q10:
select result: {a: X, b:Y}
from node U, U.value X, U.next* V, V.value Y

The first query constructs for every value x of the a attribute a new node $f(x)$ with two outgoing edges: one to a leaf holding the value x, and the other to the node $f(y)$. Thus, the edges from $f(x)$ to $f(y)$ in the output graph materialize the implicit graph given by the binary relation R. The second query uses the regular expression $next^*$ to compute the transitive closure on this graph.

Key Applications

See applications for semi-structured data.

Cross-References

▶ Semi-Structured Data
▶ Semi-Structured Data Model

▶ XML
▶ XML Tree Pattern, XML Twig Query
▶ XPath/XQuery

Recommended Reading

1. Abiteboul S, Quass D, McHugh J, Widom J, Wiener J. The Lorel query language for semistructured data. 1996. http://www-db.stanford.edu/lore/.
2. Buneman P, Davidson S, Suciu D. Programming constructs for unstructured data. In: Proceedings of the Workshop on Database Programming Languages; 1995.
3. Deutsch A, Fernandez M, Florescu D, Levy A., Suciu D. A query language for XML. In: Proceedings of the 8th International World Wide Web Conference; 1999. p. 77–91.
4. Fernandez M, Florescu D, Kang J, Levy A, Suciu D. Catching the boat with Strudel: experience with a Website management system. In: Proceedings of the ACM SIGMOD International Conference on Management of Data; 1998. p. 414–25.
5. Papakonstantinou Y, Abiteboul S, Garcia-Molina H. Object fusion in mediator systems. In: Proceedings of the 22th International Conference on Very Large Data Bases; 1996. p. 413–24.

Semi-supervised Learning

Sugato Basu
Google Inc, Mountain View, CA, USA

Synonyms

Semi-supervised classification

Definition

In machine learning and data mining, supervised algorithms (e.g., classification) typically learn a model for predicting an output variable (e.g., class label for classification) from some supervised training data (e.g., data instances annotated with both features and class labels). These algorithms use various techniques of increasing the

network is determined by how quickly individual sensor nodes drain their energy resources. This line of research examines techniques for increasing network lifetime through careful energy management or harvesting.

2. *Communication reliability.* Wireless communication is notoriously unpredictable, and sensor networks are often deployed in heavily obstructed environments. As such, being able to reliably retrieve sensed data, while still being energy-efficient, is a significant challenge.

3. *Spatio-temporal awareness.* Information generated by sensors is useful only when it is associated with a spatio-temporal context that indicates when and where the data was generated. To do this, sensor nodes need accurate positioning and time synchronization technologies. Unfortunately, these nodes are often deployed in locations (indoors, or in foliage) where the Global Positioning System (GPS) is unavailable.

4. *Programmability.* Sensor networks have a very large number of potential applications, and there is a dire need for novel programming paradigms that hide the complexity imposed by the previous challenges, yet promote reusable and easily understood systems.

5. *Security and Data integrity.* Because sensors may be deployed in unattended and harsh environments, there is critical need for software that ensures the security of the data generated by the sensors, and the overall integrity or correctness of the data.

The following paragraphs summarize research in these areas.

Energy-awareness. Perhaps most widely studied, research in improving network lifetime has proceeded on two fronts. There has been exploratory work on technologies for harvesting energy from light sources, from vibration, or by using mobility to find a nearby power source. This line of research aims to renew the supply of energy at a node. More extensively studied is the direction in which careful algorithms and systems design techniques are used to conserve energy. Specifically, it has been experimentally

determined that communication and sensing are among the primary contributors to energy usage. To conserve communication energy, researchers have explored techniques like *data aggregation* [6, 11] where sensor readings are processed and possibly compressed within the network before being transmitted. In addition, techniques employed at various software layers (medium access, routing, and application) that turn the radio off or put the node to seek during periods of inactivity have also been studied. To conserve sensing energy, researchers have proposed exploiting correlations between sensors, or using one sensor to trigger another higher-energy sensor.

Communication reliability. Wireless packet transmissions are susceptible to losses or corruption due to environmental noise and collisions from concurrent transmissions. To overcome losses due to packet corruption, researchers have explored careful coding techniques that allow for packet reconstruction using relatively small amounts of information. Another line of research has explored the construction of high-quality routing paths which attempt to reduce the likelihood of corruption [15]. To ensure reliable end-to-end delivery, researchers have explored the combined use of hop-by-hop recovery and end-to-end retransmissions. To reduce packet loss due to contention, a line of work has explored *congestion control* techniques that dynamically adapt a node's sending rate to avoid saturating the channel capacity [13]. These techniques essentially sample the current channel conditions, and adjust node sending rates according to feedback control laws that ensure stability and efficiency.

Spatio-temporal awareness. When a sensor node is deployed, it faces two important questions: "Where am I?" and "What is the time?" Without answers to these questions, the data generated by sensor networks becomes meaningless. Since GPS cannot be assumed to work well in the kinds of obstructed environments that sensor networks are likely to be deployed in, researchers have explored a wide variety of methods for *localization* and *time synchronization*. For the former problem, a variety of devices (such as ultrasound, radio, lasers) have been used to estimate the distances between two nodes, and a variety of techniques ranging from multi-lateration and least-squares regression to multidimensional scaling have been used to take these estimates and place nodes in a coordinate system [9]. For the latter problem, a similar approach has been followed: methods to locally synchronize clocks between neighboring nodes use message transmission and processing latencies and carefully compensate for clock drift and skew [12]; a second class of methods attempts to synchronize clocks across the entire network while minimizing error accumulation at every hop.

Programmability. Programming these tiny sensors because of the platform constraints imposed by form factor and lifetime requirements: energy, processor, memory, and wireless communication bandwidths are all constraints that affect application development. The literature on programming sensor networks has focused on how to relieve the burden of the programmer in dealing with these constraints. Starting from seminal work on event-driven operating systems [5], researchers have moved on to higher-level abstractions that simplify programming. These include specialized abstractions for programming individual nodes such as virtual machines [10], state machines, neighborhood communication abstractions, and so on. More recently, researchers have turned towards specialized programming languages for expressing the behavior of the network as a whole [8]. An important thread in this line of research is the application of database techniques to program sensor networks: this thread is discussed below.

Security and data integrity. More recently, research in sensor networks has turned towards techniques to ensure the integrity of data produced by the sensors and transmitted across the network. Fundamentally, sensors can produce erroneous results, and the data transmitted by sensors can be inadvertently or maliciously corrupted while in transit. Techniques to address these problems have examined encryption and authentication mechanisms for medium access [7] and routing, as well as statistical techniques for detecting outliers in sensor data or for being able to quantify the confidence attributable to a received result.

Data-Centric Techniques in Sensor Networks

An important development in the history of sensor networking has been the emergence of *data-centricity* – the use of programming abstractions that specify attributes or characteristics of data, rather than the nodes at which data may be found. Two views of data-centricity in sensor networks have emerged: the *database view* which views the network as a virtual relational table, and the networking view which views the network as a distributed hash table [14].

More prominently explored, research on the database view has been motivated by three challenges: first, the data being generated by sensors is continuously changing; second, because of quantization effects and variable spatial density, these data are inherently approximate; and third, given the energy constraints, it is infeasible to extract large volumes of data from the network. These constraints have motivated research on approximate, aggregate, continuous queries, on the virtual relational table resulting from data generated by the various sensors. This setting provides ample scope for exploring query optimization techniques. These optimization techniques exploit user-specified bounds on result error, correlations among different types of sensors, or models of the sensor field in order to reduce communication and sensing cost.

The networking view has explored one-shot point and enumeration queries by modeling the sensor network as a distributed hash table. This approach uses random or locality preserving hashing of generated sensor data to a geographic location in a two-dimensional coordinate system. The data are stored at the corresponding location. This approach enables the construction of innovative distributed indexing structures, such as, for example, those that support multidimensional range queries. Overall, this approach tends to have higher communication costs except in regimes with relatively high query rates (such as, for example, queries generated by programs running within the network).

Key Applications

The following are some examples of potential application areas for sensor networks.

1. *Atmospheric.* In this application, sensors measure the concentration of pollutants or carbon dioxide, as well as other atmospheric conditions (temperature, humidity, wind direction and speed). These dense measurements can give a much clearer picture of local variations in atmospheric conditions.
2. *Habitats.* Sensors can be used to measure light, temperature, humidity, and photosynthetically active radiation across a forest floor or a forest canopy transect. This gives a detailed picture of spatio-temporal variations in these quantities, and can help biologists understand the effect of these variations on local distributions of plant and animal life.
3. *Water.* Sensor nodes deployed on buoys on the surface of the lake or near the seashore and measuring temperature gradients and chlorophyll beneath the lake's surface can help marine biologists study the dynamics of plankton populations. These populations can significantly affect marine life and thereby have huge economic impact.
4. *Soil.* A network of sensors deployed in soil can measure temperature, light, humidity and contaminant flow. Such measurements can aid precision agriculture for improving crop yields and ensuring better land management.
5. *Man-made structures.* Integrity and energy-expenditure of structures. Sensors deployed on buildings and measuring light and temperature conditions, as well as ambient vibrations, can help understand (and control) energy usage in largebuildings as well as assess the integrity of these structures. Similar sensor networks can also be used to measure activity in seismically active areas (e.g., volcanos).

Future Directions

In the first 5–8 years of its existence, sensor networks were driven by a technological push.

Advances in miniaturization made it possible to envision networks of small devices, and these advances prompted many of the research directions described above. Now that the potential of sensor networks is well established, the next phase of research has to deliver reliable, manageable systems that provide ease of programmability. Thereafter, advances in sensor networks will be driven by experiences obtained from large-scale deployments.

Cross-References

▶ Continuous Queries in Sensor Networks
▶ Data Acquisition and Dissemination in Sensor Networks
▶ Data Aggregation in Sensor Networks
▶ Data Compression in Sensor Networks
▶ Data Estimation in Sensor Networks
▶ Data Fusion in Sensor Networks
▶ Data Storage and Indexing in Sensor Networks
▶ Database Languages for Sensor Networks
▶ Query Optimization in Sensor Networks

Recommended Reading

1. Akyildiz I, Su W, Sankarasubramaniam Y, Cayirci E. A survey on sensor networks. IEEE Commun Mag. 2002;40(8):102–14.
2. Asada G, Dong T, Lin F, Pottie G, Kaiser W, and Marcy H. Wireless integrated network sensors: low power systems on a chip. In: Proceedings of the European Solid State Circuits Conference; 1998.
3. Culler DE, Hong W. Wireless sensor networks - introduction. Commun ACM. 2004;47(6):30–3.
4. Estrin D, Govindan R, Heidemann J. Embedding the internet: introduction. Commun ACM. 2000;43(5):38–41.
5. Hill J, Szewczyk R, Woo A, Hollar S, Culler D, Pister K. System architecture directions for networked sensors. SIGPLAN Not. 2000;35(11):93–104.
6. Intanagonwiwat C, Govindan R, Estrin D. Directed diffusion: a scalable and robust communication paradigm for sensor networks. In: Proceedings of the 6th Annual International Conference on Mobile Computing and Networking; 2000. p. 56–67.
7. Karlof C, Sastry N, Wagner D. TinySec: link-layer encryption for tiny devices. In: Proceedings of the 2nd International Conference on Embedded Networked Sensor Systems; 2004. p. 162–75.
8. Kothari N, Gummadi R, Millstein T, Govindan R. Reliable and efficient programming abstrac-

tions for wireless sensor networks. In: Proceedings of the SIGPLAN Conference on Programming Language Design and Implementation; 2007. p. 10–3.

9. Langendoen K, Reijers N. Distributed localization in wireless sensor networks: a quantitative comparison. Technical University, Delft, Technical Report PDS-2002-003. 2002.
10. Levis P, Culler D. Maté: a tiny virtual machine for sensor networks. In: Proceedings of the 10th International Conference on Architectural Support for Programming Languages and Operating Systems; 2002. p. 85–95.
11. Madden S, Franklin MJ, Hellerstein JM, Hong W. TAG: tiny AGgregate queries in ad-hoc sensor networks. In: Proceedings of the 5th USENIX Symposium on Operating System Design and Implementation; 2002. p. 131–46.
12. Maróti M, Kusy B, Simon G., Lédeczi Á.. The flooding time synchronization protocol. In: Proceedings of the 2nd International Conference on Embedded Networked Sensor Systems; 2004. p. 39–49.
13. Rangwala S, Gummadi R, Govindan R, Psounis K. Interference-aware fair rate control in wireless sensor networks. In: Proceedings of the ACM SIGCOMM Symposium on Network Architectures and Protocols; 2006.
14. Woo A, Madden S, Govindan R. Networking support for query processing in sensor networks. Commun ACM. 2004;47(6):47–52.
15. Woo A, Tong T, Culler D. Taming the underlying challenges of reliable multihop routing in sensor networks. In: Proceedings of the 1st International Conference on Embedded Networked Sensor Systems; 2003.

Sequenced Semantics

Michael H. Böhlen[1,2] and Christian S. Jensen[3]
[1]Free University of Bozen-Bolzano, Bozen-Bolzano, Italy
[2]University of Zurich, Zürich, Switzerland
[3]Department of Computer Science, Aalborg University, Aalborg, Denmark

Definition

Sequenced semantics make it possible to generalize a query language statement on a nontemporal database to a temporal query on a corresponding temporal, interval time-stamped database by applying minor syntactic modifications to the statement that are independent of the particular statement. The semantics of such a generalized statement is consistent with considering the temporal database as being composed of a sequence of nontemporal database states. Sequenced semantics takes into account the interval timestamps of the argument tuples when forming the interval timestamps associated with result tuples, as well as permits the use of additional timestamp-related predicates in statements.

Key Points

A question that has intrigued temporal database researchers for years is how to systematically generalize nontemporal query language statements, i.e., queries on nontemporal databases, to apply to corresponding temporal databases. A prominent approach is to view a temporal database as a sequence of nontemporal databases. Then a nontemporal statement is rendered temporal by applying it to each nontemporal database, followed by integration of the nontemporal results into a temporal result. Sequenced semantics formalizes this approach and is based on three concepts: S-reducibility, extended S-reducibility, and interval preservation. These topics are discussed in turn.

The ensuing examples assume a database instance with three relations:

Employee

ID	Name	VTIME
1	Bob	$5-8$
3	Pam	$1-3$
3	Pam	$4-12$
4	Sarah	$1-5$

Salary

ID	Amt	VTIME
1	20	$4-10$
3	20	$6-9$
4	20	$6-9$

S

Bonus

ID	Amt	VTIME
1	20	$1-6$
1	20	$7-12$
3	20	$1-12$

S-Reducibility

S-reducibility states that the query language of the temporally extended data model must offer, for each query q in the nontemporal query language, a *syntactically similar temporal query* q^t that is its natural generalization, i.e., q^t is snapshot reducible to q, and q^t is syntactically identical to $S_1\ qS_2$. The goal is to make the semantics of temporal queries easily understandable in terms of the semantics of the corresponding nontemporal queries. The strings S_1 and S_2 are independent of q and are termed *statement modifiers* because they change the semantics of the entire statement q that they enclose.

In the following examples, statements are prefixed with the modifier *SEQ VT* [2]. This modifier tells the temporal DBMS to evaluate statements with sequenced semantics in the valid-time dimension. These examples illustrate that S-reducible statements are easy to write and understand because they are simply conventional SQL statements with the additional prefix *SQL VT*. Writing statements that compute the same results, but without using statement modifiers, can be very difficult [3]:

```
SEQ VT SELECT * FROM EMPLOYEE;
SEQ VT
SELECT ID FROM EMPLOYEE AS E
WHERE NOT EXISTS (SELECT * FROM SALARY
  AS S WHERE E.ID = S.ID);
```

The first query returns all *Employee* tuples together with their valid time – this corresponds to returning the content of *Employee* at each state. The second query determines the time periods when an employee did not get a salary. It returns $\{\langle 3,\ 1\text{–}3\rangle, \langle 3,\ 4\text{–}5\rangle, \langle 3,\ 10\text{–}12\rangle, \langle 4,\ 1\text{–}5\rangle\}$. Conceptually the enclosed statement is evaluated on each state of the database. Computationally, the interval 6–9 is subtracted from the interval 4–12 to get the intervals 4–5 and 10–12.

Extended S-Reducibility

S-reducibility is applicable only to queries of the underlying nontemporal query language and does not extend to queries with explicit references to time. Consider the following queries:

```
SEQ VT
SELECT E.ID
FROM Employee AS E, Salary AS S
WHERE E.ID = S.ID
AND DURATION (VTIME(E)) >DURATION
(VTIME(S));
SEQ VT
SELECTE.ID,VTIME(S)????????, VTIME(E)
FROMEmployeeAS E, Salary AS S
WHEREE.ID = S.ID;
```

The first query constrains the temporal join to tuples in *Employee* with a valid time that is longer than the valid time of the salary tuple it shall be joined with. This condition cannot be evaluated on individual nontemporal relation states because the timestamp is not present in these states. Nevertheless, the temporal join itself can still be conceptualized as a nontemporal join evaluated on each snapshot, with an additional predicate. The second query computes a temporal join as well, but also returns the original valid times. Again, the semantics of this query fall outside of snapshot reducibility because the original valid times are not present in the nontemporal relation states.

DBMSs generally provide predicates and functions on time attributes, which may be applied to, e.g., valid time, and queries such as these arise naturally. Applying sequenced semantics to statements that include predicates and functions on time offers a higher degree of orthogonality and wider ranging temporal support.

Interval Preservation

Coupling snapshot reducibility with syntactical similarity and using this property as a guideline for how to semantically and syntactically embed temporal functionality in a language is attractive. However, S-reducibility does not distinguish between different relations if they are snapshot equivalent. This means that different results of an S-reducible query are possible: the results will be snapshot equivalent, but will differ in how the result tuples are timestamped. As an example, consider a query that fetches and displays the content of the *Bonus* relation. An S-reducible query may return the result $\{\langle 1, 20, 1\text{--}6\rangle, \langle 1, 20, 7\text{--}12\rangle, \langle 3, 20, 1\text{--}12\rangle\}$. If Bob received a 20K bonus for his performance during the first half of the year and another 20K bonus for his performance during the second half of the year and Pam received a 20K bonus for her performance during the entire year, this is the expected result. This is also the result supported by the three tuples in the example instance displayed above. However, S-reducibility does not distinguish this result from any other snapshot equivalent result. With S-reducibility a perfectly equivalent result would be $\{\langle 1, 20, 1\text{--}12\rangle, \langle 3, 20, 1\text{--}6\rangle, \langle 3, 20, 7\text{--}12\rangle\}$.

Interval preservation settles the issue of which result should be favored out of the many possible results permitted by S-reducibility. When defining how to timestamp tuples of query results, two possibilities come to mind. Results can be coalesced. This solution is attractive because it defines a canonical representation for temporal relations. A second possibility is to consider lineage and preserve, or respect, the timestamps as originally entered into the database [1]. Sequenced semantics requires that the default is to preserve the timestamps – being irreversible, coalescing cannot be the default.

Cross-References

▶ Nonsequenced Semantics
▶ Snapshot Equivalence
▶ Temporal Coalescing
▶ Time Interval
▶ Valid Time

Recommended Reading

1. Böhlen MH, Busatto R, Jensen CS. Point- versus interval-based temporal data models. In: Proceedings of the 14th International Conference on Data Engineering; 1998. p. 192–200.
2. Böhlen MH, Jensen CS, Snodgrass RT. Temporal statement modifiers. ACM Trans Database Syst. 2000;25(4):48.
3. Snodgrass RT. Developing time-oriented database applications in SQL. San Francisco: Morgan Kaufmann; 1999.

Sequential Patterns

Jianyong Wang
Tsinghua University, Beijing, China

Synonyms

Frequent subsequences

Definition

A *sequence database* $D = \{S_1, S_2,...,S_n\}$ for sequential pattern mining consists of n input sequences (where $n \geq 1$), and an *input sequence* $S_i = \langle e_{i1}, e_{i2}, ..., e_{im}\rangle (1 \leq i \leq n)$ is an ordered list of m events (where $m \geq 1$). Each *event* $e_{i_j} (1 \leq i \leq n, 1 \leq j \leq m)$ is a nonempty set of items. Given two sequences, $S_a = \langle e_{a1}, e_{a2}, ..., e_{ak}\rangle$ and $S_b = \langle e_{b1}, e_{b2}, ..., e_{bl}\rangle$, if $k \leq l$ and there exist integers $1 \leq x_1 < x_2 < ... < x_k \leq l$ such that $e_{a1} \subseteq e_{b_{x1}}, e_{a2} \subseteq e_{b_{x2}}, ..., e_{ak} \subseteq e_{b_{xk}}$, S_b is said to *contain* S_a (or equivalently, S_a is said to be contained in S_b). The number of input sequences in D that contain sequence S is called the *support* of S in D, denoted by $sup^D (S)$. Given a user-specified minimum support threshold *min_sup*, S is called a *sequential pattern* (or a *frequent subsequence*) in D if $sup^D (S) \geq min\_sup$. If there exists no proper supersequence of a sequential pattern S with the same support as S, S is called a *closed sequential pattern* (or a *frequent closed subsequence*) in D. Furthermore, a sequential pattern S is called a *maximal sequential pattern* (or a *frequent maximal subsequence*) if it is not contained in any other sequential pattern. The problems

S

of *sequential pattern mining, closed sequential pattern mining,* and *maximal sequential pattern mining,* are to find all frequent subsequences, all frequent closed subsequences, and all frequent maximal subsequences from input sequence database D, respectively, given a user-specified minimum support threshold *min_sup.*

Historical Background

Similar to association rule mining, sequential pattern mining was initially motivated by the decision support problem in retail industry and was first proposed by Rakesh Agrawal and Ramakrishnan Srikant in [1]. Later on, it was applied to other domains. Some recent research work further validated its utility in various applications, such as identifying outer membrane proteins, automatically detecting erroreous sentences, discovering block correlations in storage systems, identifying copy-paste and related bugs in large-scale software code, API specification mining and API usage mining from open source repositories, frequent subsequence-based XML document clustering, sequence-based XML query pattern mining for effective caching, and Web log data mining.

In the seminal paper on sequential pattern mining, three algorithms were introduced [1]. Among these algorithms, AprioriSome and DynamicSome were proposed for mining maximal sequential patterns, while AprioriAll was designed for mining all sequential patterns. The same authors later generalized the sequential pattern mining problem by allowing time constraints, sliding time window, and taxonomies, and proposed a new algorithm, GSP [10]. The inefficiency of AprioriAll mainly stems from its computationally expensive data transformation operation which transforms each transaction to a set of frequent itemsets in order to find sequential patterns. As GSP overcomes the drawbacks of AprioriAll, it is much faster than AprioriAll. In [4], Jiawei Han et al. proposed the FreeSpan algorithm, which shows better performance than GSP. In [15], Mohammed J. Zaki adopted the vertical data representation

for sequential pattern mining and devised an efficient algorithm, SPADE, which fully exploits the lattice search techniques and some join operations. Another state-of-the-art sequential pattern mining algorithm is PrefixSpan, which was proposed by Jian Pei et al. [8]. It adopts a projection-based, sequential pattern growth approach to avoiding the traditional candidate-generation-and-test paradigm, thus improves the algorithm efficiency. In [3], Jay Ayres et al. designed another sequential pattern mining algorithm, SPAM. This algorithm integrates a depth-first search strategy with some effective pruning techniques, uses a vertical bitmap data representation, and can incrementally output frequent subsequences in an online fashion.

All the preceding algorithms except AprioriSome and DynamicSome mine the complete set of sequential patterns. One problem with sequential pattern mining is that it may generate too many redundant patterns, which also impedes the algorithm efficiency. One popular solution to this problem is to mine closed sequential patterns only, which usually leads to not only a more compact yet complete result set but also better efficiency. In [14], Xifeng Yan and Jiawei Han presented the CloSpan approach, which incorporates several effective pruning methods into the PrefixSpan framework and achieves much better performance than PrefixSpan. Recently, another closed sequential pattern mining algorithm, BIDE, was proposed in [12] by Jianyong Wang et al. It integrates a new pattern closure checking scheme and a new pruning technique with the PrefixSpan framework, and is both runtime and memory efficient.

Foundations

The biggest challenge faced by sequential pattern mining is the combinatorial explosion problem. To alleviate this problem, researchers have tried various ways. In the following some factors which may have an impact on the efficiency of a sequential pattern mining (or closed sequential pattern mining) algorithm are summarized. These factors mainly include the data representation format, pattern enumeration framework (i.e., search strategy), search space pruning techniques, and pattern closure checking scheme.

Sequence Data Format

The input sequence data can be represented in two alternative formats. The *horizontal representation* is a natural bookkeeping of the input sequences. Each sequence consists of an ordered list of events, while each event is recorded as a list of items (which are supposed in most algorithms to be sorted according to a certain order, say the lexicographical order). AprioriAll, GSP, PrefixSpan, CloSpan, and BIDE are typical examples adopting the horizontal representation. Note that AprioriAll needs to first find the frequent itemsets and transform each event to the set of frequent itemsets contained in the event in order to find sequential patterns. To compute the support of a subsequence, the horizontal format based algorithms need to scan the database, which is computationally expensive. To assist support counting, these algorithms devise some special mechanisms. For example, GSP introduces the hash-tree data structure, while PrefixSpan, CloSpan, and BIDE use a projection-based approach to shrink the part of database that needs to be scanned.

In the *vertical representation*, the database is represented as a set of items, where each item is recorded as a set of pairs of sequence identifier (SID) and event identifier (EID) containing the item. SPADE and SPAM use the vertical format. With the vertical representation, the support-counting can be performed using simple join operations with temporal ordering constraint. To improve the efficiency of support counting, SPAM proposes to use vertical bitmaps to represent the sequence database. Each item is converted to a bitmap, which has a bit for each event in the database. If an event contains an item, the corresponding bit regarding the event and item is set to one; otherwise, it is set to zero. Based on the transformed bitmap sequence representation, efficient support counting can be easily achieved using bitwise AND operations of bitmaps.

Search Strategy

Given an input sequence database D and a minimum support threshold *min_sup*, the set of sequential patterns are deterministic, and can be organized into a lexicographic frequent sequence tree structure. Suppose *min_sup*=2, database D contains three input sequences, $\langle$(A B D)(B C D)(A)$\rangle$, $\langle$(B) (A B E) (B)$\rangle$, and $\langle$(A B) (B C D)$\rangle$ (here the commas separating each pair of adjacent events in the same sequence are omitted), respectively, and there exists a lexicographic ordering among the set of distinct items A $\leq$B $\leq$C $\leq$D $\leq$E. For a prefix subsequence S_p, it can be extended in two ways, namely, sequence-extension and itemset-extension. The *sequence-extension* extends S_p by a new event containing a single item, while *itemset-extension* adds a new item to the last event of S_p and the new item must be lexicographically larger than any item of the last event of S_p. Assume both sequence-extension and itemset-extension are performed in lexicographic ordering and itemset-extension is performed before sequence-extension for the same prefix sequence. Then, the lexicographic sequence tree structure of the frequent subsequences in the running example is shown in Fig. 1. Note that each node in the tree shows a sequential pattern and its corresponding support (i.e., the number after the colon), and all the patterns at the same level have the same length (namely, they contain the same number of items). An edge which links a parent node P at level k to a child node C at level (k+1) indicates that the pattern at C is directly extended from the prefix pattern at node P, either by sequence-extension or by itemset-extension.

Once the sequential patterns are organized into a lexicographic tree structure, one can choose a tree traversal strategy for sequential pattern mining. The two popular search strategies are breadth-first search and depth-first search. In the *breadth-first search* method, frequent subsequences are mined in a level-wise manner, that is, before mining patterns with length (k+1), one needs to first mine all patterns with length k. In contrast, a *depth-first search* method traverses the sequence tree in depth-first order. GSP adopts the breadth-first search strategy to enumerate the

S

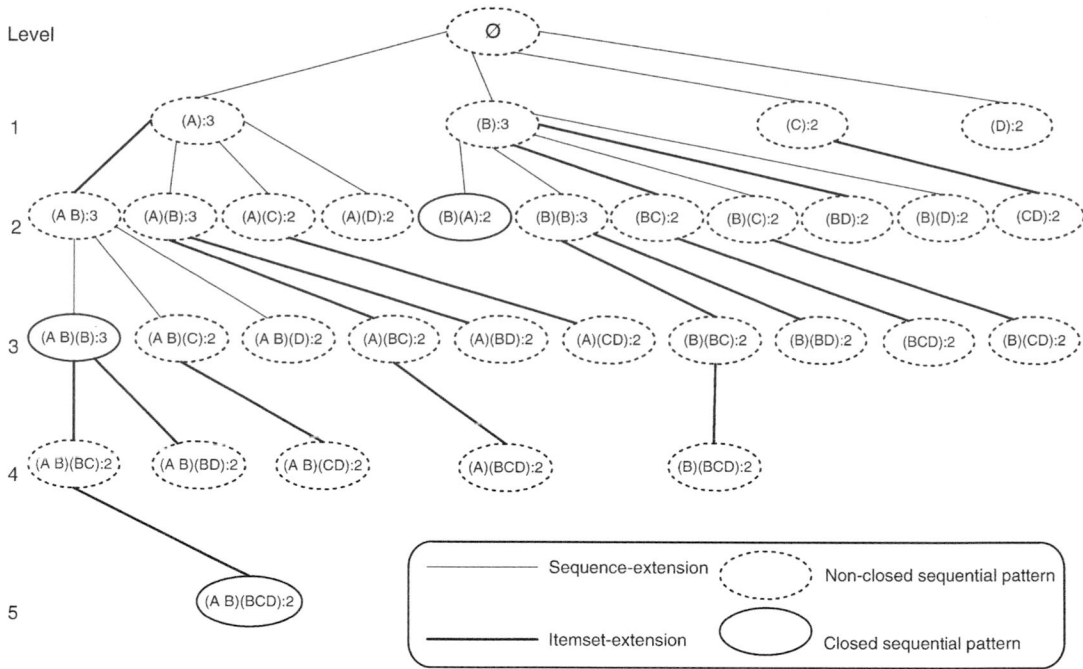

Sequential Patterns, Fig. 1 The lexicographic sequence tree structure of the frequent subsequences in the running example

sequential patterns, PrefixSpan, SPAM, CloSpan, and BIDE choose the depth-first search paradigm, while SPADE supports both breadth-first search and depth-first search methods.

Search Space Pruning

One of the most crucial optimization considerations to improve the efficiency of a data mining algorithm is to devise some effective search space pruning techniques. Based on some heuristics, if some parts of the search space are already known to be futile in generating sequential patterns, they should be found and pruned as quickly as possible. Perhaps the most well-known property used for designing pruning methods in frequent pattern mining is the *Apriori* property (also known as the downward closure property or anti-monotone property). It states in the sequence mining setting that all the subsequences of a frequent sequence must be also frequent, or equivalently, a sequence must be infrequent if it contains an infrequent subsequence. All the existing sequential pattern mining algorithms have exploited the Apriori property in different ways.

From Fig. 1 one can see that only three out of the 31 frequent subsequences are closed, namely, (B)(A):2, (AB)(B):3, and (AB)(BCD):2, and many subtrees in Fig. 1 contain no closed sequential pattern. An efficient closed sequential pattern mining algorithm should avoid traversing the subtrees containing no closed patterns, which leaves room to further prune the search space. Both CloSpan and BIDE adopt some optimization techniques. The pruning methods proposed in CloSpan are listed as follows.

- *Common prefix pruning*: if there exists a common prefix, all sequences beginning with a proper subsequence of this prefix cannot be closed.
- *Partial-order pruning*: if an item "a" always occurs before item "b" in all sequences, any sequence beginning with "b" cannot be closed.
- *Equivalent projected DB pruning*: given two subsequences, s and s', where s is a proper subsequence of s' and they have equivalent projected database, any sequence beginning with s cannot be closed.

The concepts of common prefix, partial-order, and equivalent projected database can be found in [7]. The BIDE algorithm proposes a single but effective pruning technique called *BackScan* search space pruning. Given a prefix S_p, if $\exists\ i$ (i is a positive integer and is no greater than the length of S_p) and there exists any item that appears in each of its ith semimaximum periods, S_p can be safely pruned. The interested readers are referred to [12] for more details.

Pattern Closure Checking Scheme

The optimization methods proposed in CloSpan and BIDE are very effective in pruning the unpromising parts of the search space, however, they cannot assure that each discovered sequential pattern is closed. For closed sequential pattern mining algorithms, one still needs to devise some methods to check if a sequential pattern is closed or not. In CloSpan, all the candidate closed sequential patterns are maintained in a tree data structure. CloSpan eliminates the non-closed patterns in a post-processing phase and adopts the hashing technique to accelerate the pattern closure checking. When the number of candidate sequential patterns is large, the pattern tree structure may consume non-trivial space. In [12], BIDE adopts a so-called *BI-Directional Extension* closure checking scheme. One big advantage of this new scheme is that it avoids maintaining the set of candidate closed sequential patterns, and thus saves space. If there exists no *forward-S-extension* item, *forward-I-extension* item, *backward-S-extension* item, nor *backward-I-extension* item with respect to a prefix sequence S_p, S_p is a closed sequence, otherwise, S_p must be non-closed. The definitions of a forward-S-extension item, a forward-I-extension item, a backward-S-extension item, and a backward-I-extension item can be found in [8].

Key Applications

In recent years sequential pattern mining witnessed many applications, which roughly fall into the following categories.

Frequent Subsequence-Based Classifier

In [9], the authors used an efficient implementation of generalized suffix tree to mine a set of frequent subsequences with a minimum length constraint, and built the rule-based classifier and SVM classifier based on the discovered frequent subsequences. Their performance results demonstrate that the frequent subsequence-based classifier achieves high accuracy in identifying outer membrane proteins. Recently the authors of [11] proposed a method to build associative classification rules from frequent subsequences returned by a variant of the PrefixSpan algorithm. Their performance study shows that sequence-based classification rules are very helpful in automatically detecting erroreous sentences.

Operating System and Software Engineering

In [5, 6], the authors adopted the CloSpan algorithm to mine closed sequential patterns, which have been shown very useful in discovering block correlations in storage systems and identifying copy-paste and related bugs in large-scale software code. In [7, 13], the BIDE algorithm was used to discover the set of frequent closed subsequences with the purpose of API specification mining and API usage mining from open source repositories.

Frequent Subsequence-Based XML Data Management

Sequential pattern mining has also been widely applied in semi-structured data management. In this application, an XML document is first converted to a sequence instead of a tree structure. Then some kinds of constrained frequent subsequences are mined, which can be exploited to accelerate XML query or XML document clustering. The performance results in [2] demonstrate that the frequent subsequence-based XML clustering algorithm XProj achieves better clustering quality than previous algorithms.

Web Log Data Mining

Some researchers have also applied sequential pattern mining algorithms in mining salient patterns (e.g., contiguous sequential patterns) from Web log data.

Experimental Results

Some experimental results on sequential pattern mining can be found in [3, 8, 10, 15], which compares the efficiency of some typical sequential pattern mining algorithms including GSP, SPADE, PrefixSpan, and SPAM, while [14, 12] present the performance study of two closed sequential pattern mining algorithms, CloSpan and BIDE.

Data Sets

The IBM synthetic dataset generator can generate sequence datasets and can be found from the link of http://www.cs.rpi.edu/~zaki/software/. The popularly used real sequence datasets include some Web log data, and protein sequence datasets.

URL to Code

The code for PrefixSpan and CloSpan algorithms can be found from the Illini Mine portal, http://dm1.cs.uiuc.edu/protected/im, the code for SPADE algorithm can be downloaded from the link of http://www.cs.rpi.edu/~zaki/software/, while the code for BIDE algorithm can be traced from http://dbgroup.cs.tsinghua.edu.cn/wangjy/.

Cross-References

▶ Apriori Property and Breadth-First Search Algorithms
▶ Closed Itemset Mining and Nonredundant Association Rule Mining
▶ Frequent Graph Patterns
▶ Frequent Itemsets and Association Rules
▶ Frequent Partial Orders
▶ Pattern-Growth Methods

Recommended Reading

1. Agrawal R, Srikant R. Mining sequential patterns. In: Proceedings of the 11th International Conference on Data Engineering; 1995.
2. Aggarwal CC, Ta N, Wang J, Feng J, Zaki MJ. XProj: a framework for projected structural clustering of XML documents. In: Proceedings of the 13th ACM SIGKDD International Conference on Knowledge Discovery and Data Mining; 2007.
3. Ayres J, Gehrke J, Yiu T, Flannick J. Sequential pattern mining using a bitmap representation. In: Proceedings of the 8th ACM SIGKDD International Conference on Knowledge Discovery and Data Mining; 2002.
4. Han J, Pei J, Mortazavi-Asl B, Chen Q, Dayal U, Hsu MC. FreeSpan: frequent pattern-projected sequential pattern mining. In: Proceedings of the 6th ACM SIGKDD International Conference on Knowledge Discovery and Data Mining; 2000.
5. Li Z, Chen Z, Srinivasan S, Zhou Y. C-Miner: mining block correlations in storage systems. In: Proceedings of the 3rd USENIX Conference of on File and Storage Technologies; 2004.
6. Li Z, Lu S, Myagmar S, Zhou Y. CP-Miner: finding copy-paste and related bugs in large-scale software code. IEEE Trans Softw Eng. 2006;32(3):176–92.
7. Lo D, Khoo SC SMArTIC: towards building an accurate, robust and scalable specification miner. In: Proceedings of the 14th ACM SIGSOFT International Symposium on Foundations of Software Engineering; 2006.
8. Pei J, Han J, Mortazavi-Asl B, Pinto H, Chen Q, Dayal U, Hsu MC. PrefixSpan: mining sequential patterns efficiently by prefix-projected pattern-growth. In: Proceedings of the 17th International Conference on Data Engineering; 2001.
9. She R, Chen F, Wang K, Ester M, Gardy JL, Brinkman FSL. Frequent-subsequence-based prediction of outer membrane proteins. In: Proceedings of the 9th ACM SIGKDD International Conference on Knowledge Discovery and Data Mining; 2003.
10. Srikant R, Agrawal R Mining sequential patterns: generalizations and performance improvements. In: Advances in Database Technology, Proceedings of the 5th International Conference on Extending Database Technology; 1996.
11. Sun G, Liu X, Cong G, Zhou M, Xiong Z, Lee J, Lin CY. Detecting erroreous sentences using automatically mined sequential patterns. In: Proceedings of the 45th Annual Meeting of the Association for Computational Linguistics; 2007.
12. Wang J, Han J, Li C. Frequent closed sequence mining without candidate maintenance. IEEE Trans Knowl Data Eng. 2007;19(8):1042–56.
13. Xie T, Pei J. Data mining for software engineering. In: Proceedings of the 12th ACM SIGKDD International Conference on Knowledge Discovery and Data Mining; 2006.

14. Yan X, Han J, Afshar R CloSpan: mining closed sequential patterns in large databases. In: Proceedings of the 2003 SIAM International Conference on Data Mining; 2003.
15. Zaki MJ. SPADE: an efficient algorithm for mining frequent sequences. Mach Learn. 2001;42(1/2): 31–60.

Serializability

Bettina Kemme
School of Computer Science, McGill University, Montreal, QC, Canada

Synonyms

Transactional consistency in a replicated database

Definition

While transactions typically specify their read and write operations on logical data items, a replicated database has to execute them over the physical data copies. When transactions run concurrently in the system, their executions may interfere. The replicated database system has to isolate these transactions. The strongest and most well-known correctness criterion for replicated databases is one-copy-serializability. A concurrent execution of transactions in a replicated database is one-copy-serializable if it is "equivalent" to a serial execution of these transactions over a single logical copy of the database.

Main Text

A transaction is a sequence of read and write operations on the data items of the database. A read operation of transaction T_i on data item x is denoted as $r_i(x)$ and a write operation on x as $w_i(x)$. A transaction T_i either ends with a commit c_i (all operations succeed) or with an abort a_i (whereby all effects on the data are undone before the termination).

A replicated database consists of a set of database servers $A, B, \ldots$, and each logical data item x of the database has a set of physical copies $x^A, x^B, \ldots$ where the index refers to the database server on which the copy resides. *Replica Control* translates each operation $o_i(x), o_i \in \{r, w\}$ of a transaction T_i on logical data item x into physical operations $o_i(x^A), o_i(x^B)$ on physical data copies. Given a set of transactions $\mathcal{T}$, a replicated history RH describes the execution of the physical operations of transactions in the replicated database. For simplicity, the following discussion only considers histories where all transactions commit. A database server A executes the subset of physical operations of the transactions in $\mathcal{T}$ performed on copies residing on A. The local history RH^A describes the order in which these operations occur. For simplicity, a local history is assumed to be a total order. RH is the union of all local histories with some additional ordering. In particular, if a transaction T_i executes $o_i(x)$ on logical data item x before $o_i(y)$ on logical data item y, and RH^A contains physical operation $o_i(x^A)$ and RH^B contains $o_i(y^B)$, then $o_i(x^A) <_{RH} o_i(y^B)$.

As an example, given $T_1 = w_1(y)w_1(x)$ and $T_2 = r_2(y)w_2(x)$ on logical data items x, and database servers A and B, both having a copy of both x and y, the local histories could be

$$RH^A : w_1(y^A)r_2(y^A)w_1(x^A)w_2(x^A)c_1c_2$$
$$RH^B : w_1(y^B)w_1(x^B)w_2(x^B)c_2c_1$$

The replicated history RH is the union of these two local histories plus the ordering of $r_2(y^A) <_{RH} w_2(x^B)$.

Using this notation, the following defines one-copy-serializability for the case that replica control uses ROWA (read-one-write-all approach), i.e., where each read operation is performed on one copy while write operations are performed on all copies of the data item. Failures are ignored. In this restricted case, *conflict equivalence* can be exploited. Two physical operations o_i and o_j conflict, if they are from two different transactions, access the same data copy, and at least one is a write operation.

S

Definition 1 A replicated history RH over a set of transactions $\mathcal{T}$ in a replicated system with servers $A, B, \ldots$ is one-copy-serializable if it is conflict-equivalent to a serial history H of $\mathcal{T}$ over the logical data items. This means that if $o_i(x^A), o_j(x^A) \in RH$ and the operations conflict, then $o_i(x) <_H o_j(x) \in H$ if and only if $o_i(x^A)$ is executed before $o_j(x^A)$ at server A.

Using conflict equivalence, one can easily determine whether RH is one-copy-serializable. For each local history RH^A, the serialization graph $SG(RH^A)$ has each committed transaction as node and contains an edge from T_i to T_j if $o_i(x^A)$ is executed before $o_j(x^A)$ and the two operations conflict. The serialization graph $SG(RH)$ is then the union of the local serialization graphs.

Theorem 1 *A replicated history RH over a set of transactions $\mathcal{T}$ and database servers $A, B, \ldots$ following the ROWA strategy is one-copy-serializable if and only if its serialization graph $SG(RH)$ is acyclic.*

The example history above is one-copy-serializable because its serialization graph contains only an edge from T_1 to T_2, i.e., in all local histories, and for any conflict between T_1 and T_2, T_1's operation is ordered before T_2's operation.

As soon as node failures are considered or both read and write operations only access a subset of copies, conflict equivalence is not appropriate anymore because it might miss catching conflicts at the logical level. For that purpose, one can define one-copy-serializability based on view equivalence which observes which data versions a read operation accesses and in which order write operations occur.

Cross-References

▶ Concurrency Control for Replicated Databases
▶ Consistency Models for Replicated Data
▶ Replica Control
▶ Replicated Database Concurrency Control

Recommended Reading

1. Bernstein PA, Hadzilacos V, Goodman N. Concurrency control and recovery in database systems. Reading: Addison Wesley; 1987.

Serializable Snapshot Isolation

Alan Fekete
University of Sydney, Sydney, NSW, Australia

Synonyms

SerializableSI; SSI

Definition

Serializable Snapshot Isolation is a multi-version concurrency control approach that shares many features of Snapshot Isolation and, in addition, ensures that all executions of the system have the property of serializability. A transaction T that operates under Serializable Snapshot Isolation (like a transaction in Snapshot Isolation) never observes any effects from other transactions that overlap T in duration; instead T sees values as if it were operating on a private copy or snapshot of the database, reflecting all other transactions that had committed before T started. Serializable Snapshot Isolation allows reads to occur without delay or blocking caused by concurrent updates, and also updates are never blocked by concurrent readers, so Snapshot Isolation often gives the transactions better throughput than traditional concurrency control based on two-phase locking. Serializable Snapshot Isolation guarantees that every execution is serializable, by aborting certain transactions if they could lead to anomalies in the execution. Variants of the algorithm differ in exactly which transactions to abort, and in how the necessary information is managed within the DBMS engine.

Historical Background

In 1995, Berenson et al. [1] showed that Snapshot Isolation could permit executions that were not serializable. In 2005, Fekete et al. [3] published a theory to show when a set of transactions is guaranteed to run serializably on a DBMS platform that uses Snapshot Isolation. Cahill et al. [2] introduced the term Serializable Snapshot Isolation and proposed a mechanism to achieve it. Cahill's technique reused much of the mechanism of Snapshot Isolation but also tracked dependencies between some transactions within the lock manager; using this information, the system caused certain aborts (either at the time when locks were requested or later when transactions completed) that would prevent all the non-serializable executions. Cahill described and evaluated prototype implementations of Serializable Snapshot Isolation within the BerkeleyDB and InnoDB platforms. Cahill also showed the essential correctness property (that all executions are serializable) using the theory from [3]. An implementation in the PostgreSQL platform was done by Ports and Grittner, published in 2012 [7], and included in the deployed releases of PostgreSQL since version 9.1. The design of Ports and Grittner tracks some extra dependencies, compared to Cahill's design, and uses some extra conditions to reduce the cases of unnecessary abort. Han et al. [4] have recently proposed some alternative physical structures and latching approaches, to reduce contention and improve the performance of Serializable Snapshot Isolation on multi-core hardware.

Revilak et al. [8] suggested two alternative implementation approaches for Serializable Snapshot Isolation, both of which maintain the complete graph of dependencies between transactions and then use a certification check when the transaction completes. The performance of these techniques and Cahill's own design were compared by Revilak and also by Jung et al. [5]. Another implementation technique, called the Serial Safety Net, was recently described by Wang et al. [9]. This is a certification mechanism that can be added to a variety of concurrency control algorithms, even to some that do not keep multiple versions. When Wang's method is applied to Snapshot Isolation, the combination is a design that has fewer unnecessary aborts. A concurrency control proposal from Lomet et al. [6] also guarantees serializable executions along with the snapshot property where the reads done by each transaction reflect exactly the writes by some set of other transactions. Lomet assigns transaction time stamps to match the serialization order, instead of the time stamps following the temporal sequence of commits as in Snapshot Isolation and its extensions.

Scientific Fundamentals

A concurrency control technique must handle requests for reading and writing data items that are stored by the DBMS engine; each request comes from a particular transaction. A request could be either performed immediately or delayed for a while, or the requesting transaction might be aborted. Serializable Snapshot Isolation, like Snapshot Isolation, uses a multi-version mechanism, and a request to read an item might (but does not necessarily) return a value that had been written in some earlier transaction rather than the most recent value. Just as with Snapshot Isolation, each version is labeled with a commit time stamp that indicates when the writer transaction is completed. When any transaction begins, it gets a start time stamp from the same global clock or sequence generator as used for the commit time stamps. Any request by the transaction to read an item x will not be delayed; instead it returns the value from whichever version of x has the highest commit time stamp that is less than the reading transaction's start time stamp (an exception to this policy is if the reader had earlier written to x, in that case, the reader sees its own write). The engine also makes sure that there are no lost updates: each item's version sequence is not corrupted, so that a new version is created by a write, but only if the most recent version comes from a transaction which already committed, before the new one's writer started.

To understand the workings of Serializable Snapshot Isolation, it helps to consider the multi-version serialization graph (MVSG) for a schedule; we use the definition of Weikum and Vossen [10]. In this notation, we refer to the version of item x written by transaction T_i as x_i, and each operation is subscripted by the index of the transaction that performs the operation and has an argument indicating the version of the item involved. Thus, a write by T_i to the item x is described as $w_i[x_i]$, and a read by T_j that observes this version is shown as $r_j[x_i]$. With this notation, MVSG is defined for a given schedule and a version order $\ll$ which relates every pair of versions of the same data item. For Serializable Snapshot Isolation, as for other forms of Snapshot Isolation, the version order is always taken as the order of the commit time stamps associated with those versions, which is the same as the temporal sequence of the commit of the transactions that wrote the versions. That is, one defines $x_i \ll x_j$ when T_i commits before T_j. MVSG has nodes for the transactions, and there is an edge from T_i to T_j in the following three circumstances: (i) the schedule contains $r_j[x_i]$ for some item x, (ii) $x_i \ll x_j$ and the schedule contains $r_k[x_j]$ for some x and k, or (iii) the schedule contains $r_i[x_k]$ and $x_k \ll x_j$, for some x and k. The first and second of these circumstances are referred to as direct dependencies from T_i to T_j, and the third of these circumstances is described as an anti-dependency from T_i to T_j. Informally, an anti-dependency occurs when T_i must be serialized before T_j because T_i did not observe an effect of T_j.

Serializable Snapshot Isolation concurrency control prevents any cycle arising in the MVSG. Because any direct dependency goes from one transaction to another with a higher commit time stamp, there cannot be any cycle that involves direct dependencies alone. So the focus of Serializable Snapshot Isolation is in preventing any anti-dependency edge being part of a cycle. Theory proved in [3] shows that, in fact, any cycle will have two consecutive anti-dependency edges, and there are additional constraints on the relationships among the start and commit time stamps of the transactions involved. There are variations of the detailed mechanism among proposals for Serializable Snapshot Isolation, and these variants are about how and where the system tracks the existence of anti-dependencies and about when a transaction is forced to abort in order to prevent the occurrence of patterns of anti-dependencies that risk becoming part of a cycle in MVSG.

An implementation of Serializable Snapshot Isolation needs to detect when an anti-dependency from T_i to T_j has occurred. One possibility is that the write of version x_j occurs first in time, and then a read is performed by T_i that returns an earlier version x_k. This is easy to detect, because the internal processing of the read will be examining the versions (including x_j) to choose the appropriate one to return. At this point, it is apparent that there is an anti-dependency. The case where the read $r_i[x_k]$ occurs first in time, and then $w_j[x_j]$ occurs, is harder to detect and typically involves inserting a lock on item x into a lock manager; the lock is in a special mode called SIREAD, which does not block other locks but simply tracks that the read has happened, so that when a later write occurs on x, appropriate anti-dependency information can be extracted. How should the anti-dependencies be recorded by the engine? In the implementation in PostgreSQL, the transaction block keeps a list of the incoming anti-dependencies and another list of the outgoing ones. To ensure Serializable Snapshot Isolation, that is, to prevent any cycles in MVSG, one will abort a transaction if one ever reaches a state where some transaction has both incoming and outgoing anti-dependencies; in fact, additional checks are made, concerning the relationship between the start and commit time stamps of the transactions concerned, and some unnecessary aborts are avoided. There are also issues with garbage collection for the locks and dependencies, discussed in [7].

Key Applications

The Ports and Grittner implementation of Serializable Snapshot Isolation is used in PostgreSQL (since version 9.1) as the concurrency control mechanism for "SET ISOLATION LEVEL SERIALIZABLE."

Future Directions

More research is needed to understand the impact of different implementation details on the performance of algorithms in the Serializable SI family. In particular, there seem to be significant trade-offs between the run-time effort of tracking dependencies, the frequency of checks for aborting, and the rate of unnecessary aborts; it is not yet clear which implementation would be the best choice overall for a system seeking to ensure serializable executions.

Cross-References

► Concurrency Control: Traditional Approaches
► Multiversion Serializability and Concurrency Control
► Serializability
► Snapshot Isolation
► SQL Isolation Levels

Recommended Reading

1. Berenson H, Bernstein PA, Gray J, Melton J, O'Neil EJ, O'Neil PE. A critique of ANSI SQL isolation levels. In: Proceeding of the ACM SIGMOD Conference on Management of Data; 1995. p. 1–10.
2. Cahill MJ, Röhm U, Fekete AD. Serializable isolation for snapshot databases. ACM Trans Database Syst. 2009;34(4):1–42.
3. Fekete A, Liarokapis D, O'Neil E, O'Neil P, Shasha D. Making snapshot isolation serializable. ACM Trans Database Syst. 2005;30(2):492–528.
4. Han H, Park S, Jung H, Fekete A, Röhm U, Yeom HY. Scalable serializable snapshot isolation for multicore systems. In: Proceedings of the 30th International Conference on Data Engineering; 2014. p. 700–11.
5. Jung H, Han H, Fekete A, Röhm U, Yeom HY. Performance of serializable snapshot isolation on multicore servers. In: Proceedings of the 18th International Conference on Database Systems for Advanced Applications; 2013. p. 416–30.
6. Lomet DB, Fekete A, Wang R, Ward P. Multi-version concurrency via timestamp range conflict manage-
ment. In: Proceedings of the 28th International Conference on Data Engineering; 2012. p. 714–25.
7. Ports DRK, Grittner K. Serializable snapshot isolation in postgresql. Proc VLDB Endow. 2012;5(12):1850–61.
8. Revilak S, O'Neil PE, O'Neil EJ. Precisely serializable snapshot isolation (PSSI). In: Proceedings of the 27th International Conference on Data Engineering; 2011. p. 482–93.
9. Wang T, Johnson R, Fekete A, Pandis I. The serial safety net: efficient concurrency control on modern hardware. In: Proceedings of the 11th International Workshop on Data Management on New Hardware; 2015. p. 8.
10. Weikum G, Vossen G. Transactional information systems: theory, algorithms, and the practice of concurrency control and recovery. Morgan Kaufmann; 2002.

Service Component Architecture (SCA)

Allen Chan
IBM Toronto Software Lab, Markham, ON, Canada

Synonyms

SCA

Definition

The Service Component Architecture (SCA) [1] is a collaborative effort driven by a number of software vendors in the Open SOA (OSOA) [2] Collaboration group to facilitate the building of applications and systems based on service-oriented architecture. The final specification for SCA Version 1.0 was available as of March 21, 2007.

SCA is a set of specifications in the area of service composition, assembly, protocol bindings and policy definitions, where the Service Data Objects (SDO) [3] specification is used to specify how service data can be specified and manipulated. SDO provides a uniform access pattern for heterogeneous data sources, such as XML or relational databases. Although SCA and SDO can work independently of each other, they are often used together to provide a full end-to-end framework for defining SOA applications and systems.

Key Points

Conceptually, SOA is an architecture principle to enable software applications to be exposed as services. However, since SOA itself does not dictate how these services will be packaged or assembled, some of the benefits of SOA such as reuse, manageability and scalability become unpredictable. SCA can speed up SOA adoption by using the SCA Assembly Model to define how services can be declared, implemented and connected to each other.

The basic building block for SCA is an SCA Component, which can be used to declaratively describe the business *services* exposed by an *implementation*, and declare dependency on other business services as *references*. In addition, *bindings* can be applied to any *services* or *references* to describe how a client application can invoke an existing service or how an external service can be accessed, respectively.

SCA also supports a recursive composition model to support the creation of a composite SCA Component. Another aspect of the SCA Assembly Model is the SCA Policy Framework [2], which provides a way to capture the nonfunctional aspects of an SOA system.

The SCA specification provides a framework for the implementation of scalable and manageable SOA systems, such as the enterprise service bus (ESB).

Cross-References

▶ Enterprise Service Bus
▶ Service-Oriented Architecture

Recommended Reading

1. Open SOA Collaboration. http://www.osoa.org/.
2. SCA specification, final version 1.0. http://www.osoa.org/display/Main/Service+Component+Architecture+Specifications.
3. SDO specification, final version 2.1. http://www.osoa.org/display/Main/Service+Data+Objects+Specifications.
4. Spring framework http://static.springframework.org/spring/docs/2.0.x/reference/index.html.

Service-Oriented Architecture

Serge Mankovski
CA Labs, CA Inc., Thornhill, ON, Canada

Synonyms

SOA

Definition

Service Oriented Architecture is a conceptual model for integration of software systems where system function is performed by coordinated invocation of services. In this model term service refers to *significant atomic* computational activity that can be invoked over a computer network. Service computational activity is significant in the sense that it is in order of several magnitudes more complex than a function invocation and atomic in the sense that it is a smallest element of functional decomposition in this model.

Historical Background

Historically IT systems are built in a competitive environment where each vendor is trying to develop best possible solutions to their customers and, at the same time, trying to build even more complex systems in attempt to serve any customer need and prevent customers looking for solutions from another vendor. Logic of this process led to development of IT systems that either did not have any means for integration with other systems, or at most had interfaces necessary for integration between within the brand. At the same time companies using IT systems to conduct business were trying to integrate systems in attempt to streamline operations, increase utilization of IT asset and achieve business critical functionality. These competing forces were shaping landscape IT for several decades. Over time there were a number of successful attempts to build large scale integration of IT systems, but more often than not, these integrated solutions themselves were becoming silos in its own right. A notable example of this process was emergence of electronic data interchange (EDI) system that was under development from mid 1960s almost at the same time when DARPA started work leading to development of the Internet. By the time when Tim Berners-Lee developed a first web browser in 1990, EDI was already well established as a successful model of integration of IT functionality across multiple companies.

It is perhaps not possible to pinpoint when exactly SOA way of thinking started. Impact and acceptance of SOA became more evident after successes standardization of technologies that were necessary within the SOA model. Experience and lessons learned by the industry from several decades of development, deployment and operation of various integration architectures, emergence of XML as a universal data interchange format, emergence of new open standards, and tremendous success of Internet have build technological foundation for SOA adoption. Broad adoption started when critical mass of standards comprising SOA stack has matured. At the same time Open Source Development phenomena made implementations of the standards broadly available. It removed barriers created by proprietary implementations and lowered barrier to entry for new users of the technology. Growth in scope, diversity and rapid pace of change in business requirements and demand business agility created awareness of SOA benefits in business community.

Foundations

SOA as an integration architecture is concerned with all aspects of interactions between IT systems. SOA postulates that a basic element of SOA architecture is service.

SOA system is comprised of a number of *services* deployed over a computer network. *Service* is a basic building element of SOA. Notion of a service is similar to notion of object in object oriented programing, but it is more coarse-grained. For example, a service can represent an important function of an IT system or even entire system all together. Each service has an *interface*. Service interface is *metadata*, or data about data, needed to *invoke* service operation.

In respect to invocation SOA distinguishes two roles – Service Consumer and Service Producer. Service Consumer is a system invoking service and Service Producer performs the service. Service Producer invokes service by message to an instance of Service Producer. Upon receiving an invocation conforming to the service interface, Service Producer executes requested operation using parameters provided by the invocation. Service execution might also include invocation of other services within the SOA system and hence Service Producer can also play role of Service Consumer in respect to another service. When Service Producer completes execution of the operation it replies to Service Consumer with an acknowledgement of successful completion or data containing results of the invocation or indication of fault. This reply from Service Producer to Service Consumer indicates completion of service execution.

It is important to note that service interface is an abstraction that does not take into account details of the network data transport needed to invoke the service or details of service implementation. Use of the interface abstraction allows for definition of service interactions at the higher level of abstraction than in any other integration architecture. In particular, it allows for definition of service interactions at design time and achieves high degree of flexibility regarding service implementation, location, time, and transport protocol needed for service invocation.

Since interface is separated from service instance it is necessary to associate interface with a service instance in a process that is called *service binding*. SOA allows to perform service binding at any time of SOA system life time. In particular, it can be done at run time by means of *service discovery* and *service lookup*. Service discovery is a process of discovery of services capable to perform a necessary function. Service lookup is a process of finding a service *end point reference* based on a service name, name of a service interface, or the interface metadata itself. End point reference is a piece of metadata that must contain a protocol dependent service address along with optional parameters and session identifier.

Notion of service interface is important from the software engineering point of view because allows for separation of the function performed by service from implementation of service itself. This constitutes a good software engineering practice leading to more robust system design. It also enables very powerful notion of service *orchestration*. Service orchestration is a process of delivering a *composite service* function by coordinated invocation of other services. It is usually done within an orchestration engine that executes the service in accordance with the process definition describing logic and sequence of invocation of orchestrated services along with the necessary *data transformation*. Data transformation is an activity of modifying syntax of the data exchanged between the services to accommodate their interface requirements.

SOA as a conceptual model accommodates wide variations. Any of the architectural concepts highlighted above can be omitted, except concept of service. This is perhaps why it is called service-oriented architecture.

In respect to services there are at least two major types of services:

1. SOAP services use WS-* stack of Web Service Standards. These services are based on a number of international standards developed with W3C and OASIS standardization committees and make extensive use of XML.
2. RESTful services emerged from the world of Open Source. They make use of HTTP protocol and became "standard-de-facto" for web based services and mush-ups.

Within SOA, there is a wide degree of variation in respect to use of metadata. SOAP services make extensive use of metadata: XML Schema for message syntax, WSDL and Schema for interface definition, WSDL for binding, UDDI for lookup, BPEL for orchestration. RESTful services use HTTP verbs for defining actions and use HTTP for data transport. They do not have formal definition of data syntax beyond URL syntax, no notion of binding, lookup, discovery and orchestration and hence no metadata associated with them.

In respect to invocation there are three types of invocation:

1. *Synchronous* invocation. This form of invocation passes control of execution from Service Consumer to Service Producer and it does not return to Service Consumer until service is completed. This type of invocation is common within both SOAP and RESTful services.
2. *Asynchronous* invocation. In this form of invocation Service Consumer does not need to wait for Service Producer to complete service invocation. Service Consumer can carry on performing its own function, but it requires capability on the part of the Service Consumer to receive message indicating completion of the service. This type of invocation is often done by use of a messaging system. This type of invocation is more common among SOAP services.
3. *Enterprise Services Bus (ESB) based* invocation. In this form of invocation Enterprise Service Bus performs function of discovery, lookup, binding, messaging, data transformation and orchestration. ESB-based invocation can be performed synchronously or asynchronously, but in this case service invocation can be done using abstract Service Producer interface. Service invocation is passed to ESB that performs Service Producer lookup and binding and invocation. If necessary, ESB performs data transformation. If requested service is a composite service, ESB performs necessary orchestration.

Variation in respect to service discovery, lookup and binding range from systems fully bound at design time to systems using run-time semantic-based mechanisms employing artificial intelligence methods and techniques. Systems performing mission critical function tend to drift towards design time binding. SOA systems aiming to accommodate high rate of changes tend to drift towards sophisticated run-time binding.

There is a wide variation in use and purpose of the SOA systems themselves. WS-* based service architecture tend to be used for enterprise integration. They often use ESB as a back-bone carrying majority of the business related data. It becomes a focal point where enterprise policies can be enforced and formal audit necessary for proving business compliance can be conducted. Because of this highly visible position of SOA systems within enterprise it gives rise to notion of *SOA Governance*. SOA Governance is an ongoing activity within an enterprise maximizing leverage of the SOA infrastructure for business purposes. SOA Governance has two distinctive aspects. One aspect refers to ensuring that all aspects of an enterprise functioning within SOA are performing their functions in expected manner by means of enforcing and auditing compliance with business policies, practices. In this aspect SOA provides means for automated support and tractability of decisions and actions performed within enterprise. Automated decision and action support within SOA is achieved by use of service orchestration. SOA Orchestration provides automation of business processes, automatic routing of documents, enforcement of timely document processing, notification of non-compliance, and change management. Tractability of decision and actions is achieved by retaining of service invocation data within the SOA infrastructure. It allows for cross-referencing and correlation of logging data retained with the services and allows reconstructing entire picture of business activity at any point of time. It ensures that at any point of time there is a means of checking if business activities were performed in accordance with law, regulations and in adherence to best practices.

Another meeting of SOA Governance relates to operation of the SOA system itself. Services within SOA have a certain degree of freedom and can change independently from each other as long as interfaces remain unchanged. However they are not completely independent because it is often not possible make them completely independent and there is still some degree of dependence between the services. This requires some level of control over the degree in which services can vary. SOA Governance makes sure that any change within the system does not destabilize or jeopardize business function performed by the system. It is accomplished by imposing policies to restrict degree of changes in behavior of services, ensure that services continue to perform within established service levels, managing deployment of new services and maintaining operation of the existing services. This from of governance also uses the same automation and tractability infrastructure as the other one.

Key Applications

Enterprise Application Integration, Business Process Optimization.

Future Directions

Deployment of SOA without changing business processes does not produce the same level of return on investment as if deployment is accompanied by changes in the business processes tailored to take advantage of the SOA system. On the other hand changes in business processes trigger changes in the SOA system. In the future, it would be necessary to develop a methodology for SOA deployment. This methodology would have to cover both technical and business aspects as well as provide foundation for understanding of economic impact and, ultimately, quantify return on investment associated with SOA deployment.

Cross-References

▶ Business Process Execution Language
▶ Enterprise Application Integration
▶ Enterprise Service Bus

- ▸ OASIS
- ▸ SOAP
- ▸ W3C
- ▸ Web Services

Recommended Reading

1. OASIS Reference Model for Service Oriented Architecture, http://www.oasis-open.org/committees/download.php/19679/soa-rm-cs.pdf.
2. Erl T. Service-Oriented Architecture (SOA): Concepts, Technology, and Design. Upper Saddle River: The Prentice-Hall Service Oriented Computing Series; 2005.
3. Pulier E, Taylor H. Understanding Enterprise SOA. Greenwich: Manning; 2005.
4. W3C Web Services Glossary, http://www.w3.org/TR/ws-gloss/.

Session

Sameh Elnikety
Microsoft Research, Redmond, WA, USA

Synonyms

Database interaction

Definition

A database session is sequence of interactions between a client and a database server. The session captures the state of the client's in-flight SQL commands.

Key Points

The session state may contain database objects, such as table cursor, or temporary relations that are accessible only within the session. For efficiency, some database engines maintain session state per connection rather than per client. In this case, it is called connection state.

A client expects to see the effects of its previous updates to the database. This concept is called session consistency [1] and is illustrated in the following example. A client issues a transaction to buy a book. Then, it sends a subsequent transaction to see the list of ordered books. Session consistency requires the list to contain that book which the client bought. Session consistency is trivial to implement in a centralized database system, but becomes harder in a distributed database system [2].

Cross-References

- ▸ Connection
- ▸ Strong Consistency Models for Replicated Data

Recommended Reading

1. Daudjee K, Salem K. Lazy database replication with ordering guarantees. In: Proceedings of the 20th International Conference on Data Engineering; 2004. p. 424–35.
2. Krikellas K, Elnikety S, Vagena Z, Hodson O. Strongly consistent replication for a bargain. In: Proceedings of the 26th International Conference on Data Engineering; 2010. p. 52–63.

Shared-Disk Architecture

Patrick Valduriez
INRIA, LINA, Nantes, France

Definition

In the shared-disk architecture, only the disks are shared by all processors through the interconnection network. The main memory is not shared: each processor exclusive (non-shared) access to its main memory. Each processor-memory node is under the control of its own copy of the operating system. Since any processor can cache the same disk page, a cache coherency mechanism is necessary.

Key Points

Shared-disk requires a cache coherency mechanism which allows different nodes to cache a consistent disk page. This function is hard to support and requires some form of distributed lock management. The most notable parallel database system which uses shared-disk is Oracle, with an efficient implementation of a distributed lock manager for cache consistency.

Shared-disk has a number of advantages: lower cost, good extensibility, availability, load balancing, and easy migration from centralized systems. The cost of the interconnection network is significantly less than with shared-memory since standard bus technology may be used between processor nodes. Given that each processor has enough main memory, interference on the shared disk can be minimized. Thus, extensibility can be better, typically up to a hundred processors. Since memory faults can be isolated from other nodes, availability can be very good. Load balancing is relatively easy as a query at any node can access all data on the shared disks. Finally, migrating from a centralized system to shared-disk is relatively straightforward since the data on disk need not be reorganized.

However, shared-disk suffers from complexity and potential performance problems. It requires distributed database system protocols, such as distributed locking and two-phase commit which are complex. Furthermore, maintaining cache consistency can incur high communication overhead among the nodes. Finally, access to a shared-disk is a potential bottleneck.

Cross-References

▶ Parallel Data Placement
▶ Parallel Query Processing
▶ Query Load Balancing in Parallel Database Systems

Shared-Memory Architecture

Patrick Valduriez
INRIA, LINA, Nantes, France

Synonyms

Shared-everything

Definition

In the shared-memory architecture, the entire memory, i.e., main memory and disks, is shared by all processors. A special, fast interconnection network (e.g., a high-speed bus or a cross-bar switch) allows any processor to access any part of the memory in parallel. All processors are under the control of a single operating system which makes it easy to deal with load balancing. It is also very efficient since processors can communicate via the main memory.

Key Points

Shared-memory is the architectural model adopted by recent servers based on symmetric multiprocessors (SMP). It has been used by several parallel database system prototypes and products as it makes DBMS porting easy, using both inter-query and intra-query parallelism.

Shared-memory has two advantages: simplicity and load balancing. Since directory and control information (e.g., lock tables) are shared by all processors, writing database software is not very different than for single-processor computers. In particular, inter-query parallelism is easy. Intra-query parallelism requires some parallelization but remains rather simple.

S

Load balancing is also easy to achieve since it can be achieved at run-time by allocating each new task to the least busy processor.

However, shared-memory has three problems: cost, limited extensibility and low availability. The main cost is incurred by the interconnection network which requires fairly complex hardware because of the need to link each processor to each memory module or disk. With faster processors, conflicting accesses to the shared-memory increase rapidly and degrade performance. Therefore, extensibility is limited to a few tens of processors, typically up to 16 for the best cost/performance. Finally, since memory is shared by all processors, a memory fault may affect several processors thereby hurting availability. The solution is to use duplex memory with a redundant interconnect which makes it more costly.

Cross-References

▸ Parallel Data Placement
▸ Parallel Query Processing
▸ Query Load Balancing in Parallel Database Systems

Shared-Nothing Architecture

Patrick Valduriez
INRIA, LINA, Nantes, France

Synonyms

Distributed architecture

Definition

In the shared-nothing architecture, each node is made of processor, main memory and disk and communicates with other nodes through the interconnection network. Each node is under the control of its own copy of the operating system and thus can be viewed as a local site (with its own database and software) in a distributed database system. Therefore, most solutions designed for distributed databases such as database fragmentation (called partitioning in parallel databases), distributed transaction management and distributed query processing may be reused.

Key Points

As opposed to symmetric multiprocessor (SMP), shared-nothing is often called massively parallel processor (MPP). Many research prototypes and commercial products have adopted the shared-nothing architecture because it has the best scalability. The first major parallel DBMS product was Teradata which could accommodate a thousand processors in its early version in the 1980s. Other major DBMS vendors, except Oracle, have provided shared-nothing implementations.

Shared-nothing has three main advantages: low cost, high extensibility, and high availability.

The cost advantage is better than that of shared-disk which requires a special interconnection network for the disks. By easing the smooth incremental growth of the system by the addition of new nodes, extensibility can be better (in the thousands of nodes). With careful partitioning of the data on multiple disks, almost linear speedup and linear scale up could be achieved for simple workloads. Finally, by replicating data on multiple nodes, high availability can be also achieved.

However, shared-nothing is much more complex than either shared-memory or shared-disk.

Higher complexity is due to the necessary implementation of distributed database functions for large numbers of nodes, in particular, data placement. Load balancing is more difficult to achieve because it relies on the effectiveness of database partitioning. Unlike shared-memory and shared-disk, load balancing is decided based on data location and not the actual load of the system. Furthermore, the addition of new nodes in the system presumably requires reorganizing the database to deal with the load balancing issues.

Cross-References

▶ Parallel Data Placement
▶ Parallel Query Processing
▶ Query Load Balancing in Parallel Database Systems

Side-Effect-Free View Updates

Yannis Velegrakis
Department of Information Engineering and Computer Science, University of Trento, Trento, Italy

Definition

A view is an un-instantiated relation. The contents of its instance depend on the view query and the instances of the base tables. For that reason, an update issued on the view cannot be directly applied on the view instance. Instead, it has to be translated into a series of updates on the base tables so that when the view query is applied again on the modified base table instances, the result of the view update command will be observed on the view instance. Unfortunately, it is not always possible to find an update translation such that the change observed on the view instance is the one and only the one specified by the view update command. When this happens for a view update translation, the translation is said to have no *side-effects*. To fully exploit the updateability power of views, it is desired to be able to find update translations that have no side-effects.

Historical Background

Updates on the views were introduced almost simultaneously with views. Their importance has been recognized by Codd himself. In fact, 1 of

the 12 rules that Ed Codd [2] introduced to define what a real relational database is, was referring to the ability of the views to be updateable. In particular, the sixth rule was:

> All views that are theoretically updatable must be updatable by the system.

The term *theoretically updateable* is referring to the ability of finding side-effect-free translations of the view updates.

Foundations

The problem of side-effect-free updates is based on the problem of *Updates through views*. Consider the case of a view V with a definition query Q_V on an instance I, and an update request U on the view instance $Q_V(I)$. The expected updated view instance is $U(Q_V(I))$. Since the view has no an independent instance, what is needed is to find the update W that needs to be performed on the instance I, such that the query Q_V on the updated instance W(I) gives a view instance $Q_V(W(I))$ that is equal to $U(Q_V(I))$.

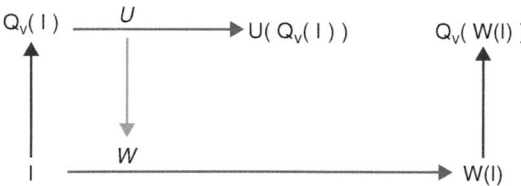

The update W is referred to as the *translation* of the view update U, and is said to have no side-effects if $U(Q_V(I)) = Q_V(W(I))$.

To better realize the problem of side-effect-free view updates, consider a database instance consisting of the three tables of Fig. 1.

Suppose that a view V_1 is defined on top of these three tables through the following view query:

```
select *
from Personnel P, Teaching T,
    Schedule S
where P.Employee = T.Professor and
T.Seminar = S.Course
```

Personnel

Department	Employee
CS	Smith
EE	Smith
Philosophy	Kole

Teaching

Professor	Equipment	Seminar
Smith	Projector	Programming
Smith	Projector	Databases
Smith	Laser	Physics
Kole	Microphone	Databases

Schedule

Course	Room
Programming	10
Databases	10
Databases	23

Side-Effect-Free View Updates, Fig. 1 Three base tables

The instance of the view will be the relation illustrated in Fig. 2.

Consider now an update command on this view that requests the deletion of the tuple t_d:[EE, Smith, Smith, Projector, Databases, Databases, 10]. Tuple t_d appears in the view instance due to the join of the three tuples [EE, Smith], [Smith, Projector, Databases] and [Databases, 10] of the base tables Personnel, Teaching and Schedule. Deletion of any (or all) of these tuples will achieve the desired result of deleting tuple t_d from the view. However, any such deletion will have additional effects in the view instance. For instance, the removal of tuple [EE, Smith] from Personnel will also eliminate the view tuples that are immediate before and after t_d. Similar observations can be made for the tuples in the other two base relations. In fact, for the particular update, it can be shown that there is no change that can be made on the base tables to achieve •

the desired tuple deletion without any additional changes, i.e., side-effects, in the view instance.

Side-effects are not observed only on deletions but also on insertions. For instance, consider the update command that requests the insertion of tuple t_i:[Economy, Smith, Smith, Projector, Databases, Databases, 10] in V_1. For the appearance of t_i in the view V_1 to be justified, tuples [Economy, Smith], [Smith, Projector, Databases] and [Databases, 10] need to exist in the instances of Personnel, Teaching, and Schedule, respectively. The last two are already there, but not the first. The translation of the insert command on the base tables will insert [Economy, Smith] in Personnel. Unfortunately, due to the value "Smith" in its attribute *Employee*, tuple [Economy, Smith] will be able to join with every other tuple of table Teaching that has value "Smith" in the attribute *Professor*. This will introduce additional tuples in the V_1 instance that the insert command did not request.

Base tables, i.e., tables that have been defined through the "create table" command, have standalone instances, thus, update commands on them can be implemented without side-effects by simply modifying their materialized instance accordingly. If views are to be used as any other table, a view needs to show the same behavior as base tables. This means that update commands need to have translations that generate no side-effects in the view instance. It would have been really surprising for an application or a user that is not aware that a relation she is interacting is actually a view, to request the deletion (or insertion) of a tuple and then see additional tuples disappearing from the view (or appearing in it).

To cope with the view update translation side-effects one option is to leave the burden to the database administrator who defines the view. During the view definition, the administrator is responsible to specify not only the view query but also how exactly each update is translated to updates on the base tables [7] and make sure that side-effects will not occur. The drawback of this option is that it requires a lot of knowledge and experience from the administrators. If the administrator determines that for a given view update there is no translation that has no side-

V_1

Department	Employee	Professor	Equipment	Seminar	Course	Room
CS	Smith	Smith	Projector	Programming	Programming	10
CS	Smith	Smith	Projector	Databases	Databases	10
CS	Smith	Smith	Projector	Databases	Databases	23
EE	Smith	Smith	Projector	Programming	Programming	10
EE	Smith	Smith	Projector	Databases	Databases	10
EE	Smith	Smith	Projector	Databases	Databases	23
Philosophy	Kole	Kole	Microphone	Databases	Databases	10
Philosophy	Kole	Kole	Microphone	Databases	Databases	23

Side-Effect-Free View Updates, Fig. 2 The view V_1 instance

effects, she can make the view not to accept this kind of updates, or allow the side-effects to happen if she believed that this is the semantically correct behavior.

Instead of letting the administrator deciding whether an update should be allowed or not, an alternative solution is to develop methods to perform this test automatically. Based on this idea, Keller [4] developed five criteria to characterize the correctness of a view update translation. The first of these criteria requires the translation to have no side-effects. A consequence of this is that keys of the base relations have to appear in the views, i.e., cannot be projected out. This reduces the cases in which side-effects may appear in the views, but does not completely eliminate them. Keller studied the different choices that exist when translating updates on select, project, select-project and select-project-join views, and provided algorithms for update translation for each case. These algorithms are guaranteed to respect his five criteria. In the same spirit a special view definition language was recently introduced that guarantees views are bidirectional, thus, every update on the view is uniquely translated to updates on the base tables [11].

For a given update on the view, there may be more than translations. Which one to be used is a decision that can be provided by the database ad-ministrator at the moment of view definition [5]. Alternatively one could have predefined ways to propagate updates for specific classes of queries, or can use provenance tracing techniques to identify at run time how an update on the view will have to be translated to base table updates [8].

Dayal and Bernstein [3] introduced the notion of the *view-trace* and the *view-dependency* graph. They are graphs that model the dependencies between the attributes of the base tables as determined by the schema and the view definitions. Through them one can determine whether there is an update translation that has no side-effects. For each update on the view, either a unique side-effect-free translation is found and is applied, or the update is not allowed to occur. This approach eliminates the need of an administrator involvement, but cannot be applied in cases in which side-effect generated translations are allowed to occur if they are semantically meaningful.

A different method to determine the kind of updates a view can accept without generating side-effects is through the *constant complement*. Two views are considered complementary if given the state of each view there is a unique corresponding database state. This means that when the instance of one of these views changes (due to an update) while the instance of the other is kept constant, then there is a unique database

S

instance from which the instances of the two views are generated. In other words, the correct translation of the view update is unique [1]. Unfortunately, given a view V, finding its view complement has been shown to be NP-complete even for views with very simple view definition queries.

View updates may be handled differently depending on the application requirements. There are situations in which view updates should not be allowed at all. If they are allowed, since there are typically more than one translations, there are same criteria for selecting the right translation. The lack of side-effects followed by the minimum number of changes that need to take place. These require reasoning with the view query at the schema level, which has all the complexity issues mentioned in the previous paragraphs. The loose approach is to let the updates to go on and be translated and then check the resulted instance for side-effects and translation correctness [9]. If a translation is found that generates side-effects the whole update execution can be dropped. Testing for all the possible translation alternatives may be time and resource consuming, and for this, summarization techniques have been used to estimate the possible side-effects [10].

But what would happen in cases in which an update on the view is absolutely necessary, as is the case of an application that can access the database only through a view interface without being aware of the fact that the relation it is accessing is actually a view and with the need to perform updates on it as it would have done if it was a base table? An idea proposed by Kotidis et al. [6] is the following. When an update command is issued on the view, the change in the view instance must be exactly the one described by the update. However, any change on the base table should not take place unless it is implied by the semantics of the view query and the update command. For instance, the deletion of the tuple [EE, Smith] from Personnel as a translation of the delete command for the view tuple [EE, Smith, Smith, Projector, Databases, Databases, 10] mentioned above, would have implied that the reason that the view tuple is deleted is that Smith stopped being affiliated with the EE department.

However, neither the semantics of the update command, nor the semantics of the view query imply something like that. Similar claims can be done for tuples [Smith, Projector, Databases] and [Databases, 10] of tables Teaching and Schedule, respectively. For the specific view tuple deletion, the claim is that no change should be observed in the instances of the three base tables, but tuple [EE, Smith, Smith, Projector, Databases, Databases, 10] will be removed from the view instance. This behavior will only be possible if one can accept views whose instances are not exclusively determined by the results of their view queries on the base tables, but also from the update commands that have been issued on them.

Key Applications

Achieving side-effect-free updates on the views is of great importance for systems that provide access to their data through views, but at the same time need to hide from their users or the applications that use the system the fact that they are dealing with views and not actual relations.

Cross-References

▶ Updates Through Views

Recommended Reading

1. Bancilhon FB, Spyratos N. Update semantics of relational views. ACM Trans Database Syst. 1981;6(4):557–75.
2. Codd EF. Is your DBMS really relational? Computer-World. 1985.
3. Dayal U, Bernstein P. On the correct translation of update operations on relational views. ACM Trans Database Syst. 1982;8(3):381–416.
4. Keller AM. Algorithms for translating view updates to database updates for views involving selections, projections, and joins. In: Proceedings of the 4th ACM SIGACT-SIGMOD Symposium on Principles of Database Systems; 1985. p. 154–63.
5. Keller AM. Choosing a view update translator by dialog at view definition time. In: Proceedings of the 12th International Conference on Very Large Data Bases; 1986. p. 467–74.

6. Kotidis Y., Srivastava D., Velegrakis Y. Updates through views: a new hope. In: Proceedings of the 22nd International Conference on Data Engineering; 2006.
7. Rowe LA, Shoens KA. Data abstractions, views and updates in Rigel. In: Proceedings of the ACM SIGMOD International Conference on Management of Data; 1979. p. 71–81.
8. Fegaras L. Propagating updates through XML views using lineage tracing. In: Proceedings of the 26th International Conference on Data Engineering; 2010. p. 309–20.
9. Wang L, Juang M, Rundensteiner EA, Mani M. An optimised two-step solution for updating XML views. In: Proceedings of the 14th International Conference on Database Systems for Advanced Applications; 2008. p. 19–34.
10. Peng Y, Choi B, Xu J, Hu H, Bhowmick SS. Side-effect estimation: a filtering approach to the view update problem. IEEE Trans Knowl Data Eng. 2014;26(9):2307–22.
11. Bohannon A, Pierce BC, Vaughan JA. Relational lenses: a language for updatable views. In: Proceedings of the 25th ACM SIGACT-SIGMOD-SIGART Symposium on Principles of Database Systems; 2006. p. 338–47.

Signature Files

Mario A. Nascimento
Department of Computing Science, University of Alberta, Edmonton, AB, Canada

Definition

A signature file allows fast search for text data. It is typically a very compact data structure that aims at minimizing disk access at query time. Query processing is performed in two stages: filtering, where false negatives are guaranteed to not occur but false positives may occur, and, query refinement, where false positives are removed.

Historical Background

Efficient and effective text indexing is a well-known and long-standing problem in information retrieval. While inverted files are a de facto standard for text indexing, in the early days, its storage overhead was not acceptable for larger datasets. In addition, accessing an inverted file on disk may require a relatively large number of (expensive) disk seeks. The main motivation for signature files is to allow fast filtering of text using a linear scan of the signature file for finding text segments that may contain the queried term(s). Given that the found segments may be false positives, a refinement step is required before the final correct answer is returned. The main compromise in signature files lies in how to build signatures for terms and for text segments that allow low storage overhead, fast disk access, and minimizes the ratio of false positives.

Foundations

Let T be a text to be indexed that is divided into nonoverlapping blocks T_i each containing b contiguous terms, e.g., strings. For each block T_i a binary signature $S(T_i)$ is built. Assume a hashing function $h(.)$ that takes as an argument a term and returns a signature of length B. The signature for a block T_i can then be obtained by performing a bitwise-OR of the signatures of the terms in that block, typically after excluding the terms in the stoplist. The signature file for T is then the set of signatures for its blocks T_i. A query term Q is also mapped into a signature $h(Q)$. At query time each block signature $S(T_i)$ is compared to the query's signature $h(Q)$. Denoting the bitwise-AND operator by "&" one can show that if $S(T_i)$ & $h(Q) = h(Q)$ then block T_i may contain the query term Q and is considered a candidate answer. However, in order to guarantee that only correct answers are returned, all candidate answers must be refined to ensure that they do contain query term Q. The reason for such false candidates is that when a block signature is built, a particular bit string matching the query's signature may appear as an incidental combination of different signatures.

Consider the sample text T: "To be, or not to be: that is the question" (punctuation marks can be ignored without loss of generality). Assume

Signature Files, Table 1 Sample term signatures

Term	H (term)
To	100,100
Be	011000
or	010010
not	101,000
that	001100
is	010001
the	100,001
question	000110

Signature Files, Table 2 Block signatures for "To be, or not to be: that is the question"

Block # (i)	Terms in the block (T_i)	Block signature $S(T_i)$
1	To be	111,100
2	Or not	111,010
3	To be	111,100
4	That is	011101
5	The question	100,111

further that a hashing function $h(.)$ is such that it hashes all the terms in this text as shown in Table 1.

The resulting block signatures will depend on the value chosen for b. Table 2 shows the resulting blocks T_i and their respective block signatures $S(T_i)$ assuming $b = 2$ and the hashing signatures shown in Table 1.

If $Q =$ "question," then T_5 satisfies the criterion $S(T_5) \,\&\, h(Q) = h(Q)$. In this particular case, T_5 indeed contains Q, and the text is selected as a true positive answer. Consider now the case where $Q =$ "to." In this *case* $(S(T_1) \,\&\, h(Q)) = (S(T_3) \,\&\, h(Q)) = (S(T_5) \,\&\, h(Q)) = h(Q)$. During query refinement the actual blocks T_1, T_3, and T_5 need to be read from disk and inspected. While T_1 and T_3 do contain the query term Q, thus being true positives, block T_5 does not, i.e., it is a false positive. This illustrates the major drawback of signature files. While one can safely discard a block T_i if $S(T_i) \,\&\, h(Q) \neq h(Q)$, the converse is not true otherwise. Without the query refinement step the query's answer is prone to contain false positives, which is typically not acceptable. Several factors need to be considered when aiming at minimizing

the probability of false positives, for instance, the length of the produced signatures (B), the number of terms per text block (b), and the number of bits randomly set in the signatures (which is denoted by n). For instance, for an optimal selection of $n = B \ln(2)/b$, then the false-positive probability is $1/2^n$ [1].

There are other important issues to be considered in addition to minimizing the false-positive probability and thus the overhead of query refinement, when using signature files. In principle the query signature could be also be the result of a bitwise-OR between several terms, hence in principle leading to the possibility of querying for phrases. Unfortunately, phrase queries are not well supported by signature files. This is due to the fact that the bitwise-OR operation used to produce the block and query signatures does not preserve any notion of ordering among the terms in the block. Therefore, two blocks with very different term ordering will have the exactly same block signature, which is an obvious problem. Another related problem occurs when terms in a phrase query occur in the boundaries of blocks. For instance, consider if one searches for $Q =$ "not to," which would yield $h(Q) = 101,100$. Using the block signature illustrated in Table 2, one would find that while T_1 and T_2 would be candidate blocks, all would be considered false positives in the query refinement, and one would be unable to find that the queried phrase is indeed in T. It would be missed altogether for being in the boundaries of the constructed text blocks.

Although less popular, negation queries, where one searches for a document not containing a query term, can also be answered using signature files. Recall that in the discussion above if $S(T_i)$ had at least the same bits set as $h(Q)$, then T_i could be an answer, pending further checking. If a block T_i does not have at least the same bits as $h(Q)$ set, it is definitely an answer to the negation query. However, if it has those same bits set, it may still be an answer because those bits may have been set independently due to other terms. Thus, in this case further checking is required as well. As an illustration consider the term $Q =$ "writer" and assume that $h(Q) = 100,010$. Since $S(T_i) \,\&\, h(Q) \neq h(Q)$

for $i = 1$, 3, and 4, one can be sure that those blocks do not contain Q, and therefore, up to that point, T could be an answer to query Q. However the fact that $S(T_i)$ & $h(Q) = h(Q)$ for $i = 2$ and 5 does not necessarily mean that those blocks contain Q. In fact, only upon checking the text of the blocks one could verify that they do not and consequently classify the text as satisfying the query.

It should be clear by now that the crucial issue in dealing with signature files is to minimize the overhead of query refinement. The more blocks needed to be further check the closer the performance will be to that of reading the whole dataset, which is obviously not desirable. This issue can be addressed by using lengthier signatures possibly through more complex hashing schemes. Due care is needed though as excessively long signatures detract from the claimed low storage overhead yielded by the signature files.

Despite its limitations above, signature files offer the possibility of very efficient search on disk. Given that the only operation necessary is to read and compare binary signatures in a linear fashion, no relatively expensive disk seeks are necessary during the file scan. Furthermore, the inspection of the signature blocks can be implemented very efficiently in memory. Updating the indexed texts can also be carried out with low overhead in a fairly straightforward way.

Bit-Slice Signature File

Even though the framework presented above is effective, fairly efficient, and relatively simple to implement, it can be further optimized by exploring the layout of the signature file. The typical layout is to have all block signatures written contiguously in a file. At query time, each and every block is read from the disk and compared to the query signature, regardless of which bits are set in the same. The bit-slice signature aims at reducing disk access per query by exploiting the fact that likely few bits are set in the query signature and in the text blocks.

In order to illustrate the idea, take Table 2 and transpose it, i.e., instead of having one row per signature, let us have one row per bit, i.e., the

Signature Files, Table 3 Bit slices for the block signatures in Table 2

Bit # (i)	Bit slice (S_i)
1	11,101
2	11,110
3	11,110
4	10,111
5	01011
6	00011

jth row will contain the jth bit of every block signature, ordered from most to less significant. Table 3 shows the resulting table after "transposing" Table 2.

Notice that when checking whether $S(T_i)$ & $h(Q) = h(Q)$, the idea is to look for bits set in corresponding positions. For instance, when searching for a block that contains Q = "or" one needs to inspect the second and fifth bits since $h(Q) = 010010$. Using the bit slice S_2 one can infer that blocks T_1, T_2, T_3, and T_4 have the second bit set, and therefore they could contain Q if they also have the fifth bit set. Similarly using S_5 one can infer that blocks T_2, T_4, and T_5 could contain Q if their second bit is also set. If S_2 & S_5 is computed, one obtains the index to the blocks that need be verified. Interestingly enough, this is equivalent to performing the intersection of the two sets of terms obtained by inspecting each bit individually. In this case S_2 & $S_5 = 01010$ and hence blocks T_2 and T_4 are candidate blocks. Note that as before, candidate blocks need to be further refined as false positives are still possible.

While the query processing using plain signature files would require a linear scan of the whole signature, using bit-slice signature this is typically not the case. The trade-off to be considered is that now one needs to perform (relatively expensive) random disk access in both the bit-slice file and the signature files. As well, the storage overhead for the bit-slice file, nearly as large as the signature file itself, needs to be taken into account. Just like the more straightforward case, a number of parameters need to be considered to produce efficient bit-slice signature files.

Signature Trees

As an alternative to simple flat signature files one can also arrange the signatures in a hierarchical balanced tree structure, similar to a B^+-tree. Once again, the idea is to trade the few disk seeks required by a linear scan of the signature file by additional seeks that would allow pruning, hence not reading large portions of the signatures.

Assume that a given text has already been broken into blocks T_i and their signatures $S(T_i)$ obtained as discussed above. A signature tree (or S-tree as it is called in [4]) can be constructed as follows. A set of block signatures are clustered together and stored in a disk page. The number of signatures per page depends primarily on the signature size and the page size. Each entry in such a node is a signature $S(T_i)$ and points to the actual text block T_i, which, as before, will be needed to check for false positives. Note that at this point it would be trivial to process a query by simply traversing all constructed nodes, which would function as a regular signature file. Fortunately, it is possible to avoid reading many of these blocks if a tree structure is used on top of the block signature nodes.

A tree can be constructed by creating an upper layer of nodes that will point to the first layer (which will become leaves of the tree). Each entry in a non-leaf node points to a leaf node containing entries for signatures $S(T_i)$, $S(T_j)$, ...$S(T_k)$, thus its entry signature will be the bit string $(S(T_i)$ | $S(T_j)$ |...| $S(T_k))$, where "|" denotes the bitwise-OR operator. The same reasoning can be applied recursively replacing the leaf nodes with the nodes of the current upper level. The sample tree depicted in Fig. 1, assuming an eight-bit block signature and a disk page that can fit two signatures, illustrates the result of this process using the set of signatures at the leaf level. For instance, the entry 0011 1110 in the root node points to a node containing signatures 0011 0100 and 0011 1010 (indeed 0011 0100 | 0011 1010 = 0011 1110). Similarly the entry 0011 0100 points to the leaf node containing signatures 0010 0100 and 0001 0100.

Figure 1: A sample signature tree. The actual block signatures are at leaf level and point to the respective text block (not shown for simplicity)

An important issue is how to cluster signatures in the leaf nodes and, similarly, how to group nodes under a single entry in the upper levels of the tree and so on and so forth until the root node. Since one of the possible criteria for this clustering task is tightly related to how the query is processed, the latter is discussed first.

Query processing starts by traversing the signature tree down from the root node choosing which subtree to traverse based on the probability that the subtree contains a candidate block. Assume that a node N has m (signature) entries N_i and it is pointed to by a parent node P under an entry with signature $P_i = (N_1$ | N_2 |...| $N_m)$. The query starts with P_i being each of the root entries. If P_i & $h(Q) \neq h(Q)$, then it is certain that the subtree pointed by that entry cannot contain a candidate block and it is discarded. Given the way the tree is constructed, the bits set are propagated from the leaf nodes up to the root entries; therefore if a given block in that subtree contained $h(Q)$, the root of that subtree entry would necessarily contain $h(Q)$. All that is needed now is to repeat the same reasoning using as the new root the node pointed to by the candidate root entry. Note that all candidate subtrees need be traversed and, when finally reaching the leaf level, query refinement step is still required.

As an illustration consider the signature tree in Fig. 1 and assume $h(Q) = 1000\ 0100$. Starting from the entries in the root node one can discard the upper subtree in the figure since 0011 1110 & $h(Q) \neq h(Q)$. The lower subtree needs to be traversed given that 1101 1100 & $h(Q) \neq h(Q)$. However, at that point and the expense of one single disk access, half of the signatures can be safely discarded, illustrating the pruning power of signature trees. Of the two subtrees only the one pointed by first entry (1000 1100) needs to be read. Of the two block signatures found in the corresponding leaf nodes, only the one corresponding to signature 1000 0100 needs to be retrieved for the mandatory refinement step.

Given the query processing reasoning above, if too many bits are set per block signature, the entries in the non-leaf nodes will quickly have too many bits set and therefore be unable to help pruning the traversal of the tree. If too many subtrees are traversed, the savings of not reading all signatures is bound to be offset by the additional disk seeks. Clearly the less bits are set higher up in the tree the more selective the traversal will be. This provides a criterion for cluster signatures within a node. Let $W(S_i)$ be defined as a function that returns the number of bits set in signature S_i. Clearly, for any pair of signatures S_i and S_j, $W(S_i \mid S_j) \geq W(S_i) + W(S_j)$. Thus the driving criteria for clustering signatures together are to minimize the value of $W(.)$ over the bitwise-OR'ed signatures. Just as in the case of plain signature files and bit-slice signature files, a number of parameters have to be set in order to obtain efficient signature trees.

Key Applications

Text indexing and search

Cross-References

▶ B+-Tree
▶ Disk
▶ Hash Functions
▶ Inverted Files
▶ Text Indexing and Retrieval

Recommended Reading

1. Baeza-Yates RA, Ribeiro-Neto BA. Modern information retrieval. New York: ACM Press/Addison-Wesley; 1999.
2. Faloutsos C. Access methods for text. ACM Comput Surv. 1985;17(1):49–74.
3. Zobel J, Moffat A, Kotagiri R. Inverted files versus signature files for text indexing. ACM Trans Database Syst. 1998;23(4):453–90.
4. Deppish U. S-tree: a dynamic balanced signature index for office retrieval. In: Proceedings of the 9th Annual International ACM SIGIR Conference on Research and Development in Information Retrieval; 1986. p. 77–87.
5. Frakes WB, Baeza-Yates RA. Information retrieval data structures & algorithms. Upper Saddle River: Prentice-Hall; 1992.
6. Witten IH, Moffat A, Bell TC. Managing gigabytes: compressing and indexing documents and images. 2nd ed. San Francisco: Morgan Kaufman; 1999.

Similarity and Ranking Operations

Michael Huggett
University of British Columbia, Vancouver, BC, Canada

Synonyms

Association; Correlation; Matching; Ordering; Proximity; Relevance

Definition

Similarity and ranking operations are fundamental to *information searching*, in which a user generates a *query* phrase of one or more words that reflects an *information need*. The query is used to find related items that satisfy that need.

Similarity operations quantify the *resemblance* or *alikeness* between two information objects. An *information object* is a conceptual unit, most typically described as a *document*, but also taking the form of a term, phrase, paragraph, page, section, chapter, article, book, or script, etc. Information objects may be printed or digital.

Similarity judgments may be subjective (performed by a user) or algorithmic (performed by a computer). Depending on the method of evaluation, alikeness judgments may be semantic (i.e., within the meaning of a document), structural (i.e., within parts of speech, position in a document, or pattern of links between documents), or statistical (i.e., within correlations between document attributes). Statistical methods that model the distribution of terms in a corpus are most common in information retrieval (IR).

In typical IR systems, documents are preprocessed to extract representative *keyword* terms. The terms of a user's query are then compared with the keywords of each document to find the best matches. Given a document that satisfies an information need, its keywords may themselves be used to query for related documents.

S

Ranking operations are based on the numeric result of a similarity measure: given a query representing an information need, the documents that are scored as most similar to the query are most highly ranked. Rank is typically presented to the user as a list of document descriptors (esp. titles) sorted monotonically in decreasing order of similarity score.

Historical Background

The history of information retrieval is parallel to and largely separate from that of database research, since the goal of IR is to find semantically related information in an arbitrary corpus of unstructured documents, whereas the traditional relational database model searches within specified value ranges in pre-defined fields. Information retrieval produces "best guess" matches for a given keyword query, whereas relational databases return items that are *true* of a (typically boolean) query statement. Information retrieval is uncertain and probabilistic; its similarity and ranking operations are vital to this distinction.

Notions of similarity and ranking are fundamental to every-day information needs, as intelligent agents (e.g., people, animals) define target objects (e.g., food) that meet certain goals (e.g., survival), and then seek objects in the environment that meet these goals. As such, the topics of similarity and ranking have been well-discussed by cognitive scientists who study the structure of knowledge (e.g., Tversky 1975), and before them by philosophers from the Enlightenment (e.g., Hume) back to Ancient Greece (e.g., Aristotle).

By the start of the Information Age in the mid-20th century, discrete mathematics was already the basis of information retrieval research. Proposed techniques included the comparison of attribute lists between documents, called *association* (Yule 1912); ranking as a process of ordinal measurement (Stevens, 1946); *distance metrics* that quantify the differences between strings, originally used for error detection (Hamming 1950); the automatic indexing of documents based on the statistical distribution of their terms (Luhn 1957); and the weighting of

terms to improve retrieval accuracy (Maron and Kuhns 1960).

Beginning in the 1960s, these techniques were refined and combined into the *associative* and *probabilistic* approaches that form the basis of current practice-although earlier explorations can claim to have produced the first operational information retrieval system (e.g., Goldberg's *Statistical Machine* as the first electronic system, c. 1927, and Tillett's *QUEASY* as the first system on a general-purpose computer, c. 1953).

The associative approach was consolidated and evaluated exhaustively in the first viable product of modern information retrieval: the SMART system (Salton 1966), which continues to be much imitated. The probabilistic approach thereafter established a solid theoretical foundation that had been lacking in earlier work, and over a series of refinements improved retrieval accuracy to become a standard benchmark (Robertson and Sparck 1976). Some researchers believe the probabilistic approach represents the future of information retrieval.

Foundations

Similarity refers primarily to alikeness based on shared attributes; it is sometimes described as *association*, *relatedness* or *relevance*. Computational similarity operates by comparing lists of representative terms derived from information objects. Comparisons are either between a query and document, or between two documents. In the latter case, if a user has an exemplar document in hand that represents an information need, that exemplar may be used to "give me more like this one".

Automatic similarity and ranking operations require information objects in digital form; paper documents may be digitized using optical character recognition (OCR). Similarity comparisons are necessarily automatic in systems that index and organize large corpora of documents. Some systems anticipate significant human interaction, but may not be practical since human similarity valuations are slower, and inter-evaluator consistency is poor.

Information retrieval systems typically encode documents as *term vectors*. Each document in a corpus is assigned a term vector. Words in the document are typically *stemmed* to pool together related words such as *stemmer, stemmed, stemming*, etc. under a single unique root term, e.g., *stem*. Each unique term is assigned a specific cell in the term vector that contains a numerical value. *Binary vectors* show a '1' in a cell if the document contains the term, '0' otherwise. *Weighted vectors* use either integer values to record the number of times each term appears in the document, or real-number values that express the degree to which the term is representative of the document. All the term vectors in the same corpus use the same (binary or weighted) numbering scheme.

All the term vectors in the same corpus are configured the same way. Two types of configuration are typical. The first type reserves a cell for each unique term that appears in the corpus, and numerical values are assigned only to cells for terms that appear in the document. In a large corpus, document vectors can contain many zeroes, and thus be inefficiently *sparse*.

The second type of configuration involves the use of document *keywords*: a small set of discriminating terms from within the document that represent its content. Keywords are fundamental to *indexing*, whose goal is to identify terms that best identify a document with respect to other documents in a corpus. Using a term-weighting algorithm, a score is calculated for each term in each document, and each document is assumed to be best described by its highest-scoring terms. These terms are chosen as the document's keywords and low-scoring terms are ignored.

The use of such automatically-extracted keywords provides some advantages. Similarity scores can be calculated more efficiently with a small number of representative keywords than by using all the terms in a document. Keywords also provide a human-readable summary of document content without requiring that all documents be read and evaluated manually. Keywords require less space for term storage: a document term vector is a compact mapping from each keyword to its numerical value. Weighted keywords are

standard in IR; more complicated methods than this seldom justify the additional complexity and difficulty (Cleverdon, Mills, and Keen 1966).

Ranking refers to any function that follows the *Probability Ranking Principle*, which states that IR systems are most effective when the documents that they retrieve in response to a query are ordered in decreasing probability of relevance to the user's information need. Ranking is always performed as a result of some user-generated query.

Associative Similarity

In associative similarity, document term vectors define a metric space: a corpus with n unique keywords can be described as a Euclidean n-space, in which each document vector describes a point, line, or hyperplane. Together, the document vectors of a corpus comprise a matrix subject to algebraic methods for space division and summarization. Associative similarity is also commonly known as the *vector space model*.

The associative approach finds documents most similar to a query by matching the terms of the query vector against all of the document term vectors in the corpus. The matching process typically uses an index to retrieve candidate document vectors. The documents that generate the highest similarity scores are assumed to be those that best meet the user's information need as represented by the query. Documents are then presented to the user in a ranked list sorted in decreasing order of similarity score.

There are several associative vector methods that can be applied to binary and weighted document vectors to generate a similarity score. The simplest method counts the number of found terms in common between two binary vectors: this produces a *simple matching coefficient*, which is useful under the assumption that *any* matching is important, for example when two objects are assumed to be incomplete descriptions of each other. In practice, this assumption proves to be coarse and inaccurate, and most similarity operations normalize the simple matching coefficient by the lengths of the

S

Similarity and Ranking Operations, Table 1 Four popular methods for calculating the similarity between two objects

Binary vectors		Weighted vectors	
Jaccard $\dfrac{\|X \cap Y\|}{\|X \cup Y\|}$	Dice $\dfrac{2\|X \cap Y\|}{\|X\| + \|Y\|}$	Jaccard $\dfrac{(\omega_X, \omega_Y)}{\sum \omega_X^2 + \sum \omega_Y^2 - \sum (\omega_X \omega_Y)}$	Dice $\dfrac{2(\omega_X, \omega_Y)}{\sum \omega_X^2 + \sum \omega_Y^2}$
Cosine $\dfrac{\|X \cap Y\|}{\sqrt{\|X\|^2 \times \|Y\|^2}}$	Overlap $\dfrac{\|X \cap Y\|}{\min(\|X\|, \|Y\|)}$	Cosine $\dfrac{(\omega_X, \omega_Y)}{\sqrt{\sum \omega_X^2 \sum \omega_Y^2}}$	Overlap $\dfrac{(\omega_X, \omega_Y)}{\min(\sum \omega_X^2, \sum \omega_Y^2)}$

The binary vectors use a simple count of matching keywords, given their attribute vectors X and Y. The weighted vectors hold term weights calculated by an indexing algorithm; (w_X, w_Y) indicates the inner product of the vector weights, and Σw^2_A indicates the sum of squares of all weights of vector A

input vectors. For weighted vectors, the inner product of two vectors is used as the equivalent of the simple matching coefficient, and sum-squares of the vectors are used for normalization. Table 1 shows binary and weighted versions of four common similarity measures. The weighted cosine method is especially popular, and represents a geometrically accurate interpretation of correlation.

In practice, these methods are virtually interchangeable, since all their scores increase monotonically given the same input data. However, it is notable that for a given method and query vector, multiple target documents with significantly different term vector compositions may generate the same similarity score. Furthermore, the effective domain of similarity scores in n-dimensional space can vary dramatically between different similarity functions. Such results suggest that different similarity operations may be more or less appropriate for corpora with different types of term distribution [2].

Probabilistic Retrieval

Probabilistic retrieval is a decision-theoretic process based on the idea that for each query, documents should be retrieved if they maximize the probability that they are *relevant* to the query, $P(relevance|document)$. Whereas associative similarity measures are largely ad hoc, probabilistic retrieval is based on an explicit theoretical framework. The intuition is that users will find documents relevant if the documents have a certain distribution of attributes, and that the probability of relevance depends on how

closely a document's distribution of attributes matches the distribution sought by the user.

Since the probabilistic model depends on the presence or absence of terms, documents are represented with binary term vectors, writing a '1' in each cell that a given term is present, '0' otherwise. In its simplest form, the model assumes that errors of false-positive choices (deemed relevant when irrelevant) and false-negative choices (deemed irrelevant when actually relevant) are negligible.

For a given document represented by term vector X, probability of relevance $P(R)$ and probability of irrelevance $P(\bar{R})$, the document is judged relevant if $P(R|X) > P(\bar{R}|X)$, i.e., if the following discriminant function $g(X)$ is greater than 1:

$$g(X) \frac{P(R|X)}{P(\bar{R}|X)}$$

Using Bayes' Law, this is rewritten as:

$$g(X) \frac{P(X|R)\,P(R)}{P(X|\bar{R})\,P(\bar{R})} \approx \frac{P(X|R)}{P(X|\bar{R})}$$

since $P(R)$ and $P(\bar{R})$ are constant for each document. The calculation of $P(X|R)$ and $P(X|\bar{R})$ depend on the probabilities of each term in the document. For each term, the following table is used, with N the number of documents in the corpus, R the number of documents relevant to the query, n the number of documents that contain the term, and r the number of relevant documents that contain the term:

	Relevant	Non-relevant	
documents w/ term	r	$n-r$	n
documents w/o term	$R-r$	$N-n-(R-r)$	$N-n$
	R	$N-R$	N

Thus for each binary term x_i in X that matches a term in the query, the relevance function is re-written as a ratio of relevant to non-relevant portions:

$$g(X) \approx \frac{P(X|R)}{P(X|\overline{R})} = \frac{\prod_i P(x_i|R)}{\prod_i P(x_i|\overline{R})}$$
$$= \sum_i x_i \log \frac{r_i/(R-r_i)}{(n-r_i)/(N-n-R+r_i)}$$

This simplifies to $g(x) \approx \sum_i x_i \log \frac{N-n_i}{n_i}$ in the absence of relevance information, and approximates further to $g(x) \approx \sum_i x_i \log \frac{N}{n_i}$ for $n_i \ll N$, noting that the log factor in this last equation is identical to the common form for *inverse document frequency* (IDF). Once $g(X)$ is calculated, documents can be ranked by their relevance score.

The crucial factor in probabilistic retrieval is how to estimate *relevance*. The problem is that R is initially unknown, since there are no retrieved documents. To gain a notion of what may actually satisfy a user's information need, a common technique has been to estimate term relevance based on distributions of terms in the corpus. One approach is to simply use term IDF scores, and to assume that all query terms have an equal probability of appearing in relevant documents. Another approach is to use a mixture of two Poisson distributions to characterize the distribution of each term among relevant and irrelevant documents in the corpus.

Once the model has made some retrievals based on these relevance estimates, the user can provide *relevance feedback* by stating explicitly whether a retrieved document is relevant or irrelevant. This feedback is used to refine the model's term weight parameters incrementally. Relevance feedback is a direct approach to system personalization, but may not be practical with large corpora, or with systems with many (perhaps anonymous) users with different information needs. Relevance feedback also imposes

a *burden of decision* that some users may wish to avoid, and its use reminds us that information retrieval is not a purely objective science: it is subject to often ill-defined information need. Users themselves may not be aware of their goal or able to describe their need, other than by a vague feeling of relevance.

Although probabilistic retrieval has been described here in its simplest form, further developments have addressed the troublesome assumption of *term independence* endemic to information retrieval, and accuracy of relevance scores has been improved by incorporating parameters for document length and *term frequency*.

Related Areas

Information retrieval deals primarily with textual information objects, although the advent of digital images and videos has led to techniques for quantifying the alikeness of non-textual media. Traditional textual similarity and ranking operations may be applied to non-textual media if they can first be interpreted into textual-symbolic form, such as by using machine-learning techniques to generate keyword descriptions of photographs. Whatever the source media, if symbolic (esp. alphanumeric) descriptive attributes can be extracted in a robust, consistent manner, then the media can be compared based on those attributes.

Alternatively, where many users share a common information space, they may add their own *tags* to information objects. Tagging is a form of *social filtering* where semantic relations between objects emerge from the collective valuations of a group of interested individuals. The resulting *tag cloud* acts as an organic index, preferentially retrieving objects strongly represented by a conjunction of query terms.

The World Wide Web, with its search engines, is by far the most popular information retrieval system: a text-based information medium with many millions of users and linked pages. Correspondingly, it has become an increasingly popular domain for information retrieval research and development. Although the term-based similarity methods discussed above can be used to compare the content of documents, other effective methods have taken advantage of the semantic collaboration that gives the Web its structure: the many Web authors who link to pages that they consider relevant to their own. Web ranking algorithms (such as HITS and PageRank) observe which pages are highly-connected *hubs* and *authorities*; these pages are retrieved preferentially for a given query.

It is often useful to calculate the similarity between documents of a corpus, whether to find documents related to an exemplar document, or as a prelude to clustering operations. Since calculating similarity scores on demand in a large corpus can be expensive (particularly where corpus-wide sums are necessary, as with *TF*IDF*), pairwise similarity scores between documents may be stored in memory for later rapid retrieval. One approach is to create a similarity matrixtriangular if the similarity property is assumed to be commutative, but twice as large otherwise.

Another common approach is to store relations between documents in an *associative similarity network* (ASN). Documents act as the nodes of a network; links between pairs of documents represent relatedness, and are typically weighted with a real-valued similarity score. As many remotely-related documents could be linked with trivial weights, nodes are only linked for similarity scores above some threshold. In addition to fast nearest-neighbour retrieval, an ASN also provides opportunities for interactive navigation through the network, following a *semantic gradient descent* moving toward clusters of documents that better support a user's information need. ASNs can be seen as an example of the *cluster hypothesis*, which states that similar documents will be relevant to the same query: after a document has been retrieved with a term-based query, other documents similar to that document are already linked and immediately available for further inspection.

Key Applications

Classification; Thesaurus construction; Recommendation systems; Similarity (nearest-neighbour) search; Summarization and Information Filtering.

Cross-References

▶ BM25
▶ Classification
▶ Clustering for Post Hoc Information Retrieval
▶ Dimensionality Reduction
▶ Index Creation and File Structures
▶ Indexing and Similarity Search
▶ Multimedia Information Retrieval Model
▶ Probability Ranking Principle
▶ Relevance Feedback for Text Retrieval
▶ Stemming
▶ Term Weighting
▶ TF*IDF
▶ Web Information Retrieval Models

Recommended Reading

1. Harman D. Ranking algorithms. In: Frakes WB, Baeza-Yates R, editors. Information retrieval: data structures & algorithms, chap. 14. Upper Saddle River: Prentice-Hall; 1992. p. 363–92.
2. Jones WP, Furnas GW. Pictures of relevance: a geometric analysis of similarity measures. J Am Soc Inf Sci. 1987;38(6):420–42.
3. Rasmussen EM. Clustering algorithms. In: Frakes WB, Baeza-Yates R, editors. Information retrieval: data structures & algorithms, chap. 16. Upper Saddle River: Prentice-Hall; 1992. p. 419–42.
4. Salton G, McGill M. An Introduction to Modern Information Retrieval. New York: McGraw-Hill; 1983.
5. van Rijsbergen CJ. Information Retrieval. London: Butterworths; 1979.

Simplicial Complex

Andrew U. Frank
Vienna University of Technology, Vienna, Austria

Synonyms

Cell complex; CW complex; Polyhedron

Definition

A simplicial complex is a topological space constructed by gluing together dimensional simplices (points, line segments, triangles, tetrahedrons, etc.).

A simplicial complex K is a set of simplices k, which satisfies the two conditions:

1. Any face of a simplex in K is also in K
2. The intersection of any two simplices in K is a face of both simplices (or empty)

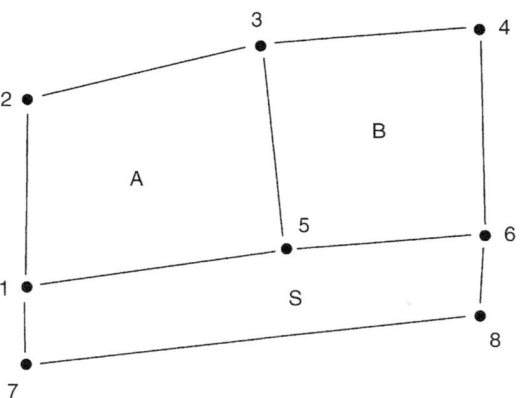

Simplicial Complex, Fig. 1 Cadastral parcels provide an example of a simplicial complex

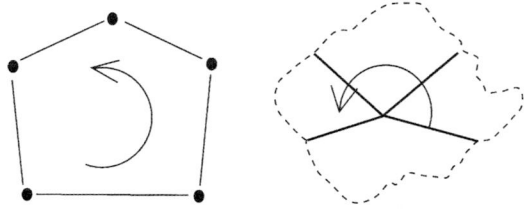

Simplicial Complex, Fig. 2 The two consistency checks: following the line segments around a face and following the line segments around a point

Historical Background

Raster (field) or vector (object) are the two dominant conceptualizations of space. Applications focusing on object with 2 or 3 dimensional geometry structure the storage of geometry as points, lines, surfaces, and volumes and the relations between them; a classical survey paper discussed the possible approaches mostly from the perspective of Computer Aided Design (CAD) where individual physical objects are constructed [1].

The representation of geographic information, e.g., maps, introduces consistency constraints between the objects; consider the sketch of a few cadastral parcels (lots) and the adjoining street (Fig. 1). Land, in this case 2 dimensional space, is divided into lots, such that the lots do not overlap and there are no gaps between them; this is called a partition (definition next section). Corbett [2] proposed to check that a sequence of line segments around a face closes and that the left neighbor of line segment and the right neighbors of the following line segment around a point is the same face; these two conditions are dual to each other (Fig. 2). This duality is the foundation of the DIME (dual independent map encoding) schema to store 2D line geometry for areas.

Every line of a graph, which represents a partition, is related to a start and an end point and to two adjacent faces (Fig. 3). Such data structures were typical for the 1980s; implemented originally with network and later relational DBMS. They did not perform acceptably fast with large Geographic Information System data, mostly because geometric operations do not translate to database operations directly (the so-called impedance mismatch of record oriented programming and tuple oriented database operations [3]), most obvious when checking geometric consistency. As late as 1985, all commercial programs to compute the overlay of two partitions, which is one of the most important operations in geographic information processing, failed.

In 1986, Frank observed that simplicial (and possibly cell) complexes enforced exactly the consistency constraints required by the large class of applications that manage geometry as 2D or 3D partitions [4]. A commercial implementation became available, designed concurrently by Herring (then with Intergraph). Alternative approaches to manage the geometry of partitions without explicit representation of topology and to reconstruct topology when required were often

Simplicial Complex, Fig. 3 An UML object diagram for a database schema for partitions

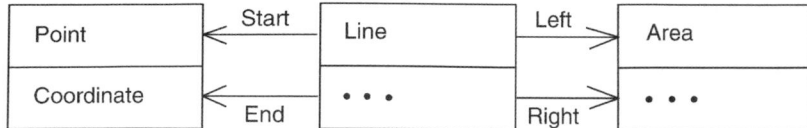

used, but cause difficulties, because of the fundamental limitations of approximative numerical processing.

Foundations

Topology, specifically the theory of homotopy, provides the mathematical theory to program geometric operations. Homotopy captures the notion that multiple metric (coordinative) descriptions of a single geometry may be different but represent "essentially" the same geometry. Fig. 1 can be transformed continuously to Fig. 4 but not to Fig. 5.

Homotopy creates equivalence classes for geometric figures. Many applications are interested in exactly these equivalence classes and benefit from the achieved abstraction that leaves out imprecisions caused, e.g., by measurements or approximative numerical processing.

Topology studies the invariants of space under continuous (homeomorphic) transformations, which preserve neighborhoods. Algebraic topology, also called combinatorial topology [5], studies invariants of spaces under homotopy with algebraic methods. The perspective of point set topology, which sees geometric figures as (infinite) sets of points is not practical for programming and the discretization of geometry achieved through algebraic topology is crucial: the unmanageable infinite sets are converted into countable objects, namely points, lines between points, and faces bounded by the boundary lines. Algebraic topology studies different "spaces" like Figs. 4 and 5 (both are embedded in ordinary 2D space, but the embedding is not in focus in algebraic topology).

The complexity of operations on arbitrary cells of a partition can be reduced by forcing a triangulation; all elements are then convex! Fig. 1 is a cell complex and the corresponding simplicial complex is Fig. 6.

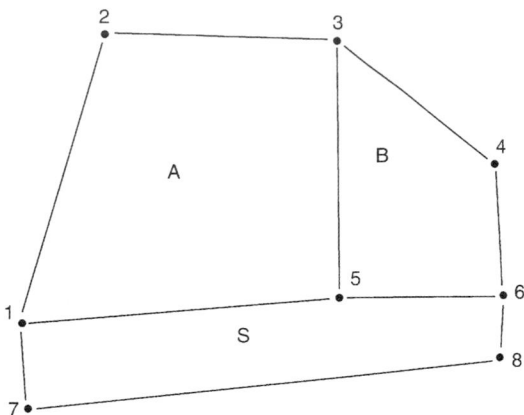

Simplicial Complex, Fig. 4 A deformed, but homotopic, copy of Fig. 1

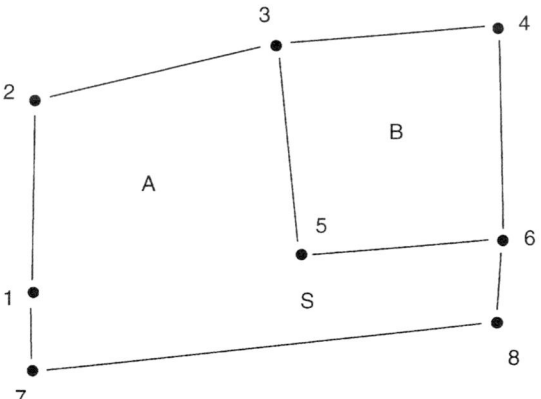

Simplicial Complex, Fig. 5 Metric is preserved, but the figure is not homotopic to Fig. 1, because elements are missing

Algebraic topology studies simplices and their relations: A simplex is the simplest geometric figure in each dimension. A zero dimensional simplex (0-simplex) is a point, a one dimensional simplex (1-simplex) is a straight line segment, a two dimensional simplex (2-simplex) is a triangle, a three dimensional simplex (3-simplex) a tetrahedron, etc. $n + 1$ points in general position define an n-simplex. Each n-simplex consists of (is bounded) by $(n + 1)$ $(n - 1)$-simplexes: a line

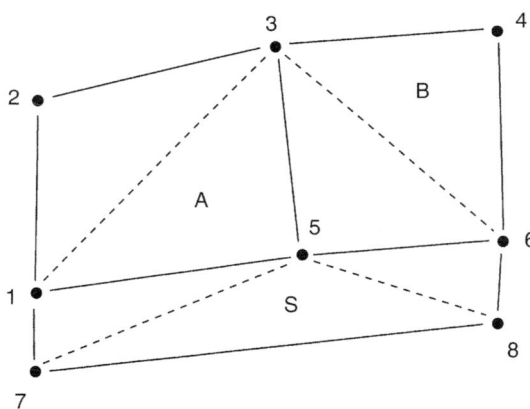

Simplicial Complex, Fig. 6 The geometry of Fig. 1 triangulated

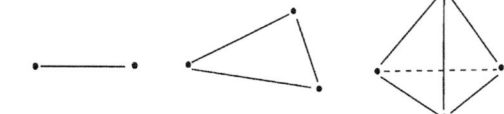

Simplicial Complex, Fig. 7 The simplices of 0, 1, 2, and 3 dimensions

(1-simplex) is bounded by 2 0-simplices (points), etc. Simplices can be oriented; the oriented 1-simplex 2–3 (in Fig. 1) is different from the oriented 1-simplex 3–2 (Fig. 7).

A k-simplicial complex K is a complex in which at least one simplex has dimension k and none a higher dimension. A homogeneous (or pure) k-complex K is a complex in which every simplex with dimension less than k is the face of some higher dimension simplex in K. For example, a triangulation is a homogeneous 2-simplicial complex, a graph is a homogeneous 1-simplicial complex. Homogeneous simplicial complexes are models of partitions of space and used therefore to model geographic spatial data. Whitehead gave for so-called CW-complexes a slightly more general, more categorical definition mostly used in homotopy theory.

Four operations are important for simplicial complexes: the *closure* of a set of simplices S is the smallest complex containing all the simplices; it contains all the faces of every simplex in S. The *star* of a set of simplices S is the set of simplices in the complex that have simplices in S as faces. The *link* of a set of simplices S is a kind of boundary around S in the complex. The *skeleton* of simplicial complex K of dimension k is the subcomplex of faces of dimension k-1 in K.

Simplicial complexes can be represented as chains, which are lists of the ordered simplices included in the complex. Chains can be written as polynomials with integer factors for the simplices included in the complex, e.g., the 2-chain of the 2-complex in Fig. 1 is $K = 1\,A + 1\,B + 1\,S$.

The boundary operator δ applied to a k-simplex gives the set of k-1-simplices, which form the boundary of the simplex; for example, the boundary of a 1-simplex gives the two 0-simplices, which are start and end point of the line, one taken with positive, the other with negative orientation. The boundary operator is applied to a chain by applying it to every oriented simplex in the chain. The boundary of a closed simplicial complex is 0; in general, the boundary of the boundary is 0.

$$\delta A = l_{12} + l_{23} + -l_{15}$$
$$\delta\,(\delta A) = \delta l_{12} + \delta l_{23} + \delta l_{35} - \delta l_{15}$$
$$= p_r - p_2 + p_2 - p_3 - p_s - p_r + p_s$$
$$= 0$$

The boundary operator is important to deduce the topological 4- and 9-intersection (Egenhofer) relations between two subcomplexes, of the same complex [6, 7]. Chains and boundary operator are easy to implement with list operators and often it is sufficient to generalize the code for operations on polynoms.

The theory of simplicial complexes can be generalized to cell complexes. Cells are homomorph to simplices, but can have arbitrary form; a 2-cell can have an arbitrary number of nodes in its boundary.

From an application point of view, it is often important that objects do not overlap and all of space is accounted for. The concept of a *partition* captures this idea; a partition of a space S is a set of subsets of the space, such that

- All subsets cover all of space (jointly exhaustive): $\bigcup_i s_i = s$
- No two subsets overlap (pairwise disjoint): $s_i \cap s_j = \emptyset$ for $i \neq j$.

These two properties are sometimes abbreviated as JEPD.

Partitions are changed by the Euler operations, *glue* and *split*, which maintain the Euler characteristic of the surface; the Euler characteristic is computed as $\chi = V{-}E + F$, where V is the number of nodes (vertices), E is the number of edges and F is the number of faces. From Fig. 1 with $\chi = 8{-}10 + 3 = 1$ merging two parcels obtains Fig. 8 with $\chi = 8{-}9 + 2 = 1$ or Fig. 9 where parcel A is split into parcel C and D with $\chi = 10{-}13 + 4 = 1$.

Consistency of these operations is difficult to check in cellcomplexes if "islands" occur as in Fig. 10, which is realistic for many application areas. The problem is avoided by triangulation and therefore simplicial complexes are an effective representation for maintainable geometric data describing partitions.

Simplicial complexes are triangulations of 2 dimensionalspace; they contain more objects than a partition represented as cells, but operations to maintain consistency in a triangulation are faster and simpler to program. The representation of a simplicial or cell complex requires the explicit representation of the boundary and converse co-boundary relation. The schema used initially (Fig. 3) contains redundancy (which is used in Corbett's tests for consistency) and is therefore difficult to maintain. Popular today are schemes with half edges (Fig. 11), where a half-edge points to the starting node and the corresponding other half edge or quad edges [8] (Fig. 12), where each quad-edge points to the next quad-edge and either a boundary node or face; in a quad-edge structure, the boundary graph and its dual are maintained in a well-defined algebra with a single operation *splice*. For example, tak-

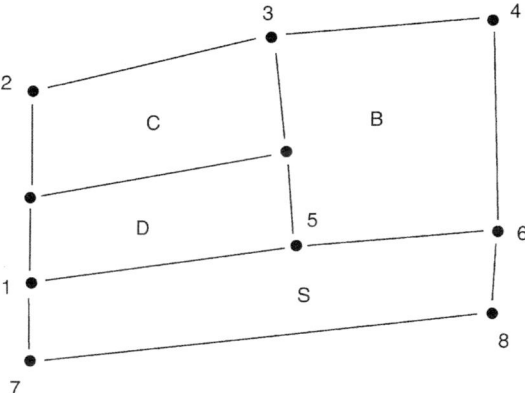

Simplicial Complex, Fig. 9 A subdivided in C and D

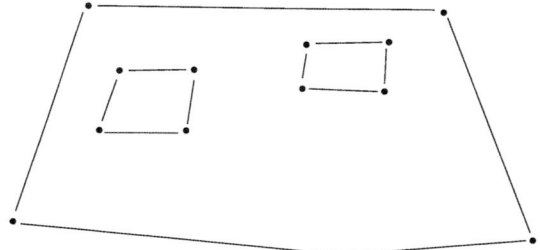

Simplicial Complex, Fig. 10 A parcel with "islands"

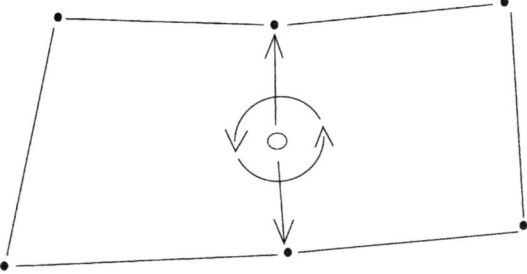

Simplicial Complex, Fig. 11 Two half edges, pointing to adjacent nodes

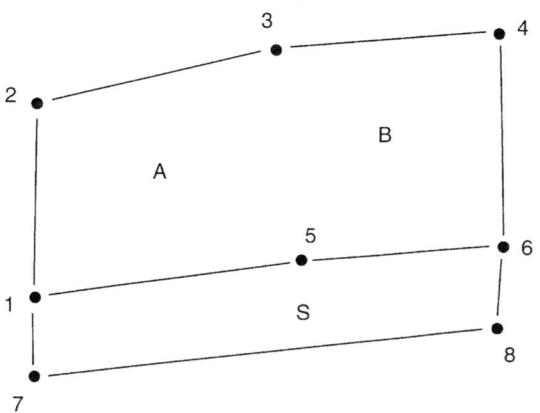

Simplicial Complex, Fig. 8 A and B merged

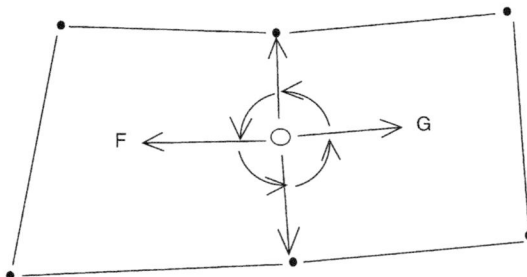

Simplicial Complex, Fig. 12 Four quad edges give one edge and point to adjacent nodes and faces

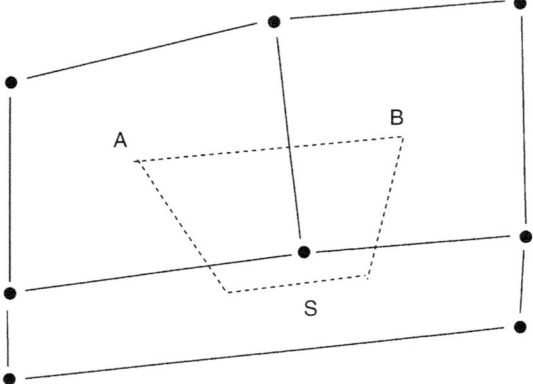

Simplicial Complex, Fig. 13 The dual graph of Fig. 1 (*dashed*) shows the neighbor relations

ing Fig. 1 as a boundary graph (primal) the dual is Fig. 13, which shows adjacency between faces.

> Quad edges represent efficiently without redundancy a much larger universe, namely partitions of orientable manifold. The Euler operations *glue* and *split* can be efficiently implemented and maintain a simplicial or cell complex. The geometry can be represented as generalized maps, for which efficient implementation using relational databases has been reported [9].

Key Applications

Many applications include geometric descriptions of objects; Computer Aided Design for mechanical and civil engineering are important, but also Geographic Information Systems, with many special applications like Utility Mapping for cities, Cadastral Maps to show ownership of land, but also car navigation systems, are popular examples.

Management of partitions is central for Geographic Information Systems (GIS); 2D partitions are wisely used for land ownership parcels, soil types, etc. Increasingly 3D models of cities and buildings are built to produce visualizations for virtual trips. Town planning applications expect that changes in 3D models over time can be visualized, which requires 4 (3 spatial plus one temporal) dimensions.

Management of the geometry of partitions of 3D space is important for CAD (Computer Aided Design), used for architecture, civil engineering but also mechanical engineering. Image processing intended to produce 3D representations of the environment is using hierarchically structured partitions and needs effective operations to subdivide these.

A generalizable approach to storing and maintaining geometry in a database integrates for many application areas the treatment of geometric data with other data. Approaches based on the theory of simplicial or cell complexes are now available as plug-ins to convert general purpose DBMS to spatial databases. They replace earlier systems where geometric data was managed in proprietary file structures and the connection between geometry and descriptive data established only in the application program.

Future Directions

Besides efforts to enhance the performances of implementations three major research goals stand out:

1. Efficient solutions for 3D data; required for example to build 3D city models and to construct operations for consistently updating these [10]
2. Generalization to *n*-dimensions to include temporal data, especially 2 and 3 dimensional geometry and time required to include time related data, movement and, in general, processes in CAD and GIS applications [11]

3. Hierarchical structures to have partitions at one level of resolution (e.g., countries of the world) and then allow subdivision (e.g., regions, departments, counties, towns) [12]

A fully general application independent, n-dimensional and hierarchical representation that supports Euler operations effectively within data stored in a database is the implied goal of research in the first decade of the twenty-first century.

Cross-References

▶ Geographic Information System
▶ Topological Data Models
▶ Topological Relationships

Recommended Reading

1. Requicha A. Representation for rigid solids: theory, methods and systems. ACM Comp Surv. 1980;12(4):437–64.
2. Corbett JP. Topological principles in cartography. Washington, DC: Bureau of the Census, US Department of Commerce; 1979.
3. Härder T. New approaches to object processing in engineering databases. In: Proceedings of the 1986 International Workshop on Object-Oriented Database Systems; 1986.
4. Frank AU, Kuhn W. Cell graph: a provable correct method for the storage of geometry. In: Proceedings of the 2nd International Symposium on Spatial Data Handling; 1986.
5. Alexandrov PS. Combinatorial topology volumes 1, 2 and 3. Mineola/New York: Dover Publications, Inc.; 1960.
6. Egenhofer M, Herring JR. A mathematical framework for the definition of topological relationships. In: Proceedings of the 4th International Symposium on Spatial Data Handling; 1990.
7. Egenhofer MJ, Franzosa RD. On the equivalence of topological relations. Int J Geogr Inf Syst. 1995;9(2):133–52.
8. Guibas LJ, Stolfi J. A language for bitmap manipulation. ACM Trans Grap. 1982;1(3):191–214.
9. Lienhardt P. Extensions of the notion of map and subdivision of a three-dimensional space. In: Proceedings of the 5th Annual Symposium on Theoretical Aspects of Computer Science; 1988.
10. Thompson RJ. Towards a rigorous logic for spatial data representation. Doctoral thesis, Delft, NCG; 2007.
11. Sellis T, et al., editors. Spatiotemporal databases: the chorochronos approach. LNCS, vol. 2520. Berlin: Springer; 2003.
12. Timpf S. Hierarchical structures in map series. Technical University Vienna, Vienna, Ph.D thesis; 1998.
13. Levin B. Objecthood: an event structure perspective. In: Proceedings Chicago Linguistic Society, vol. 35; 1999. p. 223–247.

Singular Value Decomposition

Yanchun Zhang[1] and Guandong Xu[2]
[1]Victoria University, Melbourne, VIC, Australia
[2]University of Technology Sydney, Sydney, Australia

Synonyms

Latent semantic indexing; Principle component analysis; SVD transformation

Definition

The SVD definition of a matrix is illustrated as follows [1]: For a real matrix $A = [a_{ij}]_{m \times n}$, without loss of generality, suppose $m \geq n$ and there exists SVD of A (shown in Fig. 1):

$$A = U \begin{pmatrix} \Sigma_1 \\ 0 \end{pmatrix} V^T = U_{m \times m} \sum\nolimits_{m \times n} V_{n \times n}^T$$

where U and V are orthogonal matrices $U^T U = I_m$, $V^T V = I_n$. Matrices U and V can be respectively denoted as $U_{m \times m} = [u_1, u_2, \ldots, u_m]_{m \times m}$ and $V_{n \times n} = [v_1, v_2, \ldots, v_n]_{n \times n}$, where u_i, $(i = 1, \ldots, m)$ is a m-dimensional vector $u_i = (u_{1i}, u_{2i}, \ldots, u_{mi})^T$ and v_j, $(j = 1, \ldots, n)$ is a n-dimensional vector $v_j = (v_{1j}, v_{2j}, \ldots, v_{nj})^T$. Suppose rank (A) = r and the single values of A are diagonal elements of Σ as follows:

Singular Value Decomposition, Fig. 1
SVD transformation of matrix and its approximation

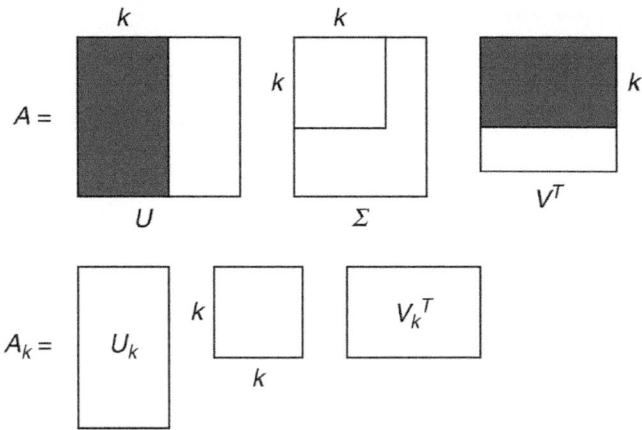

$$\sum = \begin{bmatrix} \sigma_1 0 \cdots 0 \\ 0\sigma_2 \ddots \vdots \\ \vdots \ddots \ddots 0 \\ 0 \cdots 0\sigma_n \end{bmatrix} = diag\,(\sigma_1,\sigma_2,\cdots\sigma_m)\,,$$

where $\sigma_i \geq \sigma_{i+1} > 0$, for $1 \leq i \leq r-1$; $\sigma_j = 0$, for $j \geq r+1$, that is

$$\sigma_1 \geq \sigma_2 \geq \cdots \sigma_r \geq \sigma_{r+1} = \cdots = \sigma_n = 0$$

For a given threshold ε $(0 < \varepsilon < 1)$, choose a parameter k such that $(\sigma_k - \sigma_{k+1})/\sigma_k \geq \varepsilon$. Then, denote $U_k = [u_1, u_2, \ldots, u_k]_{m \times k}$, $V_k = [v_1, v_2, \ldots, v_k]_{n \times k}$, $\sum_k = diag(\sigma_1, \sigma_2, \ldots, \sigma_k)$, and

$$A_k = U_k \sum\nolimits_k V_k T$$

As known from the theorem in algebra [1], A_k is the best approximation matrix to A and conveys main and latent information among the processed data. This property makes it possible to find out the underlying semantic association from original feature space with a dimensionality-reduced distance computational cost, in turn, is able to be used for latent semantic analysis.

Key Points

Singular Value Decomposition (SVD) algorithm could be considered as a useful means for appli-cations of data engineering and knowledge discovery, such as Web search, image and document retrieval and Web data mining. For example, finding the closely relevant pages to a given page can be carried out by manipulating SVD operation on a constructed page source to reveal the latent linkage relationships from them [3]. In order to avoid high cost of similarity computations and keep the minimum loss of linkage information contained in the feature space, SVD algorithm is applied to not only reduce the dimensionality of original feature, which leads to less com-putational costs, but also capture the semantic similarity among web pages, which is unseen intuitively. The algorithm is working as follows: (i) construct a web page space A (page source) for the given page u from link topology on the web. The page source is represented as a directed graph with edges indicating hyperlinks and nodes standing for web objects (shown in Fig. 2); (ii) since the topological relationships amongst the page source is expressed in a linkage matrix, manipulating SVD results in decomposition of original feature space $A = U_{m \times m} \Sigma_{m \times n} V_{n \times n}^T$; (iii) From the SVD theorem, the best approxima-tion matrix A_k contains main linkage information among the pages, and makes it possible to filter those irrelevant pages; (iv) by selecting a thresh-old of similarity, the relevant pages in page source to the given page are eventually found.

While SVD algorithm is usually used in con-ventional latent semantic analysis (LSA) tech-niques, some variants of LSA have been pro-

Singular Value Decomposition, Fig. 2 Page source structure for the given page u

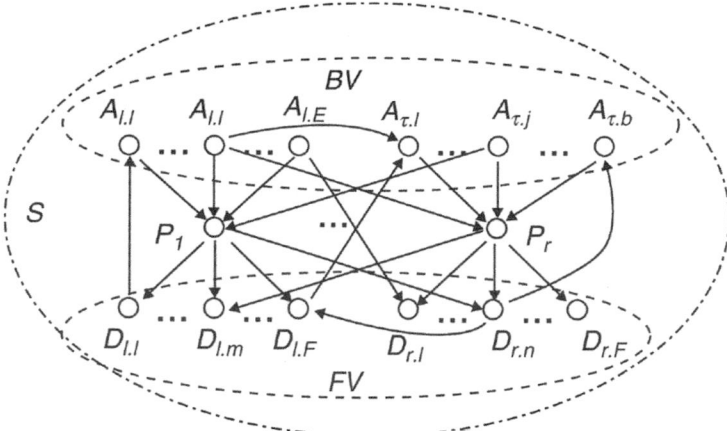

posed recently in the context of Web information processing and text mining. Apart from the difference at theoretical formulation, the common characteristics of these methods are to map the original feature space, which is utilized to model the co-occurrence observation, into a new dimensionality-reduced feature space, and maintain the maximum approximation of the original feature distance with the converted feature space. For example, Probabilistic Latent Semantic Analysis (PLSA) model is an representative of such kinds of approaches [2]. For instance, [3] proposed a PLSA-based Web usage mining approaches for Web recommendation. In this collaborative recommendation scheme, user task-oriented access patterns are extracted from Web log files, in turn, are used to predict user's likely interested Web content via referring the navigational preference of other users, who exhibit like-minded access task.

Cross-References

▶ Database Clustering Methods
▶ Principal Component Analysis

Recommended Reading

1. Datta B. Numerical linear algebra and application. Pacific Grove: Brooks/Cole Publishing Company; 1995.
2. Hofmann T. Latent semantic models for collaborative filtering. ACM Trans Inf Syst. 2004;22(1): 89–115.
3. Zhang Y, Yu JX, Hou J. Web communities: analysis and construction. Berlin: Springer; 2006.

Skyline Queries and Pareto Optimality

Peng Peng[1] and Raymond Chi-Wing Wong[2]
[1] Alibaba, Yu Hang District, Hangzhou, China
[2] Department of Computer Science and Engineering, The Hong Kong University of Science and Technology, Clear Water Bay, Kowloon, Hong Kong

Synonyms

Pareto optimal tuples

Definition

Given two d-dimensional points p and q where d is a positive integer, p is said to *dominate* or *Pareto-dominate* q, denoted by $p < q$, if p is better than or equal to q on all dimensions and p is better than q on at least one of the d dimensions. Given a set D of d-dimensional points and a point p in D, p is said to be a *skyline point* in D if p is not dominated by any other points in D. A *skyline query* is to find all skyline points in D.

Each dimension can be numeric or categorical. If a dimension is numeric, all values in this dimension are totally ordered. For any two values in the dimension, one value is more preferable than the other value. One example of a numeric dimension is the price of a product where a smaller value is more preferable. Another example of a numeric dimension is the hotel class where a higher value is more preferable. If a dimension is categorical, the ordering on the values in this dimension is more complicated. One example is airline. Some users prefer one airline A to another airline B but do not have any inclination to prefer one airline to another airline (or vice versa). Besides, some other users prefer airline B to airline A but still do not have any inclination to prefer one airline to another airline (or vice versa). No matter whether each dimension is numeric or categorical, based on the preferences on all values in the dimension, the skyline query can determine all skyline points.

For example, consider that a user would like to find a hotel near to a beach by considering two criteria, namely, the distance to a beach and the price. Figure 1a shows eight hotels, namely, a, b, $\ldots$, h. In this figure, hotel a dominates hotel d because hotel a is closer to a beach than hotel d and it is cheaper than hotel d. But, hotel a does not dominate hotel f because hotel f is closer

to a beach than hotel a. Besides, hotel a is not dominated by any hotels in this example. Hotel a is a skyline point. In this example, the set of all skyline points is $\{a, b, c\}$.

Historical Background

The skyline query was studied for traditional processing system in 1975 [1]. The term "skyline points" found in the skyline query is originally named as the term "Pareto optimal" commonly used in Economics and Engineering due to the Italian economist *Vilfredo Pareto*. Since Borzsony et al. [2] revisited it in 2001 and named the term as "skyline points," the skyline query has been studied comprehensively in the literature of databases, and the main focus is to study how to compute skyline points efficiently on very large relational databases [3, 4]. The reason why the term is named as "skyline points" is that in a two-dimensional space where smaller values are more preferable in each dimension, the visualization of the points together with the skyline points (or Pareto optimal) resembles to the real-life skyline view in a city. To illustrate, consider the same running example. Figure 1b is the same as Fig. 1a, but it includes two lines for each skyline point (i.e., a, b, and c) which can separate the space into two partitions. One partition is the one near to the origin and the other partition is the one containing all non-skyline points. The former partition can be regarded as a number of buildings

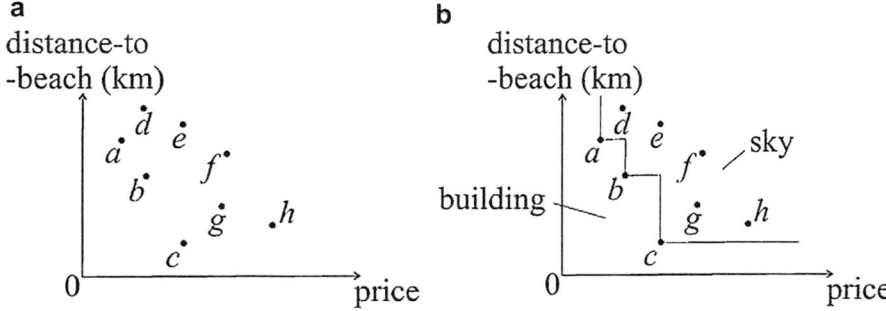

Skyline Queries and Pareto Optimality, Fig. 1 A running example

with different heights in the city and the latter partition can be regarded as a sky containing some stars each representing a non-skyline point. The lines introduced in the figure can be regarded as a skyline in the city.

Skyline is one of the most useful queries in multi-criteria decision analysis. It does not require users to give any utility function which a well-known database query called *top-k query* needs. It is shown that for any linear utility function, the point with the greatest utility value based on this function can be found in the set of skyline points. Thus, skyline points can be regarded as candidates of the traditional top-1 query. However, since there is no need to specify any utility function in a skyline query, there can be a vast number of skyline points in a dataset.

Scientific Fundamentals

Most existing studies about skyline queries are to find an efficient way of computing skyline points.

The earlier studies about skyline queries focused on skyline queries on datasets with numeric attributes only. The following studies additionally took categorical attributes into consideration for skyline queries.

First, we consider the datasets with numeric attributes only. In the following, we assume that a smaller value is preferable in each numeric attribute.

A naive method of computing skyline points for these datasets is the *block-nested loop* (*BNL*) algorithm which is to compare each point in the dataset with all other points in the dataset to see whether it is dominated by these other points and return all points which are not dominated by other points. However, this method is not efficient since it needs to reread the dataset multiple times in order to find all skyline points.

One improved skyline algorithm [1] was based on a divide-and-conquer strategy. The idea is to firstly divide the entire set of points into a number of partitions, compute all skyline points based on each separate partition, and then find all skyline points based on all partitions by filtering out all points found in each separate partition which are

dominated by other points found in other partitions. Although this method can speed up to find skyline points, it is not *progressive*. A method is said to be progressive if the method returns a subset of the answer before the algorithm ends and keeps updating the answer to the final answer until the algorithm terminates.

Another improved skyline algorithm is the sort first skyline (SFS) algorithm [3] which is progressive. In this algorithm, each point in the dataset is assigned a score which is computed based on the monotonic function on its dimensional values. This algorithm makes use of an interesting property that if a point p dominates another point p', the score of p must be smaller than that of p'. Based on this property, an efficient SFS algorithm was designed. Initially, the algorithm initializes a variable S, which stores the current skyline set, to an empty set. Then, it sorts all points in ascending order of their scores. Next, it iteratively processes each point in this ordering by comparing it with all points in S to check whether it is dominated by any point in S. If no, it is inserted into S. The final set S corresponds to the final skyline point set.

There are many other methods for skyline queries like Bitmap [5], Index [5], Nearest Neighbor [6], and Bound-and-Bound (BBS) [4]. The most influential one is the BBS method which is progressive and IO-optimal. In this algorithm, all points in the dataset are indexed by an R*-tree. Besides, in this algorithm, a min-heap is maintained to keep a list of nodes in the R*-tree where the key of each node is the minimum distance between a reference point (usually, the origin) and this node. Since each node in an R*-tree could be represented by a minimum bounding rectangle (MBR), the algorithm can make use of the geometry property of this MBR for pruning some unnecessary nodes during the execution of the algorithm, resulting in an efficient algorithm.

Next, we consider the datasets including categorical attributes in addition to numeric attributes. Some existing studies [7] focused on finding skyline points when values in each categorical attribute follow a partial ordering for any user. Some other studies [8] focused on finding skyline points when values in each categorical attribute follow different partial orderings for different users.

Some existing studies [9, 10], developed based on skyline queries, studied how to reduce the output size of a skyline query. One kind of queries is called a *k-representative skyline* query [9, 10]. A k-representative skyline query [9, 10] outputs a set of *k* "representative" points from a set of the skyline points. Different papers have different definitions about the concept of "representative" points. [9] proposed to find *k* points which dominate the greatest number of points as "representative" points. [10] proposed to find *k* points such that each of these points is far away from the other points.

Key Applications

Skyline queries are usually used for multi-criteria decision analysis where the utility function of a user is unknown. A famous example of skyline queries in the literature of databases is the hotel booking application which is shown in our running example that a traveler plans to book a hotel near to a beach for vacation. All hotels shown in the answer of the skyline query are all possible candidates for a user with any utility function.

Recently, some research studies focus on how to utilize skyline points for decision-making to serve the end purpose of business people. One example is how to find competitive products [11] and profitable products [12] among all possible products in the pool for decision-making.

Future Directions

One future direction of the skyline queries is based on some variants of the concept of "dominance," which can help users formulate their preferences in a richer manner. One representative example is *subspace dominance* [4].

Another future direction of skyline queries is to study different types of representative skylines, attempting to reduce the size of output points by utilizing more information which could be easily obtained from users.

The other future direction of skyline queries is to study the skyline queries on different types of databases such as *dynamic databases* [4], *distributed databases* [13], and *uncertain databases* [14]. More discussions on the future work of the Pareto optimal and skyline queries can be found in [15].

Data Sets

nba dataset: http://www.basketballreference.com
color dataset: http://kdd.ics.uci.edu
Household dataset: http://www.ipums.org

Cross References

► Top-K Queries

Recommended Reading

1. Kung H-T, Luccio F, Preparata FP. On finding the maxima of a set of vectors. J ACM. 1975;22(4): 469–76.
2. Borzsony S, Kossmann D, Stocker K. The skyline operator. In: Proceedings of the 17th international conference on data engineering, 2001. Heidelberg: IEEE; 2001. p. 421–30.
3. Chomicki J, Godfrey P, Gryz J, Liang D. Skyline with pre-sorting. In: Proceedings of the 19th International Conference on Data Engineering; 2003. p. 717–9.
4. Papadias D, Tao Y, Fu G, Seeger B. An optimal and progressive algorithm for skyline queries. In: Proceedings of the ACM SIGMOD International Conference on Management of Data; 2003. p. 467–78.
5. Tan K, Eng P, Ooi B. Efficient progressive skyline computation. In: Proceedings of the 27th International Conference on Very Large Data Bases; 2001.
6. Kossmann D, Ramsak F, Rost S. Shooting stars in the sky: an online algorithm for skyline queries. In: Proceedings of the 28th International Conference on Very Large Data Bases; 2002.
7. Chan C-Y, Eng P-K, Tan K-L. Stratified computation of skylines with partially-ordered domains. In: Proceedings of the 21st International Conference on Data Engineering; 2005.
8. Wong RC-W, Pei J, Fu AW-C, Wang K. Mining favorable facets. In: Proceedings of the 13th ACM SIGKDD International Conference on Knowledge Discovery and Data Mining; 2007.
9. Lin X, Yuan Y, Zhang Q, Zhang Y. Selecting stars: the k most representative skyline operator. In:

S

Proceedings of the 23rd International Conference on Data Engineering; 2007. p. 86–95.

10. Tao Y, Ding L, Lin X, Pei J. Distance-based representative skyline. In: Proceedings of the 25th International Conference on Data Engineering; 2009. p. 892–903.

11. Wan Q, Wong RC-W, Ilyas IF, Ozsu MT, Peng Y. Creating competitive products. In: Proceedings of the 35th International Conference on Very Large Data Bases; 2009.

12. Wan Q, Wong RC-W, Peng Y. Finding top-k profitable products. In: Proceedings of the 27th International Conference on Data Engineering; 2011.

13. Balke W-T, Güntzer U, Zheng JX. Efficient distributed skylining for web information systems. In: Advances in Database Technology, Proceedings of the 9th International Conference on Extending Database Technology; 2004. p. 256–73.

14. Bartolini I, Ciaccia P, Patella M. The skyline of a probabilistic relation. Knowl Data Eng IEEE Trans. 2013;25(7):1656–69.

15. Chomicki J, Ciaccia P, Meneghetti N. Skyline queries, front and back. ACM SIGMOD Rec. 2013;42(3):6–18.

Snapshot Equivalence

Christian S. Jensen[1] and Richard T. Snodgrass[2,3]
[1]Department of Computer Science, Aalborg University, Aalborg, Denmark
[2]Department of Computer Science, University of Arizona, Tucson, AZ, USA
[3]Dataware Ventures, Tucson, AZ, USA

Synonyms

Temporally weak; Weak equivalence

Definition

Informally, two tuples are *snapshot equivalent* or *weakly equivalent* if all pairs of timeslices with the same time instant parameter of the tuples are identical.

Let temporal relation schema R have n time dimensions, $D_i, i = 1, \ldots, n$, and let $\tau_i, i = 1, \ldots, n$ be corresponding timeslice operators, e.g., the valid timeslice and transaction timeslice operators. Then, formally, tuples x and y are snapshot equivalent if

$$\forall t_1 \in D_1 \ldots \forall t_n \in D_n \left(\tau_{t_n}^n \left(\ldots \left(\tau_{t_1}^1 (x) \right) \right) \ldots \right) = \tau_{t_n}^n \left(\ldots \left(\tau_{t_1}^1 (y) \right) \ldots \right) \right)$$

Similarly, two relations are snapshot equivalent or weakly equivalent if at every instant their snapshots are equal. Snapshot equivalence, or weak equivalence, is a binary relation that can be applied to tuples and to relations.

Key Points

The notion of weak equivalence captures the information content of a temporal relation in a point-based sense, where the actual timestamps used are not important as long as the same timeslices result. For example, consider the two relations with just a single attribute: {(a, [3,9]} and {(a, [3,5]), (a, [6,9])}. These relations are different, but snapshot equivalent.

Both "snapshot equivalent" and "weakly equivalent" are being used in the temporal database community. "Weak equivalence" was originally introduced by Aho et al. in 1979 to relate two algebraic expressions [1, 2]. This concept has subsequently been covered in several textbooks. One must rely on the context to disambiguate this usage from the usage specific to temporal databases. The synonym "temporally weak" does not seem intuitive – in what sense are tuples or relations weak?

Cross-References

- ▸ Point-Stamped Temporal Models
- ▸ Temporal Database
- ▸ Time Instant
- ▸ Timeslice Operator
- ▸ Transaction Time
- ▸ Valid Time
- ▸ Weak Equivalence

Recommended Reading

1. Aho AV, Sagiv Y, Ullman JD. Efficient optimization of a class of relational expressions. ACM Trans Database Syst. 1998;4(4):435–54.
2. Aho AV, Sagiv Y, Ullman JD. Equivalences among relational expressions. SIAM J Comput. 1979;8(2):218–46.
3. Gadia SK. Weak temporal relations. In: Proceedings of the 4th ACM SIGACT-SIGMOD Symposium on Principles of Database Systems; 1985. p. 70–7.
4. Jensen CS, Dyreson CE. editors. A consensus glossary of temporal database concepts – February 1998 version. In: Etzion O, Jajodia S, Sripada S, editors. Temporal databases: research and practice. LNCS 1399. Springer; 1998. p. 367–405.

Snapshot Isolation

Alan Fekete
University of Sydney, Sydney, NSW, Australia

Synonyms

Row-versioning; SI

Definition

Snapshot Isolation is a multi-version concurrency control approach that is widely used in practice. A transaction T that operates under Snapshot Isolation never observes any effects from other transactions that overlap T in duration; instead T sees values as if it were operating on a private copy or snapshot of the database, reflecting all other transactions that had committed before T started. In Snapshot Isolation, the system will not allow both of two transactions to commit if they overlap in duration and modify the same data item. Snapshot Isolation prevents many well-known anomalies (such as Lost Updates and Inconsistent Reads) that are also prevented by serializability, but it does not guarantee that all executions will be serializable. Snapshot Isolation allows reads to occur without delay or blocking caused by concurrent updates, and also updates are never blocked by concurrent readers, so Snapshot Isolation often gives the transactions better throughput than traditional concurrency control based on two-phase locking.

Historical Background

The idea of providing a multi-version concurrency control algorithm based on reading from a private snapshot has been widespread since at least 1982; Chan et al. [7] provided a detailed discussion of how this could work, considering many practical issues. However, in this early period, the idea was only considered for read-only transactions. That is, any transaction that included writes was treated with a different concurrency control, in order to ensure serializability.

In 1995, Berenson et al. [3] introduced the term Snapshot Isolation and explained the algorithm for transactions that include write operations. They did so in the context of a paper about ambiguities in how the SQL Standard defines isolation. As well as describing the algorithm, this paper showed that it could allow non-serializable executions, though it did not allow any of the particular erroneous phenomena that had been considered in the SQL Standard. Also in 1995, Oracle 7 introduced the use of the algorithm when transactions request "Isolation Level Serializable" [14]; previous releases of Oracle had not provided any transaction-level consistency but rather they had each SQL statement within a transaction see a different state of the database. In 1999, Oracle obtained a US patent (number 5870758) on their approach.

Considerable research has been devoted to understanding the properties of executions allowed by Snapshot Isolation. Bernstein et al. [4] showed how to reason about whether transactions preserve individual integrity constraints when run with Snapshot Isolation. In 2005, Fekete et al. [12] published a theory to show which sets of transactions are guaranteed to run serializably on a DBMS platform that uses Snapshot

S

Isolation (and therefore that every possible integrity constraint is preserved); some of these checks were automated by Jorwekar et al. [15]. Several papers [4, 11] look at issues that arise when some transactions run with Snapshot Isolation, while other transactions use other concurrency control techniques. An abstract characterization of the isolation provided by Snapshot Isolation was suggested by Adya [1].

Fekete et al. [11] started the agenda of considering how to modify application programs (based on analysis of the data access patterns) so that one would know that all executions on a Snapshot Isolation platform will be serializable. Alomari et al. [2] compare the performance of a number of techniques for this purpose.

The issues of storing replicated data in sites each using Snapshot Isolation have been studied extensively. Different approaches are explored by many researchers [5, 6, 8, 10, 16, 19, 20, 23]. These proposals vary in overall system architecture, how transactions are coordinated, and even in what consistency condition is provided. The use in multi-tier systems of database servers that offer SI raises similar but more complex issues [18, 24].

Foundations

Like other concurrency control techniques, Snapshot Isolation (hereafter abbreviated SI) responds to requests from clients for reading and writing data items that are stored by the DBMS engine. Each request could be either performed immediately or delayed for a while, or the requesting transaction might be aborted. SI is a multi-version mechanism, so when performing a request to read an item, the engine might return a value other than the current value of the item (that is, it might return a value that had been written in some earlier transaction, not necessarily the value written by the transaction that most recently altered the item).

The SI mechanism is defined by two properties. The first property (which explains the term "Snapshot") determines which value is to be returned in a read operation. If transaction T reads a data item x, then the value returned by the system is whichever value was written by the most recently committed transaction, among all the transactions that wrote the item x and also committed before T started. There is one exception to this rule: if T itself writes the item x, and later requests to read x, then T will see the value it wrote itself. The second property of SI is sometimes called "First Committer Wins"; it requires the engine to prevent the situation where there are two concurrent transactions (transactions T and U are concurrent if their duration overlaps, i.e., there is some intersection between the interval from the start of T to its completion and the interval from the start of U to its completion) that both write to the same item and that both commit.

To illustrate the way SI controls concurrency, consider the sequence of operations shown in Schedule 1. In presenting this and later schedules, this entry uses the standard notation where each operation has a subscript that indicates which transaction performs the operation, so r_i is a read within T_i. As well as subscripts on the operations, each data item has versions that are indicated by subscripts, where the version of item x produced by transaction T_i is represented as x_i. Thus, when T_5 writes x, this is indicated by $w_5[x_5]$, and if T_3 later does a read of x that returns the value that was written by T_5, one includes $r_3[x_5]$ in the schedule. (Readers might mistakenly think that the subscript indicates the version order; thus, they think that x_2 represents the second version of x. This is not the case in the notation used here.) The usual notation is extended by representing the start of transaction T_i as b_i (while the completion of the transaction is c_i in the case of a commit or a_i when the transaction aborts).

$$b_1 r_1 [x_0] r_1 [y_0] c_1 b_2 w_2 [x_2] a_2 b_3 r_3 [x_1] r_3 [y_1]$$
$$w_3 [x_3] b_4 r_4 [x_1] r_4 [y_1] w_3 [y_3] r_3 [x_3] c_3 c_4$$

$$(1)$$

In Schedule 1, notice that the snapshot used by T_4 includes the changes made by T_1 (which committed before T_3 started) but not those made by T_2 which aborted nor those made by T_3 which is concurrent with T_4. In particular, when T_4 reads

x, it sees the version x_1 which was written by T_1, even though there is already a more recent version x_3. In this schedule, one also can find an example where T_3 reads x twice, having modified the item in between; the second time T_3 reads x; it sees its own version.

SI is an attractive concurrency control mechanism for many reasons. It usually performs well, and in particular it does not suffer from the delays that can reduce throughput in locking-based concurrency control. For example, if a large slow transaction T is reading many items, in order to calculate some complicated statistics, traditional two-phase locking takes read locks on many items and holds these while T is running; through this long period, other transactions which want to change those items will be blocked if locking is used for concurrency control. Under SI, in contrast, the large slow transaction T does not take read locks, and concurrent transactions can update the items.

The database literature has identified a number of anomalies that can occur from uncontrolled concurrency. For example, the Inconsistent Read phenomenon happens when a transaction T sees some but not all of the changes made by another transaction U. Under SI, this can't happen: if U committed before T starts, then the effects of U are all in the snapshot used when T reads, while if U is still running when T starts, then none of the changes made by U are in the snapshot. For example, in Schedule 1 the snapshot used when T_4 reads includes both changes made by T_1 (to x and also to y) and none of the changes made by T_3. Another famous phenomenon is Lost Update. An example of this occurs when two transactions both read an item, and both produce new values that increment what they read; if both these transactions commit, the final value of the item will be incremented by one instead of by two (as would happen in a serial, non-interleaved execution). If the database platform is using SI, the First Committer Wins property will prevent Lost Update (as one of the transactions will be required to abort). For example, in Schedule 2, T_2 is not allowed to commit (since T_1 and T_2 are concurrent and both have written the same item x, the property says that they cannot both commit).

$$b_1 r_1 [x_0] w_1 [x_1] b_2 c_1 r_2 [x_0] w_2 [x_2] a_2 \qquad (2)$$

Despite preventing the well-known concurrency control anomalies, SI does not ensure that all executions are serializable. Schedule 3 shows an anomaly called "Write Skew."

$$b_1 r_1 [x_0] r_1 [y_0] b_2 r_2 [x_0] r_2 [y_0] w_1 [x_1] w_2 [y_2] c_1 c_2 \qquad (3)$$

In Schedule 3, the First Committer Wins property is not effective, because the concurrent transactions do not have any item that both of them write (T_1 writes x and T_2 writes y). The lack of serializability in this schedule can result in data corruption. For example, suppose x and y are data items that represent the balance in two different bank accounts, and suppose a business rule requires the sum of the balances to be positive. Suppose initial values are $x_0 = 100$ and $y_0 = 200$, and T_1 is reducing x by 150 (aborting if it sees insufficient funds in the combined balance), while T_2 reduces y by 175 (again, aborting if there is not enough in the total of the balances). One sees that each transaction, run alone, preserves the business rule; however the Schedule 3 is possible with SI and yet it produces a final state where x is -50 and y is 25, violating the integrity of the data according to the business rule.

Because many developers think that correct isolation ought to be what SI does (namely, a transaction does not see any effects of concurrent transactions), it is worth explaining why this is not so. Correct isolation ("serializable" execution) means that the outcome is just like in a serial or batch execution. In a batch execution, between any pair of transactions, one will come first, and so the other will see its effects. Thus, if neither of two transactions sees the other, this is not like a batch execution. The Write Skew example shows that two transactions may each decide to take some action like removing money from a bank account, where it is acceptable for one to make the change, but not when the other has already done so. When neither sees the other, they might both make the change and commit.

The Schedule 3 is not serializable, and this can be proved because it has a multi-version serialization graph with a cycle. In the literature, there are several variant definitions of multi-version serialization graph (MVSG) for a schedule; this entry uses the one in the text by Weikum and Vossen [22]. MVSG is defined for a given schedule and a version order $\ll$ which relates every pair of versions of the same data item. When working with SI, the version order is always taken as the order of the commits of the transactions that wrote the versions; that is, one defines $x_i \ll x_j$ when T_i commits before T_j. MVSG has nodes for the transactions, and there is an edge from T_i to T_j in the following three circumstances: (i) the schedule contains $r_j[x_i]$ for some item x, (ii) $x_i \ll x_j$ and the schedule contains $r_k[x_j]$ for some x and k, or (iii) the schedule contains $r_i[x_k]$ and $x_k \ll x_j$, for some x and k. It turns out that for understanding the behavior of SI, it is important to pay attention to particular edges in the MVSG: those which go between concurrent transactions. Call such an edge vulnerable, and draw it with a dashed line in the multi-version serialization graph. Notice that the First Committer Wins rule means that there can never be a vulnerable edge between two transactions if there is some data item to which both transactions write. The Snapshot property means that if there is an edge from T_i to T_j because of an operation $r_j[x_i]$, then T_i must have committed before T_j started and so the edge is not vulnerable. Thus, the only vulnerable edges arise from conflicts where one transaction reads an item which the concurrent transaction writes. The MVSG for the Schedule 3 above is in Fig. 1.

Even though SI can allow executions that are not serializable, these executions are not observed often. There are some sets of application programs which never give rise to a non-serializable execution when running with SI as the concurrency control mechanism. For example, many of the standard benchmark suites, such as TPC-C [13], generate only serializable schedules. It can

be proved [8] that in any schedule allowed by SI, if there is a cycle in the MVSG, then the cycle contains two consecutive vulnerable edges. Given a set of transactions T_1, T_2, etc., one can draw a static dependency graph SDG, which is a directed graph whose nodes are transactions, with an edge from T_i to T_j if it is possible to find a schedule h with some of these transactions, so that MVSG(h) has an edge from T_i to T_j. Furthermore, one says that the edge in SDG is vulnerable if there is a schedule h where the edge in MVSG(h) is vulnerable. Note that MVSG(h) depends on the schedule h which shows how the transactions interleave, but SDG can be found from the set of separate transactions. Because for any schedule h, MVSG(h) is a subset of SDG, it follows that if SDG has no cycle with consecutive vulnerable edges, then MVSG(h) also has no cycle with consecutive vulnerable edges, and so h is serializable. Thus, a set of transactions will always interleave in serializable executions under SI, provided that SDG does not have any cycle with consecutive vulnerable edges.

As an example, consider the transactions in Fig. 2. For these transactions, the SDG is shown in Fig. 3. The only vulnerable edges in this SDG are from T_1 to T_2 and from T_1 to T_3. The edge from T_2 to T_1 is not vulnerable because T_1 has no write operations (and under SI, a vulnerable edge can only come from a read-to-write conflict), and similarly T_3 to T_1 is not vulnerable. The edges between T_2 and T_3 are not vulnerable (in either direction) because both transactions write the item x, and so the First Committer Wins property of SI prevents these transactions both committing if they are concurrent. Thus, there are no consecutive vulnerable edges at all in SDG, and so every execution of T, T_2, and T_3 will be serializable when they run on a platform using

$$T_1 \;=\; r[x]\,r[y]\,r[z]$$

$$T_2 \;=\; r[x]\,r[y]\,w[x]\,w[y]$$

$$T_3 \;=\; r[x]\,r[z]\,w[x]\,w[z]$$

Snapshot Isolation, Fig. 2 Transactions with every execution serializable

Snapshot Isolation, Fig. 1 MVSG for Schedule 3

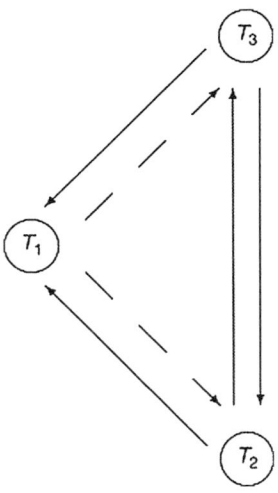

Snapshot Isolation, Fig. 3 SDG for Transactions from Fig. 2

SI for concurrency control. To use these ideas in practice, one needs to deal with application code that contains parameterized SQL statements and complicated control flow; the techniques needed are discussed in [8, 9].

What can the database administrator do if they have an application which will run on a platform where SI is the concurrency control mechanism, and yet the application is made up of transactions that are not guaranteed to have serializable executions on such a platform? The natural approach is to alter the application code, without changing the meaning of each transaction, so that the changed transactions are certain to execute serializably. This means changing programs so as to make some edges from the SDG be not vulnerable. Two classes of techniques are known to change transactions T_i and T_j where the edge from T_i to T_j is vulnerable. One can change both transactions so that they will not be allowed to run concurrently. One way to do this is to materialize the conflict on an item in the database by creating a new table called, say, Conflict and including in both T_i and T_j an update of a particular row in this table. Another form of conflict introduction explicitly sets a lock if some lock manager is available. Alternatively, one can sometimes leave T_j unaltered and introduce an identity write into T_i. That is, perform "SET x=x" in an UPDATE statement which is added to the code of T_i to affect whichever data item (row) is the one which T_i reads and T_j writes. These different techniques can lead to substantial differences in performance, depending on characteristics of the application code and the platform.

Rather than accepting the Snapshot Isolation mechanism as given, with the risk of sometimes having non-serializable executions for some applications, it is also possible for the platform developer to modify the algorithms, so that serializable execution is guaranteed. The term serializable Snapshot Isolation is used, for a concurrency control mechanism that follows the rules of SI but goes some extra checks to certify a transaction, before allowing the transaction to commit.

In conclusion, SI is a concurrency control mechanism that has many attractive features. It usually gives quite good throughput, since a read operation is never delayed by other transactions that are changing the data, and updates are not delayed when other transactions have read the data they want to change. The outdated versions that are used in SI, to respond to read requests, are often available anyway, because they are kept to support rollback recovery. SI prevents many bad executions; it can't suffer from Lost Update or Inconsistent Read or Phantoms. The way SI works is easy to understand, and indeed many articles have just assumed that being "isolated" means "not seeing any changes made by concurrent transactions" (as happens in SI). However, SI does not enforce that every execution will be serializable. Developers and users need to be aware that when SI is used, it is possible that transactions can interleave in ways that make the data invalid according to some business rule which is obeyed by every transaction running alone.

Key Applications

Snapshot Isolation is used as a concurrency control mechanism in a wide range of common platforms. For example, Microsoft SQL Server (since 2005) offers it when a user chooses to invoke "SET TRANSACTION ISOLATION LEVEL SNAPSHOT." It is similarly available as a separate isolation level in Interbase and Oracle Berkeley DB. Oracle Database uses SI for concurrency control when the client chooses "SET ISOLATION LEVEL SERIALIZABLE" [14] even though SI does allow non-serializable executions; SAP HANA has the same behavior. PostgreSQL (since version 9.1) uses Snapshot Isolation for "SET ISOLATION LEVEL REPEATABLE READ." Several large organizations have provided Snapshot Isolation in internal-use distributed data management systems, for example, Google's Percolator [17] and Yahoo!'s Omid [13]. In this setting, the pragmatics of timestamp allocation become important [9]. SI is very useful in managing replicated data. One can combine individual databases which use SI to act transparently as a global one-copy database. This is easier than to combine traditional locking databases to provide one-copy serializability. Many research prototypes combine SI with consistent replication; however, these ideas are not widely used in practice yet.

Future Directions

Research on SI continues on a number of aspects. There are many issues that arise when data are replicated between sites, some or all of which use SI rather than traditional locking for local concurrency control. Many different systems have been designed and evaluated but no clear winner has yet emerged, so research is continuing. A related issue arises in multi-tier systems, where SI may be used in the database tier, but cache management in the application tier may use a different approach to concurrency, and these may interact badly. Another topic that needs more understanding is the performance of SI, especially on multi-core platforms and when the mechanism must be distributed over multiple sites.

Cross-References

▶ Concurrency Control: Traditional Approaches
▶ Consistency Models for Replicated Data
▶ Multiversion Serializability and Concurrency Control
▶ Replication for Scalability
▶ Serializability
▶ Serializable Snapshot Isolation
▶ SQL Isolation Levels

Recommended Reading

1. Adya A. Weak consistency: a generalized theory and optimistic implementations for distributed transactions (PhD thesis). Technical Report MIT/LCS/TR-786, Laboratory for Computer Science, Massachusetts Institute of Technology, Cambridge, MA, USA, 1999.
2. Alomari M, Fekete A, Röhm U. Performance of program modification techniques that ensure serializable executions with snapshot isolation DBMS. Inf Syst. 2014;40(Mar):84–101.
3. Berenson H, Bernstein PA, Gray J, Melton J, O'Neil EJ, O'Neil PE. A critique of ANSI SQL isolation levels. In: Proceedings of the ACM SIGMOD International Conference on Management of Data; 1995. p. 1–10.
4. Bernstein AJ, Lewis PM, Lu S. Correct execution of transactions at different isolation levels. IEEE Trans Knowl Data Eng. 2004;16(9):1070–81.
5. Bornea M, Hodson O, Elnikety S, Fekete A. One-copy serializability with snapshot isolation under the hood. In: Proceedings of the 27th International Conference on Data Engineering; 2011. p. 625–36.
6. Chairunnanda P, Daudjee K, Özsu MT. ConfluxDB: multi-master replication for partitioned snapshot isolation databases. Proc VLDB Endow. 2014;7(11):947–58.
7. Chan A, Fox S, Lin W-TK, Nori A, Ries DR. The implementation of an integrated concurrency control and recovery scheme. In: Proceedings of the ACM SIGMOD International Conference on Management of Data; 1982. p. 184–91.
8. Daudjee K, Salem K. Lazy database replication with snapshot isolation. In: Proceedings of the 32nd International Conference on Very Large Data Bases; 2006. p. 715–26.
9. Du J, Elnikety S, Zwaenepoel W. Clock-SI: snapshot isolation for partitioned data stores using loosely synchronized clocks. In: Proceedings of the 32nd Symposium on Reliable Distributed Systems; 2013. p. 173–84.
10. Elnikety S, Zwaenepoel W, Pedone F. Database replication using generalized snapshot isolation. In: Proceedings of the 24th Symposium on Reliable Distributed Systems; 2005. p. 73–84.

11. Fekete A. Allocating isolation levels to transactions. In: Proceedings of the 24th ACM SIGACT-SIGMOD-SIGART Symposium on Principles of Database Systems; 2005. p. 206–15.

12. Fekete A, Liarokapis D, O'Neil E, O'Neil P, Shasha D. Making snapshot isolation serializable. ACM Trans Database Syst. 2005;30(2): 492–528.

13. Ferro DG, Junqueira F, Kelly I, Reed B, Yabandeh, M. Omid: lock-free transactional support for distributed data stores. In: Proceedings of the 30th International Conference on Data Engineering; 2014. p. 676–87.

14. Jacobs K. Concurrency control: transaction isolation and serializability in SQL92 and Oracle7. Technical Report A33745 (White Paper), Oracle Corporation; 1995.

15. Jorwekar S, Fekete A, Ramamritham K, Sudarshan S. Automating the detection of snapshot isolation anomalies. In: Proceedings of the 33rd International Conference on Very Large Data Bases; 2007. p. 1263–74.

16. Lin Y, Kemme B, Patiño-Martínez M, Jiménez-Peris R, Armendáriz-Iñigo J. Snapshot isolation and integrity constraints in replicated databases. ACM Trans Database Syst. 2009;34(2):11.

17. Peng D, Dabek F. Large-scale incremental processing using distributed transactions and notifications. In: Proceedings of the 9th USENIX Symposium on Operating System Design and Implementation; 2010.

18. Perez-Sorrosal F, Patiño-Martínez M, Jiménez-Peris R, Kemme B. Elastic SI-Cache: consistent and scalable caching in multi-tier architectures. VLDB J. 2011;20(6):841–65.

19. Plattner C, Alonso G. Ganymed: scalable replication for transactional web applications. In: Proceedings of the ACM/IFIP/USENIX 5th International Middleware Conference; 2004. p. 155–74.

20. Schenkel R, Weikum G. Integrating snapshot isolation into transactional federation. In: Proceedings of the International Conference on Cooperative Information Systems; 2000. p. 90–101.

21. Transaction Processing Performance Council. TPC Benchmark C Standard Specification, Revision 5.0. 2001. http://www.tpc.org/tpcc/

22. Weikum G, Vossen G. Transactional information systems: theory, algorithms, and the practice of concurrency control and recovery. Los Altos: Morgan Kaufmann; 2002.

23. Wu S, Kemme B. Postgres-R(SI): combining replica control with concurrency control based on snapshot isolation. In: Proceedings of the 21st International Conference on Data Engineering; 2005. p. 422–33.

24. Zellag K, Kemme B. Consistency anomalies in multi-tier architectures: automatic detection and prevention. VLDB J. 2014;23(1):147–72.

Snippet

Marcus Herzog
Vienna University of Technology, Vienna, Austria

Synonyms

Capsule; Flake; Macro; Mini; Module; Web widget

Definition

A snippet is a chunk of reusable source code. In the context of Web programming, a snippet refers to a chunk of reusable HTML source code, along with all relevant resources such as style sheets and scripts applied within the context of the snippet. In the context of Web information extraction, a snippet is a subset of the available information items that can be extracted from the Web page.

Key Points

The term snippet originates from the domain of text editors, where snippets refer to chunks of source code which can be organized for copy-and-paste usage. Snippet management allows for viewing, editing, sorting, and storing snippets in a repository of reusable source code fragments. The overall goal of snippets is to ease the process of writing code by reducing the manual effort to type in source code and to reuse existing lines of code.

Snippets can be classified according to the complexity of the interaction process: static, dynamic, and scriptable snippets. A static snippet is a fixed chunk of text that can be inserted at the cursor position. This operation is similar to a cut-and-paste operation well known from text editors. Dynamic snippets contain some dynamic elements which are filled in on insertion of the snippet into the main document. Scriptable snippets take this dynamic concept one step further by not only allowing for filling in placeholders, but by providing means to compute the values of placeholders, e.g., by applying a transformation operation on a placeholder value.

In Web programming, snippets are often used when assembling a Web page from preexisting building blocks. This is very popular in constructing social network home pages or other types of personal Web 2.0 applications such as blogs. In this context snippets are often referred to as, e.g., Web widgets, minis, or flakes, depending on the framework in which the snippet is programmed. In Web programming a snippet is already more like a mini application which can be reused in the context of a Web application, e.g., a portal such as iGoogle or MyYahoo.

In Web data extraction [1], the concept of snippet is used to refer to a particular part of the Web page which is extracted and transformed into an information item. Here the emphasis is on the reuse of existing data or content which is transformed into a presentation-independent representation, e.g., XML document format. The goal is to reuse existing data in the context of new applications which assemble data snippets from various sources and provide additional value by relating the content extracted from these independent sources.

Recommended Reading

1. Baumgartner R, Flesca S, Gottlob G. Visual web information extraction with lixto. In: Proceedings of the 27th International Conference on Very Large Data Bases; 2001. p. 119–28.

Snowflake Schema

Konstantinos Morfonios[1] and Yannis Ioannidis[2]
[1]Oracle, Redwood City, CA, USA
[2]University of Athens, Athens, Greece

Synonyms

Snowflake join schema

Definition

A *snowflake schema* has one "central" table whose primary key is compound, i.e., consisting of multiple attributes. Each one of these attributes is a foreign key to one of the remaining tables, which may, in turn, have some of its non-key attributes each be a foreign key to yet another, different table. This continues recursively with the remaining tables, until they are exhausted, forming chains or trees of foreign key dependencies rooted at the "central" table, i.e., each table in the schema (except the "central" table) is pointed to by exactly one such foreign key. (In the above, without loss of generality, we make the assumption that all tables except the "central" table have simple primary keys. This is usually the case in almost all practical situations, and as for efficiency, these keys are often generated, *surrogate keys*.)

Key Points

Many data warehouses (see definitional entry for "▶ Data Warehouse") that represent the multidimensional conceptual data model in a relational fashion [1, 2] store their primary data as well as the data cubes derived from it in snowflake schemas, as an alternative to *star schemas*. As in star schemas, the "central" table and the remaining tables of the definition above correspond, respectively, to the *fact table* and the *dimension tables* that are typically found in data warehouses. Each *fact* (tuple) in the fact table consists of a set of numeric *measures*, comprising the objects of analysis, and a set of *dimensions*, which uniquely determine the set of measures. The remaining tables store the attributes of the aforementioned dimensions at different levels of granularity.

Unlike star schemas, snowflake schemas can explicitly capture hierarchies in the dimensions, with each table in each chain (or tree path) of foreign key dependencies corresponding to one level of one such hierarchy. For instance, dimension Store in the example below contains values at different levels of detail, forming the hierarchy Street→City→State. On the contrary, star schemas capture all levels of a hierarchical dimension in a single, de-normalized table. Starting from a star schema (usually in second normal form), one may generate the corresponding snowflake schema (usually in third normal form at least) by normalization, decomposing the dimensions into multiple tables. Accordingly, star schemas lend themselves to simpler and usually faster queries, while snowflake schemas are easier to maintain and require less space.

For example, consider a data warehouse of a retail chain with many stores around a country. The dimensions may be the products sold, the stores themselves with their locations, and the dates, while the numeric measures may be the number of items and the total monetary amount corresponding to a particular product sold in a particular store on a particular date. The relevant snowflake schema, with the product, store, and date dimensions normalized, is shown below, where SalesSummary is the fact table, primary keys are in italics, and each attribute of the fact table primary key as well as each non-key 'Id' attribute of the other tables is a foreign key.

```
SalesSummary(ProductId, StoreId, DateId,
NumOfItems, TotalAmount)

Product(ProductId, ProdName,
ProdDescr, CategoryId, UnitPrice)

Category(??CategoryId, CategoryDescr)

Store(??StoreId, StreetId)

Street(??StreetId, Street, CityId)

City(??CityId, City, StateId)

State(??StateId, State)

Date(??DateId, Date, MonthId)

Month(??MonthId, Month, YearId)

Year(???YearId, Year)
```

Cross-References

► Cube Implementations
► Data Warehouse
► Dimension
► Hierarchy
► Measure
► Multidimensional Modeling
► Star Schema

Recommended Reading

1. Chaudhuri S, Dayal U. An overview of data warehousing and OLAP technology. ACM SIGMOD Rec. 1997;26(1):65–74.
2. Kimball R, Ross M. The data warehouse toolkit: the complete guide to dimensional modeling. New York: Wiley; 2002.

SOAP

Eric Wohlstadter
University of British Columbia, Vancouver, BC, Canada

Definition

SOAP [1] is an application-level protocol standard used to transport messages in distributed systems. The standard was defined and is maintained by the XML Protocol Working Group of the World Wide Web Consortium. SOAP is commonly used in the context of Web services. SOAP messages are encoded using XML and intended to carry XML encoded application data.

Key Points

SOAP provides a standard to separate infrastructure related data from application data for XML based messages. SOAP messages are known as "envelopes," which contain both a header, for infrastructure data, and a body for application data. The infrastructure which handles messages for applications is referred to as a "SOAP node." This role is commonly filled by some middleware platform. The SOAP protocol dictates the rules for the proper processing of messages by nodes on behalf of applications; this includes processing of header information and handling of faults.

The header processing rules are designed to make it easy to interpose network intermediaries between the sender and receiver of messages. The SOAP specification mentions that these intermediaries could be used for purposes such as "security services, annotation services, and content manipulation services." The specification of header information used by specific kinds of intermediaries is left to other specifications commonly known as the WS-* proposals.

The SOAP specification is intended to be extensible so that different rules for message processing can be described in further specifications. These rules are called "message exchange patterns." SOAP provides details of patterns for simple synchronous and asynchronous message exchange, which can be used for the purpose of remote procedure calls. The specification mentions but does not provide details for other more stateful patterns such as conversational exchanges and peer-to-peer message routing.

When a SOAP node is unable to process a message, an error message, called a SOAP fault is issued. Several descriptive fault types are provided by the specification as well as the conditions under which each type should be used.

SOAP provides the foundation of a Web services stack. SOAP messages are commonly layered on top of the Hypertext Transfer Protocol (HTTP). This tends to make SOAP services easier to deploy behind network firewalls; although, some critics have argued this is an abuse of HTTP. Since XML messages tend to be much larger than their binary counterparts, SOAP provides guidelines for using a binary encoding of a SOAP message body.

SOAP was originally intended as the "Simple Object Access Protocol," and its designers intended it to be used with traditional distributed object technologies such as remote method invo-

cation. When SOAP became popular for Web services the acronym was dropped because Web service interfaces are agnostic as to whether object-oriented implementations are used.

Cross-References

▶ RMI
▶ W3C
▶ Web Services

Recommended Reading

1. SOAP. Version 1.2, part 1: messaging framework (2nd ed). W3C Recommendation. http://www.w3.org/TR/soap12-part1/.

Social Applications

Maristella Matera
Politecnico di Milano, Milan, Italy

Synonyms

Collaborative software; Web 2.0 applications

Definition

Web applications, characteristic of the Web 2.0, that allow users to interact with other users by creating and sharing new content.

Key Points

The advent of the Web 2.0 has empowered the Web clients, thus providing users with richer and more complex interaction capabilities. The development of communication and interaction tools has therefore emerged, giving rise to computer-mediated communication and to collaborative approaches to content creation, based on new applications, such as social networking, file sharing, instant messaging, and blogs – just to mention a few.

The advantage from the user experience perspective is that Web users do not play only the role of passive actors accessing information, but they become creators of the contents published by Web applications.

Very often, such new applications also foster the creation of *online communities*, i.e., groups of people that interact via Web-based communication media and cooperatively create contents.

Cross-References

► Visual Interaction
► Web 2.0/3.0

Social Influence

Milad Eftekhar
University of Toronto, Toronto, ON, Canada

Synonyms

Information cascade; Information diffusion; Information spread; Innovation diffusion

Definition

Social influence is the study of individuals being affected by their peers. The subject studies how one's ideas, beliefs, or characteristics are influenced and formed by their family, friends, colleagues, acquaintances, etc. These influences in a large scale lead to so called information diffusion (aka information cascade) that explores

The author currently works at Google Inc., Mountain view, CA

the reactions of network entities against new objects and ideas as a result of the social influence they receive from their peers. The topic has been a popular subject of study in different fields including psychology, sociology, economics, and computer science.

Information diffusion explores how and to what extent a new object, called innovation, diffuses through societies. Innovations are ideas, information, products, behaviors, cultures, emotions, viruses, or other objects that are "perceived as new by an individual or other unit of adoption" [13]. Adopting a particular health behavior like obesity or happiness in a community, adoption of an instant messenger among the students of a university, switching from a product or a service to another, supporting specific political parties in an election, and participating in political uprisings in unsteady societies are only a few examples disclosing the importance of the topic. Information diffusion studies how, why, and when the entities in a society adopt an innovation and motivate their friends and peers (thanks to social influence) to adopt and diffuse it through the network.

Historical Background

The concept of information diffusion has been around for decades. It was mathematically modeled in 1978 by the 2005 Nobel prize winner Thomas Schelling [14]. Information diffusion has been one of the hot topics in computer science for the last two decades [2–4, 9, 11].

Scientific Fundamentals

Several models have been proposed to analyze innovation (information) diffusion in a network. These range from a simple model where there is a single innovation (one choice) that individuals can either accept or reject (binary decision) without changing their decision later on to more complex models where there are several innovations competing and individuals can adopt more complicated strategies (not just a binary decision of accept/reject) against them. In the next two sections, we focus on the basic models. We move forward by describing more complex models in the "Extensions" section.

The basic models of diffusion mainly differ in their definition of acceptance procedure. In other words, these models introduce different procedures for individuals to adopt an innovation. Two general categories are introduced: (1) threshold models and (2) cascade models.

A social network is mainly characterized by a graph in which nodes represent individuals in the network and edges represent the relationship between these individuals. Depending on the network and the application, there is a relationship between two individuals if they are friends, colleagues, coauthors, neighbors, etc. In various situations, the graph can be directed or undirected and weighted or unweighted. It is undirected if the relationship between any pair of individuals is reciprocal and directed otherwise. Moreover, we can define a weight for each edge in the graph to represent the influence individuals exert on each other. The weight of a directed edge (a, b) is defined as the extent of influence of a on b.

The Threshold Model

Let $G = (V, E)$ represent the social graph where V contains all nodes and E contains all edges in the social network. For each node $v \in V$, let the *threshold value* of v, denoted by θ_v, be a real number between 0 and 1 and f_v be a function from V to $[0, 1]$.

Assume there is an innovation at hand and the goal is to analyze how it is adopted by the nodes in the graph (individuals in the society). A node is called active if it adopts the innovation. Let X_v denote the active neighbors of v. Under the threshold model, node v adopts the innovation if and only if

$$f_v(X_v) \geq \theta_v \text{ [9]}$$

The function f can be defined differently based on the application we study. It can be as simple as the sum of the weight (influence) of edges from active neighbors of v to v or as complicated as *any* arbitrary function with an input of a set of nodes and an output of a real number between 0 and 1. The former case is

called the linear threshold model where node v adopts the innovation iff

$$f_v(X_v) = \sum_{\{u | (u,v) \in E, u \text{ is active}\}} w_{uv} \geq \theta_v$$

and the latter case is called the general threshold model.

The Cascade Model

The second category of models are defined as cascade models [9]. Instead of defining acceptance threshold for nodes, these models introduce success probabilities on edges. The main idea is as follows: as soon as a node adopts the innovation, it attempts once to convince its neighbors to do so. With a specified probability, each attempt leads to a success. These probabilities may depend on several factors including both sides of this activation attempt and the history of previous attempts. Formally, $g_u(v, X)$ is the activation probability when u adopts the innovation because of v's attempt, while X contains nodes that previously tried to persuade u and failed. Similar to the threshold models, in the cascade model, function g can be as simple as a constant value p or as complex as any arbitrary probability function. The former case is called the "independent cascade model," whereas the latter is called the "general cascade model."

The cascade model utilizing functions g with diminishing returns characteristics is proved to be equivalent to the general threshold models with submodular functions [12].

Extensions

The first extension is on the number of options a user has. Consider the scenario where there exist two companies competing to maximize the spread of their influence in the network. A user has an option to choose any of these competitors' products. Immorlica et al. [8] propose a model in that there exist two products (innovation) that a user may choose: the old one and the new one. A user has three options: (1) keep the old product, (2) replace it with the new product, or (3) hold both by paying extra, say L units of cost. Depending on the value of L, we observe scenarios where the new product becomes dominant in the network, or it dies out quickly, or scenarios where users choose to utilize both.

In another extension, Hartline et al. [7] study the problem where users pay differently to acquire the same innovation. To illustrate, consider a market example. We can analyze how revenue changes if the same product is sold to different users with different prices. The key application for this model is to identify the optimal price each user should pay (based on users' incentives to buy the product) to increase revenue.

Discovering the Underlying Graph

Not always the underlying social graph is given. It is important, therefore, to utilize available information to realize the influences between users and undermine the graph of relationships in a social network over which information cascades. Gomez et al. utilize the adoption time for each node in the network to discover who influenced whom [6]. Given the time when each node in the network adopts the innovation, we want to identify the paths of diffusion and the underlying graph of influence in this network. The key idea is based on this assumption that the probability of diffusion between a pair of nodes decreases by the difference between the times they adopt the innovation. In other words, if the difference in the adoption time of u and v is large, it is unlikely that v has been influenced by u. Two models have been considered to capture this assumption: an exponential model $P(u, v) \propto e^{-\frac{\Delta_{u,v}}{\alpha}}$ and a power-law model $P(u, v) \propto \frac{1}{\Delta_{u,v}^{\alpha}}$ where $P(u, v)$ is the probability of diffusion between u and v, $\Delta_{u,v}$ is the difference between the adoption time of nodes u and v, and α is the models' parameter. The goal is to create the optimal network which matches the infection times well (with the highest probability). Unfortunately this problem is NP-hard. Therefore, an approximation algorithm is proposed to estimate the best graph [6].

In another work, Leskovec et al. [10] study the cases where an individual can influence his or her friends in a positive or negative way. A technique has been proposed to determine the sign of the links in a given social network.

Key Applications

Influential Individuals

As previously discussed, communication and interpersonal relationships play a principle role in spreading innovations among members of a community. To maximize this spread, wise selection of the first adopters is crucial. The problem of identifying the most influential set of k people in a social network (the "seed set" or the "first adopters") has received an extensive attention. The most influential seed set S is the seed set with the maximum *final influence*, that is, the number of people that will eventually adopt the innovation if S contains all members that initially adopt it.

The problem has been studied extensively [1–3, 9, 11]. It has been proven that this problem is NP-hard for most models studied, even for very simple special cases [9]. Furthermore, there are models for which even approximating the optimal value within a factor of $n^{1-\epsilon}$ is NP-hard (n: the number of individuals) [12]. Therefore, several heuristic and greedy algorithms are proposed to approximate the best solution for simpler models [2, 3, 9].

Identifying the k most influential nodes is one of the well-known problems in this field of study. Since this problem is NP-hard, several heuristic and greedy algorithms have been proposed ranging from the algorithm of randomly choosing k nodes to heuristic approaches of k central nodes, k high out-degree nodes, and degree-discount [2] and finally to the greedy algorithm of hill climbing [3, 9]. It is shown that if S^* contains the k most influential nodes of the network and S contains the k nodes discovered by the greedy algorithm, $f(S) \geq (1 - 1/e) \times f(S^*)$ in which f is the final influence.

Some approaches have been proposed to improve the performance of the greedy algorithm [1, 2, 11]. Leskovec et al. take advantage of submodularity characteristics of the influence diffusion function in order to avoid expanding all candidate sets for diffusion analysis [11]. In their experiments they achieved 700 times speed up compared to the standard greedy algorithm. Chen et al. propose methods to improve the performance of the greedy algorithm for the independent [2] and weighted cascade models [1]. In the independent cascade model, the activation probability p is fixed; thus, the active edges can be predetermined by tossing a coin (with probability p and $1 - p$) before diffusion takes place. A new graph is created by removing all non-activated edges. The final influence of a given seed set is easily calculated by enumerating the number of nodes reachable from the seed set in the new graph [2]. A heuristic approach has been introduced as well – that is called degree discount – for cases where the activation probability p is small. The intuition is based on this observation that ignoring indirect influences of each node to multi-hop neighbors produces a negligible error in the estimation of the final influence when p is small.

In [1], the maximum influence path between all pairs of nodes is calculated, and linear relationships are generated between the activation probability of node pairs. Using these relationships, this approach finds the incremental influence that each node exerts on other nodes of the graph and chooses the node with the highest value as the new seed. The authors have shown that this approach is scalable and more efficient compared to the original greedy algorithm if the maximum number of connected nodes (direct or indirect) to each node is significantly smaller than the size of the graph.

Influential Groups

Traditionally, mass advertising is one of the most prevalent methods to target people. Many companies deliver their advertisements to their potential customers through TV commercials, newspaper ads, billboards, etc. The audience of each of these media constitute a group: e.g., the group of people who see a billboard, the group of people who read a newspaper ad, etc. Once a company decides to advertise, there are different media (i.e., different groups of people) it can target.

Given a fixed budget, clearly it is not possible to advertise to all existing groups. Thus, a question arises here: how to distribute the advertising budget to different groups in order to maximize revenue. Eftekhar et al. [4] address this question

by exploring the problem of identifying top-k influential groups. In other words, they study what k groups an advertiser should target to maximize its revenue (final influence). Note that as a result of advertising to a group, some members of the group become convinced to adopt the innovation. As a result of the social influence between these members and their neighbors, some new users adopt the innovation. The propagation continues until no new user adopts the innovation.

Two models have been proposed: fine-grained diffusion model (FGD) and coarse-grained diffusion model (CGD). The FGD model consists of two phases: (1) translating a mass advertisement (an advertisement to a group) into individual adopters and (2) utilizing traditional individual-based diffusion models (e.g., threshold or cascade models) to simulate diffusion in the network starting by these individual adopters. The CGD model provides a highly faster platform. Under CGD, a graph (we call it the *group graph*) is created to represent the group structure of the network: each node represents a group and edges determine the aggregate social influence between members of groups. The higher the weighs of edges between users in groups A and B, the higher the weight of the edges connecting A and B in the group graph.

It has been shown that the problem of identifying k influential groups under both models is NP-hard. The good news is that the final influence function in both models is submodular and a greedy algorithm provides a $(1 - 1/e)$ approximation bound. Please refer to [4] for more details on the final influence function, the diffusion process under FGD and CGD, the submodularity property, and the algorithm.

This paradigm shift from individual targeting to group targeting provides several advantages. It is in-line with a significant portion of advertising practices that are utilizing mass advertising techniques. It provides a platform under which the problem can be solved more efficiently (over 10 million times speed up compared to individual targeting problem). Furthermore, targeting influential groups rather than influential individuals can broaden the final influence under reasonable circumstances. The intuition is that by targeting influential groups rather than individuals, the diffusion process starts by more individuals (the initial adopters) who are less influential. Experiments show that, under reasonable assumptions, more individual adopters who are less influential may lead to higher final revenue than less individual adopters who are more influential.

Bursty Subgraphs

Information burst refers to a situation where an unusual amount of activities is performed on a specific topic. For example, a large number of email exchanges – significantly higher than the average – related to a topic in a short period of time or in a particular location specifies an instance of information burst.

This unusual activity can be instigated by external sources (e.g., an earthquake) and/or social activities between users (e.g., political chatter associated to an uprising). Clearly, social influence plays a critical role in the formation of burst and bursty subgraphs. There usually exist some users that initialize a movement in the network by posting messages. Some of their neighbors become influenced and join their peers in sharing the messages with other users. This creates a cascade and a burst in the network and forms bursty subgraphs.

Identifying subgraphs (subgraphs of a social graph) that contribute to the burst of information on a given topic is of vast interest. These subgraphs display different facets of the topic and the characteristics of people who are active on the topic. Moreover, these subgraphs are valuable targets for advertising purposes.

SODA is an algorithm to identify the bursty subgraphs for a given topic [5]. To locate these subgraphs, SODA takes both the activity of users and the social connections into account. It identifies each node as bursty or non-bursty by considering two main constraints. First, the state of each node should match its activity level on the given topic. That is, users who have posted at a rate higher than the average should be identified as bursty with a higher probability. Second, the state of neighbors should be close to each other. That is, the probability that a user is recognized as bursty is higher if its neighbors are recognized

as bursty. The latter is due to the concept of social influence between neighbors.

SODA is an iterative algorithm. The bursty state of nodes are identified when the whole process converges. See [5] for more details. Experiments on large real-world datasets show that convergence is usually achieved very fast within 10–20 iterations.

Datasets

- DBLP dataset retrieved at Oct 21, 2012, available at http://csng.cs.toronto.edu/Data/dblp.xml.zip. "The dataset is a network of coauthorship between scientists publishing papers or articles in computer science conferences or journals indexed by DBLP by Oct 21, 2012." "This graph contains about 800 thousand nodes and 6.3 million directed weighted edges. Each conference or journal (e.g., KDD, PVLDB) is a group. The DBLP dataset contains about 3200 groups" [4].
- Condensed matter collaborations 2005 dataset available at http://www-personal.umich.edu/~mejn/netdata/cond-mat-2005.zip. The dataset is "a network of coauthorship between scientists posting preprints on the Condensed Matter E-Print Archive" between Jan 1, 1995, to Mar 31, 2005. The graph contains over 40,000 nodes and 350,000 directed weighted edges.

URL to Code

- Java codes implementing the traditional threshold model, the FGD and CGD diffusion models, and the algorithms to identify influential individuals and influential groups [4] are available at http://www.cs.toronto.edu/~milad/KDD13_codes.html

Cross-References

▶ Social Networks

Recommended Reading

1. Chen W, Wang C, Wang Y. Scalable influence maximization for prevalent viral marketing in large-scale social networks. In: Proceedings of the 16th ACM SIGKDD International Conference on Knowledge Discovery and Data Mining; 2010. p. 1029–38.
2. Chen W, Wang Y, Yang S. Efficient influence maximization in social networks. In: Proceedings of the 15th ACM SIGKDD International Conference on Knowledge Discovery and Data Mining; 2009. p. 199–208.
3. Domingos P, Richardson M. Mining the network value of customers. In: Proceedings of the 7th ACM SIGKDD International Conference on Knowledge Discovery and Data Mining; 2001. p. 57–66.
4. Eftekhar M, Ganjali Y, Koudas N. Information cascade at group scale. In: Proceedings of the 19th ACM SIGKDD International Conference on Knowledge Discovery and Data Mining; 2013. p. 401–9.
5. Eftekhar M, Koudas N, Ganjali Y. Bursty subgraphs in social networks. In: Proceedings of the 6th ACM International Conference on Web Search and Data Mining; 2013. p. 213–22.
6. Gomez-Rodriguez M, Leskovec J, Krause A. Inferring networks of diffusion and influence. ACM Trans Knowl Discov Data (TKDD). 2012;5(4):1–37.
7. Hartline J, Mirrokni VS, Sundararajan M. Optimal marketing strategies over social networks. In: Proceedings of the 17th International World Wide Web Conference; 2008. p. 189–98.
8. Immorlica N, Kleinberg J, Mahdian M, Wexler T. The role of compatibility in the diffusion of technologies through social networks. In: Proceedings of the 8th ACM Conference on Electronic Commerce; 2007. p. 75–83.
9. Kempe D, Kleinberg J, Tardos E. Maximizing the spread of influence in a social network. In: Proceedings of the 9th ACM SIGKDD International Conference on Knowledge Discovery and Data Mining; 2003. p. 137–46.
10. Leskovec J, Huttenlocher D, Kleinberg J. Predicting positive and negative links in online social networks. In: Proceedings of the 19th International World Wide Web Conference; 2010. p. 641–50.
11. Leskovec J, Krause A, Guestrin C, Faloutsos C, VanBriesen J, Glance N. Cost-effective outbreak detection in networks. In: Proceedings of the 13th ACM SIGKDD International Conference on Knowledge Discovery and Data Mining; 2007. p. 420–29.
12. Nisan N, Roughgarden T, Tardos E, Vazirani VV. Algorithmic game theory, chapter 24. Cambridge/New York: Cambridge University Press; 2007.
13. Rogers EM. Diffusion of innovations. New York: Simon and Schuster; 2010.
14. Schelling T. Micromotives and macrobehavior. Nueva York: W.W. Norton & Company; 1978.

Social Media Analysis

Michael Mathioudakis
Université de Lyon, CNRS, INSA-Lyon, LIRIS, UMR5205, F-69621, France

Synonyms

Social media analytics; Social media mining

Definition

Social media analysis is the process of extracting knowledge from data that originate on social media platforms. As a field of study, it appeared in the early 2000s with the explosion of user-generated content on the web. It can be viewed as a special case of data mining.

Social media data, the object of analysis, include digital content generated by social media users (e.g., text, photos, and videos), user demographics (e.g., gender, age, location of residence), social relationships (e.g., the social network of users), and information about user activity (e.g., the actions performed by individual users on the social media platform, as well as related metadata, such as the time and location of user activity).

Social media analysis is used for a wide range of purposes, from scientific to commercial. Even though it would be difficult to list all related data mining tasks, one can distinguish among them a few broad themes that have emerged, such as emerging topic detection, sentiment analysis, and community detection. Depending on its goal, the analysis makes use of the appropriate subset of social media data and calls for different techniques, possibly borrowed from other subdisciplines of data mining, such as **text mining** or **social network analysis**. Moreover, depending on the purpose of the analysis, social media data might be analyzed in conjunction with data from other sources. For example, social media analysis for the purposes of targeted advertising might take the purchase history of users into account.

Historical Background

As a field of study, social media analysis appeared in the early 2000s. It followed the explosion of user-generated content on the web – a phenomenon enabled by technologies that allow the dynamic update of web page content, thereby empowering users not only to consume content on static web pages but also to produce content and interact with each other. **Social media**, in particular, are web platforms that host and facilitate the creation of user-generated content. They include blogging and microblogging platforms, online social networks, multimedia sharing websites, and question-answering communities and forums.

Blogging platforms were among the earliest web platforms to host user-generated content, and content published on blogs (or "weblogs") became the focus of early social media analysis. For example, Marlow [12] highlights the social character of blogging, with social ties evidenced by linking and commenting activity, and provides linking activity statistics from about 30,000 blogs over 3 years (2001–2003). Marlow suggests that methods from social sciences, and social network analysis in particular, can be used to quantify the *authority* of bloggers based on link activity – an approach that was followed by later works. Kumar et al. [10] study linking activity among blogs across time and highlight the appearance and dissolution of *dense communities*, evidenced by temporally focused linking activity among a relatively small number of blogs. Adamic and Glance [1] focus on linking activity among political blogs during the 2004 US presidential election and highlight the reflected division between liberal and conservative bloggers.

Social media continue to have a large presence on the web until the time of this writing (2014). Research on user-generated content seeks to provide systematic solutions to analysis tasks, such as emerging topic detection, sentiment analysis, community detection, and identification of authoritative users.

Scientific Fundamentals

Social media analysis encompasses a wide variety of tasks that are applied on datasets of various origins, format, and content. This section describes a typical social media dataset (section "Data"), as well as analysis tasks (section "Data Analysis Tasks") that have been studied in the literature.

Data

Social media platforms differ in the kind of user activity they host and the information they disclose about their users' activity, as per each platform's terms and policies. Therefore, data used for the purposes of social media analysis differ from case to case, depending on the platform(s) from which those data originate.

Even though it would be difficult to provide a single general description for all possible instances of social media datasets, one can argue that data from one social media platform can typically fit into relations of the following scheme:

```
users(id, name, gender, age, location,
      <other attributes>)

connections(users: idA,users: idB,
      connection type,
      <other attributes>)

actions(action type,users: id, time,
      location, content,
      <other attributes>)
```

The three relations above contain, respectively, information about *users* of the social media platform in consideration, the social *connections* between them, and their *actions* on the platform. Such information is typically made available by the social media platforms through an API.

Specifically, each entry of relation **users** contains information about a single user of the platform. Each user is associated with an id, typically a custom character string that serves as a unique identifier for the user on the platform. Other information includes the real name and gender of the user, their age, location of residence (e.g., city and country), and possibly other demographic information (e.g., the user's profession). For example, the entry

```
('jane_doe', 'Jane Doe', female, 42,
    New York City)
```

refers to a user named "Jane Doe," with id "jane doe," of female gender and age 42, who is based in New York City.

Each entry of relation **connections** represents a social connection between two users of the platform. Such connections are declared explicitly by users and can be either directed or undirected. For example, the entry

```
('jane_doe', 'john_doe', friends)
```

indicates that the two users associated with identifiers "jane_doe" and "john_doe" have declared themselves as friends on the platform.

Finally, relation **actions** can be thought to represent a log of the activity on the platform. Each entry records one action of one user, as well as information about that action, such as the time it took place and the location of the user at that time (possibly in the form of a pair of geographic coordinates). For example, the entry

```
(message, 'jane_doe', 2014-06-15T21:05:
30EST, (40.757778, -73.985833),
'Hello!')
```

indicates that user "jane_doe" generated a message with "Hello!" as its text, timestamp 2014–06-15 T21:05:30EST and geographic coordinates (40.757778, −73.985833) pointing to Times Square, NY.

Data Analysis Tasks

Social media analysis is a special case of data mining, applied on datasets that describe user activity on social media platforms. The following are examples of data analysis tasks that have been studied in the literature.

Emerging topic detection. Attention of social media users shifts dynamically as users comment on emerging events and news stories. The change of attention is reflected in the sudden increase in the rate of appearance of different pieces of text, single keywords, or phrases. Given the enormous volume of social media activity, identifying the focus of online discussion is a challenging task. A typical approach to identify emerging topics is to track the frequency of small sets of keywords (individual keywords, n-grams, or phrases) in recent user-generated documents and detect sudden surges of their values. Detection of such "bursts" is treated as signal that the related keywords are used in online discussions about an emerging topic. The approach was followed by Glance et al. in [7] to detect emerging topics in the blogosphere and by Alvanaki et al. in [3] and Mathioudakis and Koudas in [13] to detect emerging topics on Twitter.

Identification of authoritative users. As it is often the case in many social settings, some social media users enjoy a status of authority among other users, with their activity attracting significantly more attention than others. In traditional web settings, where the amount of web links that pointed to a web page was used as evidence for its authority, the authority of web pages was quantified via graph theoretic measures on the link graph. Such measures have been extended to and studied in social media settings. As an example, Bouguessa et al. [6] discuss how graph theoretic measures can be applied in the context of question-answering portals like Yahoo! Answers to quantify the authority of different users across questions of different categories.

Community detection. Interactions between social media users give rise to a network structure, with dense areas in that structure indicating the existence of communities of users. That observation was made early by Kumar et al. in [10], who documented the dynamic formation and dissolution of temporary communities of bloggers, as evidenced by their linking activity. Later, in [15], Papadopoulos et al. provide an extensive comparison of community detection techniques in social media that demonstrates how graph analysis techniques can be used on networks that arise from different kinds of social media activity.

Sentiment analysis. Due to the large volume of social media activity, manual inspection of all content of interest is next to impossible in many scenarios of data analysis. In such cases, one looks for ways to succinctly summarize the available data. For example, the campaign team of a high-profile politician running for office might be interested to know the general sentiment (positive, negative, or neutral) expressed by online users in comments about their candidate but have hardly the time to read all comments about him/her. That's a task that falls within the scope of sentiment analysis. For examples of sentiment analysis applied over social media data, see work by Wang et al. [16] that describes a system to perform sentiment analysis focused on the US presidential campaign or by Bollen et al. [4] that associates online mood with various socioeconomic events (e.g., elections and national holidays).

Other tasks. The list of social media analysis tasks discussed above contains only some of the tasks that are discussed in the literature. Examples of other related tasks include discovery of high-quality content [2], prediction of user activity [11], and tag recommendations in folksonomies [9].

Key Applications

Social media analysis finds applications in a range of domains, from scientific and commercial. The following are three application examples:

Market research. Social media analysis can be used to track online users' opinions about products and brands. For example, see Jansen et al. [8] for a study of microblogging activity on Twitter that measures the volume of messages which contain comments about a brand and analyze their linguistic structure and expressed sentiment.

Stock market prediction. Social media analysis can be used to track online users' mood, which in turn can be used as a signal to predict the behavior of the stock market. For example, see Bollen et al. [5] for a related study of microblogging activity on Twitter. The study supports the hypothesis that mood reflected on social media predicts changes of the stock market.

Social sciences. Observing social media activity, we essentially observe human activity at a much larger scale and detail than ever before. Social media analysis can thus be used to extract insights about various social phenomena. As an example, see Mendoza et al. [14] for a study on the spread of false rumors under an emergency situation. Specifically, the authors study Twitter activity during the Chilean earthquake of 2010 and identify messages that spread false rumors about the event. One positive finding of the study was that false rumors were questioned more by online users than true pieces of information.

Cross-References

▸ Interactive Analytics in Social Media
▸ Social Influence
▸ Social Media Analytics
▸ Social Networks
▸ Structure Analytics in Social Media
▸ Temporal Analytics in Social Media
▸ Text Analytics in Social Media

Recommended Reading

1. Adamic LA, Glance N. The political blogosphere and the 2004 US election: divided they blog. In: Proceedings of the 3rd International Workshop on Link Discovery; 2005.
2. Agichtein E, Castillo C, Donato D, Gionis A, Mishne G. Finding high-quality content in social media. In: Proceedings of the 2008 International Conference on Web Search and Data Mining; 2008. p. 183–94.

3. Alvanaki F, Michel S, Ramamritham K, Weikum G. EnBlogue: emergent topic detection in web 2.0 streams. In: Proceedings of the ACM SIGMOD International Conference on Management of Data; 2011. p. 1271–4.
4. Bollen J, Mao H, Pepe A. Modeling public mood and emotion: twitter sentiment and socio-economic phenomena. In: Proceedings of the 5th International AAAI Conference on Weblogs and Social Media; 2011.
5. Bollen J, Mao H, Zeng X. Twitter mood predicts the stock market. J Comput Sci. 2011;2(1):1–8.
6. Bouguessa M, Dumoulin B, Wang S. Identifying author- itative actors in question-answering forums: the case of yahoo! Answers. In: Proceedings of the 14th ACM SIGKDD International Conference on Knowledge Discovery and Data Mining; 2008. p. 866–874.
7. Glance N, Hurst M, Tomokiyo T. BlogPulse: automated trend discovery for blogs. In: Proceedings of the WWW Workshop on the Blogging Ecosystem Aggregation, Analysis and Dynamics; 2004.
8. Jansen BJ, Zhang M, Sobel K. Twitter power: tweets as electronic word of mouth. J Am Soc Inf Sci Technol. 2009;60(11):2169–88.
9. Jäschke R, Marinho L, Hotho A, Thieme LS, Stumme G. Tag recommendations in folksonomies. In: Proceedings of the 11th European Conference on Principles and Practice of Knowledge Discovery in Databases; 2007.
10. Kumar R, Novak J, Raghavan P, Tomkins A. On the Bursty evolution of blogspace. World Wide Web. 2003;8(2):159–78.
11. Lerman K, Hogg T. Using a model of social dynamics to predict popularity of news. New York: ACM; April 2010.
12. Marlow C. Audience, structure and authority in the weblog community. In: Proceedings of the International Communication Association Conference; 2004.
13. Mathioudakis M, Koudas N. TwitterMonitor: trend detection over the twitter stream. In: Proceedings of the ACM SIGMOD International Conference on Management of Data; 2010. p 1155–1158.
14. Mendoza M, Poblete B, Castillo C. Twitter under crisis: can we trust what we RT? In: Proceedings of the 1st Workshop on Social Media Analytics; 2010. p. 71–79
15. Papadopoulos S, Kompatsiaris Y, Vakali A, Spyridonos P. Community detection in social media. Data Min Knowl Disc. 2011;24(3): 515–554.
16. Wang H, Can D, Kazemzadeh A, Bar F. A system for real-time twitter sentiment analysis of 2012 us presidential election cycle. In: Proceedings of the 50th Annual Meeting of the Association for Computational Linguistics; 201

Social Media Analytics

Sihem Amer-Yahia[1,2], Alexandre Termier[3], and Behrooz Omidvar-Tehrani[4]
[1]CNRS, Univ. Grenoble Alps, Grenoble, France
[2]Laboratoire d'Informatique de Grenoble, CNRS-LIG, Saint Martin-d'Hères, Grenoble, France
[3]LIG (Laboratoire d'Informatique de Grenoble), HADAS team, Université Joseph Fourier, Saint Martin d'Hères, France
[4]Interactive Data Systems Group, Ohio State University, Columbus, OH, USA

Definition

Social media analytics is the science of developing models and algorithms in order to understand users and user-generated content in social media. Different angles have been developed in social media analytics. When user-generated content is analyzed, it is referred to as text analytics. When interactions between users are studied, it is referred to as structure analytics. The introduction of the time dimension in social media analytics is referred to as temporal analytics. Finally, the analysis could be done by involving the analyst in an interactive process that is referred to as interactive analytics. We describe each one of those axes.

Text Analytics in Social Media

Sihem Amer-Yahia (CNRS/LIG, France), Sofiane Abbar (Qatar Computing Research Institute, Qatar), Noha Ibrahim (LIG, France)

Structure Analytics in Social Media
Sihem Amer-Yahia (CNRS/LIG, France), Mahashweta Das (HP Labs, USA), Gautam Das (UT Arlington, USA), Saravanan Thirumuruganathan (UT Arlington, USA), Cong Yu (Google Research NYC, USA)

Temporal Analytics in Social Media
Sihem Amer-Yahia (CNRS/LIG, France), Themis Palpanas (Univ. Paris Descartes, France), Mikalai Tsytsarau (Univ. of Trento, Italy), Sofia Kleisarchaki (LIG, France), Ahlame Douzal (LIG, France), Vassilis Christophides (Technicolor Labs, France)

Interactive Analytics in Social Media
Sihem Amer-Yahia (CNRS/LIG, France), Alexandre Termier (LIG, France), Behrooz Omidvar Tehrani (LIG, France)

Social Media Harvesting

Yang Yang
Center for Future Media and School of Computer Science and Engineering, University of Electronic Science and Technology of China, Chengdu, Sichuan, China

Definition

Social media harvesting is the practice of collecting various types of user-generated content (e.g., tweets, blogs, image, video clips, POI, tags, users) from social media sharing platforms (e.g., Twitter, YouTube, Instagram, Pinterest) and mining useful knowledge (e.g., customer sentiment, social trending topics) to support research practice and make business decisions.

Historical Background

In big data era, driven by the advance of fast Internet, massive storage device, pervasive digital camera, and social media sharing platform, an ever-growing amount of media contents have been generated, consumed, and distributed over the Web. It was recognized that harvesting social media has become a promising topic in both academic and industry areas. Quite a bunch of popular research topics have been recognized, e.g., social content analysis/recognition/search, event modeling, sentiment analysis, social healthcare, smart city, user profiling, etc. Companies and research institutions have made great endeavors in collecting social media data. Different from existing data analytics based on well-organized data, social media harvesting and analytics aims to explore "gold" mines of intelligence from unreliable and incomplete user-generated data, which poses more challenges on traditional analytics methodology.

One of the most successful social media harvesting and analytics projects is NUS-Extreme-Tsinghua (NExT) [1] – joint led by the National University of Singapore and Tsinghua University. Their main focus is on gathering and analyzing contents that are not currently accessible from the trditional search engines. Such big social media data enables innovative applications and cutting-edge research to be realized and advanced, such as user profiling [2], social event summarization [3], event detection/visualization/tracking [4] and location estimation [5]. However, conducting research on the aforementioned topics is confronted with the following four key challenges according to the work by Chua et al. in [6]: (1) collecting relevant data from various aspects; (2) cope with the ever-growing multimedia data; (3) detect-, visual-, and track-related topics; and (4) derive predictive results to support high-level applications.

Key Applications

Event Detection/Visualization/Tracking
In social media era, event detection, visualization, and tracking have become one of the hottest topics. It can facilitate different types of applications, such as marketing and social security. During the last two decades, several techniques have been devised. For instance, in

S

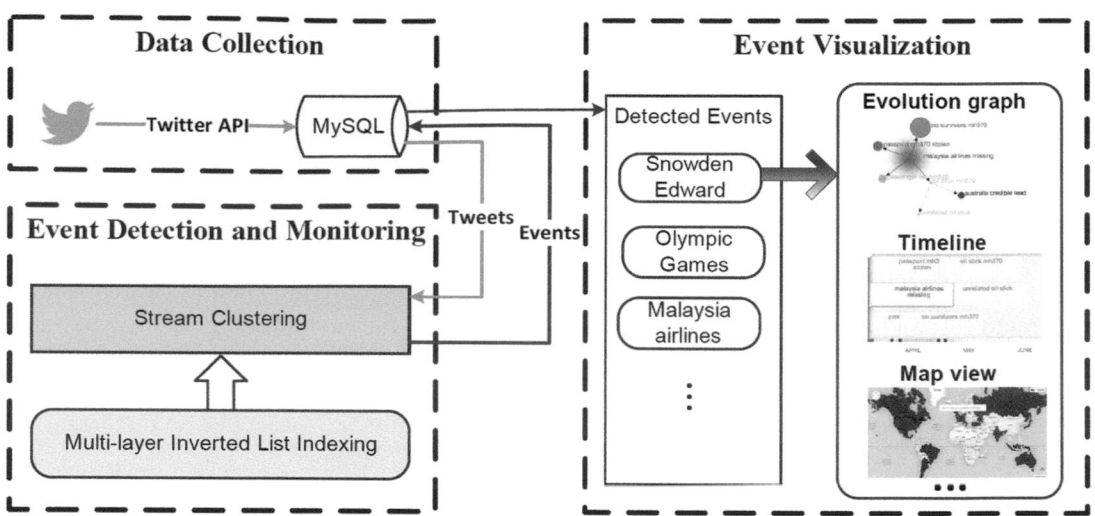

Social Media Harvesting, Fig. 1 System architecture of EventEye

[8], a live system called EventEye is developed. The overall system architecture is illustrated in Fig. 1. As illustrated, a multilayer inverted list (MIL) indexing structure is designed to support efficient event manipulation, including event maintaining and event identifying. The traditional single-pass incremental clustering algorithm is extended by taking into account four newly defined event evolution operations to detect events and capture their evolutions over time. The proposed incremental clustering algorithm enables the event detection and monitoring process in real time, which does not require any prior settings such as sliding window size. Users can browse current events or search historical events at any time. To expatiate an event, different event visualizations are adopted, including evolution graph, timeline, map view, word cloud, and representative tweets and images. Especially, to visualize the evolution of an event, all its relevant events and the associated evolving relationships are illustrated by an evolution graph, where each node represents an event, and the edge between any two nodes indicates their evolving relationship.

Tweets are collected via Twitter API and stored in the database. A stream clustering algorithm is applied on the collected tweets to detect events and monitor their evolutions. Four event operations are defined for the clustering algorithm including creation, absorption, split, and merge, to capture the event evolution patterns. With the aid of these four operations, the evolutions of the events are recorded in real time. Specifically, for each newly arrived tweet e, its nearest neighbor E_{NN} (the event that has largest similarity between e among all the existing events) is first found. If the similarity between e and E_{NN} is below a predefined threshold θ, a singleton event containing only e is created; otherwise, e is absorbed by E_{NN}. After the absorption, if the radius (the smallest similarity from the tweets in E_{NN} to the cluster center) of E_{NN} is smaller than a given threshold, E_{NN} is split into two new events. Once the split is conducted, the newly split events may be merged with nearby events if the merged event satisfies the radius threshold θ. As for the efficiency consideration, MIL indexing structure is proposed as the first event indexing structure to support large-scale dynamically evolving event maintenance and facilitate efficient event search. MIL is a multilayer indexing structure, which organizes events in different layers guided by different information-specific levels. It inspects most relevant event lists on the lowest layer to avoid exhaustive accesses to longer event lists on the upper layer. After events are detected, users can browse current events or search the historical events at any time. Five views are used to visualize a specified event, including timeline, evolution graph, map view, word cloud, and representative tweets and images.

Trending Topics Summarization

Microblogging services have revolutionized the way people exchange information. Confronted with the ever-increasing numbers of social events and the corresponding microblogs with multimedia contents, it is desirable to provide visualized summaries to help users to quickly grasp the essence of these social events for better understanding. Bian et al. [3] proposed a novel multimedia social event summarization framework to generate holistic visualized summary from the microblogs with multiple media types. The flowchart is demonstrated in Fig. 2. Specifically, the proposed framework comprises three stages: removal of irrelevant data, cross-media subevent discovery, and multimedia summary generation. First, a data cleansing approach is devised to automatically eliminate those irrelevant/noisy images. An effective spectral filtering model is exploited to estimate the probability that an image is relevant to a given event. In the second stage, a novel cross-media probabilistic model, termed Cross-Media-LDA (CMLDA), is proposed to jointly exploit the microblogs of multiple media types for discovering subevents. The CMLDA model not merely well explores and exploits the intrinsic correlations among different media types but also simultaneously characterizes both the general distribution and the subevent-specific distribution from the microblog data of various media types for reinforcing the subevent discovery process. Besides, this step could also handle the noise of the input data and remove those microblog examples from the next summarization step. Finally, based on the cross-media distribution knowledge of all the discovered subevents, a holistic visualized summary is generated for the social events by pinpointing both the representative textual and visual samples in a joint fashion. In particular, by utilizing the cross-media distributions of

microblog text, we specify three criteria, namely, coverage, significance, and diversity to measure the summarization capability of individual textual samples. Then, a greedy algorithm is devised for identifying the representative microblog texts based on the combination of the three criteria. For visual summarization, we employ the cross-media knowledge of the subevents as the prior knowledge for ranking the visual samples and selecting the most representative ones. In order to improve the descriptive power and the diversity of viewpoints, the images within a subevent are first partitioned into groups via spectral clustering. Then, for each group, we apply a manifold algorithm with the cross-media prior knowledge as initial ranking scores to identify the top-ranked image as representative. It is remarkable that both the textual and visual summarization processes utilize the cross-media knowledge of the discovered subevents and thus are intrinsically connected to reinforce each other.

User Profiling

User profiling is the process of characterizing user interests and/or attributes using multiple sorts of digital data, such as text, image, video, POI, audio, etc. Such knowledge can be used to better understand user behavior/personality and support existing applications, e.g., recommendation, online shopping, etc.

Social curation service (SCS) is a new type of emerging social media platform, where users can select, organize, and keep track of multimedia contents they like. Geng et al. [2] took advantage of this great opportunity and proposed a user profiling approach based on an example of SCSs and Pinterest. First, multimedia contents curated by users, i.e., the images in bundles as well as the associated user interest description like bundle names and tags, are collected. Due to user curation, the collected data are of high quality and focused according to the user interest. Second, an automatic ontology construction method is proposed to structuralize the curated images onto an ontology. The construction is done by pruning an expert ontology, i.e., Wikipedia category, to the desired user interest in a specific domain, such

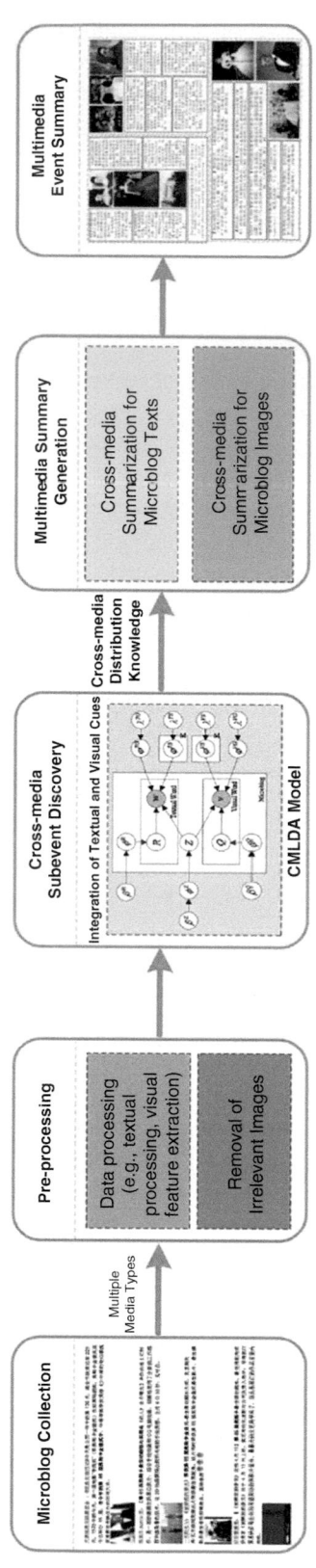

Social Media Harvesting, Fig. 2 Flowchart of multimedia summarization for social events

as fashion. Third, based on the constructed ontology, content-based models are learned to generate ontological user profiles, which are more comprehensive and personalized than the traditional text-based profiles. In particular, a novel multitask convolutional neural network (mtCNN) is proposed to leverage both the relatedness of sibling items of user interest and the cutting-edge advances in high-performance visual modeling. Fourth, a low-rank recovery framework is devised to further refine the generated user profiles by the ontological profile models, exploiting the rich user-level, bundle-level, and content-level social relations offered by social curations. Therefore, the resultant user profiles are expected to retain (a) the interest of user, (b) the interest of user-curated bundles, and (c) the semantic affinities with respect to the ontology, supporting effective fundamental social media applications such as recommendation.

Recommended Reading

1. http://next.comp.nus.edu.sg/
2. Geng X, Zhang H, Song Z, Yang Y, Luan H, Chua T-S. One of a kind: user profiling by social curation. In: Proceedings of the 22nd ACM International Conference on Multimedia; 2014. p. 567–76.
3. Bian J, Yang Y, Zhang H, Chua T-S. Multimedia summarization for social events in microblog stream. IEEE Trans Multimed. 2015;17(2):216–28.
4. Cai H, Yang Y, Li X, Huang Z. What are popular: exploring twitter features for event detection, tracking and visualization. In: Proceedings of the 23rd Annual ACM Conference on Multimedia Conference; 2015. p. 89–98.
5. Cao J, Huang Z, Yang Y. Spatial-aware multimodal location estimation for social images. In: Proceedings of the 23rd Annual ACM Conference on Multimedia Conference; 2015. p. 119–28.
6. Chua T-S. The multimedia challenges in social media analytics. In: Proceedings of the 3rd International Workshop on Socially-Aware Multimedia; 2014. p. 17–8.
7. Gao Y, Wang F, Luan H, et al. Brand data gathering from live social media streams. In: Proceedings of the International Conference on Multimedia Retrieval. ACM; 2014. p. 169.
8. Cai H, Tang Z, Yang Y, et al. EventEye: monitoring evolving events from tweet streams. In: Proceedings of the 22nd ACM International Conference on Multimedia; 2014. p. 747–8.

Social Networks

Felix Schwagereit[1] and Steffen Staab[2]
[1]University of Koblenz-Landau, Koblenz, Germany
[2]Institute for Web Science and Technologies – WeST, University of Koblenz-Landau, Koblenz, Germany

Definition

A social network is a social structure made of *actors*, which are discrete individual, corporate or collective social units like persons or departments [19] that are tied by one or more specific types of *relation* or interdependency, such as friendship, membership in the same organization, sending of messages, disease transmission, web links, airline routes, or trade relations. The actors of a social network can have other attributes, but the focus of the social network view is on the properties of the relational systems themselves [19]. For many applications social networks are treated as graphs, with actors as nodes and ties as edges. A *group* is the finite set of actors the ties and properties of whom are to be observed and analyzed. In order to define a group it is necessary to specify the network boundaries and the sampling. *Subgroups* consist of any subset of actors and the (possible) ties between them.

The science of social networks utilizes methods from general network theory and studies real world networks as well as structurally similar subjects dealing e.g., with information networks or biological networks.

Historical Background

The science of social network analysis comprises methods from social sciences, formal mathematical, statistical and computing methodology [19]. The first developments of scientific methods were empirically motivated and date back to the late nineteenth century. Jacob Moleno developed methods to facilitate the understanding of friendship patterns within small groups in the 1920s and 1930s. Other pioneers in the field of social networks were Davis, who studied social circles of women in an unnamed American city and Elton Mayo, who studied social networks of factory workers. Many of the current formal concepts (e.g., density, span, connectedness) had been introduced in the 1950s and 1960s as ways to describe social structures through measures. Another important milestone was an experiment Stanley Milgram conducted in 1967. In Milgram's experiment, a sample of US individuals were asked to reach a particular target person by passing a message along a chain of acquaintances. The average length of successful chains turned out to be about five intermediaries or six steps of separation.

Early research on social networks was limited to small networks with up to a few hundred actors, which could be examined visually. With increased computational power for data acquisition and management, networks may now comprise several millions of actors.

Foundations

Types

The simplest type of network consists of only one set of actors and one relation representing one type of ties between the actors. More complex networks can be composed of different types of actors (*multi-mode*) and different relations (*multi-relational*). Furthermore the actors and ties between them can have assigned properties, which are mostly numerical. Ties can have a direction, which makes the network a directed graph. Figure 1 shows a selection of network types [12]. Network (i) is a directed network in which each

edge has a direction; (ii) is an undirected network with only one type of actors; (iii) is a network with several types of actors and relations; (d) shows a network with different weights for actors and ties.

Of special interest in science of social networks are *bipartite graphs* [12] which contain actors of two types and ties connecting only actors of different types. They are called *affiliation networks* because they are suitable to express the membership of people (one type of actors) in groups (the second type of actors).

Notation

The common notation for social networks is the *sociometric notation* [19]. Simple social networks with one relation and only one group of actors (like the one shown above) are represented as a matrix, called *sociomatrix* or *adjacency matrix*. For one relation X, let $\mathbf{X}$ be the corresponding matrix. This matrix has g rows and g columns. The value at position x_{ij} denotes whether there exists a tie from the ith element of the social network to the jth element. An example sociomatrix for the social network (a) in Fig. 1 is shown in Table 1. For more complex networks, like multi-mode and/or multi-relational social networks, tensors may be used instead of matrices [18].

Measures

Measures have been developed in order to formalize local and global properties for social networks. Local and global properties of social networks describe ego-centric properties of individual actors and socio-centric properties of the network as a whole, respectively. Furthermore, subsets of actors (subgroups) can

Social Networks, Fig. 1
(**a–d**) Types of social networks

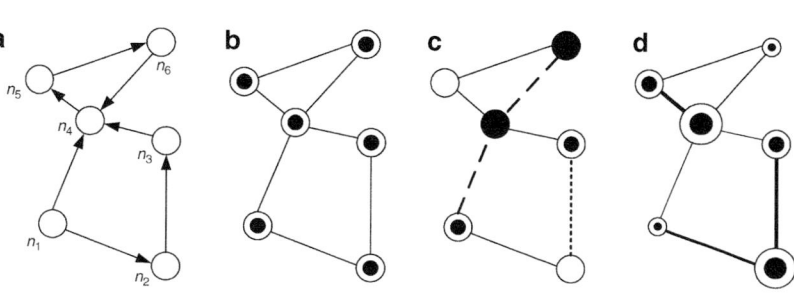

Social Networks, Table 1 Sociomatrix for network (a) in Fig. 1

	n_1	n_2	n_3	n_4	n_5	n_6
n_1	–	1	0	1	0	0
n_2	0	–	1	0	0	0
n_3	0	0	–	1	0	0
n_4	0	0	0	–	1	0
n_5	0	0	0	0	–	1
n_6	0	0	0	1	0	–

be determined. The following paragraphs contain an outline of several basic concepts.

Socio-Centric Properties

In order to compare different social networks in size and structure the following basic measures have been established.

- *Number of Actors: g*
- *Number of Ties: m*
- *Mean Standarized Degree (Density)*: $z = \frac{\sum C_{\mathrm{D}}(n_i)}{g(g-1)}$
- *Mean Actor-Actor Distance / Characteristic Path Length*: $l = \frac{1}{\frac{1}{2}g(g+1)}\sum_{i \geq j} d(n_i, n_j)$
- *Diameter*: is the longest Distance between all pairs of nodes of a given network. The distance $d(n_i, n_j)$ between a pair of nodes n_i and n_j in the network is the length of the geodesic (which is the shortest path between the two nodes).

Ego-Centric Properties

The identification of the "most important" or "prominent" actor was one of the primary goals of social network analysis [19]. Therefore various measures were developed to quantify "importance" of actors and subgroups for a given social network. The following measures can be calculated for simple undirected graphs of social networks.

- *Actor Degree Centrality* is the count of the number of ties to other actors in the network. The relevance of this measure is based on the assumption that an actor, which has more connections than other actors can be con-

sidered more active and therefore important. The actor degree centrality is calculated from sociomatrix $\mathbf{X}$ as follows:

$$C_{\mathrm{D}}(n_i) = \sum_j x_{ij}$$

- *Actor Closeness Centrality* is the degree to which an individual is close to all other individuals in a network (directly or indirectly). Therefore an actor is central if it can quickly (that means by relying on so few mediators as possible) interact with all other actors. The index of actor closeness centrality is:

$$C_{\mathrm{C}}(n_i) = \left[\sum_{j=1, j \neq i}^{g} d(n_i, n_j) \right]^{-1}$$

- where $d(n_i, n_j)$ is the length of the geodesic of actor i and actor j. To allow comparisons between different networks actor closeness can be standardized:

$$C_{C}^*(n_i) = \frac{g-1}{\left[\sum_{j=1, j \neq i}^{g} d(n_i, n_j)\right]}$$

- *Actor Betweenness Centrality* is the degree to which an individual lies between other individuals in the network. Therefore it is based on the assumption that all other actors lying in between have a certain amount of control on the interaction relying on them. So the betweenness of an actor is higher if more of the possible interactions rely on it as mediator. In order to calculate betweeness centrality two other measures are needed: g_{jk}, the number of geodesics linking two actors j and k; as well as $g_{jk}(n_i)$, which is the number of geodesics linking two actors that contain the actor i:

$$C_{\mathrm{B}}'(n_i) = \sum_{j<k} g_{jk}(n_i) / g_{ik}$$

- For comparisons the measure can be normalized:

$$C_{\mathrm{B}}'(n_i) = C_{\mathrm{B}}(n_i) / [(g-1)(g-2)/2]$$

S

Subgroups

In most social networks actors organize themselves in subgroups or cliques, which have their own values, sub-cultures, and structures. Therefore several methods to define and recognize certain kinds of subgroups were developed [19].

- A *Clique of size k* is a subgroup consisting of *k* many actors which are all adjacent to each other.
- An *n-Clique* is a subgroup with the property that the distance (length of the geodesic) between all actors is no greater than *n* and there is no actor with a distance equal or less than *n* outside the n-clique. An n-clique with $n = 1$ is equal to a normal clique.
- A *k-Core* is a subgroup with each actor is adjacent to at least *n* other actors in the subgroup.
- A *Cluster* is a subgroup consisting of actors which are similar to each other. The similarity (structural equivalence) of two actors can be defined with criteria like euclidean distance or correlation based on vectors of a sociomatrix. Similarity based clusters in undirected networks are usually created by using agglomerative or divisive hierarchical clustering methods [14]. For clustering directed networks methods like directed spectral clustering [7] can be used. In general graph theory there exist methods for partitioning graphs which can also be applied to graphs of social networks. One of these methods is the min-max cut algorithm which pursues the goal of minimizing the similarity between subgraphs while maximizing the similarity within each subgraph. Other clustering approaches are based on methods for finding densely connected subgroups by the calculation of special clustering coefficients or by comparing the number of connections within a subgroup with the number of connections to outside actors.

Topological Properties

Small-World Topology

The small-world model [20] is a well studied distribution model of actors and ties, since it has interesting properties and features. Due to the fact that networks often have a geographical component to them it is reasonable to assume that geographical proximity will play a role in deciding which actors are connected. So in a small-world network each actor is connected to actors in its near neighborhood. Other connections between more distant actors (long-range connections) are infrequent and have a low probability. The probability for each actor of having a degree *k* follows a power law $p_k \sim k^{-\alpha}$ with α as constant scaling exponent. Despite the fact that long-range connections occur only sporadicly the diameter of small-world networks is exponentially smaller than their size, being bounded by a polynomial in $\log g$, where *g* is the number of nodes. In other words, there is always a very short path between any two nodes [8].

The discovery that real world social networks might have small-world characteristics explains the importance of this model. So it can be observed that the chain of social acquaintances required to connect one arbitrary person to another arbitrary person anywhere in the world is generally short. This concept gave rise to the famous phrase six degrees of separation after a 1967 small-world experiment by Stanley Milgram. Academic researchers continue to explore this phenomenon. A recent electronic small-world experiment [5] at Columbia University showed that about five to seven degrees of separation are sufficient for connecting any two people through e-mail. Other applications of the small-world model are investigations of iterated games, diffusion processes or epidemic processes [12].

Creation of Networks

Artificially generated graphs allow comparison with real datasets and by analyzing and comparing their properties they give insights into the inner structure of social networks. They also allow for the generation of (overlay) network structures on top of existing information structures.

Several procedures are known to generate social networks from scratch. A *Poisson random graph* is the simplest way to construct a social network. This is simply done by connecting each pair of actors with the probability of p. The result of this procedure is a network with a Poisson degree distribution ($p_k = \frac{\lambda^k}{k!}e^{-\lambda}$). Since this distribution is unlike the highly skewed power-law distributions of real world networks other methods have been proposed [12].

One of the important methods is known as *preferential attachment* [1]. In this model, new nodes are added to a pre-existing network, and connected to each of the original nodes with a probability proportional to the number of connections each of the original nodes already had. I.e., new nodes are more likely to attach to hubs than peripheral nodes or in other words the "rich-get-richer". Statistically, this method will generate a power-law distributed small-world network (that is, a scale-free network).

Since there is evidence that the preferential attachment model does not show all the properties real world networks obey, like increasing of the average degree and shrinking of the diameter on growing of a network, other models have been proposed [9]. The *Community Guided Attachment*, which is based on a decomposition of actors into a nested set of subgroups, such that the difficulty of forming new links between subgroups increases with the size of the subgroups. In the *Forest Fire Model* new actors are attached to the network by burning through existing ties in epidemic fashion.

Key Applications

Distributed Information Management
Social routing allows to route efficiently in peer-to-peer networks without knowledge about the global network structure. This routing with local knowledge can be achieved by regarding the network as a social network and exploiting several properties of social networks like small-world characteristics [8, 10].

Information Replication in information networks can improve scalability and reliability. By performing social network clustering on these structures prefetching of content can be improved [15].

Information Extraction
Name disambiguation is a technique for distinguishing person names in unsupervised information frameworks (e.g., web pages), where unique identifiers can not be assumed [2].

Ontology Extraction methods can be performed on social network structures like communities and their folksonomies. This approach is based on the assumption that individual interactions of a large number of actors might lead to global effects that could be observed as semantics [11].

Social Recommendations
Social networking portals like Xing or LinkedIn allow users to express their relationships to other users and to provide personal information. This social network can be used e.g., for finding a short path to persons in special positions by identifying the geodesic to them [16].

Filtering, recommendations and inferred trust can be improved by taking into account the social networks all relevant actors are involved. So e.g., the trustworthiness of Bob can be inferred from a social network by Alice even if both are not directly known to each other [6].

Viral marketing is the strategy to let satisfied customers distribute advertisements (e.g., video clips) by recommendation or forwarding to other potential customers they know. Viral marketing campaigns are usually started by sending the advertisements to actors holding central positions in social networks in order to facilitate a rapid distribution [13, 17].

Future Directions

For the future of the social network science many areas remain insufficiently explored [12]. Many properties of social networks have been studied in the past decades. But the scientific community is still lacking the whole picture which shows what the most important properties for each application are. Especially generalized propositions (e.g., "Are more centralized organizations more efficient?") about the structure of social networks need further verification across a large number of networks [19]. Another important direction of future research is to improve the understanding of the dynamics in and the evolution of social networks [3]. In order to archive this new and more sophisticated models of social networks have to be developed. New kinds of data including more complex structures and new properties of actors or relations demand further generalization of current models. An example of these more complex structures are multiple relations which connect more than two actors.

Data Sets

- Enron Email dataset (http://www.cs.cmu.edu/~enron/ and http://www.enronemail.com/) contains about 600,000 Email messages belonging to 156 users. It was made public during the legal investigation concerning the Enron corporation.
- The Internet Movie Data Base (IMDB) (http://www.imdb.com/interfaces/) is a collection of data about movies (about 400,000) and actors (about 900,000). Especially the affiliation network of the co-appearance of actors in the same movie is subject of several studies. (cf. "The Oracle of Bacon" http://oracleofbacon.org/)
- Digital Bibliography & Library Project (DBPL) collects the bibliographic information on major computer science journals and proceedings (currently about 950,000 articles). Similar to the IMDB the co-authorship can be used to generate affiliation networks.(datasethttp://dblp.uni-trier.de/xml/)
- Southern Woman Dataset, which was collected in the 1930s is published in the classical study of Davis [4], a pioneer of social network analysis. It contains the attendance at 14 social events by 18 women in an unnamed US city.

URL to Code

Tools and Libraries
(cf. http://www.insna.org/software/index.html):

- Jung: http://jung.sourceforge.net/
- Pajek: http://vlado.fmf.uni-lj.si/pub/networks/pajek/default.htm
- UCINET: http://www.analytictech.com/ucinet/ucinet.htm

Conference Series

- International Sunbelt Social Network Conferences: http://www.insna.org/sunbelt/index.html

Journals

- Social Networks: http://www.innsa.org/pubs/connections/index.html
- CONNECTIONS: http://www.insna.org/indexConnect.html
- Journal of Social Structure: http://www.cmu.edu/joss/

Cross-References

▶ Biological Networks
▶ Cluster and Distance Measure
▶ Clustering Overview and Applications
▶ Graph
▶ Hierarchical Clustering
▶ Web Characteristics and Evolution

Recommended Reading

1. Barabási AL, Albert R. Emergence of scaling in random networks. Science. 1999;286:509–12.
2. Bekkerman R, McCallum A. Disambiguating Web appearances of people in a social network. In: Proceedings of the 14th International World Wide Web Conference; 2005. p. 463–470.
3. Berners-Lee T, Hall W, Hendler J, Shadbolt N, Weitzner DJ. Creating a science of the Web. Science. 2006;313(5788):769–71.
4. Davis A, Gardner BB, Gardner MR. Deep South. The University of Chicago Press; 1941.
5. Dodds PS, Muhamad R, Watts D. An experimental study of search in global social networks. Science. 2003;301(5634):827–9.
6. Golbeck J, Hendler JA. Inferring binary trust relationships in web-based social networks. ACM Trans Internet Technol. 2006;6(4):497–529.
7. Huang J, Zhu T, Schuurmans D. Web communities identification from random walks. In: Proceedings of the Joint European Conferences on Machine Learning

and European Conference on Principles and Practice of Knowledge Discovery in Databases; 2006.

8. Kleinberg J. Navigation in a small world. Nature. 2000;406:845.

9. Leskovec J, Kleinberg J, Faloutsos C. Graph evolution: densification and shrinking diameters. ACM Trans Knowl Discov Data. 2007;1(1):2.

10. Löser A, Staab S, Tempich C. Semantic social overlay networks. IEEE J Sel Areas Commun. 2007;25(1): 5–14.

11. Mika P. Social networks and the semantic Web: Springer; 2007.

12. Newman MEJ. The structure and function of complex networks. SIAM Rev. 2003;45(2):167–256.

13. Richardson M, Domingos P. Mining knowledge-sharing sites for viral marketing. In: Proceedings of the 8th ACM SIGKDD International Conferences on Knowledge Discovery and Data Mining; 2002. p. 61–70.

14. Scott J. Social network analysis: a handbook: Sage; 2000.

15. Sidiropoulos A, Pallis G, Katsaros D, Stamos K, Vakali A, Manolopoulos Y. Prefetching in content distribution networks via web communities identification and outsourcing. World Wide Web J. 2008;11(1):39–70.

16. Staab S, Domingos P, Mika P, Golbeck J, Ding L, Finin TW, Joshi A, Nowak A, Vallacher RR. Social networks applied. IEEE Intell Syst. 2005;20(1): 80–93.

17. Subramani MR, Rajagopalan B. Knowledge-sharing and influence in online social networks via viral marketing. Commun ACM. 2003;46(12):300–7.

18. Sun J, Tao D, Faloutsos C. Beyond streams and graphs: dynamic tensor analysis. In: Proceedings of the 12th ACM SIGKDD Internationla Conferences on Knowledge Discovery and Data Mining; 2006. p. 374–83.

19. Wasserman S, Faust K. Social network analysis. Cambridge: Cambridge University Press; 1994.

20. Watts DJ, Strogatz SH. Collective dynamics of "small-world" networks. Nature. 1998;393(6684):440–2.

Software Transactional Memory

Keir Fraser
University of Cambridge, Cambridge, UK

Definition

Software transactional memory (STM) is a method of concurrency control in which shared-memory accesses are grouped into transactions which either succeed or fail to commit in their entirety. STM provides applications programmers with an alternative to mutual-exclusion locks which avoids many of the latter's pitfalls, including risk of deadlock, unnecessary serialization, and priority inversion. Many STMs are themselves implemented using lock-free programming methods, although this is not a hard-and-fast rule.

Key Points

A software transactional memory (STM) is a software library or programming language feature which provides application programmers with an interface for allocating and accessing shared-memory variables [3]. These variables are accessible in a concurrency-safe manner without resorting to classical concurrency-management techniques such as mutual exclusion. This is achieved by grouping accesses into transactions which execute in isolation and then atomically succeed or fail in their entirety.

The application programmer chooses when transactions should start and end, rather like choosing when to acquire and release mutexes in a conventional multi-threaded program, to ensure consistency of application data structures. The STM implementation is responsible for ensuring that transactions execute in isolation and coomit atomically. Thus transactional memory guarantees the same ACID properties as classical database transactions, with the exception of durability.

The benefits of a transactional interface to shared memory are numerous. Traditional mutexes, when used conservatively, can lead to unnecessary serialization of operations that do not otherwise conflict. When a finer-grained approach is taken, involving multiple locks with individually smaller scope, the programmer must take care to avoid subtle deadlock scenarios. STM is perhaps the most promising of the proposed lock-free techniques which eschew traditional mutual exclusion and hope to enable the average programmer to implement scalable multi-threaded applications in mainstream languages [1]. Hence, although still in its infancy and an ongoing topic of research, STM is being viewed eagerly by an industry looking for salvation from the complexity of optimizing for modern multi-core systems [2].

Cross-References

▶ Performance Analysis of Transaction Processing Systems
▶ Transaction

Recommended Reading

1. Fraser K, Harris T. Concurrent programming without locks. ACM Trans Comput Syst. 2007;25(2);1–61.
2. Saha B, Adl-Tabatabai A, Hudson R, Minh C, Hertzberg B. McRT-STM: a high performance software transactional memory system for a multi-core runtime. In: Proceedings of the 11th ACM SIGPLAN Symposium on Principles and Practice of Parallel Programming; 2006. p. 187–97.
3. Shavit N, Touitou D. Software transactional memory. In: Proceedings of the ACM SIGACT-SIGOPS 14th Symposium on the Principles of Distributed Computing; 1995. p. 204–13.

Software-as-a-Service (SaaS)

Chandra Krintz
Department of Computer Science, University of California, Santa Barbara, CA, USA

Synonyms

Browser-based software; Internet application or service; Web-scale application; World wide web (WWW) application or service

Definition

Software-as-a-service (SaaS) describes Internet-accessible applications that are hosted and managed by an application provider. SaaS applications support multiple isolated, concurrent users, each of whom can customize and use their instance of the application via a web browser. Providers make SaaS applications available for free (with advertisements), via subscription, and on a pay-per-use basis.

Overview

Software-as-a-service (SaaS) represents a transformation in how software is sold, distributed, and used by consumers. Historically, software was licensed and shipped to consumers on physical media (disks, CD-ROMs) for installation on locally maintained hardware. This process was costly, time-consuming, delayed bug fixes and software improvements, and limited technology uptake. In addition, the software and the update process required testing and specialization for each target consumer device and system. Exacerbating these problems, consumers of enterprise software were required to buy, build, and maintain complex information technology (IT) infrastructures.

SaaS significantly simplifies the process of software delivery and evolution through the use of a single application deployment that the vendor manages and maintains using distributed computational resources that the vendor controls. SaaS applications are multiplexed across multiple users by providing each with a customizable instance that they access via a web browser from any device. Users are isolated from others via system-level virtual machine, operating system, or programmatic mechanisms. SaaS providers can update, test, scale, and evolve SaaS applications quickly, without impacting user experience significantly. To enable this, providers typically rely on recent advances in scalable web services and cloud computing (PaaS and IaaS) to implement and host applications. Most consumers pay for commercial SaaS applications via subscription or on a pay-per-use basis; free use is typically coupled with advertisements. Moreover, SaaS enables vendors to collect information on application use to guide marketing and engineering activities. Examples of popular SaaS applications include

Salesforce Customer Relationship Management, Google Mail and Drive, Microsoft 365, and Intuit QuickBooks, among many others.

Cross-References

▶ Cloud Computing
▶ Infrastructure-as-a-Service (IaaS)
▶ Platform-as-a-Service (PaaS)

Software-Defined Storage

Kaladhar Voruganti
Advanced Development Group, Network Appliance, Sunnyvale, CA, USA

Synonyms

SDS; Storage virtualization; Virtualized storage

Definition

The decoupling of the control and data plane in a storage controller where the control software manages commodity hardware, and the control plan receives it storage management directives in the form of high level policies.

Main Text

Traditionally, compute servers, networking devices, and storage controllers had their own specialized hardware designs. However, with the emergence of cloud architectures, the same server hardware is being used as compute server, networking device, and also as the storage device (the compute part of storage controller). The cloud architectures are known as hyper-scalar architectures where a cloud operating system manages thousands of servers in racks. The cloud operating systems use software to control whether to use a server as a compute, network, or storage device. The compute, networking, or storage software is put in virtual containers (VMs) and executed on the underlying hyper-scalar server hardware. There is a slight degradation in performance due to running the software on top of a hypervisor, but for most applications this performance degradation is not an issue. The cloud OS (orchestrator software) decides where to instantiate the different types of VMs to both improve application performance and also the overall system resource utilization. Provisioning agility is another key reason for the adoption of the software defined paradigm in the data centers because the cloud OS software can quickly spin up new compute, networking, and storage VMs in the hyper-scalar server racks. Another key tenet of software-defined storage is that the control plane receives storage provisioning and management directives from the users in terms of policies (APIs) and the control plane translates those policies into underlying low-level storage management commands. Thus, the software-defined storage approach brings both provisioning agility and management at scale. Finally, by decoupling the storage control software and the storage media, this approach allows one to use commodity storage media, and thus, it takes away the high hardware price margins that are typically associated with storage controllers.

Recommended Reading

1. Hollis C. The VMware perspective on software defined storage. White Paper. 2015.

Solid State Drive (SSD)

Kaladhar Voruganti
Advanced Development Group, Network Appliance, Sunnyvale, CA, USA

Synonyms

Flash disk drive; Solid state disk

Definition

Solid-state disk device that is made up of non-volatile memory and is accessed using traditional SCSI disk protocols.

Main Text

Regular hard disk drives (HDD) have both a computer complex (CPU, memory) and also mechanical components like arms and platters, whereas a SSD is a solid state drive with no mechanical components. SSDs also contain a computer complex. SSDs can be made up of different types of nonvolatile memories like flash, phase change memory (PCM), spin-torque memory (STT-MRAM), and nano-RAM (NRAM). Furthermore, there are many variants of flash such as single cell flash (SLC) or multiple cell flash (MLC) or 3-D NAND flash. Since SSDs do not have any mechanical arms to move, SSDs provide better random read/write performance than HDDs. The different types of SSDs provide different performance, density (impacts capacity), power usage, and endurance trade-offs. Different SSD vendors are advocating different types of nonvolatile memories. SSDs can be accessed using the traditional block oriented SCSI protocols or via faster memory-based access protocols like NVM Express (NVMe) and more recently also using key-value store interfaces. Regular HDDs still provide more cost-effective capacity than SSDs, and thus, are used by those applications that are not latency sensitive (backup, archival, and content depots).

SSDs are fundamentally changing how storage systems are getting designed. New types of SSD-based storage controllers, host (application server) side storage alternatives, and caching devices are being designed that cater to performance sensitive applications. These new systems use different types of logging, indexing, data layout, reliability mechanisms, because they are not burdened with supporting legacy HDDs. However, traditional HDD based storage controllers are also leveraging SSDs in novel ways to provide hybrid solutions that optimize on both performance and capacity dimensions.

Recommended Reading

1. Kasavajhala V. SSD vs HDD price and performance study. Dell Technical White Paper, 2011 May.

Sort-Merge Join

Jingren Zhou
Alibaba Group, Hangzhou, China

Synonyms

Merge join

Definition

The sort-merge join is a common join algorithm in database systems using sorting. The join predicate needs to be an equality join predicate. The algorithm *sorts* both relations on the join attribute and then *merges* the sorted relations by scanning them sequentially and looking for qualifying tuples.

Key Points

The sorting step groups all tuples with the same value in the join attribute together. Such groups are sorted based on the value in the join attribute so that it is easy to locate groups from the two relations with the same attribute value. Sorting operation can be fairly expensive. If the size of the relation is larger than the available memory, external sorting algorithm is required. However, if one input relation is already clustered (sorted) on the join attribute, sorting can be completely avoided. That is why the sort-merge join looks attractive if any of the input relations is sorted on the join attribute.

The merging step starts with scanning the relations R and S and looking for matching groups from the two relations with the same attribute value. The two scans start at the first tuple in each relation. The algorithm advances the scan of R as long as the current R tuple has an attribute value which is less than that of the current S tuple.

Similarly, the algorithm advances the scan of S as long as the current S tuple has an attribute value which is less than that of the current R tuple. The algorithm alternates between such advances until an R tuple R and an S tuple S with $R.r = S.s$. The join tuple $\{R,S\}$ is added to result.

There could be several R tuples and several S tuples with the same attribute value as the current tuples R and S. That is, several R tuples may belong to the current R group since they all have the same attribute value. The same applies to the current S group. Every tuple in the current R group joins with every tuple in the current S group. The algorithm them resumes scanning R and S, beginning with the first tuples that follow the group of tuples that are just processed.

When the two relations are too large to be held in available memory, one improvement is to combine the merging step of external sorting with the merging step of the join if the number of buffers available is larger than the total number of sorted runs for both R and S. The idea is to allocate one buffer page for each run of R and one for each run of S. The algorithm merges the runs of R, merges the runs of S, and joins (merges) the resulting R and S streams as they are generated.

Algorithm 1: Sort-Merge Join: $R \bowtie_{R,r=S,s} S$
```
// sorting step
Sort the relation R on the attribute r;
Sort the relation S on the attribute s;

// merging step
R = first tuple in R;
S = first tuple in S;
S' = first tuple in S;
while R ≠ eof and S' ≠ eof do
    while R.r < S'.s do
    |   R = next tuple in R after R;
    end
    while R.r > S'.s do
    |   S' = next tuple in S after S';
    end
    S = S';
    while R.r == S'.s do
    |   S = S';
    |   while R.r == S.s do
    |   |   add {R, S} to result;
    |   |   S = next tuple in S after S;
    |   end
    |   R = next tuple in R after R;
    end
    S' = S;
end
```

Cross-References

► Evaluation of Relational Operators

Recommended Reading

1. Mishra P, Eich MH. Join processing in relational databases. ACM Comput Surv. 1992;24(1):63–113.

Space-Filling Curves

Mohamed F. Mokbel[1] and Walid G. Aref[2]
[1]Department of Computer Science and Engineering, University of Minnesota-Twin Cities, Minneapolis, MN, USA
[2]Purdue University, West Lafayette, IN, USA

Synonyms

Distance-preserving mapping; Linearization; Locality-preserving mapping; Multi-dimensional mapping

Definition

A space-filling curve (SFC) is a way of mapping the multi-dimensional space into the one-dimensional space. It acts like a thread that passes through every cell element (or pixel) in the multi-dimensional space so that every cell is visited exactly once. Thus, a space-filling curve imposes a linear order of points in the multi-dimensional space. A D-dimensional space-filling curve in a space of N cells (pixels) of each dimension consists of $N^D - 1$ segments where each segment connects two consecutive D-dimensional points. There are numerous kinds of space-filling curves (e.g., Hilbert, Peano, and Gray). The difference between such curves is in their way of mapping to the one-dimensional space, i.e., the order that a certain space-filling curve traverses the multi-dimensional space. The quality of a space-filling curve is measured by its ability in preserving the locality (or relative distance) of multi-dimensional points in the mapped one-dimensional space. The main idea is that any two D-dimensional points that are close by in the D-dimensional space should be also close by in the one-dimensional space.

Key Points

Space-filling curves are discovered by Peano [3], where he introduces a mapping from the unit interval to the unit square. Hilbert [1] generalizes the idea to a mapping of the whole space. Following the Peano and Hilbert curves, many space-filling curves are proposed, e.g., [4]. Space-filling curves are classified into two categories: recursive space-filling curves (RSFC) and non-recursive space-filling curves. An RSFC is an SFC that can be recursively divided into four square RSFCs of equal size. Examples of RSFCs are the Peano SFC, the Gray SFC, and the Hilbert SFC. For the past two decades, recursive space-filling curves have been considered a natural method for locality-preserving mappings. Recursive space-filling curves are special case of fractals [2]. Mandelbrot [2], the father of fractals, derived the term fractal from the Latin adjective fractus. The corresponding Latin verb frangere means "to break" or "to fragment." Thus, fractals divide the space into a number of fragments, visiting the fragments in a specific order. Once a fractal starts to visit points from a certain fragment, no other fragment is visited until the current one is completely exhausted. By dealing with one fragment at a time, fractal locality-preserving mapping algorithms perform a local optimization based on the current fragment.

Cross-References

▶ High-Dimensional Indexing
▶ Space-Filling Curves for Query Processing
▶ Spatial Indexing Techniques

Recommended Reading

1. Hilbert D. Ueber stetige abbildung einer linie auf ein flashenstuck. Math Ann. 1891;38:459–60.
2. Mandelbrot BB. Fractal geometry of nature. New York: W. H. Freeman; 1977.
3. Peano G. Sur une courbe qui remplit toute une air plaine. Math Ann. 1890;36(1):157–60.
4. Sagan H. Space filling curves. Berlin: Springer; 1994.

Space-Filling Curves for Query Processing

Mohamed F. Mokbel[1] and Walid G. Aref[2]
[1]Department of Computer Science and Engineering, University of Minnesota-Twin Cities, Minneapolis, MN, USA
[2]Purdue University, West Lafayette, IN, USA

Synonyms

Distance-preserving mapping; Linearization; Locality-preserving mapping; Multidimensional mapping

Definition

Given a query Q, a one-dimensional index structure I (e.g., B-tree), and a set of D-dimensional points, a space-filling curve S is used to map the D-dimensional points into a set of one-dimensional points that can be indexed through I for an efficient execution of query Q. The main idea is that space-filling curves are used as a way of mapping the multidimensional space into the one-dimensional space such that existing one-dimensional query processing and indexing techniques can be applied.

Historical Background

Although space-filling curves were discovered in 1890 [14], their use in query processors has emerged only in the last two decades as it is mainly motivated by the emergence of multidimensional applications. In particular, space-filling curves have been used as a mapping scheme that supports spatial join algorithms [13], spatial access methods [2, 7], efficient processing of range queries [1, 6], and nearest-neighbor queries in [8]. Numerous algorithms are developed for efficiently generating different space-filling curves that include recursive algorithms for the Hilbert SFC [4, 15], recursive algorithms for the Peano SFC [15], and table-driven algorithms for the Peano and Hilbert SFCs [4]. The clustering and mapping properties of various space-filling curves have been extensively studied in the literature (e.g., see [10, 12]).

Foundations

Mapping Scheme

Figures 1 and 2 give examples of two- and three-dimensional space-filling curves with grid size (i.e., number of points per dimension) 8 and 4, respectively. Space-filling curves are classified into two categories: *recursive* space-filling curves (RSFC) and *non-recursive* space-filling curves.

An RSFC is an SFC that can be recursively divided into four square RSFCs of equal size. Non-recursive space-filling curves include the Sweep SFC (Figs. 1a and 2a), the Scan SFC (Figs. 1b and 2b), the Diagonal SFC (Fig. 1f), and the Spiral SFC (Fig. 1g). Recursive space-filling curves include the Peano SFC (Figs. 1c and 2c), the Gray SFC (Figs. 1d and 2d), and the Hilbert SFC (Figs. 1e and 2e). Table 1 gives the first 16 visited points for the Peano, Gray, and Hilbert space-filling curves.

The Peano SFC

The Peano SFC (Figs. 1c and 2c) is introduced by Peano [14] and is also termed Morton encoding, quad code, bit-interleaving, N-order, locational code, or Z-order. The Peano SFC is constructed recursively as in Fig. 3. The basic shape (Fig. 3a) contains four points in the four quadrants of the space. Each quadrant is represented by two binary digits. The most significant digit is represented by its x position, while the least significant digit is represented by its y position. The Peano SFC orders space quadrants in ascending order (00, 01, 10, 11). Figure 3b contains four blocks of Fig. 3a at a finer resolution and is

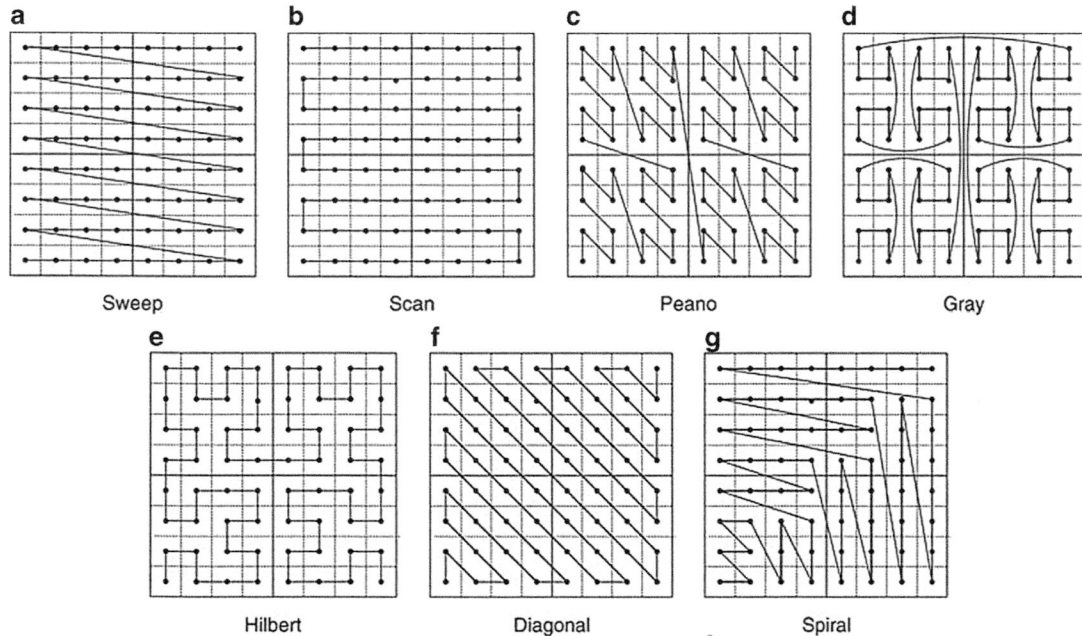

Space-Filling Curves for Query Processing, Fig. 1 Two-dimensional space-filling curves

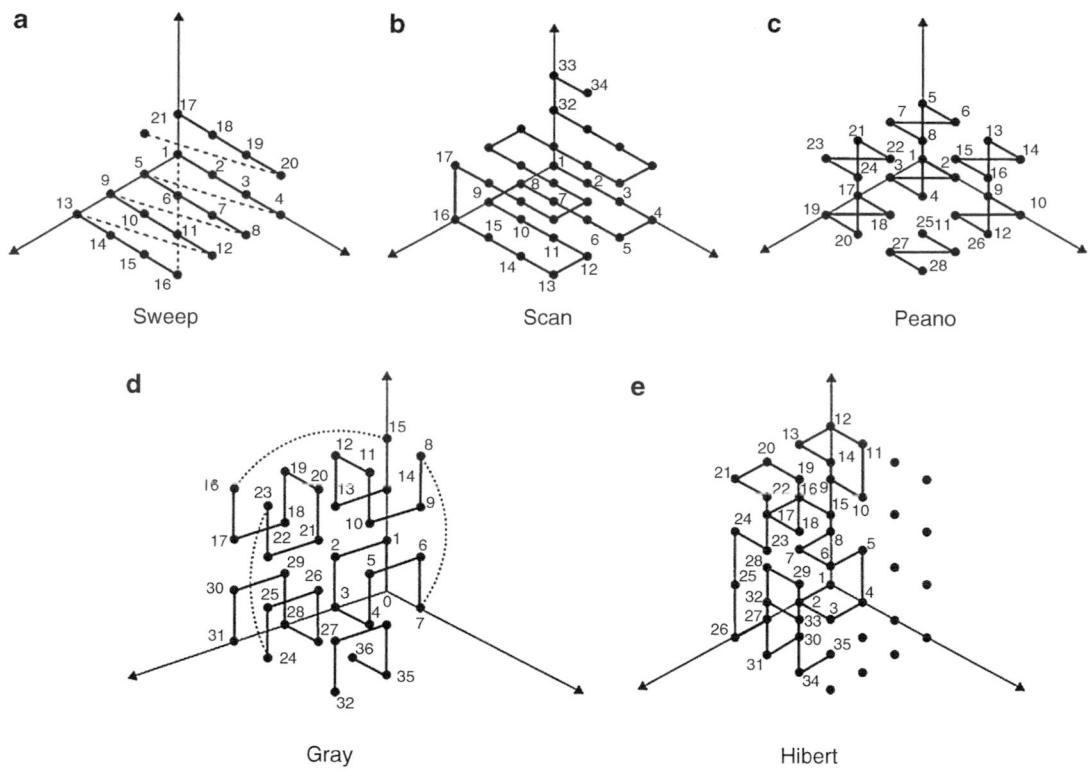

Space-Filling Curves for Query Processing, Fig. 2 Three-dimensional space-filling curves

Space-Filling Curves for Query Processing, Table 1
The first 16 traversed points by two-dimensional Peano, Gray, and Hilbert space-filling curves

Point	Peano	Gray	Hilbert	Point	Peano	Gray	Hilbert
0	(0,0)	(0,0)	(0,0)	8	(2,0)	(3,3)	(2,2)
1	(0,1)	(0,1)	(0,1)	9	(2,1)	(3,2)	(3,2)
2	(1,0)	(1,1)	(1,1)	10	(3,0)	(2,2)	(3,3)
3	(1,1)	(1,0)	(1,0)	11	(3,1)	(2,3)	(2,3)
4	(0,2)	(1,3)	(2,0)	12	(2,2)	(2,0)	(1,3)
5	(0,3)	(1,2)	(3,0)	13	(2,3)	(2,1)	(1,2)
6	(1,2)	(0,2)	(3,1)	14	(3,2)	(3,1)	(0,2)
7	(1,3)	(0,3)	(2,1)	15	(3,3)	(3,0)	(0,3)

visited in the same order as in Fig. 3a. Similarly, Fig. 3c contains four blocks of Fig. 3b at a finer resolution.

The Gray SFC

The Gray SFC (Figs. 1d and 2d) uses the Gray code representation [5] in contrast to the binary code representation as in the Peano SFC. Figure 4

gives the recursive construction of the Gray SFC. The basic shape (Fig. 4a) contains four points in the four quadrants of the space. The Gray SFC visits the space quadrants in ascending order according to the Gray code (00, 01, 11, 10). Figure 4b is constructed by having the first and fourth blocks as those of Fig. 4a, while the second and the third blocks are the rotation of the blocks in Fig. 4a by 180°. Similarly, Fig. 4c is constructed from two blocks of Fig. 4b at a finer resolution and two blocks of the rotation of Fig. 4b by 180°.

The Hilbert SFC

Figure 5 gives the recursive construction of the Hilbert SFC. The basic block of the Hilbert SFC (Fig. 5a) is the same as that of the Gray SFC (Fig. 4a). The basic block is repeated four times at a finer resolution in the four quadrants, as given in Fig. 5b. The quadrants are visited in their gray order. The second and third blocks in Fig. 5b have the same orientation as in Fig. 5a. The first block

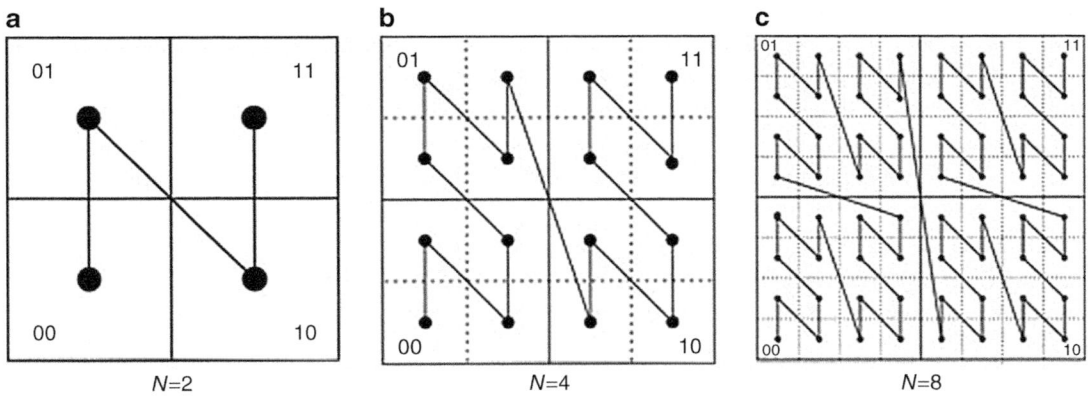

Space-Filling Curves for Query Processing, Fig. 3 The Peano SFC

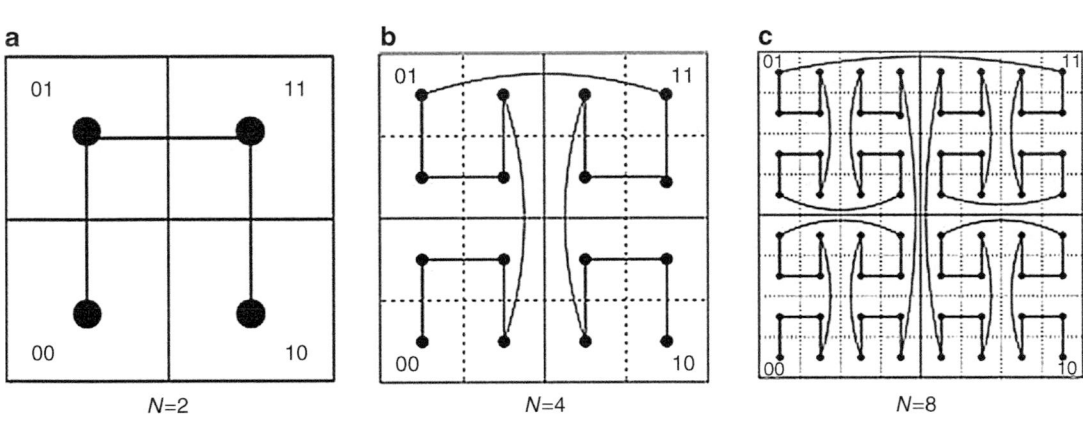

Space-Filling Curves for Query Processing, Fig. 4 The Gray SFC

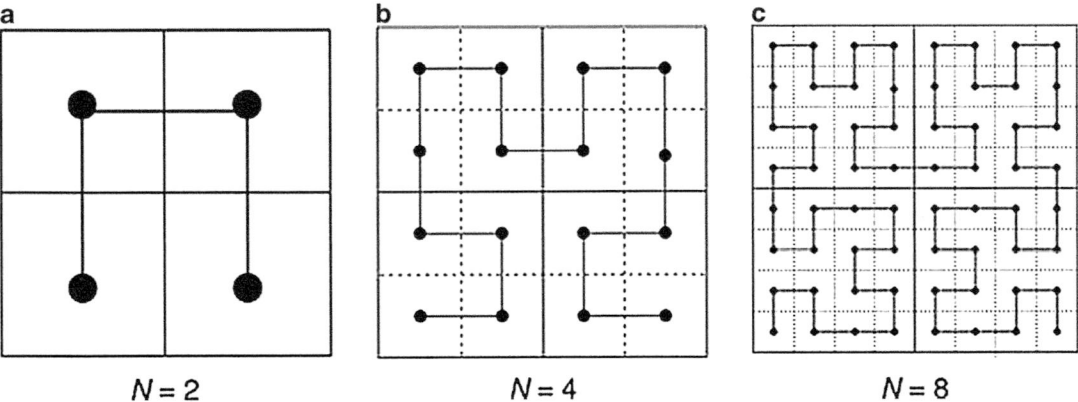

Space-Filling Curves for Query Processing, Fig. 5 The Hilbert SFC

is constructed from rotating the block of Fig. 5a by 90°, while the fourth block is constructed by rotating the block of Fig. 5 by −90°. Similarly, Fig. 5a is constructed from Fig. 5b.

Segment Types

A space-filling curve consists of a set of segments. Each segment connects two consecutive multidimensional points. Five different types of segments are distinguished, namely, *Jump*, *Contiguity*, *Reverse*, *Forward*, and *Still*. A *Jump* segment in an SFC is said to happen when the distance, along any of the dimensions, between two consecutive points in the SFC is greater than one. Similarly, a *Contiguity* segment in an SFC is said to happen when the distance, along any of the dimensions, between two consecutive points in the SFC is equal to one. On the other side, a segment in an SFC is termed a *Reverse* segment if the projection of its two consecutive points, along any of the dimensions, results in scanning the dimension in decreasing order. Similarly, a segment in an SFC is termed a *Forward* segment if the projection of its two consecutive points, along any of the dimensions, results in scanning the dimension in increasing order. Finally, a segment in an SFC is termed a *Still* segment when the distance, along any of the dimensions, between the segment's two consecutive points in the SFC is equal to zero. Closed formulas to count the number of *Jump*, *Contiguity*, *Reverse*, *Forward*, and *Still* segments along each dimension can be found in [10].

Irregularity

An optimal locality-preserving space-filling curve is one that sorts multidimensional points in ascending order for all dimensions. However, in reality, when a space-filling curve attempts to sort the points in ascending order according to one dimension, it fails to do the same for the other dimensions. A good space-filling curve for one dimension is not necessarily good for the other dimensions. In order to measure the mapping quality of a space-filling curve, the concept of *irregularity* has been introduced as a measure of goodness for the order imposed by a space-filling curve [11]. Irregularity introduces a quantitative measure that indicates the non-avoidable reverse order imposed by space-filling curves for some or all dimensions. Irregularity is measured for each dimension separately and gives an indicator of how a space-filling curve is far from the optimal. The lower the irregularity, the better the space-filling curve. The irregularity is formally defined as: For any two points, say P_i and P_j, in the D-dimensional space with coordinates

$(P_i.u_0, P_i.u_1, \ldots, P_i.u_{D-1})$, $(P_j.u_0, P_j.u_1, \ldots, P_j.u_{D-1})$, respectively, and for a given space-filling curve S, if S visits P_i before P_j, an irregularity occurs between P_i and P_j in dimension k iff $P_j.u_k < P_i.u_k$. Closed formulas to count the number of irregularities for various space-filling curves can be found in [11].

Key Applications

Preprocessing for Multidimensional Applications: Multimedia Databases, GIS, and Multidimensional Indexing

Mapping the multidimensional space into the one-dimensional domain plays an important role in applications that involve multidimensional data. Multimedia databases, geographic information systems (GIS), QoS routing, and image processing are examples of multidimensional applications. Modules that are commonly used in multidimensional applications include searching, sorting, scheduling, spatial access methods, indexing, and clustering. Considerable research has been conducted for developing efficient algorithms and data structures for these modules for one-dimensional data. In most cases, modifying the existing one-dimensional algorithms and data structures to deal with multidimensional data results in spaghettilike programs to handle many special cases. The cost of maintaining and developing such code degrades the system performance. Mapping from the multidimensional space into the one-dimensional domain provides a preprocessing step for multidimensional applications. The preprocessing step takes the multidimensional data as input and outputs the same set of data represented in the one-dimensional domain. The idea is to keep the existing algorithms and data structures independent of the dimensionality of data. The objective of the mapping is to represent a point from the D-dimensional space by a single integer value that reflects the various dimensions of the original space. Such a mapping is called a locality-preserving mapping in the sense that, if two points are near to each other in the D-dimensional space, then they will be near to each other in the one-dimensional space.

Network-Attached Storage Devices (NASDs)

Writing efficient schedulers is becoming a very challenging task, given the increase in demand of such systems. Consider the case of network-attached storage devices (NASDs) [3] as a building block for a multimedia server. NASDs are smart disks that are attached directly to the network. In a multimedia server, a major part of a NASD function goes toward fulfilling the real-time requests of users. This involves disk and network scheduling with real-time constraints, possibly with additional requirements like request priorities, and quality-of-service guarantees. NASD requirements can be mapped in the multidimensional space and a SFC-based scheduler is used. The type of space-filling curve used in NASD scheduling is determined by its requirements. For example, in NASD, if reducing the number of requests that lose their deadlines is more important than increasing the disk or network bandwidth, then the real-time deadline dimension of the scheduling space will be favored. As a result, a space-filling curve with intentional bias is favored.

Multimedia Disk Scheduling

Consider the problem of disk scheduling in multimedia servers [9]. In addition to maximizing the bandwidth of the disk, the scheduler has to take into consideration the real-time constraints of the page requests, e.g., as in the case of video streaming. If clients are prioritized based on quality-of-service guarantees, then the disk scheduler might as well consider the priority of the requests in its disk queue. Writing a disk scheduler that handles real-time and QoS constraints in addition to maximizing the disk bandwidth is challenging and a hard task. Scheduler parameters can be mapped to space dimensions and an SFC-based scheduler is used. The reader is referred to [9] to get more insight about the applicability of the irregularity in multimedia disk schedulers.

Future Directions

Future directions for space-filling curves include (i) exploiting new multidimensional applications that can make use of the properties of space-filling curves, (ii) analyzing the behavior of various space-filling curves in high-dimensional space, (iii) providing automated modules with the ability of choosing the appropriate space-filling curve for a given application, and (iv) developing new space-filling curves that are tailored to specific applications.

Cross-References

▶ High-Dimensional Indexing
▶ Space-Filling Curves
▶ Spatial Indexing Techniques

Recommended Reading

1. Faloutsos C. Gray codes for partial match and range queries. IEEE Trans Softw Eng. 1988;14(10): 1381–93.
2. Faloutsos C, Rong Y. Dot: a spatial access method using fractals. In: Proceeding of 7th International Conference on Data Engineering; 1991. p. 152–59.
3. Gibson G, Nagle D, Amiri K, Butler J, Chang FW, Gobioff H, Hardin C, Riedel E, Rochberg D, Zelenka J. File server scaling with network-attached secure disks. In: Proceeding of 1997 ACM SIGMETRICS International Conference on Measurement and Modeling of Computer System; 1997. p. 272–84.
4. Goldschlager LM. Short algorithms for space-filling curves. Softw Prac Exp. 1981;11(1):99–100.
5. Gray F. Pulse code communications. US Patent 2632058; 1953.
6. Jagadish HV. Linear clustering of objects with multiple attributes. In: Proceeding of the ACM SIGMOD International Conference on Management of Data; 1990. p. 332–42.
7. Kamel I, Faloutsos C. Hilbert r-tree: an improved r-tree using fractals. In: Proceeding of the 20th International Conference on Very Large Data Bases; 1994. p. 500–09.
8. Liao S, Lopez MA, Leutenegger ST. High dimensional similarity search with space-filling curves. In: Proceeding of the 17th International Conference on Data Engineering; 2001. p. 615–22.
9. Mokbel MF, Aref WG, El-Bassyouni K, Kamel I. Scalable multimedia disk scheduling. In: Proceeding

S

of the 20th International Conference on Data Engineering; 2004. p. 498–509.

10. Mokbel MF, Aref WG, Kamel I. Analysis of multidimensional space-filling curves. GeoInformatica. 2003;7(3):179–209.

11. Mokbel MF, Aref WG. Irregularity in high-dimensional space-filling curves. Distrib Parallel Databases. 2011;29(3):217–38.

12. Moon B, Jagadish HV, Faloutsos C, Salz J. Analysis of the clustering properties of Hilbert space-filling curve. IEEE Trans Knowl Data Eng. 2001;13(1): 124–41.

13. Orenstein JA Spatial query processing in an object-oriented database system. In: Proceeding of the ACM SIGMOD International Conference on Management of Data; 1986. p. 326–36.

14. Peano G. Sur une courbe qui remplit toute une air plaine. Math Ann. 1890;36:157–60.

15. Witten IH, Wyvill B. On the generation and use of space-filling curves. Softw Prac Exp. 1983;3: 519–25.

SPARQL

Lei Zou
Institute of Computer Science and Technology, Peking University, Beijing, China

Synonyms

SPARQL Protocol and RDF Query Language; RDF query language

Scientific Fundamentals

Semantic Web, Query Language

Definition

SPARQL (a recursive acronym for **SPARQL** **P**rotocol **a**nd **RDF** **Q**uery **L**anguage), proposed by the World Wide Web Consortium (W3C), is a structured query language that retrieves and manipulates data stored in RDF (Resource Description Framework) format. SPARQL can be used to express queries across diverse data sources, whether the data is stored natively as RDF or viewed as RDF via middleware. The results of SPARQL queries can be result sets or RDF graphs.

Historical Background

SPARQL 1.0 became an official W3C Recommendation on 15 January 2008. On 26 March 2013, the SPARQL Working Group has produced a new W3C Recommendation SPARQL 1.1 that introduces more features to 2008 version. SPARQL is emerging as the de facto RDF query language. Prior to SPARQL, there were some other popular RDF query languages, such as RQL, SeRQL, TRIPLE, RDQL, and so on. The comprehensive comparison between these languages was studied in [5].

Foundations

Syntax

The syntax of SPARQL contains five parts: prefix declarations, result clause, dataset definition, query pattern, and query modifiers. Note that prefix declarations, dataset definition, and query modifiers are optional according to SPARQL syntax. The following example illustrates a simple SPARQL statement.

Example 1 "Finding who is the developer of Tabulator." We issue the following SPARQL query, which includes four parts: prefix declarations, result clause, dataset definition, and query pattern.

```
#Prefix declarations
PREFIX foaf: <http://xmlns.com/foaf/0.1/>
PREFIX doap:  <http://usefulinc.com/ns/doap#>
#Result clause
SELECT ?name
#Dataset definition
FROM
<http://www.w3.org/People/Berners−Lee/card>
#Query pattern
WHERE
{ <http://dig.csail.mit.edu/2005/ajar/ajaw/data#Tabulator>
doap:developer ?a.
  ?a foaf:name ?name
}
```

IRIs, a subset of RDF URI references that omits spaces, in SPARQL queries are absolute, such as ⟨http://dig.csail.mit.edu/2005/ajar/ajaw/data#Tabulator⟩. However, we can use relative IRIs and prefixed names. For example, we define prefix "foaf" as ⟨http://xmlns.com/foaf/0.1/⟩. The relative IRI "foaf:name" will be resolved to produce the absolute ⟨http://xmlns.com/foaf/0.1/name⟩.

The result clause identifies the variables to appear in the query results, such as *?name*. The query pattern in SPARQL defines the graph pattern to match against the data graph. In the dataset definition, a SPARQL query specifies the dataset to be used for matching by using the FROM clause. Analog to SQL, SPARQL can also specify some modifiers, such as ORDER BY, OFFSET, LIMIT, and so on. The complete SPARQL syntax is specified in W3C SPARQL Recommendation document (https://www.w3.org/TR/rdf-sparql-query/).

Semantic

Assume there are pairwise disjoint infinite sets I, B, and L, where I denotes IRI and B and L denote blank nodes and literals, respectively. An RDF dataset is a collection of triples, each of which is denoted as $t(subject, property, object) \in (I \cup B) \times I \times (I \cup B \cup L)$. A triple can be naturally seen as a pair of nodes connected by a named relationship. Hence, an RDF dataset can be represented as a graph where subjects and objects are vertices, and triples are edges with property names as edge labels. Note that there may exist more than one property between a subject and an object, that is, multiple edges may exist between two vertices in an RDF graph.

WHERE clause in SPARQL provides the graph pattern to match against the data graph, where a basic graph pattern (BGP) is the building block. A general graph pattern can be recursively defined based on BGP. Thus, we first define BGP as follows:

Definition 1 (Basic Graph Pattern) A *basic graph pattern* is a connected graph, denoted as $Q = \{V(Q), E(Q)\}$, such that (1) $V(Q) \subseteq (I \cup L \cup V_{Var})$ is a set of vertices, where I denotes IRI, L denotes literals, and V_{Var} is a set of variables, (2) $E(Q) \subseteq V(Q) \times V(Q)$ is a set of edges in Q, and (3) each edge e in $E(Q)$ either has an edge label in I (i.e., property) or the edge label is a variable.

A match of BGP over RDF graph is defined as a partial function μ from $V(Q)$ to the vertices in RDF graph. Formally, we define the match as follows:

Definition 2 (BGP Match) Consider an RDF graph G and a connected query graph Q that has n vertices $\{v_1, \ldots, v_n\}$. A subgraph M with m vertices $\{u_1, \ldots, u_m\}$ (in G) is said to be a *match* of Q if and only if there exists a *function* μ from $\{v_1, \ldots, v_n\}$ to $\{u_1, \ldots, u_m\}$ ($n \geq m$), where the following conditions hold:

1. if v_i is not a variable, $\mu(v_i)$ and v_i have the same URI or literal value ($1 \leq i \leq n$);
2. if v_i is a variable, there is no constraint over $\mu(v_i)$ except that $\mu(v_i) \in \{u_1, \ldots, u_m\}$;
3. if there exists an edge $\overrightarrow{v_i v_j}$ in Q, there also exists an edge $\overrightarrow{\mu(v_i)\mu(v_j)}$ in G; furthermore,

$\overrightarrow{\mu(v_i)\mu(v_j)}$ has the same property as $\overrightarrow{v_i v_j}$ unless that the label of $\overrightarrow{v_i v_j}$ is a variable.

The set of matches for Q over RDF graph G is denoted as $[\![Q]\!]_G$, based on which we return variable bindings that are defined in the result clause. Example 2 shows the answers of a BGP SPARQL query.

Example 2 "Find all movies directed by Stanley Kubrick and report their movie names." The sample RDF dataset from IMDB and the SPARQL query are given as follows. We also illustrate the corresponding RDF and query graphs in Fig. 1.

Subject	Predicate	Object
mdb:film/2014	rdfs:label	"The Shining"
mdb:film/2014	movie:director	mdb:director/8476
mdb:film/2014	movie:actor	mdb: actor/29704;
mdb:film/2014	movie:relatedBook	bm:0743424425
mdb:director/8476	movie:director_name	"Stanley Kubrick"
mdb:film/2685	movie:director	mdb:director/8476
mdb:film/2685	rdfs:label	"A Clockwork Orange"
mdb:film/424	movie:director	mdb:director/8476
mdb:film/424	rdfs:label	"Spartacus"
mdb:actor/29704	movie:actor_name	"Jack Nicholson"
mdb:film/1267	movie:actor	mdb:actor/29704
mdb:film/1267	rdfs:label	"The Last Tycoon"
bm:books/0743424425	dc:creator	bm:persons/Stephen+King
bm:persons/Stephen+King	rdfs:label	"Stephen King"

```
SELECT ?moviename
WHERE {
?m rdfs:label ?moviename . ?m movie:director ?d .
?d movie: director_name ''Stanley Kubrick''.
}
```

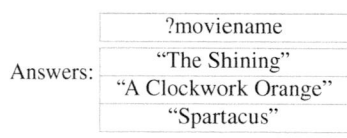

Answers:

?moviename
"The Shining"
"A Clockwork Orange"
"Spartacus"

Formally, a general graph pattern in SPARQL is defined as follows:

Definition 3 (Graph Pattern) A *graph pattern* in SPARQL is defined as follows:

1. if P is a BGP, P is a graph pattern;
2. if P_1 and P_2 are both graph patterns, P_1 AND P_2, P_1 UNION P_2, and P_1 OPTIONAL P_2 are all graph patterns;

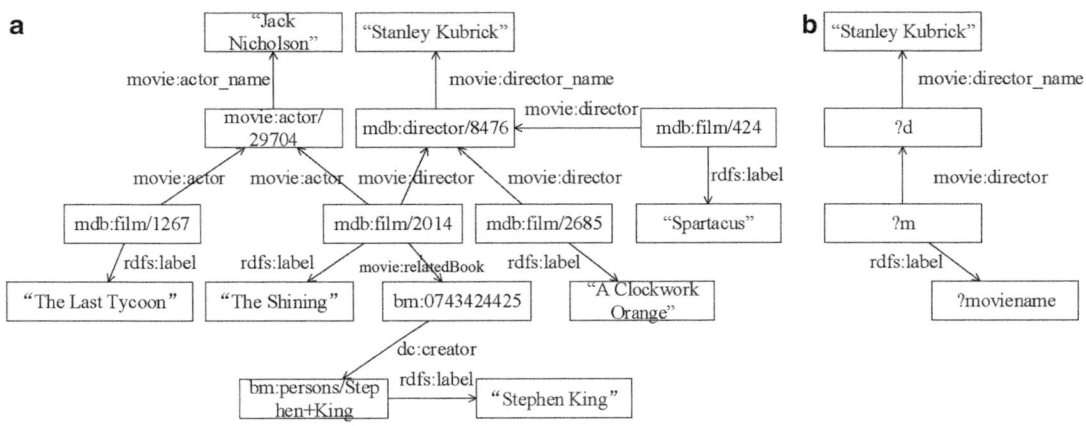

SPARQL, Fig. 1 RDF graph and SPARQL query graph

3. If P is a graph pattern and R is a SPARQL built-in condition, then the expression (P FILTER R) is a graph pattern.

A SPARQL built-in condition is constructed using the variables in SPARQL, constraints, logical connectives ($\neg$, $\wedge$,$\vee$), inequality symbols ($\leq$,$\geq$,$<$, $>$), the equality symbol ($=$), unary predicates like bound, isBlank, and isIRI, plus other features [8]. The complete list is given in [4]. We formally define the answers of SPARQL based on BGP matches.

Definition 4 (Compatibility) Given two BGP queries Q_1 and Q_2 over RDF graph G, μ_1 and μ_2 define two matching functions from vertices in Q_1 (denoted as $V(Q_1)$) and Q_2 (denoted as $V(Q_2)$) to the vertices in RDF graph G, respectively. μ_1 and μ_2 are *compatible* when for all $x \in V(Q_1) \cap V(Q_2)$, $\mu_1(x) = \mu_2(x)$, denoted as $\mu_1 \sim \mu_2$; otherwise, they are not compatible, denoted as $\mu_1 \nsim \mu_2$.

Definition 5 (SPARQL Matches) Given a SPARQL query with graph pattern Q over a RDF graph G, a set of matches of Q over G, denoted as $[\![Q]\!]_G$, is defined recursively as follows:

1. If Q is a BGP, $[\![Q]\!]_G$ is defined in Definition 2.
2. If $Q = Q_1\ AND\ Q_2$, then $[\![Q]\!]_G = [\![Q_1]\!]_G \bowtie [\![Q_2]\!]_G = \{\mu_1 \cup \mu_2\,|\,\mu_1 \in [\![Q_1]\!]_G \wedge \mu_2 \in [\![Q_2]\!]_G \wedge (\mu_1 \sim \mu_2)\}$
3. If $Q = Q_1\ UNION\ Q_2$, then $[\![Q]\!]_G = [\![Q_1]\!]_G \cup [\![Q_2]\!]_G = \{\mu\,|\,\mu \in [\![Q_1]\!]_G \vee \mu \in [\![Q_2]\!]_G\}$

4. If $Q = Q_1\ OPT\ Q_2$, then $[\![Q]\!]_G = ([\![Q_1]\!]_G \bowtie [\![Q_2]\!]_G) \cup ([\![Q_1]\!]_G \backslash [\![Q_2]\!]_G) = \{\mu_1 \cup \mu_2\,|\,\mu_1 \in [\![Q_1]\!]_G \wedge \mu_2 \in [\![Q_2]\!]_G \wedge (\mu_1 \sim \mu_2)\} \cup \{\mu_1\,|\,\mu_1 \in [\![Q_1]\!]_G \wedge (\forall\mu_2 \in [\![Q_2]\!]_G) \wedge (\mu_1 \nsim \mu_2)\}$
5. If $Q = Q_1\ Filter\ F$, then $[\![Q]\!]_G = \Theta_F([\![Q_1]\!]_G) = \{\mu_1\,|\,\mu_1 \in [\![Q_1]\!]_G \wedge \mu_1\ satisfies\ F\}$

The following example illustrates SPARQLs with "OPTIONAL." More discussions about SPARQL query semantics can be found in [8].

Example 3 "Report all movie names directed by Stanley Kubrick and their related book names if any."

```
SELECT ?moviename ?bookauthor
WHERE {
?m rdfs:label ?moviename. ?m movie:director ?d.
?d movie:director_name ''Stanley Kubrick''.
OPTIONAL {?d movie:relatedBook ?book.
         ?book dc:creator ?author.
         ?author rdfs:label ?bookauthor.
         }
}
```

Answers:

	?moviename	?bookauthor
	"The Shining"	"Stephen King"
	"A Clockwork Orange"	–
	"Spartacus"	–

Key Applications

Knowledge Graph Application

SPARQL is emerging as the de facto RDF query language, which plays a fundamental role in knowledge graph applications, that is, accessing and manipulating RDF data in knowledge graph applications.

Graph Database

RDF triple store is a special kind of graph databases, some of which are implemented based on relational approaches (e.g. [1–3, 6, 10, 11]) while others are graph native stores (e.g. [9, 12]). RDF triple stores are used in many knowledge graph applications.

Future Directions

SPARQL 1.1 Recommendation introduces more features, such as property path, to enable more powerful data access and management capability over RDF datasets. These new plug-in features require efficient implementations.

Cross-References

▶ Resource Description Framework
▶ RDF Stores

Recommended Reading

1. Abadi DJ, Marcus A, Madden S, Hollenbach K. Sw-store: a vertically partitioned DBMS for semantic web data management. VLDB J. 2009;18(2): 385–406.
2. Bornea MA, Dolby J, Kementsietsidis A, Srinivas K, Dantressangle P, Udrea O, Bhattacharjee B. Building an efficient RDF store over a relational database. In: Proceedings of the ACM SIGMOD International Conference on Management of Data; 2013. p. 121–32.
3. Broekstra J, Kampman A, van Harmelen F. Sesame: a generic architecture for storing and querying RDF and RDF schema. In: Proceedings of the 1st International Semantic Web Conference; 2002. p. 54–68.
4. T. W. S. W. Group. Sparql 1.1 overview. https://www.w3.org/TR/rdf-sparql-query/, 21 March 2013.
5. Haase P, Broekstra J, Eberhart A, Volz R. A comparison of RDF query languages. ISWC. 2004: 502–17.
6. Neumann T, Weikum G. The RDF-3X engine for scalable management of RDF data. VLDB J. 2010;19(1):91–113.
7. Özsu MT. A survey of RDF data management systems. Frontiers of Computer Science. 2016;10(3):418–432.
8. Pérez J, Arenas M, Gutierrez C. Semantics and complexity of SPARQL. ACM Trans. Database Syst. 2009;34(3).
9. Udrea O, Pugliese A, Subrahmanian VS. Grin: a graph based RDF index. In: Proceedings of the 22nd AAAI Conference on Artificial Intelligence; 2007. p. 1465–70.
10. Weiss C, Karras P, Bernstein A. Hexastore: sextuple indexing for semantic web data management. Proc VLDB Endow. 2008;1(1):1008–1019.
11. Wilkinson K. Jena property table implementation. Technical Report HPL-2006-140, HP Laboratories Palo Alto, 2006.
12. Zou L, Mo J, Chen L, Özsu MT, Zhao D. gstore: answering SPARQL queries via subgraph matching. Proc VLDB Endow. 2011;4(8):482–493.

Sparse Index

Mirella M. Moro[1] and Vassilis J. Tsotras[2]
[1]Departamento de Ciencia da Computaçao, Universidade Federal de Minas Gerais – UFMG, Belo Horizonte, MG, Brazil
[2]University of California-Riverside, Riverside, CA, USA

Synonyms

Non-dense index

Definition

Consider a tree-based index on some numeric attribute A of a relation R. If an index record (of the form <search-key, pointer>) is created for *some* of the values that appear in attribute A, then this index is *sparse*.

Key Points

Tree-based indices are built on numeric attributes and maintain an order among the indexed search-key values. Hence, they provide efficient access to the records of a relation by attribute value. Consider for example an index built on attribute *A* of relation *R*. The leaf pages of the index contain *index-records* of the form *<search-key, pointer>*, where search-key corresponds to a value from the indexed attribute *A* and *pointer* points to the respective record in the indexed relation *R* with that attribute value. If not all distinct values that appear in *R.A* also appear in index records, this index is *sparse*, otherwise it is called *dense*.

A sparse index needs a way to access even the relation records with values that do not directly appear in the index. Hence it is required that the indexed relation is *ordered* according to the values of the indexed attribute *A*; in this way the relation order can be used to access values not directly indexed by the sparse index.

Tree-indices are further categorized by whether their search-key ordering is the same with the relation file's physical order (if any). If the search-key of a tree-based index is the same as the ordering attribute of a (ordered) file then the index is called *primary*. An index built on any non-ordering attribute of a file is called *secondary*. Hence a primary index is also sparse whereas a secondary index should also be dense.

A dense index is typically larger than a sparse index (since all search-key values are indexed) and thus requires more space. It also needs to be updated for every relation update that involves the attribute value being indexed.

Cross-References

► B+-Tree
► Indexed Sequential Access Method
► Primary Index

Recommended Reading

1. Elmasri R, Navathe SB. Fundamentals of database systems. 6th ed. Reading: Addisson-Wesley; 2010.
2. Manolopoulos Y, Theodoridis Y, Tsotras VJ. Advanced database indexing. Dordecht: Kluwer; 2000.
3. Silberschatz A, Korth H, Sudarshan S. Database system concepts. 6th ed. New York: McGraw-Hill; 2010.

Spatial and Spatiotemporal Data Models and Languages

Markus Schneider
University of Florida, Gainesville, FL, USA

Definition

A *data model* provides a formalism consisting of a notation for describing data of interest and of a set of operations for manipulating these data. It abstracts from reality and provides a generalized view of data representing a specific and bounded scope of the real world. In the context of databases, a data model describes the organization, that is, the structure, of a database. In the context of complex objects like video, genomic, and multimedia objects, a data model describes a type system consisting of data types, operations, and predicates. Spatial and spatio-temporal data models are of this second kind. A *spatial data model* is a data model defining the properties of and operations on static objects in space. These objects are described by *spatial data types* like *point* (for example, representing the locations of cities in the US), *line* (for example, describing the ramifications of the Nile Delta), and *region* (for example, depicting school districts). Operations on spatial data types include, for instance, the geometric *intersection*, *union*, and *difference* of spatial objects, the computation of the *length* of a line or the *area* of a region, the test whether two spatial objects *overlap* or *meet*, and whether one object is *north* or *southeast* of another object. A *spatio-temporal data model* is a data model representing the temporal evolution of spatial objects over time. These evolutions can be discrete, that is, they happen from time to time (for example, the change of the boundary of a land parcel) or continuous, that is, they happen permanently and smoothly (for example, the devastating trajectory of a hurricane). In the continuous case, one speaks about *moving objects* and represents them by *spatio-temporal data types* like *moving point* (for example, recording the route of a cell phone user), *moving line* (for example, representing the boundary of a tsunami), and *moving region* (for example, describing the motion of an air polluted cloud). Operations on spatio-temporal data types comprise, for instance, the spatio-temporal *intersection*, *union*, and *difference* of moving objects, the computation of the *trajectory* of a moving point as a line object, the determination of the *location* of a moving object at a particular time, the calculation of a moving object during a given set of intervals, and the test whether a moving point *enters* or *crosses* a moving region. *Spatial* and *spatio-temporal query languages* enable the user to query databases enhanced by these concepts.

Historical Background

The interest to store geometric data in databases began in the late 1970s. Due to the increasing success of relational databases, the first approach has been to decompose a spatial object recursively into its constituent parts until they can be stored in tables. For example, this approach decomposes a polygon into its set of segments. Each segment is decomposed into a pair of points. A point is decomposed into a pair of two float numbers. Float numbers are a DBMS data type and can be stored in a table. This approach has revealed a number of fundamental drawbacks. Since all lines and polygons are decomposed into their constituent parts scattered as tuples over a relation, a spatial object is not treated as an entity or unit but only *corresponds* to a collection of tuples. Since this approach is based on standard domains and has no concept of spatial data types, it cannot provide and support any meaningful geometric operations. A more detailed discussion can be found in [1].

Classical research on time-varying geometric data has focused on *discrete* changes of spatial objects over time. For example, cadastral applications deal with the management of land parcels whose boundaries can change from time to time due to specific legal actions such as splitting, merging, or land consolidation. Political boundaries can suddenly disappear, as the reunification of West and East Germany shows. Different approaches have been proposed to model these discrete changes. One of them is to enhance *temporal databases* [2] with spatial data types. Each discrete change leads to a new stored snapshot with a modified spatial object in the temporal database. Another approach [3] keeps a single version of each spatial object only but annotates each of its components (for instance, a point or a segment) with a temporal element indicating the period of validity or existence of this component. Hence, discrete changes of a spatial object are registered within the object.

Foundations

Spatial data types form the basis of a large number of data models and query languages for spatial data. They are extensively leveraged by *spatial databases* [4] and embedded as attribute data types into their data models, that is, in the same way as standard data types such as *integer*, *real*, and *string*. The geometric types are designed as *abstract data types*, that is, the internal structure of a spatial object is hidden from the user, and its features can only be retrieved by (abstract) operations on this object. In this manner, they provide a high-level view of geometric data. One can distinguish the older generation of *simple* spatial data types and the newer generation of *complex* spatial data types, depending on the spatial complexity the types are able to model. In the two-dimensional space, simple spatial data types only provide simple object structures like single points, continuous lines, and simple regions (Fig. 1a–c). However, from an application perspective, simple spatial data types have turned out to be inadequate abstractions for spatial applications since they are insufficient to cope with the variety and complexity of geographic reality.

Spatial and Spatiotemporal Data Models and Languages, Fig. 1 Examples of a simple point object (**a**), a simple line object (**b**), a simple region object (**c**), a complex point object (**d**), a complex line object (**e**), and a complex region object (**f**)

From a formal perspective, they are not closed under the geometric set operations *intersection*, *union*, and *difference*. This means that these operations applied to two simple spatial objects can produce a spatial object that is *not* simple. Complex spatial data types solve these problems. They provide universal and versatile spatial objects and are closed under geometric set operations. They allow objects with multiple components, region components that may have holes, and line components that may model ramified, connected geometric networks (Fig. 1d–f).

As an example, in a relational setting, states, cities, and rivers are represented in the following relations:

- states(*sname*: *string*, *area*: *region*)
- cities(*name*: *string*, *population*: *integer*, *location*: *point*)
- rivers(*name*: *string*, *route*: *line*)

Queries can then be formulated by employing operations and predicates on spatial attribute values within an extended standard database query language such as SQL, leading to *Spatial SQL* [5]. Assume that the following operations and predicates are available:

area:	region	→ real
inside:	point × region	→ bool
intersection:	line × region	→ line
length:	line	→ real
meet:	region × region	→ bool

The predicates *inside* and *meet* represent *topological relationships* [6] that characterize the relative position between spatial objects. The operation *length* is a numerical function computing the length of a *line* object. The operation *intersection* computes the part of a *line* object intersecting a *region* object.

One can now pose queries: What is the total population of the cities in France?

- *select sum (c.pop) as total*
- *from cities as c, states as s*
- *where c.location inside s.area and s.name = "France"*

Compute the part of the river Rhine that is located within Germany and determine its length.

- *select intersection (r.route, s.area) as rhine,*
- *length (intersection(r. route, s.area)) as len*
- *from rivers as r, states as s*
- *where r.name = "Rhine" and s.name = "Germany"*

Make a list that shows for each state the number of its neighbor states, and their total area.

- *select s.name, count(*), sum(area(t.area))*
- *from states as s, states as t*
- *where s.area meet t.area*
- *group by s.name*

Spatio-temporal data types enable the user to describe the dynamic behavior of spatial objects over time. The dynamic behavior refers to the continuous change of the locations of spatial objects over time. That is, the spatial objects move, and they are therefore called *moving objects*. They are stored in special spatio-temporal databases called *moving objects databases* [7]. In the same way as spatial data types, spatio-temporal data types are also designed as abstract

S

data types and embedded as attribute types into a DBMS data model. Spatio-temporal data types are available for moving points (type *mpoint* for short), moving lines (*mline*), and moving regions (*mregion*). In case of moving regions, one can also represent the change of their extent and shape over time. Conceptually, a moving point is a function $f : time \rightarrow point$, a moving line is a function $f : time \rightarrow line$, and a moving region is a function $f : time \rightarrow region$. For example, for a moving region this means that at each time instant an object of type *region* has to be returned. Geometrically, moving objects correspond to the three-dimensional shapes. In case of a moving point it is a three-dimensional line (Fig. 2a) and in case of a moving region it is a volume (Fig. 2b). One can distinguish moving objects databases that model and query the history of movement for spatio-temporal analysis [8, 9] and moving objects databases that model, predict, and query current and future movement [10, 11]. In the latter case, location updates require a balancing of update costs and imprecision [12] and introduce the feature of uncertainty [13].

As an example, consider relations describing the movements of airplanes or storms:

- *flight*(*id*: *string*, *from*: *string*, *to*: *string*, *route*: *mpoint*)
- *weather*(*id*: *string*, *kind*: *string*, *area*: *mregion*)

One can pose queries by employing operations and predicates on spatiotemporal attribute values within an extended standard database query language such as SQL, leading to *Spatio-Temporal Query Language (STQL)* [14]. Assume that the following operations and predicates are available:

deftime:	*mpoint*	$\rightarrow$ *periods*
Disjoint:	*mpoint* × *mregion*	$\rightarrow$ *bool*
distance:	*mpoint* × *mpoint*	$\rightarrow$ *mreal*
Inside:	*mpoint* × *mregion*	$\rightarrow$ *bool*
intersection:	*mpoint* × *mregion*	$\rightarrow$ *mpoint*
meet:	*point* × *region*	$\rightarrow$ *bool*
min:	*mreal*	$\rightarrow$ *real*
trajectory:	*mpoint*	$\rightarrow$ *line*

The function *deftime* returns the set of time intervals when a moving point is defined. The *spatio-temporal predicate* [7, 14, 15] *Disjoint* checks whether a moving point and a moving region are disjoint for some period. The function *distance* computes the distance between two moving points and is a real-valued function of time, captured here in a data type *mreal* for *moving reals*. The spatio-temporal predicate *Inside* tests whether a moving point is located inside a moving region for some period. The operation *intersection* returns the part of a moving point whenever it lies inside a moving region, which is a moving point again. The topological predicate *meet* checks whether a point object is located on the boundary of a region object. The function *min* yields the minimal value assumed over time by a moving real.

One can now pose queries: Find all flights from Frankfurt that are longer than 5,000 kms.

- *select id*
- *from flight*
- *where from* = *"FRA" and length (trajectory(route)) > 5000*

Retrieve any pairs of airplanes, which, during their flight, came closer to each other than 500 m.

Spatial and Spatiotemporal Data Models and Languages, Fig. 2 Examples of a moving point object (**a**) and a moving region object (**b**)

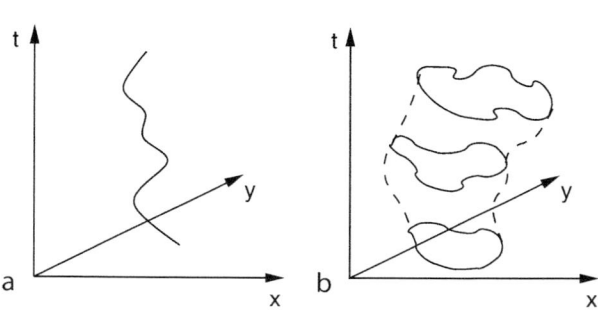

- *select f.id, g.id*
- *from flight as f, flight as g*
- *where f.id <> g.id and min(distance(f.route, g.route)) < 0.5*

At what time was flight TB691 within a snowstorm with id RS316?

- *select deftime(intersection(f.route, w.area))*
- *from flight as f, weather as w*
- *where f.id = "TB691" and w.id = "RS316"*

Which are the planes that ran into a hurricane and had to traverse it?

- *select f.id, w.id*
- *from flight as f, weather as w*
- *where w.kind = "hurricane" and*
- *f.route Disjoint >> meet >> Inside >> meet >> Disjoint w.area*

The term *Disjoint >> meet >> Inside >> meet >> Disjoint* is a spatio-temporal predicate that is composed of a *temporal sequence* of the basic spatio-temporal predicates *Disjoint* and *Inside* as well as the topological predicate *meet*. The *temporal composition operator* is indicated by the symbol >>. The query above searches for a spatio-temporal pattern in which a plane is disjoint from a hurricane for some period, then meets the boundary of the hurricane at a time instant, is inside the hurricane for some period, meets the boundary of the hurricane again at a time instant, and is disjoint again from the hurricane for some period. The alternating sequence of topological predicates that hold for some period or for some time instant is characteristic for composite spatio-temporal predicates.

Key Applications

Spatial data models containing spatial data types, operations, and predicates are a universal and general concept for representing geometric information in all kinds of spatial applications. They have found broad acceptance in spatial extension packages of commercially and publicly available database systems as well as in geographic information systems. Further, all applications in the geosciences (for example, geography, hydrology, soil sciences) as well as many applications in government and administration (for example, cadastral applications, urban planning) already benefit from them. Independent studies have shown that about 80 % of all data have spatial features (like geometric attributes) or a spatial reference (like an address). Thus, it is not surprising that independent international studies have predicted that geoinformation technology will belong to the most important and promising technologies in the future, besides biotechnology and nanotechnology.

The usage of moving objects in databases and especially in geographic information systems is still in its infancy since it is a relatively new technology. But increasingly, applications like location management, GPS-equipped PDAs, phones, and vehicles, navigation systems, RFID-tag tracking, sensor networks, hurricane research, and national security show interest in moving objects databases.

Cross-References

- ▶ Spatial Data Types
- ▶ Spatiotemporal Trajectories
- ▶ Temporal Database

Recommended Reading

1. Schneider M. Spatial data types for database systems – finite resolution geometry for geographic information systems, LNCS, vol. 1288. Berlin: Springer; 1997.
2. Tansel AU, Clifford J, Gadia S, Jajodia S, Segev A, Snodgrass RT, editors. Temporal databases: theory, design, and implementation. Benjamin/Cummings; 1993.
3. Worboys MF. A unified model for spatial and temporal information. Comput J. 1994;37(1):25–34.
4. Shekar S, Chawla S. Spatial databases: a tour. Englewood Cliffs: Prentice-Hall; 2003.
5. Egenhofer MJ. Spatial SQL: a query and presentation language. IEEE Trans Knowl Data Eng. 1994;6(1):86–94.

6. Schneider M, Behr T. Topological relationships between complex spatial objects. ACM Trans Database Syst. 2006;31(1):39–81.
7. Güting RH, Schneider M. Moving objects databases. San Francisco: Morgan Kaufmann; 2005.
8. Erwig M, Güting RH, Schneider M, Vazirgiannis M. Spatio-temporal data types: an approach to modeling and querying moving objects in databases. Geoinformatica. 1999;3(3):265–91.
9. Güting RH, Böhlen MH, Erwig M, Jensen CS, Lorentzos NA, Schneider M, Vazirgiannis M. A foundation for representing and querying moving objects. ACM Trans Database Syst. 2000;25(1): 1–42.
10. Sistla AP, Wolfson O, Chamberlain S, Dao S. Modeling and querying moving objects. In: Proceedings of the 13th International Conference on Data Engineering; 1997 p. 422–32.
11. Sistla AP, Wolfson O, Chamberlain S, Dao S. Querying the uncertain position of moving objects. In: Etzion O, Jajodia S, Sripada S, editors. Temporal databases: research and practice, LNCS, vol. 1399. Berlin: Springer; 1998. p. 310–37.
12. Wolfson O, Chamberlain S, Dao S, Jiang L, Mendez G. Cost and imprecision in modeling the position of moving objects. In: Proceedings of the 14th International Conference on Data Engineering; 1998. p. 588–96.
13. Trajcevski G, Wolfson O, Hinrichs K, Chamberlain S. Managing uncertainty in moving objects databases. ACM Trans Database Syst. 2004;29(3): 463–507.
14. Erwig M, Schneider M. Developments in spatio-temporal query languages. In: Proceedings of the IEEE International Workshop on Spatio-Temporal Data Models and Languages; 1999. p. 441–9.
15. Erwig M, Schneider M. Spatio-temporal predicates. IEEE Trans Knowl Data Eng. 2002; 14(4):1–42.

Spatial Anonymity

Panos Kalnis and Gabriel Ghinita
National University of Singapore, Singapore, Singapore

Synonyms

Anonymity in location-based services;Privacy-preserving spatial queries;Spatial k-anonymity

Definition

Let U be a user who is asking via a mobile device (e.g., phone, PDA) a query relevant to his current location, such as "find the nearest betting office." This query can be answered by a Location Based Service (LBS) in a public web server (e.g., Google Maps, MapQuest), which is not trustworthy. Since the query may be sensitive, U uses encryption and a pseudonym, in order to protect his privacy. However, the query still contains the exact location, which may reveal the identity of U. For example, if U asks the query within his residence, an attacker may use public information (e.g., white pages) to associate the location with U. Spatial k-Anonymity (SKA) solves this problem by ensuring that an attacker cannot identify U as the querying user with probability larger than $1/k$, where k is a user-defined anonymity requirement. To achieve this, a centralized or distributed anonymization service replaces the exact location of U with an area (called Anonymizing Spatial Region or ASR). The ASR encloses U and at least $k - 1$ additional users. The LBS receives the ASR and retrieves the query results for *any* point inside the ASR. Those results are forwarded to the anonymization service, which removes the false hits and returns the actual answer to U.

Historical Background

The embedding of positioning capabilities (e.g., GPS) in mobile devices has triggered several exciting applications. At the same time, it has raised serious concerns [1] about the risks of revealing sensitive information in location based services (LBS). An untrustworthy LBS may use public knowledge to relate a set of query coordinates to a specific user, even if the user-id is removed. In practice, users are reluctant to access a service that may disclose their political/religious affiliations or alternative lifestyles. Furthermore, users might be hesitant to ask innocuous queries such as "find the restaurants in my vicinity" since, once their identity is revealed, they may face unsolicited advertisements. Spatial k-Anonymity (SKA) aims at solving this problem.

k-Anonymity [15] has been used in relational databases for publishing census, medical and voting registration data (often called microdata). A relation satisfies *k*-anonymity if every tuple is indistinguishable from at least $k-1$ other tuples with respect to a set of quasi-identifier (QI) attributes. QIs are attributes (e.g., date of birth, gender, zip code) that can be linked to publicly available data to identify individuals. In the context of location based services, the *k*-Anonymity concept translates as follows: given a query, an attack based on the query location must not be able to identify the query source with probability larger than $1/k$.

A straightforward method is to pick $k-1$ random users and forward *k* independent queries (including the real one) to the LBS. This method achieves SKA because the query could originate from any client with equal probability $1/k$. However, depending on the value of *k*, a potentially large number of locations are transmitted and processed by the LBS. Also, the exact locations of *k* users are revealed, which is undesirable in many applications.

Most of the existing work adopts the framework of Fig. 1a, which assumes a trusted server, called *anonymizer*. Users access the anonymizer through a secure connection and periodically report their position. A querying user *U* sends his location-based query to the anonymizer, which removes the user-id and transforms the location of *U* through a technique called cloaking. Cloaking hides the actual location by an *anonymizing*

spatial region (*k*-ASR or ASR), which is an area that encloses *U*, as well as at least $k-1$ other users. The anonymizer then sends the ASR to the LBS, which returns a set of *candidate* results that satisfy the query condition for any possible point inside the ASR.

Figure 1b presents an example, where Bob asks for the nearest betting office. Bob forwards his request to the anonymizer, together with his anonymity requirement *k*. Assuming that $k=3$, the anonymizer generates a 3-ASR (shaded rectangle) that contains Bob and two other users U_1, U_2 (the anonymizer knows the exact locations of all users). Then, it sends this 3-ASR to the LBS, which finds all betting offices that can be the nearest-neighbor (NN) of any point in the 3-ASR (the LBS does not know where Bob is). This candidate set (i.e., $\{p_1, p_2, p_3, p_4\}$) is returned to the anonymizer, which filters the false hits and forwards the actual NN (in this case p_2) to Bob. Even if an attacker knows the location of Bob and the other users, she can only ascertain that the query originated from Bob with probability 1/3.

The privacy of user locations has also been studied in the context of related problems. *Probabilistic Cloaking* [2] does not apply the concept of SKA; instead, the ASR is a closed region around the query point, which is independent of the number of users inside. Given an ASR, the LBS returns the probability of each candidate result satisfying the query, based on its location with respect to the ASR. Kamat et al. [10] propose a model for sensor networks and

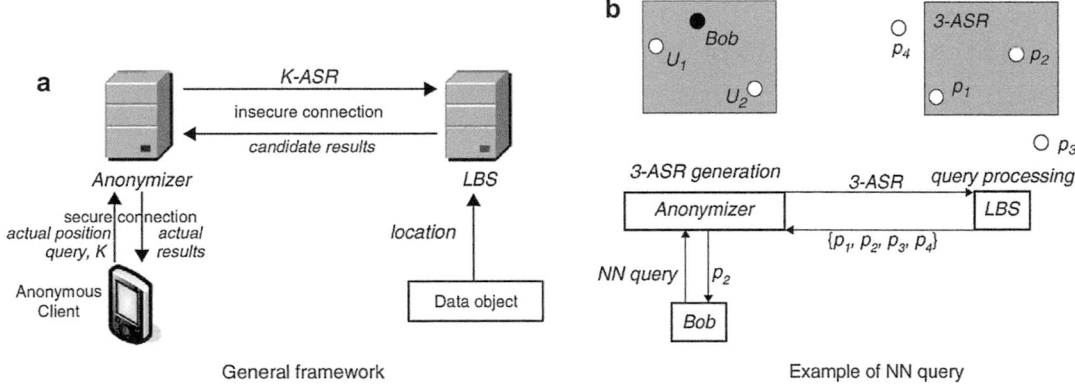

Spatial Anonymity, Fig. 1 Framework and example for spatial *k*-anonymity

examine the privacy characteristics of different sensor routing protocols. Hoh and Gruteser [8] describe techniques for hiding the trajectory of users in applications that continuously collect location samples.

Foundations

In order to solve the SKA problem, the following assumptions about the capabilities of the attacker apply:

1. The attacker intercepts the ASR, which implies that either the LBS is not trustworthy, or the communication channel between the anonymizer and the LBS is not secure.
2. The attacker knows the cloaking algorithm used by the anonymizer. This is common in the security literature where algorithms are typically public.
3. The attacker can obtain the current locations of all users. This assumption is motivated by the fact that users may often issue queries from locations (e.g., home, office), which may be identified through physical observation, triangulation, telephone catalogs etc. However, it is difficult to model the exact amount of knowledge an attacker can gain. Therefore, the third assumption dictates that the anonymization method should be provably secure under the worst-case scenario.
4. The attacker uses current data, but not historical information about movement and behavior

patterns of particular clients (e.g., a user often asking a particular query at a certain location or time). Therefore, SKA is defined only for *snapshot*, but not for continuous queries.

Given these assumptions, a spatial cloaking algorithm is said to be *secure*, if for any U and any k the probability of identifying U as the querying user is at most $1/k$. In addition to being secure, spatial cloaking should be efficient and effective. *Efficiency* means that the CPU and I/O cost of generating the ASR (at the anonymizer) should be minimized for better scalability and faster service. *Effectiveness* refers to the area of the ASR, which should also be minimized. Specifically, a large ASR incurs high processing overhead (at the LBS) and network cost (for transferring numerous candidate results from the LBS to the anonymizer). The rest of this article discusses several representative algorithms for the SKA problem.

In *Clique Cloak* [5], each query defines an axis-parallel rectangle whose centroid lies at the user location and whose extents are Δx, Δy. Figure 2 illustrates the rectangles of three queries located at U_1, U_2, U_3, assuming that they all have the same Δx and Δy. The anonymizer generates a graph where a vertex represents a query: two queries are connected if the corresponding users fall in the rectangles of each other. Then, the graph is searched for cliques of k vertices and the minimum bounding rectangle (MBR) of the corresponding user areas, forms the ASR sent to the LBS. Continuing the example of Fig. 2, if

Spatial Anonymity, Fig. 2
Example of *Clique Cloak*

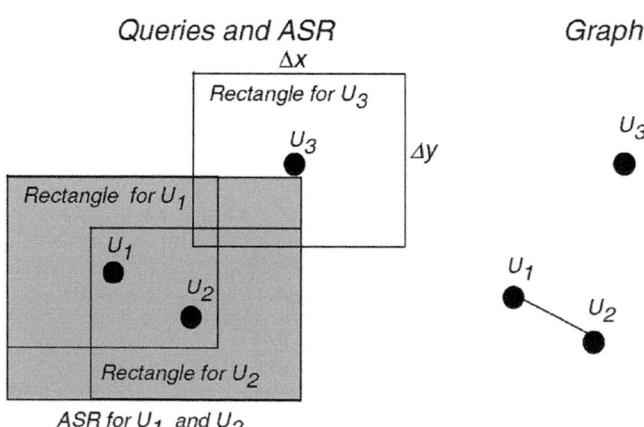

$k = 2$, U_1 and U_2 form a 2-clique and the MBR of their respective rectangles is generated so that both queries are processed together. On the other hand, U_3 cannot be processed immediately, but it has to wait until a new query (generating a 2-clique with U_3) arrives. *Clique Cloak* allows users to specify a temporal interval Δt such that, if a clique cannot be found within Δt, the query is rejected. Therefore, *Clique Cloak* may affect the quality of service, as some queries may be delayed or completely rejected. The algorithms discussed next do not suffer from this drawback.

Simply generating an ASR that includes k clients is not sufficient for SKA. Consider for instance an algorithm, called *Center Cloak* in the sequel, that given a query from U, finds his $k - 1$ closest users and sets the ASR as the MBR that encloses them. In fact, a similar technique is proposed in [4] for anonymization in peer-to-peer systems (i.e., the ASR contains the query issuing peer and its $k - 1$ nearest nodes). However, by construction, the querying user U is often closest to the ASR center. Thus, a simple "center-of-ASR" attack would correctly guess U with probability that far exceeds $1/k$, especially for large k values.

Nearest Neighbor Cloak (NN-Cloak) [9] is a randomized variant of *Center Cloak*, which is not vulnerable to the *center-of-ASR attack*. Given a query from U, *NN-Cloak* first determines the set S_0 containing U and his $k - 1$ nearest users. Then, it selects a random user U_i from S_0 and computes the set S_1, which includes U_i and his $k - 1$ nearest neighbors (NN). Finally, *NN-Cloak* obtains $S_2 = S_1 \cup U$. This step is essential, since U is not necessarily among the NNs of U_i. The ASR is the MBR enclosing all users in S_2. Figure 3 shows an example of *NN-Cloak*, where U_1 issues a query with $k = 3$. The two NNs of U_1 are U_2, U_3, and $S_0 = \{U_1, U_2, U_3\}$. *NN-Cloak* randomly chooses U_3 and issues a 2-NN query, forming $S_1 = \{U_3, U_4, U_5\}$. The 3-ASR is the MBR enclosing $S_2 = \{U_1, U_3, U_4, U_5\}$. It is not vulnerable to the *center-of-ASR attack* since the probability of U being near the center of the ASR is at most $1/k$ (due to the random choice).

In another approach called *Casper* [12], the anonymizer maintains the locations of the clients using a pyramid data structure, similar to a Quad-tree [14], where the minimum cell size corresponds to the anonymity resolution. Once the anonymizer receives a query from U, it uses a hash table on the user-id pointing to the lowest-level cell c where U lies. If c contains enough users (i.e., $|c| \geq k$), it becomes the ASR. Otherwise, the horizontal c_h and vertical c_v neighbors of c are retrieved. If the union of c with c_h or c_v contains at least k users, the corresponding union becomes the ASR. Else, *Casper* retrieves the parent of c and repeats this process recursively. Figure 4 shows an example. Cells are denoted by the coordinates of their lower-left and upper-right points. Assume a query q with $k = 2$. If q is issued by U_1 or U_2, the ASR is cell $\langle (0,2),(1,3) \rangle$. If q is issued by U_3 or U_4, the ASR is the union of cells $\langle (1,2),(2,3) \rangle \cup \langle (1,3),(2,4) \rangle$. Finally, if q is issued by U_5, the ASR is the entire data space.

Interval Cloak [7] is similar to *Casper* in terms of both the data structure used by the anonymizer (i.e., a Quad-tree) and the cloaking algorithm.

Spatial Anonymity, Fig. 3
Example of *NN-Cloak*

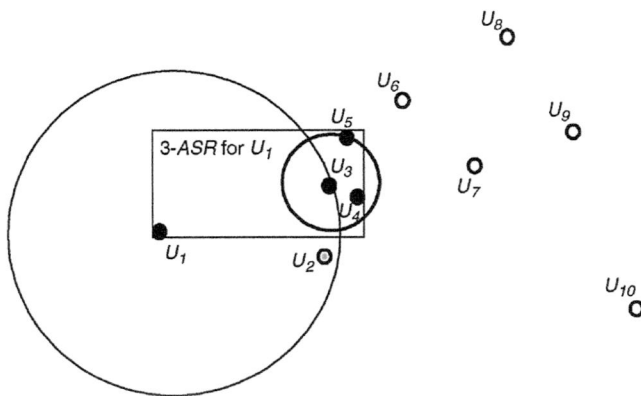

Recommended Reading

1. Beresford AR, Stajano F. Location privacy in pervasive computing. IEEE Pervasive Comput. 2003;2(1):46–55.
2. Cheng R, Zhang Y, Bertino E, Prabhakar S. Preserving user location privacy in mobile data management infrastructures. In: Proceeding of the 6th Workshop on Privacy Enhancing Technologies; 2006. p. 393–412.
3. Chow C-Y, Mokbel MF. Enabling private continuous queries for revealed user locations. In: Proceedings of the 10th International Symposium on Advances in Spatial and Temporal Databases; 2007. p. 258–75.
4. Chow C-Y, Mokbel MF, Liu X. A peer-to-peer spatial cloaking algorithm for anonymous location-based services. In: Proceedings of the 14th ACM International Symposium on Geographic Information Systems; 2006. p. 171–8.
5. Gedik B, Liu L. Location privacy in mobile systems: a personalized anonymization model. In: Proceedings of the 23rd International Conference on Distributed Computing Systems; 2005. p. 620–9.
6. Ghinita G, Kalnis P, Skiadopoulos S. PRIVE: anonymous location-based queries in distributed mobile systems. In: Proceedings of the 16th International World Wide Web Conference; 2007. p. 371–80.
7. Gruteser M, Grunwald D. Anonymous usage of location-based services through spatial and temporal cloaking. In: Proceedings of the 1st International Conference on Mobile Systems, Applications and Services; 2003. p. 31–42.
8. Hoh B, Gruteser M. Protecting location privacy through path confusion. In: Proceedings of the 1st International Conference on Security and Privacy for Emerging Areas in Communication Networks; 2005.
9. Kalnis P, Ghinita G, Mouratidis K, Papadias D. Preventing location-based identity inference in anonymous spatial queries. IEEE Trans Knowl Data Eng. 2007;19(12):1719–33.
10. Kamat P, Zhang Y, Trappe W, Ozturk C. Enhancing source-location privacy in sensor network routing. In: Proceedings of the 23rd International Conference on Distributed Computing Systems; 2005. p. 599–608.
11. Khoshgozaran A, Shahabi C. Blind evaluation of nearest neighbor queries using space transformation to preserve location privacy. In: Proceedings of the 10th International Symposium on Advances in Spatial and Temporal Databases; 2007. p. 239–57.
12. Mokbel MF, Chow CY, Aref WG. The new Casper: query processing for location services without compromising privacy. In: Proceedings of the 32nd International Conference on Very Large Data Bases; 2006. p. 763–74.
13. Moon B, Jagadish HV, Faloutsos C. Analysis of the clustering properties of the Hilbert space-filling curve. IEEE Trans Knowl Data Eng. 2001;13(1):124–41.
14. Samet H. The design and analysis of spatial data structures. New York: Addison-Wesley; 1990.
15. Sweeney L. k-Anonymity: a model for protecting privacy. Int J Uncertain Fuzziness Knowl Based Syst. 2002;10(5):557–70.

Spatial Data Analysis

Michael F. Goodchild
University of California-Santa Barbara, Santa Barbara, CA, USA

Synonyms

Geographical analysis; Geographical data analysis; Spatial analysis

Definition

Methods of data analysis perform logical or mathematical manipulations on data in order to test hypotheses, expose anomalies or patterns, or create summaries or views that expose particular traits. Data often refer to specific locations in some space. To qualify as spatial, the locations must be known and must affect the outcome of the analysis. While many spaces might be relevant, including the space of the human brain or the space of the human genome, the history of spatial data analysis is dominated by location in geographic space, in other words location on or near the surface of the Earth. Thus, geographical and spatial are often essentially synonymous. More formally, spatial data analysis can be defined as a set of techniques devised for the manipulation of data whose outcomes are not invariant under relocation of the objects of interest in some space. The term *exploratory spatial data analysis* (ESDA) describes an important subset that emphasizes real-time interaction, the creation of multiple views of data, the search for patterns and anomalies, and the generation of new hypotheses as opposed to the formal testing of existing ideas.

The term *spatial data mining* describes another important subset that emphasizes the analysis of very large volumes of spatial data.

Historical Background

Berry and Marble [2] made one of the earliest efforts to assemble a systematic review of methods of spatial data analysis, drawing on a literature that had accumulated for many decades. Their interest was sparked in large part by what later became known as the Quantitative Revolution in Geography, a paradigm shift that originated at the University of Washington in the late 1950s and spread rapidly as the original group of graduate students found faculty positions. Bunge [3] summarized the core concept: that the analysis of patterns of phenomena on the Earth's surface could lead to a set of formal theories about the behavior of human and natural systems, and that the discovery of such theories would put the discipline of geography on a sound scientific footing. Substantial progress was made in the 1960s, particularly in the study of patterns of settlement and economic activity, and in the study of such physical phenomena as meandering rivers and stream channel networks.

Beginning in the 1960s, the development of geographic information systems (GIS) provided a major impetus, by creating a simple structure in which methods of spatial data analysis could be implemented. By the 1980s, GIS had become a popular and rapidly growing software application, with a flourishing industry and tools to enable spatial data analysis, along with the necessary techniques for data acquisition, editing, and display. Today, GIS is often portrayed as an engine for spatial data analysis, and many new techniques have been added to what are now literally thousands of methods. GIS finds application in virtually all disciplines that deal with the surface and near-surface of the Earth, ranging from ecology and geology to sociology and political science [7]. It is extensively used in logistics, in planning and public decision making, in military and intelligence applications, and in the management of utility networks.

While the use of computers to perform spatial data analysis was already well established in the 1960s, ESDA emerged rather later, when the graphics and interactive capabilities of computers had advanced sufficiently. By the early 1990s, researchers were developing novel ways of linking multiple views using the windowing techniques that emerged at that time, and exploiting the high-resolution graphics that became available on standard personal computers. Today, interactive tools inspired by ESDA are widely available in GIS products, and more specialized software is also available (see, for example, GeoDa, http://geoda.uiuc.edu).

Interest in spatial data mining has grown in the past decade, driven in part by the increasing availability of very large volumes of spatial data. For example, it is now routine to capture the location and time of use of credit and debit cards, and to apply sophisticated algorithms in an effort to detect fraudulent use. Heavy use of spatial data analysis is made by intelligence agencies, based on software that can examine telephone and email traffic and detect references to places.

Foundations

Several approaches have been devised for organizing the thousands of techniques that qualify as spatial data analysis. Perhaps the commonest, represented by several recent textbooks and by the organization of some GIS user interfaces, is based on a taxonomy of spatial data types. Very broadly, one can capture variation within a space using either raster or vector structures; a raster structure is created by dividing the space into discrete, regularly shaped elements and describing the contents of each, while vector structures describe each feature present in the space as either a point, line, area, or volume, with associated attributes.

Tomlin [9] and others have systematized the analysis of raster data in schemata described as map algebras, image algebras, or cartographic modeling, and several GIS have adopted these schemata in their user interfaces. In one such schema the analysis of raster data are described

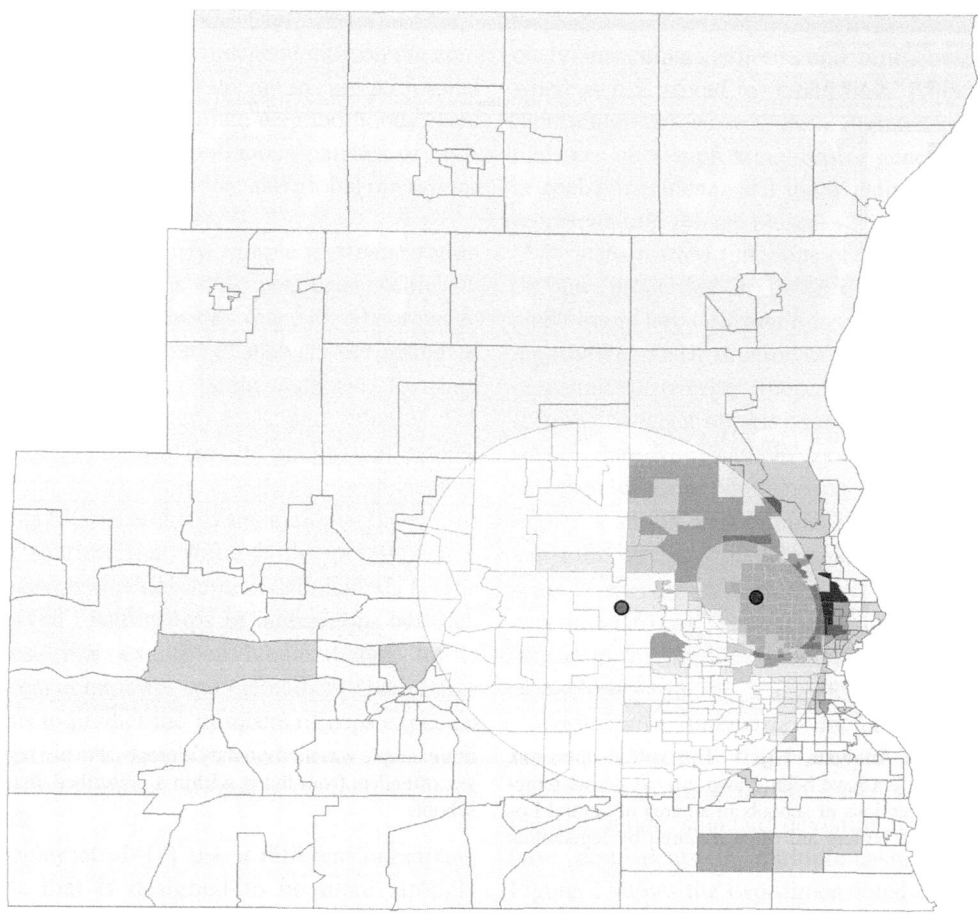

Spatial Data Analysis, Fig. 2 Two-dimensional equivalents of the mean and standard deviation. The larger ellipse shows the dispersion of the white population of Milwaukee around its centroid; the smaller ellipse shows the greater concentration of the city's black population. The map shows percent black, using 1990 data by census tract

Many important applications have derived from the need to understand the mechanisms of disease, and particularly its transmission within human populations. The work of Dr. John Snow on cholera [8] is often cited as the seminal example, but today methods of spatial data analysis are routinely used to scan data on such diseases as cancer, searching for anomalous clusters and thus for potential causal mechanisms. Spatial data analysis has been central to the study of outbreaks of new diseases such as West Nile virus and SARS.

Spatial data analysis has also been central to the study of landscape change, and related phenomena of urban sprawl, deforestation,

desertification, and habitat destruction. Such analyses are often based on snapshots of landscape obtained from Earth-orbiting satellites, and can form the basis for sophisticated models of landscape change that can be used to investigate the future effects of management alternatives.

Transportation applications are also particularly rich. Methods of spatial data analysis are routinely used to model traffic patterns, and to evaluate planning options, including new roads, mass transit, and congestion pricing. The possibility of real-time tracking of vehicles using GPS has recently given this field new impetus.

Future Directions

The insights that can be obtained from spatial data analysis are limited by its essentially cross-sectional nature - the need to draw inferences from snapshots obtained at one point in time. It is difficult, for example, to ascribe cause when no information is available about change through time. Thus there is great interest in the development of an improved suite of methods for *spatiotemporal* data analysis. In the past, the lack of suitable data has been a major impediment, but today vast new sources are becoming available as the result of developments in satellite remote sensing, GPS tracking, and Internet-based data sharing.

GIS owes much of its original stimulus to the paper map, which is of necessity flat. At global scales, analysis based on flattened or *projected* views of the Earth's surface can be misleading, and there is therefore strong interest in developing methods of spatial data analysis for the Earth's curved surface. This interest has been stimulated in part by the recent emergence of *virtual globes*, including Google Earth.

Cross-References

▶ Geographic Information System
▶ Spatial Data Mining
▶ Spatial Data Types
▶ Spatial Operations and Map Operations

Recommended Reading

1. Bailey TC, Gatrell AC. Interactive spatial data analysis. New York/Harlow: Longman Scientific & Technical; 1995.
2. Berry BJL, Marble DF. Spatial analysis: a reader in statistical geography. Englewood Cliffs: Prentice Hall; 1968.
3. Bunge W. Theoretical geography. Gleerup: University of Lund; 1966.
4. Fotheringham AS, Brunsdon C, Charlton M. Geographically weighted regression: the analysis of spatially varying relationships. Chichester/Hoboken: Wiley; 2002.
5. Haining RP. Spatial data analysis: theory and practice. Cambridge/ New York: Cambridge University Press; 2003.
6. Johnson S. The ghost map: the story of London's most terrifying epidemic and how it changed science, cities, and the modern world. New York: Riverhead Books; 2006.
7. Longley PA, Goodchild MF, Maguire DJ, Rhind DW. Geographic information systems and science. New York: Wiley; 2005.
8. O'Sullivan D, D.J U. Geographic information analysis. Hoboken: Wiley; 2003.
9. Tomlin CD. Geographic information systems and cartographic modeling. Englewood Cliffs: Prentice Hall; 1990.

Spatial Data Mining

Shashi Shekhar[1], Zhe Jiang[2], James Kang[3], and Vijay Gandhi[3]
[1]Department of Computer Science, University of Minnesota, Minneapolis, MN, USA
[2]University of Alabama, Tuscaloosa, AL, USA
[3]University of Minnesota, Minneapolis, MN, USA

Synonyms

Co-locations; Hotspots; K-primary-route summarization; Location prediction; Spatial autocorrelation; Spatial data analysis; Spatial decision trees; Spatial outliers; Spatial statistics; Ring shaped hotspots

Definition

Spatial data mining [1–3] is the process of discovering nontrivial, interesting, and useful patterns in large spatial datasets. The most common spatial pattern families are co-locations, spatial hotspots, spatial outliers, and location predictions.

Figure 1 gives an example of a spatial hotspot pattern (in the green circle) detected by SaTScan [4] from 250 cholera cases (shown by red points) that occurred near Broad Street in London, 1854. Notice that discovering spatial hotspots here is a nontrivial process due to the irregular size and special shape of the pattern. In addition, not all incidents contribute to the hotspot (e.g., red points

Spatial Data Mining,
Fig. 3 An example of
spatial co-location patterns
(Source: Huang et al. [7])

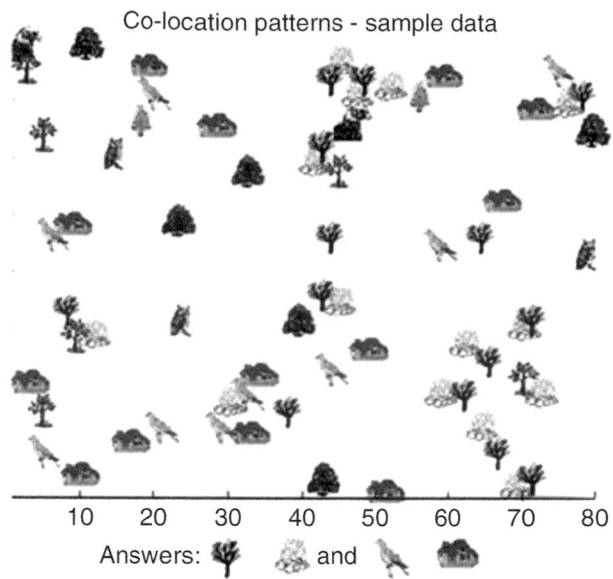

Spatial Data Mining,
Fig. 3 An example of
spatial co-location patterns
(Source: Huang et al. [7])

endangered species using maps of vegetation, water bodies, climate, and other related species. Figure 4 gives an example of a dataset used in building a location prediction model for red-winged blackbirds in the Darr and Stubble wet-lands on the shores of Lake Erie in Ohio, USA. This dataset consists of nest location, distance to open water, vegetation durability, and water depth maps. Classical prediction methods may be ineffective in this problem due to the presence of spatial autocorrelation. Spatial data mining tech-niques that capture the spatial autocorrelation of nest location such as the Spatial Autoregression Model (SAR) [5] and Markov Random Fields based Bayesian Classifiers (MRF-BC) are used for location prediction modeling. A comparison of these methods is discussed in [8].

Another problem similar to the above is the discovery of a model to infer a thematic map of different classes from the maps of other ex-planatory spatial features. For example, natu-ral resource management researchers may build models to map wetland distributions on the earth surface using remote sensing imagery. Classical prediction methods such as decision trees may be limited in this problem due to the presence of spatial autocorrelation in a target thematic class map. This may lead to salt-and-pepper noise, i.e., pixels classified differently from all neigh-

boring pixels. To address this limitation, focal-test-based spatial decision tree whose tree nodes not only test local information but also focal (neighborhood) information has been proposed [9]. Figure 4 gives an example of inputs and outputs of a decision tree and a spatial decision tree for wetland mapping in Chanhassen, MN, USA. The input feature maps consist of high-resolution (3×3 m) aerial photos (including R, G, B, near-infrared bands) collected in 2003 and 2008. The output of a decision tree (Fig. 5c) has lots of salt-and-pepper noise (e.g., black pixels highlighted by an ellipse). In contrast, the predic-tion of a spatial decision tree (Fig. 5d) is with less salt-and-pepper noise. More details on these two methods are discussed in [9].

Data summarization is an important topic in data mining for finding a compact representation of a dataset. In *spatial network activity summa-rization* [10], we are given a spatial network and a collection of activities (e.g., pedestrian fatality reports, crime reports), and the goal is to find k shortest paths that summarize the activities. This problem is important for applications where observations occur along linear paths such as roadways, train tracks, etc. For example, transportation planners and engineers need tools to assist them in identifying which frequently used road segments/stretches pose

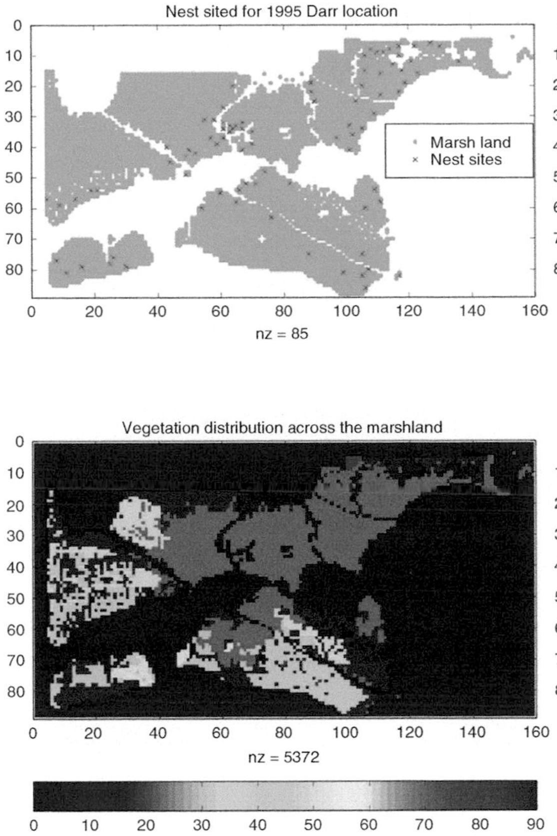

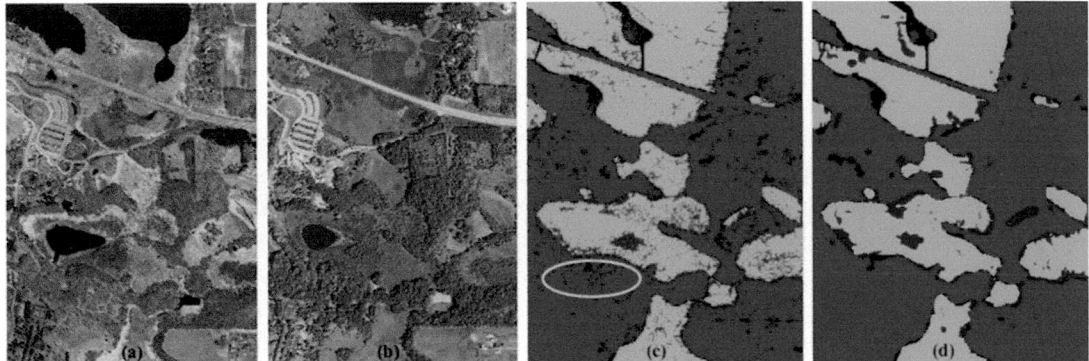

Spatial Data Mining, Fig. 4 (**a**) Learning dataset: the geometry of the Darr wetland and the locations of nests. (**b**) The spatial distribution of distance to open water. (**c**) The spatial distribution of vegetation durability over the marshland. (**d**) The spatial distribution of water depth (Source: Shekhar et al. [8])

Spatial Data Mining, Fig. 5 (**a**) Aerial photos in 2005. (**b**) Aerial photos in 2008. (**c**) Predicted class map by a traditional decision tree (*red*, true dry land; *green*, true wetland; *black*, false wetland; *blue*, false dry land). (**d**) Predicted class map by a spatial decision tree (Source: Jiang et al. [9], best viewed in color)

risks for pedestrians and consequently should be redesigned. Figure 6 shows a case study on pedestrian fatalities on a street network. As can be seen, the classical K-means approach, either Euclidean distance or network distance, cannot fully capture these network activities. In contrast, KMR can fully capture the linear patterns. For instance, the blue group and summary path

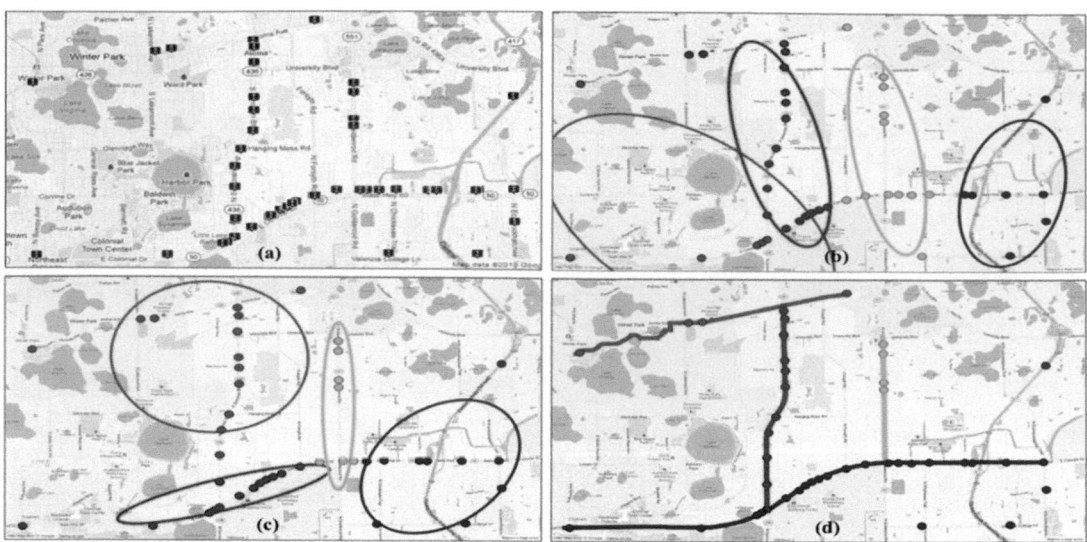

Spatial Data Mining, Fig. 6 (a) Forty-two pedestrian fatalities on a road network in Orlando, FL. (b) Crime-Stat K-means output (Euclidean space). (c) CrimeStat K-means output (network space). (d) K-main routes (Source: Oliver et al. [10], best viewed in color)

capture the activities on the arterial road that were split across three groups in K-means.

Spatial hotspots are areas where events or activities inside the areas are significantly more than outside. Examples of spatial hotspots can be concentration of crime events in a city or outbreaks of a disease. Hotspot patterns have properties of clustering as well as anomalies from classical data mining. However, hotspot discovery [11] remains a challenging area of research due to variation in shape, size, density of hotspots, and underlying space (e.g., Euclidean or spatial networks such as roadmaps). Additional challenges arise from the spatiotemporal semantics such as emerging hotspots, displacement, etc.

The *scan statistic* is a statistic test used to detect clusters in a point process. A *spatial scan statistic* is a generalization of scan statistics in high-dimensional spatial point processes. It uses a window with a predefined shape (e.g., circle) and varying sizes to scan the study area and computes a test statistic called *likelihood ratio*. The likelihood ratio is the ratio of likelihood of the alternative hypothesis (higher activity level inside the window) over the likelihood of the null hypothesis (same activity level inside and outside). A p-value indicating statistical significance is estimated through Monte Carlo simulations.

One popular scan statistics is SaTScan which detects significant circular hotspots. However, on network space (e.g., street networks), hotspot patterns may have linear shapes (e.g., routes) rather than circular shapes. [12] introduces a significant route discovery algorithm to detect significant linear hotspots from network activities. One example is in Fig. 7.

Key Applications

Spatial data mining and the discovery of spatial patterns have applications in a number of areas. Detecting spatial outliers is useful in many applications of geographic information systems and spatial databases, including the domains of public safety, public health, climatology, and location-based services. As noted earlier, for example, spatial outlier applications may be used to identify defective or out of the ordinary (i.e., unusually behaving) sensors in a transportation system (e.g., Fig. 1). Spatial co-location discovery is useful in ecology in the analysis of animal and plant habitats to identify co-locations of predator-prey species, symbiotic species, or fire events with fuel and ignition sources. Location prediction may provide applications toward predicting the cli-

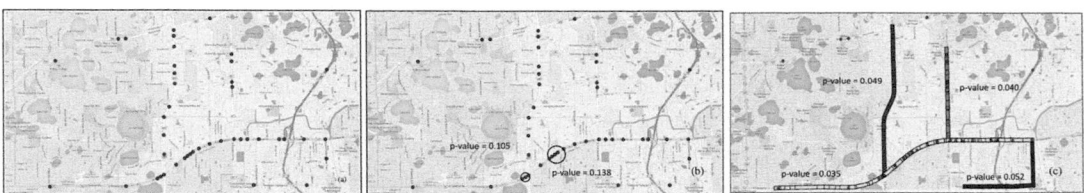

Spatial Data Mining, Fig. 7 (**a**) Pedestrian fatalities occurring on arterials in Orlando, FL, from 2000 to 2009. (**b**) Circular hotspots detected by SaTScan. (**c**) Linear hotspots detected by the significant route discovery algorithm in [12] (Source: Oliver et al. [12], best viewed in color)

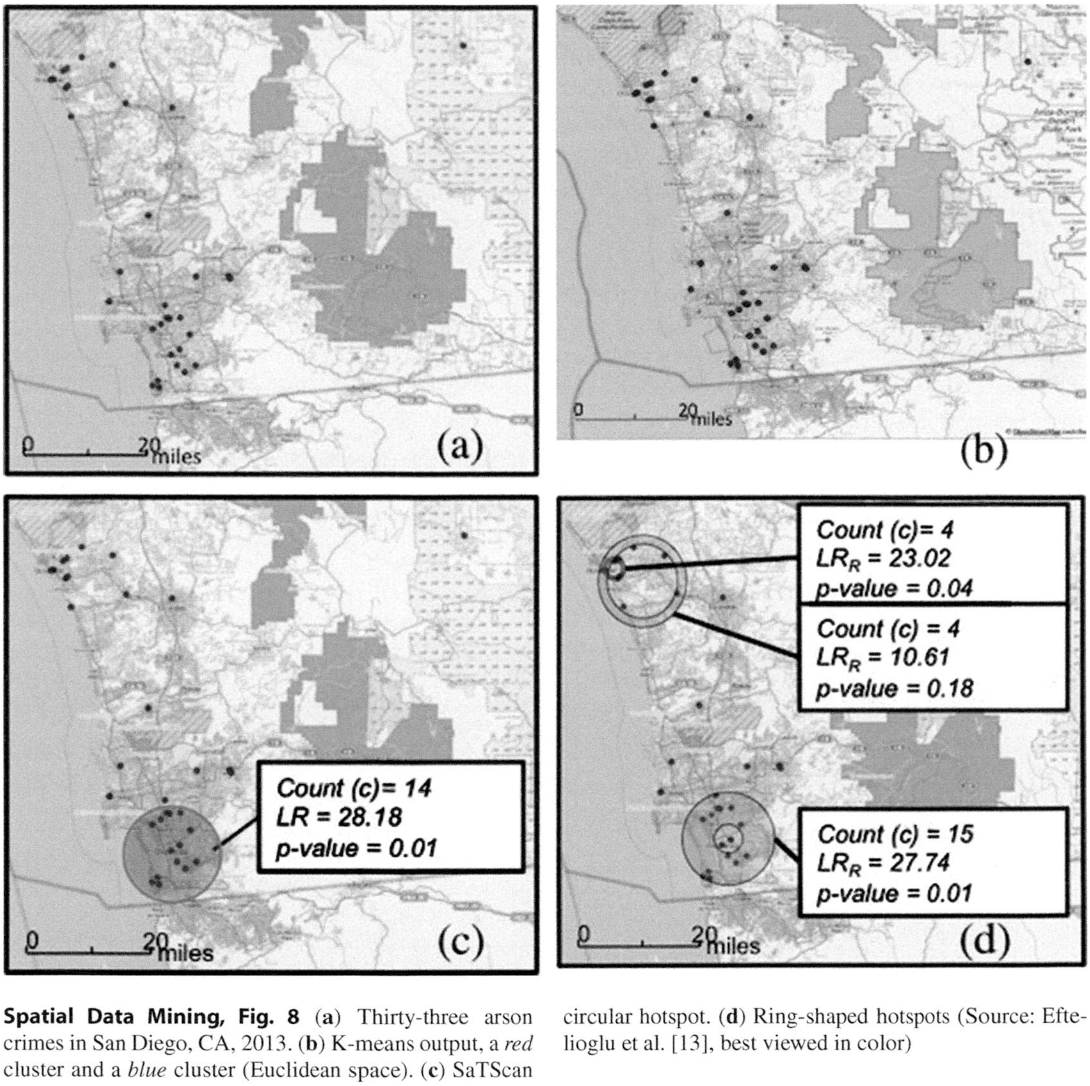

Spatial Data Mining, Fig. 8 (**a**) Thirty-three arson crimes in San Diego, CA, 2013. (**b**) K-means output, a *red* cluster and a *blue* cluster (Euclidean space). (**c**) SaTScan circular hotspot. (**d**) Ring-shaped hotspots (Source: Efte-lioglu et al. [13], best viewed in color)

matic effects of *El Nino* on locations around the world. Finally, identification of spatial hotspots can be used in crime prevention and reduction, as well as in epidemiological tracking of disease.

Future Directions

In this chapter, we have presented several major achievements in spatial data mining research.

Current research is mostly concentrated on mining spatial data in the Euclidean space. Developing spatial statistical models and spatial data mining methods on the network space (e.g., street maps, river networks) are still largely unexplored. These problems are important in many problems such as water monitoring in river networks and crime analysis on street maps. Networks pose new challenges due to the unique dependency structure, directionality, and distance metric. Future research in this area should be encouraged, e.g., how to generalize scan statistics on spatial networks. In addition, ring-shaped hotspot detection [13] is also an interesting and challenging problem. For example, Fig. 8 shows the arson crimes in San Diego, CA, in 2013. K-means clustering algorithm simply partitions the crime incidents into two clusters (in red and green colors). Circular hotspot detection algorithm, i.e., SaTScan, detects only one hotspot pattern (in green color). In contrast, ring-shaped hotspot detection algorithm finds two ring hotspots (in green) which are statistically significant. Another future direction is mining spatiotemporal data such as the outbreak of diseases and moving objects. Involving the time dimension adds new challenges such as temporal autocorrelation and temporal non-stationarity.

Cross-References

▶ Data Mining
▶ Geographic Information System
▶ Spatial Network Databases
▶ Spatiotemporal Data Mining

Recommended Reading

1. Shekhar S, Chawla S. A tour of spatial databases. Englewood-Cliffs: Prentice Hall; 2003.
2. Miller HJ, Han J. Geographic data mining and knowledge discovery. 2nd ed. Boca Raton: CRC Press; 2009.
3. Zhou X, Shekhar S, Ali R. Spatiotemporal change footprint pattern discovery: an inter-disciplinary survey. WIREs Interdiscip Rev: Data Min Knowl Disc(DMKD), 4, 1, 1–23, 2014.
4. Kulldorff M. A spatial scan statistic. Commun Stat-Theory Methods. 1997;26(6):1481–96.
5. Cressie NA. Statistics for spatial data. Rev ed. New York: Wiley; 1993.
6. Kou Y, Lu CT, Chen D. Algorithms for spatial outlier detection. In: Proceedings of the 3rd IEEE International Conference on Data Mining; 2003. p. 597–600.
7. Huang Y, Shekhar S, Xiong H. Discovering co-location patterns from spatial datasets: a general approach. IEEE Trans Knowl Data Eng. 2004;16(12):1472–85.
8. Shekhar S, Schrater P, Vatsavai R, Wu W, Chawla S. Spatial contextual classification and prediction models for mining geospatial data. IEEE Trans Multimed. (special issue on Multimedia Databases). 2002;4(2):174–88.
9. Jiang Z, Shekhar S, Zhou X, Knight J, Corcoran J. Focal-test-based spatial decision tree learning: a summary of results. In: Proceedings of the 13th IEEE International Conference on Data Mining; 2013. p. 320–9.
10. Oliver D, Shekhar S, Kang J, Laubscher R, Carlan V, Bannur A. A K-main routes approach to spatial network activity summarization. IEEE Trans Trans Knowl Data Eng. 2014;26(6):1464–78.
11. US Department of Justice – Mapping and Analysis for Public Safety report. Mapping crime: understanding hot spots. 2005. http://www.ncjrs.gov/pdffiles1/nij/209393.pdf.
12. Oliver D, Shekhar S, Zhou X, Eftelioglu E, Evans MR, Zhuang Q, Kang JM, Laubscher R, Farah C. Significant route discovery: a summary of results. In: Proceedings of the 8th International Conference on Geographic Information Science; 2014. p. 284–300.
13. Eftelioglu E, Shekhar S, Kang JM, Farah CC. Ring-shaped hotspot detection. IEEE Trans Knowl Data Eng. 2016;28(12):3367–81.
14. Longley PA, Goodchild M, Maquire DJ, Rhind DW. Geographic information systems and science. Chichester: Wiley; 2005.
15. Mamoulis N, Cao H, Cheung DW. Mining frequent spatio-temporal sequential patterns. In: Proceedings of the 5th IEEE International Conference on Data Mining; 2005. p. 82–9.
16. Shekhar S, Lu CT, Zhang P. A unified approach to detecting spatial outliers. GeoInformatica. 2003;7(2):139–66.
17. Shekhar S, Zhang P, Huang Y, Vatsavai R, Kargupta H, Joshi A, Sivakumar K, Yesha Y. Trend in spatial data mining. In: Data mining: next generation challenges and future directions. AAAI/MIT Press; 2003.
18. Solberg AH, Taxt T, Jain AK. A Markov random field model for classification of multisource satellite imagery. IEEE Trans Geosci Remote Sens. 1996;34(1):100–13.
19. Shekhar S, Evans M, Kang J, Mohan P. Identifying patterns in spatial information: a survey of methods. Wiley Interdiscip Rev: Data Min Knowl Disc. 2011;1(3):193–214.

20. Shekhar S, Gunturi V, Evans MR, Yang K. Spatial big-data challenges intersecting mobility and cloud computing. In: Proceedings of the 11th ACM International Workshop on Data Engineering for Wireless and Mobile Access; 2012. p. 1–6.

Spatial Data Types

Markus Schneider
University of Florida, Gainesville, FL, USA

Synonyms

Geometric data types

Definition

Data types are a well known concept in computer science (for example, in programming languages or in database systems). A *data type* defines a set of homogeneous values and the allowable operations on those values. An example is a type *integer* representing the set of 32-bit integers and including operations such as addition, subtraction, and multiplication that can be performed on integers. *Spatial data types* or *geometric data types* provide a fundamental abstraction for modeling the geometric structure of objects in space as well as their relationships, properties, and operations. They are of particular interest in *spatial databases* [4, 8, 12] and *Geographical Information Systems* [4]. One speaks of *spatial objects* as values of spatial data types. Examples are two-dimensional data types for *points* (for example, representing the locations of lighthouses in the U.S.), *lines* (for example, describing the ramifications of the Nile Delta), *regions* (for example, depicting air-polluted zones), *spatial networks* (for example, representing the routes of the Metro in New York), and *spatial partitions* (for example, describing the 50 states of the U.S. and their exclusively given topological relationships of adjacency or disjointedness) as well as

three-dimensional data types for *surfaces* (for example, modeling the shape of landscapes) or *volumes* (for example, representing urban areas). Operations on spatial data types include *spatial operations* like the geometric *intersection, union,* and *difference* of spatial objects, *numerical operations* like the *length* of a line or the *area* of a region, *topological relationships* checking the relative position of spatial objects to each other like *overlap, meet, disjoint,* or *inside,* and *cardinal direction relationships* like *north* or *southeast*.

Historical Background

In the late 1970s, the interest to store geometric data into databases arose. The success and efficiency of relational database technology for standard applications, which is rooted in its simple data model, its high-level query languages, and its well understood underlying theory, has led to many proposals to transfer this technology directly to geometric applications and to explicitly model the structure of spatial data as relations (tables). The consequence is that the user conceives spatial data in tabular form, just the same as standard data, and that a spatial object is represented by several or even many tuples. An example of such a relation schema is *RelName*(*id*: integer, x_1: integer, y_1: integer, x_2: integer, y_2: integer, *type*: string, <other information>) where $x_1, y_1, x_2,$ and y_2 are the coordinates of a point or a line segment. The flag *type* indicates whether a tuple describes a point, a single line, a line segment of a line, or a line segment of a polygon. The value *id* denotes the object identifier.

This approach has revealed a number of fundamental drawbacks. Since all lines and polygons are decomposed into a set of line segments (tuples) scattered over a relation, a spatial object is not treated as an entity or unit but only *corresponds* to several tuples. This is different compared to values of standard data types. A second drawback is that the approach forces the user to model complex spatial objects in flat, independent relations. Since the representations of spatial data occurs on a very low level and is exclusively based on standard domains like integers, strings, and reals (while the user has originally intended to deal with points, lines, or polygons (regions)), an adequate treatment of spatial data is impeded. Although the facilities of the query language of a DBMS are available, they are only of limited use. Since such a language is based on standard domains and has no concept of spatial data types, it cannot provide and support any meaningful geometric operations. A more detailed discussion can be found in [9].

Foundations

The numerous deficiencies of the approach of modeling spatial data as relations, have resulted in the assessment that this approach is unsuitable to manage spatial data in a clean and efficient manner, and that a high-level view of spatial objects is essential. This has led to the design of spatial data types that are represented as *abstract data types*, thus provide such a high-level view, and can be used as attribute data types in a database schema in the same way as standard data types like *integer, float*, or *string*. That is, the internal structure of a spatial object is hidden from the user, and its features can only be retrieved by (abstract) operations on this object.

One can distinguish different kinds of spatial data types. *Universal spatial data types* either only provide a single generic spatial data type called *spatial*, and therefore do not consider the dimensionality and shape of spatial objects, or they provide the types *spatial_, spatial_1, spatial_2*, and *spatial_3* and thus, consider the dimensionality but not the shape of spatial objects [6]. Another conceptual model for spatial data types is based on mathematical abstractions called *point sets*. The user is supplied with the concept that each spatial object consists of an infinite set of points that can be described by finite means. The approach in [7] introduces a type POINT-SET for point sets together with a collection of geometric operations. Aspects like dimensionality and shape of an object are not considered. A further approach of modeling spatial objects is that of using *half planes*

[13], where each half plane is defined by a *half plane segment*. A half plane segment uniquely determines a straight line which is given by an inequality, passes this segment and forms the one-sided boundary of a half plane. For constructing a polygonal region, an appropriately arranged sequence of intersection operations (conjunction of inequalities) defined on half planes is employed. This concept is the precursor of so-called *constraint spatial databases*. Most popular and fundamental abstractions of spatial objects fall into the category of *structure-based spatial data types*. These data types organize space into points, lines, regions, surfaces, volumes, spatial partitions, spatial networks, and similarly structured entities. Thus, this approach considers the structural shape and spatial extent of spatial objects, that is, their geometry. Spatial data types for points, lines, and regions have, for example, been considered in [1, 5, 6, 9, 11, 14], for surfaces and volumes in [10], for spatial partitions in [3], and for spatial networks in [12].

Structure-based spatial data types have prevailed and form the basis of a large number of data models and query languages for spatial data. They have also found broad acceptance in spatial extension packages of commercially and publicly available database systems, as well as in Geographical Information Systems. One can distinguish the older generation of *simple* spatial data types and the newer generation of *complex* spatial data types, depending on the spatial complexity the types are able to model. In the two-dimensional space, simple spatial data types only provide simple object structures like single points, continuous lines, and simple regions (Fig. 1a–c). However, from an application perspective, simple spatial data types have turned out to be inadequate abstractions for spatial applications, since they are insufficient to cope with the variety and complexity of geographic reality. From a formal perspective, they are not closed under the geometric set operations intersection, union, and difference. This means that these operations applied to two simple spatial objects can produce a spatial object that is not simple. Complex spatial data types solve these problems. They provide universal and versatile spatial objects

Spatial Data Types, Fig. 1 Examples of a simple point object (**a**), a simple line object (**b**), a simple region object (**c**), a complex point object (**d**), a complex line object (**e**), and a complex region object (**f**)

and are closed under geometric set operations. They allow objects with multiple components, region components that may have holes, and line components that may model ramified, connected geometric networks (Fig. 1d-f).

Even more complex structure-based spatial data types are spatial networks and spatial partitions. They are the essential components of maps. A *spatial network* (Fig. 2a) can be viewed as a spatially embedded graph which consists of a set of point objects representing its nodes and a set of line objects describing the geometry of its edges. Examples are highways, rivers, public transport systems, power lines, and phone lines. A *spatial partition* (Fig. 2b) is a set of region objects together with the topological constraint that any two regions either meet or are disjoint. The neighborhood relationship is of particular interest here since region objects may share common boundaries. Examples are states, school districts, crop fields, and land parcels. Both in spatial networks and in spatial partitions, their components (line objects, region objects) are annotated with thematic data like state name, unemployment rate, and parcel id.

Spatial operations manipulate spatial objects. They take spatial objects as operands and return either spatial objects or scalar values (like Boolean or numerical values) as results. One can classify them into the following categories:

- *Spatial predicates returning Boolean values.* A spatial relationship is a relationship between two or more spatial objects. A *spatial predicate* compares two spatial objects with respect to some spatial relationship and thus conforms to a binary relationship returning a Boolean value. Spatial predicates can be classified into three subcategories. *Topological predicates* characterize the relative position

of spatial objects towards each other and are preserved under topological transformations such as translation, rotation, and scaling; they do not depend on metric concepts like distance. Examples are the well known predicates *equal, disjoint, coveredBy, covers, overlap, meet, inside,* and *contains* between two simple regions. *Metric predicates* use measurements such as distances. For example, the predicates *in_circle* and *in_window* test if a spatial object is located within the scope of a predefined circle or rectangle. *Directional predicates* like *north* or *southeast* compare the cardinal direction of a target object with respect to a reference object.

- *Spatial operations returning numbers.* These operations compute metric properties of spatial objects and return a number. Examples are the operations *area* and *perimeter* computing the corresponding values of a region object, the operation *length* calculating the total length of a line object, the operation *diameter* determining the largest distance between any two object components, the operation dist computing the minimal distance between two spatial objects, and the *operation cardinality* yielding the number of components of a spatial object.

- *Spatial operations returning spatial objects.* These operations return spatial objects as results and can be subdivided into object construction operations, which construct new objects from existing objects, and object transformation operations, which transform one or more spatial objects into a new spatial object. The *object construction operations* include, for example, the *geometric set operations union, intersection,* and *difference,* which satisfy closure properties, the operation *convex_hull,* which constructs the smallest

S

convex region (polygon) enclosing a finite collection of points, the operation *boundary*, which returns the boundary of a region object as a line object, the operation *box*, which determines the minimal, axis-parallel rectangle (called *minimal bounding box* or *rectangle*) that bounds a spatial object, and the operation *components*, which extracts the vertices of a line object. Examples of *object transformation operations* are the operation *extend*, which takes a spatial object *s* and a real number *r* as operands and creates a polygonal region that is a spatial extension of *s* with distance *r* from *s* (also known as *buffer zoning*), the operation rotate, which rotates a spatial object around a point, and the operation translate, which moves a spatial object by a defined vector.

- *Spatial operations on spatial networks and spatial partitions*. An important operation on spatial networks is the *shortest_path* operator. It computes the route or path of minimum distance between a source and a destination. An important operation on spatial partitions is the *overlay* operation. It takes two spatial partitions modeling different themes as operands, lays them transparently on top of each other, and combines them into a new spatial partition by intersection. A large collection of other operations is available for both kinds of structures.

A brief example illustrates the embedding of a spatial data type into a relation schema and the posing of a spatial query. Consider the map of the 50 states of the USA. Besides its thematic attributes like name and population, each state is also described by a geometry which is a region. Cities can be represented as points, that is, one is here interested in their location and not so much in their extent. As thematic attributes, one could be interested in their name and population. In the following two relation schemas, the spatial data types *point* and *region* are used in the same way as attribute data types as standard data types.

- *states(sname: string, spop: integer, territory: region)*
- *cities(cname: string, cpop: integer, loc: point)*

A query could ask for all pairs of city names and state names where a city is located in a state. This can then be formulated as a *spatial join*:

- *select cname, sname*
- *from cities, states*
- *where loc inside territory*

The term *inside* is a topological predicate testing whether a point object is located inside a region object.

Key Applications

Spatial data types are a universal and general concept for representing geometric information in all kinds of spatial applications. Hence, they are not only applicable to a few key applications. In principle, all applications in the geosciences (for example, geography, hydrology, soil sciences) and Geographical Information Systems, as well as many applications in government and administration (for example, cadastral application, urban planning), can benefit from them. Independent studies have shown that about 80% of all data have spatial features (like geometric attributes) or a spatial reference (like an address). Thus, it is not surprising that independent international studies

Spatial Data Types, Fig. 2
Examples of a spatial network (**a**) and a spatial partition (**b**)

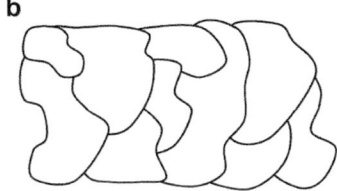

have predicted that geoinformation technology will belong to the most important and promising technologies in the future, besides biotechnology and nanotechnology.

Cross-References

▶ Cardinal Direction Relationships
▶ Dimension-Extended Topological Relationships
▶ Simplicial Complex
▶ Spatial Operations and Map Operations
▶ Three-Dimensional GIS and Geological Applications
▶ Topological Relationships

Recommended Reading

1. Clementini E, Di Felice P. A model for representing topological relationships between complex geometric features in spatial databases. Inf Sci. 1996; 90(1–4):121–36.
2. Egenhofer MJ. Spatial SQL: a query and presentation language. IEEE Trans Knowl Data Eng. 1994;6(1):86–94.
3. Erwig M, Schneider M. Partition and conquer. In: proceedings of the third international conference on spatial information theory; 1997. p. 389–408.
4. Güting RH. An introduction to spatial database systems. VLDB J. 1994;3(4):357–99.
5. Güting RH, Schneider M. Realm-based spatial data types: the ROSE algebra. VLDB J. 1995;4(2):243–286.
6. Güting RH Geo-relational algebra: a model and query language for geometric database systems. In: Advances in Database Technology, Proceedings of the 1st International Conference on Extending Database Technology; 1988. p. 506–27.
7. Manola F, Orenstein JA. Toward a general spatial data model for an object-oriented DBMS. In: Proceedings of the 12th International Conference on Very Large Data Bases; 1986. p. 328–35.
8. Rigaux P, Scholl M, Voisard A. Spatial databases - with applications to GIS. San Francisco: Morgan Kaufmann Publishers; 2002.
9. Schneider M. Spatial data types for database systems - finite resolution geometry for geographic information systems, vol. LNCS 1288. Berlin/New York: Springer; 1997.
10. Schneider M, Weinrich B An abstract model of three-dimensional spatial data types. In: Proceedings of the 12th ACM International Symposium on Geographic Information Systems; 2004. p. 67–72.
11. Schneider M, Behr T. Topological relationships between complex spatial objects. ACM Trans Database Syst. 2006;31(1):39–81.
12. Shekar S, Chawla S. Spatial databases: a tour. Upper Saddle River: Prentice-Hall; 2003.
13. Scholl M, Voisard A. Thematic map modeling. In: Proceedings of the 1st International Symposium on Advances in Spatial Databases; 1989. p. 167–90.
14. Worboys MF, Bofakos P. A canonical model for a class of areal spatial objects. In: Proceedings of the 3rd International Symposium on Advances in Spatial Databases; 1993. p. 36–52.
15. Worboys MF, Duckham M. GIS: a computing perspective. Boca Raton: CRC press; 2004.

Spatial Datawarehousing

Alejandro A. Vaisman[1] and Esteban Zimányi[2]
[1]Instituto Tecnológico de Buenos Aires, Buenos Aires, Argentina
[2]CoDE, Université Libre de Bruxelles, Brussels, Belgium

Definition

Business intelligence (BI) systems collect large amounts of data, transform them to a form that can be used to analyze organizational behavior, and store them in a common repository called a data warehouse (DW). A DW is usually designed following the multidimensional model, which represents data as facts that can be analyzed along a collection of dimensions, composed of levels conforming aggregation hierarchies.

Over the years, spatial data have been increasingly used in many application domains, like public administration, transportation networks, environmental systems, and public health, among others. Spatial data can represent geographic objects (e.g., mountains, cities), geographic phenomena (e.g., temperature, precipitation), and even data located in other spatial frames such as a human body or a house.

Similarly to conventional databases, spatial databases are typically used for operational applications, rather than to support data analysis tasks.

Spatial DWs (SDWs) emerged as a combination of spatial database and DW technologies, to provide sophisticated data analysis, visualization, and manipulation capabilities. Thus, *we define an SDW as a DW that is capable of representing, storing, and manipulating spatial data.* Usually, to represent the extent of a spatial object, spatial data types are used. A typical SDW query can thus take advantage of the operations associated with spatial data types. An example of such query is "Total number of ethnic restaurants in a radius of 10 blocks from my current position, aggregated by type of food."

Historical Background

Online analytical processing (OLAP) comprises a set of tools and algorithms that allow efficiently querying DWs [22]. Building on this concept, Rivest et al. [16] introduced the concept of Spatial Online Analytical Processing (SOLAP), a paradigm aimed at being able to explore spatial data by drilling on maps in an OLAP fashion. Stefanovic et al. [18] and Bédard et al. [2] classified spatial dimension hierarchies in (a) non-geometric, (b) geometric to non-geometric, and (c) fully geometric. Dimensions of type (a) can be treated as any descriptive dimension. In dimensions of types (b) and (c), a geometry is associated with members of the hierarchies. Malinowski and Zimányi [10] also proposed a model for SDW where a dimension level is spatial if it is represented as a spatial data type (e.g., point, region), and spatial levels are related to each other through topological relationships (e.g., contains, overlaps). A critical point in spatial dimension modeling is the problem of multiple dependencies, meaning that an element in one level can be related to more than one element in a level above it in the hierarchy. Jensen et al. [9] addressed this issue, proposing a model that supports different kinds of dimension hierarchies, most remarkably multiple hierarchies in the same dimension, although not considering geometric components. Malinowski and Zimányi [11] also proposed a model for SDW supporting multiple aggregation paths.

Spatial measures are characterized in two ways in the literature, namely, (a) measures representing a geometry, which can be aggregated along the dimensions and (b) a numerical value, using a topological or metric operator. Most proposals support option (a), either as a set of coordinates [2, 4, 10, 16] or a set of pointers to geometric objects [18].

Typically, SOLAP accounts only for discrete spatial data. Sophisticated GIS-based data analysis requires accounting also for continuous spatial data, denoted continuous fields. Continuous fields (or fields, for short) describe physical phenomena that change continuously in time and/or space, like temperature, land elevation, land use, and population density, frequently used in human geography. Conceptually, a field can be represented as a function that assigns to each point in space a value of a domain, for example, integer for altitude. In order to extend SOLAP capabilities to handle fields, spatial DW systems were proposed. In one of the first approaches, Shanmugasundaram et al. [17] proposed a data cube representation that deals with continuous dimensions, based on the idea of using the known data density to calculate aggregate queries without accessing the data. Ahmed and Miquel [1] discussed the importance of modeling multidimensional structures for field-based data and analyze how either cell values or interpolation methods can be used for inferring values at non-sampled points. Vaisman and Zimányi [19] presented a conceptual model for SOLAP that supports dimensions and measures representing continuous fields and characterized multidimensional queries over fields. They defined a field data type, a set of associated operations, and a multidimensional calculus supporting this data type. Gómez et al. [8] proposed physical data structures for implementing this set of operators. Further, Bimonte and Miquel [3] introduced a multidimensional model that supports measures and dimension as continuous field data, although they do not consider the fields as OLAP cubes, neither integrate the continuous and discrete models with an unique set of OLAP operators. Finally, in [5–7], the authors proposed a closed generic algebra over spatiotemporal continuous fields, independent of the underlying field representation and how this idea can be used for an integrated analysis of spatial and OLAP data.

Scientific Fundamentals

Spatial Dimensions and Facts

A *spatial object* corresponds to a real-world entity for which an application needs to store spatial characteristics. Spatial objects consist of a descriptive component and a spatial component. The descriptive component is represented using traditional data types, such as integer, string, and date; it contains general characteristics of the spatial object. For example, an object representing a state may be described by the state's name, population, and capital. The spatial component defines the extent of the object in the space of interest. Several spatial data types can be used to represent the spatial extent of real-world objects. Some of these data types are point, representing zero-dimensional geometries, denoting a single location in space (e.g., a village in a country); line, which represents one-dimensional geometries denoting a set of connected points defined by a continuous curve in the plane and can be used to represent, for instance, a road in a road network; PointSet, which represents sets of points, like tourist points of interest; or LineSet, representing sets of lines, for example, a road network. Spatial data types have a set of associated operations, like topological operations (e.g., intersects, disjoint, overlaps, contains), numeric operations (e.g., length, area, distance), and other ones, like boundary, buffer, intersection, etc.

An SDW is composed of the same elements as nonspatial ones, namely, dimensions, hierarchies, and facts. A *spatial level* is a level for which the application needs to store spatial characteristics. This is captured by its geometry, which is represented using a spatial data type. For example, a level city, in a geography dimension, can be spatial, and a level category in a dimension product is normally nonspatial. A *spatial attribute* is an attribute whose domain is a spatial data type, for example, an attribute elevation can be a continuous field representing the elevation of a country in a dimension level country. A *spatial hierarchy*

is a hierarchy that includes at least one spatial level. Spatial hierarchies can combine spatial and nonspatial levels. Finally, a *spatial dimension* is a dimension that includes at least one spatial hierarchy. The relationship between spatial levels is defined by topological constraints, the most usual one is the CoveredBy relationship, which indicates that the geometry of a child member is covered by the geometry of a parent member in the dimension hierarchy. In other words, spatial dimensions are standard dimensions referring to some geographical element (like cities or roads).

A *spatial fact* is a fact that relates several levels, two or more of which are spatial. A spatial fact may also have a topological constraint that must be satisfied by the related spatial levels.

Facts, whether spatial or not, are quantified by *measures* that are aggregated along a hierarchy. Measures can be numeric or *spatial*, where the latter are represented by a geometry. Note that numeric measures can be calculated using spatial operations, such as distance and area. Measures require the specification of the function used for aggregation. By default it is assumed sum for numerical measures and spatial union for spatial measures.

Types of Spatial Hierarchies

The types of hierarchies present in nonspatial data warehouses also apply to spatial ones. This way, we have:

- *Balanced spatial hierarchies*, for example, when a city is related to exactly one state in a geography hierarchy through a covering topological constraint.
- *Generalized spatial hierarchies*, which contain multiple exclusive paths that share some levels, for example, a road segment can be related either to district or to a company in charge of its maintenance.
- *Ragged spatial hierarchies*, for example, when a city can be related either to state or to a country.
- *Alternative spatial hierarchies*, composed of several nonexclusive spatial hierarchies sharing some levels.

S

- *Nonstrict spatial hierarchies*, which have at least one many-to-many relationship, for example, when a lake may belong to more than one city.

Spatiality and Measures

Spatial measures are measures represented by a geometry. For example, a spatial measure may represent the geometry (a line) of the part of a highway segment belonging to a county. Various kinds of aggregation functions for spatial data have been defined. Spatial measures allow richer analysis than nonspatial measures do. For example, we can analyze the locations of road accidents taking into account the various insurance categories (full coverage, partial coverage, and so on) and the client data, using a schema including a spatial measure representing the locations of accidents. We can use, for example, the default aggregate function (the spatial union), to roll up to the InsuranceCategory level in order to display the accident locations corresponding to each category, aggregated and represented as a set of points. Other aggregation functions can also be used for this, such as the center of *n* points.

Trajectory Data Warehouses

SDWs focus on the analysis of *static objects*, that is, the spatial features of these objects do not change (or change exceptionally) across time. However, many applications require the analysis of the so-called *moving objects*, that is, objects that change their position in space and time. Mobility analysis can be applied, for example, in traffic management, which requires to analyze traffic flows to capture their characteristics. Other applications aim at tracking the position of persons recorded by the electronic devices they carry, like smartphones, in order to analyze their behavior. Mobility data can be captured as a collection of sequences of an object's positions along their itineraries. Since such sequences can be very long, they are often processed by dividing them into segments. For instance, the movement of a car can be segmented with respect to the duration of the time intervals in which it stops at a certain location. These segments of

movement are called *trajectories*, and they are the unit of interest in the analysis of movement data. Extending DWs to cope with mobility data leads to *trajectory data warehouses* (TDW) [21]. Thus, *TDWs are DWs that contain data about the trajectories (either as facts or dimensions) of moving objects*. Such trajectories are typically analyzed in conjunction with other spatial data, for instance, a road network or continuous field data like elevation.

Mobility data can be represented as a collection of data types that capture the evolution over time of base types (e.g., integers, reals) and spatial types. These types are referred to as temporal types, that have an associated set of operations, like *duration*, *length*, and so on. In a TDW, trajectory segments are typically represented as facts, associated with dimensions. For example, each road segment can be related to a location dimension, a truck dimension, and a car dimension, allowing to analyze the movement of cars and trucks along different roads and highways. However, trajectories can also be represented as dimensions. In this case, each element in a dimension is the trajectory or a moving object [20].

Key Applications

Nowadays, organizations need sophisticated GIS-based DSSs to analyze their data with respect to geographic information. Modern DSSs must be able to support the analysis of spatial data not only represented in an alphanumeric way (addressing queries like "Total sales of cars in California") but also complex queries involving geometric components (like "Total sales in all villages crossed by the Mississippi river within a radius of 100 km around New Orleans"). Moreover, navigation of the results using typical OLAP operations like roll-up or drill-down is also required. Geographic information usually comes in the form of maps, and it is normally organized in different thematic layers. Some of these layers can represent discrete data (e.g., political division of a country, location of places of interest), while other ones can be continuous data, representing temperature, precipitation, altitude, etc. In this way, complex analysis can be performed. For example, the comprehensive management of airports requires information not only about the flights of the different airlines (i.e., alphanumerical information) but also about air routes (i.e., discrete spatial information) and weather conditions, like pressure, humidity, convective activity, and visibility, among others (i.e., continuous spatial information).

Moreover, the analysis of *historical data* is crucial for strategic management and decision-making in different kinds of organizations, from commercial companies to governmental or civil entities. This also applies to spatial data. For example, the management on the *evolution of geographical reserves* of a country over time requires the integration of alphanumeric records containing data about each zone (e.g., date of creation of the reserve, description of flora and fauna), as well as spatiotemporal data describing the evolution of the boundaries of each zone zone across time, satellite images of precipitation, temperature, and so on. As another example, *cadastral* and/or *land use applications* also require accounting for spatiotemporal data, since parcels of land (or land plots) can be split or merged. As a final example, in *traffic analysis* we may need to account for the positions of moving points, which represent pedestrians moving in a city or cars in a highway.

Future Directions

Unlike in the early days of DSSs, the proliferation of useful data nowadays exceeds the boundaries of the organizations. Moreover, the data to be incorporated into analysis are very complex and of different kinds and may involve images, geographic features, satellite maps, and so on. For example, a correlation study between a certain disease and environmental factors, such as air pollution and electromagnetic radiation emitted by telecommunication antennas, requires working with data from hospital records, social networks, pollution data, and maps reporting intensity distribution of radio-frequency fields. As we can see, most of these data are spatial and can be available in different representations. Although at present time data access is facilitated by the advance of the Internet and of the collaborative networks, the great amount and high complexity of such data makes their manipulation very difficult for end users. Therefore, future research in SDW must address ways to facilitate the seamless integration of such an overwhelming amount and variety of data.

Cross-References

▶ Data Warehouse Maintenance, Evolution, and Versioning
▶ Data Warehousing Systems: Foundations and Architectures
▶ Field-Based Spatial Modeling
▶ Geographic Information System
▶ Spatial Data Types
▶ Spatiotemporal Data Warehouses
▶ Spatiotemporal Trajectories
▶ Topological Data Models
▶ Topological Relationships

Recommended Reading

1. Ahmed TO, Miquel M. Multidimensional structures dedicated to continuous spatiotemporal phenomena. In: Proceedings of the 22nd British National Conference on Databases; 2005. p. 29–40.
2. Bédard Y, Merrett T, Han J. Fundaments of spatial data warehousing for geographic knowledge discovery. In: Miller H, Han J, editors, Geographic data mining and knowledge discovery. London: Taylor & Francis; 2001. p. 53–73.
3. Bimonte S, Miquel M. When spatial analysis meets OLAP: multidimensional model and operators. Int J Data Wareh Min. 2010;6(4):33–60.
4. Bimonte S, Tchounikine A, Miquel M. Towards a spatial multidimensional model. In: Proceedings of the 8th ACM International Workshop on Data Warehousing and OLAP; 2005. p. 39–46.
5. Gómez L, Gómez S, Vaisman A. Analyzing continuous fields with OLAP cubes. In: Proceedings of the 14th ACM International Workshop on Data Warehousing and OLAP; 2011. p. 89–94.
6. Gómez L, Gómez S, Vaisman A. A generic data model and query language for spatiotemporal OLAP cube analysis. In: Proceedings of the 15th Interna-

S

tional Conference on Extending Database Technology; 2012. p. 300–11.
7. Gómez L, Gómez S, Vaisman A. Modeling and querying continuous fields with OLAP cubes. Int J Data Wareh Min. 2013;9(3):22–45.
8. Gómez L, Vaisman A, Zimányi E. Physical design and implementation of spatial data warehouses supporting continuous fields. In: Proceedings of the 12th International Conference on Data Warehousing and Knowledge Discovery; 2010. p. 25–39.
9. Jensen CS, Klygis A, Pedersen TB, Timko I. Multidimensional data modeling for location-based services. In: Proceedings of the 10th ACM Symposium on Advances in Geographic Information Systems; 2002. p. 55–61.
10. Malinowski E, Zimányi E. Representing spatiality in a conceptual multidimensional model. In: Proceedings of the 12th ACM Symposium on Advances in Geographic Information Systems; 2004. p. 12–21.
11. Malinowski E, Zimányi E. Logical representation of a conceptual model for spatial data warehouses. GeoInformatica 2007;11(4):431–57.
12. Malinowski E, Zimányi E. Advanced data warehouse design: from conventional to spatial and temporal applications. Berlin/Heidelberg: Springer; 2008.
13. Mendelzon A, Vaisman A. Time in multidimensional databases. In: Rafanelli M, editor. Multidimensional databases: problems and solutions. Hershey: Idea Group; 2003. p. 166–99.
14. Parent C, Spaccapietra S, Zimányi E. Conceptual modeling for traditional and spatio-temporal applications: the MADS approach. Berlin: Springer; 2006.
15. Pedersen T, Jensen C, Dyreson C. A foundation for capturing and querying complex multidimensional data. Inf Syst. 2001;26(5):383–423.
16. Rivest S, Bédard Y, Marchand P. Toward better support for spatial decision making: defining the characteristics of spatial on-line analytical processing (SOLAP). Geomatica 2001;55(4):539–55.
17. Shanmugasundaram J, Fayyad U, Bradley P. Compressed data cubes for OLAP aggregate query approximation on continuous dimensions. In: Proceedings of the 5th ACM SIGKDD International Conference on Knowledge Discovery and Data Mining; 1999. p. 223–32.
18. Stefanovic N, Han J, Koperski K. Object-based selective materialization for efficient implementation of spatial data cubes. IEEE Trans Knowl Data Eng. 2000;12(6):938–58.
19. Vaisman A, Zimányi E. A multidimensional model representing continuous fields in spatial data warehouses. In: Proceedings of the 17th ACM SIGSPATIAL Symposium on Advances in Geographic Information Systems; 2009. p. 168–77.
20. Vaisman A, Zimányi E. Data warehouses: next challenges. In: Aufare MA, Zimányi E, editors. Tutorial Lectures of the First European Business Intelligence Summer School, eBISS 2011; Paris. Lecture Notes in Business Information Processing 96. Springer; 2011. p. 1–26.
21. Vaisman A, Zimányi E. Trajectory data warehouses. In: Renso C, Spaccapietra S, Zimányi E, editors. Mobility data: modeling, management, and understanding. New York: Cambridge University Press; 2013. p. 62–82.
22. Vaisman A, Zimányi E. Data warehouse systems: design and implementation. Berlin: Springer; 2014.

Spatial Indexing Techniques

Yannis Manolopoulos[1], Yannis Theodoridis[2], and Vassilis J. Tsotras[3]
[1] Aristotle University of Thessaloniki, Thessaloniki, Greece
[2] University of Piraeus, Piraeus, Greece
[3] University of California-Riverside, Riverside, CA, USA

Synonyms

Spatial access methods

Definition

A spatial index is a data structure designed to enable fast access to spatial data. Spatial data come in various forms, the most common being points, lines, and regions in n-dimensional space (practically, $n = 2$ or 3 in geographical information system (GIS) applications). Typical "selection" queries include the *spatial range query* ("find all objects that lie within a given query region") and the *spatial point query* ("find all objects that contain a given query point"). In addition, multidimensional data introduce spatial relationships (such as overlapping and disjointness) and operators (e.g., nearest neighbor), which need to be efficiently supported as well. Example queries are the *spatial join query* ("find all pairs of objects that intersect each other") and the *nearest neighbor query* ("find the five objects nearest to a given query point"). It should be noted that traditional indexing approaches (B^+-trees, hashing, etc.) are not appropriate for indexing spatial data; the basic reason is the lack of total ordering, which is an inherent characteristic in a multidimensional space. As a result, specialized access methods are necessary.

Historical Background

Many applications (VLSI, CAD/CAM, GIS, multimedia) need to represent, store, and manipulate spatial data types, such as points, lines, and regions in n-dimensional space. Although the representation of this type of data may be straightforward in a traditional database system (e.g., a two-dimensional point may be represented as a pair of x- and y- numeric values), spatial relationships (e.g., overlapping) and operators (e.g., nearest neighbor) need to be efficiently supported as well. These spatial relationships and operators have led to a variety of interesting and more complex queries like spatial joins, nearest neighbors, etc. As a result, specialized access methods have been proposed in order to quickly answer the above complex queries, as well as spatial range/point queries.

Given the characteristics of spatial data, for each spatial operator the query object's geometry needs to be combined with each data object's geometry. Nevertheless, the processing of complex geometry representations, usually polygons, is very expensive in terms of CPU cost. For that reason, the object geometries are approximated (typically by minimum bounding rectangles (MBRs)), and these approximations are then stored in underlying indices while the actual geometry is stored separately. As a result, a two-step procedure is involved during query processing, consisting of a *filter step* and a *refinement step*. The question that arises is how the object approximations (MBRs) are organized in order to answer the hits and the candidates, i.e., the result of the filter step.

Various spatial indices have been proposed in the literature and can be divided in two categories: indices designed for multidimensional points and indices for multidimensional regions. Examples in the first category are the LSD-tree [7], the grid file [10], the hB-tree [9], the buddy tree [14], and the BV-tree [3]. The major representatives in second category are the R-tree [5] and the quadtree [2] and their variants.

Given the complexity of the indexing problem and the different requirements of the multiple applications that index spatial data, it is not clear which the best index is. Nevertheless, R-tree implementations have found their way into commercial DBMSs. This is mainly due to their simplicity and ease of implementation (their structure is an adaptation of the B^+-tree for spatial data), as well as their robust performance for many applications.

Foundations

This section first discusses indices for multidimensional points, while the description of major indices for multidimensional (non-point) regions follows.

Indices for Multidimensional Points

The LSD-tree (local split decision tree), proposed in [1], maintains a catalog that separates space in a collection of (non-equal sized) disjoint subspaces using the extended k-d tree structure. New entries are inserted into the appropriate bucket. When an overflow happens, then the bucket is split, and the information about the partition line (split dimension and split position) is stored in a directory. Thus the overall structure of the LSD-tree consists of data buckets and a directory tree. The directory tree is kept in main memory until it grows more than a threshold; then a sub-tree is stored in an external catalogue in order for the whole structure to remain balanced (an example appears in Fig. 1). Inserting a new entry (point) in the LSD-tree is straightforward since nodes are disjoint. However, the target node may overflow due to an insertion; a split procedure then takes place.

The LSD-tree is a *space-driven structure*, i.e., it decomposes the complete workspace. Other members of this family include the grid file [10] and the hB-tree [9]. On the other hand, *data-driven structures* only cover those parts of the

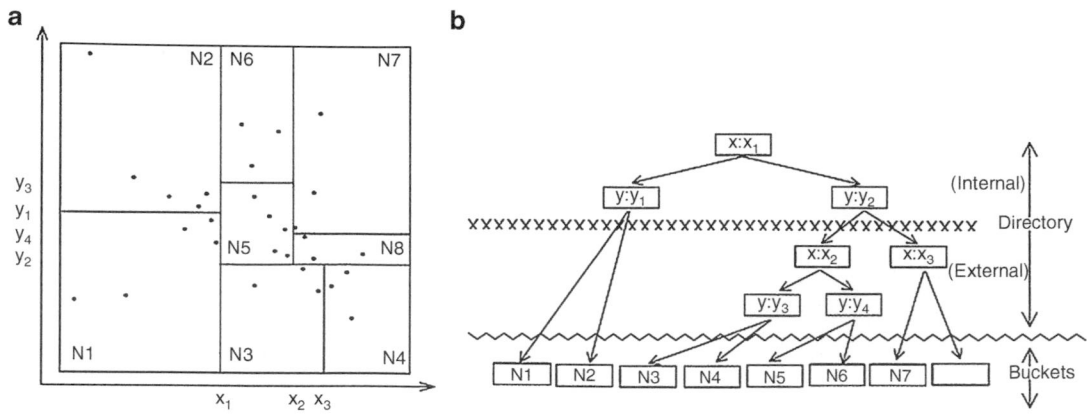

Spatial Indexing Techniques, Fig. 1 The LSD-tree

workspace that contain data objects. Examples are the buddy tree [14] and the BV-tree [3].

The grid file is an access method comprising of two separate parts: (i) the *directory* and (ii) the *linear scales*. The grid file imposes a grid on the indexed multidimensional attribute space. Each cell in this grid corresponds to one data page. The data points that "fall" inside a given cell are stored in the cell's corresponding page. Each cell must thus store a pointer to its corresponding page. This information is stored in the grid file's *directory*. The information of how each dimension is divided (and thus how data values are assigned to cells) is kept in the *linear scales*. The grid file can be thought as a multidimensional extension of hashing. As a result, exact match queries take only two disk accesses, one for the directory and one for the data page.

Indices for Multidimensional Regions

As with point indexing, two different approaches (data driven and space driven) have been proposed for indexing regions as well. The main representatives are the R-tree [5] and the quadtree, [2, 13], which were later followed by dozens of variants. In the sequel, the two structures are presented in detail. The reader is referred to a recent exhaustive survey [4] for further reading on their variants.

R-trees were originally proposed [5] as a direct extension of B$^+$-trees in n-dimensional space.

The data structure is a height-balanced tree that consists of intermediate and leaf nodes. A leaf node is a collection of entries of the form (o_id, R) where o_id is an object identifier, used to refer to an object in the database, and R is the minimum bounding rectangle (MBR) approximation of the data object. An intermediate node is a collection of entries of the form (ptr, R) where ptr is a pointer to a lower level node of the tree and R is a representation of the minimum rectangle that encloses all MBRs of the lower-level node entries. Let M be the maximum number of entries in a node and let $m \leq M/2$ be a parameter specifying the minimum number of entries in a node. An R-tree satisfies the following properties: (i) every leaf node contains between m and M entries unless it is the root; (ii) for each entry (o_id, R) in a leaf node, R is the MBR minimum bounding rectangle approximation of the object represented by o_id; (iii) every intermediate node has between m and M children unless it is the root; (iv) for each entry (ptr, R) in an intermediate node, R is the smallest rectangle that completely encloses the rectangles in the child node; (v) the root node has at least two children unless it is a leaf; and (vi) all leaves appear at the same level. As an example, Fig. 2 illustrates several minimum bounding rectangles (MBRs) m_i and the corresponding R-tree built on these rectangles (assuming maximum node capacity $M = 3$).

In order for a new entry E to be inserted into the R-tree, starting from the root node, the child

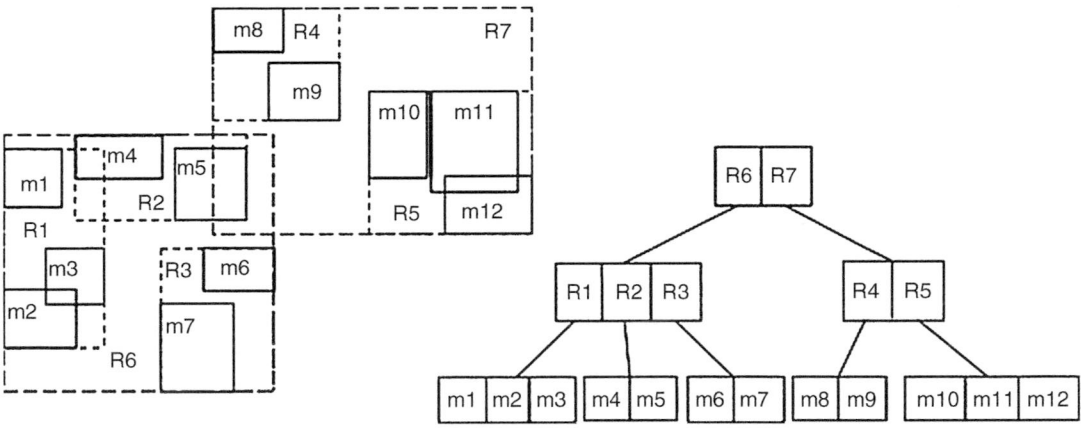

Spatial Indexing Techniques, Fig. 2 The R-tree

that needs minimum enlargement to include E is chosen (ties are resolved by choosing the one with the smallest area). When a leaf node N is reached, E is inserted into that, probably causing a split if N is already full. In such a case, the existing entries together with E are redistributed in two nodes (the current and a new one) with respect to the minimum enlargement criterion. In the original paper [6], three alternatives were proposed in order to find the two groups: an exhaustive, a quadratic-cost, and a linear-cost split algorithm. The processing of a point or range query with respect to a query window q (which could be either point or rectangle, respectively) is straightforward: starting from the root node, several tree nodes are traversed down to the leaves, depending on the result of the overlap operation between q and the corresponding node rectangles. When the search algorithm reaches the leaf nodes, all data rectangles that overlap the query window q are added to the answer set. Regarding k-nearest-neighbor queries, [12] proposed customized branch-and-bound algorithms for R-trees.

After Guttman's proposal, several researchers proposed their own improvements on the basic idea. Roussopoulos and Leifker [11] proposed the *packed* R-tree for bulk loading data in an R-tree. Objects are first sorted in some desirable order (according to the low-x value, low-y value, etc.) and then the R-tree is bulk loaded from the sorted file and R-tree nodes are packed to

capacity. Note that the above techniques allow node "overlapping": MBRs of different nodes can overlap. Since no disjointness is guaranteed, during a search multiple paths of the R-tree may be traversed. An efficient variation, namely, the R^+-tree, was proposed by Sellis et al. [15]. To preserve disjointness among node rectangles, the R^+-tree uses a "clipping" technique that duplicates data entries when necessary. However, the penalty is an (possibly high) increase in space demand due to the replication of data, which, in turn, degenerates search performance. Generally speaking, clipping techniques are ideal for point queries because a single path should be traversed, while range queries tend to be expensive, when compared with the overlapping techniques.

Later, Beckman et al. [1] and Kamel and Faloutsos [8] proposed two R-tree-based methods, the R*-tree and the Hilbert R-tree, respectively, which are currently considered to be the most efficient members of the R-tree family in terms of query performance. The R*-tree uses a rather complex but more effective grouping algorithm to split nodes by computing appropriate area, perimeter, and overlap values, while the Hilbert R-tree actually stores Hilbert values at the leaf level and ranges of those values at the upper levels, similarly to the B^+-tree construction algorithm. In addition, a "lazy" split technique is followed, where overflow entries are evenly distributed among sibling nodes, and only when all those are full, a new node (hence, split) is created.

The *region quadtree* [2] is probably the most popular member in the quadtree family. It is used for the representation of binary images, that is, $2^n \times 2^n$ binary arrays (for a positive integer n), where a "1" ("0") entry stands for a black (white) picture element. More precisely, it is a degree four tree with height n, at most. Each node corresponds to a square array of pixels (the root corresponds to the whole image). If all of them have the same color (black or white), the node is a leaf of that color. Otherwise, the node is colored gray and has four children. Each of these children corresponds to one of the four square sub-arrays to which the array of that node is partitioned. It is assumed here that the first (leftmost) child corresponds to the upper left sub-array, the second to the upper right sub-array, the third to the lower left sub-array, and the fourth (rightmost) child to the lower right sub-array, denoting the directions NW, NE, SW, and SE single ended, respectively. Figure 3 illustrates a quadtree for an 8×8 pixel array. Note that black (white) squares represent black (white) leaves, whereas circles represent internal nodes (also, *gray* ones).

Region quadtrees, as presented above, can be implemented as main memory tree structures (each node being represented as a record that points to its children). Variations of region quadtrees have been developed for secondary memory. *Linear region quadtrees* [8] are the ones used most extensively. A linear quadtree representation consists of a list of values where there is one value for each black node of the pointer-based quadtree. The value of a node is an address describing the position and size of the corresponding block in the image. These addresses can be stored in an efficient structure for secondary memory (such as a B^+-tree). There are also variations of this representation where white nodes are stored too or variations which are suitable for multicolor images. Evidently, this representation is very space efficient, although it is not suited to many useful algorithms that are designed for pointer-based quadtrees. The region quadtree [2] is probably the most popular linear implementations are the fixed length (FL), the fixed depth (FD), and the variable length (VL) linear implementations [13]. Techniques for computing various kinds of geometric properties have also been developed. Connected component labeling, polygon coloring, and computation of various types of perimeters fall in this category. Finally, many operations on images have been developed, for example, point location, set operations on two or more images (intersection, union, difference, etc.), window clipping, linear image transformations, and region expansion.

Other region quadtree variants have appeared in the literature mainly for indexing non-regional data. MX quadtrees are used for storing points seen as black pixels in a region quadtree. PR quadtrees are also used for points. However, points are drawn from a continuous space, in this case. MX-CIF quadtrees are used for small rectangles. Each rectangle is associated with the quadtree node corresponding to the smallest block that contains the rectangle. PMR quadtrees are used for line segments. Each segment is stored in the nodes that correspond to blocks intersected by the segment. A detailed presentation of these and other region quadtree variants is given in [8].

Spatial Indexing Techniques, Fig. 3 The quadtree

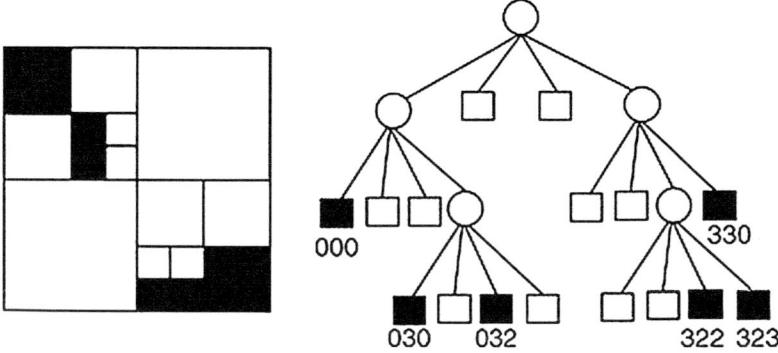

Key Applications

Geographic information systems (GIS) deal extensively with the management of two- and three-dimensional spatial data. For example, a map typically contains point objects (locations of interest), line objects (road segments, highways, rivers, etc.), as well as region objects (lakes, forests, etc.). GIS use spatial indexing as a means to provide fast access to large amounts of spatial data.

Multimedia systems manage multimedia objects like images, text, audio, video, etc. A typical query in such systems is the *similarity* query (i.e., find objects that are similar to a query object according to some measure). To answer these queries, each multimedia object is abstracted by a set of multidimensional points (features). These multidimensional points are then indexed by a spatial index. Similarly, the query object is represented by a multidimensional point. The similarity query is then answered as a nearest neighbor query (i.e., find the nearest neighbor(s) to the point that represents the query).

The World Wide Web has also provided new applications for geographic-related queries and thus spatial indexing. Users can now find maps, driving directions, etc. through specialized web sites that typically offer the ability to perform spatial queries.

Location-based services provide querying capabilities based on the location of the user (e.g., "find the cheapest gas station within 5 miles of my car"). As the user moves in space, the results of the queries change. Such queries typically have a spatial component, and spatial (and spatiotemporal) indexes are used to provide fast response.

CAD systems use spatial objects to store surfaces and bodies of design objects (for e.g., the wings or the wheels of an airplane). Typical spatial queries involve the proximity of spatial objects, their overlap, etc. Related queries (but mainly in the two-dimensional space) are also relevant for VLSI design systems; here the layout of a chip involves various rectangular regions and overlap and proximity queries are of importance.

Computer games also involve many spatial searches. In such an environment, players move around in a three-dimensional space and need to be able to see parts of (partially hidden) objects, various triggers are initiated if a player passed over them, or an explosion needs to identify the nearby objects that are affected. Spatial indexing is used to improve such query response.

Medical imaging also involves large amounts of two- and three-dimensional spatial data. Consider, for example, X-rays or magnetic resonance imaging (MRI) brain scans. Again, proximity, overlap, and related spatial queries are of interest.

Experimental Results

In general, for every presented method, there is an accompanying experimental evaluation in the corresponding reference.

Datasets

A large collection of real spatial datasets, commonly used for experiments, can be found at *R-tree portal* (URL: http://www.rtreeportal.org/).

URL to Code

R-tree portal (see above) contains the code for most common spatial and spatiotemporal indexes, as well as data generators and several useful links for researchers and practitioners in spatiotemporal databases. Similarly, the spatial index library [6] provides a general framework for developing various spatial indices (URL: http://dblab.cs.ucr.edu/spatialindexlib).

Cross-References

► B+-Tree
► Geographic Information System
► Grid File (and family)
► Nearest Neighbor Query
► Quadtrees (and Family)
► R-Tree (and Family)
► Spatial Join

Recommended Reading

1. Beckmann N, Kriegel H-P, Schneider R, Seeger B. The R*-tree: an efficient and robust access method for points and rectangles. In: Proceedings of the ACM SIGMOD International Conference on Management of Data; 1990. p. 322–31.
2. Finkel RA, Bentley JL. Quad Trees: a data structure for retrieval on composite keys. Acta Informatica. 1974;4(1):1–9.
3. Freeston MA. General solution of the n-dimensional B-tree problem. In: Proceedings of the ACM SIGMOD International Conference on Management of Data; 1995. p. 80–91.
4. Gaede V, Guenther O. Multidimensional access methods. ACM Comput Surv. 1998;30(2)·170–231.
5. Guttman A. R-trees: a dynamic index structure for spatial searching. In: Proceedings of the ACM SIGMOD International Conference on Management of Data; 1984. p. 47–57.
6. Hadjieleftheriou M, Hoel E, Tsotras VJ. SaIL: a spatial index library for efficient application integration. GeoInformatica. 2005;9(4):367–89.
7. Henrich A, Six H-W, Widmayer P. The LSD tree: spatial access to multidimensional point and non point objects. In: Proceedings of the 15th International Conference on Very Large Data Bases; 1989. p. 43–53.
8. Kamel I, Faloutsos C. Hilbert R-tree: an improved R-tree using fractals. In: Proceedings of the 20th International Conference on Very Large Data Bases; 1994. p. 500–09.
9. Lomet DB, Salzberg B. The hB-tree: a multiattribute indexing method with good guaranteed performance. ACM Trans Database Syst. 1990;15(4):625–58.
10. Nievergelt J, Hinterberger H, Sevcik KC. The grid file: an adaptable symmetric multikey file structure. ACM Trans Database Syst. 1984;9(1):38–71.
11. Roussopoulos N, Leifker D. Direct spatial search on pictorial databases using packed R-trees. In: Proceedings of the ACM SIGMOD International Conference on Management of Data; 1985. p. 17–31.
12. Roussopoulos N, Kelley S, Vincent F. Nearest neighbor queries. In: Proceedings of the ACM SIGMOD International Conference on Management of Data; 1995. p. 71–9.
13. Samet H. The design and analysis of spatial data structures. Addison-Wesley; Reading, MA, 1990.
14. Seeger B, Kriegel H-P. The Buddy-tree: an efficient and robust access method for spatial database systems. In: Proceedings of the 16th International Conference on Very Large Data Bases; 1990. p. 590–601.
15. Sellis T, Roussopoulos N, Faloutsos C. The R$^+$-tree: a dynamic index for multidimensional objects. In: Proceedings of the 13th International Conference on Very Large Data Bases; 1987. p. 507–18.

Spatial Join

Nikos Mamoulis
University of Hong Kong, Hong Kong, China

Definition

The spatial   join is one of the core operators in spatial database systems. Efficient spatial join evaluation is important, due to its high cost compared to other queries, like spatial selections and nearest neighbor searches. A binary (i.e., pairwise) spatial join combines two datasets with respect to a spatial predicate (usually overlap/intersect). A typical example is "find all pairs of cities and rivers that intersect." For instance, in Fig. 1 the result of the join between the set of cities $\{c_1, c_2, c_3, c_4, c_5\}$ and rivers $\{r_1, r_2\}$, is $\{(r_1, c_1), (r_2, c_2), (r_2, c_5)\}$.

The query in this example is a spatial intersection join. In the general case, the join predicate could be a combination of topological, directional, and distance spatial relations. Apart from the intersection join, variants of the distance join have received considerable attention, because they find application in data analysis tasks (e.g., data mining, clustering). Given two sets R and S of spatial objects (or multidimensional points), and a distance function $dist()$, the e-distance join (or else similarity join [1]) returns

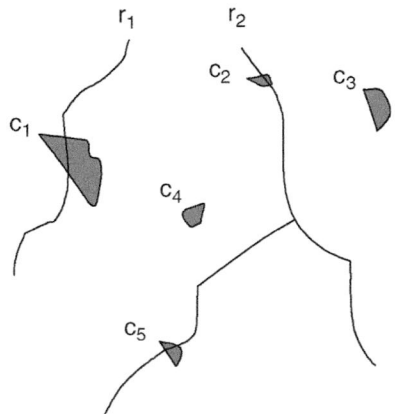

Spatial Join, Fig. 1 Graphical example of a spatial intersection join

the pairs of objects $\{(r, s) : r \in R, s \in S, dist(r, s) \leq e\}$. A closest pairs query [2] returns the set of closest pairs $CP = \{(r, s) : r \in R, s \in S\}$, such that $dist(r, s) \leq dist(r', s')$, for all $r' \in R, s' \in S : (r', s') \notin CP$.

Historical Background

The first spatial join methods [3, 4] assume that both inputs are indexed by some spatial access method (e.g., R-trees). The latest spatial join techniques do not rely on pre-existing indices [5–11]. Such situations may arise when at least one input is an intermediate result of a preceding operator. Consider for instance the query "find all rivers of width larger than 20 m, which intersect a forest." If there is a large percentage of narrow rivers, it might be natural to process the selection part of the query before the spatial join. In such an execution plan, even if there exists a spatial index on rivers, it is not employed by the join algorithm.

Table 1 classifies the spatial join techniques according to the assumption they make on pre-existing indices for the joined inputs. Methods of the first column can be applied only when both inputs are indexed (e.g., two relations Forests and Rivers are joined with respect to a spatial predicate). The second column includes algorithms suitable when only one input is indexed by an R-tree (e.g., Forests are joined with Rivers wider than 20 m). Join algorithms in the last column can be used in cases when both inputs are not indexed (e.g., Forests that intersect some City are joined with Rivers wider than 20 m). Most of the spatial join techniques focus on the filter step of the query. The refinement step (i.e., testing the exact geometry of objects against the join predicate) is applied independently of the algorithm, used for the filter step to the pairs that pass it, afterwards.

Foundations

Early Spatial Join Algorithms
Most early spatial join algorithms apply transformation of objects in order to overcome difficulties due to their spatial extent and dimensionality. The first known spatial join algorithm [4] uses a grid to regularly divide the multidimensional space into small blocks, called pixels, and uses a space-filling curve (z-ordering) to order them. Each object is then approximated by the set of pixels intersected by its MBR, i.e., a set of z-values. Since z-values are one-dimensional, the objects can be dynamically indexed using relational index structures like the B^+-tree. The spatial join is then performed in a sort-merge fashion. The performance of the algorithm depends on the granularity of the grid; larger grids can lead to finer object approximations, but also increase the space requirements. Other approaches transform the MBRs of the objects into higher dimensional points and use k-d-trees or grid-files to index the points. The join is then performed by the use of these data structures in a similar way as relational multi-attribute joins.

The R-Tree Join
R-tree Join (RJ) [2], often referred to as *tree matching or synchronous traversal*, computes the spatial join of two relations provided that they are both indexed by R-trees [5]. RJ synchronously traverses both trees, starting from the roots and

Spatial Join, Table 1 Classification of spatial join methods

Both inputs are indexed	One input is indexed	Neither input is indexed
Transformation to z-values and use of B-trees [4]	Indexed nested loops	Spatial hash join [7]
Synchronized tree traversal [3]	Seeded tree join [8]	Partition-based spatial merge join [11]
	Build a second R-tree and match it with the existing [10, 11]	Size separation spatial join [6]
	Sort and match [10]	Sweeping-based spatial join [5]
	Slot-index spatial join [9]	

following entry pairs which intersect. Let n_R, n_R be two directory (non-leaf) nodes of the R-trees that index relations R and S, respectively. RJ is based on the following observation: if two entries $e_i \in n_R$ and $e_j \in n_S$ do not intersect, there can be no pair (o_R, o_S) of intersecting objects, where o_R and o_S are under the sub-trees pointed by e_i and e_j, respectively. A simple pseudo-code for RJ that outputs the result of the filter spatial join step (i.e., outputs pairs of objects whose MBRs intersect) is given in Fig. 2. The pseudo-code assumes that both trees have the same height, yet it can be easily extended to the general case by applying range queries to the deeper tree when the leaf level of the shallow tree is reached.

Figure 3 illustrates two datasets indexed by R-trees. Initially, RJ is run taking the tree roots as parameters. The qualifying entry pairs at the root level are (A_1, B_1) and (A_2, B_2). Notice that since A_1 does not intersect B_2, there can be no object pairs under these entries that intersect. RJ is recursively called for the nodes pointed by the qualifying entries until the leaf level is reached, where the intersecting pairs (a_1, b_1) and (a_2, b_2) are output.

Two optimization techniques can be used to improve the CPU speed of RJ [2]. The first (search space restriction) reduces the quadratic number of pairs to be evaluated when two nodes n_R, n_S are joined. If an entry $e_R \in n_R$ does not intersect the MBR of n_S (that is the MBR of all entries contained in n_S), then there can be no entry $e_S \in n_S$, such that e_R and e_S overlap. Using this fact, space restriction performs two linear scans in the entries of both nodes before RJ, and prunes out from each node the entries that do not

intersect the MBR of the other node. The second technique, based on the plane sweep paradigm [15], applies sorting in one dimension in order to reduce the cost of computing overlapping pairs between the nodes to be joined. Plane sweep also saves I/Os compared to nested loops, because consecutive computed pairs overlap with high probability.

Algorithms that Do Not Consider Indexes

The most straightforward and intuitive algorithm that can be used to join two relations that are not indexed is the Nested Loops Join. This method can be applied for any type of joins (spatial, non-spatial) and condition predicates (topological, directional, distance, etc.). On the other hand, nested loops is the most expensive algorithm, since its cost is quadratic to the size of the relations (assuming that R and S have similar sizes). In fact, evaluation can be performed much faster. Spatial join algorithms for non-indexed inputs process the join in two steps; first the objects from both inputs are preprocessed in some data structures, and then these structures are used to quickly match objects that cover the same area. The algorithms differ in the data structure they use and the way the data are preprocessed.

Spatial Hash Join

The Spatial Hash Join (HJ) [9] has common features with the relational hash-join algorithm. Set R is partitioned into K buckets, where K is decided by system parameters, such that the expected number of objects hashed in a bucket will fit in memory. The initial extents of the buckets are determined by sampling. Each object is inserted into the bucket whose bounding box is enlarged the least after the insertion. Set S is hashed into buckets with the same extent as R's buckets, but with a different insertion policy; an object is inserted into all buckets that intersect it. Thus, some objects may go into more than one bucket (*replication*), and some may not be inserted at all (*filtering*). The algorithm does not ensure partitions of equal number of objects from R, as sampling cannot guarantee the best possible slots. Equal sized

```
function RJ(Node n_R, Node n_S)
  for each e_i ∈n_R
    for each e_j ∈n_S,such that ei.MBR ∩ e_j.MBR ≠ θ
      if n_R is a leaf node then /* n_S is also a leaf node* /
        output (e_i.ptr,e_j.ptr); /* a pair of object-ids
          passing the filter step */
        else /* n_R, n_S are directory nodes */
      RJ(e_i.ptr,e_j.ptr); /* run recursively for the nodes
        pointed by intersecting entries* /
```

Spatial Join, Fig. 2 The R-tree Join (RJ) Algorithm

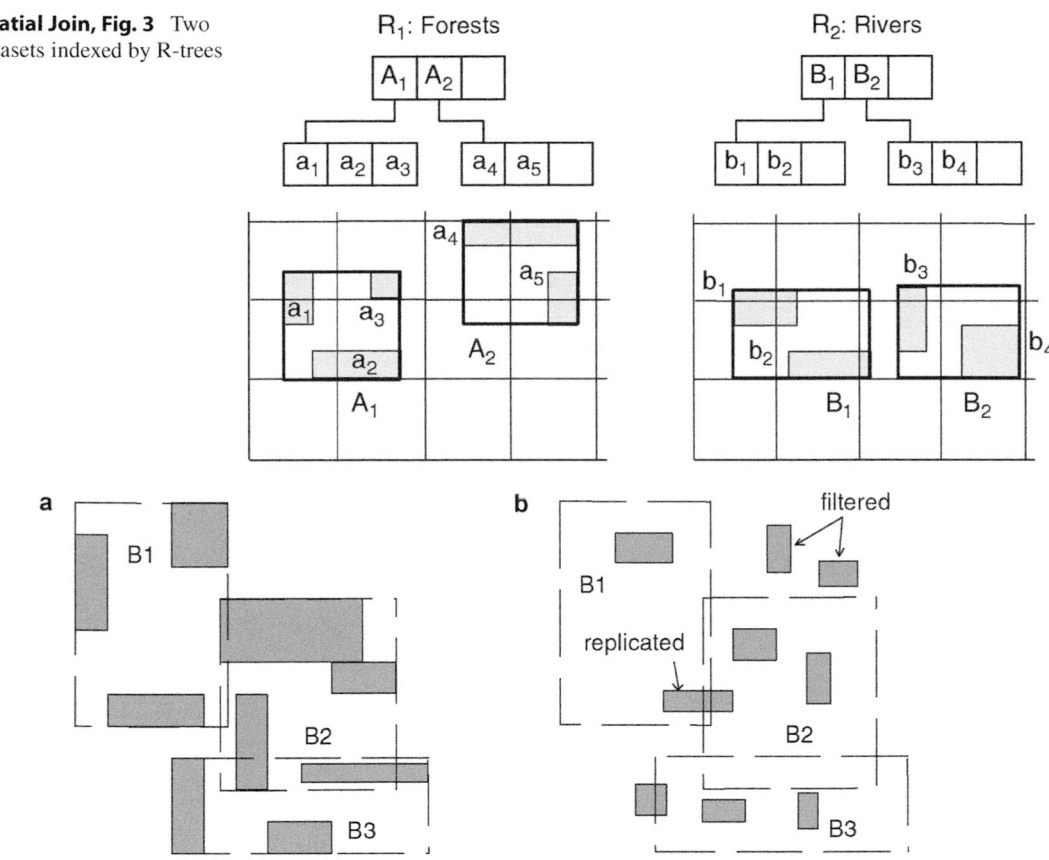

Spatial Join, Fig. 3 Two datasets indexed by R-trees

Objects from set *R* in three partition buckets   Filtering and replication of objects from set *S*

Spatial Join, Fig. 4 The partitioning phase of HJ algorithm

partitions for *S* cannot be guaranteed in any case, because the distribution of the objects in the two datasets may be totally different. Figure 4 shows an example of two datasets, partitioned using HJ.

After hashing set *S* into buckets, the two bucket sets are joined; each bucket B_i^R from *R* is matched with the corresponding bucket B_i^S from *S* that covers the same spatial region. For this phase, a single scan of both sets of buckets is required, unless for some pairs of buckets none of them fits in memory. If one bucket fits in memory, it is loaded and the objects of the other bucket are matched with it in a nested-loops fashion. If none of the buckets fits in memory, an R-tree is dynamically built for one of them, and the bucket-to-bucket join is executed in an indexed nested-loop fashion.

Partition Based Spatial Merge Join

Partition-based Spatial Merge Join (PBSM) [14] is also based on the hash join paradigm. The space, in this case, is regularly partitioned using an orthogonal grid, and objects from both datasets are hashed into partitions corresponding to grid cells, replicating wherever necessary. Figure 5a illustrates a regular space partitioning incurred by PBSM and some data hashed into the partitions. Objects hashed into the same partitions are then joined in memory using plane sweep. If the data inserted in a partition do not fit in memory, the algorithm recursively repartitions the cell into smaller parts and redistributes the objects. Since data from both datasets may be replicated, the output of the algorithm has to be sorted in order to remove pairs reported more than once.

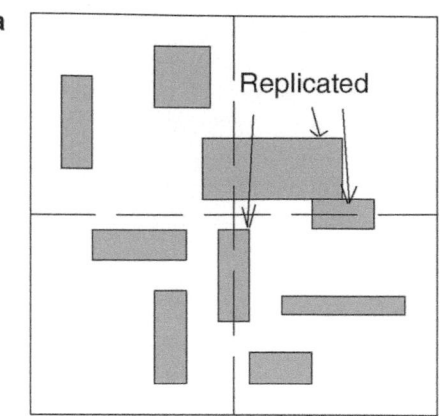

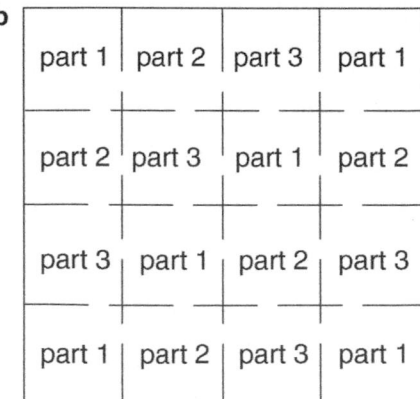

Four partitions and a set of hashed objects   A spatial hash function

Spatial Join, Fig. 5 Regular partitioning by PBSM

Spatial Join, Fig. 6 Size
separation spatial join

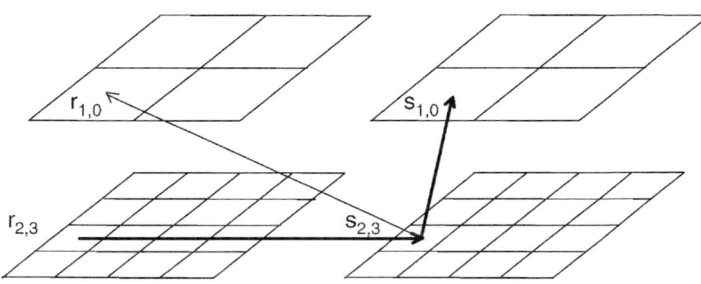

When the data to be joined are skewed, some partitions may contain a large percentage of the hashed objects, whereas others very few objects, rendering the algorithm inefficient. In order to evenly distribute the data in the partitions and efficiently handle skewed data, a spatial hash function is introduced. The cells of the grid are assigned to partitions according to this function and the space covered by a partition is no longer continuous, but consists of a number of scattered tiles. Figure 5b shows such a (round-robin like) spatial hash function.

Size Separation Spatial Join

Another algorithm that applies regular partitioning, like PBSM, but avoids object replication is Size Separation Spatial Join (S^3J) [6]. S^3J uses a hierarchical space decomposition. L partition layers of progressively larger resolution are introduced; the layer at level l partitions the space into $4l$ cells. A rectangle is then assigned to the topmost layer where it is not intersected by a grid line. This method achieves separation of the data according to their size. The rectangles in each layer are then sorted according to the Hilbert value of their MBRs center. A synchronized scan of the layer files is finally performed and the rectangles from dataset R in a partition at level l are joined with all partitions of dataset S that intersect it at levels $0, \ldots, l$. A partition from S is joined with partitions from R at levels $0, \ldots, l-1$. The Hilbert values of the data inside a layer determine the order of the join, avoiding scanning a partition more than once. Figure 6 shows two partition layers of both datasets. Partition $r_{2,3}$ is joined with $s_{2,3}$ and $s_{1,0}$, and partition $s_{2,3}$ is joined with $r_{1,0}$.

S^3J also maintains a dynamic spatial bitmap which, after partitioning the first set, indicates the cells at each layer that contain at least one rectangle, or cover same area with cells at other layers that contain at least one rectangle. This bitmap can be used during the partitioning of the second set to filter entries that cannot intersect any rectangle of the first set. If a rectangle from set S is to be hashed into a partition cell and the bitmap entry of the cell is zero, the hashed rectangle is filtered out.

Scalable Sweeping-Based Spatial Join

The Scalable Sweeping-based Spatial Join (SSSJ) [1] is a relatively simple algorithm that is based on plane sweep. Both datasets are sorted according to the lower bound of their projection on an axis (e.g., the x-axis), and some variant of plane sweep (e.g., the *forward-sweep* algorithm described before) is applied to compute the intersection pairs. SSSJ is based on the square-root rule: the expected number of rectangles in a dataset R that intersect the sweep line is $\sqrt{|R|}$, where $|R|$ is the total number of rectangles in R. SSSJ initiates an internal memory plane sweep algorithm. If it runs out of memory, i.e., the rectangles intersected by the sweep line do not fit in memory, the space is dynamically partitioned by stripes parallel to the sorted axis, the rectangles are hashed into the stripes, and plane sweep is recursively executed for each stripe.

Single-Index Join Methods

Methods in this class can be applied when one input is not indexed. Such situations often arise when processing complex queries, where another operator precedes the spatial join. Notice that in this case index-based methods cannot directly be applied, because the intermediate result is not supported by any index. Also, algorithms that consider non-indexed inputs could be expensive. All single-index join methods were proposed after RJ, and they assume that the indexed input is supported by an R-tree. Most of them build a second structure for the non-indexed input and match it with the existing tree.

Indexed Nested Loops Join

In accordance to the equivalent algorithm for relational joins, the Indexed Nested Loops Join (INLJ) applies a window query to the existing R-tree for each rectangle from the non-indexed set. This method can be efficient only when the non-indexed input is very small. Otherwise, the large number of selection queries can incur excessive computational overhead and access a large number of index pages.

Seeded Tree Join

Let R be a dataset indexed by an R-tree and S be a non-indexed dataset. The Seeded Tree Join algorithm (STJ) [10] builds an R-tree for S, using the existing for R as a seed, and then applies RJ to match them. The rationale behind creating a *seeded* R-tree for the second input, instead of a normal R-tree, is the fact that if the new tree has similar high-level node extents with RA, this would lead to minimization of overlapping node pairs during tree matching. Thus, the seeded tree construction algorithm creates an R-tree which is optimal for the spatial join and not for range searching. The seeded tree construction is divided into two phases: the *seeding* phase and the *growing* phase. At the seeding phase, the top k levels (k is a parameter of the algorithm) of the existing R-tree are copied to formulate the top k levels of the new R-tree for S. The entries in the lowest of these levels are called slots. After copying, the slots maintain the copied extent, but they point to empty (null) sub-trees. During the growing phase, all objects from S are inserted into the seeded tree. A rectangle is inserted under the slot that contains it, or needs the least area enlargement. Figure 7 shows an example of a seeded tree structure. The top $k = 2$ levels of the existing R-tree are copied to guide the insertion of the second dataset.

Spatial Join, Fig. 7 A seeded tree

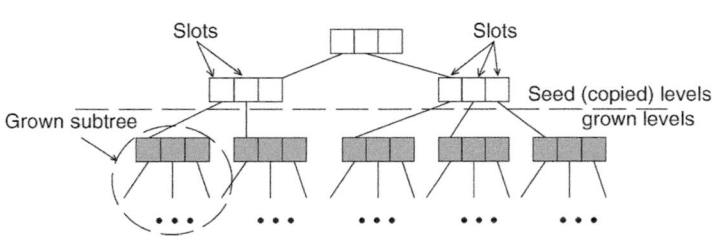

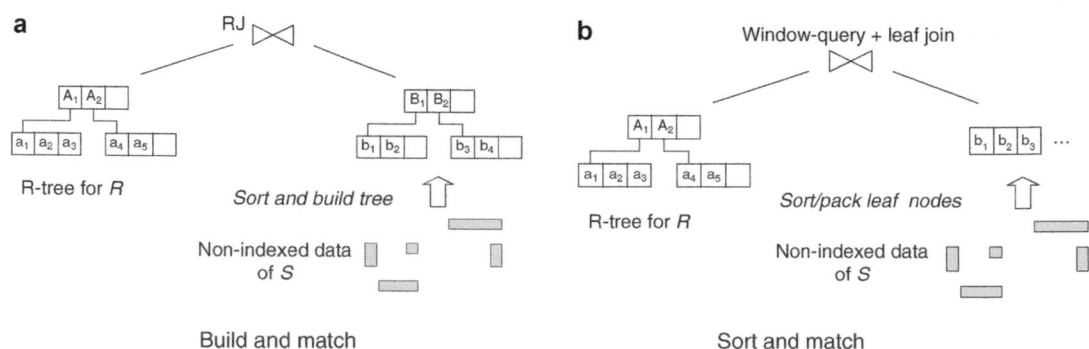

Spatial Join, Fig. 8 Algorithms based on bulk-loading

Build and Match

Building a packed R-tree using bulk loading can be much more efficient in terms of both CPU time and I/O than constructing it incrementally. Moreover, packed R-trees have a minimum number of nodes and height, and could be very efficient for range queries and spatial joins. The Build and Match (BaM) method [13, 14] first builds a packed R-tree for the non-indexed dataset S and then joins it with the existing tree of R, using RJ.

Sort and Match

Sort and Match (SaM) [13] is an alternative of BaM, which avoids building a whole R-tree structure prior to matching. The algorithm employs an R-tree bulk-loading technique [8] to sort the rectangles from the non-indexed dataset S but, instead of building the packed tree, it matches each in-memory created leaf node with the leaf nodes from the R-tree of R that intersect it, using the structure of the tree to guide search. For each produced leaf node n_L at the last phase of STR, a window query using the MBR of n_L is applied on R's tree, in order to identify the leaf nodes there that intersect n_L. Plane sweep is then applied to match n_L with the qualifying leaves of R's tree. The matching phase of SaM is expected to be efficient, as two consecutive produced nodes will be close to each other with high probability, and there will be good utilization of the LRU buffer. Graphical examples of BaM and SaM are shown in Fig. 8.

Slot Index Spatial Join

The Slot Index Spatial Join [9] is a hash-based spatial join algorithm, appropriate for the case where only one of the two joined relations is indexed by an R-tree. It uses the existing R-tree to define a set of hash buckets. If K is the desired number of partitions (tuned according to the available memory), SISJ will find the topmost level of the tree such that the number of entries there is larger than or equal to K. These entries are then grouped into K (possibly overlapping) partitions called *slots*. Each slot contains the MBR of the indexed R-tree entries, along with a list of pointers to these entries. Figure 9 illustrates a three-level R-tree (the leaf level is not shown) and a slot index built over it. If $K = 9$, the root level contains too few entries to be used as partition buckets. As the number of entries in the next level is over K, they are partitioned in nine (for this example) slots. The grouping policy used by SISJ is based on the R*-tree insertion algorithm. After building the slot index, all objects from the non-indexed relation are hashed into buckets with the same extents as the slots. If an object does not intersect any bucket it is filtered; if it intersects more than one buckets it is replicated. The join phase of SISJ loads all data from the R-tree under a slot and joins them (in memory) with the corresponding hash-bucket from the non-indexed dataset (in a similar way as HJ).

Comparison of Spatial Join Algorithms

Since indexes can facilitate the spatial join operation, algorithms (like RJ) that are based on

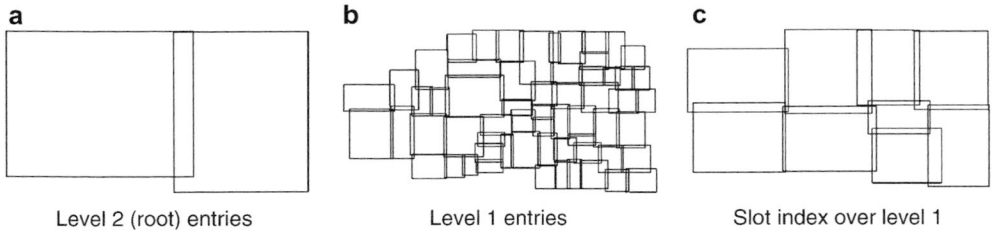

Spatial Join, Fig. 9 Entries of an R-tree and a slot index built over them

existence of indexes are typically more efficient compared to methods that do not rely on indexes. For example, RJ (which uses two R-trees) is expected to be more efficient than SISJ (which uses one R-tree), which is expected to be more efficient than HJ (which does not use trees).

RJ is the most popular index-based algorithm due to its efficiency and the fact that R-trees are becoming the standard access method in spatial database systems. Empirical and analytical studies have shown that the most efficient single-index methods are SISJ and SaM. Finally, conclusive results cannot be drawn about the relative performance of methods that do not consider indexes. S^3J is expected to be faster than PBSM and HJ when the datasets contain relatively large rectangles and extensive replication occurs in HJ and PBSM. On the other hand, this method uses sorting which is more expensive than hashing, in general. SSSJ is also based on sorting, thus it could be more expensive than hash-based methods. Furthermore, sort-based methods do not favor pipelining and parallelism of spatial joins. On the other hand, the fact that PBSM uses partitions with fixed extents makes it suitable for processing multiple joins in parallel, since the space partitions (and the local joins for them) can be assigned to different processors.

Key Applications

Spatial Database Systems
The spatial join is a core operation of Spatial Database Management Systems [4].

Geographic Information Systems
A fundamental operator in GIS is map overlay. Given two thematically different maps of the same region (e.g., elevation and political), this operator produces a join map that emphasizes on the overlaps of objects from both joined maps. Spatial join is the database operator used to produce this output.

Data Mining
In many applications that handle high dimensional data, an important analysis operation is to discover groups of objects that are close to each other in the multidimensional space. Examples of such mining tasks include "find stocks with similar movements" (time-series databases) and "find pairs of similar images" (multimedia databases). This type of clustering of complex data objects can be performed with the help of spatial joins with distance predicates [7]. Simply speaking, the original objects (e.g., images) are approximated by high dimensional feature vectors, and a spatial self-join is applied on this space to derive pairs of nearby objects, which are then postprocessed to larger groups (clusters).

Cross-References

- ► Hash Join
- ► Index Join
- ► Join
- ► Join Order
- ► Nested Loop Join
- ► R-Tree (and Family)
- ► Spatial Indexing Techniques

Recommended Reading

1. Arge L, Procopiuc O, Ramaswamy S, Suel T, Vitter JS. Scalable sweeping-based spatial join. In: Proceedings of the 24th International Conference on Very Large Data Bases; 1998. p. 570–81.
2. Brinkhoff T, Kriegel H-P, Seeger B. Efficient processing of spatial joins using r-trees. In: Proceedings of the ACM SIGMOD International Conference on Management of Data; 1993. p. 237–46.
3. Corral A, Manolopoulos Y, Theodoridis Y, Vassilakopoulos M. Closest pair queries in spatial databases. In: Proceedings of the ACM SIGMOD International Conference on Management of Data; 2000. p. 189–200
4. Güting RH. An introduction to spatial database systems. VLDB J. 1994;3(4):357–99.
5. Guttman A. R-trees: a dynamic index structure for spatial searching. In: Proceedings of the ACM SIGMOD International Conference on Management of Data; 1984. p. 47–57.
6. Koudas N, Sevcik KC. Size separation spatial join. In: Proceedings of the ACM SIGMOD International Conference on Management of Data; 1997. p. 324–35.
7. Koudas N, Sevcik KC. High dimensional similarity joins: algorithms and performance evaluation. IEEE Trans Knowl Data Eng. 2000;12(1):3–18.
8. Leutenegger ST, Edgington JM, Lopez MA. Str: a simple and efficient algorithm for R-tree packing. In: Proceedings of the 13th International Conference on Data Engineering; 1997. p. 497–506.
9. Lo M-L, Ravishankar CV. Spatial hash-joins. In: Proceedings of the ACM SIGMOD International Conference on Management of Data; 1996. p. 247–58.
10. Lo M-L, Ravishankar CV. The design and implementation of seeded trees: an efficient method for spatial joins. IEEE Trans Knowl Data Eng. 1998;10(1):136–52.
11. Mamoulis N, Papadias D. Slot index spatial join. IEEE Trans Knowl Data Eng. 2003;15(1):211–31.
12. Orenstein JA. Spatial query processing in an object-oriented database system. In: Proceedings of the ACM SIGMOD International Conference on Management of Data; 1986. p. 326–36.
13. Papadopoulos A, Rigaux P, Scholl M. A performance evaluation of spatial join processing strategies. In: Proceedings of the 6th International Symposium on Advances in Spatial Databases; 1999. p. 286–307.
14. Patel JM, DeWitt DJ. Partition based spatial-merge join. In: Proceedings of the ACM SIGMOD International Conference on Management of Data; 1996. p. 259–70.
15. Preparata FP, Shamos MI. Computational geometry – an introduction. Springer; 1985.

Spatial Matching Problems

Cheng Long[1] and Raymond Chi-Wing Wong[2]
[1] School of Electronics, Electrical Engineering and Computer Science, Queen's University Belfast, Kowloon, Hong Kong
[2] Department of Computer Science and Engineering, The Hong Kong University of Science and Technology, Clear Water Bay, Kowloon, Hong Kong

Definition

A matching is a mapping from the elements of one set to the elements of another set such that each element in one set is mapped to at most one element in another set. For example, assume two sets of objects $P = \{p_1, p_2, p_3\}$ and $O = \{o_1, o_2, o_3\}$. Then, $\{(p_1, o_1), (p_2, o_2), (p_3, o_3)\}$ is a matching with three pairs, but $\{(p_1, o_1), (p_1, o_2)\}$ is not a matching since p_1 is involved in two pairs. In general, the number of possible matchings is exponential to the cardinality of P and O; e.g., if $|P| = |O| = n$, there are $n!$ matchings with n pairs. Usually, among all possible matchings, the aim is to find one that optimizes/satisfies a certain *criterion*.

Let $c(p, o)$ be the *cost* of matching $p \in P$ with $o \in O$. *Optimal matching* [13] minimizes the *sum* of the costs of all pairs. *Bottleneck matching* [7] minimizes the *maximum* cost of any pair. *Fair matching*, also known as the *stable marriage problem*, returns a matching in which the following conditions cannot hold at the same time: (i) some $p \in P$ prefers some $o' \in O$ over the element o to which p is already matched (i.e., $c(p, o') < c(p, o)$), and (ii) o' also prefers p over the element p' to which o' is already matched (i.e., $c(p, o') < c(p', o')$). It has been proved that, when $|P| = |O|$, it is always possible to solve fair matching [6]. A comprehensive survey on matching problems can be found in [3].

In *spatial matching*, objects (or elements) have a *location* and the cost of matching $p \in P$ and $o \in O$ is based on the (Euclidean) distance $dist(p, o)$ between them [4, 8, 10, 12, 14].

Some examples include matching between access points and mobile devices, between emergency facilities (e.g., hospitals) and users, between parking slots and drivers etc. Even non-spatial applications may involve spatial matching based on distances; e.g., a matching between jobs and applicants, where each job/applicant is represented by a vector of attribute values.

Historical Background

There is extensive literature on algorithms for (non-spatial) matching problems. Currently, the time complexity of the best solution is: (i) for fair matching, $O(|P| \cdot |O|)$ [6], (ii) for optimal matching, $O((|P| + |O|)^3)$ [3], and (iii) for bottleneck matching, $O((|P| + |O|)^{2.5} \log^{0.5}(|P| + |O|))$ [5]. However, the straightforward application of these algorithms to spatial matching suffers from several drawbacks. First, most existing solutions are based on a materialized bipartite graph between P and O which has $O(|P| + |O|)$ vertices and $O(|P| \cdot |O|)$ edges. This limits their scalability since the space cost of maintaining the bipartite graph is prohibitive for large datasets. Second, the algorithms for general matching prob-

lems ignore the spatial context, which could enhance performance.

Scientific Fundamentals

The above versions of matching problems have corresponding counterparts in the spatial domain. For illustration we use the layout of Fig. 1a and the pairwise distances between P and O shown in Fig. 1b. *Spatial fair matching* (SFM) is the spatial version of the stable marriage problem. The matching $M_1 = \{(p_1, o_2), (p_2, o_1), (p_3, o_3)\}$ in Fig. 2a M_1 is fair. On the other hand, $M_2 = \{(p_1, o_1), (p_2, o_2), (p_3, o_3)\}$ in Fig. 2b is not fair because p_2 and o_1 would prefer each other to their current matches (o_2 and p_1, respectively) since $dist(p_2, o_1) < dist(p_2, o_2)$ and $dist(p_2, o_1) < dist(p_1, o_1)$. Wong et al. [14] propose four algorithms for SFM, among which the fastest has time complexity $O((|P| + |O|) \log(|P| + |O|))$.

Spatial optimal matching (SOM) refers to the optimal matching problem in the spatial context. The three matchings M_1, M_2 and M_3 of Fig. 2 have total costs (i.e., sum of distances of all pairs) 15, 14 and 15, respectively. It can be verified that among all matchings with three pairs, M_2 has the smallest total cost, and is the solution of this SOM. Hou et al. [8] model SOM as a *minimum-cost flow* problem and solve it by using *R-trees* and an adaptation of the *successive shortest path algorithm* [2].

Spatial bottleneck matching (SBM) is the spatial version of the bottleneck matching problem. The maximum cost of any pair in M_1, M_2 and M_3 is 10, 7 and 6, respectively. It can be verified that among all matchings with 3 pairs, M_3 has the lowest maximum cost and constitutes the solution of this SBM. Efrat et al. [4] extend the

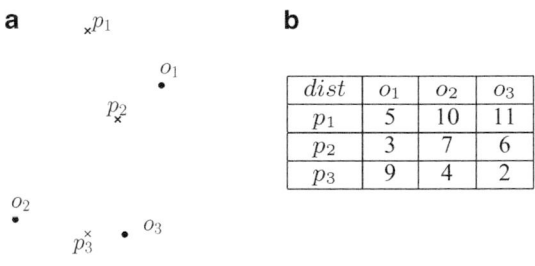

Spatial Matching Problems, Fig. 1 A running example. (**a**) Spatial layout. (**b**) Pairwise distances

$dist$	o_1	o_2	o_3
p_1	5	10	11
p_2	3	7	6
p_3	9	4	2

Spatial Matching Problems, Fig. 2 Spatial matching problems. (**a**) M_1 (SFM). (**b**) M_2 (SOM). (**c**) M_3 (SBM)

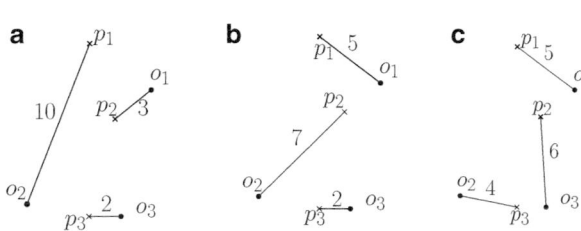

Threshold algorithm [3], originally designed for general bottleneck matching, by utilizing geometrical properties. *Swap-Chain* [10] generates an optimal solution of SBM by repeatedly re-matching pairs so that the maximum cost decreases, until no such adjustment is possible.

Key Applications

Spatial matching is useful in the following applications: emergency facility allocation (e.g., between fire stations/hospitals/police stations and users), profile matching (e.g., between jobs and applicants where both jobs and applicants are represented by vectors/multi-dimensional points), parking slot matching (e.g., between parking slots and drivers) and wireless networks (e.g., between access points and mobile devices).

Future Directions

Currently, most existing studies assume a *static* environment where the objects have fixed locations. One interesting direction refers to spatial matching problems in *dynamic* environments, where the objects move and the goal is to continuously monitor the matching changes (i.e., similar to [12] for SOM). Another direction concerns spatial matching problems with alternative criteria (e.g., *rank-maximal matching* [9], *pareto-optimal matching* [1] and *leximin-optimal matching* [11]).

Data Sets

The datasets used by [14] (for SFM) and by [10] (for SBM) can be found at the R-tree portal (http://www.rtreeportal.org/).

URL to Code

The code for SFM used by [14] can be found at http://www.cse.ust.hk/~raywong/code/quota.zip. The code for SBM used by [10] can be found at

http://www.cse.ust.hk/~raywong/code/spm-mm.zip.

Cross-References

▶ Nearest Neighbor Query
▶ R-Tree (and Family)

Recommended Reading

1. Abraham D, Cechlarova K, Manlove D, Mehlhorn K. Pareto optimality in house allocation problems. In: Proceedings of the 15th International Symposium on Algorithms and Computation; 2004. p. 3–15.
2. Ahuja RK, Magnanti TL, Orlin JB. Network flows: theory, algorithms, and applications. Englewood Cliffs: Prentice Hall; 1993.
3. Burkard RE, Dell'Amico M, Martello S. Assignment problems. Philadelphia: Society for Industrial Mathematics; 2009.
4. Efrat A, Itai A, Katz MJ. Geometry helps in bottleneck matching and related problems. Algorithmica. 2001;31(1):1–28.
5. Gabow HN, Tarjan RE. Algorithms for two bottleneck optimization problems. J Algorithms. 1988;9(3):411–17.
6. Gale D, Shapley L. College admissions and the stability of marriage. Am. Math. Mon. 1962;69: 9–15.
7. Gross O. The bottleneck assignment problem. Santa Monica: The Rand Corporation; 1959.
8. Hou UL, Yiu ML, Mouratidis K, Mamoulis N. Capacity constrained assignment in spatial databases. In: Proceedings of the ACM SIGMOD International Conference on Management of Data; 2008.
9. Irving RW, Kavitha T, Mehlhorn K, Michail D, Paluch K. Rank-maximal matchings. In: Proceedings of the 15th Annual ACM-SIAM Symposium on Discrete Algorithms; 2004. p. 68–75.
10. Long C, Wong RC-W, Yu PS, Jiang M. On optimal worst-case matching. In: Proceedings of the ACM SIGMOD International Conference on Management of Data; 2013.
11. Mehlhorn K. Assigning papers to referees. In: Automata, languages and programming, Rhodes; 2009. p. 1–2.
12. Mouratidis K, Mamoulis N. Continuous spatial assignment of moving users. VLDB J. 2010;19(2): 141–60.
13. Munkres J. Algorithms for the assignment and transportation problems. J Soc Ind Appl Math. 1957;5(1):32–8.
14. Wong RC-W, Tao Y, Fu AW-C, Xiao X. On efficient spatial matching. In: Proceedings of the 33rd International Conference on Very Large Data Bases; 2007.

Spatial Network Databases

Betsy George[1] and Shashi Shekhar[2]
[1]Oracle (America), Nashua, NH, USA
[2]Department of Computer Science, University of
Minnesota, Minneapolis, MN, USA

Synonyms

Spatial graph databases

Definition

Spatial network databases render support for
spatial networks by providing the necessary
data model, query language, storage structure,
and indexing methods. Spatial networks can
be modeled as graphs where nodes are points
embedded in space. One characteristic that
distinguishes a spatial network database is the
primary focus on the role of connectivity in
relationships rather than the spatial proximity
between objects. These databases are the kernel
of many important applications, including
transportation planning; air traffic control;
water, electric, and gas utilities; telephone
networks; urban management; utility network
maintenance, and irrigation canal management.
The phenomena of interest for these applications
are structured as a spatial graph, which consists
of a finite collection of the points (i.e., nodes), the
line-segments (i.e., edges) connecting the points,
the location of the points and the attributes of
the points and line-segments. For example, a
spatial network database storing a road network
may store road intersection points and the road
segments connecting the intersections (Fig. 1).

Foundations

Data Model of Spatial Networks

This section presents techniques related to
the data modeling of spatial networks. The
database design involves three steps, namely

conceptual modeling, logical modeling and
physical modeling.

- *Conceptual Data Model:* The purpose of conceptual modeling is to adequately represent
the data types, their relationships and the associated constraints. The Entity Relationship
(ER) model, widely used in conceptual modeling, does not offer adequate features to capture
the spatial semantics of networks. The most
critical feature of spatial networks, namely
the connectivity between objects can be expressed, using a graph framework. At the
conceptual level, the pictogram enhanced ER
(PEER) model [13] can be used. Figure 2
shows a PEER diagram for a spatial network.
In a spatial graph, vertices represent road intersections and edges represent road segments.
A path represents a street and consists of a
series of edges.

- Labels and weights can be attached to vertices
and edges to encode additional information
such as names and travel times. Two edges
are considered to be adjacent if they share a
common vertex.

- Modifications to the spatial network model
have been proposed to make it more suitable in
the context of some applications. For example,
a simple node-edge network model might not
be adequate to represent all features of a transportation network [9]. To address such limitations, various enhanced models have been
proposed. One such model is the transportation data model (UNETRANS) that organizes
the data model as three layers, namely (i)
a reference network layer that represents the
topological structure of the network, (ii) a
route features layer that defines more complex
features such as routes from the elements
of the reference network layer, and (iii) the
events layer that represents events such as
traffic signs [2].

- *Logical Data Model:* In the logical modeling
phase, the conceptual data model is implemented using a commercial database management system. Among the various implementation models such as hierarchical, network,
relational, object-relational data models and

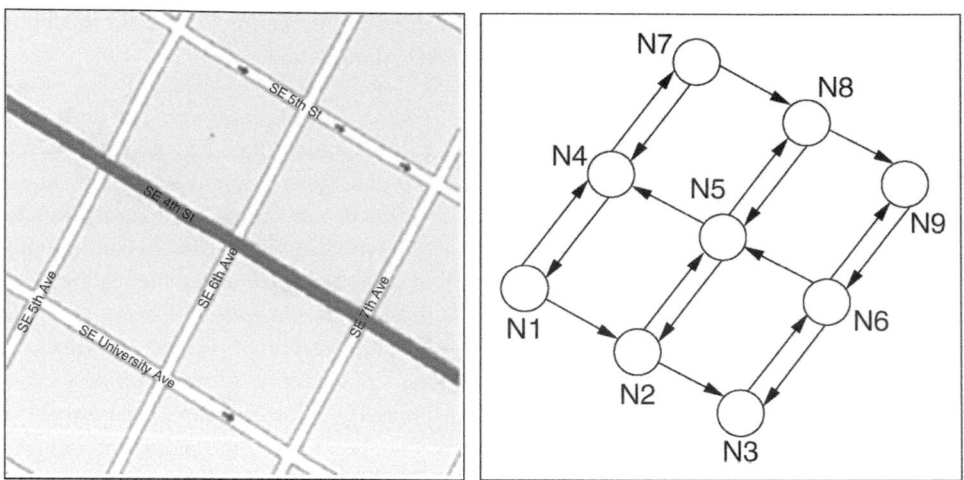

Spatial Network Databases, Fig. 1 A road map and its spatial network representation. (**a**) A road map (**b**) Spatial Graph Representation (Source for Fig. 1(a): http://maps.yahoo.com)

Spatial Network Databases, Fig. 2 A PEER diagram for spatial graph for a road network

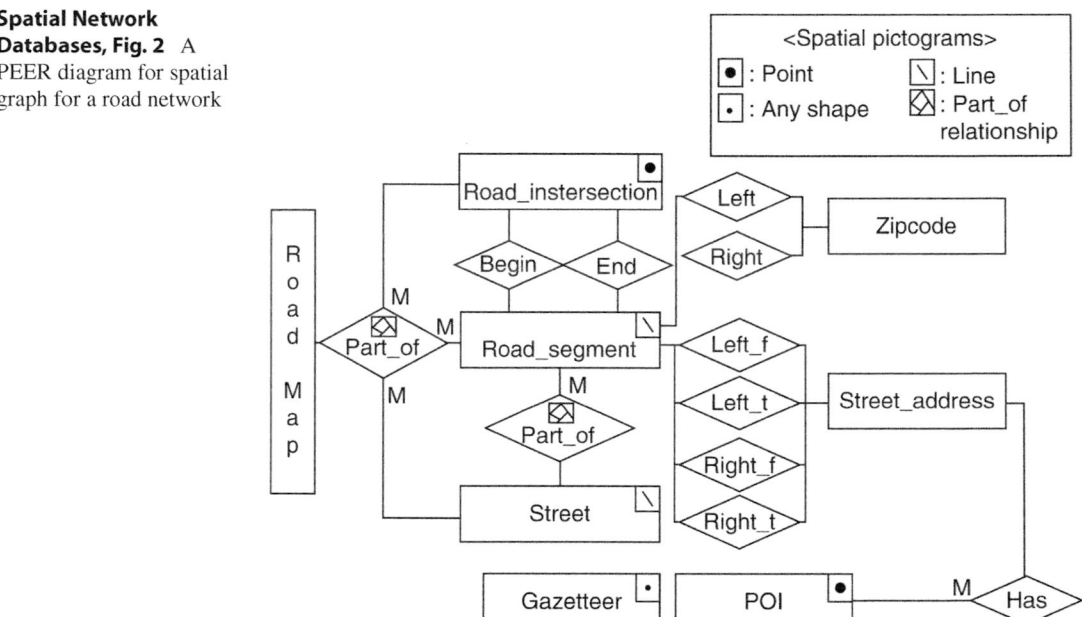

object-oriented models, the object-relational model has been gaining popularity in the representation of spatial applications. To model spatial network databases, graphs can be embedded into object-relational models. Shekhar and Chawla [11] lists some common graph operations used by spatial network applications, using a high-level object oriented notation that employs three fundamental classes in graphs, namely, Graph, Vertex, and Edge.

Models such as GraphDB [5], which allow additional data types such as path, have also been proposed. A path class explicitly stores paths or routes in a graph, which contains the list of edges and nodes. In this model, an operator called "rewrite" can apply transformations to subsequences of heterogeneous sequences such as paths.

- *Physical Data Model:* The physical data modeling phase deals with the actual implementation of the database application. Issues related to storage, indexing and memory management are addressed in this phase. Very often, queries that are posed on a network database such as a road map, involve route finding. This means the database must provide adequate support for network computations such as finding shortest paths. Figure 3 shows three representations of a graph. Adjacency-matrix and adjacency list are two well-known data structures used for implementing road networks represented as graphs [11]. In an adjacency-matrix, the rows and columns of a matrix represent the vertices of the graph. A matrix entry can be either 1 or 0, depending on whether there is an edge between the two vertices as shown in Fig. 3b. An adjacency list (shown in Fig. 3c) consists of an array of pointers. Each element of the array represents a vertex in the graph and the pointer points to a list of vertices that are adjacent to the vertex. Directed graphs can be implemented

in the relational model using a pair of relations, one for the nodes and the other for the edges. The "Node" (R) and the "Edge" (S) relations are shown in Fig. 3d and a denormalized representation is shown in Fig. 3e. The denormalized representation of a node table contains the coordinates of the node, a list of its successors and a list of its predecessors. This representation is often used in shortest path computations.

- A spatial access method called the Connectivity-Clustered Access Method (CCAM) was proposed in [12], which clusters the vertices of the graph based on graph partitions, thus providing an ordering based on connectivity.

- *Graph Algorithms*: Frequent queries on a spatial network involve operations such as shortest path, nearest neighbor search, range search and closest pairs [10]. "Shortest" path algorithms find the least cost path between two nodes in a given graph. The cost of the path could be based on network distance, travel

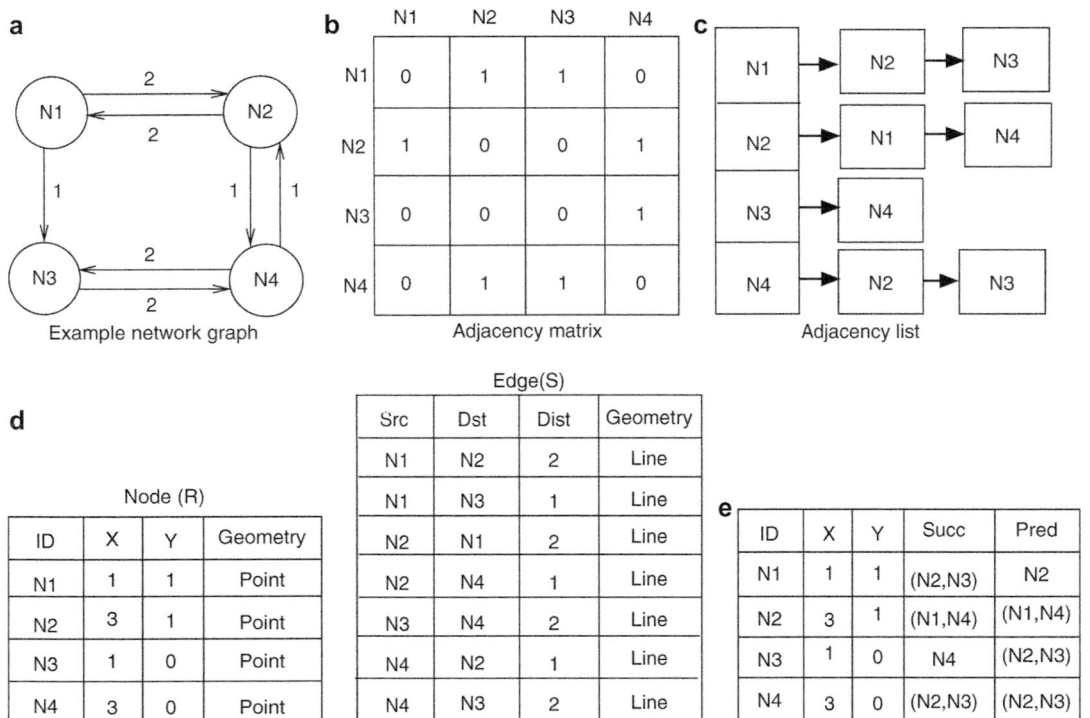

Spatial Network Databases, Fig. 3 Three different representations of a graph time or a user specified factor. Examples of popular shortest path algorithms are Dijkstra's algorithm and A* search. Nearest neighbor search algorithms find the point(s) closest to a given query point. Traditionally, the closest point was determined based on the Euclidean distances, which did not consider the network connectivity of the objects. However, in practice, trajectories of objects are usually constrained by an underlying spatial network such as a road network and hence algorithms that find nearest neighbors and closest pairs that consider network connectivity are critical in a spatial network database. Query processing algorithms that find nearest neighbors and closest pairs have been proposed [7, 10].

- *Turn Restrictions:* Turn restrictions are frequently encountered in road networks and they can affect the traversal in the network. A physical model that does not consider turn restrictions can lead to the computation of routes that are not entirely feasible. Turns have been modeled using a turn table where each turn restriction is represented as a row in the table that references the two associated edges [9]. Another proposed method to represent turn restrictions is node expansion [1]. The node that corresponds to a junction is expanded to a subgraph where permissible turns are represented as edges. This technique can lead to a substantial increase in the size of the network, which adversely affects the performance. Another method involves the transformation of the road network to a line graph where the edges in the original network are mapped to vertices in the line graph and the turns are represented as edges in the line graph [15].

A representation, consisting of a junction table, edge table and turn table was proposed in [6]. Every junction is represented as a row in the junction table. A row corresponding to a junction stores the edges that converge at the junction and the junctions connected to the given junction. The edge table stores edge identifiers and the junction where the edge originates (from-junction). A tuple in the turn table corresponds to a junction in the network. Each tuple consists of

a junction identifier, and a triplet (turn identifier, first edge-id, last edge-id) corresponding to each turn associated with the given junction.

Figure 4 illustrates the representation of turn restrictions in a road network. Figure 4a shows a part of a road network around a junction *j1* where the edges *e1*, *e2*, *e3* and *e4* meet. The curved arrows indicate the permitted turns at the junction. For example, a turn is allowed from edge *e1* to edge *e2*. Figures 4b-4d show the edge, junction and turn tables respectively, corresponding to turn *t1* in the example *e4*) and the junctions connected to it (*j2*, *j3*, *j4*, and *j5*). The turn table shows the permitted turns at junction *j1* and the edges that participate in each turn. For example, turn *t1* represents a turn from edge *e1* to edge *e2* as illustrated by the "*first edge id*" and "*last edge id*" entries in the turn table in Fig. 4d.

Key Applications

Location-Based Services

Spatial network databases are indispensable for any location-based service that involves route based queries [14]. Location-based services (LBS) provide the ability to find the geographical location of a mobile device and subsequently provide services based on that location. Spatial network databases play a key role in providing efficient query-processing capabilities such as finding the nearest facility (e.g., a restaurant) and the shortest path to the destination from a given location. Route-finding queries typically deal with route choice (shortest route to a given destination), destination choice (the nearest facility from the given location) and departure time choices (the time to start the journey to a destination so that the travel time is minimized). Though a significant amount of work has been done to find best routes and destinations, the problem of computing the best time to travel on a given route (time choice) needs further exploration.

Emergency Planning

One key step in emergency planning is to find routes in a road network to evacuate people from disaster-stricken areas to safe locations in the

Spatial Network Databases, Fig. 4 Representation of turn restrictions (Adapted from [6])

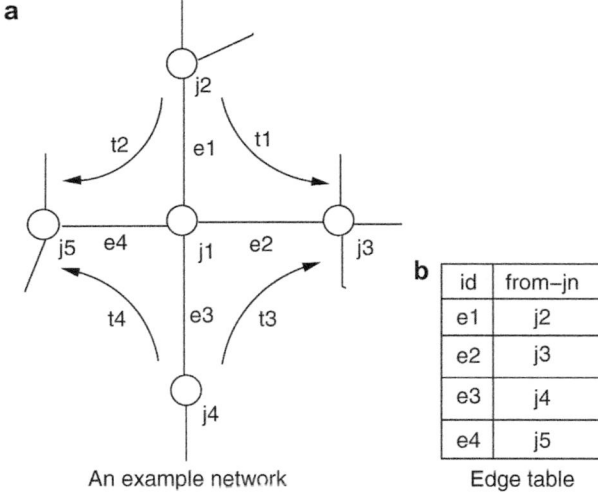

a An example network

b Edge table

id	from–jn
e1	j2
e2	j3
e3	j4
e4	j5

c Junction table

id	edge1	junc1	edge2	junc2	edge3	junc3	edge4	junc4
j1	e1	j2	e2	j3	e3	j4	e4	j5

d Turn table

id	Turn id1	First edge id1	Last edge id1	Turn id2	First edge id2	Last edge id2	Turn id3	First edge id3	Last edge id3	....
j1	t1	e1	e2	t2	e1	e4	t3	e3	e2	

least possible time. This requires finding shortest routes from disaster areas to destinations. In metropolitan-sized transportation networks, manual computation of the required routes is almost impossible, making digital road maps integral to the efficient computation of these routes.

Future Directions

A significant fraction of queries that are posed on a road network involves finding the shortest path between a pair of locations. Travel times on the road segments very often depend on the time of day due to varying levels of congestion, thus making the shortest paths also time-dependent. Road networks need to be modeled as spatio-temporal networks to account for this time-dependence. Various models such as time-expanded networks [8] and time-aggregated graphs [4, 3] are being explored in this context. A time expanded graph represents the time-dependence by copying the network for every

time instant whereas in time aggregated graphs, the time-varying attributes are aggregated over edges and nodes.

Cross-References

► Graph
► Graph Database
► Road Networks

Recommended Reading

1. Anez J, de la Barra T, Perez B. Dual graph representation of transport networks. Transp Res. 1996;30(3):209–16.
2. Curtin K, Noronha V, Goodchild M, Grise S. AR-CGIS transportation model (UNETRANS), UNETRANS data model reference, 2003.
3. George B, Shekhar S. Time-aggregated graphs for modeling spatio-temporal networks - an extended abstract. In: Proceedings of the 25th International Conference on Conceptual Modeling; 2006 p. 85–99.

4. George B, Shekhar S. Spatio-temporal network databases and routing algorithms: a summary of results. In: Proceedings of the 10th International Symposium on Advances in Spatial and Temporal Databases; 2007. p. 460–77.

5. Guting RH GraphDB: modeling and querying graphs in databases. In: Proceedings of the 20th International Conference on Very Large Data Bases; 1994.

6. Hoel EG, Heng WL, Honeycutt D. High performance multimodal networks. In: Proceedings of the 9th International Symposium on Advances in Spatial and Temporal Databases; 2005.

7. Jensen CS, Kolar J, Pederson TB, Timko I. Nearest neighbor queries in road networks. In: Proceedings of the 11th ACM International Symposium on Advances in Geographic Information System; 2003.

8. Kohler E, Langtau K, Skutella M. Time-expanded graphs for flow-dependent transit times. In: Proceedings of the 10th Annual European Symposium on Algorithms; 2002.

9. Miller HJ, Shaw SL. GIS-T data models, geographic information systems for transportation: principles and applications. Oxford: Oxford University Press; 2001.

10. Papadias D, Zhang J, Mamoulis N, Tao Y. Query processing in spatial network databases. In: Proceedings of the 29th International Conference on Very Large Data Bases; 2003.

11. Shekhar S, Chawla S. Spatial databases: a tour. Englewood Cliffs: Prentice Hall; 2002.

12. Shekhar S, Liu DR. CCAM: a connectivity-clustered access method for networks and network computations. IEEE Trans Knowl Data Eng. 1997;9(1): 102–19.

13. Shekhar S, Vatsavai R, Chawla S, Burke TE. Spatial pictogram enhanced conceptual data models and their translation to logical data models. In: Proceedings of the International Workshop on Integrated Spatial Databases, Digital Maps, and GIS; 1999.

14. Shekhar S, Vatsavai R, Ma X, Yoo J. Navigation systems: a spatial database perspective. In: Schiller J, Voisard A, editors. Location-based services. San Francisco: Morgan Kaufmann; 2004 .Chapter 3.

15. Winter S. Modeling costs of turns in route planning. GeoInformatica. 2002;6(4):345–61.

Spatial Operations and Map Operations

Michel Scholl[1] and Agnès Voisard[2]
[1]Cedric-CNAM, Paris, France
[2]Fraunhofer Institute for Software and Systems Engineering (ISST), Berlin, Germany

Synonyms

Layer Algebra; Map Algebra; Theme Algebra

Definition

Map operations refer to the operations that an end user performs on maps stored in a database. Map information is stored according to themes (for instance, cities, roads, or population), sometimes called layers in the GIS terminology. The maps considered here are stored in a vector format – as opposed to a raster format such as a grid of pixels – and can be 1 dimensional (e.g., a network of roads for a navigation system), 2 dimensional (e.g., a map of land-use for regional planning activities), or 2.5 dimensional if the elevation at certain locations is considered (for instance, the height of a building in an architecture project). A map is made of what is often called *geographic objects*. A geographic object (for instance, a city) has two parts, an alphanumeric one (e.g., its name and population) and a spatial one (e.g., a polygon), usually called *spatial object*. The alphanumeric attributes of a geographic object constitute its description. Map operations may use operations on spatial objects, commonly referred to as *spatial operations*. When describing the structure of a map – its description and its spatial part – together with its associated operations, one refers to a *map model*. The same applies to the structure and behavior of the spatial part of the geographic objects, leading to a *spatial model*.

Historical Background

When data are stored in a database, it is accessed through a query language, such as SQL. In the case of maps, however, SQL needs to be extended in order to consider their spatial component. With the emergence of GIS in the 80s, and also because maps are particular "non standard" entities, a set of operations to manipulate them was proposed by database researchers to describe these operations at a high level of abstraction, i.e., without considering SQL details. The idea was to define, at a conceptual level, unary or binary operations on maps that possibly take other arguments (alphanumerical or spatial) and that return maps, hence the term *algebra* often used in this context. Such operations also need spatial operations (geometric or topological) such as the intersection of polygons. Many proposals for lists of spatial operations were made. [8] is one of the first attempts to describe general map operations from a GIS view point. [7] proposed an extensible query language to the designer of geographic databases, independent of any underlying database model. The geo-relational algebra [2] was a pioneer approach, proposing an algebra based on the relational model that encompassed spatial operations. The SpatialSQL language [1] includes a list of spatial operations to be eventually used in conjunction with SQL. Operations on thematic layers were also proposed (e.g., [4]). The ROSE algebra [3] is a rich approach based on the relational model that allows extensible sets of functions. The OGIS Standard for SQL [5] from the Open GIS Consortium (OGC) focuses on spatial operations to be integrated in SQL and proposes an exhaustive list of such operations. Most current commercial approaches such as Oracle or ArcGIS from ESRI offer data types and operations that are OGIS compliant.

Foundations

This part focuses on end-user map operations that are performed in a database. In current applications, maps are usually stored in a relational database extended to abstract spatial data types. A kernel of elementary operations that can be combined in order to answer complex queries is presented here. The following list is coming from [2, 6].

Map Model A map M is defined as a set of geographic objects: $M = \{g\}$ where a geographic object g is defined as follows: $g = \{<A>, S\}$, where $<A>$ is a list of alphanumerical attributes and S a spatial attribute.

Spatial Model The spatial attribute S of a geographic object corresponds to its associated geometric part (it also has topological relationships with other objects). It can be simple (for instance, one polygon for a lake) or complex, i.e., made of many parts (for instance, many polygons for a given country and its islands). A

spatial attribute has a certain type and can be 0-, 1-, or 2-dimensional. Note that dimensions are often not mixed in a spatial attribute. An entity with a spatial type is usually called a spatial object and the referential used is the Euclidean plane.

The basic types of spatial objects that are usually considered are:

POINT	a 0-dimensional spatial object
LINE	a 1-dimensional spatial object made of segments
REG	a 2-dimensional spatial object made of polygons

as well as sets of these objects.

Spatial Operations

The operations presented below are primitives on spatial objects that are used in map operations. In the following, their signature is used for their short description (i.e., the type of arguments that these operations take and the one that they return). The spatial operations are presented here according to four groups: spatial predicates, spatial extractions, set operations, and geometric operations. Other classifications based for instance on the types of arguments are also sensible. Note also that the list given below is not exhaustive.

Map Operations

The operations described below constitute a common set of operations on maps. They are illustrated using the map of the 12 districts of Berlin, Germany, or a subset of it in some cases, and using the following schemas:

- **District** (Name:STRING, Population:NUM, Area:REG)
- **CityDivision** (Name:STRING, Area:REG)

Other Operations

The list of operations given above is not exhaustive but it corresponds to a kernel of common general operations on thematic maps. They are typical database operations performed in a GIS. Other GIS operations that are not detailed here include classification, zoom in, zoom out, as well as operations on layers stored in a raster form.

Group 1: Spatial Predicates

In the following, BOOL represents a Boolean value (true or false) and REG is an abbreviation for the REGION type.

Different	Tests whether two spatial objects are different in the plane
	Possible signatures: $POINT \times POINT \rightarrow BOOL$, $LINE \times LINE \rightarrow BOOL$, $REG \times REG \rightarrow BOOL$
Equal	Tests whether two spatial objects are the same (i.e., have the same value in the plane)
	Possible signatures: $POINT \times POINT \rightarrow BOOL$, $LINE \times LINE \rightarrow BOOL$, $REG \times REG \rightarrow BOOL$
Intersects	Tests whether two spatial objects intersect
	Possible signatures: $LINE \times LINE \rightarrow BOOL$, $LINE \times REG \rightarrow BOOL$, $REG \times REG \rightarrow BOOL$
Inside/Outside	Tests whether a spatial object is inside/outside a given region
	Possible signatures: $POINT \times REG \rightarrow BOOL$, $LINE \times REG \rightarrow BOOL$, $REG \times REG \rightarrow BOOL$
Adjacent	Tests whether two spatial objects are adjacent (i.e., have a common boundary)
	Possible signatures: $LINE \times REG \rightarrow BOOL$, $REG \times REG \rightarrow BOOL$

Group 2: Spatial Extractions

These operators transform their spatial input into another type of spatial object.

Intersection	Returns the intersection of two spatial objects
	Possible signatures: $LINE \times LINE \rightarrow \{POINT\}$, $LINE \times LINE \rightarrow LINE$, $LINE \times REG \rightarrow \{POINT\}$, $LINE \times REG \rightarrow LINE$, $REG \times REG \rightarrow \{POINT\}$, $REG \times REG \rightarrow LINE$, $REG \times REG \rightarrow REG$
Voronoi	Returns the Voronoi diagram of a region. Note that the returned set of regions has the particularity that the regions do not overlap
	Signature: $\{POINT\} \times REG \rightarrow \{REG\}$
Closest	Returns the closest spatial object from a given object, taken from a set
	Possible signatures: $POINT \times \{POINT\} \rightarrow POINT$, $POINT \times \{LINE\} \rightarrow LINE$, $POINT \times \{REG\} \rightarrow REG$, $LINE \times \{POINT\} \rightarrow POINT$, $LINE \times \{LINE\} \rightarrow LINE$, $LINE \times \{REG\} \rightarrow REG$, $REG \times \{POINT\} \rightarrow POINT$, $REG \times \{LINE\} \rightarrow LINE$, $REG \times \{REG\} \rightarrow REG$

Group 3: Set Operations

The following operations are common operations on sets of objects. Note again that a set only contains entities of the same type.

SetUnion	Returns the union of two sets of spatial objects, in the mathematical sense
	Possible signatures: $\{POINT\} \times \{POINT\} \rightarrow \{POINT\}$, $\{LINE\} \times \{LINE\} \rightarrow \{LINE\}$, $\{REG\} \times \{REG\} \rightarrow \{REG\}$

SetDifference	Returns the difference of two sets of spatial objects	
$\{■ ▲ ●\} - \{●\} \longrightarrow \{■ ▲\}$	Possible signatures:	$\{POINT\} \times \{POINT\} \rightarrow \{POINT\}, \{LINE\} \times \{LINE\} \rightarrow \{LINE\}, \{REG\} \times \{REG\} \rightarrow \{REG\}$
SetIntersection	Returns the intersection of two sets of spatial objects	
$\{■ ▲ ●\} \cap \{■ ▲ ◇\} \longrightarrow \{■ ▲\}$	Possible signatures:	$\{POINT\} \times \{POINT\} \rightarrow \{POINT\}, \{LINE\} \times \{LINE\} \rightarrow \{LINE\}, \{REG\} \times \{REG\} \rightarrow \{REG\}$

Group 4: Geometric Operations

In the following, NUM represents a numerical value.

Convex Hull	Returns the region (polygon) that encompasses all the points given in the argument set	
	Signature:	$\{POINT\} \rightarrow REG$
Center	Returns the center of a set of points	
	Signature:	$\{POINT\} \rightarrow POINT$
Min/Max Distance	Returns the minimal (respect. maximal) distance between two spatial objects	
	Possible signatures:	$POINT \times LINE \rightarrow NUM, LINE \times LINE \rightarrow NUM, POINT \times REG \rightarrow NUM, LINE \times REG \rightarrow NUM, REG \times REG \rightarrow NUM$
Length	Returns the length of a line	
$d1 + d2 + d3 = d$	Signature:	$LINE \rightarrow NUM$
Perimeter	Returns the perimeter of a region. In case the region is composed of many polygons, it returns the sum of their respective perimeters	
$d1 + d2 + d3 = d$	Signature:	$REG \rightarrow NUM$
Area	Returns the area of a region (total area if it is composed of many polygons)	
$a1 + a2 + a3 = a$	Signature:	$REG \rightarrow NUM$

S

Map Projection	Returns a map having a list of attributes given as argument and an unchanged spatial part	Signature:	$map \times <A> \rightarrow map$, where $<A>$ is a collection of alphanumerical attributes

Reinickendorf 245, Mitte 324, Spandau 225, Charlottenburg-Wilmersdorf 315 × District.name → Reinickendorf, Mitte, Spandau, Charlottenburg-Wilmersdorf

Map Selection	Returns a map whose geographic objects satisfy the selection criteria given as argument (e.g., population greater than 300 thousand inhabitants)	Signature:	$map \times selection\text{-}criteria\ (<A>) \rightarrow map$, where $selection\text{-}criteria\ (<A>)$ is a predicate on one or many alphanumerical attributes

Reinickendorf 245, Mitte 324, Spandau 225, Charlottenburg-Wilmersdorf 315 × DISTRICT.Population > 300 → Mitte 324, Charlottenburg-Wilmersdorf 315

Spatial Selections	– Windowing (or region query):	Returns the map made of the original map whose objects intersect the region given as argument (often, a rectangle)
	– Clipping:	Returns the part of the map that is exactly in the region given as argument
	Signature:	$map \times REG \rightarrow map$
Map Overlay		Generates a new map from two (overlaid) maps, for instance a map of the former city division in Berlin and the map of districts. It uses the intersection operation on spatial objects. It creates new geographic objects as can be seen in the center of the newly created map. This operation is also called a spatial join in the database terminology
	Signature:	$map \times map \rightarrow map$
	Note:	For legibility reasons, the following map of districts does not include their population.

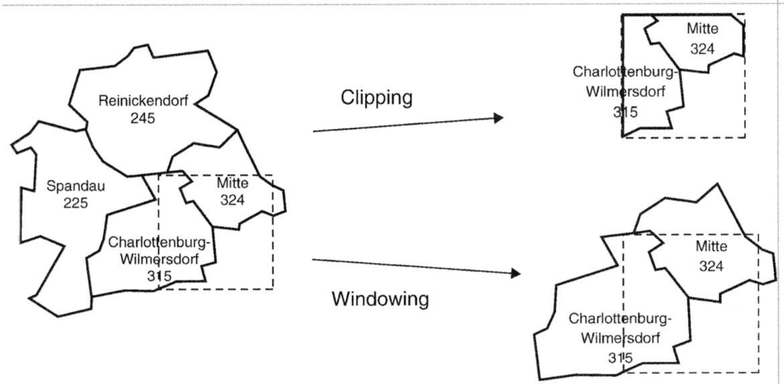

S

Map Union

Returns a map made of the two arguments

Signature:

$$map \times map \to map$$

Former East Berlin · Former West Berlin · Pankow · Lichtenberg · Marzahn Hellersdf. · Treptow-Köpenick · Reinicken-dorf · Mitte · Kreuzberg · F'hain · Tempel-hof · Neukölln · Schöneberg · Spandau · Charlottbg.-Wilmersdorf · Steglitz-Zehlendorf

Fusion

Performs the geometric union of the spatial part of geographic objects that belong to the same map. Note the sum of the population in the example

Signature:

$$map \to map$$

Reinickendorf 245 · Mitte 324 · Spandau 225 · Charlottenburg-Wilmersdorf 315 · Northwest districts 1109

Key Applications

Users likely to eventually use these map operations are end-users who need to get information from existing maps, for instance for the purpose of planning or geo-marketing. However, this conceptual approach – which moves away from implementation details – is targeted towards spatial database application designers who need to design appropriate application environments. Such environments should be easy to use, extensible, and adaptable to various application needs. They should, moreover, offer an efficient operation processing, however, this aspect is not handled by the conceptual approach presented here.

The key applications of this area concern thematic map manipulation (by the census bureau, city planners, local administrators, transportation managers, and so on) in order to perform statistics and analysis on data having a spatial dimension.

Cross-References

▶ Geographic Information System
▶ Semantic Modeling for Geographic Information Systems
▶ Spatial Data Types

Recommended Reading

1. Egenhofer MJ. Spatial SQL: a query and presentation language. IEEE Trans Knowl Data Eng. 1994;6(1): 86–95.
2. Güting RH. Geo-relational algebra: a model and query language for geometric database systems. In: Advances in Database Technology, Proceedings of the 1st International Conference on Extending Database Technology; 1988. p. 506–27.
3. Güting RH, Schneider M. Realm-based spatial data types: the ROSE algebra. VLDB J. 1995;4(2):243–86.
4. Hadzilacos T, Tryfona N. Logical data modelling for geographical applications. Int J Geogr Inf Sci. 1996;10(2):179–203.
5. Open GIS Consortium. OpenGIS® Geographic objects implementation specification. 2007.
6. Rigaux P, Scholl M, Voisard A. Spatial databases – with application to GIS. Chapter 3. San Francisco: Morgan Kaufmann/Elsevier; 2001.
7. Scholl M, Voisard A. Thematic map modeling. In: Proceedings of the 1st International Symposium on Advances in Spatial Databases; 1989. p. 167–90.
8. Tomlin D. A map algebra. In: Proceedings of the Harvard Computer Graphic Conference; 1983.

Spatial Queries in the Cloud

Ablimit Aji[1], Hoang Vo[2], and Fusheng Wang[3]
[1]Analytics Lab, Hewlett Packard, Palo Alto, CA, USA
[2]Computer Science, Stony Brook University, Stony Brook, NY, USA
[3]Stony Brook University, Stony Brook, NY, USA

Definition

Spatial queries in the cloud refer to processing of spatial queries on a distributed and interconnected network of computers that provide computation, storage, and resource management capabilities elastically in large scale. Resources in the cloud can be allocated on demand, and customers only pay for what they use. Cloud offers a number of query processing infrastructure and services ranging from parallel spatial database systems to MapReduce-based systems. Common spatial queries of interest include range queries, joins, and k-nearest neighbor queries.

Historical Background

Support of high-performance queries on large volumes of spatial data becomes increasingly important in many application domains, including geo-spatial problems in numerous fields, location-based services, and emerging scientific applications that are increasingly data and compute intensive. Past research efforts fall into three major directions toward improving spatial query performance: (i) algorithmic development on single-threaded spatial processing techniques, (ii) indexing or approximation approaches, and (iii) parallel processing of spatial queries.

Cloud computing has become cheaper, reliable, scalable, and widely available. The rapid advancement of hardware and software technologies, such as multi-core and many-core architectures, parallel programming environments, virtualization, and data center architecture, has matured cloud computing, and software service providers are moving their applications to the cloud. This allows applications to process large-scale data in parallel at low cost by utilizing a large number of cloud computing resources on demand.

MapReduce has become the prime choice for large-scale parallel data processing tasks as it is easy to program, robust to failure, and scalable. All the major cloud service providers offer MapReduce-based data processing platforms and services. Consequently, how to utilize MapReduce to process spatial queries at large scale is the focus of data-intensive spatial analytics.

Scientific Fundamentals

Most spatial query processing algorithms employ *filter and refine*, a two-step query processing strategy. Specifically, during the *filter* step, spatial objects are approximated with simplified representations such as minimum bounding regions (MBRs), and an initial query processing is performed on the MBRs. Then, during the *refinement* step, the filtered results are further refined with accurate geometric operations to generate final result.

This filter-and-refine strategy also forms the basis of spatial query parallelization at a higher level. Specifically, the underlying dataset is organized into *partitions* that can be selectively processed by applying filtering. Most often the spatial partitions take the form of axis-parallel rectangular regions (hyper-rectangles in multidimensional case). While the spatial partitions do not have to be strictly rectangular, this simple representation is easy to compute and has a very small storage footprint. Such rectangular spatial partitioning is also referred as *tiling*.

Spatial Range is a very simple but common spatial query type that returns objects whose spatial extents lie within an input range. It is also commonly referred to as *containment query*. In a MapReduce-based system, each mapper reads a portion of the dataset and performs the containment operator on each object to check if it is contained within the query range. Such brute-force approach requires an I/O intensive scan of the entire dataset regardless of the query range.

Both Hadoop-GIS [2] and SpatialHadoop [4] avoid the scan operation by utilizing a global region index for partition level data pruning. While spatial data is being loaded into the distributed file system (HDFS), both systems logically partition the space into nonintersecting rectangular regions, and records are distributed across the cloud based on their locations in the spatial coordinate system. Figure 1 illustrates an example of such global region index. In this figure, the larger rectangular *regions* represent the logical global spatial partitions. During the query processing phase, upon receiving a spatial range, both systems only scan the HDFS blocks whose region boundaries intersect with the query range. Hadoop-GIS stores the global region index on the HDFS, whereas SpatialHadoop stores it in the main memory on master node. Hadoop-GIS further partitions each global region into smaller *tiles* for within-region query processing.

Ray et al. [8] propose a system called Niharika that utilizes a relational spatial data management system on a single node and provides data partitioning, dynamic scheduling, and query processing capabilities on top of a cluster of nodes. Niharika adds a virtual *partition id* column to the spatial data table and maps this partition id to an actual spatial partition. To process a range query, the original SQL query is appended with relational filter conditions that correspond to the spatial partitions. After receiving the modified query, each single-node RDBMS engine evaluates the relational filter condition to eliminate portions of the dataset that do not contribute to the result. Finally, each node evaluates the query predicate over the remaining records to produce the final output.

Nishimura et al. [7] propose a multidimensional data infrastructure MD-HBase for location-based services (LBS), which require

Spatial Queries in the Cloud, Fig. 1 Global region index and local partition index for spatial range queries

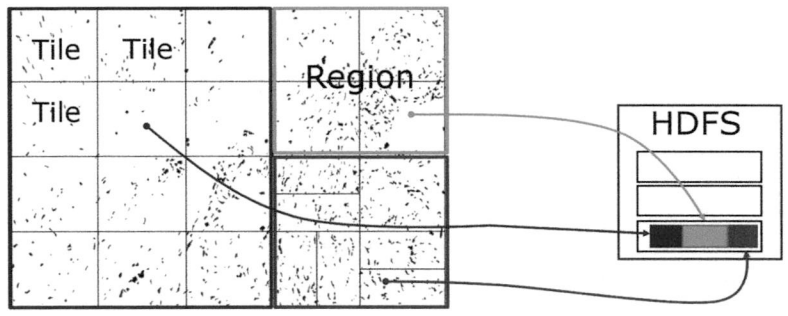

Spatial Queries in the Cloud, Fig. 1 Global region index and local partition index for spatial range queries

real-time query processing capabilities. MD-HBase uses HBase as the backend key-value store and extends it with multidimensional indexes to support spatial queries. MD-HBase splits the space into partitions using a tri-based approach and maps those partitions to physical storage buckets. Spatial partitions are linearized into one-dimensional ranges by applying Z-order, and the 1D ranges are materialized to HBase. To process a range query, MD-HBase transforms the query range into a 1D range by applying Z-order. Then, it sends the 1D range query to HBase to retrieve all the partitions which possibly contribute to the query result. A longest common prefix naming further prunes the intermediate 1D ranges that will not contribute to the final result.

Spatial Join is an important query type that combines two or more datasets with a given spatial predicate. A typical example is to *find all cities that are crossed by a river* that joins *city* and *river* datasets using the predicate *crosses*. In the cloud, most spatial join algorithms partition the datasets into manageable units that can be processed independently and distribute those units to a large number of nodes for parallel processing.

A general MapReduce framework, Hadoop-GIS [2], supports spatial join processing as follows. Input datasets are partitioned into tiles during the data loading phase, and each record is assigned a unique *partition id*. The spatial join is implemented as a MapReduce query operator processed in three steps. (i) *Map step*: each mapper, after applying some user-defined function or filter operation, emits the records with their *partition id* as the key along with a tag to indicate which dataset each record belongs to. (ii) *Shuffle*

step: records are sorted and shuffled through the cloud to group the records that have the same key (same *partition id*), and the intermediate results are materialized into the local disk of each node. (iii) *Reduce step*: each reducer is assigned a single partition, and a spatial join algorithm, such as plane-sweep or index-based join, is used to process the partition. The join algorithm can be in-memory or disk based depending on the size of the partition. During the runtime, Hadoop-GIS constructs an in-memory R*-Tree for each dataset in a partition and uses those indexes to process the query.

SpatialHadoop [4] first scans the main memory global partition index to find overlapping partitions from each dataset – a global partition level join. Then each pair of overlapping partitions are processed together to answer the query – a local within-partition join. The global join effectively eliminates irrelevant partitions and avoids processing partition pairs that do not contribute to the result. The local join uses a spatial index to process the query and writes the partial results to the HDFS. The local spatial indexes are read from the HDFS, and they are initially constructed during the loading step.

Niharika [8] mainly relies on a spatial DBMS engine that resides on each node. To process a spatial join, the coordinator process modifies the original SQL query to instrument a relational join predicate, and it uses the virtual *partition id* column to indicate the partitions that need to be processed together. Consequently, each node receives a slightly different query and only processes specific partitions. The worker database engines evaluate the query as a standard SQL query and return the results. The coordinator then aggregates the results from the worker DBMS engines and produces the final output.

A hybrid system Parallel Secondo [5] couples a single-node spatial database system with Hadoop at the engine level to run spatial queries in parallel. Similar to Hadoop, Parallel Secondo employs a master-slave architecture. It provides a parallel data model and its own parallel file system (PSFS) to facilitate query parallelization and data exchange between nodes. Parallel Secondo implements two algorithms for executing spatial joins – Distributed Hadoop Join (DHJ) and Secondo Distributed Join (SDJ). DHJ closely resembles the spatial join processing in Hadoop-GIS [2]. Initially, the datasets are partitioned and distributed to the Secondo slave nodes. Query processing has three steps. In the first step, each Secondo slave node retrieves the records in its corresponding partitions and tags them with a dataset label. It then transforms the records into key-value pairs that Hadoop can recognize and emits them to the HDFS. In the second step, Hadoop shuffles and sorts the records. In the third step, Secondo slave instances pull data from HDFS and perform the join operation on each partition. Secondo relies on several special operators to correctly handle the data distribution and query processing. Rather than relying on the HDFS to distribute intermediate data, SDJ uses PSFS to avoid the data transformation overhead.

Spatial Data Skew is a major problem for partition-based parallel spatial join processing. The irregular distribution of objects combined with a distribution-oblivious partition schema could result in a skewed partitioning where some partitions contain substantially more records than an average partition. Such data skew causes load imbalance and affects system efficiency. For example, in Fig. 2, P_3 and Q_3 contain considerably more objects compared to the rest, leading to long-running tasks.

Aji et al. [1] propose a cost-based partition assignment strategy that assigns partitions to nodes based on the partition sizes. A greedy bin-packing algorithm is used to group partitions into buckets that have similar processing cost, and each bucket is assigned to a worker node for processing. For the same problem, Eldawy and Mokbel [4] take

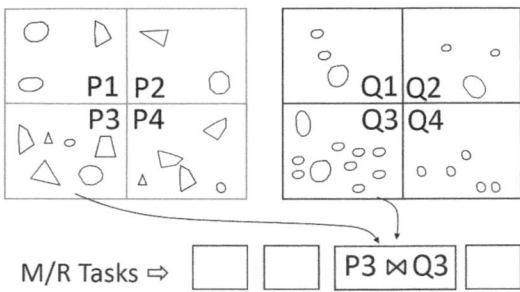

Spatial Queries in the Cloud, Fig. 2 Spatial join query processing with MapReduce

a skew-avoidance approach in which partitions are generated in a way that a skewed partitioning is less possible. In this approach, an R^+-Tree index is constructed on the input dataset during a preprocessing step. Then the lowest level index page boundaries are used as the partition boundaries. The constraint of maximal leaf page size in R^+-Tree ensures that each partition contains at most c objects, thus effectively avoids highly skewed data partitioning. Ray et al. [8] recursively partition the dataset into smaller tiles and use Hilbert space-filling curve to bucket the tiles into actual spatial partitions. While this approach effectively avoids skew to certain degree, it is difficult to achieve perfect load balancing with a priori partition assignment that does not take into account the heterogeneity of the nodes in the cloud. Ray et al. [8] propose several strategies that utilize information such as node processing capacity and partition workload to dynamically assign partitions to each node. Lu and Güting [5] take a simple approach that partitions the dataset into larger number of small grids and recreates final spatial partitions by grouping multiple grids with a hash function.

Since spatial objects have extents, partitioning them could result in *boundary objects* that cross boundaries. For example, in Fig. 3 the round object crosses the partition boundary indicated by the dotted line. Consequently, the partition independence assumption is violated, and the existence of boundary objects complicates the effective parallelization of spatial queries. Hadoop-GIS [2] takes a *replicate-and-prune* approach to resolve the boundary object problem.

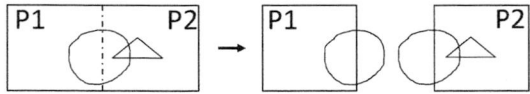

Spatial Queries in the Cloud, Fig. 3 Boundary objects and replication strategy

Specifically, a boundary object is replicated to all the spatial partitions that intersect it. A duplicate elimination step is initiated right after the query processing step to remove objects that appear multiple times. Aji et al. [2] argue that such approach is efficient as the fraction of boundary objects is very small.

Nearest Neighbor is a complex spatial query type that has several variations, such as k-nearest neighbors (kNN), reverse nearest neighbors (RNN), and maximum reverse nearest neighbors (MaxRNN). Most algorithms use an index structure to reduce the search space and computation overhead.

Akdogan et al. [3] propose a MapReduce-based algorithm that constructs a flat spatial index –Voronoi diagram (VD) – to process nearest neighbor queries. This algorithm sorts input points by a particular dimension and splits them into HDFS blocks. Each mapper builds a partial VD on the input block and outputs the result. A single reducer receives the partial VDs and merges them into a complete VD that can be used for nearest neighbor search. As the main cost of query processing comes from constructing the VD, even though query processing is not fully parallelized, it still improves query performance.

Aji et al. [1] propose two algorithms for an application scenario where the query is processed over two sets of objects and the cardinality of one set is much smaller than the other. One typical example is *for each residential house find the nearest Walmart store*; the cardinality of the *Walmart* set is much lower than the *house* set. This example can also be considered as a special case of nearest neighbor join (kNN join) in which $k = 1$. Both algorithms in Aji et al. [1] use a replication strategy to parallelize nearest neighbor queries. Specifically, the larger dataset is partitioned and distributed over HDFS, and mappers replicate the smaller dataset to each

node. Each reducer builds an in-memory index structure (VD or R-Tree) on the smaller dataset and processes the query over the larger dataset utilizing the index.

Eldawy and Mokbel [4] propose a three-step approach for kNN queries. In the initial step, for a query point q, the partition that contains q is processed using the local partition index to produce the initial k neighbors. In the correction step, a circle is drawn using the distance between the k-th neighbor and the query point q as the radius. If the circle does not intersect with other partition boundaries, all the nearest neighbors lie within the processed partition, and the current results are reported as final query output. Otherwise, the query enters a refinement step, and the partitions that intersect with the circle are also processed to produce the correct results.

Zhang et al. [9] considers a relatively complex scenario in which two datasets are joined together with a kNN as a join predicate. Zhang et al. propose an exact algorithm, Hadoop Block R-tree Join (H-BRJ), and an approximate algorithm, Hadoop-based zkNN Join (H-zkNNJ), to process kNN join. H-BRJ is an improved version of a brute-force algorithm. The brute-force algorithm partitions dataset R and S into n blocks of $(R_1, R_2, \ldots R_n)$ and $(S_1, S_2, \ldots S_n)$ and replicates them over all nodes to generate all possible n^2 block combinations of the partitioned dataset at the end of the map phase. During the reduce phase, it computes the local k-nearest neighbors for each block combination and outputs the result than can be further post-processed to answer the kNN query. H-BRJ improves the baseline by using R-Tree-based search for the local kNN search. H-zkNNJ further improves the query result by introducing random shifting and data partitioning, and it is implemented with three MapReduce jobs. First, it generates several shifted copies of original datasets and samples a fraction of those copies to compute optimal partition statistics. Second, it uses the partition statistics that are computed in the previous step to carefully partition the dataset to ensure redundancy and computes the local kNN for each partition. Last, it aggregates the local kNN results to generate final query results.

S

Lu et al. [6] propose a Voronoi diagram-based partitioning algorithm which further improves the performance of kNN join. The general idea is to carefully partition one dataset and find a minimal combination of partitions for which a local kNN can be calculated. The algorithm includes one preprocessing step and two MapReduce steps. The preprocessing step selects a set of pivot objects from one input dataset R. The first MapReduce step takes the selected pivots and both input datasets R and S and retrieves the nearest pivot for each object in both datasets. It computes the distance between the object and the pivot and assigns the object to the closest pivot. The output of this step is a partitioning of R that is generated by using the Voronoi diagram of selected pivots from the previous step. It also collects some statistics about each partition of R. The second MapReduce step, using the statistics collected in the previous step, computes a subset of S for each subset of R and performs the kNN join between those subsets.

Nishimura et al. [7] propose a key-value store-based kNN algorithm for real-time LBS applications that uses a Z-order-based spatial partitioning. A kNN query involves two steps: subspace search expansion and subspace scan. The first step incrementally expands the search region by inserting partitions into a priority queue with minimum distance from the candidate partition to the query point as the sorting criterion. The subspace scan step retrieves a partition from the top of the priority queue and computes the distance from the query point to each object in that partition. Then, it inserts the objects to a candidate result queue with the computed distance as the sorting criterion. The search process stops when the distance from query point to the k-th object in the queue is less than the distance to the closest unscanned partition.

Key Applications

Geographic Information System
Geographic information systems use cloud and MapReduce technologies to cope with the ever-increasing amount of spatial data. Large datasets are often preprocessed and cleaned using MapReduce-based ETL systems before they are loaded into the GIS.

Analytical Medical Imaging
Biomedical research professionals perform spatial analysis to study microanatomic objects and their spatial relationships. MapReduce-based systems and cloud technologies are used to manage large datasets and run large-scale queries.

Location-Based Services
Location-based systems constantly run spatial queries to provide effective service to the users. Query types include ranges, joins, and most importantly nearest neighbor queries.

Future Directions

Experimental Results

Datasets
OpenStreetMap is a largest open source mapping dataset available online, and it can be found at http://www.openstreetmap.org.

TIGER dataset is provided by United States Census Bureau, and readers can find detailed information at https://www.census.gov/geo/maps-data/data/tiger.html.

TCGA is a pathology imaging dataset that derived from high-resolution images by algorithmically segmenting microanatomic objects from the images. https://tcga-data.nci.nih.gov/tcga/

URL to Code

Hadoop-GIS – http://github.com/hadoop-gis
  SpatialHadoop – http://spatialhadoop.cs.umn.edu
  SECONDO – http://dna.fernuni-hagen.de/Secondo.html/index.html
  GIS Tools for Hadoop by ESRI – https://github.com/Esri/spatial-framework-for-hadoop

Cross-References

- ▶ Cloud Computing
- ▶ MapReduce
- ▶ Nearest Neighbor Query
- ▶ Range Query
- ▶ Spatial Datawarehousing
- ▶ Spatial Join

Recommended Reading

1. Aji, A, Wang, F, Saltz, JH. Towards building a high performance spatial query system for large scale medical imaging data. In: Proceedings of the 20th International Conference on Advances in Geographic Information Systems; 2012. p. 309–18.
2. Aji, A, Wang F, Vo H, Lee R, Liu Q, Zhang X, Saltz J. Hadoop GIS: a high performance spatial data warehousing system over mapreduce. Proc VLDB Endowment. 2013;6(11):1009–20.
3. Akdogan A, Demiryurek U, Banaei-Kashani F, Shahabi C. Voronoi-based geospatial query processing with mapreduce. In: Proceedings of the 2010 IEEE 2nd International Conference on Cloud Computing Technology and Science; 2010. p. 9–16.
4. Eldawy A, Mokbel MF. SpatialHadoop: a MapReduce framework for spatial data. In: Proceedings of the 31st International Conference on Data Engineering; 2015.
5. Lu J, Guting RH. Parallel secondo: boosting database engines with hadoop. In: Proceedings of the 18th IEEE International Conference on Parallel and Distributed Systems; 2012. p. 738–43.
6. Lu W, Shen Y, Chen S, Ooi BC. Efficient processing of k nearest neighbor joins using mapreduce. Proc VLDB Endowment. 2012;5(10):1016–27.
7. Nishimura S, Das S, Agrawal D, Abbadi AE. MD-HBase: a scalable multi-dimensional data infrastructure for location aware services. Proceedings of the 12th IEEE International Conference on Mobile Data Management; 2011. p. 7–16.
8. Ray S, Simion B, Brown AD, Johnson R. A parallel spatial data analysis infrastructure for the cloud. In: Proceedings of the 21st ACM SIGSPATIAL International Conference on Advances in Geographic Information Systems; 2013. p. 284–93.
9. Zhang C, Li F, Jestes J. Efficient parallel kNN joins for large data in MapReduce. In: Proceedings of the 15th International Conference on Extending Database Technology; 2012. p. 38–49.

Spatial-Keyword Search

Kian-Lee Tan
Department of Computer Science, National
University of Singapore, Singapore, Singapore

Synonyms

Spatial-keyword query processing; Top-k spatial-keyword search

Definition

Consider a set of spatial-textual objects where each object consists of a spatial location and a textual description. A spatial-keyword search retrieves objects of interest based on both spatial proximity to the query location and the textual relevance to the query keywords. Typically, the spatial proximity is based on distance, while the textual similarity is measured using an information retrieval model such as the cosine similarity or the language model. Query answers are usually ranked based on a linear combination of the two aspects. The goal of spatial-keyword query processing is to return the relevant answers while minimizing the processing cost.

Historical Background

Web content has traditionally been queried using keyword search. More recently, the web has taken on a new dimension – the spatial dimension. On one hand, the prevalence of GPS-enabled smartphones and social network systems has enabled users to generate, share, and access geo-tagged content (e.g., tweets, photos). On the other hand, social network- and location-based applications contribute large amount of spatial-textual content (e.g., POIs, reviews, check-ins). This has prompted a new class of applications that offer better user experience through spatial-keyword search that finds relevant content in the neighborhood of the user. For example, Google

Spatial-Keyword Search, Fig. 1 An example of spatial-keyword search scenario

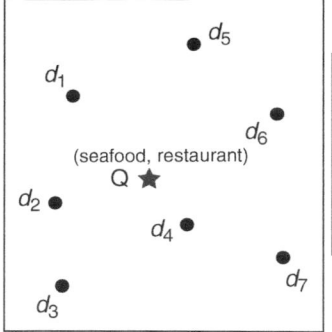

d_1	(pizza 0.6), (restaurant 0.4)
d_2	(seafood 0.9), (restaurant 0.8)
d_3	(seafood 0.2), (pizza 0.5)
d_4	(noodle 0.7), (seafood, 0.2)
d_5	(spicy 0.8), (noodle 0.5), (restaurant 0.6)
d_6	(spicy 0.4), (restaurant 0.5)
d_7	(seafood 0.1), (restaurant 0.3)

Maps allows users to retrieve POIs near them, Foursquare facilitates search for neighboring venues and restaurants, and Tinder helps users to discover other nearby users with mutual interests.

Figure 1 illustrates an example of a spatial-keyword search for a corpus with seven geo-documents d_1, ..., d_7. Each document is associated with a location and a collection of keywords together with their relevance scores. In the figure, we have a spatial-keyword query Q with keywords "seafood restaurant" issued at the location marked by the star. Spatial-keyword query comes in several flavors. The Boolean kNN query retrieves the top-k objects that are near the user's current location; these objects must contain the keywords in the query. In our example, a top-1NN query would return d_2. On the other hand, the Boolean range query searches for answers that are within a certain search region around the user's current location; the answers must contain the keywords in the query. In our example, a bounding box centered at the star would retrieve all answers within the box only. Yet another class of commonly used query, the top-k NN query, returns the top-k answers with the highest ranking scores; the scores are computed based on a function that considers both the textual relevance of the documents to the query and their spatial relevance to the user's location. In our running example, document d_2 is likely to be a better match to Q than document d_4 because d_2 is near the query location, and it is highly relevant to the query keywords; in comparison, d_4 does not match the query keywords well, though it is closer to Q than d_2. However, it is possible for d_4 to be the most relevant top 1NN answer (even

though d_4 has fewer matching keywords than d_2) if spatial distance is given a higher weight.

Scientific Fundamentals

The key challenge in processing spatial-keyword queries is to keep the processing time low to meet real-time requirements of end users. To support efficient processing of these queries for large-scale geo-document corpuses, a variety of indexes have been developed.

Early works essentially develop integrated data structures, which have both a spatial and a textual component in order to exploit both spatial and textual pruning. One such scheme is the IR-tree [1], which augments an R-tree with an inverted file. In an IR-tree, a single R-tree is used to index textual documents based on the spatial attribute (i.e., location) associated with the documents. Each R-tree node is further augmented with an inverted file that captures the textual summaries of all documents whose associated locations fall inside the node's MBR. Figure 2 illustrates the MBRs of the R-tree for our running example, as well as the corresponding IR-tree. As shown in Fig. 2, the spatial objects are inserted into an IR-tree with node capacity of 2. Each tree node is augmented with an inverted file indicating the existence of keywords in that node. In this simple database, R_1 represents the entire spatial region which contains a total of 12 objects. So the inverted file at the root node indexes these 12 documents. Likewise, the leaf node R_5 containing two objects d_5 and d_6 will point to the inverted file for these two objects.

With IR-tree, a spatial-keyword search can prune away the search space that will not contribute any answer to the query. In our example, based on the query location, assuming a top 1NN search, R_3, will be searched first, R_7 will be examined, resulting in the retrieval of d_4. Next, R_6 will be searched, leading to the retrieval of d_2. Depending on the scoring function, either d_2 or d_4 will be the answer picked. The left sub-tree of the root node (covering region R_2) will not be traversed at all, and hence the search space is significantly reduced.

Several variations of the IR-tree have been developed. In [2], the IR-tree organizes the inverted file component differently – instead of managing each inverted file for every node separately, the scheme utilizes one integrated inverted file for all the nodes. The DIR-tree [1] takes into account both the spatial and textual proximity when inserting an object. This can lead to better performance when the spatial keywords are correlated locally. Yet another scheme, the CIR-tree [1], clusters the textual documents into groups with similar textual descriptions. This optimization incurs the overhead of storing additional summarized content for each cluster in the inverted file. An experimental study on these variations [3] shows that they perform equally well.

The IR-tree-based schemes may incur high processing cost for query keywords that are frequent. In such cases, a large number of inverted files have to be accessed, resulting in high I/O cost. To improve the performance, some schemes solely utilize spatial indexes for pruning [4, 5]. These schemes partition the geo-tagged documents by keywords, and a spatial index is maintained for each keyword. In this way, only the indexes corresponding to the query keywords need to be examined.

One such scheme is the spatial inverted index (S2I) [4] which employs aggregated R-trees (aR-trees) as the underlying index structure for spatial relevance evaluation. Given a spatial-keyword query, the aR-trees related to the keywords are retrieved. Query processing starts from the most relevant documents in each aR-tree, accesses them in nondecreasing order of distance to the query location, and aggregates the partial scores from multiple keywords/trees. The algorithm terminates when sufficient (say k) best results are found to be better than the upper-bound score of unvisited documents. S2I, however, may incur large number of random accesses on tree nodes in order to combine the partial aggregates across different aR-trees.

Spatial-Keyword Search,
Fig. 2 IR-tree example

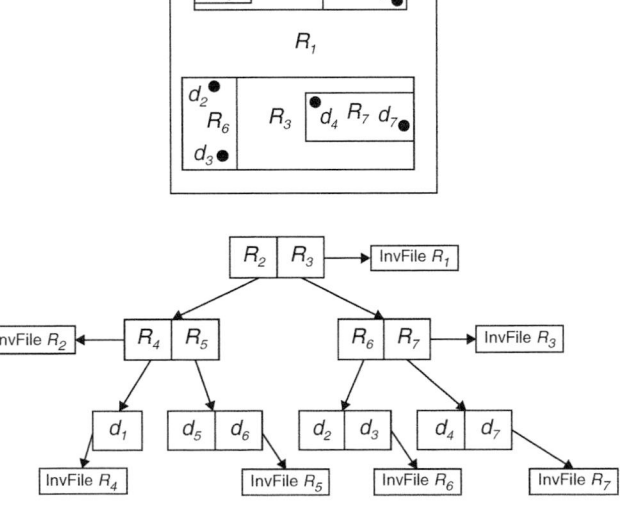

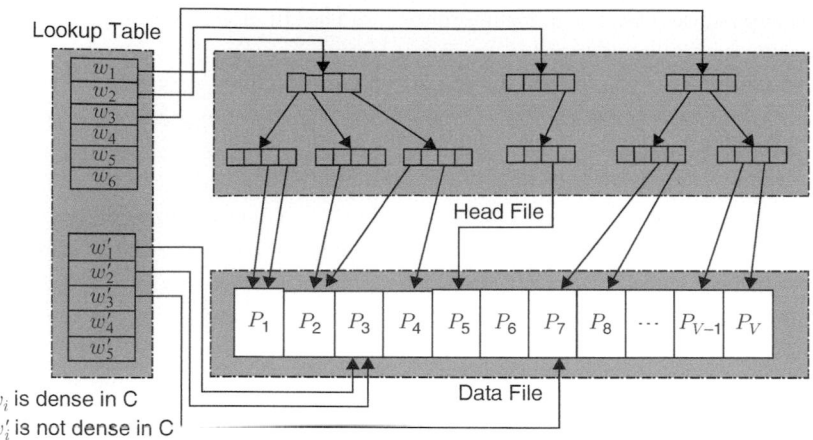

Spatial-Keyword Search, Fig. 3 Index structure of I^3

To facilitate efficient score aggregations, the integrated inverted index (I3) [5] adopts the quadtree as the indexing structure for each keyword. By enforcing the same spatial partition mechanism for all the keywords, the score aggregation across different trees can be efficiently derived. Figure 3 illustrates an example of the index design of I3. For infrequent keywords with a small number of associated documents, the inverted lists are stored in a page. Otherwise, a quadtree is built for the keyword, and each tree node is augmented with summary information to facilitate pruning. The leaf nodes of all the quadtrees are stored in the same data file.

Figure 4 shows two inverted lists for keywords "spicy" and "restaurant" derived from our running example database. Each inverted list consists of a set of cells for the keywords based on the space decomposition. In this example, each page contains at most two tuples. A keyword cell $<w_i, C_j>$ is stored in the inverted list of w_i only if the keyword w_i is neither empty nor dense in cell C_j. If $<w_i, C_j>$ is dense, it contains more than two tuples and will be split into smaller keyword cells. For example, "restaurant" is contained in three documents $\{d_4, d_8, d_7\}$ in cell C_4. Thus, the keyword cell <restaurant, C_4> is dense and is split into three smaller keyword cells: <restaurant, C_4>; <restaurant, C_4>; and <restaurant, C_4>. The query processing is top-down, beginning from the root cell of each related quadtree down to the child cells. This is the main

difference with S2I. Each cell is a candidate search space, and its upper-bound score can be calculated by summing the spatial relevance score and textual relevance score. The spatial relevance score is measured by the minimum distance from the cell to the query point, while the textual relevance is the aggregation score of different keywords. The algorithm terminates if the upper-bound scores of all the remaining cells are smaller than the k-th result found.

In [6], Zhang et al. modeled the spatial-keyword search as the well-known top-k aggregation problem [7].

The reformulation turns out to be fairly straightforward. Given a m-keyword query, each geo-tagged document D in the dataset can be modeled as a m + 1-tuple $(x_1, x_2, \ldots, x_{m+1})$ where x_i $(1 \le i \le m)$ is the textual relevance score between D and the i-th query keyword, and x_{m+1} is the spatial proximity score between D and the query location. By aggregating the values of each attribute using any monotonic aggregation function (e.g., linear combination of the textual and spatial attributes), we can effectively obtain the top-k answers to the spatial-keyword query. Such an approach has two key advantages. First, there is a wealth of work on top-k aggregation problem that can be tapped upon to solve the spatial-keyword query. Second, the approach requires only inverted index for query processing. With the TA scheme [7], the inverted lists corresponding to the attributes have

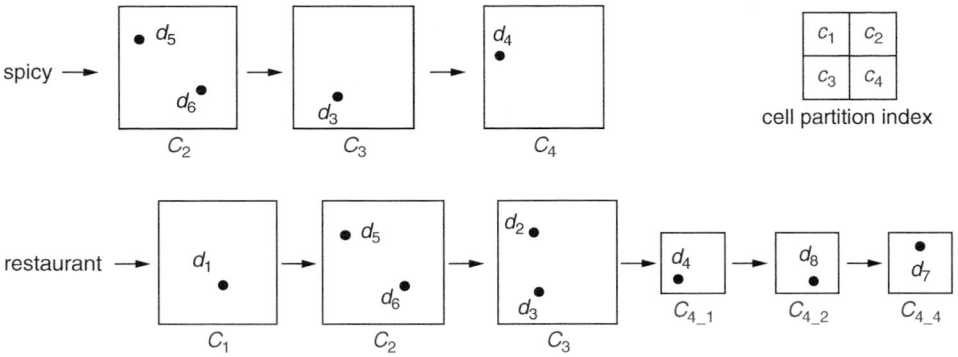

Spatial-Keyword Search, Fig. 4 Illustration for "spicy" and "restaurant"

Spatial-Keyword Search, Fig. 5 An example of top-k aggregation for query "seafood restaurant"

ψ_t(seafood) | (d$_2$, 0.9) | (d$_3$, 0.2) | (d$_4$, 0.2) | (d$_7$, 0.1) | (d$_1$, 0.0) | (d$_5$, 0.0) | (d$_6$, 0.0)

ψ_t(restaurant) | (d$_2$, 0.8) | (d$_5$, 0.6) | (d$_6$, 0.5) | (d$_1$, 0.4) | (d$_7$, 0.3) | (d$_3$, 0.0) | (d$_4$, 0.0)

ψ_s | (d$_4$, 0.8) | (d$_2$, 0.7) | (d$_1$, 0.6) | (d$_3$, 0.55) | (d$_6$, 0.5) | (d$_5$, 0.47) | (d$_7$, 0.42)

to be sorted based on the relevance scores of the attributes. Consider our running example with query keyword "seafood restaurant." Each document can be modeled as a 3-D vector (x_1, x_2, x_3) where x_1 is D's textual relevance to the keyword "seafood," x_2 is D's textual relevance to the keyword "restaurant," and x_3 is the spatial proximity between D and the query. Figure 5 shows the sorted lists for the TA algorithm. The algorithm essentially scans these lists concurrently and computes the aggregated scores of the documents that are top of each list. This may incur random access to probe for values of other attributes of these objects. These documents are ordered based on their scores. Meanwhile, a threshold value, T, is computed for documents whose aggregated values have not been fully derived. This threshold value is computed as an upper bound based on the large known scores in all the lists. The process continues with documents whose scores have not been fully computed until k answers are reported. Variants of the TA scheme, such as CA algorithm and RCA algorithm, that aim to exploit sequential access have been proposed and shown to be more efficient than the TA scheme [6].

Key Applications

There are many applications of spatial-keyword queries. Some of these include:

- *Location-based services.* Users can explore places of interests or services around the neighborhood of their locations that match their query keywords. For example, a Chinese restaurant selling peking duck within 300 m of a user is a more relevant answer to a user's query for "peking duck" than another restaurant selling the same item that is 5 km away.
- *Social news exploration.* Queries such as "dengue fever" are also distance sensitive as users may be eager to know whether there are places around their neighborhood that may be dengue hotspots (i.e., with high number of reported cases of dengue fever).
- *Location-based social networks.* There are increasingly many social network applications where locations play an important role. For example, a user looking for friends that share some common interests would prefer answers

to be ordered by the distance from his/her current location.

Future Directions

Traditionally, the user is assumed to be at a fixed location when the query is issued and the distance is based on the Euclidean distance. However, today's users always access the Internet while on the move. For example, a user may be driving his family out for dinner, and as he drives, he is searching for a "Chinese restaurant" with "peking duck" on its menu. Such continuous queries pose two key challenges. First, the context is on a road network. Second, the query is a "moving" query – while the textual part of the query remains the same (Chinese restaurant peking duck), the spatial component changes at a fast rate. For the former, the distance has to be based on the road network, and hence existing solutions based on the Euclidean space are no longer applicable. For the latter, the naive solution of repeatedly issuing the query as the user moves is unacceptable as this is bandwidth and computation inefficient. Some works have been done in these directions. In [8, 10], continuous spatial-keyword queries are studied. Each query is associated with a safe region under which the answers will not change. This means that as long as the user stays within the safe region, no further processing is required. However, once the user steps out of the safe region, the query has to be recomputed at the last known location, and the safe region has to be refreshed. On the other hand, in [9, 10], schemes for processing spatial-keyword queries on road networks are studied. However, designing efficient solutions to support these scenarios for a large-scale user base in real-time remains a challenge.

Another promising direction that requires further research is the development of location-aware publish/subscribe systems. In this context, subscriptions registered by subscribers and messages published by publishers include both spatial information and textual descriptions. For example, a Groupon subscriber in a shopping mall (location) may be interested in "Jordan shoes discount," and a Groupon message from a branded shoe store (location) may have the following promotion "Jordan running shoes up to 50% off." Clearly, messages should be delivered to relevant subscribers whose subscriptions match the message in terms of both the spatial and textual content. In our example, the system should send the message to relevant subscribers who are in the vicinity of the store. While some initial works have been done [11–13], the key challenge remains to develop a high-performance location-aware publish/subscribe system that can support millions of subscriptions and filter messages in real time.

Cross-References

▶ Continuous Query
▶ Inverted Files
▶ R-Tree (and Family)

Recommended Readings

1. Cong G, Jensen CS, Efficient DW. Retrieval of the top-k most relevant spatial web objects. Proc VLDB Endow. 2009;2(1):337–48.
2. Li Z, Lee K, ZHeng B, Lee W, Lee D, Wang X. IR-tree: an efficient index for geographic document search. IEEE Trans Knowl Data Eng. 2011;23(4):585–99.
3. Chen L, Cong G, Jensen CS, Wu D. Spatial keyword query processing: an experimental evaluation. Proc VLDB Endow. 2013;6(3):217–28.
4. Rocha-Junior JB, Gkorgkas O, Jonassen S, Nørvag K. Efficient processing of top-k spatial keyword queries. In: Proceedings of the 12th International Symposium on Spatial and Temporal Databases; 2011. p. 205–22.
5. Zhang D, Tan KL, Tung AKH. Scalable top-k spatial keyword search. In: Proceedings of the 16th International Conference on Extending Database Technology; 2013. p. 359–70.
6. Zhang D, Chan CY, Tan KL. Processing spatial keyword query as a top-k aggregation query. In: Proceedings of the 34th Annual International ACM SIGIR Conference on Research and Development in Information Retrieval; 2014. p. 355–64.
7. Fagin R, Lotem A, Naor M. Optimal aggregation algorithms for middleware. J Comput Syst Sci. 2003;66(4):614–56.
8. Huang W, Li G, Tan KL, Feng J. Efficient safe-region construction for moving top-k spatial keyword queries. In: Proceedings of the 21st ACM Interna-

tional Conference on Information and Knowledge Management; 2012. p. 932–41.

9. Rocha-Junior JB, Nørvag K. Top-k spatial keyword queries on road networks. In: Proceedings of the 15th International Conference on Extending Database Technology; 2012. p. 168–79.

10. Guo L, Shao J, Aung HH, Tan KL. Efficient continuous top-k spatial keyword queries on road networks. GeoInformatica. 2015;19(1):29–60.

11. Hu H, Liu Y, Li G, Feng J, Tan KL. A location-aware publish/subscribe framework for parameterized Spatio-textual subscriptions. In: Proceedings of the 31st International Conference on Data Engineering; 2015. p. 711–22.

12. Chen L, Cong G, Cao X, Tan KL. Temporal spatial-keyword top-k publish/subscribe. In: Proceedings of the 31st International Conference on Data Engineering; 2015. p. 255–66.

13. Guo L, Zhang D, Li G, Tan KL, Bao Z. Location-aware pub/sub system: when continuous moving queries meet dynamic event streams. In: Proceedings of the ACM SIGMOD International Conference on Management of Data; p. 843–57.

Spatiotemporal Data Mining

Nikos Mamoulis
University of Hong Kong, Hong Kong, China

Synonyms

Data mining in moving object databases

Definition

The extraction of implicit, nontrivial, and potentially useful abstract information from large collections of spatio-temporal data are referred to as spatio-temporal data mining. There are two classes of spatio-temporal databases. The first category includes timestamped sequences of measurements generated by sensors distributed in a map and temporal evolutions of thematic maps (e.g., weather maps). The second class is moving object databases that consist of object trajectories (e.g., movements of cars in a city). A trajectory can be modeled as a sequence of (p_i, t_i) pairs, where p_i corresponds to a spatial location and t_i is a timestamp. The management and analysis of spatio-temporal data has gained interest recently, mainly due to the rapid advancements in telecommunications (e.g., GPS, cellular networks, etc.), which facilitate the collection of large datasets of object locations (e.g., cars, mobile phone users) and measurement sequences (e.g., sensor readings). Mining tasks for moving object databases include detection and prediction of traffic jams, analyzing the movement behavior of animals, clustering or classification of moving objects according to their direction and/or speed, and identification of trends that associate the movement/speed of objects to their destination. In addition, from databases of measurement sequences, spatial relationships between correlated or anticorrelated sequences can be extracted (e.g., "sensors within 10 m from each other produce similar readings with high probability") or build classification models to detect abnormal combinations of sensor readings. The analysis of spatio-temporal databases is challenging due to the vast amount of collected data, and the complexity of novel mining tasks. Special issues include the fuzzy and implicit nature of spatio-temporal relationships between objects, the complex geometry of spatial objects, the varying temporal nature of events (instantaneous vs. durable), the variability of spatio-temporal data (moving objects, evolution of spatial events or phenomena, etc.), and the multiple (spatial and temporal) resolution levels of abstraction.

Historical Background

Data mining became a core field of database research in the 1990s [6]. Initial research focused on mining tasks (association analysis, classification, and clustering) applied on relational databases or transactional data that record sets of items purchased together. Two parallel streams of research were born soon after the first papers; data mining for temporal and spatial data. Temporal data mining focused on the extraction of sequential patterns from ordered transactional data or event sequences. Clustering and classification of time series, a classic problem in statistics, also

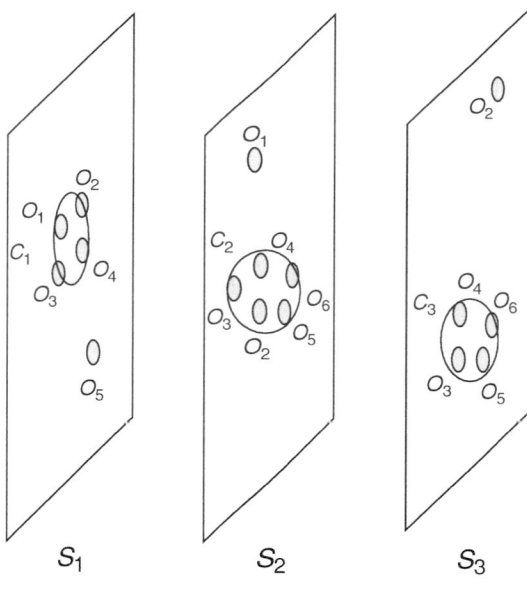

Spatiotemporal Data Mining, Fig. 2 Example of a moving cluster

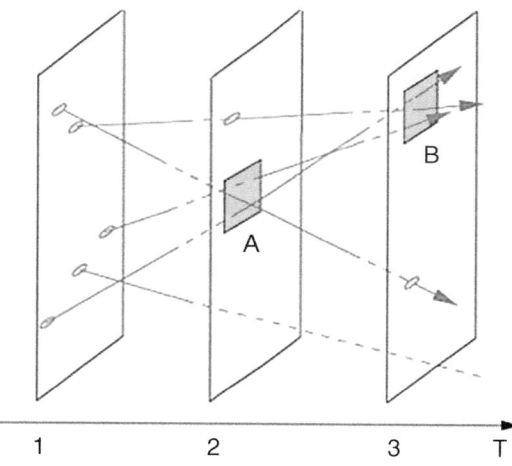

Spatiotemporal Data Mining, Fig. 3 Identifying areas of high object density

from [5]) shows a graphical example of such dense areas (*A* in timestamp 2 and *B* in timestamp 3), assuming that at least three objects must exist in an area of one square unit for the area to be considered dense. The discovery of dense regions is done after defining a space-time 3D grid and merging neighboring dense cells in this grid. Note that this problem is different to classic clustering, where objects are grouped based on the whole history (or trend) of their movement.

Classification and Prediction

Classification of trajectories is usually performed by nearest neighbor (NN) classifiers [12]. Given a trajectory $\vec{s}$ of unknown label and a database $\mathcal{D}$ of labeled samples, such a classifier (i) searches in $\mathcal{D}$ for the k most similar time series to $\vec{s}$ and (ii) gives $\vec{s}$ the most popular label in the set of k returned time series. NN classifiers, like clustering algorithms, rely on an appropriate similarity function between trajectories. Depending on the application, one of the distance functions used for clustering (as discussed above) can be used.

A related task to classification is predicting the future movement of an object given its past locations. Regression models for one-dimensional time series can be extended for (multidimensional) moving object trajectories. The movement of objects is approximated by (mixtures of) functions, which in turn are used for prediction. Given the recent past movement of an object [10] propose a methodology that computes a recursive motion function, a concise form that captures a large number of movement types (e.g., polynomials, ellipses, sinusoids, etc.). A recursive function differs from classic regression functions in that it relates an object's location to those of the recent past.

Pattern Extraction

There is limited work on extraction of patterns from spatio-temporal databases, which has been treated as a generalization of pattern mining in time series data. For example, [11] studied the discovery of frequent patterns related to changes of natural phenomena (e.g., temperature changes) in spatial regions. The locations of objects or the changes of natural phenomena over time are converted to categorical values. For instance, the map can be divided into spatial regions and replace the location of the object at each timestamp, by the region-id where it is located. Similarly, the change of temperature in a spatial region can be modeled as a sequence of temperature values. Continuous domains of the resulting time series data are discretized, prior to mining. In the case of multiple moving objects (or time series), trajectories are typically concatenated to a single long sequence. Then, an algorithm that discovers frequent subsequences in a long sequence (e.g., [13]) is applied.

In many applications, the movements obey periodic patterns, i.e., the objects follow the same routes (approximately) over regular time intervals. Objects that follow approximate periodic patterns include transportation vehicles (buses, boats, airplanes, trains, etc.), animal movements, mobile phone users, etc. For example, Bob wakes up at the same time and then follows, more or less, the same route to his work every day. Periodic patterns can be thought of as (possibly noncontiguous) sequences of object locations that reappear in the movement history periodically.

Formally, let S be a sequence of n spatial locations $\{l_0, l_1, \ldots, l_{n-1}\}$, representing the movement of an object (e.g., Bob) over a long history. Let $T \ll n$ be an integer called *period* (e.g., day, week, month). A *periodic segment* s is defined by a subsequence $l_i\, l_{i+1}\, \ldots\, l_{i+T-1}$ of S, such that i modulo $T = 0$. Thus, segments start at positions $0, T, \ldots, \left(\left\lfloor \frac{n}{T} \right\rfloor - 1\right) \cdot T$, and there are exactly $m = \left\lfloor \frac{n}{T} \right\rfloor$ periodic segments in S. A *periodic pattern* P is defined by a sequence $r_0\, r_1\, \ldots\, r_{T-1}$ of length T, such that r_i is either a spatial region or $*$. The *length* of a periodic pattern P is the number of non-$*$ regions in P. A segment s^j is said to *comply with* P, if for each $r_i \in P$, $r_i = *$ or s_i^j is *inside* region r_i. The *support* of a pattern P in S is defined by the number of periodic segments in S that comply with P. The same symbol P is used to refer to a pattern and the set of segments that comply with it. Let $min\_sup \leq m$ be

a positive integer (*minimum support*). A pattern P is *frequent*, if its support is larger than min_sup. Patterns for which the regions r_i are too sparse are not interesting; therefore, a constraint is imposed to the density of these regions. Let S^P be the set of segments that comply with a pattern P. Then each region r_i of P is *valid* if the set of locations $R_i^P := \left\{ s_i^j \mid s^j \in S^P \right\}$ form a *dense cluster*.

The discovery of partial periodic patterns from a long trajectory has been studied in [9]. This process is performed in two phases. First, S is divided into T spatial datasets, *one for each offset* of the period T. Specifically, locations $\{l_i, l_{i+T}, \ldots, l_{i+(m-1)\cdot T}\}$ go to set R_i, for each $0 \leq i < T$ (m is the length of S). A spatial clustering algorithm is applied to discover clusters in each R_i. These clusters define periodic patterns of length 1. Figure 4a shows the spatial datasets obtained after decomposing the trajectory of an object in three consecutive days (periods). A different symbol is used to denote locations that correspond to different periodic offsets and different colors are used for different segment-ids. Observe that a dense cluster r in dataset R_i corresponds to a frequent pattern, having $*$ at all positions and r at position i. Figure 4b shows examples of five clusters discovered in datasets R_1, R_2, R_3, R_4, and R_6. These correspond to five 1-patterns (i.e., $r_{11}*****$, $*r_{21}****$, etc.).

In the second phase, [9] extend the Apriori algorithm [6] to identify longer patterns level

Spatiotemporal Data Mining, Fig. 4 Locations and regions per periodic offset

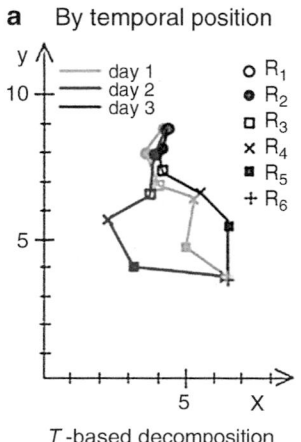

a By temporal position

T-based decomposition

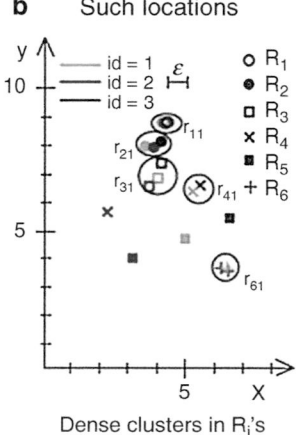

b Such locations

Dense clusters in R_i's

**Spatiotemporal Data
Mining, Fig. 5**
Subsequence
approximation

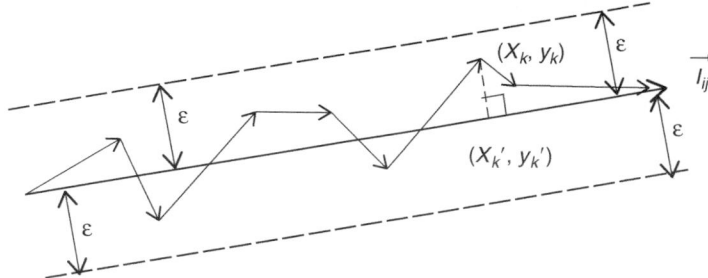

by level. In specific, pairs $\langle P_1, P_2 \rangle$ of frequent $(k-1)$-patterns with their first $k-2$ non-* regions in the same position and different $(k-1)$th non-* position create candidate k-patterns, the supports of which are counted at the next pass of the data sequence.

There has also been research on spatio-temporal pattern mining, where trajectories are regarded as sequences of locations (without giving any importance to the timestamps). In this case, the objective is to extract route-patterns of moving objects irrespectively to their speed. In this spirit, [2] define spatio-temporal patterns as sequences of line segments that form frequently followed routes by moving objects. A line simplification algorithm is used to approximate subsequences of trajectories by line segments. Figure 5 illustrates a subsequence s_{ij} which is approximated by a line segment $\overrightarrow{l_{ij}}$, such that the maximum distance of any point from s_{ij} to its projection on $\overrightarrow{l_{ij}}$ is at most ε. Simplified line segments of subsequences are clustered to form spatial regions (pattern elements) that can approximate a large number of subsequences. If a region approximates more than *min_sup* subsequences, then it forms a frequent movement pattern of length 1. A generalized frequent movement pattern is an m-length ordered sequence of pattern elements that is supported by (i.e., approximates) more than *min_sup* subsequences. The enumeration of frequent patterns is performed in two phases; first, the frequent 1-patterns are identified with the help of the clustering algorithm that is based on line simplification; then, longer patterns are found by employing a substring tree which compresses overlapping sequences of pattern elements.

Key Applications

Traffic Analysis

A motivating application of spatio-temporal data mining is to predict and analyze the causalities of traffic phenomena. Clustering or pattern extraction can help towards this purpose, since common routes that pass through the same map locations are likely the cause of traffic. Analyzing the causes of such clusters can lead to better transportation design for a city map.

Studying the Movement Behavior of Animals

With the help of GPS technology, the movements of animals can be tracked and analyzed. Identifying clusters or movement patterns can help in understanding the behavior of animals, such as the formulation and maintenance of herds, motion trends based on weather conditions, etc.

Video Analysis

The identification of objects and their movement behavior in video scenes is also an important application of spatio-temporal data mining. In this case, patterns of movement behavior can be extracted and analyzed. For example, from a soccer game, one can extract the movement style of players. Or, from a martial arts video, one can analyze movement sequences of body parts and relate them to the objective of the subject.

Cross-References

▸ Data Mining
▸ Geometric Stream Mining

▶ Spatial and Spatiotemporal Data Models and Languages
▶ Spatial Data Mining
▶ Spatiotemporal Data Warehouses
▶ Spatiotemporal Trajectories
▶ Temporal Data Mining

Recommended Reading

1. Berndt D, Clifford J. Using dynamic time warping to find patterns in time series. In: Proceedings of the AAAI-94 Workshop on Knowledge Discovery in Databases; 1994.
2. Cao H, Mamoulis N, Cheung DW. Mining frequent spatio-temporal sequential patterns. In: Proceedings of the 5th IEEE International Conference on Data Mining; 2005. p. 82–89.
3. Das G, Gunopulos D, Mannila H. Finding similar time series. In: Advances in Knowledge Discovery and Data Mining, 1st Pacific-Asia Conference; 1997. p. 88–100.
4. Gaffney S, Smyth P. Trajectory clustering with mixtures of regression models. In: Proceedings of the 5th ACM SIGKDD International Conference on Knowledge Discovery and Data Mining; 1999. p. 63–72.
5. Hadjieleftheriou M, Kollios G, Gunopulos D, Tsotras VJ. On-line discovery of dense areas in spatiotemporal databases. In: Proceedings of the 8th International Symposium on Advances in Spatial and Temporal Databases; 2003. p. 306–324.
6. Han J, Kamber M. Data mining: concepts and techniques. San Francisco: Morgan Kaufmann; 2000.
7. Kalnis P, Mamoulis N, Bakiras S. On discovering moving clusters in spatio-temporal data. In: Proceedings of the 9th International Symposium on Advances in Spatial and Temporal Databases; 2005. p. 364–381.
8. Lee J-G, Han J, Whang K-Y. Trajectory clustering: a partition-and-group framework. In: Proceedings of the ACM SIGMOD International Conference on Management of Data; 2007. p. 593–604.
9. Mamoulis N, Cao H, Kollios G, Hadjieleftheriou M, Tao Y, Cheung D.W. Mining, indexing, and querying historical spatio-temporal data. In: Proceedings of the 10th ACM SIGKDD International Conference on Knowledge Discovery and Data Mining; 2004. p. 236–245.
10. Tao Y, Faloutsos C, Papadias D, Liu B. Prediction and indexing of moving objects with unknown motion patterns. In: Proceedings of the ACM SIGMOD International Conference on Management of Data; 2004. p. 611–622.
11. Tsoukatos I, Gunopulos D. Efficient mining of spatiotemporal patterns. In: Proceedings of the 7th International Symposium on Advances in Spatial and Temporal Databases; 2001. p. 425–442.
12. Vlachos M, Gunopulos D, Kollios G. Discovering similar multidimensional trajectories. In: Proceedings of the 18th International Conference on Data Engineering; 2002. p. 673–684.
13. Zaki MJ. Spade: an efficient algorithm for mining frequent sequences. Mach Learn. 2001;42(1/2): 31–60.

Spatiotemporal Data Types

Ralf Hartmut Güting
Fakultät für Mathematik und Informatik,
Fernuniversität Hagen, Hagen, Germany
Computer Science, University of Hagen, Hagen,
Germany

Synonyms

Data types for moving objects; Spatiotemporal data types

Definition

Abstract data types to represent time-dependent geometries, in particular continuously changing geometries, or *moving objects*. The most important types are *moving point* and *moving region*.

Key Points

A *moving point* represents an entity for which only the time-dependent position is of interest. A *moving region* describes an entity for which the time-dependent location as well as the shape and extent are relevant. For example, moving points could represent people; vehicles such as cars, trucks, ships, or airplanes; or animals; moving regions could be hurricanes, forest fires, spread of epidemic diseases, etc. Moving point data may be captured by GPS devices or RFID tags; moving region data may result from processing sequences of satellite images, for example. Geometrically, moving points or moving regions exist in a 3D (2D + time) space if the movement is modeled within the 2D plane; for moving points this can be easily extended to 4D (3D + time).

Beyond the most relevant types of moving point and moving region, to obtain a closed system, there are related time-dependent data types such as real-valued functions or time-dependent Boolean values. To have a uniform terminology, these types are also called *moving real* and *moving bool*, respectively. Static spatial data types, such as *point*, *line*, or *region*, and standard data types are also needed. The data types include suitable operations such as:

trajectory: *mpoint* → *line*	Projection of a moving point into the plane
inside: *mpoint* × *mregion* → *mbool*	When is a moving point inside a moving region
distance: *mpoint* × *point* → *mreal*	Distance between a moving and a static point

The basic concept of spatiotemporal data types is developed in [2]. A comprehensive design of types and operations can be found in [4]. Implementation of data types and operations is described in [1, 3], respectively.

Cross-References

▶ Moving Objects Databases and Tracking

References

1. Cotelo Lema JA, Forlizzi L, Güting RH, Nardelli E, Schneider M. Algorithms for moving object databases. Comput J. 2003;46(6):680–712.
2. Erwig M, Güting RH, Schneider M, Vazirgiannis M. Spatiotemporal data types: an approach to modeling and querying moving objects in databases. GeoInformatica. 1999;3:265–91.
3. Forlizzi L, Güting RH, Nardelli E, Schneider M. A data model and data structures for moving objects databases. In: Proceedings of the ACM SIGMOD Conference; Dallas, TX, USA. 2000. p. 319–30.
4. Güting RH, Böhlen MH, Erwig M, Jensen CS, Lorentzos NA, Schneider M, Vazirgiannis M. A foundation for representing and querying moving objects in databases. ACM Trans Database Syst. 2000; 25:1–42.

Spatiotemporal Data Warehouses

Yufei Tao[1] and Dimitris Papadias[2]
[1]Chinese University of Hong Kong, Hong Kong, China
[2]Department of Computer Science and Engineering, Hong Kong University of Science and Technology, Kowloon, Hong Kong, Hong Kong

Synonyms

Spatio-temporal online analytical processing; Spatio-Temporal OLAP

Definition

Consider N regions $R_1, R_2, \ldots, R_N$ and a time axis consisting of discrete timestamps $1, 2, \ldots, T$, where T represents the total number of recorded timestamps (i.e., the length of history). The position and area of a region R_i may vary along with time, and its extent at timestamp t is denoted as $R_i(t)$. Each region carries a set of *measures* $R_i(t).ms$, also called the *aggregate data* of $R_i(t)$. The measures of regions change asynchronously with their extents. In other words, the measure of R_i $(1 \leq i \leq N)$ may change at a timestamp t (i.e., $R_i(t).ms \neq R_i(t-1).ms$), while its extent remains the same (i.e., $R_i(t) = R_i(t-1)$), and vice versa.

A *spatio-temporal data warehouse* stores the above information, and efficiently answers the *spatio-temporal window aggregate* query, which specifies an area q_R and a time interval q_T of continuous timestamps. The goal is to return the aggregated measure $Agg(q_R, q_T, f_{agg})$ of all regions that intersect q_R during q_T, according to some distributive aggregation function f_{agg}, or formally:

$$Agg\left(q_R,\ q_T,\ f_{agg}\right) = f_{agg}$$
$$\left\{R_i(t).ms \mid R_i(t) \text{ intersects } q_R \text{ and } t \in q_T\right\}.$$

If q_T involves a single timestamp, the query is a *timestamp query*; otherwise, it is an *interval query*. An example of a timestamp window aggregate query is "find the total number of mobile users in the city center at 12 p.m." The query will summarize the number of users (measures) in all regions intersecting q_R = "city center" at q_T = "$12_{\text{P.M.}}$"

Historical Background

The motivation behind spatio-temporal data warehouses is that many spatio-temporal applications require summarized results, rather than information about individual objects. As an example, traffic supervision systems monitor the number of cars in an area of interest, instead of their ids. Similarly, mobile phone companies use the number of phone-calls per cell in order to identify trends and prevent potential network congestion. Although summarized results can be obtained using conventional operations on individual objects (i.e., accessing every single record qualifying the query), the ability to manipulate aggregate information *directly* is imperative in spatio-temporal databases due to several reasons. First, in some cases personal data should not be stored due to legal issues. For instance, keeping historical locations of mobile phone users may violate their privacy. Second, the individual data may be irrelevant or unavailable, as in the traffic supervision system mentioned above. Third, although individual data may be highly volatile and involve extreme space requirements, the aggregate information usually remains fairly constant for long periods, thus requiring considerably less space for storage.

A considerable amount of related research has been carried out on *data warehouses* and

OLAP (on line analytical processing) in the context of relational databases. The most common conceptual model for data warehouses is the multi-dimensional data view. In this model, each *measure* depends on a set of *dimensions*, e.g., *region* and *time*, and thus is a value in the multi-dimensional space. A dimension is described by a domain of values (e.g., days), which may be related via a hierarchy (e.g., day-month-year). Fig. 1 illustrates a simple case, where each cell denotes the measure of a region at a certain timestamp. Observe that although regions are 2-dimensional, they are mapped as one dimension in the warehouse.

The *star schema* [6] is a common way to map a data warehouse onto a relational database. A main table (called *fact table*) F stores the multi-dimensional array of measures, while auxiliary tables $D_1, D_2, \ldots, D_n$ store the details of the dimensions. A tuple in F has the form $<D_i[].key, M[]>$ where $D_i[].key$ is the set of foreign keys to the dimension tables and $M[]$ is the set of measures. OLAP operations ask for a set of tuples in F, or for aggregates on groupings of tuples. Assuming that there is no hierarchy in the dimensions of the previous example, the possible groupings in Fig. 1 include: (i) group-by Region and Time, which is identical to F, (ii)-(iii) group-by Region (Time), which corresponds to the projection of F on the *region-* (*time-*) axis, and (iv) the aggregation over all values of F which is the projection on the origin (Fig. 1 depicts these groupings for the aggregation function *sum*). The

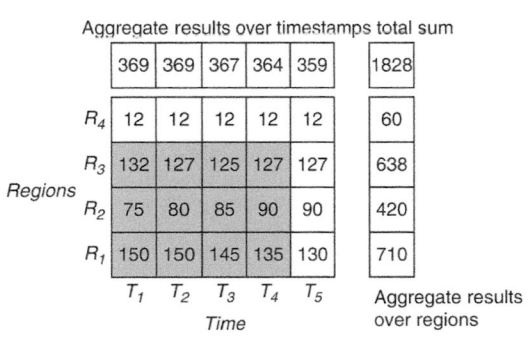

Aggregate results over timestamps total sum

	T_1	T_2	T_3	T_4	T_5	
	369	369	367	364	359	1828
R_4	12	12	12	12	12	60
R_3	132	127	125	127	127	638
R_2	75	80	85	90	90	420
R_1	150	150	145	135	130	710

Regions

Time

Aggregate results over regions

Fact table

Spatiotemporal Data Warehouses, Fig. 1 A data cube example fact table together with all possible combinations of group-bys composes the *data cube* [2]. Although all groupings can be derived from F, in order to accelerate query processing some results may be pre-computed and stored as *materialized views*.

A detailed group-by query can be used to answer more abstract aggregates. In the example of Fig. 1, the total measure of all regions for all timestamps (i.e., 1828) can be computed either from the fact table, or by summing the projected results on the *time* or *region* axis. Ideally, the whole data cube should be materialized to enable efficient query processing. Materializing all possible results may be prohibitive in practice as there are $O(2^n)$ group-by combinations for a data warehouse with n dimensional attributes. Therefore, several techniques have been proposed for the view selection problem in OLAP applications [1, 4]. In addition to relational databases, data warehouse techniques have also been applied to spatial [3, 10] and temporal [8] databases. All these methods, however, benefit only queries on a predefined hierarchy. An ad-hoc query not confined by the hierarchy, such as the one involving the gray cells in Fig. 1, would still need to access the fact table, even if the entire data cube were materialized.

Foundations

The next discussion describes several solutions to implementing a spatio-temporal data warehouse, assuming summation as the underlying aggregate function f_{agg}. Extensions to other aggregation functions (e.g., count, average) are straightforward.

- *Using a 3D aggregate R-tree*

The problem of a spatio-temporal window aggregate search can be regarded as a multi-dimensional aggregate retrieval in the 3D space (the spatial dimensions plus a time dimension) and solved using an aggregate R-tree (aR-tree). The aR-tree [5, 9] is similar to a conventional R-tree, where each node also stores summarized information about the regions in each sub-tree. Whenever the extent or measure of a region changes, a new 3D box is inserted in a 3D version of the aR-tree, called the *a3DR-tree*. Using the example of Fig. 1, four entries are required for R_1: one for timestamps 1 and 2 (when its measure remains 150) and three more entries for the other timestamps. A spatio-temporal window aggregate query can also be modeled as a 3D box, which can be processed on the a3DR-tree, following the strategy of solving a range aggregate query on an aR-tree [5, 9].

The problem with this solution is that it creates a new box duplicating the region's extent, even though it does not change. Since the measure changes are much more frequent than extent updates, the *a3DR-tree* incurs high redundancy. The worst case occurs for static regions: although the extent of a region remains constant, it is still duplicated at the rate of its measure changes. Bundling the extent and aggregate information in all entries significantly lowers the node fanout and compromises query efficiency, because more nodes must be accessed to retrieve the same amount of information. Note that redundancy incurs *whenever the extent and measure changes are asynchronous*, i.e., the above problem also exists when a new box is spawned because of an extent update, in which case the region's measure must be replicated.

- *Using a data cube*

Following the traditional data warehouse approach, it is possible to create a data cube, where one axis corresponds to time, the other to regions, and keep the measure values in the cells of this two-dimensional table (see Fig. 1). Since the spatial dimension has no one-dimensional order, the table can be stored in the secondary memory ordered by time, and a B-tree index can be created to locate the pages containing information about each timestamp. The processing of a query employs the B-tree index to retrieve the pages (i.e., table columns) containing information about q_T; then, these regions (qualifying the temporal condition) are scanned sequentially and the measures of those satisfying q_R are aggregated.

Even if an additional spatial index on the regions exists, the simultaneous employment of both indexes has limited effect. Assume that first a window query q_R is performed on the spatial index to provide a set of ids for regions that qualify the spatial condition. Measures of these regions must still be retrieved from the columns corresponding to q_T (which, again, are found through the B-tree index). However, the column storage does not preserve spatial proximity, and hence the spatially qualifying regions are expected to be scattered in different pages. Therefore, the spatial index has some effect only on very selective queries (on the spatial conditions). Furthermore, recall that pre-materialization is useless, since the query parameters q_R and q_T do not conform to pre-defined groupings.

- *The aggregate R-B-tree*

Since (i) the extent and measure updates are asynchronous and (ii) in practice, measures change much more frequently than extents (which may even be static), the two types of updates should be managed independently to avoid redundancy. This implies the deployment of two types of indexes: (i) a *host index*, which is an aggregate spatial or spatio-temporal structure managing region extents, and (ii) numerous *measure indexes* (one for each entry of the host index), which are aggregate temporal structures storing the values of measures during the history. Figure 2 shows a general overview of the architecture [13]. Given a query, the host index is first searched, identifying the set of entries that qualify the spatial condition. The measure indexes of these entries are then accessed to retrieve the timestamps qualifying

the temporal conditions. Since the number of records (corresponding to extent changes) in the host index is very small compared to the measure changes, the cost of query processing is expected to be low.

The following discussion explains an instantiation of the architecture for solving window aggregate queries when the underlying data regions are static. The instantiation leads to a structure, called the *aggregate R- B-tree (aRB-tree)*. It adopts an aR-tree as the host index, where an entry r has the form $< r$.MBR, r.aggr, r.pointer, r.btree>; r.MBR and r.pointer have the same semantics as a normal R-tree, r.aggr keeps the aggregated measure about r over the *entire history*, and r.btree points to an aggregate *B-tree* which stores the detailed measure information of r at concrete timestamps. Figure 3b illustrates an example using the data regions of Fig. 3a and the measures of Fig. 1. The number 710 stored with R-tree entry R_1, equals the sum of measures in R_1 for all 5 timestamps (e.g., the total number of phone calls initiated at R_1). The first leaf entry of the B-tree for R_1 (1, 150) indicates that the measure of R_1 at timestamp 1 is 150. Since the measure of R_1 at timestamp 2 is the same, there is no a special entry, but this knowledge is implied from the previous entry (1, 150). Similarly, the first root entry (1, 445) of the same B-tree indicates that the aggregated measure in R_1 during time interval [1, 3] is 445. The topmost B-tree stores aggregated information about the whole space, and its role is to answer queries involving only temporal conditions (similar to that of the extra row in Fig. 1).

To illustrate the processing algorithms, consider the query "find the number of phone-calls initiated during interval $q_T = [1, 3]$ in all cells intersecting the window q_R shown in Fig. 3a." Starting from the root of the R-tree, the algorithm

Spatiotemporal Data Warehouses, Fig. 2 A multi-index architecture for indexing spatio-temporal data warehouses

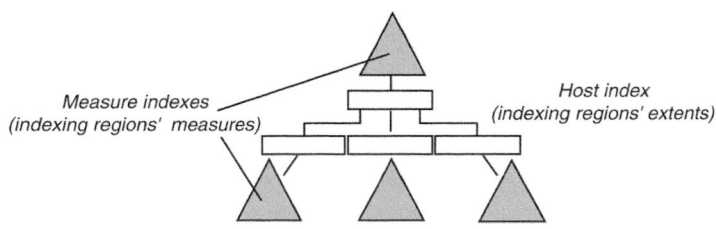

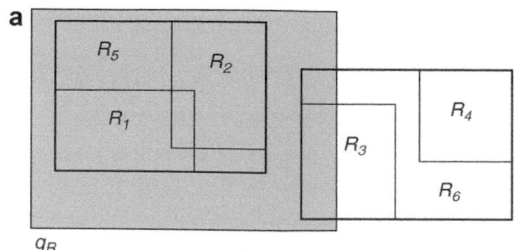

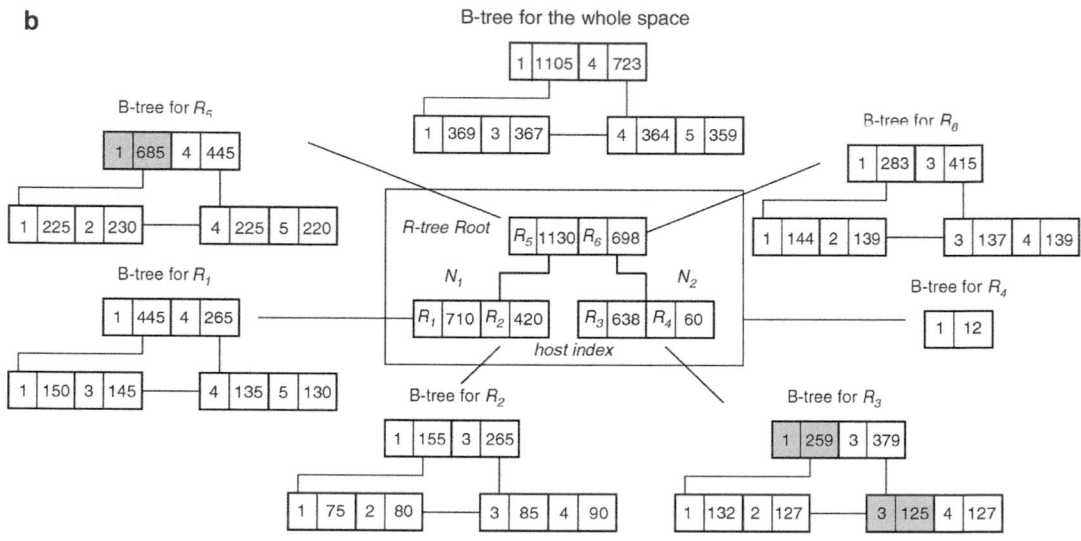

Spatiotemporal Data Warehouses, Fig. 3 A solution to static regions

visits the B-tree of R_5 since the entry is totally contained in q_R. The root of this B-tree has entries (1,685), (4,445) meaning that the aggregated measures (of all data regions covered by R_5) during intervals [1, 3], [4, 5] are 685 and 445, respectively. Hence, the contribution of R_5 to the query result is 685. The second root entry R_6 of the R-tree partially overlaps q_R, so its child node is visited, where only entry R_3 intersects q_R, and thus its B-tree is retrieved. The first entry of the root (of the B-tree) suggests that the contribution of R_3 for the interval [1, 2] is 259. In order to complete the result, the algorithm will have to descend the second entry and retrieve the measure of R_3 at timestamp 3 (i.e., 125). The final result equals $685 + 259 + 125$, which corresponds to the sum of measures in the gray cells of Fig. 3b.

Key Applications

Traffic Control

Traffic control systems require summarized information about areas of interest instead of the concrete vehicle ids. Furthermore, in most cases, approximate aggregation [11] is sufficient for tasks such as traffic jam detection and shortest path computation.

Mobile Computing

Measures such as the number of phone-calls per cell can help identify trends, prevent potential network congestion and achieve load balancing in mobile computing applications.

Sensor Systems

Spatio-temporal data warehouses can collect and store readings from geographically distributed sensors. As an example consider a pollution monitoring system, where the readings from several sensors are fed into a warehouse that arranges them in regions of similar or identical values. These regions should then be indexed for the efficient processing of queries such as "find the areas near the center with the highest pollution levels yesterday."

Future Directions

Tao and Papadias [12] discuss spatial aggregation techniques, i.e., when both the regions and their measures are static. Tao et al. [14] present a sketch-based aggregation technique that avoids counting the same object twice (*distinct counting problem*) during the computation of aggregates. A survey of spatio-temporal aggregation techniques can be found in [7].

Experimental Results

[13] contains an extensive set of spatio-temporal aggregation techniques for static and dynamic regions, and experimental (as well as analytical) comparisons.

Data Sets

Common benchmark datasets can be found at:

www.rtreeportal.org

Cross-References

► B+-Tree
► Data Warehouse
► Online Analytical Processing
► R-Tree (and Family)
► Star Schema

Recommended Reading

1. Baralis E, Paraboschi S, Teniente E. Materialized view selection in a multidimensional database. In: Proceedings of the 23th International Conference on Very Large Data Bases; 1997. p. 156–65.
2. Gray J, Bosworth A, Layman A, Pirahesh H. Data cube: a relational aggregation operator generalizing group-by, cross-tabs and subtotals. In: Proceedings of the 12th International Conference on Data Engineering; 1996. p. 152–9.
3. Han J, Stefanovic N, Koperski K. Selective materialization: an efficient method for spatial data cube construction. In: Advances in Knowledge Discovery and Data Mining, 2nd Pacific-Asia Conference; 1998. p. 144–58.
4. Harinarayan V, Rajaraman A, Ullman J. Implementing data cubes efficiently. In: Proceedings of the ACM SIGMOD International Conference on Management of Data; 1996. p. 205–16.
5. Jurgens M, Lenz H. The Ra*-tree: an improved R-tree with materialized data for supporting range queries on OLAP-data. In: Proceedings of the International Workshop on Database and Expert Systems Applications; 1998. p. 186–91.
6. Kimball R. The data warehouse toolkit. New York: Wiley; 1996.
7. Lopez I, Snodgrass R, Moon B. Spatiotemporal aggregate computation: a survey. IEEE Trans Knowl Data Eng. 2005;17(2):271–86.
8. Mendelzon A, Vaisman A. Temporal queries in OLAP. In: Proceedings of the 26th International Conference on Very Large Data Bases; 2000. p. 242–53.
9. Papadias D, Kalnis P, Zhang J, Tao Y. Efficient OLAP operations in spatial data warehouses. In: Proceedings of the 7th International Symposium on Advances in Spatial and Temporal Databases; 2001. p. 443–59.
10. Stefanovic N, Han J, Koperski K. Object-based selective materialization for efficient implementation of spatial data cubes. IEEE Trans Knowl Data Eng. 2000;12(6):938–58.
11. Sun J, Papadias D, Tao Y, Liu B. Querying about the past, the present and the future in spatio-temporal databases. In: Proceedings of the 20th International Conference on Data Engineering; 2004. p. 202–13.
12. Tao Y, Papadias D. Range aggregate processing in spatial databases. IEEE Trans Knowl Data Eng. 2004;16(12):1555–70.
13. Tao Y, Papadias D. Historical spatio-temporal aggregation. ACM Trans Inf Syst. 2005;23(1): 61–102.
14. Tao Y, Kollios G, Considine J, Li F, Papadias D. Spatio-temporal aggregation using sketches. In: Proceedings of the 20th International Conference on Data Engineering; 2004. p. 214–25.

S

Spatiotemporal Interpolation Algorithms

Peter Revesz
University of Nebraska-Lincoln, Lincoln, NE,
USA

Synonyms

Moving objects interpolation; Spatiotemporal approximation; Spatiotemporal estimation

Definition

Spatiotemporal interpolation is the problem of estimating the unknown values of some property at arbitrary spatial locations and times, using the known values at spatial locations and times where measurements were made. In spatiotemporal interpolation the estimated property varies with both space and time, with the assumption that the values are closer to each other with decreasing spatial and temporal distances.

Spatiotemporal interpolation is used in spatiotemporal databases, which record spatial locations and time instances together with other attributes that are dependent on space and time. For example, a spatiotemporal database may record the sales of houses in a town. The house sales database records the location, usually as the address of the house from which an (x, y) location can be easily found, by correlating the address with a map of the town, the calendar date when the sale occurred, the area, and the sale price of the house.

Spatiotemporal interpolation is also used in moving objects databases to estimate the trajectory of moving objects.

Historical Background

Spatiotemporal interpolation is a generalization of spatial interpolation by the addition of a temporal dimension. While spatial interpolation is well investigated [1], spatiotemporal interpolation in general, including interpolation of the trajectory of moving objects, is a relatively new area [2, 3].

Spatiotemporal interpolation problems were either assumed to be just a sequence of spatial interpolation problems, called the *time slices approach*, or they were assumed to be as easily handled as adding one more spatial dimension, called the *extension approach*. Li and Revesz [4] pointed out problems with the time slices and the extension approaches and proposed the *spatiotemporal product* (*ST product*) approach. Gao and Revesz [5] described several *adaptive spatiotemporal interpolation* approaches that improve the ST product method.

Moving objects carrying positioning devices, thus recording their positions in arbitrary spatial locations along with their corresponding timestamps, provide data for moving object databases. In moving objects databases, a variety of spatiotemporal interpolation methods can be applied, e.g., linear interpolation [1], which is the most commonly adopted, and polynomial functions and splines [6–9].

Foundations

There are several problems with the simple spatiotemporal interpolation approaches, like the time slices approach and the extension approach. Among these problems the following need to be mentioned.

Poor Estimation Accuracy: The time slices approach is not accurate in general. For example, a spatiotemporal interpolation problem is to estimate the price of houses based on sales data from sold houses in a town. To solve this problem, the time slices approach would first select just the houses that were sold in one short time slice, for example, on a single day or in a single week. Then the time slices approach would do a spatial interpolation on the selected data. The accuracy of this method is often poor, because there are simply not too many houses sold on a single day or in a single week to cover the town dense enough for accurate interpolation. Many town subdivisions may not have a single house sold on just one day.

Lack of Continuous Time: The time slices approach cannot deal with continuous time.

Non-invariance to Scaling: The extension approach handles a k-dimensional spatial and one-dimensional temporal interpolation problem as if it were a k + 1-dimensional spatial interpolation problem. Li and Revesz [4] pointed out that the extension approach can also lead to problems because of the *non-invariance to scaling* of many spatial interpolation algorithms. Scaling invariance means that if the unit of measurement changes in one dimension, then the estimated value does not change. While many spatial interpolation algorithms are scaling invariant if the units change in each spatial dimension, few spatial interpolation algorithms are scaling invariant if the units change only in one dimension. For example, the inverse distance weighting (IDW) [10] spatial interpolation method is non-invariant to scaling in only one dimension. Such methods are difficult to use because it is difficult to decide what is the right unit of time given certain units of spatial distance. For example, IDW gives a different house price estimate if the unit of time is days than if the unit of time is weeks, even if all houses are always sold on Mondays. That is a strange phenomenon that makes IDW inherently awkward to use.

The spatiotemporal product (or ST product) approach [4] improves the time slices method. Using the house price estimation problem, the ST product approach can be explained as follows. For each house that is sold several times according to the sales spatiotemporal database, the ST product approach first estimates its price based on a simple linear temporal interpolation between two consecutive sales. For example, if a house was sold for $200,000 on July 29, 2000 and was sold for $250,000 on July 1, 2002, then one can estimate that the price of the house rose $500 for each of the 100 weeks between the two sales.

After these temporal interpolations, a spatial interpolation is done for each week like in the time slices method. However, since a large percentage of the values for each week are filled in, the ST product method uses a considerably greater density of houses in the spatial interpolation part than the time slices method does. Hence, in general, the ST product method is more accurate than the time slices method. Further, the ST product method can be applied without an exact calculation of the temporal interpolation values by using a temporal parameter t. Then the ST product method yields an interpolation function of x, y, and t for any (x, y) location and time instance t, even in continuous time.

The spatiotemporal accuracy is increased further by *adoptive spatiotemporal interpolation* [5]. To explain the adoptive method, suppose $R(x, y, t, w)$ is a spatiotemporal relation where (x, y) is the location, t is the time instance, and w is the measured value. The key idea behind the adaptive method is that to interpolate the value w at any location (a, b) at time c, where a, b, and c are constants, there are two main choices:

1. *Spatial projection + 1D temporal interpolation*: Select from R the records with $x = a$ and $y = b$. Then use any 1D temporal interpolation method on the selected records.
2. *Temporal projection + 2D spatial interpolation*: Select from R the records with $t = c$. Then use any 2D spatial interpolation method on the selected records.

When the spatial and temporal projections of R both contain enough number of records to do an interpolation at location (a, b) and time c, then one could choose either (1) or (2) as an option. These two options may give different interpolation values. Which one of the two estimates is more reliable? The *choice-based adaptive interpolation method* decides that question based on the *relationship strength measures* for space and time. Intuitively, the larger the relationship strength measure for space (or time), the more reliable is the estimation using a spatial (or temporal) interpolation. These measures are denoted by the following symbols:

- $S(a,b,c)$ – the relationship strength measure for space at location (a, b) and time c
- $T(a,b,c)$ – the relationship strength measure for time at location (a, b) and time c

Both measures are localized; hence, the spatial relationship strength can be larger than the temporal relationship strength at some locations and times, while the reverse may be true at a different location and time within the same problem. The choice-based adaptive method works as follows whenever there is a choice:

$$
\begin{cases}
\text{Spatialinterpolaton} & \text{if } S\,(a,b,c) > T\,(a,b,c) \\
\text{Temporal interpolation} & \text{if } S\,(a,b,c) \\
& > T\,(a,b,c)
\end{cases}
$$

(1)

Both the ST product and the choice-based adaptive methods first interpolate the missing values for the (a, b) locations for which several measurements at different times are known. While the ST product method always preferred to do a temporal interpolation, the adaptive method at some spatiotemporal locations will prefer to do a spatial interpolation. This is the only reason why the two interpolation results may disagree. Finally, for both methods the non-measurement locations, that is, the locations that do not appear in the original data set, need to be estimated using some 2D spatial interpolation.

Gao and Revesz [5] suggested several spatial and temporal relationship strength measures. A simple measure is to use the inverse of the variance of the measured values in the spatial neighborhood of (a, b), which is defined as the nearest k other (a, b) locations for some integer k, as the spatial relationship strength. Similarly, one can use the inverse of the variance of the two measures at the earliest time following c, and latest time preceding c, as the temporal relationship strength. As an alternative to the choice-based adaptive interpolation approach, [5] also proposed a linear combination adaptive interpolation approach, which gives a weighted linear sum of the two estimates. The weights depend on the relative magnitudes of the spatial and temporal relationship strength measures.

Trajectory of Moving Objects: The estimation of the trajectory of moving objects can also be viewed as a spatiotemporal interpolation problem. Applying the above described ideas, in each sensor location, the moving object can be either sensed with value one or not sensed and assigned a value zero. However, there are more advanced interpolation techniques, which associate each moving object's approximated position with an uncertainty factor [11–13].

Key Applications

Spatiotemporal interpolation has a growing number of applications, in areas such as the following:

1. *Real Estate Analysis*: In an experiment with a database of house sales in the town of Lincoln, Nebraska, USA, the ST product method had less than a 9 % mean absolute error in estimating house prices [4]. The surprising feature of the estimation was that nothing special was known about the houses except their locations, sizes in square feet, and prices. The estimation could be easily improved with a visit to the houses to check their conditions. The house price estimation can tell whether some given houses in given years were assessed too high or too low taxes.

2. *Weather Analysis*: Weather is a spatiotemporal phenomenon that shows several cyclical patterns. The ST product spatiotemporal interpolation method was used for groundwater level and drought analysis. Carbon dioxide concentration over time is another application area.

3. *Epidemiology*: Spatiotemporal interpolation was also used to predict the spread of epidemics, for example, the spread of the West Nile virus in the continental USA [14].

4. *Forest Fires*: The spread of forest fires can be predicted similar to the spread of epidemics.

5. *Traffic Accident Report*: In a traffic accident report, spatiotemporal interpolation may be used to estimate the trajectories of the vehicles that were involved in the accident. An accurate estimation may be important to decide what caused the accident and to get insurance payments.

6. *Air Pollution and Health*: Concentrations of air pollution, such as ozone and fine particulate matter $PM_{2.5}$, are usually measured at

monitoring site locations and at fixed time instances. Spatiotemporal interpolation algorithms are used to estimate or predict the pollutant concentrations at unmeasured locations and times. Correlating the spatiotemporal interpolated pollution data with spatiotemporal health data, such as melanoma prevalence, enables the discovery of novel associations between pollutants and their health effects [15–18].

Future Directions

There are still several problems about the relationship of spatiotemporal interpolation and prediction of spatiotemporal phenomena. Spatiotemporal interpolation seems to work best if one is interested in a location that is in the middle of the space and in a time that is in the middle of the time interval recorded. Predictions are at future times; hence, they use specialized algorithms. For example, voting prediction is a research area in itself and has many specialized techniques different from the spatiotemporal interpolation approaches. For example, when one is trying to predict the outcome of an election in a voting district A, one cannot use the time slices, the ST product, and the adoptive interpolation approaches because there are no spatial neighbors in general. There are some exceptions to this rule. For example, if the neighboring voting districts have all already reported their election results while the votes in district A are still being (re)counted, then one can predict the outcome in voting district A using the results of the neighboring districts [19].

While voting problems are almost always prediction problems, on occasion, analysis of past election results may be also interesting, for example, to identify and help settle possible election fraud cases in the past. Predicting other spatiotemporal behaviors besides voting, for example, the number of customers in a town who would buy a certain product, is also an important area of interest.

Spatiotemporal databases can be conveniently represented using *constraint databases* [20].

Constraint databases allow convenient handling of different types of interpolation data [21].

Cross-References

▶ Constraint Databases
▶ Moving Object
▶ Moving Objects Databases and Tracking

Recommended Reading

1. Davis PJ. Interpolation and approximation. Mineola: Dover; 1975.
2. Güting R, Schneider M. Moving objects databases. Los Altos: Morgan Kaufmann; 2005.
3. Revesz PZ. Introduction to databases: from spatiotemporal to biological. New York: Springer; 2010.
4. Li L, Revesz PZ. Interpolation methods for spatiotemporal geographic data. J Comput Environ Urban Syst. 2004;28(3):201–27.
5. Gao J, Revesz P. Adaptive spatiotemporal interpolation methods. In: Proceedings of the 1st International Conference on Geometric Modeling, Visualization, and Graphics; 2005. p. 1622–5.
6. Hadjieleftheriou M, Kollios G, Tsotras VJ, Gunopulos D. Efficient indexing of spatiotemporal objects. In: Advances in Database Technology, Proceedings of the 8th International Conference on Extending Database Technology; 2002. p. 251–68.
7. Koubarakis M, Sellis TK, Frank AU, Grumbach S, Güting RH, Jensen CS, Lorentzos NA, Manolopoulos Y, Nardelli E, Pernici B, Schek H-J, Scholl M, Theodoulidis B, Tryfona N, editors. Spatio-temporal databases: the CHOROCHRONOS approach, LNCS 2520. Berlin/New York: Springer; 2003.
8. Ni J, Ravishankar CV. Indexing spatio-temporal trajectories with efficient polynomial approximations. IEEE Trans Knowl Data Eng. 2007;19(5):663–78.
9. Yu B, Kim SH, Bailey T, Gamboa R. Curve-based representation of moving object trajectories. In: Proceedings of the 8th International Database Engineering and Applications Symposium; 2004. p. 419–25.
10. Shepard DA. A two-dimensional interpolation function for irregularly spaced data. In: Proceedings of the 23rd ACM National Conference; 1968. p. 517–24.
11. Kuijpers B, Othman W. Trajectory databases: data models, uncertainty and complete query languages. In: Proceedings of the 11th International Conference on Database Theory; 2007. p. 224–38.
12. Pfoser D, Jensen C.S. Capturing the uncertainty of moving-object representations. In: Proceedings of the 6th International Symposium on Advances in Spatial Databases; 1999. p. 111–32.
13. Trajcevski G, Wolfson O, Zhang F, Chamberlain S. The geometry of uncertainty in moving objects

S

databases. In: Advances in Database Technology, Proceedings of the 8th International Conference on Extending Database Technology; 2002. p. 233–50.

14. Revesz PZ, Wu S. Spatiotemporal reasoning about epidemiological data. Artif Intell Med. 2006;38(2):157–70.

15. Li L, Zhang X, Holt JB, Tian J, Piltner R. Estimating population exposure to fine particulate matter in the conterminous U.S. using shape function-based spatiotemporal interpolation method: a county level analysis. Int J Comput. 2012;1(4):24–30.

16. Li L, Zhang X, Piltner R. A spatiotemporal database for ozone in the conterminous U.S. In: Proceedings of the 13th International Symposium on Temporal Representation and Reasoning, Budapest Hungary; 2006. p. 168–76.

17. Li L, Zhang X, Piltner R. An application of the shape function based spatiotemporal interpolation method on ozone and population exposure in the contiguous U.S. J Environ Inf. 2008;12(2):120–8.

18. Losser T, Li L, Piltner R. A spatiotemporal interpolation method using radial basis functions for geospatiotemporal big data. In: Proceedings of the 5th International Conference on Computing for Geospatial Research and Application; 2014.

19. Gao J, Revesz P. Voting prediction using new spatiotemporal interpolation methods. In: Proceedings of the 7th International Conference on Digital Government Research; 2006. p. 293–300.

20. Kanellakis PC, Kuper GM, Revesz PZ. Constraint query languages. J Comput Syst Sci. 1995;51(1):26–52.

21. Grumbach S, Rigaux P, Segoufin L. Manipulating interpolated data is easier than you thought. In: Proceedings of the 26th International Conference on Very Large Data Bases; 2000. p. 156–65.

Spatiotemporal Selectivity Estimation

George Kollios
Boston University, Boston, MA, USA

Synonyms

Selectivity for predictive spatio-temporal queries

Definition

In spatio-temporal databases, the locations of moving objects are usually modeled as linear functions of time. Thus, the location of an object at time t is represented as $o(t) = o_s + o_v t$, where o_s is the initial location of the object at time $t = 0$ and o_v is its velocity. Given that the object moves in a d–dimensional space, $o(t)$, o_s, and o_v are d–dimensional vectors. In this setting, the selectivity estimation of spatio-temporal queries is defined as follows:

Given a database that stores the locations of moving objects and a spatio-temporal query, estimate the number of objects that satisfy the query.

There are two important types of queries in this environment: spatio-temporal window (or range) queries and spatio-temporal distance join queries. A spatio-temporal window query (*STWQ*) specifies a (static or moving) region q_S, a future time interval q_T, and a dataset of moving objects D, and retrieves all data objects that will intersect (or will be covered by) q_S during q_T. On the other hand, a spatio-temporal distance join query (*STDJQ*) assumes two datasets D_A and D_B of moving objects, a time period q_T and a distance q_d, and asks for the pairs of objects (o_a, o_b) that $o_a \in D_A$, $o_b \in D_B$ and are closer than q_d during the time interval q_T.

Historical Background

The problem of spatio-temporal selectivity estimation is related to spatio-temporal indexing and spatial selectivity estimation. In the spatio-temporal indexing problem the system must report all the objects that satisfy the query; here the system must just give an estimate on the number of such objects. In spatial selectivity estimation the main difference is that the objects are static and therefore easier to model and represent succinctly.

In particular, selectivity estimation for spatial queries is based on spatial histograms. A histogram partitions the data space into a set of buckets, and the object distribution in each bucket is assumed to be (almost) uniform. A bucket B stores the number $B.num$ of objects whose centroids fall in B, and the average extent $B.len$ of such objects. Consider a window query Q. Then, using an analysis based on uniformly distributed data [1, 2], the expected number of qualifying objects in B is approximated by $B.num \frac{I.area}{B.area}$, where $I.area$ is the area of the intersection of the query Q with bucket B and $B.area$ the area of B. The total number of objects intersecting Q is estimated by summing the results of all buckets. Evidently, satisfactory estimation accuracy depends on the degree of uniformity of objects' distributions in the buckets. An example of a histogram construction that generates nearly uniform buckets for spatial datasets appeared in [3].

Another approach uses spatial sketches to estimate the selectivity of spatial queries and in addition provides probabilistic guarantees on the quality of the estimation [4]. In particular, given a dataset D of n spatial rectangles (points are assumed to be degenerated rectangles) it is possible to create a synopsis for each dataset, which has size poly-logarithmic to n and proportional to $\frac{1}{\epsilon^2}$, and provides answers that with high probability are $\pm \epsilon$ away from the exact answer. The synopsis is based on AGMS sketches [5] extended to handle multidimensional intervals (rectangles) in poly-logarithmic space. Note that the histogram based techniques provide no guarantees on the estimation quality (e.g., in the worst case the error can be arbitrary large), but work well in most practical cases.

Foundations

To estimate the selectivity of STWQs efficiently two basic approaches have been proposed. One is based on random sampling and the other on spatio-temporal histograms. Moreover, the histogram based approach can be extended for STDJQs as well.

Selectivity Estimation for STWQ
The simplest approach is to use random sampling [6]. Given a dataset of n moving objects the method keeps a uniform random sample S of the set of moving objects that is used to estimate the result. Note that each time an object issues an update, the function that represents its location in the database must change. Therefore, a tuple

must be deleted (the one that corresponds to the old function) and a new one must be inserted. To maintain a uniform random sample in such a dynamic environment a specialized solution must be used [7, 8]. For a query Q, let r be the size of the result of Q over the random sample S. Then, the method estimates the result of Q as $r \frac{n}{|S|}$, where $|S|$ is the size of S. If the set of queries is known beforehand, then a technique that is based on stratified sampling can be applied which reduces the size of the sample and increases the accuracy of the estimator. This improved method is called Venn sampling and appeared in [9].

The other approach is based on spatio-temporal histograms. For d−dimensional moving points, the histogram is constructed in a $2d$−dimensional space, where d dimensions represent the coordinates of the moving objects at a reference time instant (e.g., $t = 0$) and the other d dimensions their velocity. Each bucket stores the number of objects and the minimum and maximum locations and velocities of these objects in each dimension. Furthermore, it is assumed that the objects are distributed uniformly inside the bucket. Based on that, a query estimation procedure is used to find the percentage of the objects in each bucket that are expected to intersect the query. Then, the (approximate) result of Q is obtained by summing the contributions of all buckets (see [6, 10, 11] for more details). The spatio-temporal histogram can also be dynamically maintained in the presence of object updates.

Selectivity Estimation for STDJQ
For estimating the size of a STDJQ between two sets of moving objects D_A and D_B, two spatio-temporal histograms are maintained, one for each set. Then, there are two approaches to estimate the size of the join. In one approach, the distance between objects is defined using the L_{max} norm, i.e., given two d−dimensional points a and b, their distance is: $dist(a, b) = \max_{i=1,2,...,d} |a.x_i - b.x_i|$. Because of the independence between the dimensions under this distance, the selectivity of the join query can be estimated using the selectivity of each dimension

independently of the others. Thus, the selectivity of a STDJQ Q is expressed as $Sel(Q) = \Pi^d_{i=1} Sel_i(Q)$. Furthermore, the $Sel_i(Q)$ can be estimated using the projections of the buckets to dimension i. More details can be found in [12].

In the other approach, the distance function is the L_2 metric (i.e., Euclidean distance), and a more complicated formula is used to estimate the selectivity of the query. The analysis is based on the following idea: consider a moving object p_1 with specific location and velocity and another object p_2 with fixed velocity. Also, assume that the location of p_2 is distributed uniformly in space. Based on these assumptions, it is possible to compute the probability that p_2 will satisfy the query. Moreover, using this probability, the selectivity of the SPDJQ can be estimated. Note that, this method can be used for both moving points and moving rectangles and can be extended to other L_p norms. The details of this technique appears in [11].

Key Applications

Selectivity estimation is used extensively in database query optimization. Therefore, the techniques developed for spatio-temporal selectivity estimation can be useful in systems that need to generate execution plans for spatio-temporal queries. Also, many application like traffic monitoring, sensor networks or mobile communications, require aggregate information about the locations of moving objects. Furthermore, getting the exact answer to these queries can be very expensive. In that case, fast approximate answers on spatio-temporal aggregation queries using the techniques discussed above is the best alternative.

Future Directions

Current techniques for spatio-temporal selectivity estimation use the assumption that objects move linearly over time. However, in many real life application this may not be a good approximation. It is an open problem how to extend the existing

techniques to handle moving objects with non-linear motion functions.

Cross-References

▶ Indexing of the Current and Near-Future Positions of Moving Objects
▶ Spatiotemporal Data Warehouses

Recommended Reading

1. Pagel B-U, Six H-W, Toben H, Widmayer P. Towards an analysis of range query performance in spatial data structures. In: Proceedings of the 12th ACM SIGACT-SIGMOD-SIGART Symposium on Principles of Database Systems; 1993. p. 214–21.
2. Theodoridis Y, Sellis T. A model for the prediction of R-tree performance. In: Proceedings of the 15th ACM SIGACT-SIGMOD-SIGART Symposium on Principles of Database Systems; 1996. p. 161–71.
3. Acharya S, Poosala V, Ramaswamy S. Selectivity estimation in spatial databases. In: Proceedings of the ACM SIGMOD International Conference on Management of Data; 1999. p. 13–4.
4. Das A, Gehrke J, Riedewald M. Approximation techniques for spatial data. In: Proceedings of the ACM SIGMOD International Conference on Management of Data; 2004. p. 695–706.
5. Alon N, Gibbons PB, Matias Y, Szegedy M. Tracking join and self-join sizes in limited storage. In: Proceedings of the 18th ACM SIGACT-SIGMOD-SIGART Symposium on Principles of Database Systems; 1999. p. 10–20.
6. Hadjieleftheriou M, Kollios G, Tsotras VJ. Performance evaluation of spatio-temporal selectivity estimation techniques. In: Proceedings of the 15th International Conference on Scientific and Statistical Database Management; 2003. p. 202–11.
7. Frahling G, Indyk P, Sohler C. Sampling in dynamic data streams and applications. In: Proceedings of the Symposium on Computational Geometry; 2005. p. 142–9.
8. Tao Y, Lian X, Papadias D, Hadjieleftheriou M. Random sampling for continuous streams with arbitrary updates. IEEE Trans Knowl Data Eng. 2007;19(1):96–110.
9. Tao Y, Papadias D, Zhai J, Li Q. Venn sampling: a novel prediction technique for moving objects. In: Proceedings of the 21st International Conference on Data Engineering; 2005. p. 680–91.
10. Choi Y-J, Chung C-W. Selectivity estimation for spatio-temporal queries to moving objects. In: Pro-

ceedings of the ACM SIGMOD International Conference on Management of Data; 2002. p. 440–51.
11. Tao Y, Sun J, Papadias D. Analysis of predictive spatio-temporal queries. ACM Trans Database Syst. 2003;28(4):295–336.
12. Sun J, Tao Y, Papadias D, Kollios G. Spatio-temporal join selectivity. Inf Syst. 2006;31(8):793–813.

Spatiotemporal Trajectories

Elias Frentzos[1], Yannis Theodoridis[1], and Apostolos N. Papadopoulos[2]
[1]University of Piraeus, Piraeus, Greece
[2]Aristotle University of Thessaloniki, Thessaloniki, Greece

Synonyms

Moving object trajectories; Spatio-temporal representation

Definition

A spatio-temporal trajectory can be straightforwardly defined as a function from the temporal $I \subseteq \mathbb{R}$ domain to the geographical space $\mathbb{R}^2$, i.e., the 2-dimensional plane. From an application point of view, a trajectory is the recording of an object's motion, i.e., the recording of the positions of an object at specific timestamps.

Generally speaking, spatio-temporal trajectories can be classified into two major categories, according to the nature of the underlying spatial object: (i) objects without area represented as moving points, and (ii) objects with area, represented as moving regions; in this case the region extent may also change with time. Among the above two categories, the former has attracted the main part of the research interest, since the majority of real-world applications involving spatio-temporal trajectories consider objects represented as points, e.g., fleet management systems monitoring cars in *road networks*.

Focusing on trajectories of moving points, while the actual trajectory consists of a curve, real-world requirements imply that the trajectory has to be built upon a set of sample points, i.e., the time-stamped positions of the object. Thus, trajectories of moving points are often defined as sequences of (x, y, t) triples:

$$T = \{(x_1, y_1, t_1), (x_2, y_2, t_2), \ldots, (x_n, y_n, t_n)\},$$

where $x_i, y_i, t_i \in \mathbb{R}$, and $t_1 < t_2 < \ldots < t_n$, and the actual trajectory curve is approximated by applying *spatio-temporal interpolation* methods on the set of sample points; among the proposed in the literature *spatio-temporal interpolation* techniques, the notion of linear interpolation has been widely adopted, given that it is fast, natural, and easy to implement (Fig. 1).

There are several techniques developed merely for the management of spatio-temporal trajectories in databases: *spatio-temporal models and languages*, *indexing of historical spatio-temporal data*, advanced query processing, and trajectory summarization techniques.

Historical Background

From a modeling perspective, the concept of spatio-temporal trajectories was introduced in some early works [6–8], which addressed the need for capturing the complete history of objects' movement. Clearly, as location data may change over time, the database must contain the whole history of this development. Thus, the DBMS should be allowed to go back in time at any particular timestamp, and to retrieve

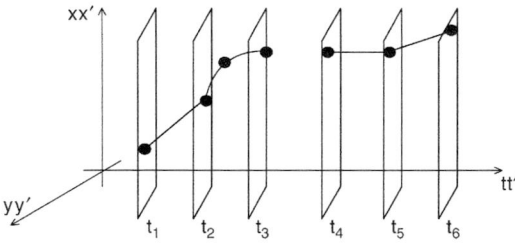

Spatiotemporal Trajectories, Fig. 1 The spatio-temporal trajectory of a moving point: *dots* represent sampled positions and *lines* in between represent alternative interpolation techniques (linear vs. arc interpolation). Unknown type of motion can be also found in a trajectory (see [t_3, t_4] time interval)

Spatiotemporal Trajectories, Table 1 Classification of spatio-temporal queries

Query type		Operation
Coordinate-based		*Overlap, inside, etc.*
Trajectory-based	Topolo-gical	*Enter, leave, cross, bypass, etc.*
	Naviga-tional	*Travelled distance, covered area, speed, heading, etc.*

the state of the database at that time. More specifically, [11] models moving points (*mpoints*) and moving regions (*mregions*) as 3-dimensional (2D space + time) or higher-dimensional entities whose structure and behavior is captured by modeling them as abstract data types. Such types and their operations for spatial values changing over time can be integrated as base (attribute) data types into an extensible DBMS. Guting et al. [11] introduced a type constructor τ which transforms any given atomic data type a into a type $\tau(a)$ with semantics $\tau(a) = time \rightarrow a$. In this way, the two aforementioned basic types, namely *mpoint* and *mregion*, may be also represented as $\tau(point)$ and $\tau(region)$, respectively. Guting et al. [11] also provided an algebra with data types (such as moving point, moving region, moving real, etc.) together with a comprehensive set of operations, supporting a variety of queries of spatio-temporal trajectory data.

On another line of research, [14] first dealt with the special requirements that spatio-temporal trajectories pose to the database engine, in terms of efficient index structures and specific query processing techniques. In particular, [14] addressed the most commonly used queries over spatio-temporal trajectories and classified them according to Table 1. Based on the observation that many query types are trajectory-based (i.e., they require the knowledge of a significant part of the moving objects' trajectory), [14] proposed the Trajectory Bundle tree (TB-tree), based on the well-known *R-tree*, considered as a seminal work in the context of *Indexing Historical Spatiotemporal Data*.

Foundations

Spatio-temporal trajectories may be queried with a variety of operators, which are mainly exten-

sions of existing spatial operators. Among them, the simple spatio-temporal range query, involving both spatial and temporal components over *R-tree* structures indexing spatio-temporal trajectories, is a straightforward generalization of the standard *R-tree FindLeaf* algorithm in the 3-dimensional space. *Nearest neighbor queries* have been also considered in the context of spatio-temporal trajectories; the algorithms over *R-trees* and variations (e.g., TB-tree) proposed in [9], are based on both depth-first and best-first *R-tree* traversals, similar to the algorithms used for nearest neighbor querying over spatial data. The proposed algorithms vary with respect to the type of the query object (stationary or moving point) as well as the type of the query result (historical continuous or not), thus resulting in four types of nearest neighbor queries.

Trajectory join has been also investigated in several papers motivated as an extension of the respective spatial operator. Bakalov et al. in [3] consider the problem of evaluating all pairs of trajectories between two datasets, during a given time interval, which, given a distance function, all distances between timely corresponding trajectory positions are within a given threshold. Then an approximation technique is used to reduce the trajectories into symbolic representations (strings) so as to lower the dimensionality of the original (3-dimensional) problem to one. Using the constructed strings, a special lower-bounding metric supports a pruning heuristic which reduces the number of candidate pairs to be examined. The overall schema is subsequently indexed by a structure based on the B-tree, requiring also minimal storage space. Another variation on the subject of joining trajectories is the closest-point-of-approach recently introduced in [2]. Closest-point-of-approach requires finding all pairs of line segments between two trajectories such that their distance is less than a predefined threshold.

The work presented in [2] proposes three approaches; the first utilizes packed R-trees treating trajectory segments as simple line segments in the $d + 1$ dimensional space, and then employs the well known *R-tree spatial join* algorithm which requires carefully controlled synchronized traversal of the two *R-trees*. The second is based on a plane sweep algorithm along the temporal dimension, while the third is an adaptive algorithm which naturally alters the way in which it computes the join in response to the characteristics of the underlying data.

Much more challenging are the so called *similarity-based* queries over spatio-temporal trajectory data. Similarity search has been extensively studied within the time series domain; consequently, techniques addressed there, are usually extended in the spatio-temporal domain, in which spatio-temporal trajectories are considered as time series. Traditionally, similarity search has been based on the Euclidean Distance between time series, nevertheless having several disadvantages which the following proposals are trying to confront. In particular, in order to compare sequences with different lengths, [12] use the Dynamic Time Warping (DTW) technique that allows sequences to be stretched along the time axis so as to minimize the distance between sequences. Although DTW incurs a heavy computation cost, it is robust against noise. Moreover, in order to reduce the effect of its quadratic complexity on large time series, a lower bounding function along with a dedicated index structure has been proposed for pruning in [12]. Longest Common SubSequence (LCSS) measure [16] matches two sequences by allowing them to stretch, without rearranging the sequence of the elements, also allowing some elements to be unmatched (which is the main advantage of the LCSS measure compared to Euclidean Distance and DTW). Therefore, LCSS can efficiently handle outliers and different scaling factors. In [5], a distance function, called Edit Distance on Real Sequences (EDR), was introduced. EDR distance function is based on the edit distance, which is the number of insert, delete, or replace operations that are needed to convert a trajectory T into Q. In the respective experimental study presented in [5], EDR was shown to be more robust than DTW and LCSS over trajectories with noise. In order to speed up the similarity search between trajectories, both [16] and [5], rely on dedicated index structures, thus achieving pruning of over 90% of the total number of indexed trajectories. However, such approaches fail to utilize the most commonly available access methods for *Indexing Historical Spatio-temporal Data* such as *R-trees*, leading to additional overhead. In order to overcome this disadvantage, [10] employ the average Euclidean distance between two trajectories as a measure of their dissimilarity, and then, using *R-trees* indexing spatio-temporal trajectories provide a series of metrics and heuristics which efficiently prune the search space. These metrics are based (i) on the observation that an upper value for the speed of the spatio-temporal trajectories can provide lower and upper bounds of the average Euclidean distance between two trajectories, and (ii), on the fact that a best-first *R-tree* traversal on the *mindist(N,q)* between a query trajectory q and a node N, provides a tighter lower bound for the average Euclidean distance. Finally, [10] provide an efficient algorithm for k-most similar trajectory search over *R-trees* indexing spatio-temporal trajectories.

In the indexing domain, a challenging line of research deals with network-constrained trajectories; this is due to the fact that the majority of the applications involving spatio-temporal trajectories deal with objects moving along *road networks* (i.e., cars, buses, trains). Following this observation, several specific access methods for objects moving in networks have been proposed; among them the most efficient is the Moving Objects in Networks tree (MON-tree) [1]. However, in order to exploit such network constrained indexes, the need for mapping the trajectories into the underlying network introduces the so called *map matching* problem. Specifically, the observation that raw trajectory positions are affected by the measurement error introduced by e.g., GPS, and, the sampling error being up to the frequency with which position samples are taken, reveals the problem of correctly matching such tracking data in an underlying map containing e.g., a *road network*. Currently, the state-of-the-art approach addressing the map matching problem is the one presented in [4], which proposes mapping the entire trajectory to candidate paths in the *road network* using the Fréchet distance, which can be illustrated as follows: suppose a human and her dog constrained to walk on two different curves, while they are both allowed to control their speed independently. Then, the Fréchet distance between the two curves is the minimal length of a leash that is necessary. The proposed global map-matching algorithms in [4] find a curve in the *road network* that is as close as possible to the given trajectory in terms of the Fréchet distance between them.

Last but not least, it is the need for compression techniques that arises due to the fact that all the ubiquitous positioning devices will eventually start to generate an unprecedented stream of time-stamped positions, leading to storage and computation challenges [13]. In this direction, [13] exploit existing algorithms used in the line generalization field, and present one *top-down* and one *opening window* algorithm, which can be directly applied to spatio-temporal trajectories. The *top-down* algorithm, named TD-TR, is based on the well known Douglas-Peucker algorithm (Fig. 2) originally used in the context of cartography. This algorithm calculates the perpendicular distance of each internal point from the line connecting the first and the last point of the polyline (line *AB* in Fig. 2) and finds the point with the greatest perpendicular distance (point *C*). Then, it creates lines *AC* and *CB* and, recursively, checks these new lines against the remaining points with the same method. When the distance of all remaining

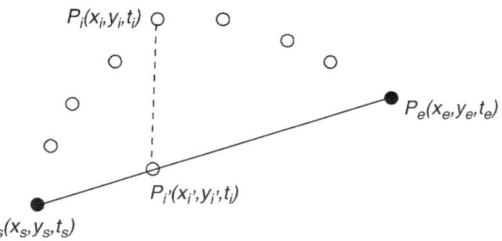

Spatiotemporal Trajectories, Fig. 3 The Synchronous Euclidean Distance (SED): The distance is calculated between the point under examination (P_i) and the point P_i' which is determined as the point on the line (P_s, P_e) the time instance t_i [MB04] [13]

points from the currently examined line is less than a given threshold (e.g., all the points following *C* against line *BC* in Fig. 2) the algorithm stops and returns this line segment as part of the new – compressed – polyline. Being aware of the fact that trajectories are polylines evolving in time, the algorithm presented in [13] replaces the perpendicular distance used in the DP algorithm with the so-called *Synchronous Euclidean Distance* (SED), which is the distance between the currently examined point (P_i in Fig. 3) and the point of the line (P_s, P_e) where the moving object would lie, assuming it was moving on this line, at time instance t_i determined by the point under examination (P_i' in Fig. 3). The experimental study presented in [13] shows that such compression techniques introduce a small and manageable error, reducing at the same time the size of the dataset under 40% of its original size.

Key Applications

- *Location-Based Services (LBS)* – Spatio-temporal trajectories are used in location-based services for determining the exact position of users, based on the map-matching solutions provided over trajectories.
- *Spatio-temporal Decision Support Systems (STDSS)* – A number of decision support tasks can exploit the presence of spatio-temporal trajectories: traffic estimation and prediction systems, analysis of traffic congestion conditions, fleet management

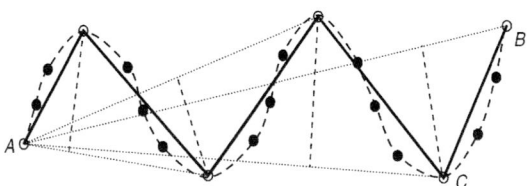

Spatiotemporal Trajectories, Fig. 2 Top-down Douglas-Peucker algorithm used for trajectory compression. Original data points are represented by *closed circles* [MB04] [13]

systems, urban and regional planers analyzing the life courses of city residents, and scientists studying animal immigration habits.

Future Directions

There are several research directions arising regarding spatio-temporal trajectory data management. For example, the problem of estimating the selectivity of a range query over historical trajectory data still remains open. More specifically, such a technique would have to deal with the *distinct counting* problem [15], which is also present in the context of *Spatio-temporal Data Warehouses*. This problem stands when an object samples its position in several timestamps inside a given query window resulting to be counted multiple times in the query result. Nevertheless, a selectivity estimation technique based on a space partitioning method, such as a histogram, would had to return the number of distinct trajectories contained inside the query region, summing the containment of several buckets; then trajectories appearing in several buckets would had to be counted only once.

Another interesting research direction appears when considering network-constraint trajectory compression; in particular, existing compression techniques do not consider that trajectories may be network-constraint, resulting in trajectories which after the compression may be invalid regarding the underline network. As such, future work should investigate on techniques which may produce compressed trajectories being still valid under the network constraints.

Experimental Results

In general, for every presented method, there is an accompanying experimental evaluation in the corresponding reference.

Data Sets

A collection of real spatio-temporal datasets, as well as links to generators for spatio-temporal trajectory data can be found at *R-tree portal* (URL: http://www.rtreeportal.org/).

Url to Code

R-tree portal (URL: http://www.rtreeportal.org/) contains the code for most common spatio-temporal indexes, as well as data generators and several useful links on spatio-temporal databases.

Cross-References

▶ Indexing Historical Spatiotemporal Data
▶ Nearest Neighbor Query
▶ Road Networks
▶ R-Tree (and Family)
▶ Spatial and Spatiotemporal Data Models and Languages
▶ Spatial Join
▶ Spatiotemporal Data Warehouses
▶ Spatiotemporal Interpolation Algorithms

Recommended Reading

1. Almeida VT, Guting RH. Indexing the trajectories of moving objects in networks. GeoInformatica. 2005;9(1):33–60.
2. Arumugam S, Jermaine C. Closest-point-of-approach join for moving object histories. In: Proceedings of the 22nd International Conference on Data Engineering; 2006. p. 86.
3. Bakalov P, Hadjieleftheriou M, Keogh E, Tsotras V. Efficient trajectory joins using symbolic representations. In: Proceedings of the 6th International Conference on Mobile Data Management; 2005. p. 86–93.
4. Brakatsoulas S, Pfoser D, Salas R, Wenk C. On map-matching vehicle tracking data. In: Proceedings of the 31st International Conference on Very Large Data Bases; 2005. p. 853–64.
5. Chen L, Özsu MT, Oria V. Robust and fast similarity search for moving object trajectories. In: Proceedings of the ACM SIGMOD International Conference on Management of Data; 2005. p. 491–502.
6. Chomicki J, Revesz P. A geometric framework for specifying spatiotemporal objects. In: Proceedings of the 6th International Workshop Temporal Representation and Reasoning; 1999. p. 41–6.

S

7. Erwig M, Güting RH, Schneider M, Varzigiannis M. Spatio-temporal data types: an approach to modeling and querying moving objects in databases. GeoInformatica. 1999;3(3):265–91.
8. Forlizzi L, Güting Nardelli E, Schneider M. A data model and data structures for moving objects databases. In: Proceedings of the ACM SIGMOD International Conference on Management of Data; 2000. p. 319–30.
9. Frentzos E, Gratsias K, Pelekis N, Theodoridis Y. Algorithms for nearest neighbor search on moving object trajectories. GeoInformatica. 2007;11(2): 159–93.
10. Frentzos E, Gratsias K, Theodoridis Y. Index-based most similar trajectory search. In: Proceedings of the 23rd International Conference on Data Engineering; 2007. p. 816–25.
11. Guting RH, Bohlen MH, Erwig M, Jensen CS, Lorentzos NA, Schneider M, Vazirgiannis M. A foundation for representing and querying moving objects. ACM Trans Database Syst. 2000;25(1):1–42.
12. Keogh E. Exact indexing of dynamic time warping. In: Proceedings of the 28th International Conference on Very Large Data Bases; 2002. p. 406–17.
13. Meratnia N, By R. Spatiotemporal compression techniques for moving point objects. In: Advances in Database Technology, Proceedings of the 9th International Conference on Extending Database Technology; 2004. p. 765–82.
14. Pfoser D, Jensen CS, Theodoridis Y. Novel approaches to the indexing of moving object trajectories. In: Proceedings of the 26th International Conference on Very Large Data Bases; 2000. p. 395–406.
15. Tao Y, Kollios G, Considine J, Li F, Papadias D. Spatio-temporal aggregation using sketches. In: Proceedings of the 20th International Conference on Data Engineering; 2004. p. 214–26.
16. Vlachos M, Kollios G, Gunopulos D. Discovering similar multidimensional trajectories. In: Proceedings of the 18th International Conference on Data Engineering; 2002. p. 673–84.

Specialization and Generalization

Bernhard Thalheim
Christian-Albrechts University, Kiel, Germany

Synonyms

Abstraction; Hierarchies; Refinement

Definition

Specialization and generalization are main principles of database modeling. Specialization is based on a refinement of types or classes to more specific ones. Generalization maps or groups types or classes to more abstract or combined ones. Typically, generalizations and specializations form a hierarchy of types and classes.

Key Points

Specialization introduces a new entity type by adding specific properties belonging to that type, which are different from the general properties of its more general type. Is-A associations specialize a type to a more specific one. Is-A-Role-Of associations consider a specific behavior of objects. Is-More-Specific-To associations specialize properties of objects of the more general type. The *student* type and the *customer* type are specializations of the *person* type. The *rectangle* type is specialized to the *square* type by adding restrictions. Different kinds of specialization may be distinguished: structural specialization which extends the structure, semantic specialization which strengthens type restrictions, pragmatic specialization which allows a separation of the different usage of objects in contexts, operational specialization which introduces additional operations, and hybrid specializations. Identification and other properties of objects of the special type can be inherited from the more general one. Methods applicable to objects of the more general one should be applicable to corresponding more special objects or specialized as well. Exceptions can be modeled by specializations. Specialization allows developers to avoid null values and to hide details from non-authorized users.

Generalization combines common features, attributes, or methods of types. It is based either on abstraction, on combination or on grouping. Generalization often tends to be an abstraction in which a more general type is defined by extracting common properties of one or more types while suppressing the differences between the subtypes. The subtypes can be virtually clustered by or generalized to or combined by a view to a general type. The *library's holding* type is a generalization of the *journal*, *book*, *preprint* and *PhD/Master thesis* types. The *occupation* type is a generalization of the *lawyer*, *merchant*, *teacher* and *banker* types. It is obtained by factoring out the commonalities among the specializations. Structural combination typically assumes the existence of a unifiable identification of all types. The *livestock* type combines the different types of farming. Generalization is represented by clusters of types. The cluster construct of the extended ER model represents common properties and abstractions. Identification of generalized objects is either inherited from the more special objects or built as an abstraction of the identification of the more special types. Generalizations often do not have their own methods.

Cross-References

▶ Semantic Data Model

Recommended Reading

1. Ter Bekke JH. Semantic Data Modeling. London: Prentice-Hall; 1992.
2. Thalheim B. Entity-Relationship Modeling - Foundations of Database Technology. Berlin/Hiedelberg/New York: Springer; 2000.

Specificity

Jovan Pehcevski[1] and Benjamin Piwowarski[2]
[1]INRIA Paris-Rocquencourt, Le Chesnay Cedex, France
[2]University of Glasgow, Glasgow, UK

Synonyms

Coverage

Definition

Specificity is a relevance dimension that describes the extent to which a document part focuses on the topic of request. In the context of structured text (XML) retrieval, a document part corresponds to an XML element.

Specificity is defined as the length ratio, typically in number of characters, of contained relevant to irrelevant text in the document part. Different Specificity values can be associated to a document part. These values are drawn from the Specificity relevance scale, which has evolved from a discrete multi-graded relevance scale to a continuous relevance scale.

Key Points

The Initiative for the Evaluation of XML Retrieval (INEX) has defined Specificity as a relevance dimension that uses values from its own relevance scale to express the extent to which an XML element focuses on the topic of request. Since 2002, different names and relevance scales were used for Specificity at INEX. It initially evolved because the relevance dimension was not sufficiently well defined, and later because the assessment procedure changed.

In 2002, Specificity was named coverage at INEX, which reflected the extent to which an XML element was focused on aspects of the information need (as represented by the INEX topic). The component coverage used a relevance scale comprising four relevance grades, from "no coverage," "too large," "too small," to "exact coverage." However, this dimension was used solely in 2002, partly because of the vagueness introduced in the terminology for its name, and partly because it has been subsequently shown that the INEX 2002 assessors did not particularly understand the notion of "too small" [1]. In particular, assessors understood "too small" as a measure of quantity while Specificity is more related to the concentration of relevant information. In 2003 and 2004, four grades were used for the Specificity relevance dimension at INEX, such

S

that the extent to which an XML element may focus on the topic of request could range from "none" (0), to "marginally" (1), to "fairly" (2), or to "highly" (3) focused. An XML element was considered relevant only if its Specificity value was greater than zero.

From 2005 onwards, a highlighting assessment procedure was used at INEX to gather relevance assessments for the XML retrieval topics. The Specificity of an XML element is automatically computed as the ratio of highlighted to fully contained text, where the relevance values that can be associated to the element are drawn from a continuous relevance scale. These values are in the range between 0 and 1, where the value of 0 corresponds to an element that does not contain any highlighted text, while the value of 1 corresponds to a fully highlighted element.

With the highlighting assessment procedure, assessors are asked to highlight all the relevant information contained by returned XML documents. This results in a reduced cognitive load on the assessor, since in this case there is no need for the assessor to explicitly associate a Specificity value to a judged element. Studies of the level of assessor agreement, which used topics that were double-judged at INEX, have shown that the use of the new highlighting procedure further increases the level of assessor agreement compared to the level of agreement observed among assessors during previous years at INEX [2, 3].

Cross-References

▶ Evaluation Metrics for Structured Text Retrieval
▶ Relevance

Recommended Reading

1. Kazai G, Masood S, Lalmas M. A study of the assessment of relevance for the INEX 2002 test collection. In: Proceedings of the 26th European Conference on IR Research; 2004. p. 296–310.
2. Pehcevski J, Thom JA. HiXEval: highlighting XML retrieval evaluation. In: Proceedings of the 4th International Workshop of the Initiative for the Evaluation of XML Retrieval; 2005. p. 43–57.
3. Trotman A. Wanted: element retrieval users. In: Proceedings of the 4th International Workshop of the Initiative for the Evaluation of XML Retrieval; 2005. p. 63–9.

Spectral Clustering

Sergios Theodoridis[1] and Konstantinos Koutroumbas[2]
[1]University of Athens, Athens, Greece
[2]Institute for Space Applications and Remote Sensing, Athens, Greece

Synonyms

Graph-based clustering

Definition

Let X be a set $X = \{\mathbf{x}_1, \mathbf{x}_2,...,\mathbf{x}_N\}$ of N data points. An *m-clustering* of X, is defined as the partition of X into m sets (*clusters*), $C_1,...,C_m$, so that the following three conditions are met:

- $C_i \neq \emptyset, i = 1,...,m$
- $\cup_{i=1}^{m} C_i = X$
- $C_i \cap C_j = \emptyset, i \neq j, i,j = 1,...,m$

In addition, the data points contained in a cluster C_i are "more similar" to each other and "less similar" to the points of the other clusters. The terms "similar" and "dissimilar" depend very much on the types of clusters the user expects to recover from X. A clustering defined as above is known as hard clustering, to distinguish it from the fuzzy clustering case.

Historical Background

The essence of clustering is to "reveal" the organization of patterns into "sensible" groups. It has been used as a critical analysis tool in a vast range of disciplines, such as medicine, social sciences, engineering, computer science, machine learning, bioinformatics, data mining and information retrieval. The literature is huge and numerous techniques have been suggested over the years. A comprehensive introduction to clustering can be found e.g., in [14]. An important class of clustering algorithms builds around graph theory. Reference [9] is one of the first efforts in this direction. Points of X are assigned to the nodes of a graph. Notions such as minimum spanning tree and directed trees have extensively been used to partition the graph into clusters (e.g., [14]). Spectral clustering is a more recent class of graph based techniques, which unravels the structural properties of a graph using information conveyed by the spectral decomposition (eigendecomposition) of an associated matrix. The elements of this matrix code the underlying similarities among the nodes (data points) of the graph (e.g., [3]). Among the earlier works on spectral clustering are [6, 12]. Fiedler [4] was one of the first to show the application of eigenvectors in graph partitioning.

Foundations

In the sequel, the simplest task of partitioning a given data set, X, into two clusters, A and B, is considered. Let $X = \{x_1, x_2, .., x_N\} \subset R^l$, where the latter denotes the l-dimensional Euclidean space. According to the previous discussion, the following preliminary steps are in order:

- Construction of a graph $G(V, E)$, where each vertex of the graph corresponds to a point x_i, $i = 1,2,...,N$, of X. It is further assumed that G is undirected and connected. In other words, there exists at least one path of edges that connects any pair of points in the graph.

- Assignment of a weight $W(i, j)$ to each one of the edges of the graph, e_{ij}, that quantifies proximity between the respective nodes, v_i, v_j in G (For notational convenience in some places i is used instead of v_i). The set of weights defines the $N \times N$ weight matrix W, also known as *affinity* matrix, with elements

$$W \equiv [W(i,j)], \quad i,j = 1,2,..,N$$

- The weight matrix is assumed to be symmetric, i.e., $W(i, j) = W(j, i)$. The choice of the weights is carried out by the user and it is a problem dependent task. A common choice is

$$W(i,j) = \begin{cases} \exp\left(-\frac{||x_i - x_j||^2}{2\sigma^2}\right), & \text{if } ||x_i - x_j|| < \epsilon \\ 0, & \text{otherwise} \end{cases}$$

- where ε is a user-defined constant and $||\cdot||$ is the Euclidean norm in the l-dimensional space.

By the definition of clustering, $A \cup B = X$ and $A \cap B = \emptyset$. Once a weighted graph has been formed, the second phase in any graph-based clustering algorithm consists of the following two steps: (i) Choose an appropriate clustering criterion for the partitioning of the graph and (ii) Adopt an efficient algorithmic scheme to determine the partitioning that optimizes the previous clustering criterion.

A clustering optimality criterion, that is in line with "common sense," is the so called *cut* [16]. If A and B are the resulting clusters, the associated *cut* is defined as:

$$cut(A, B) = \sum_{i \in A, \ j \in B} W(i,j) \quad (1)$$

Selecting A and B so that the respective *cut(A, B)* is minimized means that the set of edges,

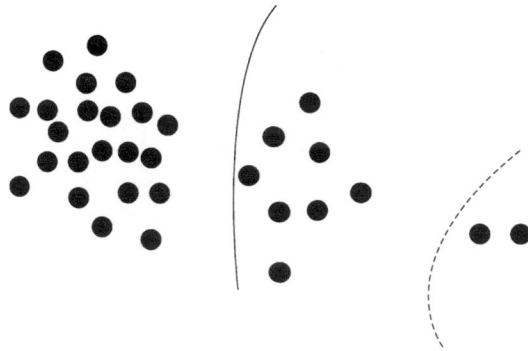

Spectral Clustering, Fig. 1 The *dotted line* indicates the clustering that is likely to favor the minimum cut criterion, while the full line indicates a more natural partitioning

connecting nodes in A with nodes in B, have the minimum sum of weights, indicating the lowest similarity between points in A and B. However, this simple criterion turns out to form clusters of small size of isolated points (least similar with the rest of the nodes). This is illustrated in Fig. 1. The minimum *cut* criterion would result in the two clusters separated by the dotted line, although the partition by the full line seems to be a more natural partitioning.

To overcome this drawback, the *normalized cut* criterion has been suggested in [12]. This is one of the most commonly used criteria in spectral clustering. The essence of this criterion is to minimize the *cut* and at the same time trying to keep the sizes of the formed clusters large. To this end, for each node, $v_i \in V$, in the graph G the index

$$D_{ij} = \sum_{j \in V} W(i, j) \qquad (2)$$

is defined. This is an index indicative of the "importance" of a node, v_i, $i = 1,2,...,N$. The higher the value of D_{ii} the more similar the ith node is to the rest of the nodes. A low D_{ii} value indicates an isolated (remote) point. Given a cluster A, a measure of the "importance" of A is given by the following index

$$V(A) = \sum_{i \in A} D_{ii} = \sum_{i \in A,\ j \in V} W(i, j) \qquad (3)$$

where $V(A)$ is sometimes known as the *volume* or the *degree* of A. It is obvious that small and isolated clusters will have a small $V(\cdot)$. The *normalized cut* between two clusters A, B is defined as

$$Ncut(A, B) = \frac{cut(A, B)}{V(A)} + \frac{cut(A, B)}{V(B)} \qquad (4)$$

Obviously, small clusters correspond to large values (close to one) for the previous ratios, since in such cases $cut(A, B)$ will be a large percentage of $V(A)$.

Minimization of the $Ncut(A, B)$ turns out to be an NP-hard task. To bypass this computational obstacle, the problem will be reshaped to a form that allows an efficient approximate solution. To this end let [1]

$$y_i = \begin{cases} \dfrac{1}{V(A)}, & \text{if } i \in A \\ -\dfrac{1}{V(B)}, & \text{if } i \in B \end{cases} \qquad (5)$$

$$\mathbf{y} = [y_1, y_2, \ldots, y_N]^T$$

where T denotes the transpose operation. After some algebraic manipulations it can be verified that

$$\left(\frac{1}{V(A)} + \frac{1}{V(B)} \right)^2 cut(A, B)$$

$$\propto \frac{1}{2} \sum_{i \in V} \sum_{j \in V} (y_i - y_j)^2 W(i, j) = \mathbf{y}^T L \mathbf{y} \qquad (6)$$

where $\propto$ denotes proportionality and

$$L = D - W, \quad D \equiv \text{diag}\{D_{ii}\}$$

is known as the *graph Laplacian* matrix and D is the diagonal matrix having the elements D_{ii} across the main diagonal. It is also easily verified that

$$\mathbf{y}^T D \mathbf{y} = \frac{1}{V(A)} + \frac{1}{V(B)} \qquad (7)$$

Combining Eqs. (4, 6 and 7) it turns out that minimizing $Ncut(A, B)$ is equivalent with minimizing

$$J = \frac{\mathbf{y}^T L \mathbf{y}}{\mathbf{y}^T D \mathbf{y}} \qquad (8)$$

subject to the constraint that $y_i \in \left\{ \frac{1}{V(A)}, -\frac{1}{V(B)} \right\}$. Furthermore, based on the respective definitions, it can be shown that

$$\mathbf{y}^T D \mathbf{1} = 0 \qquad (9)$$

where $\mathbf{1}$ is the N-dimensional vector having all its elements equal to 1. In order to bypass the computationally hard nature of the original task a relaxed problem will be solved: Eq. (8) will be minimized subject to the constraint of Eq. (9). The unknown "cluster labels," y_i, $i = 1, 2, ..., N$, are now allowed to move freely along the real axis. Let

$$\mathbf{z} \equiv D^{1/2} \mathbf{y}$$

Then Eq. (8) becomes

$$J = \frac{\mathbf{z}^T \tilde{L} \mathbf{z}^T}{\mathbf{z}} \mathbf{z} \qquad (10)$$

and the constraint in Eq. (9)

$$\mathbf{z}^T D^{1/2} \mathbf{1} = 0 \qquad (11)$$

Matrix $\tilde{L} \equiv D^{-1/2} L D^{-1/2}$ is known as the *normalized graph Laplacian* matrix. It can easily be shown that $\tilde{L}$ has the following properties

- It is symmetric, real valued and nonnegative definite. Thus, as it is known from linear algebra, all its eigenvalues are non-negative and the corresponding eigenvectors are orthogonal to each other.
- $D^{1/2} \mathbf{1}$ is an eigenvector corresponding to zero eigenvalue. Indeed,

$$\tilde{L} D^{1/2} \mathbf{1} = 0$$

- Obviously $\lambda = 0$ is the smallest eigenvalue of $\tilde{L}$, due to the non-negative definite nature of the matrix.

Furthermore, the ratio in Eq. (10) is the celebrated Rayleigh quotient for which the following hold (e.g., [5])

- The smallest value of the quotient, with respect to $\mathbf{z}$, is equal to the smallest eigenvalue of $\tilde{L}$ and it occurs for $\mathbf{z}$ equal to the eigenvector corresponding to this (smallest) eigenvalue.
- If the solution is constrained to be orthogonal to all eigenvectors associated with the j smaller eigenvalues, minimization of the Rayleigh quotient results to the eigenvector corresponding to the next smallest eigenvalue, λ_{j+1} and the minimum value is equal to λ_{j+1}.

Taking into account (i) the orthogonality condition in the constraint Eq. (11) and (ii) the fact that $D^{1/2} \mathbf{1}$ is the eigenvector corresponding to the smallest eigenvalue $\lambda_0 = 0$ of $\tilde{L}$, it follows that:
The optimal solution vector z minimizing the Rayleigh quotient in (10), subject to the constraint (11), is the eigenvector corresponding to the second smallest eigenvalue of $\tilde{L}$.

In summary, the basic steps of the spectral clustering algorithm are:

- Given a set of points $\mathbf{x}_1, \mathbf{x}_2, ..., \mathbf{x}_N$ the weighted graph $G(V, E)$ is constructed. Then, the weight matrix W is formed by adopting a similarity rule.
- The matrices D, $L = D - W$ and $\tilde{L}$ are formed. The eigenanalysis of the normalized Laplacian matrix

$$\tilde{L} \mathbf{z} = \lambda \mathbf{z}$$

- is performed and the computation of the eigenvector $\mathbf{z}_1$ corresponding to the second smallest eigenvalue λ_1 of $\tilde{L}$ is carried out. Then $\mathbf{y} = D^{-1/2} \mathbf{z}_1$ is computed.
- Finally, discretization of the components of $\mathbf{y}$ according to a threshold value takes place.

The final step is necessary since the components of the obtained solution are real-valued and our required solution is binary. To this goal, different techniques can be applied. For example, the threshold can be taken to be equal to zero. Another choice is to adopt the median value of the components of the optimum eigenvector. An alternative approach would be to select the threshold value that gives the minimum *cut*.

The eigenanalysis or spectral decomposition, as it is sometimes called, of an $N \times N$ matrix, using a general purpose solver, amounts to $O(N^3)$ operations. Thus, for large number of data points, this may be prohibitive in practice. However, for most of the practical applications the resulting graph is only locally connected, and the associated affinity matrix is a *sparse* one. Moreover, only the smallest eigenvalues/eigenvectors are required and also the accuracy is not of major issue, since the solution is to be discretized. In such a setting, the efficient Lanczos algorithm can be mobilized and the computational requirements drop down to approximately $O(N^{3/2})$.

So far, the partition of a data set into two clusters has been considered. If more clusters are expected, the scheme can be used in a hierarchical mode, where, at each step, each one of the resulting clusters is divided into two partitions. This is continued until a prespecified criterion is satisfied.

In the discussion above the focus was on a specific clustering criterion, i.e., the normalized cut, in order to present the basic philosophy behind the spectral clustering techniques. No doubt, a number of other criteria have been proposed in the related literature, e.g., [8, 13]. In [15] a review and a comparative study of a number of popular spectral clustering algorithms is presented.

Key Applications

Spectral clustering has been used in a number of applications such as image segmentation and motion tracking [13, 11], circuit layout [2], gene expression [8], machine learning [10], load balancing [7].

Cross-References

▶ Clustering Overview and Applications
▶ Graph
▶ Hierarchical Clustering
▶ Image Segmentation

Recommended Reading

1. Belikn M, Niyogi P. Laplacian eigenmaps for dimensionality reduction and data representation. Neural Comput. 2003;15(6):1373–96.
2. Chan P, Schlag M, Zien J. Spectral k-way ratio cut partitioning. IEEE Trans Comput Aided Des Integrated Circ Syst. 1994;13(9):1088–96.
3. Chung FRK. Spectral graph theory. American Mathematical Society; 1997.
4. Fiedler M. A property of eigenvectors of nonnegative symmetric matrices and its application to graph theory. Czechoslov Math J. 1975;25(100):619–33.
5. Golub GH, Van Loan CF. Matrix somputations. Baltimore: John Hopkins; 1989.
6. Hagen LW, Kahng AB. New spectral methods for ratio cut partitioning and clustering. IEEE Trans Comput Aided Des Integrated Circ Syst. 1992;11(9):1074–85.
7. Hendrickson B, Leland R. Multidimensional spectral load balancing. In: Proceedings of the 4th SIAM Conference on Parallel Processing; 1993. p. 953–61.
8. Kannan R, Vempala S, Vetta A. On clusterings- good, bad and spectral. In: Proceedings of the 41st Annual Symposium on Foundations of Computer Science; 2000. p. 367–77.
9. Ling RF. On the theory and construction of k-clusters. Comput J. 1972;15(4):326–32.
10. Ng AY, Jordan M, Weiis Y. On spectral clustering analysis and an algorithm. In: Proceedings of the 14th Conference on Advances in Neural Information Processing Systems; 2001.
11. Qiu H, Hancock ER. Clustering and embedding using commute times. IEEE Trans Pattern Anal Mach Intell. 2007;29(11):1873–90.
12. Scott G, Longuet-Higgins H. Feature grouping by relocalization of eigenvectors of the proximity matrix. In: Proceedings of the British Machine Vision Conference; 1990. p. 103–8.
13. Shi J, Malik J. Normalized cuts and image segmentation. IEEE Trans Pattern Anal Mach Intell. 2000;22(8):888–905.
14. Theodoridis S, Koutroumbas K. Pattern recognition. 4th edn. Academic Press; 2008.
15. Verma D, Meilǎ M. A comparison of spectral clustering algorithms. Technical Report, UW-CSE-03-05-01. Seattle: CSE Department/University of Washington; 2003.

16. Wu Z, Leahy R. An optimal graph theoretic approach to data clustering: theory and its applications to image segmentation. IEEE Trans Pattern Anal Mach Intell. 1993;15(11):1101–13.

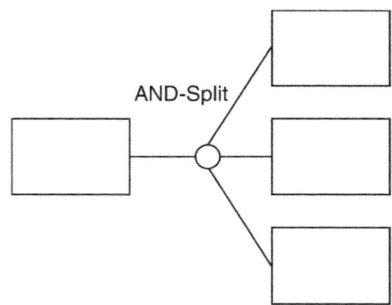

Split, Fig. 1

Split

Nathaniel Palmer
Workflow Management Coalition, Hingham, MA, USA

Synonyms

AND-split

Cross-References

► OR-Split

Definition

A point within the workflow where a single thread of control splits into two or more parallel activities.

Split Transactions

George Karabatis
University of Maryland, Baltimore Country (UMBC), Baltimore, MD, USA

Definition

The split transaction is an extended transaction model that introduces two new transaction management primitives/operations, namely, split and join. The split operation on a transaction T splits T and replaces it with two serializable transactions; each one is later committed or aborted independently of the other. The inverse of split is the join operation on a transaction T which dissolves T by joining its results with a target transaction S.

Key Points

The execution of parallel activities commences with an AND-Split and concludes with an AND-Join. For example, in a credit application process there may be a split in the workflow at which point multiple activities are completed separately (in parallel, if not simultaneously.) At an And-Split separate threads of control within the process instance are created; these threads will proceed autonomously and independently until reaching an And-Join condition. In certain workflow systems, all the threads created at an And-Split must converge at a common And-Join point (Block Structure); in other systems convergence of a subset of the threads can occur at different And-Join points, potentially including other incoming threads created from other And-split points (Fig. 1).

Key Points

The concept of split transactions was introduced by Pu, Kaiser, and Hutchinson in [3] and later elaborated in [2] to support open-ended activities such as CAD/CAM projects, engineering type of applications, and software development. The syntax of the split-transaction operation on transaction T produces two new transactions A and B and dissolves T [2, 3]:

Split Transaction (

A: (AReadSet, AWriteSet, AProcedure)

B: (BReadSet, BWriteSet, BProcedure)) where *AReadSet, AWriteSet, BReadSet,* and *BWriteSet* are sets of data items accessed by A and B. *AProcedure* and *BProcedure* are the starting points of code where A and B will begin execution. There is no need to explicitly mention T in the arguments of the operation, as by definition the operation can only be executed within the body of transaction T. The syntax for a join operation on T to join S is *join transaction (S: TID).*

The split-transaction operation can be used to commit some work of a transaction early or to distribute ongoing work among several coworkers. On the contrary, the join transaction is used to hand over and integrate results with a coworker [2]. These operations pertain to programmed transactions and to actual open-ended activities with unpredictable developments (users determine the next operation to be executed in an ad hoc manner). In the latter case, the read sets and write sets defined above are replaced with data sets on which application-specific operations (e.g., edit or compile for software development) are applied; the *AProcedure* and *Bprocedure* are replaced with *AUser* and *BUser* specifying the users who take control of A and B. Thus, the definition changes to

Split Transaction (

A: (AReadSet, AWriteSet, AUser),

B: (BReadSet, BWriteSet, BUser))

There are additional operations such as *Split-Commit* (transaction A is immediately committed, while B may be taken over by *BUser* and also *Suspend* (giving up control of a transaction) and *Accept-Join-Transaction* (a user executes this operation to accept responsibility of a joined transaction).

The main advantages of the restructuring operations (split/join transaction) on open-ended activities are:

- Adaptive recovery: committing resources that will not change

- Added concurrency: releasing committed resources or transferring ownership of uncommitted resources
- Serializable access to resources by all activities

The split and join primitives were subsequently incorporated with nested transactions to produce combined transaction models [1].

Cross-References

▶ Database Management System
▶ Distributed Transaction Management
▶ Extended Transaction Models and the ACTA Framework
▶ Serializability
▶ Transaction
▶ Transaction Management
▶ Transaction Manager

Recommended Reading

1. Chrysanthis PK, Ramamritham K. Synthesis of extended transaction models using ACTA. ACM Trans Database Syst. 1994;19(3):450–91.
2. Kaiser GE, Pu C. Dynamic restructuring of transactions. In: Elmagarmid AK, editor. Database transaction models for advanced applications. Burlington: Morgan Kaufmann Publishers; 1992. p. 265–95.
3. Pu C, Kaiser GE, Hutchinson NC. Split-transactions for open-ended activities. In: Proceedings of the 14th International Conference on Very Large Data Bases; 1988. p. 26–37.

SQL

Don Chamberlin
IBM Almaden Research Center, San Jose, CA, USA

Synonyms

SEQUEL; Structured query language

Definition

SQL is the world's most widely used database query language. It was developed at IBM Research Laboratories in the 1970s, based on the relational data model defined by E. F. Codd in 1970. It supports retrieval, manipulation, and administration of data stored in tabular form. It is the subject of an international standard named Database Language SQL.

Historical Background

Early Language Development

In June 1970, E. F. Codd of IBM Research published a paper [1] defining the relational data model and introducing the concept of data independence. Codd's thesis was that queries should be expressed in terms of high-level, nonprocedural concepts that are independent of physical representation. Selection of an algorithm for processing a given query could then be done by an optimizing compiler, based on the access paths available and the statistics of the stored data; if these access paths or statistics should later change, the algorithm could be re-optimized without human intervention. In a series of papers, Codd proposed two high-level languages for querying relational databases, called relational algebra and relational calculus (also known as Data Sublanguage Alpha) [2].

The advantages of the relational model for application developers and database administrators were immediately clear. It was less clear whether an optimizing compiler could consistently translate nonprocedural queries into algorithms that were efficient enough for use in a production database environment. To investigate this issue, IBM convened a project called System R [3] at its research laboratory in San Jose, California. Between 1973 and 1979, this project designed and implemented a prototype relational database system based on Codd's ideas, testing and refining the prototype in several customer locations. SQL was the user interface defined by the System R research project. It later became the user interface for relational database products marketed by IBM and several other companies.

The principal goals that influenced the design of SQL were as follows:

1. SQL is a high-level, nonprocedural language intended for processing by an optimizing compiler. It is designed to be equivalent in expressive power to the relational query languages originally proposed by Codd.
2. SQL is intended to be accessible to users without formal training in mathematics or computer programming. It is designed to be typed on a keyboard. Therefore it is framed in familiar English keywords and avoids specialized mathematical concepts or symbols.
3. SQL attempts to unify data query and update with database administration tasks such as creating and modifying tables and views, controlling access to data, and defining constraints to protect database integrity. In pre-relational database systems, these tasks were usually performed by specialized database administrators and required shutting down and reconfiguring the database. By building administrative functions into the query language, SQL helps to eliminate the database administrator as a choke point in application development.
4. SQL is designed for use in both decision support and online transaction processing environments. The former environment requires processing of complex queries, usually executed infrequently but accessing large amounts of data. The latter environment requires high-performance execution of parameterized transactions, repeated frequently but accessing (and often updating) small amounts of data. Both end-user interfaces and application programming interfaces are necessary to support this spectrum of usage.

The first specification of SQL was published in May 1974, in a 16-page conference paper [4] by Don Chamberlin and Ray Boyce, members of the System R project. In this paper, the language was named SEQUEL, an acronym for Structured English Query Language. The paper included a BNF syntax for the proposed language. This original paper presented only basic query features, without any facilities for data definition

S

or update. However, the basic structure of the language, including query blocks, grouping, set operations, and aggregating functions, has been consistent from this paper to the present day.

Over the course of the System R project, SE-QUEL continued to evolve based on experience gathered by users and implementers. A much more complete description of the language was published in the IBM Journal of Research and Development in November 1976 [5], including data manipulation facilities (insert, delete, and update), a more complete join facility, facilities for defining tables and views, and database administration facilities including access control, assertions, and triggers. In 1977, because of a trademark issue, the name SEQUEL was shortened to SQL.

Although SQL was designed and prototyped at IBM Research, the language was published in the open literature, and the first commercial SQL product was released by a small company called Relational Software, Inc., in 1979. This product was named Oracle, a name that was later adopted by the company, which is no longer small. The first IBM product based on SQL was called SQL/-Data System, released in 1981, followed by DB2, released in 1983 on mainframes and eventually supported on many IBM platforms. SQL has now been implemented by all major database vendors and is available in a wide variety of operating environments. In addition to commercial database products, SQL implementations include several popular open-source products such as MySQL (http://www.mysql.com) and Apache Derby (http://db.apache.org/derby/).

Standards

Shortly after the first appearance of SQL in a commercial product, an effort was made to standardize the language. Over the years, the SQL standard has contributed to the growth of the database industry by defining a common interface for use by database vendors, application developers, and tools.

The first SQL standard, named "Database Language SQL," was published by the American National Standards Institute (ANSI) in 1986 (Standard No. X3.135-1986), and an identical standard with the same name was published by the International Standards Organization (ISO) in 1987 (Standard No. ISO 9075-1987). Over the years, ANSI and ISO have cooperated to keep their respective SQL standards synchronized as they have evolved through several versions.

The original standard, often called SQL-86, occupied just under 100 pages and included only simple queries, updates, and table definitions. It was followed in 1989 by a revised standard that added several kinds of constraints for protecting the integrity of stored data. This version of the standard comprised about 120 pages and is often referred to as SQL-89 ("Database Language SQL with Integrity Enhancement").

A major revision of the SQL standard, usually called SQL-92, was published by ANSI and ISO/IEC in 1992. This version improved the orthogonality of the language, allowing expressions to be used wherever tables or scalar values are expected. SQL-92 also added several new features, including date and time datatypes, set-oriented operators such as UNION and IN-TERSECT, standard catalog tables for storing metadata, and schema-evolution features such as ALTER TABLE. SQL-92 comprised about 600 pages. A conformance test suite for SQL-92 was developed in the United States by the National Institute of Standards and Technology (NIST). After certifying several conforming products, NIST discontinued SQL conformance testing in 1996.

Between 1992 and 1999, two specialized extensions to SQL-92 were published, named Call Level Interface (CLI) and Persistent Stored Modules (PSM). CLI defines a set of functions whereby programs written in languages such as C can dynamically connect to relational databases and execute SQL statements. PSM extends SQL with assignment statements, control-flow statements, and exception handlers, making it possible to implement some database applications entirely in SQL. PSM was an attempt to standardize the procedural extensions such as PL/SQL [6] and Transact-SQL [7] that had been added to SQL by several database vendors.

The next major update of the SQL standard occurred in 1999 and is usually called SQL:1999. This new version splits the standard into several parts, incorporating CLI as Part 3 and PSM as Part 4. SQL:1999 introduced important new functionality including triggers, large objects, recursive queries, and user-defined functions. It placed major emphasis on object-relational functionality and on new features for online analytic processing (OLAP). Details of these and other recent additions to SQL are described below under "Advanced Features." The sum of all the parts of SQL:1999 exceeded 2,000 pages. Additional parts continue to be added to the SQL standard from time to time. The most recent major revision of the standard, with all its parts, comprises more than 3,600 pages.

Over the years, the SQL standard has provided a controlled framework within which the language can evolve to correct its initial limitations and to meet changing user requirements. The standard has also served to focus the industry's attention and resources, providing a common framework in which individuals and companies could develop tools, write books, teach courses, and provide consulting services. The standard has been only partially successful in making SQL applications portable across implementations; this goal has been hampered by the fact that different vendors have implemented different subsets of the standard and by the lack (since 1996) of a test suite to validate conformance of an implementation. The latest versions of the various parts of the standard can be obtained from ISO [8] or from national standards organizations such as ANSI [9]. Jim Melton, editor of the SQL standard, has also published a two-volume reference book explaining the standard in a very accessible style ([10]; see also [11]).

Foundations

Queries

SQL operates on data in the form of tables. Each table has a name and consists of one or more columns, each of which has a name and a datatype. The content of a table consists of zero or more rows, each of which has a value for each of the columns. The value associated with a given row and column may be an instance of the datatype of that column or may be a special "null" value indicating that the value is missing (not available or not applicable). SQL statements, which may be queries or updates, operate on stored tables or on tabular "views" that are derived from stored tables. The result of a query is an unnamed virtual table. The result of an update is a change to the stored data, which is visible to subsequent statements. Generally, updates can be applied to a view only if each row in the view can be mapped uniquely onto a row of a stored table. This rule makes it possible to map updates on the view to updates on the underlying table.

An SQL query consists of one or more *query blocks*. A query block consists of several clauses, each of which begins with a keyword. Some of these clauses (keywords SELECT and FROM) are required, and others (keywords WHERE, GROUP BY, and HAVING) are optional. The examples in this article use upper-case keywords and lower-case names, although SQL is a case-insensitive language. The examples are based on two tables named PARTS and SUPPLIERS. The primary key of PARTS is PARTNO and the primary key of SUPPLIERS is SUPPNO. SUPPNO also appears as a foreign key in the PARTS table (see "Database Administration" below for definitions of primary key and foreign key).

The following query block illustrates a join of two tables. Conceptually, rows from the PARTS table are paired with rows from the SUPPLIERS table according to the criterion specified in the WHERE clause (SUPPNO's must match), and the resulting row pairs are filtered by an additional condition (supplier's location must be Denver). From the surviving row pairs, the SELECT clause specifies the columns that appear in the query result (in this case, the part number and the supplier name). Many different strategies are possible for executing this query; since SQL is a nonprocedural language, choice of an execution strategy is left to an optimizing compiler:

SELECT p.partno, s.name
FROM parts p, suppliers s
WHERE p.suppno = s.suppno
AND s.location = 'Denver'

Constraints and triggers are useful for specifying and enforcing the semantics of stored data. In general, it is preferable to specify a given semantic rule by means of a constraint rather than a trigger if possible, since constraints apply to all kinds of actions and provide maximum opportunities for optimization. On the other hand, some kinds of semantic rules (e.g., "salaries never decrease") can only be specified by using triggers.

Advanced Features

Over the years, a great deal of functionality has been added to the SQL language. The full set of SQL features is far too large and complex to be explained here. The following are some of the major areas in which advanced functionality has been added to SQL:

- **Recursion:** A recursive query consists of an initial subquery that computes some preliminary results and a recursive subquery that computes additional results based on values that were previously computed. The recursive subquery is executed repeatedly until no additional results are computed. Recursion is useful in queries that search some space for an optimum result, such as "Find the cheapest combination of flight segments to travel from Shanghai to Copenhagen." Recursive queries were first defined in SQL:1999.

- **OLAP:** Online analytic processing (OLAP) is used by businesses to analyze large volumes of data to identify facts and trends that may affect business decisions. The GROUP BY clause and aggregating functions (sum, avg, etc.) of early SQL provided a primitive form of OLAP functionality, which was greatly extended in later versions of the language. For example, the ROLLUP facility enables a query to apply aggregating functions at multiple levels (such as city, county, and state). The CUBE facility enables data to be aggregated along multiple dimensions (such as date, location, and category) within a single query. The WINDOW facility allows aggregating functions to be applied to a "moving window" as it passes over a

collection of data. These facilities, and others, were introduced by SQL:1999 and enhanced in subsequent versions of the standard.

- **Functions and procedures:** Originally, SQL supported a fixed collection of functions, which grew slowly over the years. SQL:1999 introduced a capability for users to define additional functions and procedures that can be invoked from SQL statements. (In this context, a procedure is simply a function that is invoked by a CALL statement and that is not required to return a value.) The bodies of user-defined functions and procedures can be written either in SQL itself or in a host language such as C or Java.

- **Object-relational features:** Early versions of SQL could process data conforming to a fixed set of simple datatypes such as integers and strings. Over the years, a few additional datatypes such as dates and "large objects" were added. In the late 1990s, requirements arose for a more extensible type system. SQL:1999 introduced facilities for user-defined structured types and methods. These facilities support limited forms of object-oriented functionality, including inheritance and polymorphism.

- **Multimedia:** In 2000, the SQL standard was augmented by a separate but closely related standard called "SQL Multimedia and Application Packages" (ISO/IEC 13249:2000), often referred to as SQL/MM. This new standard used the object-relational features introduced by SQL:1999 to define specialized datatypes and methods for text, images, and spatial data.

- **XML-related features:** XML is an increasingly popular format for data exchange because it mixes metadata (tags) with data, making the data self-describing. The popularity of XML has led to requirements to store XML data in relational databases and to convert data between relational and XML formats. These requirements have been addressed by a facility called SQL/XML, which was introduced as Part 14 of SQL:2003 and was updated in 2006. SQL/XML includes a new XML datatype, a set of functions for converting query results into XML format, and a feature whereby SQL can invoke XQuery as a sublanguage for processing stored XML data.

Criticisms

Like most widely used programming interfaces, SQL has attracted its share of criticism. Issues that have been raised about the design of SQL include the following:

- The earliest versions of SQL lacked support for some important aspects of Codd's relational data model such as primary keys and referential integrity. These concepts were added to the language in SQL-89, along with other integrity-related features such as unique constraints and check constraints.
- The earliest versions of SQL had some ad hoc rules about how various language features could be combined and lacked the closure property because the columns of query results did not always have names. These problems were largely corrected by SQL-92.
- Null values are a complex and controversial subject. One of Codd's famous "12 rules" requires relational database systems to support a null value, defined as a representation of missing or inapplicable information that is systematic and distinct from all regular values [12]. Some writers believe that the complexity introduced by null values outweighs their benefit. However, there seems to be no method for dealing with missing data that is free of disadvantages. The SQL approach to this issue has been to support a null value and to allow database designers to specify, on a column-by-column basis, where nulls are permitted. One benefit of this approach has been that null values have proven useful in the design of various language features, such as outer join, CUBE, and ROLLUP, that have been added during the evolution of SQL.
- Unlike Codd's definition of the relational data model, SQL permits duplicate rows to exist, either in a database table or in the result of a query. SQL also allows users to selectively prohibit duplicate rows in a table or in a query result. The intent of this approach is to give users control over the potentially expensive process of duplicate elimination. In some applications, duplicate rows may be meaningful (e.g., in a point-of-sale system, a customer may purchase several identical items in the same transaction). As in the case of nulls, the SQL approach has been to provide users with tools to allow or disallow duplicate rows according to the needs of specific applications.
- Another source of criticism has been the "impedance mismatch" between SQL and the host languages such as C and Java in which it is often embedded. Exchanging data between two languages with different type systems makes applications more complex and interferes with global optimization. One approach to this problem has been the development of computationally complete SQL-based scripting languages such as PSM.

Key Applications

SQL is designed to be used in a variety of application environments.

Most SQL implementations support an interactive interface whereby users can compose and execute ad hoc SQL statements. In many cases, a graphical user interface is provided to display menus of available tables and columns and help the user to construct valid statements. These systems also usually support menu-based interfaces for administrative functions such as creating and dropping tables and views.

More complex applications usually involve use of both SQL and a host programming language. This requires a mapping between the type systems of SQL and the host language and a well-defined interface for exchanging data between the two environments. Interfaces have been defined between SQL and C, Java, and many other host languages. These interfaces fall into two major categories:

- **Embedded SQL:** In this approach, SQL statements are embedded syntactically in the host

became popular and widespread. Initially, Hive adapted a columnar data format, the RCFile [19], followed by its optimized version ORCFile [9]. At the same time, Parquet [10] emerged as a popular columnar data format. Parquet is similar to the file format used by Dremel [23] and widely used by many SQL-on-Hadoop systems. It can be used to store both tabular and nested data with arbitrary levels of nesting. It also has support for min/max indexes, also known as zone maps for skipping over large chunks of data.

In the following section, we will provide a short overview of different SQL-on-Hadoop approaches and provide a detailed discussion on a few representative systems. The interested reader can refer to [1] for a more in-depth presentation of this topic.

Overview of Representative Systems

There is a wide variety of solutions, system architectures, and capabilities in this space, with varying degree of SQL support and capabilities. The purpose of this entry is to provide a brief overview of these options and discuss various different approaches.

Before discussing various systems, it is important to underline the unique characteristics of data processing in this new ecosystem that are different than traditional relational MPP Data Warehouses. First, in the world of Hadoop and HDFS-stored data, complex data types, such as arrays, maps, structs, as well as semi-structured data, easily expressed as JSON (or BSON or ION), are more prevalent. Especially, mobile and IoT applications heavily use JSON data. Second, the users utilize user-defined functions (UDFs) widely to express their business and ETL logic, which is sometimes very awkward to express in SQL itself. Third, data is shared by many analytical frameworks and hence is not "owned" or controlled by the SQL engine. This combination of loosely coupled storage manager and query engine (or engines) imposes several challenges for query optimization and efficient query processing in general. For example, as files can be added or modified outside the tight control of a monolithic query engine, maintaining statistics about the data, or indexes, becomes challenging, if not

impossible. Another challenge for query planning and processing is that in HDFS, the structure of the sub-directories that constitute a table reflects its partitioning, and this structure can be exploited to reduce the amount of data read, through partition pruning that can happen either statically, when the predicates on the partitioning columns are known a priori, or dynamically.

Several SQL-on-Hadoop systems leverage existing relational database technology: HAWQ [12] uses large amounts of Greenplum code, and VectorH [13] uses large amounts of Actian Vectorwise code. In some cases, database files are stored in HDFS, while in other cases, database files are stored on the same physical machines as HDFS, but on a separate file system. In some cases, data is dynamically moved from Hadoop file formats to the native storage structures of the DBMS. In some cases, queries are executed by the database engine code, while in other cases, query execution is split between database engine code and native Hadoop execution engines such as MapReduce or Tez [11, 14].

HadoopDB [2] was one of the first research projects that provided SQL processing in Hadoop. The main characteristic of HadoopDB, and later its commercial offering Hadapt, was that it was replacing the file-oriented HDFS storage with DBMS-oriented storage, including column store data layouts.

Some SQL-on-Hadoop systems leverage a scalable run-time, like MapReduce or Spark. In this category, we will describe Hive and Spark SQL in some details.

Several native SQL systems use MPP architectures, including Impala [21], Drill [4], Presto [31], and Big SQL [18]. Among these, we will dive into the details of Impala and Big SQL in the following sections.

Apache Drill [4] is an open-source project which aims at providing SQL-like declarative processing over self-describing semi-structured data. Its focus is on analyzing data without imposing a fixed schema or creating tables in a catalog like Hive MetaStore.

Presto [31] is another open-source project developed by the Facebook to replace Hive. It uses a traditional MPP DBMS run-time instead of MapReduce and supports interactive analytical queries. Presto accesses data where it lives, HDFS, HBase, MySQL, or Cassandra, and provides basic federation across multiple sources.

Hive

Hive [30], an open-source project originally built at Facebook, was the first offering that provided support for an SQL-like query language, called HiveQL, and used the MapReduce run-time to execute queries. Initially, Hive compiled HiveQL queries into a series of map reduce jobs that executed various relational operations, including projection, selection, joins, and aggregations. Data movement and grouping were implemented using the MapReduce shuffle operator.

As SQL-on-Hadoop gained popularity, the MapReduce-based run-time did not provide the required response times, due to the high latency in launching map reduce jobs, and the necessary materialization of intermediate data between the various phases of the query computation. To address this issue, Hive moved to a different run-time, Tez [27], which can run DAGs as a single job, reducing the latency in launching jobs. With Tez, a complex HiveQL query can be compiled into a single Tez job. Hive-on-Tez also provides pipelining data without writing into disk between stages of a Tez job.

Hive evolved over time to provide more SQL compatible data types, and standard SQL language constructs, as well as ACID transactions. Hive now also includes a cost-based optimizer based on Apache Calcite [7] and provides various optimizations, especially targeted to avoid data shuffling. Additionally, Hive's query planner and run-time try to substantially eliminate the amount of data read from HDFS by taking advantage of the partitioning columns of the table and the predicates on these columns of the query. Through partition pruning, which takes place at run-time, Hive may skip reading entire partitions.

To address the job startup costs, live-long and process (LLAP), which consists of long-running processes on each HDFS DataNode (similar to an MPP database), was introduced in Hive 2.0. LLAP caches data in memory, supports multithreaded vectorized execution over ORCFile data, and enables interactive query processing. Clients can configure Hive to run none, partial, or all of the query in LLAP processes and Tez containers.

Apache Impala

Apache Impala (incubating) [21] is an opensource, fully-integrated MPP SQL query engine. It supports processing over a wide variety of file formats, including Parquet, RC, Avro, and CSV and compression algorithms. In addition to comprehensive SQL support, Impala offers a query language extension, based on dot notation, for querying nested data. Impala offers query processing services for data stored in four different storage managers: HDFS, HBase, Apache Kudu [22], and Amazon Simple Storage Service (S3).

Unlike other systems (often forks of Postgres), Impala is a brand-new engine written from the ground up in C++ and Java. To reduce latency, such as that incurred from utilizing MapReduce or by reading data remotely, Impala implements a distributed architecture (shown in Fig. 1) based on daemon processes that are responsible for all aspects of query execution and that run on the same machines as the rest of the Hadoop infrastructure. As shown in the figure, in addition to the daemons that run on every DataNode, Impala uses two other services: the Catalog that acts as a proxy to the Hive MetaStore Service and the Statestore that is a publish/subscribe service that monitors the health of the cluster. Every Impala daemon can accept and serve SQL requests. This provides multi-master flexibility (for high availability) but also imposes a challenge in the distributed resource management and admission control.

Figure 1 also shows the typical flow of a query. The query is received by one of the Impala daemons, a cost-based query planner that decides the distributed query execution plan, and it hands it off to the query coordinator. The coordinator sends fragments of the query to all or a subset of the other daemons in the cluster, which in turn read data stored either locally or remotely (the planner tries to minimize the need for remote reads). Data is exchanged between the participating nodes as needed, using efficient data

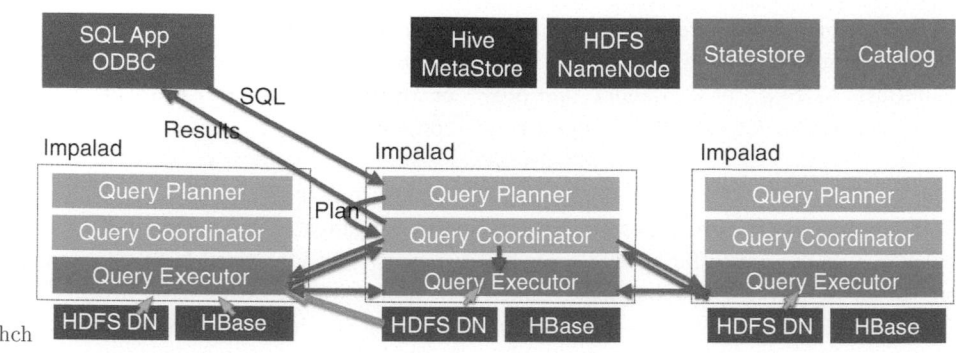

SQL Analytics on Big Data, Fig. 1 Architecture of Apache Impala

exchange operators. Eventually the query results are returned.

To perform data scans from both disk and memory at or near hardware speed, Impala uses an HDFS feature called *short-circuit local reads* to bypass the DataNode protocol when reading from local disk. Impala reads at almost full disk bandwidth and is typically able to saturate all available disks. Furthermore, *HDFS caching* allows Impala to access memory resident data at memory bus speed and also saves CPU cycles as there is no need to copy data blocks and/or checksum them.

For the query execution, Impala leverages decades of research in parallel databases. Its execution model is the traditional Volcano-style with Exchange operators [17]. Processing is performed one batch at a time: each `GetNext()` call operates over batches of rows, similar to [25]. A main characteristic of Impala's query processing is that it employs LLVM to generate code at run-time to speed up frequently executed code paths [32]. The result is performance that is on par or exceeds that of commercial MPP analytic DBMSs, depending on the particular workload.

Big SQL

IBM Big SQL [18] leverages IBM's state-of-the-art relational database technology, to process standard SQL queries over HDFS and HBase data, supporting all common Hadoop file formats, without introducing any propriety formats. Big SQL shares the same catalog and table definitions with Hive using the Hive MetaStore. Database

workers read HDFS and HBase data directly and execute relational operations.

The architecture of Big SQL is provided in Fig. 2. Database worker nodes are deployed on the same HDFS cluster, although 1-1 alignment is not required. Big SQL can work with database clusters that completely or partially overlap with the HDFS cluster. A special head node contains the database coordinator, which receives the SQL statements from the client applications and compiles them into distributed query execution plans. Big SQL also introduces a new stand-alone scheduler service which helps coordinate the distributed execution over the HDFS cluster. Big SQL scheduler assigns HDFS blocks to database workers for processing on a query by query basis. The scheduler identifies where the HDFS blocks are and decides which database workers to include in the query plan. The assignment is done dynamically at run-time to accommodate failures: scheduler uses the workers that are currently up and running. This also allows elasticity. If a new node is added to the database cluster, it can be considered immediately by the scheduler for the new queries. Similarly, if a node crashes or the cluster is scaled down, the scheduler immediately detects this change and chooses database workers for future queries accordingly.

Big SQL provides special processes, HDFS readers, which are coupled with corresponding database workers to ingest HDFS data at disk speeds. The queries are optimized using mature distributed relational optimization technology. Big SQL exploits sophisticated query rewrite transformations [28, 34] that are targeted for

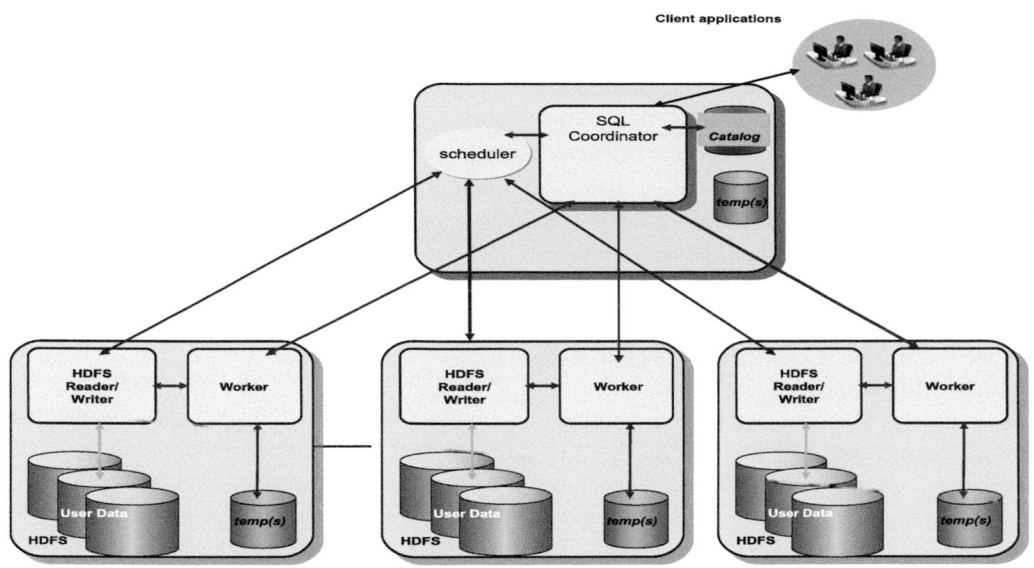

SQL Analytics on Big Data, Fig. 2 Overall architecture of Big SQL

complex nested decision support queries. It uses complex data statistics and a cost-based optimizer to choose the best query execution plan [16].

The database also provides autonomic features that optimize the allocation of various buffer pools for best memory utilization, as well as workload management tools that provide admission control for various classes of concurrent queries.

Big SQL works under YARN resource manager and allows adding and deleting nodes from its cluster, without disrupting queries, or shutting down the system. It also supports complex data types, including maps, arrays, and structs. Finally, Big SQL allows update operations over HBase tables.

Spark SQL

There are two main SQL processing engines that work on Spark: Hive on Spark and Spark SQL. Hive on Spark is an attempt at replacing MapReduce/Tez-based run-times of Hive with Spark run-time but keeping the Hive query language and Hive compiler.

Spark SQL evolved from the Shark [33] project, which replaced the MapReduce run-time of Hive [30] with Spark run-time. Shark still used

the Hive compiler and interfaces but generated Spark jobs. Shark also introduced an in-memory columnar data organization that allowed for fast execution of SQL queries.

Spark SQL is the next generation of Shark, which provides a new compiler that supports standard SQL and uses the Catalyst query optimizer [3]. Spark SQL integrates with the rest of the Spark analytics modules and facilitates easy-to-use end-to-end analytics flows. It provides a declarative DataFrame API that tightly integrates with the procedural Spark code. Its optimizer uses Scala pattern matching and provides an extensible rule-based framework. It is relatively easy to add new optimizer rules.

Spark SQL uses two interfaces, Hive and native. Hive interface supports a richer SQL subset and integrates with the Hive MetaStore. The native interface, on the other hand, only works with transient tables and an in-memory catalog, as the main target application of Spark SQL is data preparations for machine learning and other complex analytics.

Update-Optimized SQL Engines

As mentioned earlier, applications that require updates typically store their data in HBase [8], instead of HDFS, which is an append-only file system. HBase stores tables in HDFS but provides update capabilities via delta files and background compaction. HBase supports auto-sharding and fail-over technology for scaling tables across multiple servers and as such scales out to petabytes of data.

Most of the system mentioned earlier in this section can read data stored on HBase. Additionally, there are systems that provide SQL processing specifically for HBase. Two representative systems in this category are Splice Machine [29] and Apache Phoenix [5].

Both Splice Machine and Phoenix support updates, transactions, and SQL processing over HBase data, relying on HBase to execute the updates and data scans. Splice Machine leverages the Apache Derby compiler and optimizer to generate query execution plans that access HBase servers. Splice Machine even supports ACID transactions.

Phoenix provides SQL querying over HBase via an embeddable JDBC driver and converts SQL queries into execution plans composed of HBase scans. It uses coprocessors and custom filters for improved performance. Phoenix provides secondary indexes as well as basic support for joins, both of which are difficult to get with HBase.

However, these systems do not provide fast OLAP capabilities because the scans over HBase tables are quite slow. Most often, the data is transformed into a more analytical friendly format, such as Parquet, stored in HDFS and processed by one of the other SQL engines.

To bridge the gap between the fast analytic capabilities of columnar formats stored on HDFS and fast updates, Cloudera introduced the Apache Kudu storage manager [22] as an alternative to HDFS and HBase. Kudu is a storage engine that operates outside of HDFS. It partitions and replicates each partition using Raft consensus [24] and is designed for providing low-latency random accesses and update, while the data is stored in a columnar format that facilitates efficient large scans. Kudu does not provide an SQL interface. Instead, the user that has to use SQL needs to go through Impala, Drill, or Spark that have been integrated with Kudu.

Key Applications

Pretty much any vertical has needs for SQL analytics on "big data." Such systems are frequently used in financial services, telecommunications, public sector, manufacturing and retail, healthcare, insurance, and Energy and Utilities.

Cross-References

▶ Big Data Platforms for Data Analytics
▶ Data Warehousing in Cloud Environments
▶ Distributed Database Systems
▶ Parallel and Distributed Data Warehouses
▶ Parallel Database Management
▶ SQL-Based Temporal Query Languages

Recommended Reading

1. Abadi D, Babu S, Özcan F, Pandis I. SQL-on-Hadoop systems: tutorial. Proc VLDB Endow. 2015;8(12):2050–2051.
2. Abouzeid A, Bajda-Pawlikowski K, Abadi DJ, Rasin A, Silberschatz A. HadoopDB: an architectural hybrid of mapReduce and DBMS technologies for analytical workloads. Proc VLDB Endow. 2009;2(1):922–933.
3. Amburst M, Xin RS, Lian C, Huai Y, Liu D, Bradley JK, Meng X, Kaftan T, Franklin MJ, Ghodsi A, Zaharia M. Spark SQL: relational data processing in spark. In: Proceedings of the ACM SIGMOD International Conference on Management of Data; 2015.
4. Apache Drill. http://drill.apache.org/.
5. Apache Phoenix. http://phoenix.apache.org/.
6. Apache spark. https://spark.incubator.apache.org/.
7. Apache Calcite. https://calcite.apache.org/.
8. Apache HBase. https://hbase.apache.org/.
9. Apache ORC. https://orc.apache.org/.
10. Apache Parquet. https://parquet.apache.org/.
11. Bajda-Pawlikowski K, Abadi DJ, Silberschatz A, Paulson E. Efficient processing of data warehousing queries in a split execution environment. In: Proceedings of the ACM SIGMOD International Conference on Management of Data; 2011.
12. Chang L, Wang Z, Ma T, Jian L, Ma L, Goldshuv A, Lonergan L, Cohen J, Welton C, Sherry G, Bhandarkar M. HAWQ: a massively parallel processing SQL engine in Hadoop. In: Proceedings of the ACM SIGMOD International Conference on Management of Data; 2014.
13. Costea A, Ionescu A, Răducanu B, Switakowski M, Bârca C, Sompolski J, Luszczak A, Szafrański M,

de Nijs G, Boncz P. VectorH: taking SQL-on-Hadoop to the next level. In: Proceedings of the ACM SIGMOD International Conference on Management of Data; 2016.

14. DeWitt DJ, Nehme RV, Shankar S, Aguilar-Saborit J, Avanes A, Flasza M, Gramling J. Split query processing in polybase. In: Proceedings of the ACM SIGMOD International Conference on Management of Data; 2013. p. 1255–66.

15. Floratou A, Minhas UF, Özcan F. SQL-on-Hadoop: full circle back to shared-nothing database architectures. Proc VLDB Endow. 2014;7(12).1295–306.

16. Gassner P, Lohman GM, Schiefer KB, Wang Y. Query optimization in the IBM DB2 family. IEEE Data Eng Bull. 1993;16(4):4–18.

17. Graefe G. Encapsulation of parallelism in the Volcano query processing system. In: Proceedings of the ACM SIGMOD International Conference on Management of Data; 1990.

18. Gray S, Özcan F, Pereyra H, van der Linden B, Zubiri A. IBM Big SQL 3.0: SQL-on-Hadoop without compromise (2014), http://public.dhe.ibm.com/common/ssi/ecm/en/sww14019usen/SWW14019USEN.PDF

19. He Y, Lee R, Huai Y, Shao Z, Jain N, Zhang X, Xu Z. Rcfile: a fast and space-efficient data placement structure in mapreduce-based warehouse systems. In: Proceedings of the 27th International Conference on Data Engineering; 2011. p. 1199–208.

20. Hive on spark. https://cwiki.apache.org/confluence/display/Hive/Hive+on+Spark.

21. Kornacker M, Behm A, Bittorf V, Bobrovytsky T, Ching C, Choi A, Erickson J, Grund M, Hecht D, Jacobs M, Joshi I, Kuff L, Kumar D, Leblang A, Li N, Pandis I, Robinson H, Rorke D, Rus S, Russell J, Tsirogiannis D, Wanderman-Milne S, Yoder M. Impala: a modern, open-source SQL engine for Hadoop. In: Proceedings of the 7th Biennial Conference on Innovative Data Systems Research; 2015.

22. Lipcon T, Alves D, Burkert D, Cryans J-D, Dembo A, Percy M, Rus S, Wang D, Bertozzi M, McCabe CP, Wang A. Kudu: storage for fast analytics on fast data. https://kudu.apache.org/.

23. Melnik S, Gubarev A, Long JJ, Romer G, Shivakumar S, Tolton M, Vassilakis T. Dremel: interactive analysis of web-scale datasets. Proc VLDB Endow. 2010;3(1–2):330–39.

24. Ongaro D, Ousterhout J. In search of an understandable consensus algorithm. In: Proceedings of the USENIX Annual Technical Conference; 2014.

25. Padmanabhan S, Malkemus T, Agarwal RC, Jhingran A. Block oriented processing of relational database operations in modern computer architectures. In: Proceedings of the 17th International Conference on Data Engineering; 2001.

26. Presto. http://prestodb.io/.

27. Saha B, Shah H, Seth S, Vijayaraghavan G, Murthy A, Curino C. Apache Tez: a unifying framework for modeling and building data processing applications. In: Proceedings of the ACM SIGMOD International Conference on Management of Data; 2015.

28. Seshadri P, Pirahesh H, Leung TYC. Complex query decorrelation. In: Proceedings of the 12th International Conference on Data Engineering; 1996.

29. Splice machine. http://www.splicemachine.com/.

30. Thusoo A, Sarma JS, Jain N, Shao Z, Chakka P, Zhang N, Anthony S, Liu H, Murthy R. Hive – a petabyte scale data warehouse using Hadoop. In: Proceedings of the 26th International Conference on Data Engineering; 2010.

31. Traverso M. Presto: interacting with petabytes of data at Facebook. https://www.facebook.com/notes/facebook-engineering/presto-interacting-with-petabytes-of-data-at-facebook/10151786197628920.

32. Wanderman-Milne S, Li N. Runtime code generation in Cloudera Impala. IEEE Data Eng Bull. 2014;37(1):31–7.

33. Xin RS, Rosen J, Zaharia M, Franklin MJ, Shenker S, Stoica I. Shark: SQL and rich analytics at scale. In: Proceedings of the ACM SIGMOD International Conference on Management of Data; 2013.

34. Zuzarte C, Pirahesh H, Ma W, Cheng Q, Liu L, Wong K. WinMagic: subquery elimination using window aggregation. In: Proceedings of the ACM SIGMOD International Conference on Management of Data; 2003.

SQL Isolation Levels

Philip A. Bernstein
Microsoft Corporation, Redmond, WA, USA

Synonyms

Degrees of consistency; Degrees of isolation

Definition

A transaction is an execution of a well-defined set of read and write operations on shared data, which terminates with a commit operation that makes its updates permanent, or an abort operation that undoes its updates. Isolation levels define the situations in which a transaction can be affected by the execution of other transactions. In the ACID properties, isolation requires that transactions behave serializably, that is, as if they executed in a serial order with no interleaving. To obtain a serializable execution when many transactions are executing concurrently, a transaction's operations may be delayed and occasionally even rejected. This reduces the rate at which transactions execute. Users often regard this throughput reduction as unsatisfactory and therefore seek isolation levels that are less stringent than serializability, some of which are defined as part of the SQL language. These are the SQL isolation levels, which are called Read Uncommitted, Read Committed, Repeatable Read, and Serializable. They are derived primarily from the "degrees of consistency" originally presented in [3].

Key Points

The SQL isolation levels are defined in terms of the following types of interleavings, P1–P3, that each isolation allows or prohibits. The interleavings are expressed by sequences of operations, using the notation $r_1[x]$ (respectively, $w_1[x]$) to represent the execution of a read (respectively write) operation by transaction T_1 on row x.

P1. Dirty Read – A dirty read is an operation that reads a value that was written by an uncommitted transaction. In the execution "$w_1[x]\ r_2[x]$," $r_2[x]$ performs a dirty read. The problem is that transaction T_1 might abort after T_2 read row x, in which case T_2 has read a value of x that never existed.

P2. Non-repeatable Read – A transaction reads a row x, a second transaction writes into x and commits, and then the first transaction reads x again. In the execution "$w_0[x]\ r_1[x]\ w_2[x]\ c_2\ r_1[x]$," transaction T_1 has experienced a non-repeatable read, since it read one value of x before T_2 executed and a different value of x after T_2 committed.

P3. Phantom – A transaction T_1 reads a set of rows that satisfy a predicate P (such as a WHERE-clause in SQL). Before T_1 commits or aborts, a second transaction T_2 inserts, updates, or deletes rows that change the set of rows that satisfy P. Therefore, if T_1 re-executes its read, the read will return a different set of rows than it returned the first time. The rows that appear and disappear are called phantoms. This is the same situation as non-repeatable reads, except that the set of rows that T_1 retrieves is affected by T_2, not just their value. In the following execution, the row inserted by w_2 is a phantom:

- r_1[rows of the Employee table where Department = "Toy"]
- w_2[insert a new row in the Employee table where Department = "Toy"]
- r_1[rows of the Employee table where Department = "Toy"]

The SQL isolation levels are defined using the above three phenomena:

1. Read Uncommitted – All three phenomena (P1, P2, and P3) are allowed. The intent is to allow all interleavings of reads and writes by different transactions.
2. Read Committed – Dirty reads are prohibited. This ensures that each read operation reads a value that will not be undone because the transaction that wrote the value later aborts.
3. Repeatable Read – Dirty reads and nonrepeatable reads are prohibited. The intent is to ensure that transaction executions are serializable except for phantom situations that arise.
4. Serializable – All three phenomena are prohibited. The intent is to ensure that transactions are truly serializable.

Each transaction may independently define its isolation level. This creates some difficulty in how a user should interpret the levels. For example, if a transaction runs as Serializable, it may still read data that was written by transactions running (say) as Read Committed.

SQL isolation levels have been criticized because there is a gap between the definition of the phenomena they prevent and the intent of the isolation level as presented in the descriptions of (1–4) above. For details, see [1].

In principle, all transactions that perform updates should execute at the Serializable level. That way, if each transaction preserves database consistency, then each serial execution will preserve consistency. Hence a serializable execution will too. If update transactions execute at less than serializable level, then this consistency preservation guarantee is lost. Nevertheless, it is believed that most database applications execute at lower isolation levels than Serializable, typically Read Committed. This gives them higher throughput at the expense of some loss of correctness. One simulation study of transaction performance showed that transaction throughput is 2½–3 times higher with transactions executing at Read Committed level compared to Serializable level [2].

Cross-References

▶ ACID Properties
▶ Serializability
▶ Transaction

Recommended Reading

1. Berenson H, Bernstein P, Gray J, Melton J, O'Neil E, O'Neil P. A critique of ANSI SQL isolation levels. In: Proceedings of the ACM SIGMOD International Conference on Management of Data; 1995. p. 1–10.
2. Bober PM, Carey MJ. On mixing queries and transactions via multiversion locking. In: Proceedings of the 8th International Conference on Data Engineering; 1992. p. 548–56.
3. Gray J, Lorie RA, Potzulo GR, Traiger IL. Granularity of locks and degrees of consistency in a shared database. In Stonebraker M. Hellerstein J, editors. IFIP Working Conf. on Modelling in Data Base Management Systems. (3rd edn.), Reprinted in Readings in Database Systems Morgan Kaufmann, 1998; p. 175–93. 1976, p. 365–94.
4. Gray J, Reuter A. Transaction processing: concepts and techniques. Morgan Kaufmann; 1993. p. 397–403.

SQL-Based Temporal Query Languages

Michael H. Böhlen[1,2], Johann Gamper[3], Christian S. Jensen[4], and Richard T. Snodgrass[5,6]
[1]Free University of Bozen-Bolzano, Bozen-Bolzano, Italy
[2]University of Zurich, Zürich, Switzerland
[3]Free University of Bozen-Bolzano, Bolzano, Italy
[4]Department of Computer Science, Aalborg University, Aalborg, Denmark
[5]Department of Computer Science, University of Arizona, Tucson, AZ, USA
[6]Dataware Ventures, Tucson, AZ, USA

Definition

More than two dozen extensions to the relational data model have been proposed that support the storage and retrieval of time-referenced data. These models timestamp tuples or attribute values, and the timestamps used include time points, time periods, and finite unions of time periods, termed temporal elements.

A temporal query language is defined in the context of a specific data model. Most notably, it supports the specification of queries on the specific form of time-referenced data provided by its data model. More generally, it enables the management of time-referenced data.

Different approaches to the design of a temporal extension to the Structured Query Language (SQL) have emerged that yield temporal query languages with quite different design properties.

Historical Background

A number of past events and activities that included the temporal database community at large had a significant impact on the evolution of temporal query languages. The 1987 *IFIP TC 8/WG*

8.1 Working Conference on Temporal Aspects in Information Systems [7] covered topics such as requirements for temporal data models and information systems, temporal query languages, versioning, implementation techniques, as well as temporal logic, constraints, and relations to natural language.

The 1993 *ARPA/NSF International Workshop on an Infrastructure for Temporal Databases* [8] gathered researchers in temporal databases with the goal of consolidating the different approaches to temporal data models and query languages. In 1993, the influential collection *Temporal Databases: Theory, Design, and Implementation* [11] was also published. This collection describes a number of data models and query languages produced during the previous 10 years of temporal database research.

Year 1995 saw the publication of the book *The TSQL2 Temporal Query Language* [9]. TSQL2 represents an effort to design a consensus data model and query language, and it includes many of the concepts that were proposed by earlier temporal data models and query languages. In 1995, the *International Workshop on Temporal Databases* [3] was co-located with the VLDB conference.

Then, in 1996, *SQL/Temporal: Part 7 of SQL3* was accepted. This was the result of an effort aimed at transferring results of temporal database research into SQL3. The first step was a proposal of a new part to SQL3, termed SQL/Temporal, which included the PERIOD data type. In 1997, the *Dagstuhl Seminar on Temporal Databases* took place [5]. Its goal was to discuss future directions for temporal database management, with respect to both research issues and the means to incorporate temporal databases into mainstream application development.

Foundations

A discrete and totally ordered time domain is assumed that consists of time instants/points. The term (time) "period" is used to denote a convex subset of the time domain. The term (time) "interval" then denotes a duration of time, which coincides with its definition in SQL. As a running example, the temporal relation *Rental* in Fig. 1a is used, which records car rentals, e.g., customer C101 rents vehicle V1234 from time 3 to time 5. Figures 1b–e show different representations of this relation, using the strong period-based, weak period-based, point-based, and parametric model, respectively. (In all the example relation instances, the conventional attribute(s) are separated from the timestamp attribute(s) with a vertical line.) The following queries together with their intended results build on the car rental example. These will serve for illustration.

Q1: *All rentals that overlap the time period* [7,9]. Query *Q1* asks for all available information about rentals that overlap the period [7,9].

CustID VID	*T*
C102 V1245	[5,7]
C102 V1234	[9,12]

Q2: *All 2-day rentals.* This query constrains the number of time points included in a time period and teases out the difference between the use of time points versus periods.

VID	*T*
V1245	[19,20]
V1245	[21,22]

Q3: *How many vehicles have been rented?* This is an example of an ordinary query that must be applied to each state of a temporal database. The non-temporal query is an aggregation. Thus, the result at a specific time point is computed over all tuples that are valid at that time point. (Note that some query languages don't return tuples when there is no data, e.g., when the *Cnt* is 0.)

Cnt	T
1	[3,4]
2	[5,5]
1	[6,7]
0	[8,8]
1	[9,12]
0	[13,18]
1	[19,20]
1	[21,22]

Q4: *How many rentals were made in total?* This is another aggregation query; however, the aggregation is to be applied independently of any temporal information.

Cnt
5

Q5:*List all (current) rentals.* This query refers to the (constantly moving) current time. It is assumed that the current time is 5.

CustID	VID
C101	V1234
C102	V1245

Approach I: Abstract Data Types – SQL/ATD

The earliest and, from a language design perspective, simplest approaches to improving the temporal data management capabilities of SQL have simply introduced time data types and associated predicates and functions. This approach is illustrated on the *Rental* instance in Fig. 1b.

$Q1^{SQL/ATD}$: *select * from Rental where T overlaps [7,9]*

$Q2^{SQL/ATD}$: *select VID, T from Rental where duration(T) = 2*

$Q4^{SQL/ATD}$: *select count(*) as Cnt from Rental*

$Q5^{SQL/ATD}$: *select CustID, VID from Rental where T overlaps [now,now]*

The predicates on time-period data types available in query languages have been influenced by Allen's 13 period relationships [1], and different practical proposals for collections of predicates exist. For example, the overlaps predicate (as defined in the TSQL2 language) can be used to formulate Query Q1. Predicates that limit the duration of a period (Q2) and retrieve current data (Q5) follow the same approach.

Expressing the time-varying aggregation of Q3 in SQL is possible, but exceedingly complicated and inefficient. The hard part is that of expressing the computation of the periods during which the aggregate values remain constant. (This requires about two dozen lines of SQL with nested *NOT EXISTS* subqueries [10, pp. 165–166].) In contrast, counting the rentals independently of the time references is easy, as shown in Q4.

Adding a new ADT to SQL has limited impact on the language design, and extending SQL with new data types with accompanying predicates and functions is relatively simple and fairly well understood. The approach falls short in offering means of conveniently formulating a wide range of queries on period timestamped data, including temporal aggregation. It also offers no systematic way of generalizing a simple snapshot query to becoming time-varying. Shortcomings such as these motivate the consideration of other approaches.

Approach II: Folding and Unfolding – IXSQL

Another approach is to equip SQL with the ability to normalize timestamps. Advanced most prominently by Lorentzos [4, 6] in the IXSQL language, the earliest and most radical approach is

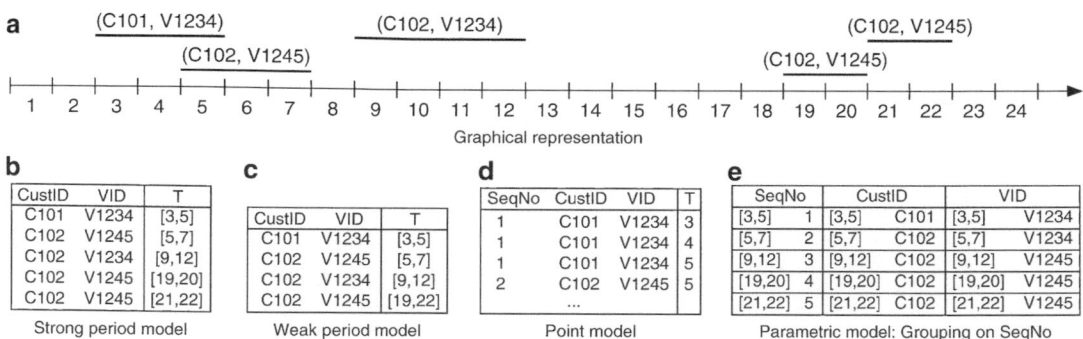

SQL-Based Temporal Query Languages, Fig. 1 Temporal relation *rental*

to introduce two functions: *unfold*, which decomposes a period-timestamped tuple into a set of point-timestamped tuples, and *fold*, which "collapses" a set of point-timestamped tuples into value-equivalent tuples timestamped with maximum periods. The general pattern for queries is then: (i) construct the point-based representation by unfolding the argument relation(s), (ii) compute the query on the period-free representation, and (iii) fold the result to obtain a period-based representation. The *Rental* relation in Fig. 1b is assumed.

Q3^{IXSQL}: *select count(*) as Cnt, T from (select * from Rental reformat as unfold T) group by T reformat as fold T*

The IXSQL formulations of Q1, Q2, Q4, and Q5 are essentially those of the ADT approach (modulo minor syntactic differences); specifically, normalization is not needed. The *fold* and *unfold* functions become useful for the temporal aggregation in Q3. The inner query unfolds the argument relation, yielding the point-based representation in Fig. 1d, on which the aggregation is computed. The *fold* function then transforms the result back into a period-stamped relation, which, however, is different from the intended result because the last two tuples are merged into a single tuple (1,[19,22]). The combination of unfolding and folding yields maximal periods of snapshot equivalent tuples and does not carry over any lineage information.

SQL with folding and unfolding is conceptually simple and offers a systematic approach to formulating at least some temporal queries,

including temporal queries that generalize non-temporal queries. It obtains the representational benefits of periods while avoiding the potential problems they pose in query formulation, since the temporal data are manipulated in point-stamped form. The *fold* and *unfold* functions preserve the information content in a relation only up to that captured by the point-based perspective; thus, lineage information is lost. This leaves some "technicalities" (which are tricky at times) to be addressed by the application programmer.

Approach III: Point Timestamps – SQL/TP

A more radical approach to designing a temporal query language is to simply assume that temporal relations use point timestamps. The temporal query language SQL/TP advanced by Toman [12] takes this approach to generalizing queries on non-temporal relations to apply to temporal relations. The point-timestamped *Rental* relation in Fig. 1d is assumed in the following.

Q1^{SQL/TP}: *select distinct a.* from Rental a, Rental b where a.SeqNo = b.SeqNo and or (b.T = 7 or b.T = 8 or b.T = 9)*

Q2^{SQL/TP}: *select SeqNo, VID, T from Rental group by SeqNo having count(T) = 2*

Q3^{SQL/TP}: *select count(*) as Cnt, T from Rental group by T*

Q4^{SQL/TP}: *select count(distinct SeqNo) as Cnt from Rental*

$Q5^{SQL/TP}$: *select CustID, VID from Rental where*
  T = now

Q1 calls for a comparison of neighboring database states. The point-based perspective, which separates the database states, does not easily support such queries, and a join is needed to report the original rental periods. The *distinct* keyword removes duplicates that are introduced if a tuple shares more than one time point with the period [7,9].

Duration queries, such as Q2, are formulated as aggregations and require an attribute, in this case *SeqNo*, that distinguishes the individual rentals. The strength of SQL/TP is in its generalization of queries on snapshot relations to queries on temporal relations, as exemplified by Q3. The general principle is to extend the snapshot query to separate database snapshots, which here is done by the grouping clause. SQL/TP and SQL are opposites when it comes to the handling of temporal information. In SQL, *time-varying* aggregation is poorly supported, while SQL/TP needs an additional attribute that identifies the real-world facts in the argument relation to support *time-invariant* aggregation (Q4).

The restriction to time points ensures a simple and well-defined semantics that avoids many of the pitfalls that can be attributed to period timestamps. As periods are still to be used in the physical representation and user interaction, one may think of SQL/TP as a variant of IXSQL where, conceptually, queries must always apply *unfold* as the first operation and *fold* as the last. To express the desired queries, an identifying attribute (e.g., *SeqNo*) is often needed. Such identifiers do not offer a systematic way of obtaining point-based semantics *and* a semantics that preserves the periods of the argument relations. The query *"When was vehicle V1245, but not vehicle V1234, rented?"* illustrates this point. A formulation using the temporal difference between the timestamp attributes does not give the expected answer {[6,7],[19,20],[21,22]} because the sequence number is not included. If the sequence number is included, the difference is effectively disabled. This issue is not only germane to SQL/TP, but applies equally to all approaches that use a point-based data model.

Approach IV: Syntactic Defaults – TSQL2

What may be viewed as syntactic defaults have been introduced to make the formulation of common temporal queries more convenient. The most comprehensive approach based on syntactic defaults is TSQL2 [9]. As TSQL2 adopts a point-based perspective, the *Rental* instance in Fig. 1c is assumed, where the periods are a shorthand representation of time points.

$Q1^{TSQL2}$: *select * from Rental where valid(Rental)*
  overlaps period '7–9'
$Q2^{TSQL2}$: *select SeqNo, VID from Rental where*
  cast(valid(Rental) as interval) = 2
$Q3^{TSQL2}$: *select count(*) as Cnt from Rental*
  group by valid(Rental) using instant
$Q4^{TSQL2}$: *select snapshot count(*) as Cnt from*
  Rental
$Q5^{TSQL2}$: *select snapshot * valid(date 'now') from*
  Rental

In TSQL2, a *valid* clause, which by default is present implicitly after the *select* clause, computes the intersection of the valid times of the relations in the *from* clause, which is then returned in the result. With only one relation in the *from* clause, this default clause yields the original timestamps as exemplified in Q1 and Q2. The cast function in Q2 maps between periods (e.g., [7−9]) and intervals (e.g., 3 days). The argument relation must be augmented by the *SeqNo* attribute (thus obtaining a relation with five tuples, as in Fig. 1b) for this query to properly return the 2-day rentals.

The default behavior of the implicit *valid* clause was designed with snapshot reducibility in mind, which shows nicely in the instant temporal aggregation query Q3. The grouping is performed according to the time points, not the original timestamps returned by *valid(Rental)*. The *using instant* is in fact the default and could be omitted (added for clarity). As TSQL2 returns temporal relations by default, the *snapshot* keyword is used in queries Q4 and Q5 to retrieve non-temporal relations.

Well-chosen syntactic defaults yield a language that enables succinct formulation of

S

common temporal queries. However, adding temporal support to SQL in this manner is difficult since the non-temporal constructs do not permit a systematic and easy way to express the defaults. It is challenging to be comprehensive in the specification of such defaults, and to ensure that they do not interact in unattractive ways. Thus, syntactic defaults lack "scalability" over language constructs.

Approach V: Statement Modifiers – ATSQL

ATSQL [2] introduces temporal statement modifiers to offer a systematic means of constructing temporal queries from non-temporal queries. A temporal query is formulated by first formulating the corresponding non-temporal query, and then prepending this query with a statement modifier that tells the database system to use temporal semantics. In contrast to syntactic defaults, statement modifiers are semantic in that they apply in the same manner to any statement they modify. The strong period-timestamped *Rental* instance in Fig. 1b is assumed in the following.

$Q1^{ATSQL}$: *seq vt select * from Rental where T overlaps [7,9]*

$Q2^{ATSQL}$: *seq vt select VID from Rental where duration(T) = 2*

$Q3^{ATSQL}$: *seq vt select count(*) as Cnt from Rental*

$Q4^{ATSQL}$: *nseq vt select count(*) as Cnt from Rental*

$Q5^{ATSQL}$: *select * from Rental*

Queries Q1 and Q2 can be formulated almost as in SQL. The *seq vt* ("sequenced valid time") modifier indicates that the semantics is consistent with evaluating the non-temporal query on a sequence of non-temporal relations, and ensures that the original timestamps are returned. Modifiers also work for queries that use period predicates, such as, e.g., Allen's relations, which cannot be used in languages of point-timestamped data models.

Query Q3 is a temporal generalization of a non-temporal query and can be formulated by prepending the non-temporal SQL query with

the *seq vt* modifier. The modifier ensures that at each time point, the aggregates are evaluated over all tuples that overlap with that time point. Query Q4 is to be evaluated independently of the time attribute values of the tuples. This is achieved by using the *nseq vt* ("non-sequenced valid time") modifier, which indicates that what follows should be treated as a regular SQL query.

A query without any modifiers considers only the current states of the argument relations, as exemplified by Query Q5. This ensures that legacy queries on non-temporal relations are unaffected if the non-temporal relations are made temporal.

Statement modifiers are orthogonal to SQL and adding them to SQL represents a much more fundamental change to the language than, e.g., adding a new ADT or syntactic defaults. The notion of statement modifiers offers a wholesale approach to rendering a query language temporal: modifiers control the semantics of any query language statement. This language mechanism is independent of the syntactic complexity of the queries that the modifiers are applied to. It becomes easy to construct temporal queries that generalize snapshot queries.

Approach VI: Temporal Expressions – TempSQL

The notion of temporal expression was originally advocated by Gadia and is supported in the TempSQL language [11, p. 28ff], which is based on the parametric data model (see Fig. 1e). Relations in TempSQL consist of tuples with attribute values that are functions from a subset of the time domain to some value domain (specified as a pair of a temporal element, a finite union of time periods, and a value). The functions in the same tuple must have the same domain. The relations are keyed. If a set of attributes is a key, then no two tuples are allowed to exist in the relation that have the same range values for those attributes. Figure 1e with the key *SeqNo* is assumed in the following.

$Q1^{TempSQL}$: *select * from Rental where [[VID]] ∪ [7,9] Ø*

$Q2^{TempSQL}$: *select VID from Rental where duration([[VID]]) = 2*

$Q3^{TempSQL}$: *select count(*) as Cnt from Rental*

$Q5^{TempSQL}$: *select * from Rental*

Queries Q1 and Q2 can be formulated using temporal expressions. If X is an expression that returns a function from time to some value domain then $[[X]]$ is a temporal expression which returns the domain of X, i.e., the time when X is true. The result of Q2 is the relation $\{\langle[19,22]$ V1245$\rangle\}$. For the aggregation query Q3, Temp-SQL automatically performs an instant temporal aggregation [11, p. 42]. A different query must be used to determine the time-invariant count in Q4. One possibility would be to formulate a query that first drops or equalizes all timestamps and then performs the above aggregation. For so-called current users, TempSQL offers built-in support for accessing the current state of a database, by assuming that the argument relations are the ordinary snapshot relations that contain the current states of the temporal relations. This is exemplified in Q5.

Temporal expressions as used in TempSQL, which return the temporal elements during which a logical expression is true, are convenient and often enable the elegant formulation of queries. Temporal expressions along with temporal elements fit well into the point-based framework. However, as of yet, little research has been done to further explore temporal expressions and to include them into query languages.

Key Applications

SQL-based temporal query languages are intended for use in database applications that involve the management of time-referenced data. Such applications are found literally in all data management application areas – in fact, virtually all real-world databases contain time-referenced data. SQL-based languages are attractive in comparison to other types of languages because SQL is used by existing database management systems.

Future Directions

While temporal query language support appears to be emerging in commercial systems, comprehensive temporal support is still not available in products.

Much research in temporal query languages has implicitly or explicitly assumed a traditional administrative data management setting, as exemplified by the car rental example. The design of temporal query languages for other kinds of data and applications, e.g., continuous sensor data, has received little attention.

Cross-References

► Allen's Relations
► Now in Temporal Databases
► Period-Stamped Temporal Models
► Point-Stamped Temporal Models
► Temporal Database
► Temporal Data Models
► Temporal Element
► Time Interval
► Time Period
► Temporal Query Languages
► TSQL2
► Valid Time

Recommended Reading

1. Allen JF. Maintaining knowledge about temporal intervals. Commun ACM. 1983;26(11):832–43.
2. Böhlen MH, Jensen CS, Snodgrass RT. Temporal statement modifiers. ACM Trans Database Syst. 2000;25(4):407–56.
3. Clifford J, Tuzhilin A, editors. Recent advances in temporal databases. In: Proceedings of the International Workshop on Temporal Databases; 1995.
4. Date CJ, Darwen H, Lorentzos N, editors. Temporal data and the relational model. Morgan Kaufmann Publishers; 2002.
5. Etzion O, Jajodia S, Sripada S, editors. Temporal databases: research and practice. Springer, Volume 1399 of lecture notes in computer science; 1998.
6. Lorentzos NA, Johnson RG. Extending relational algebra to manipulate temporal data. Inf Syst. 1988;13(3):289–96.

S

$$\text{Var}\left(\widehat{d}_{(\alpha),hm}\right) =$$

$$d_{(\alpha)}^2 \frac{1}{k}\left(\frac{-\pi\Gamma(-2\alpha)\sin(\pi\alpha)}{\left[\Gamma(-\alpha)\sin\left(\frac{\pi}{2}\alpha\right)\right]^2} - 1\right) \quad (10)$$

$$+ O\left(\frac{1}{k^2}\right).$$

- As α approaches zero, in the limit,

$$\lim_{\alpha\to 0+} -\frac{2}{\pi}\Gamma(-\alpha)\sin\left(\frac{\pi}{2}\alpha\right) = 1,$$

$$\lim_{\alpha\to 0+}\left(\frac{-\pi\Gamma(-2\alpha)\sin(\pi\alpha)}{\left[\Gamma(-\alpha)\sin(\frac{\pi}{2}\alpha)\right]^2} - 1\right) - 1. \quad (11)$$

- *The geometric mean estimator*

$$\widehat{d}_{(\alpha),gm} = \frac{\prod_{j=1}^{k}\left|x_j\right|^{\alpha/k}}{\left[\frac{2}{\pi}\Gamma\left(\frac{\alpha}{k}\right)\Gamma\left(1-\frac{1}{k}\right)\sin\left(\frac{\pi}{2}\frac{\alpha}{k}\right)\right]^k}. \quad (12)$$

$$\text{Var}\left(\widehat{d}_{(\alpha),gm}\right)$$

$$= d_{(\alpha)}^2\left\{\frac{\left[\frac{2}{\pi}\Gamma\left(\frac{2\alpha}{k}\right)\Gamma\left(1-\frac{2}{k}\right)\sin\left(\pi\frac{\alpha}{k}\right)\right]^k}{\left[\frac{2}{\pi}\Gamma\left(\frac{\alpha}{k}\right)\Gamma\left(1-\frac{1}{k}\right)\sin\left(\frac{\pi}{2}\frac{\alpha}{k}\right)\right]^{2k}} - 1\right\} \quad (13)$$

$$= d_{(\alpha)}^2\frac{1}{k}\frac{\pi^2}{12}\left(\alpha^2 + 2\right) + O\left(\frac{1}{k^2}\right). \quad (14)$$

- *The sample median estimator*

$$\widehat{d}_{(\alpha),me} = \frac{\text{median}\left\{|x_j{}^\alpha|\,|\,j=1,2,...,k\right\}}{\text{median}\{S(\alpha,1)\}^\alpha}. \quad (15)$$

The estimation variance of the sample median estimator $\widehat{d}_{(\alpha),me}$ cannot be expressed in closed-forms. Compared with the geometric mean estimator, the sample median estimator is not as accurate when the sample size k is not very large. The sample median estimator, however, is more convenient to compute.

Sample Complexity

When $\alpha = 2$, the celebrated Johnson-Lindenstrauss (JL) Lemma [9] showed that k, the required number of projections, should satisfy $k = O(\log n / \varepsilon^2)$ so that any pairwise l_2 distance among n data points can be approximated within a $1 \pm \varepsilon$ factor of the truth.

For general $0 < \alpha \leq 2$, it is proved [11] using the geometric mean estimator that the sample complexity should also be $k = O(\log n / \varepsilon^2)$. The constants can be explicitly specified.

Sampling from Stable Distributions

Sampling from a stable distribution is in general quite expensive, unless $\alpha = 2$ or $\alpha = 1$. One procedure is described in [13, Proposition 1.71.1]. A random variable W_1 is sampled from a uniform distribution on the interval $\left(-\frac{\pi}{2},\frac{\pi}{2}\right)$; and a random variable E_1 is sampled from an exponential distribution with mean 1. W_1 and E_1 are independent. Then

$$\frac{\sin(\alpha W_1)}{\cos(W_1)^{1/\alpha}}\left(\frac{\cos((1-\alpha)W_1)}{E_1}\right)^{(1-\alpha)/\alpha} \quad (16)$$

is distributed as $S(\alpha, 1)$.

Under certain reasonable regularity assumptions on the original data, it is possible to simplify the sampling procedure by replacing the α-stable distribution $S(\alpha, 1)$ with a mixture of a symmetric α-Pareto distribution (with probability $0 < \beta \leq 1$) and a point mass at the origin (with probability $1 - \beta$), i.e.,

$$\begin{cases} P_\alpha & \text{with prob. } \frac{\beta}{2} \\ 0 & \text{with prob. } 1-\beta \ , \\ -P_\alpha & \text{with prob. } \frac{\beta}{2} \end{cases} \quad (17)$$

where P_α denotes an α-Pareto variable, i.e., $\Pr(P_\alpha > t) = \frac{1}{t^\alpha}$ if $t \geq 1$, and 0 otherwise. An α-Pareto distribution has the same tail behaviors as $S(\alpha, 1)$, but it is much easier to sample from. For example, given a random variable U drawn from a uniform distribution on the unit interval $(0,1)$, then $1/U^{1/\alpha}$ follows an α-Pareto distribution.

If D random variables, r_1, r_2, ..., r_D, are sampled i.i.d. from Eq. (17), then a linear combination of r_i's is asymptotically stable, as $D \to \infty$. That is [10],

$$c_1 r_1 + c_2 r_2 + ... + c_D r_D$$

$$\Rightarrow S\left(\alpha, \beta \Gamma(1-\alpha) \cos\left(\frac{\pi}{2}\alpha\right) \sum_{j=1}^{D} |c_i^{\alpha}|\right), \tag{18}$$

provided that the data, c_1, c_2, ..., c_D, satisfy

$$\frac{\max_{1 \le i \le D} |c_i|}{\left(\sum_{i=1}^{D} |c_i^{\alpha}|\right)^{1/\alpha}} \to 0, \quad \text{as} \quad D \to \infty. \tag{19}$$

The parameter β in Eq. (17) controls the sparsity of the projection matrix. Small β values considerably reduce the processing cost for conducting random projections. β should be chosen according to the data dimension D and the prior knowledge about the data. Some synthetic and real-world data experiments in [10] indicated that, when $\alpha = 1$, using Eq. (17) with $\beta < 0.1$, achieved very similar estimation accuracy as using the exact stable distribution, even when D is not too large.

Key Applications

Stable distributions have been widely used for modeling real-world *heavy-tailed* data, arising in finance, economics, Internet traffic, computational Linguistics, and many other fields.

There have been numerous applications of stable random projections, in theoretical computer science, databases, data mining, data streams, and signal recovery [5].

Stable Random Projections for Dimension Reductions
Data mining and machine learning algorithms often assume a "data matrix" $A \in \mathbb{R}^{n \times D}$, with n rows and D columns. For many algorithms, the data matrix A is utilized only through pairwise

distances of A instead of the original data. A projection matrix $R \in \mathbb{R}^{D \times k}$ is generated by sampling each entry from i.i.d. $S(\alpha, 1)$. The projected data matrix $B = A \times R \in \mathbb{R}^{n \times k}$ contain enough information to approximately recover pairwise l_{α} distances of A. The number of projections (sample size), should satisfy $k = O(\log n / \varepsilon^2)$.

- The original data matrix A may be too large for physical memory, for example, A could be the term-by-document matrix at Web scale. Even if A may fit in memory, storing all pairwise distances of A in memory can be infeasible when $n > 10^6$. In contrast, the projected data matrix B may be small enough for the memory. Because B has only k columns, pairwise distances may be computed on demand.
- Computing all pairwise distances of A costs $O(n^2 D)$. The cost is reduced to $O(nDk + n^2 k)$ using stable random projections.
- When $\alpha = 2$, the projected data matrix B preserve not only the pairwise (squared) l_2 distances of A in expectations, but also the pairwise inner products of A in expectations. Some applications care about inner products more than distances. In databases, for example, counting the joint sizes can be viewed as computing inner products.

Stable Random Projections for Data Stream Computations
Consider the *Turnstile* model [12], which is a linear model for data streams. The input data stream $s_t = (i, I_t)$ arriving sequentially describes the underlying signal S, meaning $S_t[i] = S_{t-1}[i] + I_t$, $i = 1$ to D, where t denotes time. For example, S may represent the arriving IP addresses ($D = 2^{64}$) and $S_t[i]$ records the frequencies of IP address i. The term $\sum_{i=1}^{D} |S_t[i]|^{\alpha}$ is often referred to as the αth *frequency moment* of S_t. Due to the linearity of the *Turnstile* model, stable random projections can be applied for approximating the frequency moments.

Again, a random projection matrix $R \in \mathbb{R}^{D \times k}$ is generated by sampling each entry r_{ij} from i.i.d. $S(\alpha, 1)$. A vector x of length k is initialized so that $x_j = 0$, for $j = 1$ to k. Then for each arriving tuple $s_t = (i, I_t)$, update $x_j \leftarrow x_j + r_{ij} \times I_t$ for $j = 1$ to k.

Staged DBMS

Stavros Harizopoulos
HP Labs, Palo Alto, CA, USA

Synonyms

Staged database systems

Definition

A Staged Database Management System (DBMS) is a database software architecture that optimizes data and instruction locality at all levels of the memory hierarchy in a computer system. An additional goal of Staged DBMS is to provide a robust and efficient platform for both parallelizing and pipelining database requests. The main principle of the Staged Database System design is to organize and assign software system components into self-contained stages; database request execution is broken into stages and sub-requests are group-processed at each stage. This allows for a context-aware execution sequence of requests that promotes reusability of both instructions and data, and also facilitates development of work sharing mechanisms, which has been a key application for Staged DB; work sharing is defined as any operation that reduces the total amount of work in a system by eliminating redundant computation or data accesses. Existing database systems can be converted to staged ones by carrying over their algorithms and mechanisms, and adapt those in a platform that supports staged execution.

Historical Background

Though the Staged DBMS architecture was proposed by Harizopoulos and Ailamaki in 2003 [3], one of the earliest prototype relational database systems, INGRES [12], also consisted of four "stages" (processes) that enabled pipelining; the reason for breaking up the DBMS software was main memory size limitations. Work in staged architectures re-emerged in the early 2000s, first by Larus and Parkes as a generic programming paradigm for building server applications [7], and subsequently by Welsh et al. as a means for deploying highly concurrent internet services [13]. Initial prototypes of database systems developed at Carnegie Mellon University that followed the principles of StagedDB (Qpipe and Cordoba [4, 5]), focused on the performance benefits of work sharing. A relational engine based on the Staged DBMS design can proactively coordinate same-operator execution among concurrent queries, thereby exploiting common accesses to memory and disks as well as common intermediate result computation.

Foundations

Modern commercial DBMS are typically built as a large piece of software that serves multiple requests using a thread-based concurrency model. Queries are handled by one or more threads (or processes) that follow the query execution plan up to its completion. This model implicitly defines a query execution sequence and a resource utilization schedule in the system. Whenever a thread blocks due to an I/O, an ungranted lock request, an internal synchronization condition, or due to an expiring CPU time quantum, the thread scheduler assigns the CPU to the next runnable thread of the highest priority. This context-switching mechanism creates a logical gap in the sequence of actions the DBMS performs. While a software developer can optimize the individual steps involved in a single query's execution, she or he typically has no means of applying similar optimization techniques to a collection of multiplexed queries.

A staged database system consists of a number of self-contained software modules, each encapsulated into a *stage*. A stage is an independent server with its own queue, thread support, and resource management that communicates and interacts with the other stages through a well-defined interface. Stages accept *packets*, each carrying a query's state and private data, perform work on the packets, and may enqueue the same or newly created packets to other stages. Each stage is centered around exclusively owned (to the degree possible) server code and data. There are two levels of CPU scheduling: local thread scheduling within a stage and global scheduling across stages. The StagedDB design promotes stage autonomy, data and instruction locality, and minimizes the usage of global variables.

A stage provides two basic operations, enqueue and dequeue, and a queue for the incoming packets. The stage-specific server code is contained within dequeue. The system works through the exchange of packets between stages. A packet represents work that the server must perform for a specific query at a given stage. It first enters the stage's queue through the enqueue operation and waits until a dequeue operation removes it. Then, once the query's current state is restored, the stage specific code is executed. Depending on the stage and the query, new packets may be created and enqueued at other stages. Eventually, the stage code returns by either (i) destroying the packet (if done with that query at the specific stage), (ii) forwarding the packet to the next stage (i.e., from parse to optimize), or by (iii) enqueueing the packet back into the stage's queue (if there is more work but the client needs to wait on some condition). Queries use packets to carry their state and private data. Each stage is responsible for assigning memory resources to a query. In a shared-memory system, packets carry only pointers to the query's state and data structures (which are kept in a single copy). Each stage employs a pool of worker threads (the stage threads) that continuously call dequeue on the stage's queue, and one thread reserved for scheduling purposes (the scheduling thread). An analysis of scheduling tradeoffs in staged database systems along with a description of an initial implementation can be found in [2].

Key Applications

A key application for the Staged DBMS design has been detecting and exploiting work sharing opportunities at run-time inside a relational database engine. Traditional relational DBMS typically execute concurrent queries independently by invoking a set of operator instances for each query. To exploit common data retrievals and computation in concurrent queries, relational engines employ techniques ranging from constructing materialized views to optimizing multiple queries and sharing concurrent scans to the same table. These three techniques are briefly described next.

Materialized view selection [8] is typically applied to workloads known in advance, in order to speed up queries that contain common sub-expressions. Materialized views exploit commonality between different queries at the expense of potentially significant view maintenance costs. Tools for automatic selection of materialized views take such costs into account when recommending a set of views to create. The usefulness of materialized views is limited when the workload is not always known ahead of time or the workload requirements change frequently.

Multiple-query optimization (MQO) [10] identifies common sub-expressions in query execution plans during optimization, and produces globally-optimal plans. The detection of common sub-expressions is performed at optimization time, thus, all queries need to be optimized as a batch. In addition, to share intermediate results among queries, MQO typically relies on costly materializations. To avoid unnecessary materializations, a study described in [9] introduces a model that decides at the optimization phase which result can be pipelined and which needs to be materialized to ensure continuous progress in the system.

Shared scans allow multiple independent concurrent scans to the same table on disk to be synchronized, so that each new page fetched from disk is consumed by all scans that include the page in their range. This optimization applies to scans that can receive their input pages in any arbitrary order. Since queries interact with the buffer pool manager through a page-level interface, it requires a certain engineering effort to develop generic policies to coordinate current

S

fashion [1, 2] store their primary data as well as the data cubes derived from it in star schemas. The "central" table and the remaining tables of the definition above correspond, respectively, to the *fact table* and the *dimension tables* that are typically found in data warehouses. Each *fact* (tuple) in the fact table consists of a set of numeric *measures*, comprising the objects of analysis, and a set of *dimensions*, which uniquely determine the set of measures. The dimension tables are usually smaller than the fact table and store the attributes of the aforementioned dimensions.

For example, consider a data warehouse of a retail chain with many stores around a country. The dimensions may be the products sold, the stores themselves with their locations, and the dates, while the numeric measures may be the number of items and the total monetary amount corresponding to a particular product sold in a particular store on a particular date. The relevant star schema is shown below, where SalesSummary is the fact table, primary keys are in Italics, and each attribute of the fact table primary key is a foreign key to one of the other tables.

SalesSummary(*ProductId*, *StoreId*, *DateId*, NumOfItems, TotalAmount)
Product(*ProductId*, ProdName, ProdDescr, Category, CategoryDescr, UnitPrice)
Store(*StoreId*, Street, City, State)
Date(*DateId*, Day, Month, Year)

If one were to draw the above as a graph, with tables as nodes and foreign keys as edges, or even as an ER diagram, with the fact table as a relationship and the dimension tables as entities, the resulting image is that of a star, with the fact table in the middle, hence, the name of these schemas.

Finally, note that dimensions often consist of several attributes organized in hierarchies. For instance, dimension Store in the example above contains values at different levels of detail, forming the hierarchy Street→City→State. As also shown in the example, star schemas capture all levels of a hierarchical dimension in a single, non-normalized table. An extension of the star schema that explicitly captures hierarchies in the dimensions is the *snowflake schema*.

Cross-References

▶ Cube Implementations
▶ Data Warehouse
▶ Dimension
▶ Hierarchy
▶ Measure
▶ Multidimensional Modeling
▶ Snowflake Schema

Recommended Reading

1. Chaudhuri S, Dayal U. An overview of data warehousing and OLAP technology. ACM SIGMOD Rec. 1997;26(1):65–74.
2. Kimball R, Ross M. The data warehouse toolkit: the complete guide to dimensional modeling. 2nd ed. New York: Wiley; 2002.

State-Based Publish/Subscribe

Hans-Arno Jacobsen
Department of Electrical and Computer Engineering, University of Toronto, Toronto, ON, Canada

Definition

State-based publish/subscribe is an instance of the publish/subscribe concept. However, it is distinguished from other publish/subscribe approaches by maintaining partial matching state when processing publications, whereas, traditionally, publish/subscribe treat publications as transient and does not manage matching state. State-based publish/subscribe support the detection of composite events, event correlation and complex event processing.

Key Points

In terms of publishing, subscribing, and decoupling, state-based publish/subscribe is no different from topic-based or content-based publish/subscribe. The main difference to the other publish/subscribe approaches is that state-based publish/subscribe treats publications as non-transient. A publication is processed by the publish/subscribe system and builds up partial matching state, contributes to existing partial matching state, triggers notifications if a match is complete, or is discarded, if no matching subscription exists. This is unlike in the other publish/subscribe approaches that treat publications as transient messages, where the arriving publication is matched against the subscriptions stored with the system and forwarded towards all matching subscribers, or dropped, if no matching subscription exists. In state-based publish/subscribe, the publish/subscribe system carries state across the processing of different publications. That is, various different publications arriving over time are correlated based on conditions expressed in subscriptions to result in matching subscriptions.

The publication data model is exactly the same as in a topic-based or a content-based publish/-subscribe model, depending on the nature of the state-based approach, either topic-based, or content-based.

The subscription language model follows suite, but greatly extends the subscription language capabilities with means to express the correlation of publications. These capabilities are added to allow the application to express so called *composite subscriptions*. A composite subscription is the combination of several individual *atomic subscriptions* by means of an operator algebra that allows the developer to compose individual atomic subscriptions. An atomic subscription is a subscription in the publish/subscribe sense. It is referred to as atomic because it is matched by a single publication. A composite subscription defines a *composite event*. A composite event defines the set of events that have to occur in the specified constellation in order for the composite subscription to match. In publish/subscribe, the notion of event and publication are synonymous, while only the term composite event is used. Also, the term composite event is often used to refer to the composite subscription expression without differentiating between event and subscription.

There are large differences in the expressive power of subscription languages for specifying composite subscriptions. Common operators include the specification of composite-and (all specified events must occur in any order), composite-or (one of the events must occur), sequence operator (the specified sequence of events must occur and other events may or may not be interspersed with the sequence), and regular expression pattern (the specified pattern of events must occur). In addition, reference to time is included in many subscription languages to delimit the time a composite subscription can remain in a partial matching state before resetting to a completely unmatched state. More generally, various consumption policies attached to subscriptions express what should happen with partial matching state, as it accumulates in the system and new events arrive. For example, a consumption policy could express that a newly arriving event is correlated with the oldest or the newest event in a composite event that has accumulated a partial matching state holding many individual events already. Consumption policies are a powerful way to customize the matching behavior of state-based publish/subscribe systems.

In the state-based publish/subscribe model, the publish/subscribe matching problem is defined as follows: Given a set of subscriptions, S, and a sequence of events, E, as seen by the publish/subscribe system, determine the subscriptions in S that match under E. This formulation of the matching problem is different from the standard publish/subscribe matching problem that only looks at a single event e at a time. In state-based publish/subscribe, the matching algorithm has to manage partial matching state and correlated newly arriving events with already existing partial matching state stored in the algorithm's data structures.

State-based publish/subscribe is a fairly new sub-classification of publish/subscribe, consequently few established standards and products exist that refer to this model. However, several research projects are experimenting with the above described schemes. For example, the PADRES [?, ?] project is an example of a

Foundations

Privacy, Confidentiality, and Individual Identification

Massive databases and widespread data collection and processing offer enormous opportunities for statistical analyses, advances in the understanding of social and health problems, and benefits to society more broadly. But the explosion of computerized databases containing financial and healthcare records, and the vulnerability of databases accessible via the Internet, has heightened public attention and generated fears regarding the privacy of personal data. Identify theft and sensitive data disclosure may be just a click away from a new generation of computer users and potential intruders.

Data collected directly under government auspices or at public expense are in essence a public good; legitimate analysts wish to utilize the information available in such databases for statistical purposes. Thus, society's challenge is how to release the maximal amount of information without undue risk of disclosure of individually identifiable information. Assessing this tradeoff is inherently a statistical matter, as is the development of methods to limit disclosure risk. What distinguishes the field of statistical disclosure limitation from many other approaches to privacy protection is the ultimate goal of data access and enhanced data utility.

The term *privacy* is used both in ordinary language and in legal contexts with a multiplicity of meanings. Among these is the concept of privacy as "the right to be let alone," e.g., see Warren and Brandeis [14], and privacy in the context of data as the control over information about oneself. But privacy is personal and subjective, varies from one person to another, and varies with time and occasion depending on the context. It is even more difficult to define precisely the meaning of "privacy-preserving" with respect to databases, and the data pertaining to individual entities contained therein.

Confidentiality is the agreement, explicit or implicit, between data subject and data collector regarding the extent to which access by others to personal information is allowed. Confidential-

ity protection has meaning only when the data collector can deliver on its promise to the data provider or respondent. Confidentiality can be accorded to both individuals and organizations; for the individual, it is rooted in the right to privacy (i.e., the right of individuals to control the dissemination of information about themselves), whereas for establishments and organizations, there are more limited rights to protection, e.g., in connection with commercial secrets.

Disclosure relates to inappropriate attribution of information to a data provider or intruder, whether to an individual or organization. There are basically two types of disclosure, identity and attribute. An identity disclosure occurs if the data provider is identifiable from the data release. An attribute disclosure occurs when the released data make it possible to infer the characteristics of an individual data provider more accurately than would have otherwise been possible. The usual way to achieve attribute disclosure is through identity disclosure; first one identifies an individual through some combination of variables and then one associates with that individual values of other variables included in the released data.

Statistical disclosure limitation (SDL) is a set of techniques designed to "limit" the extent to which databases can be used to glean identifiable information about individuals or organizations. The dual goals of SDL are to assure that, based on released data, respondents can be identified only with relatively low probability, but also to release data that are suitable for non-identifiable analytical statistical purposes.

The Intruder

To protect the confidentiality of statistical data, one needs to understand what intruders or data snoopers want and how they may learn information about individuals in a database that require protection. Intruders may be those with legitimate access to databases and/or those who gain access to a database by breaking security measures designed to keep them out. In either case, one needs to distinguish among

- *Intruders with a specific target,* e.g., a friend or relative. The intruder may already know that the respondent is included in the database, will possess information about the target (e.g., height, weight, habits, income) and will search the database in order to learn additional information, e.g., drug and alcohol use.

- *Intruders in possession of data on multiple individuals whose goal is record linkage*, e.g., to build a larger database containing more individual information. Data consolidators or aggregators fit within this category.
- *Intruders without any specific target, whose goal is to embarrass the data owner.* The intruder *may* be an enemy agent or a "hacker" eager to demonstrate a capability of breaking through efforts to limit disclosure.

Data owners can be successful in protecting the confidentiality of released data if the intruder remains sufficiently uncertain about a protected target value after data release. Various authors in the SDL literature discuss confidentiality protection from the perspective of protecting against intruders or data snoopers, e.g., see [8, 12], especially those using record linkage methods [11] for attempting to identify individuals in databases.

One may consider the intruder as someone engaged in a form of a large number of statistical tests, each at significance level α associated with an effort to identify an individual in the database. The data owner needs to account for this somehow. Some of the null hypotheses will eventually be rejected whether or not they are actually false, and thus there is a problem for the intruder as well. For the data owner, simply controlling the probability of erroneously identifying each respondent, at say 1%, is not enough. To ensure that the probability that at most one out of 1,000 individuals in a sample will be identified to be less than 1%, one in fact needs to assure that the probability of identifying *each* individual is no greater than $\approx 1\,\%/1000 = 10^{-5}$.

Statistical Analysis Methods for Protecting Privacy

Matrix Masking refers to a class of SDL methods used to protect confidentiality of statistical data, transforming an $n \times p$ (cases by variables) data

matrix Z through pre- and post-multiplication and the possible addition of noise. The four most common forms of masking are:

1. *Sampling* clearly provides a measure of direct protection from disclosure provided that there is no information of which individuals or units are included in the sample. An intruder wishing to identify an individual in the sample and link that person's information to data in external files, using "key" variables such as age and geography available in both databases, needs to determine whether a record is unique in the sample, and if so, the extent to which a record that is unique in the sample is also unique in the population. For continuous variables, virtually all individuals are unique in the sample, and one needs to understand the probability that an intruder would correctly match records, e.g., in the presence of error in the key variables (e.g., see Fienberg et al. [8]). For categorical data, uniqueness corresponds to counts of "1" and various authors have shown, roughly speaking, that the probability that an individual record that is unique in the sample is also unique in the population from which the sample was drawn equals the sampling fraction, n/N, e.g., see [7]. Thus for a sample of size 2,000 drawn from a population of 200,000,000 adults the sampling fraction is 2,000/200,000,000 or 0.00001. The bottom line therefore is that sampling protects, just not absolutely.

2. *Perturbation* is an approach to data masking in which the transformation involves random perturbations of the original data, either through the addition of noise or via some form of restricted randomization. The simplest form of perturbation is the addition of noise. Common forms for the noise are observations drawn from a normal distribution with zero mean or perhaps a double exponential, also centered at zero. Someone analyzing the resulting transformed data must statistically reverse the noise addition process using methods from the literature on measurement error models - this requires release of the parameters of the noise component, e.g., the error variance in

the normal case. Other examples of perturbation include data swapping and related tabular adjustment approaches, e.g., see [9].

3. *Collapsing* is also referred to using the labels micro-aggregation and global recoding in the statistical literature [15], and *k*-anonymity in the computer science literature. In the statistical literature on tabular categorical data, collapsing across variables in a table produces a marginal table and a popular form of data release to protect confidentiality is the release of multiple marginal tables, especially when they correspond to the minimal sufficient statistics of a log-linear model. For more details, see [10].

4. *Synthetic data* are used to replace a database by a similar one, for which the *individuals* are generated through some statistical process. This can be achieved through the repeated application of data swapping, e.g., see [9], or the method known as multiple imputation, e.g., [13].

Implicit in all of these techniques is the notion that when masked data are released they can be used by responsible analysts to carry out statistical analyses so that they can reach conclusions similar to those that they would have reached had they analyzed the original data. This means that all of the details of the transformation, both stochastic and non-stochastic, must be made available to the user, a point not well understood in the computer science literature or by many statistical agencies. See the related discussion in [6]. Even when one has applied a mask to a data set, the possibilities of both identity and attribute disclosure remain, although the risks may be substantially diminished. Thus, one must still assess the extent of risk posed by the transformed data.

Putting SDL Methods to Use: Risk-Utility Tradeoff

If one is adding noise to a set of observations in order to protect confidentiality, how much noise is sufficient? And can one add too much? Clearly, too much noise will distort the data substantially and even if the details of the error variance are released the masking may impede legitimate

statistical analyses of the data. The same is true for any of the methods of SDL. Thus, one faces a tradeoff between data protection and data utility, something that one can assess formally using statistical decision analysis and depict graphically, e.g., see the chapter by Duncan et al. in [5]. For a slightly less formal approach to the tradeoff for categorical data protection through the release of multiple margins, see [10].

A crucial but relatively rarely discussed aspect of the risk-utility tradeoff involves the issue of multiplicity, introduced above. For illustration, consider a data set with information on 2,000 individuals, for each of which records the diagnostic result of an HIV test, with 1 corresponding to a positive result and 0 to a negative one. To protect the confidentiality for those individuals with positive HIV test outcomes, the data owner adds noise to each record value.

Suppose an intruder wishes to identify the individuals corresponding to the proportion ε of "1"s $(0 < \varepsilon < 1)$ and the data owner attempts to protect the records by adding i.i.d. Gaussian noise $N(0, \sigma^2)$ to each data point. Clearly σ needs to be large enough to disguise some of the 1s and make them hard to distinguish from some of the "0's. Suppose that the intruder wants to make sure that most of those identified as "1"s are indeed "1" s (otherwise the attack on the database would be unsuccessful). In statistical terms, this means that the intruder must control for the False Discover Rate (FDR), i.e., the rate of misclassified "1"s out of all those individuals labeled as being "1"s [1, 2]: FDR = [#{Misclassified " 1 " s}]/[#{All classified " 1 "s}]. Consider an intruder who decides to set the FDR at 5% by picking a threshold σt and classifying any entry as a "1' if the observed value exceeds the threshold. By elementary statistics, the number of misclassified "1's is distributed as a binomial random variable, $B(n(1 - \varepsilon), \overline{\Phi}(t))$, and the number of correctly classified "1's is distributed as $B\left(n\varepsilon, \overline{\Phi}\left(1 - \frac{1}{\sigma}\right)\right)$. Consequently, the associated $FDR \frac{n(1-\varepsilon)\overline{\Phi}(t)}{n(1-\varepsilon)\overline{\Phi}(t) + n\varepsilon\overline{\Phi}\left(t - \frac{1}{\sigma}\right)} \approx$

$\left[1 + \left(\frac{\epsilon}{1-\epsilon}\right) \frac{\overline{\Phi}\left(t - \frac{1}{\sigma}\right)}{\overline{\Phi}(t)}\right]^{-1}$, where $\overline{\Phi} = 1 - \Phi$

is the survival function of the standard normal distribution function.

Consider a high risk population where 50% of the individuals test positive for HIV, i.e., $\varepsilon = 1/2$ and suppose that the data owner chose $\sigma = 1$ as the noise variance to protect the data. To ensure that FDR $\leq 5\%$, the intruder needs to set $\left(\frac{\epsilon}{1-\epsilon}\right)\left(\frac{\overline{\Phi}(t-\frac{1}{\sigma})}{\overline{\Phi}(t)}\right) = 19$, which yields $t \approx 3.132$. The threshold is high enough so that the chance for each of the individuals exhibiting a "1" to be correctly classified as "1" is $-\Phi(2.132) \approx -\Phi(2.132) \approx 0.0165$, which seems not very large. But since $n = 2000$, the number of "0"s that are misclassified as "1"s is approximately $n (1 - \epsilon) \overline{\Phi}(3.132) \approx 2000 \times \frac{1}{2} \times 0.00087 = 0.87$, and the number of "1"s that are correctly classified as "1"s is approximately $n\epsilon\overline{\Phi}(2.132) = 2000 \times \frac{1}{2} \times 0.0165 \approx 16.5$. This says that the intruder is able to identify 17 records, out of which 16 are corrected classified!

Alternatively, one might ask about the probability that no more than k "1"s are correctly classified, i.e., $\sum_{j=0}^{k} \binom{n\epsilon}{j} p^j (1-p)^{n\epsilon-j}$, $p \equiv \overline{\Phi}(2.132)$. For $k = 0$, 3, 6, 9, the probabilities are correspondingly 5.45×10^{-8}, 5.22×10^{-5}, 2.62×10^{-3}, and 0.031. To understand the implications of these values, consider $k = 9$. This says that with probability as high as 97%, 9 or more records that are actually "1"s are correctly identified as "1"! For many this might seem to be a worrisome situation, and it raises issues associated with the efficacy of adding noise that have not appeared in the statistical literature on confidentiality protection.

Suppose that the data come from a low risk population where only 5% or 100 individuals test positive for HIV, i.e., $\varepsilon = 0.05$, and the data owner uses a similar level of noise addition for confidentiality protection, i.e., $\sigma = 1$. Then to ensure that FDR $\leq 5\%$, the intruder needs to set $\left(\frac{\overline{\Phi}(t-\frac{1}{\sigma})}{\overline{\Phi}(t)}\right) = 361$, which yields $t \approx 6.22$. Correspondingly, $\overline{\Phi}(t) = 2.5 \times 10^{-10}$ and $\overline{\Phi}\left(t - \frac{1}{\sigma}\right) = 89 \times 10^{-8}$. The expected number of "0" that are misclassified as "1"s is approximately $n (1 - \varepsilon) \overline{\Phi}(6.22) =$

$2000 \times 0.95 \times 2.49 \times 10^{-10} \approx 4.7 \times 10^{-6}$, and the number of "1" that are correctly classified as "1"s is approximately $n\varepsilon\overline{\Phi}(5.22) = 2000 \times 0.05 \times 8.95 \times 10^{-8} \approx 8.95 \times 10^{-6}$. In this case, since $n \overline{\Phi}(6.22) \ll 1$, the approximation is inaccurate and one needs to take a different approach.

In fact, the proportion of true HIV cases is so small that the example falls into the so-called *very sparse regime* studied in detail in the multiple testing literature, see for example [1, 4]. One phenomenon from that literature implies that when the noise level is relatively high, the extreme values are not necessary related to cases with positive HIV tests. Consider the following simulated data set with $n = 2,000$ cases, where 100 of them are HIV (equal to 1) and all others are non-HIV (equal to 0). The data owner adds independent standard Gaussian $N(0, 1)$ noise to each value. Figure 1 shows the result where red correspond to cases with positive HIV tests, and green correspond to cases with negative HIV tests. The red values are larger than typical green ones, but not larger than all of them. In fact, among the largest 10 values, only 2 are red, with 8 are green.

This leads us to another interesting phenomenon from the statistical literature on FDR. Let mFDR denote the minimum FDR across all possible thresholds t, mFDR $= min_{\{t\}}$ {FDR$_t$: FDR at the threshold t}. How small can mFDR be? Figure 1 shows the histogram of the mFDR values for 100 independent repetitions of the simulation experiment. More than half of the time, the mFDR value is no less than 15%, and sometimes it is as great as 50% and larger!

This simple example implies that with $\sigma = 1$, the noise level might be so large that the intruder cannot correctly identify any HIV cases. But from the perspective of the risk-utility tradeoff, one also needs to ask whether the noise level is so high that the data are no longer analytically useful. Thus one needs to ask: What is the largest noise variance that still allows for valid inferences, c.f., [1, 4]. If the number of true HIV cases is

$$m = m_n = n^{1-\beta}, \tag{1}$$

Statistical Disclosure Limitation for Data Access, Fig. 1
Top Panel: Perturbed HIV data through addition of independent draws from $N(0, 1)$. Those values associated with positive HIV tests are in *red*, and those with negative HIV tests are in *green*. *Bottom Panel*: 100 simulated mFDR values based on 100 simulation for $n = 2,000$ and $\varepsilon = .05$ and added noise from $N(0, 1)$

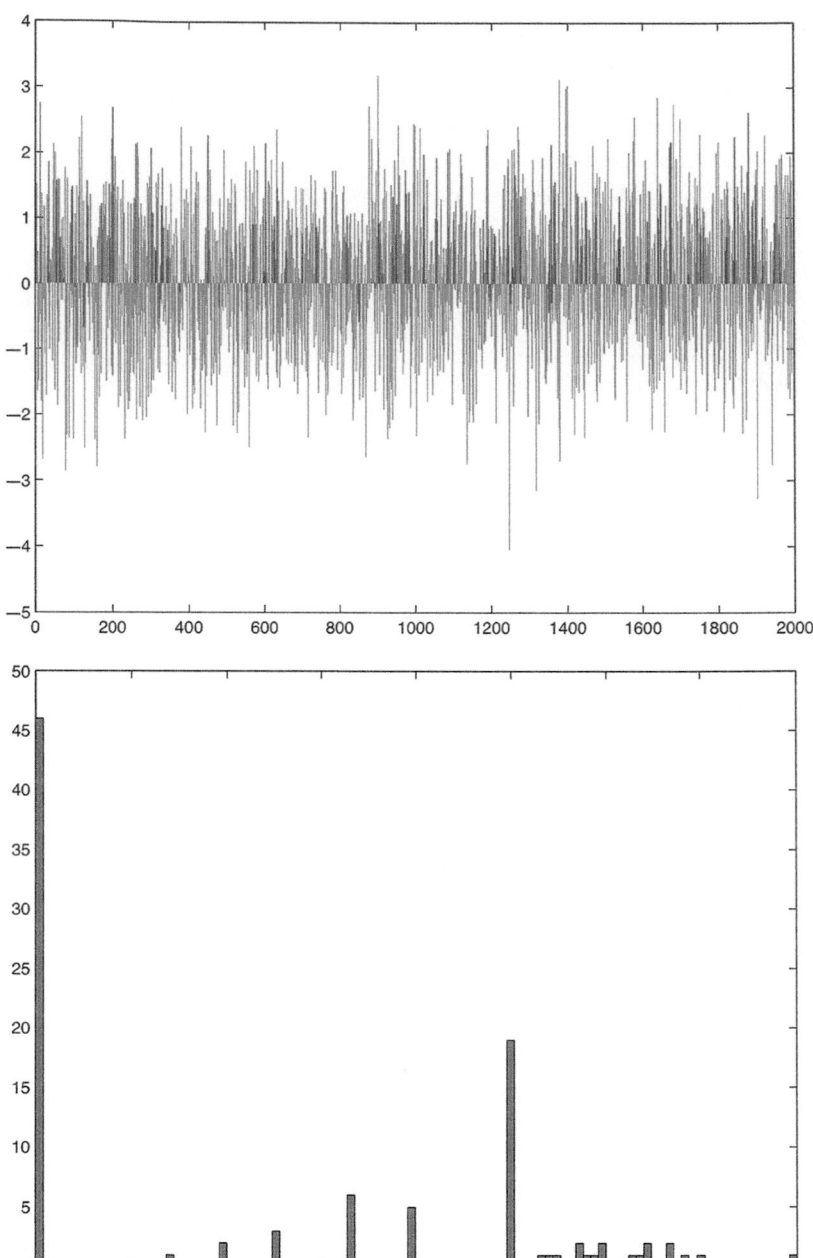

and the noise level is $\sigma = \sigma_n = \frac{1}{\sqrt{2r \log n}}$, where $0 < \beta, r < 1$ are parameters, then as n tends to ∞, there is a boundary, $r = \beta$, which separates the β-r plane into two regions: the *classifiable* region and the *non-classifiable* region; In the interior of the classifiable region, asymptotically, it is possible to isolate completely the cases with positive HIV tests from those with negative ones. In fact, there is a threshold by which one can identify that subset of the data corresponding to positive HIV tests. On the one hand, almost every "identified" HIV case has a positive HIV test and the subset includes almost all the cases with positive HIV tests. In the interior of the non-classifiable region, by contrast, such isolation of cases is impossible. In fact, given any chosen threshold, either one situation or the other occurs!

For the example of $n = 2,000$ and $m = 100$ cases with positive HIV tests. take $\beta = 1 - \log(m)/\log(n) \approx 0.3941$ in model (1). In order not to have complete isolation of cases with HIV, one should take $\sigma_n > \frac{1}{\sqrt{2\beta \log(n)}} \approx 0.409$. For $\sigma_n = 0.5$, consider a repetition of the case of $n = 2,000$ and $\varepsilon = 1/2$, as well as the case of $n = 2,000$ and $\varepsilon = 0.05$. Thus $1/\sigma = 2$ and to control the FDR at 5%, one must evaluate $\frac{\epsilon}{1-\epsilon}\left(\frac{\overline{\Phi}(t-2)}{\overline{\Phi}(t)}\right) = 19$, which yields $t = 0.54$. Correspondingly, $n(1-\epsilon)\overline{\Phi}(t) = 2,000 \times \frac{1}{2} \times 0.029 \approx 29$, and $n\epsilon\overline{\Phi}(t-2) = 2,000 \times \frac{1}{2} \times 0.54 \approx 540$. Furthermore, for $k = 0, 3, 6.9$, the probability that no more than k"1"s are correctly classified are extremely small ($<10^{-315}$). For the case of $n = 2,000$ and $\varepsilon = 0.05$. Similarly, one must similarly evaluate $\frac{\epsilon}{1-\epsilon}\left(\frac{\overline{\Phi}(t-2)}{\overline{\Phi}(t)}\right) = 19$, which yields $t = 3.62$. Correspondingly, $n(1-\varepsilon)\overline{\Phi}(t) = 2,000 \times 0.95 \times 1.47 \times 10^{-4} \approx 0.28$, and $n\varepsilon\overline{\Phi}(t-2) = 2,000 \times 0.05 \times 0.053 \approx 5$. Furthermore, for $k = 0,3,6.9$, the probability that no more than k "1"s are correctly classified are 4.5×10^{-3}, 0.22, 0.73, 0.96. Take $k = 3$. The probability that more than three "1"s are correctly classified is about 78%.

This example illustrates how the multiplicity issue arises when an intruder tries to match many records with those in a database "protected" by matrix masking. Simply protecting each individual record with high probability is not enough; there remains a substantial chance for one or more record to be vulnerable to disclosure.

Summary

The accumulation of massive data sets and the rapid development of the Internet expanded opportunities for data analyses as well as created enormous challenges for privacy protection. This brief overview of the literature on statistical disclosure limitation has stressed four categories of approaches: sampling, perturbation, collapsing, (aggregation), and the use of synthetic data. An overlooked issue in privacy protection is the notion of "multiplicity," which is present whenever one attempts to protect many records simultaneously, or a data intruder tries to match the records of multiple targets in a database simultaneously. Simply protecting each individual record with high probability does not automatically protect all records, and careful statistical measures for resolving the multiplicity issue are necessary.

Key Applications

The methods of statistical disclosure limitation outlined here are already in widespread use by government statistical agencies throughout the world. The volumes by Doyle et al. [5] and Willenborg and de Waal [15] summarize a number of the approaches and methodologies. In particular, census data released by most developed countries are protected using these methods.

Future Directions

The elaboration of approaches described here to deal with very large scale databases remains a challenge, especially in the face of demands for increased access to data and novel attacks on databases by intruders. This entry presents the first known application of ideas and results from the multiplicity literature to the problem of statistical disclosure limitation and risk-utility tradeoff. These ideas require further development and integration with the rest of the literature. And methodology dealing with the risk-utility tradeoff will clearly need to evolve in response to the evolution of intruder strategies to compromise databases.

Cross-References

▶ Individually Identifiable Data
▶ Inference Control in Statistical Databases
▶ Matrix Masking
▶ Privacy
▶ Privacy-Preserving Data Mining
▶ Randomization Methods to Ensure Data Privacy

Recommended Reading

1. Abramovich F, Benjamini Y, Donoho D, Johnstone I. Adapting to unknown sparsity by controlling the false discovery rate. Ann Stat. 2006;34(2):584–653.
2. Anderson M, William SW. Challenges to the confidentiality of U.S. federal statistics, 1910–1965. J Off Stat. 2007;23(1):1–34.
3. Benjamini Y, Hochberg Y. Controlling the false discovery rate: a practical and powerful approach to multiple testing. J Roy Statist Soc B. 1995;57(1):289–300.
4. Dalenius T. Towards a methodology for statistical disclosure control. Statist Tidskrift. 1977;5(429–444):2–1.
5. Donoho D, Jin J. Higher Criticism for detecting sparse heterogeneous mixtures. Ann Stat. 2004;32(3):962–94.
6. Doyle P, Lane JL, Theeuwes Jules JM, Zayatz LV, editors. Confidentiality, disclosure and data access: theory and practical application for statistical agencies. New York: Elsevier; 2001.
7. Fienberg SE. Confidentiality, privacy and disclosure limitation. In: Encyclopedia of social measurement, vol. 1. San Diego: Academic Press; 2005. p. 463–9.
8. Fienberg SE, Makov UE. Confidentiality, uniqueness and disclosure limitation for categorical data. J Off Stat. 1998;14(4):485–502.
9. Fienberg SE, Makov UE, Sanil AP. A Bayesian approach to data disclosure: optimal intruder behavior for continuous data. J Off Stat. 1997;13(1):75–89.
10. Fienberg SE, Makov UE, Steele RJ. Disclosure limitation using perturbation and related methods for categorical data (with discussion). J Off Stat. 1998;14(4):485–502.
11. Fienberg SE, Slavkovic AB. Preserving the confidentiality of categorical statistical databases when releasing information for association rules. Data Min Knowl Discov. 2005;11(2):155–80.
12. Hertzog TN, Scheuren FJ, Winkler WE. Data quality and record linkage techniques. New York: Springer-Verlag; 2007.
13. Lambert D. Measures of disclosure risk and harm. J Off Stat. 1993;9(2):313–31.
14. Raghunathan TE, Reiter J, Rubin DB. Multiple imputation for statistical disclosure limitation. J Off Stat. 2003;19(1):1–16.
15. Warren S, Brandeis L. The right to privacy. Harvard Law Rev. 1890;4(5):193–220.
16. Willenborg L, de Waal T. Elements of statistical disclosure control, vol. 155. New-York: Lecture Notes in Statistics Springer-Verlag; 2001.

Steganography

Radu Sion
Stony Brook University, Stony Brook, NY, USA

Synonyms

Covert communication; Information hiding

Definition

Steganography (from the greek "*steganos*" – covered) is a term denoting mechanisms for hiding information within a "cover" such that, generally, only an intended recipient will (i) have knowledge of its existence, and (ii) will be able to recover it from within its cover. In modern digital steganography applications, the cover is often a multimedia object such as an image that is minorly altered in the steganographic process. Steganographic techniques have been deployed for millenia and several primitive wartime instances are described in the Histories of Herodotus of Halicarnassus, including a case of a message tattooed on the shaven head of a slave, which, when covered with grown hair acted as an effective "cover" when traversing enemy lines.

Key Points

Steganography Versus Watermarking

A common trend of term misuse is associated with steganography. Specifically, many sources consider the term "watermarking" as equivalent. This is incorrect. There are fundamental differences, from both application perspectives and associated challenges. Steganography usually aims at enabling Alice and Bob to exchange messages in a manner as stealthy as possible, through a hostile medium where Malory could lurk. On the other hand, Digital Watermarking is deployed by a rights holder (Alice) as a court proof of rights over a Work, usually in the case when an adversary (Mallory) would benefit from using or selling that very same Work or maliciously modified versions of it. In Digital Watermarking, the actual value to be protected lies in the Works themselves, whereas pure steganography usually makes use of them as simple value "transporters." In Watermarking, Rights Assessment is achieved by demonstrating (with the aid of a "secret" known only to Alice – "watermarking key") that a particular Work exhibits a rare property ("hidden message" or "watermark"). For purposes of convincing the court, this property needs to be so rare that if one considers any other random Work "similar enough" to the one in question, this property is "very improbable" to apply (i.e., bound false-positives rate). It also has to be relevant, in that it somehow ties to Alice (e.g., by featuring the bit string "(c) by Alice"). There is a threshold determining the ability to convince the court, related to the "very improbable" assessment. This defines a main difference from steganography: from the court's perspective, specifics of the property (e.g., watermark message) are not important as long as they link to Alice (e.g., by saying "(c) by Alice") and, she can prove "convincingly" it is she who induced it to the (non-watermarked) original. In watermarking, the emphasis is on "detection" rather than "extraction." Extraction of a watermark, or bits of it, is usually a part of the detection process but just complements the process up to the extent of increasing the ability to convince in court.

Fingerprinting

In this application of steganography, license violators are "tracked" by hiding uniquely identifying "fingerprints." If the Work would then be found in the public domain, the fingerprints can then be used to assess the source of the leak.

Recommended Reading

1. Cox I, Miller M, Bloom J, Fridrich J, Kalker T. Digital watermarking and steganography. 2nd ed. Morgan Kaufmann; 2007.

Stemming

Chris D. Paice
Lancaster University, Lancaster, UK

Synonyms

Affix removal; Suffix stripping; Suffixing; Word conflation

Definition

Stemming is a process by which word endings or other affixes are removed or modified in order that word forms which differ in non-relevant ways may be merged and treated as equivalent. A computer program which performs such a transformation is referred to as a *stemmer* or *stemming algorithm*. The output of a stemming algorithm is known as a *stem*.

Historical Background

The need for stemming first arose in the field of information retrieval (IR), where queries containing search terms need to be matched against document surrogates containing index terms. With the development of computer-based systems for

IR, the problem immediately arose that a small difference in form between a search term and an index term could result in a failure to retrieve some relevant documents. Thus, if a query used the term "explosion" and a document was indexed by the term "explosives," there would be no match on this term (whether or not the document would actually be retrieved would depend on the logic and remaining terms of the query).

The first stemmer for the English language to be fully described in the literature was developed in the late 1960s by Julie Beth Lovins [11]. This has now been largely superseded by the Porter stemmer [14], which is probably the most widely used, and the Paice/Husk stemmer [12]. Stemmers have also been developed for a wide variety of other languages.

Foundations

Definitions

In an IR context, the process of taking two distinct words, phrases or other expressions and treating them as semantically equivalent is referred to as *conflation*. The two expressions need not be precisely synonymous, but they must refer to the same core concept (compare "computed" and "computable"). In this article, the term "practically equivalent" is used to mean that, *for the purposes of a particular application*, the words may as well be taken as equivalent.

The term conflation is sometimes used as though it is equivalent to stemming, but it is in fact a much broader concept, since it includes (i) cases where the strings concerned are multi-word expressions, as in "access time" and "times for access", and (ii) cases where the strings are not etymologically related, as in "index term" and "descriptor". In case (i) special string matching techniques may be used, whereas in case (ii) reference to a dictionary or thesaurus is necessary. The present account deals exclusively with the conflation of etymologically related single words.

There are various possible approaches to word conflation, including the following.

1. *Direct matching*. In this method, the character sequences of two words are compared directly, and a similarity value is computed. The words are then considered to match if their mutual similarity exceeds a predefined threshold. To give a simple example, the first six letters of the words "exceeds" and "exceeded" are the same, so these words together contain 12 matching letters out of 15. Hence, a similarity of $12/15 = 0.80$ can be computed. Use of a threshold (say, 0.70) allows a decision as to whether the words can be considered equivalent.

   With such a method, setting the threshold is problematic. Thus, the similarity between "exceeds" and "excess" is 0.62, which is below the stated threshold. However, allowing for this by lowering the threshold to 0.60 would cause "excess" and "except" (similarity 0.67) to be wrongly conflated.

2. *Lexical conflation*. In this case a thesaurus or dictionary is used to decide whether two words are equivalent. Obviously, this method can be used even for etymologically unrelated words. A problem here is obtaining a suitably comprehensive and up-to-date thesaurus, and one which explicitly lists routine variants such as plurals.

3. *Cluster-based conflation*. This method, investigated by Xu and Croft [15], involves creating clusters of practically equivalent words by analyzing the word-word associations in a large representative text corpus. Each query word is then supplemented by adding in the other words in its cluster. In contrast to method [2], the clusters created are specific to the text collection in question. However, the creation of the clusters can be very time-consuming.

4. *N-gram conflation*. In this method, each word is decomposed into a collection of N-letter fragments (N-grams), and a similarity is computed between the N-gram collections of two words; a threshold is then applied to decide whether the words are equivalent. This approach was pioneered by Adamson and Boreham [1], who used sets of *bigrams*, where $N = 2$. For example, after eliminating duplicates and sorting into order, "exceeds" can

be represented by the bigram set {ce, ds, ed, ee, ex, xc} and "exceeded" by {ce, de, ed, ee, ex, xc}. Out of 7 distinct bigrams here, 5 are shared between the two words; hence a similarity of 5/7 = 0.712 can be computed.

5. *Stemming.* Stemming refers to the removal of any suffixes (and sometimes other affixes) from an input word to produce a *stem.* Two words are then deemed to be equivalent if their stems are identical. This method is much favored because it is fast: all words can be reduced to stems on input to the system, and simple string matching used thereafter. The remainder of this article focuses on stemming in this narrow sense.

Stemming Algorithms

The most primitive type of stemming is *length truncation,* in which any word containing more than N letters is represented by its first N letters. Thus, using $N = 6$, "exceeds" and "exceeded" are both reduced to "exceed," though "excess" remains distinct. Most stemmers, however, use rules which test for specific endings which, if found, are removed or replaced.

It is possible to implement a set of stemming rules by encoding them directly as a computer program. This permits an arbitrary level of complexity in the tests and transformations used, but it makes the stemmer harder to design and modify. An alternative approach is to hold the rules in one or more tables, with a *stemming engine* designed to operate on those tables. This separation means that the same stemming engine can in principle be used with different rules on different occasions, depending on the particular stemming requirements. It also means that a given stemming engine can be used, with little or no adaptation, for a range of other languages.

In all of the stemmers to be described below, a stemming operation is subject to "acceptability" constraints. This ensures that if the ending of a word matches an ending in a table, the indicated action is only taken if relevant conditions are satisfied. These constraints vary from one stemmer to another (and sometimes from one ending to another). The Paice/Husk stemmer [12] uses a very simple constraint: an action only

proceeds if the resulting stem will contain at least two letters, including at least one vowel. Thus, "string" cannot be transformed to "str" through a hit with an "-ing" rule.

It is important to note that, for the purposes of most applications, the stem returned by a stemmer need not be an actual word of the language. The essential desideratum is that the stem should be the same for all practically equivalent words, and different for all other words.

Stemming algorithms can be classified roughly as *single-stage, multi-stage* or *iterative.* Lovins' stemmer [11] is often described as a single-stage stemmer, since it uses a single table containing all the distinct endings which are to be removed. In this table, the rules are held in decreasing order of length, and the first matching rule is the one applied. This ensures that, if the table contains the endings "-mentary," "-ment" and "-ary," the word "documentary" and "document" are both reduced to "docu." If the "-ary" ending were tested first, "documentary" would simply be stemmed to "document."

In fact, Lovins' stemmer has a second (iterative) stage, using a table of 35 "recoding rules," which can adjust the stem returned by the first stage, e.g., by replacing double final consonants by single, and making other changes – thus,

$$admission \rightarrow admis \rightarrow admis$$
$$admittance \rightarrow admitt \rightarrow admit \rightarrow admis$$

The 290 endings listed for the original Lovins' stemmer are demonstrably inadequate, but a satisfactory rule set would need to be much longer, and would show considerable structural redundancy. The Paice/Husk stemmer [12] avoids this by taking an iterative approach, where long endings are removed or transformed in a series of actions using a much shorter table containing shorter endings. When one action has been activated and completed, the table may be entered again to see if another rule will fire. Thus, some endings are removed in several stages:

$$sensibilities \rightarrow sensibility \rightarrow sensibil$$
$$\rightarrow sensibl \rightarrow sens$$

The most popular stemmer for English is that devised by Martin Porter at Cambridge University [14]. This stemmer, which was designed to reflect the linguistic structure of suffixes in English words, proceeds through five main stages. Stage 1 deals with plurals, verb inflexions, and words ending with "-y"; stages 2–4 with all the major derivational endings; and stage 5 with "-e" removal and singling of final "-ll." Acceptability constraints are based on a quantity called the "measure" of the word, which is derived by scanning the consonant/vowel pattern of each word.

To facilitate the development of stemmers, Porter developed a special language known as Snowball (see URL below).

Krovetz developed a stemmer KSTEM which uses a machine readable dictionary to decide whether a stemmed form corresponds to an acceptable root form of the original word [9]. Despite careful refining of the algorithm to allow for a range of problem cases, IR performance was not consistently better than with Porter's algorithm.

The design of a stemmer for a new language can be a labor-intensive business. An attractive alternative is to generate a set of stemming rules automatically. Thus, Bacchin et al. have developed a probabilistic approach which uses a large representative corpus to determine the optimal splitting of words into "stems" and "derivations." They showed that, in terms of retrieval performance, their approach was about as good as Porter's algorithm for a range of European languages [5].

Prefixes and Infixes

In English, stemmers are usually designed for removing suffixes from words. The removal of "intimate" prefixes such as "intro-," "pro-" and "con-" generally results in words being wrongly conflated (consider "intro-duction," "pro-duction" and "con-duction"). However, there may be a case for removing looser prefixes such as "hyper-" or "macro-." Also, prefix removal may be desirable in certain domains with highly artificial vocabularies, such as chemistry and medicine.

As explained below, there are some languages in which removal or replacement of prefixes, or even infixes, is in fact essential.

Performance and Evaluation

Since stemmers were originally developed to aid the operation of information retrieval systems, it was natural that they were first assessed in terms of their effect on retrieval performance, as well as on "dictionary compression" rates. Researchers were frustrated to find that the effects on retrieval performance for English-language material were small and often negative [10]. Removal of "-s" and other regular inflectional endings might be modestly helpful, but use of heavier stemming could easily result in a loss of performance [7]. Work by Krovetz and by Hull showed that most benefit is obtained in cases where the document or the query is short [8, 9].

Stemmers are not used only in IR systems, but in a wide range of natural language applications. A less "IR-oriented" general approach to measuring performance is to consider the number of actual stemming errors committed by an algorithm, and this forms the basis of a method developed by Paice [13]. Notice first that stemming errors are of two kinds: *understemming*, in which a pair of practically equivalent words are not conflated, and *overstemming*, in which two semantically distinct words are wrongly conflated. It is easy to see that these two types of error trade off against one another.

Paice's method makes use of a collection of distinct words (typically derived from an actual text source) which have been manually collected into groups, such that all the members of a group are practically equivalent. Two indices are computed based on a stemmer's treatment of pairs of words, which reflect the rate of understemming and of overstemming. Morerover, the resulting values are related to a baseline represented by length truncation (see above). This results in a general measure of accuracy called the "error rate relative to truncation," ERRT. Whilst this approach provides some insights into the activities of stemmers, it is unclear how such information should be used, though in future it might provide the basis for an optimization process. The use of human-defined target groups is a weak feature.

As a by-product, Paice's method yields a "stemming weight," which is the ratio of the overstemming to the understemming indices; a large stemming weight means that the stemmer is "heavy," "strong" or "aggressive." Frakes and Fox [6] present a series of other metrics related to stemming weight, as well as metrics for comparing stemmers one with another. These are all "behavior metrics," and do not relate directly to the actual accuracy of the stemming process.

Non-English Stemmers

Stemming is appropriate for most (though not all) natural languages, and appears to be especially beneficial for highly inflected languages [9]. There is neither space nor need to describe non-English stemmers here, except to note that some languages exhibit much greater structural complexity, and this warrants special approaches. Thus, a typical Arabic word consists of a root verb of three (or occasionally four or five) consonants (e.g., "k-t-b" for "to write"), into which various prefixes, infixes and suffixes are inserted to produce specific variant forms ("katabna": "we wrote" and "kitab": "book"). Some researchers have concentrated on extracting the correct root from a word [3], but Aljlayl and Frieder have demonstrated that better retrieval performance is obtained by using a simpler "light stemming" approach, in which only the most frequent suffixes and prefixes are removed [4]. Their results showed that extraction of roots causes unacceptable levels of overstemming.

Key Applications

As noted earlier, stemmers are routinely used in information retrieval systems to control vocabulary variability. They also find use in a variety of other natural language tasks, especially when it is required to aggregate mentions of a concept within a document or set of documents. For example, stemmers may be used in constructing lexical chains within a text. Stemming can also have a role to play in the standardization of data for input to a data warehouse.

Data Sets

Useful resources can be found on the two websites noted below.

URL to Code

Stemming algorithms and other resources may be obtained from the following websites:

http://www.snowball.tartarus.org/
http://www.comp.lancs.ac.uk/computing/research/
   stemming/.

Cross-References

▶ Lexical Analysis of Textual Data

Recommended Reading

1. Adamson GW, Boreham J. The use of an association measure based on character structure to identify semantically related pairs of words and document titles. Inf Process Manage. 1974;10(7/8):253–60.
2. Ahmad F, Yusoff M, Sembok MT. Experiments with a stemming algorithm for Malay words. J Am Soc Inf Sci Technol. 1996;47(12):909–18.
3. Al-Sughaiyer IA, Al-Kharashi IA. Arabic morphological analysis techniques: a comprehensive survey. J Am Soc Inf Sci Technol. 2004;55(3):189–213.
4. Aljlayl M, Frieder O. On arabic search: improving the retrieval effectiveness via a light stemming approach. In: Proceedings of the International Conference on Information and Knowledge Management; 2002. p. 340–7.
5. Bacchin M, Ferro N, Melluci M. A probabilistic model for stemmer generation. Inf Process Manage. 2005;41(1):121–37.

S

6. Frakes WB, Fox CJ. Strength and similarity of affix removal stemming algorithms. SIGIR Forum. 2003;37(1):26–30.

7. Harman D. How effective is suffixing? J Am Soc Inf Sci. 1991;42(1):7–15.

8. Hull D. A Stemming algorithms: a case study for detailed evaluation. J Am Soc Inf Sci. 1996;47(1): 70–84.

9. Krovetz R. Viewing morphology as an inference process. Artificial Intelligence. 2000;118(1/2): 277–94.

10. Lennon M, Pierce DS, Tarry BD, Willett P. An evaluation of some conflation algorithms for information retrieval. J Inf Sci. 1981;3(4):177–83.

11. Lovins JB. Development of a stemming algorithm. Mech Transl Comput Linguist. 1968;11:22–31.

12. Paice CD. Another stemmer. SIGIR Forum. 1990;24(3):56–61.

13. Paice CD. A method for the evaluation of stemming algorithms based on error counting. J Am Soc Inf Sci. 1996;47(8):632–49.

14. Porter MF. An algorithm for suffix stripping. Program. 1980;14(3):130–7.

15. Xu J, Croft WB. Corpus-based stemming using coocurrence of word variants. ACM Trans Inf Syst. 1998;16(1):61–81.

Stop-&-Go Operator

Nikos Hardavellas[1] and Ippokratis Pandis[1,2]
[1]Carnegie Mellon University, Pittsburgh, PA, USA
[2]Amazon Web Services, Seattle, WA, USA

Synonyms

Non-pipelineable operator

Definition

A Stop-&-Go operator, or non-pipelineable operator, is a relational operator which cannot produce any result tuples unless it has consumed all of its input. A typical Stop-&-Go operator is the Sort operator. The usage of Stop-&-Go operators in the query execution plan limits the degree of operator-level parallelism.

Key Points

Some relational operators need to consume their entire input before they are able to produce tuples. These operators are called Stop-&-Go or non-pipelineable operators. A typical example of a Stop-&-Go operator is the Sort operator. To sort a set of tuples, the entire input set needs to be consumed before the operator can output the tuples in sorted order. There are many Stop-&-Go operators, such as various flavors of Join and Aggregation. For example, Hash Join is a Stop-&-Go operator because the Probe phase cannot start unless the Build phase has finished. Similarly, Sort-Merge Join is a Stop-&-Go operator because the Merge phase cannot start unless the Sort phase has finished.

The Stop-&-Go operators stop the flow of tuples in a pipelined execution, so their usage limits the degree of operator-level parallelism. Thus, a query optimizer that aims to achieve high levels of operator-level parallelism may choose to replace Stop-&-Go operators with more expensive – but pipelineable – operators [1, 2] in the query execution plan.

Cross-References

▶ Hash Join
▶ Operator-Level Parallelism
▶ Pipelining
▶ Query Plan
▶ Sort-Merge Join

Recommended Reading

1. Graefe G. Encapsulation of parallelism in the volcano query processing system. In: Proceedings of the ACM SIGMOD Conference on Management of Data; 1990. p. 102–11.

2. Johnson R, Hardavellas N, Pandis I, Mancheril N, Harizopoulos S, Sabirli K, Ailamaki A, Falsafi B. To share or not to share? In: Proceedings of the 33rd International Conference on Very Large Data Bases; 2007. p. 351–62.

Stoplists

Edie Rasmussen
Library, Archival and Information Studies,
The University of British Columbia, Vancouver,
BC, Canada

Synonyms

Negative dictionary; Stopwords

Definition

Stoplists are lists of words, commonly called stopwords, which are not indexed in an information retrieval system, and/or are not available for use as query terms. A stoplist can be created by sorting the terms in a document collection by frequency of occurrence, and designating some number of high frequency terms as stopwords, or alternately, by using one of the published lists of stopwords available. Stoplists may be generic or domain specific, and are of course language specific. When a stoplist is used for indexing, as a document is added to the system, each word in it is checked against the stoplist (for example through dictionary lookup or hashing), and those which match are eliminated from further processing. In some systems, stopwords are indexed, but the stoplist is used to eliminate the words from processing when they are used as query terms.

Key Points

Hans Peter Luhn, in pioneering work on automatic abstracting, put forward the idea that certain words are too common to provide a significant discrimination value, instead contributing noise to the calculations, and should be excluded from consideration [6]. In his description of the processing needed to create Keyword-in-Context (KWIC) indexes, he described a "dictionary of insignificant words" which was to be excluded from processing. In his view, these insignificant words would include "articles, conjunctions, prepositions, auxiliary verbs, certain adjectives, and words such as "report," "analysis," "theory" and the like" [7]. This idea was incorporated in the 1960s in commercial KWIC indexes introduced by Biological Abstracts (BASIC) and Chemical Abstracts (Chemical Titles). At Biological Abstracts the number of excluded terms varied, but grew to 1,000 words, although analysis showed that 14 words were enough to prevent 80% of the entries, and the tradeoff between reduction in (printed) index size and cost of dictionary lookup became a factor as the length of the stopword list increased [2]. The use of the term "stopword" seems to come from this application, where designation of a word as a stopword stops the corresponding index entries from being printed [9]. As electronic databases became available for searching, database vendors created lists of stopwords which were not indexed or available for use in searching. The lists used by commercial systems were usually quite short; for example, in the Dialog system, the list consists of only nine words: An, And, By, For, From, Of, The, To, With [1]. Other lists which were published and used in IR research contain several hundred words; for an example see Fox [3].

There have been dual arguments put forward for the use of a stopword list, or stoplist, in building an index. The first relates to efficiencies in storage and processing. Common words follow a Bradford distribution and therefore a relatively small number of words account for a relatively large number of word occurrences. Data will vary somewhat from one corpus to another, but a typical analysis might show, for instance, that six words account for 20% of a corpus or 250–300 words for 50% [3]. Therefore, removing these words from the inverted index in a text retrieval system significantly decreases the size of an uncompressed index, though it adds to the processing time needed to create the index since a dictionary lookup or other technique is needed to identify words as stopwords when the text is processed. However, Witten et al. [10] suggest that the storage savings

S

are most obvious in an uncompressed index, and are much less significant if an appropriate compressed representation is used. Processing queries containing stopwords can also be expensive, since their frequency of occurrence results in very long lists of postings. However efficiencies in query processing can be introduced, such as sorting postings lists by term weight, so that processing can be terminated when term weights are small, as is the case with stopwords [8]. Therefore, current techniques can address to a large extent the problems associated with processing very common words in both indexes and queries.

The second rationale for using a stoplist is the claim put forward by Luhn, that these words have very little power for semantic resolution, and therefore may contribute noise rather than meaning for retrieval purposes. However, current term weighting techniques greatly reduce the contribution of common words in ranking functions, and there are many situations where an inability to use stopwords as query terms makes it difficult if not impossible to perform an effective search. There are classic examples of searches composed entirely of stopwords, such as "AT&T," "To be or not to be," or where a stopword is critical to the query, for example the "A" in "Vitamin A." In other situations the removal of stopwords makes it impossible to adequately specify the query, for example, where common words are needed to clarify the relationship between terms. One approach, as used by Google for instance, is to index stopwords but to process them in queries only when the searcher specifically requests it in the query formulation, or when the query is composed only of stopwords [4]. This allows the stopwords to be used when they would be helpful, and ignores them when they are not, but it does require some knowledge of advanced search techniques on the part of the searcher. Overall, improved storage and compression techniques, term weighting schemes, and advanced query processing techniques significantly reduce the cost of including stopwords in a text retrieval system and arguments can be made for eliminating the stopword list [8, 10].

Cross-References

▶ Index Creation and File Structures
▶ Lexical Analysis of Textual Data

Recommended Reading

1. Dialog online courses: glossary of search terms. Available at: http://training.dialog.com/onlinecourses/glossary/glossary_life.html.
2. Flood BJ. Historical note: the start of a stop list at Biological Abstracts. J Am Soc Inf Sci. 1999;50(12):1066.
3. Fox C. Lexical analysis and stoplists. In: Frakes WB, Baeza-Yates R, editors. Information retrieval: data structures and algorithms. Englewood Cliffs: Prentice-Hall; 1992. p. 102–30.
4. Google Web Search Help Center. Search basics: use of common words. Available at: http://www.google.com/support/bin/answer.py?answer=981.
5. Korfhage RR. Information storage and retrieval. Wiley: Wiley Computer Pub; 1997.
6. Luhn HP. The automatic creation of literature abstracts. IBM J Res Dev. 1958;2(2):157–65.
7. Luhn HP. Keyword-in-context index for technical literature. Am Doc. 1960;11(4):288–95.
8. Manning CD, Raghavan P, Schütze H. Introduction to information retrieval. Cambridge: Cambridge University Press; 2008.
9. Parkins PV. Approaches to vocabulary management in permuted-title indexing of Biological Abstracts. In: Proceedings of the 26th Annual Meeting on American Documentation Institute; 1963. p. 27–9.
10. Witten IH, Moffat A, Bell TC. Managing gigabytes: compressing and indexing documents and images. 2nd ed. San Francisco: Morgan Kaufmann; 1999.

Storage Access Models

Kaladhar Voruganti
Advanced Development Group, Network Appliance, Sunnyvale, CA, USA

Synonyms

Database physical layer; Database storage layer

Definition

Database management systems provide storage that can be accessed via a query language interface and that can be updated under the control of a transaction management system. A database management system can reside on top of a file system (system management storage) or on top of raw block storage (direct managed storage), or on a combination of file system and block storage (hybrid model). There are advantages and disadvantages of using these different types of underlying storage.

Historical Background

Database management systems have historically managed data on disks by themselves. Over the past few years, the management functionality in file systems has steadily improved. In order to leverage this management functionality (like data backup and recovery), and make it easier for system administrators to manage their storage infrastructure in a uniform manner, database system vendors started to architect database systems so that they can also run on top of file systems. This, in turn, has now provided users with multiple storage alternatives. These alternatives are discussed in this section.

Foundations

A database is logically organized into multiple table spaces. A table space determines the location from where a table gets its storage space. Thus, database tables are created inside table spaces. Multiple tables can exist in a table space. A table space obtains its storage from either files or a directory of a file system or multiple raw block devices. Each raw block device or file or directory that provides storage to a table space is known as a container. Multiple containers can provide storage to a single table space. Separate table spaces exist for user data, user index data, temporary data, and log data. Similarly, separate table spaces also exist for system tables and catalog tables. A table space can be classified as one of the following:

System Managed Space (File System) This type of storage container is fully under the control of the file system. The advantages of system managed space are:

- *In-time Provisioning*: Typically, database administrators use system managed storage for managing temporary storage. This ensures that space is only allocated when needed, and re-used for other purposes when it is de-allocated. This is an advantage of system managed space over database managed space, where space is pre-reserved in the containers.
- *Leverage File System Utilities:* System managed storage files can leverage backup, migration, and all of the other file system utilities. Previously, this was a major advantage for system managed storage, but over the past few years, a lot of progress has been made in building block level data management utilities that can be leveraged by a database managed system.

Database Managed Space (Raw Storage) This type of storage container is fully under the control of the database management system. The advantages of database managed space are:

- *Better Performance:* Database managed storage offers better performance because of the following reasons:
  - Unlike in system managed space, the absence of file system logging also helps with the overall performance.
  - As file systems age, the underlying managed storage system can become fragmented. The absence of file system fragmentation also helps with the performance of database managed storage.
  - The lack of an intermediate file system buffer (in addition to the database buffer) in the I/O path, and the absence of contention in the file system buffer with other non-database applications, helps to improve the

performance of database managed storage. With the recent emergence of direct I/O mechanisms, where the file system places data directly into the database buffer the disadvantages of an intermediate buffer have been reduced even in system managed storage space.

– Each database containers resource is dedicated to that container, and thus, there is no contention for storage space with other non-database applications.

• *Storage Extensibility:* Database managed storage allows for dynamic addition of more storage containers to a table space. Thus, one does not have to a priori know the maximum size of the required storage space.

• *Concurrent Access:* Some file systems put a limit on the number of concurrent accesses on a file (container). These limitations are not present in database managed containers.

Hybrid Space (Database Managed File) In this type of storage container, the file system created file is given to the database management system to manage. It is a compromise between the above two types of containers. In these containers, there is no intermediate file system buffer but the size of the container is limited to the size limits of the created file. The performance of hybrid containers is almost as good as the database managed containers, and one can leverage the conventional file system provided backup/migration utilities.

Key Applications

OLTP applications that are performance sensitive typically use system managed storage. Applications store their temporary storage in system managed containers.

Cross-References

▶ Backup and Restore
▶ Buffer Management
▶ Logging and Recovery

Recommended Reading

1. Mellish B, Aschoff J, Cox B, Seymour D. IBM ESS and IBM DB2 UDB working together. IBM Redbook, SG24-6262-00, San Jose, 2001.

Storage Area Network

Kazuo Goda
The University of Tokyo, Tokyo, Japan

Synonyms

SAN

Definition

A Storage Area Network is a network whose main purpose is to transfer data between storage devices and servers and among storage devices. The term Storage Area Network can be a synonym of the term Storage Network, but differs in that the term Storage Area Network is usually identified with a network with block-level I/O services rather than file access services. More specifically, the term Storage Area Network is often used to refer to a network with Fiber Channel technology. However, SNIA released a more general definition in which the term Storage Area Network is not connected with any specific types of network connections. Under this definition, an Ethernet-based network infrastructure for mainly connecting storage devices could also be considered a Storage Area Network. The term Storage Area Network is often abbreviated to SAN. When the term SAN is used in connection with a specific network technology X, the use of a term "X SAN" is encouraged. A SAN based on Fiber Channel technology is sometimes referred to as Fiber Channel SAN. A SAN based on TCP/IP technology is often shortened to IP SAN. Despite the original meaning, the term SAN is sometimes identified with a storage system which is also implemented using a network.

Key Points

A SAN is a general network for connecting storage devices, but as a matter of fact, currently most SANs are implemented on top of Fiber Channel technology. A typical SAN is composed of Fiber Channel switches, storage devices such as disk arrays and tape libraries, and Fiber Channel host bus adapter (HBA) cards that are installed into servers. Alternative network technologies such as iSCSI, IFCP and FCIP are used mainly in entry-level storage systems or in wide-area network connections.

Cross-References

▶ Direct Attached Storage
▶ Network Attached Storage
▶ Storage Network Architectures

Recommended Reading

1. Clark T. Designing storage area networks: a practical reference for implementing fibre channel and IP SANs. Reading: Addison-Wesley; 2003.
2. Storage Network Industry Association. The dictionary of storage networking terminology. Also available at: http://www.snia.org/.
3. Troppens U, Erkens R, Müller W. Storage networks explained. New York: Wiley; 2004.

Storage Consolidation

Hiroshi Yoshida
VLSI Design and Education Center, University of Tokyo, Tokyo, Japan
Fujitsu Limited, Yokohama, Japan

Definition

The processes of centralizing the storage infrastructure resources of multiple servers to reduce management costs, achieve better service levels, and strengthen control over data.

Key Points

In small scale IT systems, each server has its own dedicated storage infrastructure (internal disks or DAS). However, as server numbers and the amount and importance of business data stored in the storage infrastructure increases, managing such dedicated storage infrastructure resources per server becomes difficult and expensive. To solve this problem, the dedicated infrastructure resources of servers are centralized to storage infrastructure resources shared by all servers, using storage networking technologies such as SAN and/or NAS.

Once dedicated storage infrastructure resources are consolidated to SAN and/or NAS resources, storage management operations can also be consolidated and centralized. For example, data backup is performed only once for the consolidated storage instead of individually for each server. This greatly reduces the cost of storage management. Another advantage is that expensive storage solutions such as disaster recovery using replicated data in remote sites can be shared by multiple servers with consolidated storage. The result is much improved data availability that can be achieved cost-effectively. In addition, from a data management and data security viewpoint; rather than having data spread over multiple servers, often managed by multiple administrators or divisions and based on disparate policies, data stored in consolidated storage can be managed based on more consistent policies.

Cross-References

▶ Direct Attached Storage
▶ Information Loss Measures
▶ Network Attached Storage
▶ SAN File System
▶ Storage Network Architectures
▶ Storage Resource Management
▶ Storage Virtualization

Storage Devices

Kaladhar Voruganti
Advanced Development Group, Network
Appliance, Sunnyvale, CA, USA

Synonyms

CDs; DVDs; Flash; NAS servers; Optical storage; Storage controllers; Storage servers; Tape libraries; Tapes; WORM

Definition

One of the goals of database, file and block storage systems is to store data persistently. There are many different types of persistent storage devices technologies such as disks, tapes, DVDs, and Flash. The focus of this write-up is on the design trade-offs, from a usability standpoint, between these different types of persistent storage devices and not on the component details of these different technologies.

Historical Background

From a historical standpoint, tapes were the first type of persistent storage followed by disks, CDs, DVDs, and Flash. Newer types of memory technologies such as PRAM and MRAM are still in their infant stages. These newer non-volatile memory technologies promise DRAM access speeds and packaging densities, but these technologies are still too expensive with respect to cost/gigabyte.

Foundations

- *Tapes/Tape Libraries:* Tape readers/tape head, tape library, tape robot, and tape cartridge are the key components of a tape subsystem. Tapes provide the best storage packaging density in comparison to other types of persistent storage devices. Tapes do not provide random access to storage. Data on tapes can be stored either in compressed or uncompressed format. Unlike disks, tape cartridges can be easily transported between sites. Most organizations typically migrate data from older tape cartridges to newer tape cartridges once every 5 years to prevent data loss due to material degradation. One can employ disk based caches in front of tape subsystems in order to allow for tapes to handle bursty traffic. Tapes that provide Write-Once, Read Many (WORM) characteristics are also available. WORM tapes are useful in data compliance environments where regulations warrant guarantees that a piece of data has not been altered. DLT and LTO are currently the two dominant tape technologies in the market. Technology wise both these standards have minor differences. Finally, from a pure media cost standpoint, tapes are less expensive (cost per gigabyte) than disks and other forms of persistent media.

- *Disks/Storage Controllers/NAS Boxes:* Disks are the most widely used form of persistent storage media. Disks are typically accessed by enterprise level applications when they are packaged as part of the processing server box (direct attached storage model), or are part of a network attached storage box (NAS) and accessed via NAS protocols or, are packaged as part of a storage controller box and accessed via storage area network protocols (SAN). The current trend is for protocol consolidation, where the same storage controller provides support for both SAN and NAS protocols. Typically, the size of the storage controllers can vary from a few terabytes to hundreds of terabytes (refrigerator sized storage controllers). A storage controller typically consists of redundant processors, protocol processing network cards, and RAID processing adapter cards. The disks are connected to each other via either arbitrated loop or switched networks. Storage controllers also contain multi-gigabyte volatile caches. Disks are also packaged as part of laptops. There is a marked difference in the manufacturing process, and testing process between the enterprise class disks and commodity laptop class disks. Disks vary in their form factor, rotational speed, storage capacity, number of available ports, and the protocols used to access them. Currently, serial SCSI, parallel SCSI, serial ATA and parallel ATA, Fiber Channel, and SSA are the different protocols in use for accessing disks. Lower RPM and disk idle mode are new disk spin-down modes that allow disks to consume less power when they are not actively being used.

- *DVD/Juke Boxes:* DVDs and CDs are optical storage media that provide random access and WORM capabilities. Only recently, the multiple erase capacity of an individual CD, or DVD was less than the capacity of a single disk drive or tape cartridge. DVDs can store more data than a CD, and a high definition DVD can store more data than a DVD. There are numerous competing standards for CDs, DVDs and high definition DVDs, however, format agnostic DVD players and DVD writers are emerging. Usage of DVDs is more prominent in the consumer space rather than in the enterprise space. A juke box system allows one to access a library of CDs or DVDs. DVDs have slower access speeds than most types of disks.
- *Flash/SSDs/Hybrid Disks:* Flash is memory technology that has non-volatile characteristics. Flash memory has slower read times than DRAM. Moreover, it has much slower write times than DRAM. One has to perform an erase operation before one can re-use a flash memory location. One can only perform a limited number of erase operations. Thus, the number of write operations determines the Flash memory life. SLC and MLC are the two different NAND flash technologies. SLC can be erased a greater number of times, and it has faster access times than MLC based flash. NAND flash has faster write and erase times than NOR flash. NOR flash has faster read times than NAND flash. NAND flash is used to store large amounts of data whereas NOR flash is used to store executable code. People are using MLC flash in cameras and

digital gadgets, and are using SLC flash as part of solid state disks (SSDs). SSDs provide block level access interface (SCSI), and they contain a controller that performs flash wear leveling and block allocation. Hybrid disks that contain a combination of disks and Flash are emerging. Hybrid disks provide a Flash cache in front of the disk media. One typically can store meta-data or recently used data in the flash portion of hybrid disks to save on power consumption. That is, one does not have to spin-up the disk. Flash storage provide much better random access speeds than disk based storage.

Key Applications

Tapes are being used primarily for archival purposes because they provide good sequential read-/write times. Disks are the media of choice for most on-line applications. Optical media (CDs, DVDs) are popular in the consumer electronic space. Flash based SSDs are popular for those workloads that exhibit random IOs. Disks are being used in Laptops, desktops and storage servers (SANs, NAS, DAS). Tape based WORM media and content addressable based disk storage are providing WORM media capabilities in tape and disk technologies, and thus, these technologies can be used to also store compliance/regulatory data.

Cross-References

- ▶ Backup and Restore
- ▶ Direct Attached Storage
- ▶ Network Attached Storage
- ▶ Storage Area Network

Recommended Reading

1. Anderson D, Dykes J, Riedel E. More than an interface- SCSI versus ATA. In: Proceedings of the 2nd USENIX Conference on File and Storage Technologies; 2003.

2. Toigo J. Holy grail of network storage management. Englewood Cliffs: Prentice Hall; 2003.
3. Voruganti K, Menon J, Gopisetty S. Land below a DBMS. ACM SIGMOD Rec. 2004;33(1): 64–70.

Storage Grid

Kaladhar Voruganti
Advanced Development Group, Network
Appliance, Sunnyvale, CA, USA

Synonyms

Content delivery networks; Cloud computing; Data grids; Distributed databases; Peer to Peer network; Utility computing

Definition

In grid computing, storage and computing resources are geographically spread out and accessed via fast wide-area networks. Storage resources could either co-exist with computing resources, or they could exist separately from the computing resources. Databases, file systems or block storage devices can be accessed remotely across fast wide-area network based grids. A storage grid provides services for discovering storage resources, transferring data, recovering from unfinished data transfer failures, data authentication/encryption services, and data replication services for performance and availability purposes. Storage grids also typically provide the necessary mapping layers to access data from heterogeneous sources. Heterogeneity can be due to differences underlying system architectures, data formats, protocols used to access the data, and data organization. Finally, storage grids provide a global unified namespace across the resources, and one typically transfers large datasets (multi-terabytes) across the different nodes.

Key Points

Content Delivery Networks (CDNs) is a related area, where data are cached at secondary servers at various geographically distributed servers to cut down on data access latency. Peer-to-Peer networks is another related area where data are distributed across different peers in an ad-hoc manner and there is no central authority. Storage grids can be further classified based on the following criteria [1]:

Storage Grid Organization Storage grid organization deals with how resources are organized in the system. Resources can be organized in a hierarchical manner, or in a monadic manner. In hierarchical model data exists at multiple sites. Each site in turn decides which of its children sites can have access to data. In a monadic grid data exists only at a single location and everyone accesses data from that location. Data sources in a grid can also be arranged in a federated model, where each site retains independent control over the data and its participation in the grid.

Data Transport Data transport deals with transport issues, security issues, and fault-tolerance issues:

- *Transport:* Data can be transferred using protocols such as FTP and GridFTP. In addition, one could potentially employ overlay networks that provide caching functionality and control to the applications to directly control data transfer. Data can also be transferred via multiple parallel streams from one source location, and in a striped manner from multiple data source locations.
- *Security:* Authentication, encryption and authorization are the three security issues that are also applicable in grid environments.
- *Fault-Tolerance:* The primary fault-tolerance approaches are to restart a failed data transfer, or to have the ability to resume from the failed point. In some environments, if the destination node is not available, intermediate caches can

temporarily store the data and then forward the data once the unavailable node comes back on line.

Data Replication and Storage Data replication strategies can be classified in the following different ways:

- *Method:* Synchronous and asynchronous replication strategies are the two major classes of data replication mechanisms. In synchronous replication, updates are not acknowledged at the source until data has been successfully copied at the target locations. In asynchronous protocols, updates are immediately acknowledged at the source, and there is a lag in data consistency between the primary and secondary data copies.
- *Protocols:* Some grids employ open data transfer replication protocols, such as FTP or GridFTP. In open protocols, the catalog management becomes the responsibility of the application. Others employ a closed protocol which perform catalog management in an integrated manner.
- *Replication Granularity*: Data can be replicated at dataset, file, block, and database table level.
- *Replication Strategy:* Data can be replicated dynamically based on an objective function such as response time requirements, load balancing requirements or data consistency requirements, or data can be replicated statically based on an a priori schedule or on demand.

Resource Allocation and Scheduling The goal of resource allocation and scheduling is to ensure that the data are located at the appropriate site in order to meet the performance and availability goals of the application. The settings for the following parameters can be varied in this regard:

- *Process Model:* The processes can be scheduled as independent tasks, as a bag of tasks or as part of a workflow. Workflow corresponds to a sequence of tasks, whereas, a bag of tasks correspond to executing the same task on different input parameters.
- *Objective Function:* Resource allocation and task scheduling is performed based on an objective function. The objective function tries to optimize load balancing, or business profit, or application performance. The object function is assigned at the task, bag of tasks or workflow level.
- *Scope of the Scheduler:* The scheduler decides whether to replicate/migrate data/process can try to optimize the utility function at the level of an individual application or at a more global community level.
- *Types of Tasks:* Data can be migrated, replicated, cached or remotely accessed in order to satisfy application's storage requirements. Alternately, the computation process can be migrated to the location where the data exists.

Storage grids are primarily used in scientific computing environments that deal with large amounts of data. It is usually not practical to replicate all of the data, and thus, grid architecture facilitates the remote access of large datasets across wide-area networks. However, variants of storage grid architectures such as CDNs are used to transfer streaming video, and P2P networks are used to shared audio and video data. Cloud computing is the new variant of utility computing where application, storage and server resources are managed by a service provider and clients remotely access these resources.

Cross-References

- ▶ Grid and Workflows
- ▶ Grid File (and Family)
- ▶ Multimedia Resource Scheduling
- ▶ Storage Protocols
- ▶ Storage Security

Recommended Reading

1. Venugopal S, Buyya R, Ramamohan RK. A taxonomy of data grids for distributed data sharing, management, and processing. ACM Comput Surv. 2006;38(1):1–53.

Storage Management

Hiroshi Yoshida
VLSI Design and Education Center, University
of Tokyo, Tokyo, Japan
Fujitsu Limited, Yokohama, Japan

Definition

The methods and tools used to manage storage devices (disk arrays, tape libraries, etc.), storage networking devices (fiber channel switches, etc.), storage-related components inside servers (host bus adaptors, etc.), and logical objects mapped on those devices (logical units, access paths, etc.). In general, the scope of storage management is limited to the management of storage infrastructure and does not handle the data stored in the infrastructure. The functions of storage management include device management, performance management, and problem management. Those functions are usually provided as software tools.

Historical Background

Storage management technologies have developed in parallel with the evolution of storage networking.

In the early 1990s, storage devices were used as DAS devices. Even in a DAS environment, storage management functions such as storage device management were required, and those functions were provided as dedicated software tools for specific vendors and/or devices, and those tools were often bundled with the hardware.

With the evolution of storage area networks (SANs) in the late 1990s, new requirements for storage management arose:

- Initially, SAN brought significant reduction of management cost through its storage consolidation capability. The storage capacity which could be managed by storage administrators was also greatly increased, compared to DAS environments. However,

along with the growth of SAN, the number of connected devices and the amount and business importance of stored data increased, causing SANs to become more and more complex. This increased the "storage management gap," i.e., the gap between the decrease in storage hardware costs and the extreme increase in storage management costs. A reduction in the manpower required for storage management was urgently needed.

- To manage a SAN environment, both storage device management functions and network management functions, such as discovery, network configuration management, and network topology management, are necessary. In a networked environment, coping with hardware and software failures and performance problems also becomes much more difficult, compared to a DAS environment.
- SAN environments are usually constructed using products from multiple vendors. Storage management tools must cope with such multi-vendor environments. Therefore a new business model of providing software products which are independent from specific storage vendors' devices was established. This requirement also accelerated the standardization of the interfaces between storage devices and management software.

Foundations

General Classification of Storage Management Functions

Although storage management is necessary for all types of storage networking, the management of SANs is mainly discussed in the following description. This is because SANs are the most commonly used storage infrastructures and feature the most typical management require-ments. In general SAN environments, storage management is achieved by software tools running on a management server. Those tools provide functions such as device management, configuration management, performance management, and problem management.

- Device management and configuration management consists of functions to configure and to monitor storage devices (disk arrays, tape libraries, etc.), network devices (fiber channel switches, etc.), server components (fiber channel host bus adapters (HBAs), etc.), and relationship between those components. For example, disk array management provides the following functions:
- Monitoring and displaying the status of devices
- Creation/configuration/masking/mapping of logical units/logical unit numbers (LUNs)

Management of fiber channel switches and SAN configuration provides the following functions:

- Monitoring and displaying switch ports
- Collecting and displaying statistic information on switch ports
- Displaying SAN topology
- Configuration of zones
- Integrated and consistent control of multiple switches

As a SAN environment includes multiple levels of virtualization (e.g., disk arrays, volume managers, and virtualization network appliances) and access control features (zone, host affinity, and LUN mapping), the mapping of logical access paths on physical paths and the correlation between applications and physical storage devices tend to be complex. Configuration management should provide functions to visualize such mapping and correlation from the viewpoint of application and to configure multiple SAN components in a consistent manner:

- Performance management consists of functions to monitor, to analyze, and to display storage access performance based on statistic information collected from storage devices and fiber channel switches. It also includes functions to issue an alert in case that a specific parameter (e.g., device busy rate) exceeds the predefined threshold.

- Problem management consists of functions to monitor the status of storage devices as well as fiber channel switches and to notify the administrators and/or remote maintenance centers when problems such as hardware failures are detected.

"Storage management" is a very generic term, and those elemental storage management functions are sometimes named "storage x management," e.g., "storage device management" and "storage configuration management." Note that the term "storage resource management" represents a different concept from storage management mentioned here. It is usually used to indicate the functions to visualize and control usage of storage systems from the more application-aware or content-aware management viewpoint. It is described as a separate article.

Actual storage management software products include those typical functions as well as additional functions such as automation and the provision of integrated monitoring and operational views.

Elemental Technologies of Storage Management

Another aspect of understanding storage management is the technologies needed to implement storage management software. In general, the following internal functions are commonly required to implement storage management software:

- *Discovery* is a function to find storage devices in an SAN environment before knowing the topology of the SAN. When a new device such as a disk array system is connected to an SAN, it also has to be discovered.
- *Data collection* acquires necessary information, once storage devices are discovered. Information is collected through proprietary interfaces and/or standard interfaces such as SNMP (Simple Network Management Protocol) and SMI-S (Storage Management Initiative Specification). Data collection also stores that information in storage management repositories which are usually located in

S

the storage management server. Collected information includes both information on the current status of storage resources such as configuration information and historical information such as accumulated performance information on storage devices.

- *Topology management* analyzes the topology of SANs based on the collected information. Topology management handles both physical network topology which represents the relationship between physical resources such as HBA ports, switch ports, and storage device ports and logical network topology which represents the relationship between logical and/or virtualized resources such as LUNs, zones, and logical access paths.

- *Visualization* is a set of functions which display the information mentioned above on a management console and provide human interfaces to enable administrators to monitor and configure the storage resources easily.

- *Event processing* receives asynchronous events from storage resources, categorizes them based on the predefined policies, and notices administrators and/or remote maintenance centers if necessary and records those events into event log files.

- *Security management* is a set of functions which are necessary to meet the appropriate security requirements. Security management includes authorization and access control features of administrators considering multiple administrative roles, single sign-on features among multiple storage management software products, management of credentials of managed storage resources for data collection, and logging functions for auditing all operations applied to the storage infrastructure to fulfill compliance requirements.

One important topic related to storage management implementation is how the interoperability between management software products and managed storage resources is achieved. At an early stage, storage management was implemented as management tools provided by individual storage vendors. Those tools were dedicated to the storage devices of respective vendors.

Each storage vendor developed a proprietary interface and protocol which was applicable only to its storage devices and storage management software tools.

However, it became very common for datacenters to use storage devices provided by multiple vendors. Using multiple storage management tools with different looks and feels and manageability increased the management and labor costs in datacenters. This situation led to the new requirement that storage management software must be able to manage not only single vendor storage resources but multi-vendor storage resources. The goal was that every storage management software product could manage every storage resource. Standard interfaces and protocols which could be applied to the communication between management software and storage resources were crucial to achieving this goal.

Since the storage industry became aware early on of the importance of interoperability between products, standard interfaces and protocols for storage management were established as ISO standard SMI-S by the storage standards body SNIA (Storage Networking Industry Association).

Key Applications

Storage management is one of the essential features for administrative practice of computer system storage infrastructure. It allows administrators to configure, monitor, and control storage resources, particularly in SAN environments.

Storage management is also necessary as the basis of implementing higher-level management, which is described as "management applications" later. For example, cloud resource orchestration requires storage management as part of its resource management as well as server management and network management. Another example is that information management requires storage management to manage the infrastructure in which information resides. In information lifecycle management, optimal storage devices are assigned to store information in accordance with its business value. Storage management is responsible for establishing the multiple storage device tiers which meet the different service level objectives and cost requirements.

Future Directions

The following areas will become more important in terms of storage management.

Integration of Management Software Including Storage Management

Currently a huge variety of management software products are used in large datacenters and cloud service providers. Storage management software products must be integrated with those products.

To achieve consistent and automated resource management in a datacenter, storage management should be integrated with other resource management such as server and network management. This integration of resource management will provide capability such as the "provisioning" of a set of servers, storage devices, and network resources to an application. When administrators use this feature, they will not need to have detailed knowledge and skills in storage management, letting them manage storage devices in an SAN environment without regard to the SAN. This feature will be achieved through the cooperation of server management, network management, and storage management, which will configure internal connections between servers and storage devices automatically. Such integrated provisioning will greatly reduce management and labor costs in the datacenter.

Another example is integration with IT service management such as incident management, problem management, change management, release management, and configuration management. IT service management controls those management processes in a manner which is compliant to ITIL (Information Technology Infrastructure Library). In addition, configuration information needed for those management processes is stored in a CMDB (Configuration Management Database). To achieve integration with IT service management, storage management operations will have to be initiated as part of

management processes by workflow managers, and configuration information on storage resources collected by storage management will also have to be federated with CMDB.

Visualization and Optimization from a Business Viewpoint

The business value proposition brought by IT systems becomes more and more important and needs to be clearly stated. On the other hand, to achieve business risk management, the impact of IT infrastructure outages on the customer's business also has to be clearly analyzed. To fulfill such requirements, the relationship between IT resources and business applications must be visualized. The IT resources must also be optimally configured and managed to achieve the performance and availability objectives of business applications. The most important key technology is dependency mapping between business applications and resources. Another important technology is policy management which allows customers to specify the criteria of system behavior based on their business requirements. Storage management is responsible for that capability with regard to storage resources.

Establishing Framework for Management Applications

In addition to the necessity of basic management capability such as the configuring and monitoring of storage devices, the importance of higher-level "management applications" is continuously increasing. For example, such management applications include:

- Database performance management
- Resource provisioning and cloud resource orchestration
- Lifecycle management of storage resources and information lifecycle management
- Security management
- Automated management such as run book automation and autonomic management such as self-configuration

Management applications are essential for achieving integrated management and business-aware management. The amount of management applications which are available on a storage platform is a key factor in achieving strategic use of storage and stored data in an enterprise.

In general, management applications monitor and control storage resources on the level of logical resources not on physical storage device level. In the main the information directly collected with current storage management tools, from storage hardware and the control functions of storage hardware, are sometimes too detailed for management applications. For example, a management application which provides provisioning capability is only aware of LUNs, their capacities, and the zoning configurations which restrict accessibility between servers and storage resources. There are also functions which many management applications use commonly to implement their storage-related capabilities. High-level interfaces which allow management applications to handle storage resources more logically, as well as a set of common components which help the development of management applications, are required.

URL to CODE

The latest documents on SMI-S can be downloaded from the SNIA web site.

http://www.snia.org/tech_activities/standards/curr_standards/smi/

Cross-References

► Direct Attached Storage
► Information Loss Measures
► Logical Unit Number
► Storage Area Network
► Storage Consolidation
► Storage Network Architectures
► Storage Networking Industry Association
► Storage Management Initiative Specification
► Storage Protocols
► Storage Resource Management
► Storage Virtualization
► Volume

Recommended Reading

1. Cummings R. Storage Network Management, Storage Network Industry Association. 2004. Available at: http://www.snia.org/education/storage_networking_primer/stor_mngmnt/
2. Storage Network Industry Association. Storage Network Industry Association tutorials. 2014. Available at. http://www.snia.org/education/tutorials/

Storage Management Initiative Specification

Hiroshi Yoshida
VLSI Design and Education Center, University of Tokyo, Tokyo, Japan
Fujitsu Limited, Yokohama, Japan

Synonyms

SMI-S

Definition

A standard storage management interface developed by the Storage Networking Industry Association (SNIA). SNIA describes SMI-S as follows: SMI-S defines a method for the interoperable management of a heterogeneous SAN and describes the information available to a Web-Based Enterprise Management (WBEM) client from an SMI-S compliant Common Information Model (CIM) server and an object-oriented, XML-based, messaging-based interface designed to support the specific requirements of managing devices in and through SANs.

Key Points

To implement storage management, methods to retrieve configuration information on storage components such as storage devices (disk arrays, tape libraries, etc.,), network devices (fiber channel switches, etc.,), and servers (hardware such as host bus adapters and software such as drivers and volume managers) plus methods to operate those components are necessary. To manage SAN environments composed of multi-vendor products, common methods of collecting information and operations must be standardized. In the standardization of those common methods, not only the standardized interfaces and protocols but also the semantics of the information and operations, i.e., object models of managed resources, must be defined.

To achieve SAN management in multi-vendor environments, the SNIA standardized on the Storage Management Initiative Specification (SMI-S). Based on the Common Information Model (CIM) standardized by the Distributed Management Task Force (DMTF), SMI-S defines a common object model for each storage resource class as well as common methods of collecting information and operations using a Web-browser-based management framework (Web-Based Enterprise Management, WBEM).

Figure 1 shows the architecture of SMI-S. It is based on the client-server model. Managed storage components act as CIM servers, and storage management software tools act as CIM clients. CIM servers and CIM clients communicate by passing XML texts through HTTP. A SMI-S CIM server includes an HTTP server, an XML parser, an SLP agent used for discovery, and a CIM provider. The CIM provider implements actual CIM operations according to the profile defined for the corresponding storage resource class.

SMI-S includes the following storage management features and functionality:

* Hardware Devices: SMI-S providers

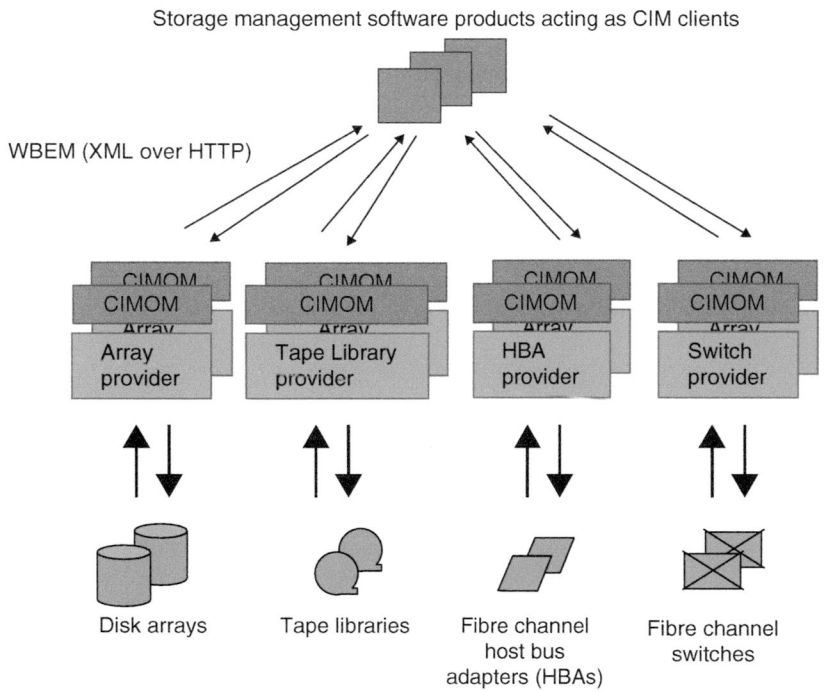

Storage management software products acting as CIM clients

WBEM (XML over HTTP)

CIMOM: CIM object manager

Storage Management Initiative Specification, Fig. 1 The architecture of SMI-S

- Fiber channel switches
- Arrays (fiber channel and iSCSI)
- NAS devices
- Tape libraries
- Host profiles (including host bus adapters)
• SMI-S Clients (software)
  - Configuration discovery
  - Provisioning and trending
  - Security
  - Asset management
  - Compliance and cost management
  - Event management
  - Data protection

The standardization effort began in 1997 and was adopted formally by the SNIA as SMI-S in 2002. It was designated as a standard by ISO/IEC in January 2007.

Cross-References

▶ Storage Management
▶ Storage Networking Industry Association

Recommended Reading

1. Storage Network Industry Association. Storage management technical specification version 1.6.0. 2012. Available at: http://www.snia.org/tech_activities/standards/curr_standards/smi/

Storage Manager

Goetz Graefe
Google, Inc., Mountain View, CA, USA

Synonyms

Disk process; Storage layer

Definition

The storage manager is a software layer within a database management system. It relies on operating system primitives for I/O, synchronization, etc. and exposes records in storage structures such as B-trees and heaps. Its principal operations are creation and removal of storage structures as well as retrieval, search, scans, insertion, update, and deletion of records. For those operations, the storage layer provides concurrency control among threads and among transactions, as well as recovery from transaction, media, and system failures. Standard implementation techniques include locking and write-ahead logging. Using a buffer pool, records are made accessible in random access memory, although the permanent storage is on block-access media, typically on disk but possibly in flash memory.

The storage layer may support data compression, "blobs" (binary large objects), bitmaps in non-unique secondary indexes, hash indexes, multi-dimensional indexes, etc. It may also manage the catalogs needed to manage its objects, or it may leave metadata management to a higher software layer. For all the supported data structures, the storage layer provides utilities such as backup and restore, bulk insertions and deletions, logical and physical consistency check, defragmentation and other forms of reorganization, etc.

Historical Background

Design and scope of the storage layer are often similar to those of the RSS (research storage system) of the System R prototype. The basic architecture of the core functions has remained unchanged, including space management for records in pages, indexing and search, concurrency control and recovery. Many specific implementation techniques have changed, e.g., introduction of multi-dimensional indexes. Another change has been the transition from write-ahead logging with recovery from a read-only log to recovery by compensation of logical actions and guaranteed exactly-once execution of physical actions, both logged and even check pointed during database recovery.

The architecture of utilities has also undergone some changes. The implementation of many storage layer functions now employs query execution and even query optimization, e.g., memory management policies or partitioning and pipelining for parallel execution. For example, creation of a new secondary index may scan existing secondary indexes rather than the primary index, or consistency check may aggregate facts gathered during a disk-order scan rather than navigate each index with many random I/O operations.

Sorting used to be a storage layer function because it was used almost exclusively for index creation, traditionally a storage layer function. Sorting can be used for many tasks, however, including complex query execution plans with merge join operations, etc., such that sort- and hash-based operations should both be part of the query processing layer.

The plethora of features and functions has led to complexity and high total cost of ownership for many installations. Extensible database management systems have not solved this problem, and the tension between "one size fits all" and "tailor-made" database systems might be resolved through factoring and mass-customization.

XML support is now available in the storage layer of many commercial database management systems, but further improvements in query processing over XML documents and databases will likely require further improvements of storage layer techniques.

In other words, as much as the storage layer may seem like a well-understood component of database technology, research, development, and competition continue unabated.

Foundations

This section describes a database system's storage layer by its external interfaces above and below, followed by internal components and data structures on disk and in memory, and completed by specific techniques for query processing, concurrency control and recovery, utilities, and catalogs.

Storage Layer Concepts

The most important services provided by a database system's storage layer revolve around indexes and records. Both of these are physical database concepts, quite different from logical database concepts such as table and row. For example, a single table may require many indexes due to redundant, secondary indexes as well as horizontal and vertical partitioning. Similarly, a single row might be represented by many records. Conversely, multiple tables may be clustered within a single index such that related information, e.g., about a purchase order and its line items, can be read, written, and buffered within a single disk page. Finally, a single record may contain information about multiple rows, e.g., a B-tree entry in a non-unique non-clustered index with a key value and a list of row identifiers.

The mapping between logical and physical concepts is usually provided by the relational layer and its query optimization component. Some implementations of the storage layer, however, provide some of this mapping, e.g., maintenance of all indexes for a row update, in order to maximize performance by minimizing the number of storage layer invocations.

Logical concepts include table, view, row, column, and domain; physical concepts include index, partition, record, and field. The word "key" is used both in the relational layer, where it restricts duplicates and null values, and in the storage layer, where it indicates index organization, sort order, or a search argument. It might be useful in some discussions to avoid the term "key" and to use "primary key," "foreign key," "index key," and "search key" instead.

Storage Layer Services

The storage layer provides access and updates for indexes and their records. Heaps can be thought of as indexes that map a record identifier to a record, with no capability to modify the record identifier. Indexes can be created and dropped. Records can be inserted, deleted, updated, scanned, and searched using an appropriate search key. Some indexes, e.g., heaps and clustered indexes, may generate unique row identifiers during insertion and possibly a new one during update. Such row identifiers can link all records representing a logical row, e.g., during retrieval using a non-clustered index or during deletion of a logical row.

All operations are part of transactions with concurrency control and recovery. The storage layer supports pre-commit for participation in coordinated commits in addition to the traditional immediate commit. The set of currently active transactions and their current state may be managed by the storage layer or by another component within the database management system.

If multiple threads invoke the storage layer at the same time, internal data structures are protected against concurrency problems using appropriate low-level synchronization.

If catalogs are managed by a higher software layer, the storage layer provides appropriate indexes and, most likely, special high-performance lock scopes and lock modes for metadata. For example, all metadata about a table may be covered by a single lock, and this lock may cover both the metadata and the data. Specifically, the weakest mode merely read-protects the metadata, whereas the strongest mode permits arbitrary changes to both metadata and data. Traditional read and write locks (shared and exclusive) imply read-protection on the metadata.

Storage Layer Requirements

A storage layer invokes two basic functions of the operating system: input/output and synchronization. Asynchronous I/O functions are valuable as otherwise they must be simulated with additional threads. Synchronization primitives that provide both shared and exclusive levels directly support the need of database management systems and their storage layers.

Storage Layer Components

The well-known components of a storage layer are the access methods (B-tree structure etc.), buffer pool, concurrency control and recovery. The less prominent components are memory management, disk space allocation, and catalogs. Asynchronous I/O could be separated into its own component, as could be latches, temporary structures (for sort- and hash-based operations), page structures (managing variable-length records), utilities for reorganization and consistency checks, backup and restore. Initial access to a database file, e.g., during recovery from a crash or when opening a database from a less-than-fully-trusted source, requires substantial logic and could be its own component, too.

Typically, multiple threads may invoke the storage layer at the same time; access and update of shared data structures within the storage layer is coordinated using latches. Those are equivalent to critical sections or locks in programming languages. Latch modes are typically shared and exclusive (read and write). Deadlock avoidance is required as latches do not participate in deadlock detection. Ordering latches based on levels is a standard technique to avoid deadlocks. Latches must not be retained while waiting for I/O to complete or while waiting for a database lock.

On-Disk Data Structures

The essential on-disk data structures are index for user data. This includes both primary, clustered indexes and secondary, non-clustered indexes on both tables and views. Additional data structures enable the essential ones: catalogs, free space management, and the recovery log. Each of those is special in their own way. For example, the recovery log is often mirrored on two devices in order to approximate the fiction of guaranteed stable storage.

Catalogs and their indexes often use the same on-disk format as user data. This is also possible for free space management, in particular if bitmap indexes are supported. The free space information is crucial during initial access to a database, i.e., system boot and recovery must be supported.

There is also the issue of free space management within each page. Each page typically consist of a page header, space for variable-length records, and an indirection vector that indicates the location of each record. A format version number in the page header enables incremental improvement of the database format without unloading and reloading entire databases. A "page LSN" (log sequence number) indicates the most recent log record pertaining to the page and is essential for exactly-once change application in modern recovery schemes.

Many database management systems and their storage layer support multiple index formats, e.g., hash indexes, heaps, multi-dimensional indexes, column storage, etc. All of them can be mapped to B-trees with reasonable efficiency in time and space, with a great savings in implementation effort. For example, implementation and testing for high scalability through a fine granularity of locking is a very substantial effort required for each storage structure.

In addition, bitmap indexes can be realized as a form of compression for non-unique non-clustered indexes. Even master-detail clustering (e.g., purchase orders and their line items) can be implemented relying merely on the sort order of B-trees and appropriate record formats. Finally, B-trees can be adapted to support "blobs" (binary large objects) for unstructured data. XML documents and other semi-structured data can be mapped to blobs plus traditional indexes for efficient search.

In-Memory Data Structures

The most prominent in-memory data structures within the storage layer are the lock manager's hash table and the buffer pool including buffers for the recovery log. Other data structures enable transaction management, checkpoints, device management, and asynchronous I/O.

In addition to the direct images of on-disk pages in the buffer pool, a storage layer may cache high-traffic data in data structures designed for fast in-memory access. Catalogs and bitmaps for free space management are obvious candidates. In order to achieve "in-memory performance" for user data and their indexes, interior B-tree nodes can be augmented in the buffer pool with in-memory pointers to child nodes also in the buffer pool, in a special form of pointer swizzling. In all cases, update propagation between cache and disk page images in the buffer pool must be ensured.

Query and Update Processing

The storage layer serves query and update processing but does not drive it. In addition to providing in-memory access to needed database pages, the storage layer speeds up query and update processing with prefetch, read-ahead, shared scans, and write-behind.

The storage layer may also provide automatic maintenance of non-clustered indexes. This design only applies to centralized systems or to parallel systems with local indexes, i.e., partitions of indexes aligned with the partitions of the table. This design forces row by-row maintenance, although index-by-index maintenance can be required for correctness (certain updates of unique indexes) or for performance (sorting large sets of changes as appropriate for each index).

In very traditional designs, the storage layer may provide the navigation from non-clustered index to clustered index during query execution, but coupling these accesses inhibits many beneficial techniques such as covering a query with a non-clustered index alone, sorting references obtained from a non-clustered index, index intersection for conjunctive predicates and index union for disjunctive predicates, joining non-clustered indexes of the same table to cover a query not covered by any one index, and joining non-clustered indexes of two tables as a form of semijoin reduction.

Concurrency Control

Concurrency control is a very important service provided by the storage layer, both latching to protect in-memory data structures from conflicting threads and locks to protect database contents from conflicting transactions. Alternatives to locking ("pessimistic concurrency control") include validation ("optimistic concurrency control") and versioning ("multi-version concurrency control"). Transactional memory may become an alternative to latching.

Locking and latching are described in detail elsewhere.

Logging and Recovery

Logging and recovery are also very important services provided by the storage layer, including transaction rollback, crash recovery, and media reconstruction. Since the units of recovery cannot be larger than the units of concurrency control, a high-concurrency system with key value locking or row-level locking cannot rely on page-based recovery provided by, for example, the file system or a network-attached storage service. The storage layer may provide log-shipping or continuous log-based replication using a "hot stand-by" database copy perpetually in recovery.

Logging and recovery are described in detail elsewhere.

Utilities

The broad term "utilities" covers all those operations needed for a complete database management system product but not directly associated with query processing and transaction processing. Examples include index creation and removal, defragmentation and other forms of reorganization, moving and partitioning data, backup and restore, statistics creation and update, consistency checks and repairs, etc. Catching up on deferred maintenance can apply to materialized views, indexes, statistics, caches, and replicas.

Utilities may be offline, online, i.e., permitting user transactions to read and modify database data while the utility is scanning or reorganizing them, or incremental, i.e., the utility operation's effects such as index creation become useful to user transactions in multiple discrete steps.

As many utility operation move data similar to a query execution plan, and since similar services are required such as memory management, partitioning and pipelining for parallel execution, etc., many utilities can be implemented using the query processing component.

Key Applications

A storage manager is useful in many applications, practically all applications that map collections of records to pages on persistent storage. This includes entertainment software such as music players, personal productivity applications such as e-mail clients, and server-side data management applications such as mail servers and database management systems. A storage manager that provides buffering and transactions for both records and large fields (such as pictures, sound tracks, videos, messages, and documents) is even more widely useful.

Future Directions

While the basics of access methods, buffer pool management, concurrency control and recovery are all well understood and documented, the need for further development continues.

Queuing

Some database management systems have integrated queuing into their feature set. It enables access patterns typical for workflow applications, electronic mail, and service-oriented architectures. Technical challenges include hotspots for both insertion and deletion as well as transaction semantics that link data records into messages and "conversations" among automatic processes and human users.

XML Support

XML is not only a message format but also a storage format, in particular for human-authored documents and in service-oriented architectures based on message passing. XML in databases creates challenges for storage, compression, fine-grained concurrency control and recovery, consistency enforcement and verification, and query processing. While initial research prototypes and even commercial implementations exist, their optimization and adaptation for large applications is not yet complete.

Transactional Memory

Forthcoming many-core processors require, for performance and power-efficiency, data structures and algorithms that permit very high degrees of parallelism as well as appropriate concurrency control among concurrent software threads. Transactional memory is a promising approach to these issues. Hardware-assisted transactional memory enables very fast execution of critical sections as well as guaranteed success based on automatic rollback and re-execution.

Self-Tuning, Self-Repair, Total Cost of Ownership

Perhaps the greatest challenge in the design and implementation of a storage layer is the total cost of ownership, i.e., the amount of human attention and trouble-shooting required to ensure the desired levels of availability, integrity, and performance. For example, can the storage layer software prevent data loss unless a user knowingly and deliberately accepts a risk? Such a storage layer would have to require, during initial deployment, that multiple storage devices with independent failures be specified, among many other things. Similarly, could the storage layer software assume all responsibility for tuning the set of indexes, or could there be a standard interface to other relevant components such as the relational layer within a relational database?

Cross-References

▶ B+-Tree
▶ B-Tree Locking
▶ Buffer Pool
▶ Transaction

Recommended Reading

1. Carey MJ, DeWitt DJ, Franklin MJ, Hall NE, McAuliffe ML, Naughton JF, Schuh DT, Solomon MH, Tan CK, Tsatalos OG, White SJ, Zwilling MJ. Shoring up persistent applications. In: Proceedings of the ACM SIGMOD International Conference on Management of Data; 1994. p. 383–94.
2. Chamberlin DD, Astrahan MM, Blasgen MW, Gray J, King WF, III Lindsay BG, Lorie RA, Mehl JW, Price TG, Putzolu GR, Selinger PG, Schkolnick M, Slutz DR, Traiger IL, Wade BW, Yost RA. A history and evaluation of system R. Commun ACM. 1981; 24(10):632–46.
3. Härder T, Reuter A. Principles of transaction-oriented database recovery. ACM Comput Surv. 1983;15(4):287–317.
4. Hellerstein JM, Stonebraker M, Hamilton JR. Architecture of a database system. Found Trends Databases. 2007;1(2):141–259.
5. Stonebraker M. Retrospection on a database system. ACM Trans Database Syst. 1980;5(2):225–40.

Storage Network Architectures

Kazuo Goda
The University of Tokyo, Tokyo, Japan

Definition

Storage Network Architecture is the conceptual structure and logical organization of a network whose main purpose is to transfer data between storage devices and servers and among storage devices. The term Storage Network is identified with such a network, but is sometimes used to refer to a storage system communicating over a network. Related terms such as Storage Area Network and Network Attached Storage are described in separate entries. Note that usages of terms related to storage network architectures may depend on contexts at times.

Historical Background

The first version of SCSI was released in 1986. SCSI then became deployed in many open systems, thus acquiring the position of the standard IO technology. However, after Fibre Channel technology was invented in the late 1990s, Fibre Channel rapidly extended its use in the market. Recent mid-range and high-end storage systems have deployed Fibre Channel as the standard IO technology. Alternative storage network technologies such as iSCSI may be available in entry-level systems. The SCSI bus technology has been replaced with Fibre Channel and iSCSI, but the SCSI protocol is still effectively used on top of such new network technologies.

S

Foundations

Inconventional IT systems, storage devices such as disk drives, disk arrays, tape drives and tape libraries are connected only to a single server. Such an IT system is often referred to as a server-centric system, in which a storage device is considered a dedicated peripheral device of a server to which the storage device is connected. Small Computer System Interface (SCSI) is the main technology of network infrastructure used in the server-centric system. Storage devices are connected together to a server by SCSI bus cables and provide block-level I/O services to the server using the SCSI protocol. The scalability of such bus technologies is rather limited. A single SCSI cable can be at most 25 m long, and allows a maximum of 15 storage devices to be connected. Storage space which a single server can accommodate is thus severely limited. A server is not able to directly access storage devices connected to another server. Instead the server has to access such storage devices indirectly through a Local Area Network (LAN). The inflexibility of interconnection thus scatters and duplicates data management functions among multiple servers. Storage management may be optimized within each server, but should be far from global optimization of the entire system. Such issues were not visible when expensive microprocessors limited the system capability. However, recent technology innovations have been decreasing the cost of high-speed microprocessors and explosively

expanding the volume of managed digital data. The system capability is then more likely to be limited by the storage system, so the issues of the conventional server-centric systems have become more obvious.

In contrast, the innovations of network technologies have given new possibilities of directly transferring data between storage devices and servers and among storage devices. All servers connected to a storage network are able to access all storage devices in the same network. Such a flexible connection helps storage devices to be consolidated in the network, thus isolating storage resource management from the server. That is, storage resources are placed and managed within a storage network, and servers are then positioned around the network. This type of IT system is often referred to as a storage-centric system. Figure 1 illustrates a server-centric system and a storage-centric system.

Storage networks have given system designers several alternatives for linking storage devices and servers. Currently available interconnections can be grouped into three architectures: Storage Area Network, Network Attached Storage and Direct Attached Storage. This categorization is widely accepted although it may not be formal. Figure 2 presents an illustrative comparison of these architectures. Separate entries give a formal definition of each storage network architecture.

A Storage Area Network, which is often abbreviated to SAN, is a network which mainly transfers data from/to storage devices. A SAN

Storage Network Architectures, Fig. 1 (a, b) A server-centric IT system and a storage-centric IT system

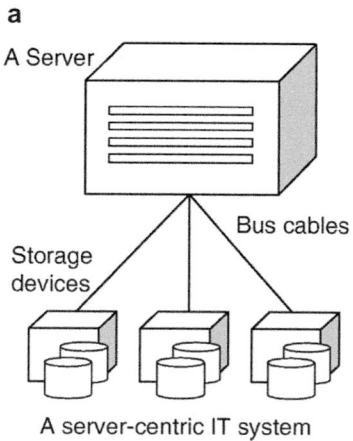

A server-centric IT system

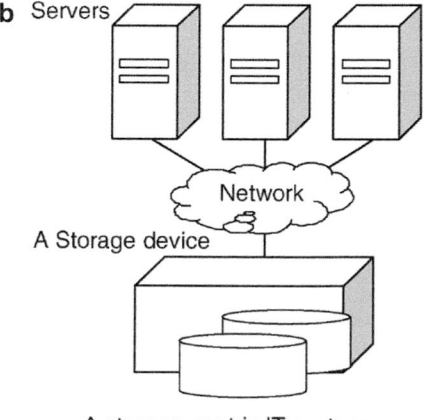

A storage-centric IT system

Storage Network Architectures, Fig. 2 (a–c) Three storage network architectures

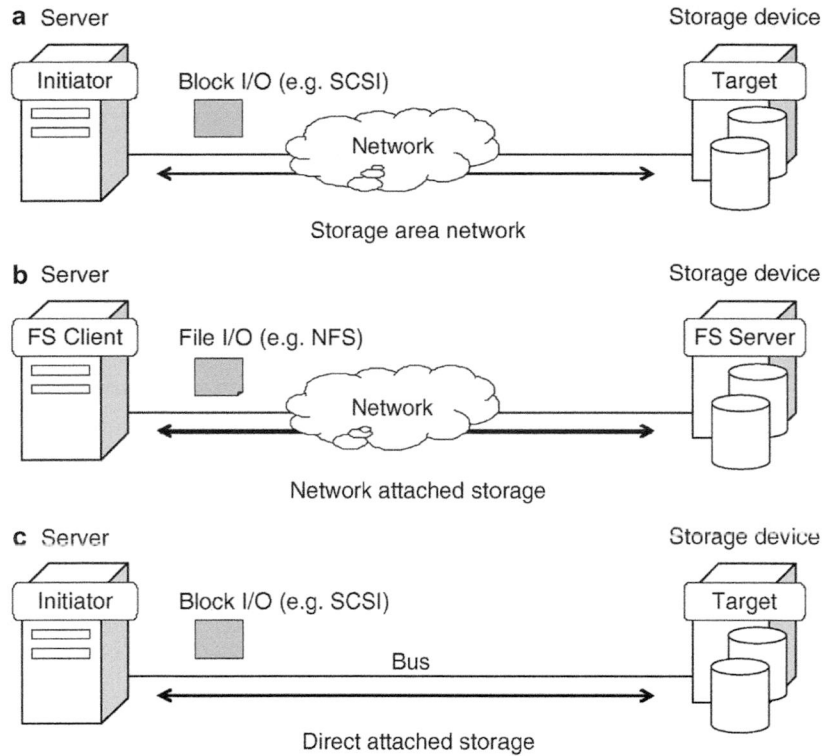

is used to connect storage devices with servers and with other storage devices. A storage system which is implemented over a network is sometimes referred to as a SAN too. A SAN provides block-level I/O services. Most SANs are implemented on top of Fibre Channel technology, although alternative technologies such as internet SCSI (iSCSI) are available for SANs. A SAN which is implemented on Fibre Channel technology is sometimes referred to as Fibre Channel SAN.

Fibre Channel is a gigabit-level network technology, which is mostly used for SANs at present. The Fibre Channel protocol stack is divided into two parts. The lower part defines fundamental network infrastructure, which can be further subdivided into four layers (FC-0-FC-3). FC-0 defines transmission media such as optical/electrical interfaces, cables and connectors. FC-1 specifies signal encoding and decoding (8b/10b conversion and 40b ordered sets) and link controls. FC-2 defines frame formats, frame transmission management (sequences and exchanges) and flow controls. FC-3 defines

common services such as multipathing, On top of these fundamental layers, the higher part (FC-4) defines protocol mappings to application protocols. When Fibre Channel is used for SANs, Fibre Channel Protocol for SCSI (FCP) maps the Fibre Channel infrastructure to the SCSI protocol. Fibre Channel is characterized by its powerful transmission capabilities such as serial transmission, low error rate and low latency. Processing of the Fibre Channel protocol is usually implemented in host bus adapter (HBA) cards at the hardware level so as to relieve servers' processors. These attributes are preferable for SANs, however, they increase the cost of Fibre Channel network devices. Fibre Channel SANs are thus deployed mainly in mid-range and larger IT systems. On the other hand, iSCSI is an approach to exploit the IP network infrastructure so that SANs can be installed and operated at much lower cost. iSCSI, which is placed on the top of the TCP/IP protocol stacks, encapsulates SCSI data in TCP/IP packets. At present, iSCSI is used for connecting servers and entry-level storage systems. Internet Fibre

Channel Protocol (iFCP) and Fibre Channel over Internet Protocol (FCIP) are alternative approaches to use TCP/IP technology for SANs. These protocols can transmit Fibre Channel frames over IP networks. In contrast to iSCSI, iFCP and FCIP are mainly used for connecting remotely distant SAN islands. Note that Fibre Channel and iSCSI are network technologies that only replace the classical SCSI bus technology. The SCSI protocol is still utilized on top of such network technologies even in today's storage-centric systems.

Network Attached Storage is a storage device that is connected to a network and provides file access services. Network File System (NFS) and Common Internet File System (CIFS) are two major protocols used for NAS networks. That is, NAS is used over IP network technology.

The history of NAS can be traced back to file sharing functions provided by operating systems. NFS and CIFS were developed for sharing files between servers in the conventional server-centric system. A file server, which exports file access services to other computers, is a type of implementation of a NAS device. Recent NAS devices are comprised of dedicated hardware and software because of the increasing demand on reliability and performance. A diskless NAS device, specifically a NAS device which has only controllers but contains no disk drives, is sometimes referred to as a NAS gateway or a NAS head. A NAS gateway/head can provide NAS clients with file access services to other storage devices that only export block-level I/O services. That is, a NAS gateway/head could be considered a service bridge between a SAN and a NAS network. Since NAS systems are based on TCP/IP technology, poor access performance is typically observed in comparison with a SAN. However, NAS systems have the strong benefit of exploiting existing IP network resources. The cost effectiveness of NAS systems has expanded their use especially in the entry-level markets.

The conventional storage system architecture, in which storage devices are connected to a single server via a SCSI bus cable, has been renamed to Direct Attached Storage after new storage network architectures such as SAN and NAS appeared. Direct Attached Storage is often abbreviated to DAS.

Key Applications

The flexible interconnection provided by storage networks enables storage devices to communicate with each other. In such a system, not necessarily all the functions need to be executed on server processors. Instead, executing some software codes on storage devices may be more efficient. Actually, storage developers are accommodating different applications into their storage devices. Below are described major storage network applications.

LAN-free Backup: in the server-centric storage architecture, a dedicated server connected to a LAN is usually responsible for creating and managing backup copies. The backup server reads data from other servers through the LAN and writes the data to the archiving storage, such as disk arrays and tape libraries, connected to the backup server. In contrast, the storage-centric architecture enables all the servers to access all the storage devices. The backup server can thus copy data directly from the source storage devices to the archiving storage device. Such LAN-free backup can be considered an approach of moving the copy traffic from the LAN toward the SAN.

Server-free Backup: server-free backup is a more advanced solution, in which storage devices or network devices connected in a storage network directly make a backup copy. Thus, the storage network does manage backup copies without any dedicated backup servers. A copy function which is incorporated in storage devices and/or network devices is often referred to as third-party copy.

Remote Replication: remote replication, which can be seen as a type of server-free backup, keeps a fresh backup copy in a remote data center. Communication between a local data center and a remote data center usually involves non-negligible latency. Two techniques are used for remote replication. Synchronous remote replication forwards a given write command to a remote storage device and then commits the command after the forwarded command is acknowledged by the remote device. This strict mode can synchronize data between the two storage devices all the time; no data would be lost even if a severe disaster damages the local storage device. However, high communication latency between the devices is likely to affect the response time of write commands, thus degrading application performance. In contrast, asynchronous remote replication commits a write command without waiting for the command to be acknowledged by the remote device. Inter-device communication latency would be invisible, but data coherence could not be guaranteed.

Data Sharing and Code Conversion: in a storage-centric environments, a storage device is shared by multiple servers. Sharing the data among multiple servers is also a natural approach. However, presentation forms of data usually depend on microprocessor architectures and operating systems. A file written by a server to a storage device may not be directly interpreted by another server. In the conventional server-centric system, the dedicated application running on the server converts data formats and character codes so that the application can interpret the data appropriately. Such code conversion facilities are being implemented in storage infrastructure. That is, when a server tries to read data stored in a storage system, the storage system converts and then exports the data to the server. Code conversion functions implemented in storage networks are deployed for downsizing from mainframes towards open systems and for file sharing between different types of machines.

SAN File system: a SAN file system is a file system which exports services for accessing files stored in storage devices connected to a SAN. A volume of a SAN file system is often shared among multiple servers. Thus, concurrency control is a key technology for a SAN file system. A SAN file system is deployed in many high-performance clusters and also in several high-end NAS systems.

Cross-References

▶ Direct Attached Storage
▶ IP Storage
▶ Network Attached Storage
▶ SAN File System
▶ Storage Area Network
▶ Storage Management

Recommended Reading

1. Benner AF. Fibre channel for SANs. New York: McGraw-Hill Professional; 2001.
2. Clark T. IP SANS: a guide to iSCSI, iFCP, and FCIP protocols for storage area networks. Reading: Addison-Wesley Professional; 2001.
3. Clark T. Designing storage area networks: a practical reference for implementing fibre channel and IP SANs. Reading: Addison-Wesley; 2003.
4. Robert W, Kembel RW, Cummings R. The fibre channel consultant: a comprehensive introduction. Tucson: Northwest Learning Association; 1998.
5. Storage Network Industry Association. The dictionary of storage networking terminology. Available at: http://www.snia.org/.
6. Troppens U, Erkens R, Müller W. Storage networks explained. London: Wiley; 2004.

Storage Networking Industry Association

Hiroshi Yoshida
VLSI Design and Education Center, University of Tokyo, Tokyo, Japan
Fujitsu Limited, Yokohama, Japan

Synonyms

SNIA

Definition

A nonprofit trade association dedicated to the development and promotion of standards, technologies, and educational services, to empower organizations in the management of information.

Key Points

The SNIA works toward these goals by forming and sponsoring technical work groups, producing a series of conferences, building and maintaining a vendor neutral technology center, and promoting activities that expand the breadth and quality of the storage and information management market. With seven regional affiliates spanning the globe, SNIA represents the voice of the storage industry on a worldwide scale.

Cross-References

▶ Storage Management
▶ Storage Management Initiative Specification

Recommended Reading

1. Storage Network Industry Association. About the SNIA, 2013. http://www.snia.org/about.

Storage of Large Scale Multidimensional Data

Bernd Reiner[1] and Karl Hahn[2]
[1]Technical University of Munich, Munich, Germany
[2]BMW AG, Munich, Germany

Synonyms

Hierarchical storage management; HSM; Multidimensional database management system; Raster data management

Definition

An identified major bottleneck today is fast and efficient access to and evaluation of high performance computing results. This contribution addresses the necessity of developing techniques for efficient retrieval of requested subsets of large datasets from mass storage devices (e.g., magnetic tape). Furthermore, the benefit of managing large spatio-temporal data sets, e.g., generated by simulations of climate models or physical experiments, with Data Base Management Systems (DBMS) will be shown. Such DBMS must be able to handle very large data sets stored on mass storage devices. This means DBMS need a smart connection to tertiary storage systems with optimized access strategies. HEAVEN (Hierarchical Storage and Archive Environment for Multidimensional Array Database Management Systems) is specifically designed and optimized for storing multidimensional array data on tertiary storage media.

Historical Background

Large-scale scientific experiments or supercomputing simulations often generate large amounts of multidimensional data sets. Data volume may reach hundreds of terabytes (up to petabytes). Typically, these data sets are permanently stored as files in an archival mass storage system, on up to thousands of magnetic tapes. Access times and/or transfer times of these kinds of tertiary storage devices, even if robotically controlled, are relatively slow. Nevertheless, tertiary storage systems are currently for the common state of the art storing such large volumes of data, because magnetic tapes are much cheaper than hard disk devices. This will also be the future trend. Furthermore, tapes are a good example for Green-IT. The generation of new data will increase extremely, because of new satellites, sensors, parameters etc. Consequently, scientists need more and more capacity for storing these large amounts of data, and tapes are well prepared for this task.

Concerning data access in the High Performance Computing (HPC) area, the main disadvantages are high access latency compared to hard disk devices and to have no random access. A major bottleneck for scientific applications is the missing possibility of accessing specific subsets of data. If only a subset of such a large data set is required, the whole file must be transferred from tertiary storage media. Taking into account the time required to load, search, read, rewind, and unload several cartridges, it can take many hours to retrieve a subset of interest from a large data set. Entire files (data sets) must be loaded from the magnetic tape, even if only a subset of the file is needed for further processing.

Furthermore, processing of data across a multitude of data sets, for example, time slices, is hard to support. Analysis of dimensions been contrary to storage patterns and requires network transfer of each required data set, implying a prohibitively immense amount of data to be shipped. Another disadvantage is that access to data sets is done on an inadequate semantic level. Applications accessing HPC data have to deal with directories, file names, and data formats instead of accessing multidimensional data in terms of area of interest and time interval. Examples of large-scale HPC data are climate-modeling simulations, cosmological experiments and atmospheric data transmitted by satellites. Such natural phenomena can be modeled as spatio-temporal array data of some specific dimensionality. Their common characteristic is that a huge amount of Multidimensional Discrete Data (MDD) has to be stored. For overcoming the above mentioned shortcomings, and for providing flexible data management of spatio-temporal data, HEAVEN (Hierarchical Storage and Archive Environment for Multidimensional Array Database Management Systems) was implemented [5].

Foundations

In order to implement smart management of large-scale data sets held on tertiary storage systems, HEAVEN combines the advantages of efficient retrieval and manipulation of data sets by using multidimensional array DBMS,

and storing big amounts of data sets on tertiary storage media. This means the DBMS must be extended with easy to use functionalities to automatically store and retrieve data to/from tertiary storage systems without user interaction. A description of related work can be found in [5].

Such intelligent concepts are implemented within the European funded project ESTEDI, and integrated into the kernel of the multidimensional array DBMS RasDaMan (Raster Data Management). RasDaMan is designed for generic multidimensional array data of arbitrary size and dimensionality. In this context, generic means that functionality and architecture of RasDaMan are not tied to particular application areas. Figure 1 depicts the architecture of HEAVEN.

One can see the original RasDaMan architecture with the RasDaMan client, RasDaMan server and the underlying conventional DBMS (e.g., Oracle, which is used as a storage and transaction manager). The additional components for the tertiary storage interface are the Tertiary Storage Manager (TS-Manager), File Storage Manager and Hierarchical Storage Management System (HSM-System). The TS-Manager and File Storage Manager are included in the RasDaMan server. The HSM-System is a conventional product like TSM/HSM (Tivoli Storage Manager) from IBM. Such an HSM-System can be seen as a normal file system with unlimited storage capacity. In reality, the virtual file system of HSM-Systems is separated into a limited cache on which the user works (load or store his data), and a tertiary storage system with robot controlled tape libraries. The HSM-System automatically migrates or stages data to or from the tertiary storage media, if necessary.

Efficient Storage of Large Multidimensional Data

For overcoming the major bottleneck, i.e., the missing possibility of accessing specific subsets of data (MDD), the tiling concept of the RasDaMan DBMS is introduced. A MDD object consists of an array of cells of some base type (e.g., integer, float or arbitrary complex types),

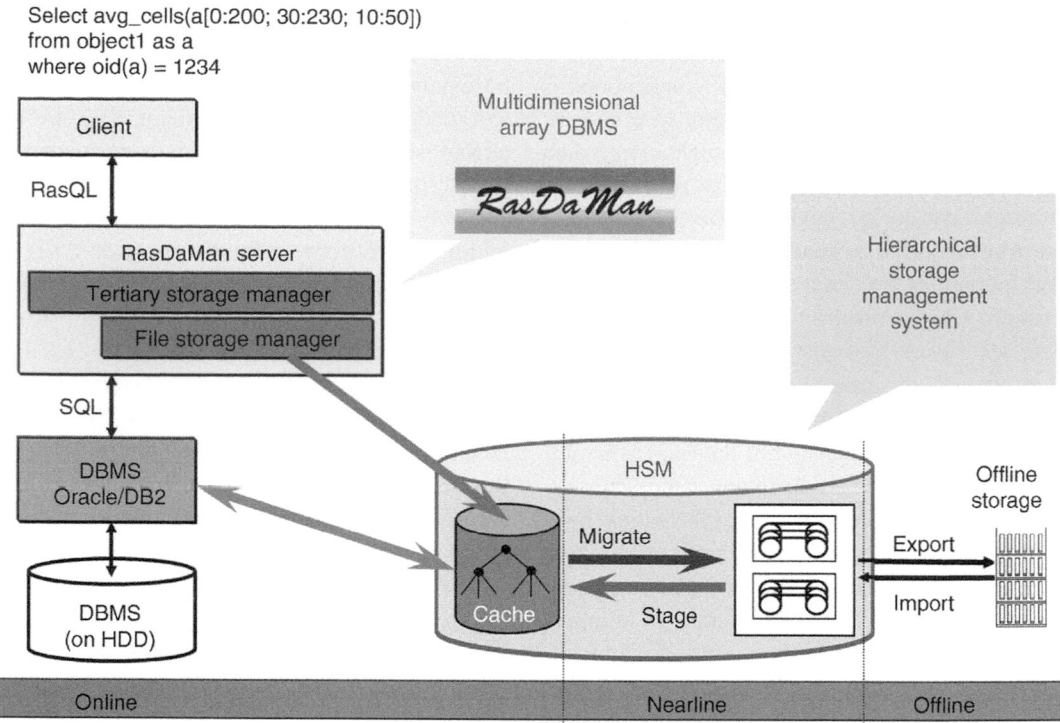

Storage of Large Scale Multidimensional Data, Fig. 1 Extended RasDaMan architecture with tertiary storage connection

which are located on a regular multidimensional grid. An often discussed approach is chunking or tiling of large data [1, 8, 2]. Basically, chunking means the subdividing of multidimensional arrays into disjoint sub-arrays. Tiling is more general than chunking, because sub-arrays don't have to be aligned or have the same size. In RasDaMan, MDD can be subdivided into regular or arbitrary tiles. Consequently one MDD object is a set of multidimensional tiles. In RasDaMan, every tile is stored as one single Binary Large Object (BLOB) in the underlying relational DBMS. This makes it possible to transfer only a subset of large MDD from the DBMS (or tertiary storage media) to client applications, because access granularity is one singe tile. Also, the problem of inefficient access to data sets stored according to their generation process order is not any longer relevant with the tiling strategy of RasDaMan. This will mainly reduce access time and network traffic. The query response time scales with the size of the query box, not any longer with the size of MDD.

Data Export to Tertiary Storage Media

The export of data sets to tertiary storage media is two-tiered. The first step is the migration of the data sets from RasDaMan to the cache area (RAID-System) of the HSM-System. Transferring data sets from the hard disk of the underlying DBMS of RasDaMan to a RAID-System is very fast. Within a second step the migration from the cache area of the HSM-System to the tertiary storage media takes place. This process is performed by the HSM-System, and does not concern the RasDaMan system regarding I/O workload. During this process RasDaMan can execute another export process or user request in parallel.

Data Retrieval

RasDaMan provides an algebraic query language RasQL, which extends SQL with powerful multidimensional operators like geometric, induced and aggregation operators. The primary benefit of such a complex query language is the minimization of data transfer between database server and client. Areas of interest can be specified with geometric operators, and complex calculations can be executed on the server side. Only the result is transferred to the client instead of the entire object [6, 7]. With this feature, RasDaMan, overcomes the mentioned shortcoming of processing data across a multitude of data sets, because RasDaMan transfers only a minimum of data to the client. Furthermore, the query language RasQL provides data access on an adequate semantic level. Users can formulate queries such as: "average temperature on the earth surface of altitude y in the area of latitude x and longitude z."

With respect to data accessibility, three well-defined areas can be differentiated, i.e., online, nearline and offline area (Fig. 1). Data sets stored online means that data sets are stored on hard disk and, therefore, access time is very fast. Data sets stored on magnetic tape and stored in robot controlled libraries are called nearline data. Access time is much higher than with online access, but the process of data retrieval is done automatically. If data sets are stored in the offline area, user interaction is necessary for retrieving data sets (tapes are not robot controlled).

The new RasDaMan tertiary storage functionality is based on the TS-Manager module. If a query is executed, the TS-Manager knows (by metadata) whether the needed data sets are stored on hard disk (online area, DBMS or HSM-Cache) or on tertiary storage media. If the data sets are held on hard disk, the query will be processed very fast. If the data sets are stored on one or more tertiary storage media, the data sets must be imported into the database system (cache area for tertiary storage data) first. The import of data sets stored on tertiary storage media is done by the TS-Manager automatically whenever a query is executed and those data sets are requested. After the import process of the data sets is done, RasDaMan can handle the data sets in the normal way. The complexity of the RasDaMan storage hierarchy is completely hidden from the user.

Techniques for Reducing Tertiary Storage Access Time

The access time for tape systems is by order of magnitude slower than for hard disk devices. It is important to use data management techniques for the efficient retrieval of arbitrary areas of interest from large data sets stored on tertiary storage devices. Hence, techniques that partition data sets into clusters based on optimized data access patterns and storage device characteristics, has to be developed. Therefore, methods for reducing tertiary storage access time, i.e., Super-Tile concept, data clustering and data caching are presented [5, 6].

Super-Tile Concept

In RasDaMan, DBMS tiles (BLOBs) are the smallest unit of data access. Typical sizes of tiles stored in RasDaMan range from 64 KB to 1 MB and are optimized for hard disk access [2]. Those tile sizes are much too small for data sets held on tertiary storage media. It is necessary to choose different granularities for hard disks and tape access because they differ significantly in their access characteristics. Hard disks have fast random access, whereas tape systems have sequential access with much higher access latency. The average access time for tape systems (20-180 s) is by order of magnitude slower than for hard disk drives (5-12 ms), whereas the difference between transfer rate is not significant [5, 10]. For this reason, HEAVEN exploits the good transfer rate of tertiary storage systems, preserving the advantages of the tiling concept. The main goal is to minimize the number of media load and search operations.

It is unreasonable to increase the RasDaMan MDD tile size, because then RasDaMan would loose the advantage of reducing transfer volumes when accessing data on HDD. The solution is to introduce an additional data granularity as provided by the so-called Super-Tile. The main goal of the new Super-Tile algorithm is a smart combination of several small MDD tiles to one Super-Tile to minimize tertiary storage access costs. "Smart" means to exploit the good transfer rate of tertiary storage devices and to preserve advantages of other concepts like data clustering. The left side of Fig. 2 visualizes one three dimensional MDD with Super-Tile and tile granularity. An algorithm for computing Super-Tiles was developed which combines tiles of spatial

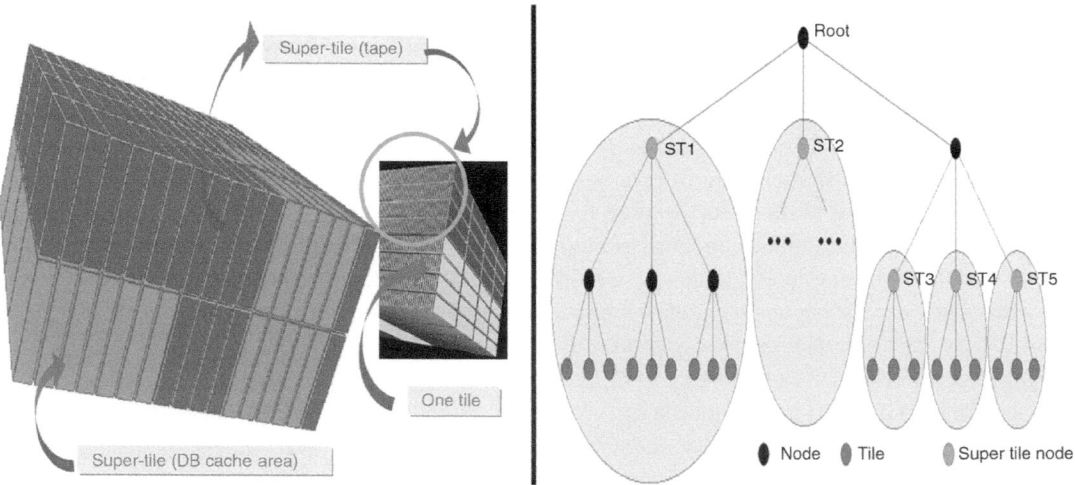

Storage of Large Scale Multidimensional Data, Fig. 2 *Left*: Visualization of one MDD with Super-Tile and Tile granularity. *Right*: Example R+ tree index of one MDD with Super-Tile nodes

neighborhood within the multidimensional object. For the realization, HEAVEN utilizes information of the RasDaMan R+ tree index [3, 5].

The creation of the multidimensional index and the index access is no performance issue compared to data retrieval and data processing. Also, the primary criticism of the R-tree, that performance problems for very many dimensions occur, is not relevant for the application field. The used scientific data from various scientific fields (e.g., climat-modeling simulations, cosmological experiments, atmospheric data, earth observation, computational fluid dynamics) does not have more than five dimensions. Therefore, the integration of advanced multidimensional index methods, e.g., the bitmap index for scientific data proposed by Rishi Sinha and Marianne Winslet [9] or the UB tree proposed by Rudolf Bayer [4], was not considered.

The conventional R+ tree index structure of the multidimensional DBMS was extended to handle such Super-Tiles stored on tertiary storage media. This means that information (whether tiles are stored on hard disk or on tertiary storage media) must be integrated into the index. Tiles of the same sub index of the R+ tree are combined into a Super-Tile and stored within a single file on tertiary storage medium (see right side of Fig. 2). Super-Tile nodes can exist on arbitrary

levels of the R+ tree. Super-Tiles are the access (import/export) granularity of MDD on tertiary storage media, which preserve the advantages of the RasDaMan tiling concept (load minimum data) and exploit the good transfer rates of tertiary storage devices. More details about determining optimal file sizes on tertiary storage media can be found in [5].

Clustering

Clustering is particularly important for tertiary storage systems where positioning time of the device is very high. The main goal is to minimize the number of search and media load operations and to reduce the access time of clusters read from tertiary storage system when subsets are needed. Clustering exploits the spatial neighborhood of tiles within data sets. Clustering of tiles according to spatial neighborhood on one disk or tertiary storage system, proceed one step further in the preservation of spatial proximity, which is important for the typical access patterns of array data, because users often request data using range queries, which implies spatial neighborhood.

The R+ tree index used to address tiles already defines the clustering of the stored MDD. With the developed Super-Tile concept intra Super-Tile clustering and inter Super-Tile clustering can be distinguished. The implemented algorithm for computing Super-Tiles maintains the predefined clustering of sub trees (of Super-Tile nodes) of the R+ tree index and achieves intra Super-Tile clustering (left side of Fig. 2). The export algorithm (export of Super-Tiles to tertiary storage) implements the inter Super-Tile clustering within one MDD. Super-Tiles of one MDD are written to tertiary storage media in the clustered order (predefined R+ tree clustering).

Caching

In order to reduce expensive tertiary storage media access, the underlying DBMS of RasDaMan is used as a hard disk cache for data sets held on tertiary storage media. The general goal of caching tertiary storage data (Super-Tile granularity) is to minimize expensive loading, rewinding and reading operations from slower storage levels (e.g., magnetic tape). In the tertiary storage version of RasDaMan, requested data sets held on tertiary storage media are migrated to the underlying DBMS of RasDaMan (Fig. 1). The migrated Super-Tiles are now cached in the DBMS. After the migration, the RasDaMan server transfers only requested tiles from the DBMS to the client application.

The tertiary storage Cache-Manager evicts data (Super-Tile granularity) from the DBMS cache area only if necessary, i.e., the upper limit of cache size is reached. A special H-LRU (HEAVEN Least Recently Used) algorithm was developed, and together with the caching component of the HSM-System a caching hierarchy (Fig. 1) was built.

Conclusion

The main goal was the realization of optimized management of large-scale data sets stored on tertiary storage systems combined with access functionality like retrieval of subsets. Therefore, a multidimensional array DBMS for optimized storage, retrieval and manipulation of large multidimensional data was introduced. In order to handle hundreds of petabytes stored on tertiary storage media, an interface was presented connecting tertiary storage systems to the multidimensional array DBMS

RasDaMan. The Hierarchical Storage and Archive Environment for Multidimensional Array Database Management Systems (HEAVEN) is specifically designed and optimized for storing multidimensional array data on tertiary storage media. For this reason, the query response time scales with the size of the query box and not with the size of the multidimensional data. This will dramatically reduce access time compared with the traditional access case.

Key Applications

HEAVEN is specifically designed for storing large-scale multidimensional array data (hundreds of terabytes) on tertiary storage media, and is optimized toward HPC. Addressed HPC areas are for example climate-modeling simulations, cosmological experiments and, atmospheric data transmitted by satellites.

Cross-References

▶ Archiving Experimental Data
▶ Data Partitioning
▶ Database Management System
▶ Disk
▶ Query Language
▶ R-Tree (and Family)
▶ Storage Management

Recommended Reading

1. Chen LT, Drach R, Keating M, Louis S, Rotem D, Shoshani A. Efficient organization and access of multi-dimensional datasets on tertiary storage. Inf Syst. 1995;20(2).
2. Furtado PA. Storage management of multidimensional arrays in database management systems, PhD Thesis, Technische Universität München. 1999.
3. Gaede V, Günther O. Multidimensional access methods. ACM Comput Surv. 1998;30(2):170–231.
4. Ramsak F, Markl V, Fenk R, Zirkel M, Elhardt K, Bayer R. Integrating the UB-Tree into a database system kernel. In: Proceedings of the 26th International Conference on Very Large Data Bases; 2000. p. 263–72.
5. Reiner B. HEAVEN – A Hierarchical storage and archive environment for multidimensional array

S

database management systems, PhD Thesis, Technische Universität München. 2005.

6. Reiner B, Hahn K, Höfling G, Baumann P. Hierarchical storage support and management for large-scale multidimensional array database management systems. In: Proceedings of the 13th International Conference on Database and Expert Systems Applications; 2002. p. 689–700.

7. Ritsch R. Optimization and evaluation of array queries in database management systems, PhD Thesis, Technische Universität München. 1999.

8. Sarawagi S, Stonebraker M. Efficient organization of large multidimensional arrays. In: Proceedings of the 10th International Conference on Data Engineering; 1994. p. 328–36.

9. Sinha RR, Winslett M. Multi-resolution bitmap indexes for scientific data. ACM Trans Database Syst. 2007;32(3).

10. Yu J, DeWitt D. Processing satellite images on tertiary storage: a study of the impact of tile size on performance. In: Proceedings of the 5th NASA Goddard Conference on Mass Storage Systems and Technologies; 1996.

Storage Power Management

Kazuo Goda
The University of Tokyo, Tokyo, Japan

Definition

Storage Power Management is a process of improving the efficiency of electric power consumption of all the concerned storage resources including storage devices, storage controllers and storage network devices. Storage Power Management may sometimes cover related equipment such as power supplies and cooling apparatuses. The definition of "the efficiency" may depend on the situation; system designers and administrators often need to balance electric power reduction against performance degradation.

Historical Background

Electric power consumption of storage devices was discussed mainly with regard to battery-operated computing environments such as laptop PCs. However, due to the rapid growth of power consumption in data centers and the increased interest in environmental issues, much attention has recently been paid to electric power consumption of enterprise-level storage systems. Energy efficiency is recognized as a new direction for research and development of storage systems.

Foundations

Hard disk drives are main components of modern storage systems. In the disk drive, a spindle motor, which rotates metal platters at high speed, consumes most of the electric power. Gurumurthi et al. [3] report that the spindle motor can account for 81% of the power consumed by the disk. For a given disk drive, the power consumption of its spindle motor P theoretically relates to the angular velocity ω as

$$P = \frac{K_e^2 \omega^2}{R}$$

where Ke is a motor voltage constant and R is a motor resistance constant. In reality, since the spindle motor rotates the platters against air drag, the angular velocity ω may has a cubic or greater effect on the power consumption P. Disk drive manufacturers have increased the rotational speed to decrease the access latency and improve the transfer rate. Disk array developers have been accommodating a number of such high-speed disk drives into a single disk array enclosure. Thus, greater electric power consumption is often seen in enterprise-level storage systems.

Many commercial disk drives have a "stand-by" mode. While a disk drive is in the stand-by mode, its head is unloaded from the platters to the ramp and its spindle motor is completely suspended. The disk drive consumes much less electric energy in the stand-by mode than in the active (currently processing read/write requests) and idle (being able to start processing read/write requests immediately) modes. However, the transition to/from the stand-by mode involves non-negligible overhead of time and electric energy.

Especially, spinning up a spindle motor to regular speed takes several to tens of seconds and consumes 10 to 100s of joules. Such a significant overhead associated with low-power modes is rarely seen in other computer components including processors and memory. Several commercial disk drives may have different low-power modes other than the stand-by mode. For example, a disk drive in a "low-rpm" mode may keep rotating the platters at lower speed with its head unloaded, accordingly consuming more power but involving less overhead.

The minimum time that the disk needs to be idle for the power saving achieved to exceed the control energy overhead is called "break-even time." The break-even time is a specific parameter of a disk drive. Assuming that the system could perfectly predict accesses that would be issued to a disk drive in the future, the system could spin down the disk drive after the disk becomes idle only if the idle period will be longer than its break-even time. In turn, the system could also spin up the drive in advance before new disk accesses are issued. This "oracle power management" gives the maximum possible energy saving for a given series of disk accesses. However, predicting the future disk accesses perfectly is impossible in reality. Alternative solutions have been studied so far.

Typical techniques of disk power management are based on idleness threshold. The simplest strategy is to spin down a disk drive to a low-power mode after a predetermined time has elapsed since the last disk access. This is based on the heuristic prediction that a disk drive is likely to continue to be idle if it has been idle for a long period. This traditional strategy is deployed in many commercial low-end disk drives, since it works effectively in end-user computing environments where the workload is dominated by interactive applications and users can accept reasonable spinning-up latency. More sophisticated techniques that try to tune the idleness threshold adaptively have been also investigated.

Dynamic Rotations Per Minute (DRPM) [3] is an attempt to exploit innovative disk drives that have the capability of dynamically changing the rotational speeds. Instead of completely spinning down the disk drive, this idea controls the rotational speeds adaptively by observing disk access performance. DRPM is helpful to balance between power and performance tradeoffs more flexibly compared with the conventional low-power modes. Gurumurthi et al. [3] validated potential benefits of DRPM techniques on a simulator. At present, disk drives that can change their rotational speeds are not commercially available but merely reported in papers.

In enterprise systems, a number of individual disk drives are incorporated into a disk array and managed by the array controller. Data layout among the disk drives is a key to power management of such systems. A disk array which has the capability of spinning down member disk drives is called Massive Array of Idle Disks (MAID). The original paper [2] of MAID investigated the caching strategy. Suppose that member disk drives of a given disk array can be divided to a small number of active disk drives and a large number of passive disk drives, and all the blocks that are exported to the server are originally located in the passive disk drives. By replicating hot blocks that are frequently accessed onto the active disks, the array can achieve long idle periods for the passive disks so that they may be spun down. Another paper [1] studied a block migration strategy called Popular Data Concentration (PDC), which clusters hot blocks onto particular disk drives in order to spin down the other drives.

This entry focuses on electric power consumption of disk drives, which are main components of modern storage systems. Other components such as RAID controllers and power supplies and other types of storage devices such as optical discs, electromagnetic tapes and solid-state memory should be more carefully studied in the future.

Key Applications

The rapid growth of electric power consumption has stimulated the economic demand of energy saving technologies. In addition, a variety of government-level restrictions and business-level standards are being considered to resolve or mitigate environmental issues. Much more attention is likely to be paid to Storage Power Management in the future.

Cross-References

► Deduplication
► Disk Power Saving
► Massive Array of Idle Disks

Recommended Reading

1. Carrera EV, Pinheiro E, Bianchini R. Conserving disk energy in network servers. In: Proceedings of the 17th Annual International Conferences on Supercomputing; 2003. p. 86–97.
2. Colarelli D, Grunwald D. Massive arrays of idle disks for storage archives. In: Proceedings of the 16th Annual International Conferences on Supercomputing; 2002. p. 1–11.
3. Gurumurthi S, Sivasubramaniam A, Kandemir M, Hubertus F. Reducing disk power consumption in servers with DRPM. IEEE Comput. 2003;36(12):59–66.
4. Hitachi Global Storage Technologies, Inc. Quietly cool. White Paper, 2004.
5. Lu YH, Micheli GD. Comparing system-level power management policies. IEEE Des Test Comput. 2001;8(2):10–9.

Storage Protection

Kazuhisa Fujimoto
Hitachi Ltd., Tokyo, Japan

Synonyms

Data protection

Definition

Storage protection is a kind of data protection for data stored in a storage system. The stored data can be lost or becomes inaccessible due to, mainly, a failure in storage component hardware (such as a hard disk drive or controller), a disastrous event, an operator's mistake, or intentional alteration or erasure of the data.

Storage protection provides the underlying foundation for high availability and disaster recovery.

Historical Background

In 1956, IBM shipped the first commercial storage that had a hard disk drive. To protect data from bit errors on disk platters, the hard disk drive commonly uses cyclic redundancy check (CRC) and an error-correcting code (ECC).

CRC and ECC cannot protect data from a whole disk failure in which an entire disk becomes inaccessible (for example, because of a disk head crash). The IBM 3990, which was shipped in the 1980s, had the replication functionality in which two identical copies of data were maintained on separate media. This approach protected data from this kind of failure. Replication functionality can be implemented in many other layers of the computer system. Most DBMS support database replication. Some file systems and Logical Volume Managers have file or volume replication functionality. Further, many storage systems and storage virtualization appliances support volume replication functionality.

RAID (Redundant Array of Inexpensive Disks) is another technology for protecting data from whole disk failure. D. Patterson et al. published a paper "A Case for Redundant Arrays of Inexpensive Disks (RAID)" in June 1988 at the SIGMOD conference [6]. This paper introduced a five level data protection scheme. The term *RAID* was adopted from this paper, but currently RAID is an acronym for *Redundant Arrays of Independent Disks*. It is noted that the patent covering RAID level 5 technology was issued in 1978 [5].

RAID level 1 is a kind of replication. RAID level 2 to 5 can reduce the capacity required to protect data against disk drive failure than replication, but it is limited to protect disk drive failure. Replication, on the other hand, can be used to protect databases, file systems and logical volume. Further replication can be used for disaster recovery, if data are replicated remotely.

Foundations

Hard disk drives commonly use Reed-Solomon code [7] to correct bit errors. Data in hard disk drives is usually stored in fixed length blocks. Controllers in hard disk drives calculate ECC for each block and record it associated with the original data. When data are read, the controller checks data integrity using ECC. CRC can be used with ECC for detecting bit errors and/or reducing the possibility of correction error.

Most DBMS support database replication with master/slave relation between the original and the replica. The master process updates and transfer it to the slave. This type of replication can provide high availability to the client of the DBMS in case of storage system failures as well as server failures. Another type of database replication is multi-master, which is mostly used to provide high performance parallel processing. Both types can be either synchronous or asynchronous replication. In synchronous replication, updates made in original are guaranteed in the replica, note there may be some delay in asynchronous replication.

Volume replication by storage system is also widely accepted as data protection. There are synchronous and asynchronous replications, the same as database replication. Asynchronous volume replication is often used for long distance remote replication. It may prevent performance degradation caused by replication delay, but could cause some data loss in case of recovery. Synchronous replication, on the other hand, may provide no data loss recovery, but may cause

performance degradation due to replication delay. Volume replication is also used within a local datacenter for online backup. Backup servers use replica volume for backup during original volume is online. To support this, a storage system can pause update delegation from original to replica volume.

RAID (Redundant Array of Independent Disks) is a set of disks from one or more commonly accessible disk subsystems, combined with a body of control software, in which part of the physical storage capacity is used to store redundant information about user data stored on the reminder of the storage capacity. The term *RAID* refers to a group of storage schemes that divide and replicate data among multiple disks, to enhance the availability of data at desired cost and performance levels. A number of standard schemes have evolved which are referred to as *levels*. Originally, five RAID levels were introduced [6], but many more variations have evolved. Currently, there are several sublevels as well as many non-standard levels. There are trade-offs among RAID levels in terms of performance, cost and reliability.

Key Applications

Storage protection is essential to achieve business continuity and legal compliance with adequate performance, cost, and reliability.

Cross-References

▶ Backup and Restore
▶ Checksum and Cyclic Redundancy Check Mechanism
▶ Continuous Data Protection
▶ Disaster Recovery
▶ Logical Volume Manager
▶ Point-in-Time Copy
▶ Redundant Arrays of Independent Disks
▶ Replication
▶ Write Once Read Many

Recommended Reading

1. ANSI. NFPA1600 standard on disaster/emergency management and business continuity programs.
2. BSI. BS25999; business continuity management.
3. Houghton A. Error Coding for Engineers. Hingham: Kluwer Academic Publications; 2001.
4. Keeton K, Santos C, Beyer D, Chase J, Wilkes J Designing for disasters. In: Proceedings of the 3rd USENIX Conferecne on File and Storage Technologies; 2004.
5. Ouchi NK. System for recovering data stored in failed memory unit. US Patent 4,092,732. 1978.
6. Patterson D, Gibson G, Katz R. A case for redundant arrays of inexpensive disks (RAID). In: Proceedings of the ACM SIGMOD International Conference on Management of Data; 1988.
7. Sweeney P. Error control coding from theory to practice. New York: Wiley; 2002.
8. http://www.sec.gov/

Storage Protocols

Kaladhar Voruganti
Advanced Development Group, Network
Appliance, Sunnyvale, CA, USA

Synonyms

ATA; CIFS; FCP; iSCSI; NFS; Parallel SCSI; Samba; SAS; SATA

Definition

The emergence of networked storage has allowed organizations to de-couple their server and storage purchasing decisions. Thus, multiple application/database servers can share storage on the same network attached storage (NAS) server, or on a block storage controller via a storage area network protocol (SAN). This is unlike direct attached storage systems (DAS) where disks are connected to an application to a database server. In DAS environments, even if one only needs to add more storage, one has to also add more servers. Various types of storage networking protocols have emerged to support both NAS and SAN systems. Currently, the same storage boxes can support both distributed file (NAS) protocols such as NFS and CIFS, and block storage (SAN) protocols such as iSCSI and FCP (fiber channel). Most of the direct attached storage systems are moving from supporting parallel SCSI protocol to supporting serial ATA (SATA) or serial SCSI (SAS) protocols. The focus of this section is to set the context with respect to when they are used, and to compare the advantages and disadvantages of these different protocols.

Historical Background

Historically the fields of storage and networking were two separate fields. With the advent of network attached storage the fields of storage and networking have now converged. Many of the existing networking protocols are now also used to carry storage payload from storage devices to the application servers.

Foundations

- *SAN Protocol Analysis:* SAN protocols transfer SCSI commands and data over a transport protocol. SAN protocol is between SCSI initiators and SCSI targets. The initiators typically reside on the host (application server), and the targets reside on a storage controller box. The SCSI initiator can be software based or it can reside in a host bus adapter card. In the SAN protocol space the comparison is primarily between Fiber Channel and iSCSI. Previously, IBM's SSA protocol competed with Fiber Channel, but has been discontinued as it was not universally adopted by all the storage controller vendors.
  - *iSCSI:* iSCSI protocol transfers SCSI blocks and commands over TCP/IP networks. iSCSI protocol allows organizations to leverage their IP networks to transfer block storage data. Until the emergence of gigabit Ethernet networks, transfer of block storage data over Ethernet/TCP/IP stack

proved to be not viable due to performance problems. Hardware TCP/IP off-load cards as well as multi-core CPU servers have alleviated the TCP/IP processing overhead. Software TCP/IP optimizations such as interrupt coalescing, and zero-copy optimizations have also reduced the CPU processing overhead. iSCSI based SANs have the following benefits over Fiber Channel based SANs:

- *Interoperability:* Until recently, Fiber Channel devices from different vendors did not always interoperate with each other. Device interoperability certification is a much more elaborate and expensive process in the fiber channel world in comparison to the IP world.
- *Distance:* IP networks have been designed to operate across large geographic distances. Fiber Channel networks have distance limitations and one typically needs channel extenders to extend their range.
- *Cost:* An organization can leverage their IP network management expertise and devices for also transferring their storage traffic.
- *Security:* The security protocols for IP networks have been well developed. Since Fiber Channel networks have been typically used behind fire-walls, the security aspects are being still developed.
- *FCP (Fiber Channel):* Sending SCSI block commands and data over the Fiber Channel protocol is known as FCP. Fiber Channel protocol stack consists of physical, data link, network, and transport layer protocol specification. Fiber Channel protocol has the following benefits in comparison to TCP/IP stack:
  - *Performance:* Fiber Channel provides better performance than TCP/IP stack due to the following reasons:
    - *Hardware Offload:* The performance of first generation iSCSI HBAs was inferior to

Fiber Channel HBAs. However, performance gap is being reduced in the newer generation iSCSI offload cards. The iSER and iWARP standards are trying to commoditize the RDMA standard for IP networks in order to bring down the cost of iSCSI offload cards. That is, these cards will be useful for additional protocols (not just iSCSI).
    - *Reservation Based Protocol:* Fiber Channel is a reservation based protocol instead of a retry based protocol. Thus, frames are not lost in Fiber Channel. The slow-start congestion control mechanism in TCP/IP is ill-suited for gigabit speed networks.

Currently, Fiber Channel protocol is the most commonly used SAN protocol in data centers.

- *Parallel* versus *Serial Protocols Analysis:* Parallel SCSI and ATA are two parallel protocols who have distance and device connectivity limitations. It is important to distinguish between the parallel SCSI transport protocol, and the SCSI block protocol which transfers data on top of the transport protocol. The physical wires for these parallel protocols are also quite wide and they make wiring a cumbersome process. Serial ATA (SATA), and Serial SCSI (SAS) have overcome the stringent distance and connectivity limitations of parallel SCSI and ATA transport mechanisms.
- *SCSI* versus *ATA Command Set Analysis:* The SCSI block protocol is transported over Fiber Channel (FCP), TCP/IP (iSCSI), and parallel SCSI, and serial SCSI (SAS) transport layers. The IDE/ATA protocol is transported over ATA and Serial ATA (SATA) transport mechanisms. It is important not to mix the block protocol with the underlying transport protocol. SCSI block and ATA block protocols have the following key differences:

– *Command Queuing at the Device:* Previously SCSI protocol allowed for the queuing of multiple commands at the SCSI device, whereas, ATA protocol did not have this functionality. Recently, tagged command queuing functionality has been added to ATA protocol. Tagged command queuing also allows the device to optimize the order in which the queued commands are executed.

– *Number of Connected Devices:* ATA protocol supported fewer devices per channel than the SCSI protocol. However, this limitation has been reduced in the serial ATA protocol with the emergence of port multipliers. Thus, 15 devices can be simultaneously supported using SATA.

– *Checksums:* ATA protocol did not initially contain support for checksums on command and data. Later versions of the protocol have added checksum support for data.

– *Hot-Plug of Devices:* ATA does not provide support for hot replacement of devices. Serial ATA has rectified this deficiency.

– *Bus Mastering:* Previous versions ATA protocol did not allow to DMA data directly from the device into memory. However, recent versions of the protocol have overcome this limitation.

• *Block* versus *File-Based NAS Protocols:* NAS protocols provide a file level abstraction to the client applications, whereas SAN protocols provide a block level abstraction. NAS protocols allow for multiple hosts to share a file system namespace, whereas, SAN protocols allow multiple hosts to share a block level (SCSI device and LUN level) namespace. Previously, Fiber Channel based SANs provided better performance than IP based NAS protocols due to the implementation differences between the IP and Fiber Channel transport protocol stacks. However, this difference is disappearing due to the arrival of TCP/IP stack offload cards. In the past, NAS systems were not as scalable as SAN systems due to the absence of clustered NAS solutions but with the advent of clustered NAS solutions, this limitation has been also overcome.

• *NAS Protocol Analysis:* NFS and CIFS are the two primary NAS protocols. NAS protocols are between NAS clients that reside on the application/database server, and a NAS server that typically is a separate box that contains a file system and manages disks. NAS clients provide a file system interface to the host applications. NFS is used primarily in the Unix/Linux environment and it is an IETF standard, whereas, CIFS is used primarily in the Windows OS environment and its management/administrative APIs are proprietary (controlled by Microsoft). Samba is a popular protocol bundle that implements many different CIFS related protocols. A Samba server can act as an open-sourced CIFS server on Unix systems that makes Unix directories appear as Windows folders to Windows clients. With the emergence of NFSv4, many of the NFS deficiencies with respect to recovery management, caching, and security have been overcome and makes NFS a competitive protocol. NFSv4 has consolidated numerous protocols such as nfs, nlm, mountd, and nsm. NFSv4 is a stateful protocol, and it introduces the concept of delegation to allow for aggressive data caching at the clients [2].

Key Applications

SANs have been typically used by applications that want block level access to storage and are performance conscious. NAS solutions have been typically used by users who are more cost conscious and want to leverage their existing IP infra-structure and IP network management experience. In SANs, Fiber Channel and iSCSI protocols are typically used to connect application servers to the storage controller boxes. Fiber Channel, SATA and SAS protocols are typically used to connect the storage controller processor/cache complex to its backend arrays/disks. In DAS environments, the server is connected to its backend storage using Parallel SCSI, ATA, SATA or SAS. In NAS environments, NFS/CIFS are used to connect the application servers to the NAS server, and SATA, Fiber Channel or SAS is used to connect the NAS processor complex to the back-end disk arrays.

Cross-References

▶ Direct Attached Storage
▶ Network Attached Storage
▶ SAN File System
▶ Storage Area Network

Recommended Reading

1. Kaladhar V, Prasenjit S. An analysis of three gigabit networking protocols for storage area networks. In: Proceedings of the 20th IEEE International Performance, Computing and Communications Conference; 2001.
2. Peter R, Li Y, Pawan G, Prasenjit S, Prashant S. A performance comparison of NFS and iSCSI for IP-networked storage. In: Proceedings of the 3rd USENIX Conference on File and Storage Technologies; 2004.

Storage Resource Management

Hiroshi Yoshida
VLSI Design and Education Center, University of Tokyo, Tokyo, Japan
Fujitsu Limited, Yokohama, Japan

Synonyms

SRM

Definition

The management of physical and logical storage resources, including storage elements, storage devices, appliances, virtual devices, disk volume, and file resources. In most cases, storage resource management is achieved by software tools which indicate and manage the storage resource utilization in a storage networking environment.

Key Points

When multiple storage devices are connected to a SAN and shared by multiple servers, the space in each device cannot be used uniformly, e.g., one device may be almost full, while another is underutilized. If the storage administrator is unaware of the correct space utilization status, inefficient resource utilization of the whole SAN environment can occur. This may cause unnecessary addition of new storage resources, or additional resources are not installed in time causing business applications to stop due to lack of required storage space.

To indicate and manage storage resource utilization, storage resource management software tools provide the following functions:

- Discovery and investigation of the utilization of all storage resources in a SAN, plus related integrated monitoring and integrated management operations
- Analysis and reporting of storage resource utilization
- Analysis and estimation of movements in resource utilization, issuing alerts, and/or execution of scripts for automation

Monitoring and indicating storage resource utilization from the business application viewpoint is an important requirement for storage resource management. To meet this requirement, storage resource management and storage configuration management, including dependency mapping between applications and storage devices, are often provided in combination.

Cross-References

▶ Information Loss Measures
▶ Logical Unit Number

▸ SAN File System
▸ Storage Consolidation
▸ Storage Management
▸ Storage Network Architectures
▸ Storage Virtualization

Recommended Reading

1. Storage Network Industry Association. Storage network industry association tutorials. 2014. Available at: http://www.snia.org/education/tutorials/.

Storage Security

Kaladhar Voruganti
Advanced Development Group, Network
Appliance, Sunnyvale, CA, USA

Synonyms

Access control; Authentication; CAS; Compliance; Data corruption; LUN masking; On-disk security; On-wire security; Port binding; Provenance; Watermarking; WORM; Zoning

Definition

The definition of storage security has many facets, and some of the key requirements are:

• Storage (and the appropriate manipulation capabilities) should only be accessible and visible to users with the appropriate permissions
• Users should be notified if their data has been tampered with or altered either intentionally or unintentionally
• Malicious users should not be allowed to access or tamper with other people's storage. If possible, the system should be able to catch malicious users.

• User's data should be physically deleted at the appropriate time. That is, it should not be present past its intended life-time
• Tracking unauthorized copying or access control delegations is also an important security concern.

Historical Background

Security has been a key computer science topic for many decades. Initially, people focused on authentication, encryption and access control for standalone systems. As distributed computing became popular, people started focusing on security within the context of network protocols. Now, with the emergence of utility computing, grid computing and cloud computing, storage is being accessed remotely. Thus, security issues pertaining to storage environment is now gaining importance. Many of the known security algorithms and techniques are now being re-used within the context of storage systems.

Foundations

The following different security mechanisms together provide the appropriate desired security functionality:

• *Storage Access Authentication:* In order to prevent unauthorized users from accessing a storage device, there is usually an authentication mechanism employed (e.g., Kerberos, CHAP) to prevent unauthorized users from accessing the storage device. If the storage device is being accessed in a client-server manner, where the client resides on a database or application server, the client-server protocol needs to employ an authentication mechanism (e.g., Kerberos, or CHAP). In cases where both the host servers and the storage devices reside behind the firewall, client authentication is typically not performed.
• *On-Wire Security:* When information flows between clients and storage servers outside

the fire-wall protected domains, information is typically encrypted and sent over the wire to prevent eavesdropping. Data are also protected by a hash key to detect data tampering. IPSec security infra-structure provides authentication, data encryption and tampering detection support at the IP layer for internet environments. Data are typically not encrypted on the wire in intranet environments behind the firewall.

- *On-Disk Encryption:* Data stored on persistent storage (disks, tapes etc) is typically not encrypted because disks usually reside in secure places. However, with the emergence of storage service providers, an organization's data are managed by others. Thus, many organizations prefer to encrypt their data before storing it at a remote location. It is very important to store and remember the keys that have been used to encrypt the data because loss of the key is as good as the loss of the data. Organizations also periodically re-encrypt data to prevent the use of brute-force approaches to break the keys being used to encrypt the data. AES, DES and 3DES are some of the encryption algorithms that are being currently used.
- *Access-Control:* Access control mechanisms have been in place and are used extensively in operating systems to control access to various resource types. Access control typically amounts to controlling read/write access for individual users, groups and everyone to files, directories of files, volumes, and storage devices. Access control mechanisms also deal with delegation of control by secondary data owners to third parties who want to access the data.
- *Compliance:* Organizations are being expected to adhere to many different types of government regulations such as:
- There should be only a single copy of a particular type of data.
- Data should be physically erased from the storage medium after so many years. That is, data cannot be logically deleted, and one should not be able to reconstruct the data from the physically deleted copy.

- Organization should be able to provide a guarantee that data has not been altered since its creation. Many organizations employ WORM (write-once, read-many) type of storage medium such as optical storage or WORM tapes or CAS systems to attain this capability.
- During audits data belonging to a certain topic should be available within a short specified period of time (for example, data should be available within 48 h). In some cases this precludes the storing of data on tape.

Data Corruption Data can be corrupted either unintentionally due to device failure or malfunction, or it could be corrupted intentionally by a malicious user or a virus. In either case, data corruption has to be detected and subsequently corrected. Data corruption can be detected at the time of data creation, or it can be checked for by a periodic checking process such as virus scans or disk scrubbing. Data corruption can dealt with in the following ways:

- *Detection:* Data corruption can be detected (not corrected) using a hash function such as MD5 or SHA-1, checksum or CRC.
- *Detection and Correction:* Data corruption can be detected and corrected using error correction codes such as Reed-Solomon.
- *Correction:* Data corruption can be only corrected using erasure correcting codes such as a variant of Reed-Solomon codes.
- *Zoning:* In storage area networks, zoning is a technique that is used to control which host port can have access to which storage device port. Zoning can be controlled at the switch firmware level (called hard zoning), by which one controls which switch input port can see which switch output port. Zoning can also be controlled at the Fiber Channel name service level (soft zoning), by which one controls which host server port can see which storage device port.
- *Port Binding:* Port binding is the process by which one controls which port can be connected to a particular switch port. That is, one explicitly specifies the address of the host

or storage device port that can exclusively transfer data to a particular switch port.

- *LUN Masking:* The process of determining which host port can access which volume on the target via which target port is called as LUN masking. Thus, using this process one can control the access of storage volumes by the different hosts.
- *Provenance and Watermarking:* With the emergence of peer to peer networks, data are getting copied at a very rapid rate often without the proper permission. In many cases, parts of the original document get copied and combined with new content. Thus, it is necessary to keep track of the trail of how data has been copied and modified from its source to its current location. In addition to keeping track of the original author and source of the data, it is also necessary to detect pirated copies of data. New digital document watermarking techniques are being developed to add hidden copyright notices to documents.
- *CAS:* Content addressable storage (CAS) systems provide a new type of data access mechanism. In CAS systems one generates a hash value out of the data content. This hash value is subsequently used to uniquely access the data. In CAS systems, one typically does not overwrite existing data but instead a new copy of the data gets created when updates are made to existing data copies. CAS systems have been used to provide WORM media capabilities for disk based systems.

Key Applications

Until recently, storage systems have been based behind company fire-walls. Thus, authentication and encryption have not been a major issue. Access control and data corruption (due to viruses or device failures) have been the major forms of security/integrity checking processes. With the emergence of government compliance regulations, data compliance has become a very major issue for most organizations. With the emergence of third party archival or storage service provider paradigms, more importance is being given to storage authentication and encryption mechanisms.

Cross-References

▶ Access Control Administration Policies
▶ Access Control Policy Languages
▶ Asymmetric Encryption
▶ Authentication
▶ Data Encryption
▶ Database Security
▶ Discretionary Access Control
▶ Homomorphic Encryption
▶ Mandatory Access Control
▶ Message Authentication Codes
▶ Network Attached Secure Device
▶ Security Services
▶ Symmetric Encryption
▶ Temporal Access Control

Recommended Reading

1. Riedel E, Kallahalla M, Swaminathan R. A framework for evaluating storage system security. In: Proceedings of the 1st USENIX Conference on File and Storage Technologies; 2002.

Storage Virtualization

Hiroshi Yoshida
VLSI Design and Education Center, University of Tokyo, Tokyo, Japan
Fujitsu Limited, Yokohama, Japan

Definition

Storage virtualization is technology to build logical storage using physical storage devices. More precisely, the Storage Networking Industry Association (SNIA) defines storage virtualization as follows:

1. The act of abstracting, hiding, or isolating the internal function of a storage (sub)system or service from applications, computer servers, or general network resources for the purpose of enabling application- and network-independent management of storage or data.
2. The application of virtualization to storage services or devices for the purpose of aggregating, hiding complexity, or adding new capabilities to lower-level storage resources.
3. Storage can be virtualized simultaneously in multiple layers of a system, for instance, to create hierarchical storage manager-like systems.

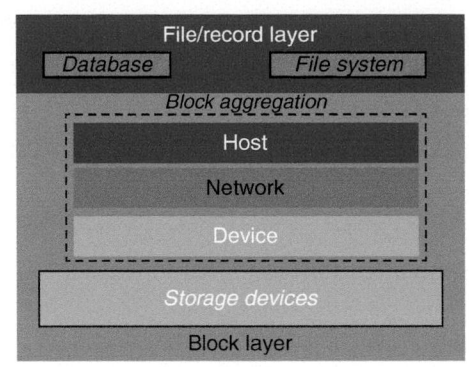

Storage Virtualization, Fig. 1 Block aggregation in shared storage model by SNIA

Key Points

Storage virtualization is used to provide an aggregation of data blocks (a logical volume) to applications running on servers. This aggregation of blocks may reside on a single storage device, may be spanned across multiple devices when it is too large to fit on a single device, may be mirrored to multiple storage devices to increase availability, or may be striped across multiple storage devices to enable parallel accesses.

Storage virtualization (block aggregation) can be implemented in various layers between the host server and the storage devices, i.e., in the host layer (software-based logical volume manager), in the network layer (switch-based virtualization appliance), and in the device layer (disk array). Figure 1 shows where block aggregation is implemented in the shared storage model defined by the Storage Networking Industry Association.

Storage virtualization technologies are now being extended to provide more capabilities. For example, multiple storage devices which reside in different locations can be virtualized as a single logical storage device with disaster recovery capability. Another example is the virtual-ization of inexpensive storage devices with high capacity together with expensive devices with high performance to form a single logical device where migration between the physical devices is done automatically. This provides a cost-effective storage system suitable for hierarchical storage management, information lifecycle management (ILM), and cloud computing.

Cross-References

▶ Information Loss Measures
▶ Storage Area Network
▶ Storage Consolidation
▶ Storage Network Architectures
▶ Storage Networking Industry Association
▶ Storage Resource Management

Recommended Reading

1. Storage Network Industry Association. Storage virtualization: the SNIA technical tutorials. 2007. Available at: http://www.snia.org/education/storage_networking_primer/stor_virt/.
2. Storage Network Industry Association. The SNIA shared storage model. 2007. Available at: http://www.snia.org/education/storage_networking_primer/shared_storage_model/.

Stored Procedure

Tore Risch
Department of Information Technology, Uppsala University, Uppsala, Sweden

Definition

Modern relational query languages such as SQL provide general programming language capabilities in addition to the statements for searching and updating the database. A stored procedure is a user program written in a query language running inside the database server. Stored procedures often include side effects that update the database. This makes it possible to define general programs using the query language. These programs are called stored procedures and are executed inside the database server.

Key Points

In SQL, the user defines stored procedures as a schema manipulation statement using as CREATE PROCEDURE statement. A so defined procedure is immediately shipped to the database server where it is compiled and stored. Stored procedures can have side effects that change the state of the database. In order to prohibit searches that change the state of the database, in SQL stored procedures cannot be called from within queries, but only from applications or general SQL interfaces. SQL provides a special EXEC statement passing to the procedure when it is called.

The advantage with stored procedures are:

- Communication time is saved in case the communication with the application program is slow or unreliable.
- Common database centered updates and computations belong naturally to the database.

A disadvantage with stored procedures is that different vendors often provide different stored procedure languages. The SQL-PSM (Persistent Stored Modules) standard defines stored procedures in SQL.

Related to stored procedures (and part of SQL-PSM) are user defined functions (UDFs), which are user defined functions to perform common side-effect free computations. UDFs are allowed as expressions in queries.

Cross-References

▶ Query Language

Stream Mining

Jiawei Han and Bolin Ding
University of Illinois at Urbana-Champaign, Urbana, IL, USA

Synonyms

Stream data analysis

Definition

Stream mining is the process of discovering knowledge or patterns from continuous data streams. Unlike traditional data sets, *data streams* consist of sequences of data instances that flow in and out of a system *continuously* and with varying update rates. They are *temporally ordered, fast changing, massive, and potentially infinite*. Examples of data streams include data generated by communication networks, Internet traffic, online stock or business transactions, electric power grids, industry production processes, scientific and engineering experiments, and video, audio or remote sensing data from cameras, satellites, and sensor networks. Since it is usually impossible to store an entire data stream, or to scan through it multiple times due to its tremendous volume, most stream mining algorithms are confined to reading only once or a small number of times using limited computing and storage capabilities. Moreover, much of stream data resides at a rather low level of abstraction, whereas analysts are often interested in relatively high-level dynamic changes, such as trends and deviations. Therefore, it is essential to develop online, multilevel, multidimensional stream mining methods. Stream mining can be considered a subfield of data mining, machine learning, and knowledge discovery.

Historical Background

There are extensive studies on stream data management and the processing of continuous queries in stream data [4]. Different from stream query processing, stream mining extracts patterns and knowledge from online stream data. It covers the topics of mining multidimensional stream statistics, frequent patterns, classification models, clusters, and outliers in online data streams. Substantial research on stream mining has appeared since only 2000, almost at the same time as the research on stream data management. However, lots of results have been generated in this line of research.

Foundations

Stream mining problems are challenging because of the following two reasons. (i) Stream data are massive, arriving with high speed, and updated frequently, so that one can neither store all the data nor scan the data repeatedly, and in the meantime, the response time is usually required to be short in applications; therefore, stream synopsis construction is popularly used to maintain a summary of stream data online using limited space without losing too much information. (ii) Stream data often evolves considerably over time, and for classification or clustering, one should often use biased sampling of the stream data to emphasize more recent behavior of the stream. In general, stream mining can be partitioned into four themes: (i) online computing multidimensional stream statistics, (ii) mining frequent items and itemsets over stream data, (iii) stream data classification, and (iv) clustering data streams.

For online computing multidimensional stream statistics, a multidimensional stream cube model was proposed by Chen et al. [6] in their study of multidimensional regression analysis of time-series data streams. A stream cube can be efficiently constructed based on (i) a tilted timeframe, where the finer granularity is used for more recent time, and the coarser granularity for more distant time, (ii) a minimal interest layer to register the minimal layer of the cube that is still of user's interest, and an observation layer for the cuboid that a user usually watches for trends or anomaly, and (iii) partial materialization that materializes the cuboids only along the popular drilling path, serving as a tradeoff between the storage space and online response time. Statstream, a statistical method for the monitoring of thousands of data streams in real time, was developed by Shasha and Zhu [15].

For mining frequent items and itemsets over stream data, one important issue is how to return approximate frequency counts for items or itemsets with limited buffer size for infinite data streams. Manku and Motwani [12] proposed Sticky Sampling algorithm and Lossy Counting algorithm for computing approximate frequency counts of items over data streams, and developed a Lossy Counting based algorithm for computing frequency counts of itemsets with the focus on system-level issues and implementation artifices. Another important issue is how to track frequent items dynamically. "Dynamically" means data streams consist of both "insertion" operations and "deletion" operations of items (imagine cars entering and exiting the parking lot). Cormode and Muthukrishnan [7] proposed algorithms for this problem using group testing and randomization techniques. Keeping track of frequent of items in such data streams arises in applications of both traditional databases and other domains, like telecommunication networks.

S

For stream data classification, the goal is to predict the class label or the value of new instances in the data stream, given some knowledge about the class membership or values of previous instances in the data stream. Since the distribution underlying the instances or the rules underlying their labeling may change over time, the class label or the target value to be predicted may change over time as well (stream data is evolving). This problem is referred to as *concept drift*. A major challenge in stream classification is how to construct highly accurate models with the existence of concept drift in stream data. Hulten et al. [10] developed an algorithm CVFDT, by integrating concept drift in time-changing data streams and a statistical measure, Hoeffding bound. Wang et al. [16] proposed an ensemble classifier to mine concept-drifting data streams. Aggarwal et al. [3] developed a k-nearest neighbor-based method for classify evolving data streams. Gao et al. [8] handled skewed distributions and proposed an ensemble-based framework that under-samples the overwhelming negative data, and repeatedly samples the scarce positive data for model construction with concept-drifting data streams.

For clustering data streams, in some applications, simply assume that the clusters are to be computed over the entire data stream, and view the stream clustering problem as a variant of single-scan clustering algorithms. For example, Guha et al. [9] gave space-efficient constant-factor approximation algorithms for the k-median problem in stream data using divide-and-conquer and randomization techniques; O'Callaghan et al. [14] proposed a k-median based stream clustering method by incrementally updating k-median centers. However, it is important to consider stream evolution over time besides the single-scan constraint and resource limitation. When stream data is evolving, the underlying clusters may also change considerably over time. If the entire data stream is used for clustering, the result is likely to be inaccurate. What's more, at one moment, users may wish to examine clusters occurring in different time periods (e.g., last week/month/year). Aggarwal et al. [2] proposed a *CluStream* framework for clustering evolving data streams by introducing a tilted timeframe, an online microclustering

maintenance, and an offline query-based macro-clustering mechanism, to achieve efficiency and high clustering quality and to provide the flexibility to compute clusters over user-defined time periods in an interactive fashion.

Aggarwal [1] provides a comprehensive survey on stream data processing and stream mining with a collection of chapters on difference issues on stream mining.

Key Applications

There are broad applications of stream mining, of which only a few examples are illustrated.

Mining Anomaly in Network Streams Stream mining has been popularly used for mining anomaly in computer network or other stream data. For example, MAIDS (Mining Alarming Incidents in Data Streams) [5] is a system that explores tilted timeframe and stream cube for mining computer network anomaly.

Computing Statistical Measures Over Time Series Data Streams Another popular application is to compute statistical measures over time series streams, such as StatStream [15].

Integration of Mobile Computing with Stream Mining Kargupta et al. [11] have developed VEDAS (Vehicle Data Stream Mining System) that allows continuous monitoring and pattern extraction from data streams generated onboard a moving vehicle.

Future Directions

There are many challenging issues to be researched further. Only a few are listed below.

Mining Sophisticated Patterns in Data Streams Due to the single-scan constraint and resource limitation, the current stream pattern mining methods are confined to simple patterns, such as single items or some limited itemsets. Tasks for mining sequential patterns [13] and structured patterns are challenging but interesting for research.

Mining Sophisticated Data Sets for Advanced Applications Text (e.g., document) streams and video streams are important real-life stream mining applications. However, it is challenging to mine such streams because it requires both sophisticated text/video analysis and real-time resource constraints. Other advanced stream mining applications include spatial data streams, financial transaction data streams, and so on.

Experimental Results

There are many experimental results reported in numerous conference proceedings and journals.

Data Sets

UCI Machine Learning Repository: http://archive.ics.uci.edu/ml/datasets.html

URL to Code

RapidMiner (previously called YALE (Yet Another Learning Environment)), at http://rapid-i.com, is a free open-source software for knowledge discovery, data mining, and machine learning. It also features data stream mining, learning time-varying concepts, and tracking drifting concept.

MassDAL (Massive Data Analysis Lab) Public Code Bank, at http://www.cs.rutgers.edu/muthu/massdal-code-index.html, is a library of routines in C and Java for stream data and other massive data set analysis, including implementations of some published algorithms for finding frequent items over stream data.

Cross-References

▶ Association Rule Mining on Streams
▶ Classification in Streams
▶ Clustering on Streams
▶ Data Mining
▶ Event Stream
▶ Frequent Items on Streams
▶ Geometric Stream Mining
▶ Stream Similarity Mining
▶ Wavelets on Streams

Recommended Reading

1. Aggarwal CC. Data streams: models and algorithms. Kluwer Academic; 2006.
2. Aggarwal CC, Han J, Wang J, Yu PS. A framework for clustering evolving data streams. In: Proceedings of the 29th International Conference on Very Large Data Bases; 2003. p. 81–92.
3. Aggarwal CC, Han J, Wang J, Yu PS. On demand classification of data streams. In: Proceedings of the 10th ACM SIGKDD International Conference On Knowledge Discovery and Data Mining; 2004. p. 503–8.
4. Babcock B, Babu S, Datar M, Motwani R, Widom J. Models and issues in data stream systems. In: Proceedings of the 21st ACM SIGACT-SIGMOD-SIGART Symposium on Principles of Database Systems; 2002. p. 1–16.
5. Cai YD, Clutter D, Pape G, Han J, Welge M, Auvil L. MAIDS: mining alarming incidents from data streams. In: Proceedings of the ACM SIGMOD International Conference on Management of Data; 2004. p. 919–20.
6. Chen Y, Dong G, Han J, Wah BW, Wang J. Multi-dimensional regression analysis of time-series data streams. In: Proceedings of the 28th International Conference on Very Large Data Bases; 2002. p. 323–34.
7. Cormode G, Muthukrishnan S. What's hot and what's not: tracking most frequent items dynamically. In: Proceedings of the 22nd ACM SIGACT-SIGMOD-SIGART Symposium on Principles of Database Systems; 2003. p. 296–306.
8. Gao J, Fan W, Han J, Yu PS. A general framework for mining concept-drifting data streams with skewed distributions. In: Proceedings of the SIAM International Conference on Data Mining; 2007.
9. Guha S, Mishra N, Motwani R, O'Callaghan L. Clustering data streams. In: Proceedings of the 41st Annual Symposium on Foundations of Computer Science; 2000. p. 359–66.
10. Hulten G, Spencer L, Domingos P. Mining time-changing data streams. In: Proceedings of the 7th ACM SIGKDD International Conference on Knowledge Discovery and Data Mining; 2001.
11. Kargupta H, Bhargava B, Liu K, Powers M, Blair P, Bushra S, Dull J, Sarkar K, Klein M, Vasa M, Handy D. VEDAS: a mobile and distributed data stream

S

mining system for real-time vehicle monitoring. In: Proceedings of the SIAM International Conference on Data Mining; 2004.

12. Manku G, Motwani R. Approximate frequency counts over data streams. In: Proceedings of the 28th International Conference on Very Large Data Bases; 2002. p. 346–57.

13. Mendes L, Ding B, Han J. Stream sequential pattern mining with precise error bounds. In: Proceedings of the 2008 IEEE International Conference on Data Mining; 2008.

14. O'Callaghan L, Meyerson A, Motwani R, Mishra N, Guha S. Streaming-data algorithms for high-quality clustering. In: Proceedings of the 18th International Conference on Data Engineering; 2002. p. 685–96.

15. Shasha D, Zhu Y. High performance discovery in time series: techniques and case studies: Springer; 2004.

16. Wang H, Fan W, Yu PS, Han J. Mining concept-drifting data streams using ensemble classifiers. In: Proceedings of the 9th ACM SIGKDD International Conferenc on Knowledge Discovery and Data Mining; 2003. p. 226–35.

Stream Models

Lukasz Golab
University of Waterloo, Waterloo, ON, Canada

Definition

Conceptually, a *data stream* is a sequence of data items that collectively describe one or more underlying signals. For instance, a network traffic stream describes the type and volume of data transmitted among nodes in the network; one possible signal is a mapping between pairs of source and destination IP addresses to the number of bytes transmitted from the given source to the given destination. A stream model explains how to reconstruct the underlying signals from individual stream items. Thus, understanding the model is a prerequisite for *stream processing* and *stream mining*. In particular, the computational complexity of a data stream problem often depends on the complexity of the model that describes the input.

Historical Background

The stream models discussed in this article were introduced in [3] and extended in [7, 8]. In addition to modeling a stream with respect to its underlying signal(s), there exist the following two related concepts. First, the stream computational model asserts that a stream algorithm must run in limited space and time and can only make a small number of passes (often only one pass) over the data [8, 6]. Furthermore, one can also model various statistical properties of a data stream, such as changes in the frequency distribution of the underlying signals or inter-arrival times of stream items (see, e.g., [9]).

Foundations

Basic Model Definitions

Consider a *data stream S* composed of individual items, $s_1, s_2, \ldots$, ordered by arrival time. Let A be a signal described by S. Assume that A is a function from a discrete and ordered domain to the range of reals; i.e., $A: [1 \ldots N] \to \mathbb{R}$. For instance, in the motivating example from above, the domain consists of IP address pairs. That is, $N = 2^{64}$ since an IP address is 32 bits long.

There are four models for representing A using individual stream items:

1. In the aggregate model, each stream item s_i corresponds to a range value for some domain value.

2. In the cash register model, each stream item s_i represents a domain value and a partial range value r_i, such that $r_i \geq 0$. Reconstructing the signal A involves aggregating all the r_i values corresponding to each domain value.

3. The turnstile model generalizes the cash register model by allowing any r_i to be negative. Thus, reconstructing the signal A involves adding/subtracting the contributions of stream items having positive/negative range values.

4. In the reset model, each stream item s_i corresponds to a range value and is understood to replace all previously reported range values for the given domain value.

Each of the four models defined above has an ordered and an unordered version. In the ordered version, stream items arrive over time in increasing order of the domain values. In the unordered version, the ordering of the domain does not correspond to the arrival order of stream items.

In terms of complexity and expressive power, the turnstile model is the most general, followed by the cash register model and the aggregate model, respectively. As a result, designing stream algorithms for the turnstile model is the most challenging. For instance, while many types of sketches have provable time and accuracy bounds in the turnstile and cash register models, *stream sampling* algorithms are typically applicable only in the cash register model [8] (effectively, the turnstile model allows deletions via negative range values; therefore, it may not be possible to maintain a fixed-size sample of the stream over time). Additionally, geometric problems over streams (e.g., estimating the diameter of a stream of points) are difficult to solve in the turnstile model since previously seen points may be deleted in the future [8]. The reset model is also quite general, and some algorithms in this model are more complex than in the other three models [8]. See [3, 7, 8] for examples of algorithms and complexity bounds for different stream models.

Examples and Extensions of Basic Models

Consider a network traffic stream S composed of IP packets. Each packet contains (among other things) the source IP address, destination IP address, and size. Define signal A_1 as a function from the source and destination address pairs to the total number of bytes exchanged by each pair (i.e., sums of sizes of all the packets sent between a given pair). Since many packets may be exchanged between two nodes and packets may arrive in random order, this example corresponds to the unordered cash register model.

For an example of an ordered cash register model, define S_2 to be the output stream of a pipelined query plan for the following query:

```
SELECT a1, a2, count(*)
FROM T
GROUP BY a1, a2
ORDER BY a1, a2
```

Define signal A_2 as a function from values of a_1 to the corresponding frequency counts (aggregated over all possible values of a_2). This example corresponds to the ordered cash register model – stream items arrive in the order of their domain values due to the *ORDER BY* clause but must be aggregated on a_1 in order to reconstruct' A_2.

Now suppose that the output stream of the above query is not ordered by a_1, but has the property that all the groups having the same value of a_1 are streamed out contiguously. This conforms to the contiguous unordered cash register model. Note that the ordered cash register model is always contiguous.

Next, suppose that stream S_3 is a preprocessed version of S, where each item s_i is a triple of the form (source IP address, destination IP address, event), where the event field denotes the start or end of a connection between two nodes. Define signal A_3 as a function from the source and destination address pairs to the total number of open connections between each pair that have not yet ended. This corresponds to the unordered turnstile model since a stream item carrying an end-of-connection event decrements the total count of open connections for the given pair of nodes; furthermore, nodes may open and close connections in arbitrary order.

In the above examples, the range of the signal corresponds to nonnegative integers. A turnstile model whose signal has a nonnegative range is said to be a strict turnstile model. For an example of a non-strict turnstile model, consider tracking the difference between the number of connections originating from two different IP addresses. Note that some sketch-based algorithms that work in the strict turnstile model do not apply in the non-strict version [8].

As in the cash register model, one may define a contiguous unordered (strict or non-strict) turnstile model; the ordered (strict or non-strict) turnstile model is always contiguous.

For an example of an aggregate model, suppose that stream S_4 is a preprocessed version of S, where each stream item denotes the total number of bytes exchanged between a given source-destination pair over a 5-min *window*. Define signal A_4 as a function from the source and destination address pairs to the total number of bytes exchanged by each pair in the window. This gives rise to an ordered aggregate model if stream items are ordered by the source and destination addresses and an unordered aggregate model otherwise. Note that S_4 may contain a concatenation of many instances of A_4, each corresponding to range values calculated over a particular 5-min *window*.

Alternatively, one can model S_4 as carrying a single signal over time, call it A_5, whose domain is the same as that of A_4, but whose range is total number of bytes exchanged by each pair in the most recent 5-min interval. This corresponds to the (unordered and noncontiguous) reset model because new items arriving on the stream (i.e., those corresponding to the most recent 5-min window) replace old items having the same domain value.

Key Applications

The unordered cash register model is appropriate for applications where the incoming stream contains multiplexed data feeds from many sources (e.g., network traffic). However, the turnstile model must be used if the input (which may be a preprocessed version of a stream originally conforming to the cash register model) includes positive and negative range values. In particular, the turnstile model can represent a signal whose range values are computed over sliding *windows*. To see this, note that values that expire from the window may be modeled as new stream items with negative range values. Stream items whose purpose is to invalidate previously arrived items are often referred to as negative tuples [1, 2, 5].

On the other hand, the aggregate model is appropriate for many types of time series data, e.g., aggregated network traffic data generated every 5 min. In some cases, the reset model may also be suitable. Moreover, the reset model is useful in applications that process locations of moving objects over time. In these applications,

moving objects (e.g., a fleet of delivery trucks) periodically report their current positions [7].

Future Directions

The four stream models described in this article can express a wide range of application scenarios. However, new *streaming applications*, such as *publish-subscribe over streams*, *XML-stream processing*, signal-oriented stream processing [4], and analysis of the results of large-scale scientific experiments, may require new models.

Cross-References

▶ Data Stream
▶ Stream Mining
▶ Stream Processing

Recommended Reading

1. Arasu A, Babu S, Widom J. The CQL continuous query language: semantic foundations and query execution. VLDB J. 2006;15(2):121–42.
2. Ghanem T, Hammad M, Mokbel M, Aref W, Elmagarmid A. Incremental evaluation of sliding-window queries over data streams. IEEE Trans Knowl Data Eng. 2007;19(1):57–72.
3. Gilbert A, Kotidis Y, Muthukrishnan S, Strauss M. Surfing wavelets on streams: one-pass summaries for approximate aggregate queries. In: Proceedings of the 27th International Conference on Very Large Data Bases; 2001. p. 79–88.
4. Girod L, Mei Y, Newton R, Rost S, Thiagarajan A, Balakrishnan H, Madden S. The case for a signal-oriented data stream management system. In: Proceedings of the 3rd Biennial Conference on Innovative Data Systems Research; 2007. p. 397–406.
5. Golab L, Özsu MT. Update-pattern aware modeling and processing of continuous queries. In: Proceedings of the ACM SIGMOD International Conference on Management of Data; 2005. p. 658–69.
6. Henzinger M, Raghavan P, Rajagopalan S. Computing on data streams. DIMACS Ser Discret Math Theor Comput Sci. 1999;50:107–18.
7. Hoffmann M, Muthukrishnan S, Raman R. Streaming algorithms for data in motion. ESCAPE. Berlin: Springer; 2007. p. 294–304.

8. Muthukrishnan S. Data streams: algorithms and applications. Found Trends Theor Comput Sci. 2005;1(2):1–67.
9. Paxson V, Floyd S. Wide-area traffic: the failure of poison modeling. IEEE/ACM Trans Netw. 1995;3(3):226–44.

Stream Processing

Michael Stonebraker[1] and Ugur Cetintemel[2]
[1]Massachusetts Institute of Technology, Cambridge, MA, USA
[2]Department of Computer Science, Brown University, Providence, RI, USA

Synonyms

Complex event processing (CEP); Data stream processing; Event stream processing (ESP)

Definition

Stream processing refers to a class of software systems that deals with processing streams of high-volume messages with very low latency. It is distinguished from business activity monitoring (BAM) or business process monitoring (BPM), in that the client of a stream processing application is often a program, rather than a human. Hence, the volume and latency requirements are often much more stringent.

Currently, stream processing is widely used in computing real-time analytics in e-trading, maintaining the state of massively multiplayer Internet games, real-time risk analysis, network monitoring, and national security applications. In the future, the declining cost of sensor technology will create new markets for this technology including congestion-based tolling on freeways and prevention of lost children at amusement parks.

Key Points

There are three main technical approaches to stream processing at the present time:

- *Custom code.* Traditionally, stream processing applications have been hand coded in a low-level programming language such as C or C++. The current trend is toward using one of the other two technologies to achieve lower development and maintenance cost.
- *Stream-oriented SQL.* Recent research activity has extended SQL with primitives for real-time operation. The main additions are the notion of real-time windows, over which SQL aggregates can be computed, facilities to perform pattern matching on sequences of messages, and primitives to deal with out-of-order data. There are now high-performance stream-oriented SQL engines from several vendors (e.g., Coral8, http://www.coral8.com and StreamBase, http://www.streambase.com).
- *Rule engines.* The final approach is to utilize a high-performance implementation of a rule engine for stream processing. These systems are descendents of the rule engines found in expert systems in the 1980s and originally specified by the artificial intelligence community in pioneering work in the 1970s. Such systems contain rich pattern matching capabilities, but must be extended with aggregation and windowing constructs. Currently, there are a variety of commercial rule engines addressing the stream processing market.

In cases where real-time processing must be combined with access to historical data, Stream-oriented SQL enjoys a natural advantage. Both real-time and historical analysis can be done in a single paradigm (SQL), whereas a rule engine must switch paradigms to access historical data. On the other hand, where very sophisticated pattern matching is the main requirement, rule engines enjoy an advantage, due to their richer pattern matching capabilities.

In either case, achieving high performance and low latency requires a collection of implementation optimizations. Extensive compilation, often to machine code, will lower message processing overhead. In addition, some systems go to great lengths to remove scheduling overhead, by precomputing what operation must be performed

S

next, and then directly calling that operation from the current one, thereby removing both scheduling and message queuing overhead. In addition, implementations based on storing real-time state in a DBMS (either a conventional disk-based one or a main memory DBMS) are not likely to be successful, because of the inherent overhead of these class of products.

Over the course of the next decade, it is expected that stream processing products will enter the mainstream, thereby complementing other system software components such as application servers and DBMSs. It is also expected that there will be a tighter integration of stream processing technologies and more traditional stored data processing technologies through systems that can deal with both models.

Cross-References

▶ Data Stream
▶ Data Stream Management Architectures and Prototypes
▶ Streaming Applications
▶ Stream-Oriented Query Languages and Operators

Recommended Reading

1. Stonebraker M, Çetintemel U, Zdonik S. The 8 requirements of real-time stream processing. ACM SIGMOD Rec. 2005;34(4):420–47.

Stream Processing on Modern Hardware

Buğra Gedik
Department of Computer Engineering, Bilkent University, Ankara, Turkey
IBM T.J. Watson Research Center, Hawthorne, NY, USA

Definition

Stream processing is a computational paradigm for on-the-fly analysis of live data at scale.

Given the ever-increasing number of online data sources, rate of streaming data, and growing demand for timely analysis, stream processing has gained an important place in today's data-driven solution architectures. A key feature of the stream processing paradigm is its amenability to parallel execution, which in turn makes stream processing an attractive domain for taking advantage of modern hardware.

To cope with the increased power consumption associated with frequency scaling, hardware manufacturers have moved toward processors and coprocessors that contain multiple cores. Such designs can provide increased computational capacity without having to work at high frequencies and thus do not suffer from high power consumption. Yet, taking advantage of such hardware requires additional effort on the software side.

In the context of stream processing systems, three major kinds of modern hardware are considered. These are general purpose multi-core processors, graphical processing units (GPUs), and field programmable gate arrays (FPGAs).

Historical Background

Stream processing systems can be divided into two broad categories, namely, *synchronous* and *asynchronous*. Synchronous streaming systems are referred to as synchronous data flow (SDF) systems [8] and can be found more often in the embedded computing domain. Asynchronous systems are referred to as *data stream processing* systems and are more prevalent in data-intensive applications [1]. This distinction is important from the perspective of taking advantage of modern hardware, as these systems are quite different in the way streaming applications are developed as well as how their supporting compile and runtime systems are designed.

SDF systems assume that the selectivity of each operator is statically declared at application development time. For instance, an operator in the SDF model will declare that it will always consume x input tuples and produce y output tuples, typically on a per port basis, where x and y are integers. The term always here means that every single invocation of the operator will behave the same way.

However, in the asynchronous model, the selectivity could depend on the contents of the data streams. For instance, an operator may produce one output tuple for each input tuple it consumes for some of the inputs and no output tuples for some others, resulting in a selectivity value between 0 and 1 (e.g., 0.5), which will be determined at runtime. In the asynchronous model, the selectivity values are not specified at development time and can change dynamically at runtime.

Another important differentiator is that the operators in the SDF model are either arbitrarily stateful, stateless, or maintain limited forms of state, such as a read-only sliding window. In contrast, most asynchronous stream processing systems support an important class of operators called *partitioned stateful* operators. Such operators keep independent state associated with each sub-stream identified by the value of a partitioning attribute.

SDF-based systems come up with an execution schedule at compile time, based on the static selectivity values of the operators. They do not need to handle rate differences, as the static schedule guarantees that no more space than the statically allocated buffers are required at runtime. Asynchronous streaming systems work by keeping buffers in between operators executed by different threads. When the buffers hit their size limit, the upstream operators are throttled down in order to handle the *backpressure*. The two models are completely different in their runtime designs, which is a direct consequence of their different programming models.

Three major kinds of parallelism are inherently present in streaming applications. An effective hardware mapping should take advantage of all these opportunities for parallelism.

- *Pipeline parallelism*: As one operator is processing a tuple, its upstream operator can process the next tuple in line, at the same time.
- *Task parallelism*: A simple fan-out in the flow graph gives way to task parallelism, where two different operators can process copies of a tuple, at the same time.

- *Data parallelism*: This type of parallelism can be taken advantage of by creating replicas of an operator and distributing the incoming tuples among them, so that their processing can be parallelized. This requires a split operation but, more importantly, a merge operation after the processing, in order to re-establish the original tuple order (if and when needed).

While pipeline and task parallelism can be applied to any operator irrespective of whether they keep state or not, data parallelism is applicable to *stateless* or *partitioned stateful* operators. Stateless operators are those that do not maintain state across tuples. Partitioned operators maintain state across tuples, but the state is partitioned based on the value of a *key* attribute. In order to take advantage of data parallelism, the streaming runtime has to modify the flow graph behind the scenes to create multiple copies of an operator. While it requires additional runtime machinery, data parallelism is not limited by the pipeline depth or task parallel width but instead only by the number of partitions in the source data.

Scientific Fundamentals

Streaming Systems on Multi-core CPUs

The parallelism inherent in streaming applications can be harnessed via the use of hardware that contain multiple processing units. The main technical challenge in this is to create a mapping from the streaming application to the parallel hardware and making this transparent to the application. This mapping has to be *safe* in the sense that the semantics of the streaming application is not changed as a result of parallel execution on the multi-core hardware. Beyond safety, the mapping needs to be *profitable*, in the sense that it should improve the performance of the streaming application. For instance, the streaming application should be able to handle an increased volume of input data as a result of the mapping to the multi-core hardware. Last, but not the least, when the characteristics of the workload processed by an application is subject to change at runtime, being able to adjust the hardware

S

resources used becomes an important capability. Streaming systems that are able to provide such dynamic mapping are called *elastic*.

Two fundamental techniques are applied for mapping a streaming application to multi-core hardware, namely, *fusion* and *fission*.

Fusion, also called superbox scheduling, aims to adjust the trade-off between the cost of scheduling and communication and the speedup achieved from pipeline and task parallelism. Operators in a subgraph that are executed by the same thread without having to go through scheduling, serialization, buffering, etc., are considered *fused*. Such operators are executed as if they are an atomic unit. This reduces the overheads associated with scheduling and communication. Yet, we also need the reverse of fusion, that is, execution of operators via different threads to take advantage of pipeline and task parallelism. As a result, finding the right fusion configuration is key in achieving good profitability. The fission optimization can be safely applied independent of the nature of the operators with respect to statefulness. Fusion decisions can be made statically at compile time (typical in SDF systems) [4] or elastically at runtime [13].

Fission, also called data parallelism, resolves a bottleneck by replicating an operator or a sequence of operators. These replicas are then assigned to different threads on a multi-core system. Fission can be applied on operators that are stateless or partitioned stateful. For stateless operators, the distribution of tuples to different replicas of the operators can be done in a flexible way, yet for partitioned stateful operators, the distribution has to make sure that sub-streams identified by partitioning keys are always sent to the same replica. The latter is often achieved via some form of hashing over the partitioning keys. For applications that require strict ordering – i.e., being able to maintain the original order of tuples after a data parallel region – a merge step is needed after data parallel execution to reorder tuples. Fission cannot be applied to arbitrarily stateful operators. Fission can be performed elastically as well [3]. Elastic fission in the presence

of partitioned stateful operators requires migrating state associated with sub-streams that are assigned to different replicas as a result of elastic scaling. Hashing techniques such as *consistent hashing* [6] can be used to minimize the amount of migration.

Streaming Systems on GPUs

The high level of hardware parallelism present in modern GPUs, coupled with the availability of flexible programming APIs used to program them, such as CUDA and OpenGL, has resulted in the popularization of general purpose programming on GPUs.

The typical approach to take advantage of GPUs for general purpose programming is to interact with them as coprocessors. The host process running on the CPU calls *kernel* functions on the GPU to take advantage of the hardware parallelism offered by the GPU. The GPU has its own memory, thus calling a kernel function on the GPU requires transferring data from the CPU memory to the GPU memory, typically via the PCI Express bus. Such transfers happen before calling the kernel function (to move data in) and after the kernel function returns (to move data out). The kernel functions run on the GPU using the single-program, multiple-data (SPMD) model. At a high level, a GPU contains a number of multiprocessors having access to a global device memory, where each multiprocessor contains many cores having access to a per-multiprocessor shared memory. Since GPUs adopt the SPMD model, a group of threads (called a *warp*) running on a GPU multiprocessor operate in lockstep. As an important implication, branching is a costly operation for GPU kernels.

The ample parallelism offered by GPUs has been successfully exploited for some of the most common data processing tasks, such as sorting and relational joins. Yet, taking advantage of GPUs in the context of stream processing is more challenging, especially with respect to profitability. One of the key characteristics of stream processing is low latency, yet the data transfers required to run GPU kernel functions require operating in batches, in order to mask the constant overheads associated with the data transfers across the CPU and the GPU memory. As such, to better take advantage of the GPU for stream processing, it is necessary to operate in mini batches. Furthermore, streaming manipulations that are more costly in terms of the computational work they do on a per data-item basis are more amenable to acceleration via the use of GPUs. Work on accelerating data streaming processing applications on GPUs has mostly focused on particular streaming operations, such as joins [7], aggregations [12], and event matching [2] to name a few.

Even though GPUs are highly data parallel, using them for running end-to-end data stream processing applications is not an approach that has been explored in depth. This is due to the highly data-dependent nature of general purpose streaming applications, which commonly perform data-dependent filtering, resulting in highly branchy code when parallelized. However, GPUs form a more suitable platform for executing SDF-based streaming applications that have a more strict programming model [14, 15].

Streaming Systems on FPGA

The ability to design custom and highly parallel hardware solutions via reconfigurable hardware has made FPGAs an attractive platform for accelerating some of today's performance critical data processing tasks [9], such as XML filtering, event matching, and sorting.

Field-programmable gate arrays (FPGAs) are semiconductor devices that are programmable. They contain programmable lookup tables used to implement truth tables for small logic circuits. They also contain memory in the form of flip-flops and small Block RAMs. A large number of these programmable blocks are connected with a programmable interconnect to implement large-scale circuits.

The typical way to program FPGAs is via hardware description languages (HALs), such as VHDL and Verilog. A more flexible, yet less performance effective, approach is to use *soft processors* that can be targeted by general purpose programming languages such as C/C++. A soft processor is an FPGA-implemented processor (typically via HALs) that has its own instruction set.

The inherent parallelism found in streaming applications and the programmability provided by FPGAs for creating custom parallel hardware solutions make direct mapping of streaming applications to hardware an attractive approach. Such FPGA-based streaming applications can either fetch their data from a CPU-based host system's main memory over the PCI Express bus or can directly consume the data from an FPGA-integrated network interface card (NIC).

Research work on mapping streaming applications to FPGAs exists both for the SDF-based streaming systems and for the asynchronous data stream processing systems. Due to the more restricted form of the SDF-based streaming applications, a complete FPGA-based hardware mapping is often possible [5]. The FPGA mappings for the asynchronous data stream processing systems often take the form of compiling SQL-based streaming queries with certain limitations (such as select-project (SP) or select-project-join (SPJ) queries with windowing) down to hardware [10, 11]. This is achieved via synthesizing FPGA circuits for individual relational streaming operators (selection, projection, windowed aggregation, grouping, join, etc.) as well as for synchronization and flow orchestration tasks.

Key Applications

High-Performance Real-Time Data Processing

Modern hardware is most commonly used for stream processing in the context of high-performance real-time data processing applications. Such applications often have very stringent latency and throughout requirements. One example is algorithmic trading in financial markets where low latency is critical. Another example is wire-speed network monitoring in cybersecurity, where high throughput is critical.

Cross-References

▸ Stream Processing

Recommended Reading

1. Carney D, Çetintemel U, Cherniack M, Convey C, Lee S, Seidman G, Stonebraker M, Tatbul N, Zdonik BS. Monitoring streams – a new class of data management applications. In: Proceedings of the Very Large Data Bases Conference; 2002. p. 215–26.
2. Cugola G, Margara A. Low latency complex event processing on parallel hardware. J Parallel Distrib Comput(JPDC). 2012;72(2):205 18.
3. Gedik B, Schneider S, Hirzel M, Wu K-L. Elastic scaling for data stream processing. IEEE Trans Parallel Distrib Syst (TPDS). 2014;25(6):1447–63.
4. Gordon MI, Thies W, Amarasinghe S. Exploiting coarse-grained task, data, and pipeline parallelism in stream programs. In: Proceedings of the 12th International Conference on Architectural Support for Programming Languages and Operating Systems; 2006. p. 151–62.
5. Hormati A, Kudlur M, Mahlke SA, Bacon DF, Rabbah MR. Optimus: efficient realization of streaming applications on FPGAs. In: Proceedings of the International Conference on Compilers, Architecture, and Synthesis for Embedded Systems; 2008. p. 41–50.
6. Karger D, Sherman A, Berkheimer A, Bogstad B, Dhanidina R. Web caching with consistent hashing. Comput Netw. 1999;31(11):1203–13.
7. Karnagel T, Habich D, Schlegel B, Lehner W, The HELLS-join: a heterogeneous stream join for extremely large windows. In: Proceedings of the 9th Workshop on Data Management on New Hardware; 2013.
8. Lee EA, Messerschmitt GD. Synchronous data flow. Proc IEEE. 1987;75(9):1235–45.
9. Müller R, Teubner J, Alonso G. Data processing on FPGAs. Proc VLDB Endow. 2009;2(1):910–21.
10. Müller R, Teubner J, Alonso G, Streams on wires – a query compiler for FPGAs. Proc VLDB Endow. 2009;21(1):229–40.
11. Sadoghi M, Javed R, Tarafdar N, Singh H, Palaniappan R, Jacobsen H-A. Multi-query stream processing on FPGAs. In: Proceedings of the 28th International Conference on Data Engineering; 2012. p. 1229–32.
12. Schneider S, Andrade H, Gedik B, Wu K-L, Nikolopoulos D. Evaluation of streaming aggregation on parallel hardware architectures. In: Proceedings of the 4th ACM International Conference on Distributed Event-based Systems; 2010. p. 248–57.
13. Tang Y, Gedik B. Auto-pipelining for data stream processing. IEEE Trans Parallel Distrib Syst (TPDS). 2013; 24(12):2344–54.
14. Udupa A, Govindarajan R, Thazhuthaveetil JM. Software pipelined execution of stream programs on GPUs. In: Proceedings of the IEEE/ACM International Symposium on Code Generation and Optimization; 2009. p. 200–9.
15. Zhang Y, Mueller F. GStream: a general-purpose data streaming framework on GPU clusters. In: Proceedings of the 2011 International Conference on Parallel Processing; 2011. p. 245–54.

Stream Reasoning

Alessandra Mileo[1], Minh Dao-Tran[2], Thomas Eiter[2], and Michael Fink[2]
[1]Insight Centre for Data Analytics, Dublin City University, Dublin, Ireland
[2]Institute of Information Systems, Vienna University of Technology, Vienna, Austria

Synonyms

Continuous reasoning; Reactive reasoning

Definition

Stream reasoning refers to inference approaches and deduction mechanisms which are concerned with providing continuous inference capabilities over dynamic data. The paradigm shift from current batch-like approaches toward timely and scalable stream reasoning leverages the natural temporal order in data streams and applies windows-based processing to complex deduction tasks that go beyond continuous query processing such as those involving preferential reasoning, constraint optimization, planning, uncertainty, non-monotonicity, non-determinism, and solution enumeration.

Historical Background

We are witnessing an unprecedented shift in the available quantity and quality of data drawn from all aspects of our lives, opening tremendous new opportunities but also significant challenges for scalable decision analytics due to its dynamicity. This makes it harder to go from data to *insight* and support effective decision-making. Such transformation of these enormous quantities of information into actionable knowledge requires new reasoning methods which are able to remove the common assumption in scalable reasoning that knowledge bases are static or evolving slowly.

Stream reasoning [1] emerged in the last few years as a new research area that aims at bridging the gap between reasoning and stream processing. Different communities have focused on complementary aspects of processing dynamic information, referred to as *stream processing* when closer to the data and as *reasoning* when closer to knowledge and event management.

In data processing, the stream processing approach materializes in the ability to answer continuous queries over low-level, high-rate input data such as those produced by (virtual or physical) sensors or social networks. Data Stream Management Systems (DSMSs) [2] considered requirements of applications where input data are unbounded and arrive in a streaming manner. Continuous queries are registered for execution when new input arrives, and various window operators are defined to target a deterministic query answer [3]. Corresponding solutions referred to as *Semantic Stream Processing* have been investigated for the Web of Data and RDF stream processing. Languages and solutions for processing Linked Stream Data (LSD) [4] are mostly based on extensions of the SPARQL query language with time stamps and operators for window-based query processing. The most widely used of such approaches are CQELS [5] and C-SPARQL [6]. In continuous query processing, incomplete information is rarely considered and not available in scalable implementations. Negation is also an important yet "expensive" construct and only rudimentary, simple forms of negation have been considered, which limits the expressive power of stream query processing approaches.

Complex event processing (CEP) emerged from publish-subscribe systems with the additional ability to express high-level events as complex patterns of incoming single events. Dealing with uncertainty is important in CEP, as information sources (usually sensors) are unreliable in general: data can be noisy or lost in communication, values can be rounded, etc. However, few works deal with this and usually use probabilistic methods [7]. A survey on CEP approaches and tools can be found in [8].

In knowledge representation and reasoning, recent works deal with streaming input, but lack scalability for high data rate. Nomonotonic reasoning formalisms serve advanced reasoning from a knowledge base that deals with incomplete and/or inconsistent information, and reasoning under changing bases such as belief revision, reasoning about actions, logic programs update, agent systems, etc. have been widely considered; however, streamed data arriving at high rate was not an issue. Recently, this has changed and initial works that study streaming have triggered investigations in this area, including ontology streams, streaming answer set programming (ASP), and incremental and reactive ASP and DSMSs with datalog [9].

The performance of these approaches is closer to program updates than to high-speed stream reasoning. This has boosted interest in *stream reasoning* as the issue of providing models, theories, and systems to bridge the gap between continuous query processing, CEP, and reasoning, paying particular attention to the aspect of scalability.

Scientific Fundamentals

In stream reasoning, input data dynamically arrive at the processing systems in form of possibly infinite streams. To deal with the unboundedness of data, such systems typically apply *window operators* to obtain snapshots of recent data. Common windows operators (also used in window-based query processing and CEP) are:

- Time-based windows: contain input data arriving within a fixed amount of time
- Tuple-based windows: contain the latest fixed number of input data

– Partition-based windows: split the input data based on different attributes and apply tuple-based windows on each substream

Based on the finite data after the application of windows, the user specifies different reasoning tasks, typically continuous queries, to get output also in terms of streams. The key aspects of stream reasoning can be expressed as a set of requirements and related methods and approaches required for stream reasoning to fulfill its promise, as outlined in the remainder of this section.

Expressivity

Stream processing systems mainly adopt operational and monotonic semantics at the logical core. The latter is less suited to produce results when data is missing and in particular not geared to deal with incorrect conclusions that must be retracted when more data is available. Such non-monotonic behavior is a key aspect in expressivity that is needed for stream reasoning.

Advanced reasoning techniques offer declarative means and model-based semantics to deal with incomplete information by (i) supplying different methods to complete data in a qualitative way and to provide different outcomes to the users, in terms of default or alternative answers and (ii) allowing to retract incorrect conclusions that were based on wrong assumptions.

Using designated, expressive language constructs, a developer can easier formulate, understand, and change the behavior of the reasoning component. Such features are extremely beneficial for reasoning on streaming data in a distributed setting, where the data can continuously change and get missing or is imprecise due to inaccurate computational models of the real world; default or rational conclusions are more useful than silence due to incomplete information.

Formal approaches of expressive reasoning to handle streaming data should comply with the following requirements:

– Ability to handle time-changing data through models that cater for abstractions

of temporally annotated data and efficient implementations of such abstractions
– Mechanisms to represent and reason about uncertainty associated to both data (missing information, noisy data) and reasoning processes (inconsistencies in deduction process, uncertain observations, propagation of uncertainty), leveraging declarative approaches to problem solving and combining quantitative and qualitative decision processes when needed
– Conflict-handling techniques for both hard and soft constraints (also referred to as preferences)
– Non-monotonic behavior in order to retract conclusions that are no longer valid

Distribution and Heterogeneity

Challenges for the integration of distributed and heterogeneous data include issues such as data provenance and trust, as well as management and combination of multiple data formats in a single formalism. The need for integration of information that comes from multiple sources motivates the usage of data models like RDF that allow semantic integration, but this is not sufficient.

Handling streaming data requires the coordination of various processing/reasoning powers through physically distributed and arbitrarily connected reasoning nodes. Each node should be capable to offer a particular form of reasoning, including, but not limited to database querying, ontological reasoning, event pattern matching, continuous queries to stream processing systems, and model building. Nodes should communicate with each other (with cycles allowed) in push- or pull-based manner, depending on the underlying supported processing/reasoning modes. Users place continuous queries at nodes providing this feature and get live results via output streams. This offers various benefits to the users, as the ability to combine potentially different reasoning mechanisms to perform sophisticated tasks that a single stream processing formalism cannot host.

For example, consider the following query posed by a tourist exploring a new city (which is similar to scenarios that stream processing approaches look at): "recommend a close DVD shop offering a movie as a birthday gift for a boy aged 12, at a reasonable price." To answer this query, realistic applications must access distributed information sources in different formalisms, including the city map, the traffic status, the user's GPS position, an ontology of the objects on the street, the websites of DVD shops with available offers, and a movie ontology. Moreover, different reasoning capabilities are needed: ontological reasoning to identify the shop from the map and to find suitable movies; reasoning about routes to find a close shop, respecting also accessibility via bike while avoiding congestion; and especially, reasoning about movie quality from price and information in the movie ontology, e.g., director, actors, etc.

At a global level, distribution requires to pay attention to synchronization and distributed processing by considering the issues of communication between nodes, global semantics of the system, and coordination of nodes to compute such a semantics. Some inspiration comes from multi-context systems [10], which is an abstract framework to interlink possibly heterogeneous, non-monotonic knowledge bases via bridge rules for belief exchange (possibly in cycles). Their semantics is given by equilibria, which are stably interlinked local models. Distributed algorithms, optimizations techniques, and a prototype system DMCS (http://www.kr.tuwien.ac.at/research/systems/dmcs/) to compute equilibria top down are available. As for data streaming, a timing concept can be attached to the belief exchange, with special management for continuous model building or query answering. The DMCS algorithms can serve as a starting point for bottom-up algorithms in a distributed streaming environment.

Scalability

There are several factors which challenge the performance of a stream processing/reasoning system. At the local level we have that (1) the high input stream rate that exceed even high computation rate; (2) the input is dynamic, that is, selectivity and distribution of incoming data change from time to time; (3) when a window slides on a stream, consecutive window snapshots may overlap; and (4) the reasoning tasks on top of the input data can have high complexity (see Expressivity above).

Current solutions in stream processing concentrate on addressing (1)–(3) on top of lightweight tasks (identified by, for example, the number of joins in queries). Traditional stream processing has efficient methods to deal with these issues. For example, adaptive evaluation adjusts computation strategies to change in dynamic input, incremental evaluation [11] avoids wasteful recomputation, and load shedding [12] reduces input while tolerating some quality of service.

Pushing for (4) faces the obstacle that solving complex reasoning tasks may take considerable time. Two observations, however, shows that advanced reasoning is still feasible:

- Often expressive queries and complex reasoning tasks need only a high-level, abstract view of data instead of low-level input (e.g., GPS positions to regions, temperature to values like warm, cold, etc.,).
- At the abstract level, changes often occur much less frequently than at the low level. A high-level reasoning algorithm has thus much more time available than a low-level algorithm.

Abstraction techniques like data filtering or summarizing are available and can be readily used.

At the global level in a distributed setting, the main challenge is that global information such as the whole routing strategy or communication branches are invisible to nodes in the system. Each node merely knows parent and child nodes, and none has a clear picture of the system status to decide about the routing strategy or to drop input/intermediate results at bottlenecks to improve global performance. To overcome this, the nodes need sophisticated methods to communicate enough information such that they can realize the need to adjust the global strategy and inform other nodes. Furthermore, a number of parameters needs a fine-grained investigation, namely, (i) the system size, i.e., the number of distributed nodes, (ii) the density and topology of the communication network, and (iii) the size of the knowledge base at each local node. The key aspect is to understand how these parameters scale while performance is still met; what is the limitation in increasing one parameter while fixing the others.

At the local and the global level, several reasoning tasks such as inconsistency checking, query answering, etc. and restrictions such as approximate semantics, syntactic restrictions, etc., are of interest. Investigating their computational complexity for various query languages will give hints on the theoretical scalability border. For example, with datalog, depending on the version, the complexity ranges from AC_0 to Σ_2^P under well-founded and answer set semantics [13].

Benchmarking and Comparison

The main stream processing/reasoning approaches have been independently proposed and few works compare them more extensively.

Linear Road [14] is currently the only complete benchmarking system for DSMSs, simulating an ad hoc toll system for motor vehicle expressways. It uses L-Rating as a single metric to measure performance, i.e., the number of expressways a system can process while meeting time and correctness constraints. Other interesting aspects for comparison are still open, e.g., expressiveness, manageable query patterns, scalability under varying static data size, number of simultaneous queries, etc.

A preliminary work [15] proposed methods to cross-compare LSD-processing engines on aspects such as functionality, correctness, and performance. They capture more basic evaluation tasks such as projection, filter, join, etc., rather than reasoning aspects, but can serve as a starting point for a more general comparison.

The lack of extensive comparison is due to several obstacles. On the practical side, comparing stream engines is nontrivial for several reasons: (i) they are based on different semantics which a benchmarking system must respect; (ii) there is pushed- vs. pull-based evaluation (also called data- vs. time-driven, eager vs. periodical evaluation): the former triggers computation when new input arrives, while the latter runs it periodically independent of input arrival; thus output may be missed (resp., repeated) when the input rate is too high (resp., slow) compared to the processing rate; (iii) the engines are fragile as any runtime deviation (e.g., processing delay) can lead to inaccurate output, especially for queries with aggregates; and (iv) some engines are black boxes.

On the theoretical side, there is no formal foundations on which existing stream processing/reasoning approaches can be captured and thus compared systematically. On both the theoretical and the practical sides, the main issues are sufficiently generic measurements and metrics, along with methods to realize them for a fair comparison.

Key Applications

With increasing numbers of data streams becoming available, the amount of observable events to be captured and processed increases dramatically, resulting in new opportunities and challenges in sensor-rich and knowledge-intensive environments. Some of the key application areas are listed in what follows.

Web of Things

The Web of Things is extremely dynamic: information is frequently changed and updated, and new data is continuously generated from a huge number of sources, often at high rate. Physical objects are made accessible as virtual sources of data streams, but the ability to select and combine them is crucial for exploiting their potential. Stream reasoning techniques that can cater for optimization and preferences help selecting and combining the best services and event sources that best contribute to a specific decision task.

Smart Cities

A smart city faces a huge volume of data that is coming from sensors monitoring the physical world in real time as well as from citizens through social streams. Processing such data streams in real time using prior knowledge of the domain (e.g., traffic, public safety) and the city (e.g., road topology) calls for stream reasoning. Traffic, public safety, healthcare, water, power grid, etc. can greatly benefit from stream reasoning providing real-time insights into collective interactions within and between each of these systems.

Social Media Analysis

The large data volume produced by users in social networks and mobile applications is part of what is called *social sensing*. The latter contains information about user contexts, including the needs, interests, relations, and activities of the users. Social media analysis aims at capturing this information and provide a characterization of the user context for context-aware applications. Social media contexts are highly dynamic but also characterized by noise and possible inconsistencies, and context properties are linked by complex relations. Stream reasoning techniques can provide mechanisms to analyze such changes, relate them with historical data, and use them to discover interesting patterns.

Location-Aware Services and Mobile Applications

Location is an important aspect of contextual knowledge that highly characterize mobile applications. Utilizing physical and virtual sensors for reasoning about the location of a user is often perturbed by noise, by imprecise or missing sensor readings as well as inconsistent information coming from different sources. Stream reasoning can help in the combination of such sources with background knowledge and provide continuous location estimation based on semantic sensor fusion.

Smart Building and Resource Optimization

When smart building is equipped with sensors for water and energy consumption, usage, and occupancy, stream reasoning techniques dealing with constraint optimization and preferential reasoning can substantially reduce cost associated with improper use of natural resources.

Emergency Management

Stream reasoning applied to dynamic planning can help in providing real-time solutions to complex problems such as actuation of lights and door locks to guidance of people in evacuation plans. This can be done by reasoning about dynamic data describing the status of an emergency and its evolution in real time (including the state of the environment and people location), combined with static knowledge about the floor plan.

Future Directions

Stream reasoning is a very young topic of research for which the requirements listed in this entry has only partially been addressed.

The lack of a unified formal foundation for advanced reasoning with streaming data hinders the potential for expressive formalisms to be used in concrete frameworks, and investigation on multiprocessing models to coordinate various reasoning posers in an advanced framework needs to be further investigated.

In terms of benchmarking, a direct cross comparison is only possible for few engines.

A formal foundation for stream reasoning does not exist yet and is a mean for methods beyond those in [15] and a key to give a reference from which existing approaches can be captured as different ways of approximation, aiding their theoretical comparison. Based on the theoretical results, properties to assess stream reasoning solutions need to be proposed and measured for practical comparison.

Efficient approaches to stream reasoning should also explore the interplay between statistical analysis and knowledge-driven inference methods to have both a quantitative perspective on streaming data patterns and a qualitative perspective on complex structural properties of events, context of validity, and logical correlations in decision processes.

Cross-References

▶ Complex Event Processing
▶ Window-Based Query Processing

Recommended Reading

1. Valle ED, Ceri S, van Harmelen F, Fensel D. It's a streaming world! Reasoning upon rapidly changing information. IEEE Intell Syst. 2009;24(6):83–9.
2. Babcock B, Babu S, Datar M, Motwani R, Widom J. Models and issues in data stream systems. In: Proceedings of the 21st ACM SIGMOD-SIGACT-SIGART Symposium on Principles of Database Systems; 2002. p. 1–16.
3. Arasu A, Babu S, Widom J. The CQL continuous query language: semantic foundations and query execution. VLDB J. 2006;15(2):121–42.
4. Phuoc DL, Nguyen-Mau HQ, Parreira JX, Hauswirth M. A middleware framework for scalable management of linked streams. J Web Sem. 2012;16(Nov):42–51.
5. Phuoc DL, Dao-Tran M, Parreira JX, Hauswirth M. A native and adaptive approach for unified processing of linked streams and linked data. In: Proceedings of the 8th International Semantic Web Conference; 2011. p. 370–88.
6. Barbieri DF, Braga D, Ceri S, Valle ED, Grossniklaus M. C-SPARQL: a continuous query language for RDF data streams. Int J Semantic Comput. 2010;4(1):3–25.
7. Wasserkrug S, Gal A, Etzion O, Turchin Y. Efficient processing of uncertain events in rule-based systems. IEEE Trans Knowl Data Eng. 2012;24(1):45–58.
8. Cugola G, Margara A. Processing flows of information: from data stream to complex event processing. ACM Comput Surv. 2012;44(3):15:1–15:62.
9. Valle ED, Schlobach S, Krötzsch M, Bozzon A, Ceri S, Horrocks I. Order matters! Harnessing a world of orderings for reasoning over massive data. Semantic Web. 2013;4(2):219–31.
10. Brewka G, Eiter T. Equilibria in heterogeneous nonmonotonic multi-context systems. In: Proceedings of the 22nd National Conference on Artificial Intelligence; 2007. p. 385–90.
11. Ghanem TM, Hammad MA, Mokbel MF, Aref WG, Elmagarmid AK. Incremental evaluation of sliding-window queries over data streams. IEEE Trans Knowl Data Eng. 2007;19(1):57–72.
12. Tatbul N, Zdonik S. Window-aware load shedding for aggregation queries over data streams. In: Proceedings of the 32nd International Conference on Very Large Data Bases; 2006. p. 799–810.
13. Dantsin E, Eiter T, Gottlob G, Voronkov A. Complexity and expressive power of logic programming. ACM Comput Surv. 2001;33(3):374–425.
14. Arasu A, Cherniack M, Galvez EF, Maier D, Maskey A, Ryvkina E, Stonebraker M, Tibbetts R. Linear road: a stream data management benchmark. In: Proceedings of the 30th International Conference on Very Large Data Bases; 2004. p. 480–91.
15. Phuoc DL, Dao-Tran M, Pham MD, Boncz PA, Eiter T, Fink M. Linked stream data processing engines: facts and figures. In: Proceedings of the 11th International Semantic Web Conference; 2012. p. 300–12.

Stream Sampling

Bibudh Lahiri and Srikanta Tirthapura
Iowa State University, Ames, IA, USA

Definition

Stream sampling is the process of collecting a representative sample of the elements of a data stream. The sample is usually much smaller than the entire stream, but can be designed to retain many important characteristics of the stream, and can be used to estimate many important aggregates on the stream. Unlike sampling from a stored data set, stream sampling must be performed online, when the data arrives. Any element that is not stored within the sample is lost forever, and cannot be retrieved. This article discusses various methods of sampling from a data stream and applications of these methods.

Historical Background

An early algorithm to maintain a random sample of a data stream is the *reservoir sampling* algorithm due to Vitter [15]. More recent random sampling based algorithms have been inspired by the work of Alon et al. [1]. Random sampling has for a long time been used to process data within stored databases - the reader is referred to [13] for a survey.

Foundations

A powerful sampling technique is *random sampling*, where random elements of the stream are selected into the sample. A random sample of a stream can be used in deriving *approximate* answers to aggregate queries such as *quantiles* [12] or *frequent elements* [11]. It can also be used in the estimation of the *selectivity of a query predicate*, which is defined as the fraction of the data items in the stream which satisfy the given user predicate. The intuition behind the above applications of random sampling is as follows. Suppose R is A uniform random sample of a data stream S. For any $A \subset S$, the size of A, $|A|$ can be estimated as $|A \cap R| \cdot |S|/|R|$. The accuracy of this estimate depends on the value of $|A \cap R|/|R|$. There are tradeoffs between the quality of the answer returned, the confidence in the answer, and the space taken by the sample.

There are two basic ways of generating a random sample of any data set - sampling without replacement and sampling with replacement. Consider a data stream with N elements and a sample size n. In random sampling with replacement, each element of the sample is chosen at random from among all N elements of the data set. It is possible that the same element is chosen more than once into the sample (though this is unlikely if the sample is much smaller than the data). A random sample without replacement is a randomly chosen subset of n elements from among all $\binom{N}{n}$ subsets of size n, thus ensuring that a data element appears no more than once in the random sample.

Reservoir Sampling

This technique, due to Vitter [15], allows the maintenance of a random sample of the stream of a particular target size in an online fashion. Suppose the objective is to maintain a random sample of n elements without replacement, from a stream of N elements, where N is not known a priori. Let the stream elements be $a_1, a_2, \ldots, a_N$.

A discussion of how to maintain a sample of a single element from the stream i.e., the case $n = 1$, can be a good point to begin. When a_1 arrives, it is always selected into the sample. For $i \geq 2$, element a_i is selected into the sample with probability $1/i$, i.e, a_2 is selected with a probability $1/2$, a_3 is selected with a probability $1/3$, and so on. Each time an element is selected into the sample, it replaces the existing element in the sample. It can be verified that the final element in the sample has an equal probability of being any of the N elements in the stream.

The above idea can be extended to maintain a random sample without replacement of size n as follows. The first n elements of the stream are (deterministically) included in the sample. For $t \geq n$, when a_{t+1} arrives, it is included in the sample with probability $n/(t+1)$. If an element is selected for inclusion in the sample, it replaces an element that is chosen uniformly at random from the currently existing elements in the sample. It is easy to verify that the resulting sample is equally likely to be any of the $n/(t+1)$ subsets of size n of the set $a_1, a_2, \ldots, a_{t+1}$. This algorithm is described in Fig. 1.

Note that if one wanted to sample n elements from a stream *with replacement*, this could be achieved by running n copies of the single element reservoir sampling algorithm. Further enhancements are possible to the algorithm in Fig. 1. In particular, instead of examining every element of the stream to see if it will be sampled, it is possible to directly generate the number of elements of the stream to be *skipped*

Input: Stream $a_1, a_2, \ldots, a_N$ where N is not known in advance; Sample Size n
Output: A sample $S[1, \ldots, n]$ of n elements chosen uniformly at random without replacement from the stream.

1. For i from 1 to n, $S[i] \leftarrow a_i$ // *The first n elements are included in the sample*

2. $t \leftarrow n$ // *t is the number of records processed so far*

3. While (there are more stream elements)

    (a) $t \leftarrow (t+1)$
    (b) Let s be a random number in the set $\{1, 2, 3, \ldots, t\}$

Stream Sampling, Fig. 1 Reservoir sampling for sampling without replacement from a stream

before the next element that will be included in the sample. This can significantly reduce the number of stream elements to be examined by the sampling algorithm. For further details, the reader is referred to [15].

Sample and Count

This is a technique pioneered by Alon et al. [1], and is based on random sampling followed by counting. This technique has been applied in the estimation of frequency dependent statistics on a data stream in very small space. To see its use in the context of estimating the frequency moments of a data stream, consider a stream $S = a_1, a_2, \ldots, a_N$, where each $a_i \in \{1, 2, \ldots, m\}$. For $1 \leq j \leq m$, let f_j denote the number of occurrences of j in stream S. For integral $k \geq 0$, the kth frequency moment of S, denoted by F_k is defined as follows.

$$ F_k = \sum_{j=1}^{m} f_j^k $$

For $k \neq 1$, computing F_k exactly on a large data stream S is provably expensive space-wise. There are lower bounds showing that such an exact computation of F_k, or even an accurate *deterministic* approximation of F_k requires $\Omega(m)$ space, in the worst case. However, a randomized approximation to F_k (for $k \geq 2$) can be found as follows. First, choose a random element a_p from S (this can be done without a knowledge of N using the reservoir sampling technique). Then maintain the count $X = |\{q : q \geq p, a_q = a_p\}|$. In other words, count the number of re-occurrences of the element a_p in the portion of the stream that succeeds a_p (including a_p). Then, the random variable $Y = N[X^k - (X-1)^k]$ is an unbiased estimator of F_k, i.e., $E[Y] = F_k$. Further, it can also be shown that the variance of Y is small. For user defined parameters $0 < \varepsilon, \delta < 1$, this can be used to generate an estimator of F_k that is within a relative error of ε with probability more than $1 - \delta$ using small space. For exact space bounds, proofs and details, the reader is referred to [1].

The sample and count technique has also been used in accurate estimation of another frequency

dependent aggregate, the *empirical entropy* of a data stream, in limited space. Consider a stream of integers, $S = a_1, a_2, \ldots, a_N$, where each $a_i \in \{1, 2, \ldots, m\}$. For $1 \leq j \leq m$, let f_j denote the number of occurrences of j in S. The empirical entropy of the stream is defined as $\sum_{j=1}^{m} -(f_j/N) \log (f_j/N)$. The entropy of a stream yields valuable information about the amount of "randomness" within the stream, and is useful in many contexts in network monitoring. Chakrabarti et al. [3] present an algorithm for estimating the entropy of a stream. Their algorithm uses the sample and count technique, and yields a provably accurate estimate of the entropy (a randomized approximate estimate) using nearly optimal space.

Distinct Sampling

In some cases, it may be necessary to compute aggregates over all the *distinct* elements in the stream. Consider a stream of tuples (i, v) where i is an item identifier and v is the value. In database query optimization, a question of interest is often just an estimate of the number of distinct values for an attribute in a relation. In network monitoring, an objective may be tracking all those sources that have contacted a large number of distinct destinations in the recent past. For computing such aggregates over all distinct identifiers, a uniform random sample will not be useful. For example, suppose a stream of 10^8 elements had 1,000 distinct identifiers, but every identifier appeared exactly once in the stream, except for the "dominant identifier", which made up the remaining $10^8 - 999$ elements of the stream. Even if a fairly large random sample of this stream is collected, say a sample of 10^4 elements, the sample is likely to contain only the dominant identifier, and any estimate from this sample is likely to be extremely inaccurate. To derive an useful sample for estimating aggregates over distinct elements in a stream, Gibbons and Tirthapura [9] introduced a technique called *distinct sampling*.

In distinct sampling, the sampling is performed with the help of a randomly chosen *hash function*, rather than through independent random choices for each element. The hash

---

Input: Stream $S = a_1, a_2, \ldots, a_N$ where $a_i \in \mathcal{I}$ and N is not known in advance; Maximum sample size n; Hash function $h : \mathcal{I} \to (0,1)$.
Output: D, a random sample of the distinct elements of the stream, chosen without replacement; The size of D is guaranteed not to exceed n.

1. Initialization: $D \leftarrow \phi$, $\ell \leftarrow 0$

2. While there are more elements in S

    (a) Let a be the next element in S
    (b) If $(h(a) < \frac{1}{2^\ell})$
        i. If $(a \notin D)$ then Add a to D
        ii. If $(|D| > n)$ then $||$ *Overflow, discard approximately half the items in D by selecting every item currently in D with a probability 1/2.*
            A. $\ell \leftarrow \ell + 1$
            B. For every $a \in D$, if $h(a) \geq \frac{1}{2^\ell}$ then discard a from D.

Stream Sampling, Fig. 2 Algorithm for maintaining a sample of all distinct elements of a stream

function h is chosen before the stream elements are observed. Given a target sample size n, the algorithm in Fig. 2 maintains a random sample of all distinct elements of a stream S of size approximately n. For example, suppose the stream of 10^8 elements had 7,500 distinct identifiers. If the target sample size was 10^4, then all distinct elements would be included in the sample. If the target sample size was 10^3, then each distinct element would be included in the sample with probability p, where p is reduced until the resulting sample fits within the target sample size.

Let the stream S be an integer sequence $a_1, a_2, \ldots, a_N$. The algorithm maintains a sample D of distinct elements, and a sampling level ℓ. At level ℓ, each distinct element is selected into the sample with probability $1/2^\ell$. For simplicity, assume a random hash function h is available such that $h(x)$ is a random real number which is uniformly chosen from the range $(0,1)$, and the outputs of the hash function on different inputs are mutually independent. The analysis of the algorithm using more practical hash functions with limited independence (and integral outputs) is presented in [9].

The above distinct sampling algorithm assumed that all elements in the stream have the same weight - in general, this may not be true. For example, a user might be interested in computing

the mean of the values received over all distinct identifiers in the stream - in this case, different elements should be weighted differently. The above algorithm was extended to the more general weighted case by Pavan and Tirthapura [14], who designed a "range-sampling" algorithm that allowed the weighted sampling problem to be reduced to the unweighted case.

Time-Decayed Sampling

The reservoir sampling (distinct sampling) algorithm maintains a sample of all elements (distinct elements) from the start of time, and such a sample is useful in computing aggregates over the entire data stream. However, in many cases the interest lies in computing *time-decayed* aggregates of data, where older elements must be discounted. For such aggregates, it is useful to have a sample where a more recent stream element has a higher probability of being included in the sample. For example, a typical *sliding-window* aggregate asks for an aggregate over all data elements that have appeared in a window of the last W elements of a stream - to answer such queries, it is useful to have a random sample of all elements within the current window. The problem with directly using any of the above algorithms to maintain a sample over a sliding window is that elements in the sample may expire, i.e., fall out of the window, and replacing them with enough new

S

sampled elements may not be possible, causing the number of elements within the sample to become too small.

The reservoir sampling algorithm was extended to sliding windows by Babcock et al. [2], who presented probabilistic guarantees on the space taken by their sampling algorithm. Gibbons and Tirthapura [10] have extended the distinct sampling algorithm to sliding windows, by sampling the stream at multiple probabilities, and maintaining a fixed number of the most recent elements at each sampling probability. It is known [4] that for computing decayed aggregates, an arbitrary time-decay function such as polynomial or exponential decay can be reduced to sliding window decay. Thus, for any stream aggregate that can be estimated well using random samples, such as quantiles, frequent elements, distinct counts and sum, the above techniques can be used to estimate time-decayed aggregates over an arbitrary decay function.

Handling Deletions

Thus far, it has been assumed that a data stream is a sequence of additions to a data set (which is so massive that it is too expensive to store it explicitly). More generally, it is necessary to deal with a stream where each element is an *operation* on the data set, which may be an addition or a deletion of one or more elements. On such a stream of add/delete operations, a direct sampling algorithm such as reservoir sampling or distinct sampling may not perform well, since elements that currently belong in the sample may be deleted by a future stream operation, leading to a sample that is too small to give useful estimates. The basic stream sampling algorithms have been extended to handle such "update streams" (sometimes called "dynamic data streams") by Ganguly [7], Cormode et al. [5], and Frahling et al. [6].

Key Applications

Estimating Query Selectivity: The selectivity of a query on a database table is defined as the ratio of the number of records that match the query to the total number of records in the table.

Suppose that the records appeared as a stream, and it was not possible to store the entire table on the disk (or disk access was too expensive). Then, a stream sampling algorithm can be used to maintain a random sample of all the records. The selectivity of the query on the random sample is an unbiased estimate of the selectivity of the query on the whole stream (i.e., the expected value of the estimate equals the actual selectivity). Of course, the larger the random sample, the more accurate is the estimate. Note that the query can be posed *after* the stream was observed, since the random sampling procedure was not sensitive to the query.

Network Monitoring: In the context of monitoring a TCP/IP network, a "network flow" is a unidirectional set of packets that arrive at the router on the same subinterface, have the same source and destination IP addresses, same transport layer protocol (TCP/UDP), same TCP/UDP source and destination ports and the same type of service (ToS) byte in the IP headers. A network monitoring tool is a software that constantly monitors the flows in a network and helps in network management tasks such as load balancing and fault management. Some network monitoring tools, for example, Random Sampled Netflow (by Cisco), Gigascope (by AT&T Research), and "Smart Sampling" (by AT&T), provide data for a subset of traffic in a router by processing only a random sample of the packet stream. Traffic sampling substantially reduces consumption of router resources while providing valuable network flow statistics.

Sensor Data Aggregation: A sensor network is a network of resource-constrained embedded devices, capable of computing and sensing, deployed to monitor environmental conditions like temperature, sound, vibration, pressure, light, etc. In a typical scenario, sensors collect readings periodically and send them to a base-station or a local cluster-head where computation/aggregation takes place. Data within a sensor network can be viewed as the union of multiple distributed streams, one per sensor node. Random samples of such distributed data streams can be used in computing key aggregates of sensor data streams, such as the mean, quantiles, frequent elements, of data. Transmitting the random samples rather than the entire observation stream can lead to savings in communication cost and hence, energy.

Cross-References

▶ AMS Sketch
▶ Data Aggregation in Sensor Networks
▶ Data Sketch/Synopsis
▶ Distributed Data Streams
▶ Frequency Moments
▶ Randomization Methods to Ensure Data Privacy
▶ Stream Mining

Recommended Reading

1. Alon N, Matias Y, Szegedy M. The space complexity of approximating the frequency moments. J Comput Syst Sci. 1999;58(1):137–47.
2. Babcock B, Datar M, Motwani R. Sampling from a moving window over streaming data. In: Proceedings of the 13th Annual ACM-SIAM Symposium on Discrete Algorithms; 2002. p. 633–4.
3. Chakrabarti A, Cormode G, McGregor A. A near-optimal algorithm for computing the entropy of a stream. In: Proceedings of the 18th Annual ACM-SIAM Symposium on Discrete Algorithms; 2007. p. 328–35.
4. Cohen E, Strauss M. Maintaining time-decaying stream aggregates. In: Proceedings of the 22nd ACM SIGACT-SIGMOD-SIGART Symposium on Principles of Database Systems; 2003. p. 223–33.
5. Cormode G, Muthukrishnan S, Rozenbaum I. Summarizing and mining inverse distributions on data streams via dynamic inverse sampling. In: Proceedings of the 31st International Conference on Very Large Data Bases; 2005. p. 25–36.
6. Frahling G, Indyk P, Sohler C. Sampling in dynamic data streams and applications. In: Proceedings of the 21st Annual Acm Symposium on Computational Geometry; 2005. p. 142–49.
7. Ganguly S. Counting distinct items over update streams. Theor Comput Sci. 2007;378(3): 211–22.
8. Gibbons P. Distinct sampling for highly-accurate answers to distinct values queries and event reports. In: Proceedings of the 27th International Conference on Very Large Data Bases; 2001. p. 541–50.
9. Gibbons P, Tirthapura S. Estimating simple functions on the union of data streams. In: Proceedings of the ACM Symposium on Parallel Algorithms and Architectures; 2001. p. 281–91.
10. Gibbons P, Tirthapura S. Distributed streams algorithms for sliding windows. Theor Comput Syst. 2004;37(3):457–78.
11. Manku GS, Motwani R. Approximate frequency counts over data streams. In: Proceedings of the 28th International Conference on Very Large Data Bases; 2002. p. 346–57.
12. Manku GS, Rajagopalan S, Lindsay BG. Random sampling techniques for space efficient online computation of order statistics of large datasets. In: Proceedings of the ACM SIGMOD International Conference on Management of Data; 1999. p. 251–62.
13. Olken F, Rotem D. Random sampling from databases – a survey. Stat Comput. 1995;5(1):43–57.
14. Pavan A, Tirthapura S. Range-efficient counting of distinct elements in a massive data stream. SIAM J Comput. 2007;37(2):359–79.
15. Vitter JS. Random sampling with a reservoir. ACM Trans Math Softw. 1985;11(1):37–57.

Stream Similarity Mining

Erik Vee
Yahoo! Research, Silicon Valley, CA, USA

Synonyms

Distance between streams; Datastream distance

Definition

In many applications, it is useful to think of a datastream as representing a vector or a point in space. Given two datastreams, along with a distance or similarity measure, the distance (or similarity) between the two streams is simply the distance (respectively, similarity) between the two points that the datastreams represent. Due to the enormous amount of data being processed, datastream algorithms are allowed just a single, sequential pass over the data; in some settings, the algorithm may take a few passes. The algorithm itself must use very little memory, typically polylogarithmic in the amount of data, but is allowed to return approximate answers.

There are two frequently used datastream models. In the *time series model*, a vector, $\overrightarrow{x}$, is simply represented as data items arriving in order of their indices: $x_1, x_2, x_3, \ldots$. That is, the value of the ith item of the stream is precisely the value of the ith coordinate of the represented vector. In the *turnstile model*, each arriving item signals an update to some component of the represented vector. So item (i,a) indicates that the value of the ith component of the vector is increased by a. For this reason, datastream items are typically written in the form $(i, x_i^{(j)})$ to indicate that this is the jth update to the ith component of the represented vector. The value of x_i is then the sum of $x_i^{(1)} + x_i^{(2)} + \ldots$ over all such updates. The update values may be negative; the special case when they are restricted to be nonnegative is sometimes called the *cash register model*.

One of the most commonly used measures for datastream similarity is the L_p distance between two streams, for $p \geq 0$. As in the standard definition, the L_p distance between points $\overrightarrow{x}, \overrightarrow{y}$ (hence, between streams representing those points) is defined to be $\sum_i |x_i^p - y_i^p|^{1/p}$. In the case that $p = 0$, the L_0 distance (sometimes called the Hamming distance) is taken to be the number of i such that $x_i \neq y_i$. For $p = \infty$, the L_∞ distance is $\max_i |x_i - y_i|$. Other measures include the Jaccard similarity, the edit distance, the earth-mover's distance, and the length of the longest common subsequence between the streams (viewed as sequences).

Historical Background

Although the earliest datastream-style algorithms were discovered some 30 years ago [11], the current resurgence of interest in datastreams began with the seminal paper of Alon et al. [2] in 1996. Implicit in their work is an algorithm for estimating the L_2 distance between streams. In 1999, Feigenbaum et al. [10] developed a datastreaming algorithm to approximate the L_1 distance between two streams. Building on this, Indyk [12] gave datastreaming algorithms to approximate the L_p distance between two datastreams, for all $p \in (0,2]$, utilizing the idea of p-stable distributions.

Later, Cormode et al. [7] demonstrated an efficient algorithm for approximating the L_0 distance (i.e., Hamming distance). Sun and Saks [15] provide lower bounds for approximating L_p, for $p > 2$ (and including $p = \infty$), showing no datastream algorithm working in polylogarithmic space can approximate the L_p distance between two streams within a polylogarithmic factor. (The bounds are even stronger for p much larger than 2).

Datar et al. [8] studied the sliding window model for datastreams, producing an algorithm that approximates the L_p distance between two windowed datastreams. Work by Datar and Muthukrishnan [9] gave an algorithm for approximating the Jaccard similarity between two datastreams in the sliding window model.

Foundations

Estimating the L_2 Distance

In their seminal paper, Alon et al. [2] provide a method for estimating F_2, the second frequency moment, of a datastream. As observed in [10, 1], this method can easily be extended to produce a datastream algorithm to approximate the L_2 distance. The ideas are briefly outlined below.

Throughout, the datastreams considered have length n. For $i = 1,2,\ldots,n$, the variable X_i is defined to be an i.i.d. (independent and identically distributed) random variable taking on the value -1 or 1 with equal probability. Of course, a datastream algorithm cannot maintain all the values of each of the random variables in memory. This will be accounted for later; for now, an algorithm is presented assuming that there is random access to these values.

The datastreams vectors are represented in the turnstile model; $(x_1, \ldots, x_n)$ denotes the accumulated values of the first stream, and $(y_1, \ldots, y_n)$ denotes the accumulated values in the second stream. The algorithm simply maintains the value of $\sum_{i=1}^{n} X_i \cdot (x_i - x_y)$. This value is straightforward to maintain: If an item $(i, x_i^{(j)})$ arrives for some i, j, the value $X_i \cdot x_i^{(j)}$ is added to it. If an item $(i, y_i^{(j)})$ arrives, the value $X_i \cdot y_i^{(j)}$ is subtracted.

The algorithm focuses on the expected value of *the square* of this quantity:

$$E\left[\left(\sum_{i=1}^{n} X_i \cdot (x_i - y_i)\right)^2\right]$$
$$= E\left[\sum_{i=1}^{n} X_i^2 \cdot (x_i - y_i)^2\right.$$
$$\left. + \sum_{i \neq j} X_i X_j \cdot (x_i - y_i)(x_j - y_j)\right]$$
$$= \sum_{i=1}^{n} (x_i - y_i)^2$$

where the last equality follows since $E[X_i] = 0$ and $X_i^2 = 1$ for all i, and all the random variables are independent. But this quantity is just the square of the L_2 distance between the two streams. Hence, the problem amounts to obtaining a good estimate of this expected value.

To do so, the above algorithm is run in parallel k times, for $k = \theta(1/\varepsilon^2)$. That is, it maintains the value $\sum_{i=1}^{n} X_i \cdot (x_i - x_y)$ for k different random assignments of the X_i. The algorithm then takes the average of their squares. For a given run t, this value is denoted $v^{(t)}$. To further ensure that the algorithm does not obtain a spurious estimate, the procedure is repeated ℓ times, for $\ell = \theta(\log(1/\delta))$. The algorithm then takes the median value over $\{v^{(1)}, v^{(2)}, \ldots, v^{(\ell)}\}$. A standard application of Chebyshev's Inequality shows that this estimates the square of the L_2 distance within a $(1 + \varepsilon)$ factor with probability greater than $1 - \delta$. (In total, this method maintains $k\ell$ values in parallel.)

Unfortunately, the procedure as described above produces and maintains values for n random variables. (In fact, due to the parallel repetitions, it actually needs $k\ell n$ random variables.) However, the technique only needed these variables to be four-wise independent. (Two-wise independence is needed for the expected value to be an unbiased estimator of the square of the L_2 distance; four-wise independence implies that the variance is small.) Hence, these fully independent random variables can be replaced with four-wise independent random variables, which is necessary for Chebyshev's Inequality to hold. These random variables can be pseudorandomly generated on the fly; the datastream algorithm thus only needs to remember a logarithmic-length seed for the pseudorandomly generated values. The full details are omitted here.

Estimating the L_p Distance: p-Stable Distributions

In 2000, Indyk [12], using many of the ideas in [2, 10], extended the results to produce datastream algorithms for approximating the L_p distance between streams, for all $p \in (0,2]$. (Feigenbaum et al. were the first to produce a datastream algorithm for L_1 distance; their technique relied on their construction of pseudorandomly generated "range-summable" variables that were four-wise independent. Although similar in flavor to the result of [2], it is somewhat more complicated). For convenience, the algorithm outlined below details the method for approximating the L_p norm of a single vector. Note, however, that in the turnstile model, it is a simple matter to produce the L_p distance between two streams (by simply negating all of the values in the second stream and finding the norm of their union). Indyk's method uses random linear projections, and relies on the notion of *p-stable distributions*.

A distribution $\mathcal{D}$ is *p-stable* if for all k real numbers $a_1, \ldots, a_k$, if $X_1, \ldots, X_k$ are i.i.d random variables drawn from distribution $\mathcal{D}$, then the random variable $\sum_i a_i X_i$ has the same distribution as $(\sum_i |a_i|^p)^{1/p} X$ for random variable X with distribution $\mathcal{D}$. There are two well-known p-stable distributions. The *Cauchy distribution*, with density function $\mu_C(x) = \frac{1}{\pi} \frac{1}{1+x^2}$, is 1-stable. The *Gaussian distribution*, with density function $\mu_G(x) = \frac{1}{\sqrt{2\pi}} e^{-x^2/2}$, is 2-stable. Although closed-form functions are not known for p-stable distributions for $p \neq 1, 2$, Chambers et al. [4] provide a method for generating p-stable random variables for all $p \in (0,2]$. Throughout the rest of this discussion, $\mathcal{D}$ denotes a p-stable distribution, for some fixed p.

The method for approximating the L_p norm of a stream will now be outlined. As previously noted, this is easily modified to give the L_p distance between two streams. Throughout, the vectors are represented as in the turnstile model, and $(z_1, \ldots, z_n)$ denotes the vector represented by the datastream. As in the previous section, the n i.i.d. random variables $X_1, \ldots, X_n$ are generated first, this time drawn from p-stable distribution $\mathcal{D}$. A brief discussion of how to reduce the number of these variables appears later.

The algorithm simply maintains the value $\sum_i X_i z_i$. Again, these values are easy to maintain: If item $(i, z_i^{(j)})$ appears for some i,j, the algorithm adds the value $X_i z_i^{(j)}$ to the sum. As in the previous section, the algorithm gains better accuracy by repeating the procedure multiple times in parallel; in this case, the algorithm runs the procedure k times in parallel, for $\mathrm{k} = \theta\left(\frac{1}{2}\log(1/\delta)\right)$. The value of $\sum_i X_i z_i$ obtained in the ℓ-th run using this procedure is denoted $Z^{(\ell)}$.

The value $Z^{(\ell)}$ is a random variable itself. Since $\mathcal{D}$ is p-stable, it is the case that $Z^{(\ell)} = X^{(\ell)} \cdot (\sum_i |z_i|^p)^{1/p}$ for some random variable $X^{(\ell)}$ drawn from $\mathcal{D}$. then the output of the algorithm is

$$\frac{1}{\gamma}\mathrm{median}\left\{|Z^{(1)}|, \ldots, |Z^{(k)}|\right\},$$

where γ denotes the median value of $|X|$, for X a random variable distributed according to $\mathcal{D}$. (The absolute value is taken for technical reasons. For instance, the median value of X is 0 when $\mathcal{D}$ is the Gaussian distribution, while the median value of $|X|$ is strictly greater than 0.) The value of the median of $\{|Z^{(1)}|, \ldots, |Z^{(k)}|\}$ is $(\sum_i |z_i|^p)^{1/p}$ times the median of $\{|X^{(1)}|, \ldots, |X^{(k)}|\}$. Hence, the above output is an approximation of $(\sum_i |z_i|^p)^{1/p}$, i.e., the L_p norm of the datastream, as needed. A more careful argument shows that this estimate is within a multiplicative factor $(1 \pm \varepsilon)$ of the true L_p norm, with probability greater than $1 - \delta$.

As in the previous section, Indyk observes that rather than storing the values of n i.i.d random variables, the values can be generated on the fly, using pseudorandom generators. The details are omitted here.

Cormode et al. [7] investigate the problem of estimating the L_0 norm. One of their key technical observations is that the L_p norm is a good approximation of the L_0 norm of the stream, for p sufficiently small. (In particular, they show the $p = \varepsilon / \log M$ is sufficient, where M

is the maximum absolute value of any item in the stream.) Thus, the Hamming distance between two streams can be approximated using the same general algorithm that was described above.

Approximating Jaccard Similarity: Min-Wise Hashing

Another useful similarity measure between two streams is their *Jaccard similarity*. Given two datastreams in the time-series model, $a_1, a_2, \ldots, a_n$ and $b_1, b_2, \ldots, b_n$ denote their respective vectors. Further, A (and B) denotes the set of distinct elements appearing in the first stream (respectively, the second stream). The Jaccard similarity between the streams is given by $|A \cap B| / |A \cup B|$.

The first explicit study of the Jaccard similarity between two streams was given by Datar and Muthukrishnan [9]. Their paper examined the sliding window model, which is discussed further in the next section. However, a datastream algorithm in the standard model was given implicitly in the work of Cohen et al. [6], although the notion of datastreams is never mentioned in the paper.

The major technical tool uses *min-wise hashing*, or min-hashing [3, 5]. For every subset A of $[n]$, the *min-hash* for A (with respect to π), denoted $h_\pi(A)$, is defined to be $h_\pi(A) = \min_{i \in A}\{\pi(i)\}$, where π denotes a permutation on $[n] = \{1, \ldots, n\}$. The wonderful property of the min-hash is that, when π is chosen uniformly at random from the set of all permutations on $[n]$, for any two subsets A,B of $[n]$, it is the case that

$$\Pr[h_\pi(A) = h_\pi(B)] = \frac{|A \cap B|}{|A \cup B|}$$

This suggests the following algorithm.

The algorithm chooses π uniformly at random from the set of permutations on $[n]$. (The fact that storing π take $\theta(n \log n)$ space will be discussed momentarily.) For the first stream, the algorithm finds the value $h_\pi(A) = \min_{i \in A}\{\pi(i)\}$, where A is the set of distinct elements occurring in the first stream. This is simple to do in a datastreaming fashion: as each new a_j appears, the algorithm updates the min value if $\pi(a_j)$ is smaller than the min seen so far. Likewise, for the second stream, the algorithm finds the value $h_\pi(B)$, where B is the set of distinct elements occurring in the second stream. From the above, the probability that the two values are equal is precisely the Jaccard similarity between the two streams.

Of course, to obtain an accurate estimate of this probability, the algorithm needs to run the procedure multiple times. In this case, it will run the procedure in parallel k times, each with an independently chosen random permutation. (Here, $k = O(\varepsilon^{-3} \log(1/\delta))$.) The value ρ is defined to be the fraction of times (out of k) that the min values for the two streams coincide. That is, if $\pi_1, \ldots, \pi_k$ are the k independently chosen random permutations, then

$$\rho = \frac{1}{k} \cdot \# \left| \left\{ j : h_{\pi_j}(A) = h_{\pi_j}(B) \right\} \right|$$

It is shown in [11] that with probability at least $1 - \delta$, the value ρ approximates the Jaccard similarity within multiplicative factor $(1 \pm \varepsilon)$.

In order for the above algorithm to be useable in a datastreaming context, it must be able to generate and store the necessary random permutations in small space. This is done using *approximately* min-wise independent hash functions. Although this introduces additional error, it can be done in small space and time. The reader is referred to [13] for more details.

Sliding Windows

In many applications, the data from streams becomes outdated or unnecessary quickly. To help understand this scenario better, researchers have proposed the *sliding window* model of datastreams. Here, the algorithm must maintain statistics (e.g., stream similarity), using only the last N items from the stream, for some N. This causes additional complications, since as each new item comes in, an old item is removed. Since memory is limited, algorithms cannot track which of these old items is disappearing. Still, there are datastream algorithms for both L_p distance and Jaccard similarity in the sliding window model.

In [8], Datar et al. define the sliding window model, and give a datastream algorithm for approximating the L_p distance between two streams (as well as several other datastream algorithms). Their technique uses what they call an *exponential histogram*. The histogram partitions the last N items (i.e., those items in the sliding window) into buckets; the last bucket may in fact contain items older than the last N. Each bucket maintains the necessary statistics for the items it contains. For instance, a bucket containing the items $a_s, a_{s+1}, \ldots, a_t$ would hold the L_p-sketch for those items. (Due to memory constraints, the bucket cannot actually maintain the values of all the items it holds.)

As new items come in, the algorithm merges old buckets to maintain the histogram structure, creating new buckets only for newly encountered items. The last bucket will eventually contain only items that do not appear in the N most recent, and will be removed from the histogram at this time. Datar et al. observe that the additional error in this windowed model, beyond that of the standard model, comes from the fact that the last bucket may contain items that are no longer in the N-item window. But the structure of the exponential histogram ensures that this error is not too large. Hence, they provide a general method for translating a wide range of datastream algorithms into windowed-datastream algorithms.

Datar and Muthukrishnan [9] study the problem of approximating the Jaccard similarity of two streams in the sliding window model. As in the non-windowed version, they use min-hashing as a primary tool. The main complication in the sliding window model is that maintaining the minimum value over a sliding window is hard. At a given time step t, the algorithm needs to know the value $\min_{i=t, \ldots, t-N+1}\{\pi_j(a_i)\}$, where π_j is a permutation chosen by the datastream algorithm in the standard model. Their solution is to maintain the value $\pi_j(a_i)$ for every *relevant* $i = t, \ldots, t-N+1$. For instance, if $\pi_j(a_i) > \pi_j(a_{i+s})$ for some $s > 0$, then the value $\pi_j(a_i)$ will never be the minimum over the sliding window at any time; hence, it may be discarded. (Here, item a_i occurs earlier than a_{i+s}, thus item a_i will move out of the window before a_{i+s})

Amazingly, with high probability, the number of relevant values that need to be maintained is at most $O(\log n)$. Hence, the standard datastream algorithm can be adapted to the sliding window model, using small space.

Lower Bounds for Stream Distance

The major technique for proving lower bounds utilize reductions from communication complexity. Here, only sketches of the very high level ideas are presented, with some of the main results cited.

An often-used communication complexity problem is DISJOINTNESS: Alice is given a set, A, and Bob is given a set, B. Neither knows what the other set is. They must communicate with each other by sending messages back and forth, until they decide whether $A \cap B$ is nonempty. (They are allowed to decide ahead of time the protocol they will use to communicate messages.) It has been shown that if the size of A and B is $\theta(n)$, the communication complexity (i.e., the number of bits that must be communicated in the worst case) is also at least $\theta(n)$ [14].

A datastream algorithm that calculates the distance between two streams can provide the basis for a communication complexity algorithm. A typical reduction gives a method for Alice to transform her set A into a datastream (without looking at set B). Likewise, the reduction gives a method for Bob to transform B into a datastream, without looking at A. Finally, the reduction guarantees that Alice's datastream and Bob's datastream are close if and only if $A \cap B$ is non-empty. Then Alice can begin running the datastream algorithm on her datastream. When it has processed her stream, the algorithm will have some memory bits indicating its current state. Alice sends a message to Bob, telling him that state. Bob can then finish running the datastream algorithm on his own datastream. If the algorithm indicates that the two streams are close, he knows $A \cap B$ is nonempty; otherwise, he knows that $A \cap B = \varnothing$ (and may communicate this to Alice in one bit). Hence, Alice and Bob have solved their communication complexity problem. Since the original communication complexity problem took at least $\theta(n)$ bits, the datastream algorithm

must also use at least this much memory. (In this case, showing that it cannot be space efficient.)

There is, of course, a great deal of technical work in providing the proper reductions; the difficulties are even greater when showing lower bounds for approximations. However, building on these ideas, Saks and Sun [15] show that approximating the L_∞ distance between two datastreams is impossible to do in sublinear space. In fact, their work shows that approximating within factor $n^{O(\varepsilon)}$ the L_p distance for any $p \geq 2 + \varepsilon$ requires space at least $n^{O(\varepsilon)}$. For p close to 2, this has very little practical implications, but the bounds become more meaningful for large p. Much simpler reductions show the impossibility of space-efficient datastream algorithms for approximating the length of the longest common subsequence between two datastreams (viewed as sequences).

Key Applications

Tracking Change in Network Traffic

The datastream algorithms outlined above allow one to take an entire day of network traffic and synopsize it using a small sketch. It is then possible to measure how different traffic is from day-to-day. Large changes in the network traffic can signal denial of service attacks or worm infestations.

Query Optimization

Most query-optimization techniques utilize data statistics to produce better plans. The L_2 norm is a useful measure for approximating join sizes, while the L_0 norm gives the number of distinct items in the stream.

Processing Genetic Data

Since genetic data consists of millions or billions of base pairs for an individual, it is useful to think of them as streams of data. The similarity of two base-pair sequences is a fundamental concept.

Data Mining

Often individual entities are represented by massive streams of data (e.g., phone calls from a large company, or IP addresses of users visiting a given web site, or items bought at a grocery store). Estimating the similarity between these streams can be a useful tool for identifying similar entities. As one example, it is possible to determine which web sites are most similar to each other, based on the IP addresses of their visitors.

Cross-References

▶ Stream Mining

Recommended Reading

1. Alon N, Gibbons P, Matias Y, Szegedy M. Tracking join and self-join sizes in limited storage. In: Proceedings of the 18th ACM SIGACT-SIGMOD-SIGART Symposium on Principles of Database Systems; 1999. p. 10–20.
2. Alon N, Matias Y, Szegedy M. The space complexity of approximating the frequency moments. In: Proceedings of the 28th ACM Symposium on Theory of Computing; 1996. p. 20–9.
3. Broder A, Charikar M, Frieze A, Mitzenmacher M. Min-wise independent permutations. In: Proceedings of the 30th ACM Symposium on Theory of Computing; 1998. p. 327–36.
4. Chambers JM, Mallows CL, Stuck BW. A method for simulating stable random variables. J Am Stat Assoc. 1976;71(354):340–4.
5. Cohen E. Size-estimation framework with applications to transitive closure and reachability. J Comput Syst Sci. 1997;55(3):441–53.
6. Cohen E, Datar M, Fujiwara S, Gionis A, Indyk P, Motwani R, Ullman J. Finding interesting associations without support pruning. In: Proceedings of the 16th International Conference on Data Engineering; 2000.
7. Cormode G, Datar M, Indyk P, Muthukrishnan S. Comparing data streams using hamming norms. In: Proceedings of the 28th International Conference on Very Large Data Bases; 2002. p. 335–45.
8. Datar M, Gionis A, Indyk P, Motwani R. Maintaining stream statistics over sliding windows. In: Proceedings of the 13th Annual ACM-SIAM Symposium on Discrete Algorithms; 2002. p. 635–44.
9. Datar M, Muthukrishnan S. Estimating rarity and similarity on data stream windows. In: Proceedings of the 10th European Symposium on Algorithms; 2002.
10. Feigenbaum J, Kannan S, Strauss M, Viswanathan M. An approximate l_1-difference algorithm for massive data streams. In: Proceedings of the 40th Annual Symposium on Foundations of Computer Science; 1999.
11. Flajolet P, Martin G. Probabilistic counting. In: Proceedings of the 24th Annual Symposium on Foundations of Computer Science; 1983. p. 76–82.
12. Indyk P. Stable distributions, pseudorandom generators, embeddings and data stream computation. In: Proceedings of the 41st Annual Symposium on Foundations of Computer Science; 2000. p. 189–97.
13. Indyk P. A small approximately min-wise independent family of hash functions. J Algorithm. 2001;38(1):84–90.
14. On the distributional complexity of disjointness. J Comput Sci Syst. 1984;2.
15. Saks M, Sun X. The space complexity of approximating the frequency moments. In: Proceedings of the 34th ACM Symposium on Theory of Computing; 2002.

Streaming Analytics

Deepak Turaga
IBM Research, San Francisco, CA, USA

Stream processing applications require the processing and analysis of continuously generated multimodal and distributed data streams. This requires a unique combination of multiple features that distinguishes streaming analytics from traditional data analysis paradigms, which are often batch and offline. These features can be summarized as follows:

In-Motion Analysis: Streaming analytics need to process data on-the-fly, as it continues to flow, in order to support real-time, low-latency analysis and to match the computation to the naturally streaming properties of the data. This limits the amount of prior data that can be accessed and necessitates one-pass, online algorithms. Several streaming algorithms are described in [1, 7].

Distributed Analysis: Data streams are often *distributed*, and/or high volume, and their large rates make it infeasible to adopt centralized solutions. Hence, the applications and analytic algorithms themselves need to be distributed.

S

High-Performance Analysis: Streaming applications require high *throughput*, low *latency*, and dynamic *scalability*. This means that analytics should be structured to exploit distributed computation infrastructures and different forms of parallelism (e.g., pipelined data and tasks). This also means that they often require joint application and system optimization.

Multimodal Analysis: Analytics often need to process streaming information across heterogeneous data sources, including *structured* (e.g., transactions), *unstructured* (e.g., audio, video, text, image), and *semistructured* data. For instance, in an application monitoring a traffic network, the data streams can include sensor readings (e.g., loop and tool-booth sensors), user-contributed text and images, traffic cameras, etc.

Loss-Tolerant Analysis: Applications need to analyze lossy data with different noise levels, statistical and temporal properties, mismatched sampling rates, etc., and hence they often need appropriate processing to transform, clean, filter, and convert data and results. This also implies the need to match data rates, handle lossy data, synchronize across different data streams, and handle various protocols. Streaming analytics need to account for these issues and provide graceful degradation of results to loss in the data.

Adaptive and Time-varying Analysis: Streaming applications are often long-running and need to adapt over time to changes in the data and problem characteristics. Hence, analytics need to support dynamic reconfiguration based on feedback, current context, and results of the analysis.

Streaming analytics therefore need to be designed to meet these requirements. There are many streaming analytic algorithms that realize different stages of the data mining and knowledge acquisition process – which include data acquisition, data pre-processing, data transformation, modeling, and evaluation. We should note that these stages are not completely disjoint from each other, and there are multiple analytics that can

be used for one or more stages. Additionally, there are also complex interactions between the analytics used in the different stages that can affect end-to-end results in various ways. For instance, the choice of data pre-processing and transformation approaches impacts the choice of the approach for the modeling step.

We now provide a brief summary of different streaming analytic algorithms partitioned into classes based on the stages of the data mining process. In section "Data Pre-processing and Transformation", we describe analytics for data pre-processing and transformation, and in section "Modeling and Evaluation", we describe analytics for modeling and evaluation.

Data Pre-processing and Transformation

Streaming analytics for data pre-processing and transformation include algorithms from the following five categories:

Descriptive statistics which are techniques focused on extracting simple, quantitative, and visual statistics from a dataset. In the context of streaming data, these statistics need to be computed in a continuous manner as tuples of a stream flow into an application. Descriptive statistics can also be computed across multiple streams to capture correlations and relationships among them. Multiple streaming algorithms have been proposed for different types of descriptive statistics including for the computation of (a) counts [19] and measures of centrality such as moments (mean, variance, skew), medians, modes, etc.; (b) distributions and histograms – including equi-width and equi-depth histograms [47, 50] and other measures such as quantiles [8,10]; (c) periodicity measures; (d) correlation and cross-correlation [52]; and (e) information theoretic measures such as entropy, Kolmogorov complexity, information gain, etc.

Sampling that includes techniques focused on reducing the volume of the input data by retaining only an appropriate subset for analysis. In the context of streaming data, this may involve discarding some tuples based on one or more criteria. Sampling techniques have been considered extensively in different areas of research including statistics, signal processing, approximation theory, and recently stream processing and mining. Streaming algorithms have been defined for (a) systematic sampling (e.g., uniform sampling [53], (b) random sampling [9], and (c) data-driven sampling, where values are chosen to be retained based on their value [53], unlike systematic and random sampling techniques that only focus on the timestamp and or the sequence number attribute associated with a tuple.

Sketches that include techniques aimed at building in-memory data structures that contain compact synopses of the temporal properties of the streaming data such that specific queries can be answered, in some cases approximately. The types of queries include frequency-based queries (e.g., related to the distribution of the data, heavy hitters etc.) and distinct value queries. Sketches that are of interest to streaming applications are those that can be built, updated, and queried efficiently, allowing the application to keep up with a high rate of data ingest. There are several *linear sketches* where each update is handled in the same way, irrespective of the history of prior updates. Linear sketches are usually implemented using a transform, often employing hash functions [18]. Example of linear sketches include the count-min sketch [17] the AMS sketch [6], the sampling sketch [49], histograms [30], and wavelet sketch [31]. Finally, while we have focused on random vector-based sketch transforms, there is also some work on deterministic sketch functions [26] for advanced types of queries.

Quantization that includes analytics focused on reducing the fidelity of individual data samples (e.g., number of bits per sample) to remove noise as well as to lower compute and memory costs of other downstream stages of a data analysis task. Examples include scalar, vector, uniform, or nonuniform techniques [56]. Specifically for streaming data, it is important that the quantization retains the temporal properties of the input data. Example of quantization

techniques that retain these temporal properties includes binary clipping [11] and moment preserving quantization [20]. Both techniques are in the family of shape-preserving quantizers, where quantizers are designed based on windows of the incoming streaming data and adaptive quantization boundaries are determined to preserve the temporal relationships between consecutive samples. Recent research in scalar quantization techniques has been focused on *compressed sensing* [12, 38] applications which perform mining tasks on data collected from distributed sensor networks. Recent research on vector quantization has focused on the development of streaming and incremental clustering methods [42,42] as the use of these techniques in continuous data processing applications is becoming more common.

Dimensionality Reduction which includes analytic algorithms that attempt to reduce the number of attributes or dimensions within each data item (i.e., attributes of a tuple) to reduce the data volume and/or to improve the accuracy of the models built on the data. Dimensionality reduction can be performed using different types of techniques including feature selection as well as transformation. Feature selection [62] techniques use information theoretic and other correlation measures to determine the optimal subset of features to retain for a particular mining problem. An exhaustive search across all subsets of available attributes is often infeasible, so practical approaches for feature selection include techniques such as feature ranking [40] and subset selection [55] [62]. Other techniques for dimensionality reduction use linear and nonlinear transformations. Linear transforms [43] such as the discrete Fourier transform [67], discrete cosine transform, the wavelet transform [43]. [41], and linear predictive coding have been used extensively for dimensionality reduction for many types of streaming data. The principal component analysis (PCA) technique can provide squared error optimal dimensionality representation of the data; however, it cannot be easily computed incrementally for streaming data. Several algorithms that approximate the PCA on streaming data have been developed and include SPIRIT [51] and

MUSCLES [65]. More recently, distributed dimensionality reduction [24, 59] techniques have also been investigated.

Modeling and Evaluation

In this section, we focus on the last two stages of the data mining process and examine streaming techniques for modeling and evaluation. These steps form the core of a mining task where automatic or semiautomatic analysis of streaming data is used to extract and create a model with previously unknown information.

In many cases for streaming applications, it is sufficient to use a methodology that combines offline modeling, where the model for a dataset is initially learned, and online evaluation, where this model is used to analyze the streaming data being processed by an application. This methodology is frequently used in stream processing settings because it can leverage many of the existing data mining algorithms devised for analyzing datasets stored in databases and data warehouses. However, it is not straightforward to satisfy the adaptive requirement on analysis that these applications have with this methodology. Thus, in this section, we examine primarily on online techniques for modeling and evaluation.

The mining process is classically divided into five different tasks. The first task, *classification*, focuses on assigning one of multiple categories or labels to each data item (tuple) from a dataset based on the values its attributes have. The second task, *clustering*, consists of assigning tuples to groups, or clusters, so that the tuples in the same cluster are more similar to each other than to those in other clusters. The third task, *Regression*, is used to estimate the relationship between one or more independent variables such as the attributes of a tuple and a dependent variable whose value must be predicted. The fourth task, *frequent pattern* and *association rule* mining, focuses on discovering recurrently occurring patterns and structures in various types of data and using this information to identify rules that capture dependencies and relationships between the occurrences of items or attributes. Finally, the last task, *anomaly detection*, aims at detecting patterns in a given data set that do not conform to an established normal behavior.

We now describe streaming analytics for each of these different tasks.

Classification The stream classification problem consists of assigning one of multiple discrete categories (or labels) to a streaming data item or tuple, based on the values its attributes have. The classification categories are predefined and known a priori. This includes both modeling, which involves supervised learning of the classifier model, and evaluation which involves applying the model to new data to obtain the predicted category.

There are several well-known classification techniques used in traditional data mining. The first group includes linear algorithms [64] that compute a linear combination of features which separates two or more classes of objects. These methods include the Fisher's linear discriminant, logistic regression, naive Bayes classifiers, and the perceptron. These are well suited for online scoring as well as for incremental learning. Other popular classification techniques [64] rely on models of different types: Gaussian mixture models, Support Vector Machines, Decision Trees and Forests, Bayesian networks, k-nearest neighbors, and hidden Markov models. These techniques are often not necessarily suitable for incremental learning in the form they were originally proposed. In addition to these techniques, there are also meta-learning procedures that build on top of these algorithms to improve the classification performance. These procedures include boosting [57] and ensemble methods [54], which leverage multiple classification models to obtain better predictive performance than could be attained from any of the individual models by themselves.

Research into online learning has led to the creation of incremental classifiers including the Very Fast Decision Trees [21] [36] algorithm for online decision tree learning. In this algorithm the node-splitting criterion is modified to account for the streaming nature of the data. There are also several algorithms that effectively take temporal locality into account; algorithms are designed to be purely single pass adaptations of conventional classification algorithms [4, 36]. Another recent innovation in on-demand stream classification methods is the use of use micro-clustering for classification purposes. This is discussed more in the clustering section.

Ensemble techniques [34, 35, 58] build and combine a collection of base algorithms (e.g., classifiers) into a joint unique algorithm (classifier). Traditional ensemble schemes for data analysis are focused on analyzing stored or completely available datasets, and examples of these techniques include bagging and boosting. In the past decade, much work has been done to develop online versions of such ensemble techniques. For example, an online version of AdaBoost is described in [23], and similar proposals are made in [63] and [45]. Extensions of this work for distributed online learning have also been recently proposed [46,61] and consist of learning agents that are linked together through a network topology in a distributed setting.

Clustering The stream clustering problem consists of assigning a set of tuples to groups, or clusters, so that the tuples in the same cluster are more similar to each other than to those in other clusters. Hence, a cluster is a group of tuples with a low distance between one another, forming a dense area of the data space or of a particular statistical distribution. To employ clustering, the tuple attributes are required to be numeric or be converted to a numeric representation ahead of time. The actual number of clusters used to represent a dataset may or may not be known a priori and may also need to be determined during the model learning step. Among the most popular of these solutions is the Lloyd's algorithm [28], a method also used in quantization, referred to more commonly as the k-means algorithm. The algorithm is a batch, gradient descent approach not well suited for incremental update.

Incremental clustering algorithms include streaming k-Means [33] the CluStream technique [2]. This is a stream processing algorithm designed to compute intermediate cluster statistics from streaming data in terms of cluster feature vectors [66] to support centroid-based stream clustering. The cluster feature vectors keep track of the first- and second-order moments of the incoming data, which are then used for stream clustering. The first- and second-order moments satisfy the additivity property and can be used to compute a host of other parameters important for clustering, for instance, a cluster's centroid and radius, making them ideal for stream clustering. The algorithm then determines microclusters based on these statistics and dynamically updates these as new data arrives. A post-processing clustering step is performed to reconstruct the actual clusters from the information stored in the microcluster snapshots. In order to ensure that the clusters yielded by the macroclustering step are comparable to offline clustering, a large number of microclusters are retained and updated.

The microcluster technique has been shown to be much more effective and versatile than the k-means-based stream clustering technique proposed earlier. CluStream has also been extended to handle a variety of other problems [7]. It has also been extended to handle cases with high-dimensional data [3] and for handling uncertain data, such as in sensor networks [5].

Regression includes a set of techniques used to estimate the relationship between one or more *independent variables* such as the attributes of a tuple and a dependent variable, possibly represented as an attribute to be added to the original tuple, whose value must be predicted. Similar to classification, regression is used for prediction and forecasting; however, in contrast with classification, the resulting prediction is also numeric (not a category). There are two main classes of regression models *parametric* models and *nonparametric* models. A parametric regression model [25] represents the relationship between the independent and dependent variables using a regression function with a finite number of (unknown) parameters. These parameters are computed during the learning step of the regression analysis. In contrast, a nonparametric regression model [60] makes no assumptions about the function that represents the relationship

Streaming Analytics

between the independent and the dependent variables.

Examples of parametric models include linear regression method that assumes a linear relationship between the independent variables and the dependent variable. The learning step consists of estimating the weights for the linear function describing this relationship by minimizing the distance between a prediction and its corresponding true value. Linear models are well suited for online learning, where the weight vector is estimated incrementally, from the streaming data. Multiple algorithms for online regression model learning have been developed, including those based on stochastic gradient descent (SGD) [27]. In this approach weights are updated using a gradient descent, where the gradient is incrementally estimated from the observed prediction error. Several enhancements [22,39,48] to online SGD-based linear regression algorithms can be accomplished by updating the learning rate or adding in safe updates.

Frequent pattern mining includes a set of analytic algorithms aimed at discovering frequently occurring patterns and structures in various types of data. Depending on the mining task and input data, a pattern may be an itemset, a collection of items that occur together; a sequence, defining a particular ordering of events; or a subgraph or a subtree, representing a common structure found in a larger one [32]. Frequent patterns provide an efficient summarization of the data from streaming sources enabling the extraction of knowledge by applying additional mining algorithms. Hence, this technique is often used by other mining tasks such as association rule mining, classification, clustering, and change detection. Neither frequent itemset nor association rule mining can be solved exhaustively, because of the combinatorial explosion in the number of possible choices.

When it comes to streaming data, multiple formulations of the problem are possible [29,37]. The first one is the entire stream approach, where frequent patterns are mined over the entire stream. Such frequent pattern can be implemented using sketch-based algorithms to identify the frequent patterns. The single pass lossy counting [44] algorithm is one example of this approach. The second formulation is the sliding window approach. As a stream changes characteristics over time, it is often desirable to determine the frequent patterns over a particular sliding window. The moment algorithm embodies this idea [16]. The third formulation is the damped window approach, where a *decay factor* is introduced such that different weights can be assigned to data from different periods of time. One example is proposed in [14] that maintains a lattice for recording the potentially frequent itemsets and their counts. Other algorithms for frequent itemset mining on streams that have been proposed [15] can be grouped according to certain characteristics based on their window management model, such as landmark or sliding; their update method, either tuple or batch; and whether they produce an exact or approximate result as well as whether they make guarantees regarding false positives and false negatives.

Anomaly detection The anomaly detection problem consists of identifying patterns in data that do not conform to *expected* behavior [13]. There are three different kinds of anomalies that usually are of interest to a streaming analysis application.

A point anomaly occurs when an individual tuple is deemed anomalous with respect to the rest of the tuples seen so far. In this case, never-before seen values and abnormally large or small values can indicate an anomaly. A contextual anomaly occurs when a tuple is deemed anomalous with respect to a particular context, but is not necessarily abnormal across all tuples seen so far. A collective anomaly occurs when a collection of related tuples is deemed anomalous with respect to the entire set of tuple seen so far.

Learning a model for anomaly detection can be done in different ways including (a) using a supervised approach where ground truth on whether a tuple is anomalous or not is provided either at an initial training period or, continuously, once it becomes known; (b) a semi-supervised approach, ground truth for a subset of tuples, typically the ones deemed normal, is provided to the algorithm; and (c) in an unsupervised manner, where no ground truth information is provided at all. Instead the algorithm relies solely on the observed tuples to determine which ones are normal and which ones are anomalies. Anomaly detection algorithms can be based on classification methods, clustering methods, nearest neighbor methods, statistical methods, information theoretic methods, and, finally, spectral methods.

So, many of the algorithms described earlier, including online classification, clustering, and descriptive statistics-based methods, may be used for streaming anomaly detection. A summary of these may be found in [7].

Conclusion

In this chapter we examined the unique requirements of streaming analytics and presented the characteristics of several stream mining analytics and algorithms for different stages of the data mining process, including data pre-processing, transformation, modeling, and evaluation.

There are several open research problems for streaming analytics – especially at the interface between online algorithms and distributed systems interface that are worth investigating. There is currently limited work on building ensembles and distributed streaming analytics and no principled approach to decompose an online-distributed large-scale learning problem such that it can be deployed as a topology/flowgraph of analytics across a distributed compute infrastructure. There is also a need to handle real-world issues of synchronization, delay, loss, and communication errors within these streaming analytics. Finally, in the presence of multiple streaming analytics, there is a need to develop ensemble approaches to handle noncooperative analytics using game theoretic approaches.

Recommended Reading

1. Aggarwal C (ed). Data streams: models and algorithms. Boston: Springer; 2007.
2. Aggarwal CC, Han J, Wang J, Yu PS. A framework for clustering evolving data streams. In: Proceedings of the 29th International Conference on Very Large Data Bases; 2003. p. 81–92.
3. Aggarwal CC, Han J, Wang J, Yu PS. A framework for high dimensional projected clustering of data streams. In: Proceedings of the 30th International Conference on Very Large Data Bases; 2004. p. 852–63.
4. Aggarwal CC, Han J, Wang J, Yu PS. On demand classification of data streams. In: Proceedings of the 11th ACM SIGKDD International Conference on Knowledge Discovery and Data Mining; 2004. p. 503–8.
5. Aggarwal CC, Yu PS. A framework for clustering uncertain data streams. In: Proceedings of the 24th International Conference on Data Engineering; 2008. p. 150–59.
6. Alon N, Matias Y, Szegedy M. The space complexity of approximating the frequency moments. In: Proceedings of the 28th Annual ACM Symposium on Theory of Computing; 1996. p. 20–9.
7. Andrade H, Gedik B, Turaga D. Fundamentals of stream processing: application design, systems, and analytics. Cambridge: Cambridge University Press; 2013.
8. Arasu A, Manku G. Approximate counts and quantiles over sliding windows. In: Proceedings of the 23rd ACM SIGACT-SIGMOD-SIGART Symposium on Principles of Database Systems; 2004. p. 286–96.
9. Ardilly P, Tillé Y. Sampling methods. Springer; 2006.
10. Babcock B, Datar M, Motwani R, O'Callaghan L. Maintaining variance and k-medians over data stream windows. In: Proceedings of the 22nd ACM SIGACT-SIGMOD-SIGART Symposium on Principles of Database Systems; 2003. p. 234–43.
11. Bagnall AJ, (Ann) Ratanamahatana C, Keogh EJ, Lonardi S, Janacek GJ. A bit level representation for time series data mining with shape based similarity. Springer Data Min Knowl Disc. 2006;13(1):11–40.
12. Boufounos P. Universal rate-efficient scalar quantization. IEEE Trans Inf Theory. 2012;58(3):1861–72.
13. Chandola V, Banerjee A, Kumar V. Anomaly detection: a survey. ACM Comput Surv. 2009;41(3).
14. Chang JH, Lee WS. Finding recent frequent itemsets adaptively over online data streams. In: Proceedings of the 9th ACM SIGKDD International Conference on Knowledge Discovery and Data Mining; 2003. p. 487–92.
15. Cheng J, Ke Y, Ng W. A survey on algorithms for mining frequent itemsets over data streams. Knowl Inf Syst. 2008;16(1):1–27.
16. Chi Y, Wang H, Yu PS, Muntz RR. Moment: maintaining closed frequent itemsets over a stream sliding window. In: Proceedings of the 4th IEEE International Conference on Data Mining; 2004. p. 59–66.
17. Cormode G, Muthukrishnan S. An improved data stream summary: the count-min sketch and its applications. J Algorithms. 2005;55(1):58–75.
18. Cormode G, Garofalakis M, Haas P, Jermaine C. Synopses for massive data: samples, histograms,

wavelets, sketches. Foundations and trends in databases series. Boston: Now Publishing; 2011.

19. Datar M, Gionis A, Indyk P, Motwani R. Maintaining stream statistics over sliding windows. SIAM J Comput. 2002;31(6):1794–813.

20. Delp E, Saenz M, Salama P. Block truncation coding. In: Al Bovik, editor. The handbook of image and video processing. Amsterdam/Boston: Academic Press; 2005. p. 661–72.

21. Domingos P, Hulten G. Mining high-speed data streams. In: Proceedings of the 6th ACM SIGKDD International Conference on Knowledge Discovery and Data Mining; 2000. p. 71–80.

22. Duchi J, Hazan E, Singer Y. An improved data stream summary: the Count-Min sketch and its applications. J Mach Learn Res. 2010;12:2121–59.

23. Fan W, Stolfo SJ, Zhang J. The application of AdaBoost for distributed, scalable and on-line learning. In: Proceedings of the 5th ACM SIGKDD International Conference on Knowledge Discovery and Data Mining; 1999. p. 362–66.

24. Fang J, Li H. Optimal/near-optimal dimensionality reduction for distributed estimation in homogeneous and certain inhomogeneous scenarios. IEEE Trans Signal Process (TSP). 2010;58(8):4339–53.

25. Fox J, editor. Applied regression analysis, linear models, and related methods. Thousands Oaks: SAGE Publications; 1997.

26. Ganguly S, Majumder A. CR-precis: a deterministic summary structure for update data streams. In: Proceedings of the International Symposium on Combinatorics; 2007. p. 48–59.

27. Gardner WA. Learning characteristics of stochastic-gradient-descent algorithms: a general study, analysis, and critique. Signal Process. 1984;6(2):113–33.

28. Gersho A, Gray RM. Vector quantization and signal compression. Boston: Kluwer Academic Publishers; 1991.

29. Giannella C, Han J, Pei J, Yan X, Yu P. Mining frequent patterns in data streams at multiple time granularities. In: Kargupta H, Joshi A, Sivakumar K, Yesha Y, editors. Data mining: next generation challenges and future directions. MIT Press; 2002. p. 105–24.

30. Gilbert A, Guha S, Indyk P, Kotidis Y, Muthukrishnan S, Strauss M. Fast, small-space algorithms for approximate histogram maintenance. In: Proceedings of the 34th Annual ACM Symposium on Theory of Computing; 2002. p. 389–98.

31. Gilbert A, Kotidis Y, Muthukrishnan S, Strauss M. Surfing wavelets on streams: one-pass summaries for approximate aggregate queries. In: Proceedings of the 27th International Conference on Very Large Data Bases; 2001. p. 79–88.

32. Goethals B. Survey on frequent pattern mining. Technical report, Helsinki institute for information technology basic research unit., 2003.

33. Guha S, Mishra N, Motwani R, OĆallaghan L. Clustering data streams. In: Proceedings of the 41st Annual Symposium on Foundations of Computer Science; 2000. p. 359–66.

34. Haipeng Z, Kulkarni SR, Poor HV. Attribute-distributed learning: models, limits, and algorithms. 2011;59(1):386–98.

35. Hansen LK, Salamon P. Neural network ensembles. 1990;12(10):993–1001.

36. Hulten G, Spencer L, Domingos P. Mining time changing data streams. In: Proceedings of the 7th ACM SIGKDD International Conference on Knowledge Discovery and Data Mining; 2001. p. 97–106.

37. Jin R, Agrawal G. An algorithm for in-core frequent itemset mining on streaming data. In: Proceedings of the 5th IEEE International Conference on Data Mining; 2005. p. 201–17.

38. Kamilov U, Goyal VK, Rangan S. Optimal quantization for compressive sensing under message passing reconstruction. In: Proceedings of the IEEE International Symposium on Information Theory; 2011. p. 459–63.

39. Karampatziakis N, Langford J. Online importance weight aware updates. In: Proceedings of the 27th Conference on Uncertainty in Artificial Intelligence; 2011. p. 392–99.

40. Kira K, Rendell L. A practical approach to feature selection. In: Proceedings of the 9th International Conference on Machine Learning; 1992. p. 249–56.

41. Lin J, Vlachos M, Keogh E, Gunopulos D. Iterative incremental clustering of data streams. In: Advances in Database Technology, Proceedings of the 9th International Conference on Extending Database Technology; 2004. p. 106–22.

42. Lughofer E. Extensions of vector quantization for incremental clustering. Pattern Recogn. 2008;41(3):995–1011.

43. Mallat S. A wavelet tour of signal processing, the sparse way. Amsterdam: Academic Press; 2009.

44. Manku GS, Motwani R. Approximate frequency counts over data streams. In: Proceedings of the 28th International Conference on Very Large Data Bases; 2002. p. 346–57.

45. Masud MM, Gao J, Khan L, Han J, Thuraisingham B. Integrating novel class detection with classification for concept-drifting data streams. In: Proceedings of the European Conference on Machine Learning and Knowledge Discovery in Databases; 2009. p. 79–94.

46. Mateos G, Bazerque JA, Giannakis GB. Distributed sparse linear regression. 2010;58(10):5262–76.

47. Matias Y, Gibbons P, Poosala V. Fast incremental maintenance of approximate histograms. In: Proceedings of the 23th International Conference on Very Large Data Bases; 1997. p. 466–75.

48. McMahan B, Streeter M. Adaptive bound optimization for online convex optimization. In: Proceedings of the International Conference on Learning Theory; 2010. p. 244–56.

49. Monemizadeh M, Woodruff DP. 1-pass relative-error lp-sampling with applications. In: Proceedings of the 21st Annual ACM-SIAM Symposium on Discrete Algorithms; 2010. p. 1143–60.

50. Motwani R, Chaudhuri S, Narasayya V. Random sampling for histogram construction. How much is enough? In: Proceedings of the ACM SIGMOD Workshop on the Web and Databases; 1998. p. 436–47.

51. Papadimitriou S, Sun J, Faloutsos C. Streaming pattern discovery in multiple time-series. In: Proceedings of the 31st International Conference on Very Large Data Bases; 2005. p. 697–708.

52. Percival D, Walden A. Spectral analysis for physical applications. Cambridge: Cambridge University Press; 1993.

53. Pharr M, Humphreys G. Physically based rendering: from theory to implementation. Burlington: Morgan Kaufmann; 2010.

54. Polikar R. Ensemble based systems in decision making. IEEE Circuits Syst Mag. 2006;6(3):21–45.

55. Russel S, Norvig P. Artificial intelligence: a modern approach. Upper Saddle River: Prentice Hall; 2010.

56. Sayood K. Introduction to data compression. Morgan Kaufmann; 2005.

57. Schapire RE, Singer Y. Improved boosting algorithms using confidence-rated predictors. Mach Learn. 1999;37(3):297–336.

58. Shinozaki T, Kubota Y, Furui S. Unsupervised acoustic model adaptation based on ensemble methods. 2010;4(6):1007–15.

59. Sugiyama M, Kawanabe M, Chui PL. Dimensionality reduction for density ratio estimation in high-dimensional spaces. Neural Netw. 2010;23(1):44–59.

60. Takezawa K, editor. Introduction to nonparametric regression. Wiley; 2005.

61. Towfic ZJ, Chen J, Sayed AH. On distributed online classification in the midst of concept drifts. Neurocomputing. 2013;112(Jul):139–52.

62. Vapnik V. Statistical learning theory. New York: Wiley; 1998.

63. Wang H, Fan W, Yu PS, Han J. Mining concept-drifting data streams using ensemble classifiers. In: Proceedings of the 9th ACM SIGKDD International Conference on Knowledge Discovery and Data Mining; 2003. p. 226–35.

64. Witten IH, Frank F, Hall MA, editors. Data mining: practical machine learning tools and techniques. 3rd ed. Amsterdam: Morgan Kauffman; 2011.

65. Yi B-K, Sidiropoulos N, Johnson T, Jagadish HV, Faloutsos C, Biliris A. Online data mining for co-evolving time sequences. In: Proceedings of the 16th International Conference on Data Engineering; 2000. p. 13–22.

66. Zhang T, Ramakrishnan R, Livny M. BIRCH: an efficient data clustering method for very large databases. In: Proceedings of the ACM SIGMOD International Conference on Management of Data; 1996. p. 103–14.

67. Zhu Y, Shasha D. Statstream: statistical monitoring of thousands of data streams in real-time. In: Proceedings of the 28th International Conference on Very Large Data Bases; 2002. p. 358–69.

Streaming Applications

Yanif Ahmad and Ugur Cetintemel
Department of Computer Science, Brown University, Providence, RI, USA

Synonyms

Continuous query processing applications; Stream-oriented applications

Definition

Streaming applications typically involve the processing of continuous data streams for the purposes of filtering, aggregation, correlation, transformation, pattern matching and discovery, and domain-specific temporal analytics. These applications often require such continuous processing to be performed with both high throughput and low latency and are able to tolerate approximate results and forego some of the persistence requirements of standard database transaction processing applications.

Key Points

A large fraction of streaming applications is monitoring oriented: they involve the tracking of events or activities to identify and act upon situations (or patterns) of interest, either manually or automatically. This so-called "sense and respond" model requires query results to be generated in real time (meaning low latency) as results lose their utility over time. As such, persistence of all the input data is often not an application requirement, unlike in traditional database applications. Thus, most input data can be simply discarded or, alternatively, asynchronously recorded for archival needs. As online data sources proliferate and consolidate, streaming applications need to deal with increasingly higher volume streams, which makes real-time operation especially challenging.

A flagship stream processing application is automated trading, which continually watches market streams (bids and asks) from financial feed providers (e.g., Reuters), evaluating sophisticated real-time patterns over them to identify arbitrage opportunities and automatically act on them. For automated trading, the desired processing latencies are in milliseconds (and continually decreasing) and the estimated peak input data rates are in 170,000 messages/s (as of July 2006), with rates roughly doubling every year [1]. Network monitoring is another application that has stringent real-time response needs under high data volumes: network elements (e.g., routers, gateways) are instrumented to log, summarize, and forward traffic data, which is then analyzed to identify and automatically respond to online security attacks (e.g., denial of service attacks) and QoS problems (e.g., SLA violations). Another early streaming application is event detection in MMORPGs, where the virtual game world is continually monitored to identify oddities, semantic bugs, and cheats.

Overall, streaming applications abound in various verticals including:

- Financial services: automated trading, market feed processing (cleaning, smoothing, and translation), smart order routing, and real-time risk management and compliance (MiFID, RegNMS)
- Government and military: surveillance, intrusion detection and infrastructure monitoring, and battlefield command and control
- Telecommunications: network management, quality of service (QoS)/service level agreement (SLA) management, and fraud detection
- Web/E-business: click-stream analysis and real-time customer experience management (CEM)
- Entertainment: online gaming (online cheat, bug detection)
- Retail and logistics: automated supply-chain management
- Healthcare: patient monitoring
- Energy: power-grid/pipeline monitoring and control

Cross-References

▶ Data Stream Management Architectures and Prototypes
▶ Stream-Oriented Query Languages and Operators

Recommended Reading

1. Options Price Reporting Authority (OPRA) Traffic Projections, http://www.opradata.com/specs/projections_2005_2006.pdf

Stream-Oriented Query Languages and Operators

Mitch Cherniack[1] and Stan Zdonik[2]
[1]Brandeis University, Wattham, MA, USA
[2]Brown University, Providence, RI, USA

Synonyms

Continuous query languages

Definition

Many research prototypes and commercial products have emerged in the new area of stream processing. All of these systems support a language for specifying queries. A fundamental difference between a stream query language and a conventional query language like SQL is that stream queries are not one-time computations, but rather, they continue to produce answers as new tuples arrive on one or more input streams. Thus, queries are registered with the system and answers continue to evolve over time. This new assumption is crucial to understanding some of the technical differences that arise in stream query languages.

Most stream query languages try to extend SQL in one way or another. The form of these extensions can be either a purely textual extension of SQL or GUI, through which users can construct dataflow diagrams that connect extended versions of relational operators. These days, many systems provide both.

The most fundamental addition to a stream query language over their relational counterparts is the notion of a window. Windows produce finite structures (i.e., tables) from infinite structures (i.e., streams). Much of the technical detail of a streaming data model revolves around the specifics of how windows are formed. This will be discussed in detail below.

One of the biggest technical challenges in the implementation of such a stream query language is the ability to produce answers with minimum latency. The latency requirement is a reaction to the kinds of applications that stream processing was invented to address. In broad terms, these applications have to do with monitoring conditions on the input streams. Typically, in this setting, the value of the answer decays quickly.

The main thrust of this article is on the technical choices that must be faced by anyone who designs a stream query language. The main concepts with examples from major systems will be illustrated. There are also many related technologies and associated languages (e.g., XML streams, temporal databases, active databases) that are not discussed in this article. The focus is on extensions to the relational model and relational languages (e.g., relational algebra, SQL) that incorporate streams.

Historical Background

There has been significant work on stream query languages in the past. The academic languages include:

CQL [3, 10] (from the STREAM project out of Stanford),
SQuAl [1, 6] (from the Aurora/Borealis project out of Brandeis, Brown and MIT), and
ESL [4, 13] (from the Atlas project out of UCLA).

The commercial languages include:

StreamSQL [11] (from Streambase),
CCL [9] (from Coral8),
EQL [7] (from Esper), and
StreaQuel [5, 8] (from Truviso).

The commercial stream query languages are, in many cases, derived from the academic languages; StreamSQL is derived from SQuAl and CQL, CCL is derived from CQL, and StreaQuel is derived from a language of the same name from the Telegraph CQ project out of UC Berkeley. The material from which information about these languages was gleaned is listed at the end. Especially in the case of commercial languages, published documentation on these languages is sometimes incomplete. Thus, there may be omissions in the descriptions of the features of these languages that follow. For example, documentation on StreaQuel is especially scant and so the description for this language is likely to be incomplete.

Foundations

Stream query languages (both academic and commercial) primarily differ in how they approach the most fundamental requirements of stream processing. These requirements include:

1. *Language Closure*: Are the language's operators closed under streams? Or does the language support operators that convert streams to relations and/or relations to streams? The approach a language takes to closure reveals a lot about how tightly integrated is the language with a relational query language such as SQL.
2. *Windowing*: Does the language support first-class windows? (I.e., are windows namable, sharable and queryable?). Or are windows internal to the definition of stateful operations. And what kinds of windows can be expressed in the language?
3. *Correlation*: Does the language provide a way to correlate tuples (events) arriving on a

stream with historical data, and with tuples arriving on a separate stream?

4. *Pattern Matching*: Does the language have a way to identify interesting subsequences of tuples on one or more streams?

Language Closure: The simplicity of the relational algebra/calculus is largely due to the closure property that says that all inputs and outputs of relational queries are relations. Beyond simplifying the type system, closure also ensures that the output of any one query can be input to another.

The various stream query languages compared here approach closure in one of two ways. Languages such as SQuAl and EQL define a language that is closed under streams. That is, every operator in these two languages accepts one or more streams as input and produces streams as output. (SQuAl includes two operators that have the side-effect of accessing a relation (*WRITESQL* and *READSQL*), but both return stream outputs.) The other languages in this list include both streams and relations in their type system. For example, CQL includes no stream-to-stream operations. Instead, operation on a stream demands that it first be converted into a relation (via windowing). Thereafter, all query operations are relation-to-relation operations as in SQL. If the desired result is a stream, the specialized operators *ISTREAM*, *DSTREAM* and *RSTREAM* produce stream outputs from relational inputs by (essentially) returning the log of changes to the relations in the order in which they occur. The other languages in this list (StreamSQL, CCL and StreaQuel) largely follow the CQL model, except that queries in these languages that contain exactly one unwindowed stream in the FROM clause are considered to be stream-to-stream. For example, in these languages, the query,

- *SELECT* *
- *FROM S*
- *WHERE p*

returns a stream if S is a stream. In CQL, this query is assumed to be a syntactic shorthand for,

- *SELECT* **ISTREAM** *
- *FROM S [∞]*
- *WHERE p*

and thus also returns a stream, but only as a result of first windowing S into a relation (see the subsection below on windowing) and then converting the result back into a stream.

Many of the query languages that include both relations and streams in their data model also define query operations that produce relations from streams and streams from relations. As mentioned previously, CQL produces streams from relations using windows, and the specialized operations *ISTREAM*, *DSTREAM* and *RSTREAM* produce relations from streams. StreamSQL, CCL and StreaQuel all include additional operations for producing relations from windows (*INSERT*, *UPDATE* and *DELETE*), which update pre-specified relations with the arrival of each tuple on a stream in the same way that the equivalent SQL operations update a relation with an individual tuple. StreaQuel and StreamSQL also support *ISTREAM* and *DSTREAM* (and in the case of StreaQuel, *RSTREAM*) operations that produce streams from relations, as in CQL. No such operations are described in the publicly available literature on CCL as of the time of this writing.

Table 1 summarizes the closure properties of the academic and commercial languages studied in this report, as well as the operations the languages support (if any) for producing relations from streams and streams from relations. Note that all stream languages support windowing as a means of producing a relation from a stream. However, some of these languages (e.g., SQuAl, ESL, EQL) are still considered to be closed under streams because their window definitions are internal to the query's operation and not output.

Windowing: All stream query languages have some form of windowing to convert infinite streams into automatically maintained, time-varying relations. But different query languages vary in whether they support first-class windows, and in terms of the features of window

Stream-Oriented Query Languages and Operators, Table 1 Closure properties of stream query languages

Features		Closed under	Operations		
			Stream-to-stream	Stream-to-relation	Relation-to-stream
Academic languages	CQL (STREAM)	Streams, Relations	No, but many queries assume use of ISTREAM	Windows	ISTREAM, DSTREAM, RSTREAM
	SQuAl (Aurora)	Streams	Default	Windows, WRITESQL	READSQL
	ESL (Atlas)	Streams	Default	Windows	–
Commercial languages	StreamSQL (Streambase)	Streams, Relations	All queries with unwindowed stream in FROM clause	Windows, INSERT, UPDATE, DELETE	ISTREAM, DSTREAM
	CCL (Coral8)	Streams, Relations	All queries with unwindowed stream in FROM clause	Windows, INSERT, UPDATE, DELETE	–
	EQL (Esper)	Streams	Default	Windows	–
	StreaQuel (Truviso)	Streams, Relations	Default (from implicit use of ISTREAM and RSTREAM)	Windows, Active Tables	?

definition that they support. Table 2 summarizes how various stream query languages support windows.

A first-class window is a window that can be named, shared and independently queries. Put simply, a query language supports first-class windows allows a window to be named and defined as the result of a statement in the query language, and subsequent queries can then access this window. Windows are not first-class if they are defined as part of a query, but are not visible outside of the execution of that query. First-class windows are typically supported in query languages that are closed under both streams and relations (i.e., CQL, StreamSQL, CCL and StreaQuel). Languages such as SQuAl, EQL and ESL that are closed under streams define windows internally within queries.

All windows are characterized as having a certain size, and advancing in some way (i.e., adding new tuples and deleting old ones as new tuples arrive on a stream). Though each language has its own syntax for specifying how to size and advance a window, all of the languages discussed looked at support windows whose size

is "row-based" (i.e., defined by the number of tuples contained in the window) or "time-based" (i.e., defined by the maximum time interval between any two tuples in the window). SQuAl and StreamSQL also support "value-based" windows over streams whose tuples are known to arrive in ascending order on some data field of the tuple. Value-based windows specify the maximum difference in value of that attribute between tuples in the same window. All languages studied in this article looked at support sliding by some query-specified amount (expressed in number of tuples, increment in time, or (in the case of SQuAl and StreamSQL), increment in value of the attribute over which the windowed stream is ordered. As well, all of the languages discussed here support "tumbling windows," which are windows that contain tuples that belong to exactly one window.

The last three columns of Table 2 show some of the window properties that are supported by some but not all of the query languages discussed here. For example, both CQL and SQuAl support a form of sampling to determine the contents of a window. In the case of CQL, a window defined with *SAMPLE* will consist of a subset of the

Stream-Oriented Query Languages and Operators, Table 2 Windowing support in stream query languages

Features		First-class	Windows (sizing)			Windows (movement)			Windows (other features)	
			Row-based	Time-based	Value-based	Sliding	Tumbling	Sampling	Top-K	Landmark
Academic languages	CQL (STREAM)	Yes	Yes	Yes	No	Yes	Yes	X% SAMPLE	–	–
	SQuAl (Aurora)	No	Yes	Yes	Yes	Yes	Yes	RESAMPLE	BSORT	–
	ESL (Atlas)	No	Yes	Yes	No	Yes	Yes	Definable with UDA's	Definable with UDA's	Definable with UDA's
Commercial languages	StreamSQL (Streambase)	Yes	Yes	Yes	Yes	Yes	Yes	–	EVICT MIN/MAX	–
	CCL (Coral8)	Yes	Yes	Yes	No	Yes	Yes	–	KEEP LARGEST/SMALLEST	Yes (KEEP FOR DURATION)
	EQL (Esper)	No	Yes	Yes	No	Yes	Yes	–	–	–
	StreaQuel (Truviso)	Yes	Yes	Yes	No	Yes	Yes	?	?	Yes

most items that have arrived on the windowed stream. In the case of SQuAl, a window defined with *RESAMPLE* will use interpolation with a user-supplied function to fill-in missing values from the windowed stream. SQuAl, StreamSQL and CCL all support "top-k" (or "bottom-k") windows, which at any point in time, contain those tuples that have arrived on the windowed stream that have the maximum (minimum) values of some specified attribute. And CCL and StreaQuel both support "Landmark Windows"; so-named because these are windows which are fixed at the start-point with an end-point that advances as tuples arrive on the windowed stream. ESL has expressive user-defined aggregate support whereby aggregates are defined with SQL statements on an internal table that specify how to initialize, increment and return a final result, and this mechanism could be used to specify each of the window types described here.

Correlation: A key component of any stream query language is its support for correlating tuples appearing on a stream with either a repository of historical data, or with the tuples appearing on another stream. As Table 3 shows, all of the query languages studied here support both forms of correlation, though in different ways.

The correlation of a tuples on a stream with historical data are expressed in most languages by allowing exactly one stream to appear in a stream query's FROM clause. Then, either periodically (e.g., CQL) or upon the arrival of each tuple on this stream (e.g., StreamSQL), the query is reevaluated using the tuple(s) that have arrived on the stream since the last time the query was evaluated. ESL, StreamSQL, CCL, EQL and StreaQuel all allow FROM clauses to include one unwindowed stream for this purpose. A CQL query containing one unwindowed stream implicitly windows that stream using the "NOW" size directive, which says to create a window with all tuples that have arrived on the stream since the last time when all queries were reevaluated. Thus, the FROM clause of a CQL query always consists solely of relations including those resulting from windowing streams. SQuAl, which has

a graphical rather than SQL-like notation, uses the operation, *READSQL* to correlate stream data with historical data.

Most query languages correlate the tuples on two streams by first windowing at least one (or in case of CQL, both) of the streams. The resulting query then is either a join of two relations as in SQL, or a join of an unwindowed stream with a relation as described in the previous paragraph. StreamSQL also has a *GATHER* operation which performs key matching to match each tuple on an input stream with the single tuple it matches on each of the other input streams.

Pattern Matching: Pattern matching is a relatively new addition to stream query languages, and is only supported in the commercial languages examined in this study (though there exist some work in the academic literature on pattern matching on streams such as [12]). Pattern matching in stream query languages can take one of two forms:

Multi-stream pattern matching resembles joins between streams in that it correlates tuples appearing on separate streams according to the order in which they arrive. For example, this form of pattern matching might identify all cases where a particular stock received a bid quote (on a BIDS stream) without a corresponding ask quote (on an ASKS stream) within some specified time period. All of the commercial stream query languages examined here (StreamSQL, CCL, EQL and StreaQuel) support this form of pattern matching.

Single-stream pattern matching looks for a sequence pattern of tuples arriving on a single stream, and typically uses a rich pattern matching language based on regular expressions to express desired patterns. For example, this form of pattern matching might identify all sequence of quotes for a particular company that resulted in an "M-pattern" whereby the stock's price rises for a time, then falls, then rises again and then falls again [2]. This form of pattern matching is part of a general SQL standard proposal put forth by Streambase, Oracle and IBM, and documented in [13].

S

Stream-Oriented Query Languages and Operators, Table 3 Correlation support in stream query languages

Features		Correlation	
		Stream-to-relation	Stream-to-stream
Academic languages	CQL (STREAM)	No, but relation-to-relation correlation with window on most recent tuples in stream (NOW) often has same effect	Must window both streams
	SQuAl (Aurora)	READSQL	JOIN
	ESL (Atlas)	JOIN with Stream, Relation in FROM Clause	Must window one of the streams
Commercial languages	StreamSQL (Streambase)	JOIN with Stream, Relation in FROM Clause	Must window one of the streams
	CCL (Coral8)	JOIN with Stream, Relation in FROM Clause	Must window one of the streams
	EQL (Esper)	JOIN with Stream, Relation in FROM Clause	Must window one of the streams
	StreaQuel (Truviso)	JOIN with Stream, Relation in FROM Clause	Must window one of the streams

Stream-Oriented Query Languages and Operators, Table 4 Pattern matching support in stream query languages

Features		Pattern matching	
		Multi-stream	Regular expression
Academic languages	CQL (STREAM)	–	–
	SQuAl (Aurora)	–	–
	ESL (Atlas)	–	–
Commercial languages	StreamSQL (Streambase)	Yes (MATCH)	Yes (PATTERN)
	CCL (Coral8)	Yes (MATCHING)	No
	EQL (Esper)	Yes (PATTERN)	Yes (PATTERN)
	StreaQuel (Truviso)	Yes (EVENT clause)	Simple (A B C)

A summary of each query language's support for pattern matching is shown in Table 4.

In short, while there are several query languages for streams with both academic and commercial roots, these languages share much in common that identify the crucial requirements of stream processing. Specifically, all of these languages are closed either under streams, or under streams and relations; all have some notion of windowing to convert streams to relations; all provide ways to correlate tuples on a stream with both historical data and tuples on other streams, and all commercial languages support some form of pattern matching to identify interesting subsequences of tuples from one or more streams.

Key Applications

The applications of stream processing revolve around low-latency monitoring of physical or virtual items or events of interest. Examples include automated trading, network security monitoring, and event detection in massively multiplayer online games.

Cross-References

▶ Continuous Query
▶ Data Stream
▶ Event and Pattern Detection over Streams

► Punctuations
► Stream Processing
► Windows

References

1. Abadi D, Carney D, Cetintemel U, Cherniack M, Convey C, Lee S, Stonebraker M, Tatbul N, Zdonik S. Aurora: a new model and architecture for data stream management. VLDB J. 2003;12(2): 120–39.
2. Anon S. Pattern matching in sequences of rows. SQL Standard Proposal. http://asktom.oracle.com/ tkyte/row-pattern-recogniton-11-public.pdf; 2007.
3. Arvind A, Shivnath B, Jennifer W. The CQL continuous query language: semantic foundations and query execution. VLDB J. 2006;15(2):121–42.
4. Bai Y, Thakkar H, Luo C, Wang H, Zaniolo C. A data stream language and system designed for power and extensibility. In: Proceedings of the international conference on information and knowledge management; 2006. p. 337–46.
5. Chandrasekaran S, Franklin M. Streaming queries over streaming data. In: Proceedings of the 28th international conference on very large data bases; 2002. p. 203–14.
6. Cherniack M. SQuAl: The Aurora [S]tream [Qu]ery [Al]gebra, Technical Report, Brandeis University; 2003.
7. Codehaus.org. Esper online documentation set, http:/ /esper.codehaus.org/tutorials/tutorials.html; 2007.
8. Conway N. An introduction to data stream query processing. Slides from a talk given on May 24, 2007. http://www.pgcon.org/2007/schedule/attachments/ 17-stream_intro.pdf; 2007.
9. Coral8 Systems, Coral8 CCL Reference Version 5.1. http://www.coral8.com/system/files/assets/pdf/ current/Coral8CclReference.pdf; 2007.
10. Jennifer W. CQL: a language for continuous queries over streams and relations. Slides from a talk given at the Database Programming Language (DBPL) Workshop, Potsdam, Germany; 2003. http://www-db. stanford.edu/~widom/cql-talk.pdf
11. Streambase Systems. StreamSQL online documentation set. http://streambase.com/ developers/docs/latest/streamsql/index.html; 2007.
12. Wu E, Diao Y, Rizvi S. High-performance complex event processing over streams. In: Proceedings of the ACM SIGMOD international conference on management of data; 2006. p. 407–18.
13. Zaniolo C, Luo C, Wang H, Bai Y, Thakkar H. An introduction to the Expressive Stream Language (ESL), Technical Report, UCLA.

Strong Consistency Models for Replicated Data

Alan Fekete
University of Sydney, Sydney, NSW, Australia

Synonyms

Copy transparency; Strong memory consistency

Definition

If a distributed database system keeps several copies or replicas for a data item, at different sites, then a replica control protocol determines how the replicas are accessed. Some replica control protocols ensure that clients never become aware that the data are replicated. In other words, the system provides the transparent illusion of an unreplicated database. Such a system is described as offering a strong consistency model. 1-copy-serializability (q.v.) is the best-known strong consistency model.

Historical Background

Early work in the 1970s investigated a range of replica control mechanisms, usually with the intention of providing transparent serializability. In the early 1980s, Bernstein and colleagues formalized the concept of 1-copy-serialiability as a consistency model [1], with a careful proof technique [2] like that for single-site serializability. Herlihy [8] extended these ideas to replicating data types with general operations (not just read and write).

1996 marked the seminal paper by Gray et al. who used some simple performance models to show the scalability barriers for different system designs [7]. This inspired research on ways to gain 1-copy serializability within a lazy single-master replica control, through restrictions on the placement of copies, and/or the ordering of update propagation [3, 4].

Since 2000, the prevalence of DBMS platforms with Snapshot Isolation concurrency control led to research on replication for these, especially by groups of researchers centered on Alonso and Kemme. The concept of 1-copy-SI was defined [10], and a proof theory was developed [9]. Variant consistency models were defined by considering different session properties [5, 6].

Foundations

A major theme in the development of distributed databases has been *transparency*, that is, the clients should not have to change if they interact with a distributed system rather than a traditional, single-site database. Transparency applies to many aspects: table naming should be the same as for a single DBMS, queries should not need rewriting even if data is fragmented between sites, etc. The shift from a single-site system to a replicated one should be seen in better quality of service, but not in altered functionality. Since a single-site database has only one copy of each data item, a transparently replicated system will provide clients with the illusion of a single copy, hiding all evidence of the replication. A scheme for replica control (q.v.) that does this can be described as providing a strong consistency model. Other replica control mechanisms do not hide the fact of replication.

There are in fact several variants among strong consistency models, because there are several different isolation models used by different DBMS platforms, and because the formal definition of isolation doesn't always capture exactly the properties of an implementation. The next paragraphs describe the main strong consistency models that have been proposed for replicated data.

Replicated Serializability. The theory of concurrency control has an established notion of

correct functionality for ACID transactions in a single-site DBMS: serializability (q.v.). This is defined by having execution equivalent to a serial (i.e., batch, non-overlapped) execution of the same transactions. When a replicated database gives to clients the transparent appearance of a single-site system with serializable transactions, one says that the replica consistency model is 1-copy-serializability (q.v.). That is, the operations of the transactions are indistinguishable from what happens if they are run serially in a database with only one site.

To illustrate the consistency model, consider a database with two logical items, x and y representing respectively the balance in the checking and savings bank accounts for a single customer. There is a single client C which submits two transactions. One client transaction $T_{1,C}$ is a transfer of two units of funds from y to x, and $T_{2,C}$ is a transaction to display the status of the customer's finances. Initial values are $x = 10$ and $y = 20$.

In (1) below is a sequence of events that might happen in a system with two sites A and B, using an eager locking-based read-one-write-all replica control. Notice how each client-submitted transaction has subtransactions at the local sites A and B; $T_{1,C}$ reads both items at site A, and then updates replicas at both sites; while $T_{2,C}$ reads the replica x^A and the replica y^B. The notation that is used in this and later examples is to indicate the event where a value 5 is written to the local replica of item x at site A, as part of transaction T_1, by $w_1[x^A,5]$. Here the subscript on the event type indicates the transaction involved, and the superscript on the item name indicates the site of the replica which is affected. The event where a client C running transaction T_3 has requested a read of the logical data item y, and the value 6 is returned, will be represented by $r_{3,C}[y,6]$. Note there is no superscript on the item since this is the client's view. Many consistency models need to refer to where transactions start and finish, so one also has events like $b_{3,C}$ for the start of transaction T_3 at client C, or $c_{3,C}$ for the commit of that transaction by the client, or indeed c_3^A for the commit of the local subtransaction of T_3 which is running at site A.

$$b_{1,C} b_1^A r_1 \left[x^A, 10\right] r_{1,C} \left[x, 10\right] b_{2,C} b_2^A r_1 \left[y^A, 20\right]$$
$$r_{1,C} \left[y, 20\right] w_1 \left[x^A, 12\right] b_1^B w_1 \left[x^B, 12\right] w_{1,C} \left[x, 12\right]$$
$$w_1 \left[y^A, 18\right] w_1 \left[y^B, 18\right] w_{1,C} \left[y, 18\right] c_1^A r_2 \left[x^A, 12\right]$$
$$r_{2,C} \left[x, 12\right] b_2^B c_1^B c_{1,C} r_2 \left[y^B, 18\right]$$
$$r_{2,C} \left[y, 18\right] c_2^A c_2^B c_{2,C} \tag{1}$$

When one hides the internal details (the events at the replicas), and only considers what the client sees, the relevant event sequence is

$$b_{1,C} r_{1,C} \left[x, 10\right] b_{2,C} r_{1,C} \left[y, 20\right] w_{1,C} \left[x, 12\right]$$
$$w_{1,C} \left[y, 18\right] r_{2,C} \left[x, 12\right] c_{1,C} r_{2,C} \left[y, 18\right] c_{2,C} \tag{2}$$

This sequence is 1-copy-serializable, because the client sees the same as in a serial execution on a single-site DBMS, in the order $T_{1,C}$ then $T_{2,C}$.

Here is a sequence that might occur in a system where site A is the primary or master site, where all updates are initially done, and site B has secondary replicas which are updated through copier transactions that lazily apply the write sets of any update transaction. The copier that transmits values produced by $T_{1,C}$ from site A to site B will be denoted by $T_\oplus^B$; notice that in contrast to T_1^B in the eager system modeled previously, $T_\oplus^B$ is not a subtransaction of any global client-submitted transaction but instead it can commit independently.

$$b_{1,C} b_1^A r_1 \left[x^A, 10\right] r_{1,C} \left[x, 10\right] r_1 \left[y^A, 20\right] r_{1,C} \left[y, 20\right]$$
$$w_1 \left[x^A, 12\right] w_{1,C} \left[x, 12\right] w_1 \left[y^A, 18\right]$$
$$w_{1,C} \left[y, 18\right] c_1^A c_{1,C} b_{2,C}$$
$$b_2^B r_2 \left[x^B, 10\right] r_{2,C} \left[x, 10\right] r_2 \left[y^B, 20\right] r_{2,C} \left[y, 20\right] c_2^B$$
$$c_{2,C} b_\oplus^B w_\oplus \left[x^B, 12\right] w_\oplus \left[y^B, 18\right] c_\oplus^B \tag{3}$$

Again hiding the internal details, and considering only what the client sees, it is

$$b_{1,C} r_{1,C} \left[x, 10\right] r_{1,C} \left[y, 20\right] w_{1,C} \left[x, 12\right]$$
$$w_{1,C} \left[y, 18\right] c_{1,C} \tag{4}$$
$$b_{2,C} r_{2,C} \left[x, 10\right] r_{2,C} \left[y, 20\right] c_{2,C}$$

This sequence is also 1-copy-serializable, since the values read are what could happen with an unreplicated system running the transactions serially in the order $T_{2,C}$ then $T_{1,C}$.

There is a detailed theory that allows one to prove that this property holds for schedules of certain read-one-write-all replica control mechanisms.

Session properties. The sequence of events shown in sequence (4) is indeed something that could happen according to the definition of serializable execution in a single-site DBMS, but it would never happen in a single site DBMS that used a concurrency control mechanism like two-phase locking. It might be very disturbing to be a client, who submits a transfer transaction $T_{1,C}$, learns that the transfer succeeded, and then uses $T_{2,C}$ to check the status of their accounts and is told that the initial balances are still unchanged. In the definition of serializability, it is enough that there exists some way to order the transactions and perform them serially; but in DBMS products, the concurrency control makes sure that the apparent serial order does not rearrange transactions unless they are actually concurrent (that is, unless they overlap). Any single-site system will not allow a situation where $T_{1,C}$ has completed, and then $T_{2,C}$ starts, but the apparent serial order has an inversion of the transaction order, with $T_{2,C}$ coming first. Thus it is often proposed to have explicit session properties [11] in a consistency model, to require that the apparent order have some relationship to what really happened. A very restrictive session requirement is *external consistency*; this means that whenever $T_{i,C}$ completes before $T_{j,D}$ starts, then the apparent serial order must contain $T_{i,C}$ ahead of $T_{j,D}$. A less restrictive property is *session consistency*, which says that the apparent order must not rearrange transactions from the same client, where one completes before the other starts. That is, whenever $T_{i,C}$ completes before $T_{k,C}$ starts, then the apparent serial order must contain $T_{i,C}$ ahead of $T_{k,C}$. But session consistency says nothing about the serialization order of $T_{i,C}$ and $T_{j,D}$ where $C \neq D$.

Replicated Snapshot Isolation. Several prominent single-site DBMS platforms provide isolation for transactions using a multiversion mechanism called Snapshot Isolation. This does not have exactly the properties of Serializability, but it avoids most of the known bad concurrency problems, and it seems to satisfy most application programmers. The key to this isolation level is that when a transaction reads an item, it sees the value which reflects all writes by other transactions which committed before the reading transaction started, but it does not see any effects of concurrent transactions.

When replicating data stored in platforms that offer Snapshot Isolation, a natural strong consistency model is to transparently appear like an unreplicated system running on the same sort of platform. This is the consistency model known as "1-copy-SI" [10]. Here is an example with concurrent client transactions $T_{3,C}$ and $T_{4,D}$, each of which reads two logical data items and increments one of them. The sequence (5) is something that a client could see in 1-copy-SI, but not in a system offering 1-copy-serializable consistency, because the values read are not the same as the serial order $T_{3,C}$ then $T_{4,D}$ (where $T_{4,D}$ would read $y = 21$), nor are they as in $T_{4,D}$ followed by $T_{3,C}$ (where $T_{3,C}$ would read x = 11).

$$b_{4,D} r_{4,D} [x, 10] r_{4,D} [y, 20] b_{3,C} r_{3,C} [x, 10]$$
$$r_{3,C} [y, 20] w_{4,D} [x, 11] w_{3,C} [y, 21] c_{3,C} c_{4,D} \tag{5}$$

As the definition of Snapshot Isolation says that a read sees the effects of all transactions that commit before the reader's transaction started, this definition implicitly includes an external consistency session property. To allow more efficient replica control algorithms, some researchers have built systems which provide more permissive consistency models. For example, in Generalized Snapshot Isolation [6], for each transaction there is a snapshot-time, which could be somewhat before the transaction starts, and the transaction's reads see exactly the effects of those transactions that committed before the reader's snapshot-time. This allows the strange inversions where a client submits a transaction, learns of its success, and then submits another transaction that does not

run in the expected state. Thus, an intermediate consistency model is Strong Session Snapshot Isolation [5], where inversions are allowed between non-concurrent transactions from different clients, but the snapshot-time for a transaction must be later than the commit of any previous transaction submitted by the same client.

Key Applications

1-copy serializability is most often provided through eager or synchronous propagation of updates among machines that are all in a single cluster. Some database engines provide options for replication internally, whereas others rely on replication tools that run along side the DBMS engine. So far, the model of 1-copy serializability with lazy propagation of updates, or the model of replicated snapshot isolation, are offered in research prototypes rather than among commercial products.

Future Directions

The algorithms known for replicated snapshot isolation may be attractive for practical use, since they can offer substantially higher performance than the algorithms for 1-copy serializability, and since programmers seem able to work with snapshot isolation in a single DBMS. Thus, this strong consistency model will probably become more widespread in the future.

Cross-References

▶ Data Replication

Recommended Reading

1. Attar R, Bernstein PA, Goodman N. Site initialization, recovery, and backup in a distributed database system. IEEE Trans Softw Eng. 1984;10(6): 645–50.

2. Bernstein PA, Goodman N. Serializability theory for replicated databases. J Comput Syst Sci. 1985;31(3):355–74.
3. Breitbart Y, Komondoor R, Rastogi R, Seshadri S, Silberschatz A. Update propagation protocols for replicated databases. In: Proceedings of the ACM SIGMOD International Conference on Management of Data; 1999.
4. Chundi P, Rosenkrantz DJ, Ravi SS. Deferred updates and data placement in distributed databases. In: Proceedings of the 12th International Conference on Data Engineering; 1996. p. 469–76.
5. Daudjee K, Salem K. Lazy database replication with snapshot isolation. In: Proceedings of the 32nd International Conference on Very Large Data Bases; 2006. p. 715–26.
6. Elnikety S, Zwaenepoel W, Pedone F. Database replication using generalized snapshot isolation. In: Proceedings of the 22nd Symposium on Reliable Distributed System; 2005. p. 73–84.
7. Gray J, Helland P, O'Neil PE, Shasha D. The dangers of replication and a solution. In: Proceedings of the ACM SIGMOD International Conference on Management of Data; 1996. p. 173–82.
8. Herlihy M. A quorum-consensus replication method for abstract data types. ACM Trans Comput Syst. 1986;4(1):32–53.
9. Lin Y, Kemme B, Patiño-Martínez M, Jiménez-Peris R, Armendáriz-Iñigo J. Snapshot isolation and integrity constraints in replicated databases. ACM Trans Comput Syst. 2009;34(2):11.
10. Plattner C, Alonso G. Ganymed: scalable replication for transactional web applications. In: Proceedings of the ACM/IFIP/USENIX International Middleware Conference; 2004. p. 155–74.
11. Terry DB, Demers AJ, Petersen K, Spreitzer M, Theimer M, Welch BB. Session guarantees for weakly consistent replicated data. In: Proceedings of the International Conference on Parallel and Distributed Information Systems; 1994. p. 140–9.

Structural Indexing

Mariano P. Consens
University of Toronto, Toronto, ON, Canada

Synonyms

Dataguide; Path index; Sketch; Structural index; Structural summary; Synopsis

Definition

Structure indexing creates summaries of the structure present in semi-structured data collections by grouping data items with similar structure, providing a mechanism to index such items. Since semi-structured data models are commonly represented by labeled graphs or trees (the XML data model being a prime example), structural indexes or summaries are naturally described as graphs where nodes represent sets of data items (called extents), and where edges represent structural relationships between the corresponding extents derived from the instance data. A concrete physical index can be created by selecting appropriate data structures to store the graph and the extents.

Structure indexing helps to find data items that satisfy structural constraints in queries by locating nodes in the structural summary graph that satisfy the query conditions (expecting far less summary nodes than data items), and then limiting query evaluation to data items in the relevant extents.

Structural summaries also provide a description of the structure present in the instance. This is in contrast with schemas, which prescribe structures that may or may not occur in an instance, but without giving an indication of the metadata that is actually present in a given collection. Note that it is possible to create a structural summary from an instance even when the instance does not conform to any schema.

Additional information can be attached to summaries (such as statistics related to nodes and relationships, and distributions of values associated with the extents), with applications in selectivity estimation and query optimization.

Historical Background

Structure indexing mechanisms were proposed as soon as the database community turned its attention to the problem of managing semi-structured data. Region Inclusion Graphs (RIG) and Region Order Graphs (ROG) [3] were proposed to help optimize the evaluation of region algebras (see [14]).

Representative Objects (RO) [10] and Dataguides [6] were motivated by a desire to describe the metadata present in semi-structured databases (modeled as labeled graphs), as well as to help in query optimization. The representative object of length 1 (1-RO) coincides with a RIG; both summaries group data items in the instance (nodes or regions, respectively) based on the label of an item. Representative objects of length k (k-RO) group instance nodes by the labels in the incoming paths of length k (or paths of arbitrary length in the case of a full representative object, or FRO). Dataguides create a summary of the path structure of a labeled graph database instance in which every label path starting at the root appears exactly once. Construction of a Dataguide is analogous to the conversion of a non-deterministic finite automaton (describing the language of the labels occurring in the instance) to a deterministic one. This construction is not unique, but a strong Dataguide is defined to be suitable as an index. For arbitrary graph data, Dataguides can in the worse case become exponentially larger than the actual instance. When instances are trees, as is the case for XML documents when links (such as IDREF) are not considered, the Dataguide size is bounded by the size of the instance.

The concept of bisimulation was introduced in the context of structural summaries by the T-index family [9]. In particular, the 1-index defines extents via partitioning the nodes of the instance using labeled bisimulation on the incoming paths. By creating a partition, bisimulation based summaries have a size bounded by the size of the instance. The F&B-Index [7] generalized the notion of bisimulation-based summaries to consider partitions created by both incoming and outgoing paths. AxPRE summaries [4] introduced a language for defining the neighborhood of interest to the bisimulation-based partitioning criteria. Depending on the axis path regular expression (AxPRE) used, all of the previously proposed bisimulation-based summaries can be defined (as well as entirely new ones).

There are also proposals that augment structural summaries with statistical information of the instance for selectivity estimation, including path/branching distribution and value distributions (e.g., XSketch [11], StatiX [5]).

Most of the existing summary proposals define all the extents using the same criteria, hence creating homogeneous summaries. These summaries are based on common element paths (in some cases limited to length k), including incoming paths (e.g., representative objects [10], dataguides [6], 1-index [9], ToXin [13], A(k)-index [8]), both incoming and outgoing paths (e.g., F&B-Index [7]), or sequences of outgoing paths (e.g., Skeleton [1]). There are also heterogeneous summaries where summary nodes adapt their criteria to query workloads (APEX [2], D(k)-index [12]), or to statistics from the instance (XSketch [11]), or where the criteria can be given explicitly (AxPRE summaries [4]).

Foundations

A suitable graph-based model for semi-structured instances is introduced below and then partition-based structural summaries are defined. The discussion then generalizes this definition to AxPRE summaries and then presents a lattice with several bisimulation-based summaries in the literature.

Semi-structured Data Example

Consider an example XML instance that consists of two RSS documents, represented as two labeled graphs in Fig. 1. RSS documents are used to encode feeds that publish frequently updated Web content such as news headlines, blog entries, and podcasts. The feeds are organized into *channels*, which in turn contain a list of *items*. A number of elements (such as *title, link, description, pubDate*) can optionally appear within channels and/or items. The items may have *content*, appearing within *groups*, that refers to multimedia files (e.g., an audio file in a podcast). The labels in the XML nodes (which are numbered for ease of reference) identify the element, and the labels in the edges the relationship (or *axis* in the XPath data model) between the XML nodes.

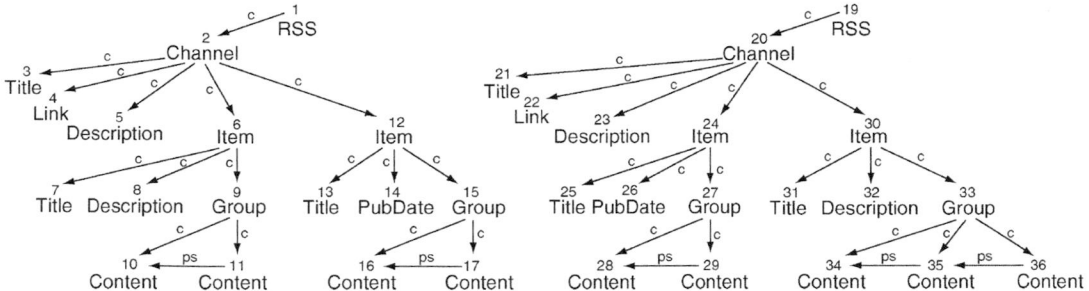

Structural Indexing, Fig. 1 Sample XML instance with two RSS feeds

The edge from node 6 to 7 labeled c means that 7 is the child of 6 (or, considering the inverse relationship, that 6 is the parent of 7). Similarly, the edge from node 29–28 labeled ps means that 28 is the preceding sibling of 29 (another XPath axis that provides information about the XML document order among siblings).

Axis Graph Definition

An *axis* graph is defined to represent XML instances. The definition can be easily applied to other semi-structure data models where instances are represented by labeled graphs.

An axis graph $\mathcal{A} = (Inst, Axes, Label, \lambda)$ is a structure where *Inst* is a finite set of nodes, *Axes* is a set of binary relations $\{E_1, \ldots, E_n\}$ in $Inst \times Inst$, *Label* is a finite set of node names, and λ is a function that assigns labels in *Label* to nodes in *Inst*. Edges in $\mathcal{A}$ are labeled by the name of the axes relations.

Structural Summary Definition

A structural summary of an axis graph $\mathcal{A}$ is another axis graph $S(\mathcal{A}) = (Sum, Axes_S, Label, \lambda_S)$ where summary nodes correspond to a partition of the nodes in $\mathcal{A}$ and summary edges are induced from the axes in $\mathcal{A}$. Each node $s \in Sum$ with $\lambda_S(s) = l$ corresponds to a subset (called extent) in a partitioning of *Inst*, such that all the nodes in the extent have the same label $l \in Label$. Also, for every edge $E_i(n, m)$ with $n, m \in Inst$ and $E_i \in Axes$ there is $E'_i(n_s, m_s)$ with $n_s, m_s \in Sum$ and $E'i \in Axes_S$, such that n_s (respectively, m_s) is the summary node that has XML node n (respectively, m) in its extent.

Sample Structural Summaries

Figure 2 shows three different structural summaries of the two RSS documents depicted in Fig. 1. The first summary, in Fig. 2a, is constructed by a partition of the XML instance nodes based solely on their labels, and as such there is a single summary node for each label in the instance. As such, the extent of the summary node s_9 labeled *description* consists of the subset {5,8,23,32} of all the XML nodes labeled *description* in the instance. Note that there are two edges labeled c incoming into the summary node s_9 since *description* appears both in *channels* and *items*.

The second summary, in Fig. 2b, is constructed from a partition that groups together XML nodes with the same ancestor elements. Note that there are two summary nodes (s_5 and s_{10}) labeled *description*, corresponding to the two possible paths of labels to the root (*channel, RSS*, and *item, channel, RSS*). Observe that, for the specific instance in the example, grouping together nodes with just the same parents would produce the same summary. However, grouping together nodes with the same children would produce a different summary (e.g., there would be two summary nodes for *item*, since the *item* nodes in the XML instance would be partitioned into the sets {6,30} and {12,24} each with the same sets of children).

The third summary, in Fig. 2c, contains an heterogeneous summary, where subsets in the partition group XML nodes with different criteria. While most the summary nodes in the figure have extents containing XML nodes that use the same criteria as above (having the

S

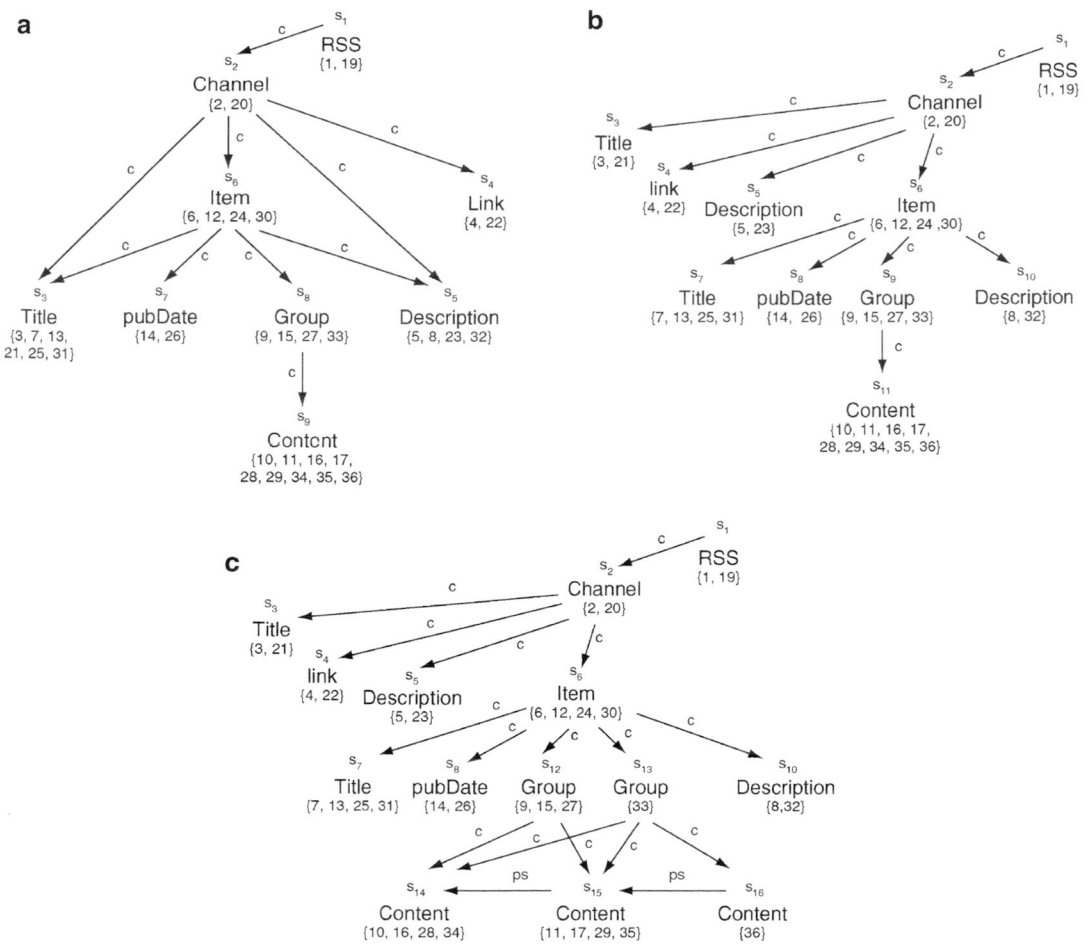

Structural Indexing, Fig. 2 Summaries of the two RSS feeds; (**a**) label, (**b**) ancestors, (**c**) heterogeneous

same ancestor elements), the summary nodes corresponding to *content* elements are partitioned according to the previous siblings (the edges labeled *ps* in the instance) and the *group* elements are partitioned according to the partition of their *content* children. So there are three summary nodes s_{14}, s_{15}, and s_{16} that group the first, second, and third *content* node siblings in the instance, and there are two summary nodes s_{12} and s_{13} corresponding, respectively, to those *group* elements that have a first and second *content* node as children, or those that have a first, second, and third *content* node as children.

AxPRE Summary Definition

An axis path regular expression (AxPRE) is a regular expression on the vocabulary of the names

of the axes relations. The AxPRE Neighborhood $N_\alpha(v)$ of an instance node $v \in Inst$ is the subgraph obtained by intersection with the prefix closed finite automaton corresponding to the AxPRE α.

A *labeled bisimulation* between two subgraphs G_1 and G_2 of an axis graph $\mathcal{A}$ is a symmetric relation $\approx$ such that for all $v \in Inst^{G_1}$, $w \in Inst^{G_2}$, $E_i^{G_1} \in Axes^{G_1}$, and $E_i^{G_2} \in Axes^{G_2}$: *(i)* if $v \approx w$, then $\lambda(v) = \lambda(w)$; *(ii)* if $v \approx w$, and $\langle v, v' \rangle \in E_i^{G_1}$, then $\langle w, w' \rangle \in E_i^{G_2}$ and $v' \approx w'$.

An AxPRE summary is a structural summary where the partition is defined as follows: two nodes $v, w \in Inst$ belong to the same partition block iff there exists a *labeled bisimulation* $\approx$ between $N_\alpha(v)$ and $N_\alpha(w)$ such that $v \approx w$.

Sample AxPRE Summaries

The summary in Fig. 2a corresponds to the AxPRE summary with an empty AxPRE, where only condition (i) in the definition of labeled bisimulation applies and the result is a partition of the instance nodes according to their labels only.

The summary in Fig. 2b corresponds to the AxPRE summary with an AxPRE p^*, which is the AxPRE that creates node neighborhoods that consists of all the ancestors of the node. As discussed earlier, for the specific instance in the example, the AxPRE could also be p, but it can not be c.

The summary in Fig. 2c is an heterogeneous summary where most summary nodes have the partition defined according to the AxPRE p^*, but the summary nodes labeled *content* use $p^*|ps^*$ and those labeled *group* use $p^*|c.ps^*$.

Summary Lattice

Figure 3 shows a lattice with relationships among several AxPRE summaries that capture bisimilarity-based proposals mentioned earlier. The lattice represents the partition refinement relationship between the summaries. Each node in the lattice of Fig. 3 corresponds to a homogeneous summary defined by the AxPRE label, with an additional textual label for the corresponding summary name. A node is also used to represent an homogeneous AXPRE summary based on the AxPRE $p^*|c.ps^*$, which applied to only some of the summary nodes in the example in Fig. 2c.

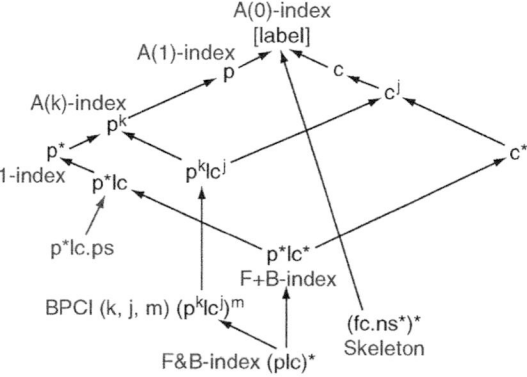

Structural Indexing, Fig. 3 Summary lattice

Key Applications

Indexing, metadata description, query processing and optimization, selectivity estimation.

Future Directions

Structure indexing techniques can be extended to support additional data models and their associated query languages, and to further refine their adaptability to different workloads.

Cross-References

▶ Query Processing and Optimization in Object Relational Databases
▶ Selectivity Estimation
▶ Semi-structured Data
▶ XML
▶ XPath/XQuery

Recommended Reading

1. Buneman P, Choi B, Fan W, Hutchison R, Mann R, Viglas S. Vectorizing and querying large XML repositories. In: Proceedings of the 21st International Conference on Data Engineering; 2005. p. 261–72.
2. Chung C-W, Min J-K, Shim K. APEX: an adaptive path index for XML data. In: Proceedings of the ACM SIGMOD International Conference on Management of Data; 2002. p. 121–32.
3. Consens MP, Milo T. Optimizing queries on files. In: Proceedings of the ACM SIGMOD International Conference on Management of Data; 1994. p. 301–12.
4. Consens MP, Rizzolo F, Vaisman AA. AxPRE summaries: exploring the (semi-)structure of XML web collections. In: Proceedings of the 24th International Conference on Data Engineering; 2008. p. 1519–21.
5. Freire J, Haritsa JR, Ramanath M, Roy P, Simeon J. StatiX: making XML count. In: Proceedings of the ACM SIGMOD International Conference on Management of Data; 2002. p. 181–91.
6. Goldman R, Widom J. Dataguides: enabling query formulation and optimization in semistructured databases. In: Proceedings of the 23th International Conference on Very Large Data Bases; 1997. p. 436–45.
7. Kaushik R, Bohannon P, Naughton JF, Korth HF. Covering indexes for branching path queries. In: Pro-

ceedings of the ACM SIGMOD International Conference on Management of Data; 2002. p. 133–44.

8. Kaushik R, Shenoy P, Bohannon P, Gudes E. Exploiting local similarity for indexing paths in graph-structured data. In: Proceedings of the 18th International Conference on Data Engineering; 2002. p. 129–40.

9. Milo T, Suciu D. Index structures for path expressions. In: Proceedings of the 7th International Conference on Database Theory; 1999. p. 277–95.

10. Nestorov S, Ullman JD, Wiener JL, Chawathe SS. Representative objects: concise representations of semistructured, hierarchial data. In: Proceedings of the 13th International Conference on Data Engineering; 1997. p. 79–90.

11. Polyzotis N, Garofalakis MN. XSketch synopses for XML data graphs. ACM Trans Database Syst. 2006;31(3):1014–63.

12. Qun C, Lim A, Ong KW. D(k)-index: an adaptive structural summary for graph-structured data. In: Proceedings of the ACM SIGMOD International Conference on Management of Data; 2003. p. 134–44.

13. Rizzolo F, Mendelzon AO. Indexing XML data with ToXin. In: Proceedings of the 4th International Workshop on the Web and Databases; 2001. p. 49–54.

14. Young-Lai M, Tompa FW. One-pass evaluation of region algebra expressions. Inform Syst. 2003;28(3):159–68.

Structure Analytics in Social Media

Sihem Amer-Yahia[1,2], Mahashweta Das[3], Gautam Das[4], Saravanan Thirumuruganathan[4,5], and Cong Yu[6]
[1]CNRS, Univ. Grenoble Alps, Grenoble, France
[2]Laboratoire d'Informatique de Grenoble, CNRS-LIG, Saint Martin-d'Hères, Grenoble, France
[3]Visa Research, Palo Alto, CA, USA
[4]Department of Computer Science and Engineering, University of Texas at Arlington, Arlington, TX, USA
[5]Qatar Computing Research Institute, Hamad Bin Khalifa University, Doha, Qatar
[6]Google Research, New York, NY, USA

Synonyms

Aggregate analytics in social media; Exploratory mining in social media; User-generated content analysis

Definition

Structure analytics in social media is the process of discovering the structure of the relationships emerging from social media use, by leveraging the rich metadata associated with items and users in online sites. It focuses on identifying the users involved, the activities they undertake, the actions they perform, and the items they create and interact with. Example items can be movies, restaurants, entities, and Web pages. The objective of structure analytics in social media is to identify interesting patterns in large amounts of user-generated content such as product reviews, rating, forums, and social media conversations and use that knowledge in subsequent actions. An example mining task is finding groups of reviewers who have similar feedback (such as high ratings) for similar (or diverse) sets of items (such as movies by the same director). Unlike unsupervised clustering approaches, groups returned by this form of analytics are meaningful due to the common structural attribute-value pairs shared by all users in each group. The core challenge lies in finding good groups from the very large space of all possible groups that results from combining the users' demographic information (such as age, gender, etc.) with the item attributes (such as genre for movies, cuisine for restaurants, etc.).

Historical Background

Social content sites that combine traditional content sites with social networking features have become increasingly prevalent. The rise of social media has necessitated the development of tools that analyze user-user and user-item interactions so as to mine interesting patterns that could be used to improve user experience. Additionally, these sites act as a medium to transmit or share information with a broad audience, as well as to create and distribute content. Previously, the analytic tasks over social media data have been fragmented and often siloed into analyzing the social network, user-item interactions, and user-generated content individually. A key feature of structure analytics is the unified treatment of all these aspects during mining.

Community detection is an example of an analytic task that only uses the topological properties of a social network (user-user interactions). It tries to identify communities which are highly connected substructures that are defined according to an objective function that formalizes the idea of groups with more intragroup than intergroup connectivity. Users in a community typically share common properties such as interests or have similar interactions within the graph. Please see [5] for a recent survey. Clustering, on the other hand, is a more generic analytic technique that groups users based on their pairwise similarity (provided by a similarity matrix). The pairwise similarity could take into account the similarity between users based on profiles and their item interactions. However, it is possible that the resulting clusters might not be structurally describable and hence not easily interpretable.

Recent years have witnessed a confluence of techniques from diverse fields like data mining, databases, machine learning, and information retrieval to process social media content. While structure analytics has similarities to well-studied topics in these fields, it is a distinct topic with unique technical challenges and techniques. Das et al. developed a general framework [2–4] for analyzing user-item interaction behavior in collaborative social content sites. The authors identified a family of analytic tasks that apply two opposing measures: similarity and diversity, to the three main components in the online sites (users, items, and user-item interactions (i.e., feedback)), and developed efficient algorithms for solving them.

Foundations

The fundamental problem in structure analytics is modeling the structure in relationship between users, items, and user-item interaction (i.e., user feedback for items) in order to allow the simultaneous analysis of how user and item attributes influence user feedback for items. User-item interactions come in different forms such as tags, numeric ratings, or detailed textual reviews. The analysis identifies patterns which are either socially structured relationships via explicit links in social networks or are behavioral relationships. The latter arises due to similarities (or dissimilarities) in user interaction with item. These patterns can be structurally expressed as a collection of units, referred to as groups. There is a great deal of flexibility in how a group is defined. The definition is critical as it plays a crucial role in determining the search space over which the analytics is performed.

The roots of structural analytics can be traced to discovery-driven exploration of OLAP data cubes. A data cube consists of two types of attributes: dimensions and measures. While dimensions consist of attributes that form a key, measures are typically numerical attributes associated with a dimension. Each cube could be considered analogous to a group of users described with attributes they share (e.g., young reviewers), while the measures determine how useful the group is (such a measure could be low ratings for all group members). Using operations such as drill-down and roll-up, an analyst can explore the cube space to identify opportunities or spot exceptions. A number of prior works [6–8] seek to aid the discovery process by automatically identifying data cubes that differ from their surrounding context or satisfy the trend specified by analyst. For example, the RELAX operator [8] identifies the coarsest granularity in which a trend (typically an anomaly such as low sales) exists.

One common way to describe a group is as a set of users all of whom share the values of a set of attributes. For example, the group {<gender, female>, <age, young>, <location, New York>} denotes the set of all young female users from New York. Another natural way of defining a group is as a set of user-item interactions (item ratings, reviews and tags, Foursquare check-in, Facebook likes, etc.) along with the set of attribute-value pairs shared among the users and the items. For example, {<genre, action>, <gender, female>, 1.5} denotes that female reviewers gave an average rating of 1.5 for war movies. Other user feedback such as tags can also be used to specify a group. For example,

{<cuisine, Thai>, <gender, male>, <age, young>, <tasty, long wait>} represents the group of young male users who tagged a Thai restaurant with tags "tasty" and "long wait."

Structure analytics also requires a measure that specifies how "interesting" a group is. It is used to summarize the group and compare it with other groups. Structure analytics returns the top-k most interesting groups based on this measure. There are a number of popular measures for a group. If the user-item interaction consists of user-item ratings, then the group can be summarized by a scalar, the variance of ratings within a group which provides a measure of the group's coherence. For user-item interactions involving tagging, the group could be summarized via the term frequency vector of the tags or even the topic mixture of the group's tags as identified using LDA [1].

The final piece of structure analytics is the optimization objective. It can be defined over a single measure – such as identify top-k groups with least variance. Alternatively, it could be based over an aggregate measure that compares a collection of groups. An example of the former is to identify top-k groups with low variance, while the latter could identify pair of similar groups who provide dissimilar feedback to similar items. The pair of groups {<genre, action>, <gender, female>, 1.5} {<genre, action>, <gender, male>, 4.5} form an interesting pattern in the aforementioned mold. In addition, constraints could be specified over the groups to ensure that the results are representative.

The two most popular ways of representing the user-item interaction information are relational and graphical. In the relational model, social media data is represented as a table containing users, items, and user-item interactions where users and items have well-defined set of attributes. In the graph-based model, the interaction is represented as an attributed bipartite graph where users correspond to a partition, items correspond to the other partition, and an annotated edge connects a user with an item based on the interaction between them. Both representations are equivalent and the choice depends on the actual analytics that is to be performed. When the interactions

are viewed as tuples in a data warehouse, the notion of group coincides with the definition of cuboids in the data cube literature. The set of all possible groups form a lattice of nodes, where the nodes correspond to groups and the edges correspond to parent/child relationships. The objective is to explore the set of nodes in the lattice and discover the interesting ones quickly. While the primary technical challenge lies in how to efficiently search the space of all possible groups (i.e., nodes in the lattice), the design of similarity and diversity (i.e., interestingness) measure is another essential factor. Additionally, constraints such as coverage (i.e., proportion of users in social media "covered" by the groups) can be specified over content. Thus, the structure analytic task can be formulated as a constrained optimization problem.

A brute-force exact algorithm to solve the structure analytic problem requires enumeration of all possible combinations of groups in order to return the most interesting pattern maximizing the mining criterion and satisfying the constraint. The number of possible patterns is exponential in the number of groups. Alternately, common heuristic techniques for solving optimization problems such as hill climbing, simulated annealing, branch and bound, etc. can be extended to handle structure analytic task.

Key Applications

The widespread proliferation of user-generated content, such as product reviews, rating, forums, microblogs, social media conversations, etc., offers unprecedented opportunities for understanding the collective opinion and preferences of a huge consumer base. Structure analytics provides the platform to analyze this knowledge to make smart consumer or business decisions. It enables users to make well-informed buying decisions effectively and efficiently, when several similar competing items are available for purchase. It helps businesses tailor campaigns to promote their content, especially when one-size-fits-all marketing strategies are inadequate, as well as improve their quality of service. Structure analytics can also effortlessly lend itself to several popular data mining applications, such as providing personalized recommendation, detecting communities, etc.

Structure analytics can greatly enhance the effectiveness of traditional recommender systems. A recommender system allows users to discover items of interest by predicting their needs based on the past behavior of similar-minded users. However, these systems require large amount of data in order to provide useful recommendations. Structure analytics can enable to discover the intrinsic structures within the social system and generate complimentary sources of information that the recommender systems can leverage in order to improve their quality of service.

Finally, structure analytics can be used to generate semantically meaningful communities as it can utilize the metadata (i.e., attributes of users and items) to organize user-generated content and implicit and explicit linkages. This provides user understandable groupings of the users or items, which are much more actionable than traditional clustering mechanisms.

Overall, structure analytics improves the utility of many existing social media applications and provides a few new applications that have the potential of engaging users more fully with the social media sites.

Future Directions

A major future direction is a better understanding of the needs of different datasets and applications to formulate relevant optimization problems and their heuristic solutions. A new avenue would be the exploration of multi-objective formulations (e.g., minimize intragroup uniformity and maximize intergroup diversity). The availability of real datasets is a real opportunity to explore Pareto-based solutions to those problems.

Cross-References

► Interactive Analytics in Social Media
► Social Media Analytics
► Temporal Analytics in Social Media
► Text Analytics

Recommended Reading

1. Blei DM, Ng AY, Jordan MI. Latent Dirichlet allocation. J Mach Learn Res. 2003;3(4/5):993–1022.
2. Das M, Amer-Yahia S, Das G, Mri CY. Meaningful interpretations of collaborative ratings. Proc VLDB Endow. 2011;4(11):1063–74.
3. Das M, Thirumuruganathan S, Amer-Yahia S, Das G, Yu C. Who tags what? An analysis framework. Proc VLDB Endow. 2012;5(11):1567–78.
4. Das M, Thirumuruganathan S, Amer-Yahia S, Das G, Yu C. An expressive framework and efficient algorithms for the analysis of collaborative tagging. VLDB J. 2014;23(2):201–26.
5. Fortunato S. Community detection in graphs. Phys Rep. 2010;486(3):75–174.
6. Sarawagi S. Explaining differences in multidimensional aggregates. In: Proceedings of the 25th International Conference on Very Large Data Bases; 1999. p. 2–53.
7. Sarawagi S, Agrawal R, Megiddo N. Discovery-driven exploration of OLAP data cubes. Springer; 1998.
8. Sathe G, Sarawagi S. Intelligent rollups in multidimensional olap data. In: Proceedings of the 27th International Conference on Very Large Data Bases; 2001. p. 531–40.

Structure Weight

Jaap Kamps[1] and Mounia Lalmas[2]
[1]University of Amsterdam, Amsterdam, The Netherlands
[2]Yahoo! Inc., London, UK

Definition

In structured text retrieval, the structure of a text component may be used to estimate the relevance of that component. This is done by associating a weight to the structure reflecting its significance when estimating the relevance of the component for a given query.

Key Points

Associating weight to the structure of a component in itself is not new, and several investigations have been reported for whole document retrieval. This entry is concerned with structure weights in the context of structured text retrieval, where the aim is to exploit the document structure to return document components, instead of whole documents.

In structured text retrieval, not all document components will trigger the same user satisfaction when returned as answers to queries. In the context of structured documents mark-up in XML, some document components, i.e., XML elements, may not be appropriate to return because they are too small, or a tag type that does not contain informative content, nested too deep in the document logical structure, or for other reasons. When ranking XML elements, their structure (size, tag type, path, depth, etc.) may prove important. The importance of the element structure is captured through a weight, which can be binary.

Using binary weights means that an element is (value one) or is not (value zero) considered for indexing and retrieval. The decision can be made by looking at the DTD (document type definition) of the collection, past relevance data, and/or the requirements of the application and user scenario. In the selective indexing strategy [3], only elements of types that were found to contain relevant content for previous query sets (relevance data) are considered. Any elements with a length size less than a given threshold can also be ignored.

Weights can be assigned to characteristics of elements, such as length, depth, location in the document logical structure, and so on. For instance, within the language modelling framework, length has been used as a normalization parameter (weight) incorporated through a prior probability in the ranking formula [2].

With statistical approaches, the weights are estimated based on training data, such as past relevance data. The weights can be determined using machine learning, and then used in the ranking function. They can also be directly calculated based on the distribution of element characteristics. For example, in [1], the distribution of tag types is used in a way similar to the binary independence retrieval model (investigating the "presence" of tags in relevant and non-relevant elements) to estimate the element weights.

Cross-References

▶ Relationships in Structured Text Retrieval
▶ XML Retrieval

Recommended Reading

1. Gery M, Largeron C, Thollard F. Probabilistic document model integrating XML structure. In: Proceedings of the 6th International Workshop of the Initiative for the Evaluation of XML Retrieval; 2007. p. 139–49.
2. Kamps J, de Rijke M, Sigurbjörnsson B. Length normalization in XML retrieval. In: Proceedings of the 27th Annual International ACM SIGIR Conference on Research and Development in Information Retrieval; 2004. p. 80–7.
3. Mass Y., Mandelbrod M. Component ranking and automatic query refinement for XML retrieval. In: Advances in XML Information Retrieval, Proceedings of the 3rd International Workshop of the Initiative for the Evaluation of XML Retrieval, 2004, Revised Selected Papers; 2005. p. 73–84.

Structured Data in Peer-to-Peer Systems

Kai-Uwe Sattler
Technische Universität Ilmenau, Ilmenau, Germany

Synonyms

Peer data management; Peer database management; P2P database

Definition

A peer database management system is a peer-to-peer (P2P) system that manages structured data. Each node in such a system maintains data which conforms to user- or application-defined structures and can be accessed and retrieved efficiently. Examples of data structures are relations as set of tuples and hierarchical or tree-organized data such as XML data and documents. In contrast to unstructured data (e.g., text and binary objects) structured data can be retrieved by specifying logical conditions and further processed by operations such as set operations, aggregations and joins.

Historical Background

One of the origins of coordinator-free distributed data management were scalable distributed data structures (SDDS). One of these approaches called LH* [7] is an extension of linear hashing to distributed nodes, and supports key search as well as parallel operations like hash joins and scans.

Distributed hashtables (DHT) [2] for managing key/object pairs in very large and changing networks were introduced around 2001, e.g., Chord, CAN or P-Grid. Based on these DHTs several approaches for supporting more database-like functionality (e.g., multi-attribute queries, join and aggregation queries) have been developed. One of the first systems was PIER [5], other examples are RDFPeers [3] as well as the work by Triantafillou and Pitoura [12].

The idea of exploiting the P2P paradigm for data integration in Peer Data Management Systems (PDMS) was first presented in 2002/2003 by systems like Edutella [8], that uses RDF/RDFS as schema language and an own RDF-based query language, and Piazza [4] which is based on XML and XQuery.

Foundations

A widely accepted classification of P2P systems distinguishes between unstructured and structured systems. In the following, techniques for managing and querying structured data in each of these two classes of systems are described.

Unstructured P2P Systems

In an unstructured P2P system peers do not maintain information about the resources managed by other peers. This means that for answering a query the request has to be forwarded to its neighbors. Despite rather simple approaches with fixed schema (e.g., filesharing systems) a typical example of an unstructured P2P system is a peer data integration system also called PDMS (Peer Data Management System).

A query in a PDMS can be issued at each peer. The peer processes the query locally but has to forward it (or portions of this query) to the neighbor peers which provide relations relevant to this query. For this purpose, the query is rewritten using the correspondence mappings. Depending on the formalism of mappings this is implemented by query unfolding (in case of GaV) or by techniques for answering queries using views (in case of LaV) known from data integration techniques. As long as the rewritten queries still contain views or subgoals instead of only base relations the query has to be further rewritten by the neighbor peer and to be forwarded to its neighbors. The remaining steps of optimization and execution follow the basic principle of distributed query processing. However, there are several ways for further optimizations, e.g., for query routing by using routing indexes or by exploiting transitivities of the correspondence paths.

Structured P2P Systems

In a structured P2P system each peer maintains information about the resources stored at other peers, e.g., in the form of a routing table. A typical example of a structured P2P system is a distributed hashtable (DHT) – a system where a logical key space is partitioned among all peers in order to manage key-object pairs $\langle k, v \rangle$.

In order to manage structured data comprising more than a key and a value component a mapping between the schema of a database relation or XML document and the $\langle k, v \rangle$ pair is needed.

In the following, three possible alternatives are presented.

In the *horizontal* or tuple-based approach, each tuple is identified and inserted into the DHT by a resource identifier (or primary key) oid. The remaining attributes A_i with the values v_i are treated as a single object:

$$\langle oid, A_1 : v_1, A_2 : v_2, \ldots, A_n : v_n \rangle \Rightarrow$$
$$\langle h(oid), [A_1 : v_1, A_2 : v_2, \ldots, A_n : v_n] \rangle$$

This data organization scheme allows to retrieve tuples by their resource identifier. Note, that the individual attribute values are still available, but cannot be accessed directly using the DHT lookup feature. In this way, the DHT can be seen as an index structure for the database relation.

If efficient access to the other attributes A_i is needed, too, additional indexes have to be constructed by inserting additional pairs $\langle [A_i : v_i], oid \rangle$ into the DHT. Instead of using the oid as value it is also possible to store the whole tuple.

Furthermore, this approach can be easily extended to more than one relation by adding a name space or relation prefix to the resource identifier. An example system following the horizontal approach is the PIER system [5] which is based on the Bamboo DHT.

An alternative approach is a *vertical* data organization where each tuple is represented by a set of triples:

$$\langle oid, A_1 : v_1, A_2 : v_2, \ldots, A_n : v_n \rangle$$
$$\Rightarrow \Big\{ \langle h(oid), A_1, v_1 \rangle,$$
$$\langle h(oid), A_2, v_2 \rangle, \ldots \langle h(oid), A_n, v_n \rangle \Big\}$$

In order to be able to query all attributes each of these triples is inserted into the DHT using the resource identifier, the attribute and eventually the value component as key, i.e., $\langle h(oid), [A_i : v_i] \rangle$, $\langle h(A_i), oid \rangle$, $\langle h(v_i), oid \rangle$. If the DHT supports prefix search, the concatenation of $A_i \# v_i$ can be used as key to allow queries both for A_i as well as $A_i = v_i$. Advantages of this vertical data organization are first, that triples with similar values are stored at the same peer or at least in the neighborhood which simplifies joins and - in combination with an order-preserving hash function - range queries, and similarity queries. Second, there is no fixed schema for a given relation. Users can extend the schema to their needs by simply adding new triples for a given tuple. The disadvantage is the required storage overhead.

This approach is not restricted to relational data. RDF data which can be represented using triples of subject, predicate, and object can be directly mapped to this scheme. Examples of systems following this idea are RDFPeers [3] based on MAAN a multidimensional extension of Chord, and UniStore [6] which is built on top of P-Grid.

Assuming a P2P system for storing and querying XML documents instead of relations a third possible approach is to exploit the idea of *path indexing* [11]. A standard technique for indexing XML documents is suffix indexing of XML paths. Given a document identified by a URI and an XML path P in this document consisting of the elements $P = e_1/e_2/ \ldots /e_n$ the following n suffixes can be derived

$$s_1 = e_1/e_2/e_3/ \ldots /e_n$$
$$s_2 = e_2/e_3/ \ldots /e_n$$
$$s_i = \ldots$$
$$s_n = e_n$$

Now, based on these subpaths the keys for the DHT are calculated and the following pairs are inserted: $\langle h(s_1), [s_1, URI] \rangle$, $\langle h(s_2), [s_2, URI] \rangle$, etc.

Note, that this requires a DHT supporting prefix search.

An important question affecting the choice of the hash function is how to fragment a relation or document collection. A viable solution has to be found between the two extreme cases:

- Distribute tuples according to their relation identifier, i.e., each relation is completely stored at exactly one peer.
- Distribute tuples according to their resource identifier (oid) such that tuples are partitioned among a set of peers. However, in this case

tuples of the same relation or with similar values of indexed attributes should be stored at neighboring peers in order to support range and nearest neighbor queries efficiently.

Appropriate fragmentation schemes are based on space-filling curves [1] and order preserving hash functions.

For querying structured data in a DHT both basic processing strategies *data shipping* and *query shipping* can be used. With data shipping the DHT is used only as a data storage: data are retrieved from the responsible peers which are identified by applying the hash functions to the value in the query predicate similarly to an index lookup or index scan in a classical DBMS.

Obviously, this also works for the path indexing approach, where the keys for the lookup operation are calculated from the path expression given in an XPath query. As an example consider a peer asking a path query *A // B /C / D // E*. By computing the longest subpath containing only the child axes (in this case *B / C / D*), the peers responsible for this key region can be identified and contacted for evaluating the remaining query locally.

This approach can be extended to process more complex queries including joins. Based on the retrieved tuples of the first relation, the matching tuples of the second relation are retrieved from the DHT by applying the hash function to the join values. However, data shipping becomes inefficient if the query operators are not very selective. In this case, very expensive scans of many peers are needed which is impractical in large-scale systems.

For *query shipping* in a DHT each peer has to provide distributed query processing capabilities rather than only the put/get operations. A query is routed to the peer that is responsible for the data addressed by the operator that is to be executed next. These peers are determined by applying the hash function to the current intermediate results. In this way, the DHT is used both as a hash table for indexing and storing tuples as well as a content-addressable network for routing tuples and/or operators by values.

This idea was exploited in PIER for implementing different join strategies. A first strategy presented in [5] is a variant of the symmetric hash join between two relations R and S. Here, each peer storing tuples of R and S scans its local data and insert these tuples into a temporary table of the DHT. The peers responsible for the key space of this table execute the probing phase by joining the received tuples locally. A second strategy developed in PIER is a fetch matches variant which can be used if one of the relation is already hashed on the join attribute. In this case, the peers storing tuples of the second relation perform a local scan and retrieve the possibly matching tuples using the get operation of the DHT.

Further improvements can be achieved by applying the idea of symmetric semijoin, i.e., perform a local projection on the join attribute of both relations, followed by a symmetric hash join and feed the results into a fetch matches join to retrieve the remaining attribute values of the original relations. Finally, bloom filters can be used, too. Here, bloom filters are created by each peer responsible for R and S and inserted into temporary DHT table. The received filters are combined and sent to all peers storing the opposite relation. The experiments presented in [5], show that these strategies help to reduce the bandwidth consumption and response time particularly in case of low selectivity values.

In case of more complex queries consisting of sequences of operators query shipping may result in multiple instances of the plan that "travel" through the network, because a single query operator might involve tuples from different peers. This approach, which was initially proposed as *mutant query plans* [9] is illustrated in Fig. 1. Given a relation $R(A, B)$ stored at peers p_1, p_2, p_3 and a relation $S(B, C)$ stored at peers p_4, p_5. Peer p_0 submits the query $\sigma_{1<A<6}(R) ^{\circledR} S$. The query is sent to $p_1...p_3$ which are identified by applying the hash function to the selection predicate. These peers evaluate the first part of the query $(\sigma_{1<A<6}(R))$ locally and replace this expression in the plan by the intermediate result. The modified plan is sent to the S-peers which are identified by applying the hash function on the B values of

S

**Structured Data
in Peer-to-Peer Systems,
Fig. 1** Mutant query plans
in query shipping

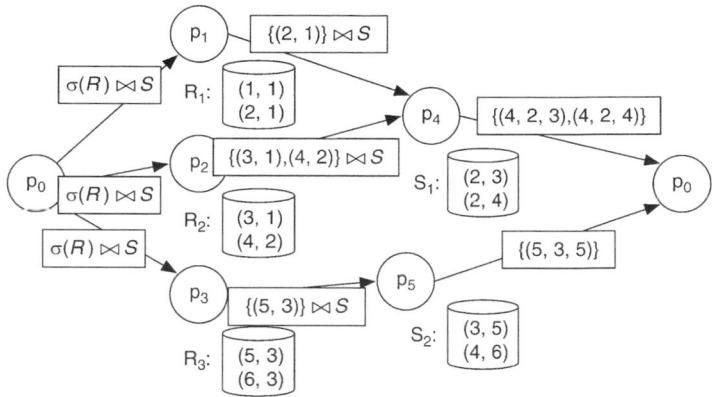

the intermediate result. Finally, p_4 and p_5 evaluate the remaining parts of the query and send the result back to the initiator.

The benefits of query shipping are exploiting computing resources of the peers as well as avoiding transfer of large datasets. However, query planning and execution is more difficult because the state of a processed query is spread over multiple peers.

Key Applications

A major application for DHT-based P2P database systems is public data management where information of a general interest, its structure and semantics is controlled by a large number of participants. Furthermore, the costs for providing the infrastructure should often be shared by the users in a fair manner. Examples of such applications are the management of public datasets in e-Sciences, e.g., genome data or data in astronomy, metadata and index data for the Semantic Web as well as specialized search engines, naming and directory services as well as social applications such as file/picture sharing, recommender systems or friend-of-a-friend networks. Note, that in these applications it is often not necessary to store the actual data itself in the DHT, but instead metadata or index data required for answering queries are publically managed.

The main application of unstructured peer data management system is data integration in large-scale, loosely-coupled scenarios.

Future Directions

Managing and querying structured data in P2P systems is a relatively new research area which raises several new challenges to established techniques, e.g., known from distributed database systems.

The first challenge is *scalability*, meaning the support of efficient query processing in networks of ten thousands or more nodes. Most of the research systems presented so far are based on simulation environments or a relatively small number of peers, e.g., in PlanetLab or similar platforms.

A second problem in P2P systems is the dynamic of the network (joining and leaving peers) as well as the unreliability of peers. Most existing approaches are best effort solutions which are unable to give guarantees wrt. result completeness, freshness or response time. Thus, an open issue is to estimate completeness of results in case of partial answers or to guarantee a certain *quality of service*.

Finally, in large P2P systems where no peer knows all other peers in the network *trustworthiness* are a further challenge. Particularly, if data are redistributed to other nodes in the network the original data producer wants to make sure that its data are not manipulated by the hosting peer. Furthermore, in order to achieve a fair balancing of load and avoiding overloaded peers the problem of rejecting requests and free riding by malicious peers has to be addressed, e.g., by incentive mechanisms.

Cross-References

▶ Distributed Join
▶ Distributed Query Processing
▶ Peer-to-Peer Data Integration
▶ Peer-to-Peer Overlay Networks: Structure, Routing and Maintenance

Recommended Reading

1. Andrzejak A, Xu Z. Scalable, efficient range queries for grid information services. In: Proceedings of the 2nd IEEE International Conference on Peer-to-Peer Computing; 2002. p. 33–40.
2. Balakrishnan H, Kaashoek MF, Karger D, Morris R, Stoica I. Looking up data in P2P Systems. Commun ACM. 2003;46(2):43–8.
3. Cai M, Frank M. RDFPeers: a scalable distributed RDF repository based on a structured peer-to-peer network. In: Proceedings of the 12th International World Wide Web Conference; 2004. p. 650–57.
4. Halevy A, Ives Z, Mork P, Tatarinov I. Piazza: data management infrastructure for semantic web applications. In: Proceedings of the 12th International World Wide Web Conference; 2003. p. 556–67.
5. Huebsch R, Hellerstein JM, Lanham N, Thau Loo B, Shenker S, Stoica I. Querying the internet with PIER. In: Proceedings of the 29th International Conference on Very Large Data Bases; 2003. p. 321–32.
6. Karnstedt M, Sattler K, Richtarsky M, Müller J, Hauswirth M, Schmidt R, John R. UniStore: querying a DHT-based universal storage. In: Proceedings of the 23rd International Conference on Data Engineering; 2007. p. 1503–04.
7. Litwin W, Neimat M-A, Schneider D. LH* - a scalable, distributed data structure. ACM Trans Database Syst. 1996;21(4):480–525.
8. Nejdl W, Wolf B, Qu C, Decker S, Sintek M, Naeve A, Nilsson M, Palmer M, Risch T. Edutella: a P2P networking infrastructure based on RDF. In: Proceedings of the 11th International World Wide Web Conference; 2002. p. 604–15.
9. Papadimos V, Maier D. Mutant query plans. Inf Softw Technol. 2002;44(4):197–206.
10. Risson J, Moors T. Survey of research towards robust peer-to-peer networks: search methods. Comput Netw. 2006;50(17):3485–521.
11. Skobeltsyn G, Hauswirth M, Aberer K. Efficient processing of XPath queries with structured overlay networks. In: Proceedings of the International Conference on Cooperative Information Systems; 2005. p. 1243–60.
12. Triantafillou P, Pitoura T. Towards a unifying framework for complex query processing over structured peer-to-peer data networks. In: Proceedings of the 1st Workshop on Databases, Information Systems, and Peer-to-Peer Computing; 2003. p. 169–80.

Structured Document Retrieval

Mounia Lalmas[1] and Ricardo Baeza-Yates[2]
[1]Yahoo! Inc., London, UK
[2]NTENT, USA - Univ. Pompeu Fabra, Spain - Univ. de Chile, Chile

Synonyms

Focused retrieval; Passage retrieval; Querying semi-structured data; Structured text retrieval; XML retrieval

Definition

Structured document retrieval is concerned with the retrieval of document fragments. The structure of the document, whether explicitly provided by a mark-up language or derived, is exploited to determine the most relevant document fragments to return as answers to a given query. The identified most relevant document fragments can themselves be used to determine the most relevant documents to return as answers to the given query.

Key Points

The aim of this entry is to clarify different terminologies that have been used to refer to or are strongly related to structured retrieval and semi-structured data.

The term "structured document retrieval," which was introduced in the early to mid 1990s in the information retrieval community, refers to "passage retrieval" and "structured text retrieval." In passage retrieval, documents are first decomposed into passages (e.g., fixed-size text-windows of words, fixed discourses such as paragraphs, or topic segments through the application of a topic segmentation algorithm). Passages could themselves be retrieved as answers to a query, or be used to rank documents as answers to the query.

Structured text retrieval is concerned with the development of models for querying and retrieving from structured text, where the structure is usually encoded with the use of mark-up languages, such as SGML, and now predominantly XML. Indeed, text documents often display structural information. For example, a scientific article will have a so-called logical structure, such as an abstract, several sections, and subsections, each of which is composed of paragraphs. A book will have a so-called layout structure, such as pages and columns.

Structured text retrieval is to be contrasted to traditional text retrieval, where the latter is concerned with the retrieval of unstructured text - so-called "raw text" or "flat text." The use of the term "structured" in "structured text retrieval" is there to emphasize the interest in the structure. Furthermore, structured text retrieval aims to exploit the available structural information to return text fragments (e.g., XML elements) as opposed to entire text documents.

The term "semi-structured" comes mainly from the database community. Traditional database technologies, such as relational databases, have been concerned with the querying and retrieval of highly structured data (e.g., from a student table, find the names and addresses of those with a grade over 80 in a particular subject). Text documents marked-up, for instance, in XML are made of a mixture of highly structured components (e.g., year, author name) typical of database records, and loosely structured components (e.g., abstract, section). Database technologies are being extended to query and retrieve such loosely structured components, called semi-structured data. Databases that support this kind of data, mainly in the form of text with mark-up, are referred to as semi-structured databases, to emphasize the loose structure of the data and use "querying data" instead of "data retrieval."

From a terminology point of view, structured text retrieval and querying semi-structured data, in terms of end goals, are the same. The difference comes from the fact that in information retrieval, the structure is added, and in database, the structure is loosened. It should, however, be pointed out that research in information retrieval and databases with respect to accessing structured text (or semi-structured data in the form of text) have been concerned, because of historical reasons, with different aspects of the access process, e.g., ranking in information retrieval versus efficiency in databases. Nowadays, there is a convergence trend between the two areas (e.g., [1]).

In the late 1990s, the interest in structured document retrieval grew significantly due to the introduction of XML in 1998, which has now became the de-facto format standard for structured documents (or structured text, semi-structured data). Research on XML retrieval was further boosted with the set-up of INEX in 2002, the Initiative for the Evaluation of XML Retrieval, which allowed researchers to compare and discuss the effectiveness of models specifically developed for XML retrieval [2]. Nowadays, XML retrieval is almost a synonym for structured document retrieval, structured text retrieval, and querying semi-structured data.

Structured document retrieval, passage retrieval, structured text retrieval, querying semi-structured data, XML retrieval, all belong to what has recently been called "focused retrieval" [3]. Focused retrieval is concerned with returning the most focused results to a given query. Such focused results include passages, XML elements, and factoid answers (e.g., London being the capital of the UK).

Cross-References

▶ Document Databases
▶ Initiative for the Evaluation of XML Retrieval
▶ Semi-Structured Data
▶ Structured Text Retrieval Models
▶ XML Retrieval

Recommended Reading

1. Amer-Yahia S, Case P, Rölleke T, Shanmugasundaram J, Weikum G. In Report on the DB/IR panel at SIGMOD 2005. ACM SIGMOD Rec. 2005;34(4):71–4.
2. Kazai G, Gövert N, Lalmas M, The FN. INEX evaluation initiative. In intelligent search on XML data, applications, languages, models, implementations, and benchmarks. New York: Springer; 2003. p. 279–93.
3. Trotman A, Geva S, Kamps J. Report on the SIGIR 2007 workshop on focused retrieval. SIGIR Forum. 2007;41(2):97–103.

Structured Text Retrieval Models

Djoerd Hiemstra[1] and Ricardo Baeza-Yates[2]
[1]University of Twente, Enschede, The Netherlands
[2]NTENT, USA - Univ. Pompeu Fabra, Spain - Univ. de Chile, Chile

Synonyms

Retrieval Models for Text Databases

Definition

Structured text retrieval models provide a formal definition or mathematical framework for querying semi-structured textual databases. A textual database contains both content and structure. The content is the text itself, and the structure divides the database into separate textual parts and relates those textual parts by some criterion. Often, textual databases can be represented as marked-up text, for instance, as XML, where the XML elements define the structure on the text content. Retrieval models for textual databases should comprise of three parts: (i) a model of the text, (ii) a model of the structure, and (iii) a query language [4]: The model of the text defines a tokenization into words or other semantic units, as well as stop words, stemming, synonyms, etc. The model of the structure defines parts of the text, typically a contiguous portion of the text called element, region, or segment, which is defined on top of the text model's word tokens. The query language typically defines a number of operators on content and structure, such as set operators and operators like "containing" and "contained by" to model relations between content and structure, as well as relations between the structural elements themselves. Using such a query language, the (expert) user can, for instance, formulate requests like "I want a paragraph discussing formal models near to a table discussing the differences between databases and information retrieval." Here, "formal models" and "differences between databases and information retrieval" should match the content that needs to be retrieved from the database, whereas "paragraph" and "table" refer to structural constraints on the units to retrieve. The features, structuring power, and the expressiveness of the query languages of several models for structured text retrieval are discussed below.

Historical Background

The storage and information and retrieval system (STAIRS), which was developed at IBM in the late 1950s, allowed querying of both content and structure. Much like today's online public access catalogues, it was used to store bibliographic data in records with fields such as *keywords* and *title*, providing structured search, but no overlapping or hierarchical structures nor full-text search. At the end of the 1980s, researchers at the University of Waterloo in Canada pursued database support for the creation of an electronic version of the Oxford English Dictionary. This resulted in a number of models for querying and manipulating content and hierarchical structure such as the parsed strings model [10], PAT expressions [15], the containment model [5], and generalized concordance lists model [7]. Similar approaches were developed elsewhere, such as the proximal nodes model [13] and the nested region model [11]. The interest in structured text retrieval models has grown since the introduction of XML in 1998 and the emergence of standard data retrieval query languages for XML data. One might argue that the structured text retrieval approaches such as PAT expressions and proximal nodes mentioned above are predecessors of XPath. The success of XML in turn has influenced the work on structured retrieval models: XIRQL was proposed in 2000 [9] as an information retrieval extension of XML query languages (XIRQL is an extension of XQL, a predecessor of XPath). More recently, in 2004, NEXI and XQuery and XPath Full-Text have been proposed as query languages for structured text retrieval, as well as examples of structured text retrieval models for the respective query languages [2, 12].

Foundations

There are several models of structured text retrieval. Since there is no consensus on how to structure a textual database, this entry addresses several modeling decisions following the taxonomy presented in [4]. The entry only addresses models for text databases and not text retrieval models in relational databases as, for instance, provided by SQL/MM. Due to the success of XML, today XML retrieval is almost a synonym for structured text retrieval, although XML retrieval only addresses the explicit, single hierarchy case below.

Explicit vs. Implicit Structure

Most models use *explicit structure*, i.e., they define unambiguously what parts of the textual database are, for instance, "sections." These models require the database to be structured explicitly and unambiguously or use terminology from markup languages: the models require the database to be well formed. This allows easy modeling of nested regions and powerful structural relationships such as the direct ancestor relationship (i.e., child and parent axis in XPath). The following query might be used in an explicit structure approach to retrieve sections that contain the word "databases":

```
section CONTAINING "databases"
```

Explicit structure is assumed by among others the proximal nodes model [13] and the full match model [2]. In systems that use *implicit structure*, however, structure is not explicitly distinguished from content. In these approaches the database is modeled as a sequence of tokens without distinguishing a word token from a markup token. A structural element should, therefore, be constructed at runtime by looking up the opening markup tokens, the closing tokens, and to return those regions starting with an opening token and ending with a closing token. The query above would then be formulated as [11] (here, the operator ".." would be pronounced as "following"):

```
("<section>" .. "</section>")
CONTAINING "databases"
```

So, the *section* element only exists at querying time. Semantically, the query is not different from a content-only query. For instance, the query("all" .. "equal") *CONTAINING* "created" retrieves regions that start with the word "all," that end with the word "equal," and that contain the word "created," matching, for instance, the phrase "all men are created equal." Nested elements or unbalanced tags are handled differently by several approaches. In the generalized concordance list (GCL) approach [7], nested sections will not be recognized by the system (instead two partially overlapped sections will be returned). In the nested region algebra approach [11] nested elements are returned properly. The approach is implemented as sgrep (structured grep) (https://www.cs.helsinki.fi/u/jjaakkol/sgrep.html). The GCL approach was recently implemented in a research system called Wumpus (http://www.wumpus-search.org).

Static vs. Dynamic Structure

The use of implicit structure also implies the use of *dynamic structure*. A system that uses a dynamic structure allows operations that define new elements or regions, i.e., elements or regions that were not previously in the database. In XQuery, this is done by element construction, but in some approaches, dynamic structure is a natural consequence of the model. As an early example of dynamic structure, consider the following bibliographic entry:

```
John Doe, "Crime," Police 6 , 2028.
```

The entry is explicitly structured by the following grammar that functions as a database schema [10]:

```
entry   := author','title','journal',
'year'.';
author  := text;
title   := ' "' text '"';
journal := text digit+ ;
year    := digit digit digit digit;
text    := ( letter | ' ' ) + ;
```

A valid database instance contains data that conforms to the grammar. The instance takes the form of a parsed string or "p-string." Note that the schema does not distinguish the author's first name(s) from his surname, but this might be done at query time by introducing a small grammar fragment *NameG* that parses the author strings into given names and surnames:

```
NameG := { name       := ( givenname ' ' )
                          + surname;
          givenname := letter +;
          surname   := letter +; }
```

The p-strings model provides a simple query language for adding additional grammar fragments. Suppose the bibliographic entry above is E, then the following query returns a p-string containing the author element with given name and surname explicitly identified.

```
(author in E) reparsed by NameG
```

The construct might be used to search for all authors with surname "Doe" that wrote a journal paper that mentions "grammar" in the title. The p-strings model uses regular expression matching as a core language primitive, and as such dynamic structure is more easily added than in, for instance, XQuery or XQuery Full-Text.

Single Hierarchy vs. Multiple Hierarchies

Although some fielded search methods use a flat structure to model text, the approaches considered here assume a hierarchical structure of the text database. The systems that use implicit structure introduced above assume a single hierarchy. Interestingly, many approaches assume multiple structural hierarchies on the same textual database. Each hierarchy might serve a different purpose. For instance, one hierarchy might represent the logical structure of the text, dividing it in chapters, sections, subsections, etc., whereas a second hierarchy might represent the layout structure in columns and pages, and a third layer might represent the results of a part-of-speech tagger, etc. Inside a single hierarchy, the structural elements are either disjoint or nested inside each other but across hierarchies elements may partially overlap, i.e., a subsection might start halfway a page and end on the following page.

Some approaches relate single views in one query [13]. An interesting approach is suggested by Alink [1], which introduces additional XPath steps (select-narrow and select-wide) that navigate from one hierarchy to another. For instance, the following XQuery Full-Text-like query fragment navigates from the paragraph elements to another hierarchy with a *verb* element that contains "killed" and to a hierarchy with a *person* element that contains "Abraham Lincoln":

```
$doc//paragraph[./select-narrow::Verb
  ftcontains
"killed" and./select-narrow::person
  ftcontains
"Abraham Lincoln"]
```

The need for multiple hierarchies is, for instance, addressed in the containment model [5] and the proximal nodes model [13]. In several publications, the hierarchies are called "standoff annotation" or "offset annotation" to stress that the structural information (or annotations) are modeled separately from the textual data.

Exact Matching Versus Ranking

Many of the early structured text retrieval models do not consider ranked retrieval results, or if they do only as an afterthought, i.e., by ranking the retrieval results using a text-only query disregarding the structural conditions in the query [5]. A simple but powerful way to take the structure of the results into account is to apply a standard information retrieval model to the retrieved content and then propagate element scores or aggregate term weights based on the text structure. In several of these approaches to ranking, propagation or aggregation is guided by weighting the paths to elements by so-called augmentation weights [9] or interpolation parameters [14], to model, for instance, that a title element is more likely to contain important information than a bibliography element. Instead of propagating or aggregating the scores from the leaf nodes, *algebraic* approaches include the ranking functionality inside each operator of the query language [2, 12]. Ranking might also include relaxation of the query's structural conditions, for instance, by relaxing complex queries stepwise to simpler queries [3].

In 2002, Fuhr and Lalmas [8] organized the first workshop of the Initiative for the Evaluation of XML Retrieval. The goal of INEX is to evaluate the quality of the retrieved results and as such the quality of the ranking provided by the system taking both content and structure into account. The initiative provides a large test-bed, consisting of XML documents, queries, and relevance judgments on the data, where the relevance judgments are human judgments that define if an XML element is relevant to the query or not. With XML databases and extensions of XML query languages becoming a de facto standard for structured text retrieval, ranking is one of the main remaining research challenges.

Key Applications

Systems based on structured text retrieval models can be applied to any problem that involves semi-structured text databases. Key applications of the approaches described in this section include managing and searching electronic dictionaries such as the Oxford English Dictionary [10, 15], managing and searching electronic journals such as the journals of the IEEE [6, 8, 12], searching stage plays such as the collected works of William Shakespeare [7], and searching hard drives for digital forensics [1].

Cross-References

► Aggregation-Based Structured Text Retrieval
► Information Retrieval Models

► Narrowed Extended XPath I
► Propagation-Based Structured Text Retrieval
► XPath/XQuery
► XQuery Full-Text

Recommended Reading

1. Alink W XIRAF: an XML information retrieval approach to digital forensics. Master's thesis, University of Twente. 2005.
2. Amer-Yahia S, Botev C, Shanmugasundaram J TeXQuery: a full-text search extension to XQuery. In: Proceedings of the 12th International World Wide Web Conference; 2004.
3. Amer-Yahia S, Lakshmanan LVS, Pandit S. FleXPath: flexible structure and full-text querying for XML. In: Proceedings of the ACM SIGMOD International Conference on Management of Data; 2004.
4. Baeza-Yates RA, Navarro G. Integrating contents and structure in text retrieval. ACM SIGMOD Rec. 1996;25(1):67–79.
5. Burkowski FJ Retrieval activities in a database consisting of heterogeneous collections of structured text. In: Proceedings of the 15th Annual International ACM SIGIR Conference on Research and Development in Information Retrieval; 1992. p. 112–24.
6. Carmel D, Maarek YS, Mandelbrod M, Mass Y, Soffer A. Searching XML documents via XML fragments. In: Proceedings of the 26th Annual International ACM SIGIR Conference on Research and Development in Information Retrieval; 2003. p. 151–8.
7. Clarke CLA, Cormack GV, Burkowski FJ. An algebra for structured text search and a framework for its implementation. Comput J. 1995;38(1):43–56.
8. Fuhr N, Gövert N, Kazai G, Lalmas M, editors. In: Proceedings of the 1st International Workshop of the Initiative for the Evaluation of XML Retrieval; 2002.
9. Fuhr N, Grossjohann K. XIRQL: a query language for information retrieval in XML. In: Proceedings of the 24th Annual International ACM SIGIR Conference on Research and Development in Information Retrieval; 2001. p. 172–80.
10. Gonnet GH, Tompa FW Mind your grammar: a new approach to modelling text. In: Proceedings of the 13th International Conference on Very Large Data Bases; 1987. p. 339–46.
11. Jaakkola J, Kilpeläinen P. Nested text-region algebra. Technical report. University of Helsinki. 1999.
12. Mihajlovic V, Blok HE, Hiemstra D, Apers PMG. Score region algebra: building a transparent XML-IR database. In: Proceedings of the International Conference on Information and Knowledge Management; 2005. p. 12–9.

13. Navarro G, Baeza-Yates RA. Proximal nodes: a model to query document databases by content and structure. ACM Trans Inf Syst. 1997;15(4):400–35.

14. Ogilvie P, Callan J. Hierarchical language models for XML component retrieval. In: Advances in XML information retrieval. Lecture notes in computer science 3493. Springer; 2005. p. 224–37.

15. Salminen A, Tompa FW. PAT expressions: an algebra for text search. Proc Complex. 1992;92:309–32.

Subject Spaces

Hans-Arno Jacobsen
Department of Electrical and Computer
Engineering, University of Toronto, Toronto,
ON, Canada

Definition

Subject spaces are a model to formalize publish/subscribe-style interactions and generalize the publish/subscribe concept. Subject spaces subsume existing publish/subscribe models, such as the channel-based, the topic-based, the type-based, and the content-based publish/subscribe models. Subject spaces go beyond these models by permitting the treatment of publications and subscriptions symmetrically, extending publications to also include expressive filter predicates, introducing the notion of selective publishing, interpreting publications and subscriptions as either stateless or stateful, and generalizing matching to encompass a wide range of possible matching semantics.

Key Points

The subject spaces model is a unifying formal framework to specify, describe and analyze the publish/subscribe concept. Subject spaces encompass existing publish/subscribe models and allow the modeling of new aspects of publish/subscribe-style interactions. Informally, a *subject space* is the set of values a publisher can publish and a subscriber can subscribe to.

The plurality of *spaces* refers to all sets of values, possibly overlapping, that publishers and subscribers publish and subscribe to in one particular instantiation of the model, respectively.

Subject spaces treat publications and subscriptions as symmetric constructs. That is, a publication must not be distinguished from a subscription and both are equally expressive. Thus, publications also contain predicates, such as range predicates, equality predicates, non-equality predicates, and string predicates found only in subscriptions in other models. The implications are that in subject spaces, both publications and subscriptions specify sets of values that can be interpreted as regions in space. Traditionally, only subscriptions specify regions in space, while publications are points in space.

In contrast to the content-based model, in subject spaces publications and subscriptions consist of predicates and attribute-value pairs, both of which are optional allowing one to model the respective counterpart in the content-based model.

Moreover, subject spaces treat publications and subscriptions as stateful entities. A publication, once published, is persisted in its associated subject space until explicitly revoked. Subsequent publish operations are updates to the previously published value. A sequence of publish operations can be interpreted as moving the published value through space. Subscriptions are interpreted in the same fashion.

Subject spaces generalize the matching of publications against subscriptions. Since both contain predicates and attribute-value pairs, a match between a publication and a subscription exists, if the attribute-value pairs of the publication satisfy the predicates of the subscription and the attribute-value pairs of the subscription satisfy the predicates of the publication. In addition, the interpretation of publications and subscriptions as regions in space enables a generalization of matching to encompass additional matching semantics such as overlap, containment, or closeness of publication and subscription regions in the space.

The symmetric treatment of publications and subscriptions enables a notion of *selective publishing*, whereby a publisher specifies predicates

S

to declaratively specify a subset of potential receivers of published messages. This is in contrast to content-based publish/subscribe, where a publication is delivered to all subscribers with matching subscriptions. The symmetric treatment of publications and subscriptions further extends the conventional interpretation of publication, as mere data points, to ranges, regions in space, or other structures. This increases the expressiveness of the subject spaces model substantially, as opposed to other publish/subscribe models.

The state-persistent nature of publications and subscriptions avoids the problem of redundant notifications. This is because subject spaces distinguish between inserting a publication and updating the previously published value. A match occurs if the publication satisfies the subscription, but not at each update of the published values. Existing content-based publish/subscribe systems cannot model this difference and are therefore susceptible to the redundant notification problem. Subject spaces can model both state-persistent and stateless publish/subscribe. The subject spaces model can be used to formalize and model publish/subscribe-style interactions. Applications of concrete realizations of the subject spaces model are similar to applications of publish/subscribe. The increased expressiveness offered by subject spaces will likely enable new applications. For a more detailed treatment of the subject see for example the work by Leung and Jacobsen [1, 2].

Cross-References

► Channel-Based Publish/Subscribe
► Content-Based Publish/Subscribe
► Publish/Subscribe
► State-Based Publish/Subscribe
► Topic-Based Publish/Subscribe
► Type-Based Publish/Subscribe

Recommended Reading

1. Leung HKY. Subject space: a state-persistent model for publish/subscribe systems. In: Proceedings of the Conference of the IBM Centre for Advanced Studies on Collaborative Research; 2002. p. 7.
2. Leung HKY, Jacobsen H-A. Efficient matching for state-persistent publish/subscribe systems. In: Proceedings of the Conference of the IBM Centre for Advanced Studies on Collaborative Research; 2003. p. 182–96.

Subspace Clustering Techniques

Peer Kröger[1] and Arthur Zimek[1,2]
[1]Ludwig-Maximilians-Universität München, Munich, Germany
[2]Department of Mathematics and Computer Science, University of Southern Denmark, Odense, Denmark

Synonyms

Bi-clustering; Co-clustering; Correlation clustering; Oriented clustering; Pattern-based clustering; Projected clustering

Definition

Cluster analysis aims at finding a set of subsets (i.e., a clustering) of objects in a data set. A meaningful clustering reflects a natural grouping of the data. In high-dimensional data, irrelevant attributes and correlated attributes make any natural grouping hardly detectable. Specialized techniques aim at finding clusters in subspaces of a high-dimensional data space.

Historical Background

While different weighting of attributes was in use since clusters were derived by hand, the problem of finding a cluster based on a subset of attributes and a specialized solution was first described 1972 by Hartigan [1]. But, triggered by modern capabilities of massive acquisition of high-dimensional data in many scientific and economic domains and the first general approaches to the problem [2–4], research focused on the problem not till 1998. The more special topic of pattern-based clustering is covered in [5]. Broad overviews are provided in several surveys [6, 7].

Scientific Fundamentals

Different Challenges: The "Curse of Dimensionality"

High-dimensional data confronts cluster analysis with several problems. A bundle of problems is commonly addressed as the "curse of dimensionality." Aspects of this "curse" most relevant to the clustering problem are: (i) In general, any optimization problem becomes increasingly difficult with an increasing number of variables (attributes) [8]. (ii) The relative contrast of the farthest point and the nearest point converges to 0 with increasing data dimensionality [9], i.e., the discrimination between the nearest and the farthest neighbor becomes rather poor in high-dimensional data spaces. In clustered data, this effect can be expected to be less expressive, but it might remain a problem in combination with the other aspects [10]. (iii) Capabilities of automated data acquisition in many application domains lead to the collection of as many features as possible in the expectation that many of these features may provide useful insights. Thus, for the task at hand in many problems, there exist typically many irrelevant attributes in a data set. Since groups of data are defined by some of the attributes only, the remaining irrelevant attributes ("noise") may heavily interfere with the efforts to find these groups. (iv) Similarly, in a data set containing many attributes, some attributes will most probably exhibit correlations among each other (in varying complexity).

Many approaches try to alleviate the "curse of dimensionality" by applying feature reduction methods prior to cluster analysis. However, the second main challenge for cluster analysis of high- dimensional data is the possibility and even high probability that different subsets or combinations of attributes may be relevant for different clusters. Thus, a global feature selection or dimensionality reduction method cannot be applied. Rather, it becomes an intrinsic problem of the clustering approach to find the relevant subspaces and to find clusters in these relevant subspaces. Furthermore, although correlation among attributes often is the basis for a dimension reduction, for many application domains, it is a main part of the interesting information what correlations exist among which attributes for which subset of objects. As a consequence of this second challenge, the first challenge (i.e., the "curse of dimensionality") generally cannot be alleviated for clustering high-dimensional data.

Different Solutions: Categories of Subspace Clustering Techniques

Subspace clustering techniques can be divided into three main families. In view of the challenges sketched above, any arbitrarily oriented subspace may be interesting for a subspace clustering approach. The most general techniques ("(arbitrarily) oriented clustering," "correlation clustering"(Note that the name "correlation clustering" relates to a different problem within the machine learning community.)) tackle this infinite search space. Yet most of the research in this field assumes the search space to be restricted to axis-parallel subspaces. Since the search space of all possible axis-parallel subspaces of a d-dimensional data space is still in $O(2^d)$, different search strategies and heuristics are implemented. Axis-parallel approaches mainly split into "subspace clustering" and "projected clustering." In between these two main fields, a group of approaches is known as "pattern-based clustering" (also "bi-clustering" or "co-clustering"). For these approaches, the search space is not necessarily restricted to axis-parallel subspaces but on the other hand does not contain all arbitrarily oriented subspaces. The restrictions on the search space differ substantially between different approaches in this group.

Axis-Parallel Subspaces

To navigate through the search space of all possible axis-parallel subspaces and to find clusters in subspaces, mainly two strategies are implemented: the top-down approach and the bottom-up approach.

Following the top-down approach, an algorithm derives a cluster approximately based on the full-dimensional space and refines the cluster by adapting the corresponding subspace based on the current selection of points. This means a lower-dimensional projection is sought for where the (iteratively refined) set of points clusters best. Thus, algorithms pursuing this approach are called "projected clustering algorithms" and, usually, assign each point to at most one subspace cluster. The first approach of this category is proposed in [3].

Bottom-up approaches start by single dimensions and search primarily for all interesting subspaces (i.e., subspaces containing clusters) as combinations of lower-dimensional interesting subspaces (often this combination is translated to the frequent item set problem and, thus, based on the Apriori property). Most of these approaches are therefore "subspace clustering algorithms" and usually can assign one point to different clusters simultaneously (i.e., subspace clusters may overlap). Their aim is to find all clusters in all subspaces. There are also "hybrid algorithms" following the projected clustering approach but allowing points to belong to multiple clusters simultaneously or, on the other hand, following the subspace clustering approach but not computing all clusters in all subspaces. The first approach in this category is proposed in [2].

In summary, approaches to axis-parallel subspace clustering handle the problem of irrelevant attributes (aspect (iii) of the "curse of dimensionality"). Bottom-up approaches, additionally, tackle mostly the problem of poor discrimination of nearest and farthest neighbor (aspect (ii)).

Pattern-Based Clustering

Pattern-based clustering algorithms seek subsets of objects exhibiting a certain pattern on a subset of attributes. In the most spread algorithms, this pattern is an additive model of the cluster, meaning each attribute value within a cluster and within the relevant subset of attributes is given by the sum of a cluster mean value and an adjustment value for the current object and an adjustment value for the current attribute. In general, covering a cluster with such an additive model is possible if the contributing attributes exhibit a simple linear positive correlation among each other. This excludes negative or complex correlations, thus restricting the general search space. Cluster objects reside sparsely on hyperplanes parallel to the irrelevant axes. Projected onto the relevant subspace, the clusters appear as increasing one-dimensional lines. In comparison to axis-parallel approaches, the generalization consists mainly in allowing the axis-parallel hyperplane to be sparse. Also the cluster in the projection subspace may remain sparse. The unifying property of all cluster members is the common pattern of correlation between attributes.

Allowing sparseness in the spatial patterns is an interesting feature of this family of approaches since this also alleviates aspects (ii) and (iii) of the "curse of dimensionality." Aspect (iv) is addressed partially.

Correlation Clustering

Correlation clustering approaches follow the most general model: Points forming a cluster can be located on an arbitrarily oriented hyperplane (i.e., subspace). These patterns occur if some attributes follow linear but complex correlations among each other (i.e., one attribute may be the linear combination of several other attributes). The main point addressed by these approaches is therefore aspect (iv) of the "curse of dimensionality." The most widespread technique is the application of principal component analysis (PCA) on locally selected sets of points. Other techniques are based on applying the Hough transform [11] to the data set. Since the Hough transform does not rely on spatial closeness of points, by using this technique, also aspect (ii) is tackled.

The first approach to this type of subspace clustering was proposed in [4]. The general model for this family of approaches is described in [12].

Key Applications

In many scientific and economic fields (like astronomy, physics, medicine, biology, archaeology, geology, geography, psychology, and marketing), vast amounts of high-dimensional data are collected. To gain the full potentials out of the gathered information, subspace clustering techniques are useful in all these domains. Pattern-based approaches are especially popular in microarray data analysis.

Future Directions

The different groups of subspace clustering techniques (subspace clustering, projected clustering, pattern-based clustering, correlation clustering) tackle different subproblems of the "curse of dimensionality." There remain challenges for each of these problems. However, as a next-generation type of approach, algorithms to tackle more and more aspects simultaneously can be expected. An open problem is the redundancy of results when similar clusters are identified in different subspaces. With respect to this problem, there are strong connections to other clustering techniques such as ensemble clustering, alternative clustering, and constrained clustering [13]. Other challenges arise when tackling the subspace clustering problem on more complex data such as dynamic data [14].

Url To Code

Many standard methods for subspace clustering are available in the open-source data mining framework ELKI [15] at http://elki.dbs.ifi.lmu.de/.

Cross-References

▶ Apriori Property and Breadth-First Search Algorithms
▶ Clustering on Streams
▶ Clustering Overview and Applications
▶ Clustering with Constraints

▶ Curse of Dimensionality
▶ Data Mining
▶ Dimensionality Reduction
▶ Feature Selection for Clustering
▶ Text Clustering

Recommended Reading

1. Hartigan JA. Direct clustering of a data matrix. J Am Stat Assoc. 1972;67(337):123–29.
2. Agrawal R, Gehrke J, Gunopulos D, Raghavan P. Automatic subspace clustering of high dimensional data for data mining applications. In: Proceedings of the ACM SIGMOD International Conference on Management of Data; 1998. p. 94–105.
3. Aggarwal CC, Procopiuc CM, Wolf JL, Yu PS, Park JS. Fast algorithms for projected clustering. In: Proceedings of the ACM SIGMOD International Conference on Management of Data; 1999. p. 61–72.
4. Aggarwal CC, Yu PS. Finding generalized projected clusters in high dimensional space. In: Proceedings of the ACM SIGMOD International Conference on Management of Data; 2000. p. 70–81.
5. Madeira SC, Oliveira AL. Biclustering algorithms for biological data analysis: a survey. IEEE/ACM Trans Comput Biol Bioinform. 2004;1(1):24–45.
6. Kriegel HP, Kröger P, Zimek A. Clustering high dimensional data: a survey on subspace clustering, pattern-based clustering, and correlation clustering. ACM Trans Knowl Discov Data (TKDD). 2009;3(1):1–58.
7. Kriegel HP, Kröger P, Zimek A. Subspace clustering. Wiley Interdiscip Rev Data Min Knowl Disc. 2012;2(4):351–64.
8. Bellman R. Adaptive control processes. A guided tour. Princeton: Princeton University Press; 1961.
9. Beyer K, Goldstein J, Ramakrishnan R, Shaft U. When is "Nearest Neighbor" meaningful? In: Proceedings of the 7th International Conference on Database Theory; 1999. p. 217–35.
10. Houle ME, Kriegel HP, Kröger P, Schubert E, Zimek A. Can shared-neighbor distances defeat the curse of dimensionality? In: Proceedings of the 22nd International Conference on Scientific and Statistical Database Management; 2010. p. 482–500.
11. Achtert E, Böhm C, David J, Kröger P, Zimek A. Global correlation clustering based on the Hough transform. Stat Anal Data Min. 2008;1(3):111–27.
12. Achtert E, Böhm C, Kriegel HP, Kröger P, Zimek A. Deriving quantitative models for correlation clusters. In: Proceedings of the 12th ACM International Conference on Knowledge Discovery and Data Mining; 2006. p. 4–13.
13. Zimek A, Vreeken J. The blind men and the elephant: on meeting the problem of multiple truths in data from clustering and pattern mining perspectives. Mach Learn. 2013;98(1–2):121–55.

S

14. Sim K, Gopalkrishnan V, Zimek A, Cong G. A survey on enhanced subspace clustering. Data Min Knowl Disc. 2013;26(2):332–97.
15. Achtert E, Kriegel HP, Schubert E, Zimek A. Interactive data mining with 3D-parallel-coordinate-trees. In: Proceedings of the ACM SIGMOD International Conference on Management of Data; 2013. p. 1009–12.

Success at n

Nick Craswell
Microsoft Research Cambridge, Cambridge, UK

Synonyms

S@n

Definition

Success at n is an information retrieval relevance measure, equal to 1 if the top-n documents contain a relevant document and 0 otherwise. When averaged across multiple queries, the success rate at n indicates how often something relevant was retrieved within the top-n.

Key Points

A system with a high success rate will be one that rarely retrieves zero relevant documents. Therefore, success rate can be employed to monitor failure. The success rate of two systems may differ even if they have the same mean precision at n, because one system has higher variance than the other. Success at n models the satisfaction of a user who does not need to see many relevant documents, is prepared to view up to n results and is disappointed by a completely irrelevant top-n.

Cross-References

▶ Precision at n
▶ Precision-Oriented Effectiveness Measures

Succinct Constraints

Carson Kai-Sang Leung
Department of Computer Science, University of Manitoba, Winnipeg, MB, Canada

Definition

Let *Item* be the set of domain items. Then, an itemset $SS_j \subseteq Item$ is a *succinct set* if SS_j can be expressed as a result of selection operation $\sigma_p(Item)$, where σ is the usual selection operator and p is a selection predicate. A powerset of items $SP \subseteq 2^{Item}$ is a *succinct powerset* if there is a fixed number of succinct sets $SS_1, ..., SS_k \subseteq Item$ such that SP can be expressed in terms of the powersets of $SS_1, ..., SS_k$ using set union and/or set difference operators. A constraint C is *succinct* provided that the set of itemsets satisfying C is a succinct powerset.

Key Points

Succinct constraints [1, 2] possess the following nice properties. For any succinct constraint C, there exists a precise "formula" – called a *member generating function* (*MGF*) – to enumerate *all* and *only* those itemsets that are guaranteed to satisfy C. Hence, if C is succinct, then C is pre-counting prunable. This means that one can directly generate precisely the itemsets that satisfy C – without looking at the transaction database and even before counting the support (or frequency) of itemsets. In other words, whether an itemset S satisfies C or not can be determined based on the selection of items from the domain. Examples of succinct constraints include $max(S.Price) \leq \$120$ and $max(S.Price) \geq \$80$, which express that the maximum price of all items in an itemset S is at most \$120 and at least \$80, respectively. The set of itemsets satisfying the first constraint $max(S.Price) \leq \$120$ can be expressed in terms of a succinct powerset: $2^{\sigma_{Price} \leq \$120(Item)}$. These itemsets can be enumerated by using only the items having prices at most

$120 – via the MGF $\{X \mid X \subseteq \sigma_{Price \leq \$120}(Item), X \neq \emptyset\}$. Similarly, the set of itemsets satisfying the second constraint $max(S.Price) \geq \$80$ can be expressed in terms of a succinct powerset formed by the set difference between two powersets: $2^{Item} - 2^{\sigma_{Price < \$80}(Item)}$. These itemsets can be enumerated by using the items having prices at least $80 (i.e., mandatory items) and other items (i.e., optional items) – via the MGF $\{Y \cup Z \mid Y \subseteq \sigma_{Price \geq \$80}(Item), Y \neq \emptyset, Z \subseteq \sigma_{Price < \$80}(Item)\}$.

Cross-References

▶ Frequent Itemset Mining with Constraints

Recommended Reading

1. Lakshmanan LVS, Leung CK-S, Ng RT. Efficient dynamic mining of constrained frequent sets. ACM Trans Database Syst. 2003;28(4):337–89. https://doi.org/10.1145/958942.958944.
2. Ng RT, Lakshmanan LVS, Han J, Pang A. Exploratory mining and pruning optimizations of constrained associations rules. In: Proceedings of the ACM SIGMOD International Conference on Management of Data; 1998. p. 13–24.

Suffix Tree

Maxime Crochemore[1,2] and Thierry Lecroq[3]
[1] King's College London, London, UK
[2] Université Paris-Est, Paris, France
[3] Université de Rouen, Rouen, France

Synonyms

Compact suffix trie

Definition

The suffix tree $S(y)$ of a nonempty string y of length n is a compact trie representing all the suffixes of the string.

The suffix tree of y is defined by the following properties:

- All branches of $S(y)$ are labeled by all suffixes of y.
- Edges of $S(y)$ are labeled by strings.
- Internal nodes of $S(y)$ have at least two children.
- Edges outgoing an internal node are labeled by segments starting with different letters.
- The segments are represented by their starting position on y and their lengths.

Moreover, it is assumed that y ends with a symbol occurring nowhere else in it (the space sign ␣ is used in the examples of the present entry). This avoids marking nodes and implies that $S(y)$ has exactly n leaves (number of nonempty suffixes).

All the properties then imply that the total size of $S(y)$ is $O(n)$, which makes it possible to design a linear-time construction of the suffix tree.

Historical Background

The first linear time algorithm for building a suffix tree of a string of length n is from Weiner [18] but it requires quadratic space: $O(n \times \sigma)$ where σ is the size of the alphabet. The first linear time and space algorithm for building a suffix tree is from McCreight [14]. It works "off-line" inserting the suffixes from the longest one to the shortest one. A strictly sequential version of the suffix tree construction was described by Ukkonen [17]. When the alphabet is potentially infinite, the optimal construction algorithms of the suffix tree can be implemented to run in time $O(n \log \sigma)$ since they imply an ordering on the letters of the alphabet. On particular integer alphabets, Farach [5] showed that the construction can be done in linear time. The minimization in the sense of automata theory of the suffix trie gives the suffix automaton. The suffix automaton of a string is also known under the name of *DAWG*, for *directed acyclic word graph*. Its

S

linearity was discovered by Blumer et al. (see [1]), who gave a linear construction (on a fixed alphabet). The minimality of the structure as an automaton is from Crochemore [2] who showed how to build with the same complexity the factor automaton of a text.

The compaction of the suffix automaton gives the compact suffix automaton (see [1]). The compaction consists in removing all internal nodes with only one child and concatenating remaining successive edge labels. A direct construction algorithm of the compact suffix automaton was presented by Crochemore and Vérin [4]. The same structure arises when minimizing the suffix tree.

The suffix array of the string y consists of both the permutation of positions on the text that gives the sorted list of suffixes and the corresponding array of lengths of their longest common prefixes (LCP) (see [9]). The suffix array of a string, with the associated search algorithm based on the knowledge of the common prefixes, is from Manber and Myers [13]. It can be built in linear time on integer alphabets (see [8, 10, 11, 15]). More on suffix arrays and their applications, mainly to bioinformatics, can be found in [16].

For the implementation of index structures in external memory, the reader can refer to Ferragina and Grossi [6].

Scientific Fundamentals

Suffix Trees

The suffix tree $S(y)$ of the string $y = \mathtt{ababbb}_\sqcup$ is presented Fig. 1. It can be seen as a compaction of the suffix trie $\mathcal{T}(y)$ of y given Fig. 2.

Nodes of $S(y)$ and $\mathcal{T}(y)$ are identified with segments of y. Leaves of $S(y)$ and $\mathcal{T}(y)$ are identified with suffixes of y. An output is defined, for each leaf, which is the starting position of the suffix in y.

The two structures can be built by successively inserting the suffixes of y from the longest to the shortest.

In the suffix trie $\mathcal{T}(y)$ of a string y of length n, there exist n paths from the root to the n leaves: each path spells a different nonempty suffix of y. Edges are labeled by exactly one symbol. The suffix trie can have a quadratic number of nodes since the sum of the lengths of all the suffixes of y is quadratic.

To get the suffix tree from the suffix trie, internal nodes with exactly one successor are removed. Then labels of edges between remaining nodes are concatenated. Edges are now labeled with strings. This gives a linear number of nodes since there are exactly n leaves, and since every internal node (called a fork) has at least two successors, there can be at most $n - 1$ forks. This also gives a linear number of edges.

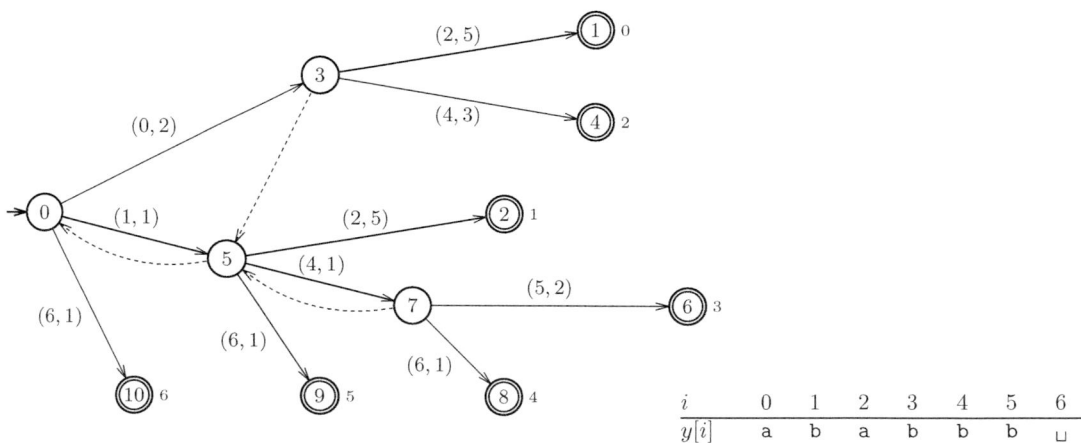

Suffix Tree, Fig. 1 Suffix tree $S(y)$ of ababbb. Nodes are numbered in the order of creation. The small numbers closed to each leaf correspond to the position of the suffix associated to the leaf

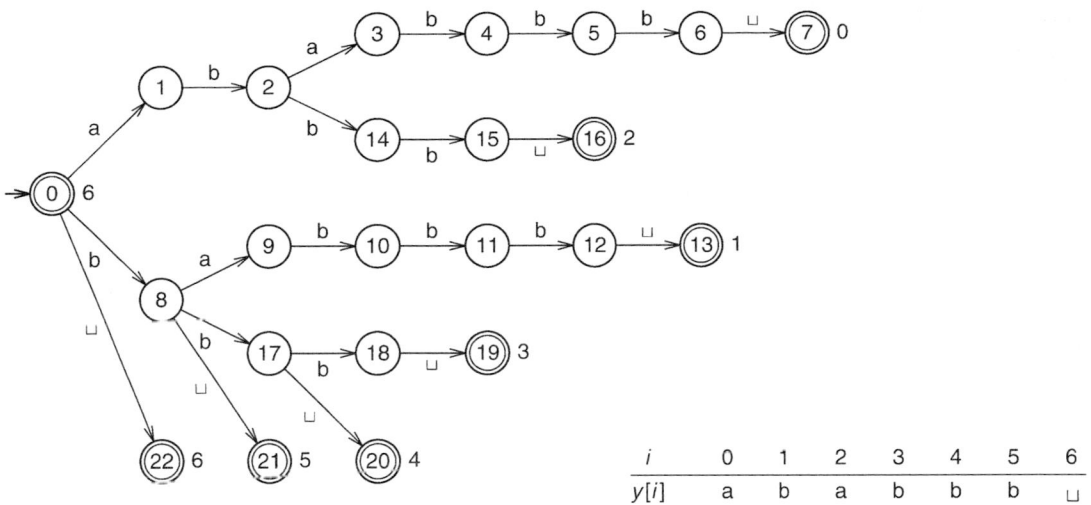

Suffix Tree, Fig. 2 Suffix trie $\mathcal{T}(y)$ of ababbb. Nodes are numbered in the order of creation. The small numbers closed to each leaf correspond to the position of the suffix associated to the leaf

Now, in order that the space requirement becomes linear, since the labels of the edges are all segments of y, a segment $y[i \, . . \, i + \ell - 1]$ is represented by the pair (i, ℓ). Thus each edge can be represented in constant space. This technique requires to have y residing in main memory.

Overall, there is a linear number of nodes and a linear number of edges; each node and each edge can be represented in constant space; thus, the suffix tree requires a linear space.

There exist several direct linear time construction algorithms of the suffix tree that avoid the construction of the suffix trie followed by its compaction.

The McCreight algorithm [14] directly constructs the suffix tree of the string y by successively inserting the suffixes of y from the longest one to the shortest one. The insertion of the suffix of y beginning at position i (i.e., $y[i \, . . \, n - 1]$) consists first in locating (creating it if necessary) the fork associated with the longest prefix of $y[i \, . . \, n - 1]$ common with a longer suffix of y: $y[0 . . n - 1], y[1 . . n - 1], \ldots,$ or $y[i - 1 . . n - 1]$. Let us call the head u the longest prefix and the tail v the remaining suffix such that $y[i \, . . \, n-1] = uv$. Once the fork p associated with u has been located, it is enough to add a new leaf q labeled by i and a new edge labeled by $(i + |u|, |v|)$ between p and q to complete the insertion of

$y[i \, . . \, n-1]$ into the structure. The reader can refer to [3] for further details.

The linear time of the construction is achieved by using a function called suffix link defined on the forks as follows: if fork p is identified with segment av, $a \in A$ and $v \in V^*$, then $s_y(p) = q$ where fork q is identified with v.

Suffix links are represented by dotted arrows on Fig. 1.

The suffix links create shortcuts that are used to accelerate head computations. If the head of $y[i - 1 . . n - 1]$ is of the form au ($a \in A, u \in V^*$), then u is a prefix of the head of $y[i \, . . \, n - 1]$. Therefore, using suffix links, the insertion of the suffix $y[i \, . . \, n - 1]$ consists first in finding the fork corresponding to the head of $y[i \, . . \, n-1]$ (starting from suffix link of the fork associated with au) and then in inserting the tail of $y[i \, . . \, n - 1]$ from this fork.

Ukkonen algorithm [17] works online, i.e., it builds the suffix tree of y processing the symbols of y from the first to the last. It also uses suffix links to achieve a linear time computation.

The Weiner, McCreight, and Ukkonen algorithms work in $O(n)$ time whenever $O(n \times \sigma)$ space is used. If only $O(n)$ space is used, then the $O(n)$ time bound should be replaced by $O(n \times \log \min\{n, \sigma\})$. This is due to the access time for a specific edge stored in each nodes. The reader can refer to [12] for specific optimized implementations of suffix trees.

For particular integer alphabets, if the alphabet of y is in the interval $[1..n^c]$ for some constant c, Farach [5] showed that the construction can be done in linear time.

Generalized suffix trees are used to represent all the suffixes of all the strings of a set of strings.

Indexes

The suffix tree serves as a full index on the string: it provides a direct access to all segments of the string and gives the positions of all their occurrences in the string. An index on y can be considered as an abstract data type whose basic set is the set of all the segments of y and that possesses operations giving access to information relative to these segments. The utility of considering the suffixes of a string for this kind of application comes from the obvious remark that every segment of a string is the prefix of a suffix of the string.

Once the suffix tree of a text y is built, searching for x in y remains to spell x along a branch of the tree. If this walk is successful, the positions of the pattern can be output. Otherwise, x does not occur in y.

Any kind of trie that represents the suffixes of a string can be used to search it. But the suffix tree has additional features which imply that its size is linear.

We consider four operations relative to the segments of a string y: the membership, the first position, the number of occurrences, and the list of positions.

The first operation on an index is the membership of a string x to the index, that is to say the question to know whether x is a segment of y. This question can be specified in two complementary ways whether we expect to find an occurrence of x in y or not. If x does not occur in y, it is often interesting in practice to know the longest beginning of x that is a segment of y. This is the type of usual answer necessary for the sequential search tools in a text editor.

The methods produce without large modification the position of an occurrence of x and even the position of the first or last occurrence of x in y.

Knowing that x is in the index, another relevant information is constituted by its number of occurrences in y. This information can differently direct the ulterior searches.

Finally, with the same assumption than previously, a complete information on the localization of x in y is supplied by the list of positions of its occurrences.

Suffix trees can easily answer these questions. It is enough to spell x from the root of $S(y)$. If it is not possible, then x does not occur in y. Whenever x occurs in y, let w be the shortest segment of y that is such x is a prefix of w and w is associated with a node p of $S(y)$. Then the number of leaves of the subtree rooted in the node p gives the number of occurrences of x in y. The smallest (respectively largest) of these leaves gives the position of the first (resp. last) position of x in y. The list of the number of these leaves gives the list of positions of x in y (see [3]).

Key Applications

Suffix trees are used to solve string searching problems mainly when the text into which a pattern has to be found is fixed. It is also used in other string-related problems such as longest repeated substring and longest common substring. It can be used to perform text compression. Word suffix trees can be used when processing natural languages in order to represent only suffixes starting after separators such as space or line feed. Gusfield [7] gives many applications of suffix trees in computational biology.

Cross-References

▶ Trie

Recommended Reading

1. Blumer A, Blumer J, Ehrenfeucht A, Haussler D, Chen MT, Seiferas J. The smallest automaton

recognizing the subwords of a text. Theor Comput Sci. 1985;40(1):31–55.

2. Crochemore M. Transducers and repetitions. Theor Comput Sci. 1986;45(1):63–86.

3. Crochemore M, Hancart C, Lecroq T. Algorithms on strings. Cambridge: Cambridge University Press; 2007.

4. Crochemore M, Vérin R. On compact directed acyclic word graphs. In: Structures in logic et computer science, LNCS, vol. 1261; 1997, p. 192–211. https://doi.org/10.1007/3-540-63246-8_12.

5. Farach M. Optimal suffix tree construction with large alphabets. In: Proceedings of the 38th IEEE Annual Symposium on Foundations of Computer Science; 1997. p. 137–43.

6. Ferragina P, Grossi R. The string B-tree: a new data structure for string search in external memory ct its applications. J Assoc Comput Mach. 1999;46(2):236–80.

7. Gusfield D. Algorithms on strings, trees and sequences. Cambridge: Cambridge University Press; 1997.

8. Kärkkäinen J, Sanders P. Simple linear work suffix array construction. In: Proceedings of the 30th International Colloquium on Automata, Languages, and Programming; 2003. p. 943–55. http://dblp.uni-trier.de/rec/bib/conf/icalp/KarkkainenS03.

9. Kasai T, Lee G, Arimura H, Arikawa S, Park K. Linear-time longest-common-prefix computation in suffix arrays and its applications. In: Proceedings of the 12th Annual Symposium on Combinatorial Pattern Matching; 2001. p. 181–92.

10. Kim DK, Sim JS, Park H, Park K. Linear-time construction of suffix arrays. In: Proceedings of the 14th Annual Symposium on Combinatorial Pattern Matching; 2003. p. 186–99.

11. Ko P, Aluru S. Space efficient linear time construction of suffix arrays. In: Proceedings of the 14th Annual Symposium on Combinatorial Pattern Matching; 2003. p. 200–10.

12. Kurtz S. Reducing the space requirement of suffix trees. Softw Pract Exper. 1999;29(13):1149–71.

13. Manber U, Myers G. Suffix arrays: a new method for on-line string searches. SIAM J Comput. 1993;22(5):935–48.

14. McCreight EM. A space-economical suffix tree construction algorithm. J Algorithms. 1976;23(2):<?pag?>262–72.

15. Nong G, Zhang S, Chan WH. Linear suffix array construction by almost pure induced-sorting. In: Proceedings of the Data Compression Conference; 2009. p. 193–202.

16. Ohlebusch E. Bioinformatics algorithms: sequence analysis, genome rearrangements, and phylogenetic reconstruction, Oldenbusch Verlag; 2013.

17. Ukkonen E. On-line construction of suffix trees. Algorithmica. 1995;14(3):249–60.

18. Weiner P. Linear pattern matching algorithm. In: Proceedings of the 14th Annual IEEE Symposium on Switching et Automata Theory; 1973. p. 1–11.

Summarizability

Arie Shoshani
Lawrence Berkeley National Laboratory,
Berkeley, CA, USA

Synonyms

Statistical correctness; Summarization correctness

Definition

Summarizability is a property that assures the correctness of summary operations over *On-Line Analytical Processing* (OLAP) databases, which are akin to Statistical Databases [10]. Such databases are generally referred to as "summary databases," and have a data model based on one or more measures defined over the cross product of *dimensions*. For example, a bookstore company may have multiple stores in many cities. Assume that there is a database containing the stores' revenues for books sold per day over the last 3 years. In such a database, "revenue" is a *measure*, and "book," "store," "day" are the dimensions that define the cross product over which the measure revenue is defined. A dimension in a summary database is said to be *summarizable* relative to a measure, if a summary statistic (sum, average, etc.) applied over the dimension produces correct results. For example, if summarization over all the books sold to obtain "total_revenues per store, per day" yields correct results, the dimension "book" is considered summarizable relative to the measure "revenue." There are certain conditions that have to hold in order to get correct results, which are discussed in the next section. Often, dimensions are organized into a *hierarchy* of categories. For example, days can naturally be organized into months, and months into years. Similarly, books can be organized by book-types (e.g., cooking, fiction, etc.). Summarization can then be applied to categories, such as summarizing over books and days to get "revenue per book-type, per store, per year." In such cases the summarizability property must apply to each category level of the category hierarchy of a dimension, for that dimension to be considered summarizable.

Historical Background

Statistical Databases, which were introduced in the 1980s [2], and OLAP databases, introduced in the 1990s [1, 3, 4] have a similar data model [1], but the issue of summarizability was not introduced until 1990 [9] and studied carefully until 1997 [7]. After that time, several authors have treated summarizability formally [5, 6, 8].

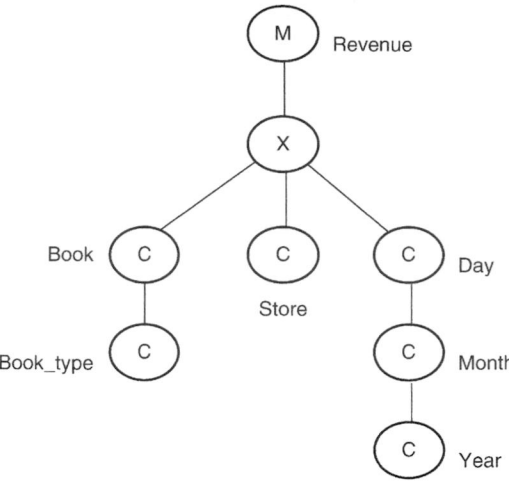

Summarizability, Fig. 1 Revenue of books sold in a particular store on a particular day

Foundations

Next, examples that violate summarizability are presented, and using these examples the conditions for summarizability are stated. There are three such conditions that must hold in order for a dimension to be summarizable. Two of the conditions refer to the category levels in a dimension, and the third is a condition of the dimension relative to the measure. Next, the basic notation used throughout this document is introduced.

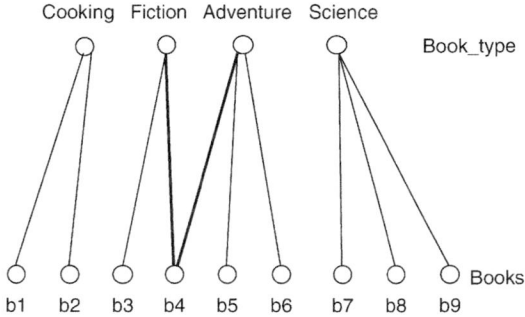

Summarizability, Fig. 2 Books organized by book_type

Notation
Consider, again, the example above of revenues per book, store, and day. Using a notation commonly used for such databases, this example database can be represented as "revenue (book, store, day)." For the category hierarchies of dimensions, the notation $[C_1 \to C_2 \to \ldots C_i \to C_{i+1}, \ldots C_n]$ is used to represent a category hierarchy of a dimension of height n, starting from the more detailed level towards higher levels. Thus, for the example above, where the two hierarchies mentioned above are over the dimensions book and day, the database will be represented as:
`"revenue ([book-> book-type],`
`store, [day -> month -> year])."`

These concepts and notation are shown graphically in Fig. 1, where the letters M, C,

and X represent Measure, Category-level, and X-product (cross-product), respectively.

The "Disjointness" Condition
Consider the revenue database above. Suppose that the book-type set is (cooking, fiction, adventure, science, etc.). Most of the books will usually belong to a single book_type, but some could be categorized under two or more types. This is shown graphically in Fig. 2, for sales in a particular day for a particular store. As can be seen, there in one book, b4, which is classified under the categories "fiction" and "adventure." If the revenues by book_type are added to generate "revenue (book-type, store, day)," the totals will be incorrect, because if the revenues for all book-types are added, the revenue for book b4 is added

twice. The reason is, of course, that book b4 belongs to two parent categories.

The disjointness condition states: given two consecutive category-levels C_low and C_high in a dimension, where [C_low -> C_high], the sets of lower-level categories elements that belong to each category element of the higher-level, must be disjoint. This condition can also be expressed as requiring that the category elements of the category levels form a "strict" hierarchy [8]. Yet another way to state this condition is to say that there must be a one-to-many relationship from C_high to C_low.

This seemingly simple observation is the source of incorrect statistics in many systems that do not enforce summarizability conditions. Under such conditions, summarization is still possible by special treatment, such as choosing to assign revenues equally to shared nodes (in the example above assigning half the revenue for book b4 to each of the "fiction" and "adventure" book types). However, this is not usually done. Note that summarization of book revenues to get, for example, "revenues (store, day)" will yield a correct result, since the category book_type is not involved.

The "Completeness" Condition

Completeness is a condition that holds if all the children of higher-level category elements exist. If some of the children are missing, then the summary to the higher level may be incorrect. Consider, for example, a database that contains "population (city, year, race, sex)." Suppose further that cities are organized by states, as shown in Fig. 3. In this database, if the population is summarized to the "state" level, the result of populations is obviously incorrect, since only populations of cities are taken into account and not populations of villages and small towns. However, if the measure was stated as "populations_in_cities" then the [city -> state] category mapping would be summarizable. This example shows that the second condition of completeness is relative to the measure semantics.

One way to overcome the "incompleteness" condition is to add instances that account for the missing elements in the category. In the

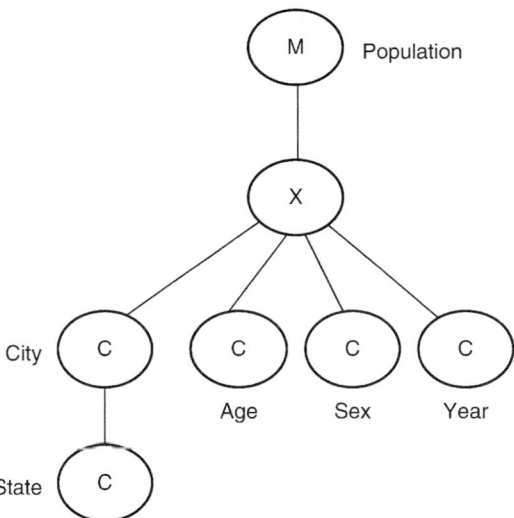

Summarizability, Fig. 3 The population database is not summarizable to the "state" level

population example, one can add for each state, in addition to the cities in the state, an instance that accounts for the population in all areas other than cities. Such a node can be labeled "other_areas" for example. If this is done, summarization to the state level would yield the correct summary population.

Another way to determine if a category level satisfies the completeness condition can often be based on external knowledge. For example, suppose that the dimension "age" in Fig. 3 is organized into age_groups: (0–10) (10–20), ..., (90–100). Is summarization to the age_group level correct in this case? It is only correct if there is external knowledge that this database does not contain people older that 100. Otherwise, a category (>100) has to be added in order to satisfy the completeness condition.

The "Measure Type" Condition

Consider the database in Fig. 3 again, where a higher category level is added to the dimension year: [year -> decade]. Obviously, population cannot be summarized to the decade level, since adding the yearly populations does not yield a meaningful measure for the decade. However, if the measure was "average population" per year (and "counts" were also recorded), the average population per decade could be calculated. Why is that? As another example, consider the book revenues database in Fig. 1. Obviously, the revenues can be summarized from days, to months, to years. However, if the measure was "number_of_unsold_books," this cannot be summarized (added) over the time dimension. The reason stems from the semantic behavior of "temporal aggregation."

In statistics the term "temporal aggregation" is used to describe the behavior of measures when aggregating over the time domain. The measures are classified into three types: "stock," "flow," and "value-per-unit." It turns out that these types behave differently when summarized over time, depending on the summary statistics used. In particular, a measure of "stock" type cannot be summed over the time domain, whereas a measure of "flow" type can be summed over the time domain. In the example discussed above, "population" is of type stock and so is "number_of_unsold_books," and therefore, they cannot be summed over the time dimension. In contrast, "book_revenues" is of type "flow" and therefore can be summed.

In general, measures of type "stock" refer to a state of the measure recorded at a particular point in time (such as inventory), while measures of type "flow" record values of events over a period of a time (such as sales). A measure of the type "value-per-unit" is similar to "stock" in that it is recorded at a particular point in time, but it has a per-unit value (such as the cost of a book). In [7] a table is given for temporal summarizability for each measure type for five common aggregation operators: min, max, sum, avg, and range. The table is reproduced in Fig. 4. As can be seem only the operator "sum" is not summarizable for the types "stock" and "value-per-unit." It turns out that "value-per-unit" is also not summarizable for non-temporal aggregation in the case of "sum," but all other cases are summarizable for non-temporal aggregation [7].

Summary

The conditions that are necessary to ensure correctness of aggregation operations over statistical and OLAP databases were presented. Such

	stock	flow	value/unit
min	yes	yes	yes
max	yes	yes	yes
sum	no	yes	no
avg	yes	yes	yes
range	yes	yes	yes

Summarizability, Fig. 4 Temporal summarizability by measure type and function type

conditions are referred to as "summarizability conditions." Summarizability conditions apply to the dimensions of multidimensional data structures and to the category hierarchies in each dimension. Such conditions depend on the summary measure type, and whether the summarization is over the time domain. Note that the summarizability of each category level in a hierarchy is independent of the others; that is, some category levels can be summarizable while others are not.

Summarizability conditions are applicable to any database system that supports aggregation operations. While these conditions were described here in the context of the OLAP data model, these conditions apply to other models that do not express the multidimensional structures explicitly. In particular, it is possible to represent an OLAP schema as a relational schema, where the semantics of multidimensionality and category hierarchies are not explicit. In order to achieve correct results in aggregation operations from such relational databases, it is necessary to identify these (multidimensionality and category hierarchies) structures, and to make sure that the summarizability conditions hold. If all the conditions do not hold, aggregation operations should be avoided and/or refused.

Key Applications

It is often claimed that statistical operations can be inaccurate for various reasons. The better-known reason is summarization over null values. For example, taking an average over a set of values where some are null, will produce the wrong result if the null elements are represented as zeros, or if they are not discounted in the computation. Summarizability conditions are just as important, but are more subtle, semantically based, and are often overlooked. It is essential that such conditions are checked in any database that provides aggregation operators, including OLAP and relational database systems.

Future Directions

The three conditions described above for ensuring summarizability are necessary conditions. However, while it is believed that these conditions are also sufficient, this was not shown formally so far. Another aspect of future work is adding annotations to schemas as to whether summarizability conditions hold, and how to automate the checking of summarizability conditions dynamically in a database as data instances are entered into (or modified in) the database. Once summarizability is made part of the data model, it is necessary to enhance the aggregation operators to avoid summarization over non-summarizable data.

Cross-References

► Dimension
► Hierarchy
► Measure
► Multidimensional Modeling
► Online Analytical Processing
► Quality of Data Warehouses
► Query Processing in Data Warehouses
► Scientific Databases

Recommended Reading

1. Agrawal R, Gupta A, Sarawagi S. Modeling multidimensional databases. In: Proceedings of the 13th International Conference on Data Engineering; 1997. p. 232–43.
2. Chan P, Shoshani A. Subject: a directory driven system for organizing and accessing large statistical databases. In: Proceedings of the 7th International Conference on Very Data Bases; 1981. p. 553–63.
3. Codd EF, Codd SB, Salley CT. Providing olap (online analytical processing) to user-analysts: an IT mandate, Codd and Associates technical report; 1993.
4. Gray J, Bosworth A, Layman A, Pirahesh H. Data cube: a relational aggregation operator generalizing group-by, cross-tabs and sub-totals. In: Proceedings of the 12th International Conference on Data Engineering; 1996. p. 152–9.
5. Hurtado CA, Mendelzon AO. Reasoning about summarizability in heterogeneous multidimensional schemas. In: Proceedings of the 8th International Conference on Database Theory; 2001. p. 375–89.
6. Hurtado CA, Gutiérrez C, Mendelzon A. Capturing summarizability with integrity constraints in OLAP. ACM Trans Database Syst. 2005;30(3):854–86.
7. Lenz H-J, Shoshani A. Summarizability in OLAP and statistical data bases. In: Proceedings of the 9th International Conference on Scientific and Statistical Database Management; 1997. p. 132–43.
8. Pedersen TB, Jensen CS. Multidimensional data modeling for complex data. In: Proceedings of the 15th International Conference on Data Engineering; 1999. p. 336–45.
9. Rafanelli M, Shoshani A. STORM: a statistical object representation model. In: Proceedings of the 2nd International Conference on Scientific and Statistical Database Management; 1990. p. 14–29.
10. Shoshani A. OLAP and statistical databases: similarities and differences. In: Proceedings of the 16th ACM SIGACT-SIGMOD-SIGART Symposium on Principles of Database Systems; 1997. p. 185–96.

Summarization

Jimmy Lin
University of Maryland, College Park, MD, USA

Synonyms

Automatic abstracting; Distillation; Report writing; Text/document summarization

Definition

Summarization systems generate condensed outputs that convey important information contained in one or more sources for particular users and tasks. In principle, input sources and system outputs are not limited to text (e.g., key frame extraction for video summarization), but this entry focuses exclusively on generating *textual* summaries from *textual* sources.

Historical Background

Summarization has a long history dating back to the 1960s, when researchers first started developing computer systems that processed natural language [6, 12]. Following a number of decades with comparatively few publications, summarization research entered a new phase in the 1990s. A revival of interest was spurred by the growing availability of text in electronic formats and later the World Wide Web. The enormous quantities of information people come into contact with on a daily basis created a need for applications that help users cope with the proverbial information overload problem. Summarization systems attempt to address this need.

Foundations

Summarization is a broad and diverse field. Traditionally, it is considered a sub-area of natural language processing, but a significant number of innovations have their origins in information retrieval. This entry is organized as follows: first, various summarization factors are discussed. Next, a tripartite processing model for summarization systems is presented, which provides a basis for discussing general issues. Finally, selected summarization techniques are briefly overviewed.

Summarization Factors

To better understand summarization, it is helpful to enumerate its many dimensions – what Sparck Jones [19] calls "factors". These factors provide a basis for understanding various automatic methods, and can be grouped into three broad categories: *input*, *purpose*, and *output*. What follows is meant to be an overview of important factors, and not intended to be exhaustive.

Input factors characterize the source of the summaries:

1. *Single* versus *multiple sources*. For example, one versus multiple reports of the same event.
2. *Genre* (categories of texts) and *Register* (different styles of writing). For example, dissertations versus blogs.
3. *Written* versus *spoken*. For example, newspaper articles versus broadcast news.
4. *Language*. Sources may be in multiple language.
5. *Metadata*. Sources may be associated with controlled vocabulary keywords, human-assigned category labels.
6. *Structure*. Source structure may be relatively straightforward (e.g., headings and sub-headings) or significantly more complex (e.g., email threads).

Purpose factors characterize the use of summaries (i.e., why they were created):

1. *Indicative* versus *informative* versus *evaluative*. Indicative summaries are meant to guide the selection of sources for more in-depth study, whereas informative summaries cover salient information in the sources at some level of detail (and is often meant to replace the original). Evaluative summaries assess the subject matter of the source and the quality of the work (e.g., a review of a movie).
2. *Generic* versus *focused*. A generic summary places equal emphasis on different information contained in the sources and provides balanced coverage. Alternatively, a summary might be focused on an information need, i.e., created to answer a question.
3. *Task*. What will the summary be used for? For example, to help write a report or to make a decision.
4. *Audience*. Whom is the summary intended for? For example, experts, schoolchildren, etc.

Output factors characterize system output (note that the input factors are relevant here also, but not repeated):

1. *Extractive* versus *abstractive*. Extractive summaries consist of text copied from the source material; typically, such approaches are based on shallow analysis. Abstractive summaries contain text that is system-generated, usually based on deeper analysis. Note that these approaches define a continuous spectrum, as many systems employ hybrid methods.

2. *Reduction, coverage, and fidelity.* Reduction, usually measured as a ratio between summary length and source length, is often inversely related to coverage, how much information of interest is preserved in the summary. The summary should also preserve source information accurately.

3. *Coherence.* Does the summary read fluently and grammatically, both syntactically and at the discourse level? For summaries not intended to be fluent prose (e.g., bullets), this factor is less important.

Input, purpose, and output factors together characterize the many dimensions of summarization and provide a basis for subsequent discussions. Note, however, that not all factors figure equally in current summarization systems – for a variety of reasons, the field has focused on some more than others.

Processing Model

Sparck Jones characterizes the process of summarization as a reductive transformation of source text to summary text through content condensation by selection and/or generalization of what is important in the source [19]. She proposes a tripartite processing model, shown in Fig. 1, that serves as a framework for understanding how various summarization techniques fit together (see also [15] for a similar model). Systems first convert source text into the source representation, which is then transformed into the summary representation. Finally, the summary representation is realized as natural language text. Note that these stages do not necessarily map to system components, as the processing model only describes abstract processing tasks. Since this model does not prescribe specific representations or particular

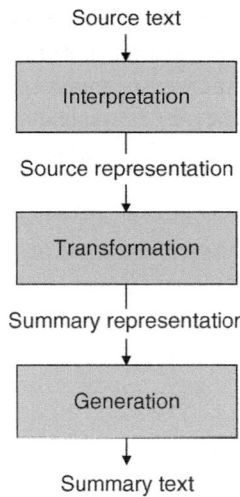

Summarization, Fig. 1 A tripartite processing model for summarization

processing methods, it is sufficiently general to describe a wide variety of summarization systems while at the same time highlighting important differences.

As previously discussed, input may come from one or multiple sources (the term "documents" is used generically, recognizing that sources may also be speech, email, etc.). Single-document summarization is challenging because simple baselines are often very difficult to improve upon. For example, since news articles are typically written in the "inverse pyramid" style (most important information first), the first sentence or paragraph makes an excellent summary. Frequently, longer documents (e.g., reports) contain "executive summaries", which nicely capture important information in the documents. Multi-document summarization faces a different set of challenges, the most salient of which is the possibility of redundant information in the sources (e.g., multiple news articles about the same event). Frequently, the redundancy is not superficially obvious, but involves paraphrase (different syntactic structures, word choices, etc.). More complex are cases where the information partially overlaps or appears contradictory (e.g., different reports of death tolls). More generally, multi-document summarization requires systems to detect similarities and differences in text.

It is generally assumed that a summarization system is provided the source text. In cases where this is assumption is not met, information retrieval techniques may be used to first select the set of documents to summarize (from a larger collection of documents). However, since most systems assume that input sources are more or less relevant to the task at hand, they may not adequately cope with imperfect retrieval results.

The use of "representation" does not necessarily imply deep linguistic analysis or processing. In fact, most extractive summarization systems adopt a "bag of words" representation at both the source and summary end – that is, text is represented as a vector that has a feature for each word. This representation makes the obviously false assumption that word occurrences are independent and ignores the rich linguistic relationships present in text. Nevertheless, extractive techniques have proven to be effective in various summarization tasks.

With extractive techniques, generation is trivial since systems simply copy material from the source. However, pure extraction often leads to problems in overall coherence of the summary – a frequent issue concerns "dangling" anaphora. Sentences often contain pronouns, which lose their referents when extracted out of context. Worse yet, stitching together decontextualized extracts may lead to a misleading interpretation of anaphors (resulting in an inaccurate representation of source information, i.e., low fidelity). Similar issues exist with temporal expressions. Note that these problems become more severe in the multi-document case, since extracts are drawn from different sources. A general approach to addressing these issues involves post-processing extracts, for example, replacing pronouns with their antecedents, replacing relative temporal expression with actual dates, etc. Such techniques, however, can not be considered purely extractive (hence the observation that most systems are, in fact, hybrid).

In general, extractive systems can be characterized as "knowledge-poor", which is contrasted against "knowledge-rich" approaches. While not synonymous, abstractive methods tend to be associated with "knowledge-rich" approaches. They involve one or more of the following: detailed linguistic analysis on source text to produce richly annotated structures, incorporation of world knowledge to support the transformation process, or generation of fluent natural language text from abstract representations.

A canonical example of abstractive summarization involves integration with information extraction (IE) systems. Information extraction concerns the automatic identification and creation of template instances from natural language text based on some pre-defined structure. For example, a template for natural disasters might contain "slots" for type, damage, death toll, etc. An IE system would analyze text sources and automatically extract information to fill these templates, in effect, populating a structured database from free text. This process can be viewed as the interpretation stage in the summarization processing model, and the templates themselves can serve as the source representation. A summarization system can then combine information from multiple templates to generate a fluent summary (e.g., [18]).

Abstractive techniques face a number of major challenges, the biggest of which is the representation problem. Systems' capabilities are constrained by the richness of their representations and their ability to generate such structures – systems cannot summarize what their representations cannot capture. In limited domains, it may be feasible to devise appropriate structures, but a general-purpose solution depends on open-domain semantic analysis. Systems that can truly "understand" natural language are beyond the capabilities of today's technology.

Finally, coherence of system-generated text is one important output factor in summarization. Coherence is usually taken to mean fluent, grammatically correct prose that "reads well". This is a tall order, mainly because coherence is very difficult to operationalize. While humans can easily identify incoherent text, they have much more difficulty defining what makes a piece of text coherent. To make matters worse, multiple arrangements of segments might be equally coherent to a human. For extractive techniques, systems must devise an ordering of extracted segments and deal with "out-of-context" issues discussed above. For abstractive techniques, generation of fluent output from an abstract representation is sufficiently difficult that it is considered another sub-area in natural language processing. Although output coherence is a requirement in both single- and multi-document summarization, the latter presents more problems (particularly for extractive systems) given the variety of sources extracts.

Overview of Selected Techniques

Due to relatively easy access to corpora, most research in summarization over the past two decades has been on written news. As most summarization systems today are primarily extractive, these methods will occupy the bulk of this discussion.

Extractive techniques first segment source text into smaller segments (sentences, paragraphs, etc.), which are then scored according a variety of features, e.g., position in the text [6], term and phrase frequencies [12], lexical chains (degree of lexical-connectedness between various segments) [1], topics present in the text [16], or discourse prominence [14]. A widely adopted approach is to use machine learning techniques to determine the relative importance of various features (the earliest example being [10]).

The features discussed above are relevant for both single- and multi-document summarization, although their relative importance varies with the task. Historically, the summarization field focused on the single-document case first, and then subsequently moved on to multi-document summarization. This move required systems to explicitly model similarities and differences in text to address redundancy, paraphrase, entailment, contradiction, and related linguistic issues. One general approach involves clustering, as exemplified by the MEAD framework [16]. Documents are first clustered to find topics present in the sources. Clusters are represented by their centroids, which are used to rank extracts (along with other features). Maximal Marginal Relevance (MMR) [7] is another effective algorithm, specifically designed for query-focused summaries (i.e., summaries that address an information need). It iteratively selects candidate segments to include in the final summary, balancing relevance and redundancy at each iteration. Redundancy is computed by content similarity between each candidate and the current summary state (using cosine similarity) – thus, candidates containing words already in the summary are penalized. Note that neither MEAD nor MMR explicitly deals with linguistic relationships such as paraphrase, but that issue has been specifically addressed in other work [8].

After scoring and selecting segments from source documents, extractive systems must decide on an ordering in the final system output. Ideally, the output should constitute a coherent piece of text. Simple baselines for ordering segments include extraction order (i.e., by score), temporal order (based on metadata or temporal expressions), and order in source document (preserving source structure). While simple to implement, these techniques frequently yield disfluent summaries. Coherence can be improved by applying computational models of content and discourse [2]. Nevertheless, text structuring is a relatively under-explored area of summarization, particularly due to difficulty in evaluation. As a final note, one possible alternative is to abandon the assumption of summaries as fluent prose, and instead present users with a bulleted list of extracts.

Although open-domain abstractive summarization using deep semantic representations is beyond the current state of the art, a variety of successful abstractive techniques operating on syntactic structures have been developed. Most of these techniques involve parsing source documents and manipulating the resulting parse trees. One popular approach involves "trimming", or removing inessential structures from the parse tree [9, 20] – for example, removing adjunct clauses that do not contribute much information. Other successful techniques include "splicing" fragments from multiple sentences (sometimes across multiple documents) – for example, embedding a simple sentence as a relative clause inside another [3, 13]. Of course, these operations are not mutually exclusive. Syntactic manipulations are particularly helpful in multi-document summarization since sentences from different sources might partially overlap, e.g., a sentence contains both redundant and new information. In this case, syntactic operations can potentially deliver the best of both worlds, by eliminating redundant information and preserving new information. However, as Sparck Jones recently noted [19], there has been comparatively little work on abstractive summarization over the last decade.

Additional Readings

Beyond this entry, a number of additional sources are recommended for further reading: slides from a tutorial presentation at SIGIR 2004 [15] provide a good starting point. Special issues of the journal *Information Processing and Management* [19] and *Computational Linguistics* [17] contain in-depth articles on selected topics. For details on specific summarization techniques, a good place to look is the online proceedings of the Document Understanding Conferences [4], an annual evaluation of summarization systems. A note on references in this entry: since a comprehensive bibliography is impossible due to space limitations, either representative early articles or recent ones are cited (in the latter case, the assumption is that the reader can trace citations backwards).

Key Applications

Summarization technology has a number of applications, many of which are outlined below:

Search result summarization Search engines typically retrieve thousands of hits (if not more) in response to a user's query. Summarization systems can provide users with an overview of results to support information seeking.

Tools for analytical support Summarization can be applied to support intelligence analysis, e.g., "prepare a report on recent insurgent activities in Basra", as well as similar activities such as investigative journalism and business intelligence.

Personal information agent A personal information agent maintains a profile of the user's interest and proactively seeks out information (e.g., retrieving and summarizing relevant news items on a continuous basis).

Accessibility assistance For example, a visually impaired person might make use of a screen reader augmented with summarization technology for greater efficiency.

Support for handheld devices Handheld devices such as cell phones and PDAs with small screens could benefit from more condensed information.

Medical applications Physicians struggle to keep current with the ever-increasing volume of medical literature. Summarization systems can be deployed to assist physicians, e.g., provide an overview of treatment options for a particular disease.

Summarization of meetings Summarization technology can be coupled with speech recognizers to automatically generate "meeting minutes".

Future Directions

Current research in summarization can be characterized by three broad trends:

Increasing linguistic sophistication Extractive techniques can benefit from richer features to characterize the appropriateness of a segment for inclusion in the summary – these features come from increasingly detailed linguistic analysis, enabled by advances in language processing technology. Of particular interest is the modeling of linguistic relations such as paraphrase, entailment, and contradiction. Separately, this task has been captured in the PASCAL recognizing textual entailment evaluations.

As discussed above, limitations of extractive methods can be addressed by incorporating abstractive techniques, e.g., manipulation of parse trees. Future developments appear to follow this trend, with increasingly richer representations (enabled by improvements in syntactic, semantic, discourse, and pragmatic analysis). In other words, abstractive summarization will likely be arrived at by successive approximations with hybrid techniques.

Exploration of different genres and domain-specific applications Recently, researchers have become interested in "informal" text – a broad genre that includes emails, conversational speech, blogs, chat, SMS messages, etc. They are important because an increasing portion of our society's knowledge is captured in these channels. Furthermore, informal text push the frontiers of summarization technology by forcing researchers to develop more general and robust algorithms.

Integration with other language processing components As technology matures, it becomes feasible to integrate summarization with other components to create more powerful applications. A few examples: integration with speech recognition to summarize TV broadcasts and meetings; integration with machine translation to summarize documents from multiple languages; integration with information retrieval and question answering to produce responses that answer complex questions.

Experimental Results

Summarization is fundamentally experimental in nature, as the effectiveness of different techniques cannot be derived from first principles. Thus, tools for assessing summary quality are critical to ensuring progress, and evaluation methods themselves represent an active area of research.

Methodologies for evaluating system output can be broadly classified into two categories: *intrinsic* and *extrinsic*. In an intrinsic evaluation, system output is directly evaluated in terms of a set of norms – for example, fluency, coverage of key ideas, or similarity to an "ideal" summary (see [19] for an overview). In partic-

ular, the last criteria has been operationalized in ROUGE [11], a commonly used automated metric that compares system output to a number of human-generated "reference" summaries. In contrast, extrinsic evaluations attempt to measure how summarization impacts some other task, for example, helping users determine if a document is relevant (see [5] and references therein). While more informative, extrinsic evaluations are much more difficult to conduct, since it often involves constructing realistic scenarios for summarization systems.

One of the most important driving forces behind summarization research is the existence of annual evaluations that provide a community-wide benchmark to assess progress. Two such evaluations are the Document Understanding Conferences [4] sponsored by the U.S. National Institute of Standards and Technology (NIST), and the NTCIR Project sponsored by Japan's National Institute of Informatics. Starting in 2008, DUC is replaced by the newly created Text Analysis Conference, also sponsored by NIST.

Data Sets

Instructions for obtaining data from the DUC and NTCIR evaluations can be found on their respective websites.

Cross-References

► Information Extraction
► Information Retrieval

Recommended Reading

1. Barzilay R, Elhadad M. Using lexical chains for text summarization. In: Proceedings of the ACL/EACL Workshop on Intelligent Scalable Text Summarization; 1997.
2. Barzilay R, Lee L. Catching the drift: probabilistic content models, with applications to generation and summarization. In: Proceedings of the 2004 Human Language Technology Conference; 2004. p. 113–20.

3. Barzilay R, McKeown KR. Sentence fusion for multidocument news summarization. Comput Linguist. 2005;31(3):297–327.

4. Document understanding conferences. http://duc.nist.gov/.

5. Dorr BJ, Monz C, President S, Schwartz R, Zajic D. A methodology for extrinsic evaluation of text summarization: does ROUGE correlate? In: Proceedings of the ACL 2005 Workshop on Intrinsic and Extrinsic Evaluation Measures for MT and/or Summarization; 2005.

6. Edmundson HP. New methods in automatic extracting. J ACM. 1969;16(2):264–85.

7. Goldstein J, Mittal V, Carbonell J, Callan J. Creating and evaluating multi document sentence extract summaries. In: Proceedings of the 9th International Conference on Information and Knowledge Management; 2000. p. 165–72.

8. Hatzivassiloglou V, Klavans JL, Eskin E. Detecting text similarity over short passages: exploring linguistic feature combinations via machine learning. In: Proceedings of the Joint SIGDAT Conference on Empirical Methods in Natural Language Processing and Very Large Corpora; 1999.

9. Knight K, Marcu D. Statistics-based summarization – step one: sentence compression. In: Proceedings of the 17th National Conference on Artificial Intelligence; 2000. p. 703–10.

10. Kupiec J, Pedersen JO, Chen F. A trainable document summarizer. In: Proceedings of the 18th Annual International ACM SIGIR Conference on Research and Development in Information Retrieval; 1995. p. 68–73.

11. Lin CY, Hovy E. Automatic evaluation of summaries using n-gram co-occurrence statistics. In: Proceedings of the 2003 Human Language Technology Conference; 2003. p. 71–8.

12. Luhn HP. The automatic creation of literature abstracts. IBM J Res Dev. 1958;2(2):159–65.

13. Mani I, Gates B, Bloedorn E. Improving summaries by revising them. In: Proceedings of the 27th Annual Meeting of the Association for Computational Linguistics; 1999. p. 558–65.

14. Marcu D. The rhetorical parsing, summarization, and generation of natural language texts. PhD Thesis, University of Toronto. 1997.

15. Radev DR. Text summarization. In: Proceedings of the 27th Annual International ACM SIGIR Conference on Research and Development in Information Retrieval; 2004.

16. Radev DR, Blair-Goldensohn S, Zhang Z. Experiments in single and multi-document summarization using MEAD. In: Proceedings of the 2001 Document Understanding Conference; 2001.

17. Radev DR, Hovy E, McKeown K. Introduction to the special issue on summarization. Comput Linguist. 2002;28(4):399–408.

18. Radev DR, McKeown K. Generating natural language summaries from multiple on-line sources. Comput Linguist. 1998;24(3):469–500.

19. Sparck JK. Automatic summarising: the state of the art. Inf Process Manag. 2007;43(6):1449–81.

20. Zajic D, Dorr B, Lin J, Schwartz R. Multi-candidate reduction: sentence compression as a tool for document summarization tasks. Inf Process Manage. 2007;43(6):1549–70.

Support Vector Machine

Hwanjo Yu
University of Iowa, Iowa City, IA, USA

Synonyms

SVM

Definition

Support vector machines (SVMs) represent a set of supervised learning techniques that create a function from training data. The training data usually consist of pairs of input objects (typically vectors) and desired outputs. The learned function can be used to predict the output of a new object. SVMs are typically used for classification where the function outputs one of finite classes. SVMs are also used for regression and preference learning, for which they are called support vector regression (SVR) and ranking SVM, respectively. SVMs belong to a family of generalized linear classifier where the classification (or boundary) function is a hyperplane in the feature space. Two special properties of SVMs are that SVMs achieve (i) high generalization (Generalization denotes the performance of the learned function on testing data or "unseen" data that are excluded in training.) by maximizing the margin (Margin denotes the distance between the hyperplane and the closest data vectors in the feature space.), and (ii) support efficient nonlinear classification by kernel trick (Kernel trick is a method for converting a linear classifier into a non-linear one by using a non-linear function to map the original observations into a higher-dimensional space; this makes a linear classification in the new space (or the feature space) equivalent to non-linear classification in the original space (or the input space).).

Historical Background

Vapnik developed the related concepts at 1979 and published an article in Russian that was translated to English at 1982. The first book introducing SVMs (written by him) were published at 1995. Since then, numerous literatures have been published including tutorials, journal articles, and books.

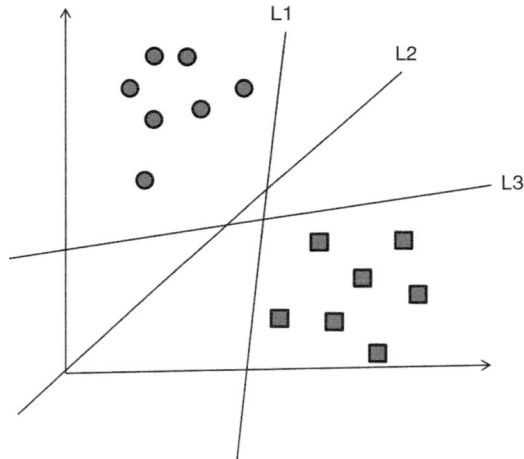

Support Vector Machine, Fig. 1 Linear classifiers (hyperplane) in two-dimensional spaces

Foundations

SVMs were initially developed for classification [1] and have been extended for regression [4] and preference learning [3, 5]. The initial form of SVMs are a binary classifier where the output of learned function is either positive or negative. A multiclass classification can be implemented by combining multiple binary classifiers using pairwise coupling method [2]. This section explains the motivation and formalization of SVM as a binary classifier, and the two key properties - margin maximization and kernel trick.

Motivation

Binary classification is to classify data objects into either positive or negative class. Each data object (or data point) is represented by a n-dimensional vector. Each of these data points belongs to only one of two classes. A *linear* classifier separates them with an "n minus 1" dimensional hyperplane. For example, Fig. 1 shows two groups of data and separating hyperplanes that are lines in a two-dimensional space. There are many linear classifiers that correctly classify (or divide) the two groups of data such as L1, L2 and L3 in Fig. 1. In order to achieve maximum separation between the two classes, An SVM pick the hyperplane so that the margin, or the distance from the hyperplane to the nearest data

point, is maximized. Such a hyperplane is likely to generalize better, meaning that the hyperplane not only correctly classify the given or training data points, but also is likely to correctly classify "unseen" or testing data points.

Formalization

The data points D in Fig. 1 (or training set) can be expressed mathematically as follows:

$$D = \left\{ \left(\vec{x}_1, y_1 \right), \left(\vec{x}_2, y_2 \right), ..., \left(\vec{x}_m, y_m \right) \right\}, \tag{1}$$

where $\vec{x}_i$ is a n-dimensional real vector, y_i is either 1 or -1 denoting the class to which the point $\vec{x}_i$ belongs. The SVM classification function $F\left(\vec{x}\right)$ takes the form

$$F\left(\vec{x}\right) \Rightarrow \vec{w} \cdot \vec{x} - b. \tag{2}$$

$\vec{w}$ is the weight vector and b is the bias, which will be computed by SVM in the training process.

First, to correctly classify the training set, $F(\cdot)$ (or $\vec{w}$ and b) must return positive numbers for positive data points and negative numbers otherwise, that is, for every point $\vec{x}_i$ in D,

$$\vec{w} \cdot \vec{x}_i - b > 0 \text{ if } y_i = 1, \text{ and}$$

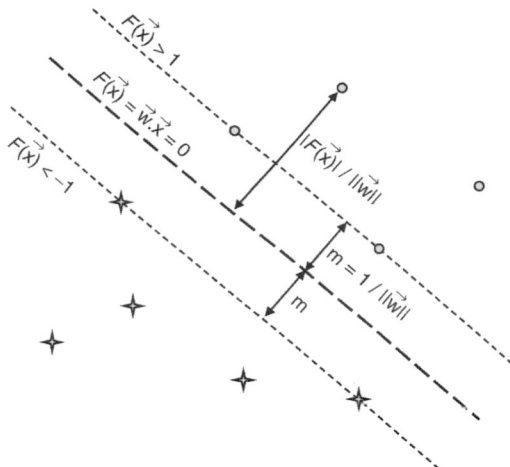

Support Vector Machine, Fig. 2 SVM classification function: the hyperplane maximizing the margin in a two-dimensional space

$$\vec{w} \cdot \vec{x}_i - b < 0 \text{ if } y_i = -1$$

These conditions can be revised into:

$$y_i \left(\vec{w} \cdot \vec{x}_i - b \right) > 0, \forall \left(\vec{x}_i, y_i \right) \in D. \quad (3)$$

If such a linear function F that correctly classifies every point in D or satisfies (3) exists, D is called *linearly separable*.

Second, F (or the hyperplane) needs to maximize the *margin*. Margin is the distance from the hyperplane to the closest data points. An example of such hyperplane is illustrated in Fig. 2. To achieve this, (3) is revised into the following (4).

$$y_i \left(\vec{w} \cdot \vec{x}_i - b \right) \geq 1, \forall \left(\vec{x}_i, y_i \right) \in D. \quad (4)$$

Note that (4) includes equality sign, and the right side becomes 1 instead of 0. If D is linearly separable, or every point in D satisfies (3), then there exists such a F that satisfies (4). It is because, if there exist such $\vec{w}$ and b that satisfy (3), they can be always rescaled to satisfy (4).

The distance from the hyperplane to a vector $\vec{x}_i$ is formulated as $\frac{|F(\vec{x}_i)|}{||\vec{w}||}$. Thus, the margin becomes

$$margin = \frac{1}{|| \vec{w} ||}. \quad (5)$$

because when $\vec{x}_i$ are the closest vectors, $F\left(\vec{x}\right)$ will return 1 according to (4). The closest vectors, that satisfy (4) with equality sign, are called *support vectors*.

Maximizing the margin becomes minimizing $|| \vec{w} ||$. Thus, the training problem in SVM becomes a constrained optimization problem as follows:

$$\text{minimize} : \frac{1}{2} || \vec{w} ||^2, \quad (6)$$

$$\text{subject to} : y_i \left(\vec{w} \cdot \vec{x}_i - b \right) \geq 1,$$

$$\forall \left(\vec{x}_i, y_i \right) \in D. \quad (7)$$

The factor of $\frac{1}{2}$ is used for mathematical convenience.

However, the optimization problem will not have a solution if D is not linearly separable. To deal with such cases, *soft margin* SVM allows mislabeled data points while still maximizing the margin. The method introduces slack variables, ξ, which measure the degree of misclassification. The following is the optimization problem for soft margin SVM.

$$\text{minimize} : L_1 (w, b, \xi) = \frac{1}{2} || \vec{w} ||^2 + C \sum_i \xi_i, \quad (8)$$

$$\text{subject to} : y_i \left(\vec{w} \cdot \vec{x}_i - b \right) \geq 1 - \xi_i,$$

$$\forall \left(\vec{x}_i, y_i \right) \in D. \quad (9)$$

$$\xi \geq 0 \quad (10)$$

Due to the ξ in (9), data points are allowed to be misclassified, and the amount of misclassification or error will be minimized as well as the margin according to the objective function (8). C is a parameter that determines the tradeoff between the margin size and the amount of error in training.

This optimization problem is called quadratic programming (QP) problem, and it is the primal form of the QP. The primal form can be changed to the following dual form using the Lagrange multipliers.

$$\text{minimize} : L_2\left(\alpha\right) = \sum_i \alpha_i - \sum_i \sum_j \alpha_i \alpha_j y_i y_j$$

$$K\left(\overrightarrow{x}_i, \overrightarrow{x}_j\right), \quad (11)$$

$$\text{subject to} : \sum_i \alpha_i y_i = 0 \quad (12)$$

$$C \geq \alpha \geq 0 \quad (13)$$

α constitute a dual representation for the weight vector such that

$$\overrightarrow{w} = \sum_i \alpha_i y_i \overrightarrow{x}_i. \quad (14)$$

$K\left(\overrightarrow{x}_i, \overrightarrow{x}_j\right)$ in (11) is originally a dot product of the two vectors, that is, $\overrightarrow{x}_i \cdot \overrightarrow{x}_j$. However, the dot product can be replaced by a non-linear kernel function, which allows the algorithm to fit the maximum-margin hyperplane in the transformed feature space. The transformed feature space is usually high dimensional, and the hyperplane (or linear classifier) in the high-dimensional space becomes non-linear in the original input space. Computing the kernel function K is often done as fast as computing a dot product. In this way, the complexity of computing a non-linear function becomes the same as that of computing a linear function in SVM. This method for computing a non-linear function is called *kernel trick*. The following are popularly used non-linear kernels.

1. Radial basis function (RBF):
   $K(a, b) = exp(-\gamma\|a - b\|^2)$
2. Polynomial: $K(a, b) = (a \cdot b + 1)^d$
3. Sigmoid: $K(a, b) = tanh(\kappa a \cdot b + c)$

Once α is learned from the dual form, the linear function F in (2) becomes the following non-linear function.

$$F\left(\overrightarrow{x}\right) \Rightarrow \overrightarrow{w} \cdot \overrightarrow{x} - b$$

$$\Rightarrow \sum_i \alpha_i y_i K\left(\overrightarrow{x}_i, \overrightarrow{x}\right) - b \quad (15)$$

Key Applications

SVMs have been widely applied for object classification and pattern recognition such as text categorization, face detection in images, and handwritten digit recognition.

Cross-References

▶ Classification

Recommended Reading

1. Burges CJC. A tutorial on support vector machines for pattern recognition. Data Min Knowl Disc. 1998;2(2):121–67.
2. Hastie T, Tibshirani R. Classification by pairwise coupling. Adv Neural Inf Proces Syst. 1998;10:507.
3. Herbrich R, Graepel T, Obermayer K, editors. Large margin rank boundaries for ordinal regression. Cambridge, MA: MIT Press; 2000.
4. Smola AJ, Scholkopf B. A tutorial on support vector regression. Technical report, NeuroCOLT2 technical report NC2-TR-1998-030. 1998.
5. Yu H. SVM selective sampling for ranking with application to data retrieval. In: Proceedings of the 11th ACM SIGKDD International Conference on Knowledge Discovery and Data Mining; 2005.

Supporting Transaction Time Databases

David Lomet
Microsoft Research, Redmond, WA, USA

Synonyms

Multi-version database; Temporal database

Definition

The temporal concepts glossary maintained at http://www.cs.aau.dk/~csj/Glossary/ defines transaction time as: "The *transaction time* of a database fact is the time when the fact is current in the database and may be retrieved." A transaction time database thus stores versions of database records or tuples, each of which has a start time and an end time, delimiting the time range during which they represent the current versions of database facts. As each version is the result of transactions, the times associated with the version are the times for the transaction starting the version (the start time) and for the transaction ending the version (the end time). These transaction times are required to agree with the serialization order of the transaction, so that the database can present a transaction consistent view of the facts being stored.

Historical Background

Postgres was the first database system that supported transaction time databases [1]. It implemented a prototype relational database system using a versioning approach for recovery. This was then augmented with a version store based on the R-tree [2]. Subsequently, the TSB-tree was introduced [3], based on the WOB-tree time splitting [4], which provided an integrated approach to storing both current and historical data.

Commercially, Oracle and Rdb [5] database systems both support multi-version concurrency control, now called snapshot isolation. Oracle subsequently added a Flashback [6] feature that permitted access to historical versions, based on saving a linear history of their recovery versions. While this permits access to historical versions, it is mostly intended as an efficient means of providing an online backup for "point-in-time" recovery, a form of "media" recovery in which the database is recovered to a point just preceding a bad user transaction.

Temporal databases have been extensively studied [7], including not only transaction time but also valid time. Included as well are bi-temporal databases, where both valid and transaction time are supported. This work has clarified the conceptual issues and provided a common vocabulary of terms for describing work in the field.

Foundations

Transaction time databases usually offer the full range of database functionality that one would expect of a database storing only current facts, which is called a current time database. In addition, a transaction time database provides access to all prior facts as represented by versions of records that existed at some prior time. Several types of functionality can be envisioned.

1. Access to database facts "as of" some past time, which are called "as of" queries
2. Access to versions of a database fact in some time range, which are called "time travel" queries
3. Access to collections of database facts in some time range, i.e., "general transaction time queries"

The functionality provided by transaction time databases is important for applications such as time series data, regulatory compliance, repeatability of scientific experiments, etc. Further, a transaction time database can provide valuable system capabilities such as snapshot isolation concurrency control, recovery from bad user transactions, and recovery from media failure.

To support transaction time functionality, one needs to change the semantics of the data manipulation operations "update" and "delete." An update creates an additional version instead of overwriting the prior state, retaining both new and old versions of the data. A delete is strictly logical, providing an end time for the previously current version. In this way, the prior version persists so that queries about the database while this version was alive can be answered.

Database systems supporting transaction time data have been built that are based on the relational data model, though this is not a fundamental limitation, simply a pragmatic choice.

Implementation Approaches

There are two generic approaches to implementing transaction time databases, which are described here.

Layered Approach

The premise of the layered approach is that application programmers cannot wait for vendors to build transaction time functionality into database products [8]. Rather, a middleware layer (MWL) is implemented that provides this functionality. The MWL processes data definition statements, adding timestamp fields to each record, processes data manipulation statements (queries plus updates) written in a language that exposes temporal functionality and translates them into equivalent ordinary SQL queries. Typically, start time and end time are added to each record of a transaction time table. The table may be organized by a clustering key that includes (user defined primary key, start time, end time). Thus, a record with a given user defined primary key is clustered next to its earlier versions, which makes "time travel" queries efficient, while implying that "as of" queries will have relatively poor performance. Grouping by time does not help much with "as of" performance, and usually compromises "time travel" performance.

Built-In Approach

The built-in approach requires the ability to modify the database engine to provide transaction time support. While a significant barrier, building transaction time support into a database system can greatly improve performance, bringing it close to current time database performance. This improved performance is the result of optimizations that are possible (i) for update, when the timestamps need to be added to versions, (ii) for storage with simple forms of version compression, and (iii) for query, because specialized indexing is possible that improves data clustering. The rest of this article discusses the issues of built-in support and how to make this support perform well.

Managing Versions

Transaction time functionality requires dealing with the multiple versions of database facts that exist to express the states of the database over time. The built-in approach has more freedom than the MWL approach in how to achieve efficiency and performance. There are a number of issues, the more important ones being:

Timestamps

Each historical version stored in a transaction time database has a begin time, at which it first became the current version, and an end time, at which it was either replaced by another version or deleted. Current versions have the usual start time, but have a special end time called "now" [9]. An MWL approach usually includes both start and end times with each version so that the SQL queries resulting from translating temporal queries are simple and efficient. With built-in support, most systems [1, 10] store only the start time with each version, the end time being derived from the start time of the subsequent version. For a deletion, this next version can be a special "delete stub" version. For a query "as of" time T, the system then looks for the version of each record with the largest start time $\leq$ T.

Storing Versions on a Page

The common approach for organizing pages for current time databases is called a slotted array. Each array element points to a record on the page. For B-trees, these records are maintained in B-tree key order. When adding temporal support to a current time database system, it is convenient to minimize the change required for current time functionality. This argues for retaining the slotted array, and back-linking versions where each version is augmented with a pointer to its preceding version. Then for each record accessed in a query, the system follows this backward chain to the first version with a timestamp $\leq$ T, the "as of" time requested.

Indexing Versions

Being able to index historical versions by time is essential to avoid increasing costs for ever earlier query times [11]. Postgres [1] used the R-tree [2] for this. The TSB-tree [3] is a more specialized index that can, with an appropriate page splitting policy [12] provide guarantees about the performance of "as of" queries. Its special feature is the introduction of a time split [4] where the time interval of a full database page is partitioned, with record versions being assigned to the resulting pages whenever their lifetimes intersect the time interval of a resulting page. Thus, a version whose lifetime intersects both resulting pages will be replicated in both pages. The result is that, when combined with ordinary B-tree key splitting, each TSB-tree page contains all versions within a key-time rectangle of the search space. This enables identifying exactly which pages can contain answers to a temporal query. Despite the need for replicating versions, the space required for the versions remains linear in the number of unique versions.

Compressing Versions

The way that versions are stored on a page and indexed makes compressing versions simple. Usually, an update changes only a small part of a record, perhaps only a single attribute. Thus, delta compression, where the compressed version represents the difference between one version and another that is adjacent in time order, can be very effective. Only the updated attribute together with location information and timestamp needs to appear in the compressed version. Backward delta's are to be preferred because this leaves the current time data uncompressed and hence unchanged, important both for compatibility and current time performance. Because time splits in the TSB-tree always replicate versions spanning the split time, and because splitting at current time is convenient, the last version in each page is always uncompressed, and this is preserved during a time split. Decompressing a version never needs information from any other page than the page upon which the version is stored.

Dealing with Timestamps

In a transaction time database, for a record identified in some way, its versions are distinguished by timestamps. The nature of timestamps, when they are chosen and included, how to optimize this process, and how to deal with user requests for transaction time are discussed below.

Nature of Timestamps

Several forms of timestamp have been used for temporal support. Some systems use transaction identifiers (XIDs) instead of time, sometimes maintaining a separate table that maps XID to time. When versioning is limited, e.g., to only support multiversion concurrency control, an active transaction's XID may be mapped to a list of transactions (their XIDs) that committed before them, hence determining which transactions have updates that should be visible to the active transaction [5]. However, for more general functionality, system time is usually used. It may need to be augmented with a sequence number because its granularity may not be sufficient to completely distinguish every transaction's updates.

When to Timestamp

A timestamp for a version must enable "as of" queries to always see a transaction consistent view of the data. This can be achieved when timestamp order agrees with the serialization order of transactions. If one chooses timestamps prior to updating, the timestamp can be added immediately to versions generated by the updating. However, early timestamp choice means that transactions that serialize differently must be aborted. Most implementations of transaction time functionality thus choose timestamps at commit, where the commit order is the same as the serialization order for the transactions. This means, however, that the timestamp is not available at time of update, and must be added later.

Lazy Timestamping

When a transaction's timestamp is determined late in the transaction, e.g., at commit time, preceding updated records need to be revisited to add the timestamp. Typically, an XID is placed in an updated record at update time, to be replaced later by the system time. Eager timestamping replaces XID with time prior to transaction commit, logging this activity as another update. This can be costly, so a lazy approach is generally preferred in which XID is replaced by time after the transaction commits. The mapping from XID to system time must be maintained persistently, at least until the timestamping is complete, to ensure that replacing XID with time can continue after a possible system crash.

Impact of User Requested Time

The SQL language supports a user's request for current (transaction) time within a query. It is essential that the user see a time that is consistent with the transaction timestamp used for updates of the transaction. Providing the user with a time for the transaction while the transaction is executing constrains the choice of timestamp when the transaction is committed [13]. A transaction must be aborted if a timestamp cannot be chosen that is consistent with the time provided to the user. To provide this, the system can exploit the fact that the user time request is usually not for the full precision of the timestamp, e.g., SQL DATE constrains only the date part of the timestamp. Further, remembering the largest timestamp on data that is seen by the transaction provides a lower bound for a possibly non-empty interval for timestamp choice.

Additional Uses

The versions maintained by a transaction time database can support a variety of other system uses.

Snapshot Isolation

Recent versions can be used to provide snapshot isolation. With snapshot isolation (the default concurrency provided by Oracle), a transaction reads not the current data (which would be used for a serializable transaction) but a snapshot (version) current as of the start of the transaction or as of the first data read by the transaction. A transaction time database keeps these versions as well as possibly older versions. Thus,

efficiently providing transaction time support also provides efficient snapshot isolation. The system may choose to garbage collect older versions more quickly when they are only used for concurrency control purposes.

Online Backup

A database backup is simply an earlier state of the database. Transaction time databases make all earlier states accessible and queryable. To use an earlier transaction time database state as a backup for, e.g., media recovery for the current state, requires two things: (i) the earlier state needs to be on a separate medium than the current state; and (ii) the media recovery log needs to include all updates from the earlier state forward to the current state and itself be on a separate device. A transaction time database system can use time splits (which it would use in a TSB-tree) to move versions to separate backup media, and it can do this incrementally as well [14].

Bad User Transactions

Occasionally, erroneous transactions commit, compromising the correctness of a database. Point-in-time recovery, where the database state is reset to an earlier time, just prior to the bad transaction, is the usual way of dealing with this. Conventional database backups used for this incur a long restore time followed by a roll forward to the just earlier time. A transaction time database lends itself greatly shortening the outage caused by this problem because earlier versions of the database are already maintained online. Oracle Flashback [6] implements point-in-time recovery in this way. One can further limit the outage by identifying exactly which transactions should be removed from the database by tracking transaction read dependencies [15].

Key Applications

In addition to the system uses just described, transaction time databases are valuable for several applications, e.g., time series analysis, repeating experiments or analysis on historical data, and auditing and legal compliance.

Future Directions

Data stream processing is a new functionality that has several important applications requiring fast reaction to sequences of events, e.g., stock market data. This data may also be stored and more carefully analyzed. It is quite natural to think about stream data as transaction time data, and ask temporal queries of it.

Cross-References

▸ Temporal Database
▸ Temporal Strata
▸ Transaction
▸ Transaction-Time Indexing

Recommended Reading

1. Stonebraker M. The design of the POSTGRES storage system. In: Proceedings of the 13th International Conference on Very Large Data Bases; 1987. p. 289–300.
2. Guttman A. R-trees: a dynamic index structure for spatial searching. In: Proceedings of the ACM SIGMOD International Conference on Management of Data; 1984, p. 47–57.
3. Lomet DB, Salzberg B. Access methods for multiversion data. In: Proceedings of the ACM SIGMOD International Conference on Management of Data; 1989. p. 315–24.
4. Easton M. Key-sequence data sets on inedible storage. IBM J Res Dev. 1986;30(3):230–41.
5. Hobbs L, England K. Rdb: a comprehensive guide. Newton: Digital Press; 1995.
6. Oracle. Oracle Flashback Technology (2005) http://www.oracle.com/technology/deploy/availability/htdocs/lashback_Overview.htm
7. Tansel U, Clifford J, Gadia SK, Segev A, Snodgrass RT. Temporal databases: theory, design, and implementation. Benjamin/Cummings; 1993.
8. Torp K, Snodgrass RT, Jensen CS. Effective timestamping in databases. VLDB J. 2000;8(4):267–88.
9. Clifford J, Dyreson C, Isakowitz T, Jensen CS, Snodgrass RT. On the semantics of "now" in databases. ACM Trans Database Syst. 1997;22(2):171–214.
10. Lomet DB, Barga R, Mokbel M, Shegalov G, Wang R, Zhu Y. Transaction time support inside a database engine. In: Proceedings of the 22nd International Conference on Data Engineering; 2006. p. 35.
11. Salzberg B, Tsotras VJ. A comparison of access methods for time-evolving data. ACM Comput Surv. 1999;31(2):158–221.
12. Becker B, Gschwind S, Ohler T, Seeger B, Widmayer P. An asymptotically optimal multiversion B-tree. VLDB J. 1996;5(4):264–75.
13. Lomet DB, Snodgrass RT, Jensen CS. Using the lock manager to choose timestamps. In: Proceedings of the International Conference on Database Engineering and Applications; 2005. p. 357–68.
14. Lomet DB, Salzberg B. Exploiting a history database for backup. In: Proceedings of the 19th International Conference on Very Large Data Bases; 1993. p. 380–90.
15. Lomet DB, Vagena Z, Barga R. Recovery from "bad" user transactions. In: Proceedings of the ACM SIGMOD International Conference on Management of Data; 2006. p. 337–46.

Symbolic Representation

Hans Hinterberger
Department of Computer Science, ETH Zurich, Zurich, Switzerland

Synonyms

Symbol graph; Symbolic graphic; Symbol plot

Definition

A written character or mark used to represent something; a letter, figure, or sign conventionally standing for some object, process, etc. (Oxford English Dictionary). Examples are the figures denoting the planets, signs of the zodiac, etc. in astronomy; the letters and other characters denoting elements, and so on. in chemistry, quantities, operations, etc. in mathematics, the faces of a crystal in crystallography.

In data visualization, the use of symbols allows the representation of multivariate data items, where each variate contributes to the symbol. The set of symbols may be displayed in an array and superimposed on coordinates to put extra information on a point plot or, if appropriate, on a geographical map.

When symbolic representations of information is used as a tool for thought or a form of communication, one distinguishes between abstract symbols where the graphical units are shapes formed by lines and areas and depictive symbols (pictograms) where the graphical units are pictorial representations of objects and scenes.

Symbolic representations in the form of tag clouds (or word clouds) have become a widely used design to visualize the frequency distribution of keyword metadata that describe the content of documents. In the particular case of website content, tag clouds have been used as a navigation aid ever since the early Web 2.0 websites and blogs.

Key Points

Symbols have considerable potential as an aid to support human cognition and as a medium for communication. To what extent symbols can successfully convey specific information and what their inherent limitations as a medium of communication are is far from being understood. There is nevertheless a useful and accumulating body of research dealing with the systematic application of symbolic representations.

One of the earliest researchers to treat symbolic representations in a theoretical context is the French cartographer Jacques Bertin. With his semiology of graphics [1], he organized the visual and perceptual elements of graphics according to the features and relations in data, distinguishing primarily between diagrams, networks, maps, and symbols.

Among the first efforts to relate visual and perceptual research to the practical problems of designing information displays was the NATO Conference on Visual Presentation of Information in 1978. The contributions to this conference are published in [2], a book that deals with basic psychological issues as well as methods of evaluation of information design, with much room given to the use of symbols.

For a classification of symbols as used for data visualization and an overview for what type of functions different symbols are used, the reader is referred to [3]. Tufte, in his series of books on information visualization, discusses with exemplary images and diagrams a wide range of different styles and techniques, good and bad, for the use of symbols; [4] is a starting point.

Cross-References

▶ Data Visualization
▶ Graph
▶ Table
▶ Thematic Map

Recommeded Reading

1. Bertin J. Semiology of graphics (trans: Berg WJ.). Madison: University of Wisconsin Press; 1983.
2. Easterby R, Zwaga H, editors. Information design, the design and evaluation of signs and printed material. London: Wiley; 1984.
3. Harris RL. Information graphics: a comprehensive illustrated reference. New York: Oxford University Press; 1999.
4. Tufte ER. The visual display of quantitative information. Cheshire: Graphics Press; 1983.

Symmetric Encryption

Ninghui Li
Purdue University, West Lafayette, IN, USA

Synonyms

Secret-key encryption

Definition

Symmetric encryption, also known as secret key encryption, is a form of data encryption where a single secret key is used for both encryption and decryption.

Key Points

Modern symmetric encryption algorithms are often classified into stream ciphers and block ciphers. In a stream cipher, the key is used to generate a pseudo-random key stream, and the ciphertext is computed by using a simple operation (e.g., bit XOR or modular addition) to combine the plaintext bits and the key stream bits. Many stream ciphers implemented in hardware are constructed using linear feedback shift registers (LFSRs). The use of LFSRs on their own, however, is insufficient to provide good security. Additional variation and enhancement are needed to increase the security of LFSRs. RC4 is the most widely-used software stream cipher and is used in popular protocols such as Secure Sockets Layer (SSL) (to protect Internet traffic) and WEP (to secure wireless networks).

A block cipher operates on large blocks of digits with a fixed, unvarying transformation. The Data Encryption Standard (DES) [1] algorithm uses blocks of 64 bits; the Advanced Encryption Standard (AES) [2] algorithm uses 128-bit blocks. When using a block cipher to encrypt a message, one needs to choose an encryption mode. Commonly used modes include the Electronic Codebook (ECB), Cipher-block chaining (CBC), Cipher feedback (CFB), Output feedback (OFB), and Counter (CTR).

Cross-References

▶ Asymmetric Encryption
▶ Data Encryption

Recommended Reading

1. Federal information processing standards publication 46–3: data encryption standard (DES), 1999.
2. Federal information processing standards publication 197: advanced encryption standard, November 2001.

Synopsis Structure

Phillip B. Gibbons
Computer Science Department and the Electrical and Computer Engineering Department, Carnegie Mellon University, Pittsburgh, PA, USA

Synonyms

Synopsis

Definition

A *synopsis structure* for a dataset S is any summary of S whose size is substantively smaller than S. Formally, its size is at most $O(|S|^{\varepsilon})$, where $|S|$ is the size (in bytes) of S, for some constant $\varepsilon < 1$.

Key Points

Synopsis structures are small, often statistical summaries of a data set. The term serves as an umbrella for any summarization structure of sufficiently small size, such as random samples, histograms, wavelets, sketches, top-k summaries, etc.

Synopsis structures are most commonly used in conjunction with data streams. The goal is to construct, in one pass over the data stream, a synopsis structure that can be used to answer any query from a prespecified class of queries. That is, at any point, a user may pose a query Q on the data stream thus far, and a (typically approximate) answer to Q must be produced using only the current synopsis structure. Two key advantages of using a synopsis structure to answer queries are that the space overhead is low (a massive dataset can be summarized using only a small amount of space) and the response time is fast (e.g., disk accesses can be avoided altogether). Moreover, in the common setting of queries comparing or aggregating over a distributed collection of streams, only the small synopsis structures (and not the massive data streams) need to be communicated between the collection points in order to answer the query.

Synopsis structures are also used within relational and XML databases, both for query optimization and for approximate query answering.

Common metrics for evaluating a synopsis structure include (i) *Coverage:* the range and importance of the class of queries supported; (ii) *Answer quality:* the accuracy and confidence of its (approximate) answers; (iii) *Space footprint:* its size, where smaller is better and often polylog space is desired (i.e., space that is $O(\log^k (|S|))$ for some constant $k \geq 1$); (iv) *Per-item processing time:* the total time to process the dataset, normalized to the number of items in the dataset; and (v) *Query time:* the time to answer a query from the synopsis structure [1].

Sometimes *sketch* is used interchangeably with synopsis structure, but more typically "sketch" is restricted to synopses based on random projections.

The term was coined by Gibbons and Matias in 1995.

Cross-References

▶ Data Stream
▶ Histogram
▶ Wavelets on Streams

Recommended Reading

1. Gibbons PB, Matias Y. Synopsis data structures for massive data sets. DIMACS Series in Discrete Mathematics and Theoretical Computer Science: External Memory Algorithms. 1999.

Synthetic Microdata

Josep Domingo-Ferrer
Universitat Rovira i Virgili, Tarragona,
Catalonia, Spain

Synonyms

Imputed data; Multiple imputation; Simulated data

Definition

Publication of synthetic – i.e., simulated – data is an alternative to masking for statistical disclosure control of microdata. The idea is to randomly generate data with the constraint that certain statistics or internal relationships of the original dataset should be preserved.

Key Points

The operation of the original proposal by Rubin [2] is next outlined. Consider an original microdata set X of size n records drawn from a much larger population of N individuals, where there are background attributes A, non-confidential attributes B and confidential attributes C. Background attributes are observed and available for all N individuals in the population, whereas B and C are only available for the n records in the sample X. The first step is to construct from X a multiply-imputed population of N individuals. This population consists of the n records in X and M (the number of multiple imputations, typically between 3 and 10) matrices of (B,C) data for the $N - n$ non-sampled individuals. The variability in the imputed values ensures, theoretically, that valid inferences can be obtained on the multiply-imputed population. A model for predicting (B,C) from A is used to multiply-impute (B,C) in the population. The choice of the model is a nontrivial matter. Once the multiply-imputed population is available, a sample Z of n' records can be drawn from it whose structure looks like the one of a sample of n' records drawn from the original population. This can be done M times to create M replicates of (B,C) values. The results are M multiply-imputed synthetic datasets. To make sure no original data are in the synthetic datasets, it is wise to draw the samples from the multiply-imputed population excluding the n original records from it.

There are other approaches to synthetic data generation based on bootstrap, Latin Hypercube sampling, Cholesky decomposition, etc. More information can be found in [1].

Cross-References

▶ Inference Control in Statistical Databases
▶ Microdata

Recommeded Reading

1. Hundepool A, Domingo-Ferrer J, Franconi L, Giessing S, Lenz R, Longhurst J, Nordholt ES, Seri G, De Wolf P-P. Handbook on statistical disclosure control. CENEX SDC Project, November 2006 (manuscript version 1.0). http://neon.vb.cbs.nl/CENEX/
2. Rubin DB. Discussion of statistical disclosure limitation. J Off Stat. 1993;9(2):461–8.

System R (R*) Optimizer

Mouna Kacimi and Thomas Neumann
Max-Planck Institute for Informatics,
Saarbrücken, Germany

Definition

The System R Optimizer is the cost-based query optimizer of System R. It pioneered several optimization techniques, including using dynamic programming for bottom-up join tree construc-tion, and the concept of interesting orderings for exploiting ordering in intermediate results. Later, it was generalized for distributed database systems in System R*.

Historical Background

System R is a database management system based on a relational data model that was proposed by E. F. Codd [4] in 1970. The system offers data independence by providing a high-level user interface through which the end user deals with data content rather than the underlying storage structures. In other words, users do not need to know how the tuples are physically stored and which access paths are available to write queries. Thus, data storage structures may change over time without users being aware of it, providing a high level of data independence and user productivity. Moreover, System R offers capabilities for database management in realistic and operational environments. Particularly, it supports multiple users concurrently accessing data, provides means for system recovery after hardware or software failures, supports different types of database use including ad hoc queries, programmed transactions and report generation. System R has been developed at the San Jose IBM Research Laboratory during three phases. First, *Phase Zero* of the project started in 1974 and ended in 1975. It involved the development of a high-level relational user language, called SQL. During this phase, a subset of SQL was implemented for one user at a time. One of the most challenging tasks of Phase Zero was the design of optimizer algorithms for efficient query processing. Second, *Phase One* took place from 1976 to 1977. It involved the development of a full-function multi-user version of System R. The multi-user prototype contained new subsystems, such as locking subsystems that prevent conflict of concurrent user accesses. Finally, *Phase Two* focused on evaluating the System R during 1978 and 1979. It was mainly composed of two parts: (i) evaluation of the system at the San Jose research Laboratory, and (ii) evaluation of the actual use of the system at a number of internal sites of IBM.

Foundations

This section is organized as follows. First, it starts by describing the role of the optimizer in processing SQL statements. Then, it describes the storage components used to access paths on the different relations. Next, it presents optimization techniques for single relations and joins in System R. Finally, it describes how these techniques are generalized to the distributed case of System R*.

SQL Query Processing

An SQL statement is composed of one or multiple query blocks depending on whether the operands of the used predicates are simple values (of the from "*column operator value*") or queries (of the from "*column operator query*"). A *query block* is represented by: (i) a *SELECT* list containing the list of items to be retrieved, (ii) a *FROM* list containing the relation(s) references and (iii) a *WHERE* tree containing the boolean combination of simple predicates specified by the user. Processing SQL statements requires four main phases namely: parsing, optimization, code generation and execution. During the first phase, each SQL statement is sent to the *Parser* to check its syntax. If no errors are detected, the optimizer component is called in the second phase. Using the System R catalogs, the optimizer verifies the existence of all the relations and columns referenced, and collects information about them. It gets from the catalog the datatype and the length of each column, and use these information to check the semantic errors and type compatibility in both expressions and predicate comparisons of the SQL statement. In addition, the optimizer obtains, from the catalogs, statistics about the referenced relations and the access paths available on each of them and use these information for access path selection process. Then, the optimizer chooses a query plan, from a tree of alternate path choices, that is a minimum cost solution. This chosen plan is a tree represented in the Access

Specification Language (ASL) [7]. In the third phase, the Code generator translates ASL trees into machine language code to finally execute the plan chosen by the optimizer in the fourth phase. When the code is executed, it accesses the Relational Storage System (RSS), via the Storage System Interface (RSI), to scan each of the query relations using the access paths chosen by the optimizer. Even though the RSS may be used for different purposes, here, we focus on its use for computing cost formulas and executing the code generated by the query processing in System R.

Relational Storage System

The Relational Storage System (RSS) provides underlying storage support for System R. Relations in the RSS are stored as a collection of tuples whose columns are physically contiguous. The storage space is logically organized into segments. Each relation resides within a single segment, however, a segment may contain one or more relations. A segment is composed of a set of equal-sized pages. Each tuple of a relation is stored within a single page and is assigned an identification of the relation to which it belongs. Note that a page may contain tuples from one or more relations.

To access tuples in a relation, an RSS scan is used. Along a given access path, a scan returns one tuple at a time. Two different types of scans can be distinguished. First, *Segment* scans find all the tuples of a given relation by examining the segment that contains the relation. All the non-empty pages of the segment are touched and the tuples belonging to the given relation are returned. Second, *Index* scans access a relation in value order using an index. An index is created on one or more columns of a relation. It is composed of one or more pages within the segment containing the relation. The pages of the index are separated from the pages containing the relation tuples. They are organized into a B-tree structure as shown in Fig. 1. Each page is a node of the B-tree that contains an ordered sequence of index entries. An entry of a non-leaf node consists of a $\prec$ *key, pointer* $\succ$ pair, where the pointer address another page in the same tree. Leaf pages contain sets of $\prec$ *key, identifier* $\succ$

System R (R*) Optimizer,
Fig. 1 Example of B-tree
index structure

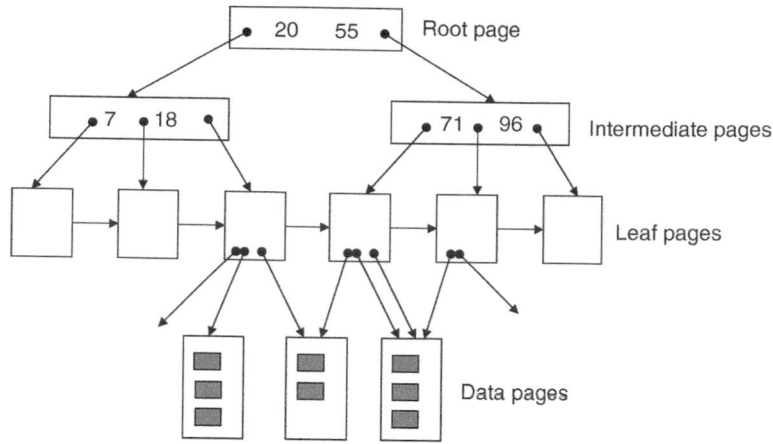

pairs, where identifier indicates the tuple that contains the corresponding key. An index scan does a sequential read along the leaf pages to get the tuple identifiers matching a given key. These identifiers are used to find and return the data tuples to the user in key value order.

When an index scan examines a relation, each page of the index is touched only once, but a data page may be touched more than once. This case may happen if a page contains tuples which are not close in the index ordering. An index is said to be clustered when the physical proximity of tuples in the same data page corresponds to the index key value. Therefore, not only each index page, but also each data page will be touched only once in a scan on that index. Additionally, to reduce the number of touched pages, starting and stopping key values can be specified when scanning tuples. Thus, only the tuples whose keys belong to the predefined interval are returned. Both index and segment scans may use a set of predicates, called also search arguments or SARGS. These predicates are of the form ("*column operator value*"). They are applied to tuples before they are returned to the RSI caller. Sargable predicates play an important role in reducing the cost by eliminating unnecessary RSI calls for tuples which can be rejected within the RSS.

Access Path Selection for Single Relations

This section describes how the optimizer chooses a query plan accessing a single relation. As presented previously, the optimizer gets from the

System R catalog the access paths available on the relation referenced by the query. The cheapest access path is obtained by evaluating the cost for each available access path, i.e., each index on the relation plus a segment scan. The optimizer formulates a cost prediction given by the following formula:

$$Cost = Page\ Fetches + W \times (RSI\ calls)$$

where *Page Fetches* represent *I/O* requirement computed by the number of index pages fetched plus number of data pages fetched. *RSI calls* indicate the predicted number of tuples returned from the RSS. This number is a good approximation of *CPU* utilization since most of System R's CPU time is spent in RSS. The parameter *W* is an adjustable weighting factor between *I/O* and *CPU*.

To find the cheapest access plan for a single relation query, the optimizer needs to examine the cheapest *unordered* access path and the cheapest access path producing tuples in each *interesting* order. *Unordered* access path may produce tuples in some order, but the order is not *interesting*. In this case, the optimizer simply chooses the cheapest access path as query plan. By contrast, a tuple order is an *interesting order* if that order is specified by the query block using GROUP BY or ORDER BY clauses. In this case, the optimizer compares the cost of producing that interesting order to the cost of the cheapest unordered path plus the cost of sorting QCARD tuples into the proper order, where QCARD represents the query cardinality. The cheapest of these alternatives is chosen as the plan for the query block. In the following, a description is given of how the query cardinality (QCARD) and RSI calls are computed by the query optimizer.

During the query processing, the optimizer gets statistics on the relations of the query and access paths available on each relation. These statistics include the cardinality of the query relations, the number of segments holding the relevant relations and the fraction of their contained pages, the number of distinct keys and the number of pages of each index. The statistics are used by the optimizer to assign a *selectivity factor F* to each predicate of the WHERE tree. This selectivity factor indicates the expected fraction of tuples which will satisfy the predicate. More details on statistics and selectivity factors are given in [8]. The optimizer computes the query cardinality (QCARD) as the product of the cardinalities of every relation in the query block's FROM list times the product of all the selectivity factors of that query block's predicates. The number of RSI calls (RSICARD) is the product of the relation cardinalities times the selectivity factors of the sargable predicates, since the sargable predicates filter out tuples without returning across the RSS interface.

Access Path Selection for Joins

A join query in SQL combines tuples from more than one relation. Two join methods have been identified as optimal or nearly optimal in most cases. For simplicity, a description of how to join two relations is given, then, it is extended to *n* relations. A two-way join involves two relations respectively called *outer* relation and *inner* relation. A predicate that relates columns of two relations to be joined is called *join predicate*. The columns referenced in a join predicate are called *join columns*. Consider the following example:

- SELECT Name, Location
- FROM STUDENT, DEPARTMENT
- WHERE STUDENT.Department=DEPARTMENT.Num

In this example, the outer and inner relations are respectively "STUDENT" and "DEPARTMENT". There is one join predicate that is "STUDENT.Departement = DEPARTMENT.Num." The join columns of this query are "STUDENT.Department" and "DEPARTMENT.Num".

The first supported join method is called *nested loops*. This method scans the outer and the inner relations in any order. The scan on the outer relation is opened and for each outer tuple obtained, a scan is opened on the inner relation to retrieve, one at a time, all the tuples of the inner relation that satisfy the join predicate. The cost of a nested loop join is computed from the costs of scans on single relations defined in the previous section and is given by the following formula:

$$C - nested - loop - join \ (path1, path2)$$
$$= C - outer \ (path1) + N \times C - innter \ (path2)$$

where *C-outer(path1)* indicates the cost of scanning the outer relation via path1, *C-inner(path2)* indicates the cost of scanning the inner relation, applying all applicable predicates, and *N* is the (product of the cardinalities of all relations R of the join so far) × (product of the selectivity factors of all applicable predicates).

The second supported join method is called *merging scans*. It requires the outer and the inner relations to be scanned in join column order. This means that join columns define *interesting* orders in addition of columns mentioned in GROUP BY and ORDER BY clauses. In case the join query contains more than one predicate, one of them is used as join predicate and the others are treated as ordinary predicates. If a relation has no index on the join column, it has to be sorted into a temporary list ordered by join column. By using ordering on join columns, the merging scan method avoids rescanning the entire inner relation for each tuple of the outer relation. The merging scan synchronizes the inner and the outer scans by matching join columns. In addition, it may take advantage of clustering on join

S

column of the inner relation. Thus, the merging scan can remember where matching join groups are located since tuples having the same values on a join column are physically close to each other. The cost of a merge scan join can be divided into the cost of actually doing the join plus the cost of sorting the outer or inner relation if required. The cost of doing the merge is given by:

$$C - merge \ (path1, path2) = C - outer \ (path1) \\ + N \times C - innter \ (path2)$$

In case the inner relation is sorted into a temporary relation, the merging scans do not scan the entire relation looking for a match. Therefore, the cost of the inner scan can be significantly reduced comparing to nested-loop joins. The cost of the inner scan, in this case, is given by the following formula:

$$C - inner \ (sorted \ list) = TEMPPAGES/N \\ + W \times RSICARD$$

where TEMPPAGES is the number of pages required to hold the inner relation. RSICARD is the number of RSI calls and W is an adjustable weighting factor between *I/O* and *CPU*.

When optimizing larger queries, these join methods are used as building blocks. The System R optimizer only considers linear join trees, where join operators may occur only on one side of other join operators. Join trees with more than two relations are therefore constructed by adding a new relation to an already existing join tree. For applying the methods described above, the join tree, which is treated like a composite relation, represents the *outer* relation and the relation being added to the join tree represents the *inner* relation. The optimizer uses a dynamic programming (DP) strategy (sketched in Fig. 2) to reduce the runtime complexity for join tree construction. It organizes the optimization by the size of the join tree, initializing the DP table with single relations and then constructing larger join trees by combining smaller join trees (S_l) with new relations (R_r). For each combination of relations the best join tree found so far is stored in the DP table, which reduces the search

```
JoinOrder (R = {R_1,...,R_n})
  for each R_i ∈ R
    dpTable[{R_i}] = R_i
  for each 1< s ≤ n ascending // size of the join tree
    for each S_l ⊂ R, R_r ∈ (R\S_l) : |S_l| = s∧dpTable[S_l] ≠ φ
      if¬(S_l can be joined with R_r) continue
      p = dpTable[S_l]⋈dpTable[{R_r}]
      if dpTable[S_l ∪ {R_r}] = φ∨ cost(p)<cost(dpTable[S_l ∪ {R_r}])
        dpTable[S_l ∪ {R_r}] = p
  return dpTable[{R_1,...,R_n}]
```

System R (R*) Optimizer, Fig. 2 Dynamic programming strategy for join tree construction

space from $O(n!)$ to $O(n2^n)$. Further, the System R optimizer considers only join trees where a join predicate between the smaller join tree and the new relation exists (i.e., it avoids cross-products). This further reduces the search space, for example to $O(n^3)$ when the relations are simply joined in a sequence.

An important aspect of this optimization strategy is the consideration of *interesting orders*. To find solutions for joining pairs of relations, the optimizer first finds access paths for each single relation in each interesting and non-interesting tuple ordering. Recall that interesting orders are defined by GROUP BY and ORDER BY clauses and also by every join column. Next, the optimizer finds the best way for joining any two relations, and starts building larger join trees. For each join tree the order of the composite result is saved to allow for merge joins that would not require sorting the composite result. As the ordering can affect later operators, a plan can only be safely pruned if a cheaper one, which satisfies the same interesting ordering, is found. After join trees for all *n* relations have been constructed, the optimizer chooses the cheapest solution that gives the required order specified by the query. Consider an example of three relations R_1, R_2, R_3 in a query and the following join predicates $R_1.x = R_2.x$ and $R_2.x = R_3.x$. Assume that the costs of nested-loop and merge scan for the subquery $\{R_1, R_2\}$ are respectively C_1 and C_2, where C_1 is lower than C_2. Intuitively, when the optimizer looks for the best plan for $\{R_1, R_2, R_3\}$, it could consider the nested-loop method to join $\{R_1, R_2\}$ since it is the cheapest alternative. However, if the optimizer considers a merge scan to join $\{R_1, R_2\}$, the composite result will be sorted on *x* which may significantly reduce the cost of the join with R_3. Thus, the optimizer has to keep track of tuple orderings that can affect the execution plans for the given query to find the optimal join tree.

R* Optimizer

The optimization algorithms described previously have been extended to efficiently process queries in a distributed database management system (R*). In such environment, data needed by queries are stored in multiple sites. Two main factors distinguish query processing in System R from processing query in System R* [6]. First, the communication delays, and second, the possibility of concurrent processing on multiple sites. These two factors raise the importance of developing an R* optimizer to deal with increasing complexity of distributed query processing.

The distribution unit in R* is a relation and each relation is stored at one site. Figure 3 shows two relations STUDENT and DEPARTMENT stored in two different sites *A* and *B*. A query

is called *distributed* if it refers to relations at sites other than the query site. The simplest form of a distributed query is a query that accesses a single relation at a remote site. To execute the query, a process at the remote site accesses the relation locally and ships the query result back to the query site. In case of a join query, the R* optimizer needs to choose a set of local and distributed parameters. The local parameters are the same as the one considered by the System R optimizer including the join method (nested-loop or merge scan), the order in which relations must be joined and the access path for each relation (index or segment scan). The distributed parameters include the choice of the join site, i.e., the site at which the join will take place, and the method for transferring a copy of the inner table to the join site, in case the inner table is not stored in the chosen join site.

R* optimizer can use different methods to transfer tuples from a site to another one. A straightforward strategy is to ship the entire relation to the join site and store it there in a

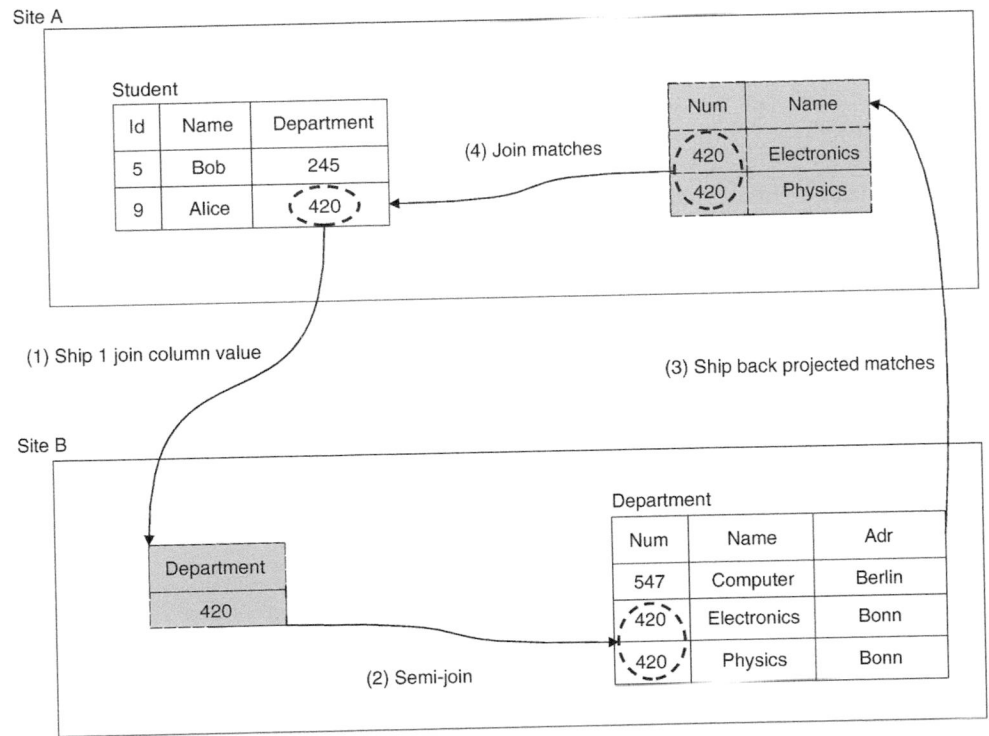

System R (R*) Optimizer, Fig. 3 Example of semijoin

temporary table. Alternatively, the R* optimizer can use other join methods such as semijoins, joins using hashing (Bloom) filters and joins using dynamically-created index. Semijoins, for example, help in reducing the number of transferred tuples by limiting the relevant domain. Specifically, only the tuples that could potentially match the join predicates are shipped to the join site. Figure 3 shows an example of semijoin procedure. First, it projects the outer relation to the join column and ships the results to the site of the inner relation. Second, it finds tuples from the inner relation that match the values received from the outer relation. Third, it ships a copy of the projected inner tuples to the join site. Last, it joins the received tuples to the outer relation.

Key Applications

The System R optimizer inspired many later optimizers, including the well known Starburst optimizer. Starburst uses a similar (though much more generalized) bottom-up constructive optimization technique, and eventually become the commercial database system DB2.

Cross-References

▶ Query Optimization (In Relational Databases)

Recommended Reading

1. Astrahan MM, Blasgen MW, Chamberlin DD, Eswaran KP, Gray J, Griffiths PP, King WF, Lorie RA, McJones PR, Mehl JW, Putzolu GR, Traiger IL, Wade BW, Watson V. System R: relational approach to database management. ACM Trans Database Syst. 1979; 1(2):97–137.
2. Chamberlin DD, Astrahan MM, Blasgen MW, Gray J, King WF, Lindsay BG, Lorie RA, Mehl JW, Price TG, Putzolu GR, Selinger PG, Schkolnick M, Slutz DR, Traiger IL, Wade BW, Yost RA. A history and evaluation of system R. Commun. ACM. 1981; 24(10): 632–46.
3. Chamberlin DD, Boyce RF. SEQUEL: A Structured English Query Language. In: Proceedings of the ACM-SIGFIDET Workshop on Data Description, Access and Control; 1974. p. 249–64.
4. Codd EF. A relational model of data for large shared data banks. Commun. ACM. 1970; 13(6):377–87.
5. Gray J. Notes on data base operating systems. In: Advanced course: operating systems. 1978. p. 393–481.
6. Lohman GM, Mohan C, Haas LM, Daniels D, Lindsay BG, Selinger PG, Wilms PF. Query processing in R*. In: Query processing in database systems. Springer; 1985. p. 31–47.
7. Lorie RA, Nilsson JF. An access specification language for a relational data base system. IBM J Res Dev. 1979; 23(3):286.
8. Selinger PG, Astrahan MM, Chamberlin DD, Lorie RA, Price TG. Access path selection in a relational database management system. In: Proceedings of the ACM SIGMOD International Conference on Management of Data; 1979. p. 23–34.

Printed by Printforce, the Netherlands